The Great Eagles

Their Evolution, Ecology and Conservation

Harpy Eagle
[With permission from Jonathan Beilby]

Golden Eagle
[With permission from Wayne Davis]

Bald Eagle
[With permission from Michelle Jean]

Philippine Hawk-eagle
[With permission from Edwin Mercado]

The Great Eagles

Their Evolution, Ecology and Conservation

Michael O'Neal Campbell

Simon Fraser University
British Columbia, Canada

CRC Press
Taylor & Francis Group
Boca Raton London New York

CRC Press is an imprint of the
Taylor & Francis Group, an **informa** business
A SCIENCE PUBLISHERS BOOK

Cover credits: Front Cover: Ger Bosma (Verreaux's Eagle)
Back Cover: Ondrej Prosicky (Changeable Hawk-eagle)

First edition published 2022
by CRC Press
6000 Broken Sound Parkway NW, Suite 300, Boca Raton, FL 33487-2742

and by CRC Press
2 Park Square, Milton Park, Abingdon, Oxon, OX14 4RN

Library of Congress Cataloging-in-Publication Data

Names: Campbell, Michael O'Neal, 1965- author.
Title: The great eagles : their evolution, ecology and conservation / Michael O'Neal Campbell, Simon Fraser University, British Columbia, Canada.
Description: First edition. | Boca Raton : CRC Press, 2022. | Includes bibliographical references and index. | Summary: "This book presents an integrated perspective (paleontological, historical, physiological, ecological, conservationist and human cultural) of eagles. Issues examined are: 21st century issues in relation to older published themes; species classifications, based on physical methods and DNA analyses; species comparisons, involving evolution, definitions and classification and ecology; and recent advances in adaptive management and other policy actions relevant to bird conservation. The book therefore targets a very wide readership from both the biological and social sciences, as well as general readers and politicians, lawyers and planners"-- Provided by publisher.
Identifiers: LCCN 2021042765 | ISBN 9781138197435 (hardcover) | ISBN 9781032112022 (paperback) | ISBN 9781315278094 (ebook)
Subjects: LCSH: Eagles. | Eagles--Evolution. | Eagles--Ecology. | Eagles--Conservation.
Classification: LCC QL696.F32 C347 2022 | DDC 598.9/42--dc23
LC record available at https://lccn.loc.gov/2021042765

ISBN: 978-1-138-19743-5 (hbk)
ISBN: 978-1-032-11202-2 (pbk)
ISBN: 978-1-315-27809-4 (ebk)

DOI: 10.1201/b21921

Typeset in Times New Roman
by Innovative Processors

Preface

Eagles are generally regarded as the kings of birds in virtually every culture across the world. Therefore, they are possibly the most important non-domesticated birds in the world, in terms of their relationships with people and are among the most recognized species, at least as much as chickens or doves. Eagles are characterized by huge or large bodies, large broad powerful wings and powerful talons and beaks, as well as their obvious strength in soaring flight and predation, and their perceived arrogant or majestic demeanor. From a human perspective, the eagles' sizes alone command respect, awe, curiosity, deification and/or even fear. Although they are a well-studied group, few studies have been made of the total relationship between eagles and people, and of eagles at the inter- and intraspecies levels, especially in the rapidly changing contexts of the 21st century. Ecologically, eagles require large living spaces and hence have wide ranges that may intrude into and conflict with human habitation. Such large birds flying over human settlements are unlikely to be ignored and may be judged positively or negatively, physically, behaviorally, or even spiritually. Eagles have been compared with lions, kings, gods, and are even believed by some to be capable of killing people or stealing children. Eagles are often perceived as dangerous, especially towards animals such as cats, small dogs, lambs and sheep. From these perspectives, eagles, like lions, tigers and bears, are often perceived as power symbols. Hence, the larger species have been symbols of political power, economic dominance and cultural supremacy. Eagles have also been associated with religious deities, and are among the principal animals seen as such, other comparative animals being lions, vultures and some smaller birds of prey such as falcons.

Since eagles are such well known and studied birds, why is another book on them important? One reason is that the twenty first century world is changing so fast that the relationship between eagles, people and the environment requires constant documentation. Many prey species and habitats are declining, and some rival predatory and scavenging species such as large vultures and medium-sized cats have experienced major population changes. Human activities, such as wars, and the development of forest and wildlife removal and rapid urbanization have proliferated throughout the ranges of most eagle species. Searching questions emerge from these engagements. Do we know enough to ameliorate all these issues? Eagles, like vultures, include some very large species and prehistorically included some even larger species, such as the huge Haast's eagle *Harpagornis moorei* (Haast, 1872). Are eagles, as large, carnivorous specialists, past their time on this earth, better adapted to the ancient world of denser prey species and more natural habitats? Can eagles adapt to the modern world of urbanization, land degradation, heavy industry, pollution and decimated prey species? What unique role if any do eagles play in the modern world, with its ever 'modernizing' trend?

This book examines answers to these questions. Six important strands of inquiry are persistent in the current literature on eagles: (1) new genetic techniques that redefine the classification of eagles, especially their evolutionary relationships with hawks and vultures, that sometimes affirm or challenge the definition of what actually is an eagle; (2) the ecological and population status of eagles, in relation to human induced landcover change; (3) the role of eagles in ecosystems and how this may be defined to support conservation efforts; (4) whether captive breeding can rescue nearly extinct species, or fundamentally change the 'wildness' and foraging abilities of eagles; (5) the negative impacts of eagle presence on human life spaces, especially on companion and agricultural animals; and (6) the historical, political, cultural and religious portrayal of eagles in societies across the world. This book researches and systematizes the vast current knowledge of all the eagle species, including their physiology, ecology, evolutionary history, and cultural and conservation studies. It attempts a distinction from earlier and less ambitious texts by adopting a total ecosystem and human relation model, including avian and mammalian competitors, disease causing agents, prehistoric and historic change and current dynamics such as the impacts of deforestation, desertification, urbanization, agricultural intensification, hunting and public attitudes. Comprehensive citations allow the reader to use this volume as a guide for further studies, for the casual reader, recreationalist, student, teacher, activist, falconer, conservationist or scientist.

The book is divided into three parts. Part 1 (Eagle Species and Classification) includes an introduction that describes the background issues necessary for the understanding of eagles, their ecology and physiology, and the field characteristics of the different species. The first two chapters consider the so-called 'true' or 'booted' eagles composed of the genera *Aquila, Hieraaetus, Spizaetus, Oroaetus, Spizastur, Nisaetus, Ictinaetus, Lophoaetus, Polemaetus* and *Stephanoaetus*. The third chapter considers the snake or serpent eagles; Subfamily *Circaetinae*, genera *Terathopius, Circaetus, Dryotriorchis, Spilornis* and *Eutriorchis*. The fourth chapter considers the sea or fish eagles of the genera *Haliaeetus* and *Ichthyophaga*. The fifth chapter considers the harpy eagles; Genera *Harpia, Morphnus, Harpyopsis, Pithecophaga* and *Buteogallus*. As some of these classifications are disputed and some authorities argue for different inter-species relations, the changing classifications are mentioned in this part. Where the species' scientific name is mentioned first, the naming authority (i.e. the name of the person or persons who gave the animal its scientific name) is placed after the scientific name. For example, the Golden eagle's scientific name is *Aquila chrysaetos* (Linnaeus, 1758). Here *Aquila* is the generic name or the name of the Genus to which the bird belongs. *Chrysaetos* is the species name. Linnaeus is the name of the person who gave the bird its classification, and the year of the naming is 1958. The name Carl Linnaeus, that of a famous geneticist and a common name for classification, is usually shortened to L. and placed after the scientific name.

Part 2 (Eagles and Their Ecological Relations) builds on the introduction of Part 1 and discusses the prehistoric evolution of relevant animal species (competitors, prey, possible predators) and landcover change and how this affects eagle ancestors and current species. This is followed by a discussion of eagles and their competitors (Chapter 6); and the relationship between eagles and the physical and biological environment, including relief, climate and seasonal changes, migration flight patterns (Chapter 7). Chapter 8 examines the relationship between eagles and cultural landscapes, touching on the topics of urbanization and rural environmental change, eagles and airplane collisions, and impacts of wind turbines and electric power lines. Chapter 9 examines the impacts of pollution on eagles, especially pesticides and heavy metals.

Part 3 (Eagles and Conservation Issues) introduces the current relationship between eagles and people. This starts with an examination of relevant human actions and cultural structures, including perception, emotional assessments, mysticism, preservation, conservation, extermination, urbanization, agricultural development, and industrialization. Chapter 10 examines the impacts of urbanization, agriculture and human hunting practices on eagles and their related competitors and prey species. Chapter 11 looks at the theoretical concepts and evidence for individualism in eagle behavior, including intra-specific behavioral differences in reaction to changing habitats. Chapter 12 examines the history of human portrayal of eagles, in symbolisms such as heraldry, statues, flags, political constructions and legends, including public attitudes and conservation policies through history and concludes with an assessment of the future of eagles. It is concluded that eagle survival into future centuries depends on their biological and ecological resilience, adaptation to environmental change and the strength, effectiveness and durability of human accepted conservation policies. The author hopes that this volume will awaken the reader to investigate further the situation of these fascinating birds in the modern world.

Michael O'Neal Campbell

Credits and Acknowledgments

Special thanks to Mike Nelson, Derek Keats, Frank McClintock, Vikas Chavan, Frank McClintock, David Nice, Pieter Verheij and Ken Behrens for their numerous pictorial contributions.

Credits for the illustrations in this book are listed below.

Part 1

Fig. P.1. Golden Eagle Talons. Permission of Michelle Jean, BC Wildlife Park
Fig. P.2. Bald Eagle Talons. Permission of Michelle Jean, BC Wildlife Park
Fig. P.3. Golden Eagle Beak. Permission of Alfred Kenneally - Unsplash
Fig. P.4. Bald Eagle Beak. Permission of Flickr
Fig. P.5. Steller's Sea Eagle Beak. Permission of Dr. Peter Schmidt.
Fig. P.6. Tawny Eagle Beak. Permission of Dr. Peter Schmidt.
Fig. P.7. Long-crested Eagle. Permission of Derek Keats
Fig. P.8. Black-breasted Snake Eagle Beak. Permission of Gustav Van Der Gryp.
Fig. P.9. Harpy Eagle Beak. Permission of Miguel "Siu" Rios.
Fig. P.10. Broad Winged Hawk Beak. This shows a smaller size than that of most eagles. Permission of Miguel "Siu" Rios.
Fig. P.11. African Fish Eagle Beak and Talons. Permission of Gustav Van Der Gryp.
Fig. P.12. Broad Winged Hawk Talons. This shows smaller form compared with those of eagles. Permission of Miguel "Siu" Rios.
Fig. P.13. Feet of an Eagle, Accipitrid vulture, Cathartid vulture and Stork
Fig. P.14. Head and Bill Profiles of Selected Eagles and Vultures
Fig. P.15. Haast's Eagle compared with Modern Eagles, including its close modern relative the Little eagle
Fig. P.16. Monotypic and Polytypic
Fig. P.17. Head and Back Feathers for a Golden Eagle. Reproduced from a photograph, with Permission of Chas Moonie
Fig. P.18. Flight Feathers of Bald Eagle in Flight. Reproduced from a photograph, with Permission of Mathew Schwartz.
Fig. P.19. Flight Feathers of the Eastern Imperial Eagle in Flight. Permission of Radovan Václav.

Chapter 1

Fig. 1.1a. Golden Eagle Perching. Permission of Paul Watts
Fig. 1.1b. Golden Eagle Perching. Permission of Linda Raymer
Fig. 1.1c. Golden Eagle Perching. Permission of Pauline Rosenberg
Fig. 1.1d. Golden Eagle Perching. Permission of Linda Raymer
Fig. 1.1e. Golden Eagle Perching. Permission of Chas Moonie
Fig. 1.1f. Immature Golden Eagle Perching. Permission of Dorothea Oldani - Unsplash
Fig. 1.1g. Golden Eagle in Flight. Permission of Linda Raymer
Fig. 1.1h. Golden Eagle in Flight. Permission of Rudolf Ernst
Fig. 1.1i. Immature Golden Eagle in Flight. Frank McClintock
Fig. 1.1j. Golden Eagle in Flight. Permission of Chas Moonie
Fig. 1.1k. Subadult Golden eagle in Flight. Permission of Chas Moonie
Fig. 1.1l. Adult (1) and Immature (2) Golden Eagles. Permission of Julian Thomas
Fig. 1.2a. Tawny Eagle Perching. Permission of Chakkrapan Na Songkhla
Fig. 1.2b. Tawny Eagle Perching. Permission of Gustav Van Der Gryp.

Fig. 1.2c. Tawny Eagle Perching. Permission of Mike Nelson-Birdtour ASIA
Fig. 1.2d. Tawny Eagle Perching. Permission of Derek Keats
Fig. 1.2e. Immature Tawny Eagle Perching. Permission of Derek Keats
Fig. 1.2f. Immature Tawny Eagle Perching. Permission of Vikas Chavan
Fig. 1.2g. Tawny Eagle in Flight. Permission of Vikas Chavan
Fig. 1.2h. Tawny Eagle in Flight. Permission of Vikas Chavan
Fig. 1.2i. Tawny Eagles in Flight. Permission of Gustav Van Der Gryp and Derek Keats
Fig. 1.2j. Tawny Eagle in Habitat South Africa. Permission of David Beadle
Fig. 1.2k. Tawny Eagle Habitat, Tanzania. Permission of Achim Mittler, Frankfurt am Main
Fig. 1.2l. Tawny Eagle Habitat Guinea Savanna. Permission of Jonathan D. Müller
Fig. 1.3a. Steppe Eagle Perching. Permission of Rob Muckley
Fig. 1.3b. Steppe Eagles Perching. Permission of Mike Nelson-Birdtour ASIA
Fig. 1.3c. Immature Steppe Eagle Perching. Permission of Frank McClintock
Fig. 1.3d. Steppe Eagle Foraging. Permission of Terrence Ang
Fig. 1.3e. Steppe Eagle in Flight. Permission of Ken Behrens
Fig. 1.3f. Immature Steppe Eagle in Flight. Permission of Terrence Ang
Fig. 1.3g. Steppe Eagle Wintering Habitat in India. Permission of Siddhartha Mukherjee
Fig. 1.3h. Steppe Eagle Habitat Asian Steppe. Permission of Ken Shono - Unsplash
Fig. 1.4a. Spanish Imperial Eagle Perching. Permission of Santiago GV
Fig. 1.4b. Spanish Imperial Eagles Perching. Permission of Wim Hoek
Fig. 1.4c. Immature Spanish Imperial Eagle Perching. Johannes Gross (j-g-fotos@web.de)
Fig. 1.4d. Immature Spanish Imperial Eagle. Credit - fsanchex
Fig. 1.4e. Spanish Imperial Eagle in Flight. Permission of Santiago Aves
Fig. 1.4f. Spanish Imperial Eagle in Flight. Permission of asanmarc
Fig. 1.4g. Immature Spanish Imperial Eagle in Flight. Permission of Frank McClintock
Fig. 1.4h. Immature Spanish Imperial Eagle in Flight. Permission of Frank McClintock
Fig. 1.4i. Spanish Imperial Eagle Perching on Nest. Permission of Tristan Bantock
Fig. 1.4j. Spanish Imperial Eagle Habitat in Spain. Permission of Ruud Kimmelaar
Fig. 1.4k. Spanish Imperial Eagle Habitat in Spain. Permission of Ruud Kimmelaar
Fig. 1.5a. Eastern Imperial Eagle Perching. Permission of Raghuvir Solanki
Fig. 1.5b. Eastern Imperial Eagle Perching. Permission of Corriplaya
Fig. 1.5c. Eastern Imperial Eagle Perching. Permission of Anshul Jain
Fig. 1.5d. Immature Eastern Imperial Eagle Perching. Permission of Ali Tellioglu
Fig. 1.5e. Immature Eastern Imperial Eagle Perching. Permission of Tanmay's Gallery
Fig. 1.5f. Immature Eastern Imperial Eagle Perching. Permission of Andy Li
Fig. 1.5g. Eastern Imperial Eagle in Flight. Permission of Radovan Václav
Fig. 1.5h. Eastern Imperial Eagle in Flight. Permission of Radovan Václav
Fig. 1.5i. Immature Eastern Imperial Eagle in Flight. Permission of Mike Nelson-Birdtour ASIA
Fig. 1.5j. Immature Eastern Imperial Eagle in Flight. Permission of Julian Thomas
Fig. 1.5k. Eastern Imperial Eagle Habitat Altai Region. Permission of BolatbekGabiden
Fig. 1.6a. Wahlberg's Eagle Perching. Permission of Mario Raimondi
Fig. 1.6b. Wahlberg's Eagle Perching. Permission of Frank McClintock
Fig. 1.6c. Immature Wahlberg's Eagle Perching. Permission of ohannesGroß (j-g-fotos@web.de)
Fig. 1.6d. Wahlberg's Eagle in Flight. Permission of Mario Raimondi
Fig. 1.6e. Wahlberg's Eagle Paler Morph. Permission of Pieter Verheij
Fig. 1.6f. Wahlberg's Eagle Habitat. Permission of Jonathan D. Müller
Fig. 1.7a. Gurney's Eagle. Permission of Christopher Hamilton
Fig. 1.7b. Gurneys Eagle Perching. Permission of Mike Nelson Birdtour ASIA
Fig. 1.7c. Immature Gurney's Eagle Perching. Permission of Davide Beadle
Fig. 1.7d. Gurney's Eagle in Flight. Permission of Raphaël Jordan
Fig. 1.7e. Gurney's Eagle in Flight. Permission of Davide Beadle
Fig. 1.7f. Gurney's Eagle in Flight. Permission of Tinyfishy
Fig. 1.7g. Immature Gurney's Eagle in Flight. Permission of Davide Beadle
Fig. 1.7h. Immature Gurney's Eagle in Flight. Tinyfishy

Fig. 1.7i. Gurney's Eagle Habitat Typical Forest in New Guinea. Permission of Zacky Ezedin
Fig. 1.8a. Wedge-tailed Eagle Perching. Permission of Brendan Shepard
Fig. 1.8b. Wedge-tailed Eagle Perching. Permission of Brendan Shepard
Fig. 1.8c. Wedge-tailed Eagle Perching. Permission of Nick Athanas
Fig. 1.8d. Immature Wedge-tailed Eagle Perching. Permission of Brendan Shepard
Fig. 1.8e. Immature Wedge-tailed Eagle Perching. Permission of Mike Nelson-Birdtour ASIA
Fig. 1.8f. Wedge-tailed Eagle in Flight. Permission of Mike Nelson-Birdtour ASIA
Fig. 1.8g. Wedge-tailed Eagle in Flight. Permission of Frank McClintock
Fig. 1.8h. Wedge-tailed Eagle Habitat and Nest. Permission of Bill Doyle
Fig. 1.8i. Wedge-tailed Eagle Habitat and Nest. Permission of Neil Edwards
Fig. 1.8j. Wedge-tailed Eagle Habitat and Nest. Permission of Bill Doyle
Fig. 1.8k. Wedge-tailed Eagle Habitat and Nest. Permission of Bill Doyle
Fig. 1.9a. Verreaux's Eagle Perching. Permission of Laura Barrio
Fig. 1.9b. Verreaux's Eagle Perching. Permission of Laura Barrio
Fig. 1.9c. Immature Verreaux's Eagle Perching. Permission of Francesco Veronesi
Fig. 1.9d. Verreaux's Eagle in Flight. Permission of Derek Keats
Fig. 1.9e. Verreaux's Eagle in Flight. Permission of Derek Keats
Fig. 1.9f. Immature Verreaux's Eagle in Flight. Permission of Alan Lee
Fig. 1.9g. Verreaux's Eagle Habitat Matobo Hills, Zimbabwe. Permission of Geoff Cooke
Fig. 1.9h. Verreaux's Eagle Habitat. Permission of Derek Keats
Fig. 1.10a. Bonelli's Eagle Perching. Permission of Juan Manuel Benjumea Fantoni
Fig. 1.10b. Bonelli's Eagle Perching. Permission of Juan Manuel Benjumea Fantoni
Fig. 1.10c. Bonelli's Eagle Perching. Permission of Pieter Verheji
Fig. 1.10d. Bonelli's Eagle in Flight. Permission of Frank McClintock
Fig. 1.10e. Bonelli's Eagle in Flight. Permission of Mike Nelson-Birdtour ASIA
Fig. 1.10f. Bonelli's Eagle in Flight. Permission of Matthew Smith
Fig. 1.10g. Bonelli's Eagle in Flight. Permission of Pieter Verheij
Fig. 1.10h. Immature Dark Morph Bonelli's Eagle in Flight. Permission of Hari K. Patibanda
Fig. 1.10i. Immature Bonelli's Eagle in Flight. Permission of Matthew Smith
Fig. 1.10j. Bonelli's Eagle Nesting. Permission of Hari K. Patibanda
Fig. 1.10k. Bonelli's Eagle Habitat. Permission of Matthew Smith
Fig. 1.11a. African Hawk-eagle Perching. Permission of Roelof van der Breggen
Fig. 1.11b. African Hawk-eagle. Permission of Hans Veth - Unsplash
Fig. 1.11c. African Hawk-eagle Perching. Permission of Derek Keats
Fig. 1.11d. African Hawk-eagle Perching. Permission of Derek Keats
Fig. 1.11e. African Hawk-eagle in Flight. Permission of Bob Burgess
Fig. 1.11f. African Hawk-eagle in Flight. Permission of Roelof van der Breggen
Fig. 1.11g Immature African Hawk-eagle in Flight. Permission of Fons Buts
Fig. 1.11h. African Hawk-eagle Habitat in Burkina Faso. Permission of Rita Willaert
Fig. 1.12a. Cassin's Hawk-eagle Perching. Permission Jeremey Roberts
Fig. 1.12b. Cassin's Hawk-eagle Perching. Permission of Bradley Hacker
Fig. 1.12c. Immature Cassin's Hawk Eagle Perching. Permission of Robert Campbell
Fig. 1.12d. Cassin's Hawk-eagle in Flight. Permission of Julian Thomas
Fig. 1.12e. Cassin's Hawk-eagle in Flight. Permission of Gil Ewing
Fig. 1.12f. Cassin's Hawk-eagle Habitat. Permission of Jonathan D. Müller

Chapter 2

Fig. 2.1a. Booted Eagle Perching. Permission of Pieter Verheij
Fig. 2.1b. Booted Eagle Perching. Permission of Derek Keats
Fig. 2.1c. Booted Eagle Light Morph Perching. Permission of Pieter Verheij
Fig. 2.1d. Booted Eagle Light Morph Perching. Permission of Hari K. Patibanda
Fig. 2.1e. Booted Eagle Dark Morph in Flight. Permission of Nick Ransdale
Fig. 2.1f. Booted Eagle Light Morph in Flight. Permission of Derek Keats
Fig. 2.1g. Booted Eagle in Flight. Permission of Gil Ewing

Fig. 2.1h. Booted Eagle Possible Breeding Display. Permission of Frank McClintock
Fig. 2.1i. Booted Eagle Habitat Varies from Open Forest to Desert. Permission of Jonathan D. Müller
Fig. 2.2a. Little Eagle Perching. Permission of David Nice
Fig. 2.2b. Little Eagle Perching. Permission of David Nice
Fig. 2.2c. Little Eagle Perching. Permission of Frank McClintock
Fig. 2.2d. Little Eagle in Flight. Permission of David Nice
Fig. 2.2e. Little Eagle in Flight. Permission of David Nice
Fig. 2.2f. Little Eagle Habitat. Permission of Benn McGuinness - Unsplash
Fig. 2.3a. Pygmy Eagle in Flight. Permission of Paul French
Fig. 2.3b. Pygmy Eagle in Flight. Permission of Nigel Voaden
Fig. 2.3c. Pygmy Eagle in Flight. Permission of John C. Mittermeier
Fig. 2.3d. Pygmy Eagle Habitat New Guinea Forest. Permission of Jay Pruett
Fig. 2.4a. Ayres's Hawk-eagle Perching. Permission of Alan Gilbert
Fig. 2.4b. Ayres's Hawk-eagle Perching and Eatong Elias Bayou
Fig. 2.4c. Immature Ayres's Hawk Eagle Perching. Permission of Antoine Marchal
Fig. 2.4d. Ayre's Hawk-eagle in Flight. Permission of Robert Davidson
Fig. 2.4e. Ayre's Hawk-eagle in Flight. Permission of Robert Davidson
Fig. 2.4f. Immature Ayres's Hawk Eagle in Flight. Permission of Feathercollector
Fig. 2.4g. Ayres's Hawk Eagle Habitat Burkina Faso. Permission of Rita Willaert
Fig. 2.4h. Ayres's Hawk-eagle Habitat. Permission of Kureng Dapel
Fig. 2.5a. Rufous-bellied Eagle. Permission of Angela Yeo
Fig. 2.5b. Rufous-bellied Eagle Permission of Angela Yeo
Fig. 2.5c. Immature Rufous-bellied Eagle. Permission of Myron Tay
Fig. 2.5d. Rufous Bellied Eagle. Permission of Mike Nelson-Birdtour ASIA
Fig. 2.5e. Immature Rufous-bellied Eagle in Flight. Permission of BP Chua
Fig. 2.5g. Rufous Bellied Hawk Eagle Habitat Sri Lanka. Permission of Genine Alyssa
Fig. 2.6a. Changeable Hawk Eagle Perching. Permission of Pieter Verheij
Fig. 2.6b. Changeable Hawk Eagle. Permission of Pieter Verheij
Fig. 2.6c. Changeable Hawk Eagle. Permission of Pieter Verheij
Fig. 2.6d. Changeable Hawk-eagle Dark Morph. Permission of Mike Nelson-Birdtour ASIA
Fig. 2.6e. Immature Changeable Hawk-eagle. Permission of Matt Trevillion
Fig. 2.6f. Immature Changeable Hawk Eagle. Permission of Frank McClintock
Fig. 2.6g. Changeable Hawk eagle. Permission of Mike (Madbirder) Nelson - Birdtour ASIA
Fig. 2.6h. Changeable or Crested Hawk Eagle. Permission of Frank McClintock
Fig. 2.6i. Changeable Hawk eagle Dark Morph. Permission of Ken Behrens
Fig. 2.6j. Changeable Hawk-eagle Habitat. Permission of Elina Sazonova from Pexels
Fig. 2.6k. Changeable Hawk-eagle Habitat. Maldare, India. Permission of Aditya-Vyas - Unsplash
Fig. 2.7a. Flores Hawk-eagle in Perching. Permission of Yovie Jehabut
Fig. 2.7b. Flores Hawk-eagle in Flight. Permission of Yovie Jehabut
Fig. 2.7c. Flores Hawk-eagle in Flight. Permission of Mu-Chi Tsai
Fig. 2.7d. Flores Hawk-eagle Habitat Lombok. Permission of KoesNadi - Unsplash
Fig. 2.8a. Mountain Hawk-eagle. Permission of H Yasui
Fig. 2.8b. Mountain Hawk Eagle Perching. Permission of Hari K Patibanda
Fig. 2.8c. Mountain Hawk Eagle. Permission of Hari K Patibanda
Fig. 2.8d. Juvenile and Adult Mountain Hawk-eagle. Permission of Skip Russell
Fig. 2.8e. Mountain Hawk Eagle. Permission of Frank McClintock
Fig. 2.8f. Immature Mountain Hawk-eagle. Permission of Mike Nelson-Birdtour ASIA
Fig. 2.8g. Mountain Hawk-eagle in Habitat. Permission of Mike Nelson-Birdtour ASIA
Fig. 2.8h. Mountain Hawk Eagle Habitat Bhutan. Permission of Yanni Shams from Pexels
Fig. 2.9a. Legge's Hawk Eagle Perching. Permission of Michael Covington
Fig. 2.9b. Immature Legge's Hawk Eagle Perching. Permission of Graham Ekins
Fig. 2.9c. Immature Legge's Hawk Eagle. Permission of Frank McClintock
Fig. 2.9d. Immature Legge's Hawk Eagle. Permission of David Beadle
Fig. 2.9e. Immature Legge's Hawk Eagle. Permission of Gil Ewing

Fig. 2.9f. Immature Legge's Hawk Eagle. Permission of Alex Aza
Fig. 2.10a. Javan Hawk-eagle Perching. Permission of Saravanan Krishnamurthy
Fig. 2.10b. Javan Hawk-eagle Perching. Permission of Mike Nelson-Birdtour ASIA
Fig. 2.10c. Javan Hawk-eagle Perching. Permission of Johnathan Beilby
Fig. 2.10d. Javan Hawk-eagle Perching. Permission of Dikky Oesin
Fig. 2.10e. Javan Hawk-eagle in Flight. Permission of Saravanan Krishnamurthy
Fig. 2.10f. Javan Hawk Eagle Habitat Rumpin Java. Permission of Tom Fisk Pexels
Fig. 2.11a. Sulawesi Hawk Eagle Perching. Permission of Peter Ericsson
Fig. 2.11b. Sulawesi Hawk-eagle Perching. Permission of Mike Nelson-Birdtour ASIA
Fig. 2.11c. Sulawesi Hawk-eagle Perching. Permission of Mike Nelson-Birdtour ASIA
Fig. 2.11d. Immature Sulawesi Hawk-eagle Perching. Permission of Mike Nelson
Fig. 2.11e. Sulawesi Hawk-eagle in Flight. Permission of Mike Nelson-Birdtour ASIA
Fig. 2.11f. Sulawesi Hawk-eagle in Flight. Permission of Mike Nelson-Birdtour ASIA
Fig. 2.11g. Sulawesi Hawk-eagle Habitat Permission of Josep Miquel Navarro
Fig. 2.12a. Philippine Hawk Eagle. Permission of Nayer Youakim
Fig. 2.12b. Philippine Hawk Eagle. Permission of Mark Jason Villa
Fig. 2.12c. Subadult Philippine Hawk-Eagle. Permission Marlz Cris.
Fig. 2.12d. Immature Philippine Hawk-eagle. Permission Edwin Mercado.
Fig. 2.12e. Philippine Hawk-eagle. Permission Marlz Cris.
Fig. 2.12f. Philippine Hawk-eagle. Permission Mike Nelson-Birdtour ASIA
Fig. 2.12g. Philippine Hawk-eagle Habitat. Permission Jeneva Stiles
Fig. 2.13a. Pinsker's Hawk Eagle Perching. Permission of Steve Young
Fig. 2.13b. Pinsker's Hawk Eagle Perching. Permission of Steve Young
Fig. 2.13c. Pinsker's Hawk Eagle. Permission of Rodrigo Layug
Fig. 2.13d. Pinsker's Hawk Eagle. Permission of Chris Chafer
Fig. 2.13e. Pinsker's Hawk Eagle. Permission of Chris Chafer
Fig. 2.13f. Pinsker's Hawk Eagle. Permission of Mike Nelson
Fig. 2.13g. Immature Pinsker's Hawk-eagle in Habitat. Permission of Steve Young
Fig. 2.14a. Blyth's Hawk Eagle Perching. Permission of Mike Nelson-Birdtour ASIA
Fig. 2.14b. Blyth's Hawk Eagle. Permission of Nick Athanas
Fig. 2.14c. Blyth's Hawk Eagle. Permission of Zaharil Dzulkafly
Fig. 2.14d. Immature Blyth's Hawk-eagle. Permission of Zaharil Dzulkafly
Fig. 2.14e. Blyth's Hawk Eagle. Permission of Mike Nelson-Birdtour ASIA
Fig. 2.14f. Blyth's Hawk Eagle. Permission of Mike Nelson-Birdtour ASIA
Fig. 2.14g. Blyth's Hawk Eagle. Habitat Malaysia. Permission of Terence Seow - Unsplash
Fig. 2.15a. Wallace's Hawk Eagle Perching. Permission of Marcus F.C. Ng
Fig. 2.15b. Wallace's Hawk Eagle Perching. Permission of Terence Ang
Fig. 2.15c. Wallace's Hawk Eagle Perching, Kinabatangan River, Sabah, Borneo. Permission of Paul Prior
Fig. 2.15d. Immature Wallace's Hawk-eagle Perching. Permission of David Koh
Fig. 2.15e. Immature Wallace's Hawk-eagle Perching. Permission of Tristan Bantock
Fig. 2.15f. Immature Wallace's Hawk-eagle Perching. Permission of Mike Nelson-Birdtour
Fig. 2.15g. Immature Wallace's Hawk-eagle Perching. Permission of Mike Nelson-Birdtour
Fig. 2.15h. Immature Wallace's Hawk-eagle Perching. Permission of Mike Nelson-Birdtour
Fig. 2.15i. Immature Wallace's Hawk-eagle in Flight. Permission of Mike Nelson-Birdtour
Fig. 2.15j. Wallace's Hawk-eagle Habitat Malaysia. Permission of Sonia Chronowska - Unsplash
Fig. 2.16a. Black Hawk-eagle Perching. Permission of Nick Athanas
Fig. 2.16b. Black-Hawk Eagle Perching. Permission of Jonathan Beilby
Fig. 2.16c. Black-Hawk Eagle Perching. Permission of Thelmå Gatuzźo
Fig. 2.16d. Black-Hawk Eagle Perching. Permission of Dave Wendelken
Fig. 2.16e. Black-Hawk Eagle In Flight. Permission of Neil Bowman
Fig. 2.16f. Black-Hawk Eagle in Flight. Permission of Nick Athanas
Fig. 2.17a. Black-and-white Hawk Eagle Perching. Permission of Jonathan Beilby
Fig. 2.17b. Black-and-white Hawk Eagle in Flight. Permission of Mike Nelson-Birdtour ASIA
Fig. 2.17c. Black-and-white Hawk Eagle in Flight. Permission of Gail Hampshire

Fig. 2.17d. Black and White Hawk-eagle Habitat - Author
Fig. 2.17e. Black-and-White Hawk Eagle Habitat. Permission of Rômulo Ferreira
Fig. 2.18a. Ornate Hawk-eagle Perching. Permission of Ian Gill
Fig. 2.18b. Ornate Hawk-eagle Perching. Permission of Ian Gill
Fig. 2.18c. Ornate Hawk-eagle Perching. Permission of Tristanba
Fig. 2.18d. Immature Ornate Hawk-eagle. Permission of Jorge Eduardo Chinchilla Arroyo
Fig. 2.18e. Immature Ornate Hawk-eagle. Permission of Xyomara Carretero-Pinzon
Fig. 2.18f. Ornate Hawk-eagle in Forested Habitat. Nick Athanas
Fig. 2.18h. Ornate Hawk-Eagle Habitat. Permission of Nadav Hurvitz from Pexels
Fig. 2.19a. Black and Chestnut eagle Perching. Permission of Luis Olarte
Fig. 2.19b. Immature Black and Chestnut Eagle. Permission of Brian and Irene Hampson
Fig. 2.19c. Black and Chestnut Eagle in Flight. Permission of Robert Davidson
Fig. 2.19d. Immature Black and Chestnut Eagle in Flight. Permission of Nick Athanas
Fig. 2.19e. Black-and-Chestnut Eagle Habitat. Permission of Robert Davidson
Fig. 2.20a. Greater spotted Eagle Perching. Permission of Vikas Chavan
Fig. 2.20b. Greater spotted Eagle Head. Permission of Quazi Sanjeed
Fig. 2.20c. Greater Spotted Eagle. Permission of Lopamudra Barman, Kolkata, India (www.flickr.comphotoslopamuidrabarman)
Fig. 2.20d. Greater spotted Eagle taking Flight. Permission of Vikas Chavan.
Fig. 2.20e. Immature Greater Spotted Eagle. Permission of Philip Roberts
Fig. 2.20f. Immature Greater Spotted Eagle. Permission of Philip Roberts
Fig. 2.20g. Greater Spotted Eagle in Flight. Permission of Mike Nelson-Birdtour ASIA
Fig. 2.20h. Greater Spotted Eagle in Flight. Permission of Kimmo Järvinen
Fig. 2.20i. Greater Spotted Eagle Habitat Russia. Permission of Max Sudyin - Unsplash
Fig. 2.21a. Lesser Spotted Eagle. Permission of Roelof van der Breggen
Fig. 2.21b. Lesser Spotted Eagle Perching. Permission of Frank McClintock
Fig. 2.21c. Lesser Spotted Eagle Perching. Permission of Frank McClintock
Fig. 2.21d. Lesser Spotted Eagle in Flight. Permission of Maarten Mortier
Fig. 2.21e. Lesser Spotted Eagle in Flight. Permission of Roelof van der Breggen
Fig. 2.21f. Lesser Spotted Eagle Habitat Ukraine. Permission of Alex-Belogub - Unsplash
Fig. 2.22a. Indian Spotted Eagle Perching. Permission of Ken Behrens
Fig. 2.22b. Indian Spotted Eagle Perching. Permission of Hari K Patibanda
Fig. 2.22c. Immature Indian Spotted Eagle Perching. Permission of Vikas Chavan
Fig. 2.22d. Indian Spotted Eagle Perching. Permission of Girish Kurudi
Fig. 2.22e. Indian Spotted Eagle in Flight. Permission of Vikas Chavan
Fig. 2.22f. Indian Spotted Eagle Habitat Open Forest India. Permission of Paul Silvan - Unsplash
Fig. 2.23a. Martial Eagle Perching. Permission of Paul Ellis
Fig. 2.23b. Martial Eagle Perching. Permission of Gustav Van Der Gryp
Fig. 2.23c. Martial Eagle Perching. Permission of Gabriel Leboff
Fig. 2.23d. Martial Eagle Perching. Permission of Frans Van Heerden - Pexels
Fig. 2.23e.Immature Martial Eagle Perching. Permission of Ken Behrens
Fig. 2.23f. Immature Martial Eagle Perching. Permission of Pieter Verheij
Fig. 2.23g. Immature Martial Eagle Perching. Permission of David Beadle
Fig. 2.23h. Martial Eagle in Flight. Permission of Frank McClintock
Fig. 2.23i. Martial Eagle in Flight. Permission of Pieter Verheij
Fig. 2.23j. Martial Eagle in Flight with Pied Crow. Permission of Roelof van der Breggen
Fig. 2.23k. Immature Martial Eagle in Flight. Permission of Theo Storm
Fig. 2.23l. Immature Martial Eagle. Permission of Rick Dunlap
Fig. 2.23m. Habitat for the Martial Eagle in South Africa. Permission of Derek Keats
Fig. 2.23n. Martial Eagle Habitat Serengeti, Tanzania. Permission of Achim Mittler.
Fig. 2.24a. Crowned Eagle Perching. Permission of Derek Keats
Fig. 2.24b. Crowned Eagle Perching. Permission of Ken Behrens
Fig. 2.24c. Juvenile Crowned Eagle Perching. Permission of John Archer-Thomson
Fig. 2.24d. Crowned eagle in Flight. Permission of Derek Keats
Fig. 2.24e. Crowned Eagle in Flight. Permission of Ken Behrens

Fig. 2.24f. Crowned Eagle in Flight. Permission of Ken Behrens
Fig. 2.24g. Crowned Eagle Habitat in Congo. Permission of Andrey Gudkov
Fig. 2.24h. Crowned Eagle habitat in Kruger National Park. Permission of Michael Kemper
Fig. 2.25a. Long-crested Eagle Perching. Permission of Frank McClintock
Fig. 2.25b. Long-crested Eagle Perching. Permission of Derek Keats
Fig. 2.25c. Long-crested Eagle Perching. Permission of Miguel Siu Rios
Fig. 2.25d. Long-crested Eagle in Flight. Permission of Frank McClintock
Fig. 2.25e. Long-crested Eagle in Flight. Permission of Derek Keats
Fig. 2.25f. Long-crested Eagle in Flight. Permission of Roelof van der Breggen
Fig. 2.25g. Long-crested Eagle Habitat. Permission of Leon Labuschagne from Pexels
Fig. 2.26a. Black Eagle Perching. Permission of Kalyan Varma kalyan@rtns.org
Fig. 2.26b. Black Eagle Perching. Permission of THEGIFT777
Fig. 2.26c. Black Eagle in Flight. Permission of Mike Nelson-Birdtour ASIA
Fig. 2.26d. Black Eagle in Flight. Permission of Nick Athanas
Fig. 2.26e. Black Eagle in Flight. Permission of Chien Lee
Fig. 2.26f. Black Eagle in Flight. Permission of Chien Lee
Fig. 2.26g. Black Eagle in Flight. Permission of Kalyan Varma kalyan@rtns.org
Fig. 2.26h. Immature Black Eagle in Flight. Permission of David Beadle
Fig. 2.26i. Black Eagle Collecting Branches for Nest. Permission of Kalyan Varma kalyan@rtns.org
Fig. 2.26j. Black Eagle Collecting Branches. Permission of Kalyan Varma kalyan@rtns.org
Fig. 2.26k. Immature Black Eagle on Nest. Permission Kalyan Varma kalyan@rtns.org
Fig. 2.26l. Black Eagle Habitat Malaysia. Permission of Kelvin Yup - Unsplash

Chapter 3.

Fig. 3.1a. Short-toed Eagle Perching. Profile. Permission of Mike Nelson-Birdtour ASIA
Fig. 3.1b. Adult Short-toed Eagle. Permission of Hari K. Patibanda
Fig. 3.1c. Immature Short-toed Eagle. Permission of Hari K Patibanda
Fig. 3.1d. Immature Short-toed Eagle Perching in Forest stand. Permission of Pieter Verheij
Fig. 3.1e. Short-toed Eagle in Flight with dark plumage. Permission of Pieter Verheij
Fig. 3.1f. Short-toed Eagle in Flight. Permission of Frank McClintock
Fig. 3.1g. Immature Short-toed Eagle in Flight. Permission of George Vittman
Fig. 3.1h. Immature Short-toed Eagle in Flight. Permission of Mike Nelson-Birdtour ASIA
Fig. 3.1i. Short-toed Snake Eagle Habitat Spain. Permission of Rudd Kimmelaar
Fig. 3.2a. Black-chested Snake Eagle Perching. Permission of Derek Keats
Fig. 3.2b. Black-chested Snake Eagle Perching. Permission of Derek Keats
Fig. 3.2c. Black-chested Snake Eagle Perching. Permission of David Beadle
Fig. 3.2d. Black-chested Snake Eagle in Flight. Permission of Gustav Van Der Gryp
Fig. 3.2e. Black-chested Snake Eagle in Flight. Permission of Derek Keats
Fig. 3.2f. Black-chested Snake-Eagle in Flight. Permission of David Beadle
Fig. 3.2g. Black-chested Snake Eagle Habitat, Serengeti Tanzania. Permission of Jérôme Christen
Fig. 3.3a. Beaudouin's Snake Eagle Perching. Permission of Neil Hilton
Fig. 3.3b. Beaudouin's Snake Eagle Perching. Permission of Jeremy Roberts
Fig. 3.3c. Beaudouin's Snake Eagle in Flight. Permission of David Beadle
Fig. 3.3c. Beaudouin's Snake Eagle in Flight. Permission of Jeremy Roberts
Fig. 3.3e. Beaudouin's Snake Eagle Habitat West Africa. Permission of David Carr
Fig. 3.3f. Beaudouin's Snake Eagle Habitat, Chad. Permission of Robert Barnes
Fig. 3.4a. Brown Snake Eagle Perching. Permission of Willie van Schalkwyk
Fig. 3.4b. Brown Snake Eagle Perching. Permission of Dubi Shapiro
Fig. 3.4c. Brown Snake Eagle Perching. Permission of Derek Keats
Fig. 3.4d. Brown Snake Eagle in Flight. Permission of Willie van Schalkwyk
Fig. 3.4e. Brown Snake Eagle in Flight. Permission of Ken Behrens
Fig. 3.4f. Immature Brown Snake Eagle in Flight. Permission of Willie van Schalkwyk
Fig. 3.4g. Brown Snake Eagle Habitat. Permission of Frank McClintock
Fig. 3.4h. Brown Snake Eagle Habitat. Permission of Rita Willaert

Fig. 3.4i. Habitat for the Brown Snake Eagle. Permission of Derek Keats
Fig. 3.5a. Southern Banded Snake Eagle Perching. Permission of Brendon White
Fig. 3.5b. Southern Banded Snake Eagle Perching. Permission of Brendon White
Fig. 3.5c. Southern Banded Snake Eagle Perching. Permission of Johan van Rensburg
Fig. 3.5d. Southern Banded Snake Eagle Perching. Permission of Nick Athanas
Fig. 3.5e. Southern Banded Snake Eagle Perching. Permission of Nick Borrow
Fig. 3.5f. Southern Banded Snake Eagle Perching. Permission of Martie De Wijn
Fig. 3.5g. Immature Southern Banded Snake Eagle Perching. Permission of Jeff Blake
Fig. 3.5h. Southern Banded Snake Eagle in Flight. Permission of Brendon White
Fig. 3.5i. Southern Banded Snake Eagle in Flight. Permission of Brendon White
Fig. 3.5j. Southern Banded Snake Eagle in Flight. Permission of Brendon White
Fig. 3.5k. Southern Banded Snake Eagle in Flight. Permission of Steve Slater
Fig. 3.5l. Southern Banded Snake Eagle Habitat. Permission of Samuel Phillips - Unsplash
Fig. 3.6a. Western Banded Snake Eagle Perching. Permission of Ken Behrens
Fig. 3.6b. Western Banded Snake Eagle Perching. Permission of Nigel Voaden
Fig. 3.6c. Western Banded Snake Eagle Perching. Permission of Frank McClintock
Fig. 3.6d. Western Banded Snake Eagle in Flight. Permission of Ken Behrens
Fig. 3.6e. Western Banded Snake Eagle in Flight. Permission of Nigel Voaden
Fig. 3.6f. Western Banded Snake Eagle Habitat Serengeti Ben Preater - Unsplash
Fig. 3.7a. Bateleur Eagle Perching. Permission of Derek Keats
Fig. 3.7b. Bateleur Eagle Perching and Flight. Permission of Gustav Van -Der Gryp
Fig. 3.7c. Immature Bateleur Eagle Perching. Permission of Peet van Schalkwyk
Fig. 3.7d. Immature Bateleur Eagle Perching. Permission of Peet van Schalkwyk
Fig. 3.7e. Immature Bateleur Eagle Perching. Permission Margaret Westrop
Fig. 3.7f. Bateleur Eagle in Flight. Permission of Photoski141
Fig. 3.7g. Bateleur Eagle in Flight. Permission of Stan
Fig. 3.7h. Bateleur Eagle in Flight. Permission of Gustav Van Der Gryp
Fig. 3.7i. Immature Bateleur Eagle in Flight. Permission of Michael Heyns
Fig. 3.7j. Immature Bateleur Eagle in Flight. Permission of Peter Gosling
Fig. 3.7k. Bateleur Habitat Okavango Delta. Permission of Fons Buts
Fig. 3.7l. Bateleur Eagle Habitat. Permission of Derek Keats
Fig. 3.8a. Philippine Serpent Eagle Perching. Permission of Steve Albano
Fig. 3.8b. Philippine Serpent Eagle Perching. Permission of Mike Nelson-Birdtour ASIA
Fig. 3.8c. Philippine Serpent Eagle Perching. Permission of Chris Chafer
Fig. 3.8d. Philippine Serpent Eagle Perching. Permission of Chris Chafer
Fig. 3.8e. Philippine Serpent Eagle Head. Permission of Chris Chafer
Fig. 3.8f. Immature Philippine Serpent Eagle. Permission of Mark Jasmin
Fig. 3.8g. Immature Philippine Serpent Eagle. Permission of Martin Alvendia
Fig. 3.8h. Immature Philippine Serpent Eagle. Permission of Mike Nelson-Birdtour ASIA
Fig. 3.8i. Immature Philippine Serpent Eagle Spreading Wings. Permission of Mike Nelson-Birdtour ASIA
Fig. 3.8j. Philippine Serpent Eagle in Flight. Permission of Mike Nelson-Birdtour ASIA
Fig. 3.8k. Philippine Serpent Eagle Habitat Forest Luzon. Permission of Med-Gadon
Fig. 3.9a. Sulawesi Serpent Eagle Perching. Permission of Mike Nelson-Birdtour ASIA
Fig. 3.9b. Sulawesi Serpent Eagle Perching. Permission of David Beadle
Fig. 3.9c. Immature Sulawesi Serpent Eagle. Permission of Mike Nelson-Birdtour ASIA
Fig. 3.9d. Sulawesi Serpent Eagle in Flight. Permission of Mike Nelson-Birdtour ASIA
Fig. 3.9e. Immature Sulawesi Serpent Eagle in Flight. Permission of Mike Nelson-Birdtour ASIA
Fig. 3.9f. Immature Sulawesi Serpent Eagle in Flight. Permission of Mike Nelson-Birdtour ASIA
Fig. 3.9g. Immature Sulawesi Serpent Eagle in Flight. Permission of Mike Nelson-Birdtour ASIA
Fig. 3.9h. Sulawesi Serpent Eagle Habitat. Permission of Josep Miquel Navarro
Fig. 3.10a. Crested Serpent Eagle Perching. Permission of Keyur Nandaniya
Fig. 3.10b. Crested Serpent Eagle. Permission of Frank McClintock
Fig. 3.10c. Crested Serpent Eagle. Permission of Pieter Verheij
Fig. 3.10d. Crested Serpent Eagle Perching. Permission of Radhika Jha

Fig. 3.10e. Crested Serpent Eagle. Permission of Frank McClintock
Fig. 3.10f. Immature Crested Serpent Eagle. Permission Frank McClintock
Fig. 3.10g. Crested Serpent Eagle in Flight. Permission of Frank McClintock
Fig. 3.10h. Crested Serpent Eagle in Flight. Permission of Pieter Verheij
Fig. 3.10i. Crested Serpent Eagle Habitat. Permission of Radhika Jha
Fig. 3.10j. Crested Serpent Eagle Habitat. Permission of Gary Ellis - Unsplash
Fig. 3.11a. Nicobar Serpent Eagle Perching. Permission of MV Shreeram
Fig. 3.11b. Juvenile Nicobar Serpent Eagle. Permission of MV Shreeram
Fig. 3.11c. Juvenile Nicobar Serpent Eagle. Permission of MV Shreeram
Fig. 3.11d. Immature Nicobar Serpent Eagle. Permission of Niranjan Sant
Fig. 3.11e. Nicobar Serpent Eagle in Flight. Permission of Shashank Dalvi
Fig. 3.11f. Nicobar Serpent Eagle in Flight. Permission of Shashank Dalvi
Fig. 3.11g. Nicobar Serpent Eagle Habitat. Permission of Sahil - Unsplash
Fig. 3.11h. Nicobar Serpent Eagle Habitat. Permission of Sahil - Unsplash
Fig. 3.12a. Andaman Serpent Eagle. Permission of Khushboo & Rahul Sharma
Fig. 3.12b. Andaman Serpent Eagle. Permission of Balaji Venkatesh Sivaramakrishnan
Fig. 3.12c. Andaman Serpent Eagle Perching. Permission of Stanislav Harvančík
Fig. 3.12d. Andaman Serpent Eagle Perching. Permission of Stanislav Harvančík
Fig. 3.12e. Andaman Serpent Eagle Perching. Permission of Anay Tamekar
Fig. 3.12f. Andaman Serpent Eagle Habitat. Permission of Lopamudra Barman, Kolkata, India (https://www.flickr.com/photos/lopamudrabarman)
Fig. 3.12g. Andaman Serpent Eagle Habitat Little Andaman. Permission of Robert Koster
Fig. 3.13a. Congo Serpent-eagle Perching. Permission of Dubi Shapiro
Fig. 3.13b. Congo Serpent Eagle Perching, Kakum Ghana. Permission of Nik Borrow
Fig. 3.13c. Congo Serpent Eagle. Permission of Johannes Ferdinand Bird-Lens.com
Fig. 3.13d. Congo Serpent Eagle in Flight. Permission of Nick Gaunt
Fig. 3.13e. Congo Serpent Eagle in Flight. Permission of Johannes Ferdinand Bird-Lens.com
Fig. 3.13f. Congo Serpent Eagle Perching in Habitat. Permission of Adam Dare
Fig. 3.13g. Congo Serpent Eagle Perching in Habitat. Permission of Ken Behrens
Fig. 3.13h. Congo Serpent Eagle Habitat. Permission of Andrey Gudkov
Fig. 3.14a. Madagascan Serpent Eagle Perching. Permission of Ken Behrens
Fig. 3.14b. Madagascar Serpent Eagle. Permission of Russell Thorstrom
Fig. 3.14c. Madagascar Serpent Eagle. Permission of Dubi Shaprio
Fig. 3.14d. Madagascar Serpent Eagle. Permission of Dubi Shaprio
Fig. 3.14e. Madagascar Serpent Eagle Perching. Permission of Russell Thorstrom
Fig. 3.14f. Immature Madagascar Serpent eagle on Nest. Permission of Russell Thorstrom
Fig. 3.14g. Madagascar Serpent Eagle in Habitat. Permission of Erwin Collaerts
Fig. 3.14h. Madagascar Serpent Eagle Habitat. Permission of Katerina Kalinina
Fig. 3.14i. Madagascar Serpent Eagle in Habitat. Credit: Theme Inn - Unsplash
Fig. 3.15a. Philippine Eagle Perching. Permission of Rodrigo Layug
Fig. 3.15b. Philippine Eagle Perching. Permission of Photo by Janssen Panizales from Pexels
Fig. 3.15c. Philippine eagle Perching. Permission of Neon Tomas BuenaflorRosell II
Fig. 3.15d. Philippine Eagle Perching. Permission of Chris Chafer
Fig. 3.15e. Philippine Eagle Perching and Displaying. Permission of William L. Lim
Fig. 3.15f. Philippine Eagle. Permission of Bram Demeulemeester
Fig. 3.15g. Philippine Eagle Nesting in Forest Habitat. Permission of Feathercollector
Fig. 3.15h. Philippine Eagle Nesting in Forest Habitat. Permission of Chris Chafer
Fig. 3.15i. Philippine Eagle Nesting in Forest Habitat. Permission of Chris Chafer
Fig. 3.15j. Philippine Eagle Habitat Luzon. Credit Aldrin William Pascua - Unsplash

Chapter 4

Fig. 4.1a. Steller's Sea Eagle Perching. Permission of Axel Blanchard on Unsplash
Fig. 4.1b. Steller's Sea Eagle Perching. Permission of Takashi Muramatsu
Fig. 4.1c. Steller's Sea Eagle Perching. Permission of Takashi Muramatsu
Fig. 4.1d. Steller's Sea Eagle in Flight. Permission of Zdeněk Macháček - Unsplash

Fig. 4.1e. Steller's Sea Eagle in Flight. Permission of Takashi Muramatsu
Fig. 4.1f. Immature Steller's Sea Eagle in Flight. Permission of Julian Thomas
Fig. 4.1g. Immature Steller's Sea Eagle in Flight. Permission of Ricardo Frantz - Unsplash
Fig. 4.1h. Steller's Sea Eagle on Nest. Permission of Russell King
Fig. 4.1i. Steller's Sea Eagle Habitat Kamchatka. Permission of Eugene Kaspersky
Fig. 4.1j. Steller's Sea Eagle Habitat. Permission of Mariya Tereshkova - Unsplash
Fig. 4.2a. White-tailed Eagle Perching. Permission of Slowmotiongli
Fig. 4.2b. White-tailed Eagle Perching. Permission of Robert Davidson
Fig. 4.2c. Immature White-tailed Eagle Perching. Permission of Robert Davidson
Fig. 4.2d. Immature White tailed Eagle Perching. Permission of Jack Seeds - Unsplash
Fig. 4.2e. White-tailed Eagle and Steller's Sea Eagle Perching. Permission of Julian Thomas
Fig. 4.2f. White-tailed Eagle in Flight. Permission of Rosemary Locock
Fig. 4.2g. White-tailed Eagle in Flight. Permission of Julian Thomas
Fig. 4.2h. White-tailed Eagle in Flight. Permission of Julian Thomas
Fig. 4.2i. Immature White-tailed Eagle in Flight. Permission of Avideus
Fig. 4.2j. White-tailed Eagle Habitat Altai Mountains. Permission of Stafichukanatoly
Fig. 4.3a. Bald Eagle Perching. Permission of Joshua J. Cotten
Fig. 4.3b. Bald Eagle Perching. Permission of Frank Cone on Prexels
Fig. 4.3c. Bald Eagle Perching. Permission of Michelle Jean
Fig. 4.3d. Immature Bald Eagle. Permission of Frank Cone - Pexels
Fig. 4.3e. Bald Eagle Adult and Immature on the Nest. Permission of Linda Raymer
Fig. 4.3f. Bald Eagle. Permission of Linda Raymer
Fig. 4.3g. Bald Eagle in Flight. Permission of Ian Dufield
Fig. 4.3h. Immature Bald Eagle in Flight. Permission of Paul Prior
Fig. 4.3i. Immature Bald eagle in Flight. Permission of Thomas Morse - Unsplash
Fig. 4.3j. Bald Eagle at Nest. Permission of Richard Lee - Unsplash
Fig. 4.3k. Bald Eagle Habitat British Columbia - Author
Fig. 4.4a. Pallas's Sea Eagle Perching. Permission of Raghuvir Solanki
Fig. 4.4b. Pallas's Sea Eagle Perching. Permission of Quazi Sanjeed
Fig. 4.4c. Pallas's Sea Eagle Perching. Permission of Mark Piazzi
Fig. 4.4d. Pallas's Sea Eagle Perching. Copyright.bryanjsmith
Fig. 4.4e. Immature Pallas's Sea Eagle Perching. Permission of Zamiruddin Faisal
Fig. 4.4f. Immature Pallas's Sea Eagle Perching. Permission of Zamiruddin Faisal
Fig. 4.4g. Immature Pallas's Sea Eagle Perching. Permission of Mark Piazzi
Fig. 4.4h. Immature Pallas's Sea Eagle in Flight. Permission of Roger Wasley
Fig. 4.4i. Pallas's Sea Eagle in Flight. Permission of Quazi Sanjeed
Fig. 4.4j. Pallas's Sea Eagle Nest and Habitat. Permission of Pete Favelle
Fig. 4.4k. Pallas's Sea Eagle on Nest. Permission of Quazi Sanjeed
Fig. 4.4l. Pallas's Sea Eagle in Grass. Permission of Lorand Szucs.
Fig. 4.4m. Pallas's Sea Eagle Habitat Kazakhstan. Permission of Dmitry Sumskoy - Unsplash
Fig. 4.5a. Madagascar Fish Eagle Perching. Permission of Mike Nelson-Birdtour ASIA
Fig. 4.5b. Madagascar Fish Eagle. Permission of Skip Russell
Fig. 4.5c. Immature Madagascar Fish Eagle Perching. Permission of Allan Hopkins
Fig. 4.5d. Immature Madagascar Fish Eagle Perching. Permission of Keith Cowton
Fig. 4.5e. Madagascar Fish Eagle in Flight. Permission of Alan Wilkinson
Fig. 4.5f. Immature Madagascar Fish Eagle Perching. Permission of Pieter Verheij
Fig. 4.5g. Juvenile and Adult Madagascar Fish Eagle. Permission of Skip Russell
Fig. 4.5h. Madagascar Fish Eagle Habitat. Permission of Rob Sall
Fig. 4.5i. Madagascar Fish Eagle Habitat. Permission of Katerina Kalinina
Fig. 4.6a. African Fish Eagle Perching. Permission of Derek Keats
Fig. 4.6b. African Fish Eagle Perching. Permission of Raghuvir Solanki
Fig. 4.6c. Adult and Immature African Fish Eagle Perching. Permission of Raghuvir Solanki
Fig. 4.6d. Immature African Fish Eagle Perching. Permission Ken Zaremba
Fig. 4.6e. Immature African Fish Eagle - Johannes Gross (j-g-fotos@web.de)

Fig. 4.6f. Immature African Fish Eagle. Permission of Brian Zarate
Fig. 4.6g. African Fish Eagle in Flight. Permission of Gustav Van Der Gryp
Fig. 4.6h. African Fish Eagle in Flight. Permission of Julian Thomas
Fig. 4.6i. Immature African Fish Eagle in Flight. Permission of Wayne Davies
Fig. 4.6j. African Fish Eagle Habitat Kruger National Park. Permission of Michael Kemper
Fig. 4.6k. African Fish Eagle Habitat Gambia River. Permission of Steve Riall
Fig. 4.7a. White-bellied Sea Eagle. Permission of Pieter Verheij
Fig. 4.7b. White-bellied Sea Eagle. Permission of Mike Nelson-Birdtour ASIA
Fig. 4.7c. Immature White-bellied Sea Eagle. Permission of Utopia 88
Fig. 4.7d. White-bellied Sea Eagle. Permission of Mike Nelson-Birdtour ASIA
Fig. 4.7e. White-bellied Sea Eagle. Permission of Frank McClintock
Fig. 4.7f. Immature White-bellied Sea Eagle. Permission of Mike Nelson - Birdtour ASIA
Fig. 4.7g. White-bellied Sea Eagle Habitat. Permission of Mike Nelson - Birdtour ASIA
Fig. 4.8a. Sanford's Sea Eagle Perching. Permission of Keith Cowton
Fig. 4.8b. Sanford's Sea Eagle in Flight. Commercial Licence from PicFair
Fig. 4.8c. Sanford's Sea Eagle in Flight. Nigel Voaden
Fig. 4.8d. Sanford's Sea Eagle Habitat. Permission of Jenny Scott
Fig. 4.9a. Lesser Fish Eagle. Permission of Mike Nelson-Birdtour ASIA
Fig. 4.9b. Lesser Fish Eagle. Permission of Praveen Pandian
Fig. 4.9c. Lesser Fish Eagle. Permission of Paul Prior
Fig. 4.9d. Immature Lesser Fish Eagle. Permission of Ken Behrens
Fig. 4.9e. Immature Lesser Fish Eagle. Permission of Gabriel Leboff
Fig. 4.9f. Lesser Fish Eagle. Permission of Andy Walker
Fig. 4.9g. Lesser Fish Eagle Habitat Malaysia. Permission Steve Douglas - Unsplash
Fig. 4.10a. Grey-headed Fish Eagle Perching. Permission of Vikas Chavan
Fig. 4.10b. Grey-headed Fish Eagle Perching. Permission of Gil Ewing
Fig. 4.10c. Grey-headed Fish Eagle Perching. Permission of Mike Nelson-Birdtour ASIA
Fig. 4.10d. Immature Grey-headed Fish Eagle Perching. Permission of Eddy Lee
Fig. 4.10e. Immature Grey-headed Fish Eagle. Permission of Frank McClintock
Fig. 4.10f. Grey-headed Fish Eagle in Flight. Permission of Mike Nelson-Birdtour ASIA
Fig. 4.10g. Grey-headed Fish Eagle in Flight. Permission of StanleyYOW
Fig. 4.10h. Grey-headed Fish Eagle in Flight. Permission of StanleyYOW
Fig. 4.10i. Immature Grey-headed Fish Eagle in Flight. Frank McClintock
Fig. 4.10j. Grey-headed Fish Eagle Habitat. Permission of Karma Talukdar - Unsplash

Chapter 5

Fig. 5.1a. Guiana Crested Eagle Perching. Permission of Jonathan Beilby
Fig. 5.1b. Guiana Crested Eagle Perching. Permission of Johannes Pfleiderer
Fig. 5.1c. Dark Morph Guiana Crested Eagle Perching. Permission of Paul Prior
Fig. 5.1d. Guiana Crested Eagle on Nest. Permission of Paul Prior
Fig. 5.1e. Immature Guiana Crested Eagle. Permission of C. McNab
Fig. 5.1f. Immature Guiana Crested Eagle on Nest. Permission of David Beadle
Fig. 5.1g. Immature Guiana Crested Eagle. Permission of Nayer Youakim
Fig. 5.1h. Immature Guiana Crested Eagle. Permission of Nayer Youakim
Fig. 5.1i. Immature Guiana Crested Eagle. Permission of Nayer Youakim
Fig. 5.1j. Guiana Crested Eagle Habitat Amazon Forest. Permission of Rômulo Ferreira
Fig. 5.2a. Harpy Eagle Perching. Permission of Chepe Nicoli
Fig. 5.2b. Harpy Eagle Perching. Permission of Anibal Pabon
Fig. 5.2c. Harpy Eagle Perching. Permission of Robert McNab
Fig. 5.2d. Harpy Eagle Collecting Twigs for Nest. Permission of Giselle Mangini
Fig. 5.2e. Harpy Eagle on Nest. Permission of Miguel "Siu" Rios
Fig. 5.2f. Immature Harpy Eagle Perching. Permission of Pia Öberg
Fig. 5.2g. Immature Harpy Eagle. Permission of Nick Athanas
Fig. 5.2h. Immature Harpy eagle Perching. Permission of Nick Athanas
Fig. 5.2i. Immature Harpy Eagle Perching. Permission of Dubi Shapiro

Fig. 5.2j. Immature Harpy Eagle Perching. Permission of Jonathan Beilby
Fig. 5.2k. Harpy Eagle in Habitat. Permission of Tristan
Fig. 5.2l. Harpy Eagle Habitat Amazon Forest. Permission of hanming_huang
Fig. 5.3a. New Guinea Harpy Eagle Perching. Permission of Gail Hampshire
Fig. 5.3b. New Guinea Harpy Eagle Perching. Permission of Johannes Pfleiderer
Fig. 5.3c. New Guinea Harpy Eagle. Permission of Mikael Bauer
Fig. 5.3d. New Guinea Harpy Eagle Habitat New Guinea. Permission of Zacky Ezedin
Fig. 5.4a. Black Solitary Eagle Perching. Permission of Peter Hawrylyshyn
Fig. 5.4b. Black Solitary Eagle Perching. Permission of Alex Pareja
Fig. 5.4c. Black Solitary Eagle. Permission of Frank McClintock
Fig. 5.4d. Black Solitary Eagle. Permission of Giselle Mangini
Fig. 5.4e. Black Solitary Eagle in Flight. Permission of Peter Hawrylyshyn
Fig. 5.4f. Black Solitary Eagle Habitat. Permission of Robert Davidson
Fig. 5.5a. Chaco Eagle Perching. Permission of Linda Bushman
Fig. 5.5b. Chaco Eagle Perching. Permission of Giselle Mangini
Fig. 5.5c. Chaco Eagle Perching. Permission of Jorge Schlemmer
Fig. 5.5d. Chaco Eagle Perching. Permission of Nayer Youakim
Fig. 5.5e. Chaco Eagle Perching. Permission of Nayer Youakim
Fig. 5.5f. Chaco Eagle in Flight. Permission of Jorge Schlemmer
Fig. 5.5g. Chaco Eagle in Flight. Permission of Jorge Schlemmer
Fig. 5.5h. Chaco Eagle Habitat, Paraguay. Permission of Katelyn Barone - Unsplash

Chapter 6

Fig. 6.1. Andean Condor Perching. Permission of Sergey Pisarevskiy
Fig. 6.2. California Condor Perching. Permission of Kojihirano
Fig. 6.3. Turkey Vulture Perching. Permission of Todd Wynia
Fig. 6.4. Black Vulture Perching. Permission of Catherine Zinsky
Fig. 6.5. Cinereous Vulture Perching. Permission of Ali Tellioglu
Fig. 6.6. Lappet-faced Vulture Perching. Permission of Sharon Verkuilen
Fig. 6.7. Griffon Vulture Perching. Permission of Laura Barrio
Fig. 6.8. Griffon Vulture Perching. Permission of Mike Nelson-Birdtour ASIA
Fig. 6.9. Rüppell's Griffon Vulture Perching. Permission of Nirav Shah - Pexels
Fig. 6.10. White Headed Vulture Perching. Permission of Leon Molenaar
Fig. 6.11. Marabou Storks compete with Eagles for Carrion. Permission of Elton Cilliers Picfair
Fig. 6.12. Osprey Perching. Permission of Frank Cone - Pexels
Fig. 6.13. Osprey in Flight Competitors with Bald eagle. Permission of Pixabay Pexels
Fig. 6.14. Black backed Jackal. Permission of Fons Buts
Fig. 6.15. Coyote. Permission of Old Fulica
Fig. 6.16. African Fish Eagle and White Headed Vulture. Permission of Michael Heyns
Fig. 6.17. Eastern Imperial Eagle and Cinereous Vulture. Permission Julian Thomas
Fig. 6.18. Steller's Sea Eagle vs Fox. Permission of Julian Thomas
Fig. 6.19. Tawny Eagle and Vultures. Permission of Roelof van der Breggen
Fig. 6.20. White-bellied Sea Eagle Chasing Eastern Osprey. Permission of Chris Chafer

Chapter 7

Fig. 7.1. Thermal Formation and Eagle Flight.
Fig. 7.2. Wing-loading Calculation. Sketched from photograph by Gustav Van Der Gryp
Fig. 7.3. Orographic Lift.
Fig. 7.4. Main Migration Bottlenecks.
Fig. 7.5. Migration through the Straits of Gibraltar.
Fig. 7.6. Migration from Vancouver Island, British Columbia, Canada
Fig. 7.7. Forest Layers
Fig. 7.8. Deforestation in Central America - Author

Fig. 7.9. Urban Sprawl and Deforestation in Central America - Author
Fig. 7.10. Urban Sprawl and Deforestation by a Central American Lake - Author

Chapter 8

Fig. 8.1. Esquimalt Victoria Bald Eagle Suburban Habitat
Fig. 8.2. Sidney Victoria British Columbia Bald Eagle Suburban Habitat with Comorants and Gulls
Fig. 8.3. Bald Eagle Suburban Habitat East Sooke, Victoria British Columbia
Fig. 8.4. Tofino Area British Columbia - Bald Eagle Habitat
Fig. 8.5. Deforestation in Central America
Fig. 8.6. Vancouver Landfill
Fig. 8.7. Airplane at 600 meters - Author
Fig. 8.8. Airplane at 1500 meters - Author
Fig. 8.9. Windfarm USA. Permission of Nico - Unsplash
Fig. 8.10. Windfarm New Zealand. Permission of Wendy Schotsmans
Fig. 8.11. Dangerous Powerlines - Author
Fig. 8.12. Dangerous Powerlines - Author
Fig. 8.13. Bald Eagle and Power lines. Permission of Sue Dietlein
Fig. 8.14. Immature Bald Eagles and Power lines. Permission of Sue Dietlein
Fig. 8.15. Western Banded Snake Eagle Perching. Permission of Frank McClintock

Chapter 12

Fig. 12.1. Habsburger Double-headed Eagle, Schönbrunn Palace in Vienna Austria. Permission of Arjan Richter
Fig. 12.2. Two Headed Eagle, Austria. Permission of Elrina753
Fig. 12.3. Two Headed Eagle. Permission of Werner Ustorf
Fig. 12.4. Eagle cameo, onyx, Octavian's elevation to Augustos. Andrew McCabe
Fig. 12.5. Philip I the Arab, 244- 249 AD. Permission of Larry Frogel
Fig. 12.6. Cassia Denarius. Romulus Sceptre, Eagle Thunderbolt Jug lituus. Permission of Andrew McCabe
Fig. 12.7. Caesar Denarius. Caesar, Eagle, Standard, Plough, Measuring Rod, Settlement Type. Permission of Andrew McCabe
Fig. 12.8. Galley Eagle Standards Praetorian Cohort type. Permission of Andrew McCabe
Fig. 12.9. Jupiter Eagle Thunderbolt Denarius. Permission of Andrew McCabe
Fig. 12.10. Thrace Koson Brutus Accensus to Gate Figure lictor. Eagle with wreath. Permission of Andrew McCabe
Fig. 12.11. Semilibral Sextans. Collateral series. Permission of Andrew McCabe
Fig. 12.12. Petillia Denarius. Eagle on angled thunderbolt. Permission of Andrew McCabe.
Fig. 12.13. Mars Eagle. Permission of Andrew McCabe
Fig. 12.14. Seleucid King Antiochus VII, 138-129 BC. Permission of Larry Frogel
Fig. 12.15. Roman Republican Coinage. Permission of Larry Frogel
Fig. 12.16. Roman Provincial Coinage Phillip 1 the Arab. Permission of Larry Frogel
Fig. 12.17. Eagle Totem Alaska. Permission of James Knitis
Fig. 12.18. Eagle Totem Alaska. Permission of James Knitis
Fig. 12.19. Eagle Totem Alaska. Permission of James Knitis
Fig. 12.20. Eagle Statue British Columbia - Author
Fig. 12.21. Eagle Totem Alaska. Permission of James Knitis
Fig. 12.22. Bald Eagle Feathers for Native American Warrior. Permission of James Knitis
Fig. 12.23. Falconry using Golden Eagles in Central Asia. Julia Volk Prexels
Fig. 12.24. Falconry using Golden Eagles in Central Asia. Julia Volk Prexels
Fig. 12.25. Crowned Eagle in Modern Falconry - Picfair Licience Purchase
Fig. 12.26. African Fish Eagle and Black-chested Buzzard-eagle at the The Falconry Centre, Hagley UK. Permission of Steve Weaver
Fig. 12.27. Bald Eagles at Hagley Bird of Prey Centre. Permission of Steve Weaver
Fig. 12.28. Golden Eagle at Hagley Bird of Prey Centre. Permission of Steve Weaver
Fig. 12.29. Steller's Sea Eagle at Royal Highland Show 2019. Permission of Christine Sutcliffe
Fig. 12.30. Steller's Sea Eagle at Royal Highland Show 2019. Permission of Christine Sutcliffe
Fig. 12.31. Verreaux's Eagle in Falconry. Permission of Jonathan Beilby

Page 623. Bald Eagle Perching on Totem Pole. Permission of Ignacio Beya Marshal

Contents

Part I

Eagle Classification, Genetics and Ecology

This part comprises four chapters that describe the world's known eagle species, their classification, genetics and ecology. The species of raptors referred to as eagles are classified in this book according to their current genetic classification, with reference to the older classificatory schemes that slowly evolved, in some cases inconclusively, to their current status. Most classifications of eagles acknowledge ancestry or at least shared ancestry with other raptors such as kites (e.g. *Milvus* kites, Lacépède, 1799) and sea eagles (*Haliaeetus*, Savigny, 1809), buzzards (e.g. *Geranoaetus*, Kaup, 1844 and *Harpyhaliaetus,* Lafresnaye, 1842, which some considered buzzards and may or may not be eagles) and Old World *Accipitrid* vultures (e.g. links between the Palm-nut vulture or Vulturine Fish eagle *Gypohierax angolensis* (Gmelin, 1788) and the sea eagles). In fact, some species now classified as eagles were previously classified as buzzards, kites, hawks or vultures, and these variable classifications are still under dispute (De Hoyo et al., 1994; Ferguson-Lees & Christie, 2001; Campbell, 2015). What follows is therefore a background for the later chapters that refer to current knowledge of eagle genetics and ecology.

Before moving to the details of the different species and genera of eagles, it is first necessary to present some preliminary issues. These include the meaning of the terms 'bird of prey', 'eagle' and 'hawk' and the criteria (as will be seen through the book, size is not the only factor for the 'eagle' designation) used to include certain birds of prey in the highly regarded club of predators termed eagles (still referred to sometimes as the kings of birds, and disproportionally present as 'National' birds, heraldic symbols and predatory representations). The terms 'predator' and 'scavenger' must also be examined, as both terms may describe some of the birds defined as eagles. Scavenging could represent a common lifestyle, or scattered events where the eagle species feeds on unattended dead animals, forces other predators off fresh kills or shares carrion with scavengers. Another important term is 'apex predator', which must be examined in terms of its application to the predatory and scavenging roles of eagles, and their relations with competitors. Finally, we need a 'map' of the complex issues considered in this book: a summary of inter-species physical differences, geographical ranges and reasons for inclusion as eagles.

What constitutes a 'Bird of Prey'or a 'raptor' is defined variously: a common definition is that it is "a bird that kills other birds or small animals for food" (Longman Dictionary of Contemporary English, 2016, 163). Of course, "small animals" may also include medium-sized mammals and reptiles, or even fish. Birds of prey may simply be defined as birds that kill and eat animals for their normal food (not on occasion, such as some species like herons, storks or hornbills may do) (Sustaita & Hertel, 2010). In this sense insect eaters, and in some cases fish eaters such as some seabirds (e.g. gulls of the Family *Laridae*, Rafinesque, 1815; and the skuas of the Family *Stercorariidae,* Gray, 1871) may also be included in this classification.

The mere ability to kill prey is an attribute shared by many species outside the Orders *Accipitriformes* (hawks and eagles), *Falconiformes* (falcons) and *Strigiformes* (owls). Other species, such as storks (*Ciconiiformes*), pelicans and herons (*Pelecaniformes*), skuas (*Stercorariidae*), larger corvids such as ravens (*Corvidae*) and even the powerful hornbills (*Bucerotiformes*) are entirely capable of killing quite large animals. These species are generally considered omnivores and sometimes scavengers. Therefore, another narrower definition would be that birds of prey are those birds which are anatomically designed to hunt, pursue and kill prey and have no other food source (e.g. an omnivorous diet), this being synonymous with the term 'raptor.' Habitual birds of prey are generally fully carnivorous, rather than omnivorous, although some such as *Accipitrid* and *Cathartid* vultures may scavenge on a full-time basis. This would include those with sharp hooked beaks and curved talons, capable of killing living animals and tearing them to pieces for food. In this sense, bird species can be divided into three groups: "*Falconiformes*, *Strigiformes* and non-raptorial species" (Csermely & Rossi, 2006, p.43). More recently, the falcons have been classified in the Order *Falconiformes*, with the remaining hawks and eagles are placed in the Order *Accipitriformes*. By this definition and modification, the group of most concern this book are the *Accipitriformes*, especially the Family *Accipitridae*, within which the various species and genera defined as eagles are situated (Campbell, 2015; Winkler et al., 2020).

Birds of prey may therefore be classified as exclusively the diurnal and nocturnal predatory species of the Orders *Falconiformes* or *Accipitriformes* (hawks, eagles, falcons and vultures) and *Strigiformes* (owls). These species are generally characterized by strong hooked beaks and talons, which are used for killing prey or in the case of vultures for ripping open carcasses, and in a few cases for predatory purposes. Brown & Amadon (1968: 17) and some other scholars (e.g. Ferguson-Lees & Christie, 2001) give four main groups of birds of prey: (1) the Accipitridae (Vieillot, 1816), including eagles, hawks, kites, ospreys and Old World *Accipitrid* vultures; (2) New World *Cathartid* vultures (which some authors place separately or near the storks (*Ciconiiformes*, Bonaparte, 1854); (3) the *Sagittariidae* (Finsch & Hartlaub, 1870) or secretary birds; and (4) falcons and caracaras, formerly part of the same order as the hawks and eagles, now with *Falconiformes* (Sharpe, 1874), separate from the *Accipitriformes* (Vieillot, 1816).

The main new development was the Order *Accipitriformes* which absorbed all the species previously in the Order *Falconiformes*,except for the falcons (Voous, 1973; Ferguson-Lees & Christie, 2001; Christidis et al., 2008; Hackett et al., 2008; Remsen, 2008; Nores et al., 2011; Campbell, 2015). However, Encyclopædia Britannica (Gill & Brown, 2020) still classifies the diurnal birds of prey within *Falconiformes*. Owls of the Order *Strigiformes* (Wagler, 1830) may be added to these groups due to their similar predatory (but mostly nocturnal) habits. Ponder and Willette (2015) argue that despite the lack of resolution on the matter, "recent systematics have aligned owls more closely with nightjars than diurnal birds of prey… Using the Sibley-Ahlquist taxonomy, the most recent addendum to the American Ornithologists' Union combines *Caprimulgiformes* with *Strigiformes*" (see also Sibley-Ahlquist Taxonomy, 2013).

Within the *Accipitriformes*, the first group, the *Accipitridae*, is very large and in terms of species classification "seldom have two authorities agreed" (Brown & Amadon, 1968: 17). Recent definitions, based on deoxyribonucleic acid (DNA) analysis have recognized the position of eagles within *Acciptridae* or *Accipitriformes* as complicated, due to disputed classifications (Campbell 2015). One idea is that birds of prey (and by later lineage eagles), are descended from primitive, less predatory species, and that kites, which are among the less predatory birds of prey, are at least partially ancestral to other species (Campbell, 2015). One group of eagles, the Sea or Fish eagles (Genera *Haliaeetus* and *Ichthyophaga*) resembles the kites in life history, some physical attributes, so-called piratical habits and displaying behavior. The Snake eagles (Genera *Circaetus*, *Dryotriorchis*, *Eutriorchis*, *Terathopius*) are also described as primitive and hence may have ancestry close to kites. The Harpy eagles (Genera *Morphus*, *Harpia*, *Harpyopsis*, *Pithecophaga*) are generally considered more specialised and genetically closer to the Genera *Buteo* (buzzards) and as the end point of a line raptorial evolution. Lastly the Booted or True eagles are generally regarded as the most highly evolved members of the related genera. The last group includes the Genera *Ictinetus*, *Aquila*, *Hieraaetus*, *Spizastur*, *Spizaetus, Nisaetus, Stephanoaetus, Oroaetus* and *Polemaetus*). The term "booted" refers to the legs, which are feathered to the toes, giving the impression of trousers, rather than the "shorts" of other eagles (Lerner & Mindell, 2005; Campbell, 2015).

The terms 'eagle' and 'hawk' must be examined in this context. Both belong, depending on the classification, within either the Family *Accipitridae* or the Order *Accipitriformes*. Hawks may include all the *Accipitridae* species or may more narrowly include only the medium-sized and small *Accipiters* (such as goshawks and sparrow hawks) (Campbell & Lack, 2013; Campbell, 2015). Eagles may be described as large hawks, or they may be in distinct groupings. Other species classified as hawks, such as kites, buzzards or the 'true hawks' (*Accipiters*) may be closely or distantly related to specific groups of eagles. Some commentators consider eagles to be any very large to large birds of prey within the *Accipitridae* or *Accipitriformes*, distinguished largely by size from the small to medium-sized birds of these groups, which are referred to as hawks. There appears to be no definitive distinction between eagles and other raptors within *Accipitridae*, except for the general concept of the superior size of eagles, and in some cases larger, stronger bills and talons and larger, more prominent heads. As argued by Lerner and Mindell (2005) the phylogeny or the physiological and evolutionary relations between birds of prey are difficult to determine and in some cases are still undetermined (Campbell, 2015). The genetic diversity of the eagle species is one difficulty, resulting in variations in the relations with other hawk species groups. In peer reviewed, scientific research journals, eagles are not a clearly defined group of closely related large birds, although they are generally defined as larger (with some overlaps) than the true hawks, buzzards and kites.

At the basic, descriptive level, eagles may be distinguished by simple definitions. An eagle can be "any of various large diurnal birds of prey (Family *Accipitridae*) noted for their strength, size, keenness of vision, and powers of flight" (Merraim-Webster, 2020) or a "large, strong bird with a curved beak that eats meat and can see very well" (Cambridge English Dictionary, 2020). Other definitions in common, non-scientific sources are cited below:

"A very large, strong bird with a beak like a hook that eats small animals, birds etc." (Longman's Dictionary of Contemporary English, 2020).

"A large bird that has very good eyesight and that kills other birds and animals for food" (Merriam Webster, 2020).

"A large bird of prey with a massive, hooked bill and long broad wings, known for its keen sight and powerful soaring flight" Family Accipitridae: several genera, in particular Aquila" (Oxford Dictionaries, 2020).

"Any of various species of very large diurnal predatory birds of the family Accipitridae of worldwide distribution, notable for keen sight and strong flight" (What does that mean Dictionary, 2016).

"Eagle is a common name for many large birds of prey of the family Accipitridae. It belongs to several groups of genera that are not necessarily closely related to each other." "Eagles are large, powerfully built birds with a heavy head and beak." "Even the smallest eagles ... have relatively larger and more evenly broad wings and more direct, faster flight..." (Wikipedia, 2020).

Typing the words "eagle" and "definition" into the Google search engine obtains the definition: "a large bird of prey with a massive, hooked bill and very broad wings, renowned for its keen sight and powerful soaring flight." The same definition appears using the Yahoo search engine.

A key issue concerns the status of the genera as either monophyletic groups (which is a taxon or group of organisms that includes an ancestral species and all descendants) or polyphyletic groups (which is group of organisms which is descended from more than one common ancestor or ancestral group; hence the members cannot be included in one taxon). For example, the study by Lerner and Mindell (2005) of the genetic characteristics of the main eagle subfamilies (Booted eagles or *Aquilinae*, Sea eagles or *Haliaeetinae*, Harpy eagles or *Harpiinae* and Snake eagles or *Circaetinae*) found that at least some of the last group were related to some Old World vultures (*Aegypiinae*), one member of the Harpy group was more related to Snake eagles, the Sea eagles were related to the *Milvinae* kites, and that within the group classified as booted eagles, some species are genetically unrelated. This tilts many genera towards polyphyletic status or polyphyly (where the last common ancestor is not a member of the current group) indicating that inclusion of a species is problematic beyond the size requirement, with the additional problem that some eagles are the same size as other birds of prey (Ferguson-Lees & Christie, 2001; Campbell, 2015).

The eagles mentioned above vary greatly in size, as can be seen from the selected species exhibited in Tables P.1 to P.8. Small species include the Little eagle (*Hieraaetus morphnoides*, Gould, 1841) and the unrelated Great Nicobar Serpent eagle (*Spilornisklossi*, Richmond, 1902). Huge species include the unrelated Steller's sea eagle (*Haliaeetus pelagicus*, Pallas, 1811) and harpy eagle (*Harpiaharpyja*, Linnaeus, 1758). Eagle species also hunt different types of prey, regardless of similarities in size. Their hunting specialization may be reflected in their anatomy. Snake hunters, such as the Brown Snake eagle (*Circaetus cinereus*, Vieillot, 1818) may have thickly padded chest feathers for protection from poisonous snakes and heavily scaled legs for protection and for grasping reptilian scales. Fish-eaters such as the Bald eagle (*Haliaeetus leucocephalus*, Linnaeus, 1758) and African Fish eagle (*Haliaeetus vocifer,* Daudin, 1800) have scaly feet for grasping fish and broad wings for lifting heavy fish out of water. These features can be seen in Figures P.1 to P.18.

Eagles that live in thick forest, such as the Harpy eagle and the Crowned eagle (*Stephanoaetus coronatus*, Linnaeus, 1766) have short wings for flying between trees, in contrast to the longer wings of the open country, soaring eagles such as the Golden eagle (*Aquila chrysaetos*, Linnaeus, 1758) and Verreaux's eagle (*A. verreauxii*, Lesson, 1830). Other features vary with little apparent connection with food. For example, some unrelated eagles have very large bills (e.g. the Philippine eagle *Pithecophaga*

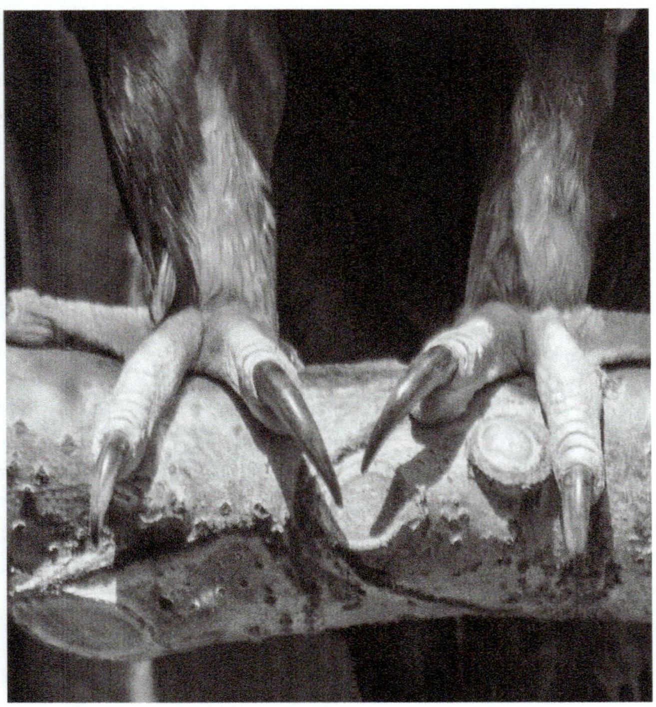

Fig. P.1. Golden eagle talons [With permission from Michelle Jean, BC Wildlife Park]

Fig. P.2. Bald eagle talons [With permission from Michelle Jean, BC Wildlife Park]

Fig. P.3. Golden eagle beak [With permission from Alfred Kenneally-Unsplash]

Fig. P.4. Bald eagle beak [With permission from Flickr]

Fig. P.5. Steller's sea eagle beak [With permission from Dr. Peter Schmidt]

Fig. P.6. Tawny eagle beak [With permission from Dr. Peter Schmidt]

Fig. P.7. Long-crested eagle [With permission from Derek Keats]

Fig. P.8. Black-breasted snake eagle Beak [With permission from Gustav Van Der Gryp]

Fig. P.9. Harpy eagle beak [With permission from Miguel "Siu" Rios]

Fig. P.10. Broad winged hawk beak. This shows a smaller size than that of most eagles [With permission from Miguel "Siu" Rios]

Fig. P.11. African fish eagle beak and talons [With permission from Gustav Van Der Gryp]

Fig. P.12. Broad winged hawk talons. This shows smaller form compared with those of eagles [With permission from Miguel "Siu" Rios]

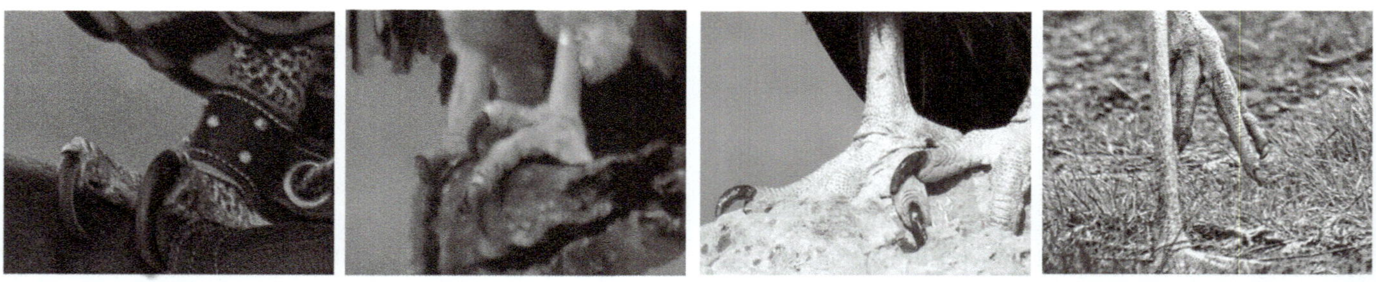

1. Crowned eagle. 2. Cinereous vulture. 3. Andean condor. 4. Marabou stork.

Fig. P.13. Feet of an eagle, Accipitrid vulture, Cathartid vulture and stork
Sources: 1. Elton Cilliers; 2. Frans Van Heerden; 3. Kojihirano; 4. Elton Cilliers.

jefferyi Ogilvie-Grant, 1897 and the Steller's Sea eagle). Other unrelated species, such as the Harpy eagles and some of the snake eagles have similar crested or prominently feathered heads. These variations further complicate the classification of eagles and create disputes among researchers concerning interspecies relations, genera classifications and ancestral links.

The largest eagle species may be considered apex predators. Wallach et al. (2015, 1453) stress that "large 'apex' predators influence ecosystems in profound ways, by limiting the density of their prey and controlling smaller 'meso-predators'." In this sense, eagles may be considered as apex predators for both birds and mammals, as they can kill all flying birds and larger mammals up to the size of wolves and small cats, and medium sized herbivores (Campbell, 2015). Huge eagles such as the Harpy, Philippine, Martial, Golden, Crowned Hawk and Verreaux's eagles may dominate their ecosystems, even competing with predatory mammals such as smaller cats (e.g. lynxes) and canids (e.g. jackals and foxes). Smaller eagle species may be considered meso-predators, or less dominant carnivores, like smaller cats and canids. The differences between apex and meso-predators may be "ambiguous" despite the dominance of the former over the latter (Wallach et al., 2015, 1453).

Fig. P.14. Head and bill profiles of selected eagles and vultures

Sources: 1. Dr. Peter Schmidt; 2. Elton Cilliers; 3. Vikas Chavan; 4. Pieter Verheij; 5. Ali Tellioglu; 6. Leon Molenaar; 7. Kojihirano; 8. Catherine Zinsky.

Fig. P.15. Haast's eagle compared with modern eagles, including its close modern relative the Little eagle.

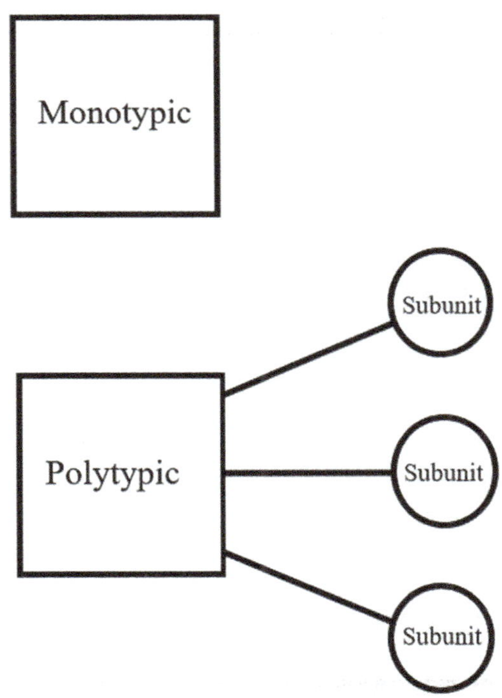

Fig. P.16. Monotypic and polytypic

Eagles may have evolved from many species, and slowly developed their similar sizes and characteristics from their ecological relations and consequent genetic adaptations. Size might be irrelevant, as the huge extinct Haast's eagle (*Hieraaetus moorei*, Haast, 1872) has been found to be related to the much smaller, modern Little eagle. Convergent evolution is an important factor for the development of eagles, as different eagle species have evolved from ancestors of different species of birds of prey, such as kites and buzzards. The convergent evolution of eagles is like that of vultures, as the latter are generally defined as descendant from two species groups: storks or similar species for the New World *Cathartid* vultures and various raptors for the Old World *Accipitrid* vultures. However, all the eagle ancestral groups are generally agreed to be birds of prey, with no links currently discerned with non-raptor species.

There is little doubt that eagles have several ancestral origins among the current and extinct birds of prey. This points to the near certainty of convergent evolution. This may be defined as the evolution of identical features in different species, Genera, Families or Orders. Structures with similar forms or functions may develop among these groups, even if these were absent from the ancestral species or groups. These analogous structures may be distinguished from homologous structures in that the latter may have a common origin with different functions. Convergent evolution is unique in that it is contrasted with divergent evolution (which involves divergent traits emerging from a single species) and the more similar parallel evolution (which refers to parallel change in two similar but independent species). The uniqueness of convergent evolution lies in the common ancestry of the contrasted descendants, compared with the differentiated ancestry of the species in the case of parallel evolution. In practice, these evolutionary processes may be difficult to discern, as the information on the ancestral groups may be incomplete, especially in cases where convergent evolution may occur at different stages in the evolution of species.

For eagles, the convergent evolution largely concerns their size, and to lesser extent the larger bills and talons and longer wings that some researchers claim distinguish eagles from other predatory birds. As such, eagles may all have a common ancestor, in which case they are monophyletic or they could be polyphyletic, which refers to ancestry from more than one common ancestor or ancestral group. In both cases, their defining characteristics could have evolved through the ecological relations of the early Age of Mammals (large prey and competitors). Consequently, any large, predatory bird with large, broad wings for soaring may be called an eagle. Unlike vultures, which have been described as a "textbook example of convergent evolution" (Del Hoyo et al., 1994) due to the plainly differing origins of the New World *Cathartid* vultures (possibly evolved from stork-like ancestors) and the Old World Accipitrid vultures (clearly related to birds of prey), there is no dispute as to the origin of eagles as within the Acciptrid group of birds of prey. Hence, the determination of eagle ancestry may be even more difficult, as narrower differences must be explored (Campbell, 2015).

Lerner and Mindell (2005) give examples of the genetic links between the main eagle Genera and other species of birds of prey, using molecular data. These groups are exhibited in Tables P.9 to P.12. The main eagle groups considered were the Booted eagles (*Aquilinae*), the Sea eagles (*Haliaeetinae*), the Harpy eagles (*Harpiinae*) and Snake eagles (Circaetinae). For the Booted eagles, which vary from most other Accipitrids by their feathered legs or tarsi, the 8 to 9 genera containing 35 to 36 species are variable in size. Five genera are described as monotypic, which refers to the genera containing only one species (as contrasted with polytypic, where the genera have two or more species). This is shown in Figure P.16. The remaining three genera, *Aquila*, *Hieraaetus* and *Spizaetus*, contain most of the species. Scholars have determined that the Booted eagle species are either monophyletic as noted by Jollie (1977b); or polyphyletic as argued by Holdaway (1994). However, these conclusions appear to have been undertaken with morphological data, rather than with molecular data, and there is some confusion on the placement of *Hieraaetus*, *Spizaetus* and *Aquila* species (Lerner & Mindell, 2005). This uncertainty precludes a clear assessment of the ancestry and relations between Booted eagles and other birds of prey (Campbell, 2015; see also Global Raptor Information Network 2021a).

Regarding the Sea eagles or Fish eagles (of the genera *Haliaeetus* and *Ichthyophaga*), a strong morphological link with some kites (Genera *Milvus* and *Haliastur*) and the Palmnut vulture or Vulturine fish eagle (Lerner & Mindell, 2005, see also Brown & Amadon, 1968; Jollie, 1977b; Holdaway, 1994; Thiollay, 1994). This relation with some kites was supported by a study using cytb sequence data by Seibold and Helbig (1996), which found a link between eight of the nine species of sea eagles and two kites of the genus Milvus. However, Lerner & Mindell (2005, 329) argue that "the relationship between the two genera of sea eagles has not been investigated with molecular sequence data and the possibility of paraphyly of the genera remains unresolved." This clearer relation (between the eagles and the non-eagle kite) contrasts with the lack of evidence of non-eagle relations for the Harpy eagles and the Snake eagles.

The few species of Harpy eagles are also considered monophyletic, as they are remarkably similar, with behavioral similarities, unfeathered tarsi and comparatively large size (Lerner & Mindel, 2005; see also Brown & Amadon, 1968; Thiollay, 1994). Few molecular links have been constructed between the Harpy eagles and other birds of prey, especially in terms of ancestry, despite some arguments of links between Harpy and Booted eagles (Lerner & Mindell, 2005; Campbell, 2015). The Snake eagles have been classified as monophyletic by Brown and Amadon (1968) and Friedmann (1950), but some studies have hinted at the polyphyletic ancestry, one reason was the failure to find relations between the Genera *Eutriorchis* and *Dryotriorchis* (Jollie, 1977b; Thiollay, 1994).

The evidence presented indicates the requirement for a comprehensive, gene-based study of eagles and related birds of prey. Similar studies have been undertaken for vultures, and while the conclusions are not conclusive, clearer results have been found that indicate the phylogenetic diversity of a polyphyletic group (Mundy et al., 1992; Sibley & Ahlquist, 1990). These results emerged from studies using biochemical nucleotide sequences of the mitochondrial cytochrome b gene which holds information on the biological relations among animals. Wink (1995, 880), note that "since also the fossil record provides no evidence that the different families within the Falconiformes had a common ancestor, 'raptors' should be a result of an evolutionary convergence between bird groups of polyphyletic origin"; the similarities "found in morphology are indeed based on convergence and not on close genetic relatedness." Studies must therefore use morphological (the study of the appearance and shape of the skeleton and body), karyological (the use of chromosomes) and DNA-based techniques (see also Sibley & Ahlquist, 1990; Mundy et al., 1992; Del Hoyo et al., 1994).

Some of the species listed in the tables below have been reclassified over the years, being moved from one subfamily or genus to another. For example, Bonelli's eagle and the African Hawk-eagle have been moved from *Hieraaetus* to *Aquila* (Helbig et al., 2005; Lerner & Mindell, 2005). The Greater, Lesser and Indian Spotted eagles were moved from their previous genus (*Aquila*) to their own genus (*Clanga*) (Clark, 2012), after being placed with the Long-Crested eagle (*Lophaetus occipitalis*) in the genus *Lophaetus* and speculation that they could be moved to *Ictinaetus* with the Black eagle (*Ictinaetus malaiensis*). The Steppe eagle and Tawny eagle were previously grouped together, now they are separate species, and according to Helbig et al. (2005) they are not even closely related.

Table P.1. Booted eagle species considered in this book (Subfamily *Aquilinae*; Genus *Aquila*)

English name	Scientific name	Mass	Length	Wingspan
True eagles	Genus *Aquila* (Brisson, 1760)	kg	cm	m
Golden eagle	*A. chrysaetos* (Linnaeus, 1758)	2.5–6.4	66–102	1.8- 2.3
Eastern Imperial eagle	*A. heliaca* (Savigny, 1809)	2.5–4.6	72–90	1.8–2.2
Spanish Imperial eagle	*A. adalberti* (C. L. Brehm, 1861)	2.5–3.5	78–82	1.8–2.1
Steppe eagle	*A. nipalensis* (Hodgson, 1833)	2.3–4.9	62–81	1.65–2.2
Tawny eagle	*A. rapax* (Temminck, 1828)	1.6–3	60–75	1.6–1.9
Verreaux's eagle	*A. verreauxii* (Lesson, 1830)	3.1–7	75–96	1.8- 2.3
Gurney's eagle	*A. gurneyi* (Gray, 1860)	3	74–86	1.7- 1.9
Bonelli's eagle	*A. fasciata* (Vieillot, 1822)	1.6–2.4	55–65	1.5
Cassin's Hawk-eagle	*A. africana* (Cassin, 1865)	0.9–1.2	50	1.2
African Hawk-eagle	*A. spilogaster* (Bonaparte, 1850)	1.2–1.6	55–65	1.30-1.6
Wedge-tailed eagle	*A. audax* (Latham, 1801)	4	1.06	2.8

Sources: Del Hoyo et al. (1994), Ferguson-Lees & Christie (2001), Lerner & Mindell (2005), GRIN.

Some scholars have argued for combining *Hieraaetus* with *Aquila*, but genetic evidence has allowed them to stay separate (Clark, 2012). Wahlberg's eagle was formerly placed in the genus *Aquila* (Stresemann & Amadon, 1979, Sibley & Monroe, 1990, Kemp, 1994a) and then *Hieraaetus* (Amadon & Bull, 1988, Ferguson-Lees & Christie, 2001). Several researchers recommend its home as in *Hieraaetus* (Helbig et al., 2005; Lerner & Mindell, 2005; Clark, 2012). Cassin's hawk-eagle was moved from the genus *Cassinaetus* (Cassin, 1865), and then into *Hieraaetus* (Kemp, 1994b), and *Spizaetus* (Voous, 1964, Amadon & Bull, 1988). Currently, the strongest arguments are for its placement in *Aquila* (Lerner & Mindell, 2005; Haring et al., 2007).

Some eagle species hover between the status of eagles and that of hawks or buzzards. For example, the genus *Harpyhaliaetus*, which contained the Crowned solitary eagle or Chaco eagle (*Harpyhaliaetus coronatus*, Lafresnaye, 1842) and the Solitary eagle (*Harpyhaliaetus solitarius*, Lafresnaye, 1842) has ceased to exist in recent classifications. Initially, the Solitary eagle and Crowned solitary eagle were considered as one species (Hellmayr & Conover, 1949), but later studies found adequate grounds to separate them in *Harpyhaliaetus* (Wetmore, 1965). Although initially *Harpyhaliaetus* was linked to the Harpy eagles (Brown, 1970), molecular studies discovered closer affinities with *Buteogallus* (especially the Great Black-hawk, *Buteogallusurubitinga*); consequently, both species are now in the genus *Buteogallus*, which mostly includes species considered to be hawks (Lerner & Mindell, 2005; Amaral et al., 2006). In another case, the Black-Chested Buzzard-eagle (*Geranoaetus melanoleucus* Vieillot, 1819) was finally not considered to be an eagle but was determined to be a buzzard (Riesing et al., 2003; Clark & Schmitt, 2006). The Black-and-White Hawk-eagle had its scientific name changed from *Spizastur melanoleucus* to

Table P.2. Booted eagle species considered in this book (Subfamily *Aquilinae*; Genera *Hieraaetus, Spizaetus*)

English name	Scientific name	Mass	Length	Wingspan
Booted eagles	Genus *Hieraaetus* (Kaup, 1844)	*kg*	*cm*	*m*
Booted eagle	*H. pennatus* (Gmelin, 1788)	6.4–1.1	42–50	1.1–1.3
Wahlberg's eagle	*H. wahlbergi* (Sundevall, 1851)	4.4–14	53–61	1.3–1.5
Ayres's Hawk-eagle	*H. ayresii* (Gurney, 1862)	0.7–1.1	44–57	1.1–1.4
Little eagle	*H. morphnoides* (Gould, 1841)	0.6–0.8	45–55	1.2–1.3
Pygmy eagle	*H. weiskei* (Reichenow, 1900)	0.5–0.8	38–48	1.1–1.3
Spizaetus eagles	Genus *Spizaetus* (Vieillot, 1816)			
Black-and-chestnut eagle	*S. isidori* (Des Murs, 1845)	-	63–74	147–170
Black Hawk-eagle	*S. tyrannus* (Wied, 1820)	0.9–1.3	58–70	135–145
Ornate Hawk-eagle	*S. ornatus* (Daudin, 1800)	1.0–1.7	56–68.5	117–142
Black-and-white Hawk- eagle	*S. melanoleucus* (Vieillot, 1816)	0.9	50–60	109–135

Sources: Del Hoyo et al. (1994), Ferguson-Lees & Christie (2001), Lerner & Mindell (2005), GRIN.

Table P.3. Booted eagle species considered in this book (Subfamily *Aquilinae*; Genera, *Nisaetus*)

English name	Scientific name	Mass	Length	Wingspan
Booted eagles	Genus *Nisaetus* (Hodgson, 1836)	kg	cm	cm
Mountain Hawk-eagle	*N. nipalensis* (Hodgson, 1836)	1.8–3.6	69–84	134–175
Changeable Hawk-eagle	*N. cirrhatus* (Gmelin, 1788)	1.2 to 1.9	60–72	127–138
Flores Hawk-eagle	*N. floris* (Hartert, 1898)	1.3 to 2.1	75–79	127–138
Sulawesi Hawk-eagle	*N. lanceolatus* (Temminck/Schlegel, 1844)	1.1–1.2	55–64	110–135
Legge's Hawk-eagle	*N. kelaarti* (Legge, 1878)	1.8–3.6	67–86	134–175
Blyth's Hawk-eagle	*N. alboniger* (Blyth, 1845)	0.83	50–58	100–115
Javan Hawk-eagle	*N. bartelsi* (Stresemann, 1924)	<1.5	56–70	110–130
Philippine Hawk-eagle	*N. philippensis* (Gould, 1863)	1.1–1.2	64–69	
Pinsker's Hawk-eagle	*N. pinskeri* (Preleuthner & Gamauf, 1998)	1.1–1.2	55–64	110–135
Wallace's Hawk-eagle	*N. nanus* (Wallace, 1868)	0.5–0.61	45–59	95–105

Sources: Del Hoyo et al. (1994), Ferguson-Lees & Christie (2001), Lerner & Mindell (2005), GRIN.

Table P.4. Booted eagle species considered in this book (Subfamily *Aquilinae*, Genera *Ictinaetus*, *Lophoaetus*, *Clanga Polemaetus*, and *Stephanoaetus*)

English name	Scientific name	Mass	Length	Wingspan
Booted eagles	Genus *Ictinaetus* (Blyth, 1843)	*kg*	*cm*	*cm*
Black eagle	*I.malaiensis* (Temminck, 1822)	1.0–1.6	70–80	164–178
	Genus *Clanga* (Adamowicz, 1858)			
Greater Spotted eagle	*C. clanga* (Pallas, 1811)	1.6–2.5	59–71	157–179
Lesser Spotted eagle	*C. pomarina* (Brehm, 1831)	1.0–2.2	61–66	146–168
Indian Spotted eagle	*C. hastata* (Lesson, 1834)	1.4–1.8	60	150
	Genus *Lophaetus*			
Long-crested eagle	*Lophaetus occipitalis* (Daudin, 1800)	0.9–1.5	53–58	112–129
	Genus *Polemaetus* (Heine, 1890)			
Martial eagle	*P. bellicosus* (Daudin, 1800)	2.2–6.5	78–96	1.9 to 2.6
	Genus *Stephanoaetus* (W.L. Sclater, 1922)			
Crowned eagle	*S. coronatus* (Linnaeus, 1766)	2.5–4.7	80–99	1.5 to 1.9

Sources: Del Hoyo et al. (1994), Ferguson-Lees & Christie (2001), Lerner & Mindell (2005).

Table P.5. Sea eagle species considered in this book (Genera *Haliaeetus,* former *Icthyophaga*)

English name	Scientific name	Mass	Length	Wingspan
Sea eagles	Genus *Haliaeetus* (Savigny, 1809)	*kg*	*cm*	*m*
White-tailed eagle	*H. albicilla* (Linnaeus, 1758)	3.1–6.9	66–94	1.8–2.5
Steller's Sea eagle	*H. pelagicus* (Pallas, 1811)	4.9–9.5	85–105	2–2.5
Bald eagle	*H. leucocephalus* (L., 1766)	3–6.3	70–102	1.8–2.3
Pallas' Fish eagle	*H. leucoryphus* (Pallas, 1771)	2–7.3	72–84	1.8–2.15
White-bellied Sea eagle	*H. leucogaster* (Gmelin, 1788)	1.8–4.5	66–90	1.8–2.2
Sanford's Sea eagle	*H.sanfordi* (Mayr, 1935)	1.1–2.7	70–90	1.7–1.85
African Fish eagle	*H. vocifer* (Daudin, 1800)	3.2–3.6	63–75	2
Madagascan Fish eagle	*H. vociferoides* (Des Murs, 1845)	2.2–3.5	60–66	1.7–1.8
Fish eagles	Genus *Icthyophaga* (Lesson, 1843)			
Lesser Fish eagle	*I. humilis* (Müller, Schlegel, 1841)	0.78–0.8	50–70	1.2–1.6
Grey-headed Fish eagle	*I. ichthyaetus* (Horsfield, 1821)	1.6–2.7	61–75	1.55–1.7

Sources: Del Hoyo et al. (1994), Ferguson-Lees & Christie (2001), Lerner & Mindell (2005). The two last species of the Genus *Icthyophaga* are now included in *Haliaeetus* as *Haliaeetus ichthyaetus* (Horsfield, 1821) and *Haliaeetus humilis* (Müller & Schlegel, 1841), with the same English names.

Table P.6. Snake Eagle Species Considered in this Book (Subfamily *Circaetinae*; Genus *Circaetus*).

English name	Scientific name	Mass	Length	Wingspan
Snake eagles	Genus *Circaetus* (Vieillot, 1816)	*kg*	*cm*	*cm*
Short-toed Snake eagle	*C. gallicus* (Gmelin, 1788)	1.2–2.3	62–67	170–185
Black-chested S. eagle	*C. pectoralis* (A. Smith, 1829)	1.2–2.3	63–68	170–185
Beaudouin's Snake eagle	*C. beaudouini* (Verreaux/Des Murs, 1862)		60–66	155–170
Brown Snake eagle	*C. cinereus* (Vieillot, 1818)	1.5–2.5	68–78	160–185
Southern Banded S. eagle	*C. fasciolatus* (Kaup, 1850)	1–1.1	55–60	119–128
Western Banded S. eagle	*C. cinerascens* (von Müller, 1851)	1.1	55–60	114
Congo Serpent eagle	*C. spectabilis* (Schlegel, 1863)		54–60	94–106
	Genus *Terathopius* (Lesson, 1830)			
Bateleur eagle	*T. ecaudatus* (Daudin, 1800)	2–2.6	55–70	186

Sources: Del Hoyo et al. (1994), Ferguson-Lees & Christie (2001), Lerner & Mindell (2005), Global Raptor Information Network (2021a). Note that the abbreviation S. eagle refers to Snake eagle.

Table P.7. Serpent eagle species considered in this book (Subfamily *Circaetinae*; Genus *Spilornis*)

English name	Scientific name	Mass	Length	Wingspan
Serpent eagles	Genus *Spilornis* (Gray, 1840)	*kg*	*cm*	*cm*
Andaman Serpent eagle	*S. elgini* (Blyth, 1863)	0.8–1.0	51–59	115–135
Great Nicobar S. eagle	*S. klossi* (Richmond, 1902)	0.5	38–42	85–95
Sulawesi serpent eagle	*S. rufipectus* (Gould, 1858)		41–54	105–120
Crested Serpent eagle	*S. cheela* (Latham, 1790)	0.4–1.8	50–74	123-155
Philippine Serpent eagle	*S. holospilus* (Vigors, 1831)	0.6–1.6	47–53	105–120
Mountain Serpent eagle	*S. kinabaluensis* (W. L. Sclater, 1919)		51–56	118–129
	Genus: *Eutriorchis* (Sharpe, 1875)			
Madagascan S. eagle	*E. astur* (Sharpe, 1875)	0.7–0.8	57–66	90–110
	Genus: *Pithecophaga* (Ogilvie-Grant, 1896)			
Philippine eagle	*P. jefferyi* (Ogilvie-Grant, 1897)	4.7–8.0	86–105	1.8–2.2

Note that the Central Nicobar serpent eagle, *S. minimus*, may be a subspecies of the Crested serpent eagle or a full-fledged species in its own right. Note that the abbreviation S. eagle refers to Serpent eagle. Sources: Del Hoyo et al. (1994), Ferguson-Lees & Christie (2001), Lerner & Mindell (2005), Global Raptor Information Network (2021a).

Table P.8. Harpy eagle species considered in this book (Subfamily *Harpiinae*: Genera *Harpia*, *Morphnus*, *Harpyopsis*, *Pithecophaga*, *Buteogallus* (formerly *Harpyhaliaetus*)

English name	Scientific name	Mass	Length	Wingspan
Harpy eagles	Genus *Harpia* (Vieillot, 1816)	*kg*	*cm*	*cm*
Harpy eagle	*H. harpyja* (Linnaeus, 1758)	4–10	86.5–107	1.8–2.2
	Genus *Morphnus* (Dumont, 1816)			
Crested eagle	*M. guianensis* (Daudin, 1800)	1.3–3.0	71–89	1.4–1.8
	Genus: *Harpyopsis* (Salvadori, 1875)			
Papuan eagle	*H. novaeguineae* (Salvadori, 1875)	1.6–2.4	75–90	1.6
	Genus: *Buteogallus* (Lesson, 1830)			
Crowned Sol. eagle	*B. coronatus* (Vieillot, 1817)	3.0	73–79	1.7–1.8
Black Sol. eagle	*B. solitarius* (Tschudi, 1844)	3.0	63–76	1.5–1.9

Sources: Del Hoyo et al. (1994), Ferguson-Lees & Christie (2001), Lerner & Mindell (2005), Global Raptor Information Network (2021a). Note: Sol. eagle refers to Solitary eagle.

Spizaetus melanoleucus. Initially, it was described as the buzzard *Buteo melanoleucus* (Vieillot, 1819) (Bierregaard et al., 2018). The name *Spizaetus melanoleucus* originally given by Louis Jean Pierre Vieillot to the Black-Chested Buzzard-eagle in 1819, and he named the Black-and-White hawk-eagle *Buteo melanoleucus* in 1816 (Bierregaard et al., 1994a,b, 2018).

Other species have had their names or classifications changed, remaining eagles in the process. For example, the Black-and-Chestnut eagle *(Spizaetus isidori)* or Isidor's eagle is sometimes placed in the monotypic (single species) genus *Oroaetus* (Bierregaard et al., 2014; BirdLife International, 2016). The species of the Genus *Nisaetus* were formerly included in *Spizaetus* (Gamauf et al., 2005). These authors suggested assigning the species of the genus *Spizaetus* to three different genera: (1) Spizaetus… in Central and South America and (2) *Nisaetus* for the Southeast to East Asian group (Gamauf et al., 2005); and (3) the African species, Cassin's hawk-eagle (*Spizaetus africanus*, later *Hieraaetus africanus*), which was moved to the genus *Aquila* (Gamauf et al., 2005). The Flores hawk-eagle (*Nisaetus floris*) was previously a subspecies (*Nisaetus cirrhatus floris*) of the Changeable or Crested hawk-eagle (*Nisaetus cirrhatus*) (Gjershaug et al., 2004; Helbig et al., 2005). Legge's Hawk-eagle (*Nisaetus kelaarti*) was previously considered as subspecies of the Mountain hawk-eagle (*N. nipalensis*) (Gjershaug et al., 2008).

Pinsker's Hawk eagle (*Nisaetus pinskeri*) was considered as subspecies of the Philippine Hawk-eagle (*N. philippensis*) as *Spizaetus philippensis pinskeri*, but molecular studies distinguished them (Gamauf et al., 2005a, 2005b; Lerner & Mindell, 2005). The Pygmy eagle is considered either *Aquila weiskei* or *Hieraaetus weiskei*, and was once a subspecies of the Little eagle, *H. morphnoides* (Lerner & Mindell, 2005). Bonelli's eagle (*Aquila fasciata*) and the African hawk-eagle (*Aquila spilogaster*) were both formerly included in the Genus *Hieraaetus* (Helbig et al., 2005). The Greater Spotted eagle (*Clangaclanga*), Lesser Spotted eagle (*Clangapomarina*) and Indian Spotted eagle (*Clangahastata*) were all placed in the Genus *Clanga*, from *Aquila* (Global Raptor Information Network, 2021a).

The Central Nicobar serpent eagle (*S. minimus*) was previously considered a subspecies of the Crested serpent eagle, (*Spilornischeela*) by several authors (Stresemann & Amadon, 1979; Amadon & Bull, 1988; Sibley & Monroe, 1990; del Hoyo et al., 1994). However, later researchers gave it full species status (Ferguson-Lees & Christie, 2001; Dickinson, 2003; Rasmussen & Anderton, 2012). Other subspecies treated as full species include Simeulue Serpent Eagle (*Spilornis abbotti*), Nias Serpent Eagle (*Spilornis asturinus*), Bawean Serpent Eagle (*Spilornis baweanus*), Natuna Serpent Eagle (*Spilornis natunensis*), Ryukyu Serpent Eagle (*Spilornis perplexus*) and Mentawai Serpent Eagle (*Spilornis sipora*) which are treated as subspecies of the Crested Serpent eagle by most authors (e.g. Del Hoyo, 1994), but as separate species by Ferguson-Lees and Christie (2001). Eagle species may have several English names, in addition to the several scientific names list above. Even after such movements, some retain multiple names. Tables P.9 to P.12 illustrate these issues.

Eagles may also be considered scavengers, in addition to their status as predators. In this regard, they may be *facultative* (part time) scavengers rather than full time, specialized *obligate* scavengers, such as vultures (Campbell, 2015). In this scavenging role, all scavengers perform an important role, removing carrion, preventing diseases and for the avian actors, leading ground scavengers to food sources. In this regard, to define scavenging, Getz (2011) gives an incisive description: "Scavenging is both a carnivorous and herbivorous feeding behavior in which the scavenger feeds on dead animal and plant material present in its habitat... Scavengers play an important role in the ecosystem by consuming the dead animal and plant material."

Table P.9. The multitude of eagle names (True eagles and Sea eagles)

Main English name	**Other English names**
True eagles	
Golden eagle.	American golden eagle
Eastern imperial eagle	Asian imperial eagle, imperial eagle.
Spanish imperial eagle	Adalbert's eagle, Iberian imperial eagle, imperial eagle
Steppe eagle	Oriental steppe eagle
Tawny eagle	Eurasian tawny eagle, Indian tawny eagle.
Verreaux's eagle	African black eagle, black eagle
Gurney's eagle	None
Bonelli's eagle	Bonelli's hawk eagle
Cassin's hawk-eagle	None
African hawk-eagle	African eagle
Wedge-tailed eagle	Eaglehawk, mountain eagle
Sea eagles	
White-tailed eagle	Erne, Eurasian sea eagle, sea eagle, white-tailed fish-eagle, white-tailed sea-eagle
Steller's sea eagle	Pacific sea eagle, Steller's eagle, Steller's fish eagle, white-shouldered eagle, white-shouldered sea eagle
Bald eagle	American eagle
Pallas' fish eagle	Band-tailed fishing eagle, long-tailed eagle, Pallas's eagle, Pallas's sea-eagle, ring-tailed fishing eagle
White-bellied sea eagle.	Gray-backed sea eagle, white-bellied fish-eagle, white-breasted fish eagle, white-breasted fish hawk, white-breasted sea eagle.
Sanford's sea eagle.	Brown fish eagle, brown sea eagle, forest fish eagle, sanford's fish eagle, solomon fish eagle, solomon sea-eagle
African fish eagle.	African fish-eagle, fish eagle, river eagle, West African river eagle
Madagascan fish eagle.	Madagascar fish-eagle, Madagascar sea-eagle
Lesser fish eagle.	Himalayan gray (or grey)-headed fishing eagle, lesser fishing eagle.
Grey-headed fish eagle.	Ceylon grey-headed fishing eagle, gray-headed fishing-eagle, gray-headed fishing eagle, greater fishing eagle, grey-headed fish-eagle, grey-headed fishing-eagle, tank eagle

Source: Global Raptor Information Network (2021a).

Eagles may be more similar in this role to the Old World *Accipitrid* vultures than to the New World *Cathartid* vultures, as the *Accipitrid* vultures have more powerful eagle-like bills, feet and wings, which are less pronounced in the *Cathartid* vultures. Eagles may compete with vultures and mammals such as canids, felids and hyenas in scavenging, whether the food sources were procured by the eagle or the other species. Eagles are strong competitors for scavengers, due to their strong bills and claws, and they may outmatch at least similarly sized vultures or even mammals as large as jackals or wolves. Eagles also match the soaring and carrion detection ability of vultures, especially as most vultures, like eagles (except for the *Cathartid* Turkey vulture) lack a sense of smell. In feeding on carrion, eagles are nonetheless at a disadvantage, as unlike vultures they lack bare skin or down-covered heads and sometimes necks, which facilitates the thrusting of heads and necks into meat, while avoiding matted blood covered feathers. The eagles also lack feet adapted to walking on the ground. The sizes of the larger *Cathartid* and *Accipitrid* vultures are given below, as these may be variably serious competitors with eagles as scavengers.

Eagles nevertheless dominate vultures at food sources, even when the latter are larger, mostly due to their more aggressive behavior and stronger talons and feet. This dominance is especially strong over the *Cathartid* vultures, which have weaker feet than those of the more eagle-like *Accipitrid* vultures. For example, Golden eagles dominate even the huge Andean and Californian condors, much more than they and other large eagles dominate the condor sized *Accipitrid* vultures, such as the Griffon, Lappet-faced and Cinereous vultures. The table below shows the sizes of the larger *Accipitrid* and *Cathartid* vultures, which may be compared with the sizes of the eagles in the tables above. These are the world's largest vultures, the species that might rival eagles in scavenging activities. The feet of large eagles are also compared to those of the *Accipitrid* and *Cathartid* vultures in Figure P.13. Those of eagles are principally for grasping and fighting, while those of the *Cathartid* vultures are principally for walking, almost like those of the carrion eating Marabou stork (*Leptoptilos crumenifer,* Lesson, 1831). The feet of the *Accipitrid*

Table P.10a. The multitude of eagle names (other Booted eagles)

Main English name	Other English names
Other Booted eagles	
Booted eagle	Booted hawk-eagle
Wahlberg's eagle	None
Ayres's hawk-eagle	Ayres's eagle
Little eagle	Australian little eagle
Pygmy eagle	New Guinea hawk-eagle
Black-and-chestnut eagle	Isidor's eagle, Isidor's crested eagle
Black hawk-eagle	Black hawk-eagle, tyrant hawk-eagle.
Ornate hawk-eagle	Barred hawk-eagle, manduyt's hawk-eagle,
Black-and-white hawk-eagle	Black and white hawk
Mountain hawk-eagle	Feather-toed hawk-eagle, Hodgson's hawk-eagle, Legge's hawk-eagle
Changeable hawk-eagle	Crested hawk-eagle, Indian crested hawk-eagle, marsh hawk-eagle
Flores hawk-eagle	Sunda hawk-eagle
Sulawesi hawk-eagle	Celebes hawk-eagle
Legge's hawk-eagle	Beautiful-crested eagle, Ceylon mountain hawk-eagle, feather-toed hawk-eagle

Source: Global Raptor Information Network (2021a).

Table P.10b. The multitude of eagle names (other Booted eagles)

Blyth's hawk-eagle	None
Javan hawk-eagle	Java hawk-eagle
Philippine hawk-eagle	Northern Philippine hawk-eagle
Pinsker's hawk-eagle	Philippine hawk-eagle, Southern Philippine hawk-eagle
Wallace's hawk-eagle	Small hawk-eagle
Black eagle	Asian black eagle, Indian black eagle
Greater spotted eagle	Large spotted eagle, spotted eagle
Lesser spotted eagle	None
Indian spotted eagle	Long-legged eagle, small Indian spotted eagle
Long-crested eagle	Long-crested hawk-eagle
Martial eagle	Martial hawk-eagle
Crowned eagle	African crowned eagle, crowned hawk-eagle

Table P.11. The multitude of eagle names (Snake eagles)

Main English name	Other English names
Snake eagles	
Short-toed snake eagle	European Snake-eagle, European Snake Eagle, Short-toed serpent-eagle
Black-chested snake eagle	Black-breasted Harrier eagle, Black-chested Harrier-eagle, Black-breasted Snake-eagle
Beaudouin's snake eagle	Beaudouin's Harrier-eagle
Brown snake eagle	Brown Harrier-eagle
Southern banded snake eagle	East African Banded Snake-eagle, East African Snake-eagle, Fasciated Snake-eagle, Fasciolated Snake-eagle
Western banded snake eagle	Banded Harrier-eagle, Banded Snake-eagle, Smaller Banded Snake-eagle, Smaller Banded Snake eagle
Congo serpent eagle	African serpent-eagle, West African serpent eagle.
Bateleur	

Source: Global Raptor Information Network (2021a).

Table P.12. The multitude of eagle names (Serpent and Harpy eagles)

Main English name	Other English names
Serpent eagles	
Andaman serpent eagle	Andaman Dark serpent-eagle
Great Nicobar serpent eagle	Great Nicobar serpent-eagle, Small serpent-eagle, South Nicobar serpent-eagle, South Nicobar serpent eagle, Southern Nicobar serpent-eagle
Sulawesi serpent eagle	Celebes serpent-eagle
Crested serpent eagle	Ceylon serpent eagle
Philippine serpent eagle	None recorded
Mountain serpent eagle	Kimabalu serpent-eagle
Madagascan serpent eagle	Long-tailed serpent-eagle, Madagascar Forest eagle
Harpy eagles	
Harpy eagle	None recorded
Crested eagle	Guiana Crested eagle, Long-tailed Harpy eagle.
Papuan eagle	New Guinea Harpy eagle, Kapul eagle, New Guinea eagle
Philippine eagle	Great Monkey-eating eagle, Great Philippine eagle, Monkey-eating eagle, Philippine Monkey-eating eagle
Crowned solitary eagle	Crowned eagle
Black Solitary eagle	Montane Solitary eagle, Solitary eagle

Source: Global Raptor Information Network (2021a).

Table P.13. The largest *Cathartid* and *Accipitrid* vultures

English name	Scientific name	Mass	Length	Wingspan
		kg	*cm*	*cm*
Andean condor	*Vulturgryphus* (Linnaeus, 1758)	8–15	100–130	2.7–3.2
California condor	*Gymnogyps californianus* (Shaw, 1797)	7–14.1	109–140	2.5–3
Cinereous vulture	*Aegypiusmonachus* (Linnaeus, 1766)	6.3–14	98–120	2.5–3.1
Himalayan griffon	*Gyps himalayensis* (Hume, 1869)	8–12	109–115	2.6–3.1
Lappet-faced vulture	*Torgostracheliotos* (Forster, 1791)	4.4–13.6	78–115	2.5–2.9
Griffon vulture	*Gyps fulvus* (Hablizl, 1783)	6.2–11.3	93–122	2.3–2.8
Cape vulture	*Gyps coprotheres* (Forster, 1798)	7–11	96–115	2.3–2.6
Rüppell's vulture	*Gyps rueppelli* (Alfred Brehm, 1852)	6.4–9	85–103	2.3–2.6

Source: Campbell (2015).

vultures are in between these two extremes. The bills of eagles and vultures are also compared in the figures above. The eagle example (Golden eagle) has the most strongly hooked bill, followed by the *Accipitrid* vulture (the Griffon vulture) and then lastly the California condor, which has the weakest bill (Fig. P.14).

Eagles are therefore exceptional animals, with multiple roles in nature and in human perceptions. However, any examination of eagles' definitions, forms and ecology must confront disputes as to which species of large birds of prey are indisputably eagles. One way to ameliorate this issue is to start with the so called 'true' eagles. This book adopts this method. Part 1 describes the different eagle species, using their feather coloration and body size and morphology for differentiation. The description is based on the feather patterns shown in Figures P.17 to P.19 below. Chapter 1 examines the true or booted eagles of the genus *Aquila*, this being followed by the description of related species and including a cursory examination of current classification schemes in succeeding chapters. Following the description of these species, Part 2 will examine their physiology and ecology more closely and assess the justification for some species' status as true eagles, and their competition with other birds and mammals. Part 3 will look at the relations between these eagles and people.

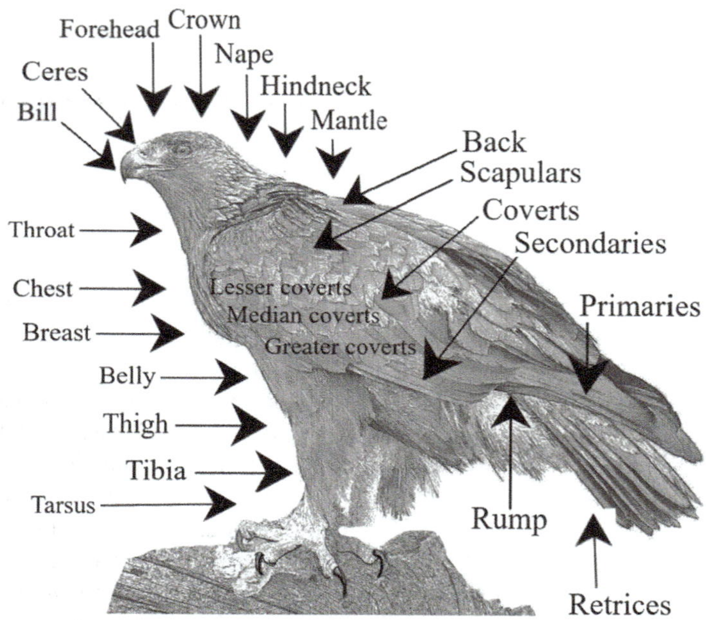

Fig. P.17. Head and back feathers of a Golden eagle [Reproduced from a photograph, with permission of Chas Moonie]

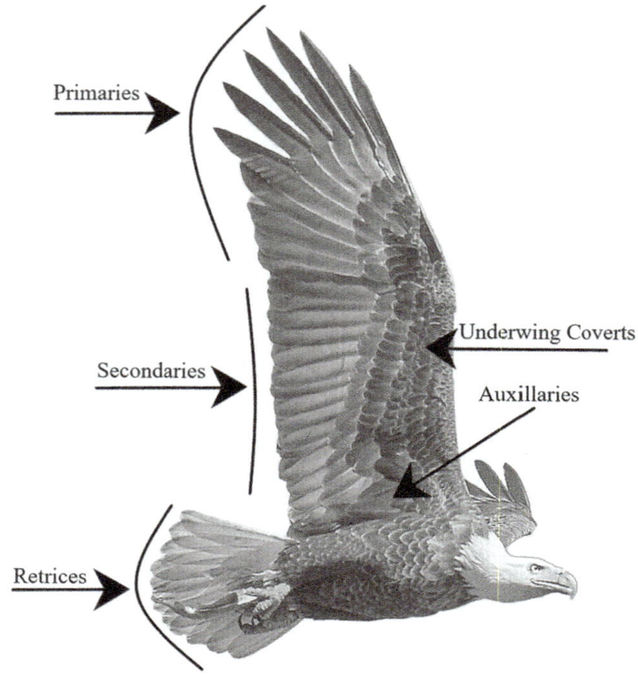

Fig. P.18. Flight feathers of Bald eagle in flight [Reproduced from a photograph, with permission of Mathew Schwartz]

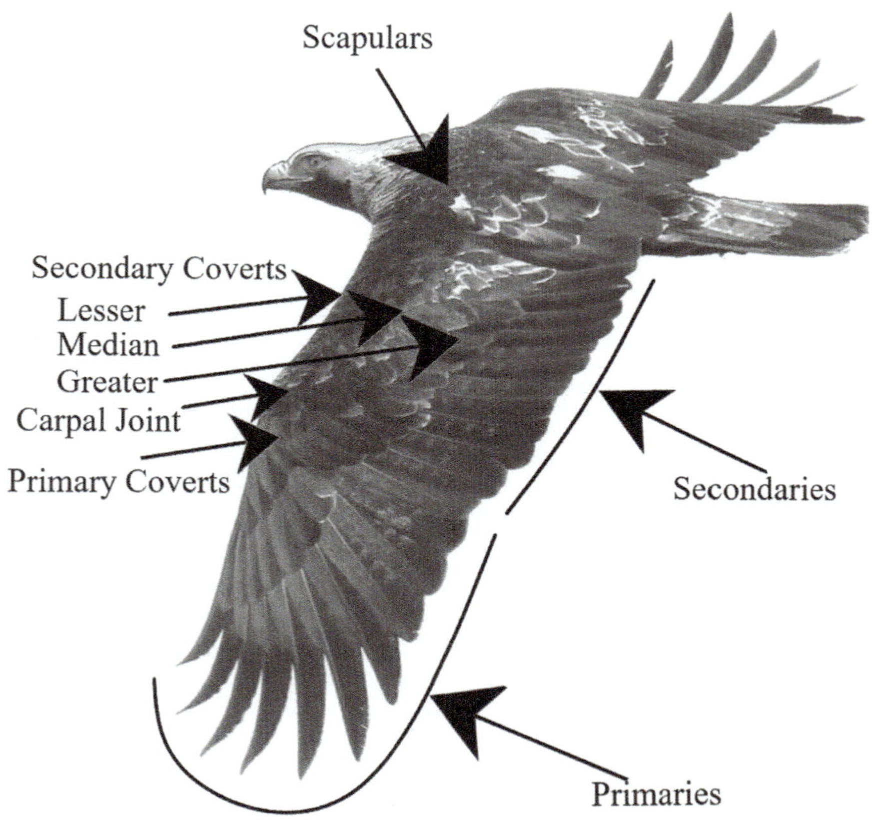

Fig. P.19. Flight feathers of the Eastern Imperial Eagle in flight [With permission from Radovan Václav]

1

Systematic List of the "True Eagles" (Subfamily *Aquilinae*; Genus *Aquila*)

1.1 Introduction

This chapter examines the forms, foraging and feeding habits, breeding patterns and statuses of the "Booted" or "True" eagles, within the subfamily *Aquilinae*. The Booted eagles are classified in the Genera *Aquila, Hieraaetus, Spizaetus, Nisaetus, Lophaetus, Clanga, Ictinaetus, Stephanoaetus, Polemaetus* and *Lophotriorchis*. These species are associated in classifications partly due to the legs being feathered to the toes, the root of the name of the Genus 'Aquila' (Brown & Amadon, 1968; Ferguson-Lees & Christie, 2001). They comprise a monophyletic group, where a monophyletic group is a group of organisms forming a clade, consisting of all the descendants from a common ancestor. According to Lerner et al. (2017: 301) there are 38 species within the *Aquilinae* or Booted eagles, these mostly within five main clades: "*Nisaetus* (10 species), *Spizaetus* (4 species), *Clanga* (3 species), *Hieraaetus* (6 species) and *Aquila* (11 species)." The Booted eagles have a wide distribution, being found on every continent except Antarctica, with the largest number of species occurring in Africa (Ferguson-Lees & Christie, 2001; Lerner et al., 2017). Throughout most parts of their ranges, the Booted eagles are a primary or dominant group of avian predators, at least as evident as other eagle species such as the harpy eagles (*Harpiinae*), sea eagles (*Haliaeetinae*), solitary eagles (*Buteoninae,* which includes the buzzard hawks *Buteo*, Lacépède, 1799) and snake eagles (*Circaetinae)* (Lerner et al., 2017).

The eagles of the Genus *Aquila* were the first to be considered eagles. The taxonomy of the Genus *Aquila* was developed by the French zoologist Mathurin Jacques Brisson in 1760. The golden eagle was floated as the type species and its close relatives were similarly designated (Jobling, 2010). This genus is cited by some as the typical genus of the true eagles and some even describe the members of this genus as the only true eagles. Other authorities place them with others as true eagles (Brown & Amadon, 1968). As pointed out by Wright (2008:1): "The English word eagle comes from Latin *aquila*. The word *aquila* or eagle, is believed to mean 'water-colored bird' and is related to the Latin *aqua* or water, as is *Aquarius*. The members of the Genus *Aquila* are medium to large-sized eagles, with no crests and tail bands. Attributes that are shared with the other booted eagles include legs feathered to the toes, longish, broad wings and tails, and strong talons. The Genus *Aquila* "has its headquarters in Eurasia and Africa, with solitary outliers in North America (a race of *A. chrysaetos*), New Guinea and Australia" (Brown & Amadon 1968: 644).

In terms of the relations between *Aquila* and the other genera of the Booted eagles, the adults of the *Aquila–Hieraaetus–Lophaetus* clade, adults generally have dark plumage, and the juvenile plumage is almost similar to that of the adult. Unlike the *Aquila* species, the *Hieraaetus* species have dark and light-colored phases, the light-colored phase distinguished by light, unmarked under-parts (Helbig et al., 2005). This is in contrast to the other booted eagles (e.g. *Spizaetus* and *Nisaetus*), which usually have adult plumage with barred underparts, and more distinct juvenile plumage with pale underparts. Current research findings indicate *Aquila* and *Hieraaetus* are polyphyletic (that is they are derived from more than one common evolutionary ancestor or ancestral group) (Helbig et al., 2005). The borders of the *Aquila* species are sometimes blurred, as some species are classified and reclassified. For example, the British Ornithologists' Union from 2005 and 2014, included both the Bonelli's eagle and the Booted eagle in *Aquila*, and Clements' Checklist (2007, 2009, 2012) put all the *Hieraaetus* species into the Genus *Aquila* from 2001 to 2009. Sangster et al. (2005, 823) writing for the British Ornithologists' Union, examined the phylogenetic studies of Wink & Seibold (1996), Wink & Sauer-Gürth (2000), Wink (2000), Roulin & Wink (2004), Wink & Sauer-Gürth (2004), Bunce et al. (2005), Helbig et al. (2005) and Lerner & Mindell (2005), and concluded that "these studies indicate that the species currently included in *Hieraaetus* and *Aquila* do not form separate monophyletic groups." Sangster et al. (2005, 823) proposed the following *Aquila* species for the Western Palearctic: Greater Spotted eagle (*Aquila clanga*), Lesser Spotted eagle (*Aquila pomarine*), Booted

Eagle (*Aquila pennata*), Golden eagle (*Aquila chrysaetos*), Verreaux's eagle (*Aquila verreauxii*), Bonelli's eagle (*Aquila fasciata*), Steppe eagle (*Aquila nipalensis*), Tawny eagle (*Aquila rapax*), Eastern Imperial eagle (*Aquila heliaca*) and Spanish Imperial eagle (*Aquila adalbertii*).

Most authorities place the *Hieraaetus* as a separate genus, with Bonelli's eagle and the African hawk-eagle placed in *Aquila* and sometimes Wahlberg's eagle is placed in *Hieraaetus* rather than in *Aquila* (or *Aquila wahlbergi*, Sundevall, 1850 rather than *Hieraaetus wahlbergi*, Sundevall, 1851). Helbig et al. (2005, 147) conducted a comprehensive analysis and phylogeny of the *Aquilini* using the "DNA sequence of one mitochondrial (cyt b) and three nuclear loci (RAG-1 coding region, LDH intron 3, and adenylate-kinase intron 5)." The results showed monophyly of the *Aquilini* (sharing a common ancestor) relative to other birds of prey, polytypic genera (with two or more different species) within *Aquilini*, such as *Spizaetus*, *Aquila*, *Hieraaetus,* were non-monophyletic and the Old World (Eurasia/Africa) *Spizaetus* and *Stephanoaetus* were the sister group of the remaining *Aquilini*. The Martial Eagle (*Polemaetus bellicosus*, Daudin, 1800) was found to be the sister species of the *Aquila/Hieraaetus/Lophaetus* clade.

Helbig et al. (2005, 148) note that the "African Wahlberg's Eagle (*wahlbergi*) has usually been included in *Aquila* (Amadon, 1982a,b; Brown and Amadon, 1968; Sibley and Monroe, 1990; Stresemann and Amadon, 1979), although its plumage dimorphisms (pale and dark morph adult) and whistling vocalizations are very atypical of that genus (Brown et al., 1982). It was recently shifted to *Hieraaetus* by Ferguson-Lees & Christie (2001) without explicit justification." Other findings by Helbig et al. (2005, 147) were that the Wahlberg's Eagle *H. wahlbergi*, formerly in *Aquila*, is part of a clade including three *Hieraaetus* species (the Booted eagle *Hieraaetus pennatus*, Gmelin, 1788; Ayres Hawk eagle *Hieraaetus ayresii*, Gurney, 1862; and the Little eagle *Hieraaetus morphnoides*, Gould, 1841).

These authors also acknowledge that "The Old World '*Spizaetus'* species should be assigned to the genus *Nisaetus* (Hodgson, 1836). *Hieraaetus fasciatus/spilogaster* are closest to *Aquila verreauxii* and should be merged with that genus."The African Long-crested eagle (*Lophaetus occipitalis*, Daudin, 1800) is the sister species of the two spotted eagles (the Greater spotted eagle, formerly *Aquila clanga*, later *Clanga clanga* (Pallas, 1811) and the Lesser spotted eagle, formerly *Aquila pomarina*, later *Clanga pomarina,* Brehm, 1831). Further, the members of Aquila (perhaps not including species moved to other genera like *Hieraaetus* but including *A. fasciata/spilogaster*) share some genetic similarities, in the nuclear and the mitochondrial cyt-B gene sequence (Helbig et al., 2005).

1.2 Golden Eagle *Aquila chrysaetos* (Linnaeus, 1758)

Physical Description

Any discussion of the genus *Aquila* must start with the Golden eagle. The word *eagle* referred specifically to the Golden Eagle in Britain before 1678. The Golden eagle is probably the best-known species of eagle in the world, partly due to its wide global range and its hunting abilities, large size and the admiration of the peoples who share its range. The name Golden eagle stems largely from the gold tinted feathers on the back of the eagle's neck. There are several subspecies, which may differ in size and physical appearance but retain the common brown plumage. These include: the nominal or 'typical' subspecies, sometimes called the European Golden eagle (*Aquila chrysaetos chrysaetos,* Linnaeus, 1758), which occurs across Europe from the northwest to central Europe, and as far as west and central Siberia and the Altai region, and into Iran. There is also *Aquila chrysaetos homeyeri* (Severtzov, 1888), a smaller darker variant that occurs in the Iberian Peninsula (Spain and Portugal). There are some individuals in North Africa from Morocco to Tunisia, and according to some, an isolated population near Ethiopia's Bale Mountains (Clouet et al., 1999) and Western Asia from Turkey to the Arabian Peninsula, Iran to the Caucasus and Kazakhstan (American Bird Census, 2015). Another subspecies is *A.c. daphanea* (Severtsov, 1888), which is even larger than *A.c. chrysaetos*, with darker plumage and a whiter tail base in juveniles, which ranges from Russian Turkestan to Manchuria to south west China and the Himalayas (Brown & Amadon, 1968).

The North American subspecies, *A.c. canadensis* (Linnaeus, 1758), is approximately the same size as the European subspecies *A.c. chrysaetos*, but is darker than both the latter subspecies and *A.c. daphanea.* Its wide range includes North America from Alaska to Mexico, and eastern and northeastern Siberia up to the Sea of Okhotsk, Altai and northern Mongolia (Brown & Amadon, 1968). The Japanese and Korean subspecies, *A.c. japonica* (Severtsov, 1888) is smaller than *A.c. canadensis*, and is distinguished by white mottling under the tail and occurs in Hondo, Japan and the Korean peninsula. There is also the east Eurasian Siberian Golden eagle (*A.c. kamtschatica*, Severtsov, 1888) with a range from western and central Siberia and eastern Altai as far as Kamchatka and the Russian Far East (Handbook of the Birds of the World, 2017). Some experts consider this subspecies to be the same as *A.c. canadensis* (Watson, 2010).

All these subspecies have broadly similar plumage. The main upper plumage is slightly variable shades of dark brown, including feathers with lighter brown tips, mostly on the nape and crown and golden feathers on the hind neck. For some individuals in "northeast United States and Scotland there is a single or pair of white scapular epaulettes" (Brown & Amadon, 1968, 664). The wing feathers are dark brown to brownish black, with greyish white or grey mottles, which may have the appearance of basal bands, especially in the inner webs. The tail feathers have bands of dark brown and dark grey of varying darkness. The upper parts are darker than the undersides. The breast feathers have darker shafts and tawny edges. The undertail and underwing coverts are rufous brown with darker mottling, and the later may be tawny at the bend of the wing. The primaries and secondaries are dark brown below, with some greyish-white shades at the carpal joint. The underside of the tail has white coloring at the base. The iris is hazel-colored, while the bill and claws are black and the feet and cere are yellow shades (Watson, 1997; Ferguson-Lees & Christie, 2001). In terms of subspecies variations, Brown and Amadon (1968, 665) note: "other races have sometimes been described, but all the races intergrade into one another, and the variation is clinal as a whole."

Juveniles are overall darker colored than adults. They have darker upper plumage than adults, including darker and slightly reddish feathers on the nape and crown. The tail is dark, almost black, with grey mottles and white coloration at the tip and tail base. The underparts are also darker than those of the adults. There is more white coloration and less rufous coloration on the undertail coverts. The outer primaries are very dark brown, as in the adult, and there is basal white on the inner primaries and secondaries, creating a white carpal joint patch. The cere and feet are yellow as in the adult, but the eye is darker. The juvenile's plumage changes into that of the adult after three to four years, manifested by the lightening of the brown colors and the reduction of the white on the wing bases (Brown & Amadon, 1968).

The Golden eagle occurs with some other similar species, hence depending on the part of the range, it may be mistaken for another species (del Hoyo et al., 1994; Campbell, 2015). Brown and Amadon (1968, 666) note that this species is "unmistakable in most of its range by reason of its dark brown color, very large size, graceful flight, and habitat usually in mountains." Other similar brown eagles which overlap in range are the Eastern imperial eagle (in Eurasia, wintering in southern Asia and East Africa), which has lighter colored feathers on the upper body; the Tawny eagle (which is smaller, and ranges in Africa and southern Asia); the smaller Lesser Spotted eagle (in Eastern Europe and South Asia, wintering to East Africa); and the smaller Greater Spotted eagle (in northern Eurasia, wintering in South Asia). The Golden eagle is more likely to confused with two species of sea eagles (the White-tailed and Bald eagles) as it is nearer their size than to smaller brown eagles. However, it differs from the adult White-tailed Sea eagle, as it lacks the latter's white tail, and the juvenile (which lacks a white tail) of that species in having a square rather than wedge shaped tail. It is like the immature Bald eagle, which has dark brown plumage and lacks the adult white head and tail, but the Golden eagle differs in having broader wings, a smaller bill, more graceful flapping flight and legs feathered to the toes. Juvenile Golden eagles also differ from the sea eagles in having large white patches on the wing carpal joint and the base of tail (Sutton et al., 1989; Grambo, 2003). The Golden eagle is a very silent bird, but rarely has a "loud clear yelping call weeeo-hyo-hyo-hyo": in display flight, it utters "mewing cries", such as "wee-o", and outside displays, "pleek or tsewk" (Brown & Amadon, 1968, 666). Illustrations of the appearance, range, and eggs of this species are shown in Figures 1.1a to 1.1n below.

Classification

The Genus *Aquila* is currently assessed as non-monophyletic, indicating they do not all have one common ancestor. In this genus, the Golden eagle is in a clade of closely related species, other closely related species being the Cassin's hawk-eagle, Bonelli's eagle, Verreaux's eagle,Wedge-tailed eagle and Gurney's eagle (Helbig et al., 2005; Lerner & Mindell, 2005). The first study used "DNA sequences from one mitochondrial and two nuclear genes" while the latter used "molecular sequences of one nuclear and two mitochondrial genes." Morphological (physical form) studies had also detected a close relationship between the Golden eagle, Gurney's eagle and Verreaux's eagle (Brown & Amadon, 1968). Other studies have postulated a close genetic relationship between the Golden eagle and several species. For example, Våli (2002) found that the Golden eagle was equally related to the Greater and Lesser Spotted eagles as to the Eastern Imperial and Steppe eagles. These studies were contrasted with that of Seibold et al. (1996) which found a significant genetic distance between the Golden eagle and the Eastern Imperial eagle, Spanish Imperial eagle, Tawny eagle, Steppe eagle, Greater Spotted eagle and the Lesser Spotted eagle.

There have also been studies of the classifications of the subspecies of Golden eagles. For example, Wink et al. (2004) and Helbig et al. (2005) discerned contrasts between the European Golden eagle (*A.c. chrysaetos*) and the North American Golden eagle (*A.c. canadensis*). These studies used DNA sequences from one mitochondrial and three nuclear genes. Wink et al. (2004) also found a close relation between the European Golden eagle and the Spanish subspecies (*A.c. homeyeri*) on the one hand, and the North American Golden eagle and the Asian Golden eagle (*A.c. daphanea*) on the other. Differences between the first two subspecies were inconsistent, hence there were suggestions that they might be joined as one subspecies, this was supported by their common range in the Iberian Peninsula. These studies used nucleotide sequences of the mitochondrial cytochrome b gene. Few differences were found between the Malian population and that of Europe, but there were significant contrasts between the European subspecies and those of Ethiopia. Wink et al. (2004) suggested there are two clades, due to the close relationship between the North American Golden eagle and the Asian Golden eagle (*A.c. daphanea*) and the Siberian Golden eagle (*A.c. kamtschatica*) (see also Orta, 1994) and the Japanese Golden eagle (*A.c. japonica*) (Global Raptor Information Network, 2021b).

Foraging

Golden eagles primarily forage in open landcover, this including grassland, scrubland, savanna or mixed vegetation on mountain terrain, slopes or flat lowlands. The Golden eagle usually soars 30 to 90 meters (98–295 feet) above ground when hunting and can go down to 7–8 meters (23–26 feet) above ground or perch while scanning the ground (Polite & Pratt, 1990) or even hunt in pairs (Kochert et al., 2002). Golden eagles employ several hunting techniques (Watson, 2010; Bodio, 2012). The "contour flight with short glide attack" is probably the commonest technique (Watson, 2010). This comprises a low glide at around 10 meters altitude along the ground contours, avoiding exposure above the skyline, usually for ambushing ground-based prey such as ground squirrels or ground breeding birds. The target is selected at very close quarters (Dekker, 1985). The "high soar with glide attack" is another method, where the eagle flies over 50 m altitude and on sighting prey it glides in an increasingly closed-wing, low-angled, accelerating dive over a kilometer, this culminating just before impact with the prey in a sudden opening of wings and spreading of tail feathers, making a loud sound as the prey is grasped (Watson, 2010). This may also start from a perch (Carnie, 1954). Another method is the "high soar with vertical stoop" for avian prey, especially large, relatively slow birds such as cranes, geese and some gamebirds (Watson, 2010). The eagle attains a height over the selected prey and then dives onto it. In some cases, prey birds may escape by increasing their flying altitude (Cameron, 1908; Meinertzhagen, 1940).

The "glide attack with tail-chase" technique is another foraging method, which involves an initial low-angled dive at the prey, followed by a chase over the ground. This commences with a low-angled stoop some distance from the quarry (Watson, 2010). This is followed by a close chase of the prey, which may be a ground bird in low flight over the ground or a running mammal prey. An example may involve a chase of grouse in low flight (Bent, 1937). There is also the "low flight with slow descent attack", where the eagle flies low and slowly swoops on the prey, usually on slower prey animals such as some mammals and reptiles (Grubac, 1988; Watson, 2010). The "low flight with sustained grip attack" is another technique, which is usually employed to attack larger ungulates such as deer or antelopes (Watson, 2010). The eagle attacks the usually running animal and grips the neck or back and may damage the vital organs, such as arteries, lungs, breathing passage or vertebra, with the attack continuing until the prey is either subdued through injury or exhaustion or escapes (Watson, 2010). There is also the "walk and grab attack", where the eagle alights and walks on the ground and attacks prey in ground cover and attempts to pull its prey out of cover (Watson, 2010). Common species hunted by this method include young antelopes or deer, some livestock and rabbits and hares (Smith & Murphy 1973; Nette et al., 1984; Palmer, 1988).

Golden eagles may hunt singly or in pairs (male and female), the latter association termed tandem hunting (Collopy, 1983; Chow, 2017). Such cooperation may be attributed to the difficulty of attacking some larger prey species (Hatch, 1968; Deblinger & Alldredge, 1996). Taking a case study of golden eagles in the Snake River Birds of Prey Area (BPA), in the Snake River Canyon in the cold desert plateaus south of Boise, Idaho, United States of America, Collopy (1983, 747) noted that "male Golden eagles attempted significantly more prey captures when solo-hunting than when tandem-hunting", the reason for this being that "males had fewer opportunities to tandem-hunt early during the nesting season when females were brooding young and not hunting."

Therefore, most of the tandem hunting occurred "late in brood-rearing when nestlings were left unattended for most of the day" (Collopy, 1983, 747). In this case study, during the tandem hunting, the eagles flew into the wind, for slow, low flight and rapid contact with prey. The male invariably flew above and led the larger female, even when returns to foraged areas occurred. Although the solo hunting areas covered the same areas with a similar flight mode and pattern, "on a per attempt basis, solo-hunting Golden eagles were significantly more successful at prey capture (29%) than those tandem-hunting (5%)"; the author notes that "These differences appear counter-intuitive, because the selective advantage of cooperative hunting behavior is assumed to be increased capture success" (Collopy, 1983, 748). This was explained by the tendency of tandem hunters to select larger and/or more elusive prey, although "no distinction between the species pursued by Golden eagles using the two hunting techniques was possible" in this study (Collopy, 1983, 748).

Golden eagles have a "strong preference for mammals"; as prey, in some cases 70% to 98% by weight of prey is mammals (Brown & Amadon, 1968). The prey species are mostly leporids (hares and rabbits), sciurids (ground squirrels, prairie dogs, marmots), birds, reptiles and fish, and in some cases carrion (Kochert et al., 2002). The prey is mostly medium sized, but they may also kill and eat larger mammals such as mule deer (*Odocoileus hemionus,* Rafinesque, 1817), in which is 1.2 to 2.1 meters long and may weigh over 50 kilograms (United States Fish and Wildlife Service, 2011). It may also kill pronghorns and Red deer (Brown & Amadon, 1968) foxes for example, the island fox (*Urocyon littoralis*, Baird, 1857) in California (Roemer et al., 2002), swift fox (*Vulpes velox*), gray fox (*Urocyon cinereoargenteus*) and island fox (*Urocyon littoralis*), coyotes (*Canis latrans*, Say, 1823), American badgers (*Taxidea taxus,* Schreber, 1778) and fishers (*Pekania pennanti*, Erxleben, 1777) in North America, Corsac fox (*Vulpes corsac, Vulpes corsac*, Linnaeus, 1768) in Mongolia, Eurasian otters (*Lutralutra*, Linnaeus, 1758) and Eurasian badgers (*Melesmeles*, Linnaeus, 1758) in Scotland, martens (*Martes foina*, Erxleben, 1777 and *Martes martes*, Linnaeus, 1758) in the French Alps, and feral pigs (*Sus scrofa domesticus* Erxleben, 1777) in California (Olendorff, 1976; Ralls & White, 1995; Ellis et al., 2000; Collins & Latta, 2009; Watson, 2010). There are a few reports of the larger wolverines, coyotes, bobcats and lynxes being taken in North America and Europe (Olendorff,1976; Bowen, 1980; Jordheim, 1980; Hornocker, 1981; Tjernberg, 1981; Neale et al., 1998; Mason, 2000; Zuskin et al., 2003; Jung et al., 2009). In some cases such large mammals comprise over five percent of the prey specimens (Hoffman et al., 2004). For example, Grubac (1988) notes that red foxes comprise 13.5% of nest remains in Sicily and 5.9% in the Republic of Macedonia. The Golden eagle has been recorded as feeding on domestic kids and lambs, although in some cases the eagle may not have killed the animals, and merely fed on carcasses. Large birds taken include geese, cranes, and gamebirds such as grouse. Gamebirds may comprise up to 40% of prey mass in areas where they are common (Brown & Amadon, 1968).

Golden eagles take carrion on occasion (Bodio, 2012; Campbell, 2015). Watson (2010) notes that carrion comprises a higher proportion of the eagles' diet in winter than in other seasons, due to the relative absence of prey species during winter (Boag, 1977; Wilmers et al., 2003; Sanchez-Zapata et al., 2010). In Scotland, dead lambs are a significant part of the Golden eagles'diet (Houston & Maddox, 1974; Brown, 1976). Dead pigs were also scavenged in Corsica and Santa Cruz Island, California (Seguin et al., 2001; Collins & Latta, 2009). Golden eagles tend to dominate both the large Andean and California condors and the smaller Turkey and Black vultures in the New World and in some cases the Griffon vultures in Eurasia and Africa (Campbell, 2015). Golden eagles also have conflicts over carrion with the larger sea eagles (White-tailed eagles in Europe, Bald eagles in North America) that share their range. The conflicts can favor either species, pitting the slightly larger size of the sea eagles against the more aggressive nature of the Golden eagle (Pettersson, 1977; Halley & Gjershaug, 2008). Golden eagles also contest with coyotes (*Canis latrans*, Say, 1823) over kills in North America, with the eagles being dominant in most encounters (Bowen, 1980; Jordheim, 1980; Jung et al., 2009).

Breeding

Nuptial displays by single birds and pairs occur in most months and comprise undulating movements, with dives and succeeding upward swoops. In soaring pairs, the male dives at the female, who then flips and presents claws; however, "grappling and whirling has not been seen" (Brown & Amadon, 1968, 667). The territories of breeding pairs vary, between 41-251 sq. km per pair (Phillips et al., 1984). Nests are on ledges of crags or in trees. These nests are at least 1.5-1.8 m wide and may be reused for each breeding season (Chow, 2017). Normally there are two eggs, which are white, spotted or blotched with browns, reddish brown and some grey markings. Fledging is after 65–80 days (Brown & Amadon, 1968) but some studies give 40 to 80 days (Chow, 2017). Watson (2010) gives 70 to 81 days in Scotland and 66 to 75 days in Idaho.

Population status

Brown and Amadon (1968, 666) note that "the Golden eagle is probably the most numerous eagle of its size in the world, by reason of the extent of its range and the huge areas of mountain country it frequents. It is the characteristic eagle of mountain country throughout its range, all others like it (*A. heliaca, A. clanga, A. pomerina* and *A. rapax*) being found more typically in plains, woodlands or swamps." The Golden eagle is not assessed as threatened (rather it is evaluated as Least Concern) by the International Union for the Conservation of Nature (IUCN) (Bird Life International, 2018). One reason is that there are estimated to be between 100,000-200,000 Golden eagles alive today; this is "very large, and hence does not approach the thresholds for Vulnerable under the population size criterion (<10,000 mature individuals with a continuing decline estimated to be >10% in ten years or three generations, or with a specified population structure) (Bird Life International, 2018). The species has a very wide range; for vulnerable status the range requirements would be less than "20,000 km^2 a declining or fluctuating range size, habitat extent/quality, or population size and a small number of locations or severe fragmentation)": Additionally, the population is stable, and "hence the species does not approach the thresholds for Vulnerable under the population trend criterion (>30% decline over ten years or three generations)" (Bird Life International, 2018).

The Golden eagle is migratory in northern parts of its range (from 55° N), for example in the Arctic migration occurs immediately the young leave the nest. Migration routes are well-defined, avoiding large water bodies. Migration peaks between mid-September and mid-October. Return migration is during March or April but is "less clearly defined and has been less studied" (Brown & Amadon, 1968, 667). In Scotland, Switzerland and southern parts of its range it is resident but moves from high to low ground in winter. During migration eagles rely on thermals or orographic lift (Campbell, 2015). Bohrer et al. (2011, 96) examined the migratory behavior and the meteorological conditions during autumn migration across eastern North America and argued that "it has not been shown previously that migration tracks are affected by species-specific specialization to a particular uplift mode" (Bohrer et al., 2011). The route choices of the Golden eagle were partly dependent on the relatively high wing-loading (mass/wing area) of the Golden eagle (7.2 kg/m^2) compared with that of the Turkey vulture (3.9 kg/m^2) (Pennycuick, 2008). Golden eagles are more muscular and have stronger flapping flight than Turkey vultures, hence they require stronger orographic lift and tended to migrate on the flight paths above the windward side of the Appalachian Mountains and ridges (see also Allen et al., 1996; Brodeur et al., 1996). The lighter, weaker Turkey vultures were able to use thermals (Bohrer et al., 2011; see also Mandel & Bildstein, 2007). Illustrations of the appearance, range, and eggs of this species are shown in Figures 1.1a to 1.1n.

Fig. 1.1a. Golden eagle perching [With permission from Paul Watts]

Fig. 1.1b. Golden eagle perching [With permission from Linda Raymer]

Fig. 1.1c. Golden eagle perching [With permission from Pauline Rosenberg]

Fig. 1.1d. Golden eagle perching [With permission from Linda Raymer]

Fig. 1.1e. Golden eagle perching. [With permission of Chas Moonie]

Fig. 1.1f. Immature Golden eagle perching [Credit: Dorothea Oldani-unsplash]

Fig. 1.1g. Golden eagle in flight [With permission from Linda Raymer]

Fig. 1.1h. Golden Eagle in flight [With permission from Rudolf Ernst]

Fig. 1.1i. Immature Golden eagle in flight [With permission from Frank McClintock]

Fig. 1.1j. Golden eagle in flight [With permission from Chas Moonie]

Fig. 1.1k. Subadult Golden eagle in flight [With permission from Chas Moonie]

Fig. 1.1l. Adult (1) and Immature (2) Golden eagles [With permission from Julian Thomas]

Fig.1.1 m. Golden eagle habitat sighting [Credit: Author]

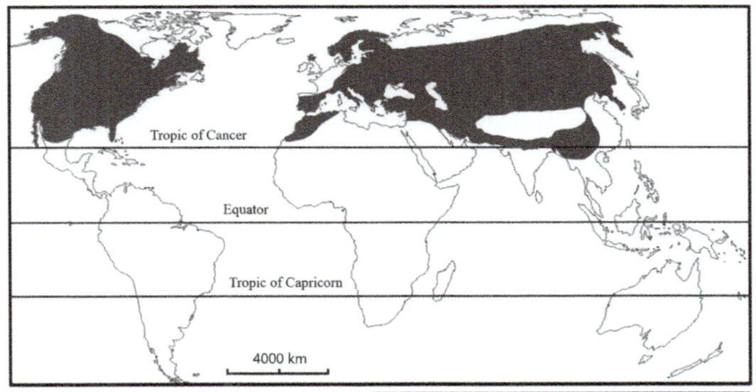

The Golden eagle's range covers most of North America and northern Eurasia, as a resident, migrant or vagrant

The Golden eagle's egg is usually light brown with varible, brownish spots.

Fig. 1.1n. Golden eagle egg and range

1.3 Tawny Eagle *Aquila rapax* (Temminck, 1828)

Physical Description

The Tawny eagle is one of the smaller members of the Genus *Aquila*. It can be confused with many other brown eagles, due to its highly variable brown plumage and wide range. However, despite its widely variable plumage color, from near white to black, the darker colored Tawny eagles generally have some tawny colored feathers on the body, distinguishing this species from other brown eagles (Clancey, 1964; Brown, 1977; Maclean, 1988). The subspecies are: *A.r. rapax* (which is the nominate or typical subspecies), *A.r. belisarius* and *A.r. vindhiana*. Two other subspecies, the Eurasian *A.r. orientalis*; and *A.r. nipalensis* have been reclassified as the Steppe eagle (*Aquila nipalen*sis) (Brown & Amadon, 1968). *A.r. belisarius* ranges in North Africa, from Morocco in the west, southwards to Nigeria and east to Ethiopia, with the range overlapping with that of *A.r. rapax* towards the equator. *A.r. vindhiana* ranges through Baluchistan and Afghanistan, to India and on to northern Myanmar and Thailand (Brown & Amadon, 1968; Knystautas, 1993). The larger, generally darker colored Steppe eagle subspecies (*A.r. orientalis* and *A.r. nipalensis*) occur in Eastern Europe and Central Asia, eastwards to southern Siberia and the Kirghiz steppes. These winter in southwest Asia and eastern Africa (*A.r. orientalis*) and in steppes in Central Asia and southwards to India (*A.r. nipalensis occurs*) (Brown & Amadon, 1968; Ferguson-Lees & Christie, 2001; Subedi, 2014). In distinguishing these from the other Tawny eagle subspecies, Shirihai (1994) notes the "Tawny Eagle Aquila rapax generally looks smaller and rather puny (and also, when perched, less elongated/horizontal) compared with Steppe Eagle *A. nipalensis*."

The Tawny eagle measures 60–75 cm (24–30 in) in length, with a wingspan of 159–190 cm (63–75 in). The weight ranges from 1.6 to 3 kg (3.5 to 6.6 lb) (Ferguson-Lees & Christie, 2001). In plumage, *A.r. rapax* is dominantly various shades of brown, with variations from very dark brown to pale rufous, with the median color reddish tawny brown (Brown & Amadon, 1968; Maclean, 1988). The tail is greyish brown with darker bars. The tail coverts are pale, and the wing quills are blackish brown. The cere and feet are yellow, and the eyes are greyish to brown. Juveniles have paler plumage (sometimes almost white), with broader pale edges and tips on the feathers of the upperparts and more streaks on the tail and white feathers at the base. The cere and feet are yellow as in the adult and the iris is dark brown (Brown & Amadon, 1968; Campbell, 1983). *A.r. belisarius* is larger, with darker brown, with less rufous coloration. *A.r. vindhiana* differs from the other two subspecies in being smaller, with an unbarred tail and is usually either lighter or darker with less reddish plumage. The two subspecies reclassified as the Steppe eagle (*A.r. orientalis* and *A.r. nipalensis*) are both darker than the Tawny eagle subspecies, and *A.r. orientalis* differs in the juvenile plumage including white upper tail coverts giving a white base tail bar, and almost white coloration to the edges of the medium and greater wing coverts, secondaries, lesser coverts and back feathers. These create whitish bars visible in flight. *A.r. nipalensis* is even darker and larger than *A.r. orientalis*. Immature birds are like *A.r. orientalis,* but possess white edges on wing coverts and secondaries (Brown & Amadon, 1968).

As noted above, the Tawny eagle may be mistaken for other species; however, Brown and Amadon (1968, 654) argue that "in most of its range it is the commonest, or the only, medium-sized brown eagle inhabiting dry grassy plains or savannah." They remark that its closest match in Africa is *Hieraaetus wahlbergi* (formerly *A. wahlbergi*), but the latter "has a lighter build and lacks the broad rounded tail, and also has different habits." (ibid.). In Europe, the Spotted eagle "differs in having round rather than oval nostrils, and lacks the rounded tail, with narrower, less rounded wing and less visible tail bars" (ibid.). Also as noted above, the Tawny eagle can generally be distinguished from the Steppe eagle by its smaller size and lighter color (Shirihai, 1994). The voice of the Tawny eagle may be written as "Kowk-Kwok" or "Kwak-Kwak" or "Kioh-Kioh". Resting females also make calls resembling "Sheep-Sheep" (Brown & Amadon, 1968, 654). Illustrations of the appearance, range, and eggs of this species are shown in Figures 1.2a to 1.2m.

Classification

The main development in the classification of the Tawny eagle is the separation of the Steppe eagle from this species, through the removal of the two subspecies *A.r. orientalis* and *A.r. nipalensis* into the new species *Aquila nipalensis*. The merger of these two species was quite recent (Vaurie, 1965; Brown & Amadon, 1968; Stresemann & Amadon, 1979). The separation was based on initial morphological, behavioral, and ecological distinctions (Brooke et al., 1972; Clark, 1992). Later molecular phylogenetic analysis supported the two-species division (Helbig et al., 2005). Further studies using the molecular sequences of one nuclear and two mitochondrial genes, indicated that the Tawny eagle may be in a monophyletic group with the Eastern imperial eagle (*A. heliaca*) (and possibly also with the Spanish imperial eagle *A. adalberti*) and the Steppe eagle (*A. nipalensis*), and they recommended that a taxonomic revision be undertaken to differentiate this group from other *Aquila* eagles (Lerner & Mindell, 2005; Global Raptor Information Network, 2021c).

Foraging

The Tawny eagle frequents open dry landcover but is also found in denser wooded savanna and woodland, at variable altitudes from sea level up to 3000 m. Brown and Amadon (1968, 654) noted that dead birds were found on Mount Everest at 26,000 feet. The Tawny eagle is often gregarious, congregating near food sources such as locust swarms, flamingo flocks or dense rat populations and even near water sources (Birdlife International, 2018). It is not deterred by human presence and may frequent farmland, suburban, urban and settled rural areas, as well as slaughterhouses (for meat discards) (Ferguson-Lees & Christie, 2001). Brown & Amadon (1968: 654) argue that the Tawny eagle "is very catholic in its tastes of food." Food sources include mammals (including rabbits), birds (including flamingos), reptiles, insects, and less commonly fish and amphibians in freshwater, and carrion through scavenging in urban areas. This species may also rob smaller raptors of their prey (Ferguson-Lees & Christie, 2001). Pirating may be practiced on other raptors as large as the Martial eagle and the Bearded vulture (*Gypaetus barbatus*), using diving attacks (Brown & Amadon, 1968). The pirating of food from the Bateleur eagle makes a major contribution to the Tawny eagle's diet in some regions (Steyn, 1982; Watson & Watson, 1987; Harrison et al., 1997, 178).

Kane et al. (2014) record substantial competition between Tawny eagles and vultures at carcasses in Africa, especially with the African White-backed vulture (*Gyps africanus*) and the Ruppell's vulture (*Gyps rueppellii*), which were observed in competition largely because they are more common than other vulture species (see also Ogada et al., 2012). The competition was termed a "producer–scrounger game" where scrounger refers to something that exists by taking from or depending on another. This was based on the biology of the competing species, as the vultures depended on thermals for flying, as flapping flight is too energy consuming for the exploration of vast areas (see also Hedenstrom, 1993; Ruxton & Houston, 2004). The smaller, lighter eagles can use the weaker thermals of the early morning for carrion detection and hence find carrion before the vultures, which can only soar with the stronger thermals of the late morning (Cone, 1962). Tawny eagles have been recorded at carcasses before the arrival of African White-backed and Ruppell's vultures. In some cases, the vultures were numerous enough to dominate the eagles and then began to feed only after the eagle had first opened the tough skin of the carcass. The relationship between the producer and the scrounger was thus fulfilled, with the eagle in the role of the producer and the vulture in the position of the scrounger (Kendall, 2013; Kane et al., 2014). Kane et al. (2014) further argue that this relationship is not because of the "abundance of raptors happening upon carcasses more often than the vultures because raptors occur at much lower densities"; the evidence for this is that "in the Masai Mara, for instance, *Gyps* species were recorded at an average density of 85.4 species per 100 km compared with 7.4 for tawny eagles" (see Virani et al., 2011).

Breeding

Aerial displays occur before and outside the breeding season. Displays include high altitude soaring, paired barking, and males diving at females, with the latter sometimes flipping over and interlocking claws with the males (Ferguson-Lees & Christie, 2001; Borello & Borello, 2004). The Tawny eagle breeds in large, platform-like nests built from sticks and sometimes animal bones, usually on the top of tall, isolated trees and less commonly, pylons (Brown & Amadon, 1968). Brown and Amadon (1968) note that the nest is small for an eagle, only 30 cm deep and one to 1.2 meters across. In Africa, the breeding season occurs from March to August in the northern range but occurs later (April to January) in the central and southern ranges and in the West (October to June). In Kenya, breeding may occur throughout the year, but mainly from May to November. In India, the breeding season lasts from November to August (Ferguson-Lees & Christie, 2001). The clutch is usually composed of one to three white eggs with brown or reddish-brown markings. Brown and Amadon (1968, 655) give a list of the breeding times in different countries, namely: Eritrea (April); Algeria (April); India (November to June, mostly December to January); Nigeria (December to February), South Africa (April to August), Zimbabwe and Malawi (April to October, mostly May to July); Sudan (September to February); Red Sea province (April); Northern Ethiopia (December); Southern Ethiopia (June); Kenya (June to October, January to February); Uganda (October to March); and Somalia (October to February). Fledging usually occurs around August, followed by migration. Fledging is usually nearly 60 days in the temperate areas and 70 days in the tropics (Brown & Amadon, 1968, 655).

Population Status

The Tawny eagle migrates southwards in August-October, to Africa through the Suez bottleneck and returns north from February to April. In West Africa, it migrates southwards in the dry season, and northwards during the wet season. East African populations move with the food supply, mostly the migratory herds of antelopes (BirdLife International, 2018).

Reports give differing opinions on the population status of this species. BirdLife International (2018) ranks the Tawny eagle as more common than other eagles of similar size and habits. The International Union for the Conservation of Nature (IUCN) (2018) states that "this species has an extremely large range, and hence does not approach the thresholds for Vulnerable under the range size criterion (extent of occurrence <20,000 km^2 combined with a declining or fluctuating range size, habitat extent/ quality, or population size and a small number of locations or severe fragmentation)." Additionally, the "decline is not believed to be sufficiently rapid to approach the thresholds for Vulnerable under the population trend criterion (>30% decline over ten years or three generations)" (IUCN, 2018). Finally, the "population size is very large, and hence does not approach the thresholds for Vulnerable under the population size criterion (<10,000 mature individuals with a continuing decline estimated to be >10% in ten years or three generations, or with a specified population structure)"; consequently, the Tawny eagle "is evaluated as Least Concern. (IUCN, 2018)." Harrison et al. (1997: 178) write that "the Tawny Eagle and the Bateleur are among the most threatened eagles in southern Africa, with centers of distribution now confined to major protected areas and sparsely inhabited regions." The populations in non-protected areas are sparse and are more numerous in northern Africa (see also Boshoff et al., 1983).

The Global Raptor Information Network (2021c) conducted a comprehensive study of the Tawny eagle in different localities and countries across its range. In Africa, the Moroccan population (*A.r. belisarius*) is described as endangered, due to "a continuous decline in both range and numbers during the second half of the 20th century," due to "human predation and deforestation" (Thévenot et al., 2003; Sayad, 2007). In Côte d'Ivoire, Mali, Burkina Faso and Niger, population declines have been recorded or hypothesized from the 1960s to the turn of the century (Thiollay, 1977, 2006a,b,c; Rondeau & Thiollay, 2004). Rondeau and Thiollay (2004) and Thiollay (2006a,b,c) conducted a comprehensive study of 7000 km of road transect surveys in 2003-2004 savannas of Burkina Faso, Mali, and Niger, duplicating similar surveys 1969-1973 by Thiollay (1977). The results showed a sharp decline in species numbers. In southern Africa, similar results have been recorded. The Tawny eagle is "now regarded as one of the most threatened eagles in southern Africa with distribution centers now confined to protected areas and sparsely inhabited regions" (Global Raptor Information Network, 2021c; see also Simmons, 1997a). It is seriously affected by poisoning and trapping in Namibia (Brown, 1991, Simmons & Brown, 2008). In Mozambique, it is affected by the decline of game and livestock resulting from local conflicts (Parker, 1999). In central Mozambique, the breeding population is estimated to be as low as 40 pairs, and only in areas with large herbivores (Parker, 2005a). In Swaziland, the Tawny eagle is endangered, going from common in the 1970s to possible breeding extinction due to poison baits, nest disturbance and habitat loss (Monadjem & Rasmussen, 2008).

In South Africa, the Tawny eagle is described as vulnerable, as it may have lost 20% of its population in the last few decades (Barnes, 2000a). Numbers recorded and estimated include 650 breeding pairs in the former Transvaal (Tarboton & Allan, 1984) and 800 pairs in the whole of South Africa (Barnes, 2000a). Declines were also recorded in the Karoo area (Boshoff et al., 1983). Reasons given include its common scavenging habits, which increase vulnerability to poisoned bait set for predators of livestock (Barnes, 2000a). Other causes of death include drownings in sheer-walled livestock reservoirs, powerline and vehicle collisions, habitat destruction, shootings and trapping (Oatley et al., 1998; Anderson et al., 1999). Wichmann et al. (2004) estimate that long term changes in rainfall patterns may also affect breeding patterns. Due to these declines, the Global Raptor Information Network (2021c) argues although it is "still categorized as a species of 'Least Concern' by BirdLife International but is now largely absent in West Africa and regarded as Threatened throughout southern Africa, hence it probably should be categorized as Near Threatened or even Vulnerable." The global population has been estimated at between 100,101 to 1,000,000 individuals (Ferguson-Lees & Christie, 2001). However, these figures may be on a declining curve (Global Raptor Information Network, 2021c).

In Asia, there are some disputes as to the extent of the Tawny eagle's range. Duckworth et al. (2008: 127) argue that "the Tawny Eagle *Aquila rapax* is usually listed as occurring widely, if rarely, in South-East Asia, primarily the northern parts"; however, in some cases "historical specimens purportedly of *A.r. vindhiana* from South-East Asia east of Myanmar were both Bonelli's eagles." Also, "interpretation of past records of '*A. rapax*' is hindered by the wide former treatment of Steppe Eagle *A. nipalensis* as a subspecies of *A. rapax*" (e.g. see Vaurie 1965). Consequently, they could "trace no acceptable record of Tawny Eagle from Indochina or Thailand: of the three records, two were certainly misidentified and the other is now unidentifiable" (Duckworth et al., 2008, 128). This was despite studies claiming vagrant status for Tawny eagle in southeast Asia (Rasmussen & Anderton, 2012; Robson, 2005). Illustrations of the appearance, range, and eggs of this species are shown in Figures 1.2a to 1.2m.

Fig. 1.2a. Tawny eagle [Credit: Chakkrapan Na Songkhla]

Fig. 1.2b. Tawny eagle perching. [Reproduced with permission from Gustav Van Der Gryp].

1.4 Steppe Eagle *Aquila nipalensis* (Hodgson, 1833)

Physical Description

The Steppe eagle, as noted earlier, was once considered to be closely related to the Tawny eagle or even part of the same species (conspecific), comprising the subspecies *A.r. orientalis* and *A.r. nipalensis*. The Steppe eagle is larger than the Tawny eagle, with darker plumage. The size range is 62–81 cm (24–32 in) in length, with a wingspan of 1.65–2.15 m (5.4–7.1 ft). Females are larger than males and may weigh 2.3–4.9 kg (5.1–10.8 lb), the latter weighing 2–3.5 kg (4.4–7.7 lb) (Ferguson-Lees & Christie, 2001). The plumage is dark overall, but there is a pale throat patch that easily distinguishes it from the Tawny eagle. Eastern birds

Fig. 1.2c. Tawny eagle perching. [Reproduced with permission of Mike Nelson-Birdtour ASIA]

Fig. 1.2d. Tawny eagle perching [With permission from Derek Keats].

Fig. 1.2e. Immature Tawny eagle perching. [Reproduced with permission from Derek Keats].

Fig. 1.2f. Immature Tawny eagle perching. [Reproduced with permission from Vikas Chavan].

Fig. 1.2g. Tawny eagle in flight [With permission from Vikas Chavan]

Fig. 1.2h. Tawny eagle in flight [With permission from Vikas Chavan]

Fig. 1.2i. Tawny eagles in flight [With permission from Gustav Van Der Gryp and Derek Keats]

Fig. 1.2j. Tawny eagle in Habitat South Africa [With permission from David Beadle]

Fig. 1.2k. Tawny eagle habitat, Serengeti Tanzania [With permission from Achim Mittler, Frankfurt am Main]

Fig. 1.2l. Tawny eagle habitat, Guinea Savanna [With permission from Jonathan D. Müller]

(formerly *A.r. nipalensis*) are larger and darker than European and Central Asian birds (*A.r. orientalis*). Juveniles are variably colored. The call is "like a crow barking, but it is rather a silent bird" (Serhal et al., 2014). Illustrations of the appearance, range, and eggs of this species are shown in Figures 1.3a to 1.3i below.

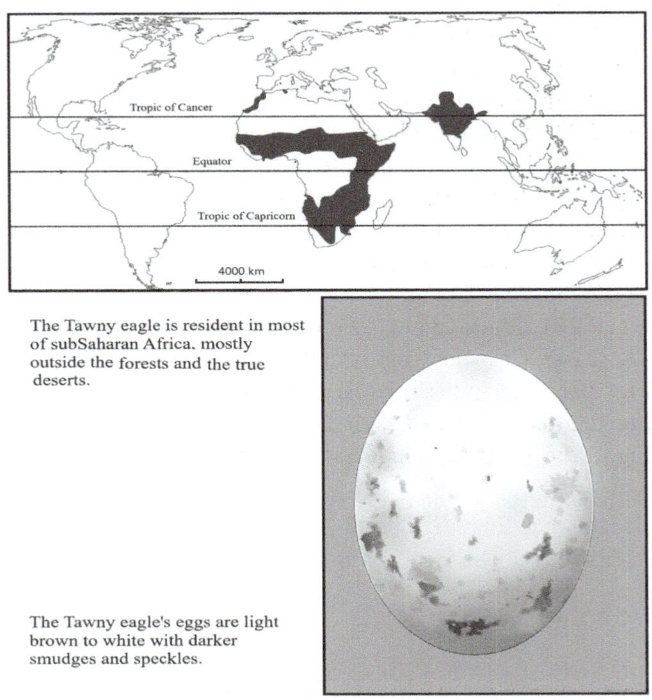

Fig. 1.2m. Tawny Eagle Range and Egg

Classification

The split of the Steppe and Tawny eagles into two separate species was initially based on the main physical differences, including morphology and anatomy, between two intra species groups, (Clark, 1992; Olson, 1994; Sangster et al., 2002). Molecular studies, each based on a very small number of genes, indicate that the species are distinct but there is disagreement over the closeness of the relationship (Global Raptor Information Network, 2021d). Olson (1994, 264) referring to the study by Clark (1992) notes that "although the Tawny Eagle *Aquila rapax* and Steppe Eagle *Aquila nipalensis* are now often treated as subspecies of a single species (*Aquila rapax*), Clark (1992) recently presented a convincing analysis of plumage, external morphology, distribution and habits, from which he concluded that "these are two very distinct species, as they were regarded by most writers in the first half of this century and before." The contesting studies that merged the two species included those of Brown and Amadon (1968), Voous (1977), Vaurie (1965), Stresemann and Amadon (1979), Amadon and Bull (1988) and Stepanyan (1990). Amadon (1982) saw the two species as members of a super-species. Clark (1992), Brooke et al. (1972) and Snow (1978) opposed this classification, using morphological, behavioral, and ecological differences as their criteria. More sophisticated analysis, using molecular phylogenetic analysis, and DNA sequences of one mitochondrial and three nuclear genes (Helbig et al. (2005), uncovered no indication that the two species are "even sister species"; "*A. rapax* is the sister species of the Imperial eagles *A. heliaca/adalberti* while *A. nipalensis* is the sister of all three" (Helbig et al., 2005, 161). Another study by Lerner and Mindell (2005), which used the molecular sequences of one nuclear and two mitochondrial genes, found evidence of a monophyletic group between *A. heliaca* (possibly *A. adalberti*) and *A. rapax* and some distance between this group and other Aquila eagles.

Foraging

The Steppe eagle is described as "sluggish, spending much of its time perched on rock piles, the ground, or telephone poles" (Global Raptor Information Network, 2021d). The diet includes small mammals, medium sized birds and carrion. Frequently, it steals food from other raptors (Brown & Amadon, 1968). Different patterns of foraging have been recorded for this species in different habitats in different countries (Global Raptor Information Network 2021d). Prey species recorded included Brandt's voles (*Lasiopodomys brandtii*) in Mongolia, termite and locust swarms and quelea colonies in Africa, ploughed ecologies in Saudi Arabia (Stagg, 1991), susliks, marmots, gerbils, hares, and occasionally nestlings and reptiles in Russia (Flint et al., 1984), Red-cheeked ground squirrels (*Spermophilus erythrogenys*), Black-bellied sandgrouse (*Pterocles orientalis*), Long-eared hedgehogs (*Hemlechinus auritus*) in Kazakhstan (Barashkova et al., 2009) and voles (*Microtus*) in Armenia (Adamian & Klem, 1999). Other prey species included termite (*Isoptera* spp.) swarms in Zambia (*Praomys natalensis*) (Brooke et al., 1972; Dowsett et al., 2008), Red-billed queleas (*Quelea quelea*) (Irwin, 1981) often competing with Lesser Spotted eagles, Steppe buzzards and Black kites in Zimbabwe (Jensen, 1972; Hartley, 1998). Similar food is taken in the Indian subcontinent, Malaysia and China (Wells, 1999; Rasmussen & Anderton, 2012; Weizhi, 2006). Some of these studies note that eagle frequently forage on the ground, for insects, small animals and carrion (Irwin, 1981; Hartley, 1998; Wells, 1999; Rasmussen & Anderton, 2012).

The wide range of predatory behaviors may be related to the variable habitats of this species. The range includes most open landcover, such as steppe, desert, semi-desert, open savanna, pastures, agricultural fields, paddy fields, grassland, and open woodland (Global Raptor Information Network, 2021d). In Asia, the Global Raptor Network (2018) notes its presence in "orchards and steppe-type fields" in Armenia (Adamian & Klem, 1999); dry landcover in Israel (Shirihai, 1996); "open woodland, wooded grassland, and floodplains, frequently in mixed raptor groups at termite swarms" (Dowsett et al., 2008); and steppes, deserts and mountains up to 2,000 m in Kazakhstan (Wassink & Oreel, 2007; Barashkova et al., 2009; Global Raptor Information Network 2021d). In Africa, the Global Raptor Network (2018) notes its presence in "floodplains, open woodland, and wooded grasslands" in Malawi (Dowsett-Lemaire & Dowsett, 2006); in large flocks over woodland savanna in Botswana (Simmons, 1997b) and Zimbabwe (Irwin, 1981; Hartley, 1998). In Zimbabwe, it is commonly associated with other raptors, such as the Lesser Spotted eagle and may form flocks near Red-billed quelea (*Quelea quelea*) colonies and insect swarms (Irwin, 1981; Global Raptor Information Network, 2021d).

Breeding

The Steppe eagle has a wide breeding range, from eastern Europe (Romania) in the west to the southern Russian and Central Asian steppes, as far east as Mongolia. Western birds from eastern Europe and Central Asian winter in Africa, while the eastern birds winter in India (Brown & Amadon, 1968). The Asian migrations pass through south-eastern Pakistan and Nepal, where large numbers are observed, sometimes up to 15.3 birds per hour during from October to November (DeCandido et al., 2001). Nests are usually placed on the ground (Gombobaatar et al., 2012). For example, in Mongolia, Gombobaatar et al. (2012) recorded the nest sites of 40 breeding pairs as located on: the ground (47.8%), rock columns (32.6%), large boulders and rock columns, cliffs (8.7%), trees (2.2%) and artificial substrates including abandoned car cabins, car tyres and artificial nest platforms (8.7%). There were no annual variations in nest site selection, nest site type, clutch size and successfully fledged juveniles. The nest is usually a large platform, composed of sticks, with a lining of twigs and small sticks, bones, feathers and in some cases even rags and dung (Bouglouan, 2020a). One to three eggs are laid, colored white with yellow brown markings. Incubation lasts for 45 days, with fledgling at about 55 to 65 days (Eagle Directory, 2017).

Population Status

The global population (the total number of the adult and immature birds at the beginning of the breeding season), of the Steppe eagle has been estimated at between 100,101 to 1,000,000 birds (Ferguson-Lees & Christie, 2001). This figure is disputed, as one other survey by BirdLife International (2009) was of 10,000 birds, although the researchers acknowledged limitations in the reliability of this survey. Other surveys by Birdlife International were 15,000 to 25,000 breeding pairs in Europe (BirdLife International/European Bird Census Council, 2000) and 5,000 to 20,000 breeding pairs (BirdLife International, 2004). Other estimates put the global population at around 37,000 breeding pairs (Karyakin et al., 2016).

The Steppe eagle is a highly migratory species, which breeds in Mongolia and China and migrates through central China, southwards and possibly south towards the Thai-Malay peninsula (Cheng, 1987; Williams, 2000; Den Besten, 2004; China Ornithological Society, 2007; Decandido et al., 2013, 21). As noted by Den Besten (2004, 9) the Steppe eagle "breeds in extreme south-eastern Europe, across central Asia, south-west Russia, east through the Kirghiz steppes and northern Kazakhstan to Transbaikalia and west Manchuria, south to the Aralsea, Tien Shan, northern Tibet and Mongolia." The Steppe eagle winters in eastern to southern Africa, southwestern Asia (the Middle East) and in south Asia (Afghanistan, the Indian subcontinent east to Myanmar). Migration starts in October, with returns from January to February (Ferguson-Lees & Christie, 2001; Den Besten, 2004).

The Steppe eagle is recorded as quite common and widespread across its range, despite recent evidence of declining populations (Yosef, 2003). However, the Global Raptor Information Network (2021d) lists this species as endangered and declining. Horvath et al. (2017) note that "This species has undergone extremely rapid population declines within its European range." These researchers further examine the shift in classification from Lower Risk/least concern (1988–2004), to Least Concern (2004–2015) to Endangered (2015 present): "The majority of its range lies outside Europe where it was not thought to be declining at a sufficiently rapid rate to approach the threshold for Vulnerable. However recent information suggests that the population outside Europe may be exposed to greater threats than was previously thought and has also undergone very rapid recent declines across much of the range. It is therefore classified as Endangered" (Horvath et al., 2017).

Evidence of population decline is recorded in Turkey and European Russia (BirdLife International, 2004) but nevertheless notes that these comprise less than 5% of the total global population); and in Israel, a major migration bottleneck, where the migrant populations have declined by 50% between 1975 and 2004 and juveniles from 30% to 2% of total migrant counts between the 1980s and 2000, possibly due to the Chernobyl nuclear disaster (Yosef & Fornasari, 2004). Other factors for the decline include frequent predation on nestlings (Ferguson-Lees & Christie, 2001), habitat loss to agriculture, human hunting and electrocution on power lines. These issues were behind the support for the change in status from Least Concern to Near Threatened (Bird & Symes, 2009).

BirdLife International (2018) gives a list of the populations in different countries. These included the European population (800–1,200 breeding pairs), Russia (2,478–3,688 breeding pairs, of which 1,176–1,895 were in European Russia), Kazakhstan (22,000–31,000 pairs). The Russian Federation had the "largest numbers"; this population included those of the Republic of Kalmykia (500-700 breeding pairs), Volgograd district (300 - 500 breeding pairs), Orenburd district (200–350 breeding pairs), Republic of Altai (400–600 breeding pairs) and Republic of Tyva (300–400 breeding pairs).

Several important breeding populations are found in the general area between Kazakhstan and Mongolia. These include the Usturt Plateau, Emba river basin, Mugodzhary mountains, Or river basin; with about 12,000 breeding pairs in Western Kazakhstan; about 9,000 in Eastern Kazakhstan; about 3,000 breeding pairs in Western Mongolia, about 2,600 breeding pairs in the Volga-Urals (between the Volga and Ural rivers northwards), about 2,000 breeding pairs in Dahuria (mostly a transboundary Russian-Mongolian population); and about 4,000 breeding pairs in Central Mongolia (Karyakin et al., 2016). Estimates for the Mongolian population range from 1,500-2,000 to 6,000-18,000 pairs (Karyakin, 2013).

The main population declines are in the western part of the range, namely eastern Europe (almost total extinction in Ukraine, Romania and Moldova) due to the development of agriculture in the steppes and direct killing of eagles (Ferguson-Lees & Christie, 2001; Meyburg & Boesman, 2013; BirdLife International, 2018). Levin and Kurkin's (2013) study in Kazakhstan, supported by other research elsewhere, found this species to be the bird of prey most often electrocuted by power lines, and in some cases by wind energy infrastructure (see also Strix, 2012; Meyburg & Boesman, 2013). Other factors for deaths in this species include pesticide poisoning, direct killings by people, habitat and suitable prey losses, nest destruction (partly from spring fires) and nestling mortality leading to low breeding success, and disturbance from agricultural activities (Karyakin et al., 2016b).

In some cases, juveniles are captured from nests and sold in western European countries (Mebs & Schmidt, 2006). Yosef and Fornadari (2004) hypothesized that the reduced total migratory population (and the proportion of juveniles among these birds) recorded migrating over Israel was a least partly due to the Chernobyl nuclear accident in 1986, and resultant radioactive contamination. In addition, the veterinary drug diclofenac used on cattle in the wintering range of Pakistan, Nepal and India, may affect this species, as it did most species of Asian vultures (Sharma et al., 2014; Campbell, 2015). This may be particularly important, as many eagles migrate westwards across Central Asia to India and Nepal during winter (Inskipp & Inskipp, 1991; Naoroji, 2006; Mallalieu, 2007, 2008). Illustrations of the appearance, range, and eggs of this species are shown in Figures 1.3a to 1.3i.

Fig. 1.3a. Steppe eagle perching [With permission from Rob Muckley]

Fig. 1.3b. Steppe eagles perching [With permission from Mike Nelson-Birdtour ASIA]

Fig. 1.3c. Immature Steppe eagle perching [With permission from Frank McClintock]

Fig. 1.3d. Steppe eagle foraging [With permission from Terrence Ang]

Fig. 1.3e. Steppe eagle in flight [With permission from Ken Behrens]

Fig. 1.3f. Immature Steppe eagle in flight [With permission Terrence Ang]

Fig. 1.3g. Steppe eagle wintering habitat in India [With permission from Siddhartha Mukherjee]

Fig. 1.3h. Steppe eagle habitat Asian Steppe [With permission from Ken Shono-unsplash]

1.5 Spanish Imperial Eagle *Aquila adalberti* (C.L. Brehm, 1861)

Physical Description

The Spanish Imperial eagle (*Aquila adalberti*) is also recorded as the Iberian imperial eagle, Spanish eagle, or Adalbert's eagle (after Prince Adalbert of Bavaria). This species averages 78–82 centimeters (31–32 in) in length and 180–210 centimeters (71–83 in) in wingspan. The weight average 2.5–3.5 kilograms (5.5–7.7 lb) (Jais, 2017). It may be distinguished from the Eastern Imperial eagle by its darker color. Compared with both the Eastern Imperial eagle and the Golden eagle, the Spanish Imperial eagle generally has a darker, blackish brown plumage that reaches the bird's throat and a paler nape and crown. It however resembles the Eastern Imperial eagle in that there is a broad white band on the shoulder and the leading edge of the wing (Thiollay et al., 1994; Ferguson & Christie, 2001). The Spanish Imperial eagle feeds mainly on rabbits, but can prey on many other animals, such as partridges, rodents, hares, pigeons, crows, ducks, foxes or—rarely, since they are not typically present in the eagle's habitat—small dogs (Cramp & Simmons, 1979; Meyburg, 1994; Seibold et al., 1996; Ferguson-Lees & Christie, 2001; Ferrer, 2001).Illustrations of the appearance, range, and eggs of this species are shown in Figures 1.4a to 1.4l below.

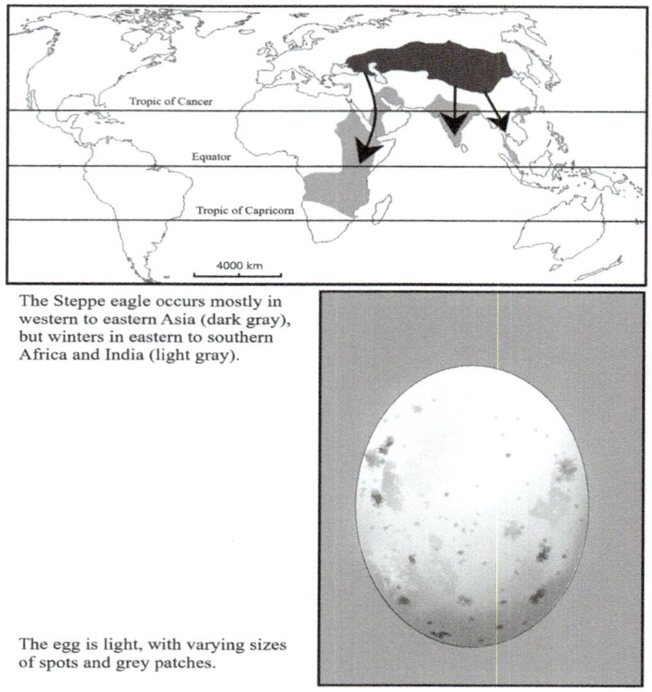

Fig. 1.3i. Steppe eagle range and egg

Classification

Older assessments considered the Spanish Imperial eagle as a subspecies of the similar Eastern Imperial eagle (Brown & Amadon, 1968; Stresemann & Amadon, 1979), but more recent research has placed it in a separate species, based on contrasts in morphology, molecular features and ecology between the Spanish and Eastern Imperial variants (Hiraldo et al., 1976; Cramp & Simmons, 1979; Gonzalez et al., 1989; Sangster et al., 2002; Meyburg, 1994; Seibold et al., 1996; Padilla et al., 1999; Helbig et al., 2005). The distinction of the two species was complicated by the fact that several studies indicated variable links with other species. For example, Helbig et al. (2005) and Lerner and Mindell (2005) used DNA sequences from mitochondrial and nuclear genes and found that *A. heliaca* and possibly *A. adalberti* were in a monophyletic group, which included *A. nipalensis* and *A. rapax*. Some studies suggested that the two species of Imperial eagle may have been parapatric (species whose habitats are separate but adjoining) within the last two centuries (Gonzales et al., 1989; Gonzalez, 1991) but other studies using the cytochrome b gene indicated no gene flow between the species since separation the Pleistocene (Seibold et al., 1996). Other studies using the nuclear genome found the separation to be more recent (Holocene or the end of the Pleistocene), as they found a male-mediated gene flow from the Eastern Imperial eagle to the Spanish Imperial eagle (Martínez-Cruz & Godoy, 2007). Seibold et al. (1996) report that "the amount of genetic differentiation between *A. heliaca* and *A. adalberti* was the same as in at least one other pair of (then) undisputed eagle species, *A.* (*Lophoaetus*) *pomarina* and *A.* (*Lophoaetus*) *clanga*. The separation between the mitochondrial lineages of the two species could have occurred slightly less than one million years ago" (Global Raptor Information Network, 2021e).

Foraging

The main foraging habitat of the Spanish Imperial eagle is woodland, marshland, open plains and mountain slopes (Valverde, 1960; González et al., 1990, Bisson et al., 2002). During the periods of its lowest numbers, it is confined mainly to mountainous areas, but more recently the population has expanded mostly in alluvial plains areas (González et al., 2008). In Morocco, it occurs in forested areas on plains and hills, especially open Cork Oak (Quercus suber L.) woodlands surrounding lowland marshes (Thévenot et al., 2003). The Eagle Directory (2018) notes that adults are mostly sedentary, while juveniles are more mobile, with movements from breeding areas, averaging 101.2 km and up to 310 km. Favored prey includes rabbits, with large birds such as flamingos taken in wetter areas, as well as small birds, reptiles and carrion (Global Raptor Information Network. 2021e).

Breeding

The Spanish Imperial eagle generally starts breeding in mid-February, but this may vary according to location (Global Raptor Information Network, 2021e). For example, in Morocco, one record describes ten clutches in late January (1), March (3), April (3), and early May (3) (Thévenot et al., 2003). The nest is usually large, composed of sticks. Trees are frequent nest locations, less commonly pylons may be used, cliffs have not been recorded (González, 1991; Global Raptor Information Network, 2021e; The Eagle Directory, 2018). Tree height and human presence are factors for nesting locations. Bisson et al. (2002: 302) note: "Our study suggests that both human persecution or disturbance and lack of adequate nesting habitat (specifically tall trees in small tree stands) appear to be important factors inhibiting the expansion of Spanish Imperial eagles."

The clutch is generally composed of 1-4 eggs, but 2 or 3 is more common. Global Raptor Information Network (2021e) lists the following averages in egg numbers: 2.42 eggs/clutch (Calderón et al., 1987); 2.47 eggs/clutch (González, 1991); 2.21 eggs/clutch (Margalida et al., 2007). Incubation is principally by the female, but the male may also incubate during food provision. The incubation period is 39-42 days (Margalida et al., 2007). Incubation may vary; in Spain it is estimated to be 44 days (Calderón et al., 1987; Margalida et al., 2007). Nestling periods may average 76.79 days (Margalida et al., 2007). Nestlings commonly kill each other, this being the commonest factor for nestling mortality (nearly 55% of nestling deaths, as noted by Margalida et al., 2007). Clutches may be replaced with new eggs if the initial clutch is lost early part in the incubation period (Global Raptor Information Network, 2021e).

Population Status

The Spanish Imperial eagle is distributed mainly in the "south-western quadrant of the Iberian Peninsula (Spain and Portugal)" (Gonzalez et al., 2008, 242). Gonzalez et al. (2008) studied the southern Spain populations between 1974 and 2004 which comprised 13 nuclei and 5 subpopulations and found that during the studied period the "population levels in all nuclei, except the one in Doñana" expanded (Gonzalez et al., 2008). A study by Bisson et al. (2002, 298) examined nest-site selection by Spanish Imperial eagles (in Doñana National Park in southwestern Spain) from 1984–1994 and found that "the presently occupied Spanish Imperial eagle territories represent the majority of the remaining suitable habitat for this endangered species in and around Doñana National Park."

Literary sources indicate that in the mid 19th century this species was abundant, with some breeding areas located near human habitation. Population decline began in the late 19th century, with populations retreating to the remote mountain ranges. Gonzalez et al. (2008) found that during their study period, breeding pairs increased in the open plains, with decreasing numbers in the mountains. The population of the main prey, rabbits has decreased, this also lowering the trophic quality of the habitat. There was also an increase in the numbers of nests in protected areas and private areas. Land ownership did not affect the breeding levels and perhaps strangely, the eagles' populations increased in non-protected areas over protected areas, even as the latter included potential habitat. The survival of this species may depend legal protection, attitude changes and habitat management (Gonzalez et al., 2008).

Other studies point to the eagle's declining populations. Höfle et al. (2008, 171) describe the Spanish Imperial eagle as "the most endangered European eagle." The Eagle Directory (2018) notes a serious decline in the eagle's population indicated by their virtual extinction as breeders in Portugal, Algeria and Morocco, leaving only Spanish birds in central and southwestern Spain, an estimated 300-400 individuals across a range of 117,000 km², with vagrants in France, the Netherlands, Senegal and Morocco. Other sources record lower population levels. For example, Gonzalez et al. (2008) estimate the Spanish population at 50 breeding pairs in the early 1960s (see also Bijleveld, 1974) and 30 pairs in 1970 (see also Simon &Geroudet 1970). Other estimates (Global Raptor Information Network, 2021e) were 60 breeding pairs in 1975 (Garzón, 1977), 125 in 1988 (Collar & Andrew, 1988), 104 in 1989 (De Juana, 1989), 146 in 1997 (Purroy 1997), 131 in 2000 (SEO BirdLife), and 175 to 200 in 2003 (González & Oria, 2003; Martí & Del Moral, 2003). The population also estimated by the BirdLife International/European Bird Census Council (2000) was 131 breeding pairs in 2000. Later a 2004 survey in Spain recorded 396 birds, which represented 198 breeding pairs (González et al., 2008). Still later, BirdLife International (2009) estimated the population of mature birds at between 300 and 400 individuals. Despite the gradual increase in population, the distribution has been described as in small clusters in Spain (González et al., 2008). Interestingly, the Spanish population has increased more outside protected areas than within, probably a reflection of changing attitudes, which have led to decreased direct persecution (González et al., op cit.).

Shooting by hunters and poisoning were historically the main factors for the near-extinction of the Spanish Imperial eagle. Later, after 1989, there were further decreases when the population of their main prey, the wild rabbit was nearly driven to extinction by the viral haemorrhagic disease (RHD). This raised major concerns among conservationists (González, 1996). Other factors for the decline of this species included habitat fragmentation from deforestation, agricultural expansion and timber cutting, and the ingestion of lead fragments from hunting bullets. Later factors included accidental electrocution and poisoning (González et al., 2007). These factors contributed strongly to the eagle's status as Vulnerable by BirdLife International and as Endangered in Spain (González & Oria, 2003).

Across Europe and North Africa, various issues have affected conservation issues. For example, Ferguson-Lees & Christie (2001) listed illegal shooting, bait poisoning, accidental captures in rabbit traps, powerline collisions and electrocutions, nesting pair disturbances and pesticide residues contributing to breeding failures. González et al. (2007) give accidental electrocution and poisoning as the main factors (47.7% and 30.7% respectively) for non-natural mortality over the period 1989 to 2004 in Spain. Regionally, their research findings indicated that electrocutions were more common in west-central region of Spain, and rarer southern Spain, while poisoning was commoner in the northern and southern regions compared to the west-central region. Illegal poison use in predator control and livestock protection was also a cited factor for eagle deaths (González et al., 2007).

Höfle et al. (2008, 171) add the danger from West Nile Virus (WNV). This was detected in some of a sample 13 wild and 18 captive eagles, using reverse transcription polymerase chain reaction (RT-PCR), immunohistochemistry and seroneutralization. The research method detected the WNV in the birds' tissues and immunohistochemistry detected WNV antigen in the Purkinje cells (a nerve cell in the cortex of the cerebellum), epithelial (lining) cells of the gizzard and duodenum, perivascular inflammatory cells, and in Kupffer-cells and hepatocytes. Serum antibodies which form in reaction against WNV "were detected in a total of five out of 21 Imperial eagles (23.8%), including free-living nestlings (two out of nine samples, 22.2%) and captive adult eagles (three out of 12 samples, 25%)" (Höfle et al. (2008, 171). The authors conclude that this gives evidence that the WNV circulates among free-living and captive Spanish Imperial eagles in South-central Spain. This is important because the area of Spain studied had no detection of WNV evidence in the previous six years (Höfle et al., 2008).

Comparing Europe and North Africa, Sayad (2007) noted that the Spanish Imperial eagle was extinct in Morocco between 1960 and 1980, largely due to human predation and deforestation. However, movements may preclude a fixed population assessment (Amezian et al., 2015). Although this species is non-migratory, there are records of juveniles flying southwards from Spain into Africa, crossing the Gibraltar Straits migratory bottleneck to winter from Morocco to Senegal, and others dispersing over Europe (Seibold et al., 1996; Ferrer, 2001; Thévenot et al., 2003; Bildstein, 2006; Weenink et al., 2011). Comparing this species with the Eastern Imperial eagle, Seibold et al. (1996) pointed out that the significant species-level genetic differences between Eastern and Spanish Imperial eagles are strong enough to rule out the reintroduction of Eastern birds into the Iberian Peninsula as a conservation solution. Possible, negative outcomes would be the Eastern birds' genetics would dominate that of the Spanish birds, or that no breeding would take place, both being unacceptable outcomes (Seibold et al., 1996). Illustrations of the appearance, range, and eggs of this species are shown in Figures 1.4a to 1.4l.

Fig. 1.4a. Spanish Imperial eagle perching [With permission from Santiago GV]

Fig. 1.4b. Spanish Imperial eagles perching [With permission from Wim Hoek]

Fig. 1.4c. Immature Spanish Imperial eagle perching [Credit: Johannes Gross (j-g-fotos@web.de)]

Fig. 1.4d. Immature Spanish Imperial eagle perching [Credit: fsanchex]

Fig. 1.4e. Spanish Imperial eagle in flight [With permission from Santiago Aves]

Fig. 1.4f. Spanish Imperial eagle in flight [With permission from asanmarc]

Fig. 1.4g. Immature Spanish Imperial eagle in flight [With permission from Frank McClintock]

Fig. 1.4h. Immature Spanish Imperial eagle in flight [With permission from Frank McClintock]

Fig. 1.4i. Spanish Imperial eagle perching on nest [With permission from Tristan Bantock]

Fig. 1.4j. Spanish Imperial eagle habitat in Spain [With permission from Ruud Kimmelaar]

Fig. 1.4k. Spanish Imperial eagle habitat in Spain [With permission from Ruud Kimmelaar]

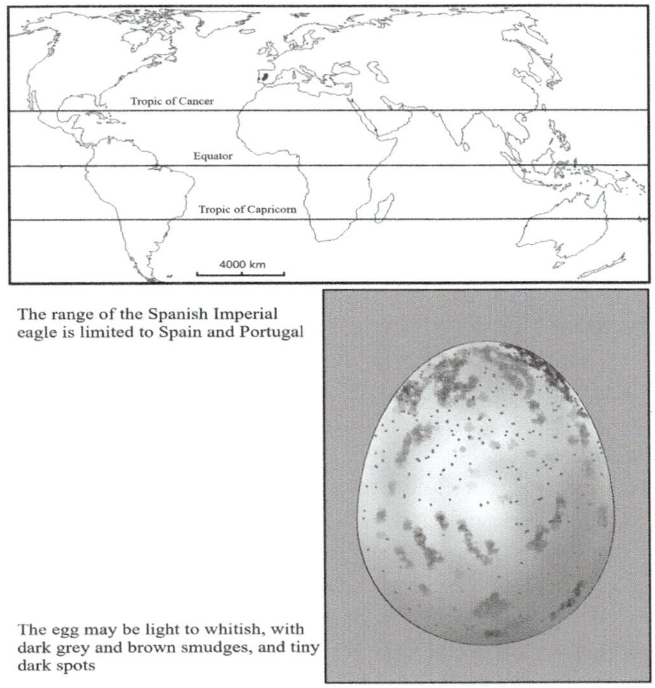

Fig. 1.4l. Spanish Imperial eagle range and egg

1.6 Eastern Imperial Eagle *Aquila heliaca* (Savigny, 1809)

Physical Description

This is "a large, very dark brown eagle with a strongly contrasting pale patch on the nape and white patch varying in extent on the shoulders" (Brown & Amadon, 1968, 656). According to Brown and Amadon (1968, 656) the adult brown with variable amounts of reddish tinging, and the "crown, nape, sides of head and neck are lanceolate pale tawny buff feathers" the visible results being a "pale patch." The forehead is streaked with black and the hind neck with central dark streaks. The tail base is grey, mottled dark brown with a black, lighter edged sub-terminal band. There is a white shoulder patch, formed from the white feathers on the scapulars. This has been referred to as an "epaulette", which in human clothing is an ornamental shoulder piece on a shirt or coat, especially on a military uniform (Brown & Amadon 1968, 656). The wing quills are black or almost black and the underparts are dark brown. The cere and feet are yellow, and the iris is hazel. This species may be distinguished from the Spanish Imperial eagle by its larger size small shoulder patch. Brown and Amadon (1968, 656) distinguishes this species from the Golden eagle in that there is a "white shoulder patch, the short hind claws and the streaked hind claw and the streaked underside of the immature bird." It is also the "largest and most majestic of the Eurasian brown eagles" (Brown & Amadon, 1968, 656).

The juvenile is lighter colored, with light brown upper parts, with fewer spots on lower back, rump and upper tail coverts. There is also brown edging on the feathers, and a pale tipped, rufous brown tail. The wing quills are blackish with lighter, lighter tipped secondaries. The reddish-brown underparts have dark-edged feathers on the throat, breast and abdomen, appearing streaked. The cere and feet are yellow, and the iris is dark brown. This juvenile plumage is less reddish than the immature Spanish Imperial eagle and more steaked below (Brown & Amadon, 1968, 656). Illustrations of the appearance, range, and eggs of this species are shown in Figures 1.5a to 1.5l.

Classification

The key issue for this species is its genetic separation from the Spanish Imperial eagle, as previously both were seen as one species with two subspecies *Aquila heliacaheliaca* (outside Spain, the current Eastern species) and *Aquila heliaca adalberti* (within Spain, the current Spanish species). Garcia (2007, 151) conducted a review of the paleontological records of the two species, considering the paleoecological environment in Eurasia during the Pleistocene (2,580,000 to 11,700 years BP) and Holocene (11,650 years BP to the present) epochs. Garcia (2007) argues that the Eastern Imperial eagle expanded its distribution during the last Pleistocenic glaciation, along the expansion of the Eurasian steppes. favoured by the expansion of the steppes (Global Raptor Information Network, 2021f).

The Spanish Imperial eagle was first recorded in the Late Pleistocene and Early Holocene, from evidence in the eastern Iberian Peninsula. Garcia (2007) notes that its distribution was linked to range of Mediterranean vegetation and the European rabbit (*Oryctolagus cuniculus*, Linnaeus, 1758). The Spanish Imperial eagle may have evolved from migrant eastern birds, which arrived in the Iberian Peninsula during the linking period between the Pleistocene and the Holocene. Such migrants, in adapting to the Mediterranean ecosystem, would have specialized in the abundant rabbit prey, which through abundance may have led the eagles into a more sedentary foraging pattern (rather than a migratory method, in search of less abundant food). Gradually, interbreeding migrants may have become genetically distinct from other eastern birds, "initiating the speciation mechanisms for sympatry or parapatry that resulted in *A. adalberti*" (Garcia, 2007, 151). Here, sympatry refers to two species in the same location, while parapatry refers to two species adjacent in range, but are almost reproductively isolated. Garcia (2007, 151) further acknowledges that "this is one possible mechanism. These findings reported here support the recent age of divergence between both taxons, and the incipient speciation supports its taxonomical considerations as a semi-species."

Martínez-Cruz & Godoy (2007) introduced a "multilocus approach and the Isolation-Migration model based on coalescence theory to estimate the time of divergence of the Spanish and Eastern Imperial eagle sister species." Here multilocus refers to the systematic sequencing of several genes or loci within the bacterial genome. A genome refers to "an organism's complete set of DNA, including all of its genes. Each genome contains all of the information needed to build and maintain that organism" (Genetics Home Reference, 2020). The findings of this research indicated that the Eastern Imperial eagle and the Spanish Imperial eagle may have diverged genetically during the Holocene or the later Pleistocene, this being perhaps later than current expectations. They also suggest there was a large population reduction at the time of the split, which may have been a fraction of the current population and the western population was significantly smaller than the eastern population.

Gene flow from the Eastern Imperial eagle to the Spanish Imperial eagle, could be detected for the nuclear genome (the complete set of DNA within the nucleus of an organism) but not for the mitochondrial genome (transmitted along the female line). Martínez-Cruz & Godoy (2007) argue that "male-mediated gene flow after divergence may explain this result, and the previously reported lower mitochondrial diversity but similar nuclear diversity in Spanish imperial eagles compared to the Eastern species." The conclusion is that the two species diverged from a common ancestor earlier than previously assumed, and there was asymmetrical gene flow after this split (Martínez-Cruz & Godoy, 2007).

This is recorded as relatively silent, but noisier during the breeding season. A barking call "owk-owk-owk" (Brown & Amadon, 1968). Another description is a deep, harsh bark owkowk, gokgok or kraw-kraw, which may be repeated rapidly several times (Ferguson-Lees & Christie, 2001; Gould & Rutgers, 1966; Schulze & Dingler, 2003). Bouglouan (2020b) gives "bark-like sounds 'rao,rao,rao' during the breeding season. Other harsh and very short calls may be heard, only one 'kaok' in flight, and rapid 'kokokoko' or 'gock-gock-gock.'"

Foraging

Birdlife International (2019) lists Forest, Shrubland, Grassland, Wetlands (inland), Artificial/Terrestrial as the main habitat. Brown and Amadon (1968) list low flat plains, normally non-mountainous areas. It forages by perching and soaring of the Eastern Imperial eagle. Food taken includes both carrion and living prey. In parts of the range, the main prey is susliks (Genus *Spermophilus*, Cuvier, 1825) which may be a factor in their distribution determines their range. Other prey taken includes snakes, gamebirds, waders (Order *Charadriiformes*, Huxley, 1867) and ducks (Order *Anseriformes*, Family *Anatidae*, Leach, 1820) (Brown & Amadon, 1968).

Breeding

Breeding starts with calling displays, paired soaring, and mutual male/female diving, and may involve rolling and locking of claws. The nest is very large, usually located in a tree in a forest, or on the forest edge or an open, treed plain. The clutch is composed of two or three eggs, colored very light brown or brownish yellow white, marked with some brown spots and with greyish markings. Egg laying takes place in mid-February to late March (Spain and southern Europe), the first half of April (Central Asia) and late April (north edge of range). Fledging takes place after more than 60 days (Brown & Amadon, 1968; Ferguson-Lees & Christie, 2001).

Population Status

The Eastern Imperial eagle occurs from east-central and south-eastern Europe in the west to central Russia, southern Siberia past Lake Baikal, the Russian Far East, Central Asia and north-western China (Xinjiang) and northern Mongolia. There are isolated populations in parts of Turkey, Cyprus, and Iran (Meyburg, 1994; Ferguson-Lees & Christie, 2001; Bakka et al., 2010; Bukreev et al., 2010; Karyakin et al., 2011; Birdlife International, 2019). The Eastern Imperial eagle is partially migratory, wintering in the northern parts of Africa, India and the Middle East, in some places migrating through the Bosphorus and Suez. Birds in the southern ranges are usually permanent residents. The Russian southward migration is mostly in groups and lasts from September to November. The northern migration is from Iran in late February to mid-March, usually by single or smaller flocks.

Birdlife International (2019) describes the Eastern Imperial eagle as sensitive to human presence, and cites this species as declining, the main threats being both direct and indirect human action. Direct actions include human action in nest robbing and illegal trade, shooting and poisoning for predatory behavior (for example, in Central Europe deliberate predator poisoning was the first cause of mortality from 2006 onwards, with 80 birds in 10 years in Hungary (Horváth et al., 2011). Indirect actions include agricultural expansion, wood and pulp plantations, utility and service lines, logging and wood harvesting, and human work and other activities. There is intensive forestry and deforestation of large trees in both mountains and lowlands. Threats include tree cutting in Russia and Bulgaria and Bulgaria (Karyakin et al., 2009a), increased human presence and infrastructural development in Central Europe and possibly other areas (Karyakin, 2011), and infrastructural and settlement network in the Hungarian Plain (Horváth, 2009). There is also the degradation of feeding habits and decline of prey species (especially the ground-squirrels of the Genus *Spermophilus*, Cuvier, 1825), electrocution by powerlines and collisions with road vehicles. Karyakin et al. (2009b) note that an average of 450 Eastern Imperial eagles died through powerline contact in the Altai region of Russia during the 2009 breeding season. The number of dead comprised one quarter of the Imperial eagle population of the region. Hunting and poisoning, and prey population reduction may have impacts along the migration routes. Interspecific competition with other eagle species also occurs, for example with the Greater Spotted eagle in the Altai region (Karyakin et al., 2009c). Illustrations of the appearance, range, and eggs of this species are shown in Figures 1.5a to 1.5l.

Fig. 1.5a. Eastern Imperial eagle perching [With permission from Raghuvir Solanki]

Fig. 1.5b. Eastern Imperial eagle perching [With permission from Corriplaya]

1.7 Wahlberg's Eagle *Aquila wahlbergi or Hieraaetus wahlbergi* (Sundevall, 1851)

Physical Description

Brown and Amadon (1968, 660) describe the Wahlberg's eagle as "a slim dark eagle, little larger than *Mivusmigrans* (the Black kite)...." It is medium-sized, about 53–61 cm long, and 437–845 g (males) to 670–1,400 g (females). The wingspan is 130–146 cm (Bouglouan, 2020c). The plumage is variable and polymorphic (with dark, intermediate and light-colored phases). The dark phase is commonest with an overall dark plumage (Bouglouan, 2020c). The adult male and female are similarly colored, with dark brown plumage, with paler coloured wing coverts, abdomen and head sides. The black primaries blend to lighter colors on

Fig. 1.5c. Eastern Imperial eagle perching [With permission from Anshul Jain]

Fig. 1.5d. Immature Eastern Imperial eagle perching [With permission from Ali Tellioglu]

Fig. 1.5e. Immature Eastern Imperial eagle perching [With permission from Tanmay's Gallery]

Fig. 1.5f. Immature Eastern Imperial eagle perching [With permission from Andy Li]

Fig. 1.5g. Eastern Imperial eagle in flight [With permission from Radovan Václav]

Fig. 1.5h. Eastern Imperial eagle in flight [With permission from Radovan Václav]

Fig. 1.5i. Immature Eastern Imperial eagle in flight [With permission from Mike Nelson-Birdtour ASIA]

Fig. 1.5j. Immature Eastern Imperial eagle in flight [With permission from Julian Thomas]

Fig. 1.5k. Eastern Imperial eagle habitat, Altai Steppe, Mountains [Credit: Bolatbek Gabiden]

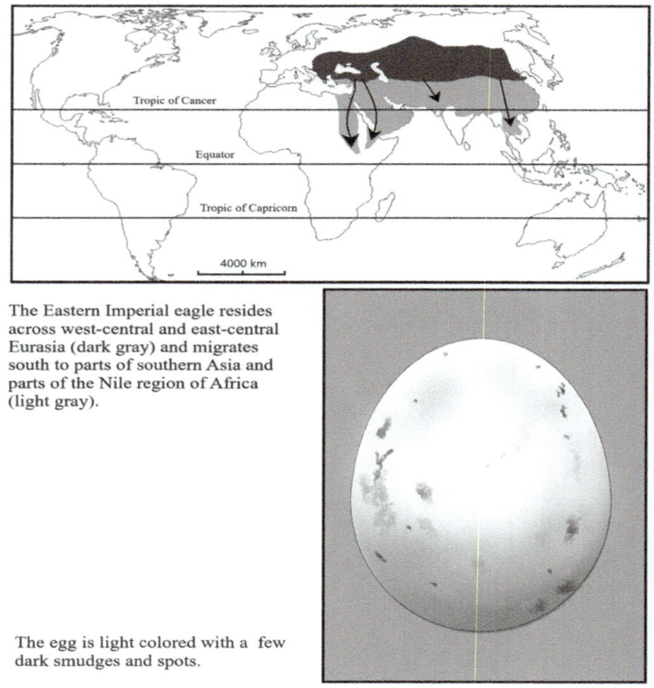

Fig. 1.5l. Eastern Imperial eagle range and map

the inner webs. The secondaries and tail are darker, the under-wing coverts are dark brown, and the underside of the flight quills and tail are silvery grey or brownish. The primaries and secondaries of the underwing have darker, blackish tips. The undertail is barred blackish. The intermediate phase is similarly colored, but the head plumage is usually darker than that of the body. The paler phase may be lighter brown, with the grey undertail and flight feathers undersides replaced with whitish color.

The pale phase is rarest, with a creamy white, dark streaked head, and dark feathers around the eyes. Head and neck are mainly creamy-white with blackish streaks on crown and around the eyes. The upperparts are dark grey-brown with lighter-edged feathers. The underparts and underwing-coverts are creamy-white with dark spots and streaks, and dark flight feathers and tail. There is a small occipital crest or nuchal crest (crest feathers in the occipital part of the back of the head) on the back of the dark brown head. The eyes are brown, the bill black, the cere yellow, and the darkly feathered legs and feet yellow. There is a substantial range in coloration from pale (even creamy or greyish white) to dark (Brown & Amadon, 1968; Bouglouan, 2020c). The juvenile is similar to the adult, but with a less visible crest, paler edged back and upper-wing covert feathers (Lachlan & Liversidge, 1978; Guggisberg, 1988; del Hoyo et al., 1994; Ferguson-Lees & Christie, 2001; Kemp & Kemp, 2005; Sinclair & Ryan, 2011).

Bouglouan (2020c) notes that the Wahlberg's eagle may be mistaken for other large birds of prey, such as the Lesser Spotted eagle, Booted eagle, Brown Snake-eagle and Black kite. However, Brown and Amadon (1968, 660) give some points that allow the identification of the Wahlberg's eagle. These include the "small, pointed crest" which "is usually visible"; "the gape only extends at maximum to the middle of the eye, whereas in lesser spotted eagle, it extends to the back of the eye." Another point is the round nostrils, which are absent in Tawny and Steppe eagles, but present in the two spotted eagles. The Tawny eagle is the most similar species, but the Wahlberg's eagle is "much slimmer when on the wing" and the tail is more square, rather than slightly rounded. This species is also "distinctly paler" (Brown & Amadon, 1968, 660). In the flight shape is also an identifier for the Wahlberg's eagle, being "very cross-shaped, with long, evenly wide wings, a slim body, and a narrow, square-ended tail… The wings are held very flat" (Brown & Amadon, 1968, 660). The dark phase of the Booted eagle may be similar but is paler. The immature plumage of the Wahlberg's eagle is similar to that of the adult, "dark parents produce dark young and vice versa", but the juvenile is paler below and the head and back are more streaked (Brown & Amadon, 1968, 660). The call is a "shrill double wail" Kleeeee-ay made during aerial displays, the other call during greetings between adults is a higher pitched Kyip-Kyip-Kyip. There is also a long "drawn yelp, reminiscent of *Larus dominicanus*, female answered by male" (Brown & Amadon, 1968, 660). Bouglouan (2020) additionally records a display call, which is "a mournful whistle "quieee-aow" given both perched and in flight." Illustrations of the appearance, range, and eggs of this species are shown in Figures 1.6a to 1.6s.

Classification

As mentioned in introduction of this chapter, genetic background of the Wahlberg's Eagle is relatively uncertain, due to its movement from the Genus *Aquila* (Stresemann & Amadon, 1979; Sibley & Monroe, 1990; Kemp, 1994a) to the Genus *Hieraaetus* (Amadon & Bull, 1988; Ferguson-Lees & Christie, 2001), these are based on analytical interpretations of the morphological and behavioral characters of this species. Using more advanced molecular phylogenetic analysis of DNA sequences from one mitochondrial and three nuclear genes, Helbig et al. (2005) indicated that this species is a sister to a monophyletic group of *Hieraaetus* species (the Booted eagle *Hieraaetus pennatus*, Gmelin, 1788; Ayres Hawk eagle *Hieraaetus ayresii*, Gurney, 1862; and the Little eagle *Hieraaetus morphnoides*, Gould, 1841). Lerner and Mindell (2005) had findings that supported these conclusions, using molecular sequences of two mitochondrial and one nuclear gene. However, the Global Raptor Information Network (2021g) argues that "the majority of raptor systematists (e.g., Wink and Sauer-Gürth, 2004; Gjershaug, 2006) and several national committees on classification and nomenclature (e.g., those in the United Kingdom and Germany) now favor merging the species formerly assigned to *Hieraaetus* into *Aquila*, and they are followed here."

Foraging

The Wahlberg's eagle is commonest in forest margins and treed savannas (Ferguson-Lees & Christie, 2001). It is rarer in denser forest and sparser open landcover and mountains, although it is recorded in the Ethiopian highlands (Brown & Amadon, 1968). The range is over sub-Saharan Africa, including south Mauritania and Senegambia, to Sudan, Eritrea and Ethiopia, through East Africa to Botswana and parts of South Africa, Angola and Namibia. It is absent from parts of the Horn of Africa and the extreme south of Africa. In the northern parts of the range, it is migrant or nomadic. A migratory species, it moves south from July to September, and northwards from February to March.

The Global Raptor Information Network (2021g) describe the Wahlberg's eagle as frequenting denser wooded landcover than the Tawny eagle, and riparian and flat landcover. Regional studies show variations in frequented landcover. For example, in Ethiopia, the main habitat is river valleys (Ash & Atkins, 2009), while in Zambia and Zimbabwe the eagle frequents a variety of woodland habitats, including miombo (woodland savanna, especially dominated by the trees of the Genera *Brachystegia*, *Julbernardia* and *Isoberlinia* (Craib & Stapf ex Holland) and is rarer in drier areas (Dowsett et al., 2008). In Malawi it also has broad habitat frequencies, including miombo and the acacia belt and woodland patches in farmland (Dowsett-Lemaire and Dowsett, 2006). In Botswana, although it has a similarly broad presence, the main habitats are tall, broad-leafed forest, in river valleys, drainage areas, Kalahari woodland patches and treed savanna areas with human habitation (Penry, 1994). In Swaziland, from a sample of 30 nests, 29 were in natural vegetation mostly near small waterbodies, while one was in a eucalyptus plantation, and the rest were in natural vegetation, mostly along drainage lines or small rivers (Monadjem & Rasmussen, 2008). The range of Wahlberg's eagle often overlaps with the Tawny eagle, but there is little competition, even having different breeding seasons when the nest is in close proximity (Irwin, 1981).

The Wahlberg's eagle forages by soaring and perching, and frequenting bushfires (Brown & Amadon, 1968). Soaring may be at about 100 m altitude (Tarboton, 1990). Nikolaus (1987) recorded large flocks in Sudan, which grouped with Tawny eagles and Black kites after rains in late March. The prey species include birds, mammals, reptiles, amphibians and even insects (Ferguson-Lees & Christie, 2001). Brown and Amadon (1968) list reptiles as the main prey (50%), followed by mammals (30%) and birds (20%). However, Tarboton (1990) ranks birds as the most frequently taken prey across the eagle's habitats. The prey may include lizards and smaller snakes, ground squirrels, hyraxes and other small mammals and game birds and poultry (Deacon & Rockingham-Gill, 2009). Foraging includes perching and soaring, and frequenting bushfires (Brown & Amadon, 1968; Global Raptor Information Network, 2021g). Various prey species have been recorded as food. These include large yellow ants (probably Camponotus, Mayr, 1861) taken from the stomach of a male specimen in Tanzania (Loveridge, 1923); small animals and prey around plows in Zambia (Osborne & Robjent, 1975; Dowsett et al., 2008); bats taken at Kasinthula in Malawi (Stephens & Blackwood, 1983); and in Zimbabwe, lizards, especially Genus *Agama* (Daudin, 1802) and plated lizards (Family *Gerrhosauridae*, Fitzinger, 1843), small mammals such as the Scrub hare (*Lepus saxatilis*, F. Cuvier, 1823) and the guineafowl (*Numididae*, Longchamps, 1842) and francolins (*Perdicinae*, Horsfield, 1821) (Irwin, 1981). Irwin (1981) notes that "in one study area over a 15-year period, mammals made up 28.3% of the diet, birds 31.5%, and reptiles 40.2%."

Breeding

The breeding season begins with displays, which involve soaring flight over breeding sites, followed by paired movements, male diving towards females and interlocking claws. Breeding times vary, generally around September to February, but may be during June to November in West Africa (Ferguson-Lees & Christie, 2001). Nests are located in tall trees, such as Acacias (*Acacia*, Mill.) and fig trees (*Ficus* L.). The nests may increase in size with time, from 75 cm across and 15 cm deep, to 60 cm deep (Brown & Amadon, 1968). Ferguson-Lees & Christie (2001) describe nests as up to 80 cm in diameter, composed of sticks. The egg is coarse, white with reddish marks at the larger end. In East Africa one egg is laid, but two eggs may be laid in the southern range. Fledging occurs 70 to 80 days after hatching (Brown & Amadon, 1968; Ferguson-Lees & Christie, 2001; Bouglouan, 2020).

Population Status

BirdLife International (2016) notes that the Wahlberg's eagle has a vast range, therefore it may not be classified as Vulnerable under the range size criterion ("extent of occurrence <20,000 km^2 combined with a declining or fluctuating range size, habitat extent/quality, or population size and a small number of locations or severe fragmentation"). Additionally, the population is large and stable, hence the species is evaluated as Least Concern. Ferguson-Lees & Christie (2001) estimated the global population of adults and immatures at the start of the breeding season at between 101,000 and 1,000,000. Although threats to the population may include accidental poisoning, human disturbance or deforestation in mixed landcover, none of these factors have been found to decimate the population (del Hoyo et al., 1994; Ferguson-Lees & Christie, 2001). There have been variable regional population estimates. For example, Simmons (1997c) estimated 50,000 pairs in southern Africa, while Parker (1999) estimated 4,000 birds in southern Mozambique, and (Parker, 2005a) estimated the breeding population in central Mozambique to be 5,000 birds. Parker (1994) estimated 200 birds in Swaziland, while Tarboton and Allan (1984) estimated the total former Transvaal population at about 18,000 birds. Monadjem & Rasmussen (2008, 1) note that "Wahlberg's eagle is the most commonly observed eagle in Swaziland… This species is not currently threatened in Swaziland" (see also Monadjem et al., 2003).

There are gaps in the knowledge of the population distribution of the Wahlberg's eagle. Meyburg (1995, 135) argued that "although Wahlberg's eagle is one of the most common eagles in Africa, its migratory movements are not properly understood." Records exist of movements of ringed birds from Zimbabwe to Rwanda, from eastern South Africa near the Kruger National Park, to Zambia, northern Sudan, and northern Zaire (Simmons, 1990, 1991). Meyburg (1995, 135) notes that "an adult female Wahlberg's eagle from northern Namibia was tracked by satellite over a total distance of 8816 km and located 104 times between 11 February and 4 November 1994. It migrated on an almost due north heading to northern Cameroon, north-eastern Nigeria and western Chad through the rain forest belt of the Congo and Zaire after the breeding season. The total trans-equatorial distance between the breeding and non-breeding ranges was 3520 km." Furthermore, the bird ranged over about 60,000 km^2 during the non-breeding season, for the six weeks between the end of April and mid-June in the three countries. For two more months it appeared to be restricted to an area of about 5000 km^2 close to Maiduguri in the savanna of north-eastern Nigeria. There was then a return migration, that lasted two weeks longer than the month-long initial movement (Meyburg, 1995, 135).

Some migration patterns are difficult to determine. For example, Steyn (1982) described the migration pattern as unclear, although it appears southern African birds migrate north across the equator. According to Brown et al. (1982) was uncertain of the non-breeding areas but surmised them to be in the northern savannas of Sudan, Chad, west to Gambia, from March to August, coinciding with rainy season. Thiollay (1975) recorded over 1200 Wahlberg's eagles during the 1974 migration in Kabalega Falls National Park, Uganda. Thiollay (1977) describes the Wahlberg's eagle's range as between 12° and 14° N in the rainy season in West Africa. Commenting on these issues, Meyburg (1995, 135) calls for more advanced research: "In view of the unlikelihood of collecting much coherent information on migration by means of ringing, wing marking, direct observation or any other methods…it is suggested that more individuals be tracked by satellite in order to better understand the migration system of different populations of this species and its dependence on various environmental factors, as well as the conservation constraints and requirements in all parts of its range." Dated literature is cited encompassing current knowledge of the population dynamics of the Wahlberg's eagle (Ulfstrand, 1960; Steyn, 1962, 1982; Brown, 1970, 1978; Smeenk, 1974; Thiollay, 1975, 1976, 1977; Anthony, 1979; Van de Weghe, 1979; Brown et al., 1982; Simmons, 1990, 1991; Auburn, 1991). Illustrations of the appearance, range, and eggs of this species are shown in Figures 1.6a to 1.6g.

1.8 Gurney's Eagle *Aquila gurneyi* (Gray, 1860)

Physical Description

This eagle has brownish black plumage, with darker and glossier feathers on the crown and shoulders. The crown and nape feathers are pointed and there is whitish mottling on the tail quills and wing quill bases, both above and below. Juveniles have a whitish to yellowish white head and underparts, and dark markings on the crown and breast. The shoulders and back are dark with lighter colored markings. The wing and tail quills are almost black, with lighter grey markings and the undersides are lighter colored. The tail is long and rounded (Brown & Amadon, 1968: 663). It may be mistaken for other species, but it is more heavily built than the Black eagle, and also "lacks the white carpal areas in the spread wing" (Brown & Amadon, 1968: 663). The Wedge-tailed eagle `has a different wedge-shaped tail, and "it is doubtful if the latter is more than a straggler to New Guinea". The immature New Guinea Harpy eagle differs from the juvenile as the latter has short, feathered legs (ibid. 663). Illustrations of the appearance, range, and eggs of this species are shown in Figures 1.7a to 1.7j.

Fig. 1.6a. Wahlberg's eagle perching [With permission from Mario Raimondi]

Fig. 1.6b. Wahlberg's eagle perching [With permission from Frank McClintock]

Fig. 1.6c. Immature Wahlberg's eagle perching [With permission from Ohannes Groß (j-g-fotos@web.de)]

Fig. 1.6d. Wahlberg's eagle in flight [With permission from Mario Raimondi]

Fig. 1.6e. Wahlberg's eagle paler morph [With permission Pieter Verheij]

Fig. 1.6f. Wahlberg's eagle habitat, open woodland and savanna [With permission from Jonathan D. Müller]

Classification

The Global Raptor Information Network (2021h) describes this species as monotypic. Lerner and Mindell (2005) used molecular phylogenetic analysis, using the DNA sequences of one nuclear and two mitochondrial genes, and found that the genus *Aquila* is not monophyletic. They found that *A. gurneyi* is part of a clade of closely related species, including *A. chrysaetos*, *A. verreauxii*, *A. audax, Spizaetus africanus* and *Hieraaetus fasciatus*. This advanced technique is used more than the morphological data study by Brown and Amadon (1968), which proposed a close link between *A. gurneyi*, *A. chrysaetos*, *A. audax* and *A. verreauxii*. Haring et al. (2007) also conducted a study based on sequences of the mitochondrial gene cytochrome b gene, which indicated *A. gurneyi* clustered with *A. africanus*, *A. fasciatus* and *A. audax*. Possibly, based on this history, the final conclusions for the position of this species may yet to be determined.

Breeding

The Global Raptor Information Network (2021h) lists no information for breeding, and Von Balen (1998) stresses the need for more research on this species and others.

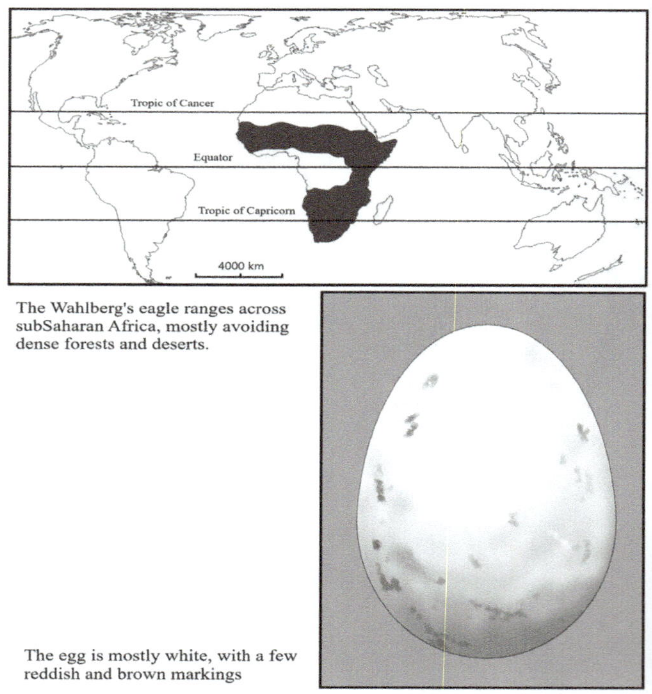

Fig. 1.6g. Wahlberg's eagle Range and Egg

Foraging

Gurney's eagle occurs frequently in dense vegetation, especially primary rainforest. However, several authors have recorded it in other habitats. For example, it has been recorded in swamp forest and cleared areas and coastal zones, foraging by soaring and perching, singly and in pairs. It may soar over forest canopy, and fly low over forest clearings (Coates, 1985; Beehler et al., 1986; Coates & Bishop, 1997; Coates, 2001). BirdLife International (2016) acknowledges that the Gurney's eagle frequents various habitats, usually below 1,000 meters elevation but up to 1,500, but the primary habitat remains primary forest. In the Moluccas, as elsewhere, it has been recorded in primary lowland and hill forest, but also coastal areas and even coconut plantations. Small islands appear to be common habitats (Rand & Gilliard, 1967). Common prey species include arboreal mammals, including possums (Family *Didelphidae*, Gray, 1821) (Pratt & Beehler, 2015).

Population Status

Gurney's eagle has a wide range through South East Asia, frequenting the Moluccas, Indonesia and New Guinea (Irham, 2012; Mittermeier et al., 2013; Reeve et al., 2014; Rheindt et al., 2014). BirdLife International (2016) notes that "population numbers are not known." In terms of population trends, BirdLife International (2016) argues "there are no data on population trends; however, the species is suspected to be in decline owing to habitat loss and degradation." Bildstein (2006) describes this species as an irruptive or local migrant. Willow (2020) classifies irruptive migration as one of three migration types: *complete,* or *obligate*, migration where the entire population of a species moves seasonally from a breeding range to a wintering range; *partial migration*, where parts of the population migrate and others stay on their breeding grounds; and *irruptive migration*, where especially nomadic species may migrate to certain areas in some years, and then migrate to other regions in other years. Willow (2020) further notes that "irruptive migration is not seasonally or geographically predictable; it is not a true form of migration. For irruptive migrants, in contrast to regular migrants, site fidelity is low, and few individuals return to the same breeding areas in successive years. Having neither a fixed breeding ground nor wintering ground, these species are the vagabonds of the bird world."

The Global Raptor Information Network (2021h) notes that this species is "little known and probably scarce throughout its fairly extensive range, but badly in need of a comprehensive population survey. Most likely, this species is declining because of habitat loss and degradation, but this has not been confirmed by actual field studies"; it is "categorized globally as 'Near Threatened' by BirdLife International, based on its apparently low numbers." Several authors have supported this position. Coates and Bishop (1997) describe this species as a rare and sparsely distributed resident of the Maluku Islands or the Moluccas. Other researchers have recorded this species in several islands and habitats in South East Asia. These include Indonesia (Rheindt & Hutchinson, 2007, both sightings at 1,200 elevation), Australia (Debus, 1998; Palliser, 2007, both near New Guinea, Global Raptor Information Network (2021h) speculates it might be resident in that area) and New Guinea, where it is resident on some islands mostly on lowland forest below 1,300 (Rand & Gilliard, 1967; Coates, 1985; Beehler et al., 1986; Coates & Peckover, 2001). It is also found near the coast of New Zealand. Also, "it has been recorded as soaring above the forests, but details of its habits are lacking" (Brown & Amadon, 1968, 663). Illustrations of the appearance, range, and eggs of this species are shown in Figures 1.7a to 1.7j.

Fig. 1.7a. Gurney's eagle [With permission from Christopher Hamilton]

Fig. 1.7b. Gurneys eagle perching [With permission from Mike Nelson Birdtour ASIA]

Fig. 1.7c. Immature Gurney's eagle perching [With permission from Davide Beadle]

Fig. 1.7d. Gurney's eagle in flight [With permission from Raphaël Jordan]

Fig. 1.7e. Gurney's eagle in flight [With permission from Davide Beadle]

Fig. 1.7f. Gurney's eagle [With permission from Tinyfishy]

Fig. 1.7g. Immature Gurney's eagle in flight [With permission from Davide Beadle]

Fig. 1.7h. Immature Gurney's eagle in flight [With permission from Tinyfishy]

Fig. 1.7i. Gurney's eagle habitat, typical forest in New Guinea [With permission from Zacky Ezedin]

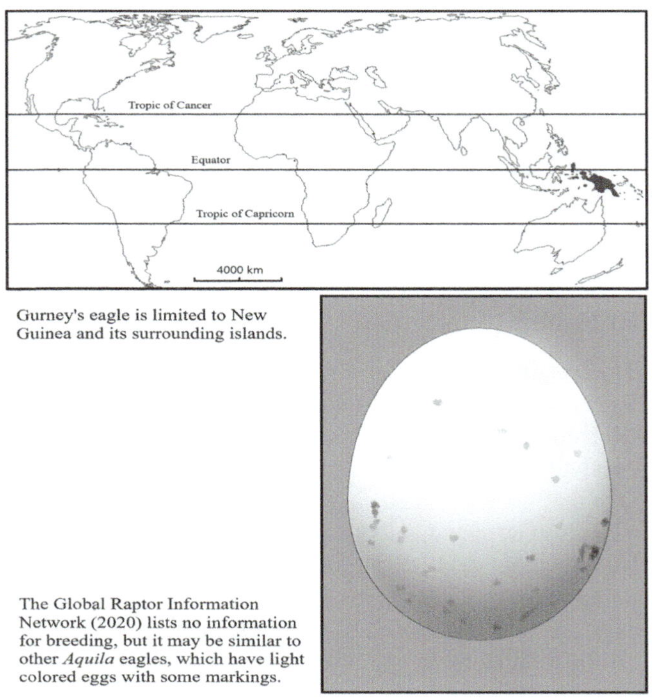

Fig. 1.7j. Gurney's eagle range and egg

1.9 Wedge-tailed Eagle *Aquila audax* (Latham, 1801)

Physical Description

The Wedge-tailed eagle is Australia's largest bird of prey. The State of Victoria Department of Environment, Land, Water and Planning (2018) describes this species as "Australia's largest raptor… which is truly a magnificent bird." According to Brown and Amadon (1968, 669) and Gill & Donsker (2019) there are two subspecies: *Aquila audax audax* (Latham, 1801) occurring mostly in Australia, southern New Guinea, and *Aquila audax fleayi* (Condon & Amadon, 1954) found mostly in Tasmania. *A.a. audax* is covered in glossy brownish black plumage, with golden reddish lanceolate nape feathers and reddish edges in the wing coverts. The eyes are brown, and the cere and feet are yellow. The juvenile has lighter colored plumage, the head and neck covered with golden rufous feathers, streaked dark on the crown, streaked paler brown on the nape and shoulders. The upperparts (back and upper wing and upper tail coverts) are covered with rufous or rufous-edged dark feathers, the wing quills are black with discernable grey mottling in the inner webs, and the tail is black. The underparts are brown, with darker streaks on the chin, throat and upper breast area, and there are dark feathers on the ear coverts, visible compared with the golden colored nape. The breast, belly and the thighs are dark brown with hidden white bases, the underwing coverts are dark brown, and the least underwing coverts at the front edge of the wing have some rufous feathers. The underwing quills are black tipped, with basal greyish brown feathers with dark mottles. The undertail feathers are black with grey mottling. *A.a. fleayi* is similar, but has light brown nape feathers, with no reddish or golden brown upper feathers. *A.a. fleayi* is the same size as *A.a. audax*, but has "heavier claws" (Brown & Amadon, 1968, 669).

The Wedge-tailed eagle's call is a shrill "I see-I see" or "pee-ya pee-ya" (Brown and Amadon, 1968, 669). These authors also give other calls: "a rolling leisurely series of emphatic whistles"; a "shrill chortling series of short calls, high pitched, followed by pee-ya"; "rapid low-toned whistling" by the male. In aggressive reaction to other members of the species the call is "a slurred, prolonged "yessirr"; the alarm cry is a "long melancholy scream"; there are also "loud mating calls of double notes with rising inflection." Finally, the voice of the female is "lower and harsher than that of the male" (ibid.). Illustrations of the appearance, range, and eggs of this species are shown in Figures 1.8a to 1.8l.

Classification

The Wedge-tailed eagle is classified in *Aquila* but was formerly classified in the separate Genus *Uroaetus* (Kaup, 1844) (Global Raptor Information Network, 2021i). However, later studies using the molecular sequences of two mitochondrial genes and one nuclear gene, showed that the genus *Aquila* is not monophyletic, and that this species is closely related to the Golden eagle, Cassin's Hawk-eagle, Bonelli's eagle, Verreaux's eagle and Gurney's eagle (Lerner & Mindell, 2005). This study was supported by Helbig et al. (2005), who used DNA sequences from one mitochondrial and two nuclear genes. More recently, Haring et al. (2007) based used sequences of the mitochondrial gene cytochrome b gene, and found that the Wedge-tailed eagle clustered with Cassin's Hawk-eagle, Gurney's eagle and Bonelli's eagle. Brown and Amadon (1968) had proposed a similar close relation between the Wedge-tailed eagle and the Golden eagle, Gurney's eagle and Verreaux's eagle, but this study was based on morphological data.

Breeding

The breeding season of the Wedge-tailed eagle begins with nuptial, aerial displays, which include the male approaching a circling female, an inward swoop and in some cases the female rolling over and presenting her claws. Calling is part of the displays. The Global Raptor Information Network (2021i) classifies this species as a solitary nester, with nests generally built in trees, mostly in eucalypts two to 73 meters off the ground (Debus, 1998), or more rarely on cliffs, rocks or the ground. Burnett et al. (1996, 277) note that "Wedge-tailed eagles are known to nest in closed forest habitats. This behaviour is particularly common in Tasmania" (see also Gaffney & Mooney, 1992) (e.g. a nest-site was recorded in an emergent Turpentine tree (*Syncarpia glomulifera* (Sm.) Nied.) on the Mt. Windsor Table and nests used for several years may be large, over two meters wide and a similar depth, with a lining of green leaves in the laying area. One to three, sometimes four white, sometimes reddish spotted and/or purplish blotched eggs are laid (Baker-Gabb, 1983). Incubation lasts 42-44 days and nestling 79-90 days (Debus et al., 2007a; Global Raptor Information Network, 2021i). Brown & Amadon (1968) note that fledging occurs 63 to 70 days after hatching. Although parents attend to the young, the female is the main provider, and the juveniles are dependent on the parents for over four months past fledging (Debus et al., 2007a). Research findings indicate regional variations in breeding. For example, one example in New Guinea found the breeding starting March with fledging in mid-October (Stronoch, 1981). In New South Wales, breeding started from early to late June or early July (Debus et al., 2007). The nesting periods in New South Wales varied between 90, 79 and 87 days (Harder, 2000; Allott et al., 2006; Debus et al., 2007a). The incubation periods varied between 42, 43 and 44 days (Harder, 2000; Allott et al., 2006; Debus et al., 2007a).

Foraging

The Wedge-tailed eagle is principally a soaring species, foraging from high altitudes. It also congregates at carrion sources. It frequents principally open country, but may occur in a wide range of landcover, from open desert to forest. Marchant & Higgins (1993) and Burnett et al. (1996, 277) emphasize the "impressive repertoire" of hunting techniques and the variety of prey taken and the habitats utilized. This study notes that "the few published accounts of Wedge-tailed eagles hunting in or at least pursuing prey into dense forests are difficult to interpret and none seem to be of rainforest hunting" of the type described in their paper (Burnett et al., 1996, 277). Studies reporting forest foraging included Mooney (1988) and Fawkner (1991), and that of Gaffney & Mooney (1992) in Tasmania, and of Smith (1984) and Czechura (1985) in Bega and south-eastern Queensland.

Most kills taken from soaring or perching are on the ground. Global Raptor Information Network (2021i) notes that "this eagle forages by low, slow quartering, high soaring, or by still-hunting from a perch. Prey is seized from the ground or occasionally from the tree canopy after a swift glide or dive; prey is rarely taken in flight. Certain prey, e.g., possums, may be removed from tree hollows." Brown and Amadon (1968) note that its predatory attack is frequently a "long slanting stoop" (ibid.). Prey taken includes rabbits, hares, and more rarely wallabies; other species taken include cats, foxes, ducks, herons, ibis, bustards, galahs and goanna lizards and lambs (the last may be taken dead or alive) (Brown & Amadon, 1968; Debus, 1998). Other prey includes road kills. It is sometimes a social feeder, with large flocks congregating at a carcass (Burnett et al., 1996, 277). There are regional variants in food taken (Global Raptor Information Network, 2021i). For example, rabbits and hares are commonly taken in the southern parts of its range, while young kangaroos and wallabies are frequently taken in the north. Opossums are a common prey in the forest. In some cases, birds are taken, such as crows, cockatoos, waterfowl, and even large cranes and bustards (Debus 1998). Pairs and social groups may cooperate in hunting and may share prey and even rob scavengers at carcasses and other raptors (Debus, 1998; Global Raptor Information Network, 2021i).

Population Status

Wedge-tailed eagles are fairly common in some localities, such as Victoria (State of Victoria Department of Environment, Land, Water and Planning, 2018). The main threat was past persecution. As noted by the State of Victoria Department of Environment, Land, Water and Planning (2018), "it has previously been believed that Wedge-tailed eagles were responsible for killing livestock, as a result, they were once legally killed in large numbers. Global Raptor Information Network (2021i) notes that "this species is still shot, trapped, and deliberately poisoned in some areas by farmers who regard it as a serious predator of sheep." Research has since established that Wedge-tailed eagles hardly ever attack healthy lambs and never fully-grown sheep." Other current threats are the clearing of habitats, poisoning of animals by pesticides and baits, the carcasses of which are later ingested by the eagles, and traffic hits on birds feeding on road kills (State of Victoria Department of Environment, Land, Water and Planning, 2018).

The Global Raptor Information Network (2021i) lists the population status of this species as "stable." Although Ferguson-Lees & Christie (2001) estimated the global population of adults and immatures at the start of the breeding season as being from 101,000 to 1,000,000 individuals, BirdLife International (2009) estimated a smaller number of 100,000 mature individuals, with an acknowledgment of insufficient data for a definitive assessment. In its range, it is a resident or an irruptive or local migrant, with some wide dispersal of juveniles from breeding areas (Baker-Gabb, 1983; Debus, 1998). One reason for its widespread distribution in Australia is the introduction of rabbits and the clearance of some forests, which allows some foraging. Declines have nevertheless been reported. The reporting rates for the Wedge-tailed eagle have declined 28% in Australia and by 15% in New South Wales (Debus, 1998). Further, the Tasmanian subspecies *A.a. fleayi* has been classified as Vulnerable by the IUCN (Garnett et al., 2011). In Tasmania, this subspecies endures habitat destruction, especially the losses of nest trees, illegal shooting, and poisoning (Mooney, 1988). Debus (2009) argues that there is insufficient federal or state support for conservation measures in Tasmania (Global Raptor Information Network, 2021i). Illustrations of the appearance, range, and eggs of this species are shown in Figures 1.8a to 1.8l.

1.10 Verreaux's or Black Eagle *Aquila verreauxi* (Lesson, 1830)

Physical Description

The Verreaux's eagle is a large eagle similar in dimensions to the Martial eagle and Golden eagle. It is 75 to 96 cm in length and weighs 3 to 4.2 kg (males) and 3.1 to 7 kg (females). The wingspan is 1.81 to 2.3 m (Ferguson-Lees & Christie, 2001). This species has totally black plumage with a small V-shaped patch of white features in the center of the back, rump and upper tail coverts, and some scapulars and some more white features on the carpal joint. These are very pronounced in flight, when viewed from above. The adult bird has brown eyes and cere, a black bill and yellow to orangish yellow eyelids and feet, which contrast strongly with the black plumage. The head is prominent with a long neck, and bill is large and strong. The juvenile has lighter colored plumage, as the crown and mantle are reddish ginger with a white streaked forehead and black cheek patches. The upper

Fig. 1.8a. Wedge-tailed eagle perching [With permission from Brendan Shepard]

Fig. 1.8b. Wedge-tailed eagle perching [With Permission from Brendan Shepard]

Fig. 1.8c. Wedge-tailed eagle perching [With permission from Nick Athanas]

Fig. 1.8d. Immature Wedge-tailed eagle perching [With permission from Brendan Shepard]

Fig. 1.8e. Immature Wedge-tailed eagle perching [With Permission from Mike Nelson-Birdtour ASIA]

Fig. 1.8f. Wedge-tailed eagle in flight [With permission from Mike Nelson-Birdtour ASIA]

Fig. 1.8g. Wedge-tailed eagle in flight [With permission from Frank McClintock]

Fig. 1.8h. Wedge-tailed eagle habitat and nest [With permission from Bill Doyle]

Fig. 1.8i. Wedge-tailed eagle habitat and nest [With permission from Neil Edwards]

Fig. 1.8j. Wedge-tailed eagle habitat and nest [With permission from Bill Doyle]

Fig. 1.8k. Wedge-tailed eagle habitat and nest [With permission from Bill Doyle]

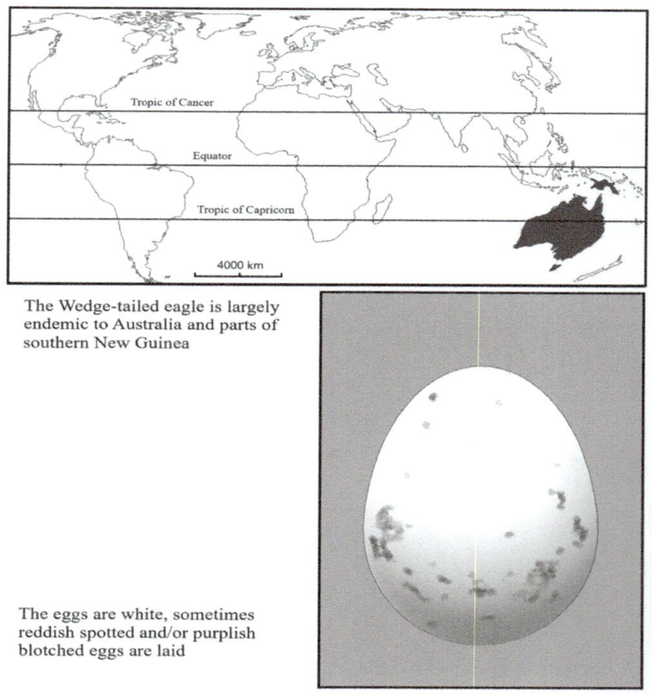

Fig. 1.8l. Wedge-tailed eagle range and egg

tail and upper wing coverts are dark with pale edges. The rump and lower back have brown spots, and the tail and wing quills are very dark brown. The underside is brown, with white feathers in the carpal joint, but the pale brown breast and lower throat contrasts with the black chest. In flight, the wings are "broadest about 2/3 along their length, and with the inner secondaries very short so that the general wing silhouette is leaf shaped" (Brown & Amadon, 1968, 672). The juvenile has a black bill, dark brown iris and yellow cere and feet. Ferguson-Lees & Christie (2001) describe this species as largely silent, but perhaps more vocal than the Golden eagle. The call is a clicking sound – "pyuck", and also a "savage bark chyow", the latter usually in alarm or danger (Brown & Amadon, 1968, 1986). Ferguson-Lees & Christie (2001) also list other calls such as long and ringing cries: whaeee-whaeee, heeeee-oh or keeooo-keeooo and in some cases screams, barks, yelps and mews (Brown & Amadon, 1986; Ferguson-Lees & Christie, 2001). Illustrations of the appearance, range, and eggs of this species are shown in Figures 1.9a to 1.9i.

Breeding

Verreaux's eagles perform nuptial displays, which include undulating flight, with male birds diving, rising and sideways rolls, diving at rolling females which may show their claws, and occasionally both birds may interlock the claws. After nuptial displays, nesting occurs on crags and less commonly in trees. The nests are broad but shallow, being up to two and half meters wide but usually less than a meter deep. One to three, usually two, eggs are laid. These are longish, chalky white tinged bluish and may have some reddish-brown markings. Laying times vary regionally. In Sudan it is December to March, in Tanzania June; in Kenya April, June, July, September; in South Africa April to July; Zimbabwe April to June; and Somalia November to December. Fledging occurs 99 days after the eggs hatch hatching (Brown & Amadon, 1968).

Foraging

The Verreaux's eagle frequents arid landcover, especially with rocky hills and crags, and is relatively rare especially in areas with more than 750 mm rainfall. Foraging is mostly over this open terrain. The Verreaux's eagle flies with the wings held in a shallow V, with the primaries turned upwards at the edge. Most prey is taken on the ground, mostly small mammals such as Rock hyrax (*Procavia capensis*, Pallas, 1766) and Yellow-spotted rock hyrax (*Heterohyrax brucei*, Gray, 1868) and dikdiks (Genus *Madoqua*, Ogilby, 1837) and mongooses (*Herpestidae*, Bonaparte, 1845). It also eats birds such as francolins (Subfamily *Perdicinae*, Horsfield, 1821), guinea fowl (*Numididae*, Longchamps, 1842), poultry (Order *Galliformes*, Temminck, 1820) and reptiles. The Verreaux's eagle also competes with other similar species such as the Martial eagle (Ferguson-Lees & Christie, 2001; Global Raptor Information Network, 2021j)).

Population Status

Population factors include the success rate of nesting, which may be up to 50% success rate. The success or otherwise of nesting has been linked to the presence of their main prey, hyraxes (Brown & Amadon, 1986). Gargett et al. (1995) further note that where there were few or no hyraxes, the majority of eagles did not even attempt to breed (66% rather than the 24% that attempt to breed in areas with hyrax presence). Breeding frequency also declines during years with higher rainfall. For example, over 80% attempted nesting when the average annual rainfall was 300 mm, but 45% of the eagles attempted nesting when the rainfall was 1,000 mm. (Gargett et al., 1995). Factors favoring the population level of the Verreaux's eagle include its main habitat (open land cover like kopje landscapes; small hills in a generally flat, arid lightly vegetated areas) which are less vulnerable to human modification than more complex landcover such as savanna. They are also less likely to take carrion than some other large eagles, as such they are less vulnerable to poisoning of carcasses for predator control (Ferguson-Lees & Christie, 2001). They may be shot as threats to small livestock (Davies, 1999). They may also be affected by human hunting of their prey, hyraxes. Case studies have indicated that some breeding eagles are resilient to these activities and may also survive close human presence and intensified activities (Simmons et al., 2007; Kruger, 2010). Illustrations of the appearance, range, and eggs of this species are shown in Figures 1.9a to 1.9i.

1.11 Bonelli's Eagle *Aquila fasciata* (Vieillot, 1822) or *Hieraaetus fasciatus* (Vieillot, 1822)

Physical Description

The Bonelli's eagle is a medium-sized eagle, one of the smallest species of the Genus *Aquila* (when classified within the Genus *Hieraaetus*, it was among the largest of that genus (Ferguson-Lees & Christie, 2001; Brown & Amadon, 1986; Naoroji & Schmitt, 2007). The adult is 55 to 74 cm long and the wingspan is from 143 to 163 cm to 180 cm, with the male at the lower end of this gradient and the female at the upper end (Naoroji & Schmitt, 2007; Dunning, 2008). The weight varies between approximately

Fig. 1.9a. Verreaux's eagle perching [With permission from Laura Barrio]

Fig. 1.9b. Verreaux's eagle perching [With permission from Laura Barrio]

Fig. 1.9c. Immature Verreaux's eagle perching [With permission from Francesco Veronesi]

Fig. 1.9d. Verreaux's eagle in flight [With permission from Derek Keats]

Fig. 1.9e. Verreaux's eagle in flight [With permission from Derek Keats]

Fig. 1.9f. Immature Verreaux's eagle in flight [With permission from Alan Lee]

Fig. 1.9g. Verreaux's eagle habitat, Matobo Hills, Zimbabwe [With permission from Geoff Cooke]

Fig. 1.9h. Verreaux's eagle habitat [With permission from Derek Keats]

1.4 to 3.03 kg (Dementiev et al., 1966; Cramp & Simmons, 1979; Ali & Ripley, 1983; Clark & Schmitt, 1999; Rasmussen & Anderton, 2012; Garcia et al., 2013).

The coloration of the adult male and female may differ slightly, with the female being browner above. The adult male is very dark brown on the upper parts, with some white partially visible feather bases, grey to greyish brown, blackish tipped tail, with several dark brown bars. The primaries are dark brown to black, with some grey on the outer webs. The underparts are lighter, with white, brown streaked cheeks, white underparts with dark streaks and white undertail coverts and thighs with dark streaks and barring. The underwing coverts are mottled white and dark, which creates a dark traverse bar at the quill bases, and the inner lining of the wing and tail quills are white to grey brown with white markings. Sometimes there is a whitish patch at the carpal joint, with a darker trailing edge. The tail has a broad black tip. The iris, cere and feet are yellow.

The juvenile is lighter colored, with a pale head, with black shaft streaks on the crown and neck. The upperparts are lighter than the adult plumage, darkish brown and the underparts are pale chestnut with black streaks, the wing quills are dark brown, with paler tips to the secondaries above. The tail quills are grey brown above, lightly banded with dark brown mottles. The underside is generally less white than that of the adult; there are white markings on the inner webs, resulting in a light patch on the carpal joint; the thighs and underwing coverts paler but more buff, The tail quills are pale, buff tinged grey and banded brown below. The eyes are brown, and the cere and feet are yellow. Brown and Amadon (1968) argue that the change from immature to adult plumage involves upper plumage darkening, lower plumage whitening, and the shaft streaks replaced by spots, over three to four years. When classified as *Hieraaetus fasciatus*, there were three subspecies: in order of size the nominate subspecies, *H.f. fasciatus*, *H.f. Renschii* and *H.f. spilogaster*. *H.f. renschii* was similar to *H.f. fasciatus*, but more darkly barred on the thighs, belly and under tail coverts, more barred remiges and tail, and more marked on the belly, thighs and crissum. *H.f. spilogaster* was blacker than *H.f. fasciatus* on the upperparts and whiter on the underparts, and with less brown on the thighs, more grey on the upper wing and a larger white patch below the wing, less black on the median and greater wing coverts and white shafts on the wing and tail quills. *H.f. fasciatus*, *H.f. renschii* and *H.f. spilogaster* are now *A.f. fasciatus*, *A.f. renschii* and the African hawk-eagle (*H.f. spilogaster*) in the new classification (Global Raptor Information Network, 2021k).

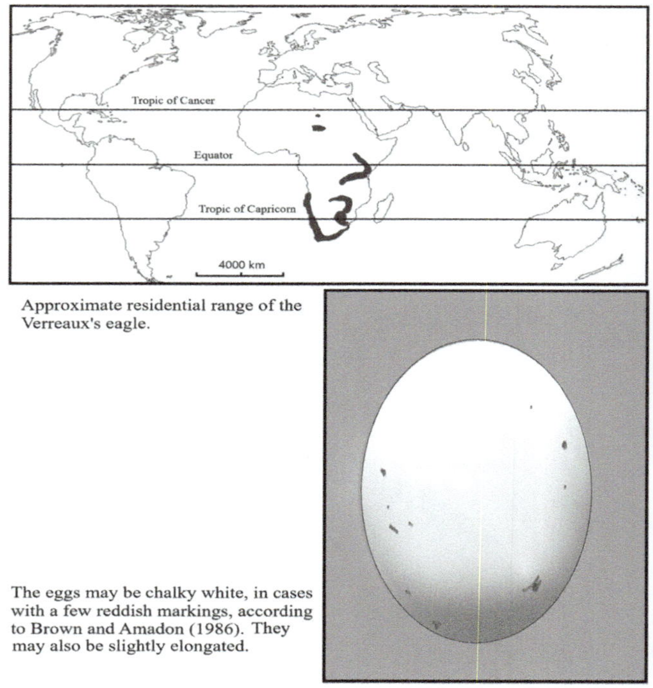

Fig. 1.9i. Vereaux's eagle range and egg

In field appearance, the Bonelli's eagle is similar in size to the Tawny eagle, slightly larger than the African hawk-eagle and larger than the Cassin's hawk-eagle (Ferguson-Lees & Christie, 2001; Dunning, 2008). The Bonelli's eagle may be distinguished by the contrasting color pattern (dark upperparts and white underparts), comparatively short wings and long tail. There is also the conspicuous grey patch on the carpal joint and for the juvenile, pale chestnut underparts (Brown & Amadon, 1968, 1986). The Bonelli's eagle is mostly silent, but its call may be "mellow fluting calls" during displays and in the vicinity of the nest, resembling a rapid, repeated "kluu-klu-klu-kluee" or "kluu-klu-klu-klu-klu" (Brown & Amadon, 1968). The excited call is a whistling "klueee-klueee"; the alarm cry is "kikiki" (ibid.). It may also give a call resembling "clek-clek-clek" or more rarely "hseeeoh". Brown and Amadon (1968, 677) also note that "its voice is generally rather lower-pitched, more fluting and melodious than some of its near relatives." Illustrations of the appearance, range, and eggs of this species are shown in Figures 1.10a to 1.10l.

Classification

As noted above, the Bonelli's eagle was previously *Hieraaetus fasciatus*, which was divided into three subspecies: *H.f. fasciatus*, *H.f. renschii* and *H.f. spilogaster*. Currently, the first two subspecies are recognized *A.f. fasciata* and *A.f. renschi*, which are parallel to the subspecies above. *H.f. spilogaster* has been reclassified as *Aquila spilogaster*, moving from *Hieraaetus fasciatus spilogaster* (Vieillot, 1822), to *Spizaetus spilogaster* (Bonaparte, 1850), to *Aquila spilogaster* (Bonaparte, 1850). Initially, Brown and Amadon (1968, 677) argued that these "differences have been considered adequate to separate it as a species by some authors, but the field habits are similar, and we prefer to regard it as a race." The Bonelli's eagle and the African hawk eagle are still considered closely related (Kemp, 1994c; Simmons, 2005). Bonelli's eagle is genetically close to the Cassin's hawk-eagle, and also to the larger golden, Verreaux's, Gurney's and Wedge-tailed eagles. The subspecies *A.f. renschi* ranges in the the Lesser Sunda Islands, distinct from the Eurasian nominate subspecies. Some researchers have suggested that this is a separate species, due to its isolation, but insufficient genetic distinctions have been found and there might have been human induced introductions into this remote location (Ferguson-Lees & Christie, 2001; Trainor et al., 2013).

Breeding

For the Bonelli's Eagle, the breeding season starts with nuptial displays, composed of undulating flight movements, steep dives and rising swoops. Nests are located in crags, and to a lesser extent in trees and rarely in buildings, less yet on buildings. The nests are large, up to two meters wide and thick, especially in crags. One to three, usually two white eggs are laid. There may be a few brown spots and streaks and lilac undermarkings (Brown & Amadon, 1968). Laying times are usually in late winter or early spring: February to April in France, January in North Africa, and December to April in India, "generally in dry seasons, but overlapping into wet periods at the equator in the tropics" and fledging is in 65 days (Brown & Amadon, 1968, 678).

Foraging

Bonelli's eagle frequents wooded mountain ranges, usually with an annual rainfall of 500 to 1140 mm. Although normally a resident species, but northern birds may migrate southwards through the Pyrenees and Asia Minor. Hunting is from soaring, and tree perching. Due to its comparatively strong feet (the hind claw is larger than that of the larger Eastern Imperial eagle), it takes a wide range of prey, including hyraxes (Order *Hyracoidea*, Huxley, 1869), medium sized rodents, small reptiles and birds such as guinea fowl (Family *Numididae*, Longchamps, 1842), herons (Family *Ardeidae*, Leach, 1820), storks (Family *Ciconiidae*, J. E. Gray, 1840), poultry (Order *Galliformes*, Temminck, 1820) and even Long-legged buzzards (*Buteo rufinus*, Cretzschmar, 1829). Notably, "its range of prey is almost as great as that of *Polemaetus bellicosus* or *Aquila chrysaetos*, eagles of twice its weight" (Brown & Amadon, 677).

Population Status

The Bonelli's eagle has a wide range, therefore according to BirdLife International (2019), this large range means that it is classified as Vulnerable, as the range is over 20,000 km^2. Additionally, the other criteria ("a declining or fluctuating range size, habitat extent/quality, or population size and a small number of locations or severe fragmentation") do not apply. BirdLife International (2019) acknowledges that the population is declining but does not classify the decline as sufficiently serious to be classified as Vulnerable. The population is also probably above 10,000 mature individuals with a slow decline. Therefore, BirdLife International (2019) classifies the Bonelli's eagle as of Least Concern.

BirdLife International (2015) describes the Bonelli's eagle's status in Europe as numbering 1,100-1,200 pairs, which equates to 2,100-2,400 mature individuals. It is further noted "Europe forms approximately 10% of the global range, so a very preliminary estimate of the global population size is 21,000-24,000 mature individuals, although further validation of this estimate is needed" (BirdLife International, 2015). Spain holds 66% of the decreasing European population, which is classified as Near Threatened within Europe. The main factors for this decline include organochlorine pesticide use in the mid-20th Century, and no subsequent population recovery. Jayaraj et al. (2016, 90) define organochlorine (OC) pesticides as "synthetic pesticides widely used all over the world. They belong to the group of chlorinated hydrocarbon derivatives, which have vast application in the chemical industry and in agriculture. These compounds are known for their high toxicity, slow degradation and bioaccumulation." The Bonelli's eagle is also persecuted by hunters and pigeon-fanciers in the western parts of its range. Juveniles are frequently killed by collisions with power lines (Rollan et al., 2010). Prey species are declining, there is increased human disturbance and poaching at nest sites, and agricultural intensification (Ferguson-Lees & Christie, 2001; Orta et al., 2016). Illustrations of the appearance, range, and eggs of this species are shown in Figures 1.10a to 1.10l.

1.12 African Hawk-eagle *Aquila africana* (Cassin, 1865)

Physical Description

The African hawk-eagle is a small-medium to medium-sized eagle, measuring 55–65 centimeters long, with a wingspan 130-160 cm and a weight of 1.15 to 1.75 kg (Ferguson-Lees & Christie, 2001; del Hoyo et al., 2014). The plumage is dark or blackish above, with white underparts heavily streaked with black feathers. The white underwing flight feathers have a black trailing edge, while the underwing coverts may be black with white spots. The juvenile plumage consists of brown upperparts and rufous rather than black coloration on the underparts. Females may be more densely marked below than males. The African hawk-eagle may be distinguished from the smaller Ayres's hawk-eagle, as in flight the latter has less white on the primary feathers, with heavier

Fig. 1.10a. Bonelli's eagle perching [With permission from Juan Manuel Benjumea Fantoni]

Fig. 1.10b. Bonelli's eagle perching [With permission from Juan Manuel Benjumea Fantoni]

Fig. 1.10c. Bonelli's eagle perching [With permission from Pieter Verheji]

Fig. 1.10d. Bonelli's eagle in flight [With permission from Frank McClintock]

Fig. 1.10e. Bonelli's eagle [With permission from Mike Nelson-Birdtour ASIA]

Fig. 1.10f. Bonelli's eagle in flight [With permission from Matthew Smith]

Fig. 1.10g. Bonelli's eagle in flight [With permission from Pieter Verheij]

Fig. 1.10h. Immature dark morph Bonelli's eagle in flight [With permission from Hari K. Patibanda]

Fig. 1.10i. Immature Bonelli's eagle in flight [With permission from Matthew Smith]

Fig. 1.10j. Bonelli's eagle nesting [With permission from Hari K. Patibanda]

Fig. 1.10k. Bonelli's eagle habitat [With permission from Matthew Smith]

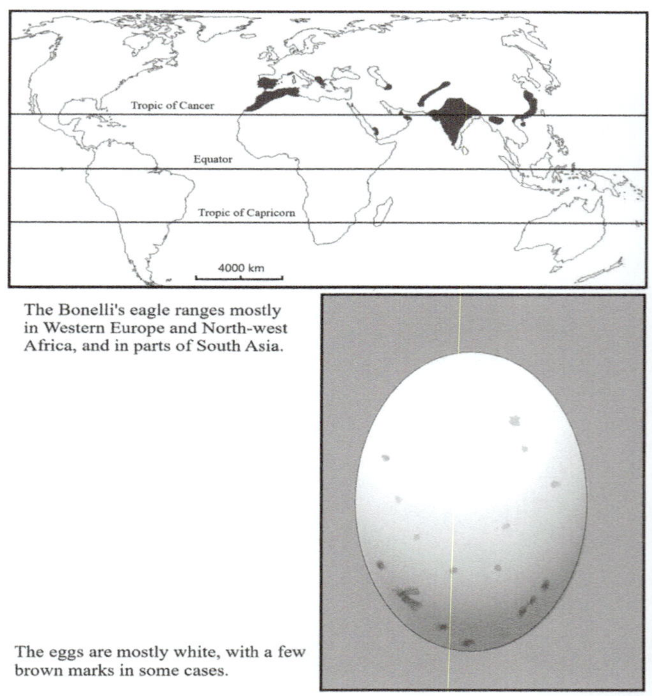

Fig. 1.10l. Bonelli's eagle range and egg

dark streaks on the underparts, and a nuchal crest (crest at the back of the neck) (Ferguson-Lees & Christie, 2001; del Hoyo et al., 2014). Bonelli's eagle is slightly larger, with less streaked white underparts and comparatively short wings and long tail, and the juvenile has a grey patch on the carpal joint and the juvenile has pale chestnut underparts (Brown & Amadon, 1968, 1986; Ferguson-Lees & Christie, 2001; Dunning, 2008). Its call may be "kluu-kluu-kluu" or a "kwee-oo" (del Hoyo et al., 1992). Illustrations of the appearance, range, and eggs of this species are shown in Figures 1.11a to 1.11i.

Classification

The African Hawk-eagle is currently usually classified as *Aquila spilogaster* (del Hoyo & Collar, 2014), was formerly placed by some in the genus *Hieraaetus*. Helbig et al. (2005) used molecular phylogenetic analyses, using DNA sequences of one mitochondrial and three nuclear genes, and recommended that *Hieraaetus spilogaster* (along with the Bonelli's eagle *Hieraaetus fasciatus*) be placed in the genus *Aquila*, and they pointed to a close relationship between these species and the Verreaux's eagle. Although the African Hawk-eagle has previously been considered as a subspecies of the Bonelli's eagle by some authors (e.g., Sinclair et al., 2002), recent research indicates species status for both eagles (e.g., Thiollay, 1994; Wink and Sauer-Gürth, 2000, 2004; Ferguson-Lees & Christie, 2001; Lerner & Mindell, 2005).

Breeding

The African Hawk-eagle builds large stick nests placed, usually placed in trees near rivers or hilly terrain, or cliffs (Steyn, 1982; Ash and Miskell, 1998; Dowsett-Lemaire, 2006; Dowsett-Lemaire and Dowsett, 2006; Dowsett et al., 2008). One or two eggs may be laid. Laying times may vary regionally but cover most months. For example, in Ethiopia, laying was recorded in December, January, February, and April (Ash & Atkins, 2009), while in Somalia laying appeared to be principally from late December to mid-February (Ash & Miskell, 1998). In Uganda, laying was recorded in September, October and November (Global Raptor Information Network, 2021l). In Angola, Malawi, Zambia, Angola and Zimbabwe, most laying records were April to August (Brooke et al., 1971; Dean, 1974; Irwin, 1981; Simmons, 1997d; Dowsett-Lemaire & Dowsett, 2006; Dowsett et al., 2008).

Foraging

The African Hawk-eagle occurs principally in dry and open woodland, or lightly treed savanna or dry forest near rivers. These habitat types may vary regionally. In Zimbabwe, it has been recorded in a wide range of woodland and savanna types, favoring wetter areas over drier locations, and sometimes hilly and riparian areas (Irwin, 1981; Penry, 1994). In Ethiopia, Eritrea and Zambia it has been recorded in light woodland and bush savanna near cliffs, hills and in some cases rivers (Smith, 1957; Dowsett et al., 2008; Ash and Atkins, 2009). In Mozambique, it frequents Acacia and mixed woodlands, and to a lesser in mopane woodland, the latter being a habitat favored generally in southern Africa (Parker, 2005a,b).

Foraging behavior is based on both perching and soaring. Prey taken includes rodents, monkeys, guineafowl, pythons (Loveridge, 1923; Robinson, 1973; Dowsett et al., 2008). Bing (2009) and Irwin (1981) found in Zimbabwe that the preys species comprised birds (74%), mammals (25%), and reptiles (1%). Most of the avian prey comprised Helmeted Guineafowl (*Numida meleagris*, Linnaeus, 1758) and some francolins (Subfamily *Perdicinae*, Horsfield, 1821). Mammals included Scrub hare (*Lepus saxatilis*, Cuvier, 1823), Slender mongoose (*Galerella sanguinea*, Rüppell, 1836) and Rock hyrax (*Procavia capensis*, Pallas, 1766) (Irwin, 1981; Middleton & Middleton, 2005).

Population Status

The African Hawk-eagle is non-migratory (Bildstein, 2006). The population is declining, largely due to habitat degradation. This includes the loss of the large trees within its range and killing by poultry farmers (Ferguson-Lees & Christie, 2001). It is classified by Birdlife International (2020) as of Least Concern, due to its large range, and large population possibly over 10,000. The population varies regionally. In West Africa across to East Africa, it is generally resident but rare, throughout the savanna belt, sometimes a local migrant (Smith, 1957; Thiollay, 1985; Grimes, 1987; Nikolaus, 1987; Gore, 1990; Elgood et al., 1994; Cheke & Walsh, 1996; Ash & Miskell, 1998; Borrow & Demey, 2001; Carswell et al., 2005; Ash & Atkins, 2009). It appears to be more common (with exceptions) in East and Southern Africa than in West Africa (Benson & White, 1957; Irwin, 1981; Pinto, 1983; Tarboton & Allan, 1984; Penry, 1994; Dean, 2000; Stevenson & Fanshawe, 2002; Parker, 2005a,b; Dowsett-Lemaire & Dowsett, 2006; Dowsett et al., 2008; Monadjem & Rasmussen, 2008). Illustrations of the appearance, range, and eggs of this species are shown in Figures 1.11a to 1.11i.

Fig. 1.11a. African Hawk-eagle perching [With permission from Roelof van der Breggen]

Fig. 1.11b. African Hawk-eagle [Credit: Hans Veth - unsplash]

Fig. 1.11c. African Hawk-eagle perching [With permission from Derek Keats]

Fig. 1.11d. African Hawk-eagle perching [With permission from Derek Keats]

Fig. 1.11e. African Hawk-eagle in flight [With permission from Bob Burgess]

Fig. 1.11f. African Hawk-eagle in flight [Credit: Roelof van der Breggen]

Fig. 1.11g. Immature African Hawk-eagle in flight [With permission from Fons Buts]

Fig. 1.11h. African Hawk-eagle habitat in Burkina Faso [With permission from Rita Willaert]

1.13 Cassin's Hawk Eagle *Aquila africana* (Cassin, 1865)

Physical Description

The Cassin's Hawk eagle is a comparatively small, long-tailed eagle. This species has been termed bicolored, due to its dark upperparts and white underparts (Brown & Amadon, 1968). The dark upperparts are dark brown, these feathers having white bases and black edges. The primaries are brown with darker tips, and there are white bases to the inner webs and two indistinct dark bars. The brown tail has three black bars and a "broad black to very dark brown sub-terminal band. The undersides and legs are white, marked with dark patches on the edges of the breast, with black underwing coverts and thigh markings. The eye, cere and feet are yellow, the bill and claws are black (Brown & Amadon, 1968).

Juvenile plumage has lighter upperparts and darker underparts. The head is cinnamon to rufous, with black steaks on the crown and throat, and dark grey brown back and upper wings, edged with lighter colors. The breast is rufous with black spots. The tail feathers are dark greyish brown, with white tips and a darker, subterminal band. The thighs are pale cinnamon and the tarsi, under white. The underwing coverts are white, with yellowish brown and tinges and blackish spots. The undertail coverts are white with cinnamon tips and dark spots. The flanks and abdomen are paler than those of the adult, with dark spots. The eye is grey brown (Brown & Amadon, 1968).

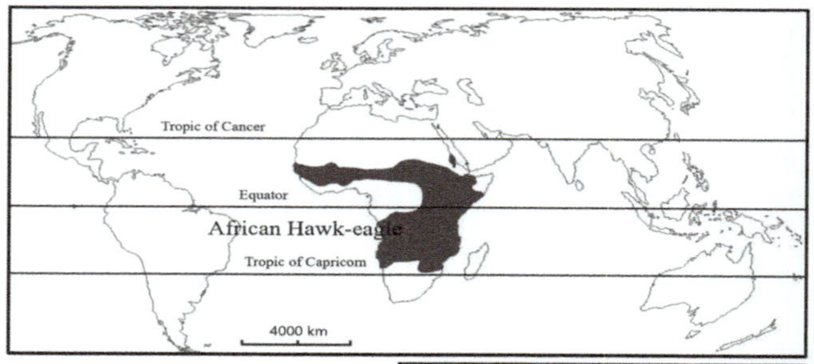

The African Hawk-eagle ranges through most of the savannas of subSaharan Africa, avoiding dense forest and desert.

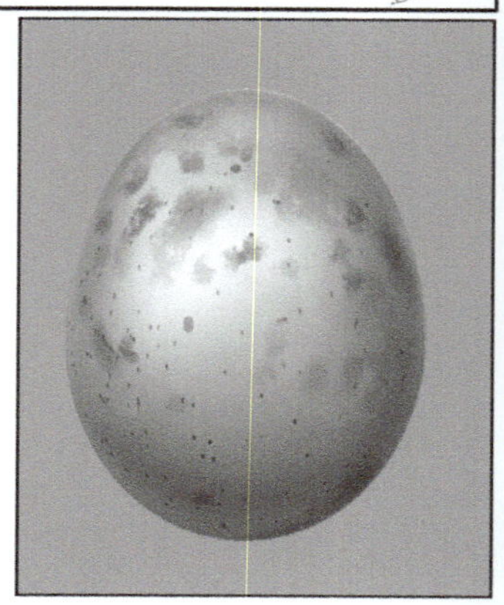

The African Hawk eagle lays eggs that may be light brown with varied markings.

Fig. 1.11i. African Hawk-eagle range and egg

Cassin's Hawk eagle may be mistaken for other, similarly sized species. For example, the similar Ayres's eagle is less barred on the wings and underparts. The current species also differs from the Congo Serpent eagle (*Circaetus spectabilis*, Schlegel, 1863 or *Dryotriorchis spectabilis*, Schlegel, 1863) in having a feathered tarsus. Ayres's eagle also differs from the Bonelli's eagle but the current species is smaller and has a comparatively longer tail. The juvenile differs in having a cinnamon-rufous head. The call is "unrecorded" (Brown & Amadon, 1968, 693). Illustrations of the appearance, range, and eggs of this species are shown in Figures 1.12a to 1.12g.

Classification

The older names of this species were *Spizaetus africanus* (Cassin, 1865), *Hieraaetus africanus* (Cassin, 1865), *Cassinaetus africanus* (Cassin, 1865) and *Aquila africanus* (Cassin, 1865) (Sibley & Monroe, 1990, 1993; del Hoyo et al., 2014; BirdLife International, 2017; Global Raptor Information Network, 2021m).

Breeding

Brown and Amadon (1968, 693) report that "nothing at all is recorded." However, Kemp & Kirwan (2020a) write that laying takes place from October to December in Ghana and Gabon and October in Uganda (Carswell et al., 2005).

Foraging

Brown and Amadon (1968, 693) note that "very little is known" about the foraging habits of this species. However, it is only rarely recorded soaring above the canopy of its forest habitat. Common prey consists of birds and squirrels caught in the canopy layer. Birdlife International (2017) records that the favored habitats are forest-subtropical/tropical moist lowland and forest-subtropical/tropical moist montane.

Population Status

Cassin's Hawk eagle has a wide range, covering most of West and Central Africa. Hence, the IUCN does not classify its status as Vulnerable, but of Least Concern, as it lacks the criterion for such a classification (small range size, with a declining or fluctuating range size, degrading habitat extent/quality, or population size and a small number of locations or severe fragmentation)

(BirdLife International, 2017). This is despite the evidence that the population appears to be decreasing, as the decline assessed to be insufficient to approach Vulnerable (>30% decline over ten years or three generations). The population is estimated to be in the range of 1,000-10,000 individuals, possibly 670-6,700 mature individuals (BirdLife International, 2017). Ferguson-Lees & Christie, 2001) note that the gradual population decline may be due to the gradual, ongoing habitat destruction. Illustrations of the appearance, range, and eggs of this species are shown in Figures 1.12a to 1.12g.

Fig. 1.12a. Cassin's Hawk-eagle perching [With permission from Jeremey Roberts]

Fig. 1.12b. Cassin's Hawk-eagle perching [With permission from Bradley Hacker]

Fig. 1.12c. Immature Cassin's Hawk eagle perching [With permission from Robert Campbell]

Fig. 1.12d. Cassin's Hawk-eagle in flight [With permission from Julian Thomas]

Fig. 1.12e. Cassin's Hawk-eagle in flight [With permission from Gil Ewing]

Fig. 1.12f. Cassin's Hawk-eagle habitat [With permission from Jonathan D. Müller]

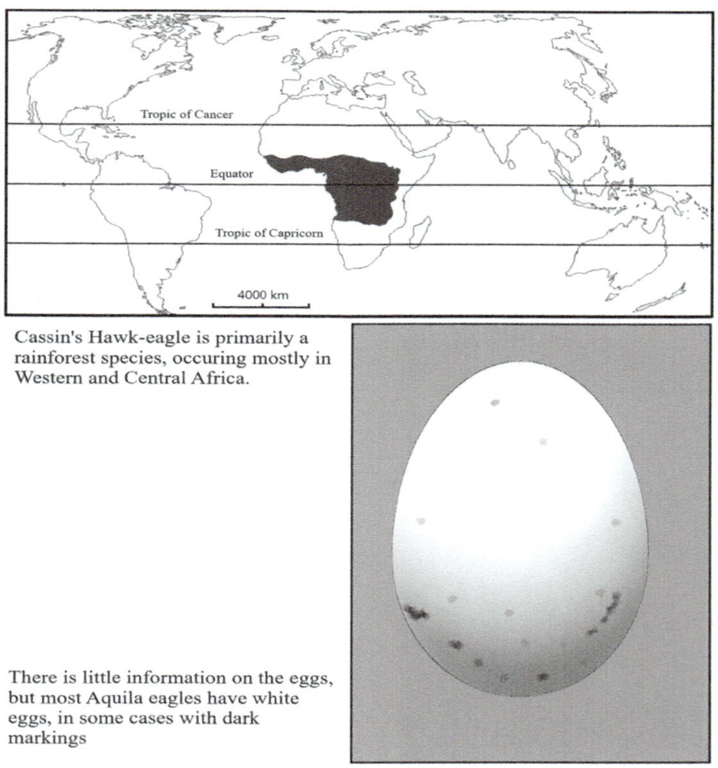

Fig. 1.12g. Cassin's Hawk-eagle range and egg

2

Systematic List of the "Booted Eagles" (Subfamily *Aquilinae*; Genera *Hieraaetus*, *Spizaetus*, *Nisaetus*, *Lophaetus*, *Clanga*, *Ictinaetus*, *Stephanoaetus*, *Polemaetus* and *Lophotriorchis*)

2.1 Introduction

This chapter examines the forms, foraging and feeding habits, breeding patterns and statuses of the genera *Hieraaetus, Spizaetus, Nisaetus, Clanga, Ictinaetus, Stephanoaetus, Polemaetus* and *Lophotriorchis* within the Subfamily *Aquilinae*. The eagles are mostly small to medium sized, similar to large or very large hawks. Classification has been a scientific issue for several years, with some species moved to *Aquila* or other genera. As argued by Brown and Amadon (1968, 674) "the genus *Hieraaetus* ... may not be a natural group ..." and "is difficult to separate from *Aquila*, *Spizastur* and *Spizaetus*." These authors further argue that there are disputes as to the classification of some of the members, as *H. africanus* has been placed in *Spizaetus*, in *Hieraaetus* or in its own genus. *H. wahlbergi* is sometimes placed with *Hieraaetus* or in *Aquila*. Some combine *Hieraaetus* with *Aquila*, "but it seems best to keep it separate on the following ensemble of characters, which admittedly do not hold for every species; size smaller; form more slender; bill smaller; legs longer, more slender; emargination on primaries deeper" (Brown & Amadon, 1968, 676).

An important issue concerns the distinction between the Genera *Aquila* and *Hieraaetus* (Clark, 2012). Clark (2012, 295) notes that there has long been a consensus on the species placement between these two genera, excepting the Wahlberg's eagle, which may be classified as *Aquila wahlbergi* or *Hieraaetus wahlbergi*. Recent studies resulted in variable findings and assessments. An important assessment was that *Aquila* and *Hieraaetus* polyphyletic (descended from more than one common evolutionary ancestor or ancestral group) (Helbig et al., 2005; Lerner & Mindell, 2005; Griffiths et al., 2007; Haring et al., 2007). For there to be monophyly (a group sharing common ancestor) efforts were made to moves species between these two genera. Other possibly weaker assessments (with few supporting, reliable studies) included the possible addition of *Hieraaetus* entirely into *Aquila* (Wink & Sauer-Gürth, 2000; Hockey et al., 2005; Sangster et al., 2005).

Clark (2012, 297) supports the separation of the two genera, that is the continued existence of the genus *Hieraaetus* rather than letting it disappear into *Aquila*, hence leaving *Aquila* and *Hieraaetus*, as monophyletic and positioned as sister taxa (Helbig et al., 2005, Lerner & Mindell, 2005; Griffiths et al., 2007; Haring et al., 2007). This classification also recognizes the monophyletic Genus *Clanga* for the spotted eagles in a clade with the genera *Lophaetus* and *Ictinaetus* (Wells & Inskipp, 2012). An important factor for the separation of *Aquila* and *Hieraaetus* concerns the fairly obvious physical differences between some species in the two groups (Brown & Amadon, 1968). Looking at these physical differences, Brown and Amadon (1968) noted that "some recent authors have combined *Hieraaetus* with *Aquila,* but it seems best to keep it separate on the following ensemble of characters, which admittedly do not hold for every species: size smaller, form more slender, bill smaller, legs longer, more slender, emargination on primaries deeper." Clark (2012, 296) adds other physical differences: compared to Genus Aquila. Genus *Hieraaetus* generally are smaller (medium-sized or large raptors, but small by eagle standards), the plumage of immature birds is more similar to that of adults, they lack the pale primary patch (light-colored primary feathers on the wings, appearing as a patch) of *Aquila* species, and generally have a longer, narrower tail of some Aquila species. Finally, the *Hieraaetus* may have more than one color morph (dimorphic or polymorphic) while *Aquila* species are monomorphic (one color morph) except for the Tawny eagle. Clark (2012, 296) acknowledges that some species have unique traits, such as "the white headlights of Booted and Ayres' Eagles and the crests of Wahlberg's and Ayres' eagles." His final list of five species for *Hieraaetus* are Ayres' eagle, Booted eagle, Little eagle, Pygmy eagle and Wahlberg's eagle. The list for *Aquila* includes 11 species: Steppe eagle, Eastern Imperial eagle, Spanish Imperial eagle, Tawny eagle, Golden eagle, African Black eagle, Wedge-tailed eagle, Gurney's eagle, Bonelli's eagle, African Hawk-eagle and Cassin's Hawk-eagle (Clark, 2012, 296).

The more recent and advanced research using molecular genetic techniques has revealed new assessments and conclusions, supporting some of these assessments, as mentioned above. At the more detailed level, molecular genetic research has determined the Genus *Hieraaetus* to be polyphyletic with the Genus *Aquila* (Helbig et al., 2004; Wink & Sauer-Gürth, 2004; Lerner & Mindell, 2005). As noted by Lerner and Mindel (2005, 339) "species in the genus *Hieraaetus* are placed in the two latest diverging clades of booted eagles and one species forms a separate early diverging clade by itself." Particularly disputed species include the African hawk-eagle, which is classified by some as *Aquila spilogaster*, but others previously placed as *Hieraaetus spilogaster* or as subspecies either *Hieraaetus fasciatus spilogaster* or *Aquila fasciatus. spilogaster*. As noted by Lerner and Mindel (2005, 339) "*Hieraaetus f. fasciatus* and *H.f. spilogaster* have been treated variously as separate species (Ferguson-Lees & Christie, 2001; Thiollay, 1994), a superspecies (Stresemann & Amadon, 1979) or subspecies (Sinclair et al., 2002)." Lerner and Mindel (2005, 342) conclude that "our findings suggest that further study with greater sampling of individuals is warranted in order to determine if *H.f. spilogaster* should be elevated to species status (*H. spilogaster*) and, if so, where the limits of its geographic distribution lie."

Other disputed species are Bonelli's eagle (*Aquila fasciata*, formerly *Hieraaetus fasciatus*) and the Rufous-bellied eagle (*Lophotriorchis kienerii*, formerly *Hieraaetus kienerii*). The Rufous-bellied eagle (*Lophotriorchis kienerii*, G. de Sparre, 1835) was removed from *Hieraaetus* and placed in the separate genus (*Lophotriorchis*, Sharpe, 1874) (Lerner & Mindell, 2005; Gjershaug, 2006). The extinct Haast's eagle was found to be closely related to the Little eagle and was classified from *Harpagornis moorei* to *Hieraaetus moorei* (Evans, 2018). Bonelli's eagle and Booted eagles were moved to *Aquila* by the British Ornithological Union (BOU) in 2005 (Sangster et al., 2005). However, in 2014, the BOU listed the booted eagle as *H. pennatus*, rather than *A. pennata* (British Ornithological Union, 2020). The pygmy eagle, or New Guinea hawk-eagle (as *H. morphnoides weiskei* used to be a subspecies of the little eagle *H. morphnoides*), but is now usually recognized as a separate species (*Hieraaetus weiskei*, Reichenow, 1900) (Gjershaug et al., 2009; Clements Checklist Team, 2012).

Other reclassifications include that of Christidis and Boles (2008), who contrastingly used the polyphyletic status (grouped together but without an immediate common ancestor) of *Aquila* and *Hieraaetus* and moved the African hawk-eagle (*Aquila spilogaster*) and Bonelli's eagle (*Aquila fasciata*) to *Aquila*, while other species originally in *Hieraaetus* and the Little eagle (formerly *Aquila morphnoides*) and Wahlberg's eagle (*Aquila wahlbergi*) within *Hieraaetus* (Christidis & Boles, 2008; The Clements Checklist, 2012). Past changes include that of The Clements Checklist between 2001 and 2009, when *Hieraaetus* was included within *Aquila* (Lepage, 2020). Later, the Clements Checklist (2009, 2012) duplicated Christidis & Boles (2008), and included Wahlberg's eagle in *Hieraaetus*, and put the booted (*Hieraaetus pennatus*, Gmelin, 1788, formerly *Aquila pennata,* Gmelin, 1788), Little (*Hieraaetus morphnoides*, Gould, 1841, formerly *Aquila morphnoides*) and Ayres's eagles (*Hieraaetus ayresii,* Gurney, 1862, formerly *Aquila ayresii*, Gurney, 1862) into *Hieraaetus* (Clements Checklist, 2007, 2009, 2012).

The Genus *Spizaetus* (Vieillot, 1816) comprises a small group of medium to large eagles, tropical forest species, ranging in the Americas mostly from Mexico to northern Argentina (Helbig et al., 2005). Some species are crested with long-tails and comparatively slender bodies. These species have "average wing lengths and proportions" slender bills, feathered legs longish toes and curved, slender talons (Brown & Amadon, 1968). Brown and Amadon (1968, 687) also note that this genus is "very similar" to the Genus *Hieraaetus* of the Old World. Differences include the greater similarity of the young to the adults, the less sharply notched primaries, the skin lining the rear edge of the lower mandible is un-feathered.

Formerly this genus included South East Asian species now included in the Genus *Nisaetus* (Hodgson, 1836) after it was found that the latter were more closely related to *Ictinaetus* (Blyth, 1843). Black-and-white hawk-eagle (*Spizaetus melanoleucus*, Vieillot, 1816) formerly *Spizasturmelanoleucus*, Vieillot, 1816) was added to *Spizaetus* by the American Ornithologists Union (Banks et al., 2006; Haring et al., 2007). The complexity of the classification and reclassification of this genus and related species has been underscored by recent research findings. Haring et al. (2007, 353) note in their study that: "The phylogenetic relationships within the New and Old World hawk-eagle assemblage (genus *Spizaetus*; Aves: *Accipitridae*) were studied using mitochondrial DNA sequences (cytochrome b, control region)." The findings for this study "clearly indicate that hawk-eagles represent a paraphyletic assemblage and thus their external similarities have to be ascribed to convergent evolution."

These assessments were indicated by clustering of the New World taxa of *Spizaetus* (including the South American species Black-and-chestnut eagle *Oroaetus isidori*, which was later classified as *Spizaetus isidori*, Des Murs, 1845). The only African member of the genus, Cassin's hawk-eagle *Spizaetus africanus*, later *Aquila Africana*, Cassin, 1865) was found in a mixed cluster of *Aquila* and *Hieraaetus* (hence it has been renamed by some authors *Hieraaetus africanus* (Cassin, 1865) and the monotypic *Cassinaetus africanus* (Cassin, 1865, in the Genus *Cassinaetus*) (see also Voous & Wijsman, 1964; Amadon & Bull, 1989; Kemp,

1994; Kemp & Kirwan, 2020a). Regarding the Asian species of *Spizaetus*, Haring et al. (2007, 353) note that "The taxa from Southeast to East Asia form a clearly separated monophyletic group." Hence, they suggest "assigning the species of the genus *Spizaetus* to three different genera: (1) *Spizaetus* (including *Oroaetus isidori*) in Central and South America and (2) *Nisaetus* for the Southeast to East Asian group. (3) The African taxon (*Spizaetus africanus*) is discussed to be included into the genus *Aquila*. Furthermore, we propose to use the former genus name *Lophotriorchis* (Sharpe, 1874), for the monotypic species *Hieraaetus kienerii*, which has an isolated phylogenetic position" (Haring et al., 2007, 353). A further reclassification was the elevation of a subspecies of the mountain hawk eagle (*Nisaetus nipalensis*, Hodgson, 1836, *Nisaetus nipalensis kelaarti*) into full species status as Legge's hawk-eagle (*Nisaetus kelaarti*, Legge, 1878) (Gjershaug et al., 2008). The Flores hawk-eagle *Nisaetus floris* (Hartert, 1898) has also been separated from the Changeable hawk-eagle (*Nisaetus cirrhatus*, Gmelin, 1788). This chapter follows the latest, accepted classifications and lists the species as above (see Table 2.1).

Table 2.1. The reclassification of some species of *Spizaetus* to *Nisaetus*

Species still in *Spizaetus*	
Black hawk-eagle	*Spizaetus tyrannus* (Wied, 1820)
Black-and-white hawk-eagle	*Spizaetus melanoleucus* (Vieillot, 1816)
Ornate hawk-eagle	*Spizaetus ornatus* (Daudin, 1800)
Black-and-chestnut eagle	*Spizaetus isidori* (Des Murs, 1845)
Species moved to *Nisaetus*	
Mountain hawk-eagle	*Nisaetus nipalensis* (Hodgson, 1836)
Legge's hawk-eagle	*Nisaetus kelaarti* (Legge, 1878)
Changeable hawk-eagle	*Nisaetus cirrhatus* (Gmelin, 1788)
Flores hawk-eagle	*Nisaetus floris* (Hartert, 1898)
Blyth's hawk-eagle	*Nisaetus alboniger* (Blyth, 1845)
Javan hawk-eagle	*Nisaetus bartelsi* (Stresemann, 1924)
Sulawesi hawk-eagle	*Nisaetus lanceolatus* (Temminck& Schlegel, 1844)
Northern Philippine hawk-eagle	*Nisaetus philippensis* (Gould, 1863)
Southern Philippine hawk-eagle	*Nisaetus pinskeri* (Preleuthner & Gamauf, 1998)
Wallace's hawk-eagle	*Nisaetus nanus* (Wallace, 1868)

2.2 Booted Eagle *Hieraaetus pennatus* (Gmelin, 1788) or *Aquila pennata*

Physical Description

The Booted eagle is a small eagle, similar in size to the Common buzzard (*Buteo buteo*, Linnaeus, 1758). In weight it is about 510–1,025 g with males in the lower range and females in the upper range. The length 45 to 55 cm, and the wingspan 110-130 cm (Kemp & Kemp, 1998; del Hoyo et al., 1998; Ferguson-Lees & Christie, 2001). There are pale and dark phases. The pale phase is *Hieraaetus pennatus pennatus*, colored mostly light grey with a darker head and flight feathers. The head and neck are colored tawny brown to tawny reddish with dark streaks. In some cases, there is a black eyebrow. The forehead and lores are whitish. The back is dark brown, and the upper tail coverts are pale brownish. The tail coverts are brown, with vague, darker barring. The lesser wing coverts are lighter brown, while the greater and primary coverts are darker brown; the lesser and greater coverts have pale edges. The primaries are blackish brown, with barring and white basal mottling on the inner webs. The secondaries and outer scapulars are lighter brown, with buff edges. The underparts are white to light brownish, with brown streaks on the chin, breast and upper flanks. The under-wing coverts and axillaries are white with black blotches, while the short axillaries are "white streaked rufous brown" (Brown & Amadon, 1968, 680). The underside and wing quills are barred dark brown and grey. The tips of the primaries and blackish brown. The eyes are brown and the cere and legs are yellow.

In the dark phase, the entire plumage is dark to mid-brown with darker streaks and grey flight feathers. Paler colors are found on the edges of the upper side feathers, tail quills and base of the primaries. The underwing coverts and the carpal joint are blackish. Brown and Amadon (1968: 680) note that in comparing this with the pale phase "the difference lies mainly in the under parts, from above or behind this phase looks very much the same as the pale phase." The immature has a more rufous underside with dark shaft stripes and the head is more rufous and the upper tail coverts are paler. The eye is also darker than in the adult. *H.p. harterti* differs from *H.p. pennatus* in being larger, with an increasing size from west to east. Similar Species. The pale phase resembles the Egyptian vulture due to the dark flight features and white underside. The dark phase may resemble Wahlberg's eagle in Africa and a dark melanistic Changeable hawk-eagle in Asia, but the current species is smaller and stockier, and has less uniformly colored plumage. The juvenileAyres's Hawk-eagle and Bonelli's eagle vary in having paler rufous underparts. The voice is a high whistle ki-kee. In display, the call is "pi-pee," pi-pi-pi-pi-pee. Also, a "longer scream kleek-kleek" (Brown & Amadon, 1968, 681). Kemp & Kemp (1998) record its call as a shrill kli-kli-kli. Illustrations of the appearance, range, and eggs of this species are shown in Figures 2.1a to 2.1j.

Classification

As noted above, the classification of this species has changed through time. For example, some have classified the Booted eagle within the genus *Aquila* (Collinson, 2006). However, some other researchers have used DNA analysis and found that the Booted eagle is in a monophyletic clade with other species including Ayres's Hawk-eagle, Wahlberg's eagle, little eagle and the Pygmy eagle, and this clade is often placed within *Hieraeetus* and the current species as *H. pennata* (Gill & Donsker, 2014; Lerner et al., 2017). The Global Raptor Information Network (2021n) notes that the classification followed on their webpage is based on that "the majority of raptor systematists (e.g., Wink & Sauer-Gürth, 2004; Gjershaug, 2006) and several national committees on classification and nomenclature (e.g., those in the United Kingdom and Germany) now favor merging the species formerly assigned to *Hieraaetus* into *Aquila*." This assessment is supported by the study of Helbig et al. (2005), which used DNA sequences from one mitochondrial and two nuclear genes, and found a link between the booted eagle, Ayres's hawk eagle, little eagle, as a monophyletic group with a sister species with Wahlberg's eagle.

The Global Raptor Information Network (2021n) lists three subspecies of the Booted eagle: *H.p. harterti*, which ranges in southwestern and central Asia; *H.p. miniscullus* ranging in Southern South Africa (Cape Province); and *H.p. pennatus* ranging from southern Europe to North Africa, to central Asia and to the Caucasus. The Global Raptor Information Network (2021n) further adds that *H.p. harterti* was merged with another form, *H.p. milvioides*, by Stresemann and Amadon (1979). One factor was that Cramp and Simmons (1979) refused to recognize *H.p. milvioides*, one reason being a perceived cline of increasing eagle size from the west to east. This led later researchers to classify the booted eagle as monotypic, until Yosef et al. (2000) described *H.p. miniscullus*. The India Biodiversity Portal (2020) also lists *H.p. pennatus*, *H.p. harterti* and *H.p. miniscullus* as subspecies.

Foraging

The Booted eagle is a generally solitary species forages and breeds in wooded landcover sometimes with hilly terrain, usually outside dense forest, and may avoid other raptor presence (Wells, 1999; Ferguson Lees & Christie, 2001). It has been recorded on mountains up to over 3,000 meters. The range of habitats is wide, including open deciduous and coniferous forest, grassland, and semi-desert to desert. Roosting birds frequent dead tree crowns of trees near nesting sites (Adamian & Klem, 1999). Generally solitary and avoids the vicinity of other raptors (Wells, 1999). There are regional differences in habitat types and foraging habitats. For example, northern birds wintering in southern Africa may frequent flat terrain and open woodland landcover (Kemp & Kemp, 1998). In southern Africa, the resident populations frequent mostly mountainous terrain, including cliff-faces and dry scrub (Clancey, 1985). The booted eagle population in northern Africa also frequents sea cliffs, variable woodlands, and human-modified landcover such as eucalyptus plantations, and palm and olive groves (Thévenot et al., 2003). In Europe, this eagle frequents a wide range of habitats, including non-forested, mixed urban and vegetation substrates in Spain (Palomino & Carrascal, 2007); open, mixed and deciduous forest in the Czech Republic (Kren, 2000); and similar environments, including some coniferous woodland in Greece (Handrinos & Akriotis, 1997). In Asia, eagles may frequent similarly wide varieties of landcover, including sparsely treed, open fields in Israel (Shirihai, 1996); forests, steppes and semi-deserts in Kazakhstan (Wassink & Oreel, 2007); dry open coniferous forest for breeding and for wintering open woodland, farmed areas and treed semi-desert in India (Rasmussen & Anderton, 2012); and rice paddylands and cleared mangrove areas in the Malay Peninsula (Ollington & Loh, 1992, Wells, 1999; The Global Raptor Information Network, 2021n). In Africa, the main habitats are open country, with light woodland and mixed forest-grassland landcover, as recorded by studies in Zambia, Malawi, Botswana and Zimbabwe (Irwin, 1981; Hustler et al., 1992; Penry, 1994; Dowsett-Lemaire & Dowsett, 2006; Dowsett et al., 2008).

López-López et al. (2016, 981) cite the example of Booted eagles in the Spanish Mediterranean area. The study covered new ground, in that it used new technology, including satellite tracking technologies "a combination of an unbiased technology (i.e., GPS telemetry), a wide geographic extension of marked birds (all over Spain), and much larger sample size than in previous works." The justification was that "detailed information about movement ecology and ranging behavior of birds is important not only from the perspective of the basic ecology, but also from the conservation point of view" and that "the Booted Eagle, likewise other forest raptors, is an adequate bioindicator of human-dominated agro-forestry in Mediterranean landscapes in which low-intensity traditional agricultural practices still persist" (López-López et al., 2016, 981). The findings indicated that eagles made long distance foraging movements, greater than 20 km from the nesting areas during the entire breeding season. These recorded movements showed that the eagles had longer breeding ranges than previously expected. The authors conclude that "hence, management measures for conservation of forest raptors based on setting restrictions around nesting sites using buffer areas of arbitrary radii clearly results in large areas of eagles' home ranges laying outside "restriction" areas" and conservation practices must be flexible enough to take into account these wide-ranging patterns (López-López et al., 2016, 981).

Breeding

Nuptial displays are "very spectacular" (Brown & Amadon, 1968, 681); several dives linked to upward swoops, with calling. The males dive at females who in turn over and present their claws to their partners. There is also a "long swift glide, with the feet sometimes hanging limp, ending in a tree or in a sudden swoop up into the air again" (Brown & Amadon, 1968, 681). It nests in trees, three to four feet across and two feet deep lined with greenery. Where there are few trees, it nests in crags (for example in Morocco). Normally, two eggs are laid, but on rare occasions only one is laid. The eggs are white with brown markings. Laying times vary. In Europe, the usual time is late April and early May. In India, it is March to June; in Morocco, June; in Asia Minor, May. Fledging is about 90 days after hatching. Nests are built from sticks, lined with green leaves usually in trees or crags, and may in some cases be a pre-existing nest of another species such as the black kite (*Milvus migrans*, Boddaert, 1783) or grey heron (*Ardea cinereal*, Linnaeus, 1758). Rearing is usually the task of the female, incubating the egg for about 45 days, and guarding the nestlings while the male does the hunting. Fledging is after 70–75 days (Ferguson-Lees & Christies, 2001).

Issam et al. (2019, 382) cite the example of nesting Booted eagles in the Chettabah forest, Algeria, which were studied during the breeding seasons of 2013, 2014 and 2015. The study covered arrivals of Booted eagles during March, finding that the mean breeding pairs' density was $1.23 \pm 0.153/\text{km}^2$, the incubation period varied from 33 to 38 days, and the frequency distribution of clutch size was 67.6% for 2 eggs, 27% for 1 and 5.4% for 3 eggs. Sixty-three eggs were monitored and 55 hatched (87%). Hatching rates for the three seasons showed differences, but these were not statistically significant (90% in 2013, 76.19% in 2014 and 84% in 2015). Issam et al. (2019, 382) note that their findings were similar to those of matching studies of the booted eagle in Spain, and that "further investigations should be conducted to improve the provision of information to support sound management interventions of this North African population" (Issam et al., 2019, 382).

Population Status

The Booted eagle occurs as a breeding resident in southern Europe, North Africa and western South Africa and Namibia, and Asia. Northern birds are migratory, moving to Sub-Saharan Africa as far south as Malawi and South Asia including India from November to February. Southern African birds are non-migratory in many cases (Global Raptor Information Network, 2021n). This is quite common through its range, but is recorded as declining largely due human actions, such as deforestation through agriculture and urban development, disturbance through human presence, direct persecution, and eradication of prey species (Ferguson-Lees & Christie, 2001).

There are regional differences in the conservation status of the booted eagle. In Europe, the population is considered to be generally stable, a summer visitor or breeding resident (BirdLife International, 2015). In Gibraltar, Bensusan et al. (2007) and (Garcia, 2007) found an increase in the numbers of birds taking part in the spring migration passage, between 1967-77 and 1990-2004 surveys, which may have resulted from increases in the Spanish populations. Similar increases were recorded between 2005 and 2006 (Garcia, 2007). In the Czech Republic, it is assessed as critically endangered, rare compared with other raptors in that country (Kren, 2000). In Croatia, it is also classified as Critically Endangered (Radovíc et al., 2003) and appears to be quire rare in Greece with a declining population (Reiser, 1905; Makatsch, 1950; Tucker & Heath, 1994; Global Raptor Information Network, 2021n). In West Africa, there appears to be declining populations in both non-protected and protected areas, based on findings of road transect surveys in the Sudano-Sahelian savannas of Burkina Faso, Mali, and Niger between 1969-1973 and replicated in 2003-2004 (Thiollay, 1977; Rondeau & Thiollay, 2004; Thiollay, 2006). In Southern Africa, the population is non-threatened, especially in protected areas (Simmons & Barnard, 2007). In Asia, Wells (1999) found declines of the eagle population as well as that of other raptor species in the Malay Peninsula, possibly due to the use of agricultural pesticides. Illustrations of the appearance, range, and eggs of this species are shown in Figures 2.1a to 2.1j.

Fig. 2.1a. Booted eagle perching [With permission from Pieter Verheij]

Fig.2.1b. Booted eagle perching [With permission from Derek Keats]

Fig. 2.1c. Booted Eagle light morph perching [With Permission from Pieter Verheij]

Fig. 2.1d. Booted Eagle light morph perching [With permission from Hari K. Patibanda]

Fig. 2.1e. Booted Eagle dark morph in flight [With Permission from Nick Ransdale]

Fig. 2.1f. Booted eagle light morph in flight [With permission from Derek Keats]

Fig. 2.1g. Booted eagle in flight [With permission from Gil Ewing]

Fig. 2.1h. Booted eagle possible breeding display [With permission from Frank McClintock]

Fig. 2.1i. Booted eagle habitat varies from open forest to desert [With permission from Jonathan D. Müller]

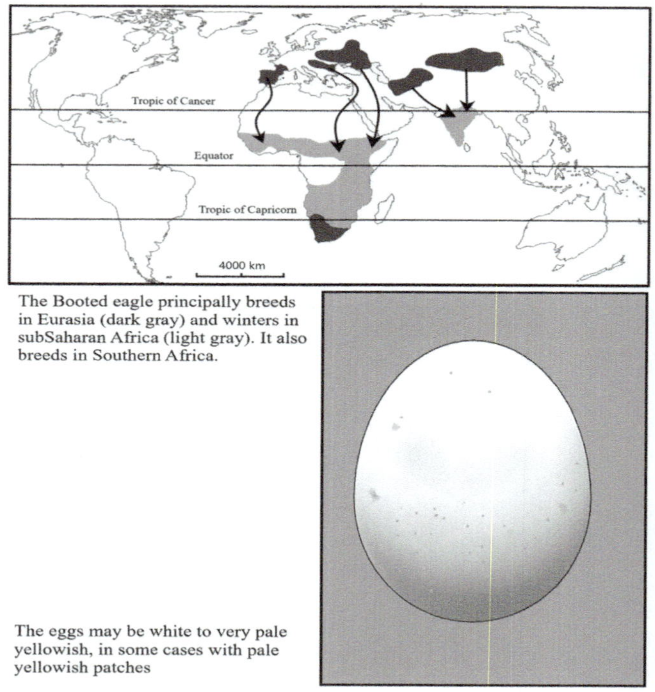

Fig. 2.1j. Booted eagle range and egg

2.3 Little Eagle *Hieraaetus morphnoides* (Gould, 1841)

Physical Description

The Little eagle is aptly named as a "small, robust eagle, appearing generally rather thickset and stumpy, about the size of the Common buzzard of Europe and Asia" (Brown & Amadon, 1968, 682). According to Brown & Amadon (1968) there were two subspecies: the nominate subspecies is *H.m. morphnoides*; and *H.m. weiskei*, but the latter has now been recognized as a separate species, the Pygmy eagle (*Hieraaetus weiskei*, Reichenow, 1900) (Gill & Donsker, 2017). The Little eagle has a light brown colored head, and darker brown feathers with light edges on the crown and crest and a white throat with dark shaft streaks. A broad black moustached streak lines the face below the eye. The upperparts are various shades of brown, from very light colored to almost black. The back rump and scapulars are brown, and the upper tail coverts are lighter with whitish tips, which creates band across the tail tip. The tail itself is greyish brown, with white mottling on the inner webs, and dark bars with lighter tips. The lesser and medium wing coverts are brown, but the greater coverts are paler with whitish tips, "producing a scaly appearance at rest" (Brown & Amadon, 1968, 682).

The Little eagle's outer primaries have blackish edges, and the inner webs are lighter basally with vague dark barring. The inner primaries are greyer, with lighter tips, and the secondaries are darker grey brown, with dark bars and also with whitish tips. The underside is white to pale reddish brown, with darker coloring on the flanks. Dark shaft streaks line the upper breast. The undertail is grey with darker barring and the under coverts are light with dark centers. The wing quills are basally white tipped and darker below. These white tips form a visible white band, joined to the coverts and a light patch at the carpal joint, contrasting with the dark tips. The eye is brown, and the cere and feet greyish white to pale yellow. For the darker phase, the upper parts are blackish brown above, and the underside is darker reddish brown with black shaft streaks, with less white in the underwings. The quill base band and the carpal joint patch is darker, rather than white. The former subspecies (*H.m. weiski*), now classified as the separate species Pygmy eagle (*H. weiski*) of New Guinea is smaller, darker above, more reddish below, with broader and darker streaks (Brown & Amadon, 1968, 682; Gill & Donsker, 2017). The Little eagle may be mistaken for the Whistling kite or eagle (*Haliastur sphenurus*, Vieillot, 1818), but the current species is much darker on the upperparts and paler on the underparts with fewer streaks and the black stripe below the eye, a crest, feathered tarsus and a larger, heavier body relative to the wings. The call is a whistle, with the first note higher than the second – "choo-oo, choo-oo" or "fee-feee-few" and some other harsher sounds which are usually made when the bird is excited (Brown & Amadon, 1968). Illustrations of the appearance, range, and eggs of this species are shown in Figures 2.2a to 2.2s.

Classification

The Little eagle is one of the eagles described above (Booted eagle, Ayres's Hawk eagle, Wahlberg's eagle and the Pygmy eagle), which may be in a monophyletic clade (Helbig et al., 2005; Gill &Donsker, 2014; Lerner et al., 2017). Lerner and Mindell (2005) also found differences between the Little and Pygmy eagles, using an analysis of molecular sequences of two mitochondrial genes and one nuclear gene. The Pygmy eagle was found to be more related to the Booted eagle than to the Little eagle, the differences between the Little eagle and Pygmy eagle being as great as that between any two full species. The conclusion was that the Pygmy eagle was a full species, and the Little eagle was monotypic (Global Raptor Information Network, 2021o).

Breeding

The Little eagle's breeding season starts with "striking, undulating displays high in the air, rising and falling as much as 200 feet, and accompanying the display with constant whistling calls" (Brown & Amadon, 1968). The male and female may glide together, or the male may fly above and dive at the female, who may role over and present the claws. These displays occur both outside and during the breeding season, especially during August. Nesting is usually in trees, and nests may be large, about 75 cm by 50 cm deep, usually in a tree. Nests are usually built high, around 5 to 45 meters altitude (Debus, 1998). Nests are built of branches and sticks, with small, leaf lined central cavities containing the eggs (Brown & Amadon, 1968). On rare occasions, Little eagle may take over the nests of other species. The breeding season varies regionally, with northern birds in Australia nesting longer (during the dry season, March to September) than those in the center and the south (August to October) (Debus, 1998). Males and females participate in nest-building. The clutch is usually 2 eggs but can be one or 3 eggs. The female is the main incubator, with the incubation period lasting 36-40 days (Bollen, 1989), and the nestling period is 54-66 days (Debus, 1998).

Debus et al. (2007) give a case study of five nests in New South Wales, Australia. These nests were in elevated locations in eucalypt woodland (*Eucalyptus calignosa*, Blakely & McKie; *Eucalyptus melliodora*, A.Cunn. ex Schaue; *Eucalyptus viminalis*, Labill.; and *Eucalyptus moluccana*, Roxb., all these species endemic to eastern and southeastern Australia). Hall (1974) also reported a similar location for a nest in *Eucalyptus dundasii* (Maiden). For the first case reported by Debus et al. (2007), nesting began on 13 August, egg-laying on 16-17 September, incubation 19 September and hatching 25 October. For the second case, copulation occurred on 8 September, incubation on 9 September, and hatching on 16 October. For the third case, nest-building began on 5 – 8 October, and later abandoned. For the fourth and fifth cases, possibly laying began in September, but only fledging was observed on 25 December and 21 December respectively (Debus et al., 2007b). Generally, clutch sizes are one to 2 eggs (Baker-Gabb, 1983), incubation periods are 37 to 39 days (Debus et al., 2007b), while some can be as long as 55 days (Baker-Gabb, 1983). Nesting successes were recorded by Debus et al. (2007) as 0.4 per eagle pair, while Baker-Gabb (1983) found 0.8 chicks fledged per territorial pair. Breeding density was 2,680 ha per eagle pair (Baker-Gabb, 1983).

Larkin et al. (2020) also studied Little eagle nests near Armidale, New South Wales, examining the parameters of the nesting trees, the surrounding woodland (≤25 m from the nest tree), and (using GIS analyses) the landscape scale (within 200-m and 2-km radii of the nest). The findings were that eagles nested ≥14 m elevation, within the canopy of eucalypts which were quite tall (>20 m tall), usually in sheltered locations, usually in mid-slope, with an aspect of north-east to southerly aspect). These were usually in woodland patches greater than 5 hectares in areas, with an average area of 76 ha). The nests were usually less than 200 m (mean 78 m) from the woodland edge, more than 10 m (mean 190 m) from an agricultural area, more than 37 meters from rural human habitation (mean 485 m), more than one kilometer from suburban development, and more distant from sealed roads (mean 832 m) than gravel roads (mean 490 m) than minor tracks (mean 291 m). The authors also found that breeding productivity in 2017–18 was 0.91 young fledged per clutch laid and 0.67 young fledged per occupied territory per year. Additionally, nesting sites were used annually for at least 3–7 years. Generally, nests were abandoned due to proximate human disturbances such as clearing, earthmoving or construction work, the death of the nest tree and pindone baiting for rabbits (*Oryctolagus cuniculus*, Linnaeus, 1758 and perhaps *Corvus tasmanicus*, Mathews, 1912). In some cases, wedge-tailed eagles and ravens (*Corvus coronoides*, Vigors & Horsfield, 1827). The authors argued most eagle nests were on private land, and greater protection would be required for "breeding habitat, nest sites and foraging habitat, woodland regeneration (especially riparian), and a buffer around established nests of ≥1 km from major developments such as urbanisation" (Larkin et al., 2020).

An older study by Debus (1984, 87) on the breeding biology of the Little eagle also found that active eagle nests were usually located in woodland patches, with a ratio of two to one egg clutches being 2:1 in favor of two eggs. The incubation period was 36 to 40 days, the nesting period 9 weeks and the post-fledging about 6 or more weeks and unlike some other studies, the males shared the incubation with the females. The finding indicated that "the breeding cycle occupied about four months from nest building to fledging, and, for the district, about five months (August-January). Fledging success was about 0.5 young per pair per year, and about one in six successful nests raised two young. Breeding density was at least one pair per 1600 ha" (Debus, 1984, 87).

Foraging

The Little eagle is a solitary, sedentary forager. Compared with the broadly similar Booted eagle, it has a calmer temperament, and more frequently it hunts from perches (Brown & Amadon, 1968). The most frequented habitats are wooded landcover, especially with hilly topography, sometimes near rivers, but it generally avoids dense forests (Olsen, 1995; Debus, 1998). When it soars, common habitats are the forest edges, otherwise it perches on the branches of living or dead trees and dives to take prey, usually on the ground rather than in flight (Beehler, 1978; Beehler et al., 1986; Debus, 1998). Common prey species are small mammals, birds, reptiles (mostly lizards) and insects. Fish may also be taken, sometimes by robbing the Whistling kite of its food (Debus, 1998). Debus (1998) argues for regional variations in prey taken. Catches in the south are principally rabbits, while northern birds take birds, and lizards are taken in more arid areas.

Examples supporting these observations include studies by Beehler et al. (1986) in New Guinea, which found that it is a bird hunter. Another study by Debus et al. (2007) in New South Wales found that the main diet at three observed nests was 78% mammals (94% of biomass), 13% birds (5% of biomass), and 4% reptiles, a Tree dragon (*Amphibolur usmuricatus*, White, 1790) (1% of biomass). The mammals consisted of rabbits (80% of biomass) and Brown hares (9%). The few birds taken included the Noisy friarbird (*Philemon corniculatus*, Latham, 1790), Common starling (*Sturnus vulgaris*, Linnaeus, 1758), Eastern bearded dragons (*Pogona barbata*, Cuvier, 1829), Peaceful dove (*Geopelia placida*, Gould, 1844), Magpie-lark (*Grallina cyanoleuca*, Latham, 1801) and White-winged chough (*Corcoraxmela norhamphos*, Vieillot, 1817) (Debus et al., 2007b). Baker-Gabb (1983) found that in Victoria, Australia, Little eagles hunted immature rabbits (72%), birds (28%), and no lizards. Hall (1974) in Western Australia observed a female eagle preying on a monitor lizard, possibly the Black-palmed rock monitor (*Varanus glebopalma*, Mitchell, 1955), and another record was of the remains of a parrot (*Barnardius zonarius*, Shaw, 1805).

Debus (1984, 87) also gives the example of Little eagles foraging and nesting in Armidale, New South Wales in 1980. Field data were supported with information from the RAOU Nest Record Scheme. The findings indicated that Little eagles hunted a wide variety of vertebrate prey species during the breeding season, as evidenced by prey remains in the studied nests. Prey remains identified at three nests consisted, by percentage of mammals (60%), birds (25%) and lizards (15%). These findings were similar to the remains found in pellets, which were 55% mammal, 25% bird and 20% lizard. Debus (1984, 87) also estimated at the prey types taken from home ranges in the six-month breeding season as 120 young lagomorphs (rabbits and hares of the Order Lagomorpha; Brandt, 1855) weighing (48 kg), 50 birds (3 kg) and 30 lizards (4 kg). Other studies have mentioned that large insects are sometimes taken (Debus et al., 2020).

Population Status

The Little eagle is considered widespread across its range, but due to population decline it is also considered vulnerable (Debus 2008, 2009; Olsen, 2009). General findings indicate a decline in population (Storr, 1991; Saunders & Ingram, 1995). However, conflicting assessments exist on the population trajectories. There may have been population increases with habitat clearance and the increased population of rabbits and starlings (Olsen, 1995; Debus, 1998); and there may also have been population declines due to European landuse in arid areas after British colonialization, but this may require some evaluation (Olsen, 1995). Some have also argued that the eagle has not suffered from direct human persecution or DDE-induced eggshell thinning (Debus, 1998). Debus et al. (2007) argue that based on existing evidence, the population of eagles may be declining. Cited evidence includes: a serious population decline in Central Tablelands in the 10 years before their research; a 14% decline of sightings between the first and second national bird atlases, and the first and second national Bird of Prey Watch Scheme (Steele, 1998); a greater than 20% decline in researched areas of eastern and southern Australian (Barrett et al., 2003); and an observed decline in the Western Australian wheat belt (Johnstone & Storr, 1998). Eagles may also experience low breeding success during droughts (Debus et al., 2007b).

Debus & Ley (2009, 76) note that "Atlas data showed a decline in the Little eagle's reporting rate over 20 years: 14% nationally, with >20% decline across the Murray–Darling Basin and eastern sheep–wheat belt, and 39% in NSW, although there was no significant decline detected in the New England Tableland bioregion" (see also Barrett et al., 2003, 2007). Other surveys of nests have found an 80–90% decline in the number of breeding pairs near Canberra, ACT (Australian Capital Territory), over the previous 15 years, which may add new information to local atlas data (Olsen & Fuentes, 2005; Olsen et al., 2009). There was also a 50% decline in the reporting rate in south-eastern NSW over the previous 30 years (Canberra Ornithologists Group, 2008). Debus & Ley (2009, 76) note that "where atlas sighting data and surveys of breeding pairs conflict, monitoring of nests more reliably indicates raptor population trends" (see also Sergio et al., 2008; Olsen et al., 2008, 2009). They add that for the Little eagle, the"objective survey data provided a case for listing as Vulnerable in the Australian Capital Territory" (see also Olsen et al., 2008). Illustrations of the appearance, range, and eggs of this species are shown in Figures 2.2a to 2.2g.

Fig. 2.2a. Little eagle perching [With permission from David Nice]

Fig. 2.2b. Little eagle perching [With permission from David Nice]

Fig. 2.2c. Little eagle perching [With permission from Frank McClintock]

Fig. 2.2d. Little eagle in flight [With permission from David Nice]

Fig. 2.2e. Little eagle in flight [With permission from David Nice]

Fig. 2.2f. Little eagle habitat [Credit: Benn McGuinness-unsplash]

2.4 Pygmy Eagle *Hieraaetus weiskei* (Reichenow, 1900) or *Aquila weiskei*

Physical Description

The Pygmy eagle, also sometimes named the New Guinea hawk-eagle is a small eagle that ranges in New Guinea, previously classified as a subspecies of the Little eagle, as *Hieraaetus morphnoides weiskei* (Reichenow, 1900) or as *Aquila morphnoides weiskei* (Reichenow, 1900) (Gjershaug et al., 2009). This is one of the world's smallest eagles, noted by some to be the smallest, measuring only 45 to 56 cm in length, with a wingspan of 112 to 126 cm (Gjershaug et al., 2009; Debus, 2017). Although like the Little eagle in some characteristics, it differs mostly in the color of its plumage. It has darker and larger streaks on the crown, nape and underparts. There are light and dark morphs, the latter perhaps ten times less common (Debus, 2017). Debus (2017) notes that the Pygmy eagle is similar in many respects to the Little eagle, with differences mostly based on the issues above. Additionally, the light morphs of the Pygmy and Little eagles show more differences than the dark morphs of both species (Debus, 2017).

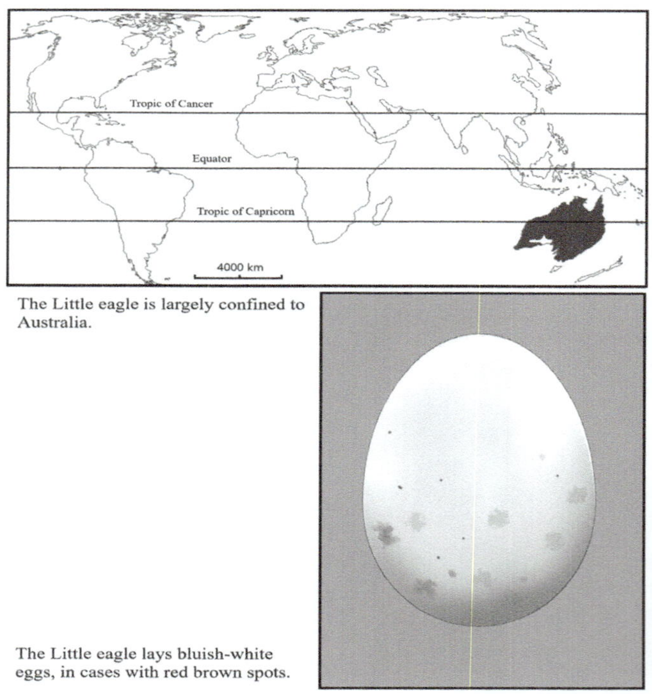

Fig. 2.2g. Little eagle range and egg

Gjershaug et al. (2009, 31) listed the main differences between the Pygmy eagles and Little eagles and broadly similar Booted eagles. These are based on 10 parameters, the first five based on the eagles' head and body are: the crown (for the Pygmy eagle, dark brown streaked; for the Little eagle, blackish; for the Booted eagle, pale rufous to cream with fine black-brown shaft streaks); crest (for the Pygmy eagle, no crest; for the Little eagle, a short crest ; for the Booted eagle, no crest); cheek (for the Pygmy eagle, dark streaked like the whole head; for the Little eagle, dark, contrasting with a lighter nape and neck-sides; for the Booted eagle, dark, darker than rest of the head); nape (for the

Pygmy eagle, dark streaked like the crown; for the Little eagle, pale rufous to cream, narrowly shaft-streaked, contrasting with darker crown and mantle; for the Booted eagle, pale rufous to cream, narrowly shaft-streaked but without contrast to crown); breast markings (for the Pygmy eagle, broad dark streaks; for the Little eagle, narrow dark streaks; for the Booted eagle, narrow dark streaks).

The remaining parameters are based on the wings and tail, comprising: the upperwing coverts (for the Pygmy eagle, no distinctive light panels; for the Little eagle, distinctive light panels; for the Booted eagle, distinctive light panels); underwing pattern in pale morph (for the Pygmy eagle, all coverts white to lesser underwing coverts rufous, greater coverts white; for the Little eagle, lesser underwing coverts rufous, white greater underwing coverts; for the Booted eagle, all underwing coverts are white); mantle (for the Pygmy eagle, dark, little difference with the tertials; for the Little eagle, lighter than the dark tertials; for the Booted eagle, lighter than the dark tertials), uppertail-coverts (for the Pygmy eagle, dark, little contrast to rump; for the Little eagle, pale buff to pale rufous, distinct from dark rump; for the Booted eagle, pale buff to pale rufous in contrast to dark rump) and tail pattern (for the Pygmy eagle, 7 to 8 dark bars with low contrast to the light bars; for the Little eagle, 7 to 8 dark bars with high contrast to the light bars; for the Booted eagle, none to 7 less distinctive dark bars) (Gjershaug et al., 2009, 31). Note that the tertials refer to the innermost flight feathers of the wing, which are attached to the humerus bone in the bird's upper arm. The voice of the Pygmy eagle is a "sip sipyeeee" usually uttered in flight, like the Little eagle and Booted eagle, and a one tone squeal (Debus, 2017).

Taxonomy

The Pygmy eagle was first comprehensively described by the German naturalist Anton Reichenow as Eutolmaetus weiskei in 1900. This species or subspecies has held several names, including 'Weiske's Eagle' which celebrates the collector Emil Weiske, who took the type specimen in Astrolabe Mountains in Jun 1898 (Parry, 2001). Other names include the 'Papuan Booted Eagle' (Debus et al., 2007b), 'Papuan Little Eagle' (Gregory, 2008), and 'New Guinea Hawk-Eagle' (Gill et al., 2020), but these were eventually superseded by Pygmy eagle (Gjershaug et al., 2009). As noted above, it was later found to be sufficiently differentiated from the Little eagle to achieve species status, a position subsequently recognized by the International Ornithologists' Union (IOC) (Gjershaug et al., 2009; Gill & Donsker, 2017).

Lerner and Mindell (2005) used molecular sequences of two mitochondrial genes and one nuclear intron, and found that the Pygmy eagle, which was then regarded as subspecies of the Little eagle, was more related to the Booted eagle, a result that was supported by Bunce et al. (2005), as part of their study of the extinct New Zealand eagle, *Aquila (Hieraaetus) moorei.* These researchers also argued that the Pygmy eagle also differed from the *Aquila* species genetically and morphologically, and also in terms of its geographical range. This was also supported by Gjershuag et al. (2009) as noted below. The alternative of *Aquila* rather than *Hieraaetus* was due to the work of raptor scientists (e.g., Wink & Sauer-Gürth, 2004; Gjershaug, 2006; Sangster et al., 2005) merging the species formerly included in the Genus *Hieraaetus* into *Aquila.*

Gjershaug et al. (2009, 24) argued that "genetic differentiation in published mitochondrial cyt-b and ND2 sequences between the nominate subspecies and *weiskei* is equivalent to that found between other accipitrid taxa at the species level, as are pairwise divergences between each Little eagle taxon and their sister species, the Booted eagle *A. pennata.*" Gjershaug et al. (2009, 32) further pointed out that "on the basis of the large number of diagnostic differences between *Aquila weiskei* and *A. morphnoides*, we recommend that *A. weiskei* should be treated as a distinct species and propose to use the English name 'Pygmy eagle' because it is the smallest eagle in the world, males being smaller even than the Little eagle." They also noted that based on their investigations, "the plumage differences between the Pygmy and Little eagles are greater than those between the Little and Booted eagles, which are widely recognized as distinct species. We therefore consider that the large number of diagnosable characters indicates that Pygmy eagle is a distinct species, regardless of which species concept is used. The large size difference between Pygmy and Little eagles (female Pygmy eagles being similar in size to Little eagle males) would probably function as a reproductive barrier should they meet". Size may also "function as an ecological isolation mechanism, as intermediate-sized hybrids could be less well-adapted" (Gjershaug et al., 2009, 34).

Current findings indicate that the Pygmy and Little eagles diverged either from the Booted eagle or from a common ancestor possibly one million years ago, evidenced by a 2.2–2.6% divergence in mitochondrial cyt-b and ND2. Using current behavioral patterns of the Booted eagle as evidence, it may be hypothesized that the vagrants of the migratory Eastern Palearctic populations of Booted eagle (or a common ancestor) which normally winter in peninsular India, may have entered New Guinea and Australia, through peninsular Malaysia, Sumatra, Java and Bali. These specimens may have diverged over time, eventually becoming the present Pygmy eagle and Little eagle. Possibly, this may also have created the extinct giant Haast's eagle *Aquila (Harpagornis) moorei* in New Zealand, which has a similar level of genetic divergence from the Booted Eagle as from the Pygmy eagle and Little eagle (Bunce et al., 2005; Gjershaug et al., 2009, 34).

Foraging

The primary habitat is primary, gallery, tropical moist and monsoon forest, including forest edges. It forages by soaring above the forest, or by low flight over the forest canopy, or perching in high trees with a view of prey movements. It has been known to attack and prey on the Grey Crow (*Corvus tristis*, Lesson & Garnot, 1827) (Coates, 1985). Other prey includes the B (*Macropygia phasianella*, Temminck, 1821) and the Mountain fruit-dove or White-bibbed fruit dove (*Ptilinopusrivoli*, Prévost, 1843). Pygmy eagles have been observed in the dense forests of Vogelkop or Bird's Head Peninsula in New Guinea and the island Buru and in the northern Moluccas. The hunting technique involves swift dives at avian prey, mostly in the canopy (Debus, 2017).

Breeding

The Pygmy eagle's breeding habits are poorly known. One member of a pair was observed performing ritual collecting of sticks and moss in the Baiyer River area in March, late in the wet season (Mackay, 1980). The eggs are white or bluish-white, unmarked, or sparsely streaked and blotched with reddish-brown (Coates, 1985).

Population Status

The Pygmy eagle is probably non-migratory (Coates, 1985). The Pygmy eagle is found mostly in Australasian region, namely New Guinea and the Moluccas. The range includes "the western part of New Guinea (Vogelkop) and the Moluccas (Halmahera, Ternate and Seram)" (Gjershaug et al., 2009, 24). It is rare but possibly resident in the Moluccas (Halmahera, Seram) (Coates & Bishop, 1997). There are isolated, possible sightings on some islands, such as on Halmahera (1989, 1994) (King, 1990; Global Raptor Information Network, 2021p), and on Seram over primary forest (1996) (Coates & Bishop, 1997). In New Guinea, this species has been recorded in the central and eastern New Guinea, such as Utakwa River region, Snow Mountains, Adelbert Mountains, Huon Peninsula, Astrolabe Mountains, and Aroa River region, at wide range of elevations from lowlands to 1,585 m (Rand & Gilliard, 1967; Coates, 1985). It is rare or absent from western Irian Jaya (Coates, 1985).

The Global Raptor Information Network (2021p) records that the population status of this species is mostly unrecorded (see also Gjershaug et al., 2009). Other authors describe it as well distributed, but rare, through its range in New Guinea and the Moluccas (Coates, 1985; Coates & Bishop, 1997). Deforestation may be a factor its rarity or declining population, but its status of Least Concern (by BirdLife International. 2016) "must be regarded as tentative"; as "there are no published population estimates for this newly recognized species, and the population category (1,000 to 10,000 total individuals) assigned here is largely speculative" (Global Raptor Information Network, 2021p). Illustrations of the appearance, range, and eggs of this species are shown in Figures 2.3a to 2.3e.

2.5 Ayres's Hawk Eagle *Hieraaetus ayresii* (Gurney, 1862) or *Aquila ayresii* (Gurney 1862)

Physical Description

The Ayres's Hawk eagle has been described as "a small, dashing raptor that is thought to be inexplicably rare throughout its range" (Hartley & Mundy, 2003, 44; see also Brown et al., 1982; del Hoyo et al., 1994; Harrison et al., 1997). This eagle measures 44-57 cm in length, with a wingspan of 106-137 cm and a weight of 685-1.05 kg (The Eagle Directory, 2012). Brown & Amadon (1968, 685) describe the adult male with white mottled, sepia upperparts, a white forehead, ash-colored tail with dark baring and a black tip. The primaries and secondaries are blackish, with a white, dark brown marked underside, and a more heavily marked breast and belly, and lighter marked thighs and vent. The brown underwing coverts are mottled white, and the underside of the flight feathers is dark with baring. The legs are white. The eye is yellow to orange, with yellow cere and feet. The beak is bluish tinted light brown, with a lighter base and black tip. Adult females are darker than males, with denser spots on the underside, with less white on the forehead and eyebrow. A melanistic phase is mostly black with white markings (Brown & Amadon, 1968). However, few descriptions exist (Hartley & Mundy, 2003; see also Brown, 1966; Brown & Davey, 1978; Brown et al., 1982; Steyn, 1982; Maclean, 1993; Del Hoyo et al., 1994; Kemp & Kemp, 1998).

Hartley & Mundy (2003, 45) note that "it is likely that plumage variation and confusion with species such as African Hawk-eagle (*H. spilogaster*), Booted eagle (*H. pennatus*), and Cassin's Hawk-eagle (*Spizaetus africanus*) have led to the Ayres's Hawk-eagle being overlooked throughout its range" (see also Brown, 1966; Ash, 1981; Clark, 1999). The immature plumage is marked by a pale buff forehead and eyebrows, pale whitish throat, dark streaked rufous crown and nape, and dark brown wings and back. The scapulars and coverts have pale edges "giving a scaly appearance"; the tail is ash brown, barred darker brown like the adult and the underside is "warm buff" with black streaks, lighter and un-streaked on the flanks, legs and tail coverts

Fig. 2.3a. Pygmy eagle in flight [With permission from Paul French]

Fig. 2.3b. Pygmy eagle in flight [With permission from Nigel Voaden]

Fig. 2.3c. Pygmy eagle in flight [With permission from John C. Mittermeier]

Fig. 2.3d. Pygmy eagle habitat, New Guinea Forest [With permission from Jay Pruett]

(Brown & Amadon, 1968, 685). The iris is grey and the cere and feet are dull greyish yellow. It differs from the Bonelli's eagle in the pale whitish eyebrows and the underside markings, as the Bonelli's eagle has lighter underparts. The immature plumage is like the African Hawk eagle, but the latter has chestnut-colored underparts, unlike the lighter buff of the Ayres's Hawk eagle. The Bonelli's eagle also lacks the pale edged upper feathers, dark quill shafts and stronger contrasts between dark and light feathers. The dark phase of the Ayres's Hawk eagle is darker than comparable species. This is "a fairly noisy eagle, emitting a high pitched "hueeeep, hip hiphiphip" or hip hiphip" (Brown & Amadon, 1968, 685). Illustrations of the appearance, range, and eggs of this species are shown in Figures 2.4a to 2.4i.

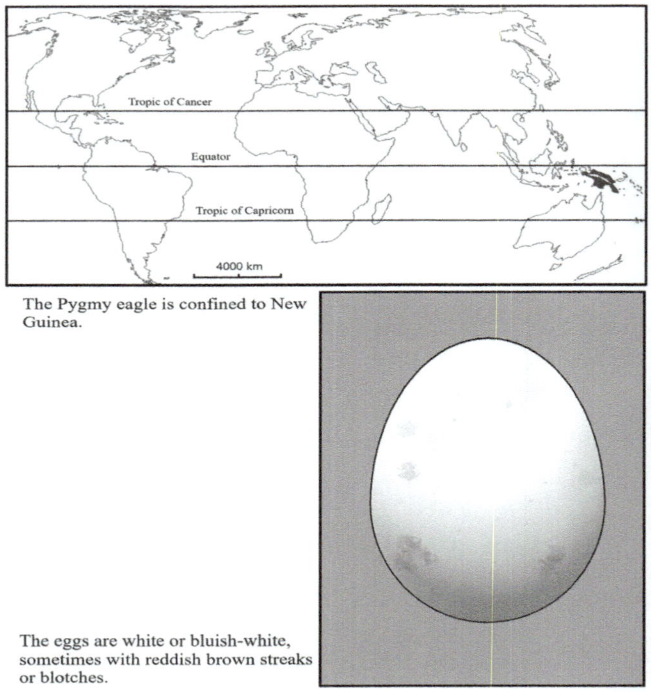

The Pygmy eagle is confined to New Guinea.

The eggs are white or bluish-white, sometimes with reddish brown streaks or blotches.

Fig. 2.3e. Pygmy eagle Range and Egg

Classification

The Ayres's Hawk eagle was classified as *Hieraaetus dubius* (Brown and Amadon, 1968; Stresemann & Amadon, 1979), but Global Raptor Information Network (2021q) states "that name apparently refers to *H. pennatus*" (see also Brooke & Vernon, 1981; Dowsett & Dowsett-Lemaire, 1993). Helbig et al. (2005), using DNA sequences from one mitochondrial and two nuclear genes, indicated that Ayres's Hawk eagle is grouped with the other smaller *Hieraaetus* species, including the Booted eagle, Little eagle and Wahlberg's eagle. Global Raptor Information Network (2021q) also argues that most raptor systematists (e.g., Wink and Sauer-Gürth, 2004; Gjershaug, 2006) and some national committees support the transfer of species formerly assigned to *Hieraaetus* into *Aquila.*

Foraging

The Ayres's Hawk eagle is mostly an aerial hunter, foraging from flight (Clark, 1999). Kemp et al. (2020a) note that the Ayres's Hawk eagle habitat can be "dense deciduous woodland, gallery forest, forest edge, savanna woodland and plantations of exotic trees; even in city suburbs." Brown and Amadon (1968) give its habitat as "open forest sometimes near rivers, and lighter vegetation such as wet savanna" with a rarer status in rainforest or open vegetation. Foraging is principally from perches for avian prey such as small birds and poultry, and occasionally arboreal mammals. Hartley & Mundy (2003, 44) point out that "little is known about this species" (Steyn, 1982; Harrison et al., 1997; Virani & Watson, 1998). They point out gaps of information on the breeding biology, with records from Kenya (Brown, 1966; Dewhurst et al., 1988) and possibly Zimbabwe (Irwin, 1981). In urban areas, Rock doves (*Columba livia*, Gmelin, 1789) and other small birds are common prey (Irwin, 1981, 1985; Lendrum, 1982; Dewhurst et al., 1988; Harrison et al., 1997).

Breeding

The breeding season begins with soaring, calling and diving into the forest near nesting sites. Nests are in tall forest trees, the nests being square-shaped with a compact form, rather than the more common basin shape, composed of branches and green leaves (Brown & Amadon, 1968, 686). Eggs are white with brown and lilac markings. The size of the clutch varies regionally; usually one egg is laid in Kenya in May to July, while in South Africa two eggs are laid in July to September. Fledgling is after 75 days. Hartley & Mundy (2003, 44) report breeding patterns from Zimbabwe, covering plumage, molt, body mass, and morphology nesting. They argue that "of nine nests, six were on well-wooded hill slopes and three were in riparian woodland. Nests were 9-15 m above ground, situated below the canopy and nest diameters were 71-129 mm. Although Ayre's Hawk-eagle is a rare species, it was not as scarce in Zimbabwe as some observers had claimed."

Population Status

BirdLife International (2017) rates the Ayres's Hawk-eagle as of Least Concern. The criterion for this assessment is the "extremely large range" and does not fulfill the other criteria for Vulnerable status ("a declining or fluctuating range size, habitat extent/quality, or population size and a small number of locations or severe fragmentation"). The population trend is stable, and the number of mature individuals is listed at 670-6,700. The Ayres's Hawk eagle species is vulnerable to habitat losses, woodland degradation and direct persecution from people one suggested reason is that it preys on racing pigeons and other birds in some regions (Hartley & Mundy, 2003; Kemp & Boesman, 2013; Global Raptor Information Network, 2021q), and presticide contamination. Illustrations of the appearance, range, and eggs of this species are shown in Figures 2.4a to 2.4i.

Fig. 2.4a. Ayres's Hawk-eagle perching [With permission from Alan Gilbert]

Fig. 2.4b. Ayres's Hawk-eagle perching and Eatong Elias Bayou

Fig. 2.4c. Immature Ayres's Hawk-eagle perching [With permission from Antoine Marchal]

Fig. 2.4d. Ayres's Hawk-eagle in flight [With permission from Robert Davidson]

Fig. 2.4e. Ayres's Hawk-eagle in flight sideview [With permission from Robert Davidson]

Fig. 2.4f. Immature Ayres's Hawk-eagle in flight [Credit: Feathercollector]

Fig. 2.4g. Ayres's Hawk-eagle habitat, Burkina Faso [With permission from Rita Willaert]

Fig. 2.4h. Ayres's Hawk-eagle habitat [With permission from Kureng Dapel]

2.6 Rufous-bellied Eagle *Lophotriorchis kienerii* (G. de Sparre, 1835) or Chestnut-bellied Hawk-eagle *Hieraaetus kienerii* (Geoffroy Saint-Hilaire, 1835)

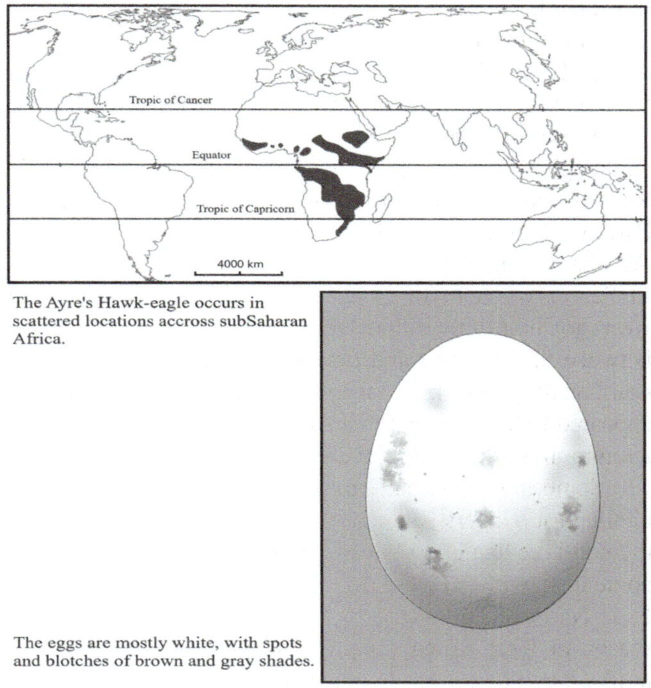

The Ayre's Hawk-eagle occurs in scattered locations accross subSaharan Africa.

The eggs are mostly white, with spots and blotches of brown and gray shades.

Fig. 2.4i. Ayres's Hawk-eagle range and egg

Physical Description

The Chestnut-bellied Hawk-eagle (*Hieraaetus kienerii*) (Lerner & Mindell, 2005) is also named the Rufous-bellied eagle or Rufous-bellied Hawk-eagle (*Lophotriorchis kienerii*, G. de Sparre, 1835) (Dickinson, 2005). Brown and Amadon (1968) list two subspecies *H.k. kienerii* and *H.k. formosus*. Mayr & Cottrell (1979) also name two subspecies the larger nominate *Lophotriorchis kienerii kienerii* ranging in India and Sri Lanka, and *Lophotriorchis kienerii formosus* ranging in Southeast Asia from Burma to Sulawesi (Mayr & Cottrell, 1979). This is a small eagle, ranging in size from 63 to 81 cm in length, with birds from Sri Lanka slightly smaller than those from the Himalayas (Rasmussen & Anderton, 2012). Eagle Directory (2012) gives its size as 46-61 cm in length, with a wingspan of 105-140 cm and a weight of 732 g. Brown and Amadon (1968) describe *kienerii* as smaller and darker, and with blacker upperparts than *formosus* (see also Ali & Ripley, 1978; Clark & Schmitt, 1993).

The general plumage is black upperparts (head, nape, mantle, back scapulars and medium and lesser wing coverts), the under parts with a white throat and chestnut belly. There may be a black hood with a short crest (Brown & Amadon, 1968; Ali & Ripley, 1978; Rasmussen & Anderton, 2012). The dark brown tail is banded with pale brown bands. The dark brown greater wing coverts and wing quills are contrasted with the paler secondaries' feathers. White feathers are on the chin, throat and upper breast, with reddish and black feathers on the chin, throat and breast. The lower breast, thighs, undertail covert, flanks and abdomen are rufous-chestnut, the last two being also covered with black streaks. Black mottling covers the chestnut underwing coverts and auxillaries. The eyes are dark brown while the cere, legs and long-toed feet are yellow (Brown & Amadon, 1968).

Immature birds have brown upperparts, pale feather edges, a white forehead line and a dark brown, grey barred, pale brown tipped tail. Neck sides and cheeks are white, with mottles and streaks, with white coloring on the throat, underparts and underwings coverts and dark brown wing quills. The eyes are brown and the legs and cere are yellow. This species may be mistaken for a few other species. The Changeable hawk-eagle or Crested hawk-eagle (*Nisaetus cirrhatus* or *Spizaetus cirrhatus*) is larger and less compact, and contrasts by having a "long flopping crest" (Brown & Amadon, 1968, 687). The juvenile Chestnut-bellied Hawk eagle differs from this species in having a shorter tail and longer wings. Unlike *N. cirrhatus*, the underwing is white, dark edged, with a black carpal joint patch. The voice is described by Brown and Amadon (1968, 687) as "unknown" and the bird as a "very silent eagle." Illustrations of the appearance, range, and eggs of this species are shown in Figures 2.5a to 2.5g.

Classification

The Rufous-bellied eagle was "traditionally considered to be a member of the genus *Hieraaetus*" (Global Raptor Information Network, 2021r, see also Dickinson, 2003). However, Lerner and Mindell (2005) and Haring et al. (2007) used the molecular sequences of mitochondrial and nuclear genes to study its genetics and did not find a close relationship between the Rufous-bellied Eagle to any other species, including the eagles of the Genus *Hieraaetus*. Global Raptor Information Network (2021r) also notes that they found it to be "genetically highly divergent from all other 'booted' eagles." The main factor was that the Rufous-bellied eagle appears to be a "morphologically specialized species with long toes, a crest, and an adult plumage different from the booted eagles" (Global Raptor Information Network, 2021r). Haring et al. (2007) found that the genetic distances between the Rufous-bellied eagle and *Hieraaetus* was sufficient for this species to be placed in a monotypic genus, *Lophotriorchis* (Sharpe 1874).

Foraging

The Rufous-bellied eagle frequents forest, and some more openhills and river valleys. It frequently soars above the canopy, diving in for small birds and mammals (Brown & Amadon, 1968). Ali & Ripley (1978) and Taher (1992) note it frequents hill forests in India. The diet is also described as including birds, such as pheasants and pigeons (Thiollay, 1994; Wells, 1999).

Breeding

Brown and Amadon (1968, 687) note that the Rufous-bellied eagle nuptial display is "undescribed but is probably similar to that of *H. pennatus* and *H. dubious*, involving vigorous vertical stoops" (Note that the species termed *H. dubious*, the Ayres's Hawk eagle, is now classified as either *Hieraaetus ayresii* (Gurney, 1862) or *Aquila ayresii* (Gurney, 1862)). Although it is a generally silent bird, during the breeding season its call is a "repeated piercing "keeee" during display flight" (Clark et al., 2020a). Eagle Directory (2012) describes its breeding season as starting with displays, circling and "sky dances that include dives and wing-shaking." The generally large nests are located on large trees, up to a meter wide and half a meter in depth (Jayaneththi, 2010; Rasmussen & Anderton, 2012). Thiollay (1994) and Robson (2000) cite the breeding season as December to March (Sri Lanka and southern India) and February (Philippines); Iqbal et al. (2011) estimate the breeding season to be April.

Population Status

BirdLife International (2018) notes that there has been no estimate of the population, "but the species is widespread and described as common in India (Western Ghats), moderately common in Sulawesi, uncommon in the Philippines, rare in Java and Burma and scarce in Nepal" (see also Del Hoyo et al., 1994). The trend of the population is "inferred to be undergoing a moderately rapid decline due to forest loss and fragmentation" (Ferguson-Lees and Christie 2001; Tracewski et al., 2016). Illustrations of the appearance, range, and eggs of this species are shown in Figures 2.5a to 2.5g.

Genus *Nisaetus* (Hodgson, 1836) and Genus *Spizaetus* (Vieillot, 1816)

These eagles (either crested or non-crested) vary in size, with comparatively slender bodies, short wings, long rounded tails, feathered legs and long talons. Genetically they have been linked to the Genera *Stephanoatus* and *Oroaetus*. The Tropical Asian members of the Genus *Spizaetus* have been moved to Genus *Nisaetus*, leaving only the South American members of *Spizaetus*

Fig. 2.5a. Rufous-bellied eagle [With permission from Angela Yeo]

Fig. 2.5b. Rufous-bellied eagle [With permission from Angela Yeo]

Fig. 2.5c. Immature Rufous-bellied eagle [With permission from Myron Tay]

Fig. 2.5d. Rufous-bellied eagle [With permission from Mike Nelson-Birdtour ASIA]

Fig. 2.5e. Immature Rufous-bellied eagle in flight [With permission from BP Chua]

Fig. 2.5f. Rufous-bellied Hawk eagle habitat, Sri Lanka [With permission from Genine Alyssa]

in the former genus. These are *S. nipalensis*, *S. philippensis*, *S. lanceolatus*, *S. bartelsi* and *S. alboniger*. These have been moved to *Nisaetus nipalensis* (Hodgson, 1836), *Nisaetus philippensis* (Gould, 1863), *Nisaetus lanceolatus* (Temminck & Schlegel, 1844), *Nisaetus bartelsi* (Stresemann, 1924) and *Nisaetus alboniger* (Blyth, 1845) respectively. *S.nanus* (later *Nisaetus nanus*, Wallace, 1868) was considered to be closely related to *S. alboniger* , and *S. cirrhatus* (later *Nisaetus cirrhatus*, Gmelin, 1788) related to *S.nipalensis*. *S. limnaeetus* (later *Nisaetus cirrhatus*, Gmelin, 1788) was seen as possibly a distinct species rather than a race. Brown and Amadon (1968, 693) argue that "the other three species of the genus, *africanus* of Africa, and *ornatus* and *tyrannus*of the New World are not obviously tied in with any of the other species of the genus" (ibid.).

More recent classifications place *S. africanus* as *Aquila africana* (Cassin, 1865), *Spizaetus ornatus* (Daudin, 1800) keeping its old classification, and *S. tyrannus* in its old designation as *Spizaetus tyrannus* (Wied, 1820). Hence, the Cassin's Hawk eagle (*Spizaetus africanus*, Cassin, 1865) is moved to *Aquila africana* (Cassin, 1865). The Global Raptor Information Network (2021m) notes that the Cassin's Hawk eagle was "previously placed in a monotypic genus, *Cassinaetus* (Cassin, 1865), in *Hieraaetus* (Kemp, 1994b), or in *Spizaetus*. The study by Lerner and Mindell (2005), based on the molecular sequences of one nuclear and two mitochondrial genes, showed that *Aquila chrysaetos*, *A. africanus*, *A. fasciatus*, *A. verreauxii*, *A. audax*, and *A. gurneyi* form a closely related clade. Their findings seem to shed the most light on the relationships of this species, so it is merged in *Aquila*, even though it shows some morphological differences from other members of that clade" (see also Voous, 1964; Amadon & Bull, 1988). The Celebes Hawk eagle (*Spizaetus lanceolatus*, Temminck & Schlegel, 1844) has been moved to be the Sulawesi Hawk-eagle (*Nisaetus lanceolatus*, Temminck & Schlegel, 1844).

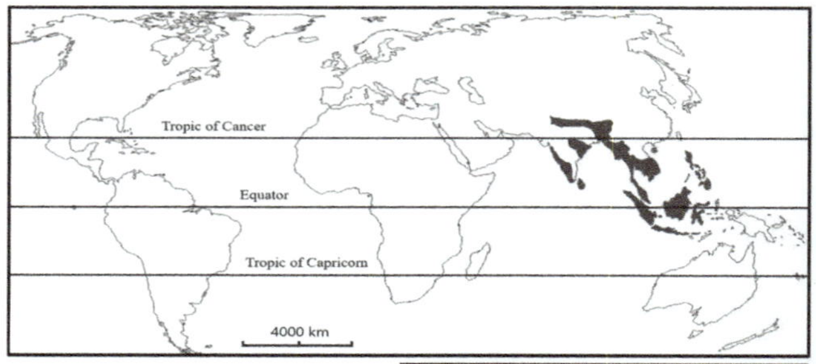

The Rufus-belied eagle has a scattered range across South east Asia.

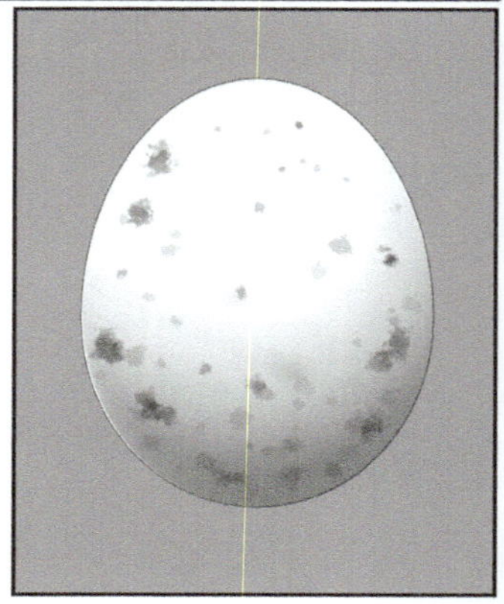

The egg is white, spotted and blotched with brown and red

Fig. 2.5g. Rufous-bellied eagle range and egg

2.7 Changeable Hawk Eagle or Crested Hawk Eagle *Nisaetus cirrhatus* (Gmelin, 1788) (formerly *Spizaetus cirrhatus*)

Physical Description

This species is a "slim brown eagle with a long tail and a prominent crest, which is prominent and sways" (Brown & Amadon, 1968, 694). There are several subspecies, the nominate subspecies, *N.c. cirrhatus* (formerly *S.c. cirrhatus*) ranging in India, *N.c. limnaetus* (formerly *S.c. limnaetus*) ranging in northern India, Malaya, Java, Sumatra, Borneo and the Philippines, *N.c. ceylanensis* (formerly *S.c. ceylanensis*) confined to Sri Lanka, *N.c. andamanensis* (formerly *S.c. andamanensis*) and *N.c. vanheurni*, confined to Simeulue Island, Indonesia (Brown & Amadon, 1968; Ferguson-Lees & Christie, 2001). The subspecies vary in size. The general size range is approximately 51 to 82 cm in length, with a wingspan of 100 to 160 cm (3 ft 3 in to 5 ft 3 in), and weight of 1.3 to 1.8 kg. Possibly, these measurements are high, representing the Flores hawk-eagle that is now considered a separate species (Ferguson-Lees & Christie, 2001; del Hoyo et al., 1999). Grimmett et al. (2016) reported lengths of up to 77 cm for *N.c. limnaeetus* in Nepal. Ali & Ripley (1978) listed average total lengths for the subspecies, in order of size: 72 cm (*N.c. cirrhatus*), 70 cm (*N.c. limnaeetus*) and 61 cm (*N.c. andamanensis*) (Ali & Ripley, 1983). *N.c. limnaeetus* in the Philippines was measured at 58.4 cm for males and 64.9 cm for females (Gamauf et al., 1998). Weights of the Philippine population (*N.c. limnaetus*) averaged at 1.36 kg (males) and 1.6 kg (females) but in some cases weights reached 1.81 kg (Brown & Amadon, 1986; Gamauf et al., 1998).

N.c. cirrhatus may be paler than the other subspecies, and has a reddish black streaked head and neck, a long black crest (the longest of all the races), brown upper parts with lighter edges to the feathers and the wing coverts, a light brown tail with three to four basal dark bars and a wider subterminal band and wing quills with dark to black tips and four or five dark bars. The long crest led some to call the subspecies the Crested Hawk eagle (Ferguson-Lees & Christie, 2001; Gamauf et al., 2005). The underparts are mostly white, with black steaks on the chin and throat, dark brown streaks on the breast, buff white barred feathers on the hind flanks and undertail coverts, dark brown and white barring on the underwing coverts, and whitish feathers with black barring on the wing quill (which also has a lighter colored patch composed of white feather bases) and tail undersides. The eye is yellow and the cere and legs are gray-brown, with a greenish tinge (Brown & Amadon, 1968). The immature plumage is lighter, with a buff head, with dark coloring on the feather centers, brown upperparts like the adult but lighter colored edgings. The underparts are brown or brownish white, streaked and spotted brown, with brown and tawny barring on the thighs and undertail coverts, and a light brown tail underside with seven dark bars, and wing quills barred with dark brown and four to five dark bars. *N.c. ceylanensis* is like *N.c. cirrhatus* with a crest but is smaller and has lighter colored plumage and only four bars on the tail.

N.c. ceylanensis is crested like the nominate subspecies. *N.c. limnaetus* has dark and light phases. The dark phase has black or dark brown plumage, in some cases there is lighter barring and a gray base to the tail. The light phase is still darker, with more spots and barring than *N.c. cirrhatus*, having brown feathers on the belly and thighs, a smaller crest (sometimes absent) and juvenile plumage being darker upperparts, lighter underparts, and light brown thighs. *N.c. andamanensis* is smaller than *N.c. cirrhatus*, has a shorter crest and darker plumage than *N.c. ceylanensis*, and is more like the pale phase of *N.c. linnaetus*. The juvenile has a white head. *N.c. floris* (now a separate species *Nisaetus floris*, Hartert, 1898) is larger than *N.c. linnaetus* and has white underparts and brown thigh markings. Gjershaug et al. (2004, 59) compared the juvenile plumage of *N.c. limnaeetus* with that of the juvenile of the former *N.c. floris* and noted that "the color of the mantle and back is darker in floris, lacking the light saddle-like band on the upperwing coverts of *limnaeetus*." *Floris* "also has distinctive white patches on the upperside of the primaries, not found in *limnaeetus*, and the dark terminal tail-bar of *floris* is broader than the other six bars (c.32 mm versus 15 mm), whereas in *limnaeetus* it is of the same width as the other bars" (Gjershaug et al., 2004, 59). *N.c. vanheurni* is the smallest subspecies, and lacks a crest and a black phase. Some observers in the past regarded *N.c. andamanensis*, *N.c. floris* and *N.c. vanheurni* as a single species, called *S. linnaetus*. Brown and Amadon (1968, 694) commented "we consider they are best regarded as more or less... races of *S. cirrhatus* to which they are in other respects similar... Alternatively, *S.c. linnaetus* may prove to be a separate species and the other races of *S. cirrhatus*."

Compared with other species, the nominate subspecies is distinctive, with the prominent crest and the long tail. Brown and Amadon (1968, 694) note that "the white underside, streaked with black, of adult plumage is unlike any other species but immature *Pernis apivorus* (European Honey buzzard) but the current species has feathered tarsus. The immature is also much paler, with a fawn-colored head, and white sparsely streaked underside." The melanistic crestless forms may resemble the melanistic forms of the Honey buzzard, but they have feathered tarsi and the less visible tail barring. The call has been compared to a "scream"; "klee – klee – ek", or a faster, rising "ki – ki – ki – ki – ki– ki – ki – ki – keeee, the last note being a "scream" (Brown & Amadon, 1968, 696). Illustrations of the appearance, range, and eggs of this species are shown in Figures 2.6a to 2.6l.

Classification

The species of the Genus *Nisaetus* were previously in the Genus *Spizaetus*, but independent molecular studies by Helbig et al. (2005) and Lerner and Mindell (2005) found that the Asian hawk-eagle species represent a different clade from the South American hawk-eagle species and should therefore be assigned to a different genus *Nisaetus* (Hodgson, 1836) (Global Raptor Information Network, 2021s). Lerner and Mindell (2005) found that *N. cirrhatus* and *N. lanceolatus* (a separate species, the Sulawesi hawk-eagle) are sister species. Gamauf et al. (2005) found a closer relationship between the western forms, *S.c. cirrhatus* and *S.c. ceylanensis* (now *N.c. cirrhatus* and *N.c. ceylanensis*), than the other subspecies, perhaps due to their proximate ranges. Amadon (1953, 1982), Stresemann and Amadon (1979), and Rasmussen and Anderton (2012) discussed *N.c. limnaeetus* as a separate species, based on morphological and vocal differences, and Rasmussen and Anderton (2012) classified *N. candamanensis* as a different species. More advanced molecular study based on two sections of the mitochondrial genome by Gamauf et al. (2005) did not support these classifications and justified *N. cirrhatus* as one species.

Foraging

The Changeable Hawk-eagle frequents dry deciduous to semi-evergreen forest, second-growth logged over forest with some clearings, mixed forest savanna and cultivated patches within the forest and is usually resident and non-migratory. *N.c. linnaetus* is common in the forests and open cultivated areas of Malaysia and Indonesia (Brown & Amadon, 1968). In Sri Lanka, Andaman Islands and Sabah it occurs mainly in tropical moist-deciduous to semi-evergreen and evergreen forest, some clearings and secondary forest and in some cases peatswamp forest, and landward zones of mangroves (Wells, 1999; Rasmussen & Anderton, 2012; Naoroji, 2006). It has been recorded in heavily logged forest in the Malay Peninsula, and even mixed agricultural landscapes with large, secluded woodland patches with emergents for nesting and roosting (Wells, 1999).

It perches on exposed branches in the canopy of trees in forests, clearings and forest edge, and soars above forest (Kennedy et al., 2000). Hunts alone, but may roost and soar in pairs (Wells, 1999). Common prey consists of small mammals and reptiles; larger mammals and domestic birds may also be taken. Most prey is taken on the ground or in trees (Brown & Amadon, 1968). Birds as large as domestic chickens may be taken and mammals up to the size of squirrels, and snakes, and lizards. Frequently hunts from exposed perches of various heights or from perches within a tree crown (Wells, 1999). In India, has been recorded ambushing large prey like parrots and monkeys (Rasmussen & Anderton, 2012). In the Malay, it is known to take domestic poultry (Robinson, 1927; Ryves, 1938; Spittle, 1952), birds like the Spotted dove (*Spilopelia chinensis*, Scopoli, 1768), snakes, and lizards. One record is of a large agamid lizard taken to the nest (Wells, 1999). In Sabah it takes similar prey, including birds, rodents and reptiles up to small monitor lizards (Genus *Varanus*, Merrem, 1820) (Sheldon et al., 2001).

Breeding

Brown and Amadon (1968) note that the Changeable Hawk-eagle has no spectacular aerial displays at the beginning of the breeding season. Nests are located in large trees, mostly composed of sticks and green leaves, about one meter across and deep, placed high in a main branch fork, sometimes near the crown center (Wells, 1999). The nests may be 18-30 m high in main branch forks, hidden within the foliage of the tree crown or on an isolated branch on a dead emergent, as nests may continue to be used after the tree dies (Wells, 1999). Common tree species used are *Shorea* species (Roxb. ex C.F. Gaertn) and *Intsiapal embanica* (Miq., 1861). One coarse, white sometimes marked reddish (mostly at the large end) egg is laid (Brown & Amadon, 1968; Global Raptor Information Network, 2021s). Global Raptor Information Network (2021s) notes differences in some eggs. One recorded egg was "plain, glossless white, sometimes with a few minute reddish-brown speckles at the broad end, and with a rough-textured surface" (Wells, 1999), another was "white with ochre stains" (Sharpe, 1879) and another was only "white" (Sharpe 1889).

Laying times vary in the different regions of Sri Lanka (December to March, especially January to February), southern India (middle November to early May, especially January to March), northeast India (for *N.c. linnaetus*, December to April, especially February to March), the Andaman Islands (January) and Borneo, Java and Sumatra (December to February, March to May, July to August, October, especially February to March and May to July) (Brown & Amadon, 1968). Breeding patterns may vary regionally. In the Indian Subcontinent, breeding takes place from January to June (*N.c. cirrhatus*) or November to June (*N.c. limnaetus*) (Rasmussen & Anderton, 2012). In the Malay Peninsula, mating has been recorded in late November-December, and laying during December-February, with the last fledglings leaving nests in late May (Edgar, 1933; Ryves, 1938; Wells, 1999). In Sabah, eggs were recorded in January 1875 (Sharpe, 1879) and 15 December 1885 (Sharpe, 1889), 17-19 February 1886, 17 January 1939 and 24 March 1939 (Sheldon et al., 2001).

Population Status

BirdLife International (2016) writes that the "global population size has not been quantified, but the species is described as common to uncommon" (see also Ferguson-Lees & Christie, 2001). Nevertheless, Ferguson-Lees and Christie (2001) estimated the global population of adults and immatures at the beginning of the breeding season, in the range of 10,001 to 100,000 individuals. The population is thought to be declining locally due to current habitat (mostly forest) destruction and increased human impacts (Ferguson-Lees & Christie, 2001). For example, in Malaysia total agricultural monocultures without landscape diversity is increasingly common, which is negative for breeding pairs of eagles that need forest clumps within agricultural lands (Wells, 1999). Global Raptor Information Network (2021s) describes this species as "the most widespread and common Asiatic hawk-eagle species, but some of the individual subspecies have small ranges and are probably under considerable threat from habitat loss." Gamauf et al. (2005) estimated their conservation and population status according to subspecies. Subspecies with small ranges are *N.c. vanheurni* with only 1,600 km^2 and *N.c. andamensis* with less than 6,475 km^2, may have comparatively small populations and may be more vulnerable. *N.c. ceylanensis* inhabits approximately 65,000 km^2 in Sri Lanka, is relatively common and not threatened (Harrison, 1999). The races *N.c. cirrhatus* and *N.c. limnaeetus* have even larger ranges and are not threatened. Hence the species is categorized as of "Least Concern"by BirdLife International (2016). Illustrations of the appearance, range, and eggs of this species are shown in Figures 2.6a to 2.6l.

2.8 Flores Hawk Eagle *Nisaetus floris* (Hartert, 1898)

Physical Description

The Flores Hawk-eagle is estimated to be the second largest hawk eagle of the Genus *Nisaetus*, after the Mountain Hawk-eagle (which is described as a "large, powerful, crested eagle" (Brown & Amadon, 1968, 697), and above the Legge's Hawk eagle and the Changeable Hawk-eagle (Ferguson-Lees & Christie, 2001; Robson & Allen, 2005). Despite the comparative lack of information on this species (Gjershaug et al., 2004, 55), the size of the Flores Hawk eagle can be inferred by comparison with other closely related species from the Genus *Nisaetus*. The larger Mountain Hawk eagle measures about 69 to 84 cm with a wingspan of 134 to 175 cm and may weigh between 1.8 and 3.5 kg (Ripley, 1982; Ferguson-Lees & Christie, 2001; Gjershaug et al, 2008). The Legge's Hawk eagle may measure 70 cm to 76 cm in length, with a wingspan of 134–175 cm and a weight of about 2 to 2.5 kg (Gjershaug et al., 2008; Harrison, 2011; Clark et al., 2020b). The Changeable hawk eagle smaller than these measurements (Ferguson-Lees & Christie, 2001).

Fig. 2.6a. Changeable Hawk eagle perching [With permission from Pieter Verheij]

Fig. 2.6b. Changeable Hawk eagle [With permission from Pieter Verheij]

Fig. 2.6c. Changeable Hawk eagle [With permission from Pieter Verheij]

Fig. 2.6d. Changeable Hawk-eagle dark morph [With permission from Mike Nelson-Birdtour ASIA]

Fig. 2.6e. Immature changeable Hawk eagle [With permission from Matt Trevillion]

Fig. 2.6f. Immature Changeable Hawk eagle [With permission from Frank McClintock]

Fig. 2.6g. Changeable Hawk eagle [With permission from Mike (Madbirder) Nelson - Birdtour ASIA]

Fig. 2.6h. Changeable or Crested Hawk eagle [With permission from Frank McClintock]

Fig. 2.6i. Changeable Hawk eagle dark morph [With permission from Ken Behrens]

Fig. 2.6j. Changeable Hawk eagle habitat [Credit: Elina Sazonova from Pexels]

Fig. 2.6k. Changeable Hawk eagle habitat, Maldare, India [With permission from Aditya-Vyas-unsplash]

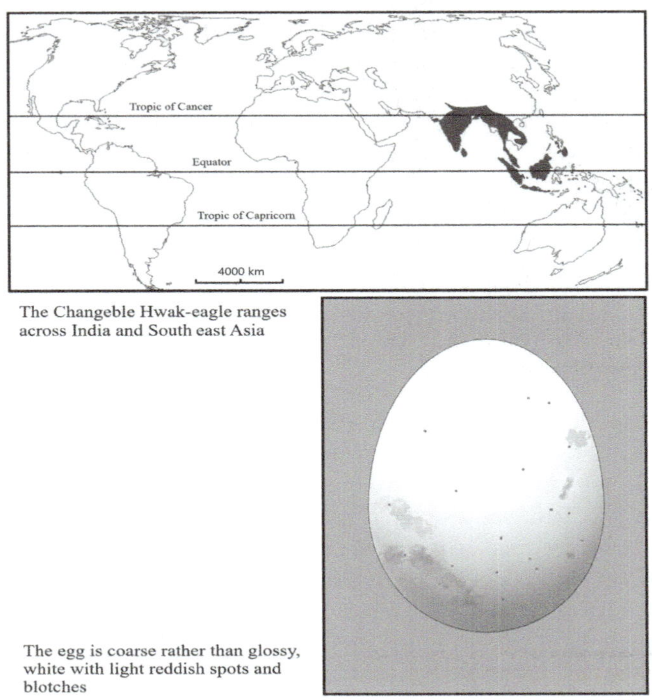

Fig. 2.6l. Changeble Hawk-eagle range and egg

The adult Flores Hawk eagle has a white head (Gjershaug et al., 2004, note that some authors wrongly depict this species with a brown head), in some cases with light fine brown streaks on the crown (see also Thiollay, 1994; Coates & Bishop, 1997). The mantle and back feathers are dark brown, and the inner vanes of the tertials, the inner parts of the outer primaries, and the outer vanes are white, these in combination creating a large white to grayish patch on the upper wing, which contrasts with the dark color of the upper wing feathers. The tail is brown with six dark bars, the outermost band being more than twice as wide as the rest, and the distance between the two outermost bands being almost twice as wide as the spaces between the other bands (Gjershaug et al., 2004, 58, see also del Hoyo et al., 2020). The uppertail-coverts and the inner tail feathers are also white, creating another white patch. The underside, including the thigh and leg feathers, is white. Gjershaug et al. (2004, 58) cite the cases of an adult eagle described by a certain Mr. Everett in South Flores (Hartert, 1898), which had "some very faint pale rufous-brown bars on the breast, sides of the breast and thighs." The juvenile plumage includes a white head with a few dark spots, a white underside, a darkish mantle and back, white patches on the upper side of the primaries, and a dark terminal tail-band twice as wide as the remaining six bands. The terminal dark tail band is more "diffuse" compared with adults, especially visible on the tail's ventral side (Gjershaug et al., 2004, 59). Illustrations of the appearance, range, and eggs of this species are shown in Figures 2.7a to 2.7e.

Classification

The main taxonomic issue for the Flores Hawk eagle is its reclassification as a full species, from its subspecies status as *Nisaetus cirrhatus floris* or *Spizaetus cirrhatus floris* (Gjershaug et al., 2004; Global Raptor Information Network, 2021t). The main issues concerned physical differences between this species and the other subspecies of *Nisaetus cirrhatus*, such as *Nisaetus cirrhatus limnaeetus.* One issue concerns its geographical range in Indonesia, which is limited to Flores, Sumbawa and Lombok and two satellite islands, Satonda near Sumbawa and Rinca near Flores (Collaerts et al., 2013, 48). For example, this creates "apparent reproductive isolation based on lack of an effective distribution barrier and apparently no significant hybridisation between *limnaeetus* and *floris*" (Gjershaug et al., 2004, 60). This mtDNA haplotype-based study found no mixing of these two subspecies, and a genetic distance of 1% implying separation during the Pleistocene. Another issue concerns the similar adult and juvenile plumage which were identical (both have white heads, unlike the other *Nisaetus* eagles, in which the head turns brown in the adult), unlike other subspecies of *N. cirrhatus* or *S. cirrhatus* which have different adult and juvenile plumage (Gjershaug et al., 2004). Gjershaug et al. (2004, 60) therefore pointed to distinct morphological plumage and genetic differences, and the lack of mixing of mtDNA haplotypes between *floris* and *limnaeetus* and concluded this indicated species status for the Flores Hawk eagle (Thiollay, 1994; Coates & Bishop, 1997).

Gjershaug et al. (2004, 60) suggested that the Flores Hawk eagle as a subspecies was "morphologically very distinct from the widespread *limnaeetus*. It has diagnostic white patches on the upper side of the inner parts of the primaries, a juvenile-like adult plumage, and it is much larger than *limnaeetus*. It is allopatric with *limnaeetus* without any known geographical overlap. The large morphological differences indicate that the two taxa are reproductively isolated. We suggest that *floris* be treated as a distinct species with the English name of Flores Hawk Eagle." This separated it from the other five subspecies: *cirrhatus*, *ceylanensis* (these two are crested), *andamanensis*, *limnaeetus* and *vanheurni* (these three are not crested) (del Hoyo et al., 1994; Gjershaug et al., 2004). This species was moved from *Spizaetus* to *Nisaetus* with the other Asian species when it was found that the Asian species of the Genus *Spizaetus* were distinct from the New World species (Helbig et al., 2005; Lerner & Mindell, 2005; Global Raptor Information Network, 2021t). See also Amadon (1953) and Inskipp et al. (1996) (e.g. Weick, 1980; del Hoyo et al., 1994; Coates and Bishop, 1997; Collaerts et al., 2013).

Breeding

The Flores Hawk eagle is a forest eagle, and breeds in forest land cover. Collaerts et al. (2013, 50) write that "pairs of Flores Hawk eagles often perform spectacular display fights over their territory and can be rather vocal." Display flights and copulation are reported on Flores in June-July, with the breeding season occurring at the same time as other islands such as Lombok (Suana et al., 2016; see also Gjershaug et al., 2004). Gjershaug et al. (2004) also estimate a territory size of 40 km^2 per pair. Verheijen (1964) reported active nests on West Flores in March, April, May, and August. Display flights and copulation were also reported observed in June and July 2003 on Flores (Gjershaug et al., 2004). This indicates breeding may occur during the dry season. Gjershaug et al. (2004) report an incident were a large tree was cut down in Mbeliling, Flores during August, and nest with a large eagle nestling was found in a nest in the tree foliage. The species identification was not definitive, but this may have indicated breeding during the dry season.

Foraging

The Flores Hawk eagle ranges and forages mostly in lowland and submontane forest up to 1,600 m elevation (BirdLife International, 2020). It is also recorded in cultivated areas, but almost always near closed or semi-closed forest, but it is uncertain if these are breeding adults or dispersing juveniles or vagrants. The extent of dispersal is also uncertain, especially if it includes flight across narrow bodies of ocean from island to island, allowing population mixing. The Flores Hawk eagle may hunt and eat small mammals, birds, and lizards, like the closely related Hawk eagle species (Coates & Bishop, 1997).

Population Status

The Flores Hawk eagle is non-migratory, but the present distribution of the species, with only possible, very local movements between closely situated islands. Collaerts et al. (2013: 50) cites this species as "one of the rarest eagles in the world" and that it is "now an Indonesian endemic only known to occur on Flores, Sumbawa and Lombok as well as on two satellite islands, Satonda near Sumbawa and Rinca near Flores."There are also records for Komodo island, with as an individual was photographed on a mangrove islet in 2011 (Coates & Bishop, 1997; Collaerts et al., 2013, 48). It has also recently been recorded on the island of Alor (Collaerts et al., 2013). There is an unconfirmed record from Paloe (Verheijen, 1961).

Collaerts et al. (2013, 48) further note that "based on the extent of suitable habitat and an estimated territory size of about 40 km^2 on Flores the population has been roughly calculated to be fewer than 100 pairs. Due to ongoing habitat destruction, the species is thought to be in rapid decline and BirdLife International (2013) has therefore ranked the species as Critically Endangered" (see also Gjershaug et al., 2004; BirdLife International, 2018). Raharjaningtrah and Rahman (2004) also estimated the population at 73-75 pairs, based on an estimated home range size of 38.5 km^2 ($n = 3$), indicating perhaps 10 pairs on Lombok, 38 pairs on Sumbawa, and 27 pairs on Flores. BirdLife International (2018) also defends the "Critically Endangered" assessment. Firstly, the Flores Hawk eagle has a very small population that is currently declining very rapidly, due to habitat clearance, with a population of estimated at 100-240 mature individuals, and possibly 20 additional pairs inhabit Alor (Collaerts et al., 2013). This indicates a total of about 150 to 360 birds. There is no data for the trend, but the main factors for the decline are hinted to be deforestation and persecution.

Persecution may take several forms. For example, Flores Hawk-eagles may be killed due to their predation on domestic fowls. In one case in April 2014, an eagle was shot in central Alor because of a generalized chicken predation threat (Verbelen, 2014). There is also the capture of eagles for the live bird trade (Gjershaug et al., 2004; Raharjaningtrah & Rahman, 2004). For example, people may journey from Java and Bali to the Lesser Sundas to participate in the live bird and taxidermy trades (Raharjaningtrah & Rahman, 2004). Birds are captured for the cagebird trade and shot for display specimens. In one case, two mounted birds were sold in Ruteng (Trainor & Lesmana, 2000; Trainor, 2002). As the species becomes rarer, there might be more shooting for collectors of preserved specimens. (Raharjaningtrah & Rahman, 2004; Global Raptor Information Network, 2021t). Trainor and Lesmana (2000) further report that in Manggarai in the western part of Flores, this eagle was formerly considered to be a 'totem' or 'empo,' i.e., a human ancestor, and therefore in the past it was not persecuted, killed or captured. Illustrations of the appearance, range, and eggs of this species are shown in Figures 2.7a to 2.7e.

Fig. 2.7a. Flores Hawk eagle perching [With permission from Yovie Jehabut]

Fig. 2.7b. Flores Hawk eagle in flight [With permission from Yovie Jehabut]

Fig. 2.7c. Flores Hawk eagle in flight [With permission from Mu-Chi Tsai]

Fig. 2.7d. Flores Hawk eagle habitat, Lombok [Credit Koes Nadi-Unsplash]

2.9 Mountain Hawk Eagle *Nisaetus nipalensis* (Hodgson, 1836), also called the Feather-toed Hawk Eagle, and formerly *Spizaetus nipalensis*

Physical Description

The Mountain Hawk eagle is a "large, powerful, crested eagle, much larger than *S. cirrhatus*" (Brown & Amadon, 1968, 697, referring to the Changeable hawk eagle, now *N. cirrhatus*) and is among the larger species of its genus (Ferguson-Lees & Christie, 2001). Brown and Amadon (1968) recognized three subspecies: *S.n. nipalensis* (ranging through continental Asia and Taiwan); *S.n. orientalis* (ranging in Japan); *S.n. kelaarti* (ranging in Sri Lanka). The last subspecies formerly also identified as *Spizaetus kelaarti* or *Nisaetus nipalensis kelaarti* is now recognized as a separate species Legge's hawk eagle *(Nisaetus kelaarti*, Legge, 1878) (Grimmet et al., 1999; Gjershaug et al., 2008; Harrison, 2011; Fam & Nijman, 2011). This was despite Brown and Amadon (1968) suggesting that *kelaarti* is "scarcely distinct." Currently, there are two subspecies: the nominate subspecies (*N.n. nipalensis*) and the Japanese subspecies (*N.n. orientalis*). Two proposed subspecies *N.n. fokiensis* and *N.n. whiteheadi* of southern China were rejected by Clark (1994) (Ali & Ripley, 1983; Ferguson-Lees & Christie, 2001). The range of *N.n. nipalensis* includes the Indian and Nepalese Himalayas eastwards into southern China to eastern China (Zhejiang) and Taiwan, southwards to northern Indochina and the northern Malay Peninsula and

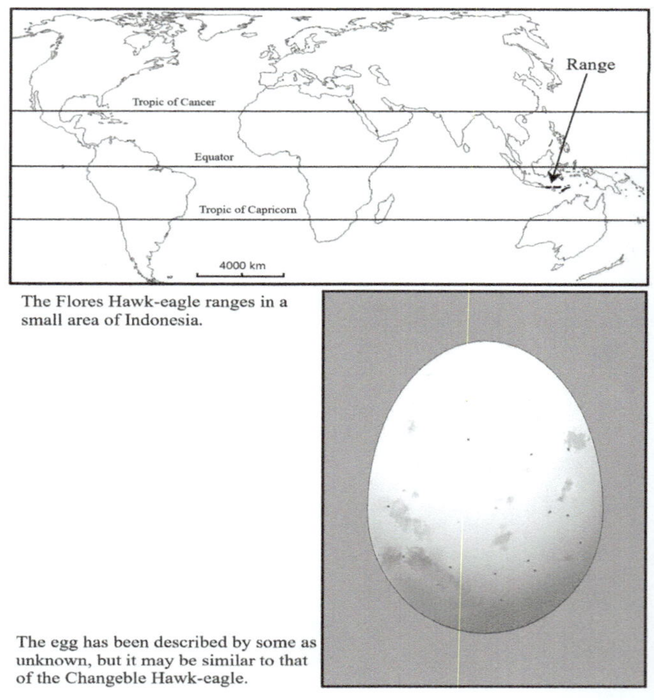

Fig. 2.7e. Flores Hawk eagle range and egg

Vietnam. The range of *N.n. orientalis*is mostly in Japan, but may include the Russian Far East, northeastern China and North Korea (Global Raptor Information Network, 2021u).

The Mountain Hawk eagle has been described as the largest species of the Genus *Nisaetus* and comparable to the large Flores hawk eagle (Ferguson-Lees & Christie, 2001; Naoroji & Schmitt, 2007; Collaerts et al., 2013). The Mountain Hawk eagle measures about 69 to 84 cm with a wingspan of 134 to 175 cm. This species has proportionally shorter wings than eagles of the open landcover, but the wings are comparatively longer than the other hawk eagles (Ripley, 1982; Ferguson-Lees & Christie, 2001). The nominate subspecies averages 72 cm in length (Gjershaug et al., 2008). One specimen from northern India was recorded with a weight of 1.83 kg, while others from the Yangtze area of east-central China weighed 2.8, 2.95 and 3.5 kg, which may represent Bergmann's rule (birds from the colder, higher latitudes are larger than the southern more tropical birds (Meiri & Dayan, 2003).

For *N.n. nipalensis* adult plumage includes a blackish crown and crest, a brown neck and back (variably with a coppery tinge) brown and white barring on the rump and upper tail coverts, a dark brown tail with dark gray barring, a wide black subterminal band and white tip, brown upper wing coverts, lighter colored median and greater coverts, blackish brown, barred wing quill, lighter brown secondaries and darker tipped feathers, and dark streaked neck sides. There is also a wide streak below the eye and behind the gape, giving a moustachial streak and a blackish streak on the throat. The underparts are primarily white and brown, with a white-barred, brown abdomen and reddish tinged lower neck and upper breast, white underwing coverts marked brown to black, and black barring on the white underside of the wing and tail quills. The eye is yellow, the feet dark yellow and the cere is grey to almost black.

The immature bird is white on the head and neck, marked with blackish steaks and brown tinged and blackish feathers forming the crest. The upperparts feathers have buffy edges, and the median wing coverts are light tawny. The wing quills are banded light and darker brown. The tail has more dark and light brown bands than that of the adult. The white underside is variably shaded brown (Brown & Amadon, 1968). The eye is pale gray and the cere and feet are yellow. *N.n. orientalis* in Japan is larger, with a shorter crest. *N.n. kelaarti* is described as Legge's Hawk eagle.

Compared with other species, the adult Mountain Hawk eagle is distinctive in having short wings with a white lining and a black trailing edge and three black bars on the secondaries, a long dark tail with four lighter bars. This species is much larger than the Changeable Hawk eagle. The Changeable Hawk eagle is slenderer, with a longer tail and narrower wings. Blyth's hawk eagle and Wallace's hawk eagles are much smaller (Brazil & Hanawa, 1991; MacKinnon et al., 2000; Ferguson-Lees & Christie, 2001; Grewal et al., 2002; Naoroji & Schmitt, 2007). The immature birds are distinctive in their white plumage. Brown and Amadon (1968, 697) describe the Mountain Hawk eagle as a "noisy eagle." The call described as "klu – weet – weet" or the display call which is a "rapid bubbling call." Illustrations of the appearance, range, and eggs of this species are shown in Figures 2.8a to 2.8i.

Classification

The Mountain Hawk eagle was previously placed in the Genus *Spizaetus*, but molecular studies by Helbig et al. (2005) found that the Asian hawk eagle species are not linked to the New World hawk eagle species and should be placed in a different genus, *Nisaetus* (Hodgson, 1836). Lerner and Mindell (2005) concurred, also indicating that Blyth's Hawk eagle and the Mountain Hawk eagle are sister species, using the molecular sequences of two mitochondrial genes and one nuclear intron. The Asian *Spizaetus* (*Nisaetus*) species are monophyletic and may be grouped into two sub-clades, one including the Mountain hawk eagle, Blyth's hawk eagle, Javan hawk eagle and Wallace's hawk eagle. Haring et al. (2007) also argued that subspecies *N.n. orientalis* may be a separate species (Brazil, 2009). Gjershaug (2006) and Gjershaug et al. (2008) argued that *N.n. kelaarti* (now Legge's hawk eagle) should be treated as a full species, based on differences in morphology, plumages, mensural characters, and vocalizations (Gjershaug et al., 2008).

Foraging

The Mountain Hawk eagle inhabits subtropical and broad-leafed forests and fir forest at higher altitudes in Bhutan; avoids pine forests (Spierenburg, 2005). It is designed forest foraging, with shortish, broad wings, longish tail and longish, powerful legs (Brazil & Hanawa, 1991; Hertel, 1995; Petty, 1998; Naoroji & Schmitt, 2007; Gjershaug et al., 2008). It frequents the forests, perching within the canopy and in some cases soaring above the canopy, these habits comparable to those of the African Crowned eagle (Brown & Amadon, 1968). The current species is a non-migratory resident, common at medium elevations (600 to 1500 meters) in south India and Sri Lanka, mountain forests in Japan and up to 2400 to 3000 meters in northern India (Brown & Amadon, 1968). Prey is commonly taken on the ground and includes mammals up to the size of a hare, large gamebirds such as the Junglefowl (*Phasianus gallus*, Linnaeus, 1758) and pheasants (Subfamily *Phasianinae*, Horsfield, 1821) and domestic poultry and lizards (Ferguson-Lees & Christie, 2001).

Breeding

The Mountain Hawk eagle performs nuptial displays, including soaring in the vicinity of the nesting area, undulating flight patterns and male and female coordinated participation. Brown and Amadon (1968) compare these displays with those of the African Crowned eagle. This species generally breeds in heavily forested montane areas up to an elevation of about 2,000 m, but the eagles may descend to altitudes during winter (Brazil, 2009). Nests are constructed in trees, both deciduous (in southern India) and conifers (in northern India and Japan). The nest is large, up to two meters across and one meter deep. Usually one egg is laid, but up to three may be laid in Japan. The coloration is pale or white with reddish tinge or small markings. Laying dates may vary according to region across the range, including Sri Lanka (December to March, most often January), northern India (February to May, usually February to March) and Japan (April). Fledging is after 80 days (Brown & Amadon, 1968). In China, a paired courtship display was recorded on 5 May 2005 in Shiwandashan National Nature Preserve, Guangxi Province (Robins et al., 2006). It breeds in April and May in Yunnan Province (Weizhi, 2006).

Population Status

The Mountain Hawk eagle is an irruptive or local migrant, with juveniles dispersing from breeding areas (Bildstein, 2006). The Mountain Hawk eagle's global population is estimated at 1,001 to 10,000 individuals, possibly closer to the latter figure, based on the number of adults and immatures at the start of the breeding season (Ferguson-Lees & Christie, 2001). This estimate included the former subspecies Legge's Hawk-eagle (*N. kelaarti*). Later estimates may be more accurate (Asai et al., 2006). This species is classified as of 'Least Concern' by BirdLife International (2016), due to its "extremely large range" which is not less than 20,000 km^2 and the slow decline (the parameters for Vulnerable classification "declining or fluctuating range size, habitat extent/quality, or population size and a small number of locations or severe fragmentation" are insufficient), the population is decreasing slowly, and is quite large. The Mountain Hawk eagle is locally common but generally less common across its range. It is seriously impacted by the wide-ranging deforestation and other types of habitat losses (Clark et al., 2015, Global Raptor Information Network, 2021u). In Japan, an eagle contracted avian influenza virus (H5N1) (Shivakoti et al., 2010). In that country, it is classified as an Endangered species in the Red Data Book of Japan (2002), and its population is considered stable (Tago et al. 2015), and does not require special conservation attention (Asai et al., 2006). In China, this species is classified as Second Class Important Protected Bird (Weizhi, 2006). Illustrations of the appearance, range, and eggs of this species are shown in Figures 2.8a to 2.8i.

Fig. 2.8a. Mountain Hawk eagle [With permission from H Yasui]

Fig. 2.8b. Mountain Hawk eagle perching [With permission Hari K Patibanda]

Fig. 2.8c. Mountain Hawk eagle [With permission from Hari K Patibanda]

Fig. 2.8d. Juvenile and adult Mountain Hawk eagle [With permission from Skip Russell]

Fig. 2.8e. Mountain Hawk eagle [With permission from Frank McClintock]

Fig. 2.8f. Immature Mountain Hawk eagle [With permission from Mike Nelson-Birdtour ASIA]

Fig. 2.8g. Mountain Hawk eagle in habitat [With permission from Mike Nelson-Birdtour ASIA]

Fig. 2.8h. Mountain Hawk eagle habitat, Bhutan [With permission from Yanni Shams from Pexels]

2.10 Legge's Hawk Eagle *Nisaetus kelaarti* (Legge, 1878)

Physical Description

Legge's Hawk eagle, formerly considered a subspecies of the Mountain Hawk eagle, is a medium-sized eagle, though generally smaller than the latter species (Ferguson-Lees & Christie, 2001; Gjershaug et al., 2008). The Legge's Hawk eagle is usually 70 cm to 76 cm in length, a wingspan of 134–175 cm and may weigh about 2 to 2.5 kg (Gjershaug et al., 2008; Harrison, 2011; Clark et al., 2020b). Clark et al. (2020b) describe this species as a "large hawk-eagle, brownish above and barred below (including remiges), with crest of varying length depending on subspecies, short wings, longish tail with dark and light bands of nearly even width, feathered tarsi and powerful feet." It differs from the Mountain Hawk eagle principally in having a "relatively much larger bill and claws and short primary projection, and that it also differs consistently in numerous plumage characters and other mensural characters" (Gjershaug et al., 2008). Primary projection refers to the text to which the primaries of the folded wing extend beyond the wing tertials (Mullarney et al., 1999).

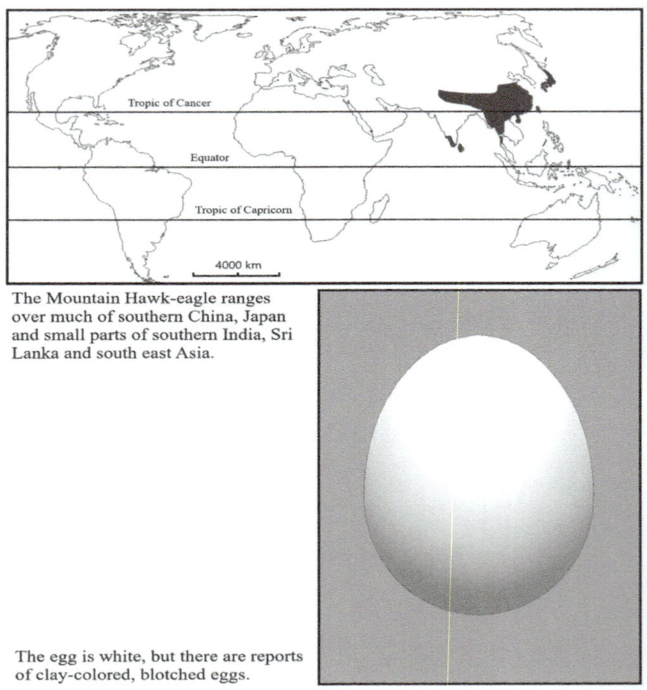

Fig. 2.8i. Mountain Hawk eagle range and egg

eBird (2020) describes this species as a "rather large, slim hawk-eagle with a white-tipped crest, streaked gray cheeks, and long feathered legs. It has strongly barred broad wings and a long tail. Unlike the otherwise very similar Crested Hawk eagle, the streaking on Legge's is restricted to the upper breast, and its breast, belly, and thighs are strongly barred. Has an upright stance when seen perched in the canopies of forested hills." The upperparts are brown, and the underparts are pale. There is barring on the undersides, on the flight feathers and tail. The breast and belly are heavily streaked. The sexes are similar. The juvenile is usually paler than adult and almost entirely whitish below; irides bluish-grey to pale yellow (Clark et al., 2020b). Illustrations of the appearance, range, and eggs of this species are shown in Figures 2.9a to 2.9g.

Classification

The Legge's Hawk eagle is monotypic. It was initially treated as a subspecies of *Nisaetus nipalensis* by some authorities (e.g. Stresemann & Amadon, 1979; Sibley & Monroe, 1990; del Hoyo et al., 1994), but later research such as that of Gjershaug (2006) and Gjershaug et al. (2008) presented evidence of full species status, such as "differences in plumage, morphology, and vocalizations, and DNA sequences between the two taxa" (Global Raptor Information Network, 2021v; see also Haring et al., 2007). This species was placed in the Genus *Spizaetus*, but comprehensive molecular studies by Helbig et al. (2005) and Lerner and Mindell (2005) found enough evidence to move the Asian hawk-eagle species to a new genus (*Nisaetus,* Hodgson, 1836), separate from the New World hawk-eagles. Haring et al. (2007) also found that *Nisaetus* species are monophyletic and are divided into two sub-clades, one including the Legge's Hawk eagle, the Mountain Hawk eagle, Blyth's Hawk eagle, Javan Hawk eagle and Wallace's Hawk eagle (Global Raptor Information Network, 2021v).

Foraging

Legge's Hawk eagle is probably sedentary, due to the relatively short wing length. It primarily frequents primary evergreen forests and remnant forest patches less than 10 m^2 in extent (Gjershaug, 2006; Gjershaug et al., 2008). It is a forest species, foraging in forest and wooded mountains. Prey taken may include small to medium-sized mammals, including hares and monkeys such as Japanese macaques (*Macaca fuscata*, Blyth, 1875), and birds (including Indian peafowl) and reptiles (Grimmett et al., 1999; Fam & Nijman, 2011; Harrison, 2011; Clark et al., 2020b). Global Raptor Information Network (2021v) notes that it "probably feeds more on mammals than the closely related Mountain Hawk eagle, based on its longer bill and hind claws" (see Wattel, 1973; Hertel, 1995; Gjershaug, 2006). It is also noted that "the shape of the wing, involving very short primary projection ("Kipp's distance"), may be an adaptation either to hunting inside forests or a lack of migratory behavior" (Kipp, 1959; Gjershaug et al., 2008).

Breeding

The breeding season is December to March in Sri Lanka (Clark et al., 2020b). It may breed in forests and also in smaller remnant forest patches (Gjershaug, 2006; Global Raptor Information Network. 2021v).

Population Status

The total population of Legge's Hawk eagle has been estimated at under 2,500 breeding individuals (Gjershaug, 2006). Gjershaug (2006) estimated about 1000 pairs in the Western Ghats and 100 pairs in Sri Lanka. These estimates were based on a hypothetical estimation of population density of one pair per 2,110 ha (Thiollay 1993) and an estimate of the existing rainforest area of 20,000 km^2 in the range in the Western Ghats (Collins et al., 1991; Gjershaug, 2006). In the Sri Lankan case, as the existing suitable rainforest was estimated at 2,135 km^2 in 1995 (Legg & Jewell, 1995). Gjershaug (2006) made the estimate of 100 pairs. Gjershaug (2006) also recommended that Legge's Hawk eagle be classified as Near Threatened but argued that for a possible future classification as Vulnerable, because of the limited population and declining numbers, as well as its specialised requirement for large areas of primary rainforest (Gjershaug et al., 2008, 2009). BirdLife International (2020) states that "This taxon is Not Recognised as a species by BirdLife International." Illustrations of the appearance, range, and eggs of this species are shown in Figures 2.9a to 2.9g.

2.11 Javan Hawk Eagle *Nisaetus bartelsi* (Stresemann, 1924) (formerly *Spizaetus bartelsi*)

Physical Description

The Javan Hawk eagle "is one of the world's least known raptors" (Nijman & Sözert, 1998, 13). This is a medium sized eagle with a blackish crest (Clark et al., 2020c). It measures 56–60 cm in length, with wingspan of 110–130 cm and unrecorded weight (probably similar to that of other similarly sized eagles) (Ferguson-Lees & Christie, 2001). There are few records of its weight. The adult plumage comprises a black, white tipped crest, dark brown upperparts, tinged rusty on the head sides and hind neck, and a brown tail with four black bands. The underparts are rusty, with a white throat with black borders, the breast is streaked with brown, and the brown belly is barred with lighter brown and the rusty colored underwing coverts are dark spotted. The

Fig. 2.9a. Legge's Hawk eagle perching [With permission from Michael Covington]

Fig. 2.9b. Immature Legge's Hawk eagle perching [With permission from Graham Ekins]

Fig. 2.9c. Immature Legge's Hawk eagle [With permission from Frank McClintock]

Fig. 2.9d. Immature Legge's Hawk eagle [With permission from David Beadle]

Fig. 2.9e. Immature Legge's Hawk eagle [With permission from Gil Ewing]

Fig. 2.9f. Immature Legge's Hawk eagle [With permission from Alex Aza]

primaries have dark and white barring below. The eye is yellow, the bill and cere are black and the toes are pale yellow (Brown & Amadon, 1968). The immature bird has a dull white to light brownish yellow head and underparts, spotted blackish on the crown. The blackish brown crest is white tipped. The upperparts are sepia brown with lighter feather margins. The brown tail has narrow, dark bars. The primaries are barred below dark and light gray, with white bases. The eye is pale yellow, and the bill is black (Brown & Amadon, 1968).

The Javan Hawk eagle may resemble several similarly sized eagles. The Changeable Hawk eagle is "a more robust bird with a relatively shorter neck and tail. Juveniles are white rather than cinnamon. Of the two morphs, only the pale morph can cause confusion, as the black one is easily excluded by its color" (Nijman & Sözert, 1998, 14). The Black eagle has a larger wingspan especially in relation to its tail, and more separated primaries in flight. The immature Black eagle is darker with no wing bars, and the flight is unstable with an upward tilt to the flight feathers (Nijman & Sözert, 1998). The Crested Serpent eagle is smaller, and its wings are held in a shallow V-shape and forwards and the adults have a white band on the underwing. The Rufous-bellied eagle is also smaller than the Javan hawk, with no barring but a similar large crest (Nijman & Sözert, 1998). The Oriental Honey-buzzard (*Pernis ptilorhyncus ptilorhyncus*, Temminck, 1821) and Barred Honey-buzzard (*Pernis celebensis*, Wallace, 1868) are much smaller, the former lacks the underparts barring and the latter has a different flight mode (deeper wing beats) (Nijman & Sözert, 1998). Illustrations of the appearance, range, and eggs of this species are shown in Figures 2.10a to 2.10g.

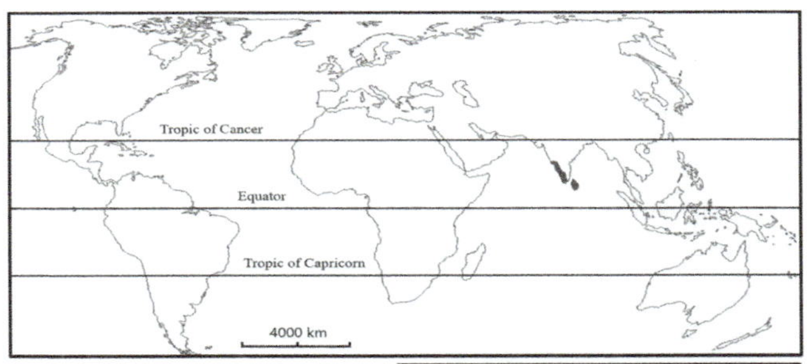

The Legge's Hawk-eagle occurs in southern India and Sri Lanka.

The egg may be white, like that of the Mountain Hawk-eagle.

Fig. 2.9g. Legge's Hawk Eagle Range and Egg

Classification

The Javan Hawk eagle is monotypic and was recognized as a separate species in the middle of the 20th century (Amadon, 1953, see also Balen et al., 2000; Global Raptor Information Network, 2021w). Initially it was placed in the Genus *Spizaetus*, but later molecular studies by Helbig et al. (2005) revealed the Asian hawk-eagle species as in a different genetic line from the New World Hawk eagle species, and should be placed in a new genus, *Nisaetus* (Hodgson, 1836). This position was independently supported by Lerner and Mindell (2005), using the molecular sequences of one nuclear and two mitochondrial genes. Haring et al. (2007) also confirmed that these Asian *Spizaetus* (*Nisaetus*) species are monophyletic, with the Javan Hawk eagle in one of two sub-clades. The Javan Hawk eagle was formerly considered to be conspecific with *N. nipalensis* (Sibley & Monroe, 1990), but the Global Raptor Information Network (2021w) notes that older references point to a probable closer link with Wallace's hawk eagle (e.g. Brown & Amadon, 1968).

Foraging

The main habitat is primary rainforest, and to a lesser extent secondary forest, tropical semi-deciduous forest, and tree plantations (Prawiradilaga, 2006). Foraging is mostly from perching, either within the canopy or on the branches of leafless trees. There is also soaring flight. Small mammals, such as including tree shrews (Order *Scandentia*, Wagner, 1855), squirrels (Family *Sciuridae*, Fischer de Waldheim, 1817), bats (Order *Chiroptera*, Blumenbach, 1779) and rats (Order *Rodentia*, Bowdich, 1821), are the main prey. In some cases, birds, snakes, and lizards are also taken (Prawiradilaga, 2006).

Breeding

The Javan Hawk eagle usually breeds from January and July and may breed in most months of the year. It is sexually mature at the age of three to four years, breeds every two years and lays only one egg. Prawiradilaga (2006, 177) describes the nest as large, usually located in a large forest tree. Most nests are located in prime undisturbed forest, but one nest was recorded in a tree plantation (Afianto, 1999). One egg is laid (usually from December to June, with a January peak (during the middle rainy season), the incubation period is 47-48 days (Sözer & Nijman, 1995), and the nestling period is 70 days long, followed by fledging (Prawiradilaga, 2006; see also Hoogerwerf, 1949; Hellebrekers & Hoogerwerf, 1967; Prawiradilaga, 1999). Both parents look after the nestling. Some juveniles stay around the parent's territory until the following breeding season. The eagles are sexually mature at the age of three or four years (Sözer & Nijman, 1995).

Population Status

The Javan Hawk eagle is endemic to the island of Java, Indonesia (Andrew, 1992, Ferguson-Lees & Christie, 2001). It is a non-migratory species (Bildstein, 2006). The global (Indonesian) population is estimated as based on the number of adults and immatures at the start of the breeding season, in the range of 101 to 1,000 individuals (Ferguson-Lees & Christie, 2001). BirdLife International (2017) estimated the number of mature birds at 300-500 individuals and puts its status as Endangered. The population is small, and its main habitat, rainforest is declining in area and increasingly fragmented, although "recent surveys showed that the population size has been underestimated in the past" (Balen et al., 2000). Although a relatively unknown species, there is increased research into its ecology. Deforestation is the main threat, followed by illegal poaching and illegal capture for trade. The Indonesian government has passed laws for its protection (Meyburg et al., 1989; van Balen, 1991; van Balen & Meyburg, 1994; Sözer & Nijman, 1995; Hapsoro et al., 1998; Setiadi et al., 1999; van Balen et al., 1999; Prawiradilaga, 2000; Afianto, 1999; Prawiradilaga et al., 2000; Kuswandono et al., 2003; Mikoyan, 2004; Widodo, 2004; Nijman et al., 2009).

Several authors rate the Javan Hawk eagle as rare and endangered and also little known to scientific observation (Meyburg et al., 1989). For example, Balen et al. (2000, 297) observe that the Javan Hawk eagle "is a threatened raptor endemic to the densely populated island of Java. Historically very little is known about its biology." They further note that there are past underestimations of the population of this species, and the breeding population "is estimated to be 137–188 pairs with a total of 600–900 birds and confirmed presence in 22 discrete forest blocks throughout Java" (Balen et al., 2000, 297). Eagles ranged through forest but were observed in isolated small forest fragments down to 3000 ha. The eagles range widely, through the dispersals of juveniles and wide generalist foraging enable adaptation to habitat fragmentation (Balen et al., 2000, 297). Prawiradilaga (2006, 177) describe this species as "globally endangered species which exclusively occupies the last remnant forests of Java island, Indonesia." These authors and others describe efforts to increase knowledge of the Javan Hawk eagle and pose solutions for its continued decline (Gjershaug et al., 2004; Nijman & van Balen, 2003; Nurwatha et al., 2004; Syartinilia & Tsuyuki, 2008; Nijman et al., 2009). Illustrations of the appearance, range, and eggs of this species are shown in Figures 2.10a to 2.10g.

Fig. 2.10a. Javan Hawk eagle perching [With permission from Saravanan Krishnamurthy]

Fig. 2.10b. Javan Hawk eagle perching [With permission from Mike Nelson-Birdtour ASIA]

Fig. 2.10c. Javan Hawk eagle perching [With permission from Johnathan Beilby]

Fig. 2.10d. Javan Hawk eagle perching [With permission from Dikky Oesin]

Fig. 2.10e. Javan Hawk eagle in flight [With permission from Saravanan Krishnamurthy]

Fig. 2.10f. Javan Hawk eagle habitat, Rumpin Java [Credit: Tom Fisk Pexels]

2.12 Sulawesi Hawk Eagle *Nisaetus lanceolatus* (Temminck & Schlegel, 1844)

Physical Description

This is among the smaller species of eagle (Brown & Amadon, 1968). Clark and Kirwan (2020) describe it as 56–64 cm long with a wingspan of 110–135 cm "a boldly marked hawk eagle, without crest, and has fairly short, broad, round-tipped wings, and longish, round-tipped tail." The adult plumage is very dark brown, with a darker crown and a purplish tinge to the back feathers. The grey tail is crossed by three pale bars, more visible below, with white tipped tail coverts. There is darker barring on the primaries, and "a suggestion of a rufous collar" (Brown & Amadon, 1968, 698). The head and breast are multi-colored, with rufous brown on the sides and breast, dark streaks on the cheeks, and broader streaks on the breast, and a medial black stripe on the white sides of the throat. There is white and black barring on the belly and underwings, while the wing quills are pale greyish (Brown & Amadon, 1968). The juvenile has a white head, with variable numbers of brown streaks on the crown. The brown back, with visible white on the inner vanes of the wing quills. The brown tail has paler bars and a lighter colored tip. The white underparts may have brown steaks on the breast, light barring on the sides and light tail bars (Brown & Amadon, 1968).

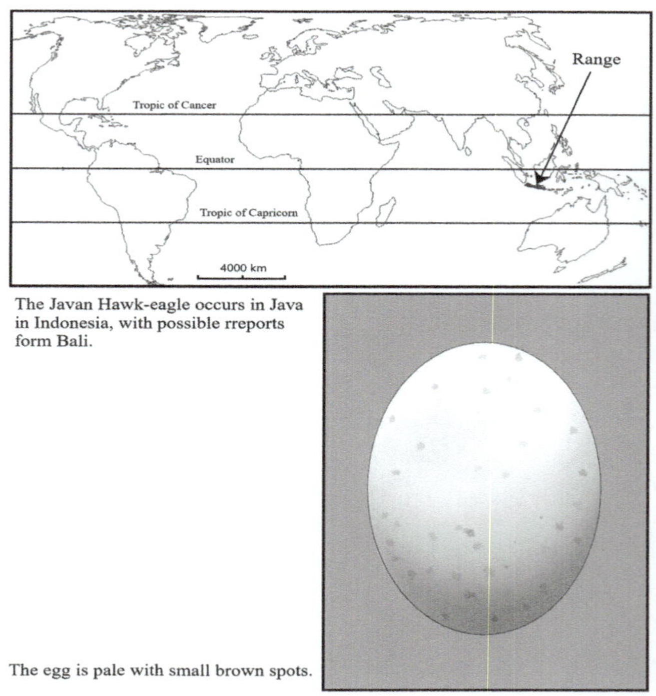

Fig. 2.10g. Javan Hawk eagle range and egg

This species may be mistaken for Barred honey buzzard (*Pernis celebensis*, Wallace, 1868) which also occurs in Sulawesi, but the eagle has a more "robust build and feathered legs" (Brown & Amadon, 1968, 700). Brown and Amadon (1968, 700) further note that "this species looks must like *S. philippensis* (which does not of course occur in its range), but this may be misleading as the tarsal feathering extends farther between the toes in *S. lanceolatus* and it does not have a crest." In this case, *S. philippensis* (*Spizaetus philippensis*) refers to the Philippine Hawk eagle, now classified as *Nisaetus philippensis* (Gould, 1863) (Global Raptor Information Network, 2021x). Illustrations of the appearance, range, and eggs of this species are shown in Figures 2.11a to 2.11h.

Classification

The Sulawesi Hawk eagle was previously placed in the genus *Spizaetus*, but as mentioned above the more advanced molecular studies by Helbig et al. (2005) differentiated the Asian hawk eagle species from the New World hawk eagle species and the latter were placed in a new genus, *Nisaetus* (Hodgson, 1836). Lerner and Mindell's (2005) independent study using the molecular sequences of one nuclear and two mitochondrial genes supported these results, and also found that the Changeable hawk eagle and the Sulawesi Hawk eagle are sister species. Two further studies, Gamauf et al. (2005b) and Haring et al. (2007), also supported the monophyletic status of the Asian *Spizaetus* (*Nisaetus*) species and found two sub-clades, with the Sulawesi Hawk eagle in one of these sub-clades.

Foraging

The Sulawesi Hawk eagle frequents "primeval forest at all elevations" soaring over the canopy and grassing clearing within the forest (Brown & Amadon, 1968, 698). The main habitat is "primary and tall secondary lowland, hill, and montane forest and forest edge. Occasionally ranges over nearby open country. Perches in tree crowns, sometimes conspicuously, and sometimes hunts in grassy clearings" (Global Raptor Information Network, 2021x; see also Coates & Bishop, 1997). Global Raptor Information Network (2021x) surmise that it feeds on mammals, birds, and reptiles, and still-hunts from a perch, like some of the other *Nisaetus* species. Prey may also be captured from the canopy (Ferguson-Lees & Christie, 2001).

Breeding

The Sulawesi Hawk eagle is a tree nester, but apparently there is "little information" on its breeding patterns (Global Raptor Information Network, 2021x). A nest was recorded in epiphytes about 20 m high on a branch of a large forest tree (Ferguson-Lees & Christie, 2001). Recently-fledged young were sighted in August, which may hint at a breeding period from May to August. The clutch may be one egg, and "the incubation and nestling periods are unrecorded" (Global Raptor Information Network, 2021x).

Population Status

The Sulawesi Hawk eagle is non-migratory, but juveniles disperse considerably from breeding sites (Bildstein, 2006). BirdLife International (2016) puts the number of mature individuals in the range 670-6,700. Ferguson-Lees and Christie (2001) estimated the global population of adults and immatures at the start of the breeding season, in the range of 1,001 to 10,000 individuals, possibly in the upper part of this range. BirdLife International (2016) also assesses this species as of Least Concern, but it was previously classified as Near Threatened (Collar et al., 1994). The justification is that the range is very large (not below 20,000 km^2) and the large range is not "combined with a declining or fluctuating range size, habitat extent/quality, or population size and a small number of locations or severe fragmentation." It is acknowledged that the population is small and decreasing, it is slower than would be evident in a Vulnerable species. The Sulawesi Hawk eagle is considered to be "rather common" (Gamauf et al., 2005, 113; see also Nurwatha et al., 2000; Thiollay & Rahman, 2002), this may be due to the wide extent of its favored habitat of forest on the Sulawesi islands. There is however substantial deforestation, and it may be shot by farmers as a predator of domestic fowl and companion animals. The distribution covers the Sulawesi and the Sula Islands, where it is a sparsely, widely distributed resident in Sulawesi, Bangka, Lembeh, Muna, Butung, Banggai Islands (Peleng, Labobo), and Sula Islands (Mangole) from lowlands to 2,300 m. The largest populations are on Sulawesi and Mangole (Coates & Bishop, 1997). Illustrations of the appearance, range, and eggs of this species are shown in Figures 2.11a to 2.11h.

2.13 Philippine Hawk Eagle or North Philippine Hawk Eagle or Northern Philippine Hawk Eagle *Nisaetus philippensis* (Gould, 1863)

Physical Description

The Philippine Hawk eagle is a medium sized, comparatively slender, crested eagle (BirdLife International, 2020). One of its defining characteristics is that it is more lightly built and visibly less powerful than the main Hawk eagle that shares its range in the Philippines, the Changeable Hawk eagle. The average length is 50 to 63 cm, with a wingspan of 105 to 125 cm and weight of

Fig. 2.11a. Sulawesi Hawk eagle perching [With permission from Peter Ericsson]

Fig. 2.11b. Sulawesi Hawk eagle perching [With permission from Mike Nelson-Birdtour ASIA]

Fig. 2.11c. Sulawesi Hawk eagle perching [With permission from Mike Nelson-Birdtour ASIA]

Fig. 2.11d. Immature Sulawesi Hawk eagle perching [With permission from Mike Nelson]

Fig. 2.11e. Sulawesi Hawk eagle in flight [With permission from Mike Nelson-Birdtour ASIA (2)]

Fig. 2.11f. Sulawesi Hawk eagle in flight [With permission from Mike Nelson-Birdtour ASIA]

Fig. 2.11g. Sulawesi Hawk eagle habitat [With permission from Josep Miquel Navarro]

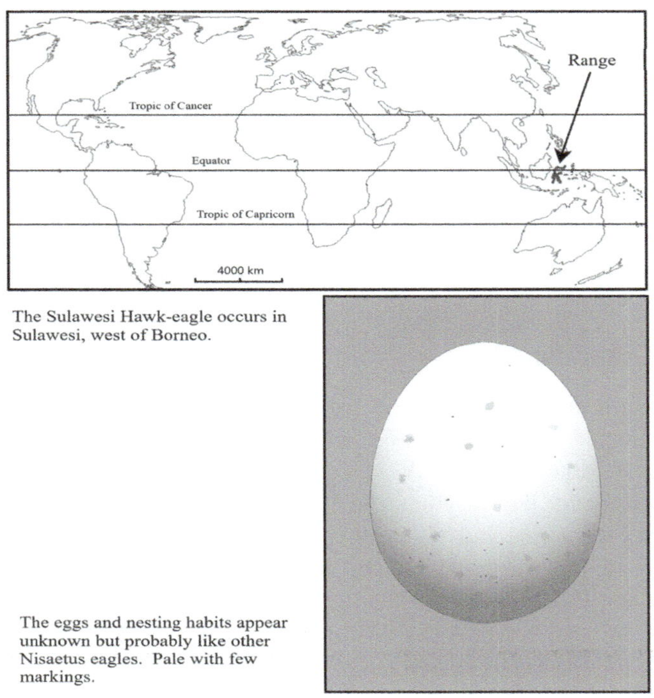

Fig. 2.11h. Sulawesi Hawk eagle range and egg

1.1-1.2 kg (Gamauf et al., 1998; Ferguson Lees & Christie, 2001). More recent descriptions may be more accurate (e.g. Gamauf et al., 1998; Ferguson Lees & Christie, 2001) than older assessments (e.g. Brown & Amadon, 1968) as in the past the Philippine Hawk eagle was combined with the former subspecies *S.p. pinskeri*, which is now recognised by some as a species, named Pinsker's hawk eagle (*Nisaetus pinskeri*), South Philippine hawk eagle or Mindanao hawk eagle (*Nisaetus pinskeri*, Preleuthner and Gamauf, 1998; BirdLife International, 2016).

The general adult plumage is dark brown with a purplish tinge (Brown & Amadon, 1968, 700). Gamauf et al. (1998, 4) describe the dominant plumage as "uniform brownish-olive", while BirdLife International (2016) describes the plumage as dark brown. Gamauf et al. (1998, 4) further describe the head as "roundish and delicate, long crest, sienna brown on the sides, with fine distinct streaks, crown with broad blackish-brown streaks." Brown & Amadon (1968, 700) state that the crown and headside feathers are edged pale brown and pointed, with a long, black crest formed from four feathers and black streaks on the throat center, and rufous feather on the throat and chest, with darker streaks. The primaries are barred darker above. The breast is ochraceous-tawny with bold black streaks and the lower belly uniformly antique brown. The leg feathering is finely barred clove brown and white (Gamauf et al., 1998, 4).

The underwing primaries have several (8 or 9) evenly spaced bars, with white bases to the primaries which form a small crescent-like shaped band. The secondaries have 7 or 8 bars on dark background. The underwing coverts are ochraceous-tawny, and finely barred ochraceous-tawny to clove brown, and some white mostly at the carpals (Gamauf et al., 1998, 4). The eyes are lemon yellow, the bill black, the cere and facial skin dark grey and the legs pale whitish yellow (Brown & Amadon, 1968, 700). The immature plumage varies with age. In the first year the head is white, with a black mask from the eyes to the cere, the head feathers turning more buffy and then darker and streaked with age. The underparts are also white. The upperparts are pale greyish, with pale-edged feathers, the tail is also grey brown. In the second year, much of the plumage becomes buffy colored, the tail developing a similar pattern to that of the adult. In the second year, more of the adult patterns emerge (Gamauf et al., 1998, 4-5).

The Philippine Hawk eagle may be mistaken for the Changeable hawk eagle (*Nisaetus cirrhatus*, Gmelin, 1788, formerly *Spizaetus cirrhatus*), but the Philippine Hawk eagle is smaller and has a contrasting crest and browner plumage (Clark et al., 2019). Gamauf et al. (1998, 5) note that "the Changeable Hawk eagle is larger with a body length of 53-68 cm. It is clearly heavier, at 1,300 to 1,900 g and more powerfully built than the Philippine Hawk eagle (see also Del Hoyo et al., 1994). Also, Gamauf et al. (2005, 3) note that "in profile the steeper forehead of the Philippine Hawk eagle gives the head a more roundish and delicate appearance compared to the rather flattened 'eagle-head' of the Changeable Hawk Eagle." The Changeable Hawk eagle may have three color morphs, the light and intermediate may be lighter than the adult Philippine Hawk eagle, while the dark morph may be darker, with most of the body blackish brown. The Pinsker's Hawk eagle or South Philippine Hawk eagle, formerly a subspecies of the Philippine Hawk eagle (*S.p. pinskeri* or *N.p. pinskeri*, and perhaps still seen as such by some commentators) has similar upperparts, but has a black streaked, white breast, a ivory to pale olive grey or dark olive buff head with fine black streaks, rather than black steaked sienna brown. The legs are heavily banded blackish and white, rather than finely barred clove brown, the underwing coverts are barred blackish and white, rather than finely barred uniformly ochraceous-tawny (Gamauf et al., 1998, 4). The Rufous-bellied Hawk eagle (*Lophotriorchis kienerii*, G. de Sparre, 1835) is smaller, with a shorter crest (Clark et al., 2019). The Barred Honey-buzzard (*Pernis celebensis*, Wallace, 1868) lacks the long crest and feathered legs (BirdLife International, 2016). The call of the Philippine Hawk eagle is a "loud, disyllabic whistle" (BirdLife International, 2016, 2020; see also Del Hoyo et al., 2014). Illustrations of the appearance, range, and eggs of this species are shown in Figures 2.12a to 2.12h.

Classification

BirdLife International (2016) notes that "*Nisaetus philippensis* and *N. pinskeri* (Del Hoyo & Collar, 2014) were previously lumped as *N. philippensis* following Haring et al. (2007), which before then was placed in the genus *Spizaetus* following Sibley and Monroe (1990, 1993)." Like the other Asian Hawk eagles, the Philippine Hawk eagle was initially placed in the Genus *Spizaetus*, but the studies of Helbig et al. (2005) and Lerner and Mindell (2005) separated these species from New World hawk eagle species and placed them in *Nisaetus* (Hodgson, 1836). The studies by Gamauf et al. (2005b) and Haring et al. (2007) named two sub-clades, one including the Philippine Hawk eagle. The Global Raptor Information Network (2021y) argues that "BirdLife International does not accept the separation of southern Philippine populations as *N. pinskeri* but designates the combined populations as Vulnerable. As treated here, the separate species should probably be classified as Endangered. Both taxa are suffering from continuing habitat loss in many parts of their respective ranges and also from hunting and trapping pressure." However, this appears to refer to the statements of BirdLife International from 2009. Currently, BirdLife International's (2020) accounts on both the North Philippine Hawk eagle and South Philippine (Pinsker's) Hawk eagle has accepted that the North Philippine Hawk eagle is separate from the South Philippine Hawk eagle.

Foraging

Brown and Amadon (1968, 700) note that the habits of this species are "practically unknown." However, Global Raptor Information Network (2021y) writes that the main habitat is forest, including both primary and secondary forest, and montane hill and lowland forest. This eagle may perch within the forest canopy but may also soar in forest areas (Kennedy et al., 2000). Foraging is from tree canopies but may also be from open landcover in forest clearings (Coates & Bishop, 1997; BirdLife International, 2016; Global Raptor Information Network, 2021y). Kennedy et al. (2000) did not record any prey species. BirdLife International (2020) writes that "it inhabits primary, selectively logged and disturbed forest, occasionally frequenting open areas, from the lowlands to lower mountain slopes, almost exclusively below 1,000 m. It appears not to tolerate much forest degradation. No migration is known, although unconfirmed reports from the migration funnel of Dalton Pass (Luzon) hint at intra-island movements."

Breeding

There are few records on the breeding patterns of this species (Kennedy et al., 2000).

Population Status

The Philippine Hawk eagle is non-migratory, although juveniles may disperse wildly from the breeding areas (Bildstein, 2006). Ferguson-Lees and Christie (2001) made an estimate that the global population of the Philippine Hawk eagle within the range of 1,001 to 10,000 individuals, but that estimate included populations now classified as South Philippine Hawk eagle or Pinsker's hawk eagle. Gamauf et al. (2005) argued that this classification of the Philippine Hawk eagle species into two different species may reduce the estimated population into two smaller populations. The new estimates were 200-220 pairs for the North Philippine Hawk eagle on Luzon and 320-340 pairs for Pinsker's hawk eagle on Mindanao (Preleuthner & Gamauf, 1998; Gamauf et al., 2005). The North Philippine Hawk eagle's detailed range covers at least 15 localities on Luzon, especially the Sierra Madre mountains and on Mindoro, even through it is rare in these locations (Preleuthner& Gamauf, 1998; Collar et al., 1999; Del Hoyo et al., 2014; BirdLife International, 2016).

The major problem for the North Philippine eagle is deforestation. Birdlife International (2020) writes that "deforestation for plantation agriculture, livestock and logging throughout its extensive, predominantly lowland range is the chief threat. In 1988, forest cover was as low as 24% on Luzon, which is likely to be an overestimate, with most lowland forest leased to logging concessions. Habitat loss is exacerbated by considerable hunting and trapping pressure." The rate of deforestation in the Philippines has been described as "very rapid in recent decades" possibly as high as 40% of total forest cover between 1970 and 1990 (Uitamo, 1999) or by 44% of closed-canopy forest between 1987 and 2002 (Walpole, 2010). Birdlife International (2020) concludes that "assuming rapid losses of primary forest over the past 56 years, and impacts from hunting and trapping pressure, it is likely that this species has experienced population declines of more than 50% over the past three generations." The area of densest population is Luzon where the population was estimated at 200-220 pairs more than 20 years ago in the late 1990s (Preleuthner & Gamauf, 1998). As there has been habitat destruction since then, it may be concluded that the population may be "a total of 400-600 mature individuals, roughly equating to 600-900 individuals" (BirdLife International, 2020). Illustrations of the appearance, range, and eggs of this species are shown in Figures 2.12a to 2.12h.

2.14 South Philippine Hawk Eagle or Pinsker's Hawk Eagle *Nisaetus pinskeri* (Preleuthner and Gamauf, 1998)

Physical Description

The Southern Philippine Hawk eagle or Pinsker's Hawk eagle (both names will be used inter-changbly) is similar to the North Philippine Hawk eagle, and if fact was once a subspecies of the former species. It is similar in size, and ecology. Fjeldså (2020) describes this species as a bird that "differs from *N. philippensis* in slightly smaller size, and somewhat paler plumage, with lower underparts densely barred." It measures 54–61 cm length, a wingspan just over 100 cm, and weighs about 1200 g. Typical hawk eagle with crest of 4–5 black feathers up to 8 cm long. This species has a crest of 4–5 black feathers up to 8 cm long (which may move according to the bird's emotions), with head feathers colored brown to dark olive-buff, and variably streaked with black (Fjeldså, 2020). The head is colored whitish brown with a pale crown, a white throat, with a black median stripe with black moustachial stripes. It also has a light breast with strong black streaks, a white lower belly with black barring. Gamauf et al. (1998, 3), the illustrations in Brown and Amadon (1968) and Weick (1980) as well as in del Hoyo et al. (1994), all show

Fig. 2.12a. Philippine Hawk eagle [With permission from Nayer Youakim]

Fig. 2.12b. Philippine Hawk eagle [With permission from Mark Jason Villa]

Fig. 2.12c. Subadult Philippine Hawk eagle [With permission from Marlz Cris]

Fig. 2.12d. Immature Philippine Hawk eagle [With permission from Edwin Mercado]

Fig. 2.12e. Philippine Hawk eagle [With permission from Marlz Cris]

Fig. 2.12f. Philippine Hawk eagle [With permission from Mike Nelson-Birdtour ASIA]

Fig. 2.12g. Philippine Hawk eagle habitat [Credit: Jeneva Stiles]

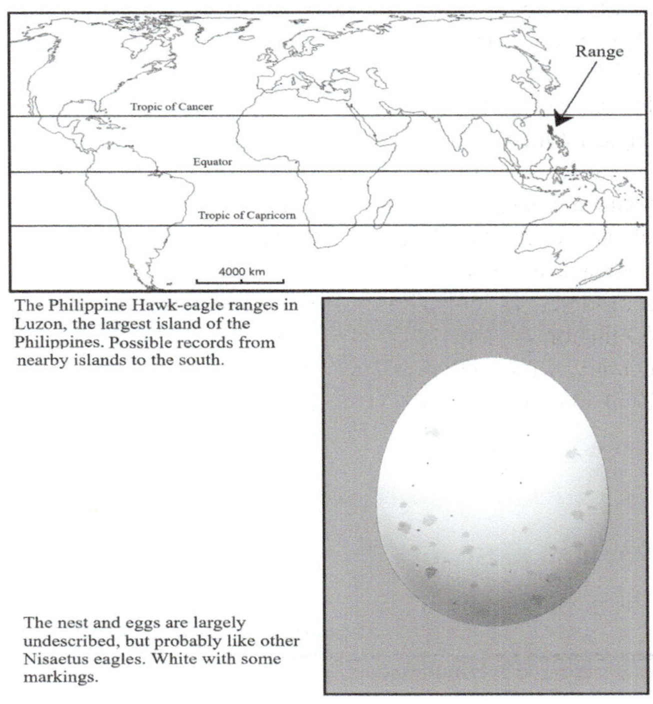

Fig. 2.12h. Philippine Hawk eagle range and egg

individuals of *S.p. philippensis*. Of course, this refers to *N.p. philippensis*. The underwing coverts are brown to blackish and white. The primaries have 8-9 regularly spaced bars, and the white bases form crescent-shaped band. There are panel, 7-8 bars on a darker on the secondaries. Males and females are similar in size and plumage. In males, the white panel is lighter than that of the female (Gamauf et al., 1998). Fjeldså (2020) notes that the juvenile is much paler than the adult, with the head and neck white except for black lores and long black crest feathers.

The South Philippine Hawk eagle may be mistaken for the North Philippine Hawk eagle, but has a black streaked, white breast, a ivory to pale olive grey or dark olive buff head with fine black streaks, rather than black steaked, sienna brown. The legs are heavily banded blackish and white, rather than finely barred clove brown, the underwing coverts are barred blackish and white, unlike the North Philippine Hawk eagle's finely-barred uniformly ochraceous-tawny lower plumage (Gamauf et al., 1998, 4). Both species have dark upperparts. It may also be mistaken for the Changeable hawk eagle, but the South Philippine Hawk eagle is smaller and has a contrasting crest and browner plumage (Del Hoyo et al., 1994; Gamauf et al., 1998; Clark et al., 2019). The Changeable Hawk eagle's three-color morphs, the light and intermediate may be lighter than the adult South Philippine Hawk eagle, while the dark morph may be darker, as most of the body plumage is blackish brown. The Rufous-bellied Hawk eagle (*Lophotriorchis kienerii*, G. de Sparre, 1835) is smaller, with a shorter crest (Clark et al., 2019). The Barred Honey-buzzard (*Pernis celebensis*, Wallace, 1868) lacks the long crest and feathered legs (BirdLife International, 2016). The call of the South Philippine Hawk eagle is a "loud, disyllabic whistle" (BirdLife International, 2016, 2020; see also Del Hoyo et al., 2014). Illustrations of the appearance, range, and eggs of this species are shown in Figures 2.13a to 2.13h.

Classification

BirdLife International (2016) notes that "*Nisaetus philippensis* and *N. pinskeri* (Del Hoyo & Collar, 2014) were previously lumped as *N. philippensis* following Haring et al. (2007), which before then was placed in the genus *Spizaetus* following Sibley and Monroe (1990, 1993)." Like the other Asian Hawk eagles, the South Philippine Hawk eagle was initially placed in the Genus *Spizaetus*, but the studies of Helbig et al. (2005) and Lerner and Mindell (2005), separated these species from New World hawk eagle species and placed them in *Nisaetus* (Hodgson, 1836). The studies by Gamauf et al. (2005b) and Haring et al. (2007) named two sub-clades, one including the South Philippine Hawk eagle. The Global Raptor Information Network (2021z) argues that "BirdLife International does not accept the separation of southern Philippine populations as *N. pinskeri* but designates the combined populations as Vulnerable. As treated here, the separate species should probably be classified as Endangered. Both taxa are suffering from continuing habitat loss in many parts of their respective ranges and also from hunting and trapping pressure." However, this appears to refer to the statements of BirdLife International from 2009. Currently, BirdLife International's (2020) accounts on both the South Philippine (Pinsker's) Hawk eagle and the North Philippine Hawk eagle has accepted that the South Philippine Hawk eagle is separate from the North Philippine Hawk eagle.

Foraging

The South Philippine Hawk eagle frequents primary, advanced second-growth, selectively logged, and disturbed forest, occasionally frequenting open areas, from the lowlands to lower mountain slopes, and in some cases montane mossy forest, usually below 1,000 m (Kennedy et al., 2000). It is therefore almost entirely a forest specialist, but frequents many types of forest, and maybe intolerant of excessive forest degradation. It is probably non-migratory, as records are from particular islands in the Philippines (Birdlife International, 2016; Global Raptor Information Network, 2021z). Hunting may be from a perch in the dense canopy, or from soaring (Kennedy et al., 2000). Kennedy et al. (2000) do not record prey species favored, but probably birds are taken (Ferguson-Lees and Christie, 2001).

Breeding

There are currently few breeding records (Kennedy et al., 2000).

Population Status

The South Philippine Hawk eagle is endemic to southern Philippine Islands (Basilan, Biliran, Bohol Mindanao, Negros, Samar, and Siquijor Islands). The principal habitat is on Mindanao. About 320-340 pairs were estimated to live on this island in the late 1990s (Preleuthner & Gamauf, 1998). Since that time, estimates of ecological change and reproductive change have led researchers to speculate on the current population. A recent population estimate is of 600-800 mature birds, which means about 900-1,200 individual birds (BirdLife International, 2016). BirdLife International (2016) classifies this species as Endangered, due to the small population, which is affected by lowland deforestation, human hunting, and trade. Illustrations of the appearance, range, and eggs of this species are shown in Figures below 2.13a to 2.13h.

Fig. 2.13a. Pinsker's Hawk eagle perching [With permission from Steve Young]

Fig. 2.13b. Pinsker's Hawk eagle perching [With permission from Steve Young]

Fig. 2.13c. Pinsker's Hawk eagle [With permission from Rodrigo Layug]

Fig. 2.13d. Pinsker's Hawk eagle [With permission from Chris Chafer]

Fig. 2.13e. Pinsker's Hawk eagle [With permission from Chris Chafer]

Fig. 2.13f. Pinsker's Hawk eagle [With permission from Mike Nelson]

Fig. 2.13g. Immature Pinsker's Hawk-eagle in habitat [With permission from Steve Young]

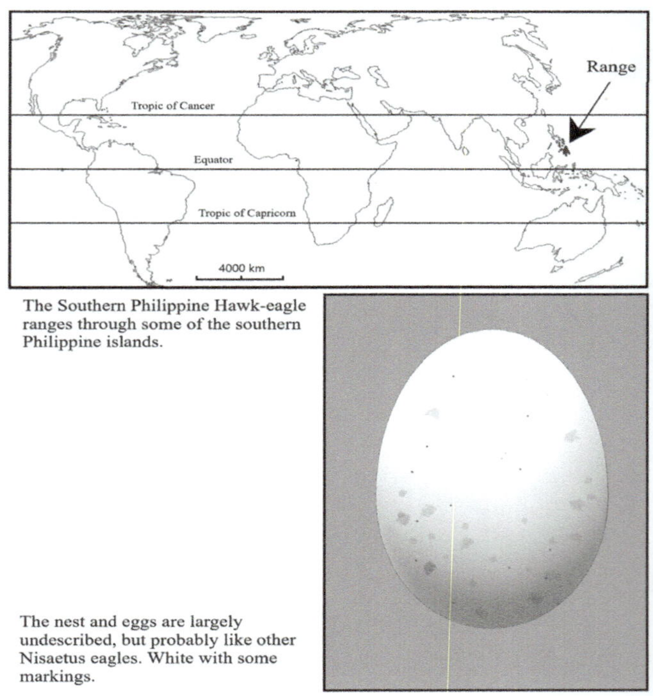

Fig. 2.13h. Pinsker's Hawk eagle range and egg

2.15 Blyth's Hawk Eagle *Nisaetus alboniger* (Blyth, 1845) (formerly *Spizaetus alboniger*)

Physical Description

The Blyth's Hawk eagle is medium-sized eagle, with a length of 51-58 cm, a wingspan: 56-70 cm and a weight of 830 g (Ferguson-Lees & Christie, 2001). Clark et al. (2020d) note that the most discerning identifier is the dark brownish-black upperparts and the strongly black-streaked white underparts. Like the other Hawk eagles, it has broad, rounded wings.

In more detail, the adult upperparts are black ("glossed with plumbeous"),with a head sided with black feathers, a white throat marked with a black line, a mottled grey tail with a lighter, central band, and a narrower basal band, with whitish coloration below. The small crest is not visible in flight but may be held erect when perched (Clark et al., 2020d). There is dark brown barring on the wing quills and coverts. The white underparts contrast with the darkly streaked breast, and the black barred wing coverts, thighs and legs. The wing quills and tail are pale grey below, barred with darker grey and marked with black tips. The plumage varies in color, especially in the proportion of the white and black coloration (Brown & Amadon, 1968, 702). The eyes are pale yellow, the bill is black fading lighter below and the toes are yellow. The immature plumage is composed of rufous brown crowns, hind necks and head sides, black, white tipped crests and brown backs, with purplish tinges to some feathers. There is whitish color to the inner edges of the wing quills. The tail is barred slightly darker. The underparts are light brown, with some feathers having a purplish sheen and/or darker tips. The back and head feathers darken to black from brown with age. The tail is less different from that of the adult, but has fewer streaks and bars, and the mottling is lighter (Brown & Amadon, 1968, 702).

The adult Blyth's Hawk-eagle may not be confused with other species: Brown and Amadon (1968, 702) note that "the handsome black and white adult is unmistakable." By contrast, the juvenile is like Wallace's Hawk eagle (*Nisaetus nanus*, Wallace, 1868, formerly Spizaetus nanus). It lacks the overall buffy brown tones of the much smaller adult Wallace's Hawk eagle. The immature Blyth's Hawk-eagle is like the Wallace's but has thinner tail barring; immatures are best distinguished by altitude (Blyth's favors higher elevations than Wallace's) and size (eBird, 2020). The call is described as a "shrill scream" (Brown & Amadon, 1968, 702). Clark et al. (2020d) describe this species as "generally silent, except in the breeding season. Commonest call is a repeated, single piercing whistle "kleeekleee". Illustrations of the appearance, range, and eggs of this species are shown in Figures 2.14a to 2.12h.

Classification

The Blyth's Hawk-eagle, along with the other members of the Genus *Nisaetus*, as mentioned above was formerly included in the Genus *Spizaetus*, prior to the molecular studies by Helbig et al. (2005) and by Lerner and Mindell (2005), which differentiated the Asian hawk eagle species from the New World hawk eagle species. One important finding of the latter study was that Blyth's Hawk eagle was that Blyth's Hawk eagle and the Mountain hawk eagle, based on the molecular sequences of two mitochondrial genes and one nuclear entron. Haring et al. (2007) also confirmed that Blyth's Hawk eagle is one of the Asian *Nisaetus* species in one of two sub-clades.

Foraging

The Blyth's Hawk eagle is a forest species, prey taken including arboreal prey. This species hunts in forest, with arboreal prey commonly taken from perches rather than from soaring flight. However, it may soar over the canopy. It may also hunt in open woodland, although higher tree densities are preferred. There are some regional variations in ecological behavior. In Malaysia, it frequents both the interior and the edge of both lowland and montane forests, including both mature and selectively logged areas. It commonly frequents steep topography, from the foot hills to the elevations of at least 1,800 m (Wells, 1999). Robinson & Seywald (1987) report nests in the Main Range (mountain range in West Malaysia, the most prominent mountain group on the Malay Peninsula), which range in location from about 400 m below the montane ecotone to minimally above it restricted by the reduced numbers of emergent trees for nesting in higher altitudes.

Global Raptor Information Network (2021aa) points out that "the diet is poorly known, but there are records of birds taking lizards and a bat. It still hunts from shaded perches inside and at the edge of forest, at interior canopy down to mid-levels, sometimes only a few meters above a quiet forest road" (see also Wells, 1999). Wells (1999) and Robinson and Seywald (1987) report a tendency to frequent forest-topped limestone karst towers, where there may be many cave-bat roosts.

Breeding

The Blyth's Hawk eagle may indulge in "noisy aerial display over forest; possible display involves undulating flights of pair; may also include soaring" (Clark et al., 2020d). In Sumatra, Holmes (1996) recorded eagles displaying during August 1986 at

Ketambe. Another displaying pair was recorded displaying and transporting nest material above the lowland and hill forests in Gunung Leuser National Park in October 1998 (Buij et al., 2006). The nest is bulky and composed of sticks, usually situated in the crown of large emergent tree in the forest. Global Raptor Information Network (2021aa) describes the nest as "a thick platform of sticks with a cup deep enough to conceal the body of the sitting parent when viewed at eye-level and placed in a main branch fork of a large emergent tree, often above the level of the surrounding canopy" (see also Wells, 1999).

The clutch size may be one egg, as Wells (1999) points out that all the reported broods have been of a single chick. Apparently, "no nests have been observed with more than one chick, suggesting clutch size is one" (Clark et al., 2020d). Global Raptor Information Network (2021aa) further notes that "the egg is apparently undescribed, and neither the incubation nor nestling periods are known. However, the total development period is no longer than four months, since a bird incubating in mid-November had a small, downy chick in December, which fledged on 17 February" (Wells, 1999). Regionally, in the Malay Peninsula, nesting and copulation have been recorded in October, with incubation from early November, and chicks appearing in late December to late April. In Pos Slim, Perak a nest was recorded during construction on 10 July 2007 (Soon, 2008). In Selangor and Pahang, there is evidence that nests with unfledged young after February were unsuccessful (Robinson & Seywald,1987; Wells, 1999). Nests are frequently located in giant legume (*Koompassia excelsa* (Becc.) Taub), which is the tallest tree of the region north from mid-Selangor, usually on slopes to the elevations just below the montane ecotone (Wells, 1999)

Population

The Blyth's Hawk eagle is a non-migratory species, but juveniles may disperse some distances from the breeding areas (Bildstein, 2006). The global population has been estimated at 1,001 to 10,000 individuals, where the counted birds are the adults and immatures at the beginning of the breeding season (Ferguson-Lees & Christie, 2001). BirdLife International (2016) assesses this species as Least Concern. The justification for this is that it has a very wide range, not under 20,000 km^2, and its occurrence is not "combined with a declining or fluctuating range size, habitat extent/quality, or population size and a small number of locations or severe fragmentation" (BirdLife International, 2016). The population is decreasing, but not fast enough to be Vulnerable or Endangered. The population size is small, but it is fairly strong, hence not Vulnerable (BirdLife International, 2016). The main factor for the population declines of the Blyth's Hawk eagle is the trend of deforestation and the increase in slash and burn agriculture throughout its range in southeast Asia (Ferguson-Lees & Christie, 2001).

Its population density varies regionally. The range includes southern Myanmar and southern Thailand through Malay Peninsula to Sumatra and nearby islands, and northern Borneo (Global Raptor Information Network, 2021aa). In the Malay Peninsula, it is a regular and uncommon to common resident, mostly frequenting hill ranges and may occur on forested outliers, including some hilly islands, including the Muntahak-Panti Ridge, eastern Johor (Wells, 1999). Although it has been recorded from Penang and Pangkor off the west coast and Tioman off the east coast, it is recorded as rare near Singapore (Jeyarajasingam & Pearson, 1999; Wells, 1999). In Sumatra, it is assumed to be resident in hilly, forested areas, despite the relative scarcity of breeding evidence (van Marle & Voous, 1988). In Sabah, it may be an uncommon resident up to 2,000 m (Sheldon et al., 2001). Possibly, this was the commonest of the *Nisaetus* eagles in Sabah (Thiollay, 1983). Illustrations of the appearance, range, and eggs of this species are shown in Figures 2.14a to 2.14h.

2.16 Wallace's Hawk Eagle *Nisaetus nanus* (Wallace, 1868) (formerly *Spizaetus nanus*)

Physical Description

The Wallace's Hawk eagle has been described as a "typical small hawk eagle" (eBird, 2020). This is one of the world's smallest eagles. This source lists its defining characteristics: A small upright crest, dark brown wings, a paler face, densely black-barred underparts, two white bands across the tail, and light brown on the inner part of the underwing. It measures 43 to 58 cm in length (46 according to BirdLife International, 2017), with a wingspan of 95 to 105 cm and a weight of 500 to 610 g (Ferguson-Lees & Christie, 2001; Clark & Kirwan, 2020). There are two subspecies *Nisaetus nanus nanus* ranging from southern Myanmar and southern Thailand through Malay Peninsula to Sumatra and Borneo; and *Nisaetus nanus stresemanni* ranging in Western Sumatra (Clark, 1994; Global Raptor Information Network, 2021ab). BirdLife International (2017) summarizes Wallace's Hawk eagle's appearance as a "small, boldly-patterned hawk eagle. Rufescent sides of head with blackish streaks, dark crest broadly tipped white, three dark bands on tail. Buffish-white base-color to underside flight feathers and warm buffish coverts with narrow dark barring,"

Regarding its detailed appearance, the small crest is more clearly visible when the bird is perched. Brown and Amadon (1968, 702) describe the adult as dark above with a "purplish cast", with the feathers, especially on the wings, tipped and barred duller,

Fig. 2.14a. Blyth's Hawk eagle perching [With permission from Mike Nelson-Birdtour ASIA]

Fig. 2.14b. Blyth's Hawk eagle [With permission from Nick Athanas]

Fig. 2.14c. Blyth's Hawk eagle [With permission from Zaharil Dzulkafly]

Fig. 2.14d. Immature Blyth's Hawk eagle [With permission from Zaharil Dzulkafly]

Fig. 2.14e. Blyth's Hawk eagle [With permission from Mike Nelson-Birdtour ASIA]

Fig. 2.14f. Blyth's Hawk eagle [With permission from Mike Nelson-Birdtour ASIA]

Fig. 2.14g. Blyth's Hawk eagle habitat, Malaysia [Credit: Terence Seow-Unsplash]

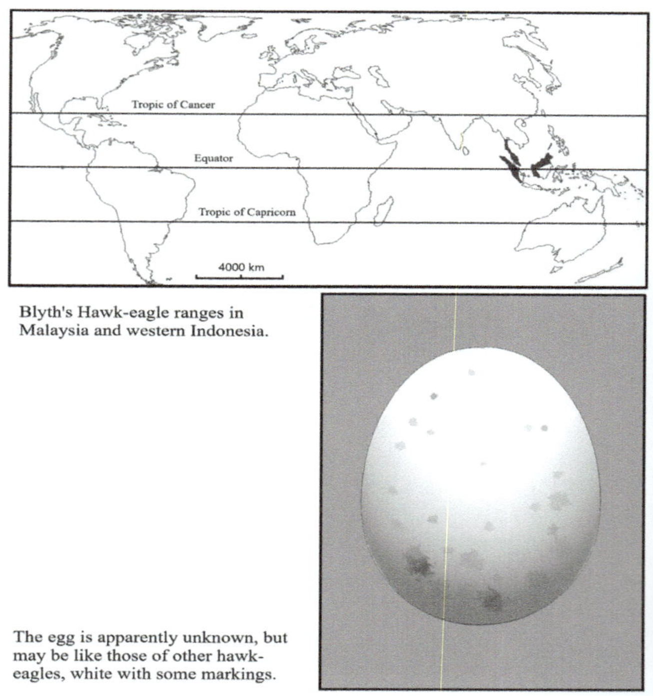

Fig. 2.14h. Blyth's Hawk eagle range and egg

the crown is streaked brown with an occipital crest of four or five black, buff tipped feathers, a slightly visible buffy brown collar between and blending into the head and the upper breast, and black streaked upper breast. The tail is banded (three black and three greyish brown bars, the last dark bar towards the tail tip broader and lighter tipped). The underparts are light brown, with a black streak on the throat, more streaks on the breast, brown and darker brown barring on the belly and legs, and black and grey barring on the tail (which is clearer than the barring above). The underwing coverts are rufous buff and lightly streaked and the wing quills are barred black and pale brown to white on the underside. The eyes and toes are yellow, while the bill is blackish. Brown and Amadon (1968, 702) argue that the plumage of *S.n. stresemanni* (now *Nisaetus nanus stresemanni*) is "unknown." The juvenile differs from the adult in having a whitish head (despite a similar crest), similarly colored underparts, with pale brown mottling and tinges, which are darker on the head and neck sides, with the coloration of the throat, legs, back lighter than those of the adult. Some of the back feathers have a purplish gloss with whitish edges. The tail is brownish grey above, with less prominent dark barring and light grey below, with a subterminal band. The juvenile of *N.n. stresemanni* has a whiter head and underparts (not pale or whitish brown) (Brown & Amadon, 1968, 703).

This species may be distinguished by its size and plumage (Ferguson-Lees & Christie, 2001). The Blyth's Hawk eagle is larger with a "narrowly white-tipped crest, blackish tail with pale grey broad, central band and narrow tip. Whitish underwing with heavy black barring on coverts and dark sides of head." The juvenile Wallace's has a pale tawny-brown head and on the underparts, while the Blyth's Hawk eagle has tail barring. Also, Wallace's Hawk eagle has light brown on the inner underwing, but the Blyth's Hawk eagle has black and white underwings (eBird, 2020). eBird (2020) notes that it is "visibly smaller than both Blyth's and Mountain hawk-eagles." It is also smaller than the Javan hawk eagle. It is similar to the immature Blyth's hawk eagle, but Brown and Amadon (1968, 703) note that the Wallace's hawk eagle is "smoother plumaged" and more "evenly barred"; image plumages of the two species are however "almost identical." Additionally, the "male *S. alboniger* are very difficult to tell from female *S. nanus*." *S. nanus* however has a shorter middle toe and the feet and beak are weaker (Brown &Amdon, 1968, 703). The Wallace's Hawk eagle has also been compared with the Jerdon's Baza (*Avicedajerdoni*, Blyth, 1842) which "can appear similar, but it has horizontal banding across the breast, whereas juvenile Wallace's has unmarked underparts, pantaloons (fully feathered legs), and a broader black tip to the tail" (eBird, 2020). By BirdLife International (2017), the call is a shrill, high-pitched yik-yee or kliit-kleeik, with rising second note. Fledged juveniles have a high-pitched, breathless call, yii-yii-yii-yii and ee-ee-ee-ee-eeee. Illustrations of the appearance, range, and eggs of this species are shown in Figures 2.15a to 2.15k.

Classification

Wallace's Hawk eagle is generally classified as including two subspecies, in the past these being *Spizaetus nanus nanus* and *Spizaetus nanus stresemanni* (Brown & Amadon, 1968). Clark & Kirwan (2019) classify the species into "*N.n. nanus* (Wallace, 1868) – S Myanmar (Tenasserim) and S Thailand through Malay Peninsula to Sumatra (including Bangka I) and Borneo" and "*N. n. stresemanni* (Amadon, 1953) – Nias I (off W Sumatra)." Further, "probably present also (race not known) on Siberut, in Mentawai Is (off W Sumatra)" (Clark & Kirwan, 2019). Like all the Asian Hawk eagles, the Wallace's Hawk eagle was initially placed in the Genus *Spizaetus*, until the studies by Helbig et al. (2005) and Lerner and Mindell (2005), cited earlier in this book (Global Raptor Information Network, 2021ab). Haring et al. (2007) also found two sub-clades, one of which included Wallace's Hawk eagle and consists of *N. alboniger*, *N. bartelsi*, *N. nanus*, and *N. nipalensis*. The Wallace's Hawk eagle was previously considered to be conspecific with Blyth's Hawk eagle because of the confusion with immature plumages of the latter species (Clark, 1994).

Foraging

The Wallace's Hawk eagle is primarily a forest species depending on the region. In the Malay Peninsula, it occurs in dry- and semi-evergreen lowland forest, undisturbed and logged forest and even in rare cases highly degraded forest, but almost invariably within range of continuous forest canopy (Wells, 1999). In Sabah, it frequents primary, logged, and riverine forest (Sheldon et al., 2001). It has also been recorded in heavily logged forest in Kalimantan and Sumatra, although one study found it in primary forest, but after logging it was absent (BirdLife International, 2017). Lowland forests are the main habitat and to a lesser extent secondary forest, in low hilly topography (Thiollay, 1985a,b). Here the Wallace's Hawk eagle generally shares its habitat with the Blyth's Hawk Eagle,the former mostly in lowland forests, and the latter in higher elevation forests (BirdLife International, 2017). Prey, mostly taken from perching, include lizards, birds, bats, frogs, and skinks (Smythies, 1981; Clark, 1994; Sheldon et al., 2001).

Breeding

The Wallace's Hawk eagle performs displays at the beginning of the breeding season (Clark & Kirwan, 2020). The nest is medium sized for an eagle's nest, built of sticks and fresh, green branchlets. Nests are usually located in a branch fork or near the base of the crown of emergents at the main canopy level. Where there is a choice, the favored nesting tree is tualang (*Koompassia excelsa* (Becc.) Taub.). Where this species is unavailable outside its range, large diptocarps (*Shorea* spp.) may be used (Wells, 1999). Most nests are located in dense forest, but Wells (1999) record a case where the eagles remained in their nest when the nesting was left isolated in open landcover when the surrounding trees were cut. There is probably one egg laid, as records indicate single chicks in the nest (Wells, 1999). The egg is not described, but clutch size is probably one, since only single nestlings have been recorded (Wells, 1999). One nest in the Malay Peninsula had a grown nestling in February and a subadult in March (Wells, 1999). Another nest was recorded in Thailand (1996-97), with an adult guard in December, an adult incubating or brooding in May and fledged juvenile nearby in August (Red Data Book Threatened Birds of Asia, 1998, in Global Raptor Information Network, 2021ab). Active nests were also recorded in September (Sabah, Smythies & Davison, 1999); April (Sarawak, Duckett, 1985); and December (Brunei, Kidd & Beales, 1977).

Population Status

The Wallace's Hawk eagle is non-migratory, with the main movements being but juveniles that disperse from breeding areas (Bildstein, 2006). Ferguson-Lees and Christie (2001) estimated the global population, based on adults and immatures at the start of the breeding season, in the range of 101 to 1,000 individuals. BirdLife International (2009) placed the population of mature birds between 2,500 to 9,999 individuals. Birdlife International (2017) classified this species as Vulnerable and estimated the number of mature individuals at 2,500-9,999. The statement reads that "The population size is preliminarily estimated to fall into the band 2,500-9,999 mature individuals. This equates to 3,750-14,999 individuals in total, rounded here to 3,500-15,000 individuals." The assessment as Vulnerable is based on the fact that "this species is suspected to be undergoing a rapid population decline owing to the widespread and on-going loss of lowland forest, which qualifies it as Vulnerable" (Birdlife International, 2017). The Global Raptor Information Network (2021ab, citing Wells, 1999) notes that "assessing the status of this species is complicated by the difficulty of separating it from the similar Blyth's Hawk eagle… in the field. However, it seems certain that populations are now generally uncommon or rare and have been seriously reduced by habitat loss in many areas… It may be near extinction in Thailand, as the result of clearing of virtually all lowland forest." Illustrations of the appearance, range, and eggs of this species are shown in Figures 2.15a to 2.15k.

Fig. 2.15a. Wallace's Hawk eagle perching [With permission from Marcus F.C. Ng]

Fig. 2.15b. Wallace's Hawk eagle perching [With permission from Terence Ang]

Fig. 2.15c. Wallace's Hawk eagle perching, Kinabatangan River, Sabah, Borneo [With permission from Paul Prior]

Fig. 2.15d. Immature Wallace's Hawk eagle perching [With permission from David Koh]

Fig. 2.15e. Immature Wallace's Hawk eagle perching [With permission from Tristan Bantock]

Fig. 2.15f. Immature Wallace's Hawk eagle perching [With permission from Mike Nelson-Birdtour]

Fig. 2.15g. Immature Wallace's Hawk eagle perching [With permission from Mike Nelson-Birdtour]

Fig. 2.15h. Immature Wallace's Hawk eagle perching [With permission from Mike Nelson-Birdtour]

Fig. 2.15i. Immature Wallace's Hawk eagle in flight [With permission from Mike Nelson-Birdtour]

Fig. 2.15j. Wallace's Hawk eagle habitat, Malaysia [Credit: Sonia Chronowska Unsplash]

2.17 Black Hawk Eagle or Tyrant Hawk Eagle *Spizaetus tyrannus* (Wied, 1820)

Physical Description

The Black Hawk eagle is a medium-sized Neotropical, forest eagle (Quintero & Jácome, 2020). eBird (2020) describes it as an "uncommon, spectacular, crested eagle of lowland tropical forest" and describes the adult as "black overall with bold white checkering under the wings and bold white bands on the tail. Immatures, although rarely seen, are browner with a white throat and crown. The short bushy crest is usually apparent only on perched birds and is not seen in flight" (Quintero & Jácome, 2020). It measures 58 to 70 cm in length, with a wingspan of 140 cm and a weight of 900 to 1300 kg. There are two subspecies: *Spizaetus tyrannus serus* and *Spizaetus tyrannus tyrannus* (Blake, 1977). The plumage for the two subspecies is similar, with largely black plumage, white spots on the underparts and barring on the wings and tail. The appearance is marked by a short broad crest, purple-tinged black plumage, white markings on the crown, thighs, underwing coverts and sometimes the throat, some concealed white feathers on the back and white barring on the upper and lower tail coverts. There are three broad greyish bars on the tail, which has a grey tip, with lighter colors on the underside of the bars. On the wing quills, there is slightly visible barring above, and more visible whitish barring below. *S.t.tyrannus* is larger with less white on the thighs and lower wing coverts (Brown & Amadon, 1968). The eye is orange yellow, the cere dark grey, the bill black and the toes orange yellow (Brown & Amadon, 1968, 703).

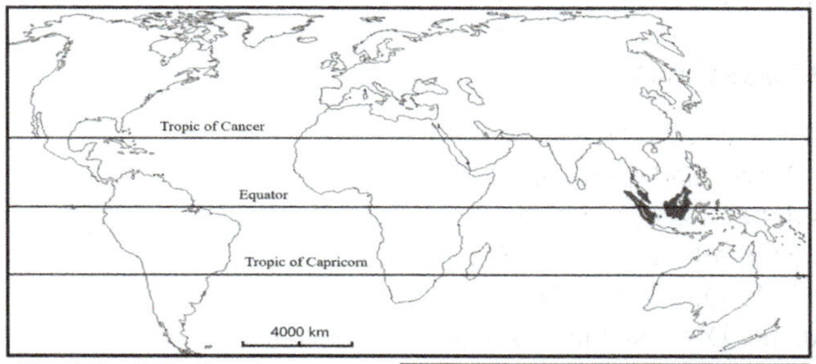

The Wallace's Hawk-eagle ranges through Malaysia and western Indonesia.

The egg is recorded as unknown, but is probably like that of the other Hawk eagles, white with markings.

Fig. 2.15k. Wallace's Hawk eagle range and egg

The juvenile is whiter on the crown and forehead, and a white line above the eye, a white tipped crest, blackish cheeks a white throat and a browner back than the black back of the adult, with white edged feathers. Further differences are black and brown breast streaks (with white midline), a blackish, mottled and barred abdomen, black and white barring on the thighs, legs and lower tail coverts, orange-brown eyes (rather than orange-yellow), yellow cere and grey feet. The Black Hawk eagle may be distinguished from the Ornate Hawk eagle (*Spizaetus ornatus*, Daudin, 1800) as in flight the wings appear narrower at the base, the tail is narrower and longer, rather than shorter and fanned. Brown and Amadon (1968, 704) describe the call as a "loud, ringing 'ker, ker, ker, ker, cu-weer." They also note that in some cases, the "cu-weer is the sole cry, like a "slightly hissing whistle." Burton (1989) describes it "whit-whit-whit-wheeeeeeeer." Illustrations of the appearance, range, and eggs of this species are shown in Figures 2.16a to 2.16i.

Classification

The Black Hawk eagle was in the past decades, regarded as a dark color phase of the Ornate Hawk eagle (Swann, 1921). The molecular studies of Helbig et al. (2005), Lerner and Mindell (2005), Gamauf et al. (2005), and Haring et al. (2007), used DNA sequences of mitochondrial and nuclear genes, and their findings indicated that the Black Hawk eagle and the Ornate Hawk eagle are not as close as previously expected. The Black Hawk eagle is less closely related to the Ornate Hawk eagle than the latter is to the Black-and-Chestnut eagle and the Black-and-White Hawk eagle. The conclusions were to retain the Black Hawk eagle and the Ornate Hawk eagle in the Genus *Spizaetus*, with the other New World hawk eagles (Black-and-Chestnut eagle and the Black-and-White Hawk eagle), separate from the Asian hawk-eagle species which were moved to the Genus *Nisaetus* (Helbig et al., 2005). Gamauf et al. (2005) also suggested placing the Black Hawk eagle in a separate, monotypic genus, *Ptenura*, due to discerned distinctions between this species and the other species in the Genus *Spizaetus*.

Foraging

The Black Hawk eagle frequents semi-open, generally low elevation (below 2000 meters) landcover. Brown and Amadon (1968, 704) argue that "little is known" of this species' feeding and foraging habits. However, Slud (1964), BirdLife International (2016), Global Raptor Information Network (2021ac) and Quintero & Jácome (2020). observe that this is a forest species, frequenting mostly lowlands and foothills. The Black Hawk eagle is also more likely than the Ornate Hawk eagle to frequent more open forests such as disturbed forest, late second-growth, forest edges and semi-open areas. Foraging is mostly from soaring flight over the canopy or in some cases open areas and perching for prey ambush (Global Raptor Information Network, 2021ac). López et al. (2012, 185) write "This handsome, crested hawk eagle is most often detected in just this fashion—as it soars high above the forest, issuing its distinctive call."

The foraging habits may vary regionally. In Guatemala, Funes et al. (1992) investigated the hypothesis that the Black Hawk eagle favors open areas over dense primary forest at Tikal National Park but found little supporting evidence of this position. In Venezuela, Naveda-Rodríguez (2007) found that most records were from the Coastal Mountain Chain and the Andes bioregions, at elevations between 700 and 1,600 m, mostly from mixed landcover (ombrophile submontane and montane forest, farms, and urban and industrial lands). In French Guiana, the Black Hawk eagle mostly occurred in more open, modified patches within mature forest stands, some with structures creating edges, both natural (inselbergs, rivers) or human constructed (clearings, roads, logging). It also frequented secondary and fragmented forests, unlike the Ornate Hawk eagles that shared habitat localities and undisturbed primary forest (Thiollay, 2007).

Prey species taken include mammals, including marmosets (Family *Callitrichidae*, Thomas, 1903), squirrels (Family *Sciuridae*, Fischer de Waldheim, 1817) and opossums (Family *Didelphidae*, Gray, 1821). It may also prey on birds such as chachalacas (Genus *Ortalis*, Merrem, 1786), macaws (Tribe *Arini*, G. R. Gray, 1840) and toucans (Family *Ramphastidae*, Vigors, 1825), and large reptiles. Hunting may be arboreal, with flight from tree to tree or pickups from the ground. In Tikal National Park, Guatemala, Funes et al. (1992) found that the main prey were mammals (68%, mostly bats (30.6%), small tree squirrels (19.3%), Woolly Opossums (*Caluromys derbianus*, Waterhouse, 1841), 10.6%) unidentified mammals (6.6%), birds (2.6%), and unidentified prey items (29.3%). Therefore, "these eagles appear to have a diet with a higher proportion of mammals than the Ornate Hawk eagle" (Global Raptor Information Network, 2021ac). In other regions, prey taken included Vermillion-crowned flycatchers (*Myiozetetessimilis*, Spix, 1825) in Costa Rica (Skutch, 1960); iguanas, vine snakes (*Oxybeli* ssp.), Black-chested jay (*Cyanocorax affinis*, Pelzeln, 1856) and a squirrel (*Notosciurus granatensis*, Humboldt, 1811) in Panama (Wetmore, 1965); lizards and squirrels and macaws in Peru (Robinson, 1994; and primates (perhaps marmosets) and bats (*Artibeus jamaicensis*, Leach, 1821) in Brazil (Barnett et al., 2015; Global Raptor Information Network, 2021ac).

Breeding

The Black Hawk eagle indulges in aerial displays during the breeding season, including paired talon grasping and roll-overs (del Hoyo et al., 1994; Hilty & Brown, 1986; Howell & Webb, 1995). The nest is large, built of sticks and vines on a branch from the main trunk of a large tree. Tree species favored included *Brosimum alicastrum* (Sw), *Bucidabuceras* (L.) observed in Guatemala, and *Manilkara chicle*n ((Pittier) Gilly) observed in El Salvador (West, 1988; Funes et al., 1992). Incubation and feeding are done by the female, while the male provides food. The clutch composed of one white and unmarked egg. The nestling period times and duration may vary regionally. The nesting period in one nest in Tikal National Park, Guatemala nest was 71 days, with the nestling hatching in May and fledging in July, and in another nest a six-week old nestling was reported in July (Funes et al., 1992). In Panama, two downy nestlings were found in February and a fledged juvenile in August in the same area three years later (Smith, 1970). In another case, Canuto (2008a) reported a pair in courtship display in July in Soberania National Park, Panama.

Population Status

The Black Hawk eagle is an irruptive or local migrant (Bildstein, 2006). The two subspecies range differently but may overlap in parts of Brazil. *Spizaetus tyrannus tyrannus* ranges through Eastern and southeastern Brazil and northeastern Argentina (Misiones Province). *Spizaetus tyrannus serus* ranges through Central Mexico through Central America to Colombia, and eastwards to the Guianas, Trinidad, northern and western Brazil, and southwards to Paraguay and northeastern Argentina (it may be recorded in west of the Andes in Ecuador and Peru) (Global Raptor Information Network, 2021ac). The global population of adults and immatures at the start of the breeding season was estimated at between 100,101 and 1,000,000 birds (Ferguson-Lees and Christie, 2001). BirdLife International (2016) estimated that the total population of mature birds is more likely in the range of 20,000 to 49,999 individuals.

BirdLife Conservation classified the Black Hawk eagle as Least Concern, as the wide range and the population status, in that the range is not less than 20,000 km^2 it does not fulfill the other criteria: "a declining or fluctuating range size, habitat extent/quality, or population size and a small number of locations or severe fragmentation." Furthermore, the population is declining but not fast enough for the classification of Vulnerable. The Black Hawk eagle is "generally regarded as more common than the Ornate Hawk eagle, possibly because it tends to occur nearer human settlements and may therefore be recorded more often"; additionally, "it may be more tolerant of human disturbance and habitat alteration than the Ornate Hawk eagle, but still doubtless suffers from habitat loss and shooting" (Global Raptor Information Network, 2021ac). Nevertheless, its population status varies. In Mexico, it was classified as endangered (Secretaría de Medio Ambiente y Recursos Naturales, 2002) and Threatened in the Los Tuxtlas region of Veracruz (1987). In El Salvador it was considered rare (Komar & Dominguez, 2001). In Ecuador, it is rare due to deforestation (Ridgely & Greenfield, 2001). It is Endangered in Paraguay (Del Castillo & Clay, 2005). It is also classified as Endangered or Threatened in parts of Brazil (Bergallo et al., 2003; Machado et al., 1998; Pacheco et al., 2007). Illustrations of the appearance, range, and eggs of this species are shown in Figures 2.16a to 2.16i.

Fig. 2.16a. Black Hawk eagle perching [With permission from Nick Athanas]

Fig. 2.16b. Black Hawk eagle perching [With permission from Jonathan Beilby]

Fig. 2.16c. Black Hawk eagle perching [With permission from Thelmå Gatuzźo]

Fig. 2.16d. Black Hawk eagle perching [With permission from Dave Wendelken]

Fig. 2.16e. Black Hawk eagle in flight [Credit: Neil Bowman]

Fig. 2.16f. Black Hawk eagle in flight [With permission from Nick Athanas]

Fig. 2.16g. Black Hawk eagle habitat [Michael Campbell]

Fig. 2.16h. Black Hawk eagle habitat [Michael Campbell]

2.18 Black-and-White Hawk-Eagle *Spizastur melanoleucus* (Vieillot, 1816) or *Spizaetus melanoleucus* (Vieillot, 1816)

Physical Description

The Black-and-White Hawk eagle is a fairly large raptor, but a small eagle, measuring 56-61 cm in length, with a wingspan and a weight of 850 g. It is a very conspicuous bird, distinguished by its white head, black crown aquiline silhouette and feathered tarsus (Brown & Amadon, 1968, 658; Tate, 2020). There are black features around the eyes, between the eye and the beak, and the crest at the back of the head. The back, mantle and barring at the tips of the flight feathers are black, and grey tipped tail has grey bars. The bill is black, the cere is reddish orange, the legs yellow and the eyes light yellow. The immature has lighter wing covert edges and one more greyish bar on the tail. The Black and White Hawk eagle is easily distinguished from other species by the white head and black crown, and the shorter tail than some other hawk eagle species (Brown & Amadon, 1968; Tate, 2020). Illustrations of the appearance, range, and eggs of this species are shown in Figures 2.17a to 2.17f.

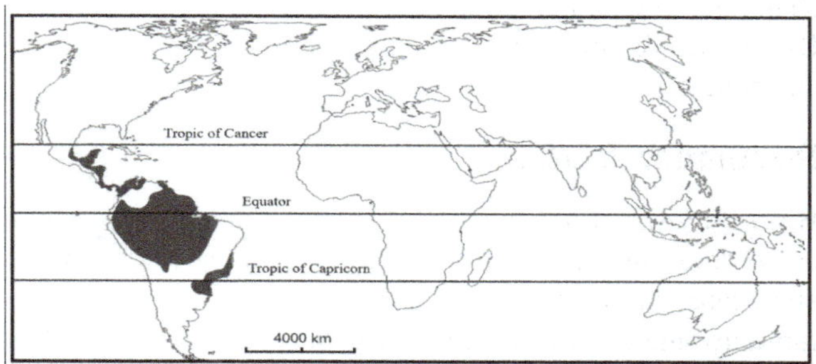

The Black eagle ranges through parts of Central America and intoColumbia, the Barzilian Amazon and Ecuador.

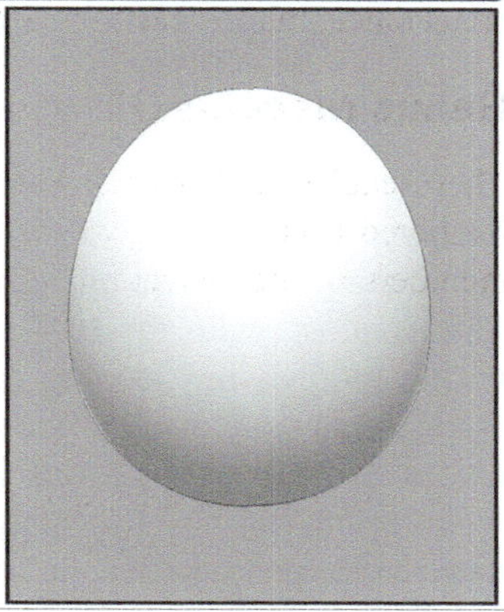

The egg is usually plain white.

Fig. 2.16i. Black Hawk eagle range and egg

Classification

Global Raptor Information Network (2021ad) points out that "the taxonomic position of this species has always been a subject of debate." It has been placed in the Genus *Spizastur* (by the majority of authors in recent times) but a few authors have placed it in the Genera *Spizaetus*, *Hieraaetus*, or *Buteo*. More recent molecular studies, such as those of Helbig et al. (2005) and Lerner and Mindell (2005), used the DNA sequences of mitochondrial and nuclear genes, and their findings indicated that *Spizastur* is closely linked to the other New World species of *Spizaetus*. The recommendation was that *Spizastur* should be merged with *Spizaetus*. Other authorities, such as the AOU Committee on Classification and Nomenclature and the AOU's South American Classification Committee agreed with this change (Banks et al., 2007; Remsen et al., 2009). Banks et al. (2007, 1112) state that "genetic studies (Helbig et al., 2005) indicate that the genus *Spizastur* should be merged into *Spizaetus*, and that the species *melanoleucus* is closely related to *S. ornatus*." They further note "delete the heading for the genus *Spizastur*; move the citation for that generic name to the synonymy of *Spizaetus*, below the citation for that name. Move the account for *S. melanoleucus* to follow that for *S. ornatus*, with the heading: *Spizaetus melanoleucus* (Vieillot) (Banks et al., 2007, 1112).

Foraging

Tate (2020) describes the Black and White Hawk eagle as a "poorly known large raptor of Neotropical forests." Brown and Amadon (1968, 688) argue that the general habits of this species are "little known." Foraging may include flight over mixed landcover, such as forest, open vegetation and rivers (Crease, 2009). Brown and Amadon (1968, 688) further note that the perching posture is "very erect and aquiline, with the flat head", but "little is known of its food" although probably birds and mammals. For example, in Argentina, prey in riverine areas included cormorants and mergansers, while in Mexico the range included non-riverine areas and presumably the prey range would be wider. Global Raptor Information Network (2021ad) gives its foraging areas as "lowlands and middle elevations, frequently edges, gaps, and heavily shaded clearings in wet gallery forest and sometimes isolated stands of wet forest remnants in savannas." Foraging involves soaring above the canopy, or flying from perch to another (Slud, 1964). It may dive into the canopy after prey (Willis, 1988). Thiollay (2007) described it as an aerial hunter in the upper canopy, in forests including disturbed areas and swamps. Records indicate that it attacks birds in fruiting trees in French Guiana. Robinson (1994) cited attacks on prey from the forest cover in Peru. Voous (1969) remarked that the large size of the claws suggests that this species preys on large mammals, but this does not coincide with its reported prey choices. It may also enter savanna (Sick, 1993).

Prey includes medium sized birds. These include toucans (Family *Ramphastidae*, Vigors, 1825), wood-quails (Genus *Odontophorus*, Vieillot, 1816), oropendolas (Genus *Psarocolius*, Wagler, 1827), caciques (Genus *Cacicus*, Lacepede, 1799) and small mammals, reptiles, and amphibians. Regional case studies have reported various type of prey. These include Aracari toucans (Genus *Pteroglossus*, Illiger, 1811) in Panama (Wetmore, 1965), Blue-headed parrots (*Pionusmenstruus*, Linnaeus, 1766), hoatzins (*Opisthocomushoazin*, Müller, 1776), Capuchin monkeys (*Sapajusapella*, Linnaeus, 1758), Squirrel monkeys (Genus *Saimiri*, Voigt, 1831), Russet-backed oropendolas (*Psarocolius angustifrons,* Spix, 1824) and Crested oropendolas (*Cacicusoseryi*, Deville, 1849) in Peru (Robinson, 1994), Brazilian merganser (*Mergusoctosetaceus*, Vieillot, 1817) in Argentina (Partridge, 1956), Little chachalacas (*Phasianus motmot*, Linnaeus, 1766) in Surinam (O'Shea, 2005) and Collared Aracari (*Pteroglos sustorquatus*, Gmelin, 1788) and Scaled pigeons (*Patagioenas speciosa*, Gmelin, 1789) in Belize (Russell, 1964). A white woodpecker (*Melanerpes candidus*, Otto, 1796) was also taken in Brazil (Olmos et al., 2006).

Breeding

The White and Black Hawk eagle's courtship display is comparatively "less elaborated" than those of the other species of *Spizaetus* (Zillio, 2017, 11). The current species' display involves "pairs gliding together, low vocal activity, chasing, and diving culminating with grasping" (Zillio, 2017, 11; see also Ferguson-Lees & Christie, 2001; Canuto, 2008b; Kohler & Rezini, 2013). This contrasts with those of the Ornate and Black Hawk eagles, which involve "elaborated flight displays including intense vocalization, diving, looping, grasping, food-touching and rolling-over" (Vannini, 1989; Ferguson-Lees & Christie, 2001). Further assessments were made of the post-display behavior of the breeding pairs of the three species mentioned above. Zillio (2017, 11) did not record "any call from Black-and-white Hawk eagle before or after copulation. Indeed, the Black and white Hawk eagle seems to be a very shy species, detected rather by visualization than vocalization." By contrast, there was "vocal communication between the pair before copulation for both Black and Ornate Hawk eagle" (Whitacre et al., 2012a,b).

Nests are large cup-shaped structures made of sticks usually on the top of tall forest emergents, for example, in Panama (Strauch, 1975) and Honduras (Anderson, 2004). The latter nest was about 1-1.5 m wide and 1 m tall, made of smaller twigs than those in nearby Ornate Hawk eagle nests (Anderson et al., 2004). These appear to serve as vantage points, due to their location high in the trees (Bierregaard, 1994; Canuto, 2008; Phillips, 2009). For example, the Honduras nest recorded by Anderson (1999) and Anderson et al. (2004) was located at the top of a 35- to 40-m tall balsamo (*Myroxylon balsamum* L.) located on a ridge. The tree in Panama was about 40 m high in the crown of a quipo tree (*Cavanillesia platanifolia* (Humb. & Bonpl.) Kunth) on a 300 m high ridge in dry forest (Strauch, 1975).

The laying dates for the Black and White Hawk eagle in Guyana are March and April (the dry season) (Brown & Amadon, 1968). One clutch of two eggs a creamy white ground color with dark brown, gray-lilac, and light brown spots (Kreuger, 1963). Observing a nest in southeast Brazil, Canuto (2008, 610) note that "In this study the observations of courtship and chick-rearing occurred during the dry season in southeast Brazil. The chick fledged at the beginning of the rainy season. No observations of breeding activities were recorded in the following year. The nest was inactive from September to November 2007." Philips (2009, 10) recorded a nest with a female incubating and feeding chick in February in the forest of Belize. Anderson et al. (2004) recorded nests on 25 January 1999 in eastern Honduras. Strauch (1975) recorded eagles building a nest on 17 September 1972 in Panama. In Guyana, eggs were collected in March and April (Amadon & Brown, 1968).

Population Status

BirdLife International (2016) classifies the Black and White Hawk eagle as Least Concern. The range is very large, over 20,000 km^2 and does not have a "declining or fluctuating range size, habitat extent/quality, or population size and a small number of locations or severe fragmentation" (BirdLife International, 2016). The population decline is also slow, and the population size appears not to be below 10,000 mature individuals. BirdLife International (2016) notes that "The population is suspected to be in decline owing to ongoing habitat destruction and human disturbance. Although it is able to tolerate some habitat degradation and human disturbance" (see also Ferguson-Lees and Christie, 2001).

In individual countries, there are different assessments. In Brazil, it is assessed as Vulnerable in the State of Rio de Janeiro (Bergallo et al., 2003), while nearby in Paraguay it is regraded as Near Threatened (del Castillo & Clay, 2005). In Ecuador, it was assessed as rare but not at risk (Ridgely & Greenfield, 2001). Further north, it is Endangered in Guatemala (Eisermann and Avendaño, 2006). In Mexico, it is classified as Endangered (Instituto Nacional de Ecología), but there are arguments for its adaptability (Gallardo del Ángel & Montejo, 2002). Zilio (2017, 2) referring the southern Atlantic Forest, Brazil, notes that the Black-and-white Hawk "is rare or at least uncommon although it could be underestimated because usually soars high and vocalizes less often than other hawk-eagles" (see also Willis, 1988; Ferguson-Lees & Christie, 2001; Menq, 2015). Illustrations of the appearance, range, and eggs of this species are shown in Figures 2.17a to 2.17f.

Fig. 2.17a. Black-and-white Hawk eagle perching [With permission from Jonathan Beilby]

Fig. 2.17b. Black-and-white Hawk eagle in flight [With permission from Mike Nelson-Birdtour ASIA]

Fig. 2.17c. Black-and-white Hawk eagle in flight [With permission from Gail Hampshire]

Fig. 2.17d. Black and White Hawk eagle habitat [Author]

Fig. 2.17e. Black-and-White Hawk eagle habitat [With permission from Rômulo Ferreira]

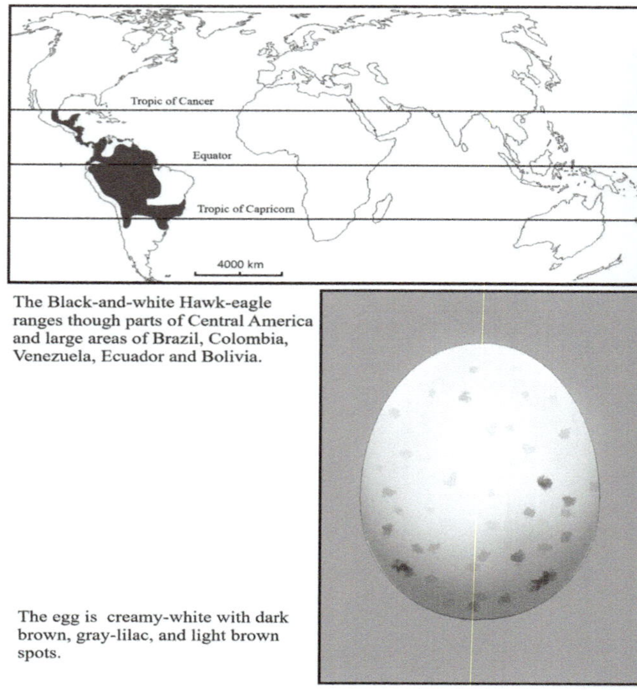

Fig. 2.17f. Black-and-White Hawk eagle range and egg

2.19 Ornate Hawk-Eagle *Spizaetus ornatus* (Daudin, 1800)

Physical Description

The Ornate Hawk eagle is a rare species that ranges across Central and South America (Howell & Webb, 1995; Van Perlo, 2009; Whitacre & Jenny, 2013; Fagan & Komar, 2016; Vallely & Dyer, 2018). It measures 56 to 68.5 cm in length, with a wingspan of 117 to 142 cm and a weight of 960 to 1,650 g (Howell & Webb, 1995; Iliff, 2020). There are two subspecies: *S.o. vicarious* and *S.o. ornatus* (Friedmann, 1935; Brown & Amadon, 1968; Ferguson-Lees & Christie, 2001). BirdLife International (2016) defines this eagle as a "large, brown-and-white hawk eagle. Black crown and occipital crest, with chestnut on the sides of the head and hindneck extending to sides of breast. Rest of underparts are white. Tail has three greyish brown bars. Yellow legs." Iliff (2020) describes the distinctive features of this species as a pointed black crest, and a black crown. These contrast with the bright chestnut coloration of the head, sides of the neck and breast. The white underparts are white, barred with black. The chin, throat and upper breast are white, sometimes black feathers on the sides of the throat and neck. There are black streaks on both the chestnut and white parts of the breast, black back feathers (with brown bases and a few white tips) and white bases of the wing quills. The upperparts are blackish to dark brown, with brown bases to the feathers which also have narrow white tips (Brown & Amadon, 1968, 704).

The wing feathers have variable shades of brown barring, and some parts of the underparts, including the wings and the primaries' tips have black and white barring. The underwing feathers are pale grey with four black stripes. The long, dark brown to blackish tail is rounded, with broad black bands. The tail is marked by a brownish grey marbled tip, and three wide bands colored whitish below. The legs are feathered, with the white streaked with black. The feet are bright yellow, and the eyes are bright orange to yellow. The black bill contrasts with the pale blue grey lores and yellow cere, which has a yellow gape edge (Brown & Amadon, 1968, 704; Iliff, 2020). The immature Ornate Hawk eagle is pale feathered, with a shorter crest, and whitish to white head, breast, and belly. There is narrow black barring on flanks and the feathered legs, and whitish underwing coverts. The barring on the body plumage and banding on the flight feathers are faint to barely visible (Iliff, 2020). The two subspecies have slight variations in adult plumage. *S.o. ornatus* has a head more strongly cinnamon-colored, with lighter ground coloration and fewer markings on the head and undersides. *S.o. vicarious* is darker in coloration, with a deeper more rufous color around the neck, and denser and darker markings in the plumage and broader tail bands (Friedmann, 1935; de Schauensee, 1950). Hence, Brown & Amadon (1968, 704) note that the *S.o. ornatus* subspecies has paler plumage than *S.o. vicarious*.

The Ornate Hawk eagle may be compared with several, possibly similar species. The defining distinction of the adult is its prominent, long crest, the chestnut coloring on the head and neck, black and white barring on the underside, the short, rounded wings and long tail (Brown & Amadon, 1968; Ferguson-Lees & Christie, 2001). The smaller Black-and-White Hawk eagle has been described as "perhaps most similar, sharing the large size and extensively white underparts" (Iliff, 2020). However, the Black-and-white Hawk eagle does not have flank and thigh barring, and has black coloration around the eye, darker upperparts, a red cere and no crest. The Black-and-white Hawk-Eagle has white feathers on the shoulder ridge, while the Ornate Hawk eagle has a dark shoulder, and broader wings and a shorter-tail (Iliff, 2020).

There are two larger crested eagles that share the Ornate Hawk eagle's range (Guiana Crested eagle and Harpy eagle), but both are "much larger and thicker-legged at rest and their heads are grayish or dark, not whitish; adult Harpy eagle has a black breastband. In flight, note the larger size and even broader wings of the two larger eagles, as well as the prominently barred tail of immature Ornate Hawk eagle (the tails of Crested and Harpy are faintly barred from below)" (Iliff, 2020). The adult Ornate Hawk eagle in flight may resemble the Black Hawk eagle but has slightly broader wings and a more rounded, longer tail (Howell & Webb, 1995). The Black Hawk eagle, although similarly sized also has darker underparts, and the immature Black Hawk eagle has a duskier breast, belly, and underwings. The immature Black-and-chestnut eagle may resemble the Ornate Hawk eagle in size, shape and the crested head, but the head is brownish, the underparts are unbarred, and the back and upperparts are whitish (Iliff, 2020). It may also resemble the much smaller, immature, Gray-bellied hawk (*Accipiter poliogaster*, Temminck, 1824), which has similar rufous coloring on the neck sides, a dark crown, and a dark back and barred flanks. However, in addition to being much smaller, the Gray-bellied hawk's taris are unfeathered and lacks a crest (Iliff, 2020). The smaller Gray-headed kite (*Leptodon cayanensis*, Latham, 1790) has some similarities in the coloration of the upper plumage (pale head and dark upperparts), but it lacks the barring on the side and thighs, feathered tarsi, crest, pale iris, and longer legs (Iliff, 2020).

The immature or juvenile Ornate Hawk eagle resembles the Black-and-White Hawk eagle but differs in having a crest and barred thighs. This juvenile may also be distinguished from the Black eagle (*Ictinaetusmalaiensis*, Temminck, 1822) by its whiter, un-streaked underside, and from the juvenile Isidor's eagle (or Black-and-Chestnut eagle, *Spizaetus isidori*, Des Murs, 1845) by its barred thighs (Brown & Amadon, 1968). Iliff (2020) note that "immatures are pale when seen from below, but the faint barring on the flanks and thighs is unlike most other pale-bellied raptors." The voice may be a "clear, thin, high pitched screaming, e.g. 'Wheeoo-whee- pee pee" or "wheeoowhee peep"; other calls include an accelerating "laughing series of notes" followed by a decelerating with "an exciting sounding set of notes that rise and fall in small crescendos" (Brown & Amadon, 1968). Illustrations of the appearance, range, and eggs of this species are shown in Figures 2.18a to 2.18i.

Classification

The Ornate Hawk eagle has two subspecies: the nominate *Spizaetus ornatus ornatus*, ranging from eastern Colombia to the Guianas, southwards and westwards to eastern Ecuador, and on to northeastern Peru, northern and eastern Bolivia and Brazil, Paraguay and northern Argentina Trinidad; and *Spizaetus ornatus vicarius*: ranging from Central Mexico southwards through Central America, western Colombia, western Ecuador and northwestern Peru (Global Raptor Information Network, 2021ae). Helbig et al. (2005), Lerner and Mindell (2005), and Haring et al. (2007), using DNA sequences of mitochondrial and nuclear genes, showed that Ornate Hawk eagle and the Black hawk eagle are less related than expected, and the Ornate Hawk eagle is closer to the Black-and-White Hawk eagle and Black-and-chestnut eagle than to the Black hawk eagle (Lerner et al., 2017; Global Raptor Information Network, 2021ae). Recent studies showed that the Ornate Hawk eagle was closely related to the Black-and-white Hawk eagle (a factor for the change in name of the genus of the latter from *Spizastur* to *Spizaetus*) and to the Black-and-Chestnut eagle (a factor for the change in name of the genus of the latter from *Oroaetus* to *Spizaetus*). The Ornate Hawk eagle and the Black-and-Chestnut eagle are sister species (Helbig et al., 2005; Haring et al., 2007; Lerner et al., 2017).

Foraging

The Ornate Hawk eagle frequents tall, dense rainforest, lowlands, and middle elevations, and in some cases the coniferous forests of southern Mexico and Belize (Crease, 2009). It soars on flat wings, but less often than the Black Hawk eagle, and usually does not perch in open landcover (Slud, 1964; Murphy, 2004). Its habitat and foraging habits may vary regionally. For example, in Panama, Canuto (2008) found that it avoided forest edges in Soberania National Park, possibly due to sensitivity to habitat fragmentation and human presence, unlike the Black Hawk eagle. In Venezuela, Naveda-Rodríguez (2007) found that most presence records were from the lowland grasslands, some areas including cattle ranches, submontane, and montane forest (between 600 and 1,500 m), meadow forest, and in some cases humid cloud forest, semideciduous forests and semideciduous riverside forest, all at varying elevations. In French Guiana, it primarily frequented primary undisturbed forest. In this it differed from the Black Hawk eagle, which more often frequented disturbed patches (lower and denser) in mature forest stands (Thiollay, 2007).

Hunting is principally within forest, between perches, with prey caught in trees and on the ground. In general, the bulk of the prey taken is birds (Iliff, 2010; Whitacre & Jenny, 2013). Prey species taken include birds, including larger birds such as curassows (Subfamily *Cracinae*, Vigors, 1825) and guans (Subfamily *Penelopinae*, Huxley, 1868), chachalacas (Genus *Ortalis*, Merrem, 1786), guans (Subfamily *Penelopinae*, Huxley, 1868), Little Blue herons (*Egrettacaerulea*, Linnaeus, 1758), domestic fowls (*Gallus gallus*, Linnaeus, 1758), New World quails (Family *Odontophoridae*, Gould, 1844) and New World parrots (Subfamily *Arinae*, Gray, 1840) (Whitacre & Jenny, 2013). Even a Black vulture (*Coragyp satratus*, Bechstein, 1793) has been recorded (Brown & Amadon, 1968, 706). Mammals may also be taken, such as the kinkajou (*Potos flavus*, Schreber, 1774) and cacomistle (*Bassariscus sumichrasti*, Saussure, 1860). Large prey weighing as much or more than the eagle may also be taken, such as the Central American agouti (*Dasyprocta punctata*, Gray, 1842), large procyonids (raccoons, coatis, kinkajousetc. of the Family *Procyonidae*, Gray, 1825) and monkeys, green iguana and the great tinamou (*Tinamus major*, Gmelin, 1789), all of which may weigh more than 4 kg (Russell, 1964; Lyon & Kuhnigk, 1985; Clinton-Eitniear et al., 1991; Robinson, 1994; Treves et al., 2001; Naveda-Rodríguez, 2004; Iliff, 2010; Whitacre & Jenny, 2013).

Other prey taken includes wood-rats on the ground, and gallinules (Genus *Gallinula*, Brisson, 1760) and wood rails (Genus *Aramides*, Pucheran, 1845) in shallow water in Peru (Robinson, 1994). Other prey included a Horned Guan (*Oreophasis derbianus*, Gray, 1844) in Mexico (Gómez de Silva, 2006); a Great curassow (*Crax rubra*, Linnaeus, 1758) and a snake in shallow water in Belize (Russell, 1964); another Great Curassow in Tikal, Guatemala (Whitacre & Jenny, 2013); a Crested guan (*Penelope purpurascens*, Wagler, 1830), Plain chachalaca (*Ortalisvetula*, Wagler, 1830), tinamou (Family *Tinamidae*G. R. Gray, 1840), Gray-headed dove (*Leptotila plumbeiceps*, Sclater & Salvin, 1868) and a leaf-nosed bat (Family *Phyllostomidae*, Gray, 1825) in Guatemala (Kilham, 1978; Lyon & Kuhnigk, 1985). Other findings indicated that the bulk of prey taken (38%) was about equal proportions of birds and mammals, species including tree squirrels (*Sciurus deppei*, Peters, 1863, and *Sciurus yucatanensis* (Allen, 1877), coatimundi (*Nasua narica*, Linnaeus, 1766), Ocellated turkey (*Meleagris ocellata*, Cuvier, 1820, Great curassow, Great tinamou, and Keel-billed toucans (*Ramphastos sulfuratus*, Lesson, 1830) (Madrid et al., 1992).

In Panama, a Ringed kingfisher was taken (Wetmore, 1965). In Peru, bird species taken included Purple gallinules (*Porphyrio martinicus*, Linnaeus, 1766) and a Gray-necked Wood rail (*Aramides cajaneus*, Müller, 1776), Brown-mantled tamarins (*Leontocebus fuscicollis*, Spix, 1823), oropendolas (Genus *Psarocolius*, Wagler, 1827) and Yellow-rumped Caciques (*Cacicuscela*, Linnaeus, 1758), Pale-winged trumpeters (*Psophialeucoptera*, Spix, 1825), macaws and parakeets at a mineral lick, domestic chickens (*Gallus gallus domesticus*, Linnaeus, 1758), hoatzins (*Opisthocomushoazin*, Müller, 1776), White-throated toucan (*Ramphastostucanus*, Linnaeus, 1758) (Robinson, 1994; Giudice, 2007). Mammals were Guianan Squirrel monkey (*Saimiri sciureus*, Linnaeus, 1758). mixed monkey troops, Silky anteaters (*Cyclopes didactylus*, Linnaeus, 1758) and wood-rats, and reptiles included Colubrid snakes (Family *Colubridae*, Oppel, 1811). In Venezuela, prey included higher proportions of mammals than birds, with rodents the commonest prey taken, including the Red-tailed squirrel (*Notosciurus granatensis*, Humboldt, 1811), Cotton rat (Genus *Sigmodon*, Say & Ord, 1825) and Red-rumped agouti (*Dasyprocta leporine*, Linnaeus, 1758). Birds taken included parrots (Genus *Pionus*, Wagler, 1832) and oropendolas. In one case, iguana (*Iguana iguana*, Linnaeus, 1758) remains were found in a nest (Naveda-Rodríguez, 2004). In Brazil, prey in one study included birds (63.5), small mammals (32.7%), and reptiles (4.1%). Birds taken, in order of frequency included tinamous (Family *Tinamidae*, Gray, 1840), macaws (Tribe *Arini*, Gray, 1840), guans (Subfamily *Penelopinae*, Huxley, 1868), a toucan, a chachalaca, and four smaller birds. Mammals taken included agoutis, opossums (*Didelphidae*, Gray, 1821), an acouchi (Genus *Myoprocta*, Thomas, 1903), and a porcupine (Family *Erethizontidae*, Bonaparte, 1845). Reptile prey included a snake and a lizard (Klein et al., 1988; 1993; Sick, 1993).

For more information on the diet of the Ornate snake eagle and detailed of the ecology of their prey species, see these references, which represent a small section of the information on this topic: (Dickey & Van Rossem, 1938; Friedmann & Foster, 1950; Russell, 1964; Wetmore et al., 1965; Kilham, 1978; Hartshorn et al., 1984; Lyon & Kuhnigk, 1985; Brown & Amadon, 1986; Trail, 1987; Klein et al., 1988; Robinson, 1988; Flatten et al., 1989, 1990; Ferrari & Ferrari, 1990; Heymann, 1990; Eitniear et al., 1991; Madrid et al., 1992; Marti et al., 1993; Peres, 1993; Rylands, 1993; Robinson, 1994; Ferguson-Lees & Christie, 2001; Treves et al., 2001; Gleason & Norconk, 2002; Boinski et al., 2003; Hilty, 2003; Greco et al., 2004; Naveda-Rodríguez, 2004; Gómez de Silva, 2006; Dornas & Pinheiro, 2007; Baur, 2008; Dunning, 2008; Wheeler, 2010; Joenck et al., 2011; Acosta-Chaves et al., 2012; Janson et al., 2012; Phillips & Hatten, 2012; Whitacre & Jenny, 2013; Bidner, 2014; Jones & Dorward, 2014; Retana et al., 2016; Duquette et al., 2017; Zilio, 2017; Cima et al., 2018; Dolotovskaya et al., 2019; Lopez et al., 2019; Iliff, 2020).

Breeding

The breeding displays include soaring, circling, short dives with half folded wings, and looping flight patterns. Displays may involve calling, gliding in tight circles, talon grabbing, occasionally touching and diving (Slud, 1964; Ffrench et al., 1991; Howell & Webb, 1995; Ferguson-Lees & Christie, 2001). Nesting is in trees (Brown & Amadon, 1968, 707). The nest is a bulky platform of sticks, lined with fresh green leaves, usually placed in a fork near the top of a large tree, or on branch jutting from the main trunk (Global Raptor Information Network, 2021ae). Average nests are about 1 to 1.25 m across and about 50 cm deep but may be smaller or considerably larger (Klein et al., 1988; Ferguson-Lees & Christie, 2001; Naveda-Rodríguez, 2004; Whitacre & Jenny, 2013). Nests are sometimes re-used in subsequent years, hence can grow in size. Nests have been recorded in the branches of large trees, such in a crotch of a large pine tree in Oaxaca, Mexico; 18.5 m in a tree in Belize (Russell, 1964); and two nests, one 20 m high on a branch 5 m off the trunk of a 30 m tall Silk Cotton tree, and another 26 m high on a 5 m branch off the canopy of a 30 m Silk Cotton tree, both in Guatemala (Lyon & Kuhnick, 1985). Madrid (1992) record nests or nesting attempts in Guatemala, averaging 19.3 to 28.3 m in Spanish cedar (*Cedrela odorata* L.), fig (Genus *Ficus* L.), Honduran mahogany (*Swietenia macrophylla*, King), a dead tree (*Calophyllum Brasiliense*, Cambess.) and a black olive tree (*Bucidabuceras*, L.). The nests were located on forks of branches jutting several meters from the main trunk. Other nests were located in Peru (21 m high, on a 32 m *Casearia javitensis* Kunth) (Giudice, 2007); in Panama (28 m high, on a 31 m *Prioriacopaifera*, Griseb.) (Brown & Amadon, 1968); in Venezuela (29 m high on a tall *Pachira insignis* ((Sw.) Savigny) (Naveda-Rodríguez, 2004); and in Brazil 37 m high on an emergent 46 m tall Hymenaeatree (Genus *Hymenaea* L. sp. (Legumniosae)) (Klein et al., 1988). The general similarity of nesting locations is evident from the widely spaced research locations (Lyon & Kuhnigk, 1985; Klein et al., 1988; Flatten et al., 1989; Flatten et al., 1990; Madrid et al., 1992; Greeney et al., 2004; Naveda-Rodríguez, 2004; Giudice, 2007; Phillips & Hatten, 2012).

The laying period begins in early March during the beginning of the dry season (Phillips & Hatten, 2012). The dates may vary regionally. For example, in Guatemala, Lyon and Kuhnigk (1985) found an active nest with an egg on 1 April 1979. The Peregrine Fund's Maya Project in the early 1990s in Guatemala, found that egg laying occurred over a 203-day span (November 5 to April, 26), the hatching period occurred over a 194-day span (December 25 to July 7), and the nestling period occurred over a 76-day span (May 12 to October 8) (Flatten, 1989, 1990; Madrid et al., 1991, 1992). The dispersal of immature birds from natal occurred over a 128-day span (March 10 to July 16). The breeding period from laying to fledging was averaged 124.6 ± 7 days (118-126 days) (Madrid et al., 1992).

The incubation period has been recorded as 44-46 days (Phillips & Hatten, 2012). Madrid et al. (1992) found the incubation period was 47, 47, and 50 days at three studied nests in Guatemala. This ensures that "the chick hatches just prior to the wet season, when avian prey is most abundant (Global Raptor Information Network, 2021ae). The ornate hawk-eagle female generally lays one white egg, which may have faint reddish brown spots (Lyon & Kuhnigk, 1985; Klein et al., 1988; Madrid et al., 1992; Ferguson-Lees & Christie, 2001; Naveda-Rodríguez, 2004; Giudice, 2007). Both males and females may participate in nest building, incubation, hunting and feeding (Madrid et al., 1992). Fledging occurred at 70-84 days in early July, the beginning of the rainy season (Phillips & Hatten, 2012). Madrid et al. (1992) noted that the nestling period in Guatemala was variable, from 66 to 93 days and averaging 79 days.

Other findings were for Panama, with an active nest on December 29, with a juvenile fledging in May (Brown and Amadon, 1968). Canuto (2008) also records paired aerial courtship displays in August. In Ecuador, a male eagle was recorded in courtship display on June 21, 1985 (Robbins & Ridgely, 1990). In Peru, an egg was laid in a nest on February 18, with a chick present on May 4 (Giudice, 2007). The well-developed chick was still in the nest on May 4. In Venezuela, active nests were recorded in February and March, with the breeding season beginning in November, hatching in March and still locally present in August (Ffrench, 1991). In Venezuela: Naveda-Rodríguez (2004) found a nest with an egg on May 31, 2001, with hatching on June 16 or 17. In Manaus, Brazil, laying was estimated in early August, with copulations on June 24 (Klein et al., 1988), and hatching between September 17-19, and fledging April 16-23.

Population Status

The Ornate Hawk eagle has a wide range, occurring from Central Mexico south through Central America to South America west of the Andes to northwestern Peru, east of the Andes across the Amazon Basin south to northern Argentina. In the Caribbean, it also occurs in Trinidad. The population has been estimated at in the range of 10,001 to 100,000, this including all adults and immatures at the beginning of the breeding season. BirdLife International (2016) describes the population as decreasing, with 13,300 to 33,300 mature individuals. Basically, this number of mature individuals equates to 20,000-50,000 individuals. BirdLife International further adds that "this species is suspected to lose 21.8-40.4% of suitable habitat within its distribution over three generations (56 years) based on a model of Amazonian deforestation" (see also Soares-Filho et al., 2006; Bird et al., 2011). There is still hope for survival, as this eagle is adaptive; "although the species is susceptible to hunting, it survives in fragmented landscapes" (BirdLife International, 2016). Therefore, it is classified as Near Threatened, but was Least Concern before 2012. BirdLife International (2016) justified this classification: "Based on a model of future deforestation in the Amazon basin, and habitat loss and persecution elsewhere within its extremely large range, it is suspected that the population of this species will decline by 25-30% over the next three generations, and it has therefore been uplisted to Near Threatened."

Fig. 2.18a. Ornate Hawk eagle perching [With permission from Ian Gill]

Fig. 2.18b. Ornate Hawk eagle perching [With permission from Ian Gill]

Fig. 2.18c. Ornate Hawk eagle perching [With permission from Tristanba]

Fig. 2.18d. Immature Ornate Hawk eagle [With permission from Jorge Eduardo Chinchilla Arroyo]

Fig. 2.18e. Immature Ornate Hawk eagle [With permission from Xyomara Carretero-Pinzon]

Fig. 2.18f. Ornate Hawk eagle in forested habitat [With permission from Nick Athanas]

Fig. 2.18g. Ornate Hawk eagle habitat [Author]

Fig. 2.18h. Ornate Hawk eagle habitat [With permission from Nadav Hurvitz from Pexels]

Global Raptor Information Network (2021ae) acknowledges these facts, stating that the Ornate Hawk-eagle is "generally fairly common in suitable habitat throughout its vast range, but at risk in many areas from habitat loss." Threats include shooting: as noted above it is considered a threat to domestic fowl. because it is regarded as a threat to poultry. The conservation classification of the Ornate Hawk eagle varies regionally. For example, in Mexico it is classified as Endangered (Winker, 1987; Semernat, 2002). In Ecuador, the population has undergone a serious decline, because of deforestation (Ridgely & Greenfield, 2001). A similar situation obtains in Brazil, were deforestation has contributed to population decline and a classification as Endangered in Minas Gerais (a state in Southeastern Brazil, which is ranked as second most populous, the third by gross domestic product (GDP), and the fourth largest by area in Brazil) (Machado et al., 1998). In Argentina, deforestation has had the same effect on the population of the Ornate Hawk eagle (Márquez Reyes et al., 2000). In Paraguay, it is classified as Vulnerable (del Castillo and Clay, 2005). In El Salvador, it was last reported in the early 1980s (Thurber et al., 1987; Komar, 2002). Illustrations of the appearance, range, and eggs of this species are shown in Figures 2.18a to 2.18i.

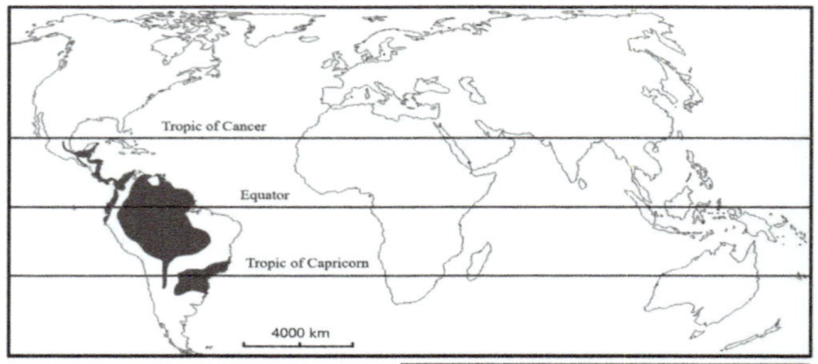

The Oranate Hawk-eagle ranges through Central America, into Colombia, Brazil, Ecuador to Venezuela and Uruguay.

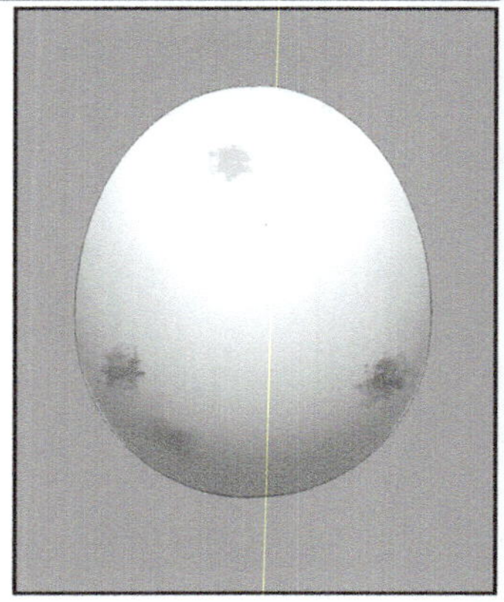

The egg may range in color from bluish-white to white with brown splotches.

Fig. 2.18i. Ornate Hawk-eagle range and egg

2.20 Black-and-chestnut Eagle or Isidor's Eagle (*Spizaetus isidori*, Des Murs, 1845, or *Oroaetus isidori*, Des Murs, 1845)

Physical Description

The Black-and-chestnut eagle is described as "a very large raptor, closely allied with the similar hawk eagles (*Spizaetus*) of the lowlands" (Bierregaard et al., 2020). Vuilleumier and Ewert (1978) list this species as the largest member of the genus *Spizaetus*, to the point that the smallest males are larger than largest female Ornate Hawk eagles. Brown and Amadon (1968, 714) describe this species as "a big, stalwart eagle, chestnut and black, with white across in flight near the ends of the broad wings." eBird (2020) describes it as a "rare, crested eagle of middle- to upper-elevation cloud forest." Aráoz (2017, 79) describe this eagle as "one of the least known Neotropical raptors throughout its entire range (see also Valdez & Osborn, 2004). However, recent studies have contributed information (Bierregaard et al., 2020). The Black-and-chestnut eagle measures 63 to 74 cm in length, with a wingspan of 147 to 166 cm (Vuilleumier & Ewert, 1978; Ferguson-Lees & Christie, 2001). Brown and Amadon (1968, 714) also described this species as larger, more robust in form, with longer less rounded wings than those of the other eagles in the Genus *Spizaetus*, when the Black-and-chestnut eagle was outside that genus as Isidor's eagle (*Oroaetus isidori*, Des Murs, 1845).

The Black-and-Chestnut eagle's most distinctive visual image is of the adult plumage, which is mostly black with chestnut coloration on the breast and belly, and the large white patches in wings and at base of tail, and chestnut in underwing (eBird, 2020). The nape feathers are long, forming a crest, and the head, throat and uppers are glossy black, with shiny shafts on the crown feathers. The crest is raised, but is depressed when the bird is excited, unlike other species such as *Morphnus, Harpia* or S*tephanoaetus* (Brown & Amadon, 1968). The throat is black. The pale grey tail has paler mottles and a black band over the terminal third. The primaries have glaucous greyish coloration on the outer webs, dull grey mottles on the inner webs, white bases and lower surfaces, and some black barring on the outer feathers. The primaries are edged with black, and the secondaries colored black above, and pale grey and black below, with black tips. The mostly dark chestnut underparts, with black streaks which are denser on the breast where they merge with the black throat. The sides are blacker, with longer black feathers on the thighs, and chestnut underwing coverts, with black-shaft streaks. The eye is orange/yellow, the bill is black with a horn-colored base and the legs are yellow (Brown & Amadon, 1968).

The juvenile plumage is different, with whitish head and body and buffy coloration bellow and gray on the back and wings (Vuilleumier & Ewert, 1978). eBird (2020) writes that the "juvenile is amazingly different with dark back, white underparts, and strongly banded wings and tail. Juvenile similar to pale morph Crested eagle, but wing and tail banding is thinner, and lacks barring on breast and belly." The crown is whitish buff or grey crown (with dark shaft streaks) and head sides, a crest with blackish feathers, which pale edges and tips, multicolored black and wing coverts (pale brownish grey and dark shaft streaks).

The marbled grey to whitish tail is tipped with white and sub-terminally banded black with three narrower bars, and almost white below. The primaries are marbled glaucous grey, lighter colored near the inner edges, with dark to black barring below. The secondaries are darker and white tipped. The underparts are white, with brown shafts streaks with pale chestnut patches on the breast sides. The eye is initially blue, but it changes to yellowish after about three months. During the transition to adult coloration, blackish feathers appear on the shoulders, chestnut breast, side and lower abdomen feathers.

Brown and Amadon (1968) give an example of specimen whose mantle changed from brownish feathers on the lower back, wing coverts and tail, to black coloration as it neared adulthood. The juvenile may be confused with the Ornate Hawk eagle, or the more similarly sized Black-and-White hawk-eagle. However, the current specie is larger with little range overlap. The call of male birds returning to the nest may be described as "p-e-e-e-eo", with a similar reply from the female, with a similar female reply. The alarm cry is "chee - chee – chee" or kee -kee – kee" (Brown & Amadon, 1968). Illustrations of the appearance, range, and eggs of this species are shown in Figures 2.19a to 2.19f.

Classification

Formerly, the Black-and-Chestnut eagle was classified as Isidor's eagle *Oroaetus isidori* (Des Murs, 1845), in the monotypic genus *Oroaetus*. Brown and Amadon (1968) noted that this genus contains one species Isidor's eagle (*Oroaetus isidori*, Des Murs, 1845). They treated this species as Black-and-chestnut eagle or Isidor's eagle (*Spizaetus isidori*, Des Murs, 1845, or *Oroaetus isidori*, Des Murs, 1845). However, Voous (1964) argued that it did not differ sufficiently from *Spizaetus* to be placed in a separate genus. Initially, Amadon (1982) moved this eagle to the Genus *Spizastur* and later to the Genus *Spizaetus* (Amadon & Bull, 1988). This move was supported by later studies, such as those by Helbig et al. (2005), Lerner and Mindell (2005), and Haring et al. (2007), which used molecular methods based on DNA sequences of mitochondrial and nuclear genes. These found that *Oroaetus* is closely related to *Spizaetus* and the Black-and-Chestnut eagle is a sister species to the Ornate Hawk eagle and the Black-and-white hawk eagle. This recommendation to place the current species in *Spizaetus* was adopted by the AOU's South American Classification Committee (Remsen et al., 2009). Most recent publications have renamed this species as the Black-and-chestnut eagle, within the genus *Spizaetus* (Vieillot, 1816), under the binomial name *Spizaetus isidori* (Des Murs, 1845). This book will adhere to this classification.

Foraging

The Black-and-Chestnut eagle's preferred habitats are dense Andean subtropical and sometimes temperate rainforest, including clearings in the forest (Valdez, 2001; Marquez & Delgado, 2010). Recorded habitats include "undisturbed wet cloud forest on slopes at middle and montane elevations, rarely lower and occasionally as high as páramos in Venezuela" (Global Raptor Information Network, 2021af). Although it frequents elevations between 1,500 and 2,800 m, it has been recorded up to 3,500 m (Fjeldså & Krabbe, 1990; Bierregaard, 1994). Foraging usually includes soaring over forest, but may also include perching inside the forest, frequent perching in tall trees, and flight in tree crowns. Davies et al. (1997) report that in Peru, this eagle frequents the less humid west-facing slopes, although these areas tend to have higher human populations, with more pressure on suitable habitats. Large prey is taken, including woolly monkeys (Genus *Lagothrix* spp. É. Geoffroy, 1812) and guans (Subfamily *Penelopinae*, Huxley, 1868) and smaller prey such as squirrels (Family *Sciuridae*, Fischer de Waldheim, 1817, *Mesosciurus*, J.A. Allen, 1915) and poultry (*Gallus gallus domesticus*, Linnaeus, 1758) (Aráoz et al., 2017; Bierregaard et al., 2020). In one case in the Andes in northwestern Ecuador in 1999, a Black-and-Chestnut eagle was observed attacking three Sickle-winged Guans (*Chamaepetes goudotii*, Lesson, 1828) (Freile & Chavez, 2004).

The ranges or former ranges of the Black-and-Chestnut eagle are increasingly modified by human activities such as agriculture and other factors for deforestation. Restrepo-Cardona et al. (2019) point to this general problem when raptors intrude into human life spaces (sometimes after people have intruded into the raptors foraging spaces) and create conflicts with people. These authors studied the diet of the Black-and-chestnut eagle in four different Andean landscapes in Colombia and found that domestic fowl were comprised 9.3%–36% in number and 12.2% to 37.1% in total biomass of prey consumed (Restrepo-Cardona et al., 2019). The suggested factor for this is habitat loss; "Furthermore, though we have no field data on this, it is very likely that deforestation occurs together with an increase in hunting pressure by people on the most important prey items of this eagle, thus exacerbating the effect of deforestation" (Restrepo-Cardona et al., 2019, 8). The main argument is that common prey items, such as medium-sized mammals including kinkajous, coatis, monkeys, and armadillos, and birds such as guans are hunted by Andean rural people for food or for their pelts and this must be controlled (see also Dickman et al., 2016; Zuluaga & Echeverry-Galvis, 2016).

Other studies point to similar results, with two case studies for Colombia. Zuluaga & Echeverry-Galvis (2016, 113) estimated an annual loss of one to two domestic fowl per local household, and 57% of the owners surveyed claiming losses. Therefore, the respondents "would prefer not to have the eagle breeding close to their houses (< 10 km) and would consider killing eagles if they preyed upon more than five domestic fowl individuals a year." Restrepo-Cardona et al. (2020) also found that in the studied rural areas, the "perception of *S. isidori* by the inhabitants was largely negative and influenced by different socio-demographic factors such as gender, chicken (*Gallus gallus*) ownership, and chicken management." Partly as a consequence of this negative, 47 eagles were shot, 16 were captured (three for illegal trafficking), and in unrelated incidents two were electrocuted on high-tension wires. They conclude that "the persecution of *S. isidori* occurs as retaliation or as a preventative measure against chicken predationand is a significant cause of mortality of this species in Colombia" (Restrepo-Cardona et al., 2020).

Breeding

Nests are "a huge structure of sticks placed high in an emergent tree" (Global Raptor Information Network, 2021af). They may measure about two meters by one meter, usually on oak (*Quercus colombiana*, Cuatrec.) or camelo tree (*Drimys winteri*, J.R. Forst. & G. Forst.). Nesting begins in February and March, "probably only one egg is laid", with fledging in 90 days (Brown & Amadon, 1968, 715). Aráoz (2017) describe one case study in montane cloud forest in the Yungas forest (Dr. Manuel Belgrano Department, Jujuy Province) in northwestern Argentina): "During the last week of May or first week of June, the female laid a single egg, and the nestling hatched on 9 July 2014. The eaglet fledged around the third week of October, although it frequently used the nest for roosting and feeding at least until the end of November. Therefore, breeding occurred during the coldest and driest months of the year, and the fledgling became independent during the rainy season (December to April)."

Population Status

The Black-and-Chestnut eagle global population size, based on the number of adults and immatures at the start of the breeding season, is 101 to 1,000 individuals (Ferguson-Lees & Christie, 2001). BirdLife International (2016) assessed this species as Endangered, with the number of mature individuals estimated at 250-999. It states that "this species has been uplisted to Endangered on the basis that its declining population is estimated to be very small, with fewer mature individuals than previously thought. The destruction of its montane forest habitat, as well as direct human persecution, are inferred to be driving a continuing decline." Global Raptor Information Network (2021af) counters that "this species may not be rare as suspected within its altitudinally narrow, but linearly extensive (6,000 km) range, but it is too poorly studied to make any informed judgment about its actual status. Despite its large size, it is difficult to detect because it occurs in such remote and rugged mountain areas." BirdLife International (2020) states that "it is thought to be rare and patchily distributed, but its status is very poorly known. The total population is precautionarily estimated to include fewer than 1,000 mature individuals (Thiollay, 1994; Ferguson-Lees and Christie, 2001). The main issues concern the loss of its primary forest habitats (Global Raptor Information Network, 2021af).

The Black-and-Chestnut eagle is non-migratory (Global Raptor Information Network, 2021af). It ranges through the Neotropical region, including the Andean slopes from northwestern Venezuela and northeastern Colombia (including the Coastal ranges of northeast Colombia (Santa Marta Mountains), northern Venezuela (Carabobo and Aragua, and the Sierra de Perijá, on the border with Columbia) and southwards through Ecuador and Peru to west-central Bolivia and northwestern Argentina (Bierregaard et al., 2020; Global Raptor Information Network, 2021af). Here it frequents middle- to upper-elevation cloud forest (eBird, 2020). In Colombia, it occurs locally, usually in mountainous areas (sometimes as high as 1,600-2,800 m) such as parts of the Cuchilla de San Lorenzo, Santa Marta Mountains, and in the western Andes near Cerro Munchique, Cauca (Hilty & Brown, 1986; see also Renjifo et al., 2002). In some areas, such as the La Planada Natural Reserve, it ranges from the upper tropical zone (1,400-2,300 m) to the subtropical zone (1,400-2,300 m) and the montane zone (2,300 and higher) (Orejuela, 1987, see also Márquez et al., 2005).

In Ecuador, the Black-and-Chestnut eagle is mostly locally uncommon in the subtropical and temperate zones of the Andes slopes, usually from higher altitudes (e.g. 1,500-3,000 m) with some occurrence from lower slopes (1,100 m) in the Cordillera de Cutucú. Overall, this species is rare in Ecuador, possibly as low as 100 pairs (Ridgely & Greenfield, 2001). BirdLife International (2020) estimates fewer than 250 mature individuals in Ecuador. Parker and Bates (1992) made a record in the upper Rio Saguayuo valley of Amboró National Park, Depto. Santa Cruz, on 17 August 1989, which at the time was the southernmost record for the Bolivian Andes. In Peru, it occurs in mountains, but is rare on the eastern slope of the Andes from 1,750-2,500 m (rarely up to 3,500 m) (Clements & Shany, 2001). In Venezuela, it is an uncommon resident in rainy, humid montane forest of the Andes, mostly above 1,200 to 2,500 m (Hilty, 2003). Birdlife International (2020) estimated 200 mature individuals in the country. Vuilleumier and Ewert (1978) described it as a visitor, in the Venezuelan páramos (páramo is an ecosystem of the regions above the continuous forest line, but below the permanent snowline) and recorded sightings at about 2,800 m. Aráoz (2017) also had a record in Argentina. Illustrations of the appearance, range, and eggs of this species are shown in Figures 2.19a to 2.19f.

Genus *Clanga* (Adamowicz, 1858)

There are three species in the Genus *Clanga*: The Greater Spotted eagle (*Clanga clanga*, Pallas, 1811), Lesser Spotted eagle (*Clanga pomarina*, C.L. Brehm, 1831), and Indian Spotted eagle (*Clanga hastata*, Lesson, 1831) (Lerner et al., 2017). These

Fig. 2.19a. Black-and-Chestnut eagle perching [With permission from Luis Olarte]

Fig. 2.19b. Immature Black-and-Chestnut eagle [With permission from Brian and Irene Hampson]

Fig. 2.19c. Black-and-Chestnut eagle in flight [With permission from Robert Davidson]

Fig. 2.19d. Immature Black-and-Chestnut eagle in flight [With permission from Nick Athanas]

Fig. 2.19e. Black-and-Chestnut eagle habitat [With permission from Robert Davidson]

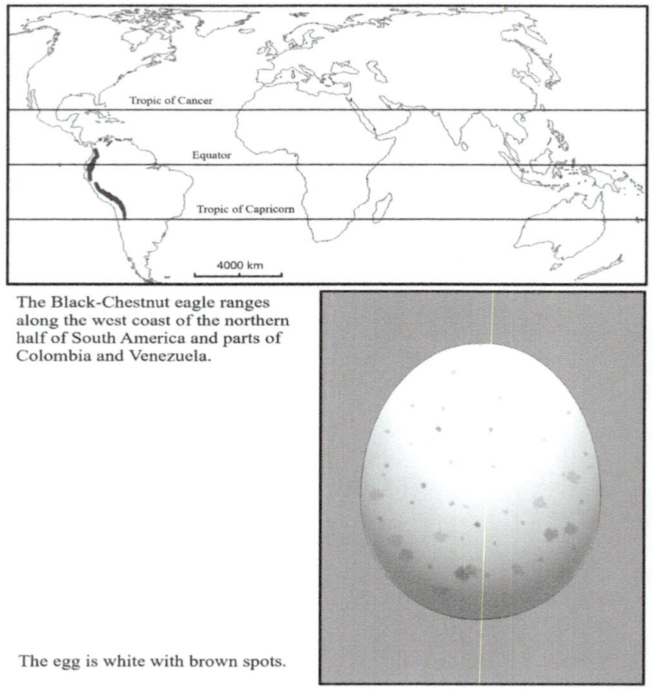

Fig. 2.19f. Black-and-Chestnut eagle range and egg

Spotted eagles were previously included in the Genus *Aquila*, but are now placed in their separate genus, *Clanga.* Lerner et al. (2017, 301), in justifying the creation of the Genus *Clanga*, argue that they applied the "recently resurrected genus *Clanga* for the spotted eagles (previously *Aquila* spp.) to resolve the paraphyly of the genus *Aquila* such that the clade including the Booted Eagle (*H. pennatus*), Little Eagle (*H. morphnoides*), Pygmy Eagle (*H. weiskei*), Ayres's Eagle (*H. ayresii*) and Wahlberg's Eagle (*H. wahlbergi*) can remain in the genus *Hieraaetus*." As used, the Biology Dictionary (2020) defines Paraphyly as descriptive of a group of animals that includes a common ancestor and some, but not all, of the descendants. Therefore, certain species within *Aquila* were moved to the Genus *Clanga* and the Genus *Hieraaetus.* Wells & Inskipp (2012) also supported moving the three spotted eagles from *Aquila* into a new genus *Aquiloides* (later *Clanga*), as the Genus *Aquila* was "too large, too unwieldy and too diverse" and "there are three distinct clades, *Aquila, Hieraaetus* and *Aquiloide*" (Clark, 2012, 295).

In some cases, as noted by Lerner & Mindell, 2005). Helbig et al. (2005) and Wink & Sauer-Gürth (2004), a clade includes the Long-crested eagle and the Spotted eagles. The Black eagle may also be included (Griffiths et al., 2007; see also Lerner et al., 2017, 308). The new *Aquiloides* (Wells & Inskipp, 2012), which was proposed for the spotted eagles, was canceled in favor of Genus *Clanga* (Adamowicz, 1858) (Gregory & Dickinson, 2012) to avoiding paraphyly in the Genus *Aquila.* Lerner et al. (2017, 310) described the *Clanga* spotted eagles as "mid-sized eagles with broad, rounded wings, shortish, rounded tails, and round nares. The secondaries of adults are longer than in juveniles, unlike *Aquila* in which the reverse is true" (see also Wells & Inskipp, 2012). They further note "large gapes but smaller, shorter bills than in *Aquila*, and the crania are long and broad; their tarsi are long, but their feet are small" and that "adults are uniform brown, and immatures have white-spotted upperwing coverts. Spotted eagles primarily give short, sharp yelping calls in series" (Lerner et al., 2017, 310; see also Parry, 2001; Rasmussen & Anderton 2012).

2.21 Greater Spotted Eagle *Clanga clanga* (formerly *Aquila clanga*) (Pallas, 1811)

Physical Description

The Greater Spotted eagle is a "moderate-sized eagle about half as big again as a Common buzzard" (Brown & Amadon, 1968: 650). It measures about 59 to 71 cm in length, with a wingspan of 157 to 179 cm, and a weight of 1.6–2.5 kg (male 1.5–1.9 kg, female 1.8–2.5 kg) (Dunning, 1992; Ferguson-Lees & Christie, 2001; Meyburg et al., 2020). The plumage is dark brown all over, with a purplish gloss to the upperparts and duller and paler underparts. The crown and nape feathers have paler edges. Sometimes the upper and undertail coverts have white edgings. The wing quills and tail are blackish, purplish brown, the primaries have blackish tips and there are slightly darker bars on the tail. The cere and feet are yellow, the eyes brown. Albinos exist, these are pale cream to golden buff coloured with dark wing coverts (Brown & Amadon, 1968). The immature birds are darker than adults and lack the pale patch on the nape. The scapulars, back and rump have broad whitish central streaks, changing to round spots on the feather tips. The upper tail coverts are also white tipped, creating a blackish bar above the tail base. The wing coverts and secondaries have whitish tips. Over four years, the whitish markings vanish, and the adult appearance appears.

The Greater Spotted eagle resembles the Tawny eagle but is slimmer, and narrower winged and more square-tailed than the latter. Albino forms of the Greater Spotted eagle may resemble the pale forms of the Tawny eagle. The Greater Spotted eagle may also resemble the Lesser Spotted eagle, but here is more dark purplish brown coloration, and more white markings on the tail base. The immature Greater Spotted eagle have a large, white spots on the back and wings, and lack a pale nape patch, which enables distinction from the immature Lesser Spotted eagle (Brown & Amadon, 1968). Meyburg et al. (2020) reports "widespread hybridization between the Greater Spotted eagle and the Lesser Spotted eagle in Europe." The Greater Spotted eagle may winter in the territory of the Indian spotted eagle and differs from the latter by its darker plumage, lighter eye color and stronger spotting in juveniles, and slightly superior size (Dunning, 1992; Ferguson-Lees & Christie, 2001; Väli & Lõhmus, 2004; Rasmussen & Anderton, 2012). Meyburg et al. (2020) writes that this species "can be quite vocal year-round, but especially in breeding season and when several birds are present." The voice is a "kluck-kluck" or "tyuck-tyuck." It is lower pitched, but more ringing than that of the Lesser Spotted eagle, but higher than the call of the Tawny eagle and the Eastern Imperial eagle (Brown & Amadon, 1968). Illustrations of the appearance, range, and eggs of this species are shown in Figures 2.20a to 2.20j.

Classification

The Greater Spotted eagle was previously included in the Genus *Aquila*. The main findings of recent research concerned the close relationship between the Greater and Lesser Spotted eagles (Global Raptor Information Network, 2021ag). One study using molecular phylogenetics analysis (using DNA sequences of one mitochondrial and two nuclear genes) found that the Lesser Spotted eagle and Greater Spotted eagle form a monophylum (a hierarchical group of organisms, which includes an ancestor species and all its descendants), including the Long-crested eagle. The recommendation was that they all be included in the Genus *Lophaetus*, including the Indian Spotted eagle. The genus would then be named *Clangus*, rather than *Clanga*, for "gender reasons" (Global Raptor Information Network, 2021ag; see also Helbig et al., 2005).

Other studies supported the merging of the Genus *Lophaetus* with the Genus *Aquila* (e.g., Wink & Sauer-Gürth, 2004). Välï's (2002) and Lerner and Mindell's (2005) molecular phylogenetic analysis study supported this position (and the close relationship between the Greater and Lesser Spotted eagles), and also favored the inclusion of the Black eagle of Asia, usually placed in the Genus *Ictinaetus*. These two studies found that the Greater and Lesser Spotted eagles were genetically distinct, but still closer than any other paired species in the studied group. As noted above, there was also hybridization between the Greater and the Lesser Spotted eagles in some places where they are sympatric (occurring together) (Lohmus & Välï, 2001; Dombrovski, 2002; Meyburg & Meyburg, 2007). As noted above later writings (Clark, 2012; Gregory & Dickinson, 2012; Wells & Inskipp, 2012; Lerner et al., 2017), justified the creation of the Genus *Clanga* for the spotted eagles.

Breeding

The Greater Spotted eagle begins the breeding season with high level soaring, and usually the male diving at the female. It breeds in wet deciduous forests near humid meadows, bogs, marshes, and other forms of wetlands up to about 1,000 m elevation (Meyburg et al., 1999). The nest is large stick construction, usually built below the canopy in a tree, more often in a deciduous than a coniferous forest, sometimes the old nest of another species with a similar nest. On rare occasions some nests may be located in low bushes in steppe areas, in some cases in East China on the ground. Laying takes place in late April or early May in Europe, and February to June (usually in April and May) in India. The clutch size usually averages two, and sometimes one or three white to greyish white eggs with some reddish-brown, grey brown or grey markings. Incubation lasts for about 42 days, with a nestling period of 63-67 days (Meyburg et al., 1999). Research findings indicate that usually only one chick survives, due to cainism (the killing of weaker or younger nestling by stronger or older siblings) (Meyburg & Pielowski, 1991).

Foraging

The Greater Spotted eagle generally frequents different types of landcover over its wide range, but usually near forest types, including "large wet forests bordering humid meadows, bogs, marshes and other wetlands including mangroves" (Meyburg et al., 2020); see also BirdLife International, 2017; Global Raptor Information Network, 2021ag). Asian habitats and wintering areas may be taiga forest, forest-steppe with wetlands, and wetlands and farmland/forested areas are preferred in winter (Brazil, 2009). During migration in Malaysia, it frequents farmed areas and even garbage dumps (Wells, 1999). Other recorded sightings include farmland, mixed farm and forest landcover, sewage farms, irrigated areas, open moorland, riparian forests in lowland and forest steppes, grassland and wetlands, in the Czech Republic, Belarus, Russia, Kazakhstan, Israel, Oman, Saudi Arabia, Eritrea, Bangladesh and the Malay Peninsula (Smith, 1957; Flint et al., 1984; Shirihai, 1996; Wells 1999; Kren, 2000; Lobley, 2007; Wassink & Oreel, 2007; Monirul & Khan 2009).

Hunting may be from perches or soaring flight. Prey species include small mammals (the Northern Water vole (*Arvicola amphibius*, Linnaeus, 1758), up to rabbits and hares), birds (including waterfowl), amphibians, lizards, snakes, frogs, small fish, carrion, and insects (Flint et al., 1984; Meyburg et al., 1999). Records exist of carrion eating (especially agriculturally poisoned rats) in Malaysia (Wells, 1999). There is also intra- and interspecies kleptoparasitism (Mallalieu, 2007). Graszynski et al. (2002) record mostly small mammals, frogs, and a few birds, while Galushin (1962) reported more birds and Glotov (1959) recorded a vast majority of mammals and few birds. Ivanowski (1996a, 1996b, 1999) reported large numbers of frogs taken in these studies in Eurasia. Karyakin et al. (2009) reported hemipodes, fish, gulls, ducks, water voles and even a Grey Heron (*Ardea cinerea*, Linnaeus, 1758) in the Altai region. In Greece, Handrinos and Akriotis (1997) recorded feeding on the Evros Delta on dead or wounded waterfowl, and even an attack on Redshanks (*Tringa tetanus*, Linnaeus, 1758). In South East Asia, food sources included dead fish, frogs and rats, domestic ducks, dead Lesser Whistling ducks (*Dendrocygna javanica*, Horsfield, 1821), and dead rails (Family *Rallidae*, Rafinesque, 1815) and bitterns (Subfamily *Botaurinae*, Reichenbach, 1850) (Round & Jukmongkol, 2003; Mallalieu, 2007). There were also kleptoparasitic attacks on other raptors, including anosprey, an Eastern Marsh harrier (*Circus spilonotus*, Kaup, 1847), a Black kite (*Milvus migrans*, Boddaert, 1783) and a juvenile Steppe eagle (Wells, 1999; Mallalieu, 2007).

Population Status

BirdLife International (2017) writes that "this species has a small population which appears to be declining owing to extensive habitat loss and persistent persecution. It is therefore listed as Vulnerable." It ranges from Central Europe through Russia to southern Siberia and northeastern China, with a few breeding populations in northern Iran and north-central India. Wintering areas are southern Europe to northeastern and eastern Africa, the Middle East, Asia Minor to northern Pakistan to southern and eastern China and Japan, to Indochina to the Malay Peninsula and Sumatra (BirdLife International, 2017). It is also rare or vagrant in many countries (Table 2.2).

The Greater Spotted eagle is long distance migrant (Bildstein, 2006; Global Raptor Information Network, 2021ag). Resident in northern and middle Eurasia, it migrates south to southern Europe, northern and eastern Africa and southwestern and southern Africa from mid-September to October, and returning late February to mid-April, usually returning later in the east ranges in Eurasia, from Finland and the Baltic States, eastwards through Siberia. Records exist from Estonia through France to Spain, from October to November (2008), and wintered in Spain and returned to Estonia in April 2009 and repeated the same journey the following season (Seeber et al., 2009). There is another record of migration crossing the Red Sea (Ash & Atkins, 2009). Dates recorded include fall migration (mid-September to the end of November, mainly in October) and spring migration (end of February to mid-April, mainly in March and the beginning of April) in Israel (Shirihai, 1996); autumn migration (September to November, peaking in October) and spring migration (mid-April to late May) in the Czech Republic (Kren, 2000); September-October (fall migration) and March-April (spring migration) in Lebanon (Ramadan-Jaradi et al., 2008); late August to late October (fall) and mostly March-April (spring) in Kazakhstan (Wassink & Oreel, 2007). In Ethiopia, migration occur between 10 October to 28 March (Ash & Atkins, 2009) and in the Malay Peninsula, the migrations are between 25 October and 5 May (Wells, 1999).

Ferguson-Lees and Christie (2001) estimated the global population (the number of adults and immatures at the start of the breeding season) in the range of 1,001 to 10,000 individuals. BirdLife International (2017) records the population as a European breeding population of 810-1,100 breeding pairs, which equates to 2,430-3,300 individual eagles. As Europe represents 25-49% of the global range, the global population size is estimated at 5,000-13,200 individuals. Rough estimates at the national levels include 100-10,000 breeding pairs and fewer migrating and wintering individuals (50-1,000 each) in China; less than 50 each migrating and wintering individuals in Taiwan and 100-10,000 breeding pairs and 50-1,000 migrating individuals in Russia (Brazil, 2009). BirdLife International (2017) notes that the Greater Spotted eagle "is suspected to have undergone at least a moderately rapid decline over the last three generations as a result of habitat loss and degradation throughout its breeding and wintering ranges, together with the effects of disturbance, persecution and competition with other predators. In Europe, the population size is estimated to be decreasing by 50-79% in 49.8 years (three generations) and by at least 20% in 33.2 years (two generations)." This species is classified as vulnerable across its range by many authors, for example, in Bulgaria: (Iankov, 2007), Korea (National Institute of Biological Resources, 2011; Choi & Park, 2012), Russia (Kalyakin & Voltzit, 2006), Ukraine (Domashevsky, 2005), Czech Republic (Mrlik, 1998), Greece (Handrinos & Akriotis, 1997), Armenia (Adamian & Klem, 1999), Thailand (Mallalieu, 2007) and the Malay Peninsula (Wells, 1999).

The Greater Spotted eagle has a much wider range distribution than the Lesser Spotted eagle, but it has a smaller population, as can be seen above it is rare in many parts of its range. Population declines have been reported through its western ranges due

Table 2.2. Population status of the Greater Spotted eagle

Country	Status	Reference
Great Britain	Vagrant (1860-1915)	Brown & Grice, 2005; Dudley et al., 2006
Netherlands	Vagrant (1800-1997)	Bijlsma et al., 2001
France	Migrant (1997-2008)	Reeber & CHN, 2009.
Hungary	Passage Migrant	Gábor, 1998
Czech Republic	Vagrant, migrant	Kren, 2000
Russia	Common to rare	Flint et al., 1984
Gibraltar	Vagrant	Garcia, 2005
Romania, Bulgaria	Breeding passage migrant/visitor	Roberts, 2000; Nankinov, 1992, 2004
Azerbaijan	Common migrant/winter resident	Patrikeev, 2004
Armenia	Rare year-round visitor, migrant	Adamian & Klem, 1999
Greece	Rare winter visitor, commonest eagle	Handrinos, 1987; Handrinos & Akriotis, 1997
Turkey	Rare winter visitor, passage migrant	Kirwan et al., 2008
Syria	Passage migrant, visitor, resident?	Baumgart et al., 2003; Murdoch & Betton, 2008
Lebanon	Irregular and scarce passage migrant	Ramadan-Jaradi et al., 2008
Israel	Uncommon passage migrant	Shirihai, 1996
Morocco	Accidental visitor, two records	Thévenot et al., 2003
Algeria	Vagrant, two modern records	Fançois, 1975, Isenmann & Maoli, 2000
Tunisia	Rare passage migrant	Isenmann et al., 2005
Libya	Vagrant three or four records	Bundy, 1976
Egypt	One record, accidental	Demey, 2006
Cameroon, Chad	Vagrant, one record, two records	Borrow & Demey, 2001
Sudan	Rare Palearctic migrant	Nikolaus, 1987; de Bont, 2009
Ethiopia, Eritrea	Rare Palearctic winter visitor	Ash and Atkins, 2009
Somalia	Palearctic migrant	Smith, 1957
Kenya, Tanzania	Rare Palearctic vagrant	Stevenson & Fanshawe, 2002
Zambia, Malawi	Palearctic migrants, tracked	Dowsett et al., 2008; Meyburg & Meyburg, 2005
Kuwait, Bahrain	Rare visitor, passage migrant	Gregory, 2005; King, 2006
Oman, UAE	Common[1], rare[2] migrant/visitor	Eriksen et al., 2003; Richardson et al., 2003
Saudi Arabia Yemen	Rare winter visitor, Yemen rarer	Ash, 1998; Porter et al.; 1996; Lobley, 2007
Kazakhstan	Rare breeding/passage migrant	Gavrilov & Gavrilov, 2005; Wassink & Oreel, 2007
Iran	Common winter visitor	Scott & Adhami, 2006
India,	Rare migrant/breeder, winter visitor	Rasmussen & Anderton, 2012
Nepal, Bhutan	Rare winter visitor/passage migrant	Inskipp & Inskipp, 1991; Spierenburg, 2005
Bangladesh	Rare winter visitor, more on coast	Monirul & Kham, 2009
Hong Kong	Rare, local winter visitor	Carey et al., 2001
Thailand, Myanmar	Winter visitor, commoner Thailand	Mallalieu, 2007; Smythies, 1986
Vietnam	Rare passage migrant	Global Raptor Information Network, 2021ag
Malay Peninsula	Sparse to common visitor/migrant	Jeyarajasingam & Pearson, 1999; Wells, 1999
Korea	Rare winter visitor	Gore & Won, 1971; Choi & Park, 2012

Source: Global Raptor Information Network (2021ag). Note: UAE is the United Arab Emirates.

to human activities such as forestry and wetland drainage, nest disturbance and robbing, shooting, and deliberate and accidental poisoning, especially zinc phosphide (Zn_3P_2) (Meyburg et al., 1999; Wells, 1999). There is also the hybridization with Lesser Spotted eagles. As noted by Väli et al. (2010, 725) "Hybridization is a significant threat for endangered species and could potentially even lead to their extinction. This concern applies to the globally vulnerable Greater Spotted eagle, a species that co-occurs, and potentially interbreeds, with the more common Lesser Spotted eagle in a vast area of Eastern Europe." In Estonia and nearby countries the range of the Lesser Spotted eagle is advancing eastward at the expense of the Greater Spotted eagle (Lõhmus & Väli, 2001; Väli, 2011). Conservation plans have been assessed and designed in Europe (Meyburg et al., 1999; Bird & Symes, 2009). Illustrations of the appearance, range, and eggs of this species are shown in Figures 2.20a to 2.20j.

Fig. 2.20a. Greater Spotted eagle perching [With permission from Vikas Chavan]

Fig. 2.20b. Greater Spotted eagle head [With permission from Quazi Sanjeed]

Fig. 2.20c. Greater Spotted eagle [With permission from Lopamudra Barman, Kolkata, India, (www.flickr.comphotoslopamuidrabarman)]

Fig. 2.20d. Greater Spotted eagle taking flight [With permission from Vikas Chavan]

Fig. 2.20e. Immature Greater Spotted eagle [With permission from Philip Roberts]

Fig. 2.20f. Immature Greater Spotted eagle 2 [With permission from Philip Roberts]

Fig. 2.20g. Greater Spotted eagle in flight [With permission from Mike Nelson-Birdtour ASIA]

Fig. 2.20h. Greater Spotted eagle in flight [With permission from Kimmo Järvinen]

Fig. 2.20i. Greater Spotted eagle habitat, Russia [Credit: Max Sudyin-unsplash]

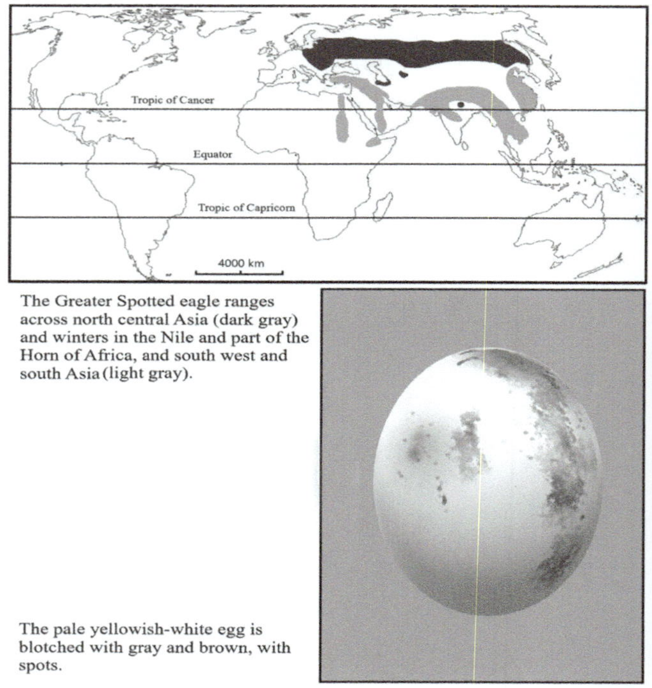

Fig. 2.20j. Greater Spotted eagle range and egg

2.22 Lesser Spotted Eagle *Clanga pomarina* (Brehm, 1831) formerly *Aquila pomarina* or *Lophaetus pomarinus*

Physical Description

The Lesser Spotted eagle is a large bird of prey, but eBird (2020) describes it as a "Medium-sized, compact eagle" which is "smaller and paler than the similar Greater Spotted Eagle." Brown & Amadon (1968) describe its appearance as that of a "small buzzard-sized eagle, differing from a buzzard in having longer wings, visibly longer in relation to the body length." It measures 55–67 cm, with a wingspan of 146–168 cm, and a weight of 1.0–2.2 kg. Its general description is of uniformly chocolate brown bird (Meyburg et al., 2016). The adult plumage is dull brown overall, but with small white markings on the upper tail coverts, which may be extensive enough in some cases to form a basal band on the tail. The wing quills are also darker than the body plumage. One distinguishing characteristic is the seventh primary which is "markedly shorter than the sixth and not deeply emarginated" (Brown & Amadon, 1968, 646). The upperwing coverts are usually paler than the flight feathers. The eye, cere and feet are yellow. The immature plumage is also brown like adult, but mostly a darker shade, with a yellow or lighter colored nape patch, and the barring on remiges and rectrices is clearer in flight. There is also a lighter basal tail band, white feathers on the primary coverts and some secondaries, forming a wing bar. The eyes are brown, and the cere and feet are yellow (Brown & Amadon, 1968; Ferguson-Lees & Christie, 2001; Meyburg et al., 2016).

The Lesser Spotted Eagle may resemble the Greater Spotted eagle, Indian Spotted eagle, Mountain Hawk eagle and Tawny eagle. Compared to the Greater Spotted eagle, the Lesser Spotted eagle usually is paler, "less purplish and less white at the base of the tail" (Brown & Amadon, 1968: 646). In flight it appears slimmer and smaller than the Greater Spotted eagle and the Tawny eagle. The contrast between the pale underwing coverts and darker flight feathers is less pronounced than in the plumage of the Greater Spotted eagle. In addition, the current species has a square tail, a less emarginated wing tip and less visible seventh primary. It is also paler and smaller than the Steppe eagle. It is paler and larger, and more uniformly colored than Wahlberg's eagle, where it winters in the latter's range. The Lesser Spotted eagle differs from the Mountain Hawk eagle by the less visible barring on remiges and tail (Meyburg et al., 2016). The immature Indian Spotted eagle also lacks the white markings on the coverts (Brown & Amadon, 1968; Ferguson-Lees & Christie, 2001; Meyburg et al., 2016).

The juvenile plumage spots on upperwing coverts are smaller and narrower than in the immature Greater Spotted eagle (Meyburg et al., 2016). The Lesser Spotted eagle also differs from other similarly sized, brown eagles, in that it frequents woodland and is less attracted to open water than the Greater Spotted eagles. Unlike the Tawny and Steppe eagles, it rarely frequents open steppe or plains, except during migration and hunting. Unlike the Greater Spotted eagle, in the Caucasus, it frequents the mountain forests up to 2,000 meters, and unlike the Greater Spotted eagle and Tawny it is in open plains and river valleys (Brown & Amadon, 1968). The normal cry is a "high yap or bark 'tjuck' (cheuk) or 'k-ek k-ek' or 'k-yeep k-yeep'" higher pitched and less vibrant than that of *A. clanga*, less deep than the Steppe and Imperial eagles. During nesting, mating or food solicitation, the male gives a whistle-like 'wiiek,' while the female gives cry 'psiiiiik' (Brown & Amadon, 1968). Illustrations of the appearance, range, and eggs of this species are shown in Figures 2.21a to 2.21g.

Classification

Formerly, the Lesser Spotted eagle was divided into two subspecies within the Genus *Aquila* (as *Aquila pomarina*, C. L. Brehm, 1831, before the development of the Genus *Clanga*) according to Brown and Amadon (1968); *A.p. pomarina* (the nominate subspecies) and *A.p. hastata* (this subspecies is now the Indian Spotted eagle, a distinct species). These authors argued that the second subspecies is similar in coloration but is larger and with longer legs. These differences have led some experts to consider it a separate species, *A. hastata*, "but to us the differences seen insufficient to warrant this" (Brown & Amadon, 1968; 646). Meyburg et al. (2016). BirdLife International (2016) wrote that "*Clanga pomarina* (del Hoyo & Collar, 2014) was previously placed in the genus *Aquila*" and *Aquila pomarina* (Sibley & Monroe, 1990, 1993) has been split into *A. pomarina* and *A. hastata* following Parry et al. (2002)." The close relationship of the Lesser Spotted eagle to the Greater Spotted eagle, is determined by the frequent hybridization of the two species (Lõhmus & Väli, 2001; Välï, 2002; Väli et al., 2010; Väli, 2011). It may also be distinguished from the Indian Spotted eagle, which has "brown vs yellow/amber iris (3); darker brown head having black shaft streaks (1); paler brown undertail coverts (1); and much larger gape (mean for females 60·1 mm vs 52·3 mm in length, 47·4 mm vs 37·4 mm in width; score at least 2)" (Meyburg et al., 2016).

Helbig et al. (2005) used molecular phylogenetic analysis of DNA sequences of one mitochondrial and two nuclear genes and found the tree spotted eagle species in a monophylum with each other and the Long-crested eagle. The recommendation from this study was the merging of these species and the Indian Spotted eagle (at that time in the Genus *Aquila* as *A. hastatus*) into the Genus *Lophaetus* "requiring the name change from *pomarina* to *pomarinus* for gender reasons" (Global Raptor Information Network, 2021ah). Other studies hinted at merging the Genus *Lophaetus* into the Genus *Aquila* (e.g., Wink & Sauer-Gürth, 2004). Lerner and Mindell (2005) while supporting these positions, also included the Black eagle in this clade, rather than in the usual placement in the Genus *Ictinaetus*. Wells and Inskipp (2012) also argued for the three spotted eagle species be placed together in the new Genus *Aquiloides*. Lerner and Mindell (2005) supported the studies mentioned above (Lõhmus & Väli, 2001; Välï, 2002; Väli et al., 2010; Väli, 2011) which showed a close relationship between the Lesser and Greater Spotted eagles, mostly through their mutual hybridization in shared ranges despite significant genetic differences (see also Lôhmus & Våli, 2001; Meyburg & Meyburg, 2007; Global Raptor Information Network, 2021ah).

Breeding

Breeding behavior begins with displays, including mutual (male and female) soaring above the breeding site. The male performs short steep undulations repetitively, with and without female presence. There is also "a winnowing display flight with wing tips rapidly vibrated" (Brown & Amadon, 1968). There is calling during mating, which occurs after the female selects a branch and the male joins her. Nests are in forest or forest edge trees, rather than in isolated trees in open country. The nests are usually in the middle part of the tree, rather than at the tree tops, the common location of Tawny eagle and Steppe eagle nests, and usually in deciduous trees. For example, in the Czech Republic, the nests were found to be in oak or beech trees, rarely in conifers (Kren, 2000). Nests are small initially (50 cm wide by 30 cm deep) but may eventually reach 90 by 60 cm. Nest materials are sticks lined with grass. The one to three, usually two eggs are white with varying amounts of reddish-brown spots and grey under-markings (Dal, 1954; Adamian & Klem, 1999; Meyburg et al., 2020). The incubation period at a nest in one study in Armenia was estimated at 43-45 days (Adamian & Klem, 1999). The laying period varies across its range. For example, April 15 to May 10 has been cited for Europe, and April 15 to July 10, but usually in May for India (Global Raptor Information Network, 2021ah). For Russia, it is cited as April or May (Flint et al., 1984). In Ukraine, laying occurred in late-April to early May, with hatching in early June, and fledging occurred in early August (Domashevsky, 2005). In the Czech Republic, breeding is between late April and early June (Kren, 2000). In Armenia, studies indicated laying beginning in late April to middle May (Dal, 1954; Adamian & Klem, 1999). Fledging occurs at the beginning of August, followed by migration (Brown & Amadon, 1968).

Foraging and Food

The Lesser Spotted eagle is mostly a forest or woodland species, but the specific ecosystem type may vary regionally (Brown & Amadon, 1968; Ferguson-Lees & Christie, 2001). For example, in the western Palearctic, it frequents moist lowland forests, sometimes near wetlands. In the Caucasian region, it frequents dry upland forest up to 2,200 m (Adamian & Klem, 1999). In Russia, it is recorded in forests and forest-steppes, river valleys, and near wet meadows (Flint et al., 1984). In the Czech Republic, it frequents any type of mature forest, mostly in highlands, sometimes lowlands, close to more open landcover such as meadows, clearings, farmland, and broad valleys (Kren, 2000). Similarly, it frequents forest patches near open landcover with agricultural lands or pastures (Roberts, 2000). Meyburg et al. (2001, 9) note that "there is little information about the species in Romania. According to the old literature the eagle has formerly been common and widespread." These authors further note that "the Lesser Spotted eagle is still widespread throughout the country and inhabits primarily deciduous woodlands in the mountains between 300 and 1600 m" (see also Meyburg, 1996 and Uhlig, 1996). In Greece, it frequents both wetlands and drier forests, generally in lowlands (100-300 m), but may range up to 1,200 m in Thessaly (Vlachos, 1989, Vlachos & Papageorgiou, 1996). In Israel, it frequents (including roosting) in woodland and more open countryside and may winter in woodland near cultivated landcover and water (Shirihai, 1996).

Wintering birds in Africa occur in any woodland areas where the annual rainfall exceeds 600 mm, probably derived savanna or denser vegetation. In some cases, it shares these contexts with Steppe eagles and Wahlberg's eagles. It may be a social bird, occurring singly or in flocks. Occurs singly or in small flocks, and extremely large concentrations are sometimes found (Dowsett-Lemaire & Dowsett, 2006). In Zambia, it has been recorded in various combinations of open woodland or wooded grassland (Dowsett et al., 2008). In Zimbabwe, it frequents densely wooded country, either singly, or in small, scattered groups, sometimes associated with Steppe eagles (Irwin, 1981; Hartley, 1998).

The Lesser Spotted eagle hunts from soaring and perching, and in some cases from ground using its relatively long legs. Prey includes small mammals such as mice, voles and small rabbits; birds such as young game birds; and lizards, large insects and frogs. Carrion is also taken, albeit rarely (Brown & Amadon, 1968). Alivizatos et al. (2020, 68) described the diet of these eagles in the Amvrakikos Wetlands National Park and found, the diet comprised 43% reptiles, 30% birds, 24% mammals, 3% amphibians and 1% arthropods. The most important prey animals were brown rat (*Rattus norvegicus*) at 19%, water snakes (*Natrix* spp.) at 11% and moorhen (*Gallinula chloropus*) at 11% of the prey composition. Alivizatos et al. (2020, 68) comment that "despite a wide availability of semi-natural and agricultural habitats around the immediate nesting areas, the species seems to have invested in foraging largely on water-dependent animals."

Dravecký et al. (2008, 1) conducted a study of the whole of Slovakia and found that prey selection was related to the range in the foothills and mountains of the Carpathian Arc. The findings indicated that the prey was composed of mammalian species (88.11%), birds (6.25) and Amphibia, Reptilia and Pisces together (4.89%). The commonest mammal prey species was the Common vole (*Microtus arvalis*, Pallas, 1778) (about 70%) followed by the European water vole (*Arvicola amphibius*, Linnaeus, 1758) (5%) and the Field vole (*Microtus agrestis*, Linnaeus, 1761). One mammal carnivore, the Least weasel (*Mustela nivalis*, Linnaeus, 1766) constituted less than one percent of the prey. Bird prey was mostly of the Order *Galliformes* (Temminck, 1820) including the Common pheasant (*Phasianus colchicus*, Linnaeus, 1758), the domesticated Jungle fowl (*Gallus gallus domesticus*, Linnaeus, 1758), Common quail (*Coturnix coturnix*, Linnaeus, 1758) and the Eurasian skylark (*Alauda arvensis*, Linnaeus, 1758) the first just above one percent, the others below one percent of prey. The commonest reptile taken was the Slowworm (a legless lizard) (*Anguis fragilis*, Linnaeus, 1758) (Dravecký et al., 2008). Similarly-sized animals are also taken in Armenia (Adamian & Klem, 1999), Greece (Vlachos, 1989, 1996), and in African countries, where winged termites may also be taken (Brooke et al., 1972; Irwin, 1981; Tarboton, 1990; Hartley, 1998; Parker, 2005a,b; Dowsett-Lemaire & Dowsett, 2006).

Population Status

The Lesser Spotted eagle is defined as a "complete long distance, trans-equatorial migrant" from mostly Eurasia, and that "except for occasional isolated individuals wintering in southeastern Europe and Asia Minor, the entire population winters in Africa" (Global Raptor Information Network, 2021ah; see also Hallmann, 1989; Pergantis & Handrinos, 1992; Simmons, 1997e; Radovíc et al., 2003; Parker, 2005a,b; Bildstein, 2006; Dravecký, 2009). Mostly non-resident in Europe, it migrates south in September to November to Central Europe, Africa and the Middle East, crossing the Mediterranean: "probably lower wing loading may enable it to cross large bodies of water more readily than other large raptors" (Brown and Amadon, 1968, 647). Knowledge on its winter quarters is scanty (ibid.). The northward migration occurs in March, and by April it reappears in the "German breeding headquarters, having travelled about 5,000 miles in six weeks" (ibid. 647).

The majority winters in eastern Africa, but some fly to westwards to Cameroon, Nigeria, and Mali, and southwards to southern Africa (Thiollay, 1977; Tarboton, 1990). For example, in Botswana, 10,000 to 15,000 Steppe and Lesser Spotted eagles were observed in woodland in Chobe National Park in January 1993 (Simmons, 1997e). In Mozambique, Parker (2005) reported sightings of over 200 birds in central Mozambique. Birds ringed or transmitter tagged have verified these transcontinental movements, and sightings conform the migrations mostly through Turkey and Greece (Benson et al., 1971; Irwin, 1981; Ristow et al., 1982; Handrinos, 1987; Meyburg et al., 1993, 1995, 2001, 2007; Baumgart et al., 1995; Handrinos & Akriotis, 1997; Vaassen & Aykurt, 2000; Dowsett & Leonard, 2001; Roxburgh & Leonard, 2007; Ramadan-Jaradi et al., 2008; Dowsett et al., 2008; Roxburgh et al., 2009). The generalized dates for southward migration are August to November, with the northward return between March and early May at the latest (Baumgart et al., 1995; Handrinos & Akriotis, 1997; Kren, 2000; Roberts, 2000; Shirihai, 1996; Dowsett-Lemaire & Dowsett, 2006; Dowsett et al., 2008; Ramadan-Jaradi et al., 2008; Ash & Atkins, 2009; Hilgerloh et al., 2009).

Poirazidis et al. (2019, 7) note that the Lesser Spotted eagle has a "stable population within its breeding range. Nonetheless, its favorable habitats (a mosaic of forest and low-vegetation forest-openings, often with water elements such as small ponds and flowing streams), are suffering from land abandonment as a result of abrupt socio-economic changes related to rural depopulation and the decline of traditional agro-silvopastoral practices" (Poirazidis et al., 2019, 7). This position may be shared by BirdLife International (2015, 2016), which describes this species as Stable and Least Concern rather than Vulnerable. One justification for this assessment is the size of the population (estimated by BirdLife International at 40,000-60,000 mature individuals, of which the European population is estimated at "16,400-22,100 pairs, which equates to 32,800-44,200 mature individuals"). As Europe contains about 73% of the global range of this species, the estimate of the global population size would be about 44,900-60,500 mature individuals. On migration routes thousands of eagles have been counted (e.g. 47,594 individuals in southern Turkey, 58,000 over the Bosporus (Fülöp et al., 2014). Another justification for the Least Concern and Stable classifications is the Lesser Spotted eagle's extremely large range, greater than 20,000 km^2 with little evidence of a declining or fluctuating range size, habitat extent/quality, or population size and a small number of locations or severe fragmentation. The issues mentioned by Poirazidis et al. (2019, 7) above may not be sufficiently wide ranging to affect the species across its entire range. The population trend is relatively unknown, and the population is estimated but neither are close to what would justify a Vulnerable designation. The main threats are listed as habitat loss, especially wet forest and meadow drainage, deforestation, and hunting (del Hoyo et al., 1994; Meyburg et al., 2014; Poirazidis et al., 2019). During migration, thousands of birds may be shot in Syria and Lebanon (del Hoyo et al., 1994). Forest management, wind energy developments, and even airport development may affect this eagle (Strix, 2012; Dravecký et al., 2015). The Lesser Spotted eagle appear to be gaining some territory in eastern Eurasia, as it hybridizes with the Greater Spotted eagle (Melniov et al., 2001). Nevertheless, in Russia it is listed in the Red Deta Books for the Moscow region and the Russian Federation (Kalyakin & Voltzit, 2006) and in Croatia it is Endangered (Radovíc et al., 2003). Illustrations

Fig. 2.21a. Lesser Spotted eagle [With permission from Roelof van der Breggen]

Fig. 2.21b. Lesser Spotted eagle perching [With permission from Frank McClintock]

Fig. 2.21c. Lesser Spotted eagle perching [With permission from Frank McClintock]

Fig. 2.21d. Lesser Spotted eagle in flight [With permission from Maarten Mortier]

Fig. 2.21e. Lesser Spotted eagle in flight [With permission from Roel of van der Breggen]

Fig. 2.21f. Lesser Spotted eagle habitat, Ukraine [Credit: Alex-Belogub-unsplash]

of the appearance, range, and eggs of this species are shown in Figures 2.21a to 2.21g.

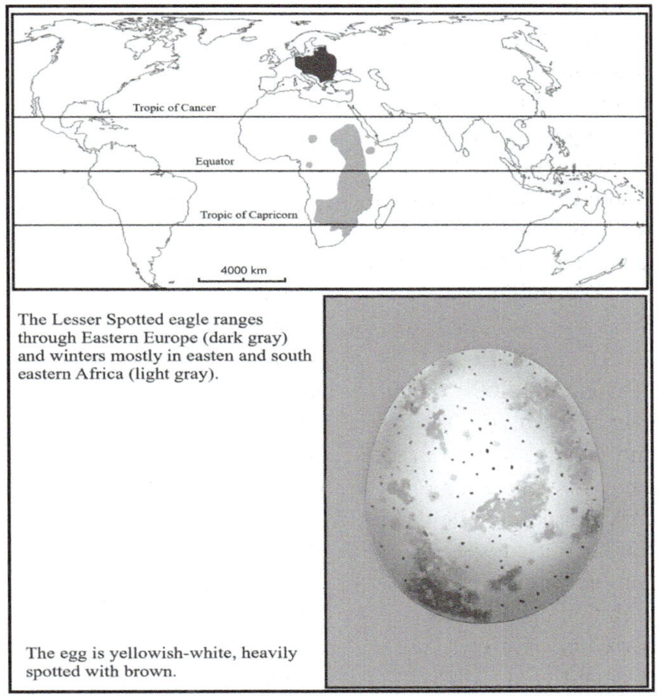

Fig. 2.21g. Lesser Spotted eagle range and egg

2.23 Indian Spotted Eagle *Clanga hastata* (Lesson, 1834)

Physical Description

The Indian Spotted eagle has been variously described as a "small eagle, slightly larger than a kite but structurally similar to other large eagles, with six splayed "fingers" at the wingtip and a pale rump" (eBird, 2020); "a stocky, medium-sized eagle with short, broad wings and a rather short tail" (BirdLife International, 2020); a "medium sized *Clanga* eagle" and a "medium sized, well-proportioned eagle, less bulky, clearly smaller and narrower-winged than the Greater Spotted eagle" (Surve et al., 2014, 220); and a "powerful predator" (BirdLife International, 2016). All these sources cite its main identifier as its smaller size compared with the Greater Spotted eagle (which may sometime require good views) and more minimal distinctions from its previous species partner, the Lesser Spotted eagle. The Indian spotted eagle is about 60 cm in length and has a wingspan of 150 cm (Parry et al, 2002; BirdLife International, 2020).

The dominant adult plumage is brown, either "rich brown overall" (ebird, 2020) or just brown (BirdLife International, 2020) with a contrast between the paler wing-coverts and darker flight feathers, both above and below, which can be identified in flight. The underwings and tail are finely barred brown on whitish to creamy white feathers. The immature plumage is paler, with light speckles on breast and belly and pale spots on the upperwing coverts, which are smaller and contrast less from the ground colour of the upperwing coverts and flight feathers, than do those of Greater Spotted Eagle (BirdLife International, 2020). Additionally, as cited above it is also "less bulky, clearly smaller and narrower-winged than the Greater Spotted eagle" (Surve et al., 2014, 220). In flight it shows "rounder wings and is lighter, slimmer and less bulky than the Greater Spotted eagle", which has "darker wing-coverts than flight feathers" (BirdLife International, 2020).

Nevertheless, many published sources point to the difficulty of identifying this species among the other spotted eagles. A noted characteristic is wide gape that has comparatively thick 'lips' that extend to the middle or behind the eye. Del Hoyo et al. (2020) point to the much larger gape compared with other comparable species (gape mean in females 60.1 mm vs 52.3 mm in length, 47.4 mm vs 37.4 mm in width). The legs also appear thin, as they are less thickly feathered than some of the other species. The nostril is round. This voice is described as a high-pitched cackle, like a laugh (BirdLife International, 2020; Del Hoyo et al., 2020; ebird, 2020, see also Grimmett et al., 1998; Robson, 2000, 2008; Parry et al., 2002; Rasmussen & Anderton, 2012; Sant et al., 2020). Illustrations of the appearance, range, and eggs of this species are shown in Figures 2.22a to 2.22g.

Classification

The "taxonomic status of the Indian Spotted eagle *Aquila (pomarina) hastata* has been an issue of dispute" (Väli, 2006, 238). The classification of this species has gone through three stages. It was first considered as a species (*Morphnus hastatus*, Lesson, 1834), after which it was classified as a subspecies of the Lesser Spotted eagle (*Aquila pomarina*, C.L. Brehm, 1831). Väli (2006, 238) considered several vital points which led to the third stage, the reclassification of this species as separate species, the Indian Spotted eagle (*Clanga hastata*, Lesson, 1834). When it was within *Aquila pomarina*, there was an important geographical separation between this subspecies (*Aquila pomarina hastata*) and the nominate subspecies (*Aquila pomarina pomarina*). The latter ranges as a breeding resident in eastern and central Europe, and in the Middle East, and is entirely migratory to Africa (Global Raptor Information Network, 2021ah; see also Hallmann, 1989; Pergantis & Handrinos, 1992; Simmons, 1997e; Radovíc et al., 2003; Parker, 2005; Bildstein, 2006; Dravecký, 2009). The former, a sedentary non-migratory subspecies ranges entirely within India (Surve et al., 2014).

Therefore, the breeding ranges of the two subspecies are separated by thousands of kilometers, complicating any study of interbreeding and hybridization, such as that between the Greater and Lesser Spotted eagles (Lohmus & Väli, 2001; Dombrovski,

2002; Meyburg & Meyburg, 2007; Meyburg et al., 2020). According to Väli (2006, 238) this prevents "any study of the reproductive barrier between them, the most important difference according to the Biological Species Concept." Väli (2006, 238) citing Parry et al. (2002), note the findings of the latter, which revealed several morphological differences, justifying the species status for *Aquila pomarina hastata*. The most obvious difference was the gape width of the mouth, which among the Spotted eagles was widest in *hastata*, medium in the Greater Spotted eagle and smallest in *A.p. pomarina*.

Molecular markers, in this case manifested by the divergence between the mitochondrial DNA lineages of the Greater Spotted eagle and *A.p. pomarina*, indicated a divergence about one million years ago. Due to the lack of such data for *hastata*, there were not clear conclusions (Seibold et al., 1996). Studies such as those of Helbig et al. (2005) and Lerner & Mindell (2005) dd not compare the two subspecies, *A.p. pomarina* and *A.p. hastata*. The study by Väli (2006) looked at the nucleotide sequence from the mitochondrial cytochrome-bgene of the Greater Spotted eagle, *A.p. pomarina* and *A.p. hastata*, and their closest monophyletic relatives, the Indian Black eagle and Long-crested eagle, and a more distant species, the Golden eagle (Helbig et al., 2005; Lerner & Mindell, 2005).

The findings were that using pairwise Kimura 2-parameter distances, cytochrome-b genes of the Greater Spotted eagle and *A.p. pomarina* differed by 1.8% while *A.p. hastata* differs from both of these eagles by 3.6%, and that the spotted eagles were separated from the other eagles. Väli (2006, 240) concluded that "phylogenetic reconstructions clearly suggest that *A.(p.) hastata* separated from *A. clanga* and *A.p. pomarina* before their differentiation." Therefore, the subspecies that became the Indian Spotted eagle is more distinct from its species partner (the Lesser Spotted eagle) than the latter is from the Greater Spotted eagle. The Indian Spotted eagle is also equidistant from the other Spotted eagles. It is estimated that the split of *A. hastata* could have happened about 1.8 million years ago. Despite the common hybridization of the Greater and Lesser Spotted eagles, assembled evidence indicated that they remain separate species.

The study by Väli (2006, using mitochondrial cytochrome-b gene), supported by Parry et al. (2002, using differences in external morphology, osteology, behavior, and clutch size) therefore justified the separate species status of the Indian Spotted eagle, and according to Global Raptor Information Network (2021ai). Before the formation of the Genus *Clanga*, the molecular phylogenetic analyses of Seibold (1994), Seibold et al. (1996), Helbig et al. (2005) and Lerner and Mindell (2005) that the three spotted eagles formed a monophylum with each other and the Long-crested eagle, and that these eagles could be merged into the Genus *Lophaetus*, and a change of the name *hastata* to *hastatus* "for gender reasons" (Global Raptor Information Network, 2021ai). Wells and Inskipp (2012) hinted that the three spotted eagle species be placed in the Genus *Aquiloide*s, although earlier studies had supported moving *Lophaetus* to *Aquila* (e.g. Wink & Sauer-Gürth, 2004; Gjershaug, 2006). However, later *Clanga* superseded *Aquiloides*. As noted above later writings (Clark, 2012; Gregory & Dickinson, 2012; Wells & Inskipp, 2012; Lerner et al., 2017) justified the creation of the Genus *Clanga* for the Spotted eagles.

Breeding

The Indian Spotted eagle's display flight includes switch-backing, wing-clapping and full loops in flight. Courtship has been observed from January, with laying in April (Sant et al., 2013). This is a tree-nesting species that frequents open landcover such as low intensity agriculture, wetlands and open forest and forest clearings year-round. The clutch ***size*** is almost always one, rarely two or three (Parry et al., 2002).

Foraging

The Indian Spotted eagle frequents lowlands, favoring open forests and forest edges, and tree patches surrounded by grassland, cultivated areas, and wooded semi urban landscapes. The prey taken is mostly mammals, most from the ground in open areas within the forest, or near denser forest stands. Other prey includes birds and frogs. In one Indian case study Prakash (1996) found mammals were the main prey, especially rats and Five-striped Palm Squirrels (*Funambulus pennantii*, Wroughton, 1905) which comprised 94% of the mammals taken. Birds taken were mainly terrestrial species (58%), such as Stone curlew (*Burhinus oedicnemus*, Linnaeus, 1758), Indian roller (*Coracias benghalensis*, Linnaeus, 1758), Grey partridge (*Francolinus pondicerianus*, Gmelin, 1789), Common myna (*Acridotheres tristis*, Linnaeus, 1766), Bustard quail (*Turnix suscitator*, Gmelin, 1789) and a Australasian pipit (*Anthus novaeseelandiae*, Gmelin, 1789). Other birds taken were marsh species (42%) such as Red-wattled lapwings (*Vanellus indicus*, Boddaert, 1783), moorhen (*Gallinula chloropus*, Linnaeus, 1758), and egrets (*Egretta* spp.). Other prey included Garden lizards (*Calotes versicolor*, Daudin, 1802) and Bull Frogs (Prakash, 1996).

Population Status

The Indian Spotted eagle is a widespread species with a fairly wide range through the Indian subcontinent (del Hoyo et al., 2020). It ranges at low densities in the lowlands of the Indian subcontinent, occurring in Nepal, India, and Myanmar (Robson, 2000; Parry et al., 2002; Rasmussen and Anderton, 2012; Tordoff et al., 2005). BirdLife International (2016) reports that despite its wide range its detailed demography is little known, hence it is assessed as Vulnerable, until further studies over its range influence a revision of this status. BirdLife International (2016) also assesses its population at extremely low density and in the band 2,500-9,999 mature individuals, possibly 3,500-15,000 individuals. The population is believed to be declining at a slow to moderate rate, mostly due to habitat change and destruction. All these assessments however require further research, as little is known about this species and its ecology (BirdLife International, 2016). For example, its presence and population status in Pakistan, Bangladesh and Cambodia is uncertain (Robson, 2008; Handschuh et al., 2011). In the last country it has been recorded in a "patchy distribution within low-lying open deciduous dipterocarp forest mosaics across northern Cambodia" (BirdLife International, 2020; see also Handschuh et al., 2011). BirdLife International (2020) further notes that "no focussed survey has been conducted to date. This raises the possibility that the species may have occurred historically across other parts of Southeast Asia, although today it has almost certainly been extirpated from much of this area due to habitat loss, hence declines are inferred in the past." There are also identification problems, as "an unknown proportion of records of this species may actually relate to Greater Spotted eagle *C. clanga*. Photographic evidence is therefore usually required to accept sightings" (BirdLife International, 2020). Illustrations of the appearance, range, and eggs of this species are shown in Figures 2.22a to 2.22g.

Genus *Polemaetus* (Heine, 1890)

The single species of the genus is a huge eagle, among the world's largest, with large, compressed bills, long pointed wings, and long toes with curved talons. Brown and Amadon (1968, 715) note that "the pattern of the immature plumage suggests a relationship to *Spizaetus*, but *Polemaetus* stands well apart from that or any other genus." The tail is shorter than those of *Spizaetus* and *Stephanoaetus*. There is only one species, the Martial eagle. Lerner (2007, 37) writes that the Genus *Polemaetus* (with its single species, the Martial eagle), was created due to the "divergent morphology" of the Martial eagle from the other booted eagles.

2.24 Martial Eagle *Polemaetus bellicosus* (Daudin, 1800)

Physical Description

The Martial eagle is the largest eagle in Africa and possibly among the two or three strongest species (Ferguson-Lees & Christie, 2001; Kemp et al., 2020b). Simmons and Brown (2004, 5) write that "the Martial eagle… is Africa's largest and one of its most

Fig. 2.22a. Indian Spotted eagle perching [With permission from Ken Behrens]

Fig. 2.22b. Indian Spotted eagle perching [With permission from Hari K Patibanda]

Fig. 2.22c. Immature Indian Spotted eagle perching [With permission from Vikas Chavan]

Fig. 2.22d. Immature Indian Spotted eagle perching [With permission from Girish Kurudi]

Fig. 2.22e. Indian Spotted eagle in flight [With permission from Vikas Chavan]

Fig. 2.22f. Indian Spotted eagle habitat open forest, India [Credit: Paul Silvan – unsplash]

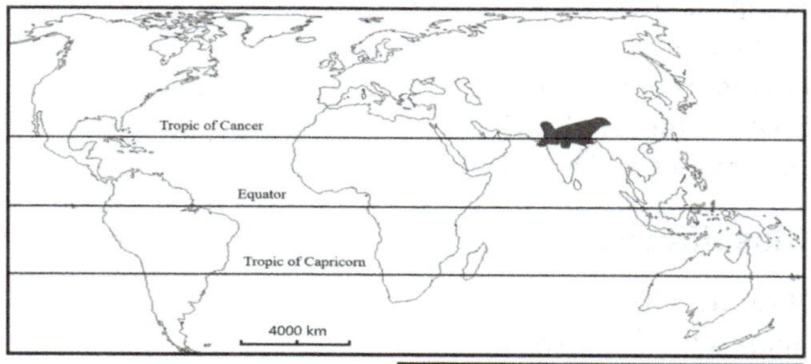

The Indian Spotted eagle ranges in northern India and slightly north of this area into Nepal and eastwards into Bangladesh.

The egg varies between white and more grayish with spots and sometimes reddish bloches at the larger end.

Fig. 2.22g. Indian Spotted eagle range and egg

powerful eagles. The female weighs ±6 kg and has a massive wingspan of ±2.6 m. Birds hunt mainly from the air and are capable of taking prey up to the size of a small antelope." Carr (1964, 108) notes that "the large, hooked beak of the Martial eagle can tear a carcass in half if it is too heavy to be carried to the nest. Its owner is the most powerful predatory bird on the open savanna." Lerner et al., 2017, 308) write of the Martial eagle, "this very large savannah eagle is characterized by unusually long, gracile tarsi, perhaps an adaptation to tall grass and/or dangerous large reptilian prey... It also has an unusually high wing loading with deep wing slots, presumably for fast gliding over its huge territory" (see also Parry, 2001). Kemp et al. (2020b) give its general description as having a "large head and long, broad blackish wings notable; upperparts, head, throat and upper breast dark brown; rest of underparts white with brown spots."

The Martial eagle measures 78 to 96 cm in length, with a relatively short tail compared with some other large species such as the Crowned eagle, reputed to be Africa's strongest eagle (Ferguson-Lees & Christie, 2001; Redman et al., 2010). The wingspan is 188 to 227 cm, which is probably about the fourth greatest among eagles, after the huge Steller's sea-eagle, the white-tailed eagle, and the wedge-tailed eagle (Mendelsohn et al., 1989; Ferguson-Lees & Christie, 2001; Redman et al., 2010). The weight range averages between 3.5 and 4.7 kg, approximately the same as the Golden eagle and the Verreaux's eagle, which makes these the heaviest of the Booted eagles, and this trio possibly the sixth heaviest eagles in the world, but some publications put the Martial eagle at 5th, behind the Steller's sea eagle, Philippine eagle, Harpy eagle and White-Tailed eagle, but ahead of the Wedge-tailed eagle, Crowned eagle, Golden eagle and Bald eagle. It must be noted that there are overlaps in the weight and other measurements of this species (Stevenson-Hamilton, 1954; Biggs et al., 1979; Mendelsohn et al., 1989; Dunning, 1992; Ferguson-Lees & Christie, 2001; Dunning, 2008; Watson, 2010; Bright et al., 2016). The Crowned eagle with its longer tail is usually cited as the longest African eagle, and after the Wedge-tailed eagle the second longest booted eagle (Dunning, 1992; Ferguson-Lees & Christie, 2001).

The adult Martial eagle has brownish grey upperparts, with pale margins on the crest feathers and light, ashy grey barring on the secondaries, scapulars and tail and darker barring on the wing quills. Lerner et al. (2017, 308) note that "adults are dark gray above with dark speckled white underparts. Brown & Amadon (1968, 715) describe the neck, throat and upper breast as colored dark grey brown like the back. The underparts covered with white feathers, spotted sub-terminally with sepia. There is sepia barring on the thighs and legs, and the inner quill linings are brown, with whitish bars, the auxillaries are sepia, and the underwing coverts mottled sepia and white. The eye is yellow, the bill and claws black and the cere and feet are blue grey (Brown & Amadon, 1968, 715; Kemp et al., 2020b).

The juvenile has dark grey upperparts, the feathers edged paler, and white underparts with two lines of dark spots on the breast sides. The underwing coverts have grey-brown and white mottling. The bars on the primaries are similarly colored to those of the adult, but with greyer coloration. The eye is yellow brown, the bill greenish grey, the cere and feet grey and the claws are black. Adult plumage is acquired through the darkening of the plumage, dark spots on the upper breast and the darkening of the throat (Brown & Amadon, 1968).

The Martial eagle is not easily mistaken for other species, due to its large size (Kemp et al., 2020b). The only similar species in flight is the Short-toed Snake eagle, but the current species has darker wings with a stronger contrast with the white spotted lower underparts and darker barred flight quills and tail. The juvenile can be mistaken for the young Crowned eagle, is duller in coloration with less grey and white and white thighs. It also has denser and less distinct wing and tail barring (Brown & Amadon, 1968). The Martial eagle is a relatively silent bird, compared with the Crowned eagle. The call is "a rather feeble" and "high pitched" whistling hweee – oh; the display call is klee – klee – klee – kloeee – kloeee – kuleee", while the young call "twip" or "quolp" during feeding (Brown & Amadon, 1968, 715). Lerner et al. (2017, 308) write that "vocalizations include a very loud, robust series of short staccato notes." Illustrations of the appearance, range, and eggs of this species are shown in Figures 2.23a to 2.23o.

Classification

The Martial eagle has been reclassified more than once. It was placed in the Genus *Hieraaetus* by Amadon (1982) and Amadon and Bull (1988), using physical attributes, mainly the osteological characters, which refer to the bone forms of the bird. According to Trail (2017) such osteological in an eagle may refer to the furcula, scapula, radius, carpometacarpus, femur, skull, sternum, pelvis (synsacrum), coracoid, humerus, ulna, tibiotarsus, and tarsometatarsus. Using more physical measurements, such as the analysis of the molecular sequences of one mitochondrial and three nuclear loci, Wink and Sauer-Gürth (2004) and Helbig et al. (2005) concluded that the Genus *Polemaetus* (of which the Martial eagle is the only species) is a sister species to a clade including the Genera *Aquila*, *Hieraaetus*, and *Lophoaetus*). In contrast, the study by Lerner and Mindell (2005), which used the molecular sequences of two mitochondrial genes and one nuclear intron, did not find a close relationship between the Martial eagle to any other raptors, not even the Booted eagles. Lerner and Mindell (2005, 341) note that "Both the Crowned hawk-eagle and the Martial eagle have been placed in monotypic genera because of their divergent morphology. Genetically these birds are also highly divergent from other booted eagles in our dataset." Lerner et al. (2017, 308) write that the Martial eagle, "with its distinct morphology, has long been recognized in the monotypic genus *Polemaetus* Heine, 1890. Its placement as sister to the clade containing *Lophaetus*, *Ictinaetus*, *Clanga*, *Hieraaetus* and *Aquila* is strongly supported here… as in other molecular studies" (see also Lerner & Mindell, 2005; Helbig et al., 2005).

Breeding

For the Martial eagle, Brown and Amadon (1968, 715) note that "no very striking display flights have been recorded." During courting it may utter a display call, when soaring or perching. Van Eeden et al. (2017, 3) report that "breeding typically occurs every second year laying a single egg, with incubation lasting 47–51 days, the fledging period lasting 90–109 days (see also Steyn, 1982). Van Eeden et al. (2017, 3) further note that "Martial eagles occur in the highest densities in the savannah biome. They are present in a wide range of habitats including open shrublands with tree cover near rivers, and open farmlands where there are trees or pylons for nesting. Birds are nearly absent from mountainous regions and grasslands void of tree cover" (see also Boshoff, 1997; Brown et al., 1982; Machange et al., 2005; Simmons, 2005; Taylor, 2015a).

Nests are usually in tall trees, near gorges and hillsides that facilitate takeoffs (Brown & Amadon, 1968). The nests are very large, sometimes two meters across and more than one meter deep but are generally smaller than those of the Crowned eagle. Materials include sticks and leafy vegetation for lining. The nest may be located in the fork of a large tree, on a cliff, or even rarely on electric power poles and pylons (Boshoff, 1993, Simmons & Brown, 2004). These locations may vary regionally. In Zambia, nests were recorded in large trees (Dowsett et al., 2008). In Zimbabwe, nests were recorded in the forks of large trees or acacia crowns (Irwin, 1981). In South Africa, some nests were recorded on human constructed structures and sometimes on cliffs in treeless areas (Boshoff, 1997).

One rounded, oval white or pale greenish-blue (sometimes with grey and brown markings) egg is laid. Eggs are large, possibly the largest of the Booted eagles (Steyn, 1983; Brown & Amadon, 1986; Beecham & Kochert, 1975; Dowsett et al., 2008). In a few cases, two eggs are recorded. For example, in Zambia in one study, 32 nests had one egg, while one nest had two eggs (Dowsett et al., 2008). In Malawi, the clutch size was one egg at four nests (Dowsett-Lemaire & Dowsett, 2006). Laying times also vary across the range. For example, late March to July (South Africa); March to July (Malawi, Zimbabwe); March, May to July, September to November (especially June or July) (Kenya); and March (Sudan). In Ethiopia, eggs were reported in nests for September, October, November, December, January (Ash & Atkins, 2009). In another study in northwest Somalia, three nests with one egg each were recorded in September, October, and November, respectively (Ash & Miskell, 1998). In Uganda, eggs were recorded in November (van Someren, 1932) and September 1944 and fledging in July 1961 (EANHS Nest Records). In Zambia, eggs were recorded in nests in March, April, May, June and July (Dowsett et al., 2008). In Malawi, eggs were recorded from April to July (Dowsett-Lemaire & Dowsett, 2006). In Namibia, laying may begin in the winter, but eggs are recorded from April to August (Simmons & Brown, 2004). In Botswana, laying was also recorded from May to August (Penry, 1994). In Zimbabwe, Irwin (1981) reported eggs from February, and March, and from April to August. In South Africa, Boshoff (1997) notes that breeding is mostly from May-November. Incubation periods may vary regionally, but usually lasts 45 days and fledging is after 99 days (Brown & Amadon, 1968; Herholdt & Kemp, 1997).

Foraging

The Martial eagle frequents open grassland, thornbush, light woodland, scattered tree savanna or more heavily wooded savannas, or riparian forests in arid landcover, and even semidesert in southern Africa, mostly below 1,500 m (Ferguson-Lees & Christie, 2001). Closed canopy forest (except for regions with clearings) and deserts are usually avoided (Brown & Amadon, 1986; Boshoff, 1997). The trees enable frequent perching, especially in the morning and evening (BirdLife International, 2018). Eagle may also perch on power pylons in the treeless areas (Verdoorn & Roth, 2000; Simmons & Brown, 2004). Foraging is from such perches or high altitude soaring (Steyn, 1982). It has been recorded in such environments across its breeding range (Irwin, 1981; Penry, 1994; Parker, 2005a,b; Dowsett-Lemaire & Dowsett, 2006; Dowsett et al., 2008; Ash & Atkins, 2009).

Prey taken includes mammals. birds and reptiles, the actual proportions varying regionally. Global Raptor Information Network (2021aj) lists eight locations across Africa, where prey choices varied. In Niger, one common prey species is the Helmeted guineafowl (*Numida meleagris*, Linnaeus, 1758) (Thiollay, 2006). In Tsavo National Park in Kenya, common prey species included dik-diks (Genus *Madoqua*, Ogilby, 1837) and bustards (Family *Otididae*, Rafinesque, 1815) both of which may weigh at least as much as the eagle itself (Smeenk, 1974). In this study of two nests, there were differences, as the pair in one nest near brushland favored in addition to dikdiks and bustards, other antelopes, gerenuks (*Litocranius walleri*, Brooke, 1879) and even Spotted eagle owls (*Bubo africanus*, Temminck, 1821). In the other nest, which was located in more open landcover, mostly wooded grassland, prey was more focussed on bustards, especially Hartlaub's bustard (*Lissotis hartlaubii*, Heuglin, 1863). The only reptile taken was the monitor lizard (*Varanus exanthematicus*, Bosc, 1792). In Zambia, prey taken include young impala (*Aepyceros melampus*, Lichtenstein, 1812) and various birds such as guineafowl (Family *Numididae*, Longchamps, 1842) and francolins (Subfamily *Perdicinae*, Horsfield, 1821) (Dowsett et al., 2008), similar to the prey taken in Botswana (Penry, 1994).

Other records were of doves, and a possible Hooded vulture predation in Tanzania (Loveridge, 1923; Global Raptor Information Network, 2021aj). Other prey comprised Denham's or Stanley's bustard (*Neotisdenhami*, Children & Vigors, 1826) in Malawi (Dowsett-Lemaire, 2006; Dowsett-Lemaire & Dowsett, 2006), and Steenbok (*Raphicerus campestris*, Thunberg, 1811) or Grey duiker (*Sylvicapra grimmia*, Linnaeus, 1758) in Zimbabwe, where one nest recorded prey proportions (54.1% gamebirds and domestic fowl, 40.5% mammals and 5.4% reptiles) (Irwin, 1981). Items gathered from one nest over a period of three years consisted of gamebirds and domestic poultry (54.1%), mammals (40.5%), and reptiles (5.4%). In northeastern South Africa, common prey included gamebirds (45%), reptiles (38%), and mammals (17%) (Boshoff et al., 1990; Tarboton & Allan, 1984). In most of these locations, other prey taken include guinea fowl (Family *Numididae*, Longchamps, 1842), francolins (Subfamily *Perdicinae*, Horsfield, 1821), bustards (Family *Otididae*, Rafinesque, 1815), hyrax (Order *Hyracoidea*, Huxley, 1869), domestic goat kids *(Capra aegagrus hircus*, Linnaeus, 1758), lambs (*Ovis aries* Linnaeus, 1758) and monitor lizards.

Naude et al. (2019, 1) using web-sourced photography, report that the most important bird prey (12%) is the Helmeted Guineafowl (*Numida meleagris*, Linnaeus, 1758), whereas the most common mammal prey species in order of frequency taken were Thompson's gazelle (*Eudorcas thomsonii*, Günther, 1884, 5%), impala (*Aepyceros melampus*, Lichtenstein, 1812, 4%), common duiker (*Sylvicapra grimmia*, Linnaeus, 1758, 4%) and banded mongoose (*Mungos mungo*, Gmelin, 1788, 4%). The commonest reptile prey taken was monitor lizards (Genus *Varanus*, Merrem, 1820, 21%). Naude et al. (2019, 1) also note that "Prey class proportions differed significantly by regions with mammalian prey being more common in eastern Africa and reptile prey being more common in southern Africa. Within South Africa, reptile prey proportion was greater in the east than in the west." Additionally, "prey composition differed between age classes, with adult eagles preying on more birds than non-adults. There was no significant difference in the average estimated prey weight by eagle age or feeding position (ground, perched, or lying)" (Naude et al., 2019, 1). Naude et al. (2019, 7) also make the interesting observation that Martial eagles that mostly rely on reptilian prey may gain more biomass for each successful hunting outcome, and those that rely on birds which are the smallest of prey classes, may require more successful hunting outcomes for suitable dietary requirements. This assessment must take cognizance of the fact that bird meat contains a higher calorific content than either mammal or reptile meat (see also Nagy et al., 1999). The monitor species taken, Africa's largest lizards include the Rock monitor (*Varanus albigularis*, Daudin, 1802), the Nile monitor (*Varanus niloticus*, Linnaeus, 1766) and the Savannah monitor (*Varanus exanthematicus*, Bosc, 1792). Because of their size, these lizards are mostly taken by female eagles by ambush (Smeenk, 1974; Boshoff & Palmer, 1980; Steyn, 1983; Tarboton & Allan, 1984; Phillips, 1995; Simmons, 2005).

Manuel (2020) used data "collected non-obtrusively by fitting five nests in the Maasai Mara and one in the Rift Valley region of Kenya with cameras that were both motion-triggered and programmed to take images every five minutes." The findings indicated found the diet of nesting Martial eagles to be mostly within four categories (gamebirds, domestic poultry, large ungulates and small ungulates), with gamebirds the commonest. Female Martial eagles selected larger, heavier prey than males. Using this logic, the male eagles were the main feeders of younger nestlings (up to the 8 to 11, with females contributing larger prey food when the nestlings are larger. Such large prey was more likely than the smaller prey to be torn to pieces, to carriage to the nest. Manuel (2020) found that the prey base of the studied eagle pairs was composed of 23 different prey species. This finding was compared with the findings of other studies, such as that of Boshoff et al. (1990), which found grassland martial eagles preying on 27 different species (birds (11), mammals (15) and reptiles (1)) using below nest collection surveys. Naude et al. (2019) used web-source photography, and found 15 bird, 23 mammal and 3 reptile species, for a total of 42 species preyed upon by Martial eagles. Hatfield (2018) found 26 different prey species preyed upon by Martial eagles in the same Maasai Mara region as that researched by Manuel (2020). Manuel (2020) note that "it is difficult to directly compare to these other studies due to habitat and methodological differences between studies. Nevertheless, the studies together corroborate each other, highlighting the importance of hyrax, mongooses and hares as well as bustards, guineafowl, francolin and spurfowl to the diet of martial eagles across southern and eastern Africa" (see also Boshoff & Palmer, 1980; Steyn, 1980; Hatfield, 2018; Naude, 2019).

Population Status

The Martial eagle is a fairly wide-ranging eagle, through the savanna landcover of West, Central, East and Southern Africa. Simmons and Brown (2004, 5) write that "although the Martial eagle probably occurs more widespread than any other eagle throughout southern Africa, it is becoming increasingly scarce in more heavily settled districts." BirdLife International (2018) reports that "it is generally scarce to uncommon or rare, but is reasonably common in some areas... It is suspected to have undergone declines in much of its range, including west Africa" (see also Ferguson-Lees & Christie, 2001; Thiollay, 2006; Cloete, 2013; Simmons, 2015; Amar & Cloete, 2017). The Martial eagle is a sedentary species, but range widely during the non-breeding season, especially juveniles after fledging (Brown et al., 1982; Steyn, 1982; Hartley, 1998; Bildstein, 2006). The recovery of ringed birds indicates some movements, but these are generally not continent-wide. Ring recoveries (examining the distance between the location the eagle was ringed and location where it was recovered) indicate the distances moved. Two examples were in Zambia, where the ring was recovered 21 years later 80 km away (Dowsett and Leonard, 2001) and in Zimbabwe, where the recovery was 65 km from the point of origin, after 307 days (Irwin, 1981).

BirdLife International (2018) states that the global population of the Martial eagle "has not been quantified but was estimated as probably 'in tens of thousands' by Ferguson-Lees and Christie (2001)". This falls within the population estimate of 10,000 to 100,000 adult and immatures at the start of the breeding season. BirdLife International (2018) lists this species as Vulnerable and also Decreasing and justifies this classification on the grounds that "it is suspected to have undergone rapid declines during the past three generations (56 years) owing to deliberate and incidental poisoning, habitat loss, reduction in available prey, pollution and collisions with power lines; threats that are likely to continue this trend into the future. Further information on trends across its large range may lead to its uplisting to Endangered in the future."

In some countries the information may be more dated, and new counts may be necessary. For example, in Western Africa, it was listed as an uncommon resident in the savannas from southern Mauritania southwards to Sierra Leone and east to Chad, Central African Republic, and southeastern Gabon (Borrow & Demey, 2001). The population in southern Africa (South Africa, Lesotho and Swaziland) was recently estimated at 800 pairs (Taylor, 2015a) of which the population in Swaziland is estimated at less than 30 birds (Parker, 1994) and the population in Namibia less than 350 pairs (Simmons, 2015). Older records put the population in Transvaal (northeastern South Africa) at 550 pairs (Tarboton & Allen, 1984; Parker (1999) estimated the population at less than 200 birds in Mozambique, but later changed this to between 200 and 1,000 birds (Parker, 2005).

There are several factors for the decline in the population, but these vary regionally (Table 2.3). These include direct shooting and trapping by farmers, indirect poisoning through poisoned carcass consumption and possibly pesticides, drowning in reservoirs, collision and electrocution on electric power lines, and habitat alteration and degradation, including local extirpation of prey species, and the loss of suitable trees for nesting (Boshoff, 1997; Oatley et al., 1998; Verdoorn & Roth, 2000; Simmons, 2005). The use of eagle parts in traditional medicine may also be a driver for hunting of eagles. It is possible that some birds may also be used in traditional medicine practices. Steyn (1982) writes that immature birds are more vulnerable than adults to direct persecution. Possibly due to high wing loading, or easily soaked feathers, the Martial eagle is the most likely large raptor to drown in steep-sided reservoirs in South Africa. These losses are serious, as one study estimated that deaths from such drownings could total 8% of the breeding population in the Northern Cape of South Africa (Anderson et al., 1999). Illustrations of the appearance, range, and eggs of this species are shown in Figures 2.23a to 2.23o.

As shown above, in individual countries, the Martial eagle varies in population and conservation status. For example, it is almost locally extirpated from parts of Niger, as well as a key prey species, Helmeted Guineafowls (Thiollay, 2006). Further east, in Central West Africa, significant population declines were reported by comprehensive, time series studies (Rondeau & Thiollay, 2004; Thiollay, 2006). In this, 7000 km of road transect surveys of raptor species were carried out in the Sudano-Sahelian savannas of Burkina Faso, Mali, and Niger, from 2003 to 2004 duplicating replicating surveys done in the same area in 1969-1973 by Thiollay (1977). The findings indicated a serious decline in eagle numbers, especially in unprotected areas. More detail can be seen in the Table 2.3.

Further east and south, similar declines have occurred (Global Raptor Information Network, 2021aj). For example, in Uganda the decline of the population in the Kampala region was sufficient for the Martial eagle to be classified as Vulnerable (Carswell et al., 2005). In Namibia, the low population estimate and indications that the population has declined by over 80% justify the assessment as Endangered (Simmons & Brown, 2004). In Botswana, it is assessed as a Bird of Concern (Hancock et al., 2007). In Malawi, environmental wildlife degradation, and human persecution from livestock predation may be driving a eagle population decline (Dowsett-Lemaire & Dowsett, 2006). In Namibia, it is Endangered, due serious population declines in the face of persecution on farmlands (Brown, 1991; Simmons and Brown, 2008). In Mozambique, it is assessed as Threatened, especially in southern and central Mozambique (Parker, 1999; Parker, 2005). In Swaziland, it is assessed as Vulnerable, possibly linked

Table 2.3. Population status of the Martial eagle

Country	Status	Sources
Sudan	Common resident	Bowen, 1926; Nikolaus, 1987
Ethiopia	Fairly common resident	Ash & Atkins, 2009
Côte d'Ivoire	Breeding resident	Thiollay, 1985
Uganda	Widespread, breeding resident	Carswell et al., 2005
Malawi	Widespread breeding resident	Dowsett-Lemaire & Dowsett, 2006
Zimbabwe	Widespread breeding resident	Irwin, 1981; Simmons & Brown, 2004
Swaziland	Widespread breeding resident	Barnes, 1998
South Africa	Widespread breeding resident	Tarboton & Allan, 1984; Anderson, 2000
Eritrea	Rare, presumed resident	Smith, 1957; Ash & Atkins, 2009
Somalia	Uncommon breeding resident	Ash & Miskell, 1998
Ghana	Uncommon resident	Grimes, 1987
Namibia	Rare resident, widespread	Simmons & Brown, 2004
Angola	Not uncommon resident	Pinto, 1983; Dean et al., 1988, Dean, 2000
Zambia	Not uncommon resident	Dowsett et al., 2008
Botswana	Fairly common to rare resident	Penry, 1994; Hancock et al., 2007
East Africa	Widespread, mostly uncommon	Stevenson & Fanshawe, 2002
Gambia	Rare resident	Gore, 1990
Togo	Breeding resident	Cheke and Walsh, 1996
Mozambique	Uncommon breeding resident	Parker, 2005; Clancey, 1996
Nigeria	Uncommon resident	Elgood et al., 1994
Niger	Uncertain	Thiollay, 1977; Newby et al., 1987
Congo	Uncertain	Rainey et al., 2009
Liberia	Vagrant	Gatter, 1997

Sources: Some of the sources in the third column are derived from Global Raptor Information Network (2021aj). In all cases the species is fairly uncommon.

Fig. 2.23a. Martial eagle perching [With permission from Paul Ellis]

Fig. 2.23b. Martial eagle perching [With permission from Gustav Van Der Gryp]

Fig. 2.23c. Martial eagle perching [With permission from Gabriel Leboff]

Fig. 2.23d. Martial eagle perching [Credit: Frans Van Heerden -Pexels]

Fig. 2.23e. Immature Martial eagle perching [With permission from Ken Behrens]

Fig. 2.23f. Immature Martial eagle perching [With permission from Pieter Verheij]

Fig. 2.23g. Immature Martial eagle perching [With permission from David Beadle]

Fig. 2.23h. Martial eagle in flight [With permission from Frank McClintock]

Fig. 2.23i. Martial eagle in flight [With permission from Pieter Verheij]

Fig. 2.23j. Martial eagle in flight with pied crow [With permission from Roel of van der Breggen]

Fig. 2.23k. Immature Martial eagle in flight [With permission from Theo Storm]

Figure 2.23l. Immature Martial eagle [With permission from Rick Dunlap]

Fig. 2.23m. Habitat for the Martial eagle in South Africa [With permission from Derek Keats]

Fig. 2.23n. Martial eagle habitat, Serengeti, Tanzania [With permission from Achim Mittler, Frankfurt am Main]

to landcover change and impacts on nesting sites (Monadjem et al., 2003; Monadjem & Rasmussen, 2008). In Southern Africa, there are several designations, including Vulnerable (Brooke, 1984; Barnes, 2000b) and Threatened (Barnes, 2000b; Simmons, 2005). Most declines and local extinctions are linked to human population increase and landcover change, hence it is rare outside protected areas (Boshoff & Vernon, 1980; Tarboton & Allan, 1984; Hustler & Howells, 1987; Brown, 1991; Boshoff, 1997; Parker, 1999). Illustrations of the appearance, range, and eggs of this species are shown in Figures 2.23a to 2.23o.

Genus *Stephanoaetus* (W.L. Sclater, 1922)

This genus comprises "very powerfully built eagles", characterized by compressed bills, short wings, long, broad tails, feather legs, short, strong toes with huge talons and large divided crests. Brown and Amadon (1968, 707) note that *Stephanoaetus* is very closely related to *Spizaetus*, of which it is a specialised offshoot." There is only one species, the Crowned eagle, which with the Martial eagle is among the largest and strongest eagles in Africa. Lerner (2007, 37) writes that, like the Genus *Polemaetus* (with its single species, the Martial eagle), the monotypic Genus *Stephanoaetus* (with its single species, the Crowned eagle) was created due to the "divergent morphology" of these species from the other booted eagles.

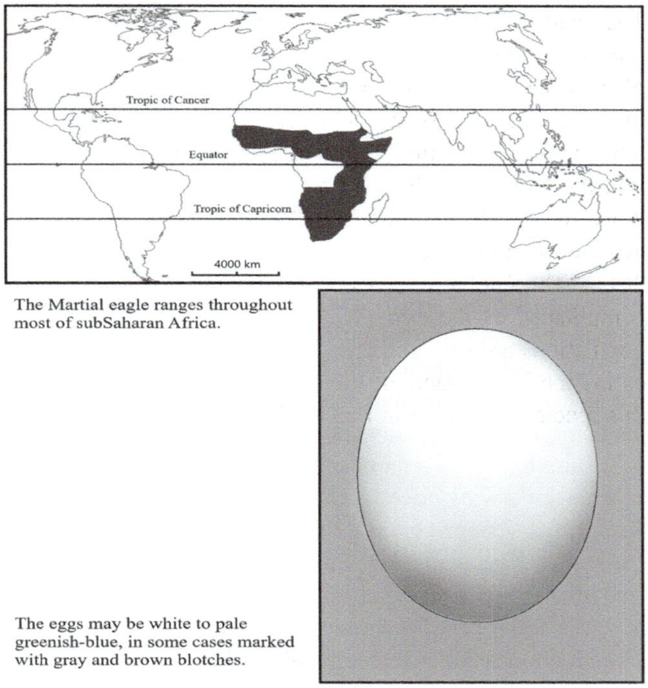

Fig. 2.23o. Martial eagle range and egg

2.25 Crowned Eagle or African Crowned Eagle or Crowned Hawk Eagle *Stephanoaetus coronatus* (Linnaeus, 1766)

Physical Description

The Crowned eagle is the largest predatory bird of the African forest and nearly matches the Martial eagle in size and perhaps is superior in strength (an estimation based on the relative sizes of the prey taken). The Crowned eagle is among those eagles mentioned by Gilliard (1958, 105), who wrote that "tropical forests around the world are the home of a small group of huge, crested, immensely powerful eagles." Due to its crested head, shortish wings (compared with the longer wings of the savanna soaring eagles like the Martial and Verreaux's eagle), extremely powerful talons and prey species (including medium to large monkeys) the Crowned eagle has been described as Africa's version of the South American Harpy eagle and Asian Philippine eagle. Brown and Amadon (1968, 708) write that "this magnificent bird is possibly the finest of all African birds of prey and probably the most powerful, though not actually the largest." Shultz (2019, 119) describe it as "a large, very dark eagle with mottled rufous breast, a distinctively barred under-belly and a crest it can erect or lay flat. This species has a large bill and fully feathered legs with powerful feet. Like many other large tropical forest raptors, its wings are short and broad, and it has a long tail. Preadults are noticeably paler than adults." Kemp et al. (2020c) describe the Crowned eagle's "long tail and short, rounded wings like a giant goshawk, with densely feathered legs, very large toes, heavy bill and comparatively small eyes."

The Crowned eagle measures 80-95 cm, with a wingspan of 151-181 cm and a weight of 2.7-4.7 kg (Shultz, 2019, 119). Other sources give varying size ranges for the Crowned eagle. Ferguson-Lees & Christie (2001) give a length of 80 to 99 cm, a weight of 3.2–4.7 kg for the females and 2.55–4.12 kg for males (see also Kemp, 1994d). Other sources give average weights of 3.64 kg (Dunning, 1992) and 3.8 kg (McGraw et al., 2006) and note that this eagle is the 9th heaviest eagle in the world (Ferguson-Lees & Christie, 2001). Some sources point to the relatively short wings of this species, roughly equal to that of the much smaller Tawny eagle (or a Short-toed Snake eagle, although they may be broader than those of the Golden eagle) (Brown, 1976). The tail is also remarked as relatively long, giving the Crowned eagle a greater overall length than the larger Martial eagle (Brown & Amadon, 1986; Ferguson-Lees & Christie, 2001). Another important point for the Crowned eagle is the relatively large and powerful talons, probably an indication of the large prey this species takes. The legs, feet and talons are larger and thicker than those of the larger Martial eagle. Two measured examples (one a museum specimen of a female, another a live, young adult female) found the hind claw to be the same as that of the larger Golden eagle, and only slightly smaller than those of the larger Harpy and Philippine eagles (Bortolotti, 1984; Dunning, 1992; Prout-Jones et al., 1997).

The Crowned eagle's distinguishing features are a dark to reddish brown crown, a black tipped double crest, reddish-brown to umber nape, neck and cheeks, bluish to black upper parts (including the upper wing coverts) and ash brown primaries with black tips, and two black bars and white bases (Brown & Amadon, 1986). The black tail has two wide, brownish bars. The underparts include a brown throat, a white and buff underside, with black barring and blotches. The wing primaries have white bases, broadly tipped with black and crossed by two black bars. the underwing quills are black with white tips and include two broken bars formed from black spots. The thighs and legs are feathered, colored with barring and black and white spots and the underwing coverts are chestnut with black spots. The eyes are pale yellow to almost white, the cere and feet are yellow ochre, the gape dull yellow and the bill and claws are black (Brown & Amadon, 1968, 707). The juvenile is lighter colored, with a white head and underparts, black spotted thighs and legs, and a reddish tinge to the breast. The upper parts are light grey to brown, with lighter edgings to the feathers, "giving a scaly appearance, especially in the upperwing coverts" (Brown & Amadon, 1968, 707; Brown & Pommery, 1984). Further juvenile characteristics are grey brown wing quills with blackish tips and two dark bars, a black tail with three lighter bars and a white tip, olive brown eyes, greyish cere and dull yellow feet and gape. The juvenile changes into the adult plumage by acquiring a darker, more reddish color on the breast, with a few black bars. The upperparts patterns stay the same, but the color darkens over the course of three years (Brown & Amadon, 1968, 707).

The adult bird is unmistakable due to exceptionally large size, the black upper parts and barred underparts, the chestnut-colored underwing coverts, and the barred wings. Juveniles may resemble the juvenile of the Martial eagle, but the current species has a much longer tail, which is more heavily barred, strong wing barring, spotted thighs and "purer grey and white plumage" (Brown & Amadon, 1968; Ferguson-Lees & Christie, 2001; Sinclair & Ryan, 2011). Brown and Amadon (1968, 708) note that "the voice of the male is always higher than that of the female, and that of the young of both sexes is also higher pitched than that of either adult"; furthermore, "this is an exceeding noisy bird, constantly drawing attention… in aerial display or calling when it is hidden inside a big forest tree." The voice varies, with a high pitched "kewee – kewee – kewee" or ki – ki – ki – ki – ki for the male during breeding display, and a lower pitched "koi – koi – koi – koi" for the female in display and perching. Nest calls for both sexes is shrill, resembling "quee – quee – quee – quee." Another male call, sometimes in reaction to stress is a softer "sheeee – oh" (Brown & Amadon, 1968, 708). Illustrations of the appearance, range, and eggs of this species are shown in Figures 2.24a to 2.24i.

Classification

The Crowned eagle was in some cases placed within the Genus *Spizaetus* (Amadon, 1982a,b; Amadon & Bull, 1988; Brown & Amadon, 1968), this assessment being based on morphological characteristics. However, more advanced molecular phylogenetics techniques, using DNA sequences of one mitochondrial and three nuclear genes, indicated that the Crowned is a sister species to the Asian hawk-eagle species (which were formerly included in the Genus *Spizaetus* but later moved to the Genus, *Nisaetus* (Helbig et al., 2005). Yet more advanced analyses examining the mitochondrial genes and the nuclear sequences found otherwise. For example, Lerner and Mindell (2005), using the molecular sequences of two mitochondrial genes and one nuclear gene, did not find a close relationship between the Crowned eagle and any of the booted species, including *Spizaetus/Nisaetus.* A related and physically similar species, now extinct was the Malagasy Crowned eagle (*Stephanoaetus mahery*, Goodman, 1994) (Goodman, 1994a,b; Wright, 1998; Feduccia, 1999; Hume & Walters, 2012).

Breeding

The breeding season is initiated by mating displays, which may occur near nests and the edges of home ranges. Displays involve steep dives and ascents, and even circles and figures of eight, with calls at the height of the ascent, which can be up to 1000 meters. Although both the male and female may perform similarly, but the female may have a lower call. The performance may include the male diving towards the female, and the female rolling and presenting claws. Nests are usually located in tall emergent trees, the larger nests being as much as two meters across and three meters deep. In some locations, certain trees are favored for nesting. Examples include the Nyika Plateau in Malawi, where many nests are in the tall forest emergents *Pouteria adolfi-friedericii* ((Engl.) *A. Meeuse*) and *Chrysophyllum gorungosanum* (Engl.), which may be over 40 meters high and emerge over the main forest canopy (Dowsett-Lemaire & Dowsett, 2006). In Zimbabwe, *Newtoniabuchananii* ((Baker) G.C.C. Gilbert & Boutique) is a common nesting tree (Irwin, 1981). In rare cases in this country, there may also be cliff nesting (Steyn, 1982). The clutch comprises one or two eggs. In one study in Zambia, some nests had one egg, while others had two eggs (Dowsett et al., 2008). In another study in Malawi, three nests contained 2 eggs each (Dowsett-Lemaire & Dowsett, 2006). In some nests when two eggs are laid, only one nestling may survive, as the weaker nestling may be killed by its sibling or outcompeted for food (Ferguson-Lees & Christie, 2001).

Laying times vary across the range, times, including September to October (South Africa), January to March, or May/July to October (Zimbabwe), October (Congo and West Africa), June to November especially August to October (Kenya), December,

March, July (Uganda). Evidence indicates that July to October is the commonest laying period for the whole range. Fledging occurs after 103 to 115 days (Brown & Amadon, 1968). Other recorded dates for eggs or nestlings in nests include March and April for grown nestlings in Liberia (Gatter, 1997), February-March for a chick in a nest in Togo (Cheke & Walsh, 1996), September and January for eggs in nests in Ethiopia (Ash & Atkins, 2009), and August to January for eggs in nests in Uganda (Brown & Britton, 1980; Skorupa, 1989; Kalina & Butynski, 1994), and nestlings July 1987 (Bataamba, 1989), and May 1989 (Carswell et al., 2005). In Angola, an incubating or brooding bird was observed in August (Dean et al., 1988), while in Zambia, eggs were reported in nests from June, to December (Dowsett et al., 2008). In Malawi, eggs were reported in March, April to June, August, and September, with one copulatory episode recorded in July (Dowsett-Lemaire & Dowsett, 2006). Other records were September in Mozambique (Parker, 2005), and scattered dates between February and November in Zimbabwe (Irwin, 1981).

Foraging

The Crowned eagle frequents mostly forests but can occur in forest patches on rocky cliffs and to a lesser extent bushy savanna, montane and lowland evergreen forest, dense woodland, and forested ravines and gorges in open savannas and thornveld (Brown & Amadon, 1968; Boshoff, 1997). These habitats may vary regionally. For example, in southern Africa, it has been recorded in montane forest and other landcover including tree plantations (Oatley et al., 1998). In Malawi, it has been reported in miombo and riparian woodland, deciduous forest, and in some cases in remnant forest near farms (Dowsett-Lemaire & Dowsett, 2006; Dowsett et al., 2008). It may also be found in evergreen montane and hilly forests (Irwin, 1981). It occurs in similar environments in Swaziland (Monadjem & Rasmussen, 2008).

Hunting techniques include perching on tall trees near clearings or waterholes, flight over the forest floor, or soaring over the canopy (Brown & Amadon, 1968). Prey taken is about 98% mammals, especially monkeys in dense forest and antelopes (e.g. the Suni, *Neotragusmoschatus*, Von Dueben, 1846 and Harvey's duiker *Cephalophus harveyi*, Thomas, 1893) and hyrax (Order *Hyracoidea*, Huxley, 1869) in more open vegetation. Other species taken include large lizards or snakes, mongooses, domestic cats or rats. The prey is generally larger than that of the Martial eagle or Verreaux's eagle (Brown & Amadon, 1968). Steyn (1982) argues that this is not the largest eagle in Africa, it is the most powerful, indicated large size of its mammalian prey, including antelopes up to six times the eagle's mass. In Zambia, it has been recorded attacking and killing monkeys, hyraxes (Genus *Dendrohyrax*, Gray, 1868), and game birds (Dowsett et al., 2008). Other studies document similar prey taken, such as hyrax, duikers (Dowsett-Lemaire, 1998), Blue monkeys (*Cercopithecus mitis*, Wolf, 1822) and Vervet monkeys (*Chlorocebus pygerythrus*, F. Cuvier, 1821) (Dowsett-Lemaire, 1983; Deacon & Rockingham-Gill, 2009), Blue Duikers (*Philantombamonticola*, Thunberg, 1789) (Irwin, 1981), klipspringers (*Oreotragus oreotragus*, Zimmermann, 1783) (Shelley & Sclater, 1897), domestic cats (Benson, 1942; Lowry, 2006). Other prey species included two species of hyrax (Rock hyrax *Procavia capensis*, Pallas, 1766; and Yellow-spotted rock hyrax *Heterohyrax brucei*, Gray, 1868), a Rock monitor lizard (*Varanus albigularis*, Daudin, 1802) and a Sharpe's Grysbok (*Raphicerus sharpie*, Thomas, 1897) (Deacon & Rockingham-Gill, 2009). There is one record of a possible attack on a human child(Stjernstedt, 1975).

The Crowned eagle's prey is mostly mammalian, with larger than average species being taken, despite the weight range being similar to those mammals taken in the same weight range of those taken by the Martial and Verreaux's eagles (Brown, 1963). In one study of prey bones, it was found that about 52% of the bones were from mammals above an adult weight of 2 kg, with some from juveniles of species with an adult weight of nearly 20 kg (Tarboton & Allan, 1984). Shultz et al., 2004) found that the average weight of prey for Crowned eagles in the rainforest community of Taï National Park in the Ivory Coast, was 5.67 kg. Ferguson-Lees & Christie (2001) noted that the Crowned eagle had the highest weight range for prey among accipitrids, with only the female Harpy eagle as a serious rival. The Crowned eagle is among the few eagles to commonly take prey weighing over 9 kg (Brown, 1976; Ginn et al., 1989; Shultz, 2002; McGraw et al., 2006; McPherson et al., 2016a,b).

Crowned eagles are also among the few eagles that take primates regularly (other such species may be the Harpy, Martial and possibly the Crested eagles (Baldellou & Henzi, 1992; Mitani et al., 2001). For example, Mitani et al. (2001, 187) in a study of Crowned eagle diets in Kibale National Park, Uganda note that "results indicate that primates form the vast majority of all prey items" (82%), and that "a nontrivial fraction of the entire primate population at Ngogo succumbs to Crowned hawk eagle predation each year." Eagles took male primates more than females (although they may be larger), and killed more Red-tailed monkeys (*Cercopithecus ascanius*, Audebert, 1799)) than Grey-cheeked mangabeys (*Lophocebus albigena*, Gray, 1850) (above expected in terms of their numerical presence) and did not discriminate between adults and young (Mitani et al., 2001, 187).

Crowned eagles have been recorded taking prey in urban contexts (McPherson et al., 2016). This study notes that a population of Crowned eagles is present within the urban greenspaces of Durban and Pietermaritzburg, in KwaZulu-Natal, South Africa, and this may create tension as the eagle may prey on domestic stock and pets. Using Time-lapse cameras at urban nest sites to document the prey composition during breeding, the findings indicated that one common prey species was a bird, the Hadeda ibis (*Bostrychia hagedash*, Latham, 1790), the occurrence of which "was several times greater than any previous diet description" (McPherson et al., 2016a,b, 383). Another common prey species was the Rock hyrax, followed by vervet monkeys. A major finding was the low percentage representing domestic stock, which comprised only 6% of the identifiable prey. The findings refuted, at least locally the common perspective that domestic dogs and cats were frequent prey. No dogs were recorded as prey, and only a few cats were taken. However, young eagles, such as juvenile and sub-adults"are most frequently identified at in situ attacks of pets, particularly toy dog breeds" (McPherson et al., 2016, 383).

Population Status

The Crowned eagle is a sedentary, but juveniles disperse from breeding areas (Oatley et al., 1998; Bildstein, 2006). Ferguson-Lees and Christie (2001) placed the global population (defined as the number of adults and immatures at the start of the breeding season) in the range of 1,001 to 10,000 birds. BirdLife International (2018) placed an estimate at 5,000-50,000 mature birds and assessed this species as Near Threatened, with a decreasing population. This assessment is justified on the grounds that the Crowned eagle is "threatened by persecution through trapping, shooting and nest destruction, competition for prey from humans, and habitat loss through deforestation." BirdLife International (2018) further comments that "this suite of widespread threats implies that the species is in moderately rapid decline, especially considering its slow reproductive rate, although the decline is currently not thought to be more severe, owing to the species tolerance of modified habitats" (Ferguson-Lees & Christie, 2001; Thomsett, 2011).

BirdLife International also commented on the ability of the Crowned eagle to resist and adapt to human presence and landcover change: "It inhabits forest, woodland, savanna and shrubland, as well as some modified habitats, such as plantations and secondary growth" and notably it "can persist in small forest fragments including urban greenspace forests" and "shows high resilience to heavy deforestation and degradation in some areas" (see also Ferguson-Lees & Christie, 2001; Dowsett-Lemaire & Dowsett, 2006; McPherson et al., 2016a). Factors for its adaptability include the "use of exotic invasive trees (especially *Eucalyptus* and *Pinus* spp.) for nesting permits (see also McPherson et al., 2016a), "dietary plasticity" in terms of prey on new prey species, especially mammals (see also Swatridge et al., 2014) and the scarcity of competitive mesopredators "which may allow it to persist in certain areas… occur at higher densities than expected" (see also Swartridge et al., 2014; Symes & Antonites, 2014; McPherson et al., 2016b).

The population and conservation status of the Crowned eagle varies regionally. For example, Parker (1994, 1999, 2005) estimated the population as possibly as high as 1000 in Mozambique and 60 in Swaziland, while Irwin (1981) estimated 60 breeding pairs in Zimbabwe and Tarboton and Allan (1984) estimated the population in northeastern South Africa at 105 pairs. Ferguson-Lees and Christie (2001) reported that it is locally common, declining, and extinct in different areas, and declining prey has impacted on its populations (Tarboton & Allan, 1984; Simmons, 2005). Hence, BirdLife International raised its classification from Least Concern to Near Threatened in 2012. Global Raptor Information Network (2021ak) report that the population of this species has "declined greatly" in Liberia and Côte d'Ivoire, due to hunting pressure on the eagles and their prey species (see also Thiollay, 1985; Gatter, 1997) and in Malawi "the range of this species outside of protected areas is doubtless decreasing as the result of habitat loss, and it has been extirpated in some parts of the country" (see also Dowsett-Lemaire and Dowsett 2006). Deforestation is also a factor for eagle species decline in Mozambique (Parker, 1999). In southern Africa, it is uncommon and considered Vulnerable, due to declines in prey species, direct persecution due to its preying on poultry, deforestation and occasional electrocution on powerlines (Tarboton & Allan, 1984; Boshoff et al., 1994; Parker, 1994; Boshoff, 1997; Oatley et al., 1998; Simmons, 2005; Monadjem & Rasmussen, 2008). Illustrations of the appearance, range, and eggs of this species are shown in Figures 2.24a to 2.24i.

Genus Lophaetus (Kaup, 1847)

These are small species, with longish, broad wings and broad tails. There is a long occipital crest, and the plumage is mostly black, with strong similarities between the adults and juveniles. Brown and Amadon (1968, 690) note that they are "generally regarded as nearest to *Spizaetus*, but perhaps a derivative of *Aquila*-like stock", with buteo-like habits (ibid. 690). This genus currently contains one species. However, there are debates surrounding the configuration of this classification. For example, Lerner et al. (2017, 301) "recommend that the Long-crested eagle (*Lophaetus occipitalis*) and the Black Eagle (*Ictinaetus malaiensis*) remain in their monotypic genera, due to their distinctive morphology", this position being at variance with the work of Helbig et al. (2005) which "suggested including *Aquila clanga* and *A. pomarina* in the genus *Lophaetus* together with *L. occipitalis*." Lerner et al. (2017, 301) add that "as *Ictinaetus malaiensis* also belongs to this clade, the correct genus name for this clade would be *Ictinaetus* Blyth, 1843, which has priority over *Lophaetus* Kaup, 1847." Based on these assessments, currently only the Long-crested eagle is within the monotypic Genus *Lophaetus* (Kaup, 1847).

Fig. 2.24a. Crowned eagle perching [With permission from Derek Keats]

Fig. 2.24b. Crowned eagle perching [With permission from Ken Behrens]

Fig. 2.24c. Juvenile Crowned eagle perching [With permission from John Archer-Thomson]

Fig. 2.24d. Crowned eagle in flight [With permission from of Derek Keats]

Fig. 2.24e. Crowned eagle in flight [With permission from Ken Behrens]

Fig. 2.24f. Crowned eagle in flight [With permission from Ken Behrens]

Fig. 2.24g. Crowned eagle habitat in Congo [Credit: Andrey Gudkov]

Fig. 2.24h. Crowned eagle habitat in Kruger National Park [With permission from Michael Kemper]

2.26 Long-crested Eagle *Lophaetus occipitalis* (Daudin, 1800)

Physical Description

The Long-crested eagle is a small, buzzard-sized eagle, with a body length of 50 to 58 cm, a wingspan of 112 to 119 cm and weight ranges of 912–1363 g for males and 1367–1523 g for females (Kemp et al., 2020d). This species has an all-black plumage Brown and Amadon (1968, 690) note that this species is "unmistakeable" due to the "all black appearance" and the "long black flopping crest, which is blown about by the wind", with white wing patches. The adult plumage is very dark brown to black plumage. There are white patches at the carpal joint (above and below), white underwing coverts, white tail base and greyish brown tarsi. The eye is golden to rufous brown, and the cere and feet are yellow. The immature plumage includes light colored tips to the neck feathers, less pronounced crest, mottled feathers. The feet and cere are pale ochre-yellow, and the iris is dark olive brown. The Long-crested eagle is a noisy species, with a loud "keee-eh" or "keee-ee-ay" or kik-kik-kik-kik-keee (Brown and Amadon (1968, 690). Illustrations of the appearance, range, and eggs of this species are shown in Figures 2.25a to 2.25h.

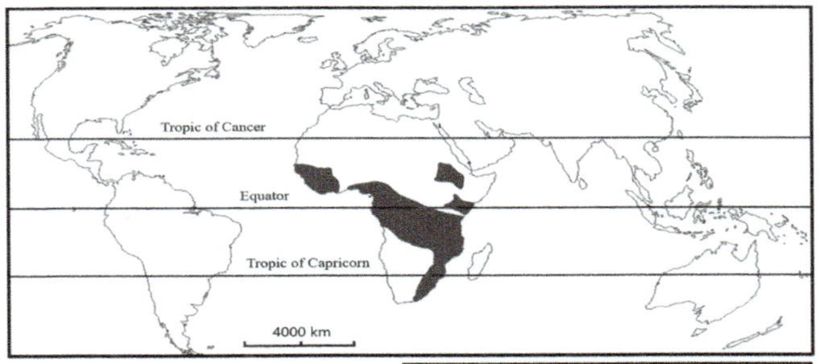

The Crowned eagle ranges through forest regions in subSaharan Africa, and into more open landcover in eastern and southern Africa.

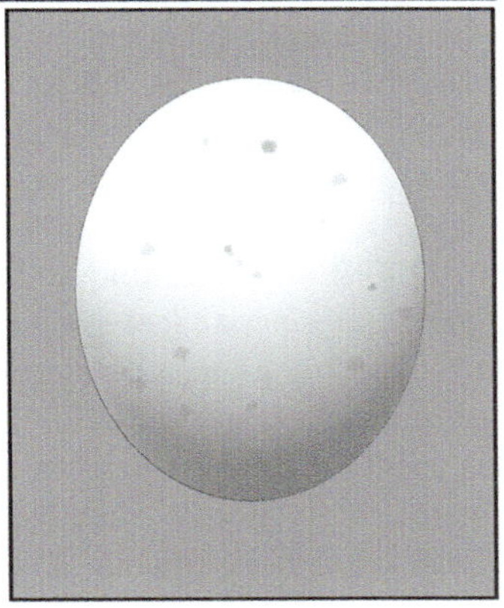

The egg is usually white, sometimes with a few reddish brown spots.

Fig. 2.24i. Crowned eagle range and egg

Classification

The Long-crested eagle was initially suggested to be linked to the Genus *Aquila* (Brooke et al., 1972), and was placed in the genus *Spizaetus* by Amadon (1982) and Amadon and Bull (1988). Later, more advanced studies using DNA sequences of one mitochondrial and three nuclear genes (see Helbig et al., 2005), established a link between the Long-crested eagle and the spotted eagles (Greater, Lesser and Indian Spotted eagles), and there was a recommendation that they all be placed in the genus *Lophaetus*. Lerner and Mindell's (2005) molecular phylogenetic study using DNA sequences of one nuclear and two mitochondrial genes, supported some of these assessments, and examined the possibly related status of the Asian Black eagle (*Ictinaetus malaiensis*, Temminck, 1822, currently in the monotypic Genus *Ictinaetus*, Blyth, 1843, not to be confused with the African Black eagle or Verreaux's eagle, *Aquila verreauxii*, Lesson, 1830). However, as noted above, the later work of Lerner et al. (2017, 301) recommended that the Long-crested eagle (*Lophaetus occipitalis*) and the Black eagle should remain in their monotypic genera, this because of their distinctive morphology compared with the spotted eagles.

Breeding

The Long-crested eagle indulges in nuptial displays above the breeding territory. Nests are located in leafy trees, commonly in river valleys. For example, in Swaziland, Monadjem and Rasmussen (2008) found that nests are usually located in tall trees, including trees in exotic plantations. The initial nest, composed of sticks and green vegetation, may be relatively small (60 cm wide and 30 cm high), but may be enlarged with time. Mostly two eggs are laid, colored dull white with shades of brown, grey and light lilac and some brown spots. In Zambia, Dowsett et al. (2008) counted 10 nests with one egg, and 14 nests with two eggs. In Malawi, Dowsett-Lemaire and Dowsett (2006) counted two nests with two eggs each. The laying periods vary across the range but are generally between July and December. Hatchlings are fledged in 55 days (Brown & Amadon, 1968). The Long-crested eagle's breeding habits may vary regionally. Global Raptor Information Network (2021al) remarks that "breeding occurs throughout the year in some areas, but is erratic, probably in response to fluctuations in rodent populations, which reflect rainfall conditions" (Irwin, 1981; Hall, 1992).

Much of the information on this species is however dated, and may require more recent additions. The nest is usually built concealed in foliage near the edge of woodland (Steyn, 1982, Hall, 1992; Kemp et al., 2020d). In Ghana, there are some records (Grimes, 1987; Wilson, 1993), while in the Democratic Republic of Congo, one young nestling was recorded early March 1942 and young flightless juvenile was recorded in late December 1949 (Vrydagh, 1949). In Eritrea, an active nest was reported in early April (Smith, 1947, 1955, 1957) while in Ethiopia Ash & Atkins (2009) found first laying dates for four nests as June, July and August. In Uganda, there were eggs reported in July (van Someren & van Someren, 1949) and in November (EANHS Nest

Records). In Angola, a downy juvenile was recorded on 8 September (Traylor, 1963), in Zambia, nesting was reported throughout the year (Dowsett et al., 2008) and in Zimbabwe most months recorded egg laying (Irwin, 1981; Baker, 2005). In Malawi, eggs were recorded in June, August and September (Dowsett-Lemaire & Dowsett, 2006). In Mozambique, there one nesting record in September (Parker, 2005). In South Africa, egg laying generally occurs between August and March (Jenkins, 1997).

Foraging

The Long-crested eagle is an adaptable bird, being "one of the eagles which prefers cultivated country and the haunts of man to uninhabited areas of bush" (Brown & Amadon, 1968, 690). This species frequents a broad range of habitats from dense forested areas, wooded savannas and moist woodlands near marshes, wetlands, and rivers. In some areas it adapts to deforestation where there are tree plantations or cultivated landcover (Hartley & Verdoorn, 2000). As it is an adaptable bird, its frequented habitats may vary regionally and locally. For example, in Ethiopia this species occurs mostly in wooded patches in wooded areas and particularly in acacia savanna (Ash & Atkins, 2009), whereas in Zambia and Malawi it appears to frequent more variable areas, such as ecotones (i.e. an area where two ecological communities interact); for example, where riparian vegetation is close to shallow wetlands and woodland patches, and forests are near swampy areas (Dowsett-Lemaire & Dowsett, 2006; Dowsett et al., 2008). In Mozambique, Botswana and Zimbabwe it occurs in different combinations of open landcover, woodlands, marshy grassland, riparian tree savanna and wetlands (Irwin, 1981; Penry, 1994; Parker, 2005).

The Long-crested eagle hunts more often from a perch than from soaring flight, mostly from treed, open landcover. For example, Dowsett et al. (2008) records still hunting from tall trees along forest or woodland margins in Zambia. Most of its time is spent on exposed perches in tall emergents, or tall trees and/or telegraph posts in open landcover, woodland patches or forest edges. Prey, mostly taken by diving from a perch, is mostly composed of small mammals, reptiles and in some cases domestic birds (Brown & Amadon, 1968, 690). In one area in Zimbabwe, most prey taken were small, diurnal rodents, especially the Vlei rat (Genus *Otomys*, F. Cuvier, 1824), and to a lesser extent the Water rat (possibly the Genus *Colomys*, Thomas & Wroughton, 1907), the four-striped grass mouse (*Rhabdomyspumilio*, Sparrman, 1784) and some shrews (Family *Soricidae*, G. Fischer, 1814) (Irwin, 1981). Irwin (1981) describes the Water rat as weighing about 1100 – 1150 g, which is about 10% of the weight of the Long-tailed eagle, and when killed by the eagle the rat is swallowed whole. In Zambia, Benon and Irwin (1967) reported that an eagle taken in 1964 had a small bird (Order *Passeriformes*, Linnaeus, 1758) in its stomach.

Population Status

The Long-crested eagle very large range, most of sub-Saharan Africa and the eastern third to half of southern Africa. This species is non-migratory, although it may disperse over breeding and habitat areas (BirdLife International, 2016). There is no evidence of regular seasonal movements of Long-crested eagles, using evidence from the atlas of southern African birds (Jenkins, 1997). However, despite its sedentary habits, it has been recorded moving between areas of varying food potential and local seasonal change (Irwin, 1981; Steyn, 1982; Gatter, 1997, Borrow & Demey, 2001). Movements may also result from human action, such as the clearing of vegetation in Uganda, which may attract Long-crested eagles (Carswell et al., 2005). Hall (1992) found, over a 16-year study in Mpulamanga, South Africa, that females were absent from their breeding ranges outside the breeding season. The main threats to the Long-crested eagle are habitat fragmentation and loss, this including the draining of wetlands, poisoning, and collisions with powerlines and vehicles (Hartley & Verdoorn, 2000). This species has prospered in some areas from the planting of exotic tree plantations, and it is also somewhat adaptable to modified habitats, so it may actually be more common in some regions than historically (Thiollay, 1985; Penry, 1994; Jenkins, 1997; Boycott & Parker, 2003; Carswell et al., 2005; Simmons, 2005; Dowsett-Lemaire & Dowsett, 2006; Hancock et al., 2007; Monadjem and Rasmussen, 2008).

Ferguson-Lees and Christie (2001) estimated the global population of adult and immature birds in the range of 10,001 to 100,000 individuals. BirdLife International (2016) rates this species as Least Concern. The justification for this assessment is the extremely large range, greater than 20,000 km^2, with little evidence of a "declining or fluctuating range size, habitat extent/quality, or population size and a small number of locations or severe fragmentation." The population trend, unlike many other eagle species in Africa, appears to be increasing, therefore is counted as Vulnerable. The latest BirdLife web source (BirdLife International, 2016) does not estimate the population but is probably above the threshold for a Vulnerable population size criterion, which is a continually declining population of less than 10,000 mature individuals. Citing Ferguson-Lees & Christie (2001), BirdLife International (2016) writes that "the population is suspected to be increasing as ongoing habitat degradation is creating new areas of suitable habitat. As a rodent eating species, it is regarded locally as beneficial and therefore is little persecuted." There are variants in rarity and local population density of this species. In West Africa, it is either a moderately common or uncommon in both forest and savanna zones, with some movements between ecosystems (Borrow & Demey, 2001). In East Africa, it is a widespread and common resident of moist woodlands and wooded savannas from sea level to 3,000 m (Stevenson & Fanshawe, 2002). Table 2.4 shows the more detailed local occurrences in western, eastern and southern Africa.

Table 2.4. Population status of the Long-crested eagle

Country	Status	Source
Gambia	Not uncommon	Gore, 1990
Liberia	Uncommon visitor	Gatter, 1997
Côte d'Ivoire	Widespread breeding resident	Thiollay, 1985
Ghana	Common breeding resident	Grimes, 1987
Togo	Possible breeding resident	Cheke & Walsh, 1996
Nigeria	Not uncommon resident	Elgood et al., 1994
Sudan	Common breeding resident	Nikolaus, 1987
Ethiopia	Common breeding resident	Ash & Atkins, 2009
Eritrea	Common breeding resident	Ash & Atkins, 2009
Somalia	Rare visitor	Ash and Miskell, 1998
Uganda	Common breeding resident	Carswell et al., 2005
Angola	Not uncommon resident	Dean et al., 1988, Dean, 2000
Zambia	Not uncommon resident, locally scarce	Dowsett et al., 2008
Malawi	Widespread breeding resident	Dowsett-Lemaire & Dowsett, 2006
Mozambique	Uncommon resident	Parker, 2005; Clancey, 1996
Namibia	Occasional visitor, vagrant	Jenkins, 1997
Botswana	Quite common to uncommon resident	Penry, 1994; Hancock et al., 2007
Zimbabwe	Widespread in forest, rarer in savanna	Irwin, 1981
Swaziland	Rare breeding resident	Monadjem & Rasmussen, 2008
South Africa	Quite common to uncommon resident	Tarboton & Allan, 1984

Throughout the cited countries, the main habitat is forest, forest edge, woodland, with some wooded savanna and savanna occurrences. The term 'not uncommon' means that it is not rare but is not numerous enough to be called common. Illustrations of the appearance, range, and eggs of this species are shown in Figures 2.25a to 2.25h.

Genus *Ictinaetus* (Blyth, 1843)

The single species of this genus is "rather large, but lightly built", characterized by a small bill, long tail, long wings, feathered legs, long, slightly curved talons and short outer toes; such peculiarities possibly adapted to hunting of tree top nesting birds (Brown & Amadon, 1968). Brown and Amadon (1968, 642) further note that "this peculiar genus is probably a much modified and specialised booted eagle, but there is some evidence in part anatomical, that it is an eagle-like kite." This is despite the existence of the feathered tarsi.

2.27 Black eagle *Ictinaetus malayensis* or *Ictinaetus malaiensis* (Blyth, 1843 or Temminck, 1822)

Physical Description

The Black eagle is an all black eagle, usually classified into two subspecies: the slightly smaller nominate subspecies *Ictinaetus malayensis malayensis* (Temminck, 1822) (ranging from Myanmar to southern China, South East Asia and Indonesia), and *Ictinaetus malayensis perniger* (Hodgson, 1836) (ranging through northern India and Nepal, and southern India and Sri Lanka) (Brown & Amadon, 1968). This is a large eagle but is comparatively lightly built. It reaches 75 cm in length, with a wingspan of 148 to 182 cm. Its weight reflects its light build, 1,000 and 1,600 g, which is half the weight of the similarly sized and partly sympatric mountain hawk-eagle. Both subspecies are black or blackish brown overall, with paler feathers on the secondaries and scapulars. In some cases, there are white markings on the upper tail coverts. The primaries have white on the bases, which creates a white patch at the carpal joint which may be visible in flight. The tail and wing quills have grey bands on the inner webs. The

Fig. 2.25a. Long-crested eagle perching [With permission from Frank McClintock]

Fig. 2.25b. Long-crested eagle perching [With permission from of Derek Keats]

Fig. 2.25c. Long-crested eagle perching [With permission from Miguel Siu Rios]

Fig. 2.25d. Long-crested eagle in flight [With permission from Frank McClintock]

Fig. 2.25e. Long-crested eagle in flight [With permission from Derek Keats]

Fig. 2.25f. Long-crested eagle in flight [With permission from Roel of van der Breggen]

Fig. 2.25g. Long-crested eagle habitat [Credit: Leon Labuschagne from Pexels]

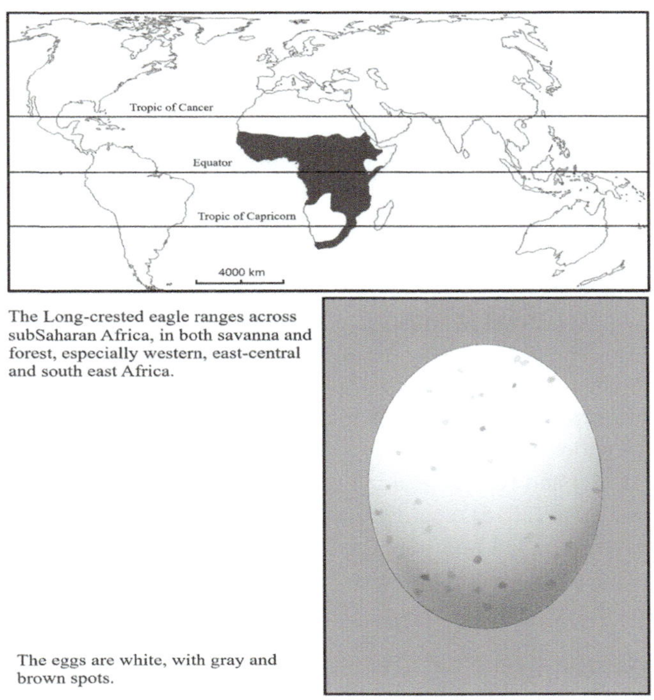

Fig. 2.25h. Long-crested eagle range and egg

eye is brown, the cere and feet are yellow (Brown & Amadon, 1968, 632; Ali & Ripley, 1978; Rasmussen & Anderton, 2012; Brazil, 2009).

The immature bird is lighter colored, generally sepia brown, with golden buff edges to the feathers on the crown, nape and some of the back feathers. The head sides and the cheeks are golden brown or buff colored. The breast is streaked with black, and the tail coverts are barred with black. The upper tail coverts have sepia and buff bands, with the one paler band at the tail base, while the underside is reddish gold. The underwing coverts and a connected patch on the upper wing at the carpal joint colored light brown. The juvenile has fewer bands than adult on the wing and tail coverts. The eyes are brown, and the cere and feet are yellow as in the adult. Both sub-species have very long primary feathers, which results in these feathers of the folded wing extending beyond the tip of the tail. Despite eagle-habits, the feet are weaker than those some other eagles. Brown and Amadon (1968) note that this species is silent, but a captive specimen once gave a loud call "ah – a" and a shrill cry. Illustrations of the appearance, range, and eggs of this species are shown in Figures 2.26a to 2.26m.

Classification

Recent studies using on DNA sequences of nuclear and mitochonrial genes (Bunce et al., 2005; Lerner & Mindell, 2005; Gjershaug, 2006; Väli, 2006), found that the Black eagle is a member of a clade, which includes the Greater Spotted and Lesser Spotted eagles (Eurasian species), and the Long-crested eagle (African species). The two Eurasian species were previously included in the Genus *Aquila*. Another study based on sequences of the mitochondrial cytochrome b gene, discovered links with Asian hawk eagles of the Genus *Nisaetus* (Haring et al., 2007). Global Raptor Information Network (2021am) observes that these results are not conclusive, and "pending further studies to clarify these relationships and deal with their taxonomic implications, the genus *Ictinaetus* is maintained here." The spelling of the second part of the binominal name varies: many authors use the term *malayensis*, but some use the original spelling *malaiensis* used by Temminck (Lebossé & Bour, 2011).

Foraging

The Black eagle nominate subspecies (*I.m. malaiensis*, Temminck, 1822) ranges across Myanmar, southern China (Yunnan, Fujian) and Taiwan, into Southeast Asia. This subspecies is usually non-migratory (Tordoff, 2002). One study in southern India found it frequented dense forests with over 50% cover (Thiollay, 1993). The other subspecies (*I.m. perniger*, Hodgson, 1836) occurs from the Himalayan foothills west through Nepal into India (the states of Himachal Pradesh and Jammu & Kashmir, the Eastern and Western Ghats) and Sri Lanka. To the west it also occurs in forested Gujarat and the Aravalli range of northwestern India (Burgess, 1937; Prater, 1940; Rao, 1968; Dharmakumarsinhji, 1985; Clark, 1994; Ferguson-Lees & Christie, 2001; Brazil, 2009; Parasharya, 2010). In these areas, it frequents variable forests, usually up to 300 meters elevation. These forests are mostly moist deciduous and evergreen lowland, hill, and montane forest (Wells, 1999).

Foraging commonly occurs by soaring low over and within tree crown and edges, clearings, and riverine forests, singly or in pairs (Coates & Bishop, 1997; Wells 1999). Its foraging habits have been described as "peculiar" which is detailed as wings held level with the back, with primaries separated; it has also been described as "harrier-like" in gliding low over hillsides (Brown & Amadon, 1968, 643). Its capacity to fly very slowly is attributed to its very long primaries, is useful in attacks on bird nests. Bird eggs and young on tree tops are frequently taken. A possible function of the comparatively less curved talons is the taking of whole nests, to take the living contents and the disengagement of the claws from the nest structure (difficult with strongly curved talons) (Brown & Amadon, 1968, 643).

One case study of the Malay Peninsula, found most birds frequented montane forest, rather than lowland forest and even over highland tea plantations (Wells, 1999). Another case study, of Sumatra found a bird flying between trees over dense understory vegetation in a pine plantation (Wells, 1999). In Sabah, it has been observed flying in or over primary, lowland, and hill forest (Sheldon et al., 2001). Prey taken also includes mammals (including bats and bonnet macaques, *Macaca radiata*, É. Geoffroy, 1812)) and reptiles (Smythies, 1968; Ali, 1986). Sheldon et al. (2001) recorded birds, bats, eggs, rodents, and lizards in Sabah. Lei et al. (2015) recorded Red and White Giant squirrel taken in China (see also Borges, 1986 on India). In one case, a Sri Lankan chick was fed only on Giant squirrels (*Ratufa macroura*, Pennant, 1769) (65%) and rats (Genus *Rattus*, Fischer de Waldheim, 1803 (35%) (Gunawardena, 2006). In another case in Perak, Malaysia, the stomach of an adult female contained remains of a rat, a bird egg, and a snake egg (Sharpe, 1888).

Breeding

Breeding begins with aerial displays, including plunging with closed wings from hundreds of meters over mountain sides. This display has been compared with those of Ayre's Hawk eagle (*Hieraaetus ayresii*, Gurney, 1862, formerly, *Hieraaetus dubius*) and Mountain buzzard (*Buteo oreophilus*, Hartert and Neumann, 1914). The acrobatics included paired displays, mutual chasing and undulating flights, near or within forests. Tree nesting, especially in the canopies of large trees is common, while crag nesting is rarer (Wells, 1999; Rasmussen and Anderton, 2012). The large nests are constructed of sticks and green leaves may be over one meter across and 60 centimeters deep (Brown & Amadon, 1968, 643; Ali & Ripley, 1978). One nest in Sri Lanka was located 25 m high in the main branches of a tree in a comparatively inaccessible location (Gunawardena, 2006). One (or rarely two) white to creamy, brown blotched, black spotted and grey marked egg is laid. Some eggs may be colored pink or reddish ochre, with darker red marks (Brown & Amadon, 1968). Laying occurs December (in southern India), November to February (India) and September to May (Assam, Java). However, "full details of the incubation, behavior and the fledgling period are lacking (Brown & Amadon, 1968, 643). Gunawardena (2006) estimated the incubation period to be 35-40 days, based on the nest cited above, and the fledging at about 60 days. Older records also exist (Buchanan, 1899; Daly, 1899; Baker, 1918).

Population Status

Lei et al. (2014, 45) note that "to date there have been relatively few studies of the Black eagle and most have focused on its breeding biology. To our knowledge, there are no specific accounts in the literature of this species in mainland China" (e.g. Soleha, 2000; Chiu & Lim, 2003; Lim & Hut, 2005; Lin, 2005; Naoroji, 2006; Lin & Lin, 2010; Samarawickrama et al., 2011). Nevertheless, the Black eagle is classified as of Least Concern (BirdLife International, 2016). The reasons for this assessment are that it has an "extremely large range", therefore it is not vulnerable, and the decreasing population is not rapid enough to be vulnerable and the population, though small, is not small enough to be vulnerable. It is nevertheless assessed as rare by many authors (Dharmakumarsinhji, 1985; Thiollay, 1993; Fernando et al., 2001; Gunawardene, 2002; Tordoff, 2002; Parasharya, 2010; Samarawickrama et al, 2011). Samarawickrama et al. (2011) note that "observing a perched Black eagle is considered to be a rare occurrence; even one of Sri Lanka's greatest ornithologists, G.M. Henry, never observed one perched. But its slow and elegant flight has caught the attention of many naturalists, including that of Henry's." Citing Henry, they quote his statement "I have never seen it perched, and the feature of its flight that has particularly struck me is the slowness with which it can sail without any wing-flapping, even when it finds itself in a deep pocket among the trees where it would seem inevitable that it must either crash or extricate itself by flapping flight. It seems to be able to avoid stalling even at what appears little more than walking pace, and certainly at very low speeds" (Henry, 1933). Deforestation is one of the more important factors for its rarity across its range. In the Malay Peninsula, its presence declines where human presence increases (Wells, 1999). In terms of population estimates, Ferguson-Lees and Christie (2001) estimated the total population, including adults and non-breeding immatures at the beginning of the breeding season, at between 10,000 to 100,000 birds. BirdLife International (2016) cites this figure, and also notes that national population sizes vary (e.g.< c.100 breeding pairs in China and c.100-10,000 breeding pairs in Taiwan) (see Brazil, 2009).

There are regional variations in occurrence. It is generally a locally common resident from northeastern Pakistan eastwards along the foothills of the Himalayas from Himachal to Bhutan, and over the South Assam Hills and Lusha Hills, the northeastern and southeastern Ghats, central India, southwards to Sri Lanka from lowlands to over 3,000 m (Rasmussen & Anderton, 2012). It is listed as fairly common or regularly distributed in Nepal and Bhutan, but rarer in Myanmar, the Malay Peninsula, Sumatra, Sabah, Bangladesh and Wallacea (Smythies, 1986; van Marle & Voous, 1988; Inskipp & Inskipp, 1991; Coates & Bishop, 1997; Jeyarajasingam & Pearson, 1999; Wells, 1999; Sheldon et al., 2001; Spierenburg, 2005). Through these areas it is either resident or a partial resident (Bildstein, 2006).

Lei et al. (2014, 48) cite authors who attempted population assessments in many Asian nations; for example, Thiollay (1998), on impacts of forest fragmentation in tropical Asia; Brazil (2009) on its rarity in eastern mainland China; Si et al. (2011), Wu et al. (2012) and Ma et al. (2012) on possible conservation actions for this species; Jiang et al. (2013) on the nationally 'Threatened' status in China; Zheng & Wang's (1998) designation as "Near Threatened"; Jiang & Luo (2012) on new designations of its conservation status such as a combination of "distribution area, life history, ecological function, anthropogenic interference and special use data"; and on Wang & Xie's (2009) non-inclusion of this species. Illustrations of the appearance, range, and eggs of this species are shown in Figures 2.26a to 2.26m.

Fig. 2.26a. Black eagle perching [With permission from Kalyan Varma – kalyan@rtns.org]

Fig. 2.26b. Black eagle perching [Credit: THEGIFT777]

Fig. 2.26c. Black eagle in flight [With permission from Mike Nelson-Birdtour ASIA]

Fig. 2.26d. Black eagle in flight [With permission from Nick Athanas]

Fig. 2.26e. Black eagle in flight [With permission from Chien Lee]

Fig. 2.26f. Black eagle in flight [With permission from Chien Lee]

Fig. 2.26g. Black eagle in flight [With permission from Kalyan Varma – kalyan@rtns.org]

Fig. 2.26h. Immature Black eagle in flight [With permission from David Beadle]

Fig. 2.26i. Black eagle collecting branches for nest [With permission from Kalyan Varma – kalyan@rtns.org]

Fig. 2.26j. Black eagle collecting branches [With permission from Kalyan Varma – kalyan@rtns.org]

Fig. 2.26k. Immature Black eagle on nest [With permission from Kalyan Varma – kalyan@rtns.org]

Fig. 2.26l. Black eagle habitat, Cameron Highland, Pahang, Malaysia [Credit: Kelvin Yup-unsplash]

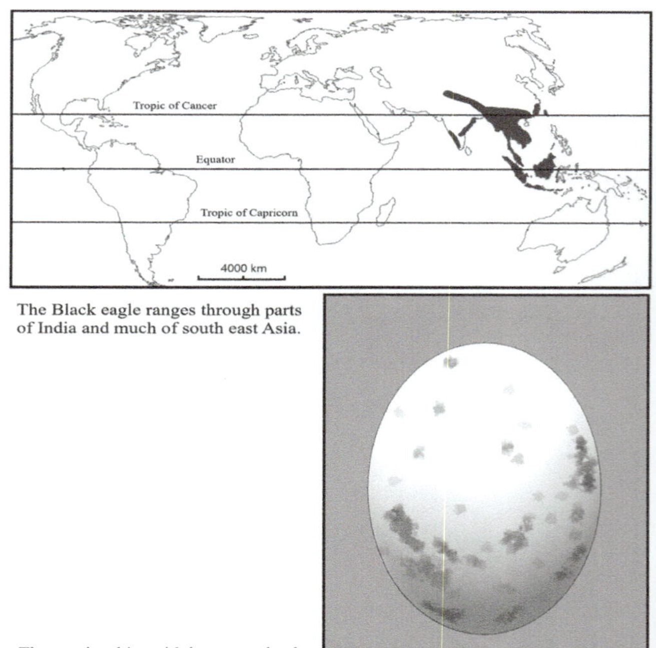

Fig. 2.26m. Black eagle range and egg

3

Systematic List of the Serpent Eagles *Circaetinae*: Genera *Circaetus*, *Eutriorchis*, *Pithecophaga*, *Spilornis* and *Terathopius*

3.1 Introduction

The Serpent eagles of the subfamily *Circaetinae* contain several genera of medium to large eagles, characterized by their broad wings and mostly snake diet (except for the Bateleur and Philippine eagles, that take more general diets); hence the name Snake eagle or Serpent eagles. The Subfamily *Circaetinae* includes five genera and sixteen species. The Genus *Circaetus* includes the Banded Snake eagle, Beaudouin's snake-eagle, Black-chested snake-eagle, Brown snake-eagle, Fasciated snake-eagle, and Short-toed eagle. The Genus *Eutriorchis* includes the Madagascar serpent-eagle, the Genus *Pithecophaga*the Philippine eagle, the Genus *Spilornis* the Andaman serpent-eagle, Crested serpent-eagle, Great Nicobar serpent-eagle, Kinabalu serpent-eagle, Philippine serpent-eagle, Sulawesi serpent-eagle, and the Genus *Terathopius* the Bateleur eagle.

Gilliard (1958, 106) writes that "the fourteen species of fierce, handsome eagles of the subfamily *Circaetinae* are found only in Eurasia, Africa and Madagascar. All have disproportionally large heads crowned with fan-like expandable crests. All feed primarily on reptiles." Gilliard (1958, 106) adds that the "best known is the Short-toed Harrier eagle" and the "most majestic of all is the Bateleur eagle." According to Brown and Amadon (1968) all the species range in tropical Africa and south Asia (*Spilornis* and *Pithecophaga*) except for the short-toed eagle, migrates between temperate Eurasia and Africa, and is resident in India. Lerner (2007, 7) describe *Circaetinae* as "Old World species feeding mainly on snakes, other reptiles and small mammals, have a reticulate pattern of heavy scales on the tarsi and relatively short toes" (citing Peters, 1931).

Lerner (2007, 34) further notes that "all of the snake eagles form a monophyletic group sister to the Old World vulture group *Aegypiinae*, except the Madagascar serpent-eagle (*Eutriorchis astur*) which is placed within the *Gypaetinae*" and there is evidence that the Philippine eagle is more closely related to the snake eagles than to the Harpy eagle group. In fact, if the "Madagascar serpent-eagle is excluded, the snake eagles form a monophyletic subfamily (*Circaetinae*) that is most closely related to some Old World vultures (*Aegypiinae*) and the Philippine eagle than to other *Accipitridae*." The vulture groups *Aegypiinae* and *Gypaetinae* are described in Table 3.1. The dynamic nature of the classification of this subfamily will be discussed further in the rest of this chapter.

Genus *Circaetus* (Vieillot, 1816)

These eagles, though fairly large birds, are small to medium sized eagles, distinguished by their large sometimes owl-like, small, crested heads and eyes, long wings, medium-length tails, longish legs and shortish, scaly toes and claws. Referring to the scaly toes, Brown and Amadon (1968, 339) note that "the heavy scalation is believed to be an adaptation to deflect the bites of serpents, which form the principal food." These authors further note that *Circaetus* is closely related to the Genus *Spilornis,* especially the Western Banded Snake eagle (*Circaetus cinerascens*, von Müller, 1851) and Southern Banded Snake eagle (*Circaetus fasciolatus*, Kaup, 1850). However, the *Circaetus* species are usually "larger, longer-winged with a simple color pattern and less developed ruff" (Brown & Amadon, 1968, 339). It is also observed that the Short-toed snake eagle (*Circaetus gallicus,* Gmelin, 1788) resembles the Congo serpent eagle (*Dryotriorchis batesi*, Sharpe, 1904 or *Dryotriorchis spectabilis*, Shelley, 1874), which is also classified as *Circaetus spectabilis* (Schlegel, 1863). The Short-toed snake eagle, Beaudouin's Snake eagle (*Circaetus beaudouini*, Verreaux & Des Murs, 1862) and Black-chested Snake eagle (*Circaetus pectoralis*, A. Smith, 1829) were also regarded as

Table 3.1. Sub-families, genera and species of the accipitrid vultures

Sub-families	Genera	Species
Gypaetinae	*Gypaetus* Storr (1784)	Bearded vulture (*Gypaetus barbatus*, Linnaeus, 1758)
	Gypohierax (Rüppell, 1836)	Palm-nut vulture (*Gypohieraxan golensis*, Gmelin, 1788)
	Neophron (Savigny, 1809)	Egyptian vulture (*Neophron percnopterus*, Linnaeus, 1758)
Aegyptiinae	*Aegypius* (Savigny, 1809)	Cinereous vulture (*Aegypius monachus*, Linnaeus, 1766)
	Gyps (Savigny, 1809)	Himalayan vulture (*Gyps himalayensis*, Hume, 1869)
		Cape vulture (*Gyps coprotheres*, Forster, 1798)
		Griffon vulture (*Gyps fulvus*, Hablizl, 1783)
		Rüppell's vulture (*Gyps rueppelli*, Alfred Brehm, 1852)
		White-rumped vulture (*Gyps bengalensis*, Gmelin, 1788)
		Indian vulture (*Gyps indicus*, Scopoli, 1786)
		Slender-billed vulture (*Gyps tenuirostris*, Hodgson (Gray, 1844)
		White-backed vulture (*Gyps africanus*, Salvadori, 1865)
	Torgos (Kaup, 1828)	Lappet-faced vulture (*Torgostrac heliotus*, Forster, 1791)
	Trigonoceps (Lesson, 1842)	White-headed vulture (*Trigonoceps occipitalis*, Burchell, 1824)
	Sarcogyps (Lesson, 1842)	Red-headed vulture (*Sarcogyps calvus*, Scopoli, 1786)
	Necrosyrtes (Gloger, 1841)	Hooded vulture (*Necrosyrtes monachus*, Temminck, 1823)

Sources: Grimmet et al. (1999), Ferguson-Lees & Christie (2001), Lerner & Mindell (2005).

conspecific, or even a superspecies (Brown & Amadon, 1968). Currently, there are six recognized species, seven if the Congo snakeeagle is included here rather than in the monotypic genus *Dryotriorchis*: Short-toed snake eagle, Black-chested Snake eagle, Beaudouin's Snake eagle, Brown Snake eagle, Southern Banded Snake eagle or fasciated Snake eagle and the Western Banded Snake eagle (Brown et al., 1982; Sibley & Monroe, 1990, 1993; Dowsett & Forbes-Watson, 1993; del Hoyo et al., 1994; Tucker & Heath, 1994; Snow & Perrins, 1998; Clark, 1999; Ferguson-Lees & Christie, 2001; Thiollay, 2007; BirdLife International, 2015).

3.2 Short-toed Snake Eagle (or Short-toed Harrier Eagle, Black-breasted Harrier Eagle, or Beaudouin's Harrier Eagle) *Circaetus gallicus* (Gmelin, 1788)

Physical Description

This is a small-medium sized eagle. Adults are 62–67 cm in length, with a 170–185 cm wingspan and, with weights ranging from 1.2–2.3 kg. Predominate features include the owl-like rounded head and the large bright yellow eyes, the mostly greyish brown upperparts and white underside, and the pale brown throat and upper breast. The tail has three to four bars on the tail and light barring under the wing (del Hoyo et al., 1994). There are slight variations among the subspecies. In the nominate subspecies (*Circaetus gallicus gallicus*, Gmelin, J.F., 1788), the male and female plumage is mostly similar. The adult has greyish brown upperparts and tail, and white underparts including the belly. There is dark barring on the flight feathers and tail, and dark primaries with lighter coloration in the inner webs and two to three lighter bars on the wing quills. The tail has three dark bars on the tail. The head is brown and white, with brown cheeks and throat, with white bases, with a white forehead and around the eye. The eye is orange yellow, the cere and feet greyish white, with a bluish, darker tipped bill.

The eastern Asian subspecies *Circaetus gallicus heptneri* is the largest subspecies, but Brown and Amadon (1968, 340) argue that the variation in size "seems more of a geographical cline of increased size from west to east than true racial variation"; other "races" or subspecies are more "distinct" and are regarded by some authorities as a separate species. The subspecies *Circaetus gallicus pectoralis* is also smaller than *C.g. gallicus* but slightly larger than the East African subspecies *Circaetus gallicus beaudiouni*, which is the smallest subspecies. As argued by Brown and Amadon (1968, 340) "as beaudouni interbreeds with pectoralis and is clearly intermediate between it and *C. gallicus*, having a dark immature plumage but adult plumage very similar to that of *C.g. gallicus*; it seems preferable to regard all three as races of *C. gallicus*." Kemp (1994) suggested that the Lesser Sundas (Southeast Asia, north of Australia) population may be subspecifically distinct from the rest of the population of *Circaetus gallicus heptneri*.

The juvenile has is paler colored, with pale reddish-brown feathers on the throat and breast and more white coloration on the nape, white undersides, edged on with light brown on the sides of the torso, and subterminal bars on the thighs and part of the underside. The eyes are yellow and the cere and feet are pale grey. The transition to the adult involves lightening and dark streaks on the breast feathers and increased barring.

There are few similar species. The Martial eagle may be mistaken for the Short-toed snake eagle, but the current species differs in having a bare tarsus, white lining on the wing and white wing coverts and breast and belly, while the Martial eagle has spots on the white underside. The juveniles of the subspecies *C.g. pectoralis* and *C.g. beaudouni* are like the adult and juvenile *C. cinereous* and the immature Bateleur eagle (*Terathopius ecaudatus*, Daudin, 1800). *C.g. pectoralis* and *C.g. beaudouni* however have more reddish underparts (especially the former), longer tails and three or four dark bars on the grey underwing. The call has been described as: "musical whistling"; in soaring a call of "huopp" or "hu – o – hu – opp" or a high-pitched "pieou – pieou, quo – quo – quo"; a call of C.g. pectoralis as a high pitched "schreee" and at nests the call is melodic, woo – dlay – ooo, weeu, weeu, weeu; the alarm cry being a sharp "puopp, puoop; and the threat display call "u – ok – ok , ue – ok – ok – ok – ok (Brown & Amadon, 1968, 349). Illustrations of the appearance, range, and eggs of this species are shown in Figures 3.1a to 3.1j.

Classification

The Short-toed snake eagle has been found to be related to the *Aegypius* clade of Accipitrid vultures, using morphological assessments (Jollie, 1976, 1977a,b; Mundy et al., 1992). Further research found that the Short-toed eagle and the Bateleur eagle formed a monophyletic clade (Wink & Sauer-Gürth, 2000). The supporting links were largely based on similarities in foraging activities, prey taken and breeding patterns (see also Sibley & Monroe, 1990). Although some regard this as a species, others have classified it as a superspecies, including *C. beaudouini* and *C. pectoralis*, which are usually seen as a subspecies of *C. gallicus* (Clark, 1999).

Breeding

Nests are usually located in trees (del Hoyo et al., 1994), sometimes in hollows, or pine or oak trees, an example being in Black (*Pinus nigra*, J.F. Arnold) or Calabrian Pines (*P. brutia*, Tenore) in the Dadia Forest Reserve of Greece (Vlachos & Papageorgiou, 1994; Shirihai, 1996). These trees may be located on south facing slopes in Europe, while in Africa, the commonest nests may be in Acacia or Euphorbia crowns and in India in thorny trees (Brown & Amadon, 1968). In some cases, nests may be situated on cliffs. In Hungary, it has been observed nesting in pine patches of oak stands and mixed deciduous forests on south-facing slopes in mountainous areas (Hadarics & Zalai, 2008). In Russia, breeding has been recorded in wetlands in the northern Novrogod District (Bakka & Kiseleva, 2009). The nests are comparatively small, compared with those of other eagles, usually under one meter in width and depth. In some cases, the nests of other species may be used, such as those of the Brown-necked raven (*Corvus ruficollis*, Lesson, 1830) in Morocco. One large, rounded, white smooth egg is laid (Vlachos & Papageorgiou, 1994; Shirihai, 1996; Adamian & Klem, 1999). Pair formation may be in April (Adamian & Klem, 1999).

Laying seasons may be late April to early June (Europe), January to March, April, November or October (India, South Africa), November (Ethiopia, Northern Nigeria), December (Somalia) or February to May, July to October (Zimbabwe). The nestling period may be 68-80 days, for example in Israel (Shirihai, 1996). The incubation period is 47 days in Greece (Vlachos & Papageorgiou, 1994) and 42 to 47 days in Israel (Shirihai, 1996). Fledging in Europe is after 70 to 75 days, and 90 days in Africa (Brown & Amadon, 1968). There is considerable regional variation in the egg-laying. For example, recorded times are 10 to 20 April (Greece, Vlachos & Papageorgiou, 1994); April to May (Israel, Shirihai, 1996); early April to mid-May (Morocco, Thévenot et al., 2003; late March to early May (Algeria, Heim de Balsac & Mayaud, 1962, and Tunisia, Isenmann et al., 2005); early May (Armenia) (Adamian & Klem, 1999); and early April (Saudi Arabia, Jennings, 1996). The breeding success may be variable, but in one example in Israel (1983 to 1985), it was one average 0.48 fledglings annually (Meir, 1988).

Foraging

The Short-toed eagle's habitats are variable, usually with some forest or tree cover, including montane areas up to 2,300 m. Brown and Amadon (1968) describe this species as a habitat generalist, frequenting open woodland, grassland and open desert, and even rocky hill sides. Adamian and Klem (1999) point out that despite its variable foraging grounds, it generally avoids dense forest. Variable foraging grounds include damp lowland forests patches of open, grassed fields and even wetlands. In the southern drier parts of the breeding range, it may frequent open, arid rocky and/or sandy terrain, with some cultivated landcover and sometimes cliffs (Adamian & Klem, 1999). Foraging areas may vary regionally (Global Raptor Information Network, 2021an). For example, in some areas it is recorded following grassland fires (Coates & Bishop, 1997), as in the Malay Peninsula (Wells, 1999). Prey is usually taken from soaring flight, usually at low elevations to a greater extent than other *Circaetus* species. It may also perch on tall trees and on large exposed bare branches (Coates and Bishop, 1997). The diet is mostly snakes, or other reptiles (del Hoyo et al., 1994). Common snakes taken include large cobras (Family *Elapidae*, F. Boie, 1827), and smaller water and grass snakes (*Serpentes* spp., Linnaeus, 1758).

In Europe, the Short-toed eagle occurs in dry rocky and mountainous areas with suitable reptilian presence (Romania, Roberts, 2000) and in Black and Calabrian Pines, and mountainous beech and mixed forest (Greece, Handrinos & Akriotis, 1997). In this region it has been recorded as taking mostly reptiles (87.2%), birds (6.9%), and rodents (5.9%) (Vlachos & Papageorgiou, 1994). Common reptile species taken included Montpelier snakes (*Malpolon monspessulanus*, Hermann, 1804) (29.4%), Grass snakes (*Natrixnatrix*, Linnaeus, 1758) (20.6%), lizards (*Lacerta* sp.) (18.7%) and European Glass lizard (*Pseudopusapodus*, Pallas, 1775 or *Ophisaurus apodus*, Mertens & L. Müller, 1928) (11.8%) (Vlachos & Papageorgiou, 1994). One bird took a European rattlesnake or Leopard snake (*Zamenis situla*, Linnaeus, 1758, or *Elaphe situla*, Engelmann et al., 1993) near Thessaloniki, Macedonia (Makatsch, 1950). In Asia, it occurs in dry, semi-open landcover during migration and breeding, usually mountainous or hilly areas with mixed woodland, bushy and open plains, in Israel (Shirihai, 1996). In Kazakhstan, it occurs in wetlands in lowland and upland steppes, and variable deserts (Wassink & Oreel, 2007). Mammals such as hares and rabbits are also taken, as are lizards, but fish is "positively refused by young birds" (Brown & Amadon, 1968, 341). In Armenia, it has been recorded with similar diets as in Europe (reptiles, fish, amphibians, small birds, and mammals, Dal, 1954). It avoids snakes with strong venom, such as the Common viper (*Viperaberus*, Linnaeus, 1758) in Europe and the Puff adder (*Bitis arietans*, Merrem, 1820) in Africa, as it lacks protection against their venom.

Population Status

The Short-toed eagle's range is Afrotropical/Indomalayan/Palearctic, covering southwestern and southern and central Europe, northwestern Africa north to the Gulf of Finland, to Iran to India and northern Mongolia and China (Xinjiang), and Africa. The northern populations are migratory to Africa (especially the southern Sahara, the Sahel zone and northern eastern Africa), India and Southeast Asia, including more rarely Sumatra, Java, and Bali, and the lesser Sunda Islands. There are resident populations in tropical Africa and India, which may intermingle with migrants (Sibley & Monroe, 1990, 1993; Tucker & Heath, 1994; Snow & Perrins, 1998; Thiollay, 2007).

Table 3.2. Migration Patterns of the Short-Toed Eagle

Location	Dates	Authors
Gibraltar	February to mid-March, declining to July	Cortes et al., 1980, Finlayson, 1992; Garcia & Bensusan, 2006; Forsman, 1999; Garcia, 2006
Romania	End of September, March	Roberts, 2000
Greece	Early September to mid-October, return from Africa early or mid-March to mid-April	Handrinos and Akriotis, 1997
Syria	Mid-September to early October, mid-March to mid-April	Baumgart et al., 1995
Lebanon	Early September to late October, early March to late April,	Ramadan-Jaradi et al., 2008
Israel	Wintering mid-August to November, February to May, summering March-April to Aug-September	Shirihai, 1996
Morocco	September to November, February to June	Thévenot et al., 2003
Egypt	February to May	Hilgerloh et al., 2009
Eritrea	September to April	Smith, 1957; Welch & Welch, 1986; Welch et al., 1992
Mauritania	September to October, February-March to mid-May	Isenmann et al., 2010
Djibouti	March	Welch & Welch, 1998
Ethiopia	August to April, outside dates May, June	Ash & Atkins, 2009
Kazakhstan	August to October, February to May	Wassink & Oreel, 2007
Armenia	October, April and departing in early October	Adamian & Klem, 1999
Tunisia	Nesting residents February/March to September/ October. Passage birds September/October and March or April (Ingelmann et al., op cit.).	Isenmann et al., 2005

For the Short-toed eagle, the periods during the first half of the year represent the spring passage northwards, while those during the second half of the year represent the autumn (fall) passage southwards. This species is migratory in some areas (mostly in the Palearctic, wintering in North Africa and Northern West Africa (there reaching the range of *C.g. beaudouni*) and eastwards to Kenya and the Indian subcontinent, moving south August to November and north February to May (Brown et al., 1982; del Hoyo et al., 1994; Snow & Perrins, 1998; Ferguson-Lees & Christie, 2001; Bildstein, 2006). Migratory routes to North and West Africa are through Gibraltar, and to East Africa through Israel and Suez to Eritrea (Brown & Amadon, 1968). During migration in the winter range, local movements relate to rainfall patterns (Adamian & Klem, 1999; Bildstein, 2006).

Local movements have been recorded in various areas (Global Raptor Information Network, 2021an). Two birds were radio tracked from France to Algeria and then on to Mali in September to October 96 (Meyburg et al., 1996, 1998). Another bird was radio tracked from France through the Mali-Mauritania border to southeast Mauritania (Meyburg et al., 1998). One bird that was ringed in Hungary was found in Syria four years later, at a location 2,050 km from its ringing site (Szitta, 2009). Meyburg et al. (1996, 1998) tracked two birds from Europe southwards, and found that the first bird, a juvenile, took 30 days to cover 4,045 km, while the second bird an adult took 20 days to cover 4,685 km, flying the autumn migration from West Central France to south-west of Niger. The passage of migratory birds to their destinations is therefore very wide ranging. The Global Raptor Information Network (2021an) gives a comprehensive list of the literature and findings on the migration patterns of this species, which is summarized in Table 3.2. Migration routes and residency/vagrancy records are displayed in Table 3.3.

Table 3.3. Migration routes

Location	Migration routes, residency	Authors
Hungary, southwards	The eastern flyway	Szitta, 2009
Italy, southwards	Apuane Alps, Arenzano, Stura Valley, avoiding Northern Tyrrhenian Sea	Ruggieri et al., 2006
Italy, southwards	Central Mediterranean flyway: Italian Peninsula, Sicily, Marettimo island, to Tunisia, narrowest point of Mediterranean	Agostini et al., 2009
Italy, northwards	Northwest Africa, Strait of Gibraltar, to Italy via Spain/France, avoiding the Central Mediterranean flyway	Agostini et al., 2009
Greece	Spring, w-NW in Thrace, from Bosphorus and/or Dardanelles; westerly direction over Lake Ismaris; Crete, spring/autumn	Handrinos & Akriotis, 1997; Handrinos, 1987; Vagliano, 1984
Indonesia	Timor in the direction of Atauro island 30 km north	Olsen & Trost, 2007
Gibraltar	Common migrant, in autumn occasional in summer	Garcia, 2008

Ruggieri et al. (2006) found juveniles migrate with adults, possibly learning the routes. Southward routes are the autumn migration and northward routes are the spring return migration.

Table 3.4. Residency/Vagrancy records for Europe

England	Vagrant, one record in 1999	Brown & Grice, 2005; Dudley et al., 2006
Netherlands	Vagrant, 9 records of 10 birds (1800-1997), May-August or October-November	Bijlsma et al., 2001
Balearic Islands	Rare or accidental on Mallorca, Menorca, Eivissa and Formentera	Anuari Ornitològic de les Balears, 2004
Russia	Rare over southern Russia, very rare in Moscow	Flint et al., 1984; Kalyakin & Voltzit, 2006
Hungary	Regular breeder, non-breeding summer visitor, March to October	Gábor, 1998; Hadarics & Zalai, 2008
Czech Republic	Vagrant, April to November, mostly in August, September, and October	Hudec, 1977; Kren, 2000
Slovakia	Regular breeder	Murin et al., 1994
Romania	Rare, local breeding migrant and passage migrant	Roberts, 2000
Bulgaria	Breeding resident, regular migrant southwards	Nankinov, 1992
Greece	Common and widespread summer visitor, wintering in Africa, September to March	Handrinos and Akriotis, 1997
Turkey	Summer migrant and passage migrant	Kirwan et al., 2008

Table 3.5. Residency/Vagrancy records for Western Asia

Azerbaijan	**Rare resident and migrant**	**Patrikeev, 2004**
Armenia	Rare resident breeder, commoner in the southeast	Adamian & Klem, 1999
Syria	Common breeder, migrants occur March to April, September to October	Baumgart, 1995
Lebanon	Summer visitor, breeder	Ramadan-Jaradi & Ramadan-Jaradi, 1999; Ramadan-Jaradi et al., 2008
Israel	Passage migrant, rare winter visitor, common summer breeder	Shirihai, 1996
Kuwait	Rare passage migrant and winter visitor	Gregory, 2005
Bahrain	Passage migrant	King, 2006
Oman	Common passage migrant and winter visitor, rare summer visitor, breeder	Eriksen et al., 2003
Saudi Arabia	Rare spring/autumn passage migrant, possible winter visitor	Jennings, 1981; Stagg, 1991
United Arab Emirates	Possible resident, rare passage migrant and winter visitor	Richardson, 1990; Richardson et al., 2003
Yemen	Rare passage migrant or winter visitor	Porter et al., 1996

Table 3.6. Residency/Vagrancy records for Eastern Asia

India	Decline in west (Gujarat), now rare, due to nest disturbance, nest site loss, deforestation	Khacher, 2003
China	Accidental in northeastern China	Brazil, 2009
Nepal	Passage migrant, possible breeder. Commoner in lowlands (tarai), also near Kakani, >2000 m	Inskipp & Inskipp, 1991
Bhutan	Vagrant, record from Kuri Chhu Valley near Lhuntse	Spierenburg, 2005
Thailand	Rare reports from northern Thailand, as migrant	Orta et al., 2020
Myanmar	Winter resident (gallicus)	Smythies, 1986
Wallacea	Rare resident in the Lesser Sundas, commoner on Flores and Timor. Possible rare migrant in winter	Kemp, 1994e; Coates & Bishop, 1997
Kazakhstan	Common breeder, passage migrant and rare resident and/or winter visitor	Wassink & Oreel, 2007
Malay Peninsula	Rare passage migrant, winter visitor, lowlands south to Singapore	Jeyarajasingam & Pearson, 1999; Wells, 1999
Iraq	Autumn passage migrant. Evidnece of breeding in occupied nests.	Ararat et al., 2011
Iran	Common summer resident	Scott & Adhami, 2006

The global population has been estimated at 10,000 to 100,000 individuals, although 52,000 individuals, with additional immatures and other non-breeding birds may be more accurate (Ferguson-Lees & Christie, 2001). Another rough estimate is 35,100-41,800 mature individuals in Europe, and a global population of 103,000-123,000 mature individuals, based on the premise that Europe forms 34% of the breeding area (BirdLife International, 2016). Variable numbers have been reported from various countries, such as Spain, which fluctuated between 1000 and 3000 in 1975, 1989, 1997 and 2003 (Garzón, 1977; De Juana, 1989; Purroy, 1997; Martí & Del Moral, 2003). During migration through the Gibraltar bottleneck, two counts recorded 1,756 in spring 2006 and 1,504 in spring 2007 (Garcia, 2008). In nearby Morocco, Thévenot et al. (1985) estimated the population at over 1,000 and assessed it as declining. In the central and eastern Mediterranean smaller numbers were recorded: in Italy, 700 breeders (Baghino et al., 2009) and in Greece about 600 to 1000 breeders (Hallmann, 1985). In eastern Europe, the estimates are much lower, for example: Bulgaria with 500-700 (2004, estimated as an increase from 50-70 pairs in 1997 and 65 pairs in 2002 (see Nankinov et al., 2004), or 270-320 breeding pairs in 2007 (see Iankov, 2008); and Croatia, about 400 to 500 pairs (Radovíc et al., 2003). The estimates for Hungary (40-50 breeding pairs, Gábor, 1998); Russia (15-20 pairs breeding in the northern Novrogod District, Bakka and Kiseleva, 2009) and Slovakia (2 breeding pairs in 2008, Danko, 2009), are much lower. In the United Arab Emirates, one estimate cited only 1-2 breeding pairs (Aspinwall & Hellyer, 2006).

Table 3.7. Residency/Vagrancy records for Africa

Canary Islands	Vagrant, seven records from Lanzarote, Gran Canaria, Tenerife, and Fuerteventura	Clarke, 2006
Morocco	Common passage migrant, rare breeding migrant, accidental winter visitor	Thévenot et al., 2003
Algeria	Widespread breeding migrant, winters southwards Passage migrants Europe to/from Africa	Isenmann and Maoli, 2000
Tunisia	Widespread, breeding migrant, in the north, few winter records	Isenmann et al., 2005
Libya	Passage migrant, casual breeder, recorded from Tripoli and coast to southern Libyan Desert	Bundy, 1976
Mauritania	Palearctic migrant, both passages; winter visitor to southwestern Mauritania, some over the summer visitor	Borrow & Demey, 2001; Isenmann et al., 2010
Ghana	Rare migrant, only savanna, dry season (winter) are *C. gallicus*, wet season are *C. beaudouinii*	Grimes, 1987
Togo	Palearctic migrant, September to April, usually northern savannas	Cheke & Walsh, 1996
Nigeria	Rare Palearctic migrant, dry season, September to March, in the northernmost savanna	Elgood et al., 1994
Sudan	Frequent Palearctic migrant/winter resident in northern acacia grassland, uncommon in the south	Bowen, 1926; Nikolaus, 1987
Eritrea	Common passage migrant, Palearctic winter visitor September to April, cross the Bab-el-Mandeb Straits	Smith, 1957; Ash & Atkins, 2009; Welch & Welch, 1992
Ethiopia	Fairly common Palearctic winter visitor (mostly west of the Rift Valley and southwards	Ash & Atkins, 2009
Uganda	Passage migrant, rare Palearctic visitor, in straggler December-February	Stevenson & Fanshawe, 2002; Carswell et al., 2005

Most reports on the Short-toed eagle indicate a stable or increasing population, hence the assessment of least concern (BirdLife International, 2016; The Global Raptor Information Network, 2021an). Some areas report population increases. For example, in Spain, it was once recorded as stable or increasing in population (Martí & Del Moral, 2003). In Gibraltar as a migration passage point, there were few changes in numbers of birds in spring passages for 1967-77 and 1990-2004 (Bensusan et al., 2007). Some areas report population declines. Hadarics and Zalai (2008) reported that the Hungarian population is slowly decreasing. In Tunisia too, the population is declining possibly due to desertification (Isenmann et al., 2005). In Mali and Niger, more specimens were recorded along survey routes in 1971–1973 (Thiollay, 1977) than found later in 2004 (Thiollay, 2006). Further declines were recorded in the Sudano-Sahelian savannas of Mali, Niger and Burkina Faso, by comparing two sets of surveys conducted in 1969-1973 and 2003-2004, using road transects in non-protected areas and national parks (Thiollay, 1977; Rondeau & Thiollay, 2004; Thiollay, 2006). The Short-toed Eagle is also a summer breeder and passage migrant through Israel, which is part of its southern breeding range and the east Mediterranean migration route to Africa (Dementiev & Gladrov, 1951; Cramp & Simmons, 1979; Shirihai, 1996; Meyburg et al., 1998; Agostini, 2002; Campora & Cattaneo, 2005; Clark et al., 2005). Darawshi (2009) gives a study in the Judean Slopes of Israel, which "is on the main western migration route in Israel, contains one of the largest Short-toed eagle concentrations in this country (estimated as 80 breeding pairs), and is known for large aggregation incidence, occurrences of different numbers of individuals together." Illustrations of the appearance, range, and eggs of this species are shown in Figures 3.1a to 3.1j.

3.3 Black-chested or Black-breasted Snake Eagle *Circaetus pectoralis* A. Smith, 1829

Physical Description

The Black-chested Snake eagle differs from the Short-toed snake eagle (which has dark blobs markings on the belly), and the Beaudouin's Snake eagle (which has a barred belly) in having unmarked belly and darker chest feathers (Clark, 1999). It is 62–67 cm in length, 170–185 cm in wingspan, 1.2–2.3 kg weight, yellow eyes, black beak and gray legs (del Hoyo et al., 1994).

Fig. 3.1a. Short-toed eagle perching profile [With permission from Mike Nelson-Birdtour ASIA]

Fig. 3.1b. Adult Short-toed eagle [With permission from Hari K. Patibanda]

Fig. 3.1c. Immature Short-toed eagle [With permission from Hari K Patibanda]

Fig. 3.1d. Immature Short-toed eagle perching in forest stand [With permission from Pieter Verheij]

Fig. 3.1e. Short- toed eagle in flight, with dark plumage [With permission from Pieter Verheij]

Fig. 3.1f. Short-toed eagle in flight [With permission from Frank McClintock]

Fig. 3.1g. Immature Short-toed eagle in flight [With permission from George Vittman]

Fig. 3.1h. Immature Short-toed eagle in flight [With permission from Mike Nelson-Birdtour ASIA]

Fig. 3.1i. Short-toed snake eagle habitat, Spain [With permission from Rudd Kimmelaar]

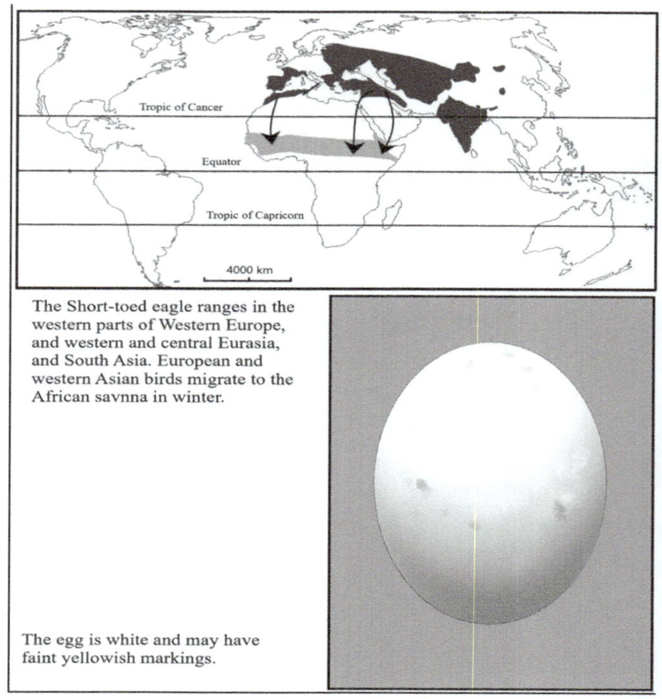

Fig. 3.1j. Short-toed snake eagle range and egg

Juveniles lack markings on underwing coverts, unlike equivalent Short-toed eagles (Clark, 1999). Illustrations of the appearance, range, and eggs of this species are shown in Figures 3.2a to 3.2h.

Classification

The Black-chested Snake eagle is a separate species, rather than a subspecies of the Short-toed snake eagle (Lerner & Mindell, 2005). It may be a superspecies with the Beaudouin's Snake eagle (Stresemann & Amadon, 1979; Amadon & Bull, 1988), but most consider the three separate species (Clark, 1999; del Hoyo et al., 2014). It may be closer to the Brown than to the Short-toed snake eagle (Lerner & Mindell, 2005), but related to the Bateleur (Wink & Sauer-Gürth, 2004).

Breeding

Nesting records exist all year across its range (Global Raptor Information Network, 2021ao), including Ethiopia, near Lake Zwai (Ash & Atkins, 2009), Somalia (Ash & Miskell, 1998), Zambia (Dowsett et al., 2008), Malawi (Dowsett-Lemaire & Dowsett, 2006), Zimbabwe (Irwin, 1981) and South Africa, where the breeding peaked during August-November (Cyrus & Robson, 1980; Steyn, 1982). One egg is laid (Dowsett et al., 2008), in trees: e.g., acacias or pine (*Pinus patula*, Schiede ex Schltdl. & Cham.) (Irwin, 1981).

Foraging

The Black-chested Snake eagle's main habitat is "open country in savanna woodlands, dwarf shrublands, thornbush, and semi-desert, often near water, avoiding mountainous and forested areas" (Global Raptor Information Network, 2021ao). Foraging is mostly from a perch or hovering, with prey largely smaller venomous snakes, lizards, rodents, frogs, and insects (Irwin, 1981; Tarboton, 1990), often grassfires (Dowsett-Lemaire & Dowsett, 2006; Dowsett et al., 2008).

Population Status

The Black-chested Snake eagle is a partial migrant (Bildstein, 2006). It ranges from eastern Sudan and Ethiopia southwards to South Africa (BirdLife International (2021). BirdLife International (2016) gives a Least concern status, with no known trend. The population may be over 10,000 (Ferguson-Lees & Christie, 2001). The range does not overlap with those of the short toed Snake eagle (western Eurasia, North Africa) or the Beaudouin's Snake eagle (western and central Africa) except due to the migration of the Short-toed snake eagle southwards. Illustrations of the appearance, range, and eggs of this species are shown in Figures 3.2a to 3.2h.

Fig. 3.2a. Black-chested snake eagle perching [With permission from Derek Keats]

Fig. 3.2b. Black-chested snake eagle perching [With permission from Derek Keats]

Fig. 3.2c. Black-chested snake eagle perching [With permission from David Beadle]

Fig. 3.2d. Black-chested snake eagle in flight [With permission from Gustav Van Der Gryp]

Fig. 3.2e. Black-chested snake eagle in flight [With permission from Derek Keats]

Fig. 3.2f. Black-chested snake eagle in flight [With permission from David Beadle]

Fig. 3.2g. Black-chested snake eagle habitat, Serengeti Tanzania [With permission from Jérôme Christen]

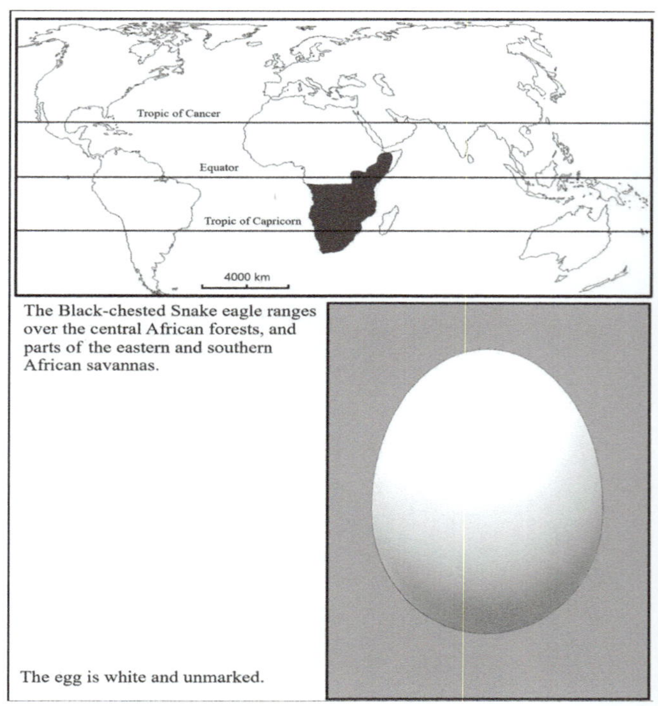

Fig. 3.2h Black-chested snake eagle range and egg

3.4 Beaudouin's Snake Eagle *Circaetus beaudouini* (Verreaux & Des Murs, 1862)

Physical Description

The Beaudouin's Snake eagle is similar in size to the Short-toed and Black-chested Snake eagles, (62–67 cm length, 170–185 cm wingspan and 1.2–2.3 kg weight) (del Hoyo et al., 1994; Ferguson-Lees & Christie, 2001). It has a grey-brown head, chest and remaining upperparts, and darkly barred white underparts, a black bill, bright yellow eyes and pale grey, unfeathered legs. Juveniles have overall dark plumage (Kemp et al., 2020e). Illustrations of the appearance, range, and eggs of this species are shown in Figures 3.3a to 3.3g.

Classification

The Beaudouin's Snake eagle is currently recognized by some as a separate species (White, 1970; Clark, 1999; Lerner & Mindell, 2005) but was formerly classified as subspecies of the Short-toed Snake eagle (Brown & Amadon, 1968; Stresemann & Amadon, 1979; Amadon & Bull, 1988; Sibley & Monroe, 1990). Possibly a superspecies with the Short-toed and Black-chested Snake eagles (Global Raptor Information Network, 2021ap).

Breeding

Breeding patterns for the Beaudouin's Snake eagle are currently little known, but with nests recorded in tree top, with one egg recorded. Isenmann et al. (2010) give a nesting on 6 April 1990 in Mauritania and Cheke & Walsh (1996) cite an eagle carrying a stick to a nest in Togo.

Foraging

The Beaudouin's Snake eagle mostly semi-open woodland, wooded savanna and more open landcover, sometimes near water bodies, but generally avoids very arid areas (Ferguson-Lees & Christie, 2001). Hunting is from perches or by hovering, and prey includes snakes, lizards and small mammals (Global Raptor Information Network, 2021ap; Kemp et al., 2020e).

Population Status

The Beaudouin's Snake eagle is usually rare or moderately common (Borrow & Demey, 2001; Kemp et al., 2020e), including in Sahelian Mauritania (Isenmann et al., 2010), Gambia (Gore, 1990), Liberia (possibly a dry season visitor and breeder) (Gatter, 1997), Côte d'Ivoire (also a slightly commoner dry season breeding visitor, in the savannas) (Thiollay, 1985), Ghana (possibly a dry season, savanna vagrant) (Harvey & Harrison, 1970; Grimes, 1987), Togo (a breeding resident (Cheke & Walsh, 1996), Nigeria (resident, but may be confused with the migratory Short-toed Snake eagle) (Elgood et al., 1994), Sudan (rare local migrant, possible breeder (Nikolaus, 1987) and Uganda and northwestern Kenya (Stevenson & Fanshawe, 2002; Carswell et al., 2005). Illustrations of the appearance, range, and eggs of this species are shown in Figures below 3.3a to 3.3g.

3.5 Brown Snake Eagle *Circaetus cinereus* (Vieillot, 1818)

Physical Description

The Brown Snake eagle measures from 66 to 78 cm, with a wingspan of 160 to 185 cm and a weight of 1.5 to 2.5 kg (Mendelsohn et al., 1989; del Hoyo et al., 1994; Borrow & Demey, 2013). The Brown Snake eagle is approximately the same size as the Bateleur eagle, but the former has a smaller wingspread (Mendelsohn et al., 1989; Borrow & Demey, 2013). The adult plumage is brown, including the wing coverts and the short tail (Ferguson-Lees & Christie, 2001). The other coloration comprises three grey tail bars, a light-colored terminal tail band, silver grey wing quill barring, and dark brown upper wing quills with white mottling on the bases of the inner webs. Some individuals have a black line above and below the eye. The wing quill underparts are silky grey, and white shafts and bars. The eye is gamboge yellow (deep saffron to mustard yellow), the cere is pale grey, the bill and claws are black, and the legs and feet greyish white. Immature plumage is paler, with a greyer head and cheeks, and there are white feather bases on the nape and belly, and lighter back feather edges which may give a "rather scaly appearance (Brown & Amadon, 1968, 346). Southern birds may show more white speckling, mostly on the head and abdomen (Steyn, 1983). The Brown Snake eagle resembles the young Bateleur and Short-toed eagles, but its coloration is darker brown. The tarsus is bare, and this differs from those of the smaller *Aquila* species, such as the Tawny eagle. The Brown Snake eagle is the largest member of the Genus *Circaetus*. Its larger head distinguishes it from other brown colored eagles in Africa. It generally has a smaller wingspread than the Bateleur, and the Black-breasted Snake eagle (Mendelsohn et al., 1989). The call is a guttural hok-

Fig. 3.3a. Beaudouin's Snake eagle perching [With permission from Neil Hilton]

Fig. 3.3b. Beaudouin's Snake eagle perching [With permission from Jeremy Roberts]

Fig. 3.3c. Beaudouin's Snake eagle in flight [With permission from David Beadle]

Fig. 3.3d. Beaudouin's Snake eagle in flight [With permission from Jeremy Roberts]

Fig. 3.3e. Beaudouin's Snake eagle habitat, West Africa [With permission from David Carr]

Fig. 3.3f. Beaudouin's Snake eagle habitat, Zakouma National Park CHAD [With permission from Robert Barnes]

hok-hok-hok, in some cases ending a kaaww, or a softer kwee-oo probably as a contact call at the nest (Ferguson-Lees & Christie, 2001). Brown & Amadon (1968) describe the call as "kok – kok – kok – kok – kaaaw" either from soaring displays or from perches; and a resting call, a throaty hok – hok or kromp – kromp. Illustrations of the appearance, range, and eggs of this species are shown in Figures 3.4a to 3.4j.

Classification

This species is part of a classification scheme for the Genus *Circaetus* by Lerner and Mindell (2005), based on the molecular sequences of two mitochondrial genes and one nuclear intron. The Brown Snake eagle in addition to the other members of this were found to be a monophyletic group linked to the *Aegypiinae* vulture group (Wink, 1995). There was also a link to the Congo Serpent eagle (Global Raptor Information Network, 2021aq).

Foraging

Thiollay (2007, 326) found that Brown Snake eagles generally hunted from perches usually in tall, isolated tree, rather than by soaring flight. This species frequented dry and closed savannah woodlands, and dense savanna and/or thorn-bush veld (with a rainfall range of 380 cm to 1270 cm) usually avoiding forest or arid grassland but may occur in dense *Brachystegia* woodland (Brown et al., 1982). The Black-breasted Snake eagle is the open country equivalent of this species (Jenkins, 1997). Brown Snake

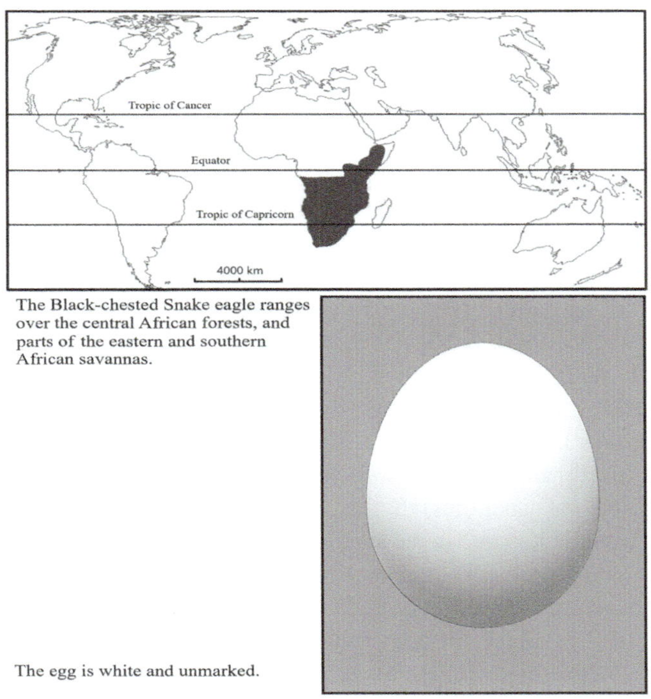

Fig. 3.3g. Beaudouin's snake eagle range and egg

eagles feed almost entirely on snakes, especially venomous species, but may also take lizards and other small vertebrates. Prey is usually taken from a glide from an exposed tree perch. It takes snakes similar to those taken by the Short-toed snake eagle, but it is a stronger bird; there may be more incidents of larger cobras and gamebirds and some mammals being taken. Unlike the Black-breasted Snake eagle, it forages and roosts singly or in pairs, but not communally.

The Brown Snake eagle's habitat and foraging habits vary regionally. In Eritrea and Ethiopia, it generally frequents thornbush and mixed tree savanna, and may occur in cultivated country (Ash & Atkins, 2009). It frequently perches on high tension pylons. In Zimbabwe it has been recorded in open woodland and sometimes near riparian fringing forest (Irwin, 1981). In Zambia, it inhabits similar environments, and here as well it avoids the open landcover which may be frequented by the Black-breasted Snake eagle (Dowsett et al., 2008). In Botswana, it frequents a wide range of woodlands, especially mixed acacia stands (Penry, 1994). Its range may include high rainfall and mature woodlands near river valleys and even in semi-desert areas (Penry, 1994). The Brown Snake eagle is considered an obligate snake hunter, taking snakes such as the Puff adder (*Bitis arietans*, Merrem, 1820), Boomslang (*Dispholidus typus*, A. Smith, 1828), Stripe-bellied Grass snake (*Psammophis subtaeniatus* W. Peters, 1882) and Black-necked Cobra (*Najamos sambica*, Peters, 1854), but it has been recorded taking a monitor lizard (Genus *Varanus*, Merrem, 1820) (Irwin, 1981). Other prey may include toads, francolins, guineafowl, domestic fowl and rats (Irwin, 1981; Steyn, 1983; Ferguson-Lees & Christie, 2001).

Breeding

Courtship begins with displays, typically including high elevation soaring, erratic patterned movements and sometimes interlocking of talons. There is also wheeling over breeding areas or high-level directional flying. The comparatively small nests are built from sticks, usually hidden within dense foliage of a tall tree or within stands of epiphytes. Nests are comparatively small (up to one meter across and 15 cm deep), usually in Euphorbia, Acacia or other savanna trees, built with small sticks and green leafy lining. One large round, white egg is laid. Laying occurs from February to May (Kenya), February, August (Sudan), September (Malawi) and April to May (Transvaal). The incubation period is 50 days. and the nestling period is 100 days. Feeding at the nest may involve the male as the main provider, bringing a single snake to the nest, which is torn to pieces to feed the young. Nests may be used only once, as this species may have nomadic tendencies (Tarboton, 1990). Fledging is after 100 to 109 days (Steyn, 1983; Ferguson-Lees & Christie, 2001).

Population Status

The global population was estimated at between 10,000 and 100,000, a fairly wide range (Ferguson-Lees & Christie, 2001). The Brown Snake eagle ranges from Senegal eastwards to northern Ethiopia and southwards to South Africa. It is a rare but resident in much of the West African Sahel and Sudan savannas, from the Atlantic coast at southern Mauritania to Liberia and eastwards along this belt to Chad and Central African Republic (Borrow & Demey, 2001). The population in this area has declined, as it was previously assessed as the commonest snake eagle in Sahelian Mauritania (Lamarche, 1988), but its population and breeding levels have since been assessed as unclear (Isenmann et al., 2010). In Ghana, Gambia and Liberia, it is rare, but more common in the inland savanna than the coastal forests, with little evidence of nesting (Grimes, 1987; Gore, 1990; Gatter, 1997). In Côte d'Ivoire, it is a more widespread breeding resident, some individuals being migrants, over the northern savannas (Thiollay, 1985). In Togo, it is a possible breeding resident. As elsewhere, it is primarily a savanna species, with some forest sightings (Cheke & Walsh, 1996). In Nigeria it is a rare resident, commonest in wooded savannas, less common or absent in forest or Sahel savanna. In its frequented habitats it is commoner than the Western banded snake eagle (Elgood et al., 1994). In Cameroon, it is found mostly in protected areas (Bobo et al., 2006). In Eastern Africa including the Horn of Africa, it may be a breeding resident in some areas. In Uganda, it is a resident breeder, considered common to the south and west of the Nile Valley. It frequents open, mixed landcover (Carswell et al., 2005). In Sudan, Eritrea and Ethiopia, it is rare breeding resident or presumed resident (for the latter two countries) (Smith, 1957; Nikolaus, 1987; Ash & Atkins, 2009). Previously commoner (Bowen, 1926; Smith, 1957). In Somalia it is rare in the south, but a vagrant in the north (Ash & Miskell, 1998).

In southern and southeast Africa, it is commoner, being widespread, sparse populations through bush and woodland vegetation, in lowlands and mountains (Stevenson & Fanshawe, 2002). For example, in Malawi, it is possibly a breeding resident, like the Black-chested Snake eagle, but frequents the woodland and lower elevations (Dowsett-Lemaire & Dowsett, 2006; Dowsett-Lemaire, 2006). In Angola and Zambia, it is fairly common, occurring in acacia and miombo woodland, in southern Angola (Pinto, 1983; Dean, 2000) and through most of Zambia (Dowsett et al., 2008). In southern Africa, it is commonest in South Africa and southern Namibia (Jenkins, 1997). In South Africa it was described by Tarboton and Allan (1984), as a "common breeding nomad." It is fairly common in Botswana and Zimbabwe (Irwin, 1981; Penry, 1994; Hancock et al., 2007), but appears rarer in Mozambique (Parker, 2005).

Regarding migratory movements, the Brown Snake eagle is generally regarded as an irruptive or local migrant, usually exhibiting local or nomadic movements, in several parts or the range in southern and West Africa (Borrow & Demey, 2001; Bildstein, 2006). An example of the wide roaming capabilities of the Brown Snake eagle is cited by Oatley et al. (1998); a bird ringed on 3 March 1974 in South Africa was recovered almost seven years later 2,100 km north in the Democratic Republic of the Congo (Oatley et al., 1998). This, however, was the only recovery of a ringed bird from the 74 birds ringed in southern Africa (Oatley et al., 1998). Although Jenkins (1997) suggested a seasonal east-west movement between northern Namibia and northern Zimbabwe, other writers disputed regular migratory movements (Parker, 2005; Dowsett et al., 2008).

Regarding conservation issues, the Brown Snake eagle may be affected by habitat destruction, the evidence for this being higher population densities in some of the protected areas (Jenkins, 1997; Simmons, 2005). Additionally, as pointed out by Rondeau and Thiollay (2004) and Thiollay (2006), compared with older work by Thiollay (1977) the populations of this species have been significantly reduced. In some areas, there are indication of increased populations. For example, in Uganda, they were rare (Jackson, 1938) but are now commoner (Carswell et al., 2005). Dowsett-Lemaire and Dowsett (2006) note that in Malawi, as the Brown Snake eagle frequents woodlands more than the Black-breasted Snake eagle, it may be more affected by deforestation. Illustrations of the appearance, range, and eggs of this species are shown in Figures 3.4a to 3.4j.

3.6 Southern Banded Snake Eagle or East African Snake Eagle or Fasciated Snake Eagle *Circaetus fasciolatus* (Kaup, 1850)

Physical Description

This is a small-medium eagle, with a large, rounded head similar to that of the above species (Kemp et al., 2020f). Average size ranges for the adults, are 55 to 60 cm in length, a wingspan of 119 to 128 cm, and a weight of 908 to 960 (Ferguson-Lees & Christie, 2001). Females are larger than males. Adult plumage consists of a greyish-brown head, including the crown and nape, the pale, greyish brown chin and throat, with a slightly darker breast, and blackish to dark brown upperparts, with blackish brown upper wing coverts with rufous tipped feathers, and dark brown upper tail coverts with some feathers white tipped. The primaries are brown, with some grey and white on the inner web bases. The secondaries are dark brown on the outer webs, greyish white on the inner webs, narrowly barred with a broader terminal band. The upper tail feathers are black with white tips, with three greyish brown bands. The underparts are paler, and the tail underside is grey with a white-tip, one broad darker band, the three thinner

Fig. 3.4a. Brown Snake eagle perching [With permission from Willie van Schalkwyk]

Fig. 3.4b. Brown Snake eagle perching [With permission from Dubi Shapiro]

Fig. 3.4c. Brown Snake eagle perching [With permission from Derek Keats]

Fig. 3.4d. Brown Snake eagle in flight [With permission from Willie van Schalkwyk]

Fig. 3.4e. Brown Snake eagle in flight [With permission from Ken Behrens]

Fig. 3.4f. Immature Brown Snake eagle in flight [With permission from Willie van Schalkwyk]

Fig. 3.4g. Brown Snake eagle habitat [With permission from Frank McClintock]

Fig. 3.4h. Brown Snake eagle habitat [With permission from Rita Willaert]

Fig. 3.4i. Habitat for the Brown Snake eagle [With permission from Derek Keats]

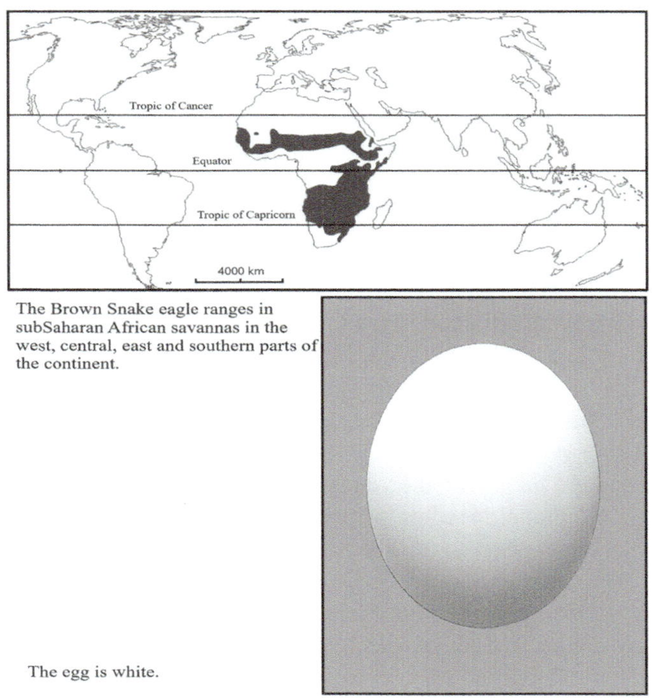

Fig. 3.4j. Brown Snake eagle range and egg

dark bands and the white underwing coverts, belly and thighs are barred brown. The underwing quills are grey, paler in the inner web bases, with a wide dark band on the tip, and several thinner bands. The eyes are greyish white to pale yellow; the bill is black and the cere and feet are light yellow (Brown & Amadon, 1968).

The immature plumage is marked by a darkly streaked face, throat and upper breast, with lighter plumage overall, mostly white and brown, with darker barring, a blackish crown with lighter streaks, a dull brownish back with paler edged feathers and brown upper tail, wing coverts and wing quills. There is paler coloring on the inner webs and thin black barring. The secondaries have black feathers with white barring on the inner webs. The tail is banded black and pale brown, with a broad black terminal band. The underside is lighter, with a paler, brown-streaked chin, throat, upper breast. The thighs and belly are white, with some brown barring. The underside of the tail is pale grey, barred with darker grey, and the underwing coverts are white, lightly barred with brown. The underside of the wing quills is grey, with black tips, white basal barring. The eye is pale yellowish grey, the cere yellow and the bill and feet are black (Brown & Amadon, 1968). The Southern Banded snake eagle may be distinguished from other similar species, due to the underside dark barring, dark bars on the white wing lining and the barred tail. The juvenile may be distinguished by its white and pale brown plumage, and with the barred wings and tail. The call is a loud "crow like koho – hoho – ho", with a softer, "mewing cry" (Brown & Amadon, 1968, 350) or "ko-ko-ko-kaw", which may be repeated in flight or when perching (repeatedly made either from a perch or while in flight (Brown et al., 1982). Illustrations of the appearance, range, and eggs of this species are shown in Figures 3.5a to 3.5m.

Classification

The Southern Banded snake eagle was formerly considered a race or subspecies of the Short-toed snake eagle (Amadon & Bull, 1988; Global Raptor Information Network, 2021ar). This species has been assessed to form a parapatric superspecies with the Western banded snake eagle, indicating a close genetic relationship. According to the molecular study of Lerner and Mindell (2005), this species is included among the snake eagles that form a monophyletic group that is linked to the Eurasian and African vulture group, *Aegypiinae*, as was detected earlier by Wink (1995) (Global Raptor Information Network, 2021ar).

Foraging

The Southern Banded snake eagle frequents forest vegetation but may also be present in open savanna (Brown & Amadon, 1968). It may also frequent coastal forests (such as coastal Brachystegia (miombo) woodland in Kenya) and inland forests near wetlands (Ferguson-Lees & Christie, 2001; Global Raptor Information Network, 2021ar). Other habitats include river ecosystems and exotic tree plantations (Kemp & Kemp, 1998) and is relatively common in coastal Brachystegia (miombo) woodland in Kenya. Hunting usually takes place from perches rather than soaring above the forest canopy. This is a shy and rather elusive species, which hides within the forest vegetation and may occur in pairs (Ash & Miskell, 1998; Parker, 2005). Prey is usually snakes, but birds, mammals, lizards, frogs and winged termites may also be taken (Brown & Amadon, 1968). Domestic poultry may be an under-researched prey item (Global Raptor Information Network, 2021ar).

Breeding

The Southern Banded snake eagle's mating season begins with calls, usually during soaring flight (Brown & Amadon, 1968). The nests are usually located in the forks of trees, are constructed by both sexes and are composed of a platform of small sticks roughly 50–70 cm wide with a smaller, inner depression lined with fresh leaves. One nest in South Africa was recorded 8-9 m high in a fork of a White Milkwood (*Sideroxylon inerme* L.) and covered with Canary-creeper growth (*Senecio* sp.) (Haagner, 2009). The single egg is laid from July to October in the northern ranges and September to October in the southern ranges (Ferguson-Lees & Christie, 2001; Cordeiro & Githiru, 2000). Eggs are white to greenish white, with reddish streaks (Brooke, 1984). Although the female is the main incubator for about 50 days, the male may participate in this activity. The male is the main food provider, with the young being fed mostly on snake parts (Ferguson-Lees & Christie, 2001).

Regional breeding patterns may be distinguished (Global Raptor Information Network, 2021ar). For example, laying takes place in August to October in Zimbabwe, and October to November in Zululand, South Africa (Brown & Amadon, 1968; Irwin 1981; Scott & Scott, 1995). Other variations include October-November for northern KwaZulu-Natal (Clancey, 1985), September and November in Mozambique (Clancey, 1996; Parker, 2005), July to October in East Africa (Kemp, 1994f) and October to November in South Africa (Clancey, 1985).

Population Status

Southern Banded snake eagle populations exist at varying densities across East Africa and South Africa, from southern Somalia to northeastern South Africa (Ferguson-Lees & Christie, 2001). Throughout its range, it may be an irruptive or local migrant (Bildstein, 2006). Some researchers speculate that it may be migratory from the southern to northern portions of its range in

July-October, especially in the case study of northern coastal Kenya where it is common during the dry season (Brown et al., 1982). The population estimates for this population are comparatively low: Ferguson-Lees and Christie (2001) estimated the global population as between 1,001 and 10,000 individuals. BirdLife International (2016) estimated the mature birds at 670-2,000 individuals.

The Southern Banded snake eagle is rare in Somalia, with most sightings in March, July, September, and October (Ash & Miskell, 1998). In East Africa, it occurs mostly in low densities in coastal forest and inland major river systems (Stevenson & Fanshawe, 2002). It may be vagrant or rare in Malawi (Hunter et al., 1977; Dowsett, 1982; Dowsett-Lemaire & Dowsett, 2006; Kemp & Kirwan, 2020b). It is a rare resident in mostly coastal, central Mozambique, and extending into the inland mountains and the Zambezi River Delta, but under serious human population pressure and deforestation (Kemp & Kirwan, 2020b). In central Mozambique, the breeding population was estimated at 200 to 500 birds, with as few as 50 birds south of the Save River, due to deforestation (Parker, 2005).In Zimbabwe, it is a rare resident of forest (Irwin, 1981; Scott & Scott, 1995), declining between the reports of Irwin (1981) and Berruti (1997).

In South Africa, it occurs mostly in forest patches and scrub, and in the east coast littoral belt and the Simangaliso Wetland Park and its surrounding forest, a protected area along the coast of South Africa's KwaZulu-Natal Province (Haagner, 2009). Berruti (1997) reported its resident status in South Africa, with immatures birds foraging for breeding areas (Barnes, 2000c). Such subadults have been reported widely (Scott & Scott, 1995; Barnes, 2000c). The Southern Banded Snake eagle ranges southwards perhaps to the Nhlabane River, east of Lesotho (Cyrus & Robson, 1980, Clancey, 1985; Berruti, 1997; Haagner, 2009). The South African population was estimated by Ginn et al. (1989) to be 80 to 100 mature individuals in 1980, 40-50 pairs in the 1980s, these estimates supported by Barnes (2000c). It was classified as rare in South Africa (Brooke, 1984) and vulnerable later by Barnes (2000c) and Chittenden (2005), mostly due to deforestation, habitat destruction and fragmentation (Berruti, 1997; Barnes, 2000c; Chittenden, 2005; Jenkins, 2008). Illustrations of the appearance, range, and eggs of this species are shown in Figures 3.5a to 3.5m.

3.7 Western Banded Snake Eagle or Smaller Banded Snake Eagle *Circaetus cinerascens* (von Müller, 1851)

Like the Southern Banded Snake eagle, the Western Banded Snake eagle has a clearly distinguishable plumage compared with the other *Circaetus* species. The distinctive features are the broad white tail and the yellow bill base. For the adult male, the

Fig. 3.5a. Southern Banded Snake eagle perching [With permission from Brendon White]

Fig. 3.5b. Southern Banded Snake eagle perching [With permission from Brendon White]

Fig. 3.5c. Southern Banded Snake eagle perching [With permission from Johan van Rensburg]

Fig. 3.5d. Southern Banded Snake eagle perching [With permission from Nick Athanas]

Fig. 3.5e. Southern Banded Snake eagle perching [With permission from Nick Borrow]

Fig. 3.5f. Southern Banded Snake eagle perching [With permission from Martie De Wijn]

Fig. 3.5g. Immature Southern Banded Snake eagle perching [With permission from Jeff Blake]

Fig. 3.5h. Southern Banded Snake eagle in flight [With permission from Brendon White]

Fig. 3.5i. Southern Banded Snake eagle in flight [With permission from Brendon White]

Fig. 3.5j. Southern Banded Snake eagle in flight [With permission from Brendon White]

Fig. 3.5k. Southern Banded Snake eagle in flight [With permission from Steve Slater]

Fig. 3.5l. Southern Banded Snake eagle habitat [With permission from Samuel Phillips - unsplash]

upperparts are slaty grey, with a grey crown, upper back and upper wing coverts (some feathers are tipped brown). The rump is brown, and the upper tail coverts are brown and white tipped (which create a whitish tail base bar). The tail is dark brown with white tips, a broad white bar and a light basal area. The cheeks are light grey, and the chin is white. The primaries are light brown above with darker tips, grading to lighter shades basally, with three black bands. The secondaries are brownish grey, shading lighter on the inner webs, with darker tips with four dark bars. The scapulars are ashy grey (Brown & Amadon, 1968). The underparts are brown, with white barring on the abdomen, thighs and undertail coverts, which may be white in some cases. The underwing coverts are brown with white barring to the wing edges. The adult female plumage is darker, with more distinct but less extensive barring. The eye is pale yellow, the bill base, cere and legs are yellow to deep yellow, and the bill tip is black.The immature bird has paler upperparts, which may be brown with white edged feathers, light colored tails with black terminal bands and hidden black bases, white dark streaked crowns, greyish white, grey brown tipped primaries, brown innerwebs, and brown secondaries with uneven, broken darker bars and darker tips (Brown & Amadon, 1968; del Hoyo et al., 1994; Kemp & Kemp, 1998; Sinclair & Ryan, 2011). The voice is a loud "kok – kok – kok– kok – kok – kok", which falls at the end, or a "mournful" "kowaaa, kowaaa" or "kooagh", which may be like some of the calls of the Bateleur eagle (Brown & Amadon, 1968, 350). Illustrations of the appearance, range, and eggs of this species are shown in Figures 3.6a to 3.6g.

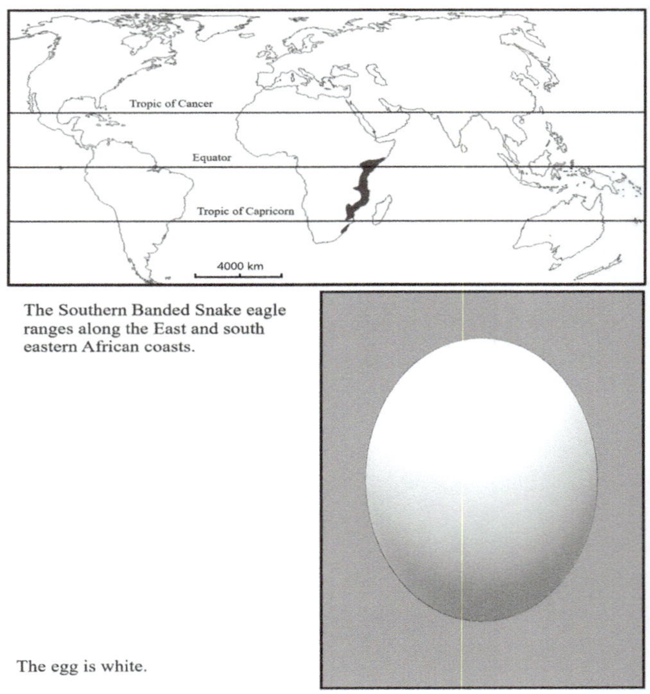

Fig. 3.5m. Southern Banded Snake eagle range and egg

Classification

The classification of the Western Banded snake eagle within the Genus *Circaetus* is the same as that of the other members of this genus.

Foraging

The Western Banded snake eagle frequents forest clearings and dense, wooded savanna, and its range may overlap with that of the Brown snake eagle (Brown & Amadon, 1968). Common forest habitats include moist riparian (riverbank) woodlands, forest edges and clearings, and savanna woodland (scattered trees in mixed grass and shrubs). Ash and Atkins (2009) characterize this eagle as secretive and elusive, identifiable by its call from soaring flight. Generally solitary, it commonly forages from perches on dead trees, surveying open landcover or ground or wetlands (Tarboton, 1990). Prey is mainly snakes, both venomous and non-venomous species. Some of its prey may be taken in trees, such as green tree snakes (Genus *Philothamnus*, A. Smith, 1847), while others such as small mammals, fish and frogs may be taken on land or water (Brown & Amadon, 1968).

The habitats and foraging habits may vary regionally. In Ethiopia, it frequents wooded landcover, foraging singly or in pairs (Ash & Atkins, 2009). In Malawi, Dowsett-Lemaire and Dowsett (2006) note that it frequents river valleys or the edge of wetlands with riparian woodland. In Zambia, it frequents riparian forest, but may also range into drier forests in the Luangwa Valley (Dowsett et al., 2008). In both Botswana and Zimbabwe, it frequents riverine forest (Irwin, 1981; Penry, 1994).

Breeding

Breeding displays include high circling flight, singly or in pairs, with loud calling like that of the Brown snake eagle, but lower. The nests are large, located in trees. One, white rough textured egg is laid, usually in January (South Sudan), February (Congo), or February, April and September (Zimbabwe and Malawi) (Brown & Amadon, 1968). Irwin (1981) reported egg-laying from December to March in the Zambezi area. The Global Raptor Information Network (2021as) notes that "there are few actual nest records from southern Africa, but some aspect of breeding activity has been reported from all times of the year" citing Herremans (1997). Nests are small, constructed from sticks, within foliage or creepers (Kemp & Kemp, 1998). In one case a nest was located in an acacia tree in a banana plantation near a village at Karonga (Benson & Benson, 1977), and another was in a patch of woodland in cultivated landcover at Nchalo (Hanmer, 1977), both in Malawi. One egg is laid (Benson & Benson, 1977; Colebrook-Robjent & Aspinwall, 1986; Tarboton, 1990; Dowsett et al., 2008; Global Raptor Information Network, 2021as).

Population Status

The Western Banded Snake eagle occurs across Sub-Saharan Africa, from Senegal eastwards to South Sudan and eastern Ethiopia, and southwards to an east to west line covering the Zambezi River, Angola, Namibia, Zimbabwe and Zambia (Global Raptor Information Network, 2021as). Ferguson-Lees and Christie (2001) have estimated the global population of adults and immatures at the beginning of the breeding season as between 1,000 and 10,000 individuals. BirdLife International (2016) estimated the global population at the lower end of the range of 10,000-100,000 individuals. Throughout its wide range, it may be an irruptive or local migrant (Bildstein, 2006). It may be locally common or uncommon, with an inconsistent and patchy distribution (Ferguson-Lees & Christie, 2001). In West Africa, it is an uncommon resident in the savanna region, stretching from the Atlantic to the Lake Chad area (Borrow & Demey, 2001). It may migrate locally during the rainy seasons (Borrow and Demey, 2001). Grimes (1987) note that it may be a breeder in West Africa, but there is little evidence of this. Presumed to breed, but there are no records from West Africa (Grimes, 1987). On a national basis, from west to east: in Gambia, it is a rare resident of riverine and open forest, with uncertain levels of breeding (Gore, 1990); in Côte d'Ivoire, it is a similarly rare resident, frequenting the wooded northern Guinean savannas and riverine forests, especially during the dry season (Thiollay, 1985); in Ghana, it has similar status in the northern savannas, possibly commoner in the wet season (Sutton, 1970; Greig-Smith, 1976; Thiollay, 1977; Cheke & Walsh, 1996); and in Nigeria, it is a rare resident of the riparian woodland of the Guinea savannas and the northern Sudan savanna (Elgood et al., 1994).

In East Africa, southeast Africa and the Horn of Africa, it is a commoner breeding resident, even in the past (Bowen, 1926) with local migratory movements, mostly in the forested patches in open landcover and riverine forest, woodland, and forest edge (Brown et al., 1982; Nikolaus, 1987; Stevenson & Fanshawe, 2002; Carswell et al., 2005). In Ethiopia, it is rare with a scattered distribution (Ash & Atkins, 2009). In Uganda, it is quite common, both as a breeding resident and intra-African migrant, frequenting the Nile Valley and nearby wetlands (Carswell et al., 2005). In Malawi, it is a common breeding resident, with local migration during the non-breeding season, and especially common near Lake Malawi (Dowsett-Lemaire & Dowsett, 2006).

In southern Africa, the population density is regionally variable. In Angola, it is rare but locally resident, observed in acacia and miombo woodlands in older and more recent research (Heinrich, 1958; Traylor, 1963; Pinto, 1983; Dean et al., 1988; Dean,

2000). In Zambia, it is commoner, frequenting riparian forest near large river systems but rarer in the drier western regions, especially the Kalahari (Dowsett et al., 2008). In Mozambique, it has been observed in arid woodland in the Zambezi Valley, possibly a breeding resident (Parker, 1995). In Namibia, several decades ago it was estimated to be as rare as 14 pairs (Brown & Hines, 1987). However, it is still recorded from Namibia (BirdLife International, 2016). In Botswana, it is a locally common resident, commonest in the Okavango Delta, but an uncertain breeder (Penry, 1994; Hancock et al., 2007). In Zimbabwe, it is quite common in the Middle Zambezi Valley and nearby areas, possibly a breeder (Irwin, 1981; Herremans, 1997).

From a conservation perspective, the Western Banded Snake eagle is under threat from human activities, that contribute to habitat loss, and also elephant damage to riparian habitats (Simmons, 2005). Rondeau and Thiollay (2004) and Thiollay (2006), found that there was a slight decline of sightings between their study in 2003-2004 and that of Thiollay (1977) in 1969-1973 in Burkina Faso, Mali, and Niger. It is classified as regionally Vulnerable in Uganda (Carswell et al., 2005), and generally impacted by deforestation in river valleys in Malawi (Dowsett-Lemaire and Dowsett, 2006) and in Namibia, where it may be impacted by riparian deforestation and African Elephants foraging near the Kavango River (Brown and Hines, 1987). In Zimbabwe, similar impacts from human action and elephants were recorded the middle Zambezi River (Parkes, 1998). Illustrations of the appearance, range, and eggs of this species are shown in Figures 3.6a to 3.6g.

Genus *Terathopius* (Lesson, 1830)

There is only one species in this genus, the Bateleur eagle. This species is characterized by very long pointed wings, comparatively short tail, short legs and the full, large "bushy" crest (Brown & Amadon, 1968, 351). The strongly scaled legs and feet are like those of the other snake eagles. Brown and Amadon (1968, 351) argue that "the Bateleur is in all ways the most specialized of the Old World Snake hawks. It is perhaps remarkable that it is the only one of them that feeds on carrion."

3.8 Bateleur Eagle *Terathopius ecaudatus* (Daudin, 1800)

Physical Description

The Bateleur eagle is the most distinctive of the snake eagles, with brightly colored plumage and exposed skin. In pointing to the unique appearance, Brown and Amadon (1968, 53) note that the Bateleur eagle "has exceptionally long wings with a greater number of secondaries (25) than possessed by any other bird of prey, and practically no tail. The Bateleur usually flies with

Fig. 3.6a. Western Banded Snake eagle perching [With permission from Ken Behrens]

Fig. 3.6b. Western Banded Snake eagle perching [With permission from Nigel Voaden]

Fig. 3.6c. Western Banded Snake eagle perching [With permission from Frank McClintock]

Fig. 3.6d. Western Banded Snake eagle in flight [With permission from Ken Behrens]

Fig. 3.6e. Western Banded Snake eagle in flight [With permission from Nigel Voaden]

Fig. 3.6f. Western Banded Snake eagle habitat, Serengeti Ben Preater - unsplash

its wing tips slightly angled backwards, bearing a resemblance to a delta wing aircraft." BirdLife International (2016): "Mid-sized, oddly-proportioned eagle, with very long pointed wings, 'tailless' appearance and bushy head. Wings held in a deep 'V' and flight fast with distinctive side to side tilting action." Adult birds average 55 to 70 cm in length with a short tail. The wingspan is around 185 cm. The average adult weight is 2 to 2.6 kg (Mendelsohn et al., 1989; del Hoyo et al., 1994; Ferguson-Lees & Christie, 2001; Buij et al., 2014; BirdLife International, 2016). The adult male plumage comprises a black head, neck, greater wing coverts and underparts, chestnut back, undertail coverts, upper tail coverts and tail, and brown median and lesser wing coverts, with lighter colored tips. The primaries are black above and white with black tips below and the underwing coverts are white. The shoulders are brownish-grey. The eye is brown, and the bare facial skin, cere and legs are red. There are paler variants, not classified as subspecies (according to Brown and Amadon, 1968) that occur in the more arid parts of the species range. These variants have pale brown rather than chestnut backs, and similarly light-colored tails (Brown & Amadon, 1968; BirdLife International, 2016).

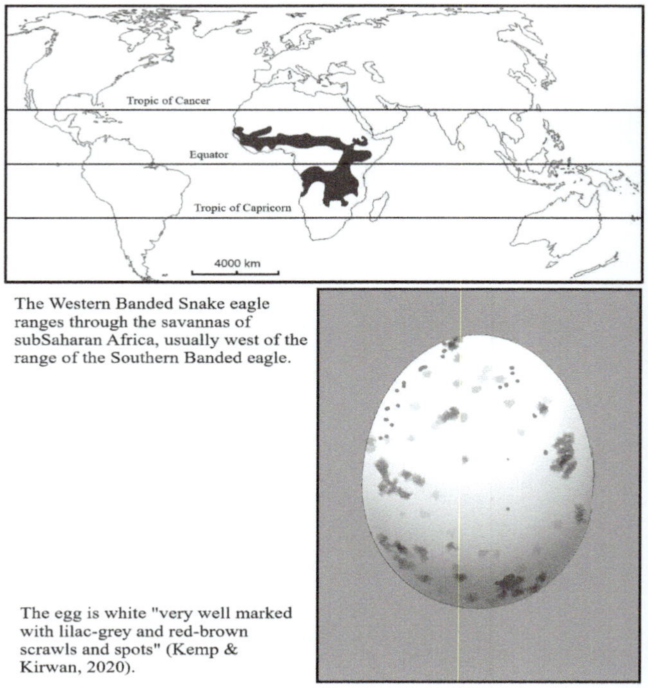

Fig. 3.6g. Western Banded Snake eagle range and egg

Adult female plumage differs in being lighter colored, with brown upper wing coverts compared with those of the male which are black on the greater coverts and brown on the medium and lesser coverts with lighter tips. The secondaries are gray and white on the inner webs, with thinner black tips on the white under of the wing quills. The paler variants may have less grey coloration on the secondaries compared with equivalent males. Females are heavier, but Brown and Amadon (1968) note that the male may have longer wings (Brown & Amadon, 1968; BirdLife International, 2016). The juvenile is entirely brown, with buff tips to most of the feathers, and a paler head and longer tail. The brown eyes, greenish blue cere and grey legs contrast with the red colors of the adult. The legs may turn red as adulthood approaches, before the brown plumage turns black, and the tail and mantle turn chestnut.

The distinctive features are the long wings and short tail (which is so short that the legs protrude beyond it in flight), the silver-white underwing (in the adult) and the dark brown underwing (in the juvenile) colors. Another distinction is the wide irregular black band on the wing edge in the male, which is thinner and less visible in the female. The juvenile has a longer tail, and unlike that of the adult, the tail hides the feet in flight. It also has a blue-grey cere, face and legs (BirdLife International, 2020). The brown color of the juvenile may resemble the Brown Snake eagle, but it is heavier with longer wings. Jackal and Augur Buzzards have a similar combination of black, white and reddish or chestnut plumage but the form of the Bateleur including the shape of the head "renders it unmistakeable" (BirdLife International, 2016). The Bateleur is also distinguished by its short tail, from which the legs protrude in flight (Newman, 1998). The calls are a harsh "chaaa – aw", and an angrier "kahkahkah" at the nest, which may be emitted with wing and hopping movements. There is also a harsh "schaaaaa", kau – kau – kau – koaggh – koaggh" during perching. Illustrations of the appearance, range, and eggs of this species are shown in Figures 3.7a to 3.7m.

Classification

The Genus *Terathopius* (Lesson, 1830) is monotypic. Research findings indicated a close relationship between the Bateleur and the Genus *Circaetus* (Kemp, 1994g). Other supportive findings suggested that the Bateleur and the short-toed eagle form a monophyletic clade, based on the nucleotide sequences in the cytochrome b gene (Wink & Sauer-Gürth, 2000). These two species have different plumage, but there is commonalities in prey taken, feeding and foraging behavior and their breeding biology (see also Sibley & Monroe, 1990). Another molecular study based on the molecular sequence from two mitochondrial genes and one nuclear intron, indicated surprisingly close relationship between the Bateleur and the Philippine eagle (Lerner & Mindell, 2005).

Foraging

The Bateleur eagle principally hunts from low soaring flight, and to a lesser extent from perches. Brown and Amadon (1968) describe the hunting flight; "it is almost invariably see gliding at a considerable speed in a straight line; it very seldom circles

Fig. 3.7a. Bateleur eagle perching [With permission from Derek Keats]

Fig. 3.7b. Bateleur eagle perching and flight [With permission from of Gustav Van-Der Gryp]

Fig. 3.7c. Immature Bateleur eagle perching [With permission from Peet van Schalkwyk]

Fig. 3.7d. Immature Bateleur eagle perching [With permission from Peet van Schalkwyk]

Fig. 3.7e. Immature Bateleur eagle perching [With permission from Margaret Westrop]

Fig. 3.7f. Bateleur eagle in flight [With permission from Photoski141]

Fig. 3.7g. Bateleur eagle in flight [With permission from Stan]

Fig. 3.7h. Bateleur eagle in flight [With permission from Gustav Van-Der Gryp]

Fig. 3.7i. Immature Bateleur eagle in flight [With permission from Michael Heyns]

Fig. 3.7j. Immature Bateleur eagle in flight [With permission from Peter Gosling]

Fig. 3.7k. Bateleur eagle habitat, Okavango Delta [With permission from Fons Buts]

Fig. 3.7l. Bateleur eagle habitat [With permission from Derek Keats]

and when it flaps its painted wings does so through a short arc. It probably could not maintain a slow air speed if it wished to, and normally flies at a speed of 3 – 50 an hour." Its habitats range across several wooded savanna and woodland associations, these including open, semi-arid Kalahari (almost desert, with grasses and shrubs) to broad-leafed woodland (Barnes, 2000c). The Bateleur eagle generally avoids dense canopied forest, and open steppe, and deserts. Dowsett et al. (2008) record the eagle foraging singly or in pairs, but in some cases larger numbers may be seen, the example being a flock of 21 recorded together near water in 1973 near Choma, Zambia.

Habitat preferences vary regionally. For example, in Ghana the Bateleur generally forages in Guinea savanna, singly or in pairs, rarely in small groups (Borrow & Demey, 2010). In Ethiopia, it frequents wooded country, but may enter more open landcover (Ash & Atkins, 2009). It frequents a wider range of habitats in Zambia, being recorded in denser woodland, wooded grassland, or open savanna, only absent from forests (Dowsett et al., 2008). In Malawi, it occupies similar landcover, but may enter cultivated landcover and large urban areas, and groups of immatures may forage together (Dowsett-Lemaire & Dowsett, 2006). In Mozambique, it frequents woodlands and savannas, mostly in pairs, but is absent from urban areas (Parker, 2005). In Botswana, it frequents all woodland with variable rainfall levels, and both tree and bush savanna, and even tarred roads (Penry, 1994). In Zimbabwe, it frequents savanna and montane grassland, generally avoiding denser woodland hilly topography (Irwin, 1981).

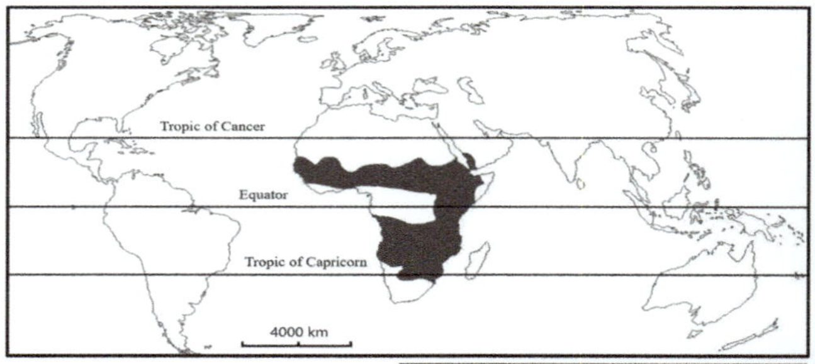

The Bateleur eagle ranges through subSaharan Africa avoiding forest and desert, and frequenting savanna landcover.

The egg is plain and white.

Fig. 3.7m. Bateleur eagle range and egg

Prey taken includes snakes, turtles, and even squirrels and hedgehogs and other small mammals. Occasionally, fish, carrion, termites and ground dwelling birds may also be taken (Brown & Amadon, 1968; Ferguson-Lees and Christie, 2001). Like other snake eagles, the coarse scutellation (scaly covering) of the legs and feet may enable the taking of snakes, by allowing a stronger grip on the scales (Global Raptor Information Network, 2021at). In Kruger National Park, South Africa, Watson (1986) notes that one third of the prey taken is of live animals, while the rest was of carrion. At carrion sources, the Bateleurs may among the first to arrive, and may be dominated by the larger vultures and sometimes the Tawny eagles (Hockey et al., 2016). In Ethiopia, an eagle was recorded taking White-winged Black Tern (*Chlidonias leucopterus*, Temminck, 1815) (Ash & Atkins, 2009). In Tsavo National Park, Kenya, one study found Dikdiks (Genus *Madoqua*, Ogilby, 1837), hares (Genus *Lepus*, Linnaeus, 1758), shrews (Family *Soricidae*, G. Fischer, 1814), and possibly Black-backed jackal (*Canis mesomelas*, Schreber, 1775), Honey badger (*Mellivora capensis*, Schreber, 1776) and Aardwolf (*Proteles cristata*, Sparrman, 1783). Birds taken included doves (Genus *Streptopelia* Bonaparte, 1855), hornbills (Family *Bucerotidae*, Rafinesque, 1815), gamebirds (*Galliformes,*Temminck, 1820) and bustards (Family, *Otididae*, Rafinesque, 1815). Other species taken included Gabar Goshawk (*Micronisus gabar*, Daudin, 1800), Striped kingfisher (*Halcyon chelicuti*, Stanley, 1814), and Crested Helmet shrike (*Prionops plumatus*, Shaw, 1809). Snakes were taken in large numbers. Road kills, including nightjars (Family *Caprimulgidae,*Vigors, 1825) and snakes were also consumed (Smeenk, 1974).

In Tanzania, an attempt to take a Honey badger (*Mellivora capensis*, Schreber, 1776) resulted in the death of both the bird and the mammal (Moreau, 1945). In Angola, prey remains in a nest included those of the Brown greater galago (*Otolemur crassicaudatus*, É Geoffroy, 1812), Greater cane rat (*Thryonomys swinderianus*, Temminck, 1827), Gambian pouched rat (*Cricetomys gambianus*, Waterhouse, 1840) and hares (Genus *Lepus*, Linnaeus, 1758) (Dean et al., 1988). In Zambia, the prey was mostly reptiles, and to a lesser extent small mammals and animals killed by bushfires. Here too, eagles may precede vultures to carcasses, and may remove the eyes of dead animals (Dowsett et al., 2008). In Zimbabwe, Irwin (1981) found in sample of 296 that the main prey was of birds (47.5%), followed by mammals (42.0%), then reptiles (8%) and fish (2.5%). The mammals had the highest biomass. The commonest species were Scrub hares (*Lepus saxatilis*, F. Cuvier, 1823) (54% of the mammals taken), game birds (22% of the birds taken), doves (21% of the birds) and glossy starlings (14%). Reptiles were less important, but carrion was important for immature birds (Irwin, 1981).

Breeding

The Bateleur is tree-nesting species (Watson, 1990; Herholdt et al., 1996; Simmons & Brown, 1997). The breeding season begins with acrobatic displays, which may include 360-degree sideways rolls, with loud calling, such as "Chaaa – aw", male diving at females, with female rolling and interlocking of claws (Brown & Amadon, 1968). The comparatively small nests (usually less than one meter across and 30 centimeters deep), composed mostly of sticks and green leaves are built in trees such as acacia (*Acacia* spp., Mill.). Other common trees are Acacia and *Diospyros mespiloformis* (Hochst. ex A. DC.) trees near watercourses (Watson, 1988; Ash & Miskell, 1998) and *Senegalia nigrescens* (Oliv.) P. J. H. Hurter) (Chittenden, 2016). This species is comparatively tolerant of people, hence may nest near human habitation and cultural landcover.

Unusually in some nesting cases, a third adult (usually male) may be present at the nesting site and may remain in proximity through the breeding cycle without incubation or rearing duties. One large, broad, oval, chalky white (red marked) is laid. Laying periods include July to November (North East Africa), January, February, April, September, October, December (Kenya), August (Zimbabwe) and September to December (West Africa) (Brown & Amadon, 1968). Egg laying in South Africa and Zimbabwe usually begins from January to March, during the rainy season (Steyn, 1982; Watson, 1986). The incubation period lasts for about 55 days, and the nestling period varies, from 101 to 194 days, comparatively long for a large eagle. Females and males share incubation (Global Raptor Information Network, 2021at). Fledging may be 90 to 120 days, but the young return to the nest after testing their wings in a first flight (Brown & Amadon, 1968; Watson, 1990; Herholdt et al., 1996).

Population Status

The Bateleur is a widely distributed species, found across sub-Saharan Africa and the Palearctic. It is recorded from Senegal in the west, eastwards to Sudan, Ethiopia, southwestern Saudi Arabia, Yemen, Southwards to Namibia and northern South Africa (Global Raptor Information Network, 2021at). Although Ferguson-Lees and Christie (2001) estimated the global population (the number of adults and immatures at the beginning of the breeding season) at over 100,000 individuals, but it is much lower now. BirdLife International (2016) estimated the total population of mature birds at "tens of thousands." Barnes (2000c) estimated the population for southern Africa, including Swaziland, at less than 700 pairs. Watson and Maritz (2000) argued that from 1980 to 2000, possibly a 75% decline population occurred in southern Africa. Parker (2005) estimated the population in central Mozambique at over 2,000 birds. For Swaziland, the estimate was only 25 pairs (Parker, 1994). For South Africa, Steyn (1982) put the population in the former Transvaal at 2,500 pairs, but later Tarboton (1990) cited 500 pairs. For Kruger National Park, over more than a decade the estimate was stable at 600 pairs (Tarboton & Allan, 1984; Watson, 1990; Simmons, 1994; Kemp et al., 2001). In the former Cape Province, the estimate was 22 breeding pairs in the early 1990s (Herholdt and de Villiers 1991), and this figure may have increased (Global Raptor Information Network, 2021at).

Current evidence indicates the population and ranges have declined, due to habitat loss, a lack of carrion, indirect persecution from the poisoning of predators such as jackals, trapping for feathers, which are used by traditional healers (Steyn 1982), and even eagles drowning in farm reservoirs (Simmons, 1997f, 2005). The populations are stable within conservation areas (such as the Kruger and Kalahari Gemsbok parks in South Africa) but declining elsewhere (Watson & Maritz, 2000). Ringing recoveries in southern Africa give evidence of eagle movements out of conservation areas, and resultant negative impacts for these birds (Oatley et al., 1998). The Bateleur has been regarded as Threatened, Vulnerable, or Endangered in different countries, and has recently been upgraded from Least Concern to Near Threatened by BirdLife International. (BirdLife International, 2016; Global Raptor Information Network, 2021at). Watson and Maritz (2000) listed the main issues for the Bateleurs status and ecologies: "Common only in larger nature reserves. Declining in West Africa (Ivory Coast) and in Sudan. Some evidence for pesticide contamination in South Africa, but levels thought to be too low for population effects. Direct and indirect persecution (shooting, trapping, poisoning) probably caused major decline elsewhere."

Throughout this region the Bateleur is an irruptive or local migrant, with the adults generally sedentary, and immature birds dispersing in large scale movements from the breeding areas (Bildstein, 2006). Oatley et al. (1998) report ringing recoveries of dispersing immatures at distances of 285 km, 77 km, 54, and 30 km from their nests. There are regional differences in the local population distributions. In West Africa, the species appears to be declining in all the countries. In Togo, the population is in decline (Cheke & Walsh, 1996). In Niger, there has been a sharp decline in tbe population the last few decades (Jensen et al., 2008). Although several were recorded in Niger in 1975 (Giraudoux et al., 1988), this species was not recorded in southeast Niger in surveys in 2002, 2003, and 2005 (Jensen et al., 2008). In Nigeria, it was recorded twice in 2000 in the northeast (Gustafsson et al., 2003). The comprehensive surveys of 2003-2004 in the Sudano-Sahelian savannas of Burkina Faso, Mali, and Nige, Rondeau and Thiollay (2004) and Thiollay (2006) report no Bateleur sightings, although these eagles were recorded in a survey of the same area in 1969-1973 (Thiollay, 1977).

Eastern and southern African birds are in variable situations. In Sudan, the number of birds sighted, and the area of the range was reduced (Nikolaus, 1987). In Uganda, sightings have increased since 1970, which may indicate a population increase (Carswell et al., 2005). In Zambia, the species has lost population, speculatively as a result of pesticides or poisons, or human presence (Dowsett et al., 2008). In Malawi, populations have survived, even in areas of high human presence (Dowsett-Lemaire & Dowsett, 2006). In Mozambique, it is found mostly in central Mozambique with relatively high populations (Parker, 2005). In Zimbabwe, it is a rare specially protected species (Hartley, 1998), possibly due to poisoning (Irwin, 1981). In Namibia, it is classified as endangered (Simmons & Brown, 2006). In Botswana, there are local movements, especially of immature birds away from their natal territory (Penry, 1994), a "Bird of Concern" (Hancock et al., 2007) and ranges widely, and feeds on carrion, which may sometimes be poisoned (Penry, 1994). In Zambia, birds are observed with greater frequency in the dry season (July to December, the non-breeding season), especially in areas of high rainfall. However, it is disputed as to whether this is due to bird movements or that they are simply less visible during the breeding season (Dowsett et al., 2008). In Swaziland, it was classified as endangered, even approaching extinction, possibly due to poisoning (Parker, 1994; Monadjem et al., 2003; Monadjem and Rasmussen, 2008).

In South Africa the Bateleur is classified as "Vulnerable" due to serious population declines and reductions over some areas such as Kalahari-Gemsbok National Park (Brooke, 1984; Herholdt et al., 1996; Simmons, 1997f; Barnes, 2000c). Conservation efforts with local education and the work of the Kalahari Raptor Project of the Raptor Conservation Group, may result in population recovery. Evidence included the colonization of farmlands in the Northern Cape Kalahari region (Watson & Maritz, 2000). Population declines in South Africa are attributed to poisoning by large-scale commercial farmers and tribal small-stock farming communities (Watson, 1986; Davies, 1988; Brown, 1991; Ferguson-Lees & Christie, 2001). Other problems are the losses of prey species, through habitat conversion to cultivation, and human disturbance and consequent nest abandonment (Steyn, 1982; Ferguson-Lees & Christie, 2001). Illustrations of the appearance, range, and eggs of this species are shown in Figures 3.7a to 3.7m.

Genus *Spilornis* (G.R. Gray, 1840)

These large to small species have roughly scaled longish legs and short toes, short talons and short crests formed from broad feathers. Brown and Amadon (1968) note that these species have limited island ranges, except for the Crested Serpent eagle, hence there is no interspecies range overlap.

3.9 Philippine Serpent Eagle *Spilornis holospilus* (Vigors, 1831)

Physical Description

The Philippine serpent eagle is a medium sized eagle, with a length of 47-53 cm, a wingspan of 105-120 cm, and weight of 0.6 to 1.6 kg (Clark & Marks, 2020). The adult plumage is mostly brownish grey, some feathers with reddish edges. The dark grey-brown crown and nape, pale sepia upper back and rufous tail coverts with four white spots on each feather, brown upper-wing coverts and upper sides of the wing quills (with some white spots) and dark brown tail with two bands (one pale brown band and a narrower band with a light tip) comprise the rest of the plumage (Brown & Amadon, 1968). The underparts are composed of a brown chin and throat, reddish brown lower breast belly and thighs and underwing coverts. Except for the throat the underside is spotted white with dark brown edges. The wing quills are dark brown to dark sepia below, marked by a grey band and another one or two narrower band close to the inner webs. The tail is dark brown with a broad grey band and tip. The eye, cere, orbital skin, legs are yellow (Brown & Amadon, 1968). Juveniles have lighter colored feathers on the head, with a white crown with some dark brown. The nape feathers are pale on the bases, with dark centers. There are buff colored edges to the brown feathers of the upperparts and the tail is pale brown, with three dark bars and a paler, buff tip. The dark brown wing quills have white bands on the inner webs. The underside is mostly white, with reddish streaks over the lower breast, and fewer streaks on the wing coverts, thighs and belly. The tail underside is grey with three dark bands, one a subterminal band. The wing quills are white basally, crossed by several dark bands and dark tipped. The eye and legs are yellow and the cere is gray blue (Brown & Amadon, 1968).

The Philippine Serpent eagle may be distinguished from serpent eagles by the clearer spots on the underparts and wings (Ferguson-Lees & Christie, 2001). The Philippine Serpent eagle may resemble the Crested Serpent eagle (*Spilornis cheela*, Latham, 1790), but it differs in having large yellow eyes and feet, facial bare skin and the pale brown tail band, which may be white in the latter species. The ranges also differ. However, Brown and Amadon (1968, 358) note that "some authors have regarded this as a race of *C. cheela*." The call is a sharp whistle, followed by a cry of "pheeeeow – pheeeeow", like that of the Crested Serpent eagle. Illustrations of the appearance, range, and eggs of this species are shown in Figures 3.8a to 3.8l.

Classification

The Philippine Serpent eagle has in some cases been assessed as a race of Crested serpent eagle (Dickinson et al., 1991) but was classified as a separate species by Ferguson-Lees and Christie (2001) and BirdLife International (2009).

Breeding

Some authors have noted that the nest and eggs are undescribed (Kennedy et al., 2000). However, these authors describe a sighting of a bird and an egg (Kennedy et al. 2000).

Foraging

Favored habitats include open country near river valleys, and denser forests (Brown & Amadon, 1968). It frequents forest canopies, searching for snakes and lizards and frequently soars high above the forest canopy and edges, clearings and river valleys (Kennedy et al., 2000).

Population Status

The Philippine serpent eagle ranges in the Philippines, from Luzon to Mindanao. It is a common endemic resident throughout this range (Kennedy et al., 2000). It is probably non-migratory (Global Raptor Information Network, 2021au). Ferguson-Lees and Christie (2001) estimated the total population at 10,000 or more breeding adults. It is still common in its wide range, and it may be more resilient and adaptable to environmental change than some other eagles (Global Raptor Information Network, 2021au). The population may be in decline due to current habitat destruction (Ferguson-Lees and Christie, 2001). However, BirdLife International (2016) notes that the very large range across the Philippines prevents its classification as Vulnerable under the range size criterion, which is a range under 20,000 km^2, in association with other factors such as declining or fluctuating range size, problems with habitat extent/quality, or small population size, few habitat locations or severe habitat fragmentation. The trend of population decline is also slower than would be required for the eagle to be considered to be Vulnerable. This would require a decline of over 30% over ten years, which might be approximately three generations. The population size is also sufficiently large, over 10,000 mature individuals, this being the required level for the vulnerability criterion. BirdLife International (2016) therefore concludes that Philippine serpent eagle should be evaluated as Least Concern. Illustrations of the appearance, range, and eggs of this species are shown in Figures 3.8a to 3.8l.

Fig. 3.8a. Philippine Serpent eagle perching [With permission from Steve Albano]

Fig. 3.8b. Philippine Serpent eagle perching [With permission from Mike Nelson-Birdtour ASIA]

Fig. 3.8c. Philippine Serpent eagle perching [With permission from Chris Chafer]

Fig. 3.8d. Philippine Serpent eagle perching [With permission from Chris Chafer]

Fig. 3.8e. Philippine Serpent eagle head [With permission from Chris Chafer]

Fig. 3.8f. Immature Philippine Serpent eagle [With permission from Mark Jasmin]

Fig. 3.8g. Immature Philippine Serpent eagle [With permission from Martin Alvendia]

Fig. 3.8h. Immature Philippine Serpent eagle [With permission from Mike Nelson-Birdtour ASIA]

Fig. 3.8i. Immature Philippine Serpent eagle spreading wings [With permission from Mike Nelson-Birdtour ASIA]

Fig. 3.8j. Philippine Serpent eagle in flight [With permission from Mike Nelson-Birdtour ASIA]

Fig. 3.8k. Philippine Serpent eagle habitat, Forest Luzon [Credit: Med-Gadon]

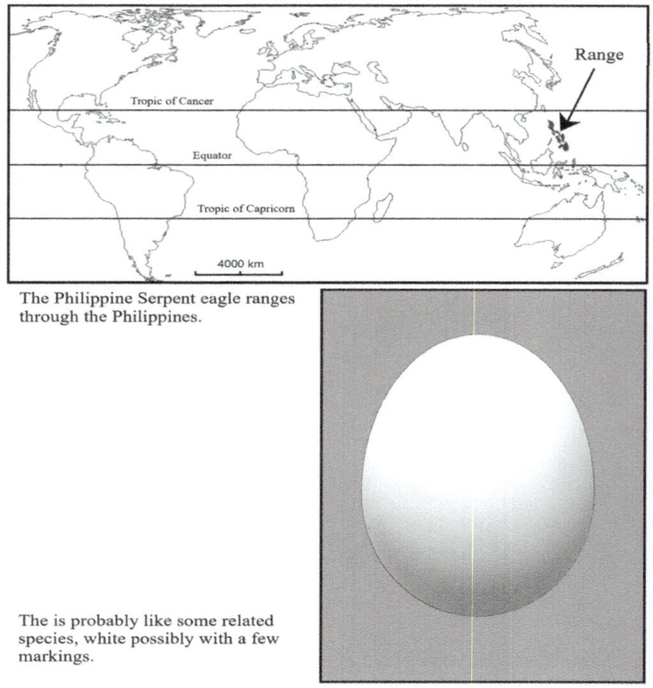

Fig. 3.8l. Philippine Serpent eagle range and egg

3.10 Sulawesi Serpent Eagle or Celebes Serpent Eagle *Spilornis rufipectus* (Gould, 1858)

Physical Description

The Sulawesi serpent eagle is a medium-sized eagle, with two subspecies *Spilornis rufipectus rufipectus* and *Spilornis rufipectus sulaensis* (Clark & Kirwan, 2020). The former, the nominate subspecies is slightly larger, darker, with less visible barring on the breast, belly and underwing coverts (Brown & Amadon, 1968). The adult has a black crown and nape, some white on largest feathers, a dark brown back, upper wing coverts and wing quills feathers, with some lighter edges on the feathers on the upper tail coverts, and white barring on the scapulars and darker bars on the flight quills. The dark brown tail has a lighter band and tip. The underparts are dark, with a dark grey to black chin and throat, and dark reddish-brown upper breast feathers, with black shaft streaks. "Chocolate" brown feathers, with white barring and spots cover the lower breast, belly and underwing coverts (Brown & Amadon, 1968, 358). The undertail feathers are grey, crossed by two darker bars and a lighter terminal band. The very dark underwing coverts combine with a wide, light colored band across the inner webs, and one or two thinner mottled grey bars on the primaries. The eyes, cere, facial skin and feet are yellow (Brown & Amadon, 1968).

The immature plumage is lighter overall, with a light creamy head and neck with variably colored back feathers (light bases and edges, and darker centers). The chin, throat and breast are white, the latter spotted dark brown. The white belly and thighs are streaked reddish brown, and the underwing coverts are reddish brown, with white mottling. The dark brown tail has a pale colored band and tip, with less visible bands. The inner wing quills are pale and the outer wing quills are dark brown, with white and dark barring on the inner webs. The undertail feathers are gray, with thinner dark bands and a wide subterminal band and a gray tip. The underwing wing quill feathers are white, with black tipped primaries and brown tipped secondaries, with fine dark barring. The eyes and legs are yellow and the cere is greenish yellow (Brown & Amadon, 1968).

The Sulawesi Serpent eagle resembles the Crested Serpent eagle, but differs in the large yellow eyes, black throat, reddish breast and dark brown underparts. Juveniles may be distinguished by their creamy, steaked head feathers. The call is described as "Fli – wi – keek" or "keek" or "kek", repeated several times (Brown & Amadon, 1968). Illustrations of the appearance, range, and eggs of this species are shown in Figures 3.9a to 3.9i.

Classification

The Global Raptor Information Network (2021av) lists two subspecies: *S.r. rufipectus* from Sulawesi and nearby islands and *S.r. sulaensis* on the Sula islands. This species was formerly assessed as a subspecies of the Crested serpent eagle (e.g., Stresemann & Amadon, 1979).

Foraging

The Sulawesi serpent eagle's natural habitat is tropical moist lowland forest (Brown & Amadon, 1968). Favored habitats include forests, from lowland to montane elevations, tall secondary forest, scrub woodland, forest edge, and lightly wooded cultivated landcover. For hunting, it frequents nearby open grassy savannas and grassland, and fly over forest canopies. It may perch on exposed elevated branches near the forest edge, and also inside the canopy. Foraging is undertaken singly, or in pairs (Coates and Bishop, 1997). Prey comprises lizards, small snakes, and more rarely rodents, sometimes near grass fires (Ferguson-Lees and Christie, 2001). It may associate with the snake-eating Woolly-necked stork (*Ciconia episcopus*, Boddaert, 1783), which is also a snake eater (Brown & Amadon, 1968).

Breeding

Evidence indicates that the breeding season lasts from January to April (Ferguson-Lees and Christie, 2001).

Population Status

The Sulawesi serpent eagle is endemic to Sulawesi and the Sula Islands in Indonesia, where it is a common and widespread endemic resident, and may be migratory (Coates and Bishop, 1997). Ferguson-Lees and Christie (2001) estimated the global population of adults and non-breeding immatures at the beginning of the breeding season, at 10,000 to 100,000 birds, based on territory size. It was regarded as a threatened species up to the 1980s, but recent findings are that it is common in some studied areas (Ferguson-Lees & Christie, 2001). Deforestation is a threat, but it seems adaptive to habitat change. It is categorized globally as a species of "Least Concern" and the population is "Stable" (BirdLife International, 2016). Illustrations of the appearance, range, and eggs of this species are shown in Figures 3.9a to 3.9i.

Fig. 3.9a. Sulawesi Serpent eagle perching [With permission from Mike Nelson-Birdtour ASIA]

Fig. 3.9b. Sulawesi Serpent eagle perching [With permission from David Beadle]

Fig. 3.9c. Immature Sulawesi Serpent eagle [With permission from Mike Nelson-Birdtour ASIA]

Fig. 3.9d. Sulawesi Serpent eagle in flight [With permission from Mike Nelson-Birdtour ASIA]

Fig. 3.9e. Immature Sulawesi Serpent eagle in flight [With permission from Mike Nelson-Birdtour ASIA]

Fig. 3.9f. Immature Sulawesi Serpent eagle in flight [With permission from Mike Nelson-Birdtour ASIA]

Fig. 3.9g. Immature Sulawesi Serpent eagle in flight [With permission from Mike Nelson-Birdtour ASIA]

Fig. 3.9h. Sulawesi Serpent eagle habitat [With permission from Josep Miquel Navarro]

3.11 Crested Serpent Eagle *Spilornis cheela* (Latham, 1790)

Physical Description

The Crested Serpent eagle is a medium sized eagle (Clark et al., 2020e). This species includes several subspecies, with some records giving 20 (Brown & Amadon, 1968). Ferguson & Christie, 2001) list 21. The adult plumage includes a black crown and nape feathers with brownish yellow edges and less visible white bases. These form a black crest with variably visible white markings. The dark brown upperparts have white markings on the lesser coverts, feathers with black quills and white tips and a white band above the tail, brown primaries and secondaries with black tips and a black band or two and white coloring on the inner webs. The dark brown chin and throat have barring, and dark and light brown barring on the upper breast. Reddish brown feathers with dark barring cover the belly, thighs and underwing and undertail coverts. White markings with black rings form barring on the thighs, underwings and undertail coverts (Brown & Amadon, 1968, 360). The tail underside is black with a white crossbar and tip, and the underside wing quills have grey bases and black tips. The primaries on the wing edges have one wide and two thinner bars. The eye, cere and facial skin and the legs are yellow, the first being the brightest and the last the dullest (Brown & Amadon, 1968).

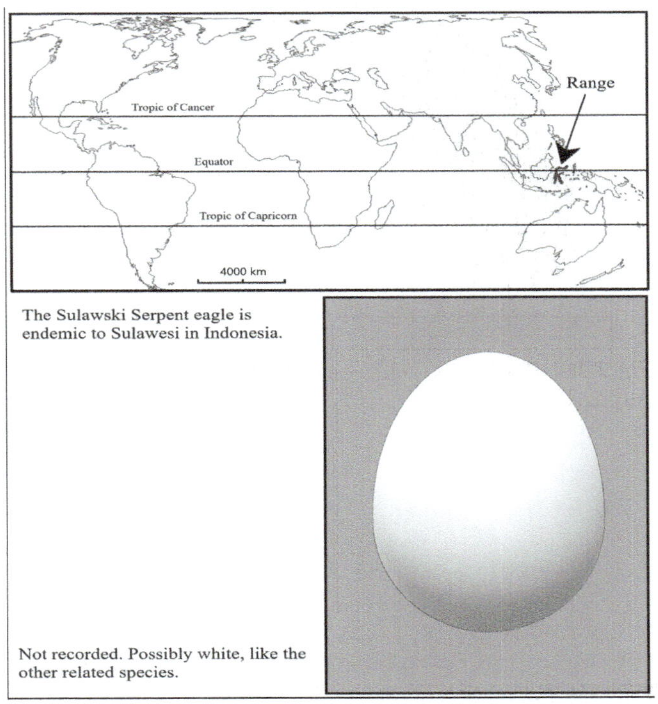

Fig. 3.9i. Sulawesi Serpent eagle range and egg

The immature plumage differs in having dark brown upperparts and a tail with white markings, and two darker bands and tip on the tail. The underparts are a brownish buff-white color, and reddish-brown streaks on the throat and chest, and reddish-brown barring on the underwing coverts, thighs, flanks and axillaries. The underside of the tail is grey with two dark bars and a dark terminal band. The wing quills have white coloring on the inner webs, and several dark bars and black tip. The eye is gray, the cere and feet are yellow (Brown & Amadon, 1968). Illustrations of the appearance, range, and eggs of this species are shown in Figures 3.10a to 3.10k.

Classification

The classification of this species is complicated. For example, Brown and Amadon (1968, 360) compare 19 subspecies with the nominate subspecies *S.c. cheela* and give plumage comparisons. Brown and Amadon (1968, 361) noted that "further study, especially in the field, may indicate that some of these races merit recognition as separate species or alternatively that some now regarded as distinct species should be merged with *S. cheela*." Ferguson-Lees and Christie (2000) argue that some species boundaries are unclear, and some subspecies may be full species. The tables below give a list of the subspecies (note that this list may not be conclusive) as defined by Brown and Amadon (1968, 360) and Ferguson-Lees and Christie (2000). Approximately, the smaller subspecies are higher up in the tables, based on the comparative description of Brown and Amadon (1968). The mention of the possible full species status of some subspecies is based on the description and analysis of Ferguson-Lees and Christie (2000). The first group are some of those found on the mainland of Asia (Table 3.8), in order of size from smallest to largest, while the second list (Table 3.9) is those found on smaller islands, and also considered possibly separate species.

Table 3.8. Comparison of Crested Serpent eagle subspecies with *S.c. cheela* (Asian Mainland)

Subspecies	Features	Range
S.c. malayensis	Darker, with more visible white markings	Thai-Malay Peninsula and northern Sumatra
S.c. melanotis	Darker, black bands and tip on the primaries, less black on the throat	India
S.c. burmanicus	Paler colored, with less barring and fewer spots	Bawean, island of Indonesia
S.c. ricketti	Paler, with less underside barring and fewer spots, more brownish color	Northern Vietnam and southern China
S.c. rutherfordi	Darker, with clearer markings	Hainan, South China Sea
S.c. pallidus	Small and pale, with reddish underparts	Northern Borneo
S.c. spilogaster	A grey throat and is dark	Sri Lanka
S.c. melanotis	Lighter colored	South India
S.c. hoya	Dark and more clearly marked	Taiwan
S.c. kinabaluensis Mountain Serpent eagle	Darkest of the subspecies, dark brown plumage and white marks on the breast, belly and wing coverts	Brunei, Indonesia, and Malaysia, island of Borneo.
S.c. batu	Lighter colored	Southern Sumatra and Batu Islands, Indian Ocean
S.c. bido	Dark colored with white markings on the breast, belly and wing coverts	Java and Bali, Indonesia
S.c. palawanensis	Reddish brown bands across the breast	Palawan, Philippines

Sources: Brown & Amadon, 1968; Mayr & Cottrell, 1979; Ferguson-Lees & Christie, 2001; Rasmussen & Anderton, 2012. Note size overlaps. *S.c. kinabaluensis* may be a separate species, the Mountain Serpent eagle, *Spilornis kinabaluensis* Sclater, WL, 1919.

This species differs from *Circaetus* or the Bateleur eagle in having a smaller head, rounded wings and a brightly colored eyes legs and feet. The call is loud, resembling scream or whistle: "Kuk – kuk – qweee – ar" or "hueeee ah" or "heeeuwheheh" (mostly in India) or "keu – keu – keu – keuiye – keeuiye" (mostly in Thailand) or "keeeeu" (in Borneo) (Brown & Amadon, 1968). Rasmussen & Anderton (2012) describe the call as "Kluee-wip-wip" the first note high and rising. Illustrations of the appearance, range, and eggs of this species are shown in Figures 3.10a to 3.10k.

The number of subspecies is debatable. Brown and Amadon (1968) lists 20 subspecies. Brown and Amadon (1968, 361) note that "further study, especially in the field, may indicate that some of these races merit recognition as separate species or alternatively that some now regarded as distinct species should be merged with *S. cheela*." Ferguson-Lees & Christie (2001) list 21, both as shown in the table above. The Global Raptor Information Network (2021aw) lists 14 "races" cited as: *S.c. batu*: southern Sumatra and Borneo; *S.c. bido*: Java and Bali; *S.c. burmanicus*, Eastern Bangladesh, southern and eastern India (Assam),

Table 3.9. Comparison of Crested Serpent eagle subspecies with *S.c. cheela (*Asian Islands)

Subspecies	Features	Range
S.c. minimus Central Nicobar serpent eagle	Light brown upperparts, a dark brown tail with greyish brown median bar and base	Central Nicobars, eastern Indian Ocean
S.c. abbotti, Simeulue serpent eagle	With dark plumage, and a black, light brown banded tail	Simeulue is an island of Indonesia
S.c. natunensis Natuna serpent eagle	Lighter colored underparts and a shorter tail	Natuna Regency, South China Sea, Natuna Sea
S.c. sipora Mentawai serpent eagle	Comparatively small and dark brown	Mentawai Islands Regency, Indonesia
S.c. baweanus Bawean serpent eagle	Narrow white band on the tail, three darker bands under the primaries	Bawean, Java Sea, Indonesia
S.c. perplexus Ryukyu serpent eagle	Smaller and lighter colored with black crown and nape, contrasting back coloration	Ryukyu Islands, Japanese islands
S.c. davisoni	Browner, with more barring on the breast	Andamans, Indian Ocean
S.c. asturinus Nias serpent eagle	Light colored above with a short black tail	Near Sumatra, Indonesia

Sources: Brown & Amadon, 1968; Mayr & Cottrell, 1979; Ferguson-Lees & Christie, 2001; Rasmussen & Anderton, 2012. Note species size overlaps. Those at the top of the table are generally smaller than those at the bottom. The subspecies recognized by some authorities, are indicated by name.

Burma, Thailand, and Indochina (except Tonkin); *S.c. cheela*, Pakistan, Kashmir, Nepal, Bhutan, Bangladesh, India (northern and western Assam); *S.c. davisoni*, Andaman Islands and possibly Nicobar Islands; *S.c. hoya*, Taiwan; *S.c. malayensis*, Southern Tenasserim, Thailand, and Malaysia; *S.c. melanotis*, India south of 25 degrees N; *S.c. palawanensis*, Philippines (Palawan); *S.c. pallidus*, Lowlands of northern Borneo; *S.c. richmondi*, southern Borneo; *S.c. ricketti*, South China (Tonkin); *S.c. rutherfordi*, China (Hainan); *S.c. spilogaster*, Sri Lanka.

Foraging

The Crested Serpent eagle frequents mostly forest vegetation, to a lesser extent savanna (Brown & Amadon, 1968; Weizhi, 2006). Favored landcover includes moderately or wooded hills, from lowlands to over 2000 meters elevation, and the edges of cultivated land and mangroves (Brazil, 2009). This species is comparatively tolerant of moderate habitat disturbance or degradation, especially if large trees survive. Regional case studies record the adaptability of this species. For example, in Malaysia it has been recorded in overgrown edge vegetation inside cultivated ecosystems and may be comparatively common in mangrove landcover (Wells, 1999). In Hong Kong, it has been recorded foraging in marshlands and wooded areas, and even over rocky hilltops hunting exposed snakes (Carey et al., 2001). It may also hunt from a perch (Clark, 1994; Pande et al., 2003; Naoroji, 2006). In China, it frequents forested areas in Yunnan Province and may perch in high trees or soar singly (Weizhi, 2006). In Sabah, Malaysia it frequents primary, logged, and swamp forest, and may intrude into clearings and tree plantations (Sheldon et al., 2001). Pendlebury & Chasen (1932) describe this as a lowland form, which frequents lower elevations than the Kinabalu Serpent Eagle, both eagles being sympatric at intermediate elevations (note that the Kinabalu Serpent eagle may be of the same species, either a subspecies (*Spilornis cheelakina baluensis*) or the Mountain serpent eagle *Spilorniskina baluensis* W.L. Sclater, 1919 (Brown & Amadon, 1968; Ferguson-Lees & Christie, 2001; Phillips & Phillips, 2011; BirdLife International, 2017).

Generally, it is a solitary or paired hunter, but it nevertheless may also forage in groups of 4 or 5, in forests and may also roost in such environments (Baker, 1914; Global Raptor Information Network, 2021aw). Brown & Amadon (1968) record the main prey taken, mostly from a tree perch as reptiles, usually tree snakes, lizards, and sometimes small mammals or birds, and carrion. Wells (1999) also record snakes, as well as lizards, frogs, toads, and more rarely small birds, usually taken from a canopy-level perch both inside forest vegetation and at the forest edge, with dives to the ground or vegetation (Wells, 1999). Other records include foraging near wet grassland in forested areas (Ueta & Minton, 1996; Lerner et al., 2005; Gokula, 2012).

Eagles have been recorded following snakes on the ground in India (Naoroji & Monga, 1983). Chia-hong Liu (2010) used radio-telemetric methods to survey eagle movements in Taiwan, the results indicating that eagles spent 98% of the day perched, and generally found food during the morning hours. The hunting technique was a perching and waiting for the appearance of the prey (Chia-hong Liu, 2010). Regional variation in prey taken have been discerned. For example, in Yunnan Province, China, the main prey was snakes, other reptiles and small mammals and birds (Weizhi, 2006). In the Malay Peninsula, the prey included smaller to larger snakes (some about 1 m long), lizards, amphibians (frogs and toads), and small, grassland birds, including weavers, munias, and pipits (Batchelor, 1959; Wells, 1999). In Sabah, Malaysia, reptiles, rodents, large insects, and crabs were recorded as prey (Sheldon et al., 2001). In at least one case, endoparasitic nematodes, including Madelinema angelae (Schmidt and Kuntz, 1971) were retrieved from the intestines of eagles (Schmidt & Kuntz, 1971; Yoshino et al., 2009).

Breeding

The breeding season of the Crested Serpent eagle starts with aerial displays, including soaring, loud calls, rolling and stooping (diving). The comparatively small nests, usually about 60 centimeters wide are constructed from sticks, with a green vegetation lining. These nests are generally in the upper levels of trees. For example, in the Malay Peninsula, Wells (1999) recorded nests in a relict mangrove tree, the lowest horizontal branch of a durian tree at a forest edge, and the main crown forks of inland forest living or dead trees. Both males and females build nests, and birds may sometimes re-use old nests (Wells, 1999). One white or reddish white, marked egg is laid (Brown & Amadon, 1968; Wells, 1999). Only females incubate the eggs, but the male guards the nest while the female forages (Gokula, 2012). There is considerable variation in laying periods according range and subspecies. For example, in India, *S.c. cheela* lays in February and March in low elevations and March to May in higher elevations. *S.c. melanotis* lays from December to March, especially February to March, and more rarely from June to September. In Sri Lanka *S.c. spilogaster* lays from January to April, most frequently February. In Myanmar, *S.c. burmanicus* lays from February to May, most frequently in March. In Java, *S.c. bido* lays from February to April, June to August and November. In Borneo, *S.c. pallidus* lays February to April, and August. In the Malay Peninsula, aerial displays from early September to early April, with nest building in January and March, nest occupation from February-May, and recent fledgings in June and late September (Wells, 1999). Fledging after 60 days (Brown & Amadon, 1968; Wells, 1999).

Population Status

The Crested Serpent eagle is a non-migratory resident throughout its range, although in some areas it occurs only in summer (Brown & Amadon, 1968; Purandare, 2002; Rasmussen & Anderton, 2012). It may however be an irruptive or local migrant, especially as juveniles dispersing from breeding areas and in some cases, it may be an altitudinal migrant (Bildstein, 2006). The global population, including all adults and immatures at the start of the breeding season, has been estimated at 100,101 to 1,000,000 individuals (Ferguson-Lees & Christie, 2001). Throughout most of its range, it is quite common, and may even be the commonest local eagle species.

BirdLife International (2020) notes that "the population size has not been quantified, but it is not believed to approach the thresholds for Vulnerable under the population size criterion (<10,000 mature individuals with a continuing decline estimated to be >10% in ten years or three generations, or with a specified population structure). For these reasons, the species is evaluated as Least Concern." Additionally, the large range is not less than 20,000 km^2 and the population trend is stable (BirdLife International, 2020). There are however regional variations. In the Malay Peninsula, deforestation and losses of tall woodland has reduced the population in some settled areas, especially in Singapore, but the species is not threatened (Wells, 1999). In Hong Kong, populations have increased possibly due to mature woodland expansion (Carey et al., 2001). In China, it has been classified as a "Second Class Important Protected Bird in China" (Weizhi, 2006). Illustrations of the appearance, range, and eggs of this species are shown in Figures 3.10a to 3.10k.

3.12 Nicobar Serpent Eagle *Spilornis klossi* (Richmond, 1902)

Physical Description

The Nicobar serpent eagle may refer to two separate species: The Great Nicobar serpent eagle also named the South Nicobar serpent eagle; and the Central Nicobar serpent eagle usually considered a subspecies of the Crested Serpent eagle as *Spilornis cheela minimus* (Amadon, 1974; Clark, 1994; Ferguson-Lees & Christie, 2001; Global Raptor Information Network, 2021ax). In the current book it refers to the former species (Global Raptor Information Network, 2021ax). The Nicobar serpent eagle is the only species on Nicobar Island and is one of the world's smallest eagles (Ferguson-Lees & Christie, 2001). Brown and Amadon (1968, 364) note that "it is small bird, scarcely larger than a big sparrow hawk." It is 38 to 42 centimeters in length and has a wingspan of 85 to 95 centimeters, and a weight of 448 grams (Clark et al., 2020f). It has a distinctive appearance, having a dark crest, yellow legs, rounded wings and a barred tail. The plumage consists of a blackish cinnamon-tipped crest, crown and nape, a brown back, a paler band between the crest and the back, dark brown upper plumage, with a few white tips on the coverts, darker wing quills, paler edges to the tail feathers, two narrow black tail bars and a broader subterminal tail bar. The underparts consist of a light brown throat and chest, a lighter belly and underwing coverts. The tail below and wing quills are paler brown, with black barring. The eyes, cere and feet are yellow. The immature plumage is very dark crown, tipped light brownish white, a brown back

Fig. 3.10a. Crested Serpent eagle perching [With permission from Keyur Nandaniya]

Fig. 3.10b. Crested Serpent eagle [With permission from Frank McClintock]

Fig. 3.10c. Crested Serpengt eagle [With permission from Pieter Verheij]

Fig. 3.10d. Crested Serpent eagle perching [With permission from Radhika Jha]

Fig. 3.10e. Crested Serpent eagle [With permission from Frank McClintock]

Fig. 3.10f. Immature Crested Serpent eagle [With permission from Frank McClintock]

Fig. 3.10g. Crested Serpent eagle in flight [With permission from Frank McClintock]

Fig. 3.10h. Crested Serpent eagle in flight [With permission from Pieter Verheij]

Fig. 3.10i. Crested Serpent eagle habitat [With permission from Radhika Jha]

Fig. 3.10j. Crested Serpent eagle habitat [With permission from Gary Ellis - unsplash]

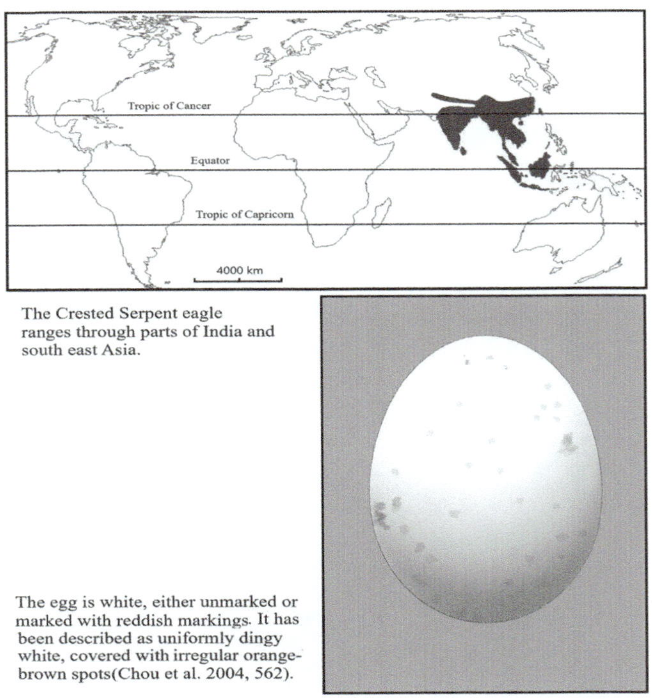

Fig. 3.10k. Crested Serpent eagle range and egg

like that of the adult, but with broader lighter feather tips. The eyes are brownish grey, the cere and facial skin is greenish yellow, and the legs are dark yellow. Illustrations of the appearance, range, and eggs of this species are shown in Figures 3.11a to 3.11i.

Classification

Global Raptor Information Network (2021ax) notes that the Nicobar Serpent eagle was "treated as a race of *S. cheela*" by Stresemann and Amadon (1979), or as conspecific with *S. minimus* by others (e.g., Amadon and Bull, 1988). Clark (1994) and Ferguson-Lees and Christie (2000) regarded it as a distinct species. Molecular genetics studies are needed to clarify the relationships of this group. Brown and Amadon (1968, 364) noted that "we have considered it as a separate species on account of its very small size and unmarked pale brown underside" (ibid.).

Breeding

Brown and Amadon (1968) write "No information."

Foraging

Brown and Amadon (1968, 364) note that the general habits of this species are "little known, but probably similar to the smaller races of *Spilornis cheela*." It is however frequently recorded in the forest canopy (Richmond, 1903; Abdulali, 1967, 1978; Sankaran, 1998), over elevations from sea level to 600 m (Ferguson-Lees & Christie, 2001; Global Raptor Information Network, 2021ax). Ferguson-Lees and Christie (2001) describe its diet as reptiles, rodents, and birds, and cite an example of the stomach contents of one bird, which comprised lizards, rats, a small bird, and an Emerald Dove (*Chalcophaps indica*, Linnaeus, 1758).

Population Status

The Nicobar Serpent eagle is classified as "Near Threatened", with a declining population trend by BirdLife International (2016). The range is Indomalayan, being endemic to Great Nicobar, Little Nicobar, and Menchal in the South Nicobar Islands group, off the Indian coast. It may not be migratory (Global Raptor Information Network, 2021ax). BirdLife International (2016) points out that "this recently recognised species is believed to have a small and declining population, although it is not as scarce as previously reported." The main ecological issues concerning this species are urbanization and increased pressure on eagle habitats, especially concerning pressure on its very small range (Stattersfield et al., 1998). Nevertheless, "its population is not regarded as severely fragmented or restricted to a few locations, hence it is classified as Near Threatened" (BirdLife International, 2016). The global population size has not been quantified, but the species is described as common (Ferguson-Lees & Christie, 2001). However, these authors estimated, without any quantitative data, that the global population would be anything from 11 to 1,000 individuals. The range on the island of Great Nicobar (with an area of 800 km²) is small, therefore there would have to be a very high population density for the numbers to reach the estimated three figure levels. Illustrations of the appearance, range, and eggs of this species are shown in Figures 3.11a to 3.11i.

3.13 Andaman Serpent Eagle *Spilornis elgini* (Blyth, 1863)

Physical Description

The Andaman Serpent eagle is a small eagle, with overall brown plumage. It is 51–59 cm in length, with a wingspan of 115–135 cm and a weight of 790–1000 grams (Clark et al., 2020g). The adult plumage is brown (including the upper wing and tail coverts), with white spots on the underside and white flecks on the upperparts. The black feathers on the crown and nape, have light brown edges and hidden bases. The wing quills are dark brown, and the primaries are darker brown, while the secondaries have white tips. The underside and the thighs are dark umber brown, spotted with white, and there is a greyish tinge to the sides of the neck. The underside wing quills are greyish brown at the bases and the inner webs, with blackish tips, crossed with several greyish white bars, wider on the primaries than on the secondaries. The upper tail feathers are brown with two to three narrow lighter bars, and white bars on the undertail coverts. The underside of the tail is black, with two to three grey bars. The eyes and feet are yellow and the cere is peach colored. The juvenile plumage is "either unknown or does not differ from the adult" (Brown & Amadon, 1968, 365). The Andaman serpent eagle may be distinguished from *S.c. davisoni* as it has a different range, is colored

Fig. 3.11a. Nicobar Serpent eagle perching [With permission from MV Shreeram]

Fig. 3.11b. Immature Nicobar Serpent eagle [With permission from MV Shreeram]

Fig. 3.11c. Immature Nicobar Serpent eagle [With permission from MV Shreeram]

Fig. 3.11d. Immature Nicobar Serpent eagle [With permission from Niranjan Sant]

Fig. 3.11e. Nicobar Serpent eagle in flight [With permission from Shashank Dalvi]

Fig. 3.11f. Central Nicobar Serpent eagle in flight [With permission from Shashank Dalvi]

Fig. 3.11g. Nicobar Serpent eagle habitat [Credit: Sahil - unsplash]

Fig. 3.11h. Nicobar Serpent eagle habitat [Credit: Sahil - unsplash]

dark brown spotted clear white below, three tail bands (as opposed to one) and the wings with several (5 to 6) rather than a few (one to three) lighter colored bars. Some authors have hinted that it may be a race of *S. cheela*, "but as it occurs on the same islands as *S.c. davisoni* with no integration, it is a species" (Brown & Amadon, 1968, 365). Brown and Amadon (1968) note that there is no recorded call. Illustrations of the appearance, range, and eggs of this species are shown in Figures 3.12a to 3.12h.

Classification

The Andaman Serpent eagle was previously classified as a subspecies of the Crested serpent eagle, but not by Stresemann and Amadon (1979). It has also been considered as conspecific with the Central Nicobar serpent eagle (*S. minimus* or *S.c. minimus*) by some authorities, such as Amadon and Bull (1988). Clark (1994) and Ferguson-Lees and Christie (2000) classified it as a distinct species. Amadon and Bull (1988) observe that the Andaman Serpent eagle is sympatric with the subspecies of the Crested Serpent eagle, *S. cheela davisoni* but they do not interbreed. The latter species may have originated from a later repopulation of the Andaman Islands. Current findings indicate that molecular genetics studies may be needed to clarify the statuses of the species and subspecies of this group of eagles (Global Raptor Information Network, 2021ay).

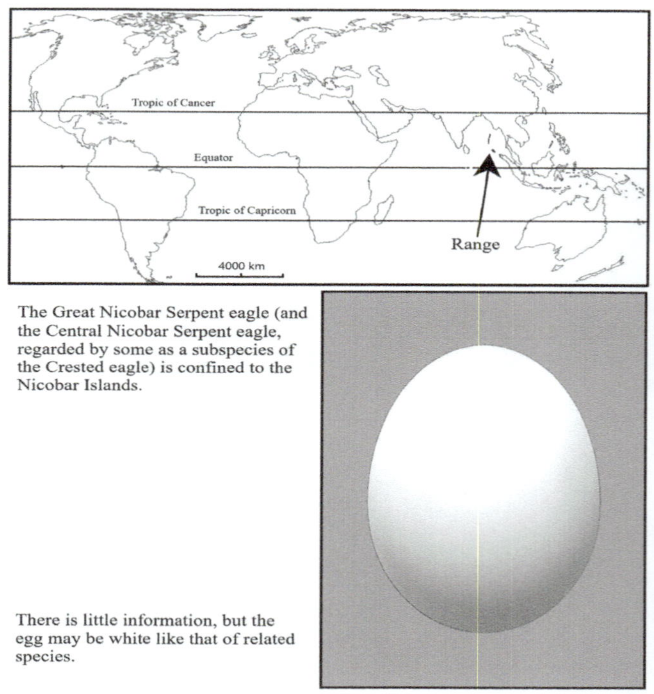

Fig. 3.11i. Nicobar Serpent eagle range and egg

Breeding

The breeding season begins with displays, which involve calling, with mutual soaring over the breeding territory. Breeding activity begins in September, with *laying* in October to December. The nest is very large usually more than 30 meters off the ground in a large tree (Clark et al., 2020g).

Foraging

Brown and Amadon (1968) note that the Andaman serpent eagle's general habitats are mostly unknown, but unlike the Crested Serpent eagle it frequents inland forest clearings, while the latter is commoner on the coast. It has however been recorded in closed canopy wet evergreen, semi-evergreen and even degraded forests, cultivated land, and wetlands (Clark et al., 2017). Commoner in inland areas, especially in inland forest clearings and hillsides with scattered trees, it nevertheless has been recorded in coastal areas, occurring from sea level to 700 meters elevation (Ali & Ripley, 1978; Koparde & Manchi, 2013; Rajeshkumar et al., 2015; BirdLife International, 2001, 2017). BirdLife International (2017) notes that it may be ecologically separated from the endemic subspecies of the Crested Serpent eagle, which frequents coastal forests on the same island (Davidar et al., 1996; Grimmett et al., 1998). Hunting is mostly from a perch (Ferguson-Lees and Christie, 2001). In addition to snakes, it has been recorded taking lizards and rats (Rao et al., 2013), as well as crabs and prawns (Pande et al., 2009).

Population Status

The Andaman Serpent eagle is endemic to the Andaman Islands, in the Bay of Bengal east of India and west of Myanmar, ranging over the North Andaman island southwards to the Little Andaman Islands (Davidar et al., 2007, Koparde and Manchi 2013, Rajeshkumar et al., 2015). Ferguson-Lees and Christie (2001) estimated the global population as in four figures (1,000+), based on a projected average density of 1 pair/13 km^2. BirdLife International (2017) estimated the population of mature individuals at 1,000-4,000.

It is considered fairly common across its range and in suitable habitat and is probably non-migratory (BirdLife International 2001; Vijayan, 2009; BirdLife International, 2017; Clark et al., 2017). Due to the small population within a limited range, and the fragmented, multi-island subpopulation structure, conservation issues may emerge (Clark et al., 2017). Human populations on the larger islands are increasing, with deforestation under agriculture, grazing, and logging, and hunting activities widespread (Stattersfield et al., 1998). Therefore, eagle habitats in the interior of the Andaman Islands are threatened which may affect the eagles' foraging ranges, and population levels. For these reasons, the Andaman Serpent eagle is listed as "Vulnerable" by BirdLife International (2017). The Global Raptor Information Network (2021ay) lists this species as "Near Threatened" with a declining population trend. Illustrations of the appearance, range, and eggs of this species are shown in Figures 3.12a to 3.12h.

Genus *Dryotriorchis* (Shelley, 1874)

This genus has only one species, Congo Serpent eagle (*Dryotriorchis spectabilis* or *Circaetus spectabilis*, Schlegel, 1863). This species is medium-sized, with a large head and eyes, short wings and a long, rounded tail. Smaller than the *Circaetus* species, the immature birds resemble the members of the Genus *Spilornis* (Brown & Amadon, 1968).

3.14 Congo Serpent Eagle *Dryotriorchis spectabilis* or *Circaetus spectabilis* (Schlegel, 1863)

Physical Description

The Congo Serpent eagle is a medium-sized, lightly built eagle species with shortish, rounded wings and a long, rounded tail. It is 54 to 60 centimeters in length, of which 25 to 27 centimeters represents the tail. The wingspan is 94 to 106 centimeters (Negro, 2008; Ferguson-Lees & Christie, 2001). The weight may be estimated around one kilogram (Kemp & Kemp, 1998; Kemp & Kirwan, 2020c). The female is slightly larger than the male. The adult plumage includes dark brown upperparts, pale brown cheeks, a reddish throat with a black streak on the throat and another on the lower cheek aligned with the mouth (giving the appearance of a moustache), a dark brown crown, mantle, and upper parts, with white bases on the nape and occipital area, and reddish brown on the collar and neck sides. The feathers on the crown are slightly crested (Negro, 2008). The tail is light brown with six dark bars, blackish upper wing coverts and less coverts, the latter tipped white, and the dark brown primaries and lighter brown inner webs which are tipped and barred with black (Ferguson-Lees & Christie, 2001; Kemp & Kirwan, 2020c).

The white underparts, undertail and underwing coverts are tinted reddish brown, with black spots, the thighs are barred sepia brown and white and the underwing has dark spots and bars. The eyes are dark brown or gray, and the cere, legs and feet are

Fig. 3.12a. Andaman Serpent eagle [With permission from Khushboo & Rahul Sharma]

Fig. 3.12b. Andaman Serpent eagle [With permission from Balaji Venkatesh Sivaramakrishnan]

Fig. 3.12c. Andaman Serpent eagle perching [With permission from Stanislav Harvančík]

Fig. 3.12d. Andaman Serpent eagle perching [With permission from Stanislav Harvančík]

Fig. 3.12e. Andaman Serpent eagle perching [With permission from Anay Tamekar]

Fig. 3.12f. Andaman Serpent eagle habitat [Credit: Lopamudra Barman, Kolkata, India (www.flickr.com/photos/lopamudrabarman)]

Fig. 3.12g. Andaman Serpent eagle habitat, Little Andaman [With permission from Robert Koster]

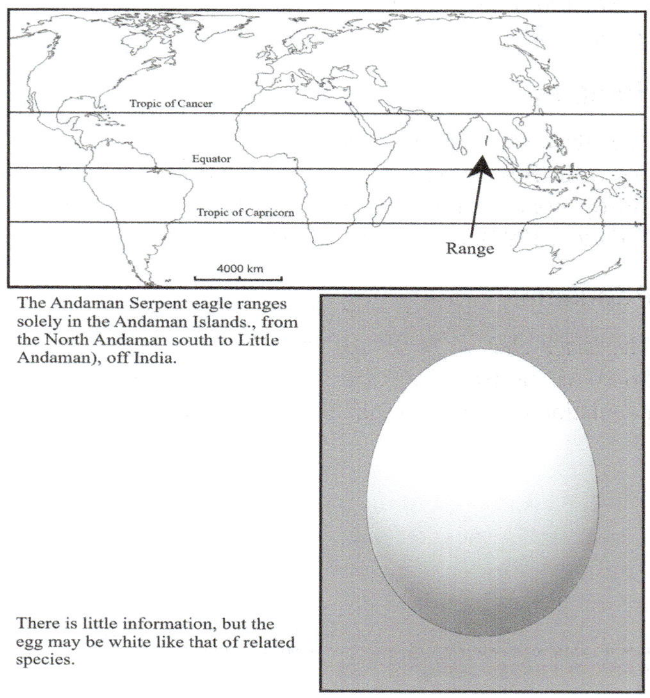

Fig. 3.12h. Andaman Serpent eagle range and egg

yellow (Brown & Amadon, 1968). Immature plumage comprises a white crown and mantle, with the terminal half of the feathers reddish brown with black tips. The underparts are spotted blackish and reddish brown. The subspecies *D.s. batesi* has browner upperparts and whiter underparts, with the fewer spots found only on the flanks, unlike *D.s. spectabilis*, which has spots on the chest (Brown & Amadon, 1968; Brown et al., 1982). The immature plumage includes a spotted chest which is reduced with age. The Congo Serpent eagle resembles Cassin's hawk eagle but has bare yellow legs. It also resembles the Black Sparrow hawk, but differs in having a reddish collar, black streaks on the lower cheeks and throat, and six tail bars (Negro, 2008; Kemp & Kirwan, 2020c). This species is vocal, with a call "cat-like mewing" and a "nasal cow – cow" (Brown & Amadon, 1968; Brown et al., 1982; Ferguson-Lees & Christie, 2001). Gatter (1997) describes its call as "a far-carrying peacock-like miaowing." Illustrations of the appearance, range, and eggs of this species are shown in Figures 3.13a to 3.13i.

Classification

The Congo Serpent eagle is classified as the two subspecies: the nominate subspecies is *Dryotriorchis spectabilis spectabilis* and *Dryotriorchis spectabilis batesi* (Sharpe, 1904; Kemp & Kirwan, 2020c). The latter was originally a separate species, *Dryotriorchis batesi*, defined by Richard Bowdler Sharpe in 1904, based on its unique, unspotted breast. Shelley (1874) moved this species from its original classification as *Astur spectabilis* (Schlegel, 1863) to the Genus *Dryotriorchis*, due to certain unique characteristics: comparatively short wings, longish tail, short crest, and oval nostrils (BirdLife International, 2016). Some have placed it in the Genus *Circaetus* as Lerner and Mindell (2005) found through a study of the molecular sequences of two mitochondrial genes and one nuclear intron, that it is close to *Circaetus*. Lerner and Mindel (2005, 343) note that "the West African serpent-eagle (*Dryotriorchis spectabilis*) falls within a clade of species in the genus *Circaetus*, suggesting that the taxonomy of these two genera should be revised." They note differences between *Circaetus* and *spectabilis*, in that "the genus *Circaetus* is not monophyletic in these analyses when the West African serpent eagle (*Dryotriorchis spectabilis*) is included" (Lerner & Mindel, 2005, 340).

Foraging

The Congo Serpent eagle frequents dense forest, and has large eyes adapted for hunting in shaded, canopied forest, but is less adapted to more open secondary forest (Ferguson-Lees & Christie, 2001). The species ecology is comparatively under researched (Demey & Ossom, 2007). The Global Raptor Information Network (2021az) describes it as "secretive and easily overlooked in lowland and medium-altitude primary rainforests" also spending "most of its time perched at mid-levels of forests and forest edges", for example in Sierra Leone (Dowsett-Lemaire & Dowsett, 2008). Physically, its large head and eyes may be adaptations hunting in the dim light of the lower layers of the closed canopy forests (Snow, 1978). The Congo Serpent eagle's close resemblance to the sympatric, bird-eating Cassin's Hawk-eagle (*Aquila africana*, Cassin, 1865), has been described as a possible ecological advantage (Negro, 2008). The argument is the Serpent eagle derives an advantage by disguise, as perhaps snakes may not hide from the bird-eating Cassin's Hawk eagle. Also, competitive from the latter bird may be less for the Serpent eagle, and less mobbing from smaller birds that may normally avoid the Cassin's Hawk eagle. Another suggestion is that the similarities of the two species may be convergent evolution in a similar context with shared ecological constraints (Global Raptor Information Network, 2021az).

Recent and historic records indicate the main prey animals are reptiles, amphibians and to a lesser extent, small mammals, mostly captured from tree woody structures and foliage, or on the ground, all mostly taken by stooping from a perch (Kemp & Kemp, 1998). For example, Olmos and Turshak (2009) saw an eagle carrying a whip snake (possibly of Genus *Psammophis*, Fitzinger, 1826). Congo Serpent eagles have been observed flying above forest canopies and clearings, but evidence indicates that most hunting is from perches within the dense forest or in cleared areas, including road and other clearings (Borrow & Demey, 2001). Gatter (1997) reports eagles perching on branches overhanging rivers, possibly tracking water reptiles.

The distribution of the Congo Serpent eagle is based on the Afrotropical forests from Sierra Leone to Cameroon in the west, to western Uganda and Zaire in the east, and southwards to northern Angola. There are regional variations in its population density (Borrow & Demey, 2001; Global Raptor Information Network, 2021az). In Liberia, it is fairly common, although elusive, frequenting primary and old secondary forest from the coast into the forested interior (Gatter, 1997). For the rest of its range, it has been recorded in dense forest (Dowsett Lemaire & Dowsett, 2008). In some cases, as in Cote de Ivoire it frequents only primary forest, avoiding even secondary forest (Thiollay, 1985). It is a rare resident in Ghanaian forests (Grimes, 1987) and in Nigeria (Elgood et al., 1994; Olmos & Turshak, 2009). It is rare and a possible breeding resident in Uganda (Carswell et al., 2005) and Angola (Pinto, 1983; Dean, 2000). Favored foods include snakes and lizards, and also toads and small mammals. Hunting is commonly done from a perch (Brown & Amadon, 1968; Ferguson-Lees & Christie, 2001).

Breeding

Brown and Amadon (1968) note that nothing is known about the breeding habits, possibly laying is from June to November. Nevertheless, some authors have published records on breeding observations. For example, in the Democratic Republic of Congo Prigogine (1971) collected three juveniles during July, mid-August, and mid-December in South Kivu (eastern Congo) and surmised that the egg-laying month in that region was January. Ferguson-Lees & Christie (2001) write that breeding may take place in October to December in Gabon and from June to November in the Democratic Republic of the Congo. Herroelen (2006), also describes some juveniles in the Royal Museum for Central Africa, which were collected from Oriental Province in northern Congo (November to December), Tshuapa District (NW Central Congo) and Mbandaka, Equateur (NW Congo) (February to March, and June), and Bolobo, Bandundu (Western Congo) (December).

Population Status

The Congo Serpent eagle has a very extensive range (2,880,000 square kilometers) including the forested parts of the territories of Sierra Leone, Guinea, Liberia, Côte d'Ivoire, Ghana, Nigeria, the Central African Republic, Democratic Republic of the Congo, Gabon, Republic of the Congo and Angola (Ferguson-Lees & Christie, 2001). *D.s. spectabilis* ranges through the upper Guinean forests from Liberia to northern Cameroon, while *D.s. batesi* ranges through the lower Guinean forests from southern Cameroon through Gabon and western Uganda and northern Angola (Brown et al., 1982). Ferguson-Lees and Christie (2001) have estimated the population could be in the five-figure range (10,000+), based on the large range and forest habitat. Thiollay (1985, 2000) classified this species as among the most threatened species in Côte d'Ivoire, due to human-induced deforestation, and its intolerance of expanding secondary habitats. Hence, it might be more common in the larger tracts of dense forest in Central Africa (Thiollay, 2000).

Due to its extensive range and some population factors the Congo Serpent eagle is assessed as Least Concern rather than Vulnerable by BirdLife International (2016). This assessment uses the criterion that vulnerability implies a range size of less than 20,000 km^2 and the additional parameters of a declining or fluctuating range size, and a general degradation of the habitat extent and/or quality, a small number of locations and pronounced habitat fragmentation, and a declining population size (BirdLife International, 2016). Current records indicate a slow population decline, but this is gradual and is too slow to be considered a threat and a factor for species vulnerability. For this designation, the decline must be greater than 30% over ten years or over three generations. The population size is still too large to be considered vulnerable, being over 10,000 mature individuals with only a minimal decline (BirdLife International, 2016). Illustrations of the appearance, range, and eggs of this species are shown in Figures 3.13a to 3.13i.

Genus *Eutriorchis* (Sharpe, 1875)

This genus has one species, the Madagascar serpent eagle, a medium sized eagle with a small crest formed from longer crown and nape feathers, short, rounded wings, a comparatively long rounded tail, and scaly legs and toes. The crown is like that of the genus *Spilornis*, and the tail is slightly longer than that of the similar Congo Serpent eagle (Ferguson-Lees & Christie, 2001).

3.15 Madagascar Serpent Eagle *Eutriorchis astur* (Sharpe, 1875)

Physical Description

The Madagascar serpent eagle is medium-sized, measuring 57 to 66 cm in length, with a wingspan of 90 to 110 cm and a weight of 700-800 g (Kemp & Kemp, 1998, 2005; Morris & Hawkins, 1998; Ferguson-Lees & Christie, 2001; Stafford et al., 2015). The adult plumage is primarily composed of a greyish-brown to earth-brown back, a paler belly, breast, and throat and overall barring, with "narrow, blackish barring in crown, nape, some ear-coverts, and wider black barring on the tertials, secondaries and primaries" (BirdLife International, 2016). The crown and nape are brown, the chin and throat are grey with black streaks, and the back and tail are grayish brown, with eight darker tail bars. The upper wing coverts are pale brown with darker barring and the wing quills are similarly colored with about seven bars. The white underparts have slight reddish coloration with dark barring. The tail underside is gray with eight dark bars, and the underwing quills are gray crossed with seven darks bars. The eye and cere yellow, and it has "long, powerful, yellowish, heavily-scaled legs" (BirdLife International, 2016).

This species is like Henst's Goshawk (*Accipiter henstii*, Schlegel, 1873), but is larger and longer tailed and lighter colored. BirdLife International (2016) further note the differences between the two species: "Differs from Henst's Goshawk *Accipiter henstii* by paler earth-brown upperparts combined with horizontal breast-barring, inconspicuous dark bars in upperparts (especially tertials and on head), rather long tail, rather bulging head-shape with slight nape crest, and call." According to Brown

Fig. 3.13a. Congo Serpent eagle perching [With permission from Dubi Shapiro]

Fig. 3.13b. Congo Serpent eagle perching Kakum, Ghana [With permission from Nik Borrow]

Fig. 3.13c. Congo Serpent eagle perching [With permission from Johannes Ferdinand Bird-Lens.com]

Fig. 3.13d. Congo Serpent eagle in flight [With permission from Nick Gaunt]

Fig. 3.13e. Congo Serpent eagle in flight [With permission from Johannes Ferdinand Bird-Lens.com]

Fig. 3.13f. Congo Serpent eagle perching in habitat [With permission from Adam Dare]

Fig. 3.13g. Congo Serpent eagle perching in habitat [With permission from Ken Behrens]

Fig. 3.13h. Congo Serpent eagle habitat [Credit: Andrey Gudkov]

& Amadon (1968, 368) the call is unrecorded, but BirdLife International (2016) describes a "loud call, a far-carrying and measured *wah...wah...wah...wah...*, often followed by a lower *rugh*." Illustrations of the appearance, range, and eggs of this species are shown in Figures 3.14a to 3.14j.

Classification

The Madagascar Serpent eagle is monotypic (Global Raptor Information Network, 2021ba). However, classification schemes have changed with time (del Hoyo et al., 2014). For example, Brown and Amadon (1968) suggested that this species and the Congo Serpent eagle might be congeneric, and that the Madagascar Serpent eagle might be related to the serpent eagles, *Spilornis*. More recently, Lerner and Mindell's (2005) study, which was based on more advanced molecular sequences of two mitochondrial genes and one nuclear intron, found that the Madagascar Serpent eagle has links with the subfamily, *Gypaetinae*, where it clusters with the Palm-nut vulture (*Gypohier axangolensis*, Gmelin, 1788), the Egyptian vulture (*Neophron percnopterus*, Linnaeus, 1758), and the Bearded vulture (*Gypaetus barbatus*, Linnaeus, 1758). The findings indicated a high degree of genetic divergence between the related species, but closer links between these species than with other accipitrine species (Lerner & Mindell, 2005).

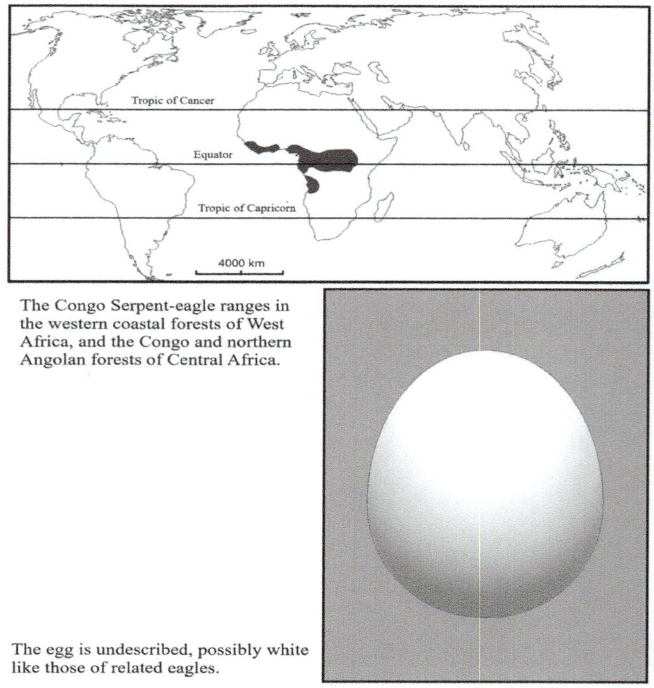

Fig. 3.13i. Congo Serpent eagle range and egg

Foraging

The Madagascar Serpent eagle frequents thick humid forests, hunting largely from perches (Brown & Amadon, 1968; Langrand, 1990; BirdLife International, 2016; Kemp & Christie, 2020). Thorstrom et al. (2003) record this species as frequenting primary rainforest and forest edges up to 1,000 m, mostly perching below the canopy. This is an elusive species (from a human observer's perspective), rarely soaring above the canopy, with their presence detected mostly by their call (Thorstrom et al., 2003; Kemp & Christie, 2020). An example, reported by Safford and Duckworth (1990) and Sheldon and Duckworth (1990) concerned the observation of a bird from the Marojely Natural Reserve on 23 September 1988, which was perched on some sub-canopy branches near a river. The bird was observed glancing at the surrounding forest floor and flying short distances between the trees (Safford & Duckworth, 1990; Sheldon & Duckworth, 1990).

The Madagascar Serpent eagle feeds mostly reptiles, especially lizards; Thorstrom & René de Roland (2000) record chameleons (*Furcifer* and *Calumna* spp.), Leaf-tailed geckos (*Uroplatus* spp.) and tree frogs. More rarely, it takes snakes and even bats, and insects (Thorstrom et al., 2003). Global Raptor Information Network (2021ba) notes that "reports by local residents of it feeding on small mammals and chickens are doubtful and probably result from confusion with Henst's Goshawk *Accipiter henstii*." Thorstrom et al. (2003) monitored two nests for prey remains; identifiable remains included 264 chameleons or geckos (85.1%), 41 frogs (13.2%), 2 snakes (0.7%), 2 bats (0.7%), and 1 insect (0.3%). As noted above, chameleons and Leaf-tailed geckos were the commomest, totally 263 (85%) of the identified items. Less common were frogs (Genera *Boophis*and *Mantidactylus*, snakes (*Stenophis* spp.), and bats (*Hipposideros commersoni*). Brown & Amadon (1968) also recorded a chameleon as prey.

Breeding

According to Brown and Amadon (1968, 368), the breeding patterns are "unknown." However, Thorstrom et al. (2003) describe two nesting attempts in which the nests were constructed of sticks and hidden in epiphytic ferns within the canopy of primary rainforest. Nests of this species are comparatively small, approximately 80 × 60 cm. There was a single, plain white egg in each of the studied nests (although other eggs recorded by these authors are described as stained). Although both adults incubated and delivered fresh twigs with leaves to the nest, the female did most of the incubation (78%) and also looked after the nestling (Thorstrom & René de Roland, 2000). The incubation period lasted 40 days at one nest, and the nestling periods were 58 and 62 days respectively (Thorstrom et al., 2003). The male brought 75% of the prey items to the nest. The female began hunting when the nestling was 18 days old. Both parents continued to bring food to the chick after fledging, and it began hunting by itself 10 days after fledging (Thorstrom et al., 2003). These author report nests that were located in primary rainforest in large epiphytic ferns (*Asplenium nidus*) surrounded by dense vines and branches. One nest was about 20 m elevation, in an isolated *Potameia capuron* (Lauraceae) tree and the other nest was at 25.2 m elevation in a *Canthium madagascariensis* tree of 35 cm (Thorstrom et al., 2003).

Population Status

The range of the Madagascan Serpent eagle is limited to Madagascar (BirdLife International, 2016). It is generally rare and occurs most frequently in the eastern rainforests (Thorstrom et al., 1995; Wilcove, 1999). The Global Raptor Information Network (2021ba) reports a population estimate of 66-1,660 individuals, points out that 250-999 mature individuals (375-1,499 individuals in total) may be more accurate. The population is recorded as declining, with climate change playing a possible role. Hints of past, denser populations may be derived from studies indicating bird captures in the southern region of Madagascar, outside the current range (Collar & Stuart, 1985; Dee, 1986; Langrand, 1990; Ferguson-Lees & Christie, 2001). Other studies recorded the main sightings as in the eastern parts of Madagascar, from the northern to southern parts of this range (Morris & Hawkins, 1998; Thorstrom et al., 2003). It is non-migratory (Bildstein, 2006).

The Madagascan serpent eagle was surmised to be very rare or even extinct during the middle of the twentieth century, as the only known specimens were taken before 1935 (Dee, 1986). By 1993, however, new specimens were recorded, some by work conducted by the Peregrine Fund (Thorstrom et al., 1995). The Global Raptor Information Network (2021ba) notes that "subsequently, studies by Russell Thorstrom, Lily René de Roland, Steve Goodman, and other researchers have shown that the Madagascar Serpent eagle is more common than formerly believed and not in imminent danger of extinction, as previously thought." However, due to its narrow range, and its vulnerability to loss of primary forests it is classified by BirdLife International (2016) as "Endangered." Illustrations of the appearance, range, and eggs of this species are shown in Figures 3.14a to 3.14j.

Genus *Pithecophaga* (Ogilvie-Grant, 2018)

The single species of this genus is a huge bird, comparable in size to the New World Harpy eagle and with the limited range of the larger Philippine islands. Like the latter bird, the tail is long and the wings are relatively short and rounded. The bill is very large "extremely compressed (equalled in this respect only by the Steller's Sea eagle)" (Brown & Amadon, 1968, 637). The legs also resemble those of the Harpy eagle, in that they are relatively short and very powerful, with thick toes and talons. The crest is composed of longer feathers on the nape and crown.

Fig. 3.14a. Madagascan Serpent eagle perching [With permission from Ken Behrens]

Fig. 3.14b. Madagascar Serpent eagle [With permission from Russell Thorstrom]

Fig. 3.14c. Madagascar Serpent eagle [With permission from Dubi Shaprio]

Fig. 3.14d. Madagascar Serpent eagle [With permission from Dubi Shaprio]

Fig. 3.14e. Madagascar Serpent eagle perching [With permission from Russell Thorstrom]

Fig. 3.14f. Immature Madagascar Serpent eagle on nest [With permission from Russell Thorstrom]

Fig. 3.14g. Madagascar Serpent eagle in habitat [With permission from Erwin Collaerts]

Fig. 3.14h. Madagascar Serpent eagle in habitat [With permission from Katerina Kalinina]

Fig. 3.14i. Madagascar Serpent eagle in habitat [Credit: Theme Inn - unsplash]

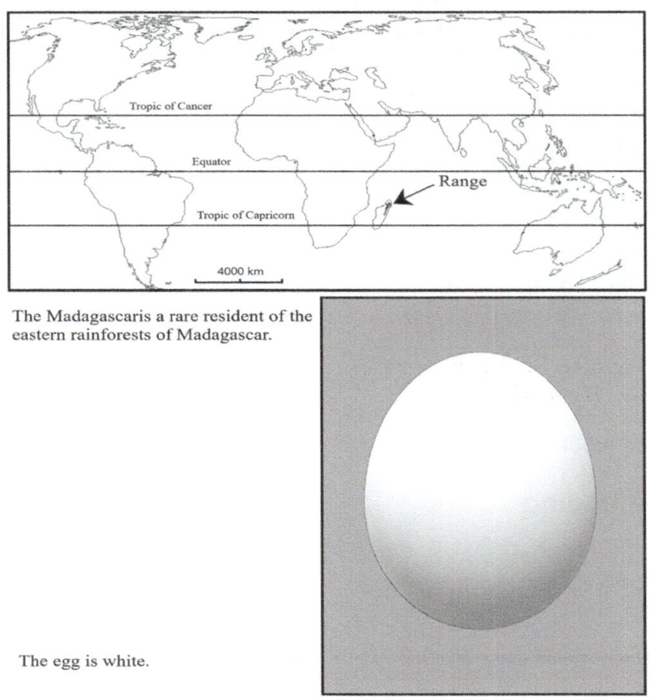

Fig. 3.14j. Madagascar Serpent eagle range and egg

3.16 Philippine Eagle, Monkey-eating Eagle or Great Philippine eagle *Pithecophaga jefferyi* (Ogilvie-Grant, 2019)

Physical Description

The Philippine eagle is a very large eagle (or large raptor by any standards). This species is considered the world's largest eagle by length and wing surface, and only slightly smaller than the Harpy eagle and Steller's Sea eagle by weight and bulk (Ferguson-Lees & Christie, 2001; Tabaranza, 2005). It may measure about 86 to 102 cm in length, with a wingspan of 184 to 220 cm (Ferguson & Christie, 2001). Some researchers have given varying lengths; for example, an analysis of some specimens from natural history collections gave averages of 95 cm for adult males and 105 cm for adult females (Gamauf et al., 1998). The tail is 42 to 50 cm long (Dupont, 1971; Ferguson-Lees & Christie, 2001). The Philippine eagle may thus be among the longest eagles in the world, at least equalling its main competitors in this regard, the female Harpy and the Steller's Sea eagles (Clark, 1994; Ferguson-Lees & Christie, 2001). Its weight may lie between 4.0 and 8.0 kg (Clark, 1994; BirdLife International, 2000, 2018; Kennedy et al., 2000). The weight of this species has also been disputed by different authorities. Some sources cite weights between 4.7 and 8.0 kg (Mearns, 1905; Seth-Smith, 1910; Ferguson-Lees & Christie, 2001). Other sources cite a narrower range, a 4.5 kg average for males and a 6.0 kg average for females (Gamauf et al., 1998). In terms of sexual dimorphisms, males are about 10% smaller than females (Ferguson-Lees & Christie, 2001; Gamauf et al., 1998).

For the adult, the crown and nape are creamy-brown to buff and steaked with black shafts (Brown & Amadon, 1968). There are black feathers on the chin and rufous streaks on the brown throat. The crest is composed of erectile and lanceolate feathers on the nape. The upper parts including the back and tail are dark brown, the feathers edged paler with white hidden bases, contributing to mottle appearance. The wing quills are black or very dark brown, with lighter colored inner webs and bases (Brown & Amadon, 1968). The underparts are white to lighter colored, with rufous streaks on the thighs and the underwing coverts. The wing quills have grey tips, with white bases, and with light wing linings. The underside tail quills are silver to grey. The eyes are bluish grey, the legs yellow and the cere greenish blue. According to Brown & Amadon (1968), juveniles are like adults. However, immature birds may have pale fringes to the upperpart (BirdLife International, 2018). The Philippine eagle's bill is one of the largest of any eagle, like that of the Steller's eagle. The bill may be over 7 cm in length from the gape (Dupont, 1971; Tabaranza, 2005). This species is unlikely to be confused with any other eagle, due to its very large size, bill size, and the contrasting dark upperparts and light underparts. The call is a "rather mellow whistle", "wheeew" which rises or descends and appears rather "weak" for a bird of this size; when excited, the call may be "a series of short, rather intense notes" (Brown & Amadon, 1968). Kennedy (1977) describes the cry as gives loud, high pitched whistling sounds. Illustrations of the appearance, range, and eggs of this species are shown in Figures 3.15a to 3.15k.

Classification

The Genus *Pithecophaga* is monotypic with no known subspecies. Brown and Amadon (1968) note that the Genus *Pithecophaga* may be related to the less specialised *Harpyopsis* of New Guinea, and less related to *Harpia* and *Morphnus* of South America. Shufeldt (1919) reported that initial studies of skeletal features in 1919 hinted at a close link with the Harpy eagle. This led to its inclusion in the subfamily Harpiinae the use of DNA sequences which identified them as unrelated (Lerner & Mindell, 2005). The Global Raptor Information Network (2021bb) notes that recent molecular studies (Lerner & Mindell, 2005), using molecular sequences from two mitochondrial genes and one nuclear intron, "indicated a previously unsuspected close relationship with the Bateleur, *Terathopius ecaudatus* of Africa." It is further noted that "*Pithecophaga* is sister to a clade of snake eagles (*Circaetinae*), instead of the clade including the genera *Harpia* (Harpy eagle), *Morphnus* (Crested eagle), and *Harpyopsis* (New Guinea Harpy eagle)." Earlier names for the Philippine eagle are the Monkey-eating eagle (due to local observations that it preyed on monkeys) and the Great Philippine eagle (due to its huge size) (Kennedy, 1977; Salvador & Ibanez, 2006).

Breeding

The breeding season begins with aerial displays, including paired soaring, male pursuit of the female, mutual presentation of talons with the male above and the female on her back, loud calling and nest building (Ferguson-Lees & Christie, 2001; Ibañez et al., 2003; Ibañez, 2007). The nesting season begins in September and October on Mindanao, but this varies according to location in its range (breeding in Mindanao and Luzon may differ) (Ibañez et al., 2003). The huge nest is built of sticks, usually in a tree at an elevation of over 30 meters. One white egg is laid, and the incubation is about 58 to 68 days, with hatching in January, and the nestling period 4 to 5 months (Brown & Amadon, 1968; Ferguson-Lees & Christie, 2001). The fledged young remain dependent on the parents for several months, and the whole breeding season may last up to two years (Kennedy, 1985; Miranda et al., 2000; Ibañez et al., 2003). The Philippine eagle has one of the longest breeding cycles of any bird of prey (Brown & Amadon, 1986; Ferguson-Lees & Christie, 2001). Eagle breeding pairs form monogamous bonds for life (BirdLife International, 2018).

Ibañez (2007) looked at the breeding biology, diet, behavior and nest characteristics of Philippine eagles in Mindanao using data from five Philippine eagle pairs nesting from 1999 to 2007. The findings indicated that the "Philippine Eagle has the longest and energetically most expensive parental investment for any birds of prey" the parameters of this assessment "nest attendance, incubation and brooding bout duration, trip duration and feeding rates during each stage of breeding for each sex." Additionally, the Flying lemur (*Cynocephalus Volans*, Linnaeus, 1758) was the main prey species taken, followed by other arboreal mammals, with at least 14 prey taxa taken and 9 species identified. There was no sign of prey base partitioning among sexes (for example, the larger females taking larger prey than males) but recorded predation on domestic animals and pets (Ibañez, 2007).

Foraging

The Philippine eagle forages in dense forest, usually below the canopy but sometimes above it, with flight usually between perches. The main habitat covers primary forest, forest edge, and logged forest (Kennedy et al., 2000). BirdLife International (2018) describes the primary habitat as dipterocarp forest, especially in areas with steep terrain, and to a lesser extent secondary growth and gallery forest, from lowlands to about 1,800 m elevation, while avoiding the open canopy forest. Salvador and Ibañez (2006) recorded nests in habitats, ranging from old growth diptocarp forest to second-growth and even highly modified areas. Due to lowland deforestation eagles have increasingly moved to mountainous areas such as the sides of steep slopes canyons, and slopes. The ranges of breeding pair can be as large as 12 to 100 km^2 (Rabor, 1971; Kennedy, 1977). The Philippine eagle evolved in an ecosystem where other large predators were absent, giving them dominant apex predator status and a specialised foraging ecology vulnerable to deforestation (Bueser et al., 2003).

Prey was initially believed to be mostly monkeys (the main monkey species occurring in its range is the Crab-eating macaque (*Macaca fascicularis*, Raffles, 1821), but later studies recorded other prey animals such as medium-sized mammals, such as monkeys, lemurs, civets, flying squirrels, fruit bats, and macaques; large snakes, monitor lizards, and even large birds, such as hornbills, as it is "reputed by natives to take large birds" (Brown & Amadon, 1968, see also Kennedy, 1985, Ibañez et al., 2003). Hunting is mostly from perches in the forest (Kennedy et al., 2000). Miranda et al. (2008) used the distribution of nests in Mindanao and suggested that a breeding pair forages over an average of 133 km^2, which includes on average of 68 km^2 of forest cover (Miranda et al., 2008). Pairs may hunt cooperatively, using distracting and rear attacks (Gonzales, 1968). Kennedy (1985) documented 218 prey items; of the 158 which were identified, mammals were the majority at 84.3%, birds were only 7.8%, and reptiles 7.8%. The commonest mammals were Flying lemurs (54.2%), Palm civets (12.0%), Flying squirrels (7.8%), Rousettefruit bat (5.4%), Long-tailed macaque (3.0%), and small bats, rats and deer (1.8%). Birds included the Rufous hornbill (6.0%), a medium-sized raptor (1.2%), and a small fledgling owl (0.6%). Reptiles included unidentified snakes (5.4%), two Philippine cobras (1.2%), and a monitor lizard (0.6%) (Kennedy, 1985).

The primary prey varies across the inhabited islands. The Philippine flying lemurs are common prey in Mindanao, but these are absent in Luzon. In Luzon, prey taken includes monkeys, birds, flying foxes, giant cloud-rats (*Phloeomys pallidus*, Nehring, 1890) and large snakes and lizards (Hurrell, 2014; Clark et al., 2015; BirdLife International, 2018). The Philippine eagle hunts from perching, or mixed perching and gliding through the lower branches of the forest. Eagles in Mindana may use gliding to hunt the elusive flying lemurs (Ferguson-Lees & Christie, 2001). Eagle pairs sometimes hunt troops of monkeys cooperatively, with one bird distracting the monkeys and the other eagle to swoop in from behind, hopefully unnoticed, for the kill (Ferguson-Lees & Christie, 2001).

Population Status

The Philippine eagle has been reported to occur primarily on four islands: Luzon, Samar, Leyte and Mindanao (Bueser et al., 2003; Salvador & Ibañez, 2006). Confined to these islands, it is non-migratory (Bildstein, 2006). It has a unique conservation status and some protections, due to its position as the national bird of the Philippines (Kennedy et al., 2000; Ferguson-Lees & Christie, 2001). Nevertheless, its status is "Critically Endangered" (BirdLife International, 2018). BirdLife International (2018) defends this classified status: "This long-lived species qualifies as Critically Endangered because it has an extremely small population, as a result of extremely rapid declines in the past three generations (56 years), owing to extensive deforestation. Recruitment to the adult population currently appears to be very low indicating that declines may continue into the future. Confirmation of trends is required and may lead to a change in status in the future." Salvador and Ibanez (2006, 171) support this designation, arguing that the Philippine eagle "is regarded as one of the most endangered eagles in the world. Habitat loss and human persecution continue to impact the species adversely. These problems are complicated further by a number of human-induced factors." Nevertheless, conservation activities have derived some positive results through sustained efforts. Recent actions include reintroductions.

Bueser et al. (2003) have estimated the population on Mindanao to be 82 to 233 pairs, with far smaller numbers on the other islands "perhaps six pairs on Samar, two on Leyte and probably very few on Luzon, giving a total population size of perhaps

90-250 pairs, or 180-500 mature individuals, roughly equating to 250-750 individuals in total" (BirdLife International, 2018). BirdLife International (2018) give examples of research that located evidence that may be "treated with caution" (Collar et al., 1999). These include the work of a Philippine Eagle Foundation survey team that located the first active Philippine eagle nest in Apayao province, Luzon (Ranada, 2015). There is also the work of Miranda et al. (2008), which was based on 1992 forest-cover data. Based on ecological parameters, this hinted at 226 mature eagles, within a total population of 350-670 birds. Using this as a nest density base model, a total of 340 pairs may be guessed across the other islands. A limitation was the lack of knowledge of the densities on the other islands (Miranda et al., 2008).

The decline of the eagle's population had previously been blamed on insufficient recruitment to the breeding population (Miranda et al., 2006), but later research blamed the poor dispersal and survival of juveniles and subadults (Miranda et al., 2008). Earlier reports noted the killing of birds during the 20th century. For example, Kennedy (1977) note that most older reports were of eagles that had been shot or captured. Thirty-five birds were lost from these causes during the period of his study. A recent example (2006), was that of a rescued bird (wounded by shot) which was rehabilitated by the Philippine Eagle Foundation, subsequently rehabilitated and released with a radio-tracking device. This bird was found dead, shot and eaten by local villagers (Salvador & Quiblat, 2009).

The breeding and release of captive birds is a recent action with some unsuccessful results. For example, the first release in Mount Apo, Mindanao 2004 failed when the bird was electrocuted nine months after the release (Block, 2004). Another released bird which was released on Mindanao in 2008 was killed by a hunter four months after the release, but this did not discourage conservation policies using this method (Salvador & Ibañez, 2006). For example, the Philippine Eagle Foundation manages the Philippine Eagle Centre in Davao City, Mindanao, and the captive-breeding and management of wild populations. The Philippine Eagle Foundation had produced 18 chicks in captivity by 2006, (Salvador & Ibañez, 2006). By 2008, the Philippine Eagle Foundation was developing a full reintroduction programme (Philippine Eagle Foundation, 2008; Salvador & Quiblatt, 2009).

The Global Raptor Information Network (2021bb) has compiled the Philippines eagle's status on the islands within its range. On Mindanao island, documented records are derived from the observations of Clemens (1907), Davidson (1934), McGregor (1907, 1921), Mearns (1905), and Seth-Smith (1910). Rabor (1965) described the population as only a few birds. Gonzales (1969) estimated 36-50 birds, based on a survey that found one eagle at Mt. Timolan, South Zamboanga and three eagles at the Kibawalan-Takalon forests, South Davao. Other estimates of eagles were 40 (Alvarez, 1970), 36 (Gonzales, 1971), 50 to 60 (Rabor, 1971) and Kennedy (1977) who cited 70 eagles based on the numbers cited by Alvarez (1970), Rabor (1971), and Gonzales (1971) and 309 to 580 eagles, based on wider, suitable habitat analysis. A recent estimates were 82-233 pairs (Bueser et al., 2003, based on 1991-1998 data on remaining forest and the mean nearest neighbor distance between breeding pairs of 12.74 km) and 200 pairs (Salvador & Ibañez, 2006), the last number being "the most recent estimate of the size of the Mindanao population, which is the best studied to date", with similar results to that of Kennedy (1977) (Global Raptor Information Network, 2021bb).

On Leyte island, records are based on individual observations, taken in the 19th century (Ogilvie-Grant, 1897), and sightings by Philippine Parks personnel in 1951, 1963, 1968, 1969, and two eagles killed in 1965 (Kennedy, 1977). In 1970, the estimate was of 8 to 10 eagles (Kennedy, 1977). On Luzon island, the historical records include an eagle killed in May 1907 (McGregor, 1907), two eagles sighted in August 1907 (Seth-Snith, 1910), one eagle captured in January 1917 (McGregor, 1918), another eagle captured in July 1926 (McGregor, 1927) and one taken, perhaps in 1933-1934 (Davidson, 1934). Rabor (1971) visited Luzon in 1959, 1960, and 1961, and reported that in some areas eagles had been unseen since 1930s (the Cordillera and Ilocos mountain ranges in northwest Luzon) but were still reported in the 1960 in other areas (northeastern Luzon, including Cagayan, Isabela, Nueva Vizcaya, and Quezon provinces). The last record in Luzon was taken in 1963 (Gonzales, 1968). Global Raptor Information Network (2021bb) states that this species is probably extirpated on Luzon. On Samar island, the recent estimate is six pairs (BirdLife International, 2018). Older records show fewer specimens. For example, Ogilvie-Grant (1897) and Davidson (1934) made few records, Rabor (1965) found no records from the previous 30 years and Rabor (1971) assessed the species as extinct in Samar, based on a 1957 research visit.

Maala (2001) notes some of the legislative pieces that have been created by the Philippine Government to protect the Philippine eagle. These include Administrative Order 235 (1970) prohibiting the harming, killing and trading of eagles; Republic Act 6147 (1970) declaring the protection of Philippine eagles; Parks and Wildlife General Administrative Order 1 (1971) regulating the preservation of the eagle; Proclamation 1732 (1978) changing the former name "monkey-eating eagle" to Philippine eagle; Republic Act 7586 (1992) creating a National Integrated Protected Areas System including eagle habitats; Proclamation 615 (1995) declaring the Philippine national bird; and Proclamation 79 (1999) declaring June 4-10 of every year as Philippine eagle week. Illustrations of the appearance, range, and eggs of this species are shown in Figures 3.15a to 3.15k.

Fig. 3.15a. Philippine eagle perching [With permission from Rodrigo Layug]

Fig. 3.15b. Philippine eagle perching [Credit: Photo by Janssen Panizales from Pexels]

Fig. 3.15c. Philippine eagle perching [With permission from Neon Tomas Buenaflor Rosell II]

Fig. 3.15d. Philippine eagle perching [With permission from Chris Chafer]

Fig. 3.15e. Philippine eagle perching and displaying [With permission from William L. Lim]

Fig. 3.15f. Philippine eagle [With permission from Bram Demeulemeester]

Fig. 3.15g. Philippine eagle nesting in forest habitat [Credit: Feathercollector]

Fig. 3.15h. Philippine eagle nesting in forest habitat [With permission from Chris Chafer]

Fig. 3.15i. Philippine eagle nesting in forest habitat [With permission from Chris Chafer]

Fig. 3.15j. Philippine eagle habitat, Luzon [Credit: Aldrin William Pascua - unsplash]

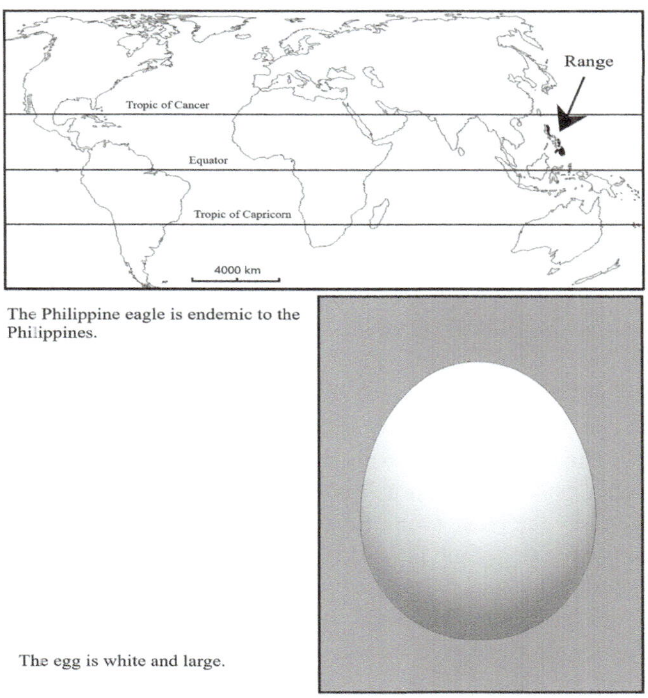

Fig. 3.15k. Philippine eagle range and egg

4

Systematic List of the Sea Eagles Subfamily *Haliaeetinae*; Genus *Haliaeetus*

4.1 Introduction

The Subfamily *Haliaeetinae* or sea eagles has one genus, *Haliaeetus* with 10 species (Lerner & Mindell, 2005). These are the African fish eagle, Bald eagle, Gray-headed fish eagle, Lesser fish eagle, Madagascar fish eagle, Pallas's fish eagle, Sanford's sea eagle, Steller's sea eagle, White-bellied sea eagle and the White-tailed eagle (refer to the Preface for the scientific names and the nomenclatural authorities). The Genus *Icthyophaga* (Lesson, 1843) comprises two species, which are included in this subfamily. The Subfamily *Haliaeetinae* is found on all continents except South America. Members of the Genus *Haliaeetus* are described by Brown and Amadon (1968) as "large to very large" eagles, with long broad wings, short to medium, rounded or wedge-like tails. Their bills are strong and large, with high to moderate compression (compressed refers to the bill being higher than wide). The legs are relatively short, feathered on the upper thighs (unlike the Genus *Aquila* eagles, where the legs are feathered to the toes) with powerful feet and talons. Brown and Amadon (1968) further note that the names 'sea eagle' and more recently 'fish eagle' are derived from the ranges of these species, which are generally near water bodies, especially coasts. Their diets are also generally focussed on fish, and to a lesser extent mammals and birds.

The genetic classification of sea eagles is complex (Wink et al., 1996). Using morphological research, some studies concluded that some *Haliaeetus* species are genetically more closely related than others. For example, Brown and Amadon (1968, 279) argue that six species probably form three "super-species." The White-tailed Sea eagle (*H. albicilla*, ranging in Eurasia) is linked to the Bald eagle (*H. leucocephalus*, ranging in North America); the African fish eagle (*H. vocifer*, ranging in Africa) is linked to the Madagascan fish eagle (*H. vociferoides*, ranging in Madagascar); and White-bellied sea eagle (*H. leucogaster*, ranging from India to Australia) is linked to Sanford's sea eagle (*H. sanfordii*, ranging in the Solomons). The last two species Pallas's fish eagle (*H. leucoryphus*, ranging in central Eurasia) and Steller's sea eagle (*H. pelagicus*) ranging in East Asia). Brown and Amadon (1968, 279) further argue that the *Haliaeetus* species are "very closely related" to those of the Genus *Ichthyophaga* and also to some kites, including scavengers of the Genus *Haliastur*, the Genus *Gypohierax* (Palm nut vulture/Vulturine Fish eagle, *Gypohierax angolensis*, Gmelin, 1788). This is argued to be a possible link with the Accipitrid Old World vultures.

Lerner and Mindel (2005, 342) using more advanced genetic research, note that Sea eagles have been classified as a monophyletic group, closely linked to the *Milvinae* kites, a point supported by the older morphological studies mentioned above (Brown & Amadon, 1968; del Hoyo et al., 1994; Wink et al., 1996). This relationship was discerned using morphological research, which found that sea eagles and Milvinae kites both possessed a basal fusion of the second and third phalanges (Holdaway, 1994), a feature not shared by the other accipitrid species. Lerner and Mindel (2005) using molecular methods argue against the findings of older similar works (e.g. Seibold and Helbig, 1996) which argue for the monophyly (sharing a common ancestor) of the sea eagles in the genus *Haliaeetus*. There is evidence, using more detailed methods, for the monophyly of the subfamily *Haliaeetinae*, but not when the genus *Haliaeetus* and genus *Ichthyophaga* are combined (Lerner & Mindel, 2005). Their findings also support a sister relationship between the sea eagles and Milvinae kites (genera *Milvus* and *Haliastur*).

Brown and Amadon (1968) argue for a division of the *Haliaeetus* eagles between those species with northern distributions (White-tailed eagle, Bald eagle and Steller's sea eagle) and species of tropical distributions (African fish eagle, Madagascan fish eagle, White-bellied sea eagle and Sanford's sea eagle) (Lerner & Mindel, 2005). Pallas's fish eagle is classified as unique, in that it is a tropical resident, while also breeding in the northern temperate region, but clustering with the northern group (see also

Seibold & Helbig, 1996). The two species, the Lesser fish eagle (*Haliaeetus humilis*, Müller & Schlegel, 1841, or *Ichthyophaga humilis*) and the Grey-headed fish eagle (*Haliaeetus ichthyaetus*, Horsfield, 1821, or *Ichthyophaga ichthyaetus*) are included with the tropical group (Lerner & Mindel, 2005, 343).

4.2 Steller's Sea Eagle *Haliaeetus pelagicus* (Pallas, 1811)

Physical Description

This is a very distinctive species in appearance. BirdLife International (2016) describes the Steller's Sea eagle as a "huge, unmistakable eagle with massive yellow bill in adults. Adults have blackish-brown plumage with white shoulders and white, wedge-shaped tail. Dark morph lacks white shoulders." The Steller's Sea eagle is one of the world's principal eagles, as it appears in the top five of the largest, strongest, longest winged, and heaviest eagles (del Hoyo et al., 1994; Ferguson-Lees & Christie, 2001). The largest of the sea eagles, this species may be the world's heaviest eagle at 5 to 9 kg, although it may be smaller than the Harpy and Philippine eagles in terms of some other parameters (Ferguson-Lees & Christie, 2001; Dunning, 2008). The larger females may range in size from 6 to 9 kg, with male weights ranging from 5 to 7 kg (Ferguson-Lees & Christie, 2001; Dunning, 2008). The total length ranges from 85 to 105 cm with male averages around 90 cm and females about 100 cm (BirdLife International, 2016). The wingspan ranges from 1.95 to 2.5 m, one of the largest among eagles. Different authors give different median or maximum wingspan; cited average (some of possibly doubtful validity) figures include 2.13 m (Ferguson-Lees & Christie, 2001), 2.2 m (Saito, 2009), and maximum wingspans of 2.44 m and even 2.7 to 2.8 meters (Palmer, 1988; National Geographic, 2015; Cincinnati Zoo & Botanical Garden, 2020).

The feathers of the upper body are dark brown to blackish brown, while those on the head are paler and greyish brown with streaked shafts (Brown & Amadon, 1968, 298). The forehead and in some cases parts of the crown have white patches. Part of the upper parts are white, these including the upper tail coverts, medium and lesser upper wing coverts, under wing and tail coverts and thighs. The eye is light yellow, the cere legs and the very large bill are yellow. This is one of the largest bills for any eagle, or even bird of prey (Ladyguin, 2000). The tail is longer than that of the White-tailed eagle and diamond-shaped (Lee, 2005). Immature birds, which attain adult plumage at about 4 years of age, are largely dark brown, with grey brown streaks on the head and neck. There are also lighter tips on the wing quills and greater wing coverts. The white tail has black mottles. The eyes are brown dirtier yellow than that of the adult. The feet are a lighter shade of yellow than those of the adult. The wing and tail feathers are "longer than those of the adult" (Brown & Amadon, 1968, 298).

For the possible Korean subspecies (*H.p. niger*) there is dark plumage all over, except for the white tail, with yellow eyes, bill, and feet, and it is "said to be slightly smaller" (Brown & Amadon, 1968, 298). Brown & Amadon (1968, 298) speculated on the status of this subspecies: "birds with the typical plumage have been obtained from Korea and *niger* may be only a color phase; it is now said to be very rare if not extinct" (ibid.). More recently, this has been revealed to be a dark morph, not a separate subspecies of species (Ferguson-Lees & Christie, 2001; Davies, 2010; Kaiser, 2010). As argued by Davies (2010), "for over a century, experts questioned whether two Steller's sea eagle species exist: one with white feathers and a darker one. But a dark, captive Steller's sea eagle in a Berlin zoo, the only living bird of its kind, has solved the mystery. Born to white feathered parents, the dark Steller's sea eagle confirms they are two variants of the same species." Kaiser (2010, 207) adds "Berlin Tierpark has a breeding female of the dark form of Steller's Sea eagle *Haliaeetus pelagicus* ("niger"). This is the first known specimen for about half a century and the first evidence that it is not a subspecies but just a colour phase."

The adult of this species may not be mistaken for any other species, due to its unique appearance; its very large size and its distinctive white and dark coloration. However, the juvenile plumage is like that of the juvenile of the White-tailed eagle, but the former has a larger, heavier bill, darker coloration and a wedge-shaped white tail (ibid.). The immature bird differs from all other immature sea eagles in having a yellow bill and 14, rather than 12, rectrices (Wink et al., 1996). The voice is a "deep toned barking cry, 'ra – ra – ra – rau – rau.' deeper and louder" than that of the White-tailed Sea eagle, "but otherwise reminiscent of it in flight a loud, gull-like call" (Brown & Amadon, 1968, 298). Illustrations of the appearance, range, and eggs of this species are shown in Figures 4.1a to 4.1k.

Classification

Brown and Amadon (1968) noted that there are possibly two species: *H.p. pelagicus* (ranging outside Korea) and *H.p. niger* (ranging in Korea) although it is also classified as monotypic. However, later studies have refuted this classification as above (Ferguson-Lees & Christie, 2001; Davies, 2010; Kaiser, 2010). The Steller's Sea eagle, being part of *Haliaeetus* is included in studies conducted over the last 30 years that have defined this genus using techniques of increasing complexity and sophistication (Wink et al., 1996). For example, starting with morphological analyses (Zimbelmann, 1992), studies moved on to allozyme analysis (Schreiber & Weitzel, 1995), and to molecular phylogenetic analyses by Wink et al. (1996), Seibold and Helbig (1996),

Fig. 4.1a. Steller's Sea eagle perching [Credit: Axel Blanchard – unsplash]

Fig. 4.1b. Steller's Sea eagle perching [With permission from Takashi Muramatsu]

Fig.4.1c. Steller's Sea eagle perching [With permission from Takashi Muramatsu]

Fig. 4.1d. Steller's Sea eagle in flight [With permission from Zdeněk Macháček – unsplash]

Fig. 4.1e. Steller's Sea eagle in flight [With permission from Takashi Muramatsu]

Fig. 4.1f. Immature Steller's Sea eagle in flight [With permission from Julian Thomas]

Fig. 4.1g. Immature Steller's Sea eagle in flight [With permission from Ricardo Frantz – unsplash]

Fig. 4.1h. Steller's Sea eagle on nest [With permission from Russell King]

Fig. 4.1i. Steller's Sea eagle habitat, Kamchatcka [With permission from Eugene Kaspersky]

Fig. 4.1j. Steller's Sea eagle habitat [With permission from Mariya Tereshkova – uunsplash]

and Lerner and Mindell (2005). These concluded that the Genus *Haliaeetus* is monophyletic with strong links to the genera *Milvus* and *Haliastur*. Lerner and Mindell (2005) linked the northern sea eagle species, the Steller's Sea eagle, White-tailed Sea eagle, Bald eagle and Pallas's fish eagle, into a clade separate from the southern species of *Haliaeetus*. The Steller's Sea eagle was found to have diverged from a genetic development that also resulted in the bald eagle and white-tailed eagle about 3 to 4 million years BP. These three species share some similar characteristics, including yellow eyes, beaks, and feet, hence differing from their next nearest relation, Pallas's fish eagle. These findings resulted from an analysis of the cytochrome b gene of mitochondrial DNA (Wink et al., 1996).

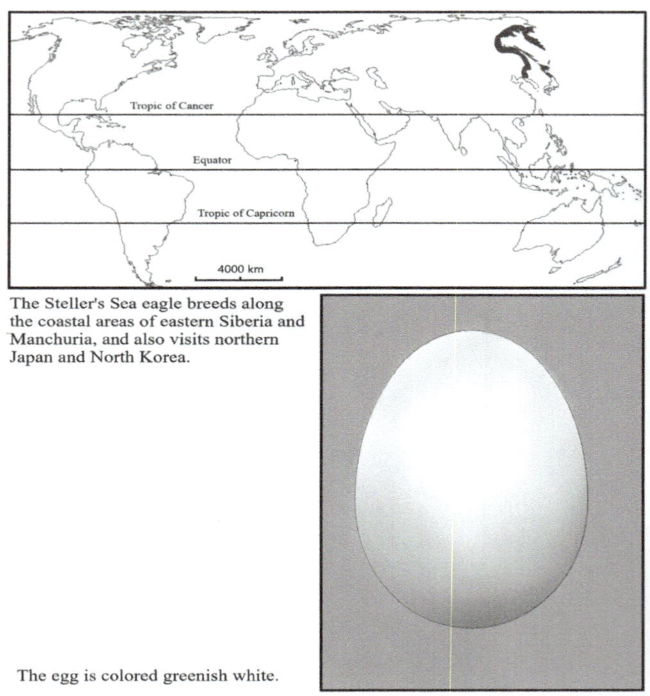

Fig. 4.1k. Steller's Sea eagle range and egg

Foraging

The Steller's Sea eagle has been called "a candidate for the title of the world's most aggressive bird" (Brown & Amadon, 1968, 298). Favored habitats and foraging spaces are rocky sea coasts, inland rivers, and along large river valleys, usually with some forest vegetation and flood plains (Shiraki, 2001). In colder areas, there may be migration to unfrozen lakes on the onset of winter (Wunderlich, 1980). Steller's sea eagles hunt most commonly from a perch in a tree or rocky ledge located 5–30 m above the water, although they may also hunt while circling 6–7 m above the water. This species dives into water feet-first to catch fish seen from the air, swoops into water from a perch or wades into shallows or from a sandbar or low promontory (Ferguson-Lees & Christie, 2001). The diet includes large fish, especially Pacific salmon and trout (*Oncorhynchus* spp.), especially Pink salmon (*Oncorhynchus gorbuscha*, Walbaum, 1792), chum salmon (*Oncorhynchus keta*, Walbaum, 1792), grayling (*Thymallus*, Linck,

1790) and three-spined stickleback (*Gasterosteus aculeatus*, Linnaeus, 1758) (Utekhina et al., 2000). Steller's sea eagles may congregate near salmon spawning areas late in the year (Ueta et al., 1999). Their diet includes dead and dying fish during the annual salmon runs, in this performing a similar ecological role in Eurasia as the Bald eagle in North America. It also forages on offal, and spends periods perching on cliffs, trees and other raised structures. fish, some weighing up to 5 kg (Ladyguin, 2000). Other fish taken include the Bering wolffish (*Anarhichas orientalis*, Pallas, 1814; Sea raven (*Hemitripterus villosus*, Pallas, 1814), smooth lumpfish (*Aptocyclus ventricosus*, Pallas, 1769), and sculpins (Genus *Myoxocephalus*, Tilesius, 1811) (Utekhina et al., 2000). Alaska pollock (*Gadus chalcogrammus*, Pallas, 1814) and cod (Genus *Gadus*, Linnaeus, 1758) are important food sources for wintering eagles in Japan (Ferguson-Lees & Christie, 2001).

The Steller's sea eagle, unlike most other *Haliaeetus* species may also prey on birds such as the ptarmigan (*Lagopus*, Brisson, 1760, specifically the Willow ptarmigan *Tetrao lagopus*, Linnaeus, 1758 and the Rock ptarmigan *Lagopus muta*, Montin, 1776), capercaillie (*Tetrao urogallus*, Linnaeus, 1758), guillemots or murres (*Alcidae*, Leach, 1820, such as the Common murres *Uria aalge*, Pontoppidan, 1763 and Thick-billed murres *Uria lomvia*, Linnaeus, 1758), geese and ducks (*Anatidae*, Leach, 1820) and Slaty-backed gull (*Larus schistisagus*, Stejneger, 1884) (Brown & Amadon, 1968; Global Raptor Information Network, 2021bc). The Steller's Sea eagle may also take birds of prey and omnivores such as Snowy owl (*Bubo scandiacus*, Linnaeus, 1758), Short-eared owls (*Asioflammeus*, Pontoppidan, 1763), Common ravens (*Corvus corax*, Linnaeus, 1758) and Carrion crows (*Corvus corone*, Linnaeus, 1758) (Utekhina et al., 2000). Ferguson-Lees & Christie (2001) cite an example of the eagle killing and eating a great albatross (Genus *Diomedea*, Linnaeus, 1758), which would be a vagrant from the southern sub-Antarctic oceans. Mammalian carnivores may also be taken, such as Arctic foxes (*Vulpes lagopus*, Linnaeus, 1758), sable (*Martes zibellina*, Linnaeus, 1758) and young seals (*Otariidae* and *Phocidae* spp.). Other species taken include the Red fox (*Vulpes vulpes*, Linnaeus, 1758), and small domestic dogs (*Canis lupus familiaris*, Linnaeus, 1758) (Brown & Amadon, 1986). Brown and Amadon (1986) recorded eagles taking seal pups. Carrion, including carcasses of seals and sea lions and deer are also eaten (Ueta et al., 2003). The Steller's Sea eagle engages with many other species during foraging activities. The Golden eagle and White-tailed eagle are common competitors, as revealed in a video where a Golden eagle defeats an immature Steller's sea eagle (Encyclopaedia Britannica, 2020). The Steller's Sea eagle is also recorded dominating the smaller eagles (Watson, 2010).

Breeding

Potapov et al. (2015, 77) note of the Steller's Sea eagle, that "due to its restricted remote range in the Russian Far East and northern Japan, it remains one of the least studied of all raptors, and almost no data are published about its behaviour in the pre-laying period." The Steller's eagle breeds after aerial displays, which involves calling and flights over breeding areas. The comparatively large nests may reach over two meters in width and four meters in depth, made principally of branches. Nests are usually located in large deciduous or coniferous trees but may also be found on crags. Laying usually begins in late April or early May. The clutch size is usually of two slightly greenish white eggs, but 1 or 3 eggs may be laid. Incubation is usually 38 to 45 days, with fledgling at 70 days (Flint et al., 1984; Brown & Amadon, 1968; BirdLife International, 2016; Global Raptor Information Network, 2021bc).

Population Status

BirdLife International (2016) notes that "this species has a small, regionally declining population as a result of habitat degradation, pollution, poisoning by lead shot, and over-fishing. It therefore qualifies as Vulnerable." BirdLife International (2016) estimates the global population at 4,600 - 5,100 birds, this including c.1,830-1,900 breeding pairs. Moores et al. (2018) describe the Steller's Sea eagle as "Red-Listed, i.e. they are the Highest Conservation Priorities" "vulnerable" and a winter visitor, with 10-99 individuals per year. Ferguson-Lees and Christie (2001) estimated the global population (adults and immatures at the start of the breeding season) at between 1,001 and 10,000 birds. They based this on 1990s estimates within the Russian breeding range, which included Kamchatka. These estimated 7,500 birds, with estimates of another 4,000 in Kamchatka and about 2,000 in Japan. This species is a partial migrant as some populations are migratory and others are sedentary (Bildstein, 2006).

The breeding areas include the Kamchatka peninsula, the coastal area close to the Sea of Okhotsk, the Amur river's lower reaches southwards to the Gorin river and on northern Sakhalin and Shantar Islands, part of and off Russia (Belopolskiy & Rogova, 1947; Nakagawa et al., 1987; Babenko et al., 1988; Lobkov, 1991; Aliskerov,1996; Shiraki, 1996; Brazil, 2009). Because of its large Russian distribution, much of literature on this species range and ecology is in Russian publications, for example: Tikhomirov, 1966. Stenchenko, 1974; Lobkov, 1978; Shibnev, 1981; Galushin, 1982; Lobkov & Neufeldt, 1986; Ladigin et al., 1991; Kondratiev et al., 1992; Ladygin, 1992, 1994, 1996, 1997; Masterov, 1992; Masterov & Zykov, 1992; Golubova & Pleshenko, 1997). Wintering areas are southern Kuril Islands and Hokkaido, Japan, and to a lesser extent in Kamchatka, the northern parts of Sea of Japan, and the coast of Okhotsk, and rarer cases in north-eastern China, and North and South Korea (Nakagawa et al., 1987; Working Group for White-tailed Eagles and Steller's Sea-eagles, 1996; Choi & Park, 2012; Global Raptor

Information Network, 2020). Potapov et al. (2010, 2012) describe reduced breeding successes for inland river populations of Magadan district, Russia, between 1991 and 2009, while the breeding success of coastal populations increased during this period.

Regarding widespread, accidental distributions, termed vagrancy, the Steller's sea eagle is less often recorded than the White-tailed eagle. One reason is that the juveniles of the former species do not engage in long-range dispersal (McGrady et al., 2003; Whitfield et al., 2009a,b). Nevertheless, vagrants have been recorded in North America and Asia, including Attu, Unimak, Simeonof, Taku River, near Juneau and Kodiak Island near Alaska, and inland Asia westwards to Beijing in China and Yakutsk in Russia, and southwards to Taiwan (Brown & Amadon, 1986; American Ornithologists Union, 1998). It may also be a very rare winter visitor in the Korean Peninsula (Lee et al., 2005; Choi & Park, 2012; Moores et al., 2018) and a rare vagrant in the islands of Hawaii (American Ornithological Union, 1998).

The Global Raptor Information Network (2021bc) writes that the Steller's Sea eagle is "generally uncommon to rare over its range, although locally common in a few areas. It has suffered population declines from the loss of habitat in Russia during the development of hydroelectric power projects, proposed large-scale coastal and offshore developments of the petrochemical industry, and logging for timber" (see also Ferguson-Lees & Christie, 2001). Factors for the decline also include human competition for fish stocks in Russia and Japan, resulting in some wintering birds moving to non-coastal areas, where many subsist on deer carrion, with lead poisoning possibilities in Japan (Kurosawa, 1998, 2000; Kim et al., 1999; Iwata et al., 2000; Saito, 2009). Eagles may also be shot (National Institute of Biological Resources, 2011). Illustrations of the appearance, range, and eggs of this species are shown in Figures 4.1a to 4.1k.

4.3 White-tailed Eagle *Haliaeetus albicilla* (Linnaeus, 1758)

Physical Description

The White-tailed Sea eagle, also referred to as the ern or erne, Gray Sea eagle, or Eurasian Sea eagle (Bishop, 1905; Gilliard, 1958; Love, 1983), has been described as "a huge, grey brown eagle with a white wedge-shaped tail, yellow bill and bare yellow tarsus" (Brown & Amadon, 1968, 292). Kamarauskaitė et al. (2020, 26) describe it as "the largest avian predator in the temperate and boreal latitudes of Europe." This huge bird is only slightly smaller than the Steller's Sea eagle. It is the largest eagle in Europe, including Western Russia, and all parts of Eurasia outside the range of the larger Steller's Sea eagle. Its huge size ranks it as possibly the fourth largest and heaviest eagle globally, after the Steller's Sea eagle, Harpy eagle and Philippine eagle (Del Hoyo et al., 1994; Ferguson-Lees & Christie, 2001). The White-tailed Sea eagle also varies in size according to range, with a north (larger sizes) to south (smaller sizes) clinal trend, and a decline in size from west to east (Love, 1983; Zhengjie, 2001; Hailer et al., 2007). The largest specimens are found in Greenland, while the smallest are found in the southern ranges, including Turkey, southern Kazakhstan and the Korea Bay area (Ferguson-Lees & Christie, 2001).

The White-tailed eagle measures 66 to 94 cm in length and 1.78 to 2.45 m in wingspan (Ferguson-Lees & Christie, 2001; National Geographic Field Guide to the Birds of North America, 2002). Ferguson-Lees & Christie (2001) entertain the possibility that this species has the largest wingspan among eagles, even greater than that of the heavier and longer Steller's Sea eagle, and rivalling that of the wedge-tailed eagle. Studies and reports by various authors (specifically Willgohs,1961; Brown & Amadon, 1968; Smárason, 2011; Krone et al., 2009a) have recorded large wingspans for this species; males at 2.26 m and females at 2.37 m in Norway, males at 2.1 m and females at 2.3 m in another location, one individual at 2.53 m in Greenland and another at 2.6 m, respectively. Although the White-tailed eagle forms a species pair with the Bald eagle, the latter is usually slightly smaller in body mass, bill and talon size, wing chord and wingspan, but a slightly longer tail and hence total length (Friedmann, 1950; Imler & Kalmbach, 1955; Love, 1983; Ferguson-Lees & Christie, 2001).

The White-tailed Sea eagle's weight varies widely, from 4 to 7 kg in females and 3.1 to 5.4 kg in males (Ferguson-Lees & Christie, 2001). Examples of size variation are specimens from Europe, which may weigh from 4.02 kg for a sample of males and 5.11 kg for a sample of females, and 4.98 kg for males and 6.06 kg for females in another European sample (Brown & Amadon, 1968; Love, 1988). Specimens from northeast China averaged 2.8 to 3.78 kg for males and 3.75 to 4.6 kg for females (Zhengjie, 2001). Some weights may rank the White-tailed Sea eagle among the world's heaviest eagles. Recorded weights have reached 7.5–8 kg for females and 6.5 kg for males, which at least for the male may exceed the maximum recorded for the male Harpy eagles and Philippine eagles (in these species the male may be comparatively smaller in relation to the female), which may generally be below 5 kg (Ferguson-Lees & Christie, 2001; Dunning, 2008). Compared with the sympatric eagle species (sharing the same range), the average female Steller's Sea eagle may weigh almost 25% more than the average White-tailed Sea eagle, considering the global average weight for the White-tailed eagle may be 5 kg (Dunning, 2008; Sutton, 2015; Nadjafzadeh et al., 2015a,b). Comparing the White-tailed Sea eagle with the other sympatric eagle species, the Golden eagle, the average European golden eagle may weigh about 11-12% less than the average European white-tailed eagle. For the non-sympatric but related Bald eagle, the average White-tailed Sea eagle may be about 10% heavier (Patterson et al., 1998; Dunning, 2008; Watson, 2010).

The plumage is brown, with yellowish-white tips to the brown feathers on the crown, nape, neck and throat. The upperparts are dark brown, the underparts are brown, the upper tail coverts are white with brown mottling and the tail is white with some brown tripped bases. Younger specimens may be darker than older (Brown & Amadon, 1968, 292). Immature birds have more mottled plumage, brown nape and crown with lighter feather tips, and brown rump and mantle feathers with reddish edges and darker tips and centers. The scapulars are dark brown, the tail has dark and light brown mottling, the wing quills are less grey (more brownish) than those of the adult and the inner secondaries have light brown mottling, the innermost being paler brown with dark tips. The underparts are brown with darker shafts and tips, and the feather edges are more reddish with more prominent white bases than the adult's feathers. The bill and eyes are brown, unlike the yellow coloration of the adult, and the cere is greenish yellow and the feet are yellow. The coloration of the Greenland specimens or the subspecies *Haliaeetus albicilla groenlandicus* is similar (Brown & Amadon, 1968, 292).

The White-tailed eagle may be readily identified and is unlikely to be mistaken for some other species, due to the huge size and/or white tail. Pallas's fish eagle is smaller and lacks the yellow bill and white tail (Brown & Amadon, 1968). The immature birds may be more comparable, but the White-tailed Sea eagle has darker plumage, a larger bill and white mottling on the tail feathers. The call is generally a series of barking calls: these may be described as "grigrigri" and krickkrickkrick, kyikyikyitykliklek yak, ending with ko koko for the male; for the female "gragragragra, krawkrawkrawkraw"; the alarm cry may be "gahgahgah or jokjokjok" (Brown & Amadon, 1968, 294) and the young may call a monotonous "veee-veee" (Love, 1983; Mecionis & Jusys, 1994; Ferguson-Lees & Christie, 2001).

Classification

Some authors recognize two subspecies of the White-tailed eagle: *Haliaeetus albicilla albicilla* (Linnaeus, 1758) occurring in Eurasia and *Haliaeetus albicilla groenlandicus* in Greenland (Brehm, 1831) (Brown & Amadon, 1968; Kampp & Wille, 1979). *H.a. groenlandicus*, thought to be endemic to Greenland and generally larger than Eurasian specimens, was classified as distinct largely on the basis of this size distinction (Salomonsen, 1979). Later authors rejected this division, and hold the species is monotypic (Cramp & Simmons, 1979; Stresemann & Amadon, 1979; Thiollay, 1994). As noted above, a size cline has been identified, increasing sizes from southeast to northwest (Ferguson-Lees & Christie, 2001).

The Global Raptor Information Network (2021bd) describe the classification of this species within *Haliaeetus*, and the links between this genus and the milvine kites of the genera *Milvus* and *Haliastur.* These assessments were based on morphological analyses by Zimbelmann (1992), and allozyme analysis by Schreiber and Weitzel (1995), and molecular phylogenetic analyses by Wink et al. (1996), Seibold and Helbig (1996), and Lerner and Mindell (2005). These studies found that the Genus *Haliaeetus* is monophyletic. Additionally, these analyses found that the northern sea eagle species, including the White-tail Sea eagle, form a clade which is distinct from the southern members of the genus. Additionally, the White-tailed Sea eagle is a sister species with the Bald eagle (Wink et al., 1996). These authors argue that these two species may have diverged genetically from the other sea eagles between 10 million and 28 million years ago, a conclusion that is supported by a more recent study using mitochondrial DNA (Hailer et al., 2007). The divergence between these two species may have occurred in the North Pacific, with the White-tailed Sea eagle expanding westwards into Eurasia and the Bald eagle moving eastwards into North America (Hailer et al., 2007). The Greenlandic white-tailed eagles have been suggested as a subspecies (*H.a. groenlandicus*), but their isolation in Greenland may not have been for sufficient time to form the requirements for classification as a subspecies. An eastern subspecies (*H.a. brooksi*) was also suggested, but there was insufficient supporting evidence for such a status, as it was possibly merely a group within the clinal variation in size and plumage color (Ferguson-Lees & Christie, 2001).

African occurrences have also been recorded. Maurer et al. (2010, 208) describe Egypt as the southernmost part of the breeding range, the species is now probably extinct in that area. The Egyptian population was isolated from the rest of the White-tailed Eagle's range and may have had some morphological differences from the other populations. Some researchers attempted to classify the Egyptian birds from the area of Lake Manzala as a distinct species (Brehm, 1856; von Heuglin, 1856). These classifications created *Haliaeetus cinereus* (von Heuglin, 1856) and *Haliaeetus funereus* (Brehm, 1856). These were both published by Brehm's father, Christian Ludwig Brehm (Brehm, 1855) as *nomina nuda.* Here nomina nuda refers to a name that is like a scientific name, but is not truly recognized as such, because it has not been published with suitable description, hence it is a 'naked name' (nomina nudem *singular* or nomina nuda *plural*) (Dubois, 2008) by Brehm's father, Christian Ludwig Brehm (Brehm, 1855). However, Sharpe (1874) placed *H. cinereus* and *H. funereus* in synonymy with *H. albicilla* (Linnaeus, 1758) (i.e. the same species) and this is the classification used until today (Maurer et al., 2010, 208).

Breeding

The White-tailed eagle's breeding season begins with aerial displays, which include diving, grappling and even some mock fighting (Brown & Amadon, 1968). The nests are in large trees, or lower on hummocks, reed beds and even sand beds or bushes.

These nests are quite large, over 1 meter high and wide, in some cases twice these measurements. The Global Raptor Information Network (2021bd) describes the nest as a "massive stick nest usually placed high in a tree, but sometimes on a rocky cliff. The nests may be one meter across and up to 2 meters deep or even much larger than this (Ussher & Warren, 1900; Willgohs,1961; Brown & Amadon, 1968; von Blotzheim et al., 1993; Ferguson-Lees & Christie, 2001), usually situated in tall tree that commands a good view of the landcover, shelter and protection from predators (Love, 1983). Cliff nests may be smaller, even only a shallow depression in soil or 1.5 m wide and a few centimeters deep (Willgohs, 1961). White-tailed eagles may also use nests built by other raptors, including black kites and common buzzards (Fischer, 1982; von Blotzheim et al., 1993).

Nests are sometimes re-used in successive years, increasing in size each year. The clutch is usually of two (sometimes 1 or 3) unmarked white to yellowish white eggs (Shirihai, 1996), which may be slightly larger in *Haliaeetus albicilla groenlandicus*. Some reports indicate that the average clutch size in Scotland is 1.56, and the average for Norway is 2.16 (Willgohs, 1961; Love, 1983). Four egg clutches were reported in Norway, Poland and Romania, and the Isle of Skye (Willgohs, 1961; Fischer, 1970; Witherby et al., 1943). Other studies have noted differing results for brood sizes: for example, mostly one egg is laid (57%), followed by two eggs (41%) and three eggs (2.5%) and four eggs (1%); some findings give an average of 1.52 eggs across the range, with a minimum of 1.1 eggs in western Germany to a maximum of 1.9 eggs in southern Kazakhstan (Willgohs, 1961; Love, 1983; von Blotzheim et al., 1993; Helander & Stjernberg, 2003). The female does most of the early brooding, with the male joining in after about a month (Willgohs, 1961; Brown & Amadon, 1968; Fischer, 1970). Fledging occurs after about 70 days, with full development of light after about 90 days. Fledging has been reported in July in Norway, and August near the Russian White Sea (Willgohs, 1961; Brown & Amadon, 1968; Flerov, 1970; Mori, 1980). After fledging, the young eagle may remain in close proximity for another month and be slowly weaned off their parents' feeding (Brown & Amadon, 1968; Mori, 1980).

The White-tailed eagle's breeding range is over northern Europe and northern Asia, this including parts of Greenland, Iceland, Great Britain, mainland western and eastern Europe, south east to the Black Sea, and eastwards through the Middle East to Kazakhstan, Russia, Mongolia, China and North Korea, and Sakhalin Island, the Kuril Islands and Hokkaido, Japan (Brown, 1969; Flerov, 1970; Bijleveld, 1974; Puscariu & Filipascu, 1975; Bogucki, 1976; Stjernberg, 1981; Love, 1983; Knystautas, 1993; Ferguson-Lees & Christie, 2001; Helander & Stjernberg, 2003; Firouz, 2005; Ehmsen et al., 2011; BirdLife International, 2016; Barkham, 2019; Global Raptor Information Network, 2021bd; Orta et al., 2020).

Regional breeding variations have been historically noted (Willgohs, 1961; Love, 1983). The breeding periods may vary regionally, usually from January to July (more narrowly February to early March) in the south ranges, and from April to September (more narrowly late April to mid May) in the northern ranges (Ferguson-Lees & Christie, 2001). For example, in European Russia, eggs are generally laid in March-April (Flint et al., 1984). However, in the Czech Republic, breeding may start earlier, in February, with the latest egg-laying at the beginning of March (Kren, 2000). In Romania, breeding activities, including courtship displays may begin in mid-January (Roberts, 2000). In Greece records indicate egg-laying beginning in early to mid-February (Reiser, 1905; Handrinos & Akriotis, 1997), and in the lower Volga river it is during February. In Germany, most egg-laying is in March, and in the subarctic of Lapland and northern Russia it is in late April to early May (Bent, 1937; Ingólfsson, 1961; Flerov, 1970). In Scotland, it is usually later than in Norway in late March to early April (Bannerman, 1959; Willgohs, 1961; Love, 1983). In Finland, laying may be in early as March, to May (Love, 1983). Older pairs may nest earlier than younger pairs (Brown & Amadon, 1968). In Asia, breeding may be January to May in Israel (Shirihai, 1996), January in Iran (when it does breed), September to February in Bangladesh (Global Raptor Information Network, 2021bd); and after March to fledging in July to August in Novosibirsk area, Siberia (Andreenkov et al., 2009).

The choice of nest sites may also vary regionally, but tall trees are generally favored for nest construction. For example, in Germany, nesting may be in poplar plantations, planted in 1950s, which were recorded as 11% of breeding pairs in 2009 (Hauff, 2009). Hauff (2009) recorded 276 nest sites and classified them according to the tree type. Most were in pine (112/40.6%), beech (78/28.3%), poplar: 34/12.3%), oak (26/9.4%) and spruce (11/4.0%). Smaller numbers were in ash (4/1.4%), larch (3/1.1%), alder (2/0.7%), lime (1/0.4%), birch (1/0.4%), willow (1/0.4%), Douglas fir (1/0.4%), elm (1/0.4%) and maple (1/0.4%) (Hauff, 2009). In Croatia, Radovic and Mikuska (2009a,b) found nests in eight tree species, including Pedunculate Oak (*Quercus robur*), Sessile Oak (*Quercus petraea*), Narrow-leafed Ash (*Fraxinus angustifolia*), White Poplar (*Populus alba*), Black Poplar (*Populus nigra*), Beech (*Fagus sylvatica*), Euro-American Poplar (*Populus* sp.), and White Willow (*Salix alba*). The average nest height was 24.75 m, with those in ash trees being highest (avg. 25.2 m; n = 14). In Romania, Goriup et al. (2007) found nests in tall trees in thick forest, sometimes located on river islands. In Greece, nests were only recorded in trees, especially in riverside forests and hill pines, and in a few cases in mixed woodlands (Handrinos & Akriotis, 1997). Depending on the height of trees, the elevation of the nest may vary. Nests recorded in Romania, on black poplars and willows were 15 to 25 m above the ground. One set of nests was 19 to 33 m above the ground (Bannerman, 1959; Hagen, 1976). In eastern Germany, nests ranged in height from 8 to 30 m (Oehme, 1961). In Croatia, the national breeding population was about 135 breeding pairs, nesting on mature pedunculate oaks, narrow-leafed ash and white poplars below 140 m above sea level and over one km from human presence (Radović & Mikuska, 2009, 156).

Nests are most often cited near to food sources. For example, in Germany, about three quarters of the nests were recorded within 3 km of fresh water (Oehme, 1961), and most were located near forests, wooded islands or promontories and even marshy ground or cultivated land. One cliff nest was recorded in Germany, on Rugen, a Baltic island. Past records also showed ground nests in the German Baltic region (Oehme, 1961; Fischer, 1970). Evans et al. (2010, 473) compared the nesting of White-tailed eagles and Golden eagles in western Scotland, where White-tailed eagles were reintroduced from 1975 onwards. Their findings indicated that "Golden and White-tailed eagles selected different habitats for nesting": more specifically White-tailed Eagles usually nested at lower altitudes than Golden eagles, and in trees in more wooded habitats closer to open water. Additionally, "all Golden Eagles in the study area nested on cliffs… whereas more than half of White-tailed Eagle nests were in trees" (Evans et al., 2010, 477). The Golden eagle's presence was unaffected by the presence of the larger White-tailed eagles. The authors note that "there were 3359 km^2 of potential White-tailed eagle nesting habitat within 25 km of existing White-tailed eagle nests, containing 54 Golden eagle territory centers, but we found no difference in change of occupancy for Golden Eagle territories close to White-tailed Eagles compared with those further away" (Evans et al., 2010, 477).

The conclusions were therefore that "White-tailed and Golden eagles appear to partition nesting habitat in the west of Scotland by altitude. This corresponds with behavior in western Norway and with the situation described in historical accounts of nest-sites in western Scotland prior to extinction of White-tailed Eagles" (Evans et al., 2010, 477). These findings were consistent with other studies that found little overlap in breeding season diet of the two eagle species in western Scotland, and possible partitioning of foraging habitat by altitude. The authors also concluded that "the likelihood of competitive exclusion is less than previously suggested." This study was compared with that of Halley (1998) in western Norway, where White-tailed eagles are more associated with coastal locations and tree sites, and Golden eagles are more commonly associated with higher elevation cliffnest-sites (Halley, 1998; Evans et al., 2010, 477).

Despite these consistent reports of tree-nesting by White-tailed eagles, some observers have recorded cliff nesting for this species. For example, Reiser (1905) entertains the possibility that cliff nesting existed in the past, especially on islands. Cliff nesting was recorded in countries and regions where trees were scarce (Brown & Amadon, 1968). For example, 92% of nests recorded in Norway were on cliffs, while 8% were in trees. In Scotland, 79% were on cliffs, and the figures for Iceland and Greenland were 77% and 100% respectively (Willgohs, 1961; Ingólfsson, 1961; Christensen, 1979; Love, 1983). These cliff nests may be higher in elevation than tree nests, sometimes over 75 meters above the nearby flat ground. Nests are only rarely recorded on the ground, low hummocks, small trees and bushes, and on sand banks or among reed beds (Brown & Amadon, 1968). One nest was recorded on a buoy near a Norwegian shipping route (Willgohs, 1961).

Two Asian examples are given by Global Raptor Information Network (2021bd). In Israel, the White-tailed eagle's breeding population is extinct, and the species is a very rare winter visitor and passage migrant, despite some Reintroduction Program efforts by the Israeli Nature and Parks Authority in the Hula Valley (Kislev, 2020). When it was a breeder, nests were usually located in tall trees such as pistachio (*Pistacia atlantica*, Desf) and eucalyptus (Genus *Eucalyptus* L'Hér.) (Shirihai, 1996). In Siberia, Karyakin et al. (2009) found most nests in the Novosibirsk district near water, except for two nests in the Karakan pine forest which were over one kilometer from water reservoirs. In Hokkaido, Japan, nests were generally located in the tallest tree in the vicinity, with the nesting tree averaging 2.8 m taller than the average tree in the immediate vicinity, with the elevation being similar to that of Europe at 16.5 to 25 m above ground (Mori, 1980). Nests were commonly in *Piceaglehnii* (F. Schmidt) and *Alnus japonica* ((Thunb.) Steud.) (Shiraki, 1994).

In Africa, Maurer et al. (2010, 208) note that "little is known concerning the biology of the now extinct Egyptian population of White-tailed eagle *Haliaeetus albicilla*, and few specimens remain in museums to assess its alleged subspecific status" in what was formerly the southernmost part of the breeding range. Citing Meinertzhagen (1930), Cramp & Simmons (1979) and Goodman & Meininger (1989), Maurer et al. (2010, 208) note that the "White-tailed Eagle *Haliaeetus albicilla* is now a rare and irregular winter visitor to Egypt… but has long been claimed as a former breeding bird at the lakes of lower Egypt"; however, "these claims are based solely on 19th-century published descriptions by the German explorers and ornithologists Martin Theodor von Heuglin (1824–76) and Alfred Edmund Brehm (1829–84), and observations by the British politician and agriculturalist, the Hon. Murray Finch-Hatton, 12th Earl of Winchilsea and seventh Earl of Nottingham (1851–98)." The study by Maurer et al. (2010) used three White-tailed eagle eggs in museums; one in the Natural History Museum (NHM), and two in Stuttgart at the Staatliches Museum für Naturkunde (SMNS) to "re-examine evidence of breeding by *H. albicilla* in Egypt and to summarize published and specimen breeding data (Maurer et al., 2010, 208).

Foraging

The White-tailed Sea eagle forages mostly near coasts, large lakes and rivers. In Europe it frequents lowland areas but frequents higher altitudes up to 2000 meters in Siberia and Central Asia. Foraging is by perching and soaring, with aerial diving for fish such as pike (*Esox Lucius*, Linnaeus, 1758), seabirds (e.g. swan *Cygnus* spp. Garsault, 1764, coots *Fulica* spp. Linnaeus, 1758, eider *Somateria* spp. Leach, 1819, ducks *Anatidae* spp. Leach, 1820 and gulls *Laridae* spp. Rafinesque, 1815) and land animals (including rodents *Rodentia* spp. Bowdich, 182 and even small deer *Cervidae* spp. Goldfuss, 1821). In some cases, the eagle may dive to capture fish. According to Brown & Amadon (1968) over 50% of the prey may be birds, followed by fish and then mammals and carrion. There are inter- and intra species variations in the food taken. For example, Nadjafzadeh et al. (2015a) found differences between the diets of German and Greenlandic White-tailed eagles. In the German population, there were seasonal shifts as the eagles responded to temporal changes in food availabilities, and there were age-related differences, indicating different diets in adults and juveniles. There were also intra-individual short-term changes in the German and Finnish but not the Greenlandic population. They conclude that "this suggests that local feeding niches of this generalist predator may vary with local food supplies, which determines the niche width (from generalist to specialist) at the individual level" (Nadjafzadeh et al., 2015a).

Zawadzka (1999, 65) note that "the White-tailed eagle preyed almost exclusively on birds and fish" this being a "specialized species" as "food taken in water habitats accounted for 95% of its diet." In this study the mean body mass of prey specimens was 578 g (Zawadzka, 1999, 65). This species frequents areas near large water bodies, such habitats combining coastal saltwater and inland freshwater, with an abundant supply of aquatic prey and large trees or more rarely cliffs for nesting. This ecological configuration is basically identical to that frequented by the White-tailed eagles North American equivalent, the Bald eagle (Love, 1983; Wink et al., 1996; Ferguson-Lees & Christie, 2001).

The White-tailed eagle actually frequent a wide range of habitats, "from tundra to desert" (Global Raptor Information Network, 2021bd), including higher elevations (1,500 to 2,300 m, especially in Central Asia and Siberia) near water (Brown & Amadon, 1968), sea cliffs, lowland islands and archipelagos, estuaries and coastal marshes, wood areas, areas of coastal human habitation or even farmland, and commercial fish farms (Ferguson-Lees & Christie, 2001; Helander & Stjernberg, 2003; Radović & Mikuska, 2009; Evans et al., 2010; Tingay & Katzner, 2010; Treinys et al., 2016). Avoided areas include those with strong human disturbances (e.g. economic and recreational development) (Love, 1983; Ferguson-Lees & Christie, 2001; Helander & Stjernberg, 2003). Deforestation, especially of large, matured trees and disturbance in wetlands reduces breeding opportunities (Rosenvald & Lõhmus, 2003; Rijn et al., 2010). This is less of a soaring species than some other eagles, and may have vulture-like, heavy flapping flight when moving or fishing (Rasmussen and Anderton, 2012).

Habitats frequented varies regionally (Orta et al., 2020; Global Raptor Information Network, 2021bd). The typical European breeding and foraging landcover is woodland/forest in floodplain areas. In Greenland, it occurs near breeding areas near the coast, and in the fiords, and near inland lakes (Christensen, 1979; Kampp & Wille, 1990; Boertmann, 1994). In the Czech Republic, it breeds and forages mostly deciduous forest and woodland near floodplains, rivers and fishponds (Kren, 2000). In Croatia, it breeds and forages in alluvial hardwood and softwood forests near floodplains, and fishponds (Radovic & Mikuska, 2009). In Hungary, it is usually present in floodplain forests near the larger rivers and fishponds (Gábor, 1998). In Asia, it frequents more open terrain. In China, it frequents more open grasslands and marshes, near open water such as lakes, reservoirs, and rivers (Weizhi, 2006). In Tibet, it frequents mostly riverine and lacustrine habitats, and areas near reservoirs (Lang et al., 2007). In Kazakhstan, it frequents forest steppes and riparian and floodplain (Wassink & Oreel, 2007). In Israel, it frequents open country, especially near wooded stands and open water, and in Mediterranean lowlands (Shirihai, 1996).

White-tailed eagles may perch on trees or crags for fish in shallow water, or a general view of possible food sources. Prey taken includes fish, birds (including waterbirds such as ducks, gulls, coots and land birds such as partridges) and mammals (hares, muskrats, susliks, which are ground squirrels of the Genus *Spermophilus*, Cuvier, 1825). It also takes carrion and offal and robs other smaller raptor species of their prey. In some cases, it associates with human fishing activities (Rasmussen & Anderton 2012; Global Raptor Information Network, 2021bd). In Eastern Europe, non-fish diets are common. For example, in Russia, it was recorded taking large birds such as geese and mammals such as dogs and foxes (Dementiev & Gladkov 1951). In Poland, it preyed on broods of White storks (Jakubiec & Peterson, 2005). In Romania, it takes coots as a key prey animal (Roberts, 2000). In Armenia, food taken included a coot (Adamian & Klem, 1999), and carrion (Dal, 1954). A wide range of prey are taken in Greece, apart from fish including medium to large waterbirds and mammals. Carrion is also taken, including sources from the vulture feeding station in Dadia Forest Reserve (Handrinos & Akriotis, 1997). In China, fish, wild ducks, and rodents have been recorded as prey (Weizhi, 2006). In Siberia, eagle nests have been recorded away from fish sources, and near dense populations of the Gray or Altai Marmot (*Marmota baibacina*, Kastschenko, 1899). Remains of this mammal were discovered in pellets and under the eagle nests (Karyakin et al., 2009). One eagle was recorded in Hawaii, preying on Laysan Albatrosses (*Phoebastria immutabilis*, Rothschild, 1893) (Zaun, 2009).

Fish prey is mostly taken by snatching from the water surface, rather than by diving (Willgohs, 1961; Love, 1983; von Blotzheim et al., 1993; Valero, 2008; Nadjafzadeh et al., 2015a,b). The main prey are fish and water birds, with a higher proportion of waterbirds taken than are taken by other *Haliaeetus* species such as the Bald eagle and Steller's sea eagle (Ferguson-Lees & Christie, 2001; Mlíkovský, 2009). The relative proportions of the two top prey items, fish and birds varies between locations (Kampp & Wille, 1979; Love, 1983; Wille & Kampp, 1983; Vrezec, 2016). Kampp & Wille (1983) studying eagles in Greenland, noted that the prey items delivered to nests were 90% fish, the remaining 10% being various birds and Arctic fox pups.

Overall, nearly 70 species of fish have been recorded as prey of the White-tailed eagle, with larger fish being commoner victims (Bergman, 1961; Love, 1983; Valero, 2008; Nadjafzadeh et al., 2015a,b; Vrezec, 2016). The Northern pike (*Esox lucius*, Linnaeus, 1758) is recorded as the commonest fish taken across the eagle's range (Fischer, 1982; Sulkava et al., 1997; Ekblad, 2004; Vrezec, 2016). The Common bream (*Abramisbrama*, Linnaeus, 1758) is another commonly taken species (Samago, 1995; Adrianova & Kashevarov, 2005; Yurko, 2016). Other fish species taken include Arctic char (*Salvelinus alpinus*, Linnaeus, 1758) and cod (Atlantic cod, *Gadus morhua* Linnaeus, 1758; and Greenland cod, *Gadus ogac*, Richardson, 1836) (Wille & Kampp, 1983; Harris et al., 2010) taken mostly in Greenland; common lumpsucker (*Cyclopterus lumpus,* Linnaeus, 1758), Atlantic wolffish (*Anarhichas lupus,* Linnaeus, 1758) taken mostly in Norway (where birds are taken more frequently) (Willgohs, 1961; Collette et al., 2015; Simpson et al., 2016). Fish were the commonest prey taken in Sweden, Belarus, Hokkaido, Japan (especially the Alaska pollock, *Gadus chalcogrammus*, Pallas, 1814) and the Kostomuksha Nature Reserve of Russia, the Danube Delta of Romania (especially the Common carp, *Cyprinus carpio*, Linnaeus, 1758, and Prussian carp, *Carassius gibelio*, Bloch, 1782) (Mori, 1980; Love, 1983; Sulkava et al.,1997; Bailey et al., 1999; Ekblad, 2004; Adrianova & Kashevarov, 2005; Tuvia & Välia, 2007; Ivanovsky, 2010; Sandor et al., 2015; Vrezec, 2016; Yurko, 2016).

Bird species recorded as prey include about 170 species. These include about 38 species of waterfowl (Order *Anseriformes*, Wagler, 1831), 42 species of shorebird, (Order *Charadriiformes*, Huxley, 1867) and any available loons, grebes, rails, herons, storks and other waders. Adult swans are among the largest species taken (Willgohs, 1961; Love, 1983; Zawadzka, 1999; Ekblad, 2004; Nolet et al., 2006; Brazil, 2010; Nadjafzadeh et al., 2015a,b; Vrezec, 2016). The Mallard ducks (*Anas platyrhynchos*, Linnaeus, 1758) is the commonest bird species taken, possibly due to its wide range and common status wetland landcover shared with eagles (Love, 1983; Dunning Jr, 2008; Vrezec, 2016). Examples of areas where mallards are commonly taken include Müritz National Park, Germany, and Augustów Primeval Forest, Poland. Eurasian coots (*Fulicaatra*, Linnaeus, 1758) are the next most taken bird species after mallards. Examples of where coots are taken include Wigry National Park, Poland and the Augustów Primeval Forest, Poland where they are 47 to 66 percent of the prey taken. Coots were also strongly taken in the Danube Delta, Romania (Fischer, 1970, 1982; Zawadzka, 1999; Sandor et al., 2015; Vrezec, 2016). Mlíkovský (2009, 35) notes that "a long-term study (1991–2001) of the food of White-tailed Sea eagles in the Svâtoj Nos wetlands at Lake Baikal, Northeastern Russia, revealed that these eagles feed predominantly on water birds, mainly ducks. Anecdotal data from the Selenga Delta in Southeastern Lake Baikal indicate that White-tailed eagles generally prefer birds as their prey in the Lake Baikal area."

The White-tailed eagle also takes grey herons (*Ardea cinerea*), common eiders (*Somateria mollissima*), loons (Genus *Gavia*, Forster, 1788), little grebes (*Tachybaptus ruficollis*, Pallas, 1764), dabbling ducks, geese and swans (Family *Anatidae*, Leach, 1820) (including waterfowl injured from hunters gunshots), larger gulls (Family *Laridae*, Rafinesque, 1815) and even a northern goshawk (*Accipiter gentilis*, Linnaeus, 1758) that mob eagles and nests of gulls and other seabirds, corvids or other accipitrids. Most gulls taken are nestlings and juveniles, e.g. in Germany and Scotland. In rare cases eggs of kittiwakes, eiders, cormorants and gulls are taken (Wolley, 1907; Uttendorfer, 1939; Rudebeck, 1950; Meinertzhagen, 1959; Willgohs, 1961; Fischer, 1970; Flerov.1970; Fischer, 1982; von Blotzheim et al., 1993).

Mammals are generally rarer as prey species (Love, 1983; Flerov, 1970; Vrezec, 2016). In Europe, rodents taken include European rabbits (*Oryctolagus cuniculus*, Linnaeus, 1758), Mountain hares (*Lepus timidus*, Linnaeus, 1758), European hares (*Lepus europaeus*, Pallas, 1778), Arctic hares (*Lepus arcticus*, Ross, 1819), muskrat (*Ondatra zibethicus*, Linnaeus, 1766) (some from "muskrat farms" in Finland) and young of Eurasian (*Castor fiber*, Linnaeus, 1758) and North American beavers (*Castor canadensis*, Kuhl, 1820) (Sulkava et al., 1997; Feldhamer et al., 2003). Smaller rodents are also taken (Flerov, 1970; Fischer, 1982; Chapman & Flux, 1990; Watson et al., 1992; von Blotzheim et al., 1993; Tannerfeldt & Angerbjörn, 1996; Samago, 1995; Halley, 1998; Probst, 2009; Whitfield et al., 2013). Young herbivores are also taken, including those of domestic sheep (*Ovis aries*, Linnaeus, 1758), goats (*Capra aegagrus hircus*, Linnaeus, 1758), reindeer (*Rangifer tarandus*, Linnaeus, 1758), red deer (*Cervus elaphus*, Linnaeus, 1758) and European roe deer (*Capreolus capreolus*, Linnaeus, 1758) and wild boar (*Sus scrofa*, Linnaeus, 1758), the last being frequently taken in Polesski Reserve, Belarus (Willgohs, 1961; Love, 1983; Marquiss et al., 2003, 2004; Yurko, 2016). In a few cases, adult deer such the roe deer are taken (Alexander & Buxton, 1986; Gorbet & Harris, 1991).

Predatory mammals taken include Arctic foxes and red foxes, mustelids up to the size of European otters (*Lutralutra*, Linnaeus, 1758) and young European badgers, Baikal seal (*Pusasibirica*, Gmelin, 1788) and Grey seal (*Halichoerus grypus*, Fabricius, 1791) (Wille & Kampp, 1983; Korpimäki et al., 1989; Larivière & Pasitschniak-Arts, 1996; Audet et al., 2002; Mlíkovský, 2009; Sidorovich et al., 2011; Arkadevich et al., 2014; Bowen, 2016). Other mammals are mostly taken as carrion, these including European bison, horses, moose and/or cattle (Sulkava et al., 1997; Selva et al., 2003; Rijn et al., 2010). Rarer takings include squids and octopuses. Examples may be taken from the Ural mountains and Hokkaido, where they may take cephalopods such as robust clubhook squid (*Onykia robusta*, Verrill, 1876) and Giant Pacific octopus (*Enteroctopus dofleini*, Wülker, 1910) (Uttendörfer, 1952). Insects have also been recorded as part of their diet, for example, Odonata species (dragonflies and damselflies) in Augustów Primeval Forest, Poland (Sherrod, 1978; Mori, 1980; Samago, 1995; Vrezec, 2016).

Population Status

The White-tailed eagle has a wide range from Greenland and Iceland to Hokkaido, Japan (the regions just south such as the Mediterranean and the middle east) and winters south to India, southwest China, Taiwan, Korea and Japan, and possibly Africa (Brown & Amadon, 1968; Forsman, 1999; Abuladze, 2013; Orta et al., 2020), but their range is mostly sparsely populated. BirdLife International (2020) down listed the White-tailed eagle from Near Threatened to Least Concern in 2005. BirdLife International (2016) justifies this reclassification, as it has an extremely large range, over 20,000 km^2 and does not fulfil all the required criteria ("a declining or fluctuating range size, habitat extent/quality, or population size and a small number of locations or severe fragmentation"). The population trend also appears to be increasing, is not Vulnerable in the sense of a serious decline in numbers. The population size may also be over 10,000 mature individuals.

The global population (defined as the number of adults and immatures at the inception of the breeding season) is estimated at 10,001 to 100,000 birds (Ferguson-Lees & Christie, 2001). BirdLife International (2016) puts the population at 20,000-49,999, and reports that the "The population is increasing locally owing to conservation measures such as protecting eyries, providing safe (non-poisoned) food and re-introductions to certain areas such as Bavaria" citing Ferguson-Lees and Christie (2001). The European breeding population, also increasing has been estimated at 9,000-12,300 breeding pairs, which means 17,900-24,500 mature individuals (BirdLife International, 2016). BirdLife International (2016) assesses that "Europe forms 50-74% of the global range, so a very preliminary estimate of the global population size is 24,200-49,000 mature individuals, although further validation of this estimate is needed. It is placed in the band 20,000-49,999 mature individuals."

As noted above, BirdLife International (2020) classifies species according the classification of the International Union for the Conservation of Nature (IUCN) Red List Categories classification; from most to least serious, Extinct (EX), Extinct In The Wild (EW), Critically Endangered (Possibly Extinct) CR (PE), Critically Endangered (Cr), Endangered (EN), Vulnerable (VU), Near Threatened (NT), Least Concern (LC), Data Deficient (DD). The down listing was due to the population recovery mostly after the early 1970s, especially in the northern and western European portions of its range. During the 19th and early 20th centuries, factors for local and regional extinction and decline included the banning of DDT and DDE, increased protection of the birds and habitats and successful reintroduction programs. Threats still exist from the wetland losses, deforestation and loss of nesting trees, human disturbances, wind turbine collisions, and prevailing poisoning, and collisions (BirdLife International, 2007).

The White-tailed eagle is a partial migrant (Bildstein, 2006), but mostly resident, with the western populations less migratory, and the northern and eastern populations mostly migrating southwestwards, after breeding, others are nomadic, altitudinal migrants or stationary (Boertmann, 1994; Adamian & Klem, 1999; Brown & Grice, 2005). Juveniles are usually more migratory and dispersive than adults (Brown & Amadon, 1968; Ferguson-Lees & Christie, 2001). The wintering range is a belt south of the breeding countries, including parts of central and southern Europe, southwest to southern Asia generally north of India, China, South Korea and Japan, and rarely, across the Bering Sea to Alaska and nearby islands (Bent, 1937; Murie, 1959; Flerov, 1970; Bijleveld, 1974; Love, 1983; Tobish & Balch, 1987; Knystautas, 1993; Ferguson-Lees & Christie, 2001; Helander & Stjernberg, 2003; Firouz, 2005; Rasmussen & Anderton, 2012; Brazil 2009; Rijn et al., 2010; Howell et al., 2014; Khan, 2015; BirdLife International, 2016; Mackinnon & Hicks, 2017). Most migrations begin late in the year, with returns early in the year. For example, in Kazakhstan the migration dates are generally September to mid-November and mid-February to May (Wassink & Oreel, 2007). In the Czech Republic, the fall migration passage is from mid-October to December, with the passage north in February to March (Kren, 2000). In Israel, the fall migration passage occurs from September to November, and in the spring from mid-February to March, while the wintering birds arrive in November-December and leave in February-March (Shirihai, 1996).

Evidence of migration is based on recorded observations with seasonal variations and ringing. For example, one eagle ringed in was recovered over 2,400 km away in Hortobágy, Hungary (Horváth, 2009). Another bird which was ringed in eastern Estonia was recovered dead in the east of the Czech Republic. Another bird was similarly found dead, ringed in Estonia and recovered in Poland (Volke & Randla, 2003). A German ringed bird was recovered in England (Brown and Grice, 2005). These recoveries illustrate both the distances and the patterns of the migrations. Serbia: A bird ringed in Belgrade on 12 May 2007 was recovered in Hungary on 9 February 2008 (Stanković & Raković, 2008).Banded Norwegian juveniles have been recovered 720 km away from the banding location in Karlstad and other locations in southern Sweden, the Wadden Sea in the Netherlands. Juveniles from Finland were recovered 520 km away in northern Norway and further away in Bulgaria. German juveniles generally remain with 50 km of nesting areas. Some however, have been traced to Italy (about 1,030 km) or Gironde, France (1,520 km) (Hagen, 1976; Stjernberg, 1981; Helander, 1985; von Blotzheim et al., 1993; Rijn et al., 2010).

Reintroductions of this species have been generally successful. The White-tailed eagle has been successfully reintroduced into the United Kingdom of Great Britain (BirdLife International, 2016). Human activities, consequent environmental change and habitat destruction and related poisoning is a major factor for the scarcity of eagles. These events have led observers to classify this species as endangered or locally extinct, and to take conservation action, including reintroductions, poaching and poison bans and conservation awareness strategies (Love, 1988; Ferguson-Lees & Christie, 2001; Krone et al., 2006; Krüger et al., 2009; BirdLife International, 2016).

Krüger et al. (2010, 710) give an example of reintroduction analyzing population growth through a demographic matrix model to analyze relevant population processes. The case study is of the White-tailed eagle population in the Federal State of Schleswig-Holstein, Germany, which has been studied from re-colonisation in 1947 and population recovery. The demographic model employed illustrated the trajectories of a recovering population for the population management of a rare species. The findings indicated that between 1947 and 1974, the eagle population was stable, and the predicted growth rate using an age-structured matrix model was −6.1% per annum, and the authors infer that "the small but stable population must have been maintained by immigration" (Krüger et al., 2009, 710). For the next period, from 1975 to 2008, the observed and predicted population growths were similar (6.7% and 4% per annum respectively). Indications were that the main factors were adult and pre-breeding survival. The authors concluded that while survival was the main factor influencing population growth, another factor that should be considered is the increasing reproduction rate during this period, hence management actions must consider the dynamics of these two issues. Management efforts must focus on the more important mortality factors, including lead poisoning and wind turbine collisions. The authors cite a model that predicted a maximum carrying capacity of 255 pairs for the Federal State of Schleswig-Holstein but argue that referring to the observed density of 1.4 pairs per 100 km^2 and habitat suitability variation, a more reasonable estimate for carrying capacity would be 122 pairs. Due to these ecological factors, they argue that the population growth should slow, when the limits of carrying capacity are reached (Krüger et al., 2009, 710).

Abuladze (2013) records the status of this species in Georgia and the Caucasus Region. In this country it is a year-round resident, and a rare passage migrant and winter visitor (Paakinen et al., 1981; Sojamo, 1982, 1983; Abuladze & Eligulashvili, 1996; Abuladze et al., 2011; Verhelst et al., 2011). The Eurasian population is increasing, and there have been local increases in Scotland and the Netherlands. In other areas, declines have been noted. For example, Abuladze et al., 2013, 37) notes that it is a "former breeder at the Black sea coastlands and in flood-lands of large rivers but extirpated as a breeding species in 1950's-1960's. In the middle of 1990's after 30-40 intervals again became to nest in Georgia. The present distribution and numbers not well-understood, but probably 3-5 pairs breed in country during last years. Further study of the White-tailed Sea eagle in Georgia is needed." This eagle was apparently commoner in the past: "A great number of works in the ornithological literature of the 19th to early 20th centuries affirm that the species was common and widespread at that time throughout the Caucasus, occurring all along the shores of the Black Sea and Caspian Sea, on large lakes, marches and in flood-lands of large rivers. The total number was unknown, but we can presume that in Caucasus between 1890's and 1920's it could be estimated at perhaps several hundred pairs, including 20-40 pairs in Georgia. The decline of the White-tailed Sea eagle's Caucasian population has been marked everywhere in the region since the 1930's" (Abuladze, 2013, 37 - 38). The population decline in the southern Caucasus is attributed to habitat loss, concerning the decline of flood-lands and forest/woodland near large wetlands, decline in large trees used for nesting, the drainage and reclamation of waterbodies and wetlands, reservoir creation, river management, pesticide use, poisoned bait and traps near animal carcasses, other forms of environmental destruction and modification, human recreation and direct killings of eagles by people (Abuladze & Eligulashvili, 1996; Abuladze, 2013).

Global Raptor Information Network (2021bd) lists several locations where the White-tailed eagle occurs. In the Americas it has been recorded in Alaska, the contiguous United States and Greenland. In Alaska, it is a breeder in the Aleutian Islands (especially the island of Attu) and has been recorded on the island of Shemya (American Ornithologists' Union, 1998). There was also an unconfirmed record from Baffin Island (Cumberland Sound), Northwest Territories, Canada (American Ornithologists' Union, 1998). A vagrant record exists for Massachusetts near Nantucket Lightship (American Ornithologists' Union, 1998). Zaun (2009, 35) reports on "the First Modern Record of the White-tailed Eagle in Hawaii" in Kilauea Point National Wildlife Refuge, Kauai. In Greenland, it is a resident breeder or intraregional migrant, commoner in West Greenland from Qaqortoq/Julianehlb to southern Aasiaat/Egedesminde (Hansen, 1979; Kampp & Wille, 1990; Boertmann, 1994) and rarer farther north, with a nest recorded at Ililissat/Jakobshavn (Bennike & Feilberg, 1982), and a few vagrants recorded northwards to Uumannaq (Salmonsen, 1967). Johnsen (1953) reported a skull in Northeast Greenland. Vagrants have been reported in Southeast Greenland (Knudsen, 1935), and possible local sightings from local people in Ammassalik, southeastern Greenland (Helms, 1926).

In Europe, this species has been recorded in the British Isles (Orta et al., 2020). In England, it may have been common up to the 10th century, but it extinct as a resident breeder by 1916 (Brown & Grice, 2005). Currently, it is a rare visitor, with a few individuals recorded between 1958 and 2005, mostly in the southeast: Norfolk, Suffolk, or Kent, mostly during the autumn or spring migration (Brown & Grice, 2005). In Scotland there has been a reintroduction program from 1959, which has created a breeding population in western Scotland (Love, 1988; Orta et al., 2020), which has been monitored by the 2006 Royal Society for the Preservation of Birds Sea Eagle Project Team and Scottish Raptor Monitoring Scheme breeding survey (Forrester & Andrews, 2007; Etheridge et al., 2008). The number of territorial pairs/young produced in Scotland increased from 1995 to 2007, from 1995 (11/7) to 2007 (42/34) (Holling and Rare Breeding Birds Panel, 2007, 2008, 2009, Etheridge et al., 2008).

In Ireland, the White-tailed eagle was named Iolar mara, and it was a common breeding resident on all Irish coasts. BirdWatch Ireland (2020) notes that "formerly the last wild pair bred in County Mayo in 1912", and there has been "mixed success in terms of breeding in Ireland since the recent reintroduction at Mountshannon and Portumna." Currently, "Killarney National Park in County Kerry offers the best chance to see the re-introduced birds. Tracking by GPS has shown that the released eagles have dispersed significantly and have been recorded in almost all Irish counties" (BirdWatch Ireland, 2020). Also, "a pair of White-tailed Eagles nesting on an island in Lough Derg, County Clare, in the spring of 2012" represented the first breeding of this species in Ireland in more than 100 years (Global Raptor Information Network, 2020).

In Northern Europe, the White-tailed eagle was formerly common in Iceland (Wille, 2003). This author reports that nearly 170 historical nest sites have been discovered, and that by the early 1800s, there were possibly 150-200 nesting pairs. However, intense persecution by people from the mid-1800s, resulted in its near extinction, except for some individuals in the northwest. When poisoned fox baits were banned in 1964, there was a gradual recovery, to perhaps 53 pairs in 2002 with 39 breeding (Wille, 2003). In Sweden, the distribution of this species is in three main regions: the Baltic coastline, northwestern Sweden (termed Lapland), and central and southern Sweden (Helander, 2003). The distinctive ecological character of the Lappland region has led some scientists to hint the Lapland population as genetically distinct (Cederberg et al., 2003). Throughout Sweden, there is evidence that the population is increasing (Helander, 2003). In Denmark, there is also evidence of increased populations, from seven pairs in 1999 to 28 in 2009 (Ehmsen et al., 2011). This was after, as noted by Herrmann et al. (2020) "at the beginning of the 20th century, as a consequence of severe persecution the White-tailed Sea Eagle faced extinction all around the Baltic Sea".

In Sweden, according to Helander (1988) the Swedish population in the early 19th century was at least about 500 pairs. After more than a century of persecution, the population declined to about 40 pairs on the Baltic Coast and a few pairs in Lapland in the 1920s (Helander, 2003). Helander (1988) estimated an increase to about 50 pairs by 1940, and by the mid-1950s probably 100 breeding pairs. The effects of the eggshell thinning contaminant, DDE, contributed to a population decline from the late 1950s into the 1960s. The DDT ban in the early 1970s allowed a population recovery, starting by 1980 until the present time. An estimated 300 territorial pairs lived in Sweden in 2000 (Helander, 2003) and the latest estimate by Herrmann et al. (2020), was of 496 territorial pairs in Sweden in 2007. In Denmark, the species even disappeared after 1911, and the same happened in Lithuania and the Kaliningrad Region of Russia." In Norway there were breeding populations, but British Birds (2006) once reported "It's a very grim case of 'we told you so'. The White-tailed Eagle *Haliaeetus albicilla* population on the SmÃÂ¸la archipelago in Norway has been decimated by an ill-advised windfarm. This year, only one eagle is expected to fledge from the windfarm site on the bird's former stronghold of SmÃÂ¸la, a set of islands about 10 km off the coast of northwest Norway."

Van den Berg & Haas (2007, 242) also write that "this spring, four additional White-tailed eagles *Haliaeetus albicilla* have been killed at the wind turbine park on Sømla in Norway, raising the total for this site to 13 victims. For the second consecutive year, the first-ever breeding pair of the Netherlands produced a young at Oostvaardersplassen, Flevoland." Van den Berg & Haas (2008, 188) noted that "the first-ever breeding pair of White-tailed eagle *Haliaeetus albicilla* for the Netherlands, raising a single young both in 2006 and 2007, built a new nest at Oostvaardersplassen, Flevoland, which contained two nestlings in April 2008." In the Netherlands, mostly birds from Norway, Sweden, and northwestern Germany are wintering vagrants, mostly subadults. Its summer presence is very rare (Bijlsma et al., 2001). Eagles have also bred at Oostvaarderplassen, Flevoland (Van Rijn et al., 2010).

In the Mediterranean area and Eastern Europe and Russia, this species is rare to vagrant. It is a vagrant on Mallorca in the Balearic Islands (López-Jurado, 2004). It is also a vagrant in the Canary Islands, with some old records from Tenerife, Gran Canaria, and Lanzarote (Clarke, 2006; Brinkley & Byrne, 2013) and possibly in Malta (Azzopardi, 1978). In Russia, the population is low and widely distributed through suitable habitat outside tundra and desert. Northern birds migrate southwards in winter, while southern populations are nonmigratory, and may be nomadic (Flint et al., 1984; Mlíkovský, 2009). In Romania, it is a rare migratory, breeding resident, generally confined to the Delta, lower Danube, and Prut, usually recorded near goose flocks and wetland forests (Roberts, 2000). In Bulgaria, it is a breeder, and is also a wintering migrant (Nankinov, 1992, 1997, 1998, 2003). In the Czech Republic it is rare, but a breeding resident, with wintering birds from more northerly parts of Europe (Kren, 2000). These are summarized in Table 4.1.

In Hungary, there are resident breeding birds and non-breeding vagrants from the north (Russia, the Baltic States and Scandinavia), the latter either in passage or wintering locally (Horváth, 2009). Breeders are mostly recorded near wetlands or in forest near large rivers, such as the Danube (Gábor, 1998; Horváth & Pinter, 2005; Hadarics & Zalai, 2008). In Croatia, the national breeding population was found to be around 135 breeding pairs (Radović & Mikuska, 2009, 156). In Turkey, it is a rare resident and winter visitor (Kirwan et al., 1999). In Greece, although fairly common in the past, it is rare, either a resident or a visitor, frequenting wetlands of Macedonia and Thrace (Handrinos & Akriotis, 1997) and more rarely, Crete (Wilde, 1980) and the Aegean islands (Handrinos, 1987). In Armenia it is a rare visitor recorded all year (Adamian & Klem, 1999), but with nesting records since 1954 (Global Raptor Information Network, 2021bd).

In Poland, by the beginning of the 20th century, the White-tailed eagle was almost extinct, with possibly less than 20 breeding pairs (Banzhaf, 1937; Tischler, 1941; Król, 1992). After 1945, when protective measures were adopted, the population began to increase. The population was estimated to have increased to 49-56 pairs by 1960 (Mizera, 2003), this number increasing to about 100 pairs in the 1960s (Król, 1992). Another population decline occurred in the 1970s, mostly due to the impact of the use of DDE effects in agriculture, but another a long-term population increase started after the ban on DDT use in 1975 (Mizera, 2003). By 2004, the increase was successful, and the breeding population was estimated as high as 1,150-1,200 pairs. The highest density was located in the Odra-Lagoon area in Poland (Hauff & Mizera, 2006). A positive estimate was that of Herrmann et al. (2009), that there were about 700-800 territorial pairs throughout Poland in 2007, a massive increase over the previous half-century. These are summarized in Table 4.1 and Table 4.2.

Table 4.1. Records for the presence of White-tailed eagles

Location	Pairs	Date	Reference
Greenland	150-170	1990-2000	Kampp & Wille, 1990; Wille, 2003
Iceland	53	2002	Skarphédinsson, 2003
Finland	200; 294	2000; 2009	Stjernberg et al., 2003; Herrmann et al., 2010
Norway	2,000	2000	Folkestad, 2003
Sweden	496	2007	Herrmann et al., 2010
Denmark	6, 17, 28	2003, 2000, 2009	Gënsbøl, 2003; Herrmann et al., 2010; Ehmsen et al., 2011
Britain	36-42	2008-2010	Etheridge et al., 2008; Holling & Rare Birds Breeding Panel, 2010
Netherlands	1	2006-2008	Bijlsma et al., 2001; Orta et al. 2013.
Germany	368, 575, 295	2000, 2007	Hauff, 2003; Hauff, 2009; Herrmann et al., 2010
Poland	<1,200, <800	2004, 2007	Hauff & Mizera, 2006; Herrmann et al., 2010
Austria	7-10	2009	Probst, 2009
Danube	210-223	2000	Schneider-Jacoby et al., 2003
Hungary	55-65, 166	1998, 2007	Gábor, 1998; Horváth, 2009
Czech Rep.	10-15	1994-2000	Kren, 2000
Slovakia	7	2008	Chavko, 2009
Romania	20	2007	Goriup et al., 2007
Bulgaria	10-15	2007	Todorov et al. (2015).
Croatia	135	2006	Radovic & Mikuska, 2009
Greece	104	1994	Handrinos & Akriotis, 1997
Belarus	20-25, 28+	1991, 2000	Ivanovsky, 1996; Ivanovsky, 2003
Estonia	80-90, 150-70	2000, 2009	Volke & Randla, 2003; Herrmann et al., 2010
Latvia	20-25, 25	2000, 2007	Lipsbergs & Bergmanis, 2003; Herrmann et al., 2010
Lithuania	50-60, 90	2000, 2007	Mecionis & Dementavicius, 2003; Herrmann et al., 2010
Ukraine	80-100	2000	Gavrilyuk and Grishchenko, 2003
Kola, Russia	30-35	2000	Ganusevich, 2003; Ptschelinzev, 2003
Kaliningrad	<20	2007	Herrmann et al., 2010
St.Petersburg	25-30	2007	Herrmann et al., 2010
Altai Krai	124-137	2008	Vazhov, 2009
Ob, Siberia	18-21	2002	Karyakin et al., 2005

Pairs refer to breeding, territorial pairs. Kola is the Kola Peninsula in the far northwest of Russia. Kaliningrad and St. Petersburg are in Northwestern Russia. Ob refers to the Ob River area in Siberian Russia. Danube refers to the Danube River Basin. Altai Krai is in central Russia.

Table 4.2. Population status of the White-tailed eagles

Country	Status	Past and Present Threats, Status Change
Greenland	Near Threatened Slow recovery	Weather, prey decline, lead poisoning, shooting (Wille, 2003).
Iceland	Near Threatened Slow recovery	Persecution, past bounties and poison baits, weather, disturbance(Skarphédinsson, 1994, 2003)
Finland	Near Threatened Slow recovery	Eggshell thinning/DDE, persecution, poisoned bait, habitat loss, disturbances (Stjernberg et al. 2003)
Norway	Near Threatened Slow recovery	Nest robbing, hunting, trapping, poisoned bait, wind turbines (Folkestad, 2003; RSPB, 2006)
Sweden	Near Threatened Slow recovery	Poisoning/past DDT, lead poisoning (Helander, 2003)
Denmark	Near Threatened Slow recovery	Persecution, extinction, slow recovery (Gënsbøl, 2003)
Ireland	Near Threatened Slow recovery	Persecution, extinction for 100 years, breeding now (Levy, 2011).
Scotland	Near Threatened Slow recovery	Common historically, extinct, now reintroduced (Smith, 2007; Yalden, 2007; 2012)
Netherlands:	Near Threatened Slow recovery	Possibly extinct, now recovering from German migrant birds (van den Berg and Orta et al. 2013).
Spain	Near Threatened Slow recovery	Rare migrant or vagrant, sightings may be reliable (Prieta, 2012)
Russia	Threatened (Red Book) Slow recovery	Recovering, widespread, rare (Russian Raptor Research and Conservation Network, 2018; Andreenkov et al., 2009)
Ukraine	Near Threatened in some areas, Slow recovery	Recovering, declining in the south (Kalyakin & Voltzin 2006; Gavrilyuk & Grishchenko, 2003)
Estonia	Near Threatened Slow recovery	Breeding resident, various threats recorded (Eagle Club, 2020)
Latvia	Near Threatened Slow recovery	Habitat loss, forestry, shooting, trapping, taxidermist sales (Lipsbergs & Bergmanis, 2003)
Lithuania	Endangered Slow recovery	Persecution, possibly extinct in mid-20th century, now recovering (Mecionis & Dementavicius, 2003)
Germany	Near Threatened Slow recovery	Persecution, lead poisoning, electrocution, disease, recovering (Krone et al., 2003; Hauff, 2003, 2009)
Austria	Near Threatened Slow recovery	Persecution, pesticides, extinction, (Probst & Peter, 2009; Probst, 2009; Wichmann & Kohler, 2009)
Poland	Least Concern Slow recovery	DDT impact, then ban, DDE, felling nest trees (Mizera, 1990, 2003)
Hungary	Near Threatened Slow recovery	Severe decline, persecution, recent recovery (Gábor, 1998; Bank et al., 2004; Horváth, 2009).
Czech Republic	Critically endangered Slow recovery	Possible breeding extinction, recent reintroduction (Kren, 2000; Rajchard & Procháka, 2009)
Romania	Near Threatened Slow recovery	Near extinction, striychnine poisoning, electrocution, late recovery (Roberts, 2000)

The RSPB refers to the Royal Society for the Protection of Birds.

In Asia, the White-tailed eagle is widespread as a winter visitor or vagrant, and rarer as a breeding resident outside Russia and Japan. For example, in Azerbaijan, it is a resident but is very rare, with a decline in nesting sites (Patrikeev, 2004). In South West Asia, it is rare as a passage migrant. In Syria, it was once fairly common as a breeder near rivers such as the Euphrates, its tributaries, the Aleppo and Orontes, but after the mid 20th century, it was extirpated (Baumgart & Stephan, 1986; Baumgart et al., 1995). In Lebanon it is a rare vagrant (Ramadan-Jaradi & Ramadan-Jaradi, 1999; Ramadan-Jaradi et al., 2008; Serhal & Khatib,

2014). In Israel it is a very rare passage migrant and winter visitor, especially in the former breeding areas of northern Israel (Shirihai, 1996). Breeding pairs were recorded in the mid-nineteenth century the Hula and Jezreel Valleys (Zahavi, 1957; Merom, 1960; Geller, 1983). These are summarized in Table 4.2. This is based on information collated by Global Raptor Information Network (2021bd) and other sources and details the conservation issue surrounding the White-tailed eagle in different countries. Although BirdLife International (2016) classifies the White-tailed eagle as Least Concern, it may be classified as Near Threatened or Critically endangered in some countries.

In Kuwait, there is one record of a White-tailed eagle (Balmer & Murdoch, 2010). In Kazakhstan it is rare, but a breeding resident, and also a passage migrant and winter visitor. It is widespread over Kazakhstan during this migration, and non-breeding immatures are also recorded widely during the summer. Individuals are recorded around water bodies such as lakes (Gavrilov & Gavrilov, 2005; Wassink & Oreel, 2007). It is common as a winter visitor, but rare as a summer visitor (Scott &Adhami, 2006). In the Indian subcontinent, the White-tailed eagle is usually a rare inter visitor, in many cases immatures in dispersal, in Afghanistan, Pakistan, Nepal, Bhutan, Assam in India and Bangladesh, but may breed locally in the last country (see Monirul & Khan, 2009; Inskipp & Inskipp, 1991; Thompson & Johnson, 2003; Rasmussen & Anderton, 2012; Spierenburg, 2005). In Central and East Asia and Siberia, it is commoner in some places such as South Korea (Gore & Won, 1971), and parts of Siberia (Andreenkov et al., 2009; Karyakin et al., 2009) but is rarer in China (Weizhi, 2006), and Tibet (Lang et al., 2007) and the Commander Islands (especially Bering Island and Medny Island) (Gibson & Byrd, 2007). There was one record of a White-tailed eagle from Hawaii (Zaun, 2009). In the North African Maghreb, it is rare or accidental visitor. For example, in Morocco, it is an accidental visitor, with one known record off the Mediterranean Coast on 20 March 1983 (Thévonot et al., 2003). In Algeria, Isenmann and Maoli (2000) report a few breeders noted in 1975 by J.P. Thomas, which were recorded later, and some wintering records. In Tunisia, it is a rare accidental visitor (De Balsac & Mayard, 1962; Isenmann et al., 2005). The Global Raptor Information Network (2021bd) gives regional population estimates for this species.

Other countries and areas where the White-tailed eagle is classified as near threatened include the Faeroe Islands, Switzerland, France, Slovakia, Slovenia, Italy, Moldova, Macedonia, Georgia, Turkey, Syria and Israel (Global Raptor Information Network, 2021bd). In other countries, there are other classifications. For example, in China it is a First Class Important Protected Bird (Weizhi, 2006). In Croatia, the breeding population is Endangered (Radovíc et al., 2003). In Armenia, it is a species of special concern (Movsesian & Ayrumian, 1987) and Endangered due to the poor nest records (Adamian & Klem, 1999). In Greece, it is classified as Critically Endangered, although it was common during the 19th century (Reiser, 1905). In South Korea, it is classified as Vulnerable (National Institute of Biological Resources, 2011). Illustrations of the appearance, range, and eggs of this species are shown in Figures 4.2a to 4.2k.

Fig. 4.2a. White-tailed eagle perching [With permission from Slowmotiongli]

Fig. 4.2b. White-tailed eagle perching [With permission from Robert Davidson]

Fig. 4.2c. Immature White-tailed eagle perching [With permission from Robert Davidson]

Fig. 4.2d. Immature White-tailed eagle perching [Credit: Jack Seeds – unsplash]

Fig. 4.2e. White-tailed eagle and Steller's Sea eagle perching [With permission from Julian Thomas]

Fig. 4.2f. White-tailed eagle in flight [With permission from Rosemary Locock]

Fig. 4.2g. White-tailed eagle in flight [With permission from Julian Thomas]

Fig. 4.2h. White-tailed eagle in flight [With permission from Julian Thomas]

Fig. 4.2i. Immature White-tailed eagle in flight [With permission from Credit Avideus]

Fig. 4.2j. White-tailed eagle habitat, Altai Mountains [With permission from Stafichukanatoly]

4.4 Bald Eagle *Haliaeetus leucocephalus* (Linnaeus, 1766)

Physical Description

The Bald eagle, one of the world's most recognizable birds, on account of its status as the national bird of the United States of America, has been described as the "typical sea eagle" (Brown & Amadon, 1968, 289). These authors list two subspecies: *Haliaeetus leucocephalus leucocephalus* (Linnaeus, 1766) ranging in the southern United States and *Haliaeetus leucocephalus washingtoniensis* (Audubon, 1827) mostly ranging in the region of Alaska. In size, the northern birds are slightly larger than those in the south. The body and tail length totals 70–102 cm, and the wingspan is between 1.8 and 2.3 m and the weight ranges from 3 to 6.3 kg (Ferguson-Lees & Christie, 2001). Like most raptors, there is strong sexual dimorphism in terms of size, with females being about 25% larger than males. The average female may weigh almost 5 kilograms, while the male may weigh about 4 kilograms (Grossman & Hamlet, 1964; Dunning, Jr., 1993; Patterson et al., 1998; Bird, 2004; Maehr & Kale, 2005). However, the Bald eagle varies in size, with the larger specimens found in the northern regions, according to the tenets of Bergmann's Rule, which holds that the species increases in size northwards from the Equator, verified by research that found that "Bergmann's rule is a valid ecological generalization for birds and mammals" (Meiri & Dayan, 2003, 331). Generalizable variations according to latitude are recorded in published findings, which are presented in Table 4.3.

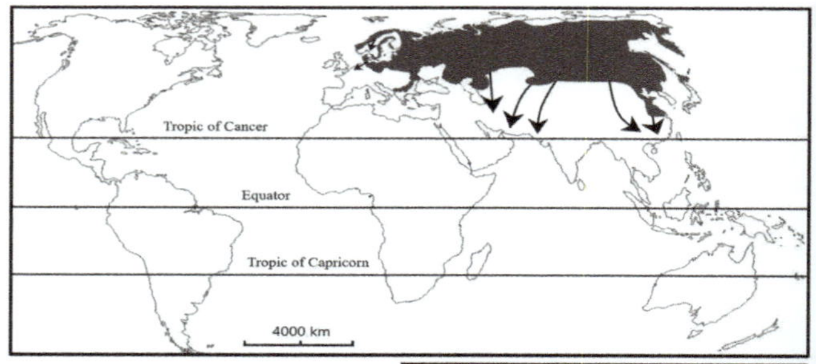

The White-tailed eagle breeds across Siberia and parts of Western Europe and winters south of this range.

The egg is dull white, but may have a few smudges from the nest.

Fig. 4.2k. White-tailed eagle range and egg

Table 4.3. Weight variation in Bald eagles

Location	Latitude	Weight (kg)	Reference
Alaska	66.160507	4.68	Brown & Amadon, 1986; Palmer, 1988
Saskatchewan	52.939900	4.23	Gerrard & Bortolotti, 1988
Montana	46.965260	4.30	Patterson et al., 1998
New Brunswick	46.5653	6.9	Wright, 1953
Arizona	34.048927	4.74	Zylo, 2012
South Carolina	33.836082	3.27	Murphy & Hope, 2006
Florida	27.994402	4.13	Maehr & Kale, 2005

The latitude is in decimal degrees north. The weight is averaged between the males and females. Extreme records were of an outsized female in Alaska weighing 7.4 kg (Gende, 2008).

The adult plumage is unmistakable throughout its range, due to the snowy white head and tail, with all other feathers dark brown to brownish black (Alderfer, 2006; Ferguson Lees & Christie, 2001). The bill, feet and eyes are bright yellow, creating an overall appearance of strong and sharp contrasts. This adult plumage is attained after 4 to 5 years (Brown & Amadon, 1968, 289). The Bald eagle's plumage is uniform for both males and females, but females are up to 25% larger than males (Bent, 1937; Brown & Amadon, 1968; Brown, 1970; Bird, 1983; Gerrard & Ingram, 1985; Godfrey, 1986; Stalmaster, 1987; Gerrard & Bortolotti, 1988; Savage, 1989; Del Hoyo et al., 1994).

The immature plumage is less unique, being overall dark brown, less blackish than the adult plumage, with some white mottling overall. For the immature birds, the feet are colored more lemon-yellow contrasting with the adult bird's canary-yellow feet. The immature birds also have eyes and cere of a more greyish shade of yellow, with a bill that is more brownish, than those of the adult (Brown & Amadon, 1968, 289). The immature may only be mistaken for an adult or immature Golden eagle, as the Golden eagle shares the entire range of the Bald eagle, but the large size of the immature Bald eagle distinguishes it from any other species. All the buzzards and vultures (except for the condors, which are larger but have un-feathered heads) are smaller, and the larger hawks are generally multicolored. These include the dark morph of the Swainson's hawk (*Buteo swainsoni*, Bonaparte, 1838), the dark morph of the Red-tailed hawk (*Buteo jamaicensis*, Gmelin, 1788), the dark morph of the Rough-legged buzzard (*Buteo lagopus*, Pontoppidan, 1763) and the dark morph of the larger Ferruginous hawk (*Buteo regalis*, Gray, 1844). However, even the latter, which is North America's largest buzzard, has a wingspan reaching only around 1.5 meters. The most similar species in appearance is the closely related and slightly smaller African fish eagle, bit this African bird does not share the same range and has a more reddish-brown plumage and a white chest and black tipped bill (Ferguson-Lees & Christie, 2001; Del Hoyo et al., 1994).

The all dark brown immature Bald eagle may be distinguished from the Golden eagle by the former bird's larger head, deeper and heavier bill and un-feathered legs below the thighs (like human shorts, while for the Golden eagle the legs are feathered to the toes, like human trousers), the blotchy white coloration on the underwing coverts, axillaries and tails, and the shorter tails of immature bald eagles (Alderfer, 2006; Bortolotti, 1984a; Bortolotti, 1984b; Dickinson, 1991; Sibley, 2003; Watson, 2010). The immature Bald eagle also has straighter edged wings which are held flat during flight, while the Golden eagle's wings are held in a flattened 'v'. The Bald eagle also has a stiffer wing beat (Sibley, 2000). The Golden eagle also differs in having a mote solid, rich brown plumage than the immature bald eagle, and in having golden colored feathers on its nape, and immature Golden eagles have white squared patches on the wings (Porter, 1981; Bortolotti, 1984c; Sibley, 2000). The related Sea eagles only rarely wander in Alaska (Ferguson-Lees & Christie, 2001). The immature White-tailed eagle may have more light brown coloration on the tail and whitish feather edgings on the dark plumage (Brown & Amadon, 1968; Love, 1983; Forsman, 1999). The immature Steller's Sea eagle is larger than the Bald eagle. The plumage is dark brown with some greyish-brown streaking on the head and the neck, with white feather bases, and light mottling on the rectrices. The tail is similarly colored to that of the immature Bald eagle, but is more wedge shaped (Masterov, 2000). The Bald eagle's call has been termed "almost gull like", the female call resembling "karkarkar" or "karkkarkkark" (Brown & Amadon, 1968, 289). Other sources have described the voice as a weak whistling "kleekkikikikik", like the call of a gull, with harsher calls from immature birds than adults (Sibley, 2000; Ferguson-Lees & Christie, 2001).

Classification

The Bald eagle is regarded as a sister species to the White-tailed Sea eagle (Wink et al., 1996). The species pair is thus a one white-headed species and one tan-headed species which are approximately even in size, with the same ecological role which may have diverged from their ancestral species between the Early Miocene (10 million BP) and the Early/Middle Oligocene (28 million BP) (Wink et al., 1996). The Bald eagle, alongside its sister species is classified as one of the eagles of the Genus *Haliaeetus*, which was found to be monophyletic, and closely linked to the kites of the Genera *Milvus* and *Haliastur* (the latter contains two species, the Brahminy kite *Haliasturindus* Boddaert, 1783, and *Haliastursphenurus* Vieillot, 1818, which together with those of the Genus *Milvus* is part of the subfamily *Milvinae*, Friedmann, 1950). This link between the sea eagles and the kites was constructed and confirmed from morphological analyses by Zimbelmann (1992), an allozyme analysis by Schreiber and Weitzel (1995), and molecular phylogenetic analyses by Seibold and Helbig (1996), Wink et al. (1996) and Lerner and Mindell (2005).

The four northern sea eagle species, the Bald eagle, White-tailed eagle, Steller's Sea eagle and Pallas's fish eagle were also classified into a clade that is distinct from the southern members of the Genus *Haliaeetus* (Wink et al., 1996; Lerner & Mindell, 2005). Two subspecies are recognized by some authors and rejected by others. The subspecies are *Haliaeetus leucocephalus leucocephalus* (Linnaeus, 1766), which is the nominate subspecies, and *Haliaeetus leucocephalus washingtoniensis* (Audubon, 1827) (Del Hoyo et al., 1994; Integrated Taxonomic Information System, 2020). The nominate subspecies occurs in the southern United States and Baja California, Mexico. The second subspecies, a larger bird occurs through the northern United States, Canada and Alaska (Del Hoyo et al., 1994). However, the Global Raptor Information Network (2021be) notes that "infraspecific variation in this species appear to be clinal, and it seems unnecessary to recognize subspecies, e.g., *washingtoniensis*, whose boundaries cannot be determined."

Foraging

Bald eagle habitats include a variety of wetlands, including coastal ecosystems, large lakes, rivers and marshes, with the main criterion being abundant fish stocks. Research findings indicate eagles predominantly frequent water bodies of water with a circumference over 11 km, with breeding commoner near lakes over 10 km^2 (Peterson, 1986). The Bald eagle forages either alone or in pairs, mostly soaring along shores and in some cases perching on high ground in such areas (Stalmaster, 1987; Global Raptor Information Network, 2021be). It may forage over the lower elevations of islands, and over lacustrine and fluviatile waters, and inshore and nearshore waters (Gibson & Byrd, 2007). Food includes live fish, ailing waterfowl, carrion and pirated food (freshly caught fish from ospreys, carrion from vultures). Communal aggregations have been observed, usually near food sources and after the nesting period. Brown and Amadon (1968, 389) note that "the greatest concentrations occur along certain rivers near the coast of Alaska, like the Chilkaty where in mid-November 3000–4000 eagles may be present along a ten-mile stretch, preying on dead and spent dog salmon." Watson et al. (1991, 492) pointed out that in their study "eagles acquired food by hunting live prey (57%), scavenging (24%), and pirating (19%)." The Global Raptor Information Network (2021be) classifies this species as a "dietary generalist, feeding primarily on fish in watered areas, but it is opportunistic and will take any mammalian or avian prey, as well as carrion, in upland areas, especially during the nonbreeding season." The Bald eagle may hunt from a perch or from flight and may take fish from Ospreys, or other fish-eating birds (Global Raptor Information Network, 2021be).

The dietary habits of the Bald eagle may be divided into two main categories; those areas where fish are the main food items, to the near exclusion of other food items, and those where the other items assume more prominence. Fish comprises the main food of the Bald eagle across its range, with at least 100 species of fish recorded in the diet (Sherrod, 1978). Regional analyses of Bald eagle diets have uncovered dietary variations. For example, Stalmaster (1987) examined 20 food habit studies across the Bald eagle's range and found that the diet of nesting eagles comprised fish (56%), birds 28%, mammals 14% and other prey 2%. In the Pacific Northwest, including Alaska, the main diet is spawning salmon, herring, flounder, pollock, and salmon (Alaska Department of Fish and Game, 2020). In Southeast Alaska, the main prey is pink salmon (*Oncorhynchus gorbuscha*, Walbaum, 1792), coho salmon (*Oncorhynchus kisutch*, Walbaum, 1792) and sockeye salmon (*Oncorhynchus nerka*, Walbaum, 1792). Possibly, Chinook salmon may be too large (12 to 18 kg) to be taken from the water and may be eaten as carrion (Armstrong, 2008). Other fish taken include Pacific herring (*Clupea pallasii*, Valenciennes in Cuvier and Valenciennes, 1847), Pacific sand lance (*Ammodytesh exapterus* Pallas, 1814) and eulachon (*Thaleichthys pacificus*, Richardson, 1836) (Armstrong, 2008). Fish as the dominant prey have been recorded by other authors (Armstrong, 2008). For example, Sidle et al. (1986) assessed fish as the main diet of eagles in the area. Using eagle stomach analysis Imler and Kalmbach (1955) decided that fish averaged 66% of the annual diet, and Ofelt (1975) found that fish comprised 78% of the prey brought to nests.

Eagles living in the Columbia River Estuary in Oregon were found to rely on fish for 90% of their dietary intake, while birds and mammals comprised 7% and 3% of the identified prey taken by eagles through one year in one case study (Watson et al., 1991). Here the commonest fish taken included large scale suckers (*Catostomus macrocheilus*, Girard, 1856) (17.3%), American shad (*Alosa (Alosa) sapidissima*, Wilson, 1811) (13%) and common carp (*Cyprinus carpio*, Linnaeus, 1758) (10.8%) (Watson et al., 1991). For eagles on the opposite side of the American continent, in Chesapeake Bay in Maryland the main food items were the American gizzard shad (*Dorosoma cepedianum*, Lesueur, 1818), threadfin shad (*Dorosoma petenense*, Günther, 1867) and white bass (*Morone chrysops*, Rafinesque, 1820) (Mersmann, 1989). Further south in Florida, common prey includes catfish, especially the brown bullhead (*Ameiurus nebulosus*, Lesueur, 1819) and to a lesser extent catfish of the Genus *Ictalurus* (Rafinesque, 1820), mullet (Family *Mugilidae*, Jarocki, 1822), trout (*Salmonidae*, Cuvier, 1816), needlefish (Family *Belonidae*, Bonaparte, 1832), and eels (Order *Anguilliformes*, Berg, 1943) (Broley, 1947; Travsky & Beauvais, 2004). In Nebraska, case studies of the Platte River and the Columbia River found the commonest prey were the American gizzard shad and the common carp, which were obtained by direct river capture, scavenged as carcasses and obtain through attacks on other fish predators (Watson, 1991; Stalmaster & Plettner, 1992). Studies by several researchers hint that Bald eagles prefer large fish (Ofelt, 1975; Delong, 1990; Kozie & Anderson, 1991; Watson et al., 1991; Jenkins & Jackman, 1994; Collins et al., 2005). Large fish selected when the choice of smaller fish was possible included species over 34 cm long from Lake Britton in California, over 35 cm from Lake Superior, 30 – 60 cm from the Columbia River estuary, most fish preyed on by eagles were estimated to measure between 30 and 60 cm and even 86 cm, and larger fish such as Pacific halibut (*Hippoglossus stenolepis*, Schmidt, 1904) and lemon sharks (*Negaprion brevirostris*, Poey, 1868), which are too heavy to be carried (the last may be taken as carcasses or when young).

In some areas, birds and mammals comprise a larger proportion of the diet, such consumption patterns at times varying locally. For example, Newsome et al. (2015, 396) report a case study of the Channel Islands, California, where "eagles on Santa Catalina Island consumed a high proportion (~60%) of marine fish and a lower proportion (25–30%) of seabirds, while their counterparts on the Northern Channel Islands consumed equal proportions (~40–45%) of these prey types." These differences were attributed to "a combination of natural and anthropogenic factors." The development of recreational fishing off Santa Catalina Island increased access to fish species there, while seabird nesting near Northern Channel Islands, may have increased the bird diet of the eagles in that area (Newsome et al., 2015, 396).

Across the Bald eagle's range, birds especially waterbirds are next commonest species groups taken after fish and may even comprise the largest portion of the diet in some areas (Watson et al., 1991; Brisbin & Mowbray, 2002). Researchers have recorded scores of bird species identified as prey for Bald eagles. For example, Collins et al. (2010) examining a historic nest site on San Miguel, found that the prey remains were 61.2% birds and only 18.6% fish, 13.6% invertebrates, 5.2% mammals, and 1.4% reptiles. They note that "remains of 297 birds (45–48 species), 90 fish (four species, two genera and seven families), seven reptiles (one species), 25 mammals (seven species), and 66 invertebrates (23 species, three genera and three families) were identified" (Collins et al., 2010, 103). The avian prey taken by eagles, as a proportion of the total diet may vary seasonally: Watson et al. (1991, 494) found that "overall there was a 16% higher proportion of birds in eagle diets during the nonbreeding period and a corresponding lower proportion of fish" despite the overall 7% figure for birds for the year.

Some findings indicate that bird species diversity is higher than that of the fish or mammals taken by eagles (Palmer, 1988; Sherrod, 1978; Collins et al., 2005). Birds taken are mostly medium-sized waterbirds and waders, such as gulls, geese, grebes, coots and herons (Watson et al., 1991; Ferguson-Lees & Christie, 2001; Travsky & Beauvais, 2004; Birdlife International, 2016). Larger birds may also be taken, such as sandhill cranes (*Antigone canadensis*, Linnaeus, 1758); Common loons (*Gaviaimmer*, Brünnich, 1764), Great black-backed gulls (*Larus marinus*, Linnaeus, 1758), snow geese (*Anser caerulescens*, Linnaeus, 1758), Emperor geese (*Anser canagicus*, Sevastianov, 1802) and Canada geese (*Branta canadensis*, Linnaeus, 1758). For example, American herring gulls (Larus smithsonianus) are among the commoner species taken in the area near Lake Superior (Kozie & Anderson, 1991). Even larger species may be attacked, such as the very large Great Blue heron (*Ardea herodias*, Linnaeus, 1758), Brown pelicans (*Pelecanus occidentalis*, Linnaeus, 1766) and even young American white pelicans (*Pelecanus erythrorhynchos*, Gmelin, 1789) taken when perching or in flight, singly or from colonies (Stalmaster, 1987; Evans & Knopf, 1993; Gill Jr. & Kincheloe, 1993; Wood et al., 1993; Vlietstra & Paruk, 1997; Ferguson-Lees & Christie, 2001; Buehler, 2020; Good, 2020).

Bald eagles may also take raptors, including hawks and owls. Only the Golden eagle among North American raptors may match the Bald eagle in size and strength (Ferguson Lees & Christie, 2001; Campbell, 2015). Raptors recorded as prey include Red-tailed hawks (*Buteo jamaicensis*, Gmelin, 1788), peregrine falcons (*Falco peregrinus*, Tunstall, 1771), northern goshawks (*Accipiter gentilis*, Linnaeus, 1758), ospreys (*Pandion haliaetus*, Linnaeus, 1758), black vultures (*Coragypsatratus*, Bechstein, 1793) and turkey vultures (*Cathartes aura*, Linnaeus, 1758) (Coleman & Fraser, 1986; Watson et al., 1993; MacDonald, 1994; Anthony et al., 2008; Cartron, 2010). Owls may also be taken, ranging from the small Western Screech-owls (*Megascops kennicottii*, Elliot, 1867) to the large Great horned (*Bubo virginianus*, Gmelin, 1788) and Snowy owls (*Bubo scandiacus*, Linnaeus, 1758). CTV Television Network (2012) produced a video detailing a Bald eagle killing a Great horned owl on a golf course in Parksville, Vancouver Island, British Columbia (Sherrod, 1978; Palmer, 1988; Collins et al., 2005; Jackman et al., 1999). Bird numbers taken as a proportion of the eagles' diet varies regionally. For example, in the Greater Yellowstone area, birds rivalled fish as a food item, both comprising about 40% of the diet (Swenson et al., 1986). The Florida Fish and Wildlife Conservation Commission (2020) writes that "Bald eagles are opportunistic foragers, feeding or scavenging on a wide variety of prey." Taking the example of Florida, the primary prey is various fish and waterfowl species. In a cited research paper by McEwan & Hirth (1980), the prey in north-central Florida comprised fish (78%, especially catfish, mostly brown bullhead, *Ameiurus nebulosus*, Lesueur, 1819), birds (17%, especially American coot, *Fulica americana*, Gmelin, 1789), mammals (3%) and amphibians and reptiles (1% together) (McEwan and Hirth, 1980). In Florida, carrion is also a common food source, including roadkills and garbage at landfills (Millsap et al., 2004).

In another example, of Prince of Wales Island, Alaska, Bald eagles may have killed and eaten two Sitka Black-tailed deer fawns (Gilbert, 2016). In the Aleutian Islands it has been recorded at beached carcasses of various marine animals such as otters, seals, sea lions and whales all year round, and seasonally at seabird colonies during summer, and at streams during the fall salmon-run and near refuse dumps and fish-processing plants during winter months (Kenyon, 1961; Sherrod et al., 1976; Gibson & Byrd, 2007). In the Shasta Reservoir area, California the diet biomass was recorded as 86.7% fish, 0.9% birds and 0.7% mammals. Common fish species taken included bass and catfish, and smaller numbers of carp, salmon or trout. Birds taken comprised a goose, dabbling ducks (*Anatinae*, Leach, 1820) and American coot (*Fulica americana*, Gmelin, 1789). Two Black-tailed Jackrabbits and two squirrels were also taken (Global Raptor Information Network, 2021be).

In Wyoming, sheep (*Ovis aries*, Linnaeus, 1758) and pronghorn antelope (*Antilocapra americana*, Ord, 1815) were recorded as 28% of prey items, while lagomorphs were 38% of prey items (Anderson & Patterson,1988). Less taken prey items included carrion from domestic cattle (*Bos taurus*, Linnaeus, 1758) and mule deer (*Odocoileus hemionus* Rafinesque, 1817), and smaller animals such as domestic cat (*Felis catus*, Linnaeus, 1758), prairie dog (Genus *Cynomys*, Rafinesque, 1817) and American Robin (*Turdus migratorius*, Linnaeus, 1766) (Global Raptor Information Network, 2021be). In Nebraska, research indicated that the diet comprised birds (75.8%), mammals (33.8%), fish (11.1%), invertebrates (0.2%), and plant materials (3.2%) (Lingle & Krapu, 1986). The commonest animal species taken from these categories were waterfowl, Cottontail rabbits (Genus *Sylvilagus*, Gray, 1867) and of the fish, carp (Family *Cyprinidae*, Rafinesque, 1815) and gizzard shad (*Dorosoma cepedianum*, Lesueur, 1818). In Texas, eagle diets comprised birds (34%), reptiles (31%) and fish (30%). Commonest species taken from these categories were American coots, soft-shelled turtles (Genus *Apalone*, Rafinesque, 1832), and freshwater catfish (Genus *Ictalurus*, Rafinesque, 1820; Genus *Noturus*, Rafinesque, 1818) (Mabie et al., 1995). Bald eagles also feed on road kills, especially that of deer and large mammals (Ortego et al., 2009).

Mammal prey taken includes rabbits and hares (Family *Leporidae*, Fischer de Waldheim, 1817), ground squirrels (*Marmotini*, Pocock, 1923), raccoons (*Procyon lotor*, Linnaeus, 1758), muskrats (*Ondatra zibethicus*, Linnaeus, 1766), beavers (*Castor canadensis*, Kuhl, 1820), Virginia opossums (*Didelphis virginiana* Kerr, 1792), nine-banded armadillos (*Dasypusnovem cinctus* Linnaeus, 1758), and even deer fawns (Buehler, 2020). Predators taken include Red foxes (*Vulpes vulpes*, Linnaeus, 1758), Arctic foxes (*Vulpes lagopus*, Linnaeus, 1758), striped skunks (*Mephitis mephitis*, Schreber, 1776) and adult bobcats (*Lynx rufus*, Schreber, 1777) (may have been scavenged) (Mersmann, 1989; Grubb, 1995; Mabie et al., 1995). Harbor seal (*Phoca vituline*,

Linnaeus, 1758) pups may also be taken (Hayward et al., 2010). It may take livestock, especially sheep (*Ovis aries*, Linnaeus, 1758) but is less likely than the Golden eagle to do so (McEneaney & Jenkins, 1983; O'Gara, 1994). Reptiles are also taken, for example turtles, such as Diamondback terrapins (*Malaclemys terrapin*, Schoepff, 1793), Common Musk turtles (*Sternotherus odoratus*, Latreille in Sonnini & Latreille, 1801) and Common Snapping turtles (*Chelydra serpentina*, Linnaeus, 1758) (Clark, 1982; Travsky & Beauvais, 2004; Ruzicka, 2006), in addition to snakes, amphibians, crustaceans such as crabs and crayfish, and mussels. Grubb and Coffey (1982) report that "a 1978 study of the winter habitat of the Bald Eagle (*Haliaeetus leucocephalus*) in the Coconino National Forest, Arizona, indicated repeated and potentially heavy use of a freshwater mussel (*Anodonta corpulenta*) in the eagles' diet."

In foraging, Bald eagles frequently encounter interspecific conflicts. Bald eagles may dominate some canids such as coyotes and foxes, and smaller birds such as vultures, gulls and corvids (Sherrod et al., 1976; Campbell, 2015). They may encounter more difficulty with bobcats (*Lynx rufus*) and domestic dogs (*Canis lupus familiaris*) (McCollough et al., 1994). The foraging and competitive abilities of the Bald and Golden eagles are rated as even or a slight advantage to the Golden eagle (Sabine & Gardner, 1987; Watson, 2010; Campbell, 2015).

Breeding

Bald eagles utilize varied landscapes for their activities throughout their range. For example, Suring (2008, 104) points out that "studies of Bald eagles in Alaska show that they consistently select areas with specific landscape features for nesting, perching, roosting and foraging. They also choose different landscape features, or habitat, for different activities." The breeding season starts with acrobatic courtship flights, including talon locking, somersaults and aerial diving. Nests, which are generally built on tall trees, rocky promontories or the ground, are located near large inland lakes and/or rivers, or the coast. Building materials are usually sticks, with smaller structures such as pine needles. These can be very large structures, up to 4 meters high and 2 to 3 meters wide. In areas with scanty wood growth, "the nest may be only a little pile of seaweed on the grounds" (Brown & Amadon, 1968, 389).

Nesting is principally near inland lakes. Nesting generally occurs in old-growth and mature stands of coniferous or hardwood trees, the important criteria being the tree's height, composition and location (usually within 200 meters of open water) (Suring, 2008). Shorter trees (e.g. 6 meters) in water, such as swamp ecosystems may (usually 16 to 38 meters) also be utilized for nesting. These nesting choices may however vary regionally. For example, in Florida, nests may be located in mangrove swamps, river and lake shorelines, pine forests, seasonally flooded woodland, hardwood swamps, and open, treed prairies and even pastureland. Favored nesting trees in Florida include slash pines (*Pinus elliottii*, Engelm.), longleaf pines (*P. palustris*), loblolly pines (*P. taeda*) and cypress trees (Family *Cupressaceae*, Bartlett). In the southern coastal areas nests may be in mangroves. In Wyoming, nests may be in cottonwoods or pines near rivers, including old-growth stands of ponderosa pines (*Pinus ponderosa*, Douglas ex C. Lawson) and riparian trees near rangeland. In Southeast Alaska, Sitka spruce (*Picea sitchensis*, (Bong.) Carr) provided 78% of the nesting trees, while hemlocks (*Tsuga*, Carrière) provided 20%. Nests are also found in trees near human constructed reservoirs stocked with fish. The nesting trees are usually tall. Examples are 28 m in Chesapeake Bay, 23 m in Florida, 27 m in the Greater Yellowstone area. Most nests are near water. Human disturbances may affect nesting eagles, nests usually located nearly 2 km from human presence. Rarely, nests are with large estuaries or forested areas within cities, like Hardtack Island on the Willamette River in Portland, Oregon or John Heinz National Wildlife Refuge at Tinicum in Philadelphia, Pennsylvania. One nest was recorded in New York City in 2010.

The Global Raptor International Network (2020) notes that the Bald eagles build "a huge nest of sticks, generally placed in a tall tree, but sometimes on a cliff ledge. Nests are used in successive years and sometimes become massive structures 3-4 m high or more." The National Eagle Center (2020) notes that "the largest recorded bald eagle nest, located in St. Petersburg, Florida, was 9.5 feet in diameter, 20 feet deep and weighed almost 3 tons." Two dull white, rough-shelled eggs, more rarely one or 3 are laid. Laying periods may vary across its range. For example, in western Canada, March to early April, in Northern Canada usually later as the inland lakes may melt later, in southern Florida, usually early November and late January. Incubation is usually around 35 days, but this is variable (Brown & Amadon, 1968; Ferguson-Lees & Christie, 2001). In the Alexander Archipelago, Alaska, nesting starts in early April, and incubation in mid-May (Robards & Hodges, 1976). In the Aleutian Islands, nesting starts in late January, egg laying from late March to May, first hatching in early May, and fledging from early July to late August (Sherrod et al., 1976, White et al., 1971). Some breeding occurs up to a month later (Gibson & Byrd, 2007). In Washington State, courtship began on 31 December, with a majority courting at the beginning of February, and incubation by the end of March and hatching March to May (Bortolotti, 1984). Fledging occurred June to August, when most nestlings were 11 weeks old. The post-fledging period averaged 29 days, (Bortolotti, 1984; Watson, 2007).

Nest sites vary regionally. In Alaska, specifically in the Alexander Archipelago, southeast Alaska, nesting was generally in the huge old growth trees (especially Sitka Spruce and Western Hemlocks) near later, some 25.5 m high (Robards & Hodges, 1976). In the Aleutian Islands, nests are mostly located on sea stacks (these are almost vertical rock columns, created by coastal wave erosion) and promontories near the sea. A few nests were located on artificial habitats, such as a crane structure, an old bulldozer set, and runway-approach platform (Byrd & Day, 1986; Sherrod et al., 1976; Byrd & Williams, 1992; Gibson & Byrd, 2007). In California, common nesting trees are Ponderosa and Sugar pines, with the vast majority of such trees selected less than two kilometers from water, associated with reservoirs, with a clear un-constructed view of open water (Lehman, 1979).

In an interesting study, Dawe (2016, 35) document an example of cooperative breeding by Bald eagles on Vancouver Island, British Columbia. In this case, a male and two females cooperated in breeding over two seasons, including nest incubation and protection. Dawe (2016) cites others who define cooperative breeding as "a reproductive system in which more than a pair of individuals show parent-like behavior towards young of a single nest or brood" (Hatchwell & Komdeur, 2000). Factors for cooperative breeding include delayed dispersal, which may be due to ecological factors such as habitat saturation or life history issues such as low mortality of adults (Kimball et al., 2003; Arnold & Owens, 1998; Emlen, 1982; Hatchwell & Komdeur, 2000). Cooperative breeding yields the benefits of teamwork (assistance in food provision, territorial defence, and knowledge exchange (Kimball et al., 2003). It is argued that such cooperative breeding is common in at least 29% of the genera and 14% of species of diurnal raptors. Reproductive pairs may include a third bird, immature or adult, male or female (Kimball et al., 2003). For the Bald eagle, cooperative breeding is rare, in some cases polyandry and polygyny (Dawe, 2016, 35). Dawe (2016) lists reports of such behavior in Bald eagles across the United States, where in most cases the third eagle was a female (Heglund & Reiswig, 1980; Fraser et al., 1983; Garcelon & Slater, 1995; Hopkins et al., 1993, 1995, 1999; Nye, 1994; Ortego et al., 2009). They maintain that the Bald eagle is still recorded as monogamous, and observers sighting third birds at nests might regard such individuals as transients (Kimball et al., 2003).

Population Status

The Bald eagle ranges across North America, from Bering Island and the Aleutian Islands, southwards to southern Florida and Baja California. Records exist from the Aleutian Islands, central Alaska (southern Brooks Range) to southern Alaska (west to Buldir in the Aleutian Islands (once on Attu) and east along the southern coast), and northern Canada (Yukon, Manitoba, Labrador, Newfoundland) south through continental United States to northern Mexico (Baja California Norte, Sonora). It winters from southern Alaska southward (American Ornithological Union, 1998). In Canada, it breeds from northern Yukon, northwestern and southern Mackenzie, northern Saskatchewan, Manitoba and Ontario, central Quebec, Labrador, and Newfoundland southwards to southern Canada, while wintering in southern Canada (American Ornithological Union, 1998).

The Bald eagle is a partial migrant (Bildstein, 2006). The Global Raptor Information Network (2021be) notes that "northern populations are migratory and disperse somewhat sporadically over most of North America." Through this range the Bald eagle, along with the Golden eagle is one of the most well-known bird species, with a wide-ranging bibliography (for example, Arent, 2007; Avery & Watson, 2009). The Handbook of the Birds of the World (2018) lists the range of *H.h. leucocephalus* as "S USA (California, Arizona E to Virginia and Florida) S to NW Mexico (Baja California and Sonora)." The other subspecies, *H.l. washingtoniensis*, ranges from the "Aleutian Is, Alaska and Canada E to Labrador and Nova Scotia, S in USA to Oregon, Utah, C Kansas and Illinois; inland Alaskan and Canadian populations winter from coastal Alaska and Canada S to USA and Mexico." One migration route has been recorded as between Canada and California (Linthicum et al., 2007). Other migrants have been recorded in Texas (Lockwood & Freeman, 2004). Generally, migrants move south in the fall (August to October, and northwards in spring (February to April) (Global Raptor Information Network, 2021be).

The Bald eagle is the only sea eagle that is endemic to North America. During a time of decline in the mid-twentieth century, its limited range was Alaska, the Aleutian Islands, northern and eastern Canada, and Florida (Bull & Farrand, Jr., 1987). After this period the population increased, to widespread status in all the states and provinces of the United States and Canada (Ferguson-Lees & Christie, 2001; Buehler, 2020). It is also accidental in the Commander Islands, Kamchatka Peninsula, and the Kolyma River region in Siberia (Global Raptor Information Network, 2021be). The Bald eagle has been recorded as a vagrant in the Caribbean (mostly Bahamas) (Norton et al., 2011), Mexico in Baja California (where it may breed) and Sonora southwards to Nayarit and Zacatecas (Howell & Webb, 1995). Records exist from Veracruz (Monroy-Ojeda et al., 2016). Breeds in Baja California (Magdalena Bay, formerly in other localities), and Sonora (Rio Yaqui) (American Ornithological Union, 1998). Other accidental or vagrant records are from Ireland (Petersen, 1987; Smiddy & O'Sullivan, 1996; Haas, 2012), Northern Ireland (British Ornithologists' Union, 2020, Smiddy & O'Sullivan, 1996; Haas, 2012); Wales (British Ornithologists' Union Records Committee, 1980, 2009; Rogers & Rarities Committee, 1980; Haas, 2012) and possibly (unreliable) England (Haas, 2012). It has been recorded on the Arctic coast of northeastern Siberia (Global Raptor Information Network, 2021be).

Regarding population levels, BirdLife International (2016) assesses this species as increasing and Least Concern. It further notes that "past declines have been attributed to intense hunting, unintentional poisonings (notably use of DDT and lead shot), and habitat destruction in combination with the loss of great herds of bison, a seasonally important food source"; however, "this species has undergone a large and statistically significant increase over the last 40 years in North America (779% increase over 40 years, equating to a 72.2% increase per decade; data from Breeding Bird Survey and/or Christmas Bird Count: Butcher and Niven, 2007)" (BirdLife International, 2016). It is not classified as Vulnerable, as its population is increasing and is very large

(>10,000), it has a very wide range (>20,000 km^2) (BirdLife International, 2016). Global Raptor International Network (2020) cites sources that put the population at 19,578 individuals in 2006.

DDT pollution in the mid-twentieth century was a major factor for the decline in the population. DDT and DDE impacts included egg losses through shell thinning (Global Raptor International Network, 2020). The result of this was that "Bald Eagle populations plummeted in the United States and southern Canadian portions of its range from the 1950s to the early 1970s" and the species was placed on the endangered list; however, when "DDT was banned for domestic use in the United States in 1972, following an earlier ban in Canada, the species enjoyed a spectacular recovery, which was also facilitated by the protection of many breeding and roosting sites, a reduction in persecution (shooting and trapping), and a decline in lead poisoning, as a result of a ban on the use of lead shot for waterfowl hunting on federal lands." This resulted in the conservation status being downgraded from Endangered to Threatened in 1995, to Least Concern (BirdLife International, 2020).

For example, in a case study of Bald eagle population recovery in the US state of Louisiana, Smith et al. (2016) note that "the population of nesting *Haliaeetus leucocephalus* (Bald eagle) in Louisiana reached precariously low numbers in the early 1970s after experiencing marked declines similar to the rest of the continental population." This study used aerial data collected from 1975 to 2008, Findings indicated that "active nests increased exponentially from 7 to 387, exhibiting a mean annual rate of increase of 11.1 ± 0.3% per year with no indications of slowing." The main factor for this increase appeared to be slight increases in reproduction. These results exceeded the objectives of the Southeastern States Bald Eagle Recovery Plan, justifying the 2007 removal of the Bald eagle from the federal list of endangered and threatened wildlife. Nickolas et al. (2016) nevertheless argue that "the continued stability and growth of the nesting population may depend on the ability of Bald Eagles to cope with increasing levels of human activity, as well as the protection and availability of current and future nesting habitats." Illustrations of the appearance, range, and eggs of this species are shown in Figures 4.3a to 4.3l.

4.5 Pallas's Sea Eagle or Pallas's Fish Eagle *Haliaeetus leucoryphus* (Pallas, 1771)

Physical Description

The plumage is brownish overall, with lighter colored feathers on the head. The crown, nape, head, neck and upper back are golden brown, while the forehead, throat, chin and fore-neck and ceres are white. The lower back, rump and lower breast, belly, undertail coverts, flanks, thighs, tail quills and upper tail coverts are darker brown while the tail is almost black with a broad central white band (Brown & Amadon, 1968, 287; Orta et al., 2020). The tail is banded dark and white. The immature plumage is

Fig. 4.3a. Bald eagle perching [With permission from Joshua J. Cotten]

Fig. 4.3b. Bald eagle perching [With permission from Frank Cone on Prexels]

Fig. 4.3c. Bald eagle perching [With permission from Michelle Jean]

Fig. 4.3d. Immature Bald eagle [With permission from Frank Cone – Pexels]

Fig. 4.3e. Bald eagle adult and immature on the Nest [thepcspud]

Fig. 4.3f. Bald eagle [thepcspud]

Fig. 4.3g. Bald eagle in flight [With permission from Ian Dufield]

Fig. 4.3h. Immature Bald eagle in flight [With permission from Paul Prior]

Fig. 4.3i. Immature Bald eagle in flight [With permission from of Thomas Morse – unsplash]

Fig. 4.3j. Bald eagle at nest [With permission from Richard Lee – unsplash]

Fig. 4.3k. Typical Bald eagle habitat, British Columbia [Author]

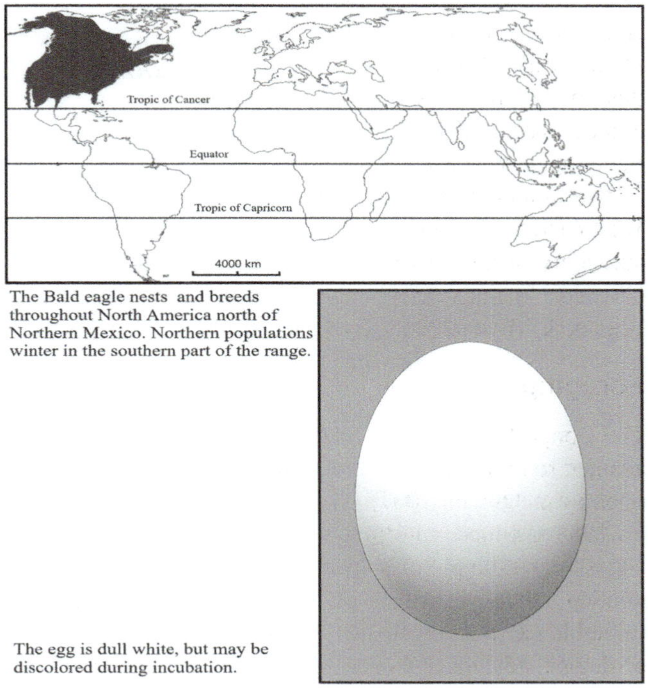

Fig. 4.3l. Bald eagle range and egg

brown, with lighter tinge, and white coloring on the secondaries' inner webs. The underside is darker, with paler tips. The tail is dark brown with white mottling and the legs are whitish, the cere is greyish and eyes brown (Brown & Amadon, 1968, 287).

There are few species that closely resemble the Pallas's Sea eagle. It can be confused with the Golden eagle, but the Golden eagle has its legs feathered to the toes, while the Pallas's Sea eagle has a bare tarsus. The Pallas's Sea eagle is also smaller than the White-tailed Sea eagle, which may be similarly brown overall (Brown & Amadon, 1968, 287). The Grey-headed Fish eagle and White-tailed eagle lack the Pallas's Sea eagle's plumage pattern of dark body coloring, light colored hood and black-and-white tail. The Grey-headed is much smaller (BirdLife International, 2018). The voice is usually a barking call, which has been described as "like the yapping of a Pekinese dog, rather hoarse", quoted as "kwokkwokkwok"; other calls are during breeding resemble Black kite (*Milvus migrans*, Boddaert, 1783) and gull calls (Brown & Amadon, 1968, 287). BirdLife International (2018) describes the voice as a "loud, guttural kha-kha-kha-kha or gao-gao-gao-gao, and sometimes high-pitched, excited yelping. Hoarse guttural continuous kook-kook-kook is commonest call." Illustrations of the appearance, range, and eggs of this species are shown in Figures 4.4a to 4.4n.

Classification

This species is monotypic. It is among the species in *Haliaeetus* that were found to be related to the milvine kites of the genera *Milvus* and *Haliastur*. These conclusions were arrived at using morphological analysis (Zimbelmann, 1992), allozyme analysis (Schreiber & Weitzel, 1995) and molecular phylogenetic analyses (Lerner & Mindell, 2005; Seibold & Helbig, 1996; Wink et al., 1996). Pallas's Sea eagle is not among the northern sea eagle species, *H. albicilla*, *H. leucocephalus*, *H. pelagicus*, and *H. leucoryphus*, that were classified as a clade distinct from the southern members of the genus (of which it is one). Due to its range being frequently away from coastal areas, some have argued that naming it 'sea eagle' is a misnomer (Beaman, 1994). This species has no close link with the other sea eagles, using mtDNA cytochrome b sequence data. It may be closer to the northern sea eagles, based on the results of morphological assessments (Wink et al., 1996).

Foraging

The Pallas's Sea eagle ranges from Central Asia, the Kyrghiz steppes, Manchuria, Kokonor (Qinghai province, China), Tibet and Mongolia, to Iraq, Pamir, North India and Myanmar. It winters south to Crimea, Iraq, Northern India and the Indo-Chinese region (Orta et al., 2020). Brown and Amadon (1968, 287) note that "the name Sea eagle is generally a misnomer for this species, for it is not a bird of sea coasts, but of large or small inland lakes and large rivers of Central Asia." Wintering grounds may reach the

Black and Caspian Seas, otherwise it is largely an inland bird. It often occurs in highlands, sometimes up to 14,000 to 15,000 feet in Tibet, "the commonest large eagle in such places" (ibid., 288). Other habitats are rivers of northern India and Myanmar.

The Global Raptor Information Network (2021bf) describes the main habitats as lakes, marshes, and large rivers in inland areas, including dry, open grassed landcover (Brazil, 2009), ranging from lowlands to altitudes of 5,000 m. It regularly perches on branches and may rest on the ground after eating (Global Raptor Information Network, 2021bf). Foraging is mostly from a perch. In some cases, carrion is taken. Like Bald eagles, it snatches fish from the surface of water. It also practices kleptoparasitism against other raptor species (Rasmussen & Anderton, 2012; Orta et al., 2020). Such competitors include ospreys, and in some cases the more dominant Himalayan vultures. Community behavior has been observed, which is more frequent than some other Sea eagles. Foods include stranded fish and disabled waterbirds, and "rarely takes fully active birds" (Brazil, 2009). Other prey taken include mammals, including rabbits and rodents, and reptiles and amphibians such as frogs, snakes and terrapins (Ferguson-Lees & Christie, 2001). Two exceptional prey incidents include record of killing greylag geese (*Anseranser*, Linnaeus, 1758), which is slightly heavier than the eagle, and a large carp weighing 6 kg or twice the eagle's weight (Wood, 1983).

Breeding

The Pallas's Sea eagle's breeding season begins with calling and soaring, and more rarely acrobatics. Nesting varies, in the northern ranges such as Tibet, most nests are built on crags, while in the southern ranges the nests are mostly located in tall trees near waterbodies, or more rarely on the ground. The older, larger nests may reach two meters in breadth and one meter deep. Nest materials are mostly dried sticks and green vegetation. Nests may be reused for many years, with materials being added each season. The coarse-surfaced, white eggs are laid in clutches of 2 to 4, averaging 2 in the northern ranges and 3 in the southern ranges. Eggs are laid earlier in the northern ranges; late March to early April in Turkestan, and May in Tibet and Transbaikal. In Russia, nesting may begin early March-April (Flint et al., 1984), while in Bhutan it may nest in September to April, during periods with low river water levels and many wintering waterbirds (Spierenburg, 2005). In the southern ranges, for example, breeding may start earlier: October to February is common in India. Incubation is nearly 40 days and fledgling after 70 to 105 days. Brown and Amadon (1968) note that in Tibet, due to the late laying, birds that have fledged must migrate within a month.

Population Status

The Pallas's Sea eagle ranges over the Indomalayan/Palearctic region (BirdLife International, 2018). This includes central and southern Asia, ranging from Iraq to Kazakhstan and Uzbekistan, eastwards to northern China and southwards to the northern parts of Indian subcontinent (note that the Indian subcontinent includes Bangladesh, Bhutan, India, Maldives, Nepal, Pakistan, Sri Lanka) and south-central China. Eagles in the northern parts of the range migrate to Iran and Afghanistan, more rarely to Europe and southwest Asia (Global Raptor Information Network, 2021bf).

The Pallas's Sea eagle has a sparse distribution over a very wide range (Ferguson Lees & Christie, 2001). The population is small and declining; hence it is classified as Endangered by the IUCN Red List (BirdLife International, 2018). The population decline is due to habitat losses, mostly the draining of wetlands and deforestation, mainly losses of the large trees that the eagle uses for perching and nesting. Other issues include the silting of lakes and the spread of water hyacinth (*Pontederia crassipes*, Mart. Eichnornia) in India, shooting, disturbance, and reduction of the prey base from hunting and overfishing and possible impacts from environmental contaminants, especially DDE (Global Raptor Information Network, 2021bf).

The Pallas's Sea eagle records various populations across its wide range. BirdLife International (2018) states that it may "breed only in northern India (apparent strongholds in Assam and Uttarakhand), Bangladesh and Myanmar, with very small numbers in Bhutan, dispersing north of the Himalayas to Kazakhstan, Russia and Mongolia in its non-breeding season (May to September)." In Europe, a vagrant was recorded in the Netherlands (Bijlsma, 2001) and may be extinct or very rare in Bulgaria (Nankinov, 1992; Nankinov et al., 2004). In Turkey, it may have been recorded in the past, but there are few reliable records that it is present today (Kirwan et al., 2008). In Asia, it appears as a vagrant in several countries. For example, it is possibly extinct and a past vagrant in Azerbaijan (Patrikeev, 2004); a recent vagrant in Oman (Eriksen et al., 2003); and may be a vagrant in the United Arab Emirates (Richardson, 1990) and Saudi Arabia (Bundy & Warr, 1980). It is also a vagrant or rare winter visitor in Kazakhstan (although formerly a breeder) (Balmer & Betton, 2007; Wassink & Oreel, 2007; Wassink, 2009) and Iran (Jennings, 1981; Scott & Adhami, 2006). It is also a rare breeder in central Asia (Flint et al., 1984).

Further east, it is generally a declining and rare breeder, passage migrant or visitor. In the Indian subcontinent, it is a declining resident, with the main range being across northern section, southern and east-central Afghanistan, across eastwards to the Himalayas and Assam and southwards to Gujarat and Bangladesh (Rasmussen & Anderton, 2012). In this zone, it is a possible breeder, but defined mostly as a winter visitor and passage migrant in the Fall and Spring (Inskipp & Inskipp, 1991). In Bhutan and Bangladesh, it is recorded as a possible resident (Spierenburg, 2005; Monirul & Khan, 2009). In China, it is a breeder in northeastern China, and also a migrant, southwards to Jiangsu (Bildstein, 2006; Brazil, 2009). The global population has been estimated at 1,001 to 10,000 individuals, possibly nearer the smaller figure (Ferguson-Lees and Christie, 2001). BirdLife

International (2018) estimates the population at 1,000-2,499. Although this species is not as well known as the Bald eagle, it has nevertheless been reported by several researchers (BirdLife International, 2001; Brazil, 2009; Gilbert & Gombobataar, 2009; Kovalenko, 2009; Sourav et al., 2011; Gilbert et al., 2014; Steele, 2017; IUCN, 2018). Illustrations of the appearance, range, and eggs of this species are shown in Figures 4.4a to 4.4n.

Fig. 4.4a. Pallas's Sea eagle perching [With permission from Raghuvir Solanki]

Fig. 4.4b. Pallas's Sea eagle perching [With permission from Quazi Sanjeed]

Fig. 4.4c. Pallas's Sea eagle perching [With permission from Mark Piazzi]

Fig. 4.4d. Pallas's Sea eagle perching [Copyright.bryanjsmith]

Fig. 4.4e. Immature Pallas's Sea eagle [With permission from Zamiruddin Faisal]

Fig. 4.4f. Immature Pallas's Sea eagle [With permission from Zamiruddin Faisal]

Fig. 4.4g. Immature Pallas's Sea eagle perching [With permission from Mark Piazzi]

Fig. 4.4h. Immature Pallas's Sea eagle in flight [With permission from Roger Wasley]

Fig. 4.4i. Pallas's Sea eagle in flight [With permission from Quazi Sanjeed]

Fig. 4.4j. Pallas's Sea eagle nest and habitat [With permission from Pete Favelle]

Fig. 4.4k. Pallas's Sea eagle on nest [With permission from Quazi Sanjeed]

Fig. 4.4l. Pallas's Sea eagle in grass [With permission from Lorand Szucs]

Fig. 4.4m. Pallas's Sea eagle habitat, Kazakhstan [With permission from Dmitry Sumskoy – unsplash]

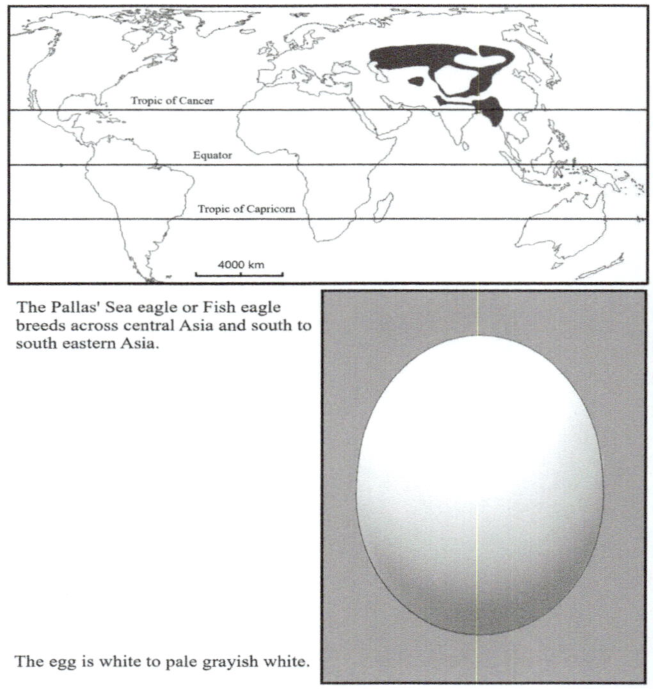

Fig. 4.4n. Pallas' Sea eagle range and egg

4.6 Madagascar Fish Eagle or Madagascan Fish Eagle *Haliaeetus vociferoides* (Des Murs, 1845)

Physical Description

The Madagascar Fish eagle or Madagascar Sea eagle is a medium-sized eagle, but a large raptor. The body and tail measure 60–66 cm in length, with a wingspan of 165–180 cm. The weight varies from 2.2 to 3.5 kg. BirdLife International (2018) gives the length as 70 to 80 cm. In common with most raptors, the female is slightly larger than the male. The plumage is mostly brown (Ferguson Lees & Christie, 2001). The head is distinct with a light coloration, and a greyish white, reddish-streaked crown and nape, and a whitish throat with brown streaks. The dark reddish-brown back and upper wing coverts have reddish borders to the feathers, while the similarly dark brown breast and belly have reddish streaks (BirdLife International, 2018). The wings are generally dark brown. The upper wing quills are dark grey, and the lesser wing coverts are dark brown, also with reddish streaks. The wing quills and greater coverts are dark grey. The white tail is rather short (BirdLife International, 2018). The eyes are brown and the cere and legs are light or whitish (Brown & Amadon, 1968; Ferguson Lees & Christie 2001; del Hoyo et al., 1994, 2014; Kemp et al., 2020g).

The immature plumage consists of browner head feathers, with a brown crown and nape streaked lighter, and paler general plumage. The back and upper tail coverts are also brown with whitish tips to the feathers, dark brown upper breast and underwing coverts with paler reddish streaks and whitish tips to the feathers. This coloration gives a "mottled appearance" (Brown & Amadon, 1968, 286). The dark greyish brown tail has whitish edges and is grey below (BirdLife International, 2018). The wing primaries are darkish brown, almost black, the secondaries lighter brown and the innermost wing feathers are lighter. The wing quills are grey, with whitish edges (Ferguson Lees & Christie, 2001; del Hoyo et al., 1994, 2014; Kemp et al., 2020g). This species may not be confused with other species, as it is the only bird of prey of this size in Madagascar, with brown and white coloration, long wings and short tail and a unique voice call. However, BirdLife International (2018) notes that this species "could only be confused with Madagascar Buzzard *Buteo brachypterus* or Madagascar Harrier-hawk *Polyboroides radiatus*, from which separated by white tail and cheeks in adult, and huge size, strong head and short tail of juvenile. Hints much the largest raptor in Madagascar." The voice is generally a loud "quay-quay or hoarse hoai – hoai" this like the African fish eagle (Brown & Amadon, 1968). Illustrations of the appearance, range, and eggs of this species are shown in Figures 4.5a to 4.5j.

Classification

Global Raptor Information Network (2021bg) classifies this species as monotypic, with no subspecies. This species is among the 4 southern sea eagle species, the other being the African fish eagle, Sanford's Sea eagle and the White-bellied sea eagle. The Madagascar fish eagle and the African Fish eagle (or African Sea eagle) are sister species (Wink and Sauer-Gürth, 2000, 2004; Lerner and Mindell, 2005), and both were formerly placed in a separate genus, *Cuncuma*. This study complemented that mentioned above that founded that *Haliaeetus* is monophyletic with a close relationship to the milvine kites of the genera *Milvus* and *Haliastur*. This was based on morphological analysis by Zimbelmann (1992), allozyme analysis by Schreiber and Weitzel (1995), and molecular phylogenetic analyses by Wink et al. (1996), Seibold and Helbig (1996), and Lerner and Mindell (2005) (Global Raptor Information Network, 2021bg). Del Hoyo et al. (1994) also describe the African fish eagle its closest relative, the two forming a distinct species pair, diverged from the rest of the genus *Haliaeetus*. Similarities between the two species include the dark bill, talons, and eyes, and the partially or totally white tails (even in the immature plumage) and at least partially white tails, even while juvenile. Like the Bald eagle and White-tailed eagle species pair, one species has a white head (African fish eagle) and the other species has a tan head (the Madagascan fish eagle) (Del Hoyo et al., 1994).

Foraging

The Madagascar Fish eagle frequents wooded areas near open water (Bamford et al., 2017; Rabarisoa et al., 1997), usually occurring singly, or in pairs, sometimes pairing outside the breeding season (Global Raptor Information Network, 2021bg). Hunting takes place near or over water, with frequent perching on tall trees and soaring over water (BirdLife International, 2018). Such areas are frequented largely because of the large trees in a shoreline ecosystem (Berkelman, 1997). Habitats are listed as including coastal habitats, islands, lakes, and rivers with nearby woodlands and large shoreline trees, and mangroves, from sea level up to 1,200 m (Langrand, 1990; Rabarisoa et al., 2003). Deeper, clearer lakes appear to be preferred (Berkelman et al., 1999a,b; Watson et al., 2000).

One study (Rand, 1936) inferred increased presence near muddy mangrove-bordered bays near the coast, where the shallow water was frequented by fish. In these areas, foraging was mostly from perches on mangrove or other trees, on the mud or sand flats that were exposed at low tide and shoreline rocks. In some cases, eagles entered native fish weirs, and removed fish before the tide declined enough for the people to remove the fish (Rand, 1936). In another study, Morris and Hawkins (1998) found the main habitat as rocky coasts, mangrove landcover, and lakes and large rivers with wooded banks, which are favored perching areas.

The main prey is fish, in one case especially two species of non-native tilapia (Langrand, 1990; Berkelman, 1997; Berkelman et al., 1999a,b). Brown and Amadon (1968) note that it differs from other Sea eagles in that it generally does not take larger fish in open water. Fish may be taken when swimming near the surface or when stranded in shallow pools, or as carrion (Brown & Amadon, 1968). The Global Raptor Information Network (2021bg) notes that it may take fish from the water surface "in a shallow dive"; food other than fish includes "crabs, young marine turtles, and occasionally birds, and they pirate prey from other bird species."Factors for prey choices include the relative abundance of the fish species, and common species include four species of Tilapia and Channa, introduced to Madagascar, and the native species *Megalops cyprinoides* and *Arius madagascarensis*. *Fistularia* spp. and *Tylosurus* spp. are also taken in the northern northwestern coastal regions (Berkelman et al., 1999a,b; Rasamoelina, 2001). Other prey recorded included young domestic ducks, and water birds such as *Anhinga melanogaster*, *Phalacrocorax africanus*, *Ardea cinerea*, *A. humbloti*, *Dendrocygna viduata*, and *Platalea alba*, which in some cases may be motivated by territorial defense (Berkelman et al., 1999a,b; Rabarisoa et al., 2003).

Breeding

The breeding season starts in May with courtship, involving soaring flight and calling, followed by nesting in trees, mostly mangroves. The large, bulky, mostly stick built nest is about 100 to 120 cm in diameter and 30 to 50 cm in depth. The dead branches used for construction may be 10 to 30 mm thick and 30 to 60 cm long (Langrand & Meyburg, 1989; Razafindramana, 1995; Watson et al., 1999). This large nest is usually located on a tree fork on or near the top of a mangrove or other large tree, or in some cases on a rocky crag or cliff (Rabarisoa et al., 2003). These authors argue that nest site selection may be conditioned by the size of the mangrove trees, with most trees selected being near low-turbidity water, where fish are more easily spotted. Although many nests are built in littoral areas, in some other areas, especially in northern to northwestern Madagascar, many eagles nest on crags on rocky islands, where vegetation may in some cases be absent (Rabarisoa et al., 2003). The commonest tree species used for nesting are *Tamarindus indica* (L. 1753) and *Adansonia* spp., while other species may also be used, such as *Broussonetia greveana* ((Baill.) C.C. Berg), *Neobeguea mahafaliensis* (J.F. Leroy), and *Colvillea racemose* (Bojer ex Hook.) have also been used (Rabirisoa et al., 2003).

Egg laying begins at the start of the rainy season. Brown and Amadon (1968, 287) noted that "other details of breeding are not recorded." However, more recent studies have found that the Madagascar Fish eagle generally nests on large trees or rock cliffs (Rabarisoa et al., 1997). Favored trees in coastal areas include *Rhizophora mucronata* and *Ceriops tagal* mangrove trees (Kemp et al., 2020g, Razafimanjato et al., 2014). One or two (one record of three eggs) eggs are laid, but usually one chick is raised, as one chick may kill the other (Watson, 1998, 1999; Watson et al., 1999; Rabarisoa et al., 2003). The eggs are plain white (Watson et al., 1999). The incubation period lasts from 37 to 43 days, and with both parents participating in the rearing, but the female taking the larger share (Watson et al., 1999). There may be replacement clutches when a clutch fails. The nestling period lasts from 78 to 89 days, and the nestlings move from the nest to the surrounding branches in about 70 days (Watson et al., 1999; Rabarisoa et al., 2003). The immature fish-eagles may return to the parental territory until the following breeding season but may disperse 14 to 28 weeks after fledging (Rafanomezantsoa, 1998; Rabarisoa et al., 2003; Global Raptor Information Network, 2021bg).

Polyandry and co-operative breeding have been reported for this species (Watson et al., 1999, Razafimanjato et al., 2017), and polyandrous breeding attempts may be more successful than pairs, as in the latter a third of breeding attempts may be unsuccessful (Watson et al., 1999; Razafimanjato et al., 2017). In cooperative polyandry and cooperative breeding, the extrapair adults are usually males (Tingay, 2000). Breeding pairs usually nest about 0.7-1.7 km apart (Watson & Razafindramanana, 1999). In one case study of the Antsalova region from 1991 to 1997, the nesting success was 0.29 fledglings per territorial pair per year (Rabarisoa, 1999).

Population Status

The Madagascar Fish eagle occurs only on the island of Madagascar, principally near inland water bodies and the western coast (Global Raptor Information Network, 2021bg). This is a non-migratory species, but juveniles may disperse widely from breeding areas (Bildstein, 2006). Langrand & Meyburg (1989) point out that "the Madagascar Sea Eagle is one of the rarest birds of prey in the world, with probably approximately 50 pairs. Its range is confined to a small strip along the west coast of the island, from Antsiranana (Diego Suarez) at its northernmost point to Belo sur Mer, about 50 km south of Morondava, 600 km further south. It is doubtful whether it occurred further south in the past." These authors describe three biotopes as its typical habitat: (1) marine rocky areas; (2) mangroves; and (3) lakes and rivers. The main threats to its population are listed as hunting, trapping and taking young from nests and degradation of the natural habitats listed above. Crucially, it is argued that the current (at time of writing) protected areas are insufficient for the survival of the species. The suggestion is for a conservation area covering the small rocky islands of the north-west and the larger lakes in the Antsalova region (Langrand & Meyburg, 1989).

BirdLife International (2018) lists the Madagascar Fish eagle as 'Critically Endangered' with a total population of about 240 birds. Threats are listed as the development of annual & perennial non-timber crops, hunting & trapping of terrestrial animals, logging & wood harvesting, fishing & harvesting aquatic resources and human work and other activities. Razafimanjato et al. (2014, 88) support this assessment with a case study. In this study, intensive field surveys "in western Madagascar from Antsiranana in the north to Manja in the south during the 2005 and 2006 breeding seasons (May–October)" recorded 287 birds, this total including 98 breeding pairs, 23 breeding trios (69 birds), 15 single adults and seven immature birds. On location observation, 128 birds (44.6%) were recorded on lakes, 116 (40.4%) in coastal areas, 103 (35.9%) in mangroves and 13 (4.5%) in estuaries), 32 (11.2%) on marine islands and 11 (3.8%) on rivers. These findings indicated an increase in population from surveys in 1995, from 222 to 287, with the number of pairs increasing from 99 to 121. It was concluded that these results, though positive did not override the impact of human persecution (hunting, collection of eggs and nestlings), overfishing and habitat destruction on the population of the Madagascar Fish eagle. More population monitoring is suggested, especially in western and northern Madagascar (Razafimanjato et al., 2014, 88).

Regional variations in populations have been recorded. For example, in the northwest coast of Madagascar, it was evidently common in the 19th century, but it has since become much rarer (Owen, 1833; Schlegel & Pollen, 1868; Rand, 1936). Later sources noted it was scarce in some locations (Van Someren, 1947). By the latter half of the 20th century, most reports commented on the rare status of the Madagascar Fish eagle. Langrand (1990) argued that the eagle was only a limited range on the west coast. Thiollay and Meyburg (1981) recording only one pair in a field survey, estimated only 10 pairs remaining. Langrand (1990) estimated the total population at 50 pairs, although this was not based on actual field work. Morris and Hawkins (1998) reported that despite its reported presence along the whole west coast of Madagascar, it was last reported from the southwest in 1975 and at time of writing it was restricted to a 600-m stretch of coastline (Morris & Hawkins, 1998). Therefore, they estimated the total population at about 100 pairs. Rabarisoa et al. (1997) documented 63 pairs between 1991 and 1995, and estimated the population at 222 adults in 1995, and later Rabarisoa et al. (2003) reported about 75 pairs. Global Raptor Information Network (2021bg) reports that "based on later surveys, the National Director of The Peregrine Fund's Madagascar Project, Lily René de Roland, estimated the total population at 350 individuals in August 2006 (L. René de Roland pers. comm.)." Tingay (2005) argued that, based on historical records, there is no evidence that the Madagascar Fish eagles range was ever wide enough to reach the east coast of Madagascar. Any such records were attributed to "vagrants, misidentified birds, or flawed locality data" (Global Raptor Information Network, 2021bg). The study therefore concluded that the present range is similar to the historical range.

Although Birdlife International (2018) has classified the Madagascar Fish eagle as "Critically Endangered" based on "an extremely small population which is probably in decline", it points that "recent data have suggested that its population is stable." Tingay (2005) cited evidence that this species is normally rare, with no real population decline. Johnson et al. (2009) supported this position, using genetic evidence that revealed low genetic diversity. These findings indicate a species with a prehistorically low population. Nevertheless, conservation issues abound regarding the possibilities of protection and extinction. Authors regarding the situation of this species as dire include Meyburg and Langrand (1985), Langrand (1990), Rabarisoa et al. (1997, 2003), Watson et al. (1999) and Watson and Rabirisoa (2000).

All these studies point to human factors as the main factors for the threatened population. Such actions include the destruction of nests and killing of nestlings, the theft of nestlings for pets and food, the shooting and trapping of adults, the tasking of eagle body parts for traditional medicine, and habitat destruction, mostly the over killing of fish and the loss of nesting and foraging structures (e.g. large trees used for nesting, extracted to make dugout canoes, and smaller trees used for firewood and construction). Conservation methods have included the encouragement of population expansion, and the rescue of the weaker chicks before the stronger chicks kill them. sibling rescue (removal of one nestling from a brood of two before the youngest dies) and control of overfishing and deforestation (O'Daniel, 1995; Watson et al., 1996; Watson & Rabarisoa, 2000; Watson et al., 2000; Johnson et al., 2009). These actions point to the gravity of the conservation situation for this species (Langrand, 1990; Berkelman, 1997; Rabarisoa et al., 1997; Watson, 1998; Berkelman et al., 1999a,b; Watson et al., 1999; Watson & Rabarisoa, 2000; Johnson et al., 2009; Safford & Hawkins, 2013; Kemp et al., 2020g; Razafimanjato et al., 2007, 2014, 2017; Bamford et al., 2017). Illustrations of the appearance, range, and eggs of this species are shown in Figures 4.5a to 4.5j.

4.7 African Fish Eagle or African Sea Eagle *Haliaeetus vocifer* (Daudin, 1800)

Physical Description

The African Fish eagle or African Sea eagle is colored like the Bald eagle, with a white head and tail and brown body plumage, but the brown feathers are more reddish in color, the mantle is darkly streaked, and the underside is from the chin to the belly is also white. The scapulars, greater coverts, upper tail coverts, wing quills and rump are also black rather than brown as in the Bald eagle. The lesser wing coverts, belly, under wing coverts and sides are chestnut colored and the wing quill undersides are

Fig. 4.5a. Madagascar Fish eagle perching [With permission from Mike Nelson-Birdtour ASIA]

Fig. 4.5b. Madagascar Fish eagle [With permission from Skip Russell]

Fig. 4.5c. Immature Madagascar Fish eagle perching [With permission from Allan Hopkins]

Fig. 4.5d. Immature Madagascar Fish eagle perching [With permission from Keith Cowton]

Fig. 4.5e. Madagascar Fish eagle in flight [With permission from Alan Wilkinson]

Fig. 4.5f. Immature Madagascar Fish eagle perching [With permission from Pieter Verheij]

Fig. 4.5g. Juvenile and adult Madagascar Fish eagle [With permission from Skip Russell]

Fig. 4.5h. Madagascar Fish eagle habitat [With permission from Rob Sall]

Fig. 4.5i. Madagascar Fish eagle habitat [With permission from Katerina Kalinina]

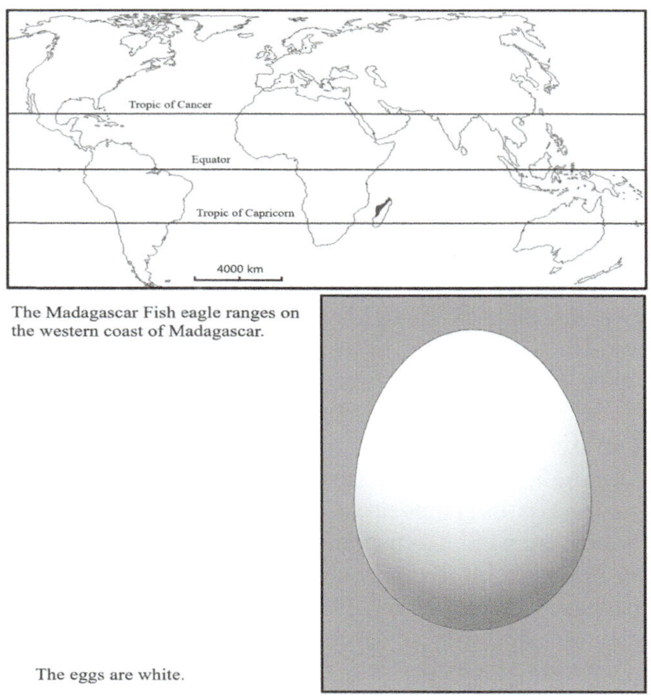

Fig. 4.5j. Madagascar Fish eagle range and egg

slate grey (Brown & Amadon, 1968). The eyes are hazel, and both the cere and some facial skin are yellow, while the feet are yellowish to slightly pinkish. The bill and claws are black (Brown & Amadon, 1968). This is a fairly large species, with a body and tail length of 63–76 cm, a wingspan of 1.8 to 2.4 meters and a weight of 1.8 to 3.6 kg (Peregrine Fund, 2020; Kemp & Marks, 2020).

The immature bird has a brown crown, with black feather tips, while the nape and feathers around the eyebrow are greyish. The upperparts are medium to dark, blackish brown, with black feather tips. The underparts are buff, the low neck, chest and belly are brown with white streaks. The immature plumage after two years changes, with a white underside, and more white feathers on the head, increased chestnut coloring on the belly, and darker patches develop on the upperparts, which transforms into the dark upperparts of the adult after two more years. The adult African Fish eagle is distinctive, and may not be mistaken for any other species, due to the white head and chestnut brown body plumage. The juvenile of this species may resemble some Snake eagles, but the current species has a light-colored nape and eyebrow and white tail feathers (Brown & Amadon, 1968). Brown and Amadon (1968) describe its voice as a yelping call, "weeah – hya – hya – hya" or a lower "quock – quock – quock", with the female giving a shriller call than the male. Illustrations of the appearance, range, and eggs of this species are shown in Figures 4.6a to 4.6l.

Classification

The African Fish eagle is monotypic. It was formerly placed with the Madagascar Fish eagle in the genus *Cuncuma*, separately from the genus *Haliaeetus*. Like the Madagascar Fish eagle, it was classified as more closely related to the southern sea eagle species, *H. leucogaster*, *H. sanfordi*, *H. vocifer*, and *H. vociferoides*, *H. vocifer* and *H. vociferoides* than to the northern species of sea eagles (Wink & Sauer-Gürth, 2000; Lerner & Mindell, 2005). Like the other species, it is classified as close to the milvine kites of the genera Milvus and Haliastur, using research conducted by Zimbelmann (1992), Schreiber and Weitzel (1995), Wink et al. (1996), Seibold and Helbig (1996), and Lerner and Mindell (2005).

Foraging

The African Fish eagle generally frequents lowland (usually below 1300 meters) forested and wooded areas near rivers, lakes and sea coasts. It is absent from the drier, northern half of Africa, being confined to the region south of a line stretching approximately from Senegal to Ethiopia, southwards to South Africa. This species is a lowland denizen, rarely occurring above 1300 meters. (Brown & Amadon, 1968). Foraging, usually from perches, takes place singly, but less commonly pairs or groups may hunt or associate in food procurement, especially in areas of food high density. Dowsett-Lemaire and Dowsett (2006) report up to 75 individuals in areas of limited resources such as declining water sources or fish sources around fishing boats. Fish comprises 90% of the diet (Brown & Amadon, 1968). Other food sources are waterbirds such as flamingos (*Phoenicopteridae* spp., Bonaparte, 1831) and ducks. Ash and Atkins (2009) reported an incident where eagles captured a Tropical Boubou or Bell shrike (*Laniarius major*, Hartlaub, 1848), a White-winged Black tern (*Chlidonias leucopterus*, Temminck, 1815) and a Little egret (*Egretta garzetta*, Linnaeus, 1766), all in flight (Ash & Atkins, 2009). Other food sources are carrion (including the remains of fish, birds and mammals) and kleptoparasitism, where the eagle robs other species such as ospreys (*Pandion haliaetus*, Linnaeus, 1758), herons and cormorants. For example, in Zambia, it is reported to attack and rob Goliath Herons of their catch (Dowsett et al., 2008), an activity also reported in Zimbabwe (Irwin, 1981). In Zambia, it was also reported feeding on antelope and fish carrion (Benson et al., 1970). In Malawi, attacks on poultry and small passerines are reported (Dowsett-Lemaire, 2006, Dowsett-Lemaire and Dowsett, 2006), while in Zimbabwe, it may take young birds from nests (Irwin, 1981). Insects, including termites are also taken, especially during droughts when the waterbodies decline or dry up (Dowsett-Lemaire, 2006).

Regional differences may be observed, but generally the habitats are near water sources (Ferguson Lees & Christie, 2001). For example, in Ghana it is reported near the large rivers, and also by large rivers, lagoons, lakes, and reservoirs (Borrow & Demey, 2010). In Eritrea and Ethiopia, it is recorded near both alkaline and freshwater lakes, and larger rivers and marshes, in some cases in groups up to 20 individuals (Ash & Atkins, 2009). In Zambia, its habitats are similar, but it is more numerous near rivers with high fish presence and less so near rivers in densely wooded areas (Benson et al., 1970; Dowsett et al., 2008). In Botswana, it occurs as elsewhere where there are lakes, rivers and reservoirs, but it is mostly resident only where there is permanent water (Parker, 2005). In Mozambique, it most commonly frequents treed waterbodies (including bays and estuaries) and wetlands with some open water, and even drying wetlands (Parker, 2005). In Zimbabwe, it frequents areas near any permanent waterbodies with fish, but is absent from those waterbodies missing fish (Irwin, 1981).

Breeding

The African Fish eagle breeding season starts with soaring, both singly and in pairs, calling and talon grappling and whirling dives. Most nests are in trees such as acacias (Genus *Acacia*, Martius, 1829), figs (Genus *Ficus*, Linnaeus) and Euphorbia species, less commonly they are on crags, shorter trees and bushes or even on the ground (Brown & Amadon, 1968). Nests, mostly constructed of sticks and green vegetation for lining, may reach 2 meters across and over 1 meter deep after annual reuse. Clutches vary from 1 to 3 or 4 eggs, but commonly 2 eggs are laid (Global Raptor Information Network, 2021bh). There are records of egg numbers from Ethiopia (1 or 2) (Ash & Atkins, 2009), Zambia (mostly 2, then 3 and 1) (Dowsett et al., 2008) and Malawi (mostly 2, and a few clutches of 3) (Dowsett-Lemaire & Dowsett, 2006). The coloration is white, with slight gloss and in some cases faint grey-brown spots.

Laying varies; in the northern ranges (West Africa, Sudan) laying generally coincides with the end of the rainy season in October; in the equatorial ranges (Ethiopia, Kenya, Uganda, northern Lake Victoria) laying generally occurs during the dry season from June to August; in the southern ranges (southern Lake Victoria) laying generally occurs from May to June; in more southerly ranges (e.g., South Africa) laying may be as late as June to September). The assessment is that there is more correlation between the dry season and the laying times, than in the equatorial ranges. Incubation is for about 42-45 days. The eggs usually hatch at different times and in some cases the smaller, younger chick may die. The nestling period is about 65/70 to 75 days, and the fledged young may stay close to the nest for two to three months before leaving (Brown & Amadon, 1968; Global Raptor Information Network, 2021bh).

Nesting for the African Fish eagle is generally late in the year, has been recorded in Mauritania (Morel & Morel, 1990; Rodwell et al., 1996; Isenmann et al., 2010); Liberia (Gatter, 1997); Ghana (Harvey & Harrison, 1970; Grimes, 1987); Ethiopia (Ash & Atkins, 2009); Somalia (Ash & Miskell, 1998); Uganda (Brown and Britton, 1980); Angola (Dean, 1974; Pinto, 1983); Zambia, here many from the first half of the year, as well as later (Dowsett et al., 2008); Malawi, here between March and September (Dowsett-Lamaire & Dowsett, 2006); South Africa, mid-year (Brown et al., 1982; Tarboton & Allan, 1984); Botswana, mid-year (Penry, 1994); and Zimbabwe, January to October (Irwin, 1981). Nesting locations include mostly baobabs or Afzelia trees, with one on a rock in Malawi (Dowsett-Lemaire, 2006); and trees in Mozambique (Douglas, 2002; Parker, 2005) and Zimbabwe (Irwin, 1981). In some cases, the nesting density is high: in one case in Zimbabwe, there were 67 nest sites on the 200 km long shoreline of Matusadona National Park (Irwin, 1981). Irwin (1981) notes that only 36 nests were occupied in one season, but 50 nests (one for each breeding pair) were available each year, which indicated a density of about 2 km of lake frontage per bird (Irwin, 1981). In South Africa, the population was sparser, as the distances between breeding pairs were estimated between 0.5 km and 11 km (Harrison et al., 1997).

Population Status

The African Fish eagle has a wide distribution across Sub-Saharan Africa, from Senegal in the west to Ethiopia in the east, and southwards to South Africa (Ferguson Lees & Christie, 2001). The total population was estimated by Brown et al. (1982) at 100,000 to 200,000 adult pairs, with an unknown or unstated number of immatures. Ferguson-Lees and Christie (2001) estimated the number of adults and immatures at the beginning of the breeding season as between 100,101 and 1,000,000 birds. individuals. BirdLife International (2016) does not give a number but notes that "the population is suspected to be stable in the absence of evidence for any declines or substantial threats" and assesses the species of "Least Concern." Regional population estimates have been given; for example, in Namibia, the population was once estimated at 600 birds (Simmons & Brown, 2009); in Mozambique, the estimate was 1,500+ (Parker, 2005); and in South Africa: the population in Kruger National Park was estimated at 128 birds (Kemp et al., 2001). It is an irruptive or local migrant (Bildstein, 2006). Immatures and subadults are more mobile than adults (Brown et al., 1982; Oatley et al., 1998).

Global Raptor Information Network (2021bg) describes the African Fish eagle "not usually persecuted by humans, or affected by habitat changes, but it may suffer from competition with humans for its food supply in some areas." Problems include impacts from DDE-induced eggshell thinning, especially in Zimbabwe (Douthwaite, 1992) and South Africa (Davies & Randall, 1989), and drownings from artificial structures, and hunting and overfishing (Thiollay, 1985). The population density varies according to region and country.

For example, in West Africa (Burkina Faso, Mali, Niger), it appears to be declining, especially in unprotected areas (Thiollay, 1977; Rondeau & Thiollay, 2004; Thiollay, 2006). It is a common or not uncommon resident in the Mauritanian wetlands (e.g., in the Senegal delta and valley, Nouakchott) (Gee, 1984; Lamarche, 1988; Rodwell et al., 1996; Isenmann et al., 2010); Gambia along the river (Gore, 1990); and in Ghana, in the north near the major rivers and near the coast (Bannerman, 1951; Grimes, 1987; Dowsett-Lemaire & Dowsett, 2008). In West Africa, it is also rare in some places, such as Liberia (Gatter, 1997); Togo, where it is an uncommon presumed breeding resident (Millet-Horsin, 1923; Cheke & Walsh, 1996); Côte d'Ivoire where it is a rare breeding resident (Thiollay, 1985); and Nigeria, where it is an uncommon resident (Elgood et al., 1994). In Central Africa, it frequents the forests of central Cameroon (Bobo et al., 2006). In Côte d'Ivoire, it was formerly abundant in coastal lagoons during the 1900s, but is now almost extinct (Thiollay, 1985). In Liberia, it was formerly common, but by 1981-1984, only 10 birds were recorded in a 380 km river stretch (Gatter, 1997).

In Northeast Africa, in Sudan, the African Fish eagle has been a common breeding resident (Bowen, 1926; Nikolaus, 1987). In Egypt, it may be a visitor, but sightings have been uncertain (Goodman & Meininger, 1989; Gantlett, 1999; Haas, 2012). In East Africa, it is common near waterbodies (Stevenson & Fanshawe, 2002). In Ethiopia, it is a fairly common breeding resident, common along rivers but rare in arid areas (Ash & Atkins, 2009). In Somalia, it is also a fairly common resident along the rivers but rarer on the coast (Gatter, 1997). In Uganda, it is a fairly common breeder on most water bodies (Carswell et al., 2005). The population is declining (Eltringham, 1975; Krueger, 1997; Carswell et al., 2005). In Eritrea, it is comparatively rare (Smith, 1957; Ash & Atkins, 2009).

In Southern Africa, it the African Fish eagle appears to be more common. Oatley et al. (1998) note that the population is concentrated on the eastern part of the region, but it is also found in the riverine and watered areas of the drier western areas. It is a common to a not uncommon resident breeder in some countries such as Angola (on the larger rivers) (Traylor, 1963; Pinto, 1983; Dean et al., 1988, Dean, 2000) and Malawi (especially near Lake Malawi and larger rivers) (Dowsett-Lemaire & Dowsett, 2006). However, in Malawi, it appears unthreatened despite fishing exploitation (Dowsett-Lemaire & Dowsett, 2006). In Zambia, it is widespread and locally common especially near wooded rivers (Benson et al., 1970; Dowsett et al., 2008). In Botswana, it is common resident of the main rivers and the Okavango Delta, and locally common elsewhere (Penry, 1994; Boshoff, 1997; Hancock et al., 2007). In Zimbabwe, it is widespread on large waterbodies (Irwin, 1981), and past population increases (Irwin, 1981; Hartley, 1998). In South Africa, it is fairly common in the Transvaal (Tarboton & Allan, 1984). In this country, the population has benefited from dam construction (Boshoff, 1997; Simmons, 2005). In Swaziland, it is not uncommon in conservation areas and waterbodies (Monadjem et al., 2003; Monadjem & Rasmussen, 2008). Dangers include overuse of water by the sugarcane industries, irrigation and pesticides in waterbodies (Monadjem & Rasmussen, 2008). It is an uncommon breeding resident in Mozambique (Parker, 2005). In Namibia, it is considered Vulnerable with few sightings, but increased with human depopulation and fish repopulation (Simmons & Brown, 2008). Illustrations of the appearance, range, and eggs of this species are shown in Figures 4.6a to 4.6l.

4.8 White-bellied Sea Eagle or White-Breasted Sea Eagle *Haliaeetus leucogaster* (Gmelin, 1788)

Physical Description

The White-bellied Sea eagle or White-breasted Sea eagle is a large, generally light-colored bird (Strange, 2000). The adult's head, neck, underside and outer third of the tail are white, while the back, upper wing and upper tail coverts and the basal area of the tail

Fig. 4.6a. African Fish eagle perching [With permission from Derek Keats]

Fig. 4.6b. African Fish eagle perching [With permission from Raghuvir Solanki]

Fig. 4.6c. Adult and immature African Fish eagle perching [With permission from Raghuvir Solanki]

Fig. 4.6d. Immature African Fish eagle perching [With permission from Ken Zaremba]

Fig. 4.6e. Immature African Fish eagle [Johannes Gross (j-g-fotos@web.de)]

Fig. 4.6f. Immature African Fish eagle [With permission from Brian Zarate]

Fig. 4.6g. African Fish eagle in flight [With permission from Gustav Van Der Gryp]

Fig. 4.6h. African Fish eagle in flight [With permission from Julian Thomas]

Fig. 4.6i. Immature African Fish eagle in flight [With permission from Wayne Davies]

Fig. 4.6j. African Fish eagle habitat, Kruger National Park [With permission from Michael Kemper]

Fig. 4.6k. African Fish eagle habitat, Gambia River [With permission from Steve Riall]

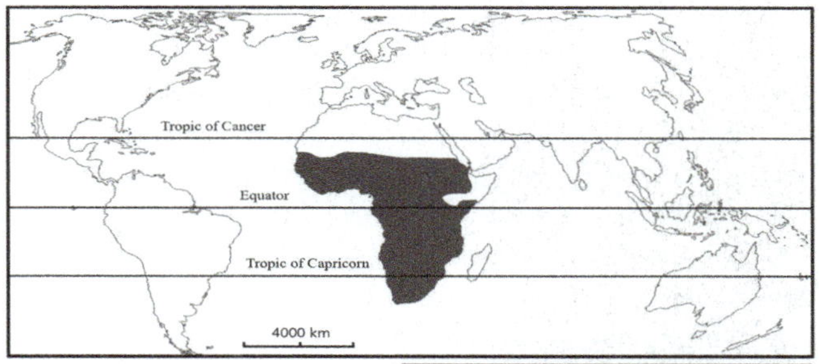

Fig. 4.6l. African Fish eagle range and egg

are grey or greyish brown. The wing quills are dark to blackish grey-brown. The eyes ate brown, the cere is bluish grey and the legs are yellowish white (Brown & Amadon, 1968). The body and tail measure 66 to 90 cm, and the wingspan is approximately 1.78 to 2.2 m. The weight varies from 1.8 to 4.5 kg (Ali, 1993; Ferguson-Lees & Christie, 2001; Hollands, 2003; Shephard et al., 2004; Zahm, 2015; Debus & Kirwan, 2020). The immature plumage over two years age has a partially white head and tail, with the rest of the plumage being brown to greyish brown. The adult plumage appears at 3.5 to 5 years (Brown & Amadon, 1968; Strange, 2000).

This species may be distinguished from other similar species by its large size. It is nonetheless "the slimmest and most graceful of the genus" (Brown & Amadon, 1968, 279). Zahm (2015) writes that "White-bellied sea eagles can sometimes be confused with the Brahminy kites or Egyptian vultures. However, … these species are much smaller than White-bellied Sea eagles. Wedge-tailed eagles can also be mistaken for White-bellied Sea eagles, but wedge-tailed eagles have feathered legs and more dark coloration than these sea eagles" (see also Einoder & Richardson, 2007; Ferguson-Lees and Christie, 2001; Shephard, et al., 2004). Immature, White-bellied Sea eagles differ from wedge-tailed eagles, in that the latter are larger, darker, longer tailed with more feathered legs. They also contrast with the Black-breasted buzzard (*Hamirostramelanosternon*, Gould, 1841), in that the latter species is smaller, with white wing patches (Marchant & Higgins, 1993; Ferguson-Lees & Christie, 2001). The Egyptian vulture in India is also whiter, smaller with a whiter back and wings (Ferguson-Lees & Christie, 2001; Campbell, 2015). The Philippine eagle in the Philippines has a crest and is larger (Kennedy et al., 2000). The immature, White-bellied sea Eagles differ from immature grey-headed fish eagles, by the former bird's browner underparts and flight feathers, and wedge-shaped tail (Del Hoyo et al., 1994; Kennedy et al., 2000). The voice may be described "ah – ah – ah" or "cank – cank – cank" or "glak – glak – glak" and a "duck like qua – qua – qua" (Brown & Amadon, 1968, 279). Strange (2000) describes the call as a loud goose-like honking, with the male giving a higher-pitched and more rapid call than the female. Illustrations of the appearance, range, and eggs of this species are shown in Figures 4.7a to 4.7h.

Classification

The White-bellied Sea eagle, like the other sea eagles, is part of the Genus *Haliaeetus*, is monophyletic and has a close link with the kites of the genera *Milvus* and *Haliastur*. As described above, these conclusions were arrived at through morphological, allozyme and molecular phylogenetic analyses by Zimbelmann (1992), Schreiber and Weitzel (1995), Wink et al. (1996), Seibold and Helbig (1996), and Lerner and Mindell (2005) (see also Jollie, 1977; Holdaway, 1994; Thiollay, 1994). The White-bellied Sea eagle is also closely related to Sanford's sea eagle, with which it is considered by some researchers as a superspecies or clade (Wink et al., 1996). Wink et al. (1996, 783) notes that "*H. vocifer* clusters with the *H. leucogaster*/*H. sanfordi* clade in a monophyletic tropical clade." The latter two species are genetically very close (Helbig et al., 2005).

The nucleotide sequences of the cytochrome b gene of the two sea eagles which were analysed in Seibold & Helbig (1996), hints that the genetic divergence of the two species is small enough (0.3%) and that the ancestors of the two species may have diverged 150,000 years ago, this recent event possibly resulting after New Guinea-based white-bellied sea eagles reached the Solomon Islands, the current home of the Sanford's Sea eagle (Wink et al., 1996; Seibold & Helbig; 1996; Wink & Sauer-Gnrth 2004). Although on account of this close relation they could be considered one species, Wink et al. (1996) argued two full species may be concluded, due to the differences in morphology and behavior.

The physical differences between this species pair (the White-bellied Sea eagle and the Sandford's Sea eagle) and those species pairs described above are similar, in that one species has a white head and the other has a dark head. Similar to the other Southern Hemisphere sea eagles, the bill and eyes are dark colored, and the talons are dark yellow. The White-bellied Sea eagle's links with the other sea eagles (apart from the Sanford's Sea eagle), may be relatively uncertain. Current findings indicate the genetic links between the White-bellied Sea eagle and other sea eagles (other than the Sanford's eagle) require more research, as the molecular data links it to the tropical sea eagles (African fish eagle and the Madagascar fish eagle), but allozyme data shows links with the northern sea eagles (Schreiber & Weitzel, 1995; Seibold & Helbig, 1996). Further links have been found using molecular methods between the White-bellied Sea eagle and the Sanford's sea eagles, with the African fish eagle and the Madagascar fish eagle and the two species of the Genus *Ichthyophaga* (Lerner & Mindel, 2005).

Foraging

The White-bellied Sea eagle frequents sea coasts and offshore islands, and inland water bodies, including estuaries, rivers, mangroves, and artificial reservoirs and swamps usually near open country rather than forest (Beehler et al., 1986; Olsen, 1995; Sheldon et al., 2001; Debus, 2008). Foraging is based on perching and soaring, sometimes in pairs, and practices intra-species kleptoparasitism, as well as on ospreys. Prey species include sea snakes, fish, herons and even small crocodiles and mammals. Its range includes India, Sri Lanka, South China, Australia, New Guinea and Bismark Archipelago (Brown & Amadon, 1968). Foraging generally involves observational perching near water bodies, singly, in pairs or small family groups (Coates & Bishop, 1997, Kennedy et al., 2000). Prey taken includes mammals (including fruit bats), birds (including gulls, cormorants, and gannets), reptiles (including sea snakes), fish (including spiny poisonous fish), and small turtles, taken from the water surface, and more rarely carrion (Coates, 1985; Debus, 1994; Kennedy et al., 2000; Coates & Peckover, 2001; Rasmussen & Anderton, 2012). Mo et al. (2017, 43) report that White-bellied Sea eagles "feed on fish, aquatic birds and turtles, with most reported observations of hunting involving fish-catching." However, there are records where they attacked birds, including Striated Heron (*Butorides striata*, Linnaeus, 1758), Silver gulls (*Chroicocephalus novaehollandiae*, Stephens, 1826), Australian ravens (*Corvus coronoides*, Vigors & Horsfield, 1827) and Little Pied cormorants (*Microcarbo melanoleucos*, Vieillot, 1817) (Mo et al., 2017). Mo et al. (2017) further note that in their study, the prey remains of this species were all birds, rather than fish, despite perceptions of the White-bellied eagle as a fish eater (see also Rose, 1973; Marchant & Higgins, 1993; Leonard, 1995; Collins & Hines, 1999; Reid & Reid, 1999; Thompson, 1999; Olsen et al., 2006a, b, 2013; Debus, 2006, 2008; Corbett & Hertog, 2011, 2012; Harford, 2012; O'Donnell & Debus, 2012; Debus et al., 2014; Roderick, 2014; Stace, 2015; O'Brien & Lacey, 2016).

Inter-species kleptoparasitism is practiced against Whistling kites (*Haliastur sphenurus*, Vieillot, 1818) in Australia (Coates, 1985; Olsen, 1995) and Ospreys and Brahminy kites (*Haliastur indus*, Boddaert, 1783) in Malaysia (Wells, 1999). Madoc (1956) reported large crabs being dropped 30-40 m onto rocks to break their shells open. Olsen et al. (2006) in an Australian case study recorded prey taken as 6.0% mammals, 47.4% birds (cormorants, grebes, ducks), 8.6% reptiles, 36.0% fish, and 1.7% invertebrates. Another Australian study (Smith, 1985) found a prey breakdown of 52% fish, 44% birds, 4% reptiles (sea snakes), and no mammals. Eckert's (1971) Australian study also found a diet of 60% birds (shearwaters, Family *Procellariidae*, Leach, 1820), 20% fish and 20% mammals (bandicoot, Order *Peramelemorphia*, Ameghino 1889, and probable carrion sources).

Other studies have mostly reported similar prey groups, such as birds, fish and mammals (and in some cases carrion, crustaceans, squid, and sea turtle hatchlings, as well as kleptoparasitism) in Hong Kong, the Malay Peninsula, Sabah and New Guinea (Taylor, 1933; Chasen, 1939; Fleay, 1948; Gilliard & Lecroy, 1966; Rand & Gilliard, 1967; Marchant & Higgins, 1993; Debus, 1998; Wells, 1999; Beehler et al., 1986, 2001; Heinsohn, 2000; Sheldon et al., 2001; Hollands, 2003; Rasmussen & Anderton 2012; Dennis & Baxter, 2006; Gopi & Pandav, 2006; Mikula et al., 2016). Olsen et al. (2006, 193) compared the foraging habits of white-bellied sea eagles and wedge-tailed eagles in a shared, Australian habitat. The findings indicated that the two species shared territories but did not share prey taken. White-bellied sea eagles took fish, water-dwelling reptiles and waterbirds. Wedge-tailed eagles took rabbits, macropods, terrestrial birds and passerines. Common prey was only the maned duck and a few rabbits. There were very few interspecies interactions, as white-bellied sea eagles foraged near lake shores, and wedge-tailed eagles foraged away from the water bodies (Olsen et al., 2006, 193). Other studies have reported conflicts between these species, over nesting sites (Spencer & Lynch, 2005).

Breeding

The breeding season of the White-bellied Sea eagle begins with soaring, calling and paired birds chasing each other. These displays may include diving, gliding and chasing while calling, and mutual copying of flight patterns and movements, as well as interlocking, talon-grappling displays with one bird in an inverted position and plunging cartwheeling dives (Lindgren, 1972; Hollands, 2003). Nests are usually located in tall trees, less commonly on rocks. The nests are built of sticks and are lined with smaller vegetation, such as twigs, leaves, grass, and seaweed. It may be located in tall dead or live trees, on a sheltered rocky cliff, on the ground when nesting is on offshore islands) and in some cases on telegraph poles and communications pylons (Wells, 1999). When in trees, nests may be located in the forks of large trees near water (Marchant & Higgins, 1993). Generally, nests are over 30 m high (Debus, 1998; Wells, 1999; Kennedy et al., 2000). Larger nests may be three meters across and two meters deep, composed of sticks and green leaves (Brown & Amadon, 1968; Dennis et al., 2012). Usually, coarse textured 2 eggs are laid, but clutches of 3 eggs may also be laid (Debus, 1998; Wells, 1999; Kennedy et al., 2000).

Laying varies according to the region of the range: August to February (India and Sri Lanka); November to December (Myanmar); December to February (Malaysia); April to June (Java), December to January (Philippines); December to March (Borneo); May to June (New Guinea); May to July (Australia, Queensland); and June to August (Australia, Northwest Cape). Generally, the incubation period lasts for about 40 days (Debus,1998). The nestling period lasts for 65-70 days in Australia (Debus, 1998), or as long as 85 days in more northern ranges (Global Raptor Information Network, 2021bi). Carey et al. (2001) note that in Hong Kong, out of two nestlings, generally only one survives, as the stronger chick may kill the other. When the nesting is unsuccessful, the pair may attempt a second laying (Favaloro, 1944). In Australia, 34% of successful nests produced two young (Debus, 1998). Fledgling occurs after 65 to 70 days, and the period of dependence after fledging may last for two or three months, or even until the next breeding season (Brown & Amadon, 1968; Corbett & Hertog, 2011; Global Raptor Information Network, 2021bi).

The success of nesting may depend on the extent of human disturbance. For example, Dennis et al. (2011a, 179), in a Kangaroo Island (South Australia) case study, assessed breeding territories against measures of relative isolation from human disturbance, in Low-, Moderate- or High-disturbance categories. The findings were that "the level of disturbance was found to significantly affect fledging outcomes, with high-disturbance territories having significantly lower fledging success... Territories with high-disturbance levels produced eggs less often, fledged fewer young, and had higher rates of nesting failure." Suggested solutions including "site-specific habitat management prescriptions" such as "buffer-zone refuge provisions, are required to minimise the effects of human activity on breeding outcomes" (Dennis et al., 2011a, 179). As most South Australian nests are on open cliffs, the authors conclude that in such areas with little visual screening over long distance "refuge dimensions should be double those prescribed elsewhere for nests in tall forest habitat" (Dennis et al., 2011a, 179).

Population Status

The White-bellied Sea eagle ranges through India and Sri Lanka eastwards to Southern China, and Southeast Asia, Philippines, Wallacea, New Guinea, to Australia and Tasmania (Dennis et al., 2011b). Throughout much of this region it is an irruptive or local migrant, with juveniles dispersing widely from breeding areas (Bildstein, 2006). In Australia, juveniles may disperse up to 3,000 km from their natal territory about four months after fledging (Debus, 1998). Coastal adults are mostly sedentary, but some breeding adults may disperse during autumn and winter (Blakers et al., 1984). The population estimates vary widely; Ferguson-Lees and Christie (2001) estimated the global population (defined as the number of adults and immatures at the start of the breeding season) in the range of 10,001 to 100,000. BirdLife International (2016) estimated 670 to 6,700 mature individuals.

Dennis et al. (2011b, 1) used a South Australian case study and estimated the population of White-bellied Sea eagle as 70 to 80 pairs in study area. These findings compared with previous studies indicated a 21.7% decline in the White-bellied Sea eagle population. The authors note that most of the studied areas were on offshore islands. Threats to the breeding and population success of the eagles were listed as "rural land-use change leading to increased human activity in the coastal landscape and corresponding levels of disturbance at nest sites; poorly conceived tourism developments; ill-timed research projects and pest plant control programs coincident with the breeding season; and inter-species conflicts between White-bellied Sea eagles with Wedge-tailed eagles"; these threats were particularly severe as the White-bellied Sea eagle is described as "disturbance-intolerant" (Dennis et al., 2011b, 1). These results complemented earlier studies that showed that the White-bellied Sea eagle's population had declined in Australia since European colonization (Dennis & Lashmar, 1996). The declining population led to the elevation of the conservation status of the eagle to Endangered in South Australia and listing on the Marine and Migratory Species Schedules of the Federal Environment Protection and Biodiversity Conservation Act, 1999.

Over its range, BirdLife International (2016) rates the White-bellied Sea eagle as of "Least Concern." Global Raptor Information Network (2021bi) describes it as "widespread and generally fairly common or common in most parts of its range. Not as common now in Australia, however, where it is categorized as Vulnerable in Victoria and Tasmania, Endangered in South

Australia, rare in the Australian Capital Territory, and proposed as Vulnerable in New South Wales" (see also Debus, 2009). The declines in the population are listed as "human disturbance, shooting, poisoning, loss of suitable breeding sites caused by the clearance of waterside forests, and possibly pesticides, especially DDT, and other pollutants" (Global Raptor Information Network, 2021bi; see also Ferguson-Lees & Christie, 2001). Another factor for population losses is the use of eggs in traditional Chinese medicine (Herklots, 1953).

Debus (1994) points out that in some areas, there may be population recoveries, due to the introduction of exotic species and the creation of rubbish dumps. Regionally, in Hong Kong, there have been some increases in population over the period 1930s to 1990s, possibly due to the decline in nest robbing and disturbances by grass cutters (Carey et al., 2001). In Thailand, there is also hunting pressure (Wells, 1999). On the Malay Peninsula, specifically on the main Singapore Island, 10 nest sites were recorded during 1987-1992, of which five would be active in a given season (Wells, 1999). In 1992-93, two pairs nested on the Selangor mainland coast between Kapar and Sekinchan, these being artificial sites. Three natural sites active until the late 1980s, were abandoned due to tree losses (Wells, 1999). Wells (1999) notes possible negative impacts of increasing tourism and possible benefits from the use of artificial nests. In Australia, it is classified as generally common, but declines have been noted in southern Australia due to habitat losses and nest site disturbances (Debus, 1998). Recent benefits include the development of inland reservoirs with Common carp (*Cyprinus carpio*, Linnaeus, 1758) (Olsen, 1995), the cancelation of DDT use in Australia, and the introduction of some prey species and development of rubbish dumps. These developments, however, have not prevented some local population declines in parts of South Australia (Debus, 1998). For these reasons it is classified as Vulnerable in Victoria and Tasmania, and Endangered in South Australia (Debus, 2009). In Tasmania, it is listed as Vulnerable under Schedule 3.1 of the Tasmanian Threatened Species Protection Act 1995. Threats include nest disturbance, nesting habitat loss, shooting, poisoning, trapping, and powerline and wind turbine collisions and environmental pollution (Threatened Species Section, DPIW, 2006; Wiersma & Richardson, 2009).

In Asia, the White-bellied Sea eagle is a resident breeder from Mumbai and northwards, and eastwards to eastern India, and into Bangladesh and Sri Lanka, coastal Southeast Asia including Myanmar, Thailand, Malaysia, Indonesia, Vietnam, Laos, Cambodia, the Philippines, southern China through New Guinea, the Bismarck Archipelago, and Australia (Acharya, 1936; Coates, 1985; Kennedy, 2000; Strange, 2000; Ferguson-Lees & Christie, 2001; Clunie, 2003; Rasmussen & Anderton, 2012; Threatened Species Section, DPIW, 2006; Dennis et al., 2011b). There are also some reports from New Zealand (Marchant & Higgins, 1993). In the northern Solomon Islands, it occurs only on Nissan Island (Hadden, 2004) and on the other islands it is replaced by Sanford's Sea eagle (Ferguson-Lees & Christie, 2001; Coates, 1985). Ranges include coastal areas, more rarely inland (e.g. sightings in the Panna Tiger Reserve in central India, almost 1,000 km from the coast) usually where with minimal human presence (Coates, 1985; Grimmett et al., 1998. Kennedy, 2000; Naoroji, 2006; Sriram et al., 2008; Shurpali, 2008; Sadvi, 2009; Department of Sustainability, Environment, Water, Population and Communities, 2011).

The White-bellied Sea eagle is quite common in some areas such as Hong Kong, and some parts of India, such as Sindhudurg and in the Ratnagiri district of Maharashtra, and Netrani Island (for the latter, especially during the 19th century) (Hume, 1876; Katdare & Mone, 2003; Katdare et al., 2004). It is rarer or more affected by human disturbance in Thailand (Ferguson-Lees & Christie, 2001), Kangaroo Island in South Australia (Dennis et al., 2011a), the Eyre Peninsula in South Australia (Dennis, 2004), Andhra Pradesh in India (Gadgil, 2004; Madhyastha, 2004; Sekhar et al., 2004). The White-bellied Sea eagle is seriously affected by DDT, as it fed in areas treated with DDT, such as swamps (Olsen et al., 1993). Illustrations of the appearance, range, and eggs of this species are shown in Figures 4.7a to 4.7h.

4.9 Sanford's Sea Eagle *Haliaeetus sanfordi* (Mayr, 1935)

Physical Description

The Sanford's Sea eagle has been compared with the White-bellied Sea eagle. Brown and Amadon (1968, 282) note that this species is "a representation of the widespread *H. leucogaster*, in which the typical adult plumage is absent and is replaced by one resembling the immature plumage." It is therefore regarded as a separate species. The Sanford's Sea eagle reaches a length of 70 to 90 cm, with a wingspan of 165 to 185 cm and a weight of 1.1–2.7 kg (Del Hoyo et al., 1994). The adult plumage is blackish to brown on the back, wings and tail, lacking brown on the primaries and tail feathers. The underparts are reddish brown (Debus et al., 2020). The head is lighter colored, appearing golden. The eyes are brown, the bill dark blackish and the legs yellowish (Brown and Amadon, 1968, 282). The immature plumage is brown, spotted greyish white above and streaked with a similar color below. The tail is long and narrow. These authors note that this species appears closer to the true eagles than the sea eagles, and in fact it frequently feeds on birds and phalangers (Genus *Phalanger*, Storr, 1780) rather than fish. The voice is high pitched, sounding as "eh – eh – eh" (Brown & Amadon, 1968, 282). Illustrations of the appearance, range, and eggs of this species are shown in Figures 4.8a to 4.8f.

Fig. 4.7a. White-bellied Sea eagle [With permission from Pieter Verheij]

Fig. 4.7b. White-bellied Sea eagle [With permission from Mike Nelson-Birdtour ASIA]

Fig. 4.7c. Immature White-bellied Sea eagle [With permission from Credit Utopia 88]

Fig. 4.7d. White-bellied Sea eagle [With permission from Mike Nelson-Birdtour ASIA]

Fig. 4.7e. White-bellied Sea eagle [With permission from Frank McClintock]

Fig. 4.7f. Immature White-bellied Sea eagle [With permission from Mike Nelson - Birdtour ASIA]

Fig. 4.7g. White-bellied Sea eagle habitat [With permission from Mike Nelson-Birdtour ASIA]

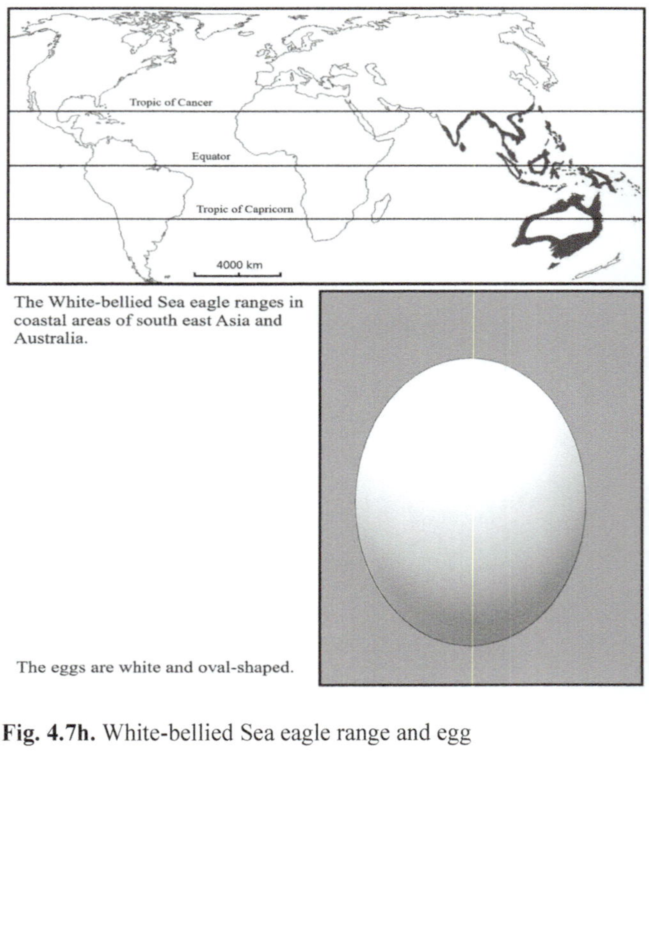

Fig. 4.7h. White-bellied Sea eagle range and egg

Classification

As a closely related species to the previous species, the White-bellied Sea eagle, all the classification issues presented for the latter species are valid for the Sanford's Sea eagle (Global Raptor Information Network, 2021bj). These include the bases of the classification; the morphological analysis by Zimbelmann (1992), the allozyme analysis by Schreiber and Weitzel (1995), and the molecular phylogenetic analyses by Wink et al. (1996), Seibold and Helbig (1996), and Lerner and Mindell (2005), which linked the Genus *Haliaeetus* to the kites of the genera *Milvus* and *Haliastur*, the distinction of the southern species (Sanford's Sea eagle, White-bellied Sea eagle, African Fish eagle and Madagascar Fish eagle), from the northern species of *Haliaeetus*.

As hinted by some researchers, there is speculation that Sanford's Sea eagle was derived from the White-bellied Sea eagle very recently, concluded by the small genetic distance (0.3%) between the two species (Wink & Sauer-Gürth, 2004; Helbig et al., 2005). Although this small genetic distance is about the grade for intra-species differences, Wink et al. (1996) put the Sanford's Sea eagle and the White-bellied Sea eagle as two full species, based on their size morphological and behavioral differences. Some authors dispute this, and still argue for the merging of the two species into one (Debus, 2006).

Foraging

The Sanford's Sea eagle mostly frequents forest and forested woodland, coastal areas, and sea shores, although it soars widely over most terrain and landcover types, usually or in pairs (Cain & Galbraith, 1956; Coates, 1985; Doughty et al., 1999; Dutson, 2011; Pikacha et al., 2012). In some areas, it is commoner along the shorelines with coconut trees than in the forested habitats (Blaber, 1990; Pikacha et al., 2012) and may be recorded in inland areas (Diamond, 1975). One old report by Mayr (1936) recorded eagles on Choiseul Bay along the coast and the dense lowland forest, and inland slopes up to 610 m. On San Cristobal and Malaita, eagles were recorded on slopes up to 1,220 m. Some records dispute the feeding habits of this species: for example, Mayr (1936) claimed that this species, like its sister species the White-bellied eagle, feeds mostly on phalangers and large birds such as pigeons, rather than the dominant fish and carrion diet of most sea eagles (Olsen et al., 2006). Some of these points are supported by Coates (1985), who reports feeding on pigeons (Genus *Columba*, Linnaeus, 1758), flying foxes (Genus *Pteropus*, Brisson, 1762), and phalangers (possums) (Superfamily *Phalangeroidea*, Thomas, 1888) (Heinsohn, 2000; Hamilton, 2004; Mikula et al., 2016). However, Debus (1994) states that the feeding and foraging of the Sanford's Sea eagle are like those of other *Haliaeetus* species, mostly fish. One bird was reported to be feeding on a shark on Fera Islands, Isabel Province, Solomon Islands. Sibley (1951) also reported two birds foraging by wading in shallow water in New Georgia.

Breeding

There is little information on the breeding patterns of the Sanford's Sea eagle (Cain & Galbraith, 1956; Petersson et al., 2016; Global Raptor Information Network, 2021bj). Ferguson-Lees and Christie (2001) report courtship in June and August, and a large stick nest placed in trees, these sometimes may be in located in mangroves. In Kolombangara island, New Georgia, one report estimates a minimum territory size of 10 km^2 in comparatively pristine coastal habitat (Buckingham et al., 1995). Near Makira or San Cristobal Island, on the Olu Malau Islands (or Three Sisters Islands), there were four pairs on a land mass of less than 20 km^2, in the 1950s (French, 1957; Petersson et al., 2016).

Population Status

The Sanford's Sea eagle has an Australasian distribution, being endemic to Papua New Guinea and the Solomon Islands (Webb, 1992; Olsen, 1994; Van Oosten & Wyant, 1999; Read, 2013). Olsen (1997, 81) describes the Solomon Island succinctly: "The Solomon Islands are a scattered double chain of 992 islands, atolls and cays just south of the Equator, extending 1646 km in a south-easterly direction from Bougainville Island in Papua New Guinea. The six major islands, Guadalcanal, New Georgia, Malaita, Santa Isabel, Choiseul and Makira, vary in length from 87 to 171 km and in width from 25 to 50 km." This eagle is a non-migratory (Bildstein, 2006). Ferguson-Lees and Christie (2001) estimated the population of adults and immatures at the beginning of the breeding season from 101 to 1,000 individuals and suggested that the population is more likely to be in the middle to upper hundreds than in thousands. BirdLife International (2016) estimates mature eagles at 250 to 999 individuals.

The population is declining, as it is affected by deforestation, and persecution (Katovai et al., 2015). Olsen (1997, 82) describes persecution: "This eagle is the largest avian predator in the islands and unlike Brahminy Kites and Ospreys, it was killed by locals because it was said to catch dogs, cats, poultry and other animals valued by these villagers. Australian Catholic priests at Vanga Point apparently shot the local eagles as part of their teachings about raising poultry and other stock. In pre-Christian times Sanford's Sea-Eagle was worshipped as a totem and therefore, not killed. This changed as values and beliefs changed" (see also Hadden, 2004). Global Raptor Information Network (2021bj) lists the principal threats as deforestation and other habitat decline or modification, overfishing, pollution mostly from logging, and in a few cases killing for food, sport and/or protection for poultry and domestic animals, such as cats and dog and perhaps competition with human fishers (Olsen, 1997; Ferguson-Lees & Christie, 2001). It is classified as Vulnerable by BirdLife International (2016) but adds the following qualification: "This species is classified as Vulnerable on the basis of a small, estimated population which is suspected to be declining. It is judged to have distinct subpopulations, some totalling more than 250 birds. If subpopulations are found to be smaller or, conversely, that movement of birds between them means that all birds are in a single population, the species would warrant uplisting to Endangered." The ecology and status of the Sanford's Sea eagle has been debated and described by several researchers (Cain & Galbraith, 1956; French, 1957; Blaber, 1990; Buckingham et al., 1995; Petersson et al., 2016). Illustrations of the appearance, range, and eggs of this species are shown in Figures 4.8a to 4.8f.

Genus *Haliaeetus* (Savigny, 1809) or Genus *Ichthyophaga*

Brown and Amadon (1968) note that the eagles of the Genus *Ichthyophaga* or Fishing eagles are related to the Sea eagles of the Genus *Haliaeetus,* but the former eagles may rely to a greater extent on a fish diet acquired by their own efforts. Lately they have been classified as Sea eagles of the Genus *Haliaeetus*. There are two species, one is the Lesser fish eagle (*Haliaeetus humilis*, Müller & Schlegel, 1841, formerly *Ichthyophaga nana* or *Ichthyophaga humilis*, Müller & Schlegel, 1841). This species is divided into two subspecies: *Haliaeetus humilis humilis*, and *Haliaeetus humilis plumbea* (the latter sometimes named the Himalayan Grey-headed fishing eagle, which may be named *Ichthyophaga humilis humili*s (S. Müller & Schlegel, 1841) and *Ichthyophaga humilis plumbea* (Jerdon, 1871), or *Ichthyophaga nana nana* (Blyth, 1842) and *Ichthyophaga nana plumbea* (Jerdon, 1871), (Clark et al., 2018; Global Raptor Information Network, 2021bj). The other species is the Grey-headed Fish eagle (*Haliaeetus ichthyaetus*, Horsfield, 1821) also named *Ichthyophaga ichthyaetus* (Horsfield, 1821) (Gill & Wright, 2006; GRIN, 2018; ITIS Report, 2018).

These two species are smaller than the other Sea eagles and are ranked as small to medium sized eagles. Their anatomy is well-adapted to fish predation, having broad wings for lift, unfeathered legs with spicules under the toes for grasping slippery fish, and long, slender, curved talons. Generally, the adult plumage is white and brown, while that of the juveniles is streaked and mottled, with paler coloration. Primary habitats are riverine areas, rather than coasts (Brown & Amadon, 1968). This may distinguish them from the larger Sea eagles such as the Bald, White-tailed and Steller's Sea eagles, which are noted to frequent coastal areas, as well as rivers and lakes (Ferguson-Lees & Christie, 2001).

Fig. 4.8a. Sanford's Sea eagle perching [With permission from Keith Cowton]

Fig. 4.8b. Sanford's Sea eagle in flight [Detail Commercial Licence from PicFair]

Fig. 4.8c. Sanford's Sea eagle in flight [Commercial Licence from PicFair]

Fig. 4.8d. Sanford's Sea eagle in flight [With permission from Nigel Voaden]

Fig. 4.8e. Sanford's Sea eagle habitat [With permission from Jenny Scott]

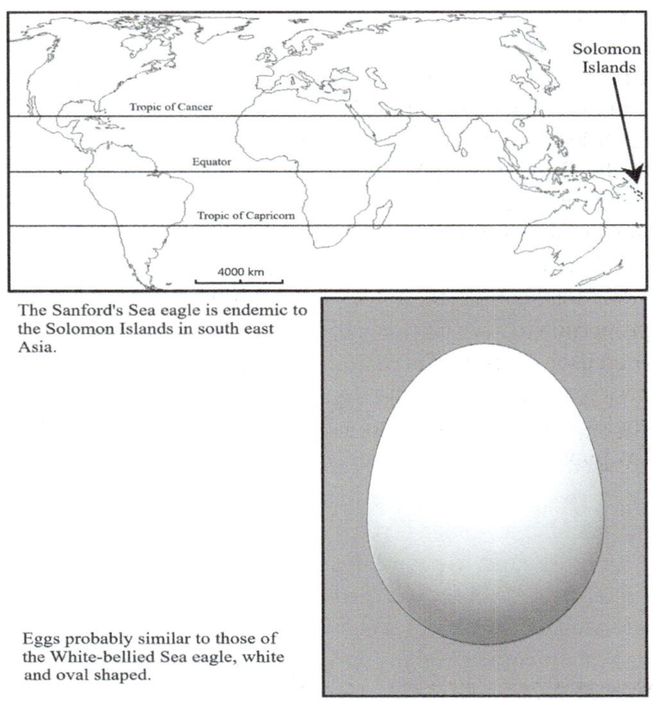

Fig. 4.8f. Sanford's Sea eagle range and egg

4.10 Lesser Fish Eagle *Haliaeetus humilis* (Müller & Schlegel, 1841) or *Icthyophaga humilis* (Müller & Schlegel, 1841)

Physical Description

The Lesser Fish eagle has two subspecies *H.h. humilis* and *H.h. plumbea* similar in plumage, despite the marginally greater size of *H.h. plumbea.* This species reaches a length of 51 to 64 cm in length and a wingspan of 120 to 123 cm and a weight of 780 to 785 g (Ferguson-Lees & Christie, 2001; Planet of Birds, 2013; Del Hoyo et al., 2014). The head and neck feathers are brownish grey, with black feather shafts, while the back, upper wing and tail coverts are brown, with darker feathers with paler tips in the tail center and a black subterminal tail band, which is mottled white and brown with a darker end underneath. The lateral tail quills are mottled brown, with white coloring for the basal tail section (Brown & Amadon, 1968). The dark brown primaries have white mottling on the "inner webs basally" (Brown & Amadon, 1968, 299). The white underparts are colored brown on the breast and underwing coverts. The wing quills contrast as dark grey brown, combined with white mottling on the inner webs on the outer primaries. The golden eyes have a brown cere. The feet are white, with a blend to yellowish grey (Clark et al., 2020h). Both the Lesser Fish eagle and the Grey-headed Fish eagle have strongly recurved talons similar to the Osprey's claws, a specialism for taking fish, which the *Haliaeetus* eagle lack (Poole, 1989).

The immature plumage is overall paler than that of the adult. It is brown above with visible white feather bases, mostly strongly on the rump. Both the primaries and the secondaries are dark brown, with the former a darker shade. The throat is brown, and the breast is darker with lighter streaks. The belly, thighs, undertail and underwing coverts are white, with brown mottling on the underwing coverts. The wing quills are brown above, and dark grey-brown below, with white mottling on the inner webs. The dark greyish-brown tail is lighter below basally, with a brown tip (Brown & Amadon, 1968; Ferguson-Lees & Christie, 2001).

This species may be confused with the Grey-headed Fish eagle, despite its significantly smaller size and brown tail. eBird (2021) distinguishes between the Lesser Fish eagle and the Grey-headed Fish eagle: "Most easily separated from larger Gray-headed Fish-Eagle in flight: note Lesser's "dirtier" tail with a less contrasting black band and more mottled underwings in adult; note overall paler underside patterning in juvenile. White belly also tends to have a more poorly-defined and jagged shape than the sharp "V" of Gray-headed." The immature of the current species also has less white streaking, fewer white feathers above, darker underwing and less white mottling at the tail base. This is a generally silent bird, but it has a cry 'hak – hak' (Brown & Amadon, 1968). Illustrations of the appearance, range, and eggs of this species are shown in Figures 4.9a to 4.9h.

Classification

The Lesser Fish eagle was formerly named *Ilcthyophaga nana* (Mees, 1967; Brown & Amadon, 1968). This species and genus are closely linked to the sea eagles (*Haliaeetus*) in general, and more closely linked to the southern species of *Haliaeetus* (Lerner and Mindell, 2005). Some taxonomic authorities place this species in the monotypic genus *Ichthyophaga*. Others place it in the genus *Haliaeetus* (del Hoyo et al., 2014; Clark et al., 2020h; Global Raptor Information Network, 2021bk). BirdLife (2016) notes that "the genus *Icthyophaga* (del Hoyo and Collar, 2014) was previously spelled *Ichthyophaga*."

Foraging

BirdLife International (2016) writes that the Lesser Fish eagle "depends on forest-fringed waterbodies with good stocks of fish. It is thought to be undergoing a moderately rapid population reduction owing to forest degradation, over-fishing and perhaps especially, pollution, and may already have a moderately small population. It is consequently classified as Near Threatened." Brown and Amadon (1968) describe this species as favoring low latitude forests (<750m, <4200 m in Nepal) with streams and sometimes open country (Brown & Amadon, 1968). Its range only rarely overlaps with that of the Grey-headed Fish eagle, but the latter species occurs more frequently near lakes and coasts, rather than wet inland forests. Hunting methods are principally based on perching and inter-perch flights, and attacks from perches rather than soaring.

The habitat types of the Lesser Fish eagle have been distinguished from those of the closely related Grey-headed Fish eagle, in that the Lesser Fish Eagle is more likely to frequent hill rivers, while the Grey-headed Fish Eagle inhabits standing and sluggish wetlands, although the two species may have substantial habitat overlaps (Fuchs et al., 2007). Common foraging habitats include perching in riverine woodland, especially on dead tree branches and on branches under living green canopies over shaded water. Other foraging includes soaring, either singly or paired (Coates & Bishop, 1997). Case studies show slight variants in habits and foraging. For example, in Malaysia, it frequents "clear-flowing rivers and broad streams cutting through tall, plains-level lowland forest, which provides perches overhanging water" (Global Raptor Information Network, 2021bk; see also Wells, 1999). In Sabah, the main habitats are "along rivers, swamps, mangroves, and in logged forests and oil palm plantations, but mainly along forest-lined rivers of the interior" (Global Raptor Information Network, 2021bk; see also Sheldon et al., 2001).

In some cases, they frequent higher elevations, the majority below 1000 meters, but some up to over 4,000 in Nepal (BirdLife International, 2020). Prey taken is mainly fish, usually taken by a pounce from a perch on overhanging branches, or from rocky mid-stream areas (Coates & Bishop, 1997; Wells, 1999; Clark et al., 2020h; Global Raptor Information Network, 2021bk).

Breeding

Brown and Amadon (1968) argue that there are no known courtship displays. Breeding occurs from November to February, but the breeding season generally lasts from March to August, especially for the birds in Northern India and Nepal, and from November and end in April in other areas. The large nests, usually smaller than the nest of the Grey-headed Fish eagle, are usually located in trees near water. The nests expand in time; recently constructed nests may be about 1 meter wide and 30 centimeters deep, eventually reaching about 120 centimeters across and 150 centimeters deep. Composing materials include sticks and green leaves. The clutch of 2 to 3, occasionally 4 eggs is white (Brown & Amadon, 1968; Clark et al., 2020h).

Population Status

The Lesser Fish eagle occurs mostly in the Indian subcontinent, especially near the foothills of the Himalayas, and south-east Asia. There are records from Gujarat, Central India and in more recent times from the Kaveri river valley in southern India (Lethaby, 2005; Naoroji, 2006). *Haliaeetus humilis humilis* occurs in the Malaysian Peninsula, Sumatra, Borneo, and Sulawesi; *Haliaeetus humilis plumbeus* occurs from Kashmir through southeast India, Nepal, and Burma towards Indochina (Global Raptor Information Network, 2021bk). The Global Raptor Information Network (2021bk) lists its distribution as including India, Kashmir and Nepal, eastwards towards southern China and Southeast Asia to the Malay Peninsula, Sumatra, Borneo, Sulawesi, and Southern Moluccas (Buru). BirdLife International (2016) lists its range as including India (limited to the Himalayan foothills and north-east, and a small population in Karnataka state in the south), Myanmar and Bangladesh, with a rare to very rare status in the following countries: Nepal, Bhutan, China (especially Hainan), Thailand, Cambodia, Laos, Peninsular Malaysia and east Malaysia, Brunei, Indonesia and the Sula islands and Buru (BirdLife International, 2016). This species is a partial migrant and altitudinal migrant (Bildstein, 2006). BirdLife International (2016) estimates the number of mature individuals at 10,000-50,000. Ferguson-Lees and Christie (2001) estimated the global population at the start of the breeding season at 1,000 to 10,000 individuals but cast doubt on a figure above 1,000.

BirdLife International (2020) lists the threats against this species as "loss of forest habitat along rivers, siltation, over-fishing and increasing human disturbance of waterways are causing widespread declines. It is also declining in Uttar Pradesh, India, partly because of pesticide use and this is presumably relevant throughout much of its range" (see also Round, 1988; Clark, 1994; Fuchs et al., 2007; Global Raptor Information Network, 2021bk). Like the closely related Haliaeetus species, there is a possibility of vulnerability to DDE-induced eggshell thinning, (Naoroji, 1997, 2006). The Lesser Fish eagle was declining in Uttar Pradesh, India partly due to pesticide use (Naoroji, 1997, 2006), but as the population is decreasing (BirdLife International, 2016), more recent research may be needed to ascertain this for the last decade. It is "Nationally critically endangered, following a prolonged decline" (Global Raptor Information Network, 2021bk; see Baral & Inskipp, 2004). In Laos, it is classified as a "Species at Risk" (Fuchs et al., 2007). A case study of Phongsaly Province, Laos, along the Nam Ou River, the Nam Khang River, the Nam Houn River, all in the Phou Dendin National Protected Area, found this area to be possibly the densest population in Laos, despite a perception of low numbers of immatures and possibly low breeding levels (Fuchs et al., 2007). Fuchs et al. (2007) (see also Thewlis et al., 1998) note that the Fish eagle is "At Risk" in Laos, and document the problems of the Fish eagles in Laos:

> Fish eagles are nationally threatened because they need forested large watercourses, among the most threatened habitats in Laos...Large rivers are fishing sites and access routes and seasonal camps for rural Lao, and so their birds are more in contact with people than are interior forest species. Fish eagles are large, easily seen and hence tempting targets: they perch for long periods in riverside trees, rather than soaring high, out of gun range.

Other problems include the lack of populations in remote areas for the repopulation of the more accessible areas where they are rare. Their nesting locations may also be exposed, increasing the likelihood of robbery. These contributed to the situation in Laos, where the population of the Lesser Fish eagle "is now reduced to a few isolated populations in the least-disturbed drainage basins (XeKong, Nam Theun, Nam Ou and perhaps others) between which dispersal may be limited. Each is vulnerable by nature of its small size" (Thewlis et al., 1998, 77).

In Myanmar, the population is concentrated in the Hukaung Valley in north Myanmar (Fuchs et al., 2007). In Thailand, the species may be rarer, as Robson (2008) documents two adult eagles. In the Malay Peninsula, the population has seriously declined (Wells, 1999), with historical records describing much larger populations (Chasen, 1939). Wells (1999) listed factors for such declines, including river siltation, which makes water too murky for fishing, tourist disturbance in national parks. In Sabah, the impact of human presence is disputed: some authors hold that human disturbance upsets the eagles' ecology (Thiollay, 1983), while others cite observations in heavily logged and burned areas (Sheldon et al., 2001). Illustrations of the appearance, range, and eggs of this species are shown in Figures 4.9a to 4.9h.

Fig. 4.9a. Lesser Fish eagle [With permission from Mike Nelson-Birdtour ASIA]

Fig. 4.9b. Lesser Fish eagle [With permission from Praveen Pandian]

Fig. 4.9c. Lesser Fish eagle [With permission from of Paul Prior]

Fig. 4.9d. Immature Lesser Fish eagle [With permission from Ken Behrens]

Fig. 4.9e. Immature Lesser Fish eagle [With permission from Gabriel Leboff]

Fig. 4.9f. Lesser Fish eagle [With permission from Andy Walker]

Fig. 4.9g. Lesser Fish eagle habitat, Malaysia [Credit: Steve Douglas – unsplash]

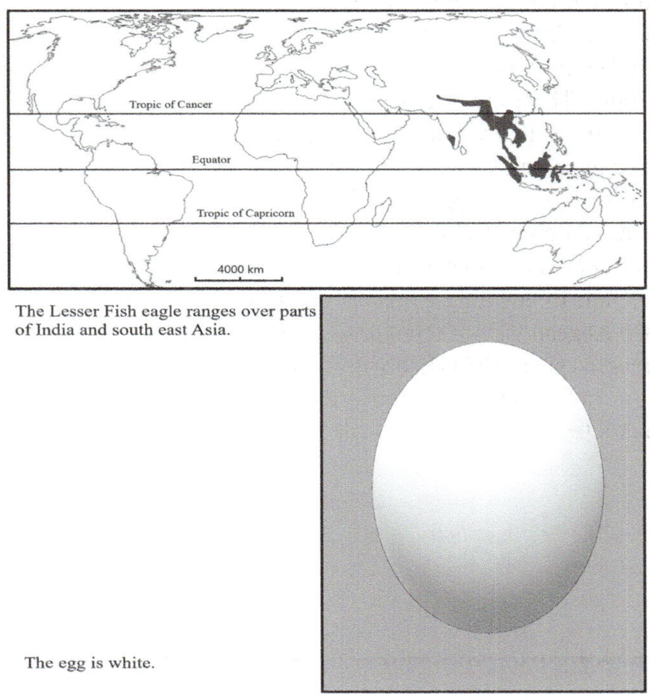

Fig. 4.9h. Lesser Fish eagle range and egg

4.11 Grey-headed Fish Eagle *Haliaeetus ichthyaetus* (Horsfield, 1821) or *Ichthyophaga ichthyaetus*

Physical Description

The Grey-headed Fish eagle is a small-medium eagle. The body length is 61–75 cm, with a wingspan of 155 to 170 cm and a weight of 1.6. to 2.7 kg (Thiollay, 1994; Ferguson-Lees & Christie, 2001). The grey-brown head and neck of the adult contrasts with the darker brown upperparts, which include darker wing and tail coverts and paler upper back feathers, with yet paler feathers on the breast. The underparts below the breast is white, the underwing coverts are dark grey-brown, the upper wing quills almost black and the underwing quills dark brown. The white tail has a dark brown subterminal band, with a paler tip. The yellow eye has a dark grey cere and the feet are light dirty yellow (Brown & Amadon, 1968). The immature plumage includes a brown head and neck, with paler streaks, dark brown back with light feather edges, brown breast streaked with white, and white, reddish brown streaked thighs. The brown wing quills have white basal parts, especially on the secondaries. The underwing coverts are brown with white mottling, The grey brown tail ends in a broad, dark brown band (Brown & Amadon, 1968). The tail has also been described as black and white with a broad dark subterminal band and white tip (Rasmussen & Anderton, 2012). Brown and Amadon (1968, 302) note that specimens from Southern India and Sri Lanka are marginally smaller, hence some authors place them in a separate subspecies (Global Raptor Information Network, 2021bl), although the current authors do not recognize this classification (Clark, 1994).

The only similar species is the Lesser Fish eagle. The Grey-headed Fish eagle may be distinguished by the white tail base and much larger size. Del Hoyo et al. (1994) note the Lesser Fish eagle has a brown tail, while the Grey-headed Fish eagle has a white tail with a black terminal band, and the former species is about half of the weight of the latter. The immature birds may be distinguished from those of the Lesser Fish eagle, by the former species' white streaked head neck and breast, darker upperparts, more basal white in the wing quills, and whitish tail base. eBird (2021) distinguishes between the Grey-headed Fish eagle and the Lesser Fish eagle: "The smaller Lesser Fish-Eagle has a less contrasting tail pattern and typically a more jagged-edged white belly patch in adults; juveniles average paler below overall." The immature Grey-headed Fish eagle may resemble the immature, White-bellied Sea eagle (*Haliaeetus leucogaster*, Gmelin, 1788), except for the square tail, which contrasts with the more wedge-shaped tail of the latter. The voice has been defined as a "loud gurgling shout" *awh – awhr* or a similar laugh to a barbill *Tockus griseus*" or a cry *cheewarr* repeated 5 to 6 times (Brown & Amadon, 1968, 302) (note barbill *Tockus griseus* is currently named *Ocyceros griseus*, Latham, 1790). It has also been described as awh-awhr and chee-warr repeated several times, or ooo-wok, ooo-wok, ooo-wok, or uh-wuk and a louder high scream, which may be emitted with ascending pitch while in flight (Ferguson-Lees & Christie, 2001; Rasmussen & Anderton, 2012). Vocalizations increase during the breeding season, sometimes even at night (Ferguson-Lees & Christie, 2001). Illustrations of the appearance, range, and eggs of this species are shown in Figures 4.10a to 4.10k.

Classification

The Grey-headed Fish eagle is monotypic. The only classification controversy concerns the occasional classification of the Sri Lankan as a separate subspecies *Ichthyophaga ichthyaetus plumbeiceps*, but it was not recognized by many authors (Clark, 1994). The Grey-headed Fish eagle is closely related to the sea eagles, *Haliaeetus*, especially with the southern species (Lerner & Mindell, 2005; Global Raptor Information Network, 2021bl). Lerner & Mindell placed the Grey-headed fish eagle in the Genus *Haliaeetus* (sea eagles), but others place it in the Genus *Ichthyophaga.*

Foraging

The Grey-headed Fish eagle frequents freshwater lakes, marshes, swamps, forest reservoirs, lagoons, forested slow-moving rivers, lowland forest, generally below 1,500 m above sea-level (Brown & Amadon, 1968; Coates & Bishop, 1997; Grimmett et al., 1998; Kennedy et al., 2000; Rasmussen & Anderton, 2012). Common foraging grounds are near water; including streams, lagoons, lakes, estuaries and coasts, usually in low altitudes below 300 meters and frequents irrigation tanks in Sri Lanka, (Ferguson-Lees & Christie, 2001; Rasmussen & Anderton, 2012; Tingay et al., 2006; Tingay et al., 2012. Visal & Mahood, 2015). It rarely leaves the close proximity of water, but unlike the Lesser Fishing eagle it may avoid smaller forest streams. Described as sluggish, most hunting is from a perch but may also be from soaring flight (Coates & Bishop, 1997). Brown & Amadon (1968) argue that its fishing skills are rated below those of the osprey or White-bellied Sea eagle. The main diet is fish, including carrion (Coates & Bishop, 1997, Kennedy et al., 2000; Ferguson-Lees & Christie, 2001; Tingay et al., 2012. Visal & Mahood, 2015). Like the Lesser Fish eagle, the talons are strongly recurved, a specialization for fish capture (Poole, 1989). One survey in Cambodia revealed a diet of water snakes (Tingay et al., 2006, 2010).

Breeding

The breeding season begins with calls. The breeding season is usually between November and May in most areas but may be from December to March in Sri Lanka and November to January in India (Ferguson-Lees & Christie, 2001). Thiollay (1994) records nests found in January to March in Burma, April in Sumatra and August in Borneo, but their breeding status was uncertain. Tingay et al. (2006) describe a case study in the PrekToal protected area of the Tonlé Sap (a water body in Cambodia), where nesting follows the flood regimes that start in September, and the eggs hatching at peak flood waters in October–November. The nest is large, over 1 meter wide and about 2 meters deep. Global Raptor Information Network (2021bl) describes this as "a massive stick nest lined with leafy twigs placed in a tree adjacent to water." Their nests are usually located near water, such as slow-moving rivers and streams, lakes, lagoons, reservoirs, marshes, swamps and coastal lagoons and estuaries (Ferguson-Lees & Christie, 2001; Tingay et al., 2006, 2010). One in Malaysia was 30 m high in the main-crown fork of an emergent tree, and another was 23 m high in a stand of dead paperbarks (*Melaleuca cajiputi*) in open swampland" (see also Wells, 1999).

The clutch is usually 2, or even 3 to 4, white eggs, "stained coarser and rougher than *Haliaeetus leucogaster*" (Brown & Amadon, 1968, 302). Common breeding times are November to June in India (occasionally October to March), November to March or December in Sri Lanka and August in Borneo (Brown & Amadon, 1968). These authors note that there are no records on fledgling. Builds a massive stick nest lined with leafy twigs placed in a tree adjacent to water. One in Malaysia was 30 m high in the main-crown fork of an emergent tree, and another was 23 m high in a stand of dead paperbarks (*Melaleuca cajiputi*) in open swampland (Wells, 1999). Clutch size is two eggs, which are white and unmarked. Incubation, foraging and fledgling feeding are shared by the male and female, incubation lasts 45 to 50 days and the fledgling period is 70 days (Ferguson-Lees & Christie, 2001; Tingay et al., 2010).

Population Status

The Grey-headed Fish eagle ranges across India and Sri Lanka, and South-East Asia through Malaysia, Western Indonesia, Greater Sundas (Java, Sumatra, Borneo, and Sulawesi), and Philippines. Over this wide range it is generally uncommon, listed as rare or uncommon in (going eastwards and southwards) Rajasthan, Uttar Pradesh, Bihar and Assam in Northern and Eastern India, North and East Sri Lanka, Nepal, Bangladesh, South Thailand, Laos, Vietnam, Cambodia and Malaysia (including Sabah) to Sumatra, Java and Sulawesi, Borneo and the Philippines (Thiollay, 1998; Wells, 1999; Ferguson-Lees & Christie, 2001; Sheldon et al., 2001; Thiollay & Rahman, 2002; Rasmussen & Anderton, 2012; Tingay et al., 2006, 2010, 2012; Choudhury, 2012; BirdLife International, 2017). In this region, it is non-migratory, although juveniles may disperse from the breeding areas (Bildstein, 2006).

Ferguson-Lees and Christie (2001) estimated the global population, including adults and immatures at the beginning of the breeding season as between 1,000 and 100,000 individuals. BirdLife International (2017) estimated the number of mature birds as 10,000-100,000. This source notes that "This species' global population is preliminarily estimated at 10,000-100,000 mature individuals on the basis that it may not exceed a five-figure total. This estimate equates to 15,000-150,000 individuals in total…" and for the trend justification there is a "moderately rapid and on-going population decline suspected on the basis of rates of habitat degradation and levels of pollution and over-fishing." Despite its widespread distribution, it is only locally common, and hence the population is relatively small and declining. For these reasons, it is categorized as "Near Threatened" (BirdLife International, 2017).

The Global Raptor Information Network (2021bl) admits that the Grey-headed Fish eagle is "a poorly studied species thought to be in recent decline in many portions of its range, possibly as the result of habitat loss (deforestation and loss of wetlands), overfishing, siltation, human disturbance, and pesticide contamination", citing Ferguson-Lees & Christie (2001) (see also Tingay et al., 2012; Visal & Mahood, 2015). Global Raptor Information Network (2021bl) nevertheless notes that some such statements are based "mostly on anecdotal evidence" (Tingay et al., 2006, 2010). These authors found a negative correlation between human presence and eagle nest site occupancy levels in Cambodia.

Thewlis et al. (1998, 77) notes that the Grey-headed Fish eagle is rare in Laos, and cites a few sightings, including one bird sighted in 1990, singles birds in 1994 and 1995, 3 birds sighted in 1992, 2 in 1993 and 2 in 1995. This study also compares the Grey-headed Fish eagle with the Lesser Fish eagle, cites the former as possibly less common, with historically low populations across its wide range (see Oustalet, 1899-1903; Delacour & Jabouille, 1931, Engelbach, 1932). However, the Thewlis et al. (1998, 77) take issue with some of the historical records of these two species: "The historical records of the two fish-eagles need cautious interpretation… Recent records come from locations and habitats similar to those where Lesser Fish-eagle was recorded." Citing Parr et al. (1996) it is noted that "Lao numbers are tiny: at Tonle Sap lake in Cambodia over 100 pairs were estimated in 1996… and the Lesser Fish-eagle has yet to be recorded from Cambodia."

Several case studies have illustrated the status of this species in southern Asia. In Cambodia, Tingay et al. (2006, 2010) found a negative correlation between human presence and eagle nest site occupancy levels. In a study of the Tonle Sap Lake in Cambodia Tingay et al. (2006) recorded 32 pairs (and 2 outside the lake area) in about 80 km^2, an area representing only 2.2% of the swamp forest surrounding Tonle Sap Lake. The findings indicated that the population around the lake is numerically important, especially due to significant declines elsewhere in nearby Thailand, Laos, and Vietnam (Poole, 2001). However, this population may be threatened by hydroelectric dam developments (which may affect flooding regimes of the Tonlé Sap) on the Mekong River, and other projects in Cambodia, Laos and Thailand (Tingay et al., 2010), and the Chinese Mekong (Poole, 2005), with few studies on the impacts of the developments on the Tonle Sap region. Water snake hunting by local people may also impact on the foraging of the Grey-headed Fish eagles (Tingay et al., 2006, 2010). In Peninsular Malaysia, there is major habitat loss over the past decades (Gibson-Hill, 1950). The population in this region was more recently estimated at only 40 pairs (Wells, 1999) with one evolving problem being tourism traffic on reservoirs. Conservation actions have been undertaken to protect the Grey-headed Fish eagle. There have been nesting surveys since 2005 at PrekToal Ramsar Site (Visal & Mahood, 2015). Suggestions include more regular monitoring of breeding areas throughout its range, protection of habitat forests, and awareness campaigns involving local residents' conservation of wetland habitats (BirdLife International, 2017). Illustrations of the appearance, range, and eggs of this species are shown in Figures 4.10a to 4.10k.

Fig. 4.10a. Grey-headed Fish eagle perching [With permission from Vikas Chavan]

Fig. 4.10b. Grey-headed Fish eagle perching [With permission from Gil Ewing]

Fig. 4.10c. Grey-headed Fish eagle perching [With permission from Mike Nelson-Birdtour ASIA]

Fig. 4.10d. Immature Grey-headed fish eagle [With permission from Eddy Lee]

Fig. 4.10e. Immature Grey-headed fish eagle [With permission from Frank McClintock]

Fig. 4.10f. Grey-headed Fish eagle in flight [With permission from Mike Nelson-Birdtour ASIA]

Fig. 4.10g. Grey-headed Fish eagle in flight [With permission from StanleyYOW]

Fig. 4.10h. Grey-headed Fish eagle in flight [With permission from StanleyYOW]

Fig. 4.10i. Immature Grey-headed Fish eagle in flight [With permission from Frank McClintock]

Fig. 4.10j. Grey-headed Fish eagle habitat [Credit: Karma Talukdar – unsplash]

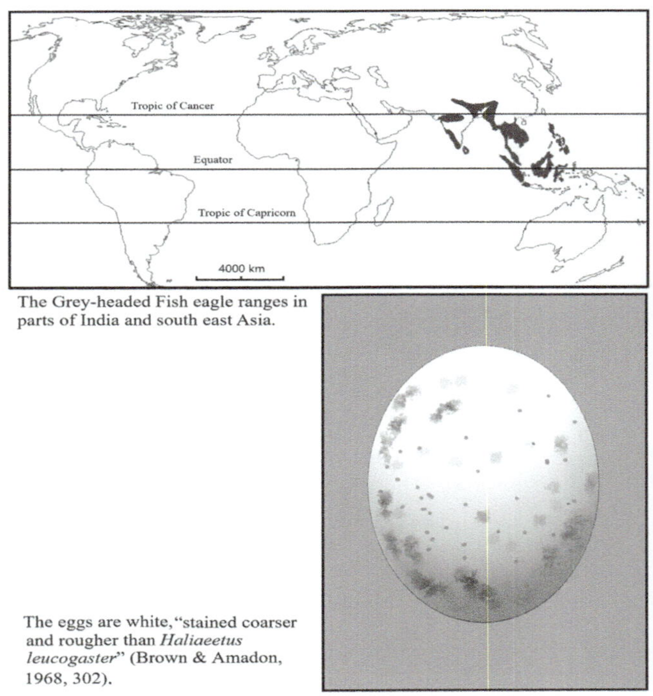

Fig. 4.10k. Grey-headed Fish eagle range and egg

5

Systematic List of the Harpy Eagles Subfamily *Harpiinae;* Genera *Harpia, Morphnus, Harpyopsis,* and *Buteogallus* (formerly *Harpyhaliaetus*)

5.1 Introduction

The Harpy eagles are among the most powerful and conspicuous eagles in the world, distinguished by their size, strength, large crests and strong relations with their local ecologies and human residents. Gilliard (1958, 105) writes that "tropical forests around the world are the home of a small group of huge, crested, immensely powerful eagles" referring of course to the family of Harpy eagles and the African and Asian eagles mentioned earlier, and the Harpy eagle listed in Table 5.1. Lerner and Mindel (2005, 343) write that "the members of the harpy eagle group (as defined by Brown and Amadon, 1968) are easily distinguished from other Accipitridae eagles by traits such as their extremely large size, with female wing-spans ranging from 1.76 to 2.01 m in length and female body weights ranging from 6 to 9 kg in *Pithecophaga* and *Harpia*"(see also Ferguson-Lees & Christie, 2001). Other shared characteristics include their primary tropical forest habitats, their mostly medium sized mammal prey (such as sloths and monkeys in Neotropical and Tropical Asian forests). However, as will be discussed in this chapter, these species are not all closely related, despite their similar ecologies and in some case anatomy. Lerner and Mindel (2005) list four species: two tropical American (Harpy eagle and Crested eagle) and two tropical Asian (Philippine eagle and New Guinea harpy eagle, or Papuan eagle), three of which are recorded in the table below.

In this book we have already considered the Philippine eagle, so this species, although commonly considered one of the Harpy eagles (but more often recently held to be related to the snake eagles) will be omitted from this chapter. Some other species such as the crowned eagle of Tropical African resemble the Harpy eagles in appearance (crested head, short wings for forest maneuverability, powerful feet, medium-sized mammal prey)and ecology (dense rainforest habitat) but are actually different birds. The two *Harpyhaliaetus* species, the Crowned Solitary eagle (formerly *Harpyhaliaetus coronatus*, now Buteogalluscoronatus (Vieillot, 1817)) and the Black Solitary eagle (formerly *Harpyhaliaetus solitarius*, now *Buteogallus solitarius*, Tschudi, 1844) (BirdLife International. 2016a,b) are not usually considered part of the Harpy eagle group and how now been placed with the buzzards of the genus *Buteogallus* species (Brown & Amadon, 1968).The Global Raptor Information Network (2021bm) notes that Brown (1970) "thought that Harpyhaliaetus is closely related to the harpy eagle group", but this perspective changed through the molecular studies of Lerner and Mindell (2005) and Amaral et al. (2006), which found that they are closer to *Buteogallus* (especially the Great Black-hawk, B. urubitinga, Gmelin, 1788) than to the genera *Harpia*, *Aquila* and *Spizaetus* (Global Raptor Information Network (2021bm). The Crowned Solitary eagle and the Black Solitary eagle are included in this chapter as they are regarded as eagles (BirdLife International. 2016a, b).

Genus *Morphnus* (Dumont, 1816, Subfamily: *Harpiinae*)

The Genus *Morphnus* is related to the Genus *Harpia*, both being within the subfamily *Harpiinae*, but the former eagle is slenderer and less powerful with an undivided, occipital crest, long secondaries and tails and un-feathered legs. The single species is the Guiana Crested eagle. Whitacre et al. (2002, 77) note concerning this species that it is comparatively little known: "Although this is the second largest of widespread Neotropical forest eagles, virtually all that is known concerning the species' nesting biology and behavior is based on a single nest" (Bierregaard, 1984).

Table 5.1. Harpy Eagle species considered in this book (Subfamily *Harpiinae*: Genera *Harpia*, *Morphnus*, *Harpyopsis* and *Buteogallus* (formerly *Harpyhaliaetus*)

English name	Scientific name	Mass (kg)	Length (cm)	Wingspan (cm)
Genus *Harpia* (Vieillot, 1816)				
Harpy eagle	*H. harpyja* (Linnaeus, 1758)	4 – 10	86.5–107	1.8- 2.2
Genus *Morphnus* (Dumont, 1816)				
Crested eagle	*M. guianensis* (Daudin, 1800)	1.3 - 3.0	71–89	1.4 -1.8
Genus: *Harpyopsis* (Salvadori, 1875)				
Papuan eagle	*H. novaeguineae* (Salvadori, 1875)	1.6–2.4	75–90	1.6
Genus: *Buteogallus* (Lesson, 1830)				
Crowned Solitary eagle	*B. coronatus* (Vieillot, 1817)	3.0	73–79	1.7–1.8
Black Solitary eagle	*B. solitarius* (Tschudi, 1844)	3.0	63–76	1.5 –1.9

Sources: Hoyo et al. (1994), Ferguson-Lees & Christie (2001), Lerner & Mindell 2005, Global Raptor Information Network (2021bm).

5.2 Guiana Crested Eagle or Crested Eagle *Morphnus guianensis* (Daudin, 1800)

Physical Description

The Guiana Crested eagle is described as "a rare, patchily-distributed, forest-dwelling eagle occurring from southern Mexico (Chiapas) to Argentina" (Araújo et al., 2015). This species is a large eagle, with a slender build, measuring 71–89 cm in length, with a wingspan of 138 to 176 cm and a weight of 1.3 to 3 kg (Ferguson-Lees & Christie, 2001; Hilty, 2002). There are two color phases, termed the normal and the dark phase (Brown & Amadon, 1968). Although historically, the dark morph was seen as a separate species *Morphnus taeniatus* (Gurney, 1879; Wetmore, 1965; Blake, 1977) it was later seen as only a dark morph (Lehmann, 1943; Gomes & Sanaiotti, 2015). For the 'normal' morph, the adult plumage comprises a dark grey to almost black crown, with lighter colored feathers on the head and neck sides, with lighter tipped feathers (Brown & Amadon, 1968). The throat and breast are grey, the lighter throat feathers with black side markings, and a dark brown to black mantle, white tipped upper wing and tail coverts, and lower back feathers. The underside is white to pale brown, with dark and reddish-brown barring. The tail is pale tipped, basally black, with three mottled grey bands. The primaries have black and mottled grey barring. Adult females are darker colored than males, on the head and breast. The bill is black, the eye grey to yellowish, and the cere, lores and skin in the eye area are blackish (Brown & Amadon, 1968). The dark morph has a black head and upper breast, a paler throat, and black and white barring in the under parts. Some are almost entirely black, covering even the barred underparts of the lighter morph. There are also intermediate morphs, colored between the normal and dark phases, which was one reason the dark morph was not given species status (Brown & Amadon, 1968). Immature individuals have a crest that is colored black with white on the base and tip. The head and underparts are white, sometimes with grey marks and marbled, with black and light grey barring on the back and wings. The barred primaries are like the adult's primaries. The pale wing coverts have grey markings. The tail is marbled grey and white, with several black bars (Brown & Amadon, 1968). Dark morph juveniles are darker brownish-gray (Ferguson-Lees & Christie, 2001).

The Guiana Crested eagle is unmistakable. Despite similarity to the Harpy eagle (with which it has a considerable range overlap), the lighter build (about half the bulk of the Harpy eagle), longer tail and a single feathered crest contrast with the Harpy eagle's heavier build, shorter tail and multi-feathered crest. The immature bird may be taken for the Ornate eagle or Isidor's eagle, but the immature Guiana Crested eagle has un-feathered, yellow legs and its back has marbled grey/white coloration. Brown and Amadon (1968: 631) argued that "when soaring, the long tail and broad, blunt wings give a curious outline." The Guiana Crested eagle is usually silent but may make a double whistling call (Ferguson-Lees & Christie, 2001). Illustrations of the appearance, range, and eggs of this species are shown in Figures 5.1a to 5.1k.

Classification

The Guiana Crested eagle is monotypic. As mentioned above, the dark morph was formerly seen as a separate species (Gurney, 1879; Wetmore, 1965; Blake, 1977; Global Raptor Information Network, 2021bn) and was originally described as a separate species, *taeniatus*, by Gurney (1879). In the past, Amadon and Bull (1988) described the Genus *Morphnus* as "barely separable" from the Genus *Harpia*. This assessment was supported by Helbig et al. (2005), who used molecular studies and found that the two genera were closer than demonstrated in other studies, and that the two genera might be united. Lerner and Mindell (2005) used molecular sequences of two mitochondrial genes and one nuclear intron, and found that the Genera *Morphnus*, *Harpia* and *Harpyopsis* are closely related into one clade. The Philippine eagle was found to be closer to the clade of snake eagles (Lerner & Mindell, 2005).

Foraging

The Guiana Crested eagle has a scattered distribution through its range, which although absent from Mexico (Vargas et al., 2006) extends across northern Guatemala southwards through Belize, Honduras, Nicaragua, Costa Rica, and Panama, and into Colombia, Venezuela, Guyana, Suriname, French Guiana, Brazil and further into Ecuador, Peru, Paraguay and Bolivia to northern Argentina (Ferguson-Lees & Christie, 2001; Vargas et al., 2006; Gomes, 2014; Gomes & Sanaiotti, 2015). The habitats are generally humid lowland rainforests, and in some cases narrow, gallery forests and forested ravines, usually below 600 m in most areas and up to 1,600 m in Andean regions (Crease, 2009). According to Brown and Amadon (1968: 631) the general habitats of the Guiana Crested eagle are "little known", except that it "inhabits heavy jungle (see also Lehmann, 1943). However, recent habitat studies have been undertaken. The habitat behavior varies regionally. For example, in French Guiana, it occurs in palm swamps and forest edges, and logged and palm swamps, with interspecific avoidance with the Harpy eagle (Ridgely & Greenfield, 2001; Thiollay, 2007). In Colombia, it occurs only in the very wet rainforest, and riverine areas (Lehmann, 1943). In Brazil, it usually frequents virgin, canopied forest (Bierregaard, 1984; Sick, 1993).

Although it shares much of its range with the Harpy eagle, the Guiana Crested eagle is the smaller species and appears to avoid conflict with the more powerful Harpy eagle by hunting smaller prey, including a larger percentage of birds (e.g. jays (Family *Corvidae*, Leach, 1820), trumpeters (Genus *Psophia*, Linnaeus, 1758), guans (Subfamily *Penelopinae*, Huxley, 1868) and cocks-of-the-rock (Genus *Rupicola*, Brisson, 1760)) as well as small mammals (e.g. capuchin monkeys (Subfamily *Cebinae*, Bonaparte, 1831), tamarins (Genus *Saguinus*, Hoffmannsegg, 1807), woolly monkeys (Genus *Lagothrix*, É. Geoffroy, 1812), opossums (Family *Didelphidae*, Gray, 1821) and kinkajous (*Potos flavus*, Schreber, 1774)). Reptiles such as snakes and lizards may also be taken, with most hunting from perching, singly or in pairs (Brown & Amadon, 1968; Bierregaard, 1984, 1994; Ferguson-Lees & Christie, 2001; Oversluijs Vasquez & Heymann, 2001; Sick, 1993; Hilty, 2003; Guyra Paraguay, 2004; Grosselet and Gutierres-Carbonell, 2007; Gomes and Sanaiotti, 2015).

Foraging behavior may vary regionally. For example, in Peru, predation on monkeys, including a Squirrel monkey (*Saimiri sciureus*, Linnaeus, 1758) (Robinson, 1994). Oversluijs Vasquez and Heymann (2001) reported two successful attacks on infant monkeys of the species Moustached tamarin (*Saguinus mystax*, Spix, 1823) and Brown-mantled tamarin (*Leontocebus fuscicollis*, Spix, 1823). In Venezuela, there was an unsuccessful attack on a Cock-of-the-Rock (*Rupicola rupicola*, Linnaeus, 1766) (Trail, 1987). In French Guiana, a Spider monkey (*Ateles paniscus*, Linnaeus, 1758) was killed by an eagle (Julliot, 1994). In Brazil, one nest contained the remains of six snakes, one frog, and eight small mammals, and a kinkajou was observed delivered to the nest, while a flock of Gray-winged Trumpeters were attacked (*Psophia crepitans*, Linnaeus, 1758) (Bierregaard, 1984). A white-faced Saki monkey (*Pithecia pithecia*, Linnaeus, 1766) was also attacked (Gilbert, 2000).

Breeding

The Guiana Crested eagle's breeding season begins in March to April, which is the transition from the dry to the wet season. The nests are very large, usually located in the central fork of large, tall trees (Crease & Tepedino, 2013). The clutch is of one to two eggs (Kreuger, 1963; Bierregaard, 1984; Whitacre et al., 2002), which may vary in color from dull white or cream-colored with no spots, or pale yellowish brown and lilac-greyish spots (Palmer, 1962; Kreuger, 1963; Brown & Amadon, 1968; Bierregaard, 1984; Kiff et al., 1989). The incubation period may be 40 to 50 days (Bierregaard, 1984). The male provides for the female for most of the first 80 days of the nestling period. The nestling period may vary; for example, 109-114 days in a Guatemalan nest, and 103-105 days at the Oklahoma City Zoo (Whitacre et al., 2002). There are similar nesting habits across the range (Bierregaard, 1984; Whitacre et al., 2002; Raine, 2007). In one case, a female Crested Eagle fed a recently fledged Harpy eagle near the Harpy's nest in the Bagre Mountains, Panama. The prey delivered comprised two snakes, one Geoffroy's Tamarin (*Saguinus geoffroyi*, Pucheran, 1845), one squirrel (*Scirurus* sp.), and one Brown-throated Tree Sloth (*Bradypus variegatus*, Schinz 1825) (Vargas et al., 2006). Whitacre et al. (2002, 80) observed that this species has a comparatively slow biological development compared with smaller, South American eagle species: "The relative speed of development of the above three species accords with their relative body sizes. Black Hawk eagles (most rapid to develop) are also smallest, with mean adult female mass of 1115 g. Ornate Hawk eagles, with intermediate rate of development, are intermediate in size, with females averaging 1450 g (Whitacre et al., 2002). Crested eagles, slowest to develop, have a mean adult female mass of about 1850 g."

Population Status

The Guiana Crested eagle may not be migratory. Ferguson-Lees and Christie (2001) tentatively estimated the total population, including adults and immatures at the beginning of the breeding season, at 1,000 to 10,000 birds. BirdLife Internation al (2017) noted that the "global population size has not been estimated, but it has been described as sparsely distributed in the south of its range" referring to Whitacre et al. (2012) and Bierregaard & Kirwan (2017). It is rare to uncommon across its range, possibly commoner in Amazonia and the Orinoco basin. Citing information reported by Tracewski et al. (2016), Soares-Filho et al. (2006) and Bird et al. (2011), BirdLife International (2017) argues that suitable habitat has declined by 16.5% over the past three generations (54 years) and may lose 29.8-47% of suitable habitat 25 to 29 population over three generations, with lower losses outside Amazonia. Population losses are from shooting and habitat loss through its range. It also may compete with people for, prey (Galetti et al., 1997). It appears historically rarer and Endangered in Central America than in South America and Near Threatened or Vulnerable in parts of South America, but overall, it is classified as Near Threatened (BirdLife International, 2017). The Global Raptor Information Network (2021bn) lists "Northern Guatemala (Petén), and Belize south locally through humid Honduras, Nicaragua, Costa Rica, and Panama to Colombia and south east of the Andes to southern Brazil, Paraguay, and extreme northeastern Argentina (Misiones); west of the Andes only in central Colombia (Serranía de Baudó)." Through this wide range it varies in population density. For example, it is rare or absent from Mexico, but there is one record from 18 December 2004 in the Reserva de la Biosfera de Monte Ázul, Chiapa, photographed by Grosselet and Gutierrez Carbonel (2007) possibly the first record for Mexico (Global Raptor Information Network, 2021bn).

In Central America, Global Raptor Information Network (2021bn) reports increased sightings, for example the lack of records north of Honduras before 1968 but argues that "this is presumably an indication of better coverage, rather than an actual range extension." In Honduras, it was rarely recorded, mostly at the individual level (Bangs, 1903; Monroe, 1968; Anderson et al., 2004), or as rare breeding resident (Bonta& Anderson, 2002). However, the Crested eagle appears to be rarer than the Harpy eagle in shared contexts, possibly due to competitive exclusion (Jones & Komar, 2008; Whitacre et al., 2017). Whitacre & Burnham (2017) notes that the Guiana Crested eagle is the largest common Neotropical raptor and the most space demanding, after the Harpy eagle. In Belize, it is a rare resident (Howell & Webb, 1995; Jones & Komar, 2008), while in Guatemala it is also rare (Ellis & Whaley, 1981; Howell &Webb, 1995). The Peregrine Fund's Maya Project found nests in Guatemala (Whitacre et al., 2002). Possibly, it is a breeding resident of the Atlantic lowlands (Eisermann & Avendaño, 2007). In Nicaragua, its distribution is uncertain as some authors do not list it as present, while others record it as probably resident especially in the Caribbean coast area (Martínez-Sánchez, 2000; Kjeldsen, 2005; Múnera-Roldán et al., 2007). In Costa Rica, it is a very rare resident, but probably historically common; it was only recently recorded outside the Sarapiquí region (Finca La Selva) and the Península de Osa (Stiles & Skutch, 1989). A few recent records exist from Finca La Selva, Rara Avis, and Braulio Carillo National Park (Jones 2004). There are also records from Tortuguero, possibly Nicaraguan visitors (Jones & Komar, 2008). In Panama, it is rare on Caribbean and Pacific slopes, and the Canal area (Ridgely & Gwynne, 1989), and some authors argue that it is commoner than supposed (Wetmore, 1965; Jones & Komar, 2008).

The Guiana Crested eagle's situation is similar in South America, mostly a rare resident. In Columbia, it is rare but frequents lowland forested areas (Hilty & Brown, 1986; Márquez et al., 2005). In Ecuador, it is a rare resident in the lowlands of eastern Ecuador below 300 m (Ridgely & Greenfield, 2001). In Bolivia, it is similarly resident in lowland forests (Pearman, 1994; Hennessey et al., 2003). In Peru is rare but has been recorded in the eastern Andean lowland gallery forests (Clements & Shany, 2001). In Venezuela, it is a widely distributed resident, albeit rare, in the lowlands and lower montane forests, especially in the Orinoco region, and the lower Rio Caura (Hilty, 2003). In Guyana, it rarely recorded in lowland forests (Braun et al., 2000). In French Guiana, it frequents primary forests, and may be commoner than the Harpy eagle in modified and degraded forests (Thiollay, 2007). In Suriname, it may be a nesting resident, but there are no nesting records, and it is rarely recorded in mature forest and some coastal areas (Haverschmidt & Mees, 1994). In Brazil, unlike in Central America (where in some areas the population appears to be increasing) the population is declining in the Mato Grosso and Rio Grande do Sul (von Ihering, 1899; Belton, 1984; Sick, 1993; Bencke, 1997). In Argentina, it is rare and resident in some areas and a visitor in others (Bertoni, 1913; Olrog, 1985; Mazar Barnett & Pearman, 2001; Pearman, 2001). In Paraguay, its presence and breeding residence is uncertain, with some inconclusive records (Hayes, 1995; Madroño-Nieto et al., 1997; Del Castillo & Clay, 2004).

In terms of population levels, it varies across the region. For example, in Guatemala it is Critically Endangered (Eisermann & Avendaño, 2006), as in Paraguay (Guyra Paraguay, 2005), while in Mexico it is Endangered (Instituto Nacional de Ecología). In Panama, it is declining due to human landuse and presence (Ridgely & Gwynne, 1989). In Colombia and Bolivia, it is classified as Near Threatened (Renjifo et al., 2002; Global Raptor Information Network, 2021bn). In Ecuador, it has been classified as Vulnerable (Ridgely and Greenfield, 2001), even outnumbered by the Harpy eagle, largely due to habitat losses and direct persecution. In French Guiana, it may be present, with few records (Thiollay, 1989, 2007). In Brazil, it appears to be rarer and possibly declining outside Amazonia, for example the Mato Grosso and Rio Grande do Sul, and areas north of these (Sick &

Teixeira, 1977; Sick, 1993; Albuquerque et al., 2006) mostly due to habitat destruction (mostly deforestation from woodcutting, pine plantation creation and charcoal production, and hydropower dams and associated flooding of the forest). Flooded forests include those near the Uruguay River. Illustrations of the appearance, range, and eggs of this species are shown in Figures 5.1a to 5.1k.

Fig. 5.1a. Guiana Crested eagle perching [With permission from Jonathan Beilby]

Fig. 5.1b. Guiana Crested eagle perching [With Permission from Johannes Pfleiderer]

Fig. 5.1c. Dark morph Guiana Crested eagle perching [With permission from Paul Prior]

Fig. 5.1d. Guiana Crested eagle on nest [With permission from Paul Prior]

Fig. 5.1e. Immature Guiana Crested eagle [With permission from C. McNab]

Fig. 5.1f. Immature Guiana Crested eagle on nest [With permission from David Beadle]

Fig. 5.1g. Immature Guiana Crested eagle [With permission from Nayer Youakim]

Fig. 5.1h. Immature Guiana Crested eagle [With permission from Nayer Youakim]

Fig. 5.1i. Immature Guiana Crested eagle [With permission from Nayer Youakim]

Fig. 5.1j. Guiana Crested eagle habitat, Amazon Forest [With permission from Rômulo Ferreira]

Genus *Harpia*, Vieillot, 1816 (Subfamily *Harpiinae*)

There is only one species in this genus. This very large species is marked by comparatively short, rounded wings with broad secondaries, long tails, divided crests un-feathered and very thick tarsi (in the larger female, these are the thickest of any bird of prey) and huge talons, which are "coarser and less acute than the booted eagles such as *Stephanoaetus* or *Oroaetus*" (Brown & Amadon, 1968, 632). By this, the authors are referring to eagles of the Genus *Stephanoaetus* (W.L. Sclater, 1922) which is represented by only one species, the Crowned eagle or Crowned hawk-eagle (*Stephanoaetus coronatus*, Linnaeus, 1766) and of the Genus *Oroaetus*, which also has one species the Black-and-chestnut eagle (usually classified not as *Oroaetus isidori* (Des Murs, 1845) but as *Spizaetus isidori* (Des Murs, 1845). Other recent information on the regional, ecological, nesting, feeding habits and conservation of the Harpy eagle can be found in Alvarez-Cordero (1996), Galetti & Carvalho-Jr (2000), Rotenberg et al. (2000); Piana & Arpia, (2002), Rettig (2002), Sanaiotti (2002), Luz (2005), Trinca et al. (2008), Muniz-Lopez (2005), Vargas-Gonzalez et al. (2006a,b), Giudice et al. (2007), Piana (2007), Crease (2009), Seymour et al. (2010), Sanchez-Lalinde et al. (2011), Vargas-Gonzalez & Vargas (2011), Aguiar-Silva et al. (2012, 2014, 2015), Muniz-Lopez et al. (2012), Vargas-Gonzalez et al. (2014), Kuniy et al. (2015). Sanaiotti et al. (2015), Sousa et al. (2015) and Watson et al. (2016).

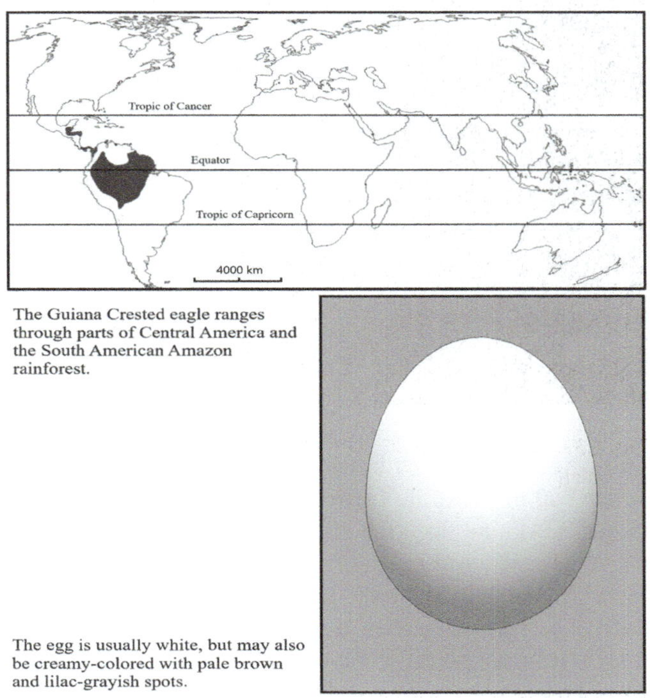

Fig. 5.1k. Guiana Crested eagle range and egg

5.3 Harpy Eagle *Harpia harpyja* (Linnaeus, 1758)

Physical Description

The Harpy eagle is a very large eagle, probably the largest or second largest in the world (after the Steller's Sea eagle, and larger than the Philippine eagle). Bates (1964, 99, 100) writes "the Giant Harpy, most powerful of the world's eagles, terrorizes mammals and birds of the forest canopy…the heaviest of all eagles." This species stands out throughout South America for its size and strength: "the largest and fiercest inhabitant of the upper canopy is certainly the harpy eagle, a most unusual bird of prey…Weighing some 20 pounds and with short, powerful wings… (Bates, 1964, 104). Aguiar-Silva (2014, 24) argues that the Harpy eagle is "the heaviest and the most powerful bird of prey in the canopy of the Neotropical rainforests." The Harpy eagle may measure from approximately 87 cm to 107 cm in length, with a wingspan of 176 to 224 cm (Wetmore, 1965; Thiollay, 1994; Howell & Webb,, 1995; Ferguson-Lees & Christie, 2001). The female is larger than male, in common with and other eagle species. Weights for females may range from 6 to 9 kg, but weights of up to 10 kg and even 12.3 kg (in captivity) have been recorded, with an average of 7.35 to 8.3 kg. The male's weight may range from 4 to 6 kg, with an average of 4.4 to 4.8 kg (O'Connor, 1984; Thiollay, 1994; Ferguson-Lees & Christie, 2001; Arent, 2007; Dunning, 2008; Trinca et al., 2008; Whitacre & Jenny, 2013; Miranda et al., 2018). The wingspan is relatively shorter compared with similarly sized or even smaller eagles, an adaptation that allows the Harpy eagle to move through dense forest (Ferguson-Lees & Christie, 2001).

The adult is ashy grey-headed, with a divided crest of white-tipped black feathers. The dark grey to slate black back feathers are edged and marbled with lighter grey, which can appear as light bars on the secondaries (Brown & Amadon, 1968). The black tail is crossed with three greyish bars and tip. The white underparts are crossed on the chest by a wide, horizontal black bar, and black barring on the on the thighs. The wing linings are also white with black markings, and black barring on the outer primaries. The light brown eyes have grey spots, while the orbit and lores are dark to blackish and the bill is bluish tinged black. The legs are creamy, pale yellow (Brown & Amadon, 1968). The immature birds have white crests and heads, with the backs being lighter brownish grey than those of the adults. Wing and tail quills are black. The black adult chest band begins with brownish feathers and darkens with age. The dark brown eyes slowly lighten with age. The legs are more yellow than those of the adult. The massive talons are the largest of any living eagle (Ferguson-Lees & Christie, 2001). The Harpy eagle is unmistakable. It is like the Guiana Crested eagle, but is much bulkier, with a divided crest and black chest band. The voice is a "loud wheeeoooooo", while the juvenile has a higher pitched whee e-e-e-e, with up to twelve repetitions, in addition to "mewing" and "croaking" sounds (Brown & Amadon, 1968, 634). Some have described the call as a weak scream (Rettig, 1978; Schulenberg, 2020).

Classification

The Harpy eagle is monotypic (Global Raptor Information Network, 2021bo). Helbig et al. (2005) used molecular studies and found that the genera *Harpia* and *Morphnus* are sister genera, close enough to entertain the possibility of merging. Lerner and Mindell (2005) used molecular sequences of two mitochondrial genes and one nuclear intron, and found that the genera *Harpia*, *Morphnus* (Crested Eagle), and *Harpyopsis* (New Guinea Harpy Eagle) are similar enough to form a defined clade, while the Philippine eagle is separate, and is closer to the clade of *Circaetinae* snake eagles.

Foraging

The Harpy eagle inhabits rainforests from Mexico, where it is almost extinct southwards into Central America, where it is also almost extinct, to South America, reaching Argentina (Ferguson-Lees & Christie, 2001; Weidensaul, 2004). It is commonest in the Amazon forest of Brazil (Camargo/Conexão Planeta, 2020).These authors write that "there are an estimated 5,000 harpy eagles in the Amazon and 300 in the Atlantic Forest, with deforestation the main threat to their survival." They mostly frequent forests below 900 m (BirdLife International, 2017), with most hunting in the below the canopy or emergent trees. Brown and Amadon (1968, 634) note that "the wild birds were never seen to fly above the trees." They may enter deforested areas, such as forest clearings and forest/savanna/pasture mosaics (Bierregaard, 1994; Rettig & Hayes, 1995; Parker et al., 1996; Álvarez-Cordero, 1996; Sigrist, 2013; Muñiz-López, 2016). Global Raptor Information Network (2021bo) writes that "Harpies can apparently survive in isolated patches of primary forest, selectively logged primary forest, and late second-growth with a few large emergent trees at least temporarily, if it can avoid persecution and if enough prey is still available."

The Harpy eagle hunts mostly from perches within the canopy and may fly from such perches to catch prey either through the canopy to another tree or to the ground (Ferguson-Lees & Christie, 2001; Global Raptor Information Network, 2021bo). Aguiar-Silva et al. (2014, 24) reports that "throughout the Harpy Eagle's entire range, 69 prey species have been documented, indicating that it can use a wide range of food resources." Miranda (2018, 1) reported 102 prey species reported in the literature and the current research in Suriname. Female eagles, being larger, take larger prey (e.g., a minimum of 2.7 kg) than males (0.5 to 2.5 kg) (Ferguson-Lees & Christie, 2001).

The Harpy eagle's main prey is arboreal mammals, such as monkeys (*Platyrrhini* species, E. Geoffroy, 1812), coatis (*Nasua* spp., Storr, 1780; *Nasuella* spp., Hollister, 1915), sloths (*Bradypus*, Linnaeus, 1758; *Megalonychidae*, Gervais 1855), opossom (*Didelphidae*, Gray, 1821) and New World porcupines (*Erethizontidae*, Bonaparte 1845) (Fowler & Cope, 1964; Rettig, 1978; Sanaiotti et al., 2001; Galetti & de Carvalho, 2000; Piana, 2007; Seymour et al., 2010; Rotenberg et al., 2012; Schulenberg, 2020). Ground species such as agoutis (*Dasyprocta*, Illiger, 1811), large birds and reptiles are also captured on the ground (Brown & Amadon, 1968). Global Raptor Information Network (2021bo) adds armadillos and deer as prey species. Bates (1964, 104) writes the Harpy preys on "monkeys, sloths, agoutis, and an occasional opossum or porcupine…the harpy displays extraordinary maneuverability in pursuing its victims through the trees. One observer watching a harpy chasing a capuchin monkey estimated that it flew at 40 to 50 miles per hour, dodging tree trunks and branches; on another occasion, a harpy was seen to roll over completely onto its back to pluck a monkey from underneath the branch on which it had sought refuge. Big as they are, these eagles are almost never seen flying above the canopy of the rain forest and rarely have they been found to hunt in the open savanna."

Aguiar-Silva et al. (2014) studying Harpy nest in the Brazilian Amazon, found 79% of the Harpy's prey was sloths (the Brown-throated sloth (*Bradypus variegatus*, Schinz 1825) (39%), and the two-toed sloth (*Choloepus didactylus*, Linnaeus, 1758) (40%)). In Guyana, monkeys were about 37% of the prey taken in one case study (Izor, 1985). In Ecuador, one study found cebid monkeys (Family *Cebidae*, Bonaparte, 1831) were 35% of remains recorded (Muñiz-López et al., 2007). In Panama, two captive-bred subadults were found to have preyed on two sloth species (52% of one eagle's prey, and 54% of the other eagle's prey), the Brown-throated sloth and Hoffmann's two-toed sloth (*Choloepus hoffmanni*, Peters, 1858) (Touchton et al., 2002). Monkeys may also be taken, such as: Tufted capuchin (*Sapajusapella*, Linnaeus, 1758), saki monkeys (Genus *Pithecia*, Desmarest, 1804), howler monkeys (Genus *Alouatta*, Lacepede, 1799), titi monkeys (Subfamily *Callicebinae*, Pocock, 1925), squirrel monkeys (*Saimiri sciureus*, Linnaeus, 1758), and spider monkeys (Genus *Ateles*, E. Geoffroy, 1806) (Izor, 1985; Ferguson-Lees & Christie, 2001). In the Pantanal (a wetland area, of flooded grasslands in the Brazilian state of Mato Grosso do Sul, and Bolivia and Paraguay) one pair fed mostly on porcupine (*Coendou prehensilis*, Linnaeus, 1758) and the agouti (*Dasyprocta azarae*, Lichtenstein, 1823) (Ubaid et al., 2011). Birds are also taken, including red-and-green macaws (*Ara chloropterus*, Gray, 1859), Aguiar-Silva (2007, 2014) curassows (Family *Cracidae*, Rafinesque, 1815) and seriemas (Family *Cariamidae*, Bonaparte, 1853). Reptiles taken include iguanas (Genus *Iguana*, Laurenti, 1768), tegus (Genus *Salvator*, Duméril & Bibron, 1839), and snakes (Suborder *Serpentes*, Linnaeus, 1758). Rarely, domestic livestock, such as chickens, lambs, goats, and pigs may be taken (Ferguson-Lees & Christie, 2001; Global Raptor Information Network, 2021bo).

There are regional variations in the prey taken, at least depending on sometimes scattered findings. For example, in French Guiana, Thiollay (1989, 2007) found that the Harpy eagle was rarer than the Crested eagle in modified including logged forest, and in primary forest. For foraging there appeared to be interspecific avoidance as they are less often found together than alone (Thiollay, 1989, 2007). In Mexico, harpies were observed preying on iguanas, and guans and curassows, but not Spider monkeys and Black Howler monkeys, which were locally common (Lowery & Dalquest, 1951). In Belize, recorded prey included spider monkeys, porcupines, opossums, gray foxes, kinkajous and coatimundis (Matola, 2006). There is one record of a Harpy eagle eating a tamandua (*Tamandua Mexicana*, Saussure, 1860) (Jones, 2002). In Panama, there are records of harpies taking as sloths, pigs, two-toed sloths, Brown-throated Tree sloths, marmosets, deer, macaws, other large birds, kinkajous, tayras, porcupines, and a gray fox (Sturgis, 1928; Wetmore, 1965; Muela et al., 2007; Springer et al., 2011).

In Brazil, foraging was commonly from tall forest trees near rivers with rapids (e.g. in Amapá) and salt banks that attracted many animals come to lick the salty earth (e.g. Mato Grosso, Pará) (Sick, 1993), and also from soaring flight over forests and nearby savannas. In Ecuador, there are records of two-toed sloths, Three-toed sloths, guans macaws, titi monkeys, Humboldt's White-fronted capuchins, Saki monkeys, Venezuelan red howler monkeys, Brown woolly monkeys, kinkajous, prehensile-tailed porcupines, Black agoutis (Galetti & de Carvalho, 2000; Muñiz-Lopez et al., 2007; Aguiar-Silva et al., 2014). Deer such as Brockett deer were also taken (Genus *Mazama* spp. Rafinesque, 1817) (Peres, 1993; Sick, 1993). In Peru, similar prey was reported, including Squirrel monkeys, Red howlers and sloths, macaws, bicolored-spined porcupines, Hoffmann's two-toed sloths, Brown-throated sloths, kinkajous, Southern tamanduas, South American coatis, Common opossums, Venezuelan red howlers, Southern naked-tailed armadillos, Night monkeys, Tufted capuchins, tayras (*Eira Barbara*, Linnaeus, 1758) and birds such as Amazon parrots and toucans (Family *Ramphastidae*, Vigors, 1825) (Eason, 1989; Sherman, 1991; Robinson, 1994; Brightsmith, 2002; Piana, 2007). In Guyana, similar species were recorded, mostly sloths, primates and smaller mammals (Blake, 1977; Fowler & Cope, 1961; Rettig, 1978; Izor, 1985). In Argentina, in addition to the general prey of monkeys, porcupines and opossums, there was a record of a margay (*Leopardus wiedii*, Schinz, 1821) (Anfuso et al., 2008). Other recent information on the regional, ecological, nesting, feeding habits and conservation of the Harpy eagle can be found in Alvarez-Cordero (1996), Galetti & Carvalho-Jr (2000), Rotenberg et al. (2000); Piana & Arpia, (2002), Rettig (2002), Sanaiotti (2002), Luz (2005), Trinca et al. (2008), Muniz-Lopez (2005), Vargas-Gonzalez et al. (2006a, b), Giudice et al. (2007), Piana (2007), Crease (2009), Seymour et al. (2010), Sanchez-Lalinde et al. (2011), Vargas-Gonzalez & Vargas (2011), Aguiar-Silva et al. (2012, 2014, 2015), Muniz-Lopez et al. (2012), Vargas-Gonzalez et al. (2014), Kuniy et al. (2015). Sanaiotti et al. (2015), Sousa et al. (2015) and Watson et al. (2016).

Breeding

Harpy eagle nests are usually in large trees such as the Silk Cotton tree (*Ceiba pentandra* (L.) Gaertn.) in Guyana. Records include nesting locations nearly 60 meters high in canopies nearly 40 meters wide, and nests 1.5 meters wide and 1.2 meters high (Brown & Amadon, 1968, 634). Global Raptor Information Network (2021bo) describes the nest as "a huge platform of sticks with a shallow central cup, placed in a high crotch of a huge emergent tree in primary forest, or partially logged forest with remaining large trees. Both parents participate in nest building." Piper (2007) describes Harpy eagle nests as often located in the crown of the kapok tree, which is among the tallest trees in its range. Due to the frequency of this location, some cultures preserved kapok trees, in the knowledge that it would help eagle breeding (Piper, 2007). One or 2 dull white eggs are laid, which may acquire brown nest stains. Sick (1993) described a fresh egg as gray with a white wash. Incubation is principally undertaken by the female, while the male eagle is the food provider. In a two-eagle clutch, one egg may hatch well before the other, and only the first-born eaglet may survive. The incubation period is generally about 56 days, and a Guyana chick fledged at 141-148 days (Rettig, 1978), but these periods may vary regionally as detailed below. Breeding maturity may be reached at 4 to 6 years of age (Rettig, 1978; Rettig & Hayes, 1995; Ferguson-Lees & Christie, 2001).

For example, in Brazil, eggs were recorded at Costanhal, near the Rio Apehu, on 27 April and 9 May (Norris, 1927). Eggs may also be laid from September and November. Sick (1993) stated that eggs are laid between September and November in Goiás, Center-West Brazil. In the Ecuadorian lowlands, however, one case study failed to find seasonal peaks in egg laying and incubation (Muñiz-Lopez, 2007). Activity dates included the period between the dry and wet season, the start of the wet season, and during the dry season, where the dry season lasts from May to November, and the rainy season from December to April. There was an interesting cases study where a pair built a nest during July-August 2003, abandoned it in August, renovated the nest in July 2004, and fledged a nestling in September (Muñiz-Lopez, 2007). In Guyana, nesting evidence included fledged juveniles near nests in January and February (Fowler & Cope, 1964), and a nest that was redeveloped during mid-May to mid-June, with two eggs laid in mid June (on 11-17 June and 19 June, respectively) and one egg hatching on 7 August, and fledging in December to January (Rettig, 1978). Rettig (1978) judged incubation periods to be 56 days, and nestling periods 141-148 days.

Various locations have been recorded for nests. Nests are located at varying distances apart, for example 3 km in parts of Guyana and Panama, 5 km in Venezuela and 7.4 km in Peru. In degraded habitats such as fragmented forests, the inter-nest distances were estimated at 25 km (Thiollay, 1994). In Brazil, locations included one, interestingly lined with sloth hair, on a limb 35 meters above ground (Norris, 1972), nests on comparatively tall trees above the canopy (e.g. ceibas *Ceiba pentandra*, (L.) Gaertn, and Brazil Nut *Bertholletia excelsa*, Humb. & Bonpl.), and nests on the crowns of *Mauritia carana* (Wallace) or *Mauritia flexuosa* (L.f.) (Sick, 1993), and a nest 18 meters high in a dead *Astronium graveolens* (Jacq.) tree in Mato Grosso do Sul near the Serra da Bodoquena National Park (Pereira & Salzo, 2006). In Ecuador, Muñiz-Lopez et al. (2007) found 11 nests in primary lowland wet forest between elevations of 206-240 meters and the trees were ceibas, *Cedrelinga cateniformis* (Ducke) Ducke, and *Huberodendron* (Ducke). In Peru, nine nests were found in very tall emergents (all but one living trees); Brazil Nut trees, four in ironwood trees (*Dipteryx micrantha*, Harms, 1926), and one in a ceiba in the Tambopata National Reserve (Giudice et al., 2007; Piana, 2007). In Guyana, two nests were located in ceibas about 40 m high (Fowler & Cope, 1964). In Misiones, northeastern Argentina, one nest was located 21 m high in hillside tree (*Enterolobium contortisiliquum* (Vell.) Morong) (Anfuso et al., 2008). Other recent information on the regional, ecological, nesting, feeding habits and conservation of the Harpy eagle can be found in Alvarez-Cordero (1996), Galetti & Carvalho-Jr (2000), Rotenberg et al. (2000); Piana & Arpia, (2002), Rettig (2002), Sanaiotti (2002), Luz (2005), Trinca et al. (2008), Muniz-Lopez (2005), Vargas-Gonzalez et al. (2006a,b), Giudice et al. (2007), Piana (2007), Crease (2009), Seymour et al. (2010), Sanchez-Lalinde et al. (2011), Vargas-Gonzalez & Vargas (2011), Aguiar-Silva et al. (2012, 2014, 2015), Muniz-Lopez et al. (2012), Vargas-Gonzalez et al. (2014), Kuniy et al. (2015). Sanaiotti et al. (2015), Sousa et al. (2015) and Watson et al. (2016).

Population Status

The Harpy eagle population has been estimated to be 10,000 to 100,000 at the start of the breeding season (Ferguson-Lees & Christie, 2001), possibly on the assumption that there is still a large population in Amazonia (Global Raptor Information Network, 2021bo). BirdLife International (2017) estimates the population at between 20,000-49,999 birds. It is further noted that the population is decreasing, "is suspected to lose 27.6-45.5% of suitable habitat within its distribution over three generations (56 years) based on a model of Amazonian deforestation" (citing Soares-Filho et al., 2006, Bird et al., 2011). Losses outside Amazonia "are judged to be likely to be lower, so the species is therefore suspected to decline by 25-29% over three generations" (BirdLife International, 2017). Throughout most of its range, the Harpy eagle is an irruptive or local migrant (Sick 1993; Alvarez-Cordero 1996; Bildstein, 2006; Muela et al., 2007).

The Harpy eagle is considered a rare species and is certainly rarer than it was 50 years ago from southern Mexico to north Argentina (Brown & Amadon, 1968; Miranda, 2015). Global Raptor Information Network (2021bo) lists its range as "Southern Mexico (formerly north to Veracruz, but now probably only in Chiapas) south locally through Caribbean Central America to Colombia east to Venezuela and the Guianas and south through eastern Bolivia and Brazil to extreme northeastern Argentina (Misiones) and Paraguay; formerly in western Colombia and western Ecuador, but now nearly extirpated in that region, except for Esmeraldas Province." Aguiar-Silva et al. (2014, 24) note that the Harpy eagle "is critically endangered in some parts of its range, mainly due to hunting pressure and habitat loss by deforestation." The Harpy eagle is commonest in Brazil and is almost extinct in Mexico and Central America (Weidensaul, 2000). Hence, the conservation of Harpy eagles is generally considered to require the protection of nesting trees, suitable habitat territories, and prey species throughout the geographic range. Miranda (2015, 70) writes that in Central and South America, Harpy eagles are threatened by retaliation for predation (imagined and real) on domestic animals, habitat loss, use as food, and by curious settlers and colonists (Trinca et al., 2008; Godoi et al., 2012; Freitas et al. 2014).

The distribution and status of the Harpy eagle varies regionally. In Mexico, it may occur only in Chiapas and to a lesser extent Oaxaca and may be a rare resident from Veracruz and eastern Oaxaca south to Central America (Howell & Webb, 1995). It may be extinct from much of its former range, including most of Veracruz and eastern Oaxaca and perhaps the western part of Chiapas (Howell & Webb, 1995). In Belize, in the past there was only one record (Russell, 1964), but it has more recently been noted as a rare, breeding resident (Jones, 2002, 2003; Jones & Komar, 2008). Rotenberg et al. (2012) documented the fledging of a juvenile, which "represents the first confirmed nesting record for Belize, and it may be near the northern limits of the breeding range of the harpy" (Global Raptor Information Network, 2021bo). In Guatemala, it is a rare resident, with scattered sightings: a record from Alta Verapaz (Land, 1970), a dead specimen (Sclater & Salvin, 1959), a mounted specimen in the Museo Nacional de Historia Natural in Guatemala City, and skeleton from a bird killed on El Petén, at the Natural History Museum of Los Angeles County (Global Raptor Information Network, 2021bo). Small populations have been recorded from northwestern Petén in Guatemala, and nearby Chiapas, Mexico and into southern Belize, and the Atlantic lowlands of nearby areas (Whitacre et al., 2002; Eisermann & Avendaño, 2007).

In Honduras, there were two historic records (Monroe, 1968). There was a photographed record in 1996, near the Rio Plátano (Global Raptor Information Network, 2021bo), another record of two eagles in 1991 in the Valle Sutawala (Anderson et al., 2004), possibly indicating rare presence as a permanent resident in the Moskitia region (Anderson et al., 2004). In Nicaragua, there were two historic records from the early 20th century, one of which is now in the American Museum of Natural History (Martínez-Sánchez & Will, 2010). There was another record from 1999, and another from 2005 in the Indio Maíz Biological Reserve near some Howler monkeys (Múnera-Roldán et al., 2007). In Costa Rica, it was formerly quite widespread, but was nearly extinct in the 1980s (Stiles, 1985; Stiles & Skutch, 1989). Other reports are from Slud (1964), Jones (2002) and May et al. (2010). In Panama, it is rare in areas where it was formerly common (Wetmore, 1965; Ridgely & Gwynne, 1989; Aparicio, 2004).

The largest populations historically and recently are in Brazil, especially in Amazonia (Carvalho, 1949; Zerries, 1962; Sick 1993; Lees, 2006; Trinca et al., 2008; Global Raptor Information Network, 2021bo). It was historically commoner outside Amazonia (Belton, 1978; Sick, 1993). In the South American countries of Colombia, Ecuador, Peru, Venezuela, Guyana, French Guiana and Suriname, it is generally a rare and local resident. In Colombia, it is mostly known in localities west and east of the Andes (Hilty & Brown, 1986; Márquez et al., 2005). In Ecuador, it is commoner in the east and northwest than in the west (Ridgely & Greenfield, 2001; Muñiz-Lopez, 2005; Muñiz-Lopez, 2007). In Peru, there are recent nesting records, which are the first in Peru (Clements & Shany, 2001; Piana, 2007). In Guyana and Venezuela, it is recorded as an uncommon resident in lowland forests (Braun et al., 2000; Hilty, 2003). In French Guiana, it is recorded as widespread in the forests (Thiollay, 2007). In Suriname, there are records from the undisturbed forests in the south, with occasional sightings in coast areas (Haverschmidt & Mees, 1994). In Paraguay, it is a rare breeding resident (Hayes, 1995; Del Castillo & Clay, 2004).

The Harpy eagle is Near Threatened across its range (Collar et al., 1992, 1994; BirdLife International, 2017), the main threats being trophy shooting, deforestation, capture of birds (particularly juveniles) for social use and hunting and trapping or terrestrial prey animals (Ferguson-Lees & Christie, 2001; Alvarez-Cordero, 1996; Lees, 2006; BirdLife International, 2017). It is endangered or local extinct in Mexico (Instituto Nacional de Ecología; Winkler, 1987); rare in Honduras (Anderson et al., 2004); declining in Panama (Ridgely & Gwynne, 1989); Near Threatened in Colombia (Renjifo et al., 2002); Vulnerable in Ecuador (Ridgely & Greenfield, 2001; Muñiz-Lopez et al., 2007); Near Threatened in Bolivia (Global Raptor Information Network, 2021bo); rare in Suriname (Haverschmidt & Mees, 1994), rare in French Guiana (Thiollay, 1989, 2007) and Critically Endangered in Paraguay (Paraguay, 2005). More information on the regional, ecological, nesting, feeding habits and conservation of the Harpy eagle can be found in Alvarez-Cordero (1996), Galetti & Carvalho-Jr (2000), Rotenberg et al. (2000); Piana & Arpia, (2002), Rettig (2002), Sanaiotti (2002), Luz (2005), Trinca et al. (2008), Muniz-Lopez (2005), Vargas-Gonzalez et al. (2006a,b), Giudice et al. (2007), Piana (2007), Crease (2009), Seymour et al. (2010), Sanchez-Lalinde et al. (2011), Vargas-Gonzalez & Vargas (2011), Aguiar-Silva et al. (2012, 2014, 2015), Muniz-Lopez et al. (2012), Vargas-Gonzalez et al. (2014), Kuniy et al.

Fig. 5.2a. Harpy eagle perching [With permission from Chepe Nicoli]

Fig. 5.2b. Harpy eagle perching [With permission from Anibal Pabon]

Fig. 5.2c. Harpy eagle perching [With permission from Robert McNab]

Fig. 5.2d. Harpy eagle collecting twigs for nest [With permission from Giselle Mangini]

Fig. 5.2e. Harpy eagle on nest [With permission from Miguel "Siu" Rios]

Fig. 5.2f. Immature Harpy eagle [With permission from Pia Öberg]

Fig. 5.2g. Immature Harpy eagle [With permission from Nick Athanas]

Fig. 5.2h. Immature Harpy eagle perching [With permission from Nick Athanas]

Fig. 5.2i. Immature Harpy eagle perching [With permission from Dubi Shapiro]

Fig. 5.2j. Immature Harpy eagle perching [With permission from Jonathan Beilby]

Fig. 5.2k. Harpy eagle perching in habitat [With permission from Tristan]

Fig. 5.2l. Harpy eagle habitat, Amazon Forest [With permission from hanming_huang]

Fig. 5.2m. Harpy eagle habitat, Central America [Author]

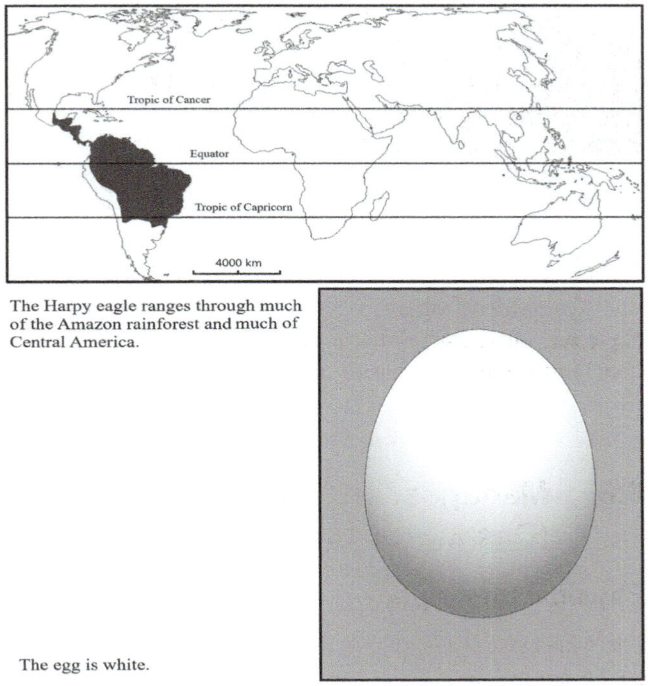

Fig. 5.2n. Harpy eagle range and egg

(2015). Sanaiotti et al. (2015), Sousa et al. (2015) and Watson et al. (2016). Illustrations of the appearance, range, and eggs of this species are shown in Figures 5.2a to 5.2n.

Genus *Harpyopsis* (Salvadori, 1875)

Eagles of this genus are large, with short, rounded wings and long rounded tails. Brown and Amadon (1968, 637) note that the only species of the genus, the Papuan eagle, is "presumably related to *Morphnus* and *Harpia* of South America."

5.4 New Guinea Eagle or Papuan Eagle or Papuan Harpy Eagle *Harpyopsis novaeguineae* (Salvadori, 1875)

Physical Description

The New Guinea eagle is a large and rare eagle endemic to New Guinea (Watson & Asoyama, 2001). Although smaller than the American Harpy eagle, it is described as a "very large, powerful eagle" (BirdLife International, 2016) with a "powerful beak" (Del Hoyo et al., 1994). The total length averages 75 to 90 cm, with a wingspan 157 cm and a weight of 1600 to 2400 grams) (Del Hoyo et al., 1994). The dominant upperpart plumage color is described as greyish brown, grey-brown or brownish grey (Brown & Amadon, 1968; Del Hoyo et al., 1994; BirdLife International, 2016; Debus et al., 2020). There is a short crest on the back of the head, termed by BirdLife International (2016) "an erectile occipital ruff", with buff tips on the head feathers and white tips on the lower back feathers (Brown & Amadon, 1968).

The adult plumage has brownish grey upper parts, with buff tips on the head feathers and white tips on the lower back feathers. There is a small crest on the back of the head. The underparts are white, with a greyish throat and breast (or pale brown upper breast), and spots on the sides in some individuals. The wings are broad and three banded (Del Hoyo et al., 1994). The brownish grey tail has a narrow white tip, with a dark bar just above it and several thinner dark bars across the rest of the tail, and dark barring on the primaries and secondaries (Brown & Amadon, 1968). The tail is long and rounded (Del Hoyo et al., 1994). There is an orange to brownish eye with a large iris and a blackish bill, with a lighter tip. The feet are orange to brownish grey. Immature birds have pale brown, lightly barred underparts rather than white (Brown & Amadon, 1968; Del Hoyo et al., 1994; Debus et al., 2020).

The Papuan eagle may not be mistaken with other similarly sized birds of prey in New Guinea, such as Gurney's eagle (*Aquila gurneyi*, Gray, 1860) and White-bellied Sea eagle or White-breasted Sea eagle (*Haliaeetus leucogaster*, Gmelin, 1788), largely due to its short, rounded short wings, long tail and yellow, bare legs (Brown & Amadon, 1968). BirdLife International (2016) notes that "in size, it is matched only by the short-tailed White-bellied Sea-eagle *Haliaeetus leucogaster* and the uniformly very dark brown Gurney's Eagle *Aquila gurneyi*. In plumage, it is similar to the smaller Long-tailed Buzzard *Henicopernis longicauda* and Doria's Hawk *Megatriorchisdoriae* but has an unstreaked breast." The voice of a zoo individual was given as "an unpleasant, high-pitched, short, vibrant cry of irritation" (Brown & Amadon, 1968). BirdLife International (2016) describe "repeated, deep, resonating calls and grunts, often at night", which can be "heard and occasionally seen in any large forest with limited shotgun hunting." Illustrations of the appearance, range, and eggs of this species are shown in Figures 5.3a to 5.3e.

Classification

The New Guinea eagle is monotypic (Global Raptor Information Network, 2021bp). One finding by Debus (1994) hinted that this eagle was distant from the Harpy or Philippine eagles, and possibly closer to Australasian forms.However, Lerner and Mindell (2005) used more advanced molecular sequences of two mitochondrial genes and one nuclear intron, and found that the genera *Harpyopsis*, along with *Harpia* (Harpy eagles), *Morphnus* (Crested eagles) are closely related and together are a clade. These genera were found to be at least only distantly related to the Philippine Eagle, which is in or close to a clade of snake eagles.

Foraging

The New Guinea eagle's favored habitat is lowland and low elevation forest, with foraging involving perching and soaring (Brown & Amadon, 1968, 637). Some findings indicate that it also occurs in gallery forest within savanna landcover, and monsoon scrub (stunted tree or shrubs, under fluctuations in rain) up to elevations over 3000 meters and clearings (Coates, 1985; Beehler et al., 1992; Debus, 1994). Peckover & Filewood (1976) described this species primarily as a montane species, despite its occurrence in suitable lowland habitat. The habitat frequented by this species varies according to landcover across its range. Beehler (1978) designated this species as a non-disturbed forest specialist, absent from non-forest landcover, usually foraging just over the canopy, an assessment shared by Beehler et al. (1986). Diamond (1972) also observed eagles flying low over the forest or perched inside the crowns of trees. Debus (1994) also described low flight over the canopy or through the forest, less commonly over clearings, and perching inside the canopy. Rand and Gilliard (1967) add new behaviors, frequenting forest edges and soaring flight over steep forested slopes or ridges, a flight pattern that "has not been reported by other observers" (Global Raptor Information Network, 2021bp). Rand and Gilliard (1967) and Diamond (1972) recorded soaring flight, while Watson and Asoyama (2001) never recorded such soaring flight. Bailey (1992) sighted an eagle in disturbed forest, and Diamond (1972) describes crepuscular (evening), or even partially nocturnal (night) foraging or calling.

In terms of prey taken, small mammals, such as wallabies (e.g. *Macropus* species, Shaw, 1790) and "piglets" have been described (Brown & Amadon, 1968, 637; Mayr & Gilliard, 1954; Rand & Gilliard, 1967; Diamond, 1972). Other species named include possums, cuscuses, giant rats and dogs, birds, lizards and snakes (Diamond, 1972; Coates, 1985; Mack & Wright, 1996; Watson & Asoyama, 2001). According to a study by Legra (2005) mammals comprised 70% of the diet, birds 23%, and reptiles 7%. One record found the remains of a cuscus (*Phalanger gymnotis*, Peters & Doria, 1875) and a Painted ringtail possum (*Pseudochirulus forbesi*, Thomas, 1887) (Mack & Wright, 1996). The local marsupial, called Kapul is a frequent kill, hence the name Kapuleagle (Mayr & Gilliard, 1954; Coates, 1977). Prey may be taken from the ground by diving from perches or pursued by movement on the ground or walking up trunks and branches. Prey may also be caught from tree cavities and hiding in hanging vegetation and epiphytes (Beehler et al., 1992; Debus, 1994; Coates & Peckover, 2001; Watson & Asoyama, 2001). One record has a New Guinea Harpy eagle chasing a monitor lizard (*Varanus indicus*, Daudin, 1802) along and up tree trunks and branches (Beehler et al., 1992).

Some researchers have speculated that the facial disc, owl-like in appearance may have auditory functions, trapping noise and facilitating the tracking of prey, especially in the evening or night (Diamond, 1972; Coates, 1977). Coates (1977) supports this surmise with a description of hunting behavior: "Perching in the canopy in its characteristic upright pose with tail curving forwards, it peers inquisitively downward. Its head is bobbed from side to side and sometime turned through peculiar angles in an attempt to get a better view. The erected ruff on its head perhaps helps it focus on slight sounds down below." One record was of an eagle attacked by a Blyth's Hornbill (*Rhyticeros plicatus*, J.R. Forst., 1781) and a Sulphur-crested Cockatoo (*Cacatua galerita*, Latham, 1790), indicates these species as prey (Mack & Wright, 1996). One record that the eagle capturing a child is false (Majnep & Bulmer, 1977).

Breeding

The breeding season begins early in the year, during the wet and dry seasons, and nesting is from April to August (Mayr & Gilliard, 1954; Coates, 1985; Debus, 1994). Nests are generally very large due to reuse, reaching 3 meters high and 3 meters deep, composed of epiphytic vegetation and located just below the crown of an emergent tree, on a large branch (Peckover & Filewood,

1976; Legra, 2005). Legra (2005) found that many nests are located on horizontal branches a few meters from the main tree trunk, which may provide space for the movement of fledging immature birds and the catching of food. The commonest trees used were *Syzygium* (*Myrtaceae*) trees, usually emergent averaging 36 meters and nearly 10 meters higher than the surrounding trees (Legra, 2005). One nest was 18.3 meters up in a tree in montane forest (Rand & Gilliard, 1967). Another nest, with one young bird was located 20 m high in a montane forest tree (Coates, 1985). One or two eggs many be laid, only one young bird was fledged from recorded nests (Coates, 1985; Legra, 2005). The chick is fed by both parents (Global Raptor Information Network, 2021bp).

Population Status

The population of the New Guinea eagle at the start of the breeding season is estimated at 1,000 to 10,000 birds (Ferguson-Lees & Christie, 2001). This species is Vulnerable to the IUCN Red List of Threatened Species (BirdLife International, 2016). However, Global Raptor Information Network (2021bp) reports that "very little is actually known about the population size or density of this species." The New Guinea eagle is endemic to Australasia, specifically New Guinea, historically ranging over savanna, lowland and mountain forest up to 2,000 m. Forests include montane forest (1,829-3,048 m in the Central Highlands, 2,896 m in the Eastern Highlands, 2,800 m in the Wau region, and 3,200 or 3932 m in New Guinea in general) and lowlands (Mayr, 1941; Mayr & Gilliard, 1954; Rand & Gilliard, 1967; Diamond 1972; Beehler 1978, 1986; Coates, 1985; Debus, 1994; Coates, 2001; Debus et al., 2020). It is non-migratory, but juveniles disperse from breeding areas (Bildstein, 2006). It is affected by hunting (for its feathers for ceremonial head-dresses) and habitat loss mostly deforestation due to logging and losses of prey species to hunting (Coates, 2001; Watson & Asoyama, 2001; Legra, 2005; Debus et al., 2020). Although Coates (1977) wrote that feather hunting was still done in the mid 20th century, more recent reports note that this practice has declined (Healey, 1990; Watson & Asoyama, 2001). Illustrations of the appearance, range, and eggs of this species are shown in Figures 5.3a to 5.3e.

Genus *Buteogallus* (Lesson, 1830); formerly Genus *Harpyhaliaetus* (Lafresnaye, 1842)

Eagles of this genus are medium-large, neotropical raptors, some of which are referred to as hawks and others as eagles. Their status as eagles appears disputed, as they were formerly in the Genus *Harpyhaliaetus* (Lafresnaye, 1842). This was considered a genus of eagles, with two species: the Solitary eagle, *Buteogallus solitarius* (Tschudi, 1844) and the Chaco or Crowned eagle, *Buteogallus coronatus* (Vieillot, 1817). Both species were found to be closely related to the Black hawks, hence the Genus *Harpyhaliaetus* was cancelled, and the two species were moved to *Buteogallus*. The two species have been researched and reported over their ranges, sometimes in comparison with other eagle species and have been found to play an eagle-like role throughout mostly forested Latin America and have attracted copious literature on their biology and ecology and those of

Fig. 5.3a. New Guinea Harpy eagle perching [With permission from Gail Hampshire]

Fig. 5.3b. New Guinea Harpy eagle perching [With permission from Johannes Pfleiderer]

Fig. 5.3c. New Guinea Harpy eagle [With permission from Mikael Bauer

Fig. 5.3d. New Guinea Harpy eagle habitat, Typical Forest of New Guinea [With permission from Zacky Ezedin]

associated prey and competitive species. Most of such studies refer to these two eagles as part of the Genus *Harpyhaliaetus*, before the shift *to Buteogallus.*

Earlier works for example, for the Solitary eagle include: Carriker (1910); Griscom (1932); van Rossem (1948); Amadon (1949); Meyer de Schauensee (1949); Sutton (1951); Shäfer & Phelps (1954); Stager (1954); Slud (1964); Wetmore (1965); Monroe (1968); Land (1970); Blake (1977); Harrison & Kiff (1977); Meyer de Schauensee & Phelps (1978); Remsen & Ridgely (1980); Ridgely (1980); Schulenberg & Parker (1981); Amadon (1982); Dunning (1982); Parker et al. (1982, 1985, 1991); Smith (1982); Hilty (1985); Olrog (1985); Stiles (1985); Thiollay (1985a, 1985b, 1985c); Weyer (1985); Clinton-Eitniear (1986); Collar (1986); Hilty & Brown (1986); Meyburg (1986); Ortiz-Crespo (1986); Ramos (1986); Kamstra (1987); Collar & Andrew (1988); Binford (1989); Stiles & Skutch (1989); Remsen & Traylor (1989); Ridgely & Gwynne (1989); Vannini (1989); Ortiz et al. (1990); Robbins & Ridgely (1990); Fjeldså & Krabbe (1990); Clinton-Eitniear (1991); Beavers et al. (1991); Best & Clarke (1991); Willard et al. (1991); Collar et al. (1992); Martinez-Gomez (1992); Tostain et al. (1992); Winker et al. (1992); Best et al. (1993); Bierregaard (1994, 1995a, b, 1998); Howell & Webb (1995); Remsen & Parker (1995); Wege & Long (1995); Urbina Torres (1996); Herzog et al. (1997); Parker (1997); Perry et al. (1997); Schaldach & Escalante-Pliego (1997); Schulenberg & Awbrey (1997); Schulenberg & Wust (1997); Winker (1997); American Ornithologists' Union (1998); Bildstein et al. (1998); Russell & Monson (1998); Donegan & Dávalos (1999); De la Peña (1999); Komar (1998); Miller & Miller (1998); and Fraga & Clark (1999).

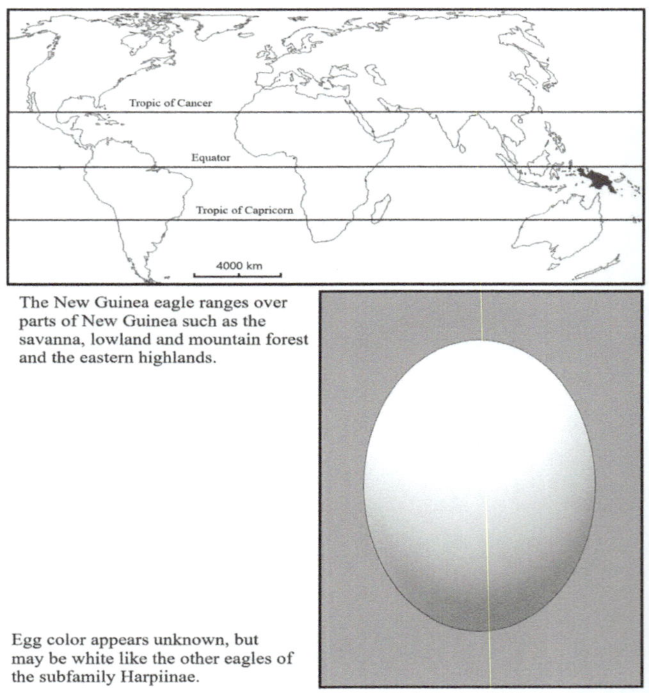

Fig. 5.3e. New Guinea eagle range and egg

More recent (post 20th century) publications include Ceballos & Márquez-Valdelamar (2000); Herzog & Kessler (2002); Iñigo-Elias (2000); Jones et al., 2000); Clements & Shany (2001); Jones & Vallely (2001); Komar & Domínguez (2001); Ridgely & Greenfield (2001); Salaman et al. (2001); Blanco Márquez (2002); Bonta & Anderson (2002); Claessens (2002); Hilty (2002); Komar (2002a, b); Naveda-Rodríguez (2002); Pereira (2002); Pérez-León (2002); Pueblo-Olivares et al. (2002); Mee et al. (2002); Peterson et al. (2003); Renjifo et al. (2002); Walker (2002); Bonta (2003); Hennessey et al. (2003); Ingels et al., 2003); Strewe & Navarro (2003); Roesler & Juhant (2004); De Silva (2005, 2005, 2006a, b; 2007a. b, c); Jones (2003, 2004, 2005, 2006); Seipke (2004); De la Peña (2005); Eisermann (2005); MacLeod et al. (2005); Márquez et al. (2005); Jones & Komar (2008); Kocum (2006); Clark (2007); Clark et al. (2006, 2007); Griffiths et al. (2007); Juhant (2007); Lerner et al., 2008; McCrary et al. (2009). Martínez-Sánchez & Will (2010). Seminario et al. (2011); Phillips (2012); Phillips & Martinez (2013); Phillips et al. (2014) and Novy & Van Putte (2016).

For the Chaco eagle, there are several similar studies that cover its ecology and eagle status, some very old and up until the early 20th century, mostly still referring to this species as *Harpyhaliaetus coronatus* across Brazil, Argentina and Paraguay. Older texts include: Kerr (1892); Dabbene (1908); Naumburg (1930); Peters, 1931; Amadon (1949); Pinto, 1978; Albuquerque (1979, 1986); Stresemann & Amadon (1979); Sick et al. (1981); Belton (1978, 1984); Nores et al. (1983); Contreras et al. (1990); Collar et al. (1992); Delhey (1992); De Lucca (1992, 1993); Dubs (1992); Chebez (1993); Kratter et al. (1993); Sick (1993, 1997); Bierregaard (1994); Ordano & Bosisio (1994); Salvador & Eroles (1994); Cardoso da Silva (1995); Hayes (1995); Wege & Long (1995); Scherer-Neto & Straube (1995); Travaini et al. (1995); Lowen et al. (1996); Bellocq et al. (1998; 2002); Moreno (1997); Ordano & Bosisio (1997); de la Peña (1978, 1987, 2006); Straube (1993); Bellocq et al. (1998); De Andrade & Andrade (1998); Gonnett & Blendinger (1998); Mazar Barnett & Kirwan (1999); Parrin et al. (1999) and Robbins et al. (1999).

More recent (post 20th century) texts include: BirdLife International (2000); Guerrero Ayuso & Arambiza Segund (2001); Mazar Barnett & Pearman (2001); Pulido Jiménez & Scheidegger Soboll (2002); Carvalho et al. (2002); Herzog & Kessler (2002); Bencke et al. (2003); Hennessey et al. (2003); Maceda et al. (2003); Pautasso (2003); Sferco & Nores (2003); Pautasso et al. (2003); del Castillo & Clay (2004; 2005); Pereyra Lobos (2004); Straube et al. (2004); Di Giacomo (2005); Albuquerque et al. (2006); Amaral et al. (2006, 2009); Barcellos & Accordi (2006); Granzinolli et al. (2006); Peyrera Lobos & Alvarado (2006); Sarasola & Maceda (2006); Sigrist (2006); Zorzin et al. (2006); Bragagnolo et al. (2007); Carvalho & Marini (2007); Capdevielle et al. (2007a, b); Clark (2007); Filho et al. (2007); Granzinolli & Motta-Junior (2007); Joenck et al. (2007); Juhant (2007); Lerner (2007); Maceda (2007a, b); Maceda et al. (2007); Pereyra Lobos et al. (2007a,b, 2008); Salvador et al. (2007); Amaral (2008); Carvalho Filho et al. (2008); Lerner et al. (2008); Pereyra Lobos et al. (2007a,b, 2011); Chiaravalloti et al. (2009); Galmes (2009a,b); Carrete et al. (2009); Lopes et al. (2009); Remsen et al. (2009); Saggese et al. (2009); Santillán et al. (2009); Sarasola et al. (2010); Vidoz et al. (2010); Pimentel & Olmos (2011); Berkunsky et al. (2012); Urios et al. (2014); Global Raptor Information Network, 2021bq, br). Alvarado et al. (2015); Barbar et al. (2016); Montalvo et al. (2015); and Monsalvo et al. (2018).

5.5 Black Solitary Eagle or Solitary Eagle *Buteogallus solitarius* (Tschudi, 1844)

Physical Description

The Black Solitary eagle is a large raptor but a medium-sized to medium-large eagle, it is 63–76 cm long, with a wingspan about 152–188 cm and a weight of 2.75 kg. The plumage is dark gray to almost black plumage, with white feathers in a broad, medial band on the tail. The eyes are dark brown, the beak is dark grey, and the cere, face and feet are yellow. The Black Solitary eagle may be mistaken for the common black hawk and great black hawk, but the eagle is much larger with a more massive body, more than double the weight and broader wings (Howell & Webb, 1995; Ferguson-Lees & Christie, 2001; Jones, 2003, 2005; Bierregaard et al., 2020). Illustrations of the appearance, range, and eggs of this species are shown in Figures 5.4a to 5.4g.

Classification

Del Hoyo et al. (2014) note that this species was once placed in the Genus *Harpyhaliaetus*. Bierregaard et al. (2020) also note that it was "previously placed in genus *Harpyhaliaetus*, but molecular studies indicate that present species is part of a clade containing three other dry-habitat members of *Buteogallus*." The Global Raptor Information Network (2021bm) writes that initially, under the work of Hellmayr and Conover (1949), the Black Solitary eagle was classified with *Harpyhaliaetus coronatus*. This classification was criticized by Wetmore (1965), who found major differences between the two species, and suggested different genera. Later molecular data found differences between the two species, but close links (not conspecific, belonging to the same species). Brown (1970) hinted that *Harpyhaliaetus* is closely related to the Harpy eagle group (*Harpiinae*, including the Genera *Harpia* Vieillot, 1816, *Morphnus* Dumont, 1816, and *Harpyopsis* Salvadori, 1875), but the later molecular studies of Lerner and Mindell (2005) and Do Amaral et al. (2006) found closer links between the Black Solitary eagle and the Genus *Buteogallus*, especially the Great Black-hawk, *Buteogallus urubitinga*, Gmelin, 1788) than to eagle genera of *Harpia*, *Aquila* and *Spizaetus* (Lerner et al., 2008).

In the current book, this species is included due to similarities with the eagle species and the fact that some sources still classify it as part of the Genus *Harpyhaliaetus*. Two subspecies have been identified, but this classification is disputed. The Global Raptor Information Network (2021bm) lists *Harpyhaliaetus solitarius solitarius* as the nominate subspecies occurring from northeastern Mexico through Central America to northwestern Argentina; and *Harpyhaliaetus solitarius sheffleri* in western Mexico. However, Sibley and Monroe (1990) cast doubt on this classification by defining this species as monotypic (a monotypic species has no subspecies). Global Raptor Information Network (2021bm) notes that "other authors have suggested that the Mexican race *sheffleri* may be weakly defined. Even if it is valid, the southern extent of its range is poorly known, as few specimens have been critically examined."

Breeding

Global Raptor Information Network (2021bm) reports that few nests have been recorded. This species nests in forest trees (Seminario et al., 2011; Phillips et al., 2014). Two nests were found in Mexico, which were large structures constructed of sticks lined with thinner sticks and fresh leaves, with some palm fronds (Harrison & Kiff 1977). One was in a Yellow Pine (*Pinus ponderosa*), and the other was in a fig tree (*Ficus* sp.). The single egg in one nest was dull white. A third nest was in Belize (Phillips, 2012). One nest in Mexico contained a nestling on 7 June 1947, and the nest with the egg was recorded on 9 May 1958 (Harrison & Kiff, 1977). A breeding pair recorded copulating in early March, near King Vulture Falls, Hidden Valley, Cayo, Belize (Jones, 2005). In June of 2011, a two-month-old nestling was recorded in a nest (Novy & Van Putte, 2016). These recorded at the breeding season starting from March to April to May, with a single egg laid and perhaps fledgling from July, but very few records exist (Novy & Van Putte, 2016). Novy and Van Putte (2016, 32) note that despite gaps in information, their study found from an observed nest that "our observations indicated that the adult male was the primary food provider early in the nesting cycle, while the female remained mostly at the nest."

Foraging

The Black Solitary eagle frequents forests in middle level and middle and higher montane altitudes, usually from 600 to 2,200 meters, more rarely it occurs on lowlands. It has been recorded in dry pine forests in the Sierra Madre Occidental of Mexico, but also occurs in moist forests of Central and South America. Most foraging is soaring or perching on treetops. Prey taken includes snakes, and to a lesser extent birds and small mammals (Ridgely & Greenfield, 2001; Hilty, 2003; Novy & Van Putte, 2016).

Population Status

The Black Solitary eagle is an irruptive or local migrant (Bildstein, 2006). BirdLife International (2020) rates this species as having "a moderately small population size which is likely to be declining owing to habitat loss and poaching. It is consequently classified as Near Threatened but may be uplisted to Vulnerable based on further evidence of its population size and trends." The population may range between 1,000 and 2499 mature individuals, including younger birds this estimated at approximately 1,500-4,000 individuals. BirdLife International further notes that "there are no population density figures and rigorous national or global population estimates are lacking." Hence, the global population is uncertain. It frequents variable landcover from western Mexico, through Central America and north western South America to extreme north-west Argentina. There are no recent reports of its presence from El Salvador, Honduras and Guyana. Throughout its range, it is rare and localized (Ferguson-Lees & Christie, 2001). There is speculation on the presence of this species in Texas (Clark, 2007). Clark (2007, 18) notes a photograph of one specimen "south of Texas, near El Cielo in southern Tamaulipas in the Sierra Madre Oriental"; however, "because of the paucity of birders in this area of Texas, the extreme rarity of the eagle, and the previous lack of information of how to distinguish it from the Common Blackhawk, an uncommon resident in the Davis Mountains, no Solitary Eagles have been reported from Texas." This might change with more accurate identification (Clark, 2007). Illustrations of the appearance, range, and eggs of this species are shown in Figures 5.4a to 5.4g.

5.6 Chaco or Crowned Eagle or Crowned Solitary Eagle *Buteogallus coronatus* (Vieillot, 1817)

Physical Description

The Chaco eagle is a large eagle similar in size or slightly larger than the Black Solitary eagle, but dissimilar in appearance. It measures 73 to 79 cm, with a wingspan of 170-183 cm and a weight of about 3 kg (Ferguson-Lees & Christie, 2001; Do,

Fig. 5.4a. Black Solitary eagle perching [With permission from Peter Hawrylyshyn]

Fig. 5.4b. Black Solitary eagle perching [With permission from Alex Pareja]

Fig. 5.4c. Black Solitary eagle [With permission from Frank McClintock]

Fig. 5.4d. Black Solitary eagle [With permission from Giselle Mangini]

Fig. 5.4e. Black Solitary eagle in flight [With permission from Peter Hawrylyshyn]

Fig. 5.4f. Black Solitary eagle habitat [With permission from Robert Davidson]

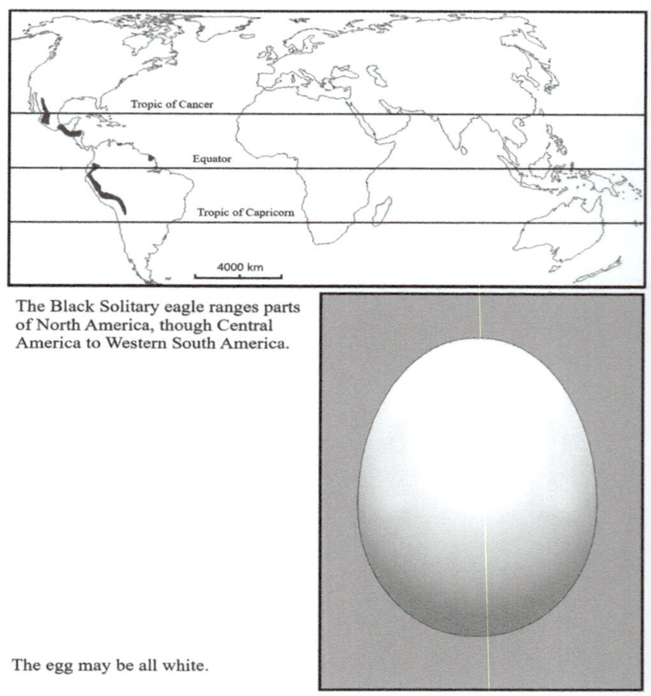

Fig. 5.4g. Black Solitary eagle range and egg

2020). Rather than very dark grey or black, the plumage is gray, with a short black tail with a wide white medial band and a white tip as well as a white tip. There is a crest composed of darker feathers. The immature bird has a dark brown crest, a creamy head and underbody, with gray streaks (Ferguson-Lees & Christie, 2001; Do, 2020). Ebird (2020) describes the Chaco eagle as a "huge imposing eagle" with a "distinctive flight shape with a short tail and very broad inner wings that taper towards the tip." Illustrations of the appearance, range, and eggs of this species are shown in Figures 5.5a to 5.5i.

Classification

Like the Black Solitary eagle, the Chaco eagle was previously (and in some cases still is) classified in the Genus *Harpyhaliaetus* (Lafresnaye, 1842) (e.g., Pereyra Lobos et al., 2011). The Chaco eagle was formerly linked with the Black solitary eagle as one species *Harpyhaliaetus solitarius* (Hellmayr & Conover, 1949), but this classification was challenged by Wetmore (1965) who argued that the differences between the two species were sufficient to place them in separate genera. Brown (1970) thought regarded *Harpyhaliaetus* as closely related to the Harpy eagle group. Sibley and Monroe (1990) saw a relationship between the two species but thought that rather than separate genera, a superspecies might be a more accurate description. Later molecular studies found a close relationship between the Chaco eagle and the Black Solitary eagle, but also found that they are not conspecific (i.e., belonging to the same species). The molecular studies of Lerner and Mindell (2005) and Amaral et al. (2006) found that both species are closer related to the Genus *Buteogallus* (especially the Great Black-hawk) than to the eagles of the genera *Harpia*, *Aquila* and *Spizaetus*.

These studies also found that the Chaco eagle is part of a clade, which includes the Savanna hawk (*Buteogallus meridionalis*, Latham, 1790) and the Great Black hawk (*Buteogallus urubitinga*, Gmelin, 1788).

One recent argument concerned the former name of the Chaco eagle, 'Crowned Eagle.' Schulenberg (2015, Proposal 681) wrote that the eagle *Buteogallus coronatus* "currently is called 'Crowned Eagle' by SACC" (note that the SACC refers to the South American Classification Committee) and that "this name has a long history in the literature on Neotropical birds (dating back at least to Meyer de Schauensee 1966)" In an argument to the South American Classification Committee, Schulenberg (2015) stated that the name 'Crowned eagle', although better than older names (e.g., Crowned Harpy eagle or Crowned Solitary eagle, see Hellmayr & Conover, 1949; Meyer de Schauensee, 1966; Brown & Amadon, 1968; Dickinson & Remsen 2013) confuses with the common name for the African species, the Crowned eagle (see Amadon & Brown, 1968; Britton, 1980; Brown et al., 1982; Borrow & Demey, 2001). Schulenberg (2015) gave two possible solutions: either change the name of the African eagle to a new name such as 'African Crowned Eagle' or 'Crowned Hawk Eagle' (see Sibley & Monroe, 1990; Zimmerman et al., 1996; Fergusson-Lees & Christie, 2001; Stevenson & Fanshawe, 2002), or change the name of the South American species "to remove the basis of the conflict" one suggestion used by some members of the SACC, 'Chaco Eagle.' The current book uses this suggestion, which has been adoped by numerous authors and researchers.

Breeding

The Chaco eagle has a large platform nest composed of stocks, which is usually situated in trees or ravines, usually in the fork of a large tree. One white, unmarked egg is laid (Sick & Telxeira, 1977). However, "its movements are poorly understood, and it may be partially crepuscular" (BirdLife International, 2016). Maceda (2007) records egg-laying in Argentina, lasting from August to October, and hatching from November-December in Argentina. The incubation period at one nest in Argentina nest lasted at least 39 or 40 days, but in this case the nest was abandoned before hatching (Di Giacomo, 2005). Generally, only the female incubates, and the male is present at the nest only when he is delivering food to the female (Global Raptor Information Network, 2021bq).

Foraging

The Chaco eagle frequents lowland, semi-open seasonal dry country, including palm-savanna, light and sparse woodland, and bushy steppes (Collar et al., 1992; BirdLife International, 2016). Do (2020) notes the Chaco eagle as ranging "from eastern Brazil and Bolivia south to central Argentina in caatinga, pampas, cerrado, and chaco." Here, caatinga refers to tropical, dry xeric shrubland and thorn forest, mostly in eastern Brazil (Santos et al., 2011). Leal et al. (2005, 701) describes the caatinga as "735,000 km^2, and "a mosaic of thorn scrub and seasonally dry forests, with more than 2000 species of vascular plants, fishes, reptiles, amphibians, birds, and mammals." The pampas refer to temperate to subtropical grasslands with dry to moist climates, found in eastern Argentina and Uruguay, and small parts of Brazil (Roesch et al., 2009; Ricard, 2019). The cerrado is a vast area of Brazil, with small areas within Paraguay and Bolivia (Oliveira & Marquis, 2002). The chaco is a hot and semiarid lowland, tropical and subtropical dry broadleaf forest, found in Paraguay, Bolivia, and Argentina (Bucher & Huszar, 1999). The Chaco eagle has also been recorded from cattle ranches in Paraguay (BirdLife International, 2016).

The Chaco eagle hunts either by perching and ambush, or "performing circular low-altitude flights approximately 100 m above the ground" (Pereyra Lobos et al., 2011, 180). Sarasola et al. (2010) record its commonest prey in La Pampa, Argentina as armadillos and snakes. The Global Raptor Information Network (2021bq) describes its diet as "mostly on armadillos and other medium-sized mammals, especially skunks, medium-sized birds, lizards, and carrion of various types." In Brazil, it was found to hunt armadillos take carrion (Sick, 1993), and even skunks (Belton, 1984). Pereyra Lobos et al. (2011, 180) further argue that "detailed data on the feeding ecology of the Crowned Eagle has been published" and gave a study of the diet in the Mendoza desert, northwestern Argentina. Pereyra Lobos et al. (2011, 182) supported these findings, as in their study "mammals were the most common prey both by number and biomass" greater than 70% of both number and biomass; nearly 50% of both number and biomass being the dwarf armadillo (*Zaedyuspichiy*, Desmarest, 1804). The next commonest prey species were reptiles, nearly 23% of the total individual prey and total biomass, with the commonest reptiles taken being snakes. Birds were a "negligible" food source (Pereyra Lobos et al., 2011, 182).

Sarasola et al. (2010, 207) found similar results in Argentina, with the prey taken being mammals (67.7%), reptiles (16.2%), birds (3.3%), fish (2.5%), and invertebrates (16.9%). They argued that this indicates that the eagles very rarely preyed on domestic livestock. This was an important finding, as "previous studies have documented that human persecution is the most important threat to the endangered crowned eagle *Harpyhaliaetus coronatus* in central Argentina, due to a local belief that crowned eagles heavily and consistently prey on livestock" (Sarasola et al., 2010, 207). Their findings indicated that only one percent of the prey taken by eagles in their study were domestic livestock, a domestic goat (*Capra aegagrus hircus*, Linnaeus, 1758). These findings therefore do not justify the persecution of this eagle on the grounds that it preys on livestock.

Population Status

This species is an irruptive or local migrant (Bildstein, 2006). Pereyra Lobos et al. (2011, 180) describe the Chaco eagle's range as "a geographic distribution restricted to southern South America, including Bolivia, Brazil, Paraguay, and Argentina" (see also Collar et al., 1992, del Hoyo et al., 1994, Ferguson-Lees and Christie, 2001). The Crowned Solitary Eagle is classified as Endangered by BirdLife International (2016), which notes that "the total population is placed in the band 250-999 mature individuals, equating to 375-1,499 individuals in total, rounded here to 350-1,500" and for the trend justification "a moderate and on-going population decline is suspected to be taking place on the basis of rates of hunting, habitat loss and persecution."

The Global Raptor Information Network (2021bq) describes the Chaco eagle as rare or uncommon throughout its range, due to habitat loss, mostly due to agricultural expansion and direct shooting. It is argued that "the available information suggests that this eagle has a very small, highly fragmented population and probably a significant and continuing decline in numbers." Canal et al. (2017, 235) argued that the Chaco eagle (they used the name Crowned Solitary Eagle) "is one of the rarest and most severely threatened birds of prey in the Neotropical region" but nevertheless found "genetic homogeneity across the sampled regions, which may be explained by a high dispersal capability of Crowned Solitary Eagles resulting in high gene flow or relatively recent population expansion." This meant that despite the "recent demographic reduction", "inbreeding was not detected" (Canal et al., 2017, 235). The results were mostly attributed to the existing connectivity between the widely dispersed populations (Canal et al., 2017).

The rarity of the Chaco eagle has spurred some conservation actions. For example, this species is legally protected in Argentina (Bellocq et al., 2002). In this country, threats include human persecution and habitat disturbance (Sarasola & Maceda, 2006; Sarasola et al., 2010; Barbar, 2016). In Uruguay, Azpiroz and Cortés (2014, 110) note that research findings indicate either extinction or vagrancy and conclude that "unambiguous records in Uruguay are still lacking. Given that the species has recently been observed at several localities in nearby Argentina and Brazil, individuals may reach Uruguay sporadically, probably dispersing juveniles" (Azpiroz & Cortés, 2014, 110). More research is required across its range to evolve effective conservation policies. Illustrations of the appearance, range, and eggs of this species are shown in Figures 5.5a to 5.5i.

Fig. 5.5a. Chaco eagle [With permission from Linda Bushman]

Fig. 5.5b. Chaco eagle perching [With permission from Giselle Mangini]

Fig. 5.5c. Chaco eagle perching [With permission from Jorge Schlemmer]

Fig. 5.5d. Chaco eagle perching [With permission from Nayer Youakim]

Fig. 5.5e. Chaco eagle perching [With permission from Nayer Youakim]

Fig. 5.5f. Chaco eagle in flight [With permission from Jorge Schlemmer]

Fig. 5.5g. Chaco eagle in flight [With permission from Jorge Schlemmer]

Fig. 5.5h. Chaco eagle habitat, Paraguay [Credit: Katelyn Barone – unsplash]

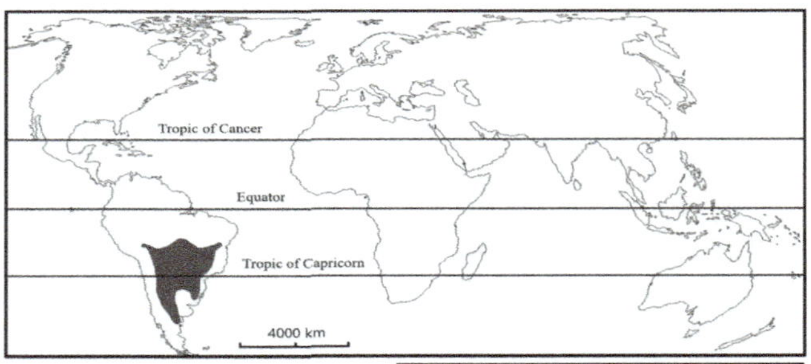

The Chaco eagle ranges across south-central South America.

The egg is white with gray spots.

Fig. 5.5i. Chaco eagle range and egg

Part II

Eagles and Their Ecological Relations

This part looks at the ecology of eagles and environmental relations, both natural and cultural. As an introduction, animals with past and present relations with eagles, as competitors, prey and even predators are examined. As eagles may be both predators and facultative scavengers, the larger megafauna that provided the carcasses upon which eagles fed in competition with vultures and mammal scavengers are also considered. Megafauna in this sense refers to the huge pachyderms, armored mammals and some cases reptiles and large cats. Any discussion of the evolution of eagles must consider the wide range of species that are classified as eagles. *Where did eagles come from, different types of birds of prey*, is a key question. Also, *why did many different types of birds of prey evolve into eagles*? Finally, *at what stage in their evolution would they be considered eagles*? Such a discussion must also consider the definition, method and characteristics of convergent evolution, as particular evolutionary factors may have conditioned different species, with different genetic relations (possibly related to buzzards, kites or even vultures) to slowly evolve into physically similar eagles.

Eagles evolved during the period termed the Age of Mammals, which followed the Age of Reptiles. The abundant mega- and meso-fauna of this period may have been a factor for the large size of some eagles. In this sense, the particular ecological scenarios may be factors for the evolution of eagles from earlier, possibly smaller forms. Eagles evolved during the period termed the Age of Mammals, which followed the evolution of birds during the preceding Age of Reptiles. Palaeontogical evidence indicates that groups ancestral to eagles developed during the Age of Mammals, and the increased numbers of smaller mammals and birds provided food for these birds. Eagles and their ancestral species may also have competed with vultures for the carcasses of the megafauna, in some cases reliant on the large carnivores that opened the carcasses of thick skinned giants for the avian scavengers. These densely inhabited ecosystems possibly facilitated the development of large avian predators and scavengers, with large broad wings for soaring, hunting and scavenging, huge, strong talons and beaks for killing prey and sharp eyesight of foraging from great heights.

A major question concerns the factors for the large size of eagles during this evolutionary period. Researchers argue that the rate of competition with other predators may be a factor for the size of eagles: the less the competition, the larger the eagle. For example, Bunce et al. (2005) point to Haast's eagle, which "represents an extreme example of how freedom from competition on island ecosystems can rapidly influence morphological adaptation and speciation." This eagle lived on New Zealand, where there were no known mammalian competitors and it ranked at the top of the food chain.

The evolution of eagles may be traced to the earliest forms of life, as birds are generally believed to have emerged from reptiles or common reptilian forms, during the past Eons, Eras and Periods (in order of duration, Eons are the longest, followed by Eras and then Periods). A brief discussion of bird evolution is useful at this stage, linked to a chronology of the evolution of life. The discoveries of the bird fossils that fuel our knowledge of bird evolution is largely based on the science of *biostratigraphy* or *paleontological stratigraphy*, which studies the fossil record of organisms in rock layers. Life began during the Precambrian supereon (including the Hadean, Archean, Proterozoic eons) Period, which endured from the Earth's genesis about 4600 million years ago, to the beginning of the Cambrian Period, about 540 million years ago. The Precambrian saw life forms evolve from unicellular to multicellular forms. The Cambrian Period, the first period of the current Phanerozoic Eon (from about 540 million years to the present), which is divided into the Paleozoic, Mesozoic and Cenozoic Eras, was the main period when life forms and species types evolved.

Animal life developed during the Cambrian period, an event referred to as the "Cambrian Explosion" especially during the Paleozoic when aquatic, multicellular forms of mollusks appeared in marine environments and unicellular animals on land. This was followed by the Age of Reptiles, during the next Era, the Mesozoic, which lasted from 252 to 66 million years BP. During the transition between the Paleozoic and the Mesozoic, during the end of the Permian Period of the Paleozoic Era and the Triassic Period of the Mesozoic Era, about 252.28 million years BP there occurred the Permian–Triassic (P–Tr) extinction event or the

Great Dying. During this, possibly the greatest recorded extinction event, about 70 percent of terrestrial vertebrates and 96 percent of marine species became extinct.

The evolution of organisms ancestral to modern birds occurred during this period; these organisms were largely predatory and may have fulfilled the ecological role of modern eagles. The first such animals were the archosaurs, which are generally considered ancestral to both birds (the *Avemetatarsalia*) and dinosaurs (*Pseudosuchia*). The former, named and described by Benton (1999) included the subgroup, *Ornithodira*, which also included the *Pterosauromorpha*, some of which were the first flying vertebrates. More developed birds, such as the Avialans (including the Archaeopteryx) appeared during the following Jurassic Period. These forms were associated with large dinosaurs, mostly the herbivorous sauropods and the carnivorous Theropods. The following Cretaceous Period witnessed the development of birds and mammals, which replaced many of the large reptiles. However, there are few recorded skeletal remains of birds from this period; those discovered are "scarce and usually difficult to interpret" (Olson & Parris, 1987). The Cretaceous–Paleogene extinction event occurred during this period, approximately about 66 million years BP), and the dinosaurs, and birds (including the toothed *enantiornithines* and *hesperornithiforms*) became extinct, basically leaving current life forms of the Cenozoic Era (66 million years BP to present) the Age of mammals (Alvarez et al., 1980; Martin et al., 1996). Larger birds and mammals developed during this period, possibly due to the extinction of the large reptiles. During the first two periods of the Cenozoic Era, the Paleogene and the Neogene, the bird species were like modern species, although some were larger. Records from the last period of the Cenozoic, the Quaternary Period (especially during the Pleistocene Epoch of the Quaternary Period, 2,588,000 to 11,700 years BP) showed many species larger than modern species.

For the evolution of eagles and their associated ecologies, the Cenozoic Era is most important era, as predatory birds and their associated ecologies evolved during this period. The key issues concern the evolution of larger predatory birds and larger mammal competitors, and the possibilities of scavenging competition with avian and mammals, on both prey animals and the megafauna. Eagles evolved within the context of megafauna and large, dominant competitors. Possibly, the large size of some eagles may be due to the large competitors in this scenario. Here, megafauna refers to the huge mammals that largely became extinct during the Quaternary extinction events, this altering the food sources of scavengers and carnivores, both aerial and terrestrial. These extinctions are currently attributed to increased human hunting; climatic change; disease; and/or an asteroid or comet, especially in Americas, northern Eurasia and Australia (Campbell, 2015).

Megafauna and associated carnivores were important for the evolution of eagles, as they also performed the role of scavenging on the carcasses of the huge mammals killed and opened by the mammalian carnivores, competing with the huge, contemporary vultures. Examples of such large carnivores that later suffered extinction in Pleistocene North America included: the dire wolf *Canis dirus* (family Canidae, which was larger and fiercer than the modern wolf (*Canis lupus*, Linnaeus 1758); the huge short faced bear (*Arctodussimus*, Cope 1897), the Florida spectacled bear (*Tremarctos floridanus*, Gidley 1928; family Ursidae), and the huge cat such as the Smilodon of the subfamily Machairodontinae, which were an extinct subfamily of the family Felidae (true cats). These lived in Asia, Africa, North America, South America, and Europe from the Miocene to Pleistocene from about 23 million to 11,000 years BP. The dominant predators and hence providers of carcasses, these huge cats included the earliest Genus *Miomachairodus* from Africa and Turkey (van den Hoek Ostende et al., 2006) and later, three groups; the huge Smilodontini, which had the longest upper canines the Machairodontini or Homotherini, including the Machairodus or Homotherium, with smaller upper canines and the Metailurini, with the smallest teeth (Martin, 1989; Barnett et al., 2005; Van Valkenburgh, 2007; Slater & Van Valkenburgh, 2008; Meloro & Slater, 2012). The Smilodon especially, was a giant apex predator, that killed mastodons, huge ground sloths, bison, camels, and wild horses; its main competitors were the American Lion and the Dire Wolf (Vanvalkenburgh & Hertel, 1993; Coltrain et al., 2004; Fennec, 2005; Christiansen & Harris, 2005).

These predators preyed on mammals much larger than eagles could kill, and hence eagles like vultures would also scavenge from the kills. Prey species for the large carnivores, included the mammoth (*Mammuthus primigenius*, Blumenbach, 1799; family Elephantidae and the American mastodon (*Mammut*, Blumenbach, 1799; family Mammutidae) and Glyptotherium (family Glyptodontidae), Megalonyx (family Megalonychidae), Genera *Pampatheriuma* and *Holmesina* (family Pampatheriidae), *Eremotherium* and *Nothrotheriops* (family Megatheriidae), and *Paramylodon* (family Mylodontidae). These, members of the Order Cingulata (SuperOrder Xenarthra), were very large, placental, vegetarian mammals, armoured, in a similar manner to modern armadillos Dasypodidae Gray, 1821. Glyptodonts, which were some of the largest, weighed as much as two tonnes and measured up to 3.5 m, with a covering of a protective shell of hundreds of 2.5 cm-thick bony plates or osteoderms or scutes. The Order Artiodactyla was also present, including Genera such as *Mylohyus* and *Platygonus* (family Tayassuidae, larger versions of modern peccaries); *Camelops* and *Hemiauchenia* (family Camelidae, related to the modern, generally smaller camels). The Genera *Navahoceros* and *Cervalces* (family Cervidae) were similar to modern deer. Other Genera were *Capromeryx*, *Tetrameryx* and *Stockoceros* of the family Antilocapridae, resembling the modern pronghorn (*Antilocapra americana* Ord, 1815); The Genera *Euceratherium*, *Bootherium*, *Oreamnosharringtoni* (family Bovidae) were similar to the modern musk ox (*Ovibosmoschatus*, Zimmermann 1780 and mountain goat (*Oreamnos americanus*, Blainville 1816).

Smaller mammals could be taken by eagles directly. These smaller mammals included primates and possibly hominins ancestral to the human species. In fact, McGraw and Berger (2013, 289) argue that "Given that raptors actively prey on living primates and the likelihood that several behavioral and life history characteristics were influenced by raptor predation, it is reasonable to expect that raptors have posed a threat to primates, possibly including hominins, throughout their evolutionary history." Gilbert et al. (2009, 421) support this position: "Recent studies suggest a large raptor such as the crowned eagle (*Stephanaoetus coronatus*) was responsible for collecting at least a portion of the primate fauna from the South African fossil site of Taung, including its lone hominin specimen… on the basis of this evidence, we conclude that at least some of the Angolan cercopithecids were most likely raptor prey and hypothesize that raptor predation has been a strong and perhaps underappreciated selective force during the course of primate evolution." Howard (1947, 12) gives the example of the ancestral change in the physical form of the Golden eagle from the Pleistocene, including a reduction in the size of the eagle linked to the extinction of larger prey species: "A comparison of fossil with recent specimens, therefore, reveals the minuteness of change which has taken place over a period of some 50,000 years or more. In the case of the eagle, we may postulate that the changes were adaptive. The strong jaws, and the musculature to operate them, which characterized the Pleistocene form, were appropriate for the day in which hit lived-the day of larger mammals. As smaller forms began to dominate the scene, the need for heavy equipment to cope with larger prey would, presumably, have diminished...the trend toward smaller size (from Pleistocene to Recent time) which, in the overall picture of the raptor group, is observable in the extinction of the larger eagles and vultures, and the increased abundance of the smaller hawks and owls."

Therefore, the largely extinct large, medium and small mammals may have been part of the reason for the evolution of the larger size of eagles, both as predators and scavengers, rivalling the size of the huge, contemporary vultures. This trend would have slowed and been reversed with the decline of the larger prey animals and predatory competitors. Eagles developed relations with other large species, that compete with eagles until today. These include the descendants of the mammalian and avian predators and scavengers. The small, medium and large herbivores descended from these mammals are an important part of the diet of modern eagles, both as live prey and carrion. Most eagles are capable of killing small to medium-sized mammals, and even if they cannot lift the carcass, they can tear it apart on the ground. In the competition for such food sources there are three main sets of competitors: felines (cats), especially the smaller species; canines (dogs), including from the smaller species such as foxes, through coyotes and jackals, to wolves; and other birds of prey, including other eagles, large vultures and smaller hawks. More recently, eagles have competed with domesticated cats and dogs, and even people for food sources and habitat access. As the subsequent chapters will show, there is a complex relationship between eagles and these species.

6

Eagles and Their Competitors

6.1 Introduction

The competition and conflicts between eagles and other avian and mammalian predators are largely driven by the size of the species involved. Regarding avian predators, eagles stand at the top of the food chain as apex predators. They exceed the largest hawks, falcons and owls in size, with mostly superior size in beaks and talons, and generally take larger prey such as large snakes (snake and serpent eagles, with especially strong and scaled feet), fish (sea eagles, which exceed ospreys), medium sized and small mammals (the Harpy eagles) and large and medium-sized birds (some of the *Aquila* and booted eagles). Nevertheless, some raptors are bold enough to challenge at least smaller eagles in predatory, nesting and migratory actions. Eagles are dominant over the *Cathartid* vultures, even the huge condors, at carcasses. They are less dominant over the larger *Accipitrid* vultures, due to the latter's more eagle-like strength and anatomy compared with the more stork-like *Cathartid* vultures. Perhaps the most serious avian rivals for eagles are other larger or similarly sized eagles.

Concerning mammals, the competitors are medium and smaller carnivores, such as canids and felines. Competition has been observed between eagles and domestic dogs, foxes, coyotes, jackals, wolves, bobcats, servals, caracals, lynxes, ocelots and even cougars and leopards. The main determinants of the outcome of such encounters, whether interspecies predation or conflicts over food, is the size and fighting technique of the parties involved. Generally, felines are described as tougher fighters than canids, possibly due to their sharp, retractile claws, their ability to climb trees, jump into the air to intercept flying birds, and their maneuverability. The larger eagles are able to kill animals as large as wolves, but they are also recorded as falling victim to these larger animals.

This chapter looks at the relations of eagles and other eagles, raptors, felines and canids. There is particular emphasis on the role of the larger eagles as indicator species and apex predators. Case studies are presented of Golden eagles in Europe, Asia and North America, Bald eagles in North America, Harpy eagles in Central and South America, and Martial, Crowned and Verreaux's eagles in Africa. The role of human activities in the development of these relations are also examined. Some species of eagle, such as the Golden eagle are much more researched than rarer and smaller species. Therefore, some species are looked at in more detail. The other non-eagle species are also examined for their ecological and physiological characteristics. Remarkable inter-continental similarities exist between different eagle species and their associated competitors. There is evidence that competitors both encourage and avoid conflict, in a continuous relationship. It is concluded that eagles are highly adaptive, able to live alongside such competition with success.

6.2 Eagle Competition with Other Eagles

An interesting study concerns the relationship between the White-tailed eagle and Golden eagle in Scotland and Norway, where important ecological differences were noted. The White-tailed eagle and the Golden eagle share ranges in Europe, with increasing overlaps due to the reintroduction of the former species into its former range. The White-tailed eagle is the larger bird, but it is primarily a fish-eater, while the Golden eagle is faster and a hunter of mammals and birds. Conflicts commonly occur when the two species attempt to compete for fish carrion and other carcasses. Historically, there was conflict between these species, with the larger bird prevailing in many cases. For example, Halley (1998, 171) notes: "In the eighteenth and nineteenth centuries, both the White-tailed Eagle *Haliaeetus albicilla* and the Golden Eagle *Aquila chrysaetos* lived in Scotland, but their ranges there were segregated, the latter's distribution seemingly restricted by competition from the White-tailed; Golden Eagles colonized the Scottish range of White-tailed Eagle following that species' extermination in the early twentieth century." In the 20th century, the reintroduction of the White-tailed eagle into Scotland may have resurrected their historic relationship with the Golden eagle. However, this author also notes a difference between the diet preferences in Scotland and Norway. In Scotland, the main diets for both species are sheep and deer carrion, while in Norway the difference is greater, with more variable diets. Here, "Although Golden Eagles are wholly dominant in direct competition, White-tailed Eagles are nevertheless able to outcompete them through having a wider spectrum of available prey" (Halley, 1998, 171).

In Scotland complications emerge as human action in the form of long-term deforestation and overgrazing, has depleted the habitat resources for both species, hence the increased possibilities for competition for scarce resources. Citing work by Love (1983), Halley (1998) argues that the historic (18th and 19th centuries) Scottish distributions of the two eagles were mostly separate (allopatric), the larger bird being mostly confined to the coast and to a lesser extent near inland freshwater bodies, while the Golden eagles mostly frequented inland ranges. The extinction of the White-tailed eagle at the beginning of the 20th century, was followed by the expansion of the Golden eagle's range into the coastal areas previously occupied by the White-tailed eagle, this occurrence leading researchers to conclude that the presence of the White-tailed eagle was the main determinant of the previous exclusion of the Golden eagle from such areas (Love, 1983; Watson, 1997; Halley, 1998). More recent research appears to support this conclusion, White-tailed eagles have been observed driving Golden eagles from breeding areas in western Scotland, and the highest density of Golden eagle populations are on Scotland's west coast and islands (Watson, 1992a,b; Watson & Rae, 1993). This contrasts with the situation in Norway, where the ranges of the two species overlap, for example along the west coast of Norway, including larger islands and some coastal fjords, with neighboring nests and overlapping home ranges, and scavenging flocks (adults and immatures) of both species attending carcasses (Gjershaug et al., 1994; Halley & Gjershaug, 2008). Adding more complication, Halley & Gjershaug (2008, 295) found in a case study in Norway that the Golden eagle the dominant species, strongly contrasting with the results of the Scottish case study: "Although smaller, Golden Eagles were strongly dominant over Sea Eagles in direct competition for carcass access… Conflicts for carcass access tended to be most escalated between Golden Eagles and least escalated between Sea Eagles, with interspecific conflicts intermediate."

Halley (1998, 172) argues that based on these results, "the situation in Scotland, where the Golden appears to be excluded from certain areas by the presence of White-tailed (competitive exclusion), cannot, therefore, be explained as reflecting an intrinsic feature of the two species' ecology." Explanations for these discrepancies may be found a comparison of the ecologies of the two case studies. The Golden eagle's diet in Norway was mostly hares (Genus *Lepus*, Linnaeus, 1758) and grouse (Subfamily *Tetraoninae*, Vigors, 1825), some other animals and livestock carcasses. The White-tailed eagle's diet was mostly fish, seabirds and waders, and to a much lesser extent, livestock carcasses, hence little dietary overlap with the Golden eagle. Halley & Gjershaug (2008) found that carrion was absent from the diet of the two species in summer, even when it was available. This is contrasted with the situation in Scotland, where the diets of the two species were found to be more similar (an overlap of over 90%), mostly carcases of sheep and deer, and rabbits (Watson et al., 1992a,b, 1993).

There was also the factor of direct competition between the two species. Golden eagles in western Scotland were found to be highly dependent on carrion, and in competition with the newly reintroduced White-tailed eagles (Watson et al., 1992b). Halley (1998, 174) adds that "It has sometimes been claimed in Britain that White-tailed Eagles are dominant over Golden Eagles: a reasonable assertion, as they are larger, heavier birds, and one which would provide a ready explanation for their displacement of Golden Eagle pairs. The opposite is, however, the case" (see also Halley & Gjershaug, 2008). The Golden eagle enjoys significant physical advantages over the larger eagle, having longer legs, a wider talon span and stronger flight, which may enable them to wing aerial fights (Cramp & Simmons, 1979; Halley, 1998).

Environmental degradation is also a factor (Halley, 1998, 176). Although the study areas of Scotland and Norway were environmentally similar, except that Norway was more forested, the Scottish environment was more deforested through intensive landuse, and local extinction of the wildlife, especially grouse and hares, and inshore fish levels (Halley, 1998, 177, see also Hunter, 1994). This depleted the resources available to each species of eagle and hence may have increased the potential for conflict. Therefore, the allopatric populations of these two eagle species may present fewer problems, as the lack of overlapping ranges lessens the potential for conflict between the two populations, but in regions with overlapping populations, the possibility of conflict may complicate conservation efforts (Halley, 1998).

A later study by Whitfield et al. (2013, 67) found that "The breeding season diets of White-tailed Eagles and Golden Eagles in western Scotland were different, and there was no evidence of competition between the two species", and the findings indicated that "Diet breadth was similar between species but diet composition was significantly different, with White-tailed Eagles taking more sheep and aquatic or coastal food items, whereas Golden Eagles took more gamebirds (*Galliformes*, Temminck, 1820), lagomorphs and other terrestrial prey. Diet overlap comparisons, between species-pairs that nested close together or far apart, rejected a hypothesis that there was competition for food, but partially supported an alternative hypothesis that diet overlap indicated abundant shared food. There was no indication that White-tailed eagles have had any long-term effect on the breeding productivity or abundance of territorial Golden Eagles."

A similar study by Sabine and Gardner (1987, 120) examines relations between Golden eagles and Bald eagles, the North American equivalent of the White-tailed eagle, near Salt Lake City, Utah, United States of America. This study is based on the wintering population of Bald eagles, about 200 to 300 of which cluster in the Rush Valley, which compete with Golden eagles, as well as two smaller buzzards, the Ferruginous hawk (*Buteo regalis*, Gray, 1844, the largest North American buzzard) and the Red-tailed hawk (*Buteo jamaicensis*, Gmelin, 1788). Sabine and Gardner (1987, 120) argue that at the time of writing, their study represented "the only reported observations of agonistic interactions between Bald Eagles and Golden Eagles or Ferruginous Hawks." One conflict recorded a conflict between one Bald eagle and two Golden eagles, over a Black-tailed jackrabbit (*Lepus californicus*, Gray, 1837). The fight involved aerial strikes with the first Golden eagle and ground tussles with the second, with the Golden eagle leaving after one minute of ground fighting. After the Bald eagle won the carcass, other Bald eagles arrived and fought over the carcasses. In another incident, a Golden eagle feeding on a jackrabbit, was driven away by a Bald eagle, which was also supplanted by a second Golden eagle, and the Bald eagle was prevented from reapproaching the feeding Golden eagle by a third Golden eagle. The Bald eagle eventually took over the kill, and all the Golden eagles left.

Sabine and Gardner (1987, 120) note that "Based on the incidents of 17 and 24 January, we believe that Bald Eagles may occasionally steal prey from Golden Eagles, which contrasts with the dominant position Golden Eagles appear to have in some Bald Eagle winter roosts" (see also Lish, 1973). The authors attribute the conflicts of the two eagle species to the seriously declining food supply, principally of jackrabbits, the main food of eagles in the Rush Valley according to older sources (Edwards, 1969; Platt, 1976). This decline was so serious that examined eagles lost body fat (Sabine & Gardner, 1987). Bald eagles also attacked Ferruginous Hawks, and the latter also attacked the eagles, as did a Red-tailed hawk (LeDuc, 1970).

6.3 Competition between Eagles and Other Avian Species

One of the more important engagements experienced by eagles is their relationship with various vulture species. This relationship is largely over carrion, as vultures are mostly obligate or full-time scavengers and eagles may interact with vultures as facultative or part-time scavengers (Campbell, 2015). To uncover the complex relationship between these different species requires a complex analysis of the published research findings on this topic. Four main issues dominate the relationship between eagles and vultures in terms of foraging competition: (1) whether the vulture is an Old World (African or Eurasian) *Accipitrid* vulture, or a New World (North or South American) *Cathartid* vulture, this largely determining the strength of the vulture; (2) the size of the respective eagle or vulture; (3) the extent to which the particular eagle species takes carrion; and (4) possibly, the aggressive nature of the eagle or vulture species.

The background of these issues is the main difference between eagle and vulture anatomy. Eagles have much stronger feet and talons, designed for grasping and killing prey, in some cases mammals or birds as large or larger than the eagle concerned. Some eagles have feet better designed for killing large fish or snakes (e.g., Sea eagles, Snake eagles or Serpent eagles). The largest eagles, such as the Harpy, Crowned and Philippine eagles, have feet and talons capable of killing large mammals such as monkeys and even canids and felines. Golden eagles are capable of killing wolves, coyotes and foxes. Vulture feet are relatively weak, mostly designed for walking. Eagles also have stronger, sharper beaks designed for killing mammals and birds, while vulture beaks are more for tearing open carcasses, or reaching into carcasses after other birds or mammals have torn open the carcass (Campbell, 2015).

Considering the first point (whether the vulture is an Old World (African or Eurasian) *Accipitrid* vulture, or a New World (North or South American) *Cathartid* vulture), the most relevant issue concerns the relatively stronger build of the latter. Old World (Accipitrid) vultures are more eagle-like than the New World (*Cathartid*) vultures. These two groups are a result of convergent evolution, the independent development of species from different ancestors, of identical or similar characteristics in species of different Orders, Families or Genera (Sibley & Ahlquist, 1990; Mundy et al., 1992; Del Hoyo et al., 1994; Wink, 1995). These issues are still under debate (Campbell, 2015). The difference between these two types of vultures are the result of the vague idea of what should be termed a vulture, which is currently defined a bird with the following attributes: a featherless or almost featherless head or head and neck, obligate or specialised scavenging habits (rather than facultative or part time scavenging), soaring flight, weaker feet than the predatory raptors, and long intestines with strong stomach acids for the digestion of carrion (Newton, 1990; Brown and Amadon, 1968; Mundy et al., 1992; Glutz von Blotzheim et al., 1971; Cramp & Simmons, 1979; del Hoyo et al., 1994). As noted in detail below, the Acciptrid vultures are usually considered to share common ancestry with eagles or other large raptors, while *Cathartid* vulture are considered either to be related to storks, to share common ancestry with storks, or to be a separate Order with links to both storks and raptors (Houde, 1988; Cracraft & Rich, 1972; Mourer-Chauvire, 1982; Sibley & Ahlquist, 1991; Wink, 1995; Ferguson-Lees & Christie, 2001).

The *Cathartid* and *Accipitrid* vultures differ on fundamental points, relevant to their physical strength and ecological relations. Rea (1983: 45) supporting the placement of the *Catheridae* with the storks (*Ciconiformes*) wrote that "there is nothing in the fossil history to unite the groups or to suggest that one might be derived from the other" and "the most inescapable conclusion from an examination of the *Cathartidae* is that they have virtually no trait, whether anatomical, physiological, cytological or behavioral, that suggests that they are related to the Falconiformes." Contrasts between the two groups of vultures include the following characteristics of the *Cathartids*: thermoregulation (like storks, they defecate on their legs and feet for cooling purposes, termed *urohidrosis*), slightly longer, webbed front toes (Feduccia, 1999); clawed but weaker feet (Krabbe & Fjeldså, 1990); different leg and pelvic muscle anatomy; the lack of a syrinx or voice box (Rea, 1983; Howell & Webb, 1995; Kemp & Newton, 2003); a relatively weaker beak, compared to other raptors (Ryser & Ryser, 1985; Krabbe and Fjeldså, 1990); oval, perforate nostrils not divided by a septum, i.e. one can see through the bill by the nostrils (Terres, 1991; Allaby, 1992) as in the Turkey vultures; and the lack of the bony brow of the raptors (Terres, 1991). Behavioral similarities between the *Cathartids* and the storks, in addition to *urohidrosis*, include wing spreading, bill clapping and bill interlocking before and during copulation, inflation of neck sacs and gender-neutral egg incubation (see also Konig, 1982; Rea, 1983; Seibold & Helbig, 1995).

The scientific studies that distinguished the *Cathartid* vultures from other raptors were at first based on physical assessments. For example, Jollie (1976) used pterylography (the study of the bird's feather arrangements), osteology (the study of bone structure) and myology (the study of muscle structure). Later studies used karyotypes, based on the number and appearance of chromosomes in the nucleus of a eukaryotic cell, and/or the set of chromosomes in a species (Ligon, 1967; de Boer, 1975; König, 1982; Rea, 1983; Seibold et al., 1993, 1995; Wink, 1994, 1995; Wink et al., 1993a,b, 1996). More advanced studies used marker genes, to uncover a "more precise phylogenetic reconstruction of relationships within and between genera, subfamilies and families of birds" (Wink, 1995: 868; see also Avise et al., 1994; Sibley, 1994) because marker genes can be easily amplified by PCR and sequenced for clearer results (Kocher et al., 1989; Hillis & Moritz, 1990; Kornegay et al., 1993; Hedges & Sibley, 1994; Meyer, 1994). Molecular phylogenetics analyses molecular differences in organisms, mostly in DNA sequences, to obtain information on the organism's evolution. Deoxyribonucleic acid (DNA) is a molecule containing the codes of the genetic markers for the development and functioning of living organisms (Sibley & Ahlquist, 1990; Avise, 1994; Sibley, 1994). The nucleotide sequence of the mitochondrial cytochrome b (cyt b) illustrates phylogenetic events, such as species divergence from ancestral forms up to 20 million years BP. This technique was used effectively during the 1990s (Edwards et al., 1991; Richman & Price, 1992; Helm-Bychowsky & Cracraft, 1993; Kocher et al., 1989; Meyer, 1994; Taberlet et al., 1992; Heidrich et al., 1995; Wink, 1994; Wink et al., 1993a,b, 1994, 1996; Seibold et al., 1993, 1995).

Four main results came from these increasingly sophisticated studies. Firstly, the placement of the Cathartids close to the *Ciconiformes* (storks) was challenged (Griffiths, 1994; Cracraft et al., 2004; Ericson et al., 2006; Gibb et al., 2007; Slack et al., 2007; Livezey and Zusi, 2007; Hackett et al., 2008). Secondly, they were classified as a subfamily of storks (Ligon, 1967; Sibley & Monroe, 1990). This, however, was criticized as an oversimplification (Griffiths, 1994; Fain & Houde, 2004). An early DNA sequence study that supported this position (Avise, 1994) was based on erroneous data and subsequently retracted (Cracraft et al., 2004; Gibb et al., 2007; Brown & Mindell, 2009). Thirdly, they were not seen as a subfamily of storks, but closer to storks than raptors (Konig, 1982; Rea, 1983; Sibley & Ahlquist, 1990; Avise et al., 1994; Seibold, 1994; Seibold et al., 1993, 1995). Fourth, the Cathartids were found to be distantly related or unrelated to either the Accipitrid raptors or the storks (Wink, 1995: 880), hinting these vultures "represent a phylogenetically inhomogenous, i.e., polyphylectic group"; descendant from different ancestral groups (see also Mundy et al., 1992; Sibley and Ahlquist, 1990; del Hoyo et al., 1994). Wink's study acknowledged the morphological data and DNA-DNA hybridisation that link the Cathartids to storks ancestrally (see Garrod, 1873; Ligon, 1967; Konig, 1982; Rea, 1983; Sibley and Ahlquist, 1990); however, the results indicated a separation between the *Cathartids* and storks (especially the Genera *Ciconia*, *Leptoptilos* and *Mycteria*) and similar genetic distances between Cathartids and Accipitrid vultures, and the latter and storks. A similar study tested the phylogeny of three species of New World vultures, 11 species of *Cathartid* vultures and other species, using 1026 nucleotides of the mitrochondrial cytochrome b gene (Siebold & Helbig, 1995). The findings indicated that the Cathartids were unrelated to the *Falconiformes*, but also not closer to the storks than to *Falconiformes*.

The *Accipitrid* vultures' evolution and classification is clearer and more agreed among scholars and researchers (Campbell, 2015). An important issue concerns their status as either a monophyletic group (descended from one common ancestor) or a polyphyletic group (descended from several ancestors). The two subfamilies of *Acciptrid* vultures, *Gypaetinae* and *Aegyptiinae* are either monophyletic (Brown & Amadon, 1968; Thiollay, 1994) or polyphyletic (the subfamily *Gypaetinae* including the genera *Gyophierax*, *Neophron*, and *Gypaetus*, separate from the sub-family Aegyptiinae) (Jollie, 1977b; Mundy et al., 1992; Seibold & Helbig, 1995). Within this group, the Cinereous vulture is possibly a sister species to the Lappet-faced vulture (Wink, 1995: 875). The Genus *Gyps* (Griffon vultures) possibly share a common ancestor. The Genera *Neophron* and *Gypaetus* (the Egyptian and Bearded vultures) were related, using advanced DNA studies (Wink, 1995), affirming the findings of studies using karyological (de Boer & Sinoo, 1984) morphological, anatomical (Jollie, 1976, 1997a,b) and embryological data sources (Thaler et al., 1986).

Another important issue concerns the relation of some *Accipitrid* vultures to some eagles, buzzards or kites. Some species (e.g. the feather-headed Bearded and Palm-nut vultures of the subfamily Gypaetinae) may be closer to other raptors than to the other *Accipitrid* vultures (Jollie, 1976, 1977a,b; Mundy et al., 1992; Lerner & Mindell, 2005; Wink, 1995). The Palm-nut vulture is sometimes surmised to be intermediate between vultures and Sea eagles (Brown & Amadon, 1968). The *Aegypiinae* is sometimes viewed as a monophyletic group, perhaps related to the *Circaetinae* Snake eagles (especially the Madagascan Serpent eagle (*Eutriorchis astur*, Sharpe, 1875) or to the *Aquilinae* eagles, while the *Gypaetinae* vultures appear to be related to the *Perninae* hawks (Wink, 1995). Close relations have been found between the Gyps/Aegypius clade of vultures and the Short-toed Snake eagle (*Circaetus gallicus*, Gmelin, 1788) (Jollie, 1976, 1977a,b; Mundy et al., 1992) and between Neophron/Gypaetus vultures and the Honey buzzard (*Pernis apivorus*, Linnaeus, 1758) (Wink, 1995).

These published findings indicate the status of the *Cathartid* and *Accipitrid* vultures, and their relationship with raptors such as eagles (Campbell, 2015). The question emerging from these complex details concerns the impact on the relationship with raptors such as eagles in foraging activities. Presumably, the *Acciptrid* vultures are closer to eagles, and this may translate into superior strength and foraging abilities. Therefore, the crucial question is whether the stronger, more eagle-like physique of the *Accipitrid* vultures translates into superior foraging capabilities and success in competition with large raptors, than the ability of similarly sized *Cathartid* vultures. Current research findings indicate that the *Accipitrid* vultures are either physically stronger or similar to similarly-sized *Cathartid* vultures. This superiority is more marked for the larger species, such as the *Accipitrid* Lappet-faced, Cinereous and Griffon vultures, which are generally stronger towards competitors than the *Cathartid* Andean and California Condors. The smaller vultures, such as the *Accipitrid* Egyptian and Hooded vultures are more similar in this respect to the similarly sized *Cathartid* Turkey and Black vultures (Campbell, 2015). Table 6.1 gives the approximate sizes of the larger vultures and the eagles likely to interact with them at carrion sites. Table 6.2 shows a summary of the research on the competition between eagles and vultures, based on the vultures' perspective. Figures 6.1 to 6.11 illustrate some of the vultures and scavengers that compete with eagles.

Table 6.1. The largest *Cathartid* and *Accipitrid* vultures

English name	Scientific name	Mass (kg)	Length (cm)	Wingspan (cm)
Cathartid vultures				
Andean condor	*Vulturgryphus* (Linnaeus, 1758)	8–15	100–130	2.7–3.2
California condor	*Gymnogyps californianus* (Shaw, 1797)	7–14.1	109–140	2.5–3
Accipitrid vultures				
Cinereous vulture	*Aegypius monachus* (Linnaeus, 1766)	6.3–14	98–120	2.5–3.1
Himalayan griffon	*Gyps himalayensis* (Hume, 1869)	8–12	109–115	2.6–3.1
Lappet-faced vulture	*Torgostracheliotos* (Forster, 1791)	4.4–13.6	78–115	2.5–2.9
Griffon vulture	*Gyps fulvus* (Hablizl, 1783)	6.2–11.3	93–122	2.3–2.8
Cape vulture	*Gyps coprotheres* (Forster, 1798)	7–11	96–115	2.3–2.6
Rüppell's vulture	*Gyps rueppelli* (Alfred Brehm, 1852)	6.4–9	85–103	2.3–2.6
White-headed vulture	*Trigonoceps occipitalis* (Burchell, 1824)	4–4.7	72–85	2.1–2.3
Eagles				
Bald eagle	*Haliaeetus leucocephalus* (Linnaeus, 1766)	4.1–5.6	70–102	1.8–2.3
Golden eagle	*Aquila chrysaetos* (Linnaeus, 1758)	4–6.4	66 -102	1.8–2.3
Martial eagle	*Polemaetus bellicosus* (Daudin, 1800)	3–6.2	78–96	1.9–2.6
Tawny eagle	*Aquila rapax* (Temminck, 1828)	1.6–3	60–75	1.6–1.9
Bateleur eagle	*Terathopiusecaudatus* (Daudin, 1800)	2–2.6	55–70	1.9
Verreaux's eagle	*Aquila verreauxii* (Lesson, 1830)	3–5.8	75–96	1.8–2.3
Steppe eagle	*Aquila nipalensis* (Hodgson, 1833)	2–4.9	62–81	1.7–2.2
Imperial eagle	*Aquila heliacal* (Savigny, 1809)	2.5–4.6	72–90	1.8–2.2
Booted eagle	*Hieraaetus pennatus* (Gmelin, 1788)	1	46	1.2
Bonelli's eagle	*Aquila fasciata* (Vieillot, 1822)	1.6–2.5	55–65	1.5–1.8
Spotted Eagle	*Clanga pomarine* (Brehm, 1831)	3.2	59–71	1.6 –1.8

Sources: del Hoyo et al. (1994, 2020); Ferguson-Lees & Christie (2001); Campbell (2015). The smaller vultures, such as the Hooded, Egyptian, Turkey and Black vultures are not posted, as they are easily dominated by their sympatric eagles. These vultures are approximately the same size as the Spotted eagle at the bottom of the table.

Table 6.2. Possible dominance hierarchy of *Cathartid* and *Accipitrid* vultures

Vulture	Notes
Lappet-faced vulture	Aggressive and possibly dominant over Golden and similarly sized eagles
Cinereous vulture	Approximately equal of the Lappet-faced vulture
Himalayan vulture	Approximately equal of the Lappet-faced vulture, but less aggressive
Griffon vulture	Smaller than Lappet-faced vulture, social but less aggressive
White-headed vulture	Aggressive, match for most eagles
Ruppell's vulture	Slightly larger than the White-headed vulture, social aggressive
Andean condor	Slightly larger than the California condor, possibly subordinate to eagles
California condor	Subordinate to Golden and Bald eagles
Hooded vulture	Larger and more aggressive than the Egyptian vulture, inferior to eagles
Egyptian vulture	Less aggressive and slightly smaller than the Hooded vulture
King vulture	Larger than Turkey vulture
Turkey vulture	Marginally dominant over solitary Black vulture, inferior to all eagles
Black vulture	Subordinate to all, but dominant in flocks

A trawl and analysis of the relevant literature may answer these issues, especially the second issue posed above: the importance of the size of the respective eagle or vulture in competitive activities. Is size more important in the case of the *Accipitrid* vulture/eagle match (as a large *Accipitrid*, eagle-like vulture may be too much of a match for a smaller eagle) than the *Cathartid* vulture/eagle match (where the weaker, more stork-like physique of the *Cathartid* vulture may not be offset by its larger size compared with the eagle)? Two hypotheses may be derived: (1) for competition between eagles and *Accipitrid* vultures, size is more important than the vulture/raptor distinction; and (2) for competition between eagles and *Cathartid* vultures, the vulture/raptor distinction is more important than size. Hence, eagles will be dominant over the largest *Cathartid* vultures, while eagles will be less dominant over the largest *Accipitrid* vultures. An inference is that the condors are weaker than the medium sized *Accipitrid* vultures, based on the relations with similarly sized eagles. The vultures are ranked top to bottom by their possible physical strength in competition.

In the trawl of research findings, the main lines were of enquiry concerned the relation between the *Cathartids* and their main competitors, the Golden eagle and Bald eagle and between the *Accipitrid* vultures, and their main competitors the Tawny eagle, Bateleur eagle, Bonelli's eagle, Booted eagle, Golden eagle, Imperial eagle, Spotted eagle, Steppe eagle and Verreaux's eagle (Table 6.1). In this survey, the *Accipitrid* and *Cathartid* vultures were divided into two groups (large and small) and related to eagles and canids, to assess their competitive interactions. For the canid species, the similarly sized jackals (Eurasia and Africa) and coyotes (Americas) were considered. These are facultative (part time) scavengers; carrion accounts for less than 5 percent of jackal diets (Houston, 1979; Loveridge & Macdonald, 2002). Larger species, which may be too large for eagle competition were omitted: these included wolves (*Canis lupus*, Linnaeus, 1758), feral dogs, Spotted Hyena (*Crocuta Crocuta*, Erxleben, 1777), Black bear (*Ursus americanus* Pallas, 1780, Brown bear (Ursus arctos, Linnaeus, 1758), cougar (*Puma concolor*, Linnaeus, 1771), African Hunting dog (*Lycaon pictus*, Temminck, 1820), lion (*Panthera leo*, Linnaeus, 1758), leopard (*Panthera pardus*, Linnaeus, 1758) and cheetah (*Acinonyx jubatus*, Schreber, 1775).

The Golden and Bald eagles are the main eagles of the Americas that interact with vultures. Other eagles such as the Harpy eagle are only rarely competitors for carrion. For example, Jonesand & Dorward (2014, 27) note that the Ornate Hawk eagle of Central and South America is a "species formerly assumed only to prey on live food items" and referring to this species within a guild of Amazonian raptors including the Black-and-white Hawk-eagle, Black Hawk-eagle, Crested eagle and Harpy eagle, "scavenging behavior or feeding on carrion appears to be very rare in this guild of species, although there are some records of species feeding on stored prey that are likely to be at least partially decomposed." They also argue that "we believe the interactions with carrion by Ornate Hawk-eagle to be the first records of possible scavenging behavior in the species" (Jones & Dorward, 2014, 27). They also refer to a finding by Springer et al. (2011) of a Harpy Eagle, feeding on a dead Three-toed Sloth (*Bradypus variegatus*) "the first report of a non-scavenging raptor repeatedly feeding for more than two days on prey" (Jones & Dorward, 2014, 27). Globally, regarding eagles "such behavior also seems to be rare or unrecorded in similar species worldwide, although (African) Crowned Eagles… reportedly rarely feed on carrion (although no specific example is given) and, after a kill, have been recorded to cache and re-visit food for several days" (Jones & Dorward, 2014, 28; see also Ferguson-Lees & Christie 2001). To eagle scavenging levels in tropical forests, camera trap surveys "may reveal opportunistic scavenging to be more widespread in tropical forest raptors than has generally been assumed" (Jones & Dorward, 2014, 27).

Fig. 6.1. Andean condor perching [With permission from Sergey Pisarevskiy]

Fig. 6.2. California condor perching [With permission from Kojihirano]

Fig. 6.3. Turkey vulture perching [With permission from Todd Wynia]

Fig. 6.4. Black vulture perching [With permission from Catherine Zinsky]

Fig. 6.5. Cinereous vulture perching [With permission from Ali Tellioglu]

Fig. 6.6. Lappet-faced vulture perching [With permission from Sharon Verkuilen]

Fig. 6.7. Griffon vulture perching [With permission from Laura Barrio]

Fig. 6.8. Griffon vulture perching [With permission from Mike Nelson-Birdtour ASIA]

Fig. 6.9. Rüppell's Griffon vulture perching [With permission from Nirav Shah - Pexels]

Fig. 6.10. White Headed vulture perching [With permission from Leon Molenaar]

Fig. 6.11. Marabou storks may compete with eagles and vultures at carrion [With permission from Elton Cilliers Picfair]

Fig. 6.12. Osprey perching [Credit: Frank Cone - Pexels]

Fig. 6.13. Osprey in flight. These birds are competitors with Bald eagles for fish [With permission from Pixabay Pexels]

Fig. 6.14. Black backed jackal [With permission from Fons Buts]

Back to the studies of Golden and Bald eagles, their competition with vultures appears to be non-predatory, even the small Black and Turkey vultures have no "major predators in North America away from their nests" and with "few published accounts of interactions of either the Black vulture (*Coragypsatratus*) or the Turkey vulture (*Cathartes aura*) with large predatory birds", there yet might be "competitive interactions due to overlap in use of carrion" (Stolen, 1996, 43; see also Townsend, 1937; Jackson, 1988a, b). However, both eagles have absolute dominance over all the *Cathartids*, encouraging leaving behavior in the latter (Stolen, 1996; Snyder & Snyder, 2000). California condors even avoid nesting where Golden eagles were common (Snyder & Snyder, 2000). In dominance at carcasses, Golden eagles are ranked above Bald eagles (Wilmers et al., 2003). Nevertheless, there are records of Bald eagles dominating vultures (Meinertzhagen, 1959; Brown & Amadon, 1968). Golden eagles dominate California condors, which normally dominate other species at carcasses (United States Fish and Wildlife Service, 1996; Snyder & Schmitt, 2002). California Condors avoided nesting where Golden Eagles were common (Snyder and Snyder, 2000). Over carcasses, eagles were dominant over condors even where the eagles were totally outnumbered. For example, in successive incidents one eagle drove 16, 14, and 12 condors from carcasses, in other cases the condors only waited in the distance and in one recorded incident a group of condors could crowd the eagles off the food. The dominance of Golden Eagles over condors was greater than that of condors over Turkey vultures (Koford, 1953).

Fig. 6.15. Coyote [Credit: Old Fulica]

In one incident, an immature Golden eagle disturbed a large, feeding flock of Turkey and Black vultures and attacked one Black vulture, causing it to regurgitate its food (Coleman & Fraser, 1986). A reintroduced California condor was killed by a Golden eagle (Meretsky et al., 2000) and Golden eagles among other predators were implicated in 18 condor deaths (Rideout et al., 2012). Oberholser (in Bent, 1937) observed that Bald eagles do not ". . . hesitate even to pursue the vultures and compel them to disgorge, when if it fails to catch the coveted morsels before they reach the ground it alights and devours them." The arrival of eagles at a carcass caused vultures to ". . . retire and patiently wait until their betters are satisfied", this being an example of strong dominance. In one incident a Bald eagle killed a vulture that "for some reason was unable completely to disgorge" (Audubon, 1937 in Bent, 1937).

Another study documented two Bald eagle flyovers that drove two flocks of vultures (seven Black and three Turkey vultures and nine Black vultures and one Turkey vulture) from carcasses; Black vultures "appeared agitated and...flew between perches and looked around more" when Bald eagles were harassing Ospreys within 100 meters (Stolen, 1996: 43). Also, when a Bald eagle arrived in the vultures' roost tree, all the Black and Turkey vultures were perched 200 meters distant (Stolen, 1996, 43). Therefore, "interactions with large predatory birds may constitute a selective pressure on Black and Turkey vulture behavior" (Stolen, 1996, 43). Eagle-vulture interactions are important at carcasses; vultures regurgitate food in response to harassment or threats from eagles and other predators (for different opinions, see also Maynard, 1881; Brown & Amadon, 1968; Jackson, 1988a,b; Ritter, 1983).

Even other smaller hawks such as the Variable hawk (*Geranoaetus polyosoma*, Quoy & Gaimard, 1824) and Southern Crested caracara (*Caracara plancus*, Miller, 1777), which are smaller than Turkey or Black vultures (interspecies average length 49–65 cm, wingspan 107–132 cm, weight 800–1600 g) are generally dominant over these vultures (Rodriguez-Estella & Rivera-Rodriguez, 1992; see also Sherrod, 1978; Glazener, 1964; Strange, 1996; Catry et al., 2008; Dwyer & Cockwell, 2011). Only Andean condors and King vultures can hold their own against these hawks and the Turkey vulture may dominate the smaller Striated caracara (*Phalcoboenus australis*, Gmelin, 1788) (Wallace & Temple, 1987).

Koford (1953) found that cougars and to a lesser extent Black bears provided carcasses, but bears were more like than cougars to be competitors at carcasses. Golden eagles could drive as many as 16 Californian condors from such feeding sources, but the condors gradually lost their fear with each attack. This study found that the dominance ranking (what might be termed the pecking order) was Golden eagles at the top, followed by condors, then turkey vultures and finally ravens, with the largest gap between eagles and condors, and the smallest gap between turkey vultures and ravens (Koford, 1953).

Wilmer et al. (2003) found that in the Greater Yellowstone Ecosystem, the dominance ranking was coyotes, Golden eagles, Bald eagles, ravens and magpies. In this area, the providers of carcasses were coyote, wolf (*Canis lupus*), cougar (*Puma concolor*), grizzly bear (*Ursus arctos*), and black bear (*U. americanus*), with the wolves and humans the top providers, and elk were the main carcasses (see also, Murie, 1940; Gese et al., 1996; Lemke et al., 1998; Crabtree & Sheldon, 1999a,b; Mech et al., 2001). There were significant foraging differences between the Golden and Bald eagles. The Golden eagle is more sedentary and solitary, with a smaller feeding radii and lacking "the degree of social interaction that enhances information transfer concerning resource locations" (Wilmer et al., 2003: 997, see also LeFrank Jr. & Clark, 1983; Heinrich, 1988; Marzluff et al., 1996; Buehler, 2000; Dall, 2002). In the study by Wilmer et al. (2003) human provided carcasses were relatively locally located and dominated by local scavengers with lower feeding radii (such as magpies), while wolf provided carcasses, which were widely dispersed and dominated by scavengers with wider feeding radii (such as Golden Eagles). Although vultures were absent, due to the winter period.

The dominance of eagles over the *Cathartid* vultures can easily be seen by their dominance over the huge condors, which are also dominant over the smaller vultures (Campbell, 2015). The dominance hierarchy among the *Cathartid* vultures is: (1) condors; (2) the King vulture; (3) *Cathartes* vultures i.e. Turkey vulture, Greater Yellow-headed vulture and Yellow-headed vulture; and (4) the Black vulture (Koford, 1953; Wallace & Temple, 1987; Lemon, 1991; Houston, 1994; Gomez et al., 1994; Muller-Schwarze, 2006; Snyder & Snyder, 2006). The difference between the *Cathartes* vultures and the Black vulture is however minimal; in some cases, the more social Black vulture is numerically and hence socially dominant (Koford, 1953; Stewart, 1978; Wallace and Temple, 1987; Houston, 1988, 1994; Graves, 1992; Buckley, 1996; Restall et al., 2006). The Andean condor has absolute and strong dominance over the generally solitary King vulture (Wallace & Temple, 1987). The latter has the strongest, most raptor-like bill of the *Cathartids* (Fisher, 1944) and being larger, is dominant over the *Carthartes* and Black vultures. It has superior ability to open carcasses than the smaller vultures (similar to the Old World tearers Lappet-faced, White-headed and Red-headed vultures), while the smaller Cathartids have the role of the pickers (similar to the Hooded and Egyptian vultures) (Schlee, 2005).

In some cases, Black vulture flocks may dominate Andean condors, both when the condor is the first arrival at a carcass (especially in mountains) and also when Black vultures were first arrivals (especially in open plains) (Mikami & Kawata, 2004; Carrete et al., 2010). Therefore, the numbers of condors are negatively related to the abundance of Black vultures (both in plains and mountains, but more so in the plains). In some cases, "although black vultures do not completely prevent the arrival of

Andean condors to carcasses, they represent serious obstacles for feeding. Thus, while dominance hierarchy at carcasses could be related to body size, carcass consumption was determined by species abundance" (Carrete et al., 2010, 390). California condors have absolute dominance over the generally solitary Turkey vultures (the latter lacks the social power of Black vultures) (Koford, 1953). Cathartes vultures have only one advantage, a sense of smell, enabling first arrival at carcasses before larger or social vultures exert dominance (Stager, 1964; Hilty, 1977; Olrog, 1985; Houston, 1986, 1988; Graves, 1992; Snyder & Snyder, 2006; Schulenberg, 2010).

The relations between eagles and *Acciptrid* vultures are more complex (Campbell, 2015). Among the *Accipitrid* vultures, the dominance hierarchies are similarly based on size and numbers, and aggressive behavior. Old World vultures may be classified morphologically from most to least dominant: (1) the dominant, large and medium sized tearers with large bills for tearing flesh (the huge Cinereous and Lappet-faced vultures, the large-medium sized Indian King and White-headed vultures); (2) the large to medium-sized pullers (Griffons), with longer necks for pulling flesh from deep within opened carcasses; and (3) the small pickers or peckers (the Hooded and Egyptian vultures) with small narrow bills for picking small morsels from carcasses (Kruuk, 1967; Konig, 1974, 1983). Kruuk (1967), Brown (1971) and Anderson and Horwitz (2008) ranked six species of African vultures according to size and dominance: (1) Lappet-faced vulture 7.5 kg; (2) White-headed vulture 5.9 kg; (3) Ruppell's vulture 6.4 kg; (4) African White-backed vulture 5.7 kg; (5) Hooded vulture 1.9 kg; and (6) Egyptian vulture 1.9 kg. The tearer White-headed vulture is usually dominant over the larger puller Ruppell's Griffon vulture.

All the larger *Accipitrid* vultures dominate the smaller Hooded and Egyptian vultures, and the slightly larger and more aggressive hooded vulture dominates the Egyptian vulture (Lack, 1971; Konig, 1974; Houston, 1975). Numbers are also relevant. The larger Cinereous and Lappet-faced vultures dominate the large griffons in one-to-one conflict, but the more social Griffons may dominate when in flocks (Konig, 1973, 1974). The tearers, like their *Cathartid* equivalents (the condors and the King vulture) are comparatively solitary foragers. Aggressive behavior and threatening postures are also important; the Lappet-faced vulture dominates the Griffons when in threatening posture, but when it is in a less threatening attitude it may be driven back, possibly indicating that hunger may be more important than fixed dominance (Konig, 1983). Based on these research findings, the two hypotheses posed earlier may be answered: (1) for competition between eagles and *Accipitrid* vultures, size is more important than the vulture/raptor distinction; and (2) for competition between eagles and *Cathartid* vultures, the vulture/raptor distinction is more important than size. Therefore, eagles will be dominant over the largest *Cathartid* vultures, while eagles will be less dominant over the largest *Accipitrid* vultures.

Tawny, Bateleur, Bonelli's, Booted, Golden, Imperial, Spotted, Steppe and Verreaux's eagles are the main competing eagles for vultures in Eurasia and Africa (Kendall, 2013; Spiegel et al., 2013). The larger vultures with tearing abilities, for example the huge Lappet-faced and Cinereous vultures, and the smaller but eagle-like White-headed vultures, have similar beak strengths as the larger eagles (Mundy, 1992; Campbell, 2015). The nearly condor-sized Lappet-faced vulture is dominant over several species of eagles in Saudi Arabia (including the Golden eagle and Imperial eagle, and the slightly smaller Bonelli's eagle, Booted eagle,Spotted eagle and Steppe eagle (Shobrak, 1996). Comparing the dominance of these huge, condor-sized larger vultures over the Eurasian Golden eagle with the dominance of the American Golden eagle over the condors in the Americas illustrates the difference in strength between the *Accipitrid* and *Cathartid* vultures is apparent (Koford, 1953; Murn et al., 2009; Stolen, 1996; Snyder & Snyder, 2000; Snyder & Schmitt, 2002; Campbell, 2009, 2015; Dwyer & Cockwell, 2011;). The Lappet-faced vulture was also found to be moderately dominant over Tawny and Bateleur eagles in the Masai Mara of southwestern Kenya, East Africa (Kendall, 2013).

The Tawny eagle and Bateleur eagle appear to have particularly strong conflicts with larger vultures such as the Lappet-faced and White-backed vultures (Spiegel et al., 2013). According to Kendall (2013), among these species there is generally hierarchy. For example, the Tawny eagle was generally dominant over the Bateleur eagle in conflicts at carrion sources, and the Lappet-faced, Ruppell's and White-backed vultures also dominated Bateleur eagles to the point of leaving behavior in the eagle. Tawny eagles were in some cases dominated by White-backed vultures (Kendall, 2013). Kendall's (2013) study in the Masai Mara National Reserve of Kenya found that the size and possibly beak strength, aggressive behavior and in some cases the social group size, may have been determinants of interspecies dominance achievement and competitive abilities, between the Bateleur eagles and Tawny eagles, and the Lappet-faced, White-headed, African White-backed, Ruppell's vulture and Hooded vultures. The weights of these species, from smallest to largest, were: Hooded vulture 1.9 kg, Bateleur 2.2 kg., Tawny eagle 2.3 kg, White-headed 4.3 kg, White-backed 5.6 kg, Lappet-faced 6.8 kg, Ruppell's 7.6 kg. Concerning beak strength's, the eagles and the White-headed and Lappet-faced vultures are high, the White-backed and Ruppell's vultures have medium strength and that of the Hooded vulture is low (the source is adapted from Mundy et al., 1992). The Lappet-faced vulture, a huge aggressive species with a large and powerful beak, can dominate the generally heavier Ruppell's vulture and smaller White-backed vultures (except when the latter more social birds were in groups) and also the Tawny eagle and the Bateleur eagle (Kendall, 2013).

Bateleur eagles were rated as among the best foragers (possibly due to flight height in foraging or visual acuity, or different wing-loading), lending them a compensatory advantage at kills. Due to their superior foraging abilities, Bateleurs were generally found in high quality landcover (high wildlife density, low human settlements) similar to the more socially dominant vultures like the Lappet-face vultures (Pennycuick, 1972; Kendall, 2013). The Tawny eagle's foraging capacity was lower, similar to that of some of the vultures, for example the White-backed and Hooded vultures. These species were commoner in low quality landcover, with more human presence and lower wildlife density. The Bateleur eagle (and White-backed, and Lappet-faced vultures) were also more often morning foragers, possibly increasing food access (Pennycuick, 1983; Kendall, 2013).

Local declines of large vultures in such African cases studies may possibly result in increased populations of Tawny and Bateleur eagles, due to the decline in competition from the vultures (Virani et al., 2011). This contrasts with the decline of nesting California Condors where Golden eagles were common (Snyder & Snyder, 2000). In some cases, eagles may attack the larger *Accipitrid* vultures, but these attacks if predatory, may usually be on nestlings. For example Mundy et al. (1986) reported numerous successful and unsuccessful attacks by Verreaux's eagles on nestlings of the huge, near condor-sized Cape Griffon vultures (see also Pitman, 1960). They point out that "Always the eagle has initiated an encounter; sometimes the vulture has reacted to the eagle (a true interaction) and at other times it has not" (Mundy et al., 1986, 61). Furthermore, "while only once was an eagle seen eating a vulture nestling, many other attacks by eagles on the nestlings and eggs of the Cape Griffon vulture… were interpreted as predation attempts. The vulture recognizes the eagle as a predator, and the plumage and behavior of feathered nestlings possibly have a camouflage function. Eagles also attack flying vultures, but these instances are probably territory defense" (Mundy et al., 1986, 61). In one case, a Verreaux's eagle killed an adult Cape Griffon vulture, but Mundy et al. (1986) argued with Pitman's (1960) opinion that this attack was predation, as the eagle did not alight on the ground and feed on the vulture. Other attacks by eagles have been recorded on Bearded and White-headed vultures (Mundy et al., 1986; Campbell, 2015).

Several eagle species (Bonelli's, Booted, Golden, Imperial, Spotted and Steppe eagles) were recorded in aggressive attacks on the huge Lappet-faced vultures in Saudi Arabia (Shobrak, 1996). These conflicts at carcasses were mostly intraspecific. For the eagles collectively, 63.3% of the conflicts were intraspecifc and 36.7% were interspecific. For Lappet-faced vultures, 71.5% of the conflicts were interspecific and 29.5% were intraspecific. Lappet-faced vultures won the majority (70.9%, losing 29.1%) of their interspecifc conflicts, while the eagles won fewer such interspecific fights (64.8%, 35.2%) (Shobrak, 1996). These results the Lappet-faced vultures as possibly at the top of the scavenging groups for this case study.

Eagles may also dominate smaller hawks, as cited above in the interactions between Bald eagles and Ferruginous hawks, and Red-tailed hawks (Sabine & Gardner, 1987, see also LeDuc, 1970). Bald eagles have frequently observed taking food from Ospreys (Howell, 1932; Bent, 1937; Blume, 1947; Brown & Amadon, 1968; Peterson, 1969; Garber, 1972; Ogden, 1975; Levenson, 1976; Yves Prévost, 1979). Although "in general, Ospreys and Bald Eagles were tolerant of each other" Yves Prévost (1979, 414) observed three types of interactions between Bald eagles and Ospreys: "(1) eagles attacking Ospreys with a fish, (2) eagles approaching Ospreys with no fish, and (3) Ospreys attacking Bald eagles." In this study, "Bald Eagles were observed trying to steal fish from Ospreys on 15 occasions...the eagle approaching to within a few meters of the Osprey while the Osprey tried to escape with violent swerves, all the time hurrying toward the shore." In such interactions the pursuits ended when the Osprey entered treed areas outside the estuary, usually after dropping the fish, and not all the attacks were successful "only 2 of the 1,793 captures observed were lost because of Bald Eagle attacks, which was less than the number of fish dropped accidentally (4)" estimated as representing 8 losses over two years in the study area. The author contrasts these findings with those of Levenson (1976), which found Ospreys lost their fish in all of the 3 eagle attacks observed. Bald eagles also attacked Ospreys without fish, and in 51% of cases this led to the Ospreys leaving the study area, and in some cases the Ospreys either attacked first (21%) or retaliated to the eagle's attack (79%). Bald eagles have also been observed attacking Northern Harriers (*Circus hudsonius*, Linnaeus, 1766) (Baldwin, 1940).

6.4 Eagles, Canids, Felids and Other Medium-Sized Animals

The mammalian facultative scavengers and mesopredators (coyotes in the Americas, jackals in Eurasia and Africa) are dominant facultative scavengers (Wilmers et al., 2003; Kendall, 2013). These similarly sized canids (coyotes and jackals) share the ranges of most eagles and may compete with both eagles and vultures for carrion. Vultures, being obligate scavengers may be present in many such interactions and participate in conflicts (Campbell, 2015). Hence, they are included in this analysis. Coyotes are the main scavengers in some wild lands in North America. Regarding their relationship with eagles, the coyote can be dominant over the Bald and Golden eagles, more so than smaller omnivores such as raccoons (*Procyon lotor*, Linnaeus, 1758) and opossums (*Didelphis virginiana*, Gray, 1821) (Wilmers et al., 2003; DeVault et al., 2011; Campbell, 2015). Any competition between coyotes and eagles, is usually at the apex of most carrion food sources, at least in the absence of wolves (*Canis lupus*, Linnaeus, 1758), cougars and black and brown bears. Coyotes are serious opponents, usually dominant over all vultures. A study by Wilmers et al. (2003) in the Greater Yellowstone Ecosystem illustrated the relationship between coyotes and avian scavengers. In winter, with vultures absent, the "primary winter scavengers in Greater Yellowstone are, in order of dominance at carcasses, coyote, Golden eagle, Bald eagle, raven and magpie" (Wilmers et al., 2003, 997). In one case, one coyote drove 3 Bald eagles of a carcass (BBC Earth, 2015). Bowen (1980, 377) recorded coyotes driving Golden eagles off carcasses, despite threatening behavior from the eagles. The author concludes that "these observations suggest that coyotes have priority of access to ungulate carcasses in Jasper National Park, although Golden eagles occasionally attempt to defend food from coyotes." Bowen (1980, 376) notes that coyotes and Golden eagles "have similar food habits", taking small mammals in summer and carrion in winter, and this may lead to conflicts between the two species, although most observations of these two species where of coyotes driving eagles off carrion.

However, Golden eagles have killed coyotes and other predators, including red fox (*Vulpes vulpes*, Hatch, 1968) (Mason, 2000). Moehrenschlager et al. (2007) also report on Golden eagle predation on swift foxes (*Vulpes velox*, Say, 1823) in Canada and kit foxes (*Vulpes macrotis*, Merriam, 1888) in Mexico, but these are smaller animals than coyotes. Mason (2000, 244) described an eagle killing an adult male coyote, and "that the talons had punctured the lungs and aorta. There were no other obvious wounds." Other reports of eagle attacks on coyotes, some successful and some unsuccessful, include those of Ford and Alcorn (1964), Wells and Bekoff (1978), Bowen (1980), Dekker (1985) and Woelfl and Woelfl (1994). MAPS Media Institute (2019) placed a video on YouTube that shows one eagle driving two coyotes off a carcass. Ford and Alcorn (1964) once reported two incidents where Golden eagles attacked coyotes, in what appeared to be predatory attacks. An important factor for conflicts between eagles and mammal scavengers and carnivores in North America may also involve the wide range of larger prey species taken by eagles (as not all carrion is derived from wolf or human hunter kills (Campbell, 2015). Eagles are capable of taking larger mammals too heavy to carry to the nest, hence large carrion pieces may be left on the ground (Cooper, 1969; McEneaney & Jenkins, 1983; Love, 1989; Deblinger & Alldredge, 1996; Phillips et al., 1996; Nybakk et al., 1999).

North American coyotes are more dominant over the *Cathartid* vultures than the Eurasian and African jackals are over the *Accipitrid* vultures, hence they may be the competitors with the American eagles, which are also strongly dominant over *Cathartid* vultires (Campbell, 2015). The coyote is dominant over the California Condor at carcasses (Koford, 1953; Snyder & Snyder, 2000). A reintroduced condor was killed by a coyote (Meretsky et al., 2000) and coyotes among other predators are implicated in 18 condor deaths (Rideout et al., 2012). They are also easily dominant over the smaller Cathartids (Personal observation, 2014). Koford (1953) recorded incidents were a coyote's approach caused up to 20 condors abandoned food sources. Coyote abundance appears partly dependent on the absence of wolves (Berger & Gese, 2007). Although wolves were abundant over most of North America at the onset of European colonisation, they were gradually exterminated by human action. Coyotes were more adaptive, and began to dominate previously wolf-dominated areas, based study areas such as Grand Teton National Park (GTNP), Wyoming, USA, and the Greater Yellowstone Ecosystem (GYE) (where wolves caused 56% of transient coyote deaths), and in Algonquin Provincial Park (APP), Ontario, Canada (Berger & Gese, 2007; Rutledge et al., 2012).

Regarding the jackals of Eurasia and Africa, the relationship with eagles is more even, with both sides prevailing. For example, Lockhart (2020) provides photographs of a Tawny eagle challenging an African jackal, which ended in a standoff without the eagle landing. Black-backed jackals (*Canis mesomelas*, Schreber, 1775) may be killed by Martial eagles (Steyn, 1983; Estes, 1992; Sillero-Zubiri et al., 2004). Boshoff et al. (1990), Kingdon (1988) and Kingdon et al. (2013) report canids in the Martial eagle's diet, including the small Bat-eared foxes (*Otocyon megalotis*, Desmarest, 1822), Black-backed jackals and the larger African Golden wolf (*Canis lupaster*, Hemprich and Ehrenberg, 1832). Even adult domestic dogs (*Canis lupus familiaris*, Linnaeus, 1758) have been killed (Govender, 2015). Conflicts are shown in Figures 6.16 to 6.20.

The Black-backed and Golden jackals in Africa and Eurasia are commonly present at carrion sources, and hence may compete with scavenging eagles; these canids also "compete heavily with vultures" (Kruuk, 1967). The relation between the jackals and the largest *Accipitrid* vultures (with a size ratio similar to that of the condors and coyotes) is more even than that between the American coyotes and the *Cathartid* condors. The larger vultures may also attack jackals at carcasses, as numerous photographs published online attest (see for example Perry, 2007). Lappet-faced vultures, one of the most aggressive vultures in Africa, have been filmed and recorded attacking and repelling jackals from carcasses (Mundy et al., 1992; Perry, 2007; Daily Telegraph, 2010). There are no records of vultures killing jackals however (Campbell, 2015). Jackals may have moderate dominance over vultures, but rarely have absolute dominance (Shimelis, 2005). Golden jackals (*Canis aureus*, Linnaeus, 1758) are noted to be aggressive competitors, and "can dominate even the largest vultures, and may hold dozens at bay while it monopolizes the remains of a kill, threatening, snapping, lunging with ears flattened and tail whipping from side to side, sometimes jumping into the air to bite at a bird that tries to land too close" (Estes, 1992, 400). Holding vultures at bay would be considered strong or moderate dominance, but as vultures may also dominate jackals, the relationship may also be termed indeterminate (Campbell, 2015).

Large eagles may also be killed by other predators. For example, Crowned eagles, possibly the strongest eagles in Africa, may be killed as prey by large carnivores. In one case, two young eagles reintroduced into the wild were killed, one by a leopard that surprised the eagle on a monkey kill and another was killed by a crocodile when feeding on a bushbuck at the water's edge (Thomsett, 2011). Another Crowned eagle was killed by a Sanje mangabey (*Cercocebus sanjei*, Mittermeier, 1986) (Jones et al.,

2006), and another was seriously injured in the attempted predation of a baboon (Thomsett, 2011). Referring to the killing by the mangabey, Jones et al. (2006, 359) argued that this was the "first ever reported observations of a hunting African crowned eagle… being killed by a primate, in the Udzungwa Mountains of south-central Tanzania. An adult female eagle launched an attack on a young Sanje mangabey *Cercocebus sanjei* who was feeding in a tree but was intercepted and bitten by an adult mangabey who was feeding nearby. The adult mangabey and the eagle then fell together 25 m to the forest floor below. The eagle subsequently died from her injuries, while the mangabey escaped and is thought to have survived."

Table 6.3 shows the dimensions of the canids. Gray wolves are the largest, and are rarely killed by eagles, but in Central Asia the Golden eagle may be used by hunters to kill wolves with variable success (Stewart, 2002; Riviera & Yiu, 2013). The larger species are within the range of the more normal prey for the larger eagles. As all the canid species are habitual facultative scavengers, all of them may conflict with eagles and vultures at carrion sources (Campbell, 2015).

Table 6.3. Canids likely to compete with eagles

Canine species	Length (cm)	Shoulder height (cm)	Weight (kg)
Gray wolf (*Canis lupus*, Linnaeus, 1758)	200	76	45
Coyote (*Canis latrans*, Say, 1823)	100–130	60	9–23
Golden wolf (*Canis lupaster*, Hemprich& Ehrenberg, 1832)	120–130	40	7–15
Side-striped jackal (*Canis adustus*, Sundevall, 1847)	99–122	35–50	6.5–14
Golden jackal (*Canis aureus*, Linnaeus, 175)	95–110	40	8–10
Black-backed jackal (*Canis mesomelas*, Schreber, 1775)	86–130	38–48	6–13

Sources: Este (1992); Heptner & Naumov (1998); Burnie & Wilson (2005); Kingdon & Hoffman (2013); Campbell (2015); Encyclopaedia Britannica, 2020a,b).

Fig. 6.16. African Fish eagle and White Headed vulture [With permission from Michael Heyns]

Fig. 6.17. Eastern Imperial eagle and Cinereous vulture [With permission from Julian Thomas]

6.5 Eagles and Domesticated Animals

Eagles frequently take domesticated animals and protected animals, a crucial issue as these actions, perceived or real may result in persecution of the eagles (Campbell, 2015, see also Cooper, 1969; McEneaney & Jenkins, 1983; Love, 1989; Deblinger &

Fig. 6.18. Steller's Sea eagle vs fox [With permission from Julian Thomas]

Fig. 6.19. Tawny eagle and vultures [With permission from Roelof van der Breggen]

Alldredge, 1996; Phillips et al., 1996; Nybakk, et al., 1999). For example, Thomson (2014) documents incidents in Campbell River, British Columbia, Canada where two young Bald eagles killed several domestic cats in some neighborhoods. Thomson (2014) notes that "According to eagle experts, a Bald eagle capturing a cat is not impossible, but highly unusual. Their main diet is fish, but they are opportunistic and will eat squirrels, ducks, seagulls, turtles and Canada geese. They will also feast on carcasses, including deer." Beckford (2017) argues that "Bald eagles readily accept livestock carrion, although some individuals may prefer live prey to carrion. Eagles are efficient predators, and they can cause severe losses of young livestock, particularly where concentrations of eagles exist. Generally, they prey on young animals, primarily sheep and goats, although they are capable of killing adults. Understandably, it is difficult to control eagles using lethal methods as they are a protected species in Florida." O'Gara (1994, 45) further adds that "Golden eagles are more likely to prey on livestock than are Bald eagles. Both species readily feed on livestock carrion and carcasses left by foxes and coyotes, although some individuals prefer live prey to carrion. Eagles are efficient predators and can cause severe losses of young livestock, particularly where concentrations of eagles exist. Generally, they prey on young animals, primarily lambs and kids, although they are capable of killing adults."

Fig. 6.20. White-bellied Sea eagle chasing Eastern Osprey [With permission from Chris Chafer]

Matchett & O'Gara (1987, 85) assert that Golden eagle predation on lambs may be partially due (either/or) to lowered populations of jack rabbits and the lambing seasons are cool or wet. Other factors include "natural prey densities, availability of carrion, weather, ranching practices, vegetative cover, and topography as well as Golden eagle age, behavior, densities and distribution" and the "Interaction of these factors in combination with economic, social and political values contributes to emotionally charged, opinionated and highly contested conflicts" (Matchett & O'Gara (1987, 85). Dated citations on these factors include Arnold (1954), Lockie & Stephen (1959), Snow (1973), Bolen (1975), Brown (1975), Tigner & Larson (1977, 1981), O'Gara (1978, 1981) and Gober & Lockhart (1979). For a case study, "such conditions were present during 1974 and 1975 on two ranches near Dillon, Montana, when eagle depredation on lambs was high." Actions taken to stop the depredation were mostly ineffective and included trapping of eagles on ranches, with the captured birds moved to areas with lower possibilities of depredation, chasing of eagles with small airplanes, firing rifles and explosive shotgun shells, use of human-like scarecrows combined with the harassment methods. Matchett & O'Gara (1987, 85) concluded that "no single predator control technique will always work in every situation, but we conclude that scarecrows, combined with harassment, offer the most feasible means of protecting lambs under range lambing conditions in southwestern Montana."

Matchett & O'Gara (1987, 85) develop their argument by noting that the "topic of Golden Eagle (*Aquila chrysaetos*) depredation on lambs of Domestic Sheep (*Ovis aries*) is controversial. Raptor enthusiasts and sheepmen are often at odds over the severity of depredations or the fact that they even occur." The Golden eagle is cited as a highly efficient predator capable of killing both wild and domestic mammals of appreciable size, including lambs, domestic calves, Mule deer, Bighorn sheep, Dall's Sheep, Mountain goats and coyotes. This argument is supported by dated literature: (McLean, 1925; Dixon, 1934; Alien, 1939; Couey, 1944; Wood, 1946; Boyer, 1948; Clawson, 1948; Kennedy, 1948; Brown & Couey, 1950; Casebeer et al., 1950; Riney, 1951; Ford & Alcorn, 1964; Glover & Heugly, 1970; Nesse et al., 1976; Tigner & Larson, 1977, 1981; Foster & Crisler, 1978, 1979; O'Gara, 1978, 1981; Wade & Livingston, 1978, 1979). The impacts of Golden eagles' depredation on livestock are most pronounced during the lambing periods, and in the past decades these had serious economic impacts. For example, estimates of losses included $87,000 in New Mexico from 1 January-1 May 1981 (Littauer & White, 1981), and $38,000 (1974) and $48,000 (1975) on two ranches in southwestern Montana (O'Gara, 1978). Matchett & O'Gara (1987, 85) cite Arnold (1954, 30), who studied incidents in Colorado, Wyoming and Texas and concluded that depredations were "spasmodic... and may vary considerably under changing local conditions."

6.6 Conclusions

This chapter has explored the relationship between eagles, other raptors, wild mammals and livestock. Eagles plainly have a wide range of connected ecologies and play an extremely important role in ecological regions. Eagles are almost always the dominant species, due to their position as the world's strongest raptors. Other species may only challenge, dominate or endanger eagles when they possess far greater size and strength. In cases, the eagle's most serious avian opponents may be other species of eagles, at the intra-species and inter-species levels, as few other birds possess the strength and armament to take on the larger eagles. The most serious mammals are obviously cats, as even medium sized cats such as bobcats and caracals may pose more serious challenges than similarly-sized canids such as coyotes, small wolves and jackals. The eagle's status as indicator species and apex predators, while ecologically beneficial for the eagle species, may also incur some negative consequences, especially when the eagles challenge human-constructed ecosystems and biological constructions, inviting ameliorative human action. This chapter has examined a few of the possible human-eagle interactions, mostly centered on livestock predation. Further human interactions will be examined in the succeeding chapters.

7

Climate, Relief, Landcover and Eagles

7.1 Introduction

This chapter examines the relation between eagles and environmental factors. These include variables such as climate temperature and rainfall, and the weather phenomena that emerge from these, such as thermals, wind and daily and seasonal weather changes, relief, topography and elevation, especially related to eagle foraging and nesting and the local, regional and global vegetation types, related to eagle habitats. These include closed forest vegetation, open rocky terrain, grassland, savanna and mixed tree-grass environments, usually used for foraging and nesting. Waterbodies, such as lakes, rivers and the sea are also considered in terms of their relationship to eagle foraging and habitats. Climatic and weather variables affect eagle life histories, in terms of their ecological relations, migration patterns and even their physical hunting behavior. Orographic lift and thermals determine soaring flights and hence foraging activities. Vegetation may also affect eagles, especially in terms of the foraging when vegetation may obscure prey observation (for example, in dense forest). Examples are taken from the global climate and vegetation types in the six continents where eagles are resident, and local ecosystems were general and unique issues may occur.

7.2 Climate, Weather and Relief

The most important climate issues for eagles concern temperature and rainfall, seasonal change and regional variation, leading to migration, impacts on prey species and foraging from the sky using thermals (Campbell, 2015). This section will look at each of these issues systematically. Daily weather issues are considered first, as such weather patterns are generalized into climate change over longer periods. The climate of an area is defined as the average weather conditions, such as precipitation, temperature, humidity, pressure, wind over a period of about 30 years. The Intergovernmental Panel on Climate Change (2019, 808) notes that "climate in a narrow sense is usually defined as the average weather, or more rigorously, as the statistical description in terms of the mean and variability of relevant quantities over a period of time ranging from months to thousands or millions of years." The World Meteorological Organization (2019) notes, regarding the duration of a climatic period of weather generalization, that "the classical period for averaging these variables is 30 years." The relevant quantities are most often surface variables such as temperature, precipitation and wind. Climate in a wider sense is the state, including a statistical description, of the climate system.

Weather on the other hand refers to the "state of the atmosphere at a given time and place with regard to temperature, atmospheric pressure, humidity, wind, cloudiness, and precipitation" (Government of Canada, 2019). Weather therefore covers short time periods, particularly the immediate period, and may be determined by variables such as local altitude, regional latitude, marine proximity and orientation of mountain barriers to winds and ocean currents. These day to day events may be stable over decades creating a climatic system, or may gradually change, resulting a slow evolution of the regional climatic system. From an ecological perspective, climatic region is determined by the particular combination of five environmental components: the atmosphere (air), hydrosphere (water), cryosphere (soil), land surface, and biosphere (plants) (Campbell, 2015).

Climate studies have emphasized three systems of classification: the Köppen system, the Thornthwaite system and the Bergeron and Spatial Synoptic Classification. The Bergeron Classification, now the most accepted system used air masses for classification, based on three letters: the first letter for moisture properties (c for dry continental air masses, and m for moist maritime air masses); the second letter for the thermal classification of the source region (A for Arctic or Antarctic, E for equatorial, M for monsoon, P for polar and T for tropical); and the third letter for the stability of the atmosphere. Air masses warmer than the ground below are designated w; and those colder than the ground below are designated k. Baum et al. (2014, 4) notes that the "Bergeron Classification, based on North American source regions, is the most popular and well-known system."

The Spatial Synoptic Classification system (SSC) is based on the Bergeron classification system and has categories: Dry Polar (similar to continental polar), Dry Moderate (similar to maritime superior), Dry Tropical (similar to continental tropical), Moist Polar (similar to maritime polar), Moist Moderate (a hybrid between maritime polar and maritime tropical), and Moist Tropical (similar to maritime tropical, maritime monsoon, or maritime equatorial). Baum et al. (2014, 4) describe this system as "perhaps the most unique and practical of all weather type schemes available for meteorological research since daily air mass calendars of seven air mass types can be accessed for download on a website for over 300 North American locations dating back to the early 1900s for most stations" (see also Sheridan, 2002). The Spatial Synoptic Classification system (SSC) is also unique as it is based on surface weather parameters recorded at a location rather than the source region. Surface weather parameters include diurnal temperature, dew point depression and dew point ranges, cloud cover and mean sea level pressure records (Sheridan, 2002; Baum et al., 2014, 4). The Köppen classification relies on average monthly values of temperature and precipitation. The common version uses five categories: A, tropical; B, dry; C, mild mid-latitude; D, cold mid-latitude; and E, polar. These can also be divided into secondary classifications, from the hottest and wettest to the coldest and/or driest, rainforest, monsoon, tropical savanna, humid subtropical, humid continental, oceanic climate, Mediterranean climate, steppe, subarctic climate, tundra, polar ice cap, and desert (Campbell, 2015).

Relief, referring to the elevation, slope and orientation of Earth's surface, is another important physical attribute of the Earth's surface that is relevant to eagle ecology. The local relief affects altitudinal climate and eagle habitat suitability and breeding mainly for cliff nesters. Relief also creates the rain shadow effects, that may create dry or desert climates, such as the American Southwest inland from the Rocky Mountains. Longitudinal mountain canyons can also create orographic windflows possibly used by soaring raptors. Large, rising blankets of hot air, termed thermals also assist in soaring flight for large, heavy, soaring raptors like eagles that may otherwise be unable to maintain long distance, high altitude flight. Relief may also influence or determine surface water flows, such as streams, rivers and lakes, which are frequented by many eagle species. Eagles frequent all the climatic regions listed above, except for the Antarctic and Arctic ice caps, tundra and pure desert. Rainforests are the wettest climatic region, have high rainfall, with definitions setting minimum normal annual rainfall between 1,750 millimeters and 2,000 millimeters. The mean monthly temperatures may exceed 18 °C throughout the year. Some of these regions are affected by monsoons, which is a season dominated by winds that bring rain for several months, mostly in tropical regions (Campbell, 2015).

Tropical savanna is a drier biome with varying proportions of grasses, shrubs and trees, mostly occurring in semiarid to semi-humid climate regions of subtropical and tropical latitudes, where the average rainfall is between 750 millimeters and 1,270 millimeters a year and temperatures above 18 °C. Savannas with heavier rainfall may have more tree cover, sometimes even clumps of woodland, while the savannas under lighter rainfall are drier, with a higher proportion of grasses. Most savannas are located in subSaharan Africa, southern Asia and northern South America, usually forming an intermediate zone between forest and desert, with the wetter savanna near the drier forest, and the drier savanna near semi-desert. Dry grassland with few or no trees (steppe in Eurasia, prairie in North America, Sahel in subSaharan Africa) may occur in the northern regions of the Americas and Eurasia, usually with a high temperate range between hot summers up to 40 °C and cold winters down to −40 °C. Deserts are the driest of these landcover types, and in fact may entirely lack vegetated landcover, due to minimal precipitation. The very low humidity contributes large seasonal and diurnal ranges in temperature; sometimes between 45 °C and 0 °C. Such regions may form in the rain shadow behind mountains, which obstruct moist winds from ocean or wetter landcover (Campbell, 2015).

7.3 Eagles, Thermals and Orographic Lift

Eagles engage with climate and daily weather in three main ways: soaring on thermals, soaring on orographic lift, and migration due to seasonal temperature change (summer, fall or autumn, winter, spring). Here soaring is defined as "the use of air movements to sustain gliding flight" (Alexander, 2003). Eagles are too heavy to main flapping flight. As argued by Murgatroyd et al. (2018, 6788) "Unlike smaller raptors, which can readily use flapping flight, large raptors are mainly restricted to soaring flight due to energetic constraints." Wing loading is defined as the ratio of the wing area to the weight of the bird and is related to gliding in the rising hot air in thermals, as "the ability to fly slowly, and turn in small circles, is critical for exploiting thermals, and therefore the wing loading is a useful indicator of a gliding bird's capacity for this activity" (Pennycuick, 2008: 266). This author also notes that the wing loading does not have any special significance for powered flapping flight. For rapid gliding flight, a high wing loading may be helpful, as it allows the rapid flight for hunting. For thermal soarers, such as foraging eagles, low wing loading is needed because they can thus fly in smaller thermals (Alexander, 2003).

Ákos et al. (2010) define thermals as "spatially and temporally localized parts of the atmosphere created by solar radiation heating the ground, typically moving upwards with a speed in the range of 1–10 m/s. The ground heats up the air nearby which rises in columns." The flight based on these thermals is defined as thermal soaring, which is "a form of flight where the flying objects use only convection currents, called thermals, to stay in the air without any additional power source (motor power in the case of airplanes or flapping of wings in the case of birds)" (Ákos et al., 2010) (Figure 7.1).

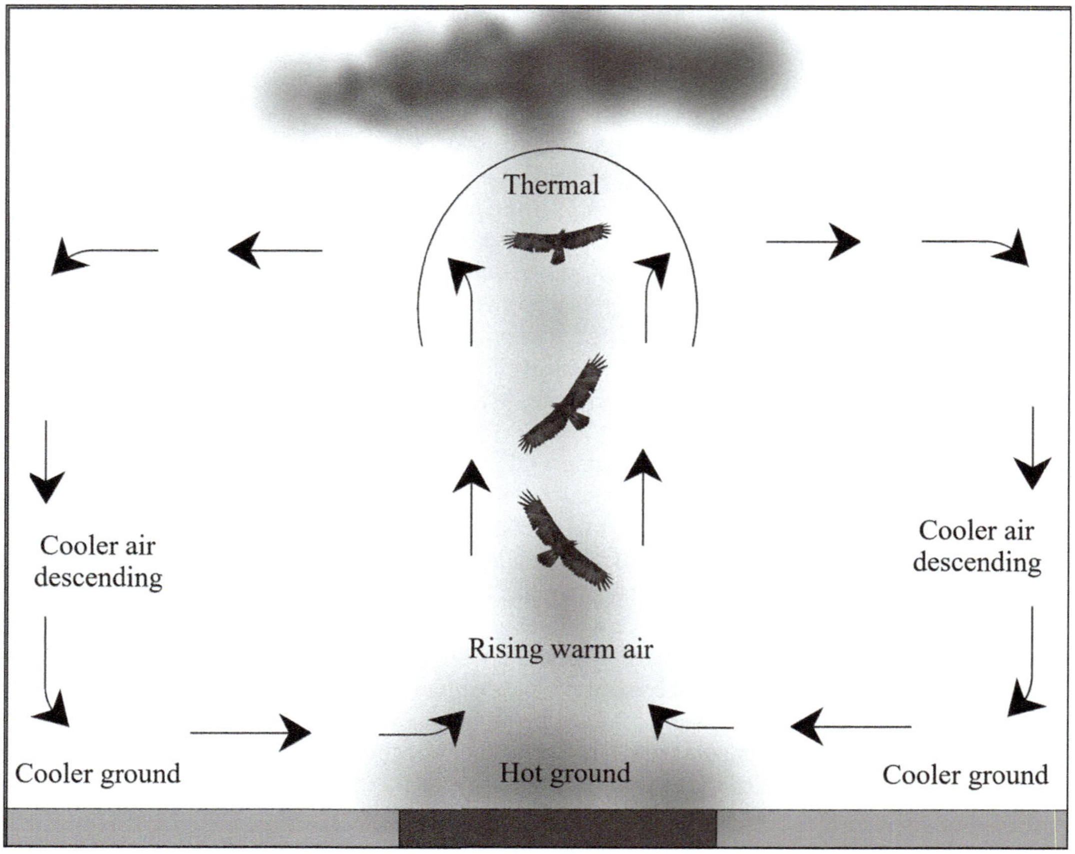

Fig. 7.1. Thermal formation and eagle flight

These thermals form through convection, defined as the transfer of heat through the movement of a fluid, either air or water (Incropera et al., 1990). The process starts when local, earth surface heating causes the air molecules over the hot surface into movement, causing them to gain energy through conduction, creating heated air that expands and rises. This heated rising air bubble is less dense than the rising air and attracts a flow of cooler air under the large, rising bubble of hot air. Fed by the cooler inward air flow, which becomes heated over the hot ground surface, the hot air bubble or thermal grows higher and larger, created this movement which is termed convection. When the thermal reaches a particular altitude it cools, creating a cool downwards flow of cooler air that re-enters the bottom of the thermal and starts the cycle again. This is a convective circulation or a thermal cell (Ahrens & Henson, 2019).

This rising hot air assists soaring flight. Eaglesare unable to maintain flapping flight for long periods, as their heavy bodies relative to the wing areas increasing the expenditure of energy. The bird weight to wing area ratio is termed the wing-loading (Brown & Amadon, 1968; Lish et al., 2016). The wing-loading is calculated by dividing the mass of the bird by the total area of the upper surface of its wings: i.e., wing loading = body mass (kg)/wing area (m^2). If a bird has a low mass but large wings, there will be a low wing loading. Where there is a large body mass and small wings there is a high wing loading. For example, Kirmse (1998) calculated the wing-loading of nine palearctic eagles and the Osprey: The Golden, Greater and Lesser Spotted eagles, Eastern Imperial, Steppe and Short-toed snake eagles, the White-tailed, Bonelli's and Booted eagles, and the Osprey. Using skyward photos of flying raptors, the calculations were based on length of the whole wing and of the handwing, plane of the whole wing and of the handwing. Here the handwing (HWI) of a bird is defined as presented by Claramunt et al. (2012), where HWI is calculated using the equation HWI = 100 × (WL − FSL)/WL, where WL (or L_w as represented by Sheard et al., 2020) is the distance on the bird's wing from the carpal joint to the tip of the longest primary feather, and FSL (or S_l as represented by Sheard et al., 2020) is the distance from the carpal joint to the tip of the first secondary feather. HWI is a standard index of flight and dispersal ability in birds. Sheard et al. (2020, 1) argue that "HWI is a strong predictor of geographical range size" and is "the strongest predictor of migration across all birds." An illustration of these measurements is shown in Figure 7.2. Tables 7.1a, b show an approximation of the findings of Kirmse (1998). The weights were derived from Weick (1980). In Table 7.1a, the wing loadings are sorted from lowest to highest.

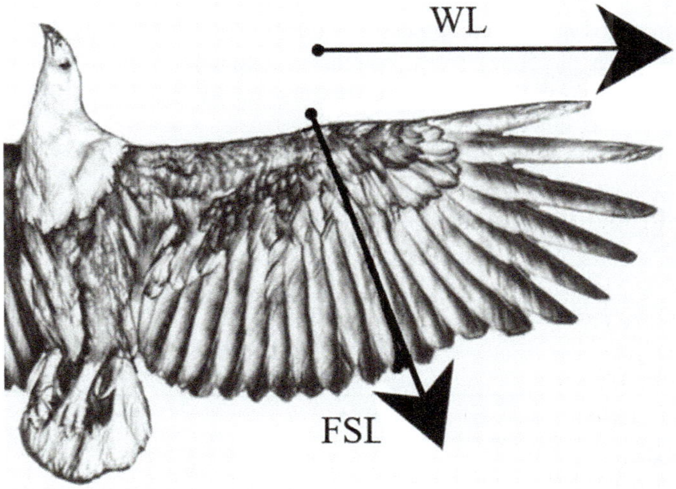

Fig. 7.2. Wing-loading calculation. Picture derived from photograph by Gustav Van Der Gryp of flying African Fish eagle.

Table 7.1a. Wing-loading of Palearctic Eagles and Osprey (Males)

Eagle species	Sex	Body mass (kg)	Wing length (cm)	Handwing (cm)	Wing loading
Booted eagle	Male	0.683	0.596	0.380	0.63
Lesser Spotted eagle	Male	1.281	0.777	0.464	0.66
Greater Spotted eagle	Male	1.800	0.791	0.507	0.73
Short-toed Snake eagle	Male	1.750	0.875	0.548	0.79
Osprey	Male	1.430	0.787	0.480	0.96
Bonelli's eagle	Male	1.712	0.753	0.485	1.03
Eastern Imperial eagle	Male	2.885	0.876	0.582	1.09
Steppe eagle	Male	2.481	0.850	0.536	1.17
White-tailed eagle	Male	4.030	1.048	0.615	1.17
Golden eagle	Male	3.850	1.005	0.620	1.29

Sources: Weick (1980); Kirmse (1998). Note that wing loading = body mass (kg)/wing area (m^2). For comparison, the Bald eagle has a wing loading of 80.94 (Kerlinger, 1989). Raptor View Research Institute (2020) notes that "wing-loading estimates for adult female Golden eagles are among the heaviest reported for any raptor."

Table 7.1b. Wing-loading of Palearctic Eagles and Osprey (Females)

Eagle species	Sex	Body mass (kg)	Wing length (cm)	Handwing (cm)	Wing loading
Lesser Spotted eagle	Female	1.677	0.820	0.490	0.79
Short-toed Snake eagle	Female	1.919	0.900	0.563	0.82
Booted eagle	Female	0.993	0.620	0.395	0.84
Greater Spotted eagle	Female	2.675	0.838	0.537	0.99
Osprey	Female	1.630	0.804	0.490	1.04
Eastern Imperial eagle	Female	3.383	0.961	0.638	1.15
Bonelli's eagle	Female	2.386	0.785	0.506	1.33
White-tailed eagle	Female	5.575	1.120	0.657	1.42
Golden eagle	Female	5.207	1.095	0.675	1.47
Steppe eagle	Female	3.562	0.898	0.566	1.50

Sources: Weick (1980); Kirmse (1998). Note that wing loading = body mass (kg)/wing area (m^2). For comparison, the Bald eagle has a wing loading of 80.94 (Kerlinger, 1989). Raptor View Research Institute (2020) notes that "wing-loading estimates for adult female Golden Eagles are among the heaviest reported for any raptor."

There are significant correlations between the variables in the columns, as shown in Table 7.2. Larger eagles normally have larger wings, but as noted in part one this is not always the case, as forest eagles such as the Harpy eagle may have comparatively short wings to allow rapid flight through the forest trees, compared with long winged species like the Golden eagle, which are open landcover soarers. However, even in such cases, the shorter-winged eagles may have broader wings to allow lift and maneuverability (Ferguson-Lees & Christie, 2001).

Table 7.2. Correlations between wingloading variables of selected eagle species.

Variables	Body mass (kg)	Wing length (cm)	Handwing (cm)	Wing loading
Body Mass (kg)	1	0.923	0.913	0.838
Wing length (cm)	0.923	1	0.970	0.686
Handwing (cm)	0.913	0.970	1	0.712
Wing loading	0.838	0.686	0.712	1

Sources: Weick (1980); Kirmse (1998).

The results in Table 7.2 show strong correlations between the variables in the columns (Tables 7.1a, b). It is to be expected that there will be a high correlation between body mass and wing length for this sample, because all the sampled eagles have a similar body and wing shape (forest eagles with shorter and possibly broader wings and heavy bodies were not included). With similarly shaped wings, the correlation between wing length and handwing is high. Although lower, there is a high correlation between body mass and wing loading, which may infer that larger eagles may have more difficulty rising in soaring flight. The correlation between wing loading and wing length is lower, possibly because handwing is an important variable for such as a relationship (long wings may be narrower or broader, and this may affect relations between the wing length and wing loading. Research on the problems of wing loading for large raptors indicates that larger birds need more wing area for lift, even for non-eagle species. For example, Houston (1983: 137) describe three weight categories of Griffon vultures. The heaviest at 12 kg, the Himalayan Griffon, is a slope soarer, that rarely uses thermals. Houston (1983: 137) argues that "wind speeds in the Himalayas are high, and the rising airflows over hill ridges provide the extremely good soaring conditions needed to support such a large soaring bird." This orographic lift is discussed later.

This heavy body and high wing-loading may be adapted to gliding flight in rising airflows. The medium weight category, the Cape, Ruppell's, Common, Indian and Slender-billed vultures (weighing about 7–8 kg, similar to the very large eagles) use thermals. "All four species have comparable body proportions, presumably a compromise between the high wing loading that is efficient in gliding flight and the lower wing loading required for good performance when soaring in weak thermals" (Houston, 1983: 138, see also Pennycuick, 1972). The two lightest Griffon species, the African White-backed and the Indian White-Rumped vultures (weighing about 5 kg each, similar to the larger eagles) forage in flat savanna lands where there are thermals, but no mountain airflows, where the choice is between laborious flapping flight (very difficult for heavier birds) and use of thermals (Houston, 1983; see also Pennycuick 1972, Houston 1975). This supports the work of Murgatroyd et al. (2018, 6788) who argue that "topography is important in determining the occurrence of soaring flight" and that "thermal soaring occurs in relatively flat areas which are likely to have good thermal uplift availability."

Body temperature is also important. Large birds generate more heat from metabolism and store more heat than small birds, especially during exercises such as flapping flight, when body heat can increase ten-fold over the resting temperature (Kleiber, 1961; Calder & King, 1974; Mahoney, 1983). It is also arguable that dark plumage can retain more heat (see Hamilton & Heppner, 1967). This, in addition to wing loading, provides reasons why large eagles avoid serious flapping flight over long distances and may remain in perches when thermal flight is not possible. It is also a factor for the large, broad wings of eagles, among the largest wing area of any birds. These provide maximum lift but are still insufficient for sufficient lift without thermals (Campbell, 2015).

Orographic lift is also useful to soaring birds (Figure 7.3). This occurs when an air mass, originating either over flat land or the ocean, moves up rising terrain, and cools as it moves up in altitude, creating strong winds that blow up the slope and turbulence over the land surface. Soaring birds use the slope winds and the turbulence created, which can lift heavy birds such the condors and large vultures. Orographic lift has been cited as a extremely useful foraging tool for some large non-eagle species such as the Himalayan Griffon vulture, Bearded vulture, California condor, and Andean condor, which live in mountain terrains (Belding, 1879; Koford, 1953; Stillman, 1967; McGahan, 1973; Wilbur, 1978; Pennycuick & Scholey, 1984; Inskipp & Inskipp, 1991; Shirihai, 1996; Adamian & Klem, 1999; Ferguson-Lees & Christie, 2001; Brandes & Ombalski, 2004; Bildstein, 2006; Wassink & Oreel, 2007; Mandel et al., 2008; Lambertucci & Ruggiero, 2013; Rivers et al., 2014).

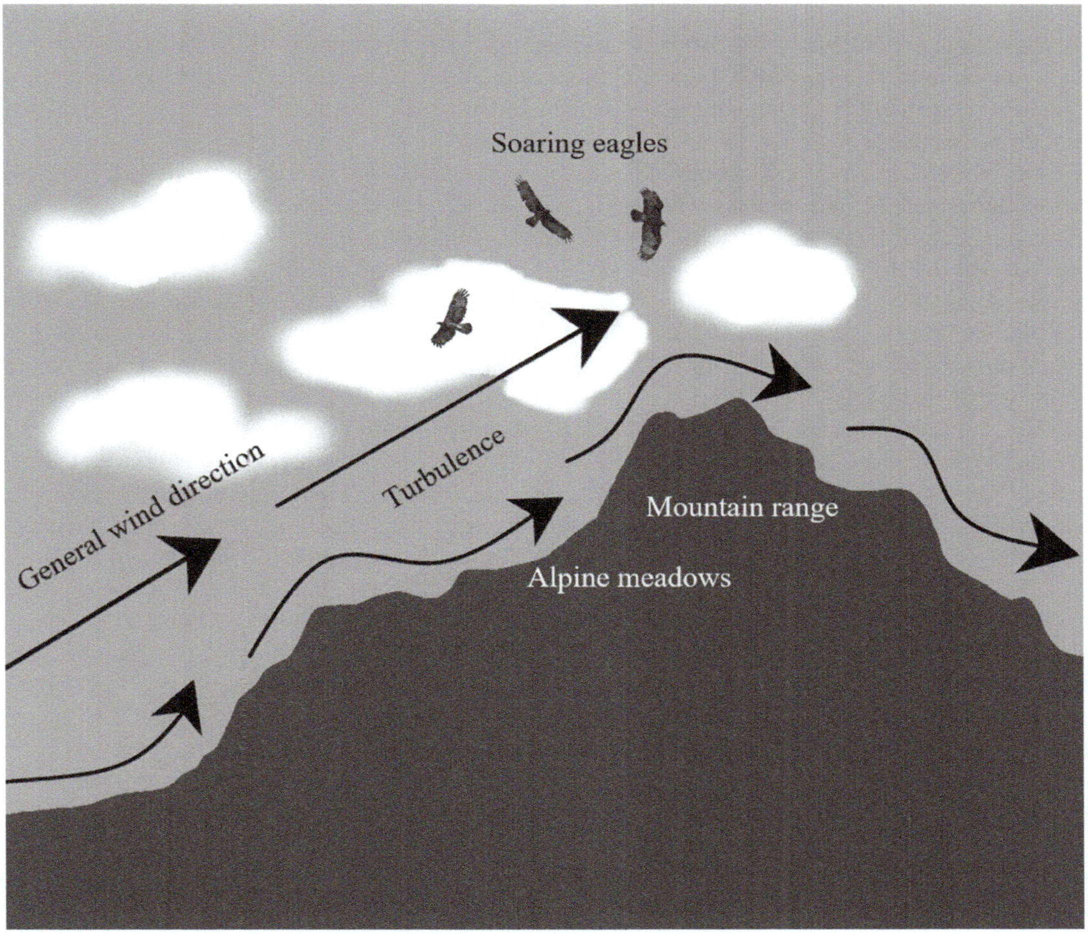

Fig. 7.3. Orographic lift

Johnston et al. (2013) point out that "complex mountain terrain is often favored by large-bodied soaring species during migration because of the enhanced opportunities for lift, i.e., slope winds, deflected winds, and thermals, compared with areas of flatter terrain (see also Kerlinger & Moore, 1989; Hedenstrom, 1993, Spaar & Bruderer, 1996; Yates et al., 2001; Mandel et al., 2008; Katzner et al., 2012; Lanzone et al., 2012). The vertically rising air, flowing up mountains, allow the heavy soaring birds to maintain energy expenditures that may be almost similar to that of resting state, allowing more tireless migration (Brodeur et al., 1996; Weimerskirch et al., 2000; Alerstam, 2001; Bildstein, 2006; Mandel et al., 2008; McIntyre et al., 2008; Bittner, 2010).

Murgatroyd et al. (2018, 6789) made a comparison between flat land thermal soaring and mountainous orographic lift soaring, for the Verreaux's eagle in the flat Sandveld and mountainous Cederberg in the Western Cape, South Africa, examining "the topographic and meteorological correlates of the two uplift-assisted flight modes (thermal soaring and orographic soaring)." Murgatroyd et al. (2018, 6796) notes that "if thermal soaring is more energetically beneficial than orographic soaring and there is better availability of this uplift type in the Sandveld, then these differences could contribute to the significantly higher breeding productivity of eagles in the Sandveld than in the Cederberg, a finding which is not yet fully understood (see also Murgatroyd et al., 2016). It is further argued that although "uplift strength can determine the speed at which a soaring bird can gain altitude" (see also Shepard et al., 2013; Spaar & Bruderer, 1996), and "thermal uplift might be beneficial for cross-country flying" it is "less useful for foraging, where eagles will need to maintain visibility of the ground" (Murgatroyd et al., 2018, 6796; see also Péron et al., 2017; Shepard et al., 2013).

In another relevant example on a non-eagle species, Hiraldo and Donázar (1990) compared nesting nuclei of Cinereous vultures, in a mountainous area and a lowland area, and found that in the mountainous area there was stronger slope lift, and vultures started flights nearer sunrise, and concluded foraging nearer sunset. In flatland, where they depended only on weaker thermals, they foraged for shorter periods each day. Vultures nesting in the mountainous area, which foraged in the flat areas began flying early in the mountainous area, and glide down to the flatland foraging area, and perch until the thermals were strong enough for soaring flight, about seven hours after sunrise. The vultures nesting in the flatland area started foraging at this time as well. Griffon vultures in nearby habitats matched this behavior. Regarding the vultures nesting in the mountains, with vultures leaving the mountain nests earliest in winter (February) and returning latest in summer (June), when the strong thermals lasted longest, and the flying periods were longest. "The arrival of vultures at the colony occurred relatively closer to sunset in summer than in winter probably because of the high temperatures in the afternoon which permit thermal lift to continue until the last hours of the day" (Hiraldo & Donázar, 1990, 131).

Ground and airmass temperatures, and associated thermals are also associated with eagle migration. Here migration refers to the regular seasonal movement of birds seeking warmer climates, mostly due to cold winters. Studies of migration have become important for many species of birds, as migratory movements may influence the survival of many species. Vidal-Mateo et al. (2015, 89) point out that "in the last decades, research on bird migration has focused on several aspects, such as flight mode, fuel deposition, predation risk, use of stopover sites, routes and detours, migration schedules, and the effect of wind" (see also Alerstam, 2011). Furthermore, in such migratory movements "birds must deal with the displacement caused by crosswinds and their behavior can vary according to different factors such as flight mode, migratory season, experience, and distance to goal areas" (Vidal-Mateo et al., 2015, 89).

Most migratory routes are from the north to regions further south, usually using north to south routes, but in some cases routes are tangential to such routes, even almost west-east or east-west. Such alternates may be due to the geomorphological factors listed above, and to the availability of food, day length and breeding. Some studies have found that migrating birds rely on the sun and stars, the earth's magnetic field, and mental maps for their clearly accurate routing. In some species, not all populations are migratory; this is termed partial migration. For eagles, migration is generally over long distances, and the variable geomorphology (elevation, slope, valleys, flatlands, ocean bodies) may affect thermal and orographic air movements and the ability of eagles to remain in migratory flight (Leshem & Yom-Tov, 1998; Campbell, 2015). Migration times and routes may be influenced and even determined by the ability of the raptors to adapt to these variables, duration of migration, energy cost, routes, flight speed and flight strategies (Maransky, 1997; Shamoun-Baranes et al, 2006; Shamoun-Baranes, 2010; Vardanis, 2011).

Shamoun-Baranese et al. (2010: 280) argue that "meteorology should be integrated into research on migration, especially when trying to understand natural variability observed in aspects like the timing of migration; migratory routes; orientation; use of stopover sites; or population trends such as effects on survival or breeding success as a result of changes in arrival time or physiological condition." Climate, manifested by longer term meteorological variables, is also important: "meteorology is also of interest at longer time scales when trying to understand the evolution of particular migratory systems" (Shamoun-Baranes et al., 2010: 280). Supporting literature targets the important topics relating to environmental variables and general bird migration. General sources include Richardson (1978, 1990), Dingle (1996), Liechti (2006), and Newton (2008). Studies of the impact of atmospheric conditions on migration onset include Shamoun-Baranes et al. (2006) and Gill et al. (2009), and those on migratory success include Erni et al. (2005) and Reilly and Reilly (2009). Other topics include flight speeds (Garland & Davis, 2002; Shamoun-Baranes et al., 2003), flight altitudes (Bruderer et al., 1995; Schmaljohann et al., 2009), flight strategy (Pennycuick et al., 1979; Spaar & Bruderer, 1997; Spaar et al., 1998; Sapir, 2009), bird stopover decisions (Åkesson & Hedenström, 2000; Dänhardt & Lindström, 2001; Schaub et al., 2004), migration phenology (Hüppop & Hüppop, 2003; Jonzen et al., 2006; Bauer et al., 2008, 2009), flight orientation and flight trajectories (Thorup et al., 2003); and the probability and/or intensity of migration (Erni et al., 2005; van Belle et al., 2007).

Wind is another important natural weather phenomenon that affects and even determines general bird migration (Klaassen et al., 2011; Mellone et al., 2011a; Safi et al., 2013). For example, Vidal-Mateo et al. (2015, 89) argue that "wind is among the most important environmental factors shaping birds' migration patterns. Birds must deal with the displacement caused by crosswinds and their behavior can vary according to different factors such as flight mode, migratory season, experience, and distance to goal areas." Wind affects bird migration, influencing speeds and flight altitudes, routes taken, foraging patterns during migration, and the timing of migrations. Vidal-Mateo et al. (2015, 89) further argue that "recently, the effect of winds upon migratory behavior has received great attention, because it is apparently the most important environmental factor influencing bird behavior in different stages of the routes…The effect of wind reaches such importance that it intervenes directly in the movement vector of the bird: birds' direction and flight speed is the sum of the wind vector and the bird's own flight vector relative to the surrounding air" (see also Alerstam, 1990; Akesson & Hedenstrom, 2000; Shamoun-Baranes et al., 2003; Green et al., 2004; Thorup et al., 2003, 2006; Kemp et al., 2010; Panuccio et al., 2010; Klaassen et al., 2011; Mellone et al., 2011a; Mateos-Rodrıguez & Liechti, 2012; Mellone et al., 2012, 2015; Liminana et al., 2013; Safi et al., 2013; Hernandez-Pliego et al., 2014; Malmiga et al., 2014; Vansteelant et al., 2015).

The difference between large raptor migration using soaring flight, and smaller birds using flapping flight is another important issue related to eagle migration. For example, Vidal-Mateo et al. (2015) base their study on the fact that while smaller birds migrate through flapping flight, larger birds such as eagles and vultures use soaring and gliding flight, the reason being that the energy expenditure required for flapping flight increases with bird size, mass and weight (see Pennycuick, 1975). Medium and large birds, with higher wing loadings may migrate by soaring, using horizontal or vertical (thermals) winds rather than active, flapping flight (Alerstam, 1990; Vidal-Mateo et al., 2015). Larger raptors may employ a strategy of using soaring and gliding, using soaring in thermals and then gliding to the next thermal updraft, while smaller raptors may use a similar strategy with more flapping (Kerlinger, 1989; Spaar & Bruderer, 1997). Using the example of two eagle species (Booted eagle and Short-toed Snake eagle) and one vulture species (Egyptian vulture) the study found that the movements of the Booted eagles and Egyptian vultures were strongly affected by tailwinds during both the southward autumn and northward spring migrations, but the short-toed eagles were less affected during the autumn migration. Crosswinds had significant impacts in all cases. Booted eagles and Egyptian vultures were found to adapt to wind patterns, partly by looping routes, adaptations that may occur in other species (Klaassen et al., 2010; Limin˜ana et al., 2012a, 2012b; Bradley et al., 2014; Drake et al., 2014). Factors for differences between the studied species were attributed to their different weights and wingloading. Booted eagles had a lower wing loading than Egyptian vultures, enabling the vultures to have a stronger glide speed (Spaar, 1997; Kirmse, 1998; Mellone et al., 2012). Vidal-Mateo et al. (2015, 89) argue that "our results highlight the effect of wind patterns on the migratory routes of soaring raptors, with different outcomes in relation to species, season, and latitude, ultimately shaping the loop migration patterns that current tracking techniques are showing to be widespread in many long-distance migrants."

Considering these variables, a common occurrence is a phenomenon termed a 'migration bottleneck'(see Figures 7.4, 7.5, and 7.6. These are areas where migrating birds may congregate, because of limiting factors (such as wide seas, deserts or high mountains) that influence them to crown into narrow habitat channels. As thermals only form over land, birds that would otherwise travel across the seas use these bottlenecks, such as narrow land links between continents (e.g., the isthmus of Panama), narrow seas that partition landmasses (e.g., Gibraltar, Falsterbo, and the Bosphorus). Other migratory links include passes between high mountains, such as the Central American migratory bottleneck and the Batumi bottleneck in the Caucasus, the latter avoiding the Black Sea and some high mountains (Maanen et al., 2001; Campbell, 2015). These areas are important in that they are good places to survey migrating birds, as all the migrating raptors may congregate at these points (Bildstein et al., 2009). Premuda et al. (2015) argue that such surveys at migration bottlenecks are particularly important as "surveying and monitoring raptor populations in their breeding areas is challenging, due to large territories, high costs in terms of funds and human resources, and potential disturbance to breeding sites" and "counts at migration watch-sites are also used to study migration behaviour and ecology, such as flying dynamics, energy management, weather effects, species interactions, roosting and feeding", the origin and destination of raptors, and their population status (see also Bednarz et al., 1990; Kjellén & Roos, 2000; Lewis & Gould, 2000; Zalles & Bildstein, 2000; Yosef & Fornasari, 2004; Agostini et al., 2007; Bildstein et al., 2009).

Another important migration bottleneck, this in the Americas, is the Mesoamerican Land Corridor. This land corridor connects North America and South America, and is the "numerically the most important raptor-migration flyway in the world" in which most birds "soared in and glided between thermals; some engaged in slope soaring" (Bildstein & Sabario, 2000, 197, 200;

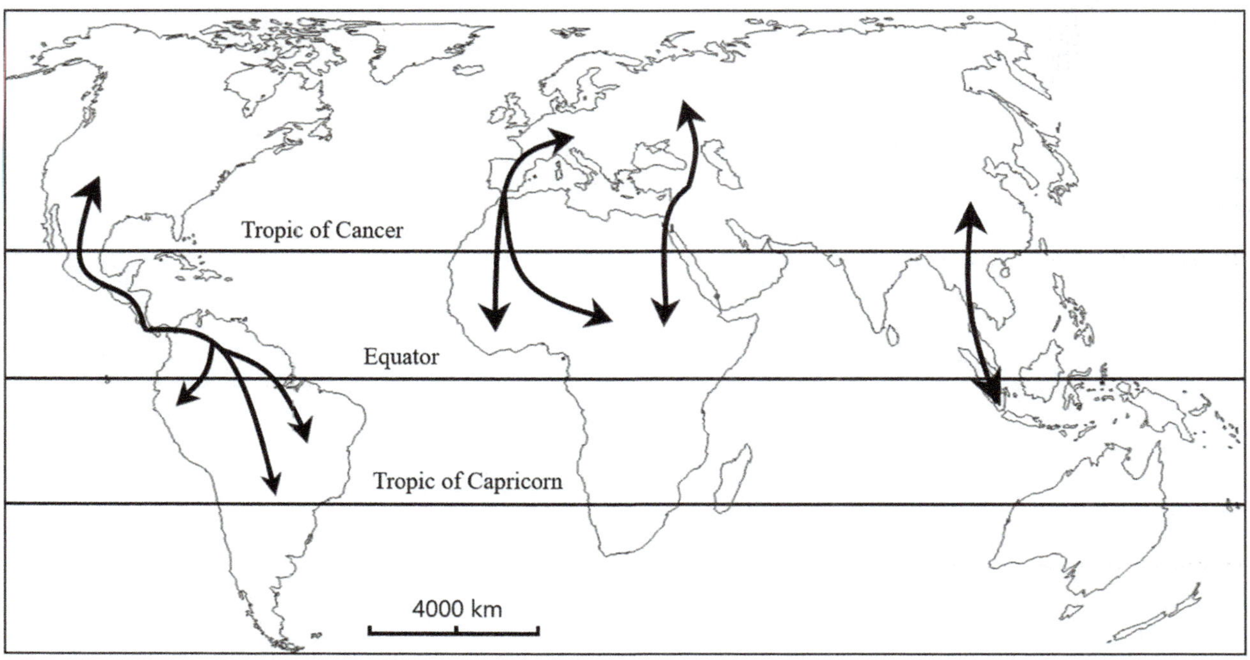

Fig. 7.4. Main migration bottlenecks

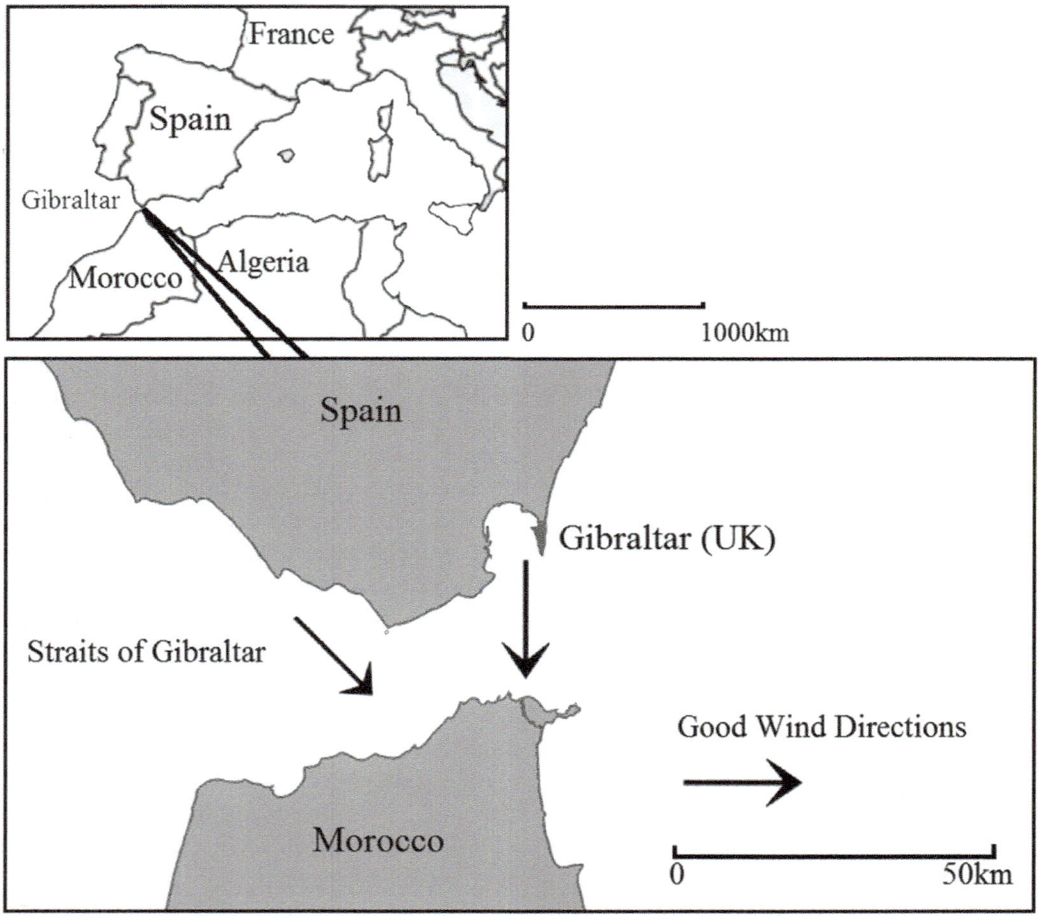

Fig. 7.5. Migration through the Straits of Gibraltar

see also Carriker, 1910, Skutch, 1945; Andrle, 1968; Smith, 1980, 1985a, 1985b; Thiollay, 1980; Stiles & Skutch, 1989; Tilly et al., 1990; Hernandez & Zook,1993; Hidalgo et al.,1995; Sutton & Sutton, 1999; Zalles & Bildstein, 2000).

Another migratory bottleneck for raptors is the Straits of Gibraltar (Bernis, 1975; Finlayson, 1992; Bensusan et al., 2007; Bildstein et al., 2009; ProgramaMigres, 2009; Onrubia et al., 2011; Martin et al., 2014, 2016; Panuccio et al., 2013, 2016, 2017). Wild Andalucia (2020) notes that at peak migration times, about 9000 soaring birds pass through the straits, including more than 3,000 Black kites, 2,000 Honey buzzards, 1,000 Booted eagles and hundreds of Short-toed eagles, Griffon vultures and dozens of smaller raptors. Computing total numbers of migrating birds for the last two years, the eagle species numbered 18,000 Short-toed eagles and 26,000 Booted eagles, less than the White stork, Black kites and Honey buzzards (Wild Andalucia, 2020). Bensusan et al. (2007, 83) record increases in the number of migrating Booted eagles, this increase in numbers corresponding "with those observed elsewhere in western Europe", additionally the "mean hourly rate of Booted Eagle arrivals during the pre-nuptial migration has risen by 65%" (Bensusan et al., 2007, 83). Describing the status of the Short-toed eagle, Bensusan et al. (2007, 87) noted that "small numbers of Short-toed eagles winter in Iberia but the majority cross the Strait to sub-Saharan winter quarters. Most of the western European population breeds in Spain, and are recorded by observations at Gibraltar, where no significant changes in numbers are recorded indicating a similarity with the population records of these two eagle species in continental Europe (Cramp & Simmons, 1979; Martí & del Moral, 2003).

Commenting on the ability of Short-toed eagles to cross the Straits of Gibraltar, Mellone (2016) notes that this species undertakes long journeys from Italy through southern France into southern Spain, the ostensible reason being their fear of crossing the wider straits between Sicily and North Africa. In particular, "Why do these eagles follow such a detoured route? The answer lies in their morphology: broad wings suited for soaring flight using thermal currents, that being absent over water, make sea-crossings very expensive from the energy point of view. Following the detoured route, eagles can avoid the costly and risky crossing of the Sicily channel (150 km over the sea), switching to the safer Strait of Gibraltar (14 km), even if the overland flight

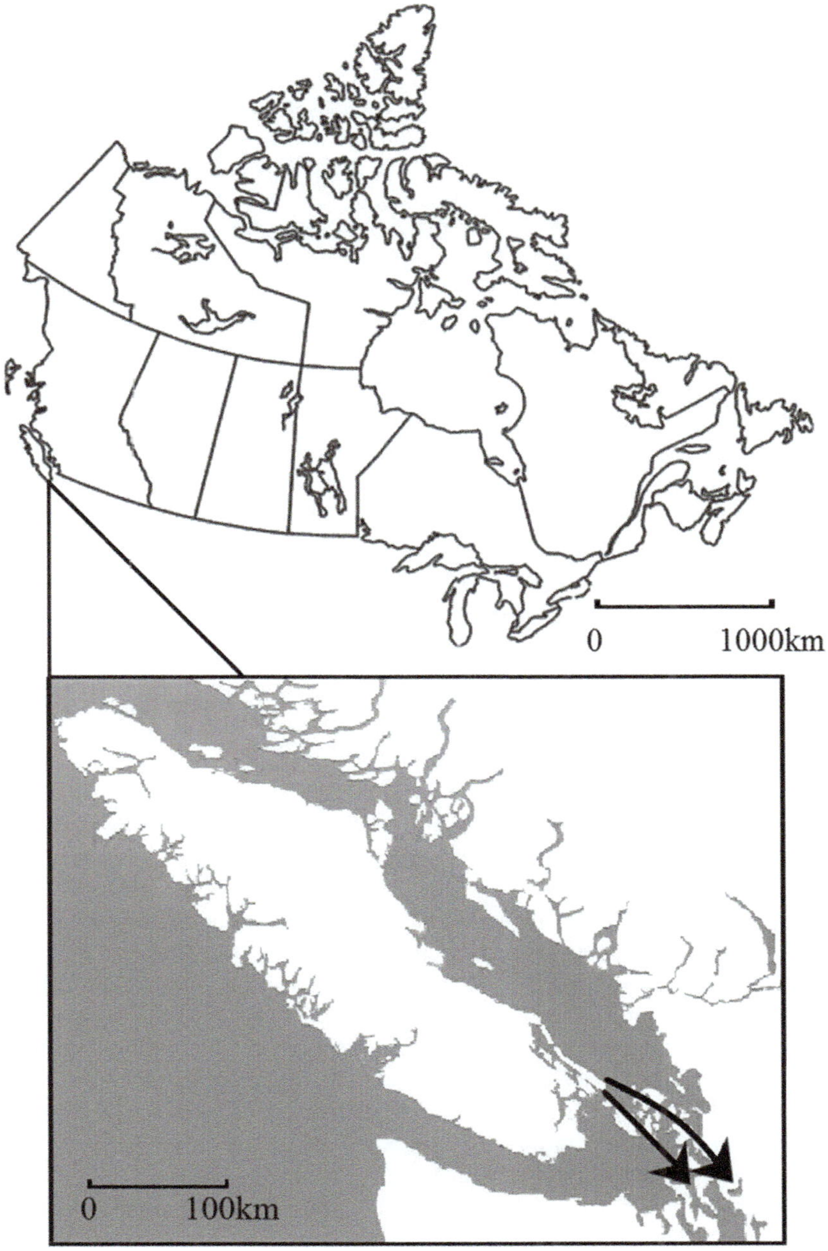

Fig. 7.6. Migration from Vancouver Island, British Columbia, Canada.

becomes much longer. Furthermore, the observation of many juveniles in mixed groups with adults during autumn suggests that these inexperienced individuals can learn the route from experienced ones."

Another migratory bottleneck for raptors is the Channel of Sicily in the central Mediterranean (Agostini, 2005). The main migration northward points are the Cap Bon Peninsula (a peninsula in far northeastern Tunisia) and western Sicily. Bird counts at the promontory at Cap Bon have been undertaken for several years (Thiollay, 1975; Giordano, 1991; Agostini et al., 1994a, 1994b; Agostini & Duchi, 1994; Agostini & Malara, 1997; Agostini & Logozzo, 1998; Agostini, 2001; Ferguson-Lees & Christie, 2001; Brichetti & Fracasso, 2003; Sara, 2003; Panuccio et al., 2004). Agostini (2005) writes that data was collected focusing on the flight behavior of raptors in the spring of 1990 and 1992 (see Agostini & Duchi, 1994; Agostini et al., 1994a, 1994b). In particular "during these observations, upon reaching the coast near the promontory, most raptors hesitated at the water barrier; they often stopped migrating and exhibited an unexpected series of movements, first disappearing across the sea, then returning to the coast, and then flying back inland." The origins of this behavior were uncertain, but possibilities included the "morphology and age of the hawk, weather…" and as "thermals are mostly absent over water, raptors cannot alternate soaring with gliding during the crossing, as they do over land" (Agostini, 2005, see also Kerlinger, 1985, 1989; Giordano, 1991; Agostini & Logozzo,

1998; Agostini, 2001; Corso, 2001; Panuccio et al., 2004). Eagles recorded in this area included Short-toed eagles and Lesser Spotted Eagles (Thiollay, 1977; Zalles & Bildstein, 2000).

It is arguable that the Straits of Gibraltar, being narrower than the route between Sicily and Tunisia, encourage more eagles to cross. Ferguson-Lees and Christie (2001) cited Thiollay's work and concluded that the "Italian population (380-415 pairs) presumably crosses by Sicilian Channel to and from Tunisia's Cap Bon, whence total of ca. 200 travelled northward on spring migration during 2-18 May 1975." Other evidence infers that most of these Italian eagles cross the Mediterranean Sea at the Strait of Gibraltar (only 14 km wide), and use a winding route to arrive in central Italy during the southward fall and northward spring migrations (see also Agostini et al., 2002a,b; Agostini et al., 2004; Baghino, 2003; Premuda, 2004). However, there are records that several eagles cross the Sicilian Channel, possibly a late fall flight of juvenile Short-toed eagles across the Sicilian Channel (Agostini, 2005, see also Agostini et al., 2004).

This heavy wing loading, that restricts long sea passage by eagles, also applies to vultures. For example, for the Griffon vulture, which has a similar wing as some of the largest eagles, Bildstein et al. (2009: 382) found that the 14-km-wide sea channel "significantly impeded the southern migration of the species into Africa, with many birds attempting repeated passage for weeks before crossing, and others not crossing at all and overwintering in Southern Spain." Flocks of vultures (never singly) usually attempted to cross the straits around mid day, when winds were light or variable, or there were stronger winds from the north or west, but cross was avoided when strong winds blew from the east or south. Possibly due to the lack of thermals over water, while crossing the vultures flapped over 10 times more often (with more flaps per flapping bout) than when during the overland flights immediately preceding the straits crossing, a rate of flapping almost impossible for them to maintain for any length of time. Bildstein et al. (2009: 382) further noted that "vultures flying over water that flapped at rates of 20 flaps or more per minute typically aborted attempted crossings and returned to Spain in intermittent flapping and gliding flight" (ibid. 382). Some exhausted vultures fell into the sea and drowned or fell onto the beaches after crossing (see also Barrios Partida, 2006).

Premuda et al. (2015) cite the example of the Apuane Alps, where recorded migrants include the Short-toed eagle, Booted eagle, Black or Yellow-billed kite and Western Honey buzzard (Premuda, 2007, Premuda et al., 2010b). The main focus was on the Short-toed eagle, a strongly migratory species (Zalles & Bildstein, 2000), with at least 90% of all eagles leaving the breeding range during the non-breeding season (Kerlinger, 1989). This species, like many large raptors avoids long sea crossings, and uses the Italian peninsula for crossings into North Africa (Agostini et al., 2002a, 2002b, 2004, 2009; Premuda, 2002, 2004a, 2004b; Baghino & Premuda, 2005, 2007; Ruggieri et al., 2006; Premuda, 2007; Agostini & Mellone, 2008; Premuda et al., 2008, 2010a, 2010b; Mellone et al., 2011a, 2011b; Baghino et al., 2012). Some birds may take the longer route through Greece at Mount Olympus to the Bosphorus (Panuccio et al., 2012).

Another migration bottleneck is located in Israel's north-south extension, which is part of the Eurasian-East African flyway and the Great Rift Valley, the second most important migration route in the world (Leshem & Yom-Tov, 1998; Krumenacker, 2013). This author estimates that 500 million birds pass over Israel during spring and autumn migration, as part of the bridge between Europe, Asia and Africa. The Lesser Spotted eagle is a common species using this flyway. In Israel there are three main north to south flyways for migrating birds, a western route near and parallel to the coastline into the northern Sinai, an eastern route to the east of the central mountain chain along the Jordan Valley, the Judean Desert, the Negev Mountains into central Sinai, and a southern route northeast to southwest through Jordan and the southern tip of Israel into the Sinai. In some areas the western and eastern routes may link with a joint route separating on in the central mountain chain (Alon et al., 2004; Leshem, 1985, 1989; Leshem & Yom-Tov, 1996, 1998). Almost the whole population of the Lesser Spotted eagle migrates along these flyways through Israel during the southward fall migration and the spring northward migration. Smaller numbers migrate through Gibraltar (e.g. one record counted 47 birds between 1998 and 2009 cross Gibraltar) (Onrubia et al., 2011), but these do not match the numbers recorded in Israel (Leshem & Yom-Tov, 1996; Shirihai, 1996; Shirihai et al., 2000, Alon et al., 2004; Meyburg et al., 2007).

Some eagle species migrate over relatively flat continental areas or use mountain passes. For example, Bohrer et al. (2011, 96) examined the migratory behavior and meteorological contexts for thermal and orographic uplift "to contrast flight strategies of two morphologically similar but behaviorally different species: Golden eagle and Turkey vulture, during autumn migration across eastern North America." The findings indicated Golden eagles primarily relied on orographic lift during migration, while only the Turkey vultures used thermal lift. As this infers species-specific specialisation to a particular uplift mode and migration paths, the authors note "it has not been shown previously that migration tracks are affected by species-specific specialisation to a particular uplift mode." In this study, the main differences were the higher wing loading (mass/wing area) of Golden eagles (7.2 kg/m^2) than the Turkey vultures (3.9 kg/m^2) (see also Pennycuick, 2008). This was attributed to the greater muscle mass and consequently stronger flapping flight of the eagle over the vulture. The Turkey vulture's lower wing-loading adapts to the weak updrafts and a scavenging or non-predatory specialisation, while the higher wing loading of Golden eagles is needed for faster flight and a predatory specialisation. These adaptations influence the migration paths frequented by Golden eagles, being above the windward side of the Appalachian Mountains and ridges, with strong orographic lift (see also Allen et al., 1996; Brodeur et al., 1996). The Turkey vultures used thermals, and even moved between scattered sources of thermal lift (Bohrer et al., 2011, see also Mandel & Bildstein, 2007).

Mojica et al. (2008) also studied the migration of subadult Bald eagles across North America, using standard tracking methods, satellite telemetry, unlike earlier studies using band recoveries and VHF radio transmitters (Harmata et al., 1985; Hunt et al., 1992; McClelland et al., 1994; Wood & Collopy, 1994; Harmata, 2002). Eagles studied in Florida have been found to be partial migrants, i.e., sub-adults migrated north in summer, up to Prince Edward Island, Canada despite many adults remaining year-round residents (Broley, 1947; Wood & Collopy, 1994). Band recoveries indicated that Florida eagles migrate as far north as Prince Edward Island, Canada in summer and returned to Florida in winter (Broley, 1947; Wood & Collopy, 1994).

Migrating eagles used both Coastal Plain and Appalachian Mountain routes on their first migration north. The mountain migrants traveled farther than coastal migrants. The eagles changed between migration routes less often on northbound and southbound movements as they aged, as the one-year-old eagles changed their choice of routes between yearly spring and fall migrations 57% of the time, 2-year-olds dropped to below 30%, and 3–5-year-olds changed to only 17% of the time (Mojica et al., 2008, 305).

The findings of this study indicated factors for Bald eagle choices for migration routes northward from Florida included the character of the "prevailing winds at time of migration, location of natal areas, or genetic pre-programming" (Mojica et al., 2008, 309; see also Broley, 1947; Hunt et al., 1992). Other factors may include "seasonally available food", especially "exothermic fish near the surface that are sensitized to seasonal temperature fluctuations" (see also Gerrard & Bortolotti, 1988), the attempt to avoid higher temperatures, access higher prey counts (Broley, 1947; Buehler et al., 1991; Wood et al., 1998; Weidensaul, 2000). The authors (Mojica et al., 2008, 306) argue that "wide variation in both departure and return dates suggests that migration behavior may not be attributed solely to prey abundance or air temperature." The main contribution of this piece was that the findings indicated that Bald eagles used migratory routes along the coasts and through the mountains, while "earlier studies in southwest and north central Florida indicated that eagles used coastal routes predominantly" (Mojica et al., 2008, 309; Broley, 1947; Wood & Collopy, 1994). Also, eagles tended to specialize with particular routes with maturity: "An increase in route fidelity as eagles age suggests that young eagles learn to migrate efficiently. As an eagle gains more experience, it may seek the same stopover sites on a particular route." Possibly, migrating eagles travelled further by using the mountain updrafts absent from the Coastal Plain (see also Heintzelman, 1975; Hunt et al., 1992).

A recent problem with wind-directed flight concerns collisions with wind turbines, discussed in more detail below. For example, Johnston et al. (2013) give the case study of Golden eagles in the eastern foothills of the Rocky Mountains in northeastern British Columbia, where there are possibilities for the development of wind turbines, but Golden eagles use such areas for migration due to the strong uplift from mountain turbulence. Johnston et al. (2013) cite numerous findings that point to the risks to raptors from such wind turbines, a risk greater than that for other birds as raptors rely on topographically channeled wind turbulence (Barrios & Rodríguez, 2004; Duinker & Greig, 2006; Drewitt & Langston, 2008; Katzner et al., 2008, 2012; Ferrer et al., 2012; Lanzone et al., 2012). Johnston et al. (2013) note that "complex mountain terrain is often favored by large-bodied soaring species during migration because of the enhanced opportunities for lift, i.e., slope winds, deflected winds, and thermals, compared with areas of flatter terrain" (Kerlinger & Moore, 1989; Hedenstrom, 1993; Spaar & Bruderer, 1996; Weimerskirch et al., 2000; Yates et al., 2001; Mandel et al., 2008; Katzner et al., 2012; Lanzone et al., 2012). This is particularly serious for rarer species like the Golden eagle (Carrete et al., 2009).

The case study by Johnston et al. (2013) for the Golden eagle, takes account of the fact that the Golden eagle migration in western North American Cordillera is concentrated along the eastern Rocky Mountains. In this area the winds generally blow perpendicularly to the travel direction (Brodeur et al., 1996; Alerstam, 2001; Bildstein, 2006, McIntyre et al., 2008). The orographic lift that results from the deflected SW winds creates a so-called migration highway, which allows a migratory soaring flight for raptors (Bittner, 2010). Based on general theory based on scattered case studies in the Northern Hemisphere, Johnston et al. (2013) argue that for two relevant points: species numbers are not necessarily related to the numbers of wind turbine fatalities; fatalities are more correlated to species-specific flight behavior, related to local weather conditions, topography, or mode of flight such as gliding, flapping and soaring (see also Barrios & Rodríguez, 2004; Hoover & Morrison, 2005; Smallwood, 2007; Lekuona & Ursua, 2007; De Lucas et al., 2008; Ferrer et al., 2012). In case studies in Spain, breeding and overwintering birds were found in some cases to be more likely to collide with wind turbines than migrants in transit to and from Africa (Barrios & Rodríguez, 2007; De Lucas et al., 2012). In Californian cases, Golden eagles and other raptors have been found to be vulnerable in areas where there are overlaps between with raptor foraging and overwintering grounds and wind turbine developments, especially as a result of bird flight patterns near the turbines (Orloff & Flannery, 1992; Hunt, 1995; Thelander & Rugge, 2001; Smallwood & Thelander, 2004, 2008; Barrios & Rodríguez, 2007; Smallwood et al., 2009; Smallwood & Karas, 2009; Smallwood & Neher, 2009).

The Johnston et al. (2013) case study of fall migration in the eastern foothills of the Rocky Mountains in northeastern British Columbia found that passage rates peaked at midday and increased with increased wind speed and temperature, juvenile eagles were more likely to pass the ridge-top area where the turbines would be located, and eagles were more likely to cross the ridges at risk height (< 150 m) considering the turbine heights with headwinds or tailwinds. During spring migration, eagles frequented the ridge-top area under eastern crosswinds. Collision risks might be higher when maximum migration rates intersected with less high-altitude lift (Duerr et al., 2012; Ainslie et al., 2013). Johnston et al. (2013) conclude that the strong SW winds during the fall migration may generate “deflected air masses that generate orographic lift but may also dissipate thermals on the ridgelines. This may cause fall-migrating eagles in the region to rely more on slope soaring than thermaling flight” and push eagles nearer wind developments (see also Duerr et al., 2012). However, the spring winds are weaker and less directional, creating thermals consistent in direction, and influence these thermals rather than the orographic lift, and “This seasonal difference may alter the relative impacts of wind developments on eagle migration, depending on the locations of the installations” (Johnston et al., 2013, see also Duerr et al., 2012).

Bald eagles may also be affected by wind turbine collisions. Mojica et al. (2016) note that the Bald eagle “is known to be vulnerable to collision with wind turbines and federal wind energy guidelines include an eagle risk assessment for new projects.” The findings of this study indicated that Bald eagle migration routes were mostly concentrated in two corridors along the Appalachian Mountains and the Atlantic Coast. In the study area, there were 3,123 wind turbines 100 m in height in the study area (1,405 turbines located in the United States and 1,718 in Canada). The findings indicated that less than 10% of the wind power classes overlapped with the areas most used by migrating eagles. The authors concluded that this finding is “encouraging because it suggests that wind energy development can still occur in the study area at sites that are most viable from a wind power perspective and are unlikely to cause significant mortality of migrating eagles.”

These turbines may not strongly affect Bald eagle populations, due to the wide range of the species. Bald eagle migration counts may be used to estimate populations and thus assess the impact of various contributions to species populations. For example, Wright (2016) compared the growth rates of two populations of Bald eagles in eastern North America. In this study, Bald eagle numbers were recorded at a springsite on Lake Ontario, separating the northern summer breeding populations and the southern breeding populations. The findings indicated that the southern breeding populations grew much faster than the northern breeding populations, based on migration numbers past the study area over the period 1991 to 2015. The conclusions of this study note that such studies along a more accurate of the population status and environmental issues surrounding Bald eagle populations over wide ranges (Wright, 2016).

Saito (2017) also records the dangers faced by the Steller’s Sea-Eagle. The Steller’s Sea eagle wintering in Hokkaido, including oil and gas developments and wind millpower plant facilities. Particularly, substantial numbers of wind mills in Northern Hokkaido are within the path of the eagles’ migration routes. Additionally, in the breeding ranges in Russia from where the migration starts, there are also industrial developments in Russia and lead-poisoning in Japan (Saito, 2017).

Eagles may also migrate in reaction to other animals, the latter moving in reaction to seasonal weather change. For example, Nilsson (2014) notes that Golden eagles in Scandinavia may migrate with the movements of the semi-domesticated reindeer (*Rangifer tarandus* L.), which is a prey species for large carnivores, this interaction providing carcasses for scavenging eagles. In Northern Sweden, the high seasonality results in seasonal migrations of reindeer migrations twice a year: from the boreal forest in the winter (the winter-feeding range) to the alpine tundra in the summer (summer feeding range, and calving area). Some eagles move with the reindeer movements in spring. Nilsson (2014, 17) however cautions that “this does not necessarily prove that eagles primarily scavenge on reindeer.”

Steppe eagles have also been recorded migrating in the Himalayan area of Asia (Den Besten, 2004). This author recorded 8,194 Steppe eagles (from which about 10,000 to 11,000 were extrapolated) migrating during fall along the Himalayas in Dharamsala, Himachal Pradesh, India. About 5,200 individuals were counted in spring (from which a total of 5,900-6,600 was estimated). Although during fall, eagles mostly migrated 9 am to 5 pm, but in the spring mostly in the morning between 9 am and 11.30 am and 4 pm and 6 pm in the afternoon. The author also recorded other eagle species such as a few Greater Spotted eagles and Imperial eagles (Den Besten, 2004). The Steppe eagle has a very wide range, breeding in south-eastern Europe, central Asia, south-west Russia, the Kirghizsteppes and northern Kazakhstan to western Manchuria, and southwards to the Aral Sea, Tien Shan, northern Tibet and Mongolia. Over this vast range, it winters southwards, in the west to eastern Africa, and in smaller proportions to southern Africa, southwest and south Asia east to Myanmar. Migratory routes include north and south of the Caspian Sea, through the Israel and Suez and the north and south ends of the Red Sea, and through Arabia to Yemen (Ferguson-Lees and Christie, 2001). Migration along the Himalayas, including Nepal and Tibet (Fleming, 1983; Christensen & Sorensen, 1989; Welch & Welch, 1991; Ellis et al., 2001).

A curious feature of the migration of the Steppe eagle is that up to 40 percent more birds are sometime recorded moving south than to the north. Den Besten (2004, 12) argues that “this would appear to be too large a difference in numbers to be explained by winter mortality alone.” He also notes that “discrepancies between spring and autumn migration of Steppe Eagles have also been described for migration bottlenecks such as Eilat, Suez and Bab-el-Mandeb at the south-western tip of the Arabian Peninsula” citing Yom-Tov (1984, 1988). It is suggested that birds may actually take roundabout routes, such as documented by Meyburg et al. (2003), where eagles wintering in southern Africa take a longer route around the Red Sea during migration, and speculatively due to the strong easterly winds between October and April which may make the return flight through the Bab-el-Mandeb onerous (Meyburg et al., 2003).

Meyburg et al. (2003) monitored several eagles that wintered in Africa and found that the return migration was around the northern tip of the Red Sea, the difference between this route and the fall migration being "probably caused by east winds that blow from October until April, making the return migration difficult via Bab-el-Mandeb. This finding should help to explain the difference in eagle numbers between spring and autumn at such migration bottlenecks as Eilat, Suez, and Bab-el-Mandeb." The eagles moving north from wintering grounds in southern Africa (different from those moving north from wintering grounds in Sudan and Ethiopia), "must make a considerable detour of over 1200 km to complete this loop. The increase in Steppe eagles overwintering in Arabia has probably contributed to the decline in the number of birds passing through Eilat in spring during recent years" (Meyburg et al., 2003, 219).

It is hypothesized that "a similar phenomenon may cause the difference between the numbers of Steppe Eagles migrating along the Himalayas in spring and autumn, with some individuals taking a more northerly route in spring to fly more directly towards breeding grounds in areas such as Mongolia" (Den Besten, 2004, 12–13). Den Besten (2004, 13) however cautions that "it is clear that more research is needed in order to find out exactly where the Steppe Eagles that migrate along the Himalayas winter, and the routes they take to return in spring." More information on such migration routes may be discerned from other writings such as Donald (1923), Fleming (1983), Christensen and Sorensen (1989), Ellis et al. (2001), Parry et al. (2002), and Welch & Welch (1991).

The Booted eagle's migration routes are from its summer residency in southern Europe to Africa, through Straits of Gibraltar and the Bosphorus (Cramp & Simmons, 1979; Martínez & Sánchez-Zapata, 1999; Thiollay, 2006; Baghino et al., 2007; Chevallier et al., 2010; Mellone et al., 2013). Mellone et al. (2013) used remote GPS tracking, based on satellite transmitters and data loggers, to track adult Booted eagles (two males and two females). Their results indicated they had difficulties crossing the 14 km straits, due to wind and rain vagaries (see also Bernis, 1980; Muñoz et al., 2010). The tracked Booted eagles wintered in the forests of West Africa, further south than some other species such as the Short-toed eagle, which primarily wintered in the Sahel (Thiollay, 2006; Chevallier et al., 2010; García-Ripollés et al., 2010, Mellone et al., 2011a; Limiñana et al., 2012a, 2012b).

In some cases, the Booted eagles may winter in southern Italy and Sicily (Baghino et al., 2007). This study found increases in the population of eagles wintering in southern Italy and Sicily, and consequently a western European origin for those species heading north during the spring migration. Baghino et al. (2007, 57) note that "the northward spring migration observed from the southern regions of Italy along the Tyrrhenian coast (towards France), validates the hypothesis of a West European origin of the individuals migrating in the spring of 2005 and can be regarded as a case of circuitous migration in the Booted Eagle" (see also Baghino & Premuda, 2005).

Premuda (2007) describes the Booted eagle as a trans-Saharan migratory species, and in some cases as a partial migrant, over wintering in Southern Europe and North Africa (see also Zalles and Bildstein, 2000; Thévenot et al., 2003; Isenmann et al., 2005), with two main population clusters in south-western France and Spain, and another in Russia and Turkey. For the western populations, migration is through the Straits of Gibraltar and to a much lesser extent the Bosphorus (Cramp & Simmons, 1979; Finlayson, 1992; Zalles & Bildstein, 2000; Thévenot et al., 2003). Possibly there is a migration route from France through to southern Italy, and perhaps the Sicilian Channel to North Africa (see also Bijleveld, 1986; Premuda & Baghino, 2004). Premuda (2007) also touts the possibility of winding migration routes along the coast from Spain to France and Italy, similar to that of the Short-toed eagle (Olioso, 1991; Dubois et al., 2001; Agostini et al., 2002a, 2002b, 2004; Kayser et al., 2003; Guillosson & Garcia, 2006). Numerous records are reported in Sicily (Iapichino & Massa, 1989; Iapichino, 1993, Ciaccio & Priolo, 1997; Lo Valvo et al., 1993; Grussu & Corso, 1997; Corso & Iapichino, 1998; Premuda & Baghino, 2004). Premuda (2007) concludes that research findings have indicated a migratory route from France to Italy (see Premuda & Baghino, 2004), but "more research is needed in the near future to improve the knowledge of this circuitous migration of the Booted eagle and the new wintering tendencies in Southern Europe."

Wahlberg's eagle is another species that migrates over open land, but it generally migrates northwards in the southern hemisphere. Its migratory patterns were unknown or speculated in the past, with possibilities of a final destination in South Sudan (Steyn, 1962, 1982). More recent studies have used tracking technologies to locate southern African birds further north in northern Zambia, northern Zaire (Congo) and northern Sudan (Simmons, 1990, 1991). Meyburg et al. (1995, 135) admit that "few connected observations of migration have been recorded" and cite the examples of Ulfstrand (1960) and Thiollay (1975), the latter recording over 1200 eagles in the Murchison Falls National Park in Uganda. Meyburg's study was the first to use satellite telemetry for tracking eagles, and the findings supported the hypothesis of "intra-African transequatorial migration" and migration through the Congo rainforest (Meyburg et al., 1995, 135). In this study, one eagle was tracked from northern Namibia 8816 km through northern Cameroon, north-eastern Nigeria, western Chad and Congo. This bird ranged over thousands of square kilometers, and the northward journey of about one month was about two weeks less than the return migration.

7.4 Eagles and Biomes

Eagle foraging and nesting is highly dependent on vegetation cover, indicated by presence frequencies in forest (e.g., species like the Harpy eagle and Crowned eagle), savanna (for example, the Martial eagle), mixed vegetation on mountains (e.g., the Golden eagle) and riverine and coastal areas (e.g. the Bald, Steller's and African Fish eagles). Eagles may also nest in forest trees and on bare cliffs. It is possible to predict or hypothesize about the conservation potential for eagles using classifications of variably classified vegetation types. There are many classifications of global vegetation patterns, which may be useful. For example, the WorldWide Fund for Nature (WWF), recognises 14 biomes or Terrestrial Ecoregions or Major Habitat Types. A biome is defined as a region of similar climatic conditions with communities of animals and plants associated with this climatic regime (Olson et al., 2001; Campbell, 2015). The biomes under this classification are listed in Table 7.3.

Table 7.3. Biomes linked to eagles

No.	Biome	Example of eagle
1	Tropical and subtropical moist broadleaf forests	Harpy eagle, Philippine eagle
2	Tropical and subtropical dry broadleaf forests	Booted eagle
3	Tropical and subtropical coniferous forests	Solitary eagle
4	Temperate broadleaf and mixed forests	Golden eagle, White-tailed eagle
5	Temperate coniferous forests	Golden eagle, Bald eagle
6	Boreal forests/taiga	White-tailed Sea-eagle
7	Tropical/subtropical grasslands, savannas, shrubs	Short-toed snake eagle
8	Temperate grasslands, savannas, and shrublands	Eastern imperial eagle, Chaco eagle
9	Flooded grasslands and savannas	Greater and Lesser Spotted eagle
10	Montane grasslands and shrublands	Wedge-tailed eagle
11	Tundra (Arctic)	Steller's Sea eagle
12	Mediterranean forests, woodlands, scrubs	Spanish Imperial eagle
13	Deserts and xeric shrublands	Tawny eagle
14	Mangrove	Bald eagle

Sources: Ferguson-Lees & Christie (2001), World Wildlife Fund (2020).

As shown above, eagles can occur in all the 14 biomes, but most occur in the forest, savanna and grassland ecosystems, especially five biomes: the Tropical and subtropical moist broadleaf forests; Biome 2, Tropical and subtropical dry broadleaf forests; Biome 7, Tropical and subtropical grasslands, savannas, and shrublands; Biome 8, Temperate grasslands, savannas, and shrublands; and Biome 10, Montane grasslands and shrublands (alpine or montane climate). They may occasionally frequent Biome 11 (tundra) and Biome 13, the deserts, possibly as migrants and occasional foragers. Biomes 1 and 2, the tropical and subtropical moist broadleaf forests and the drier tropical and subtropical dry broadleaf forests have high rainfall, and a dry season in the drier areas. Within the moist broadleaf forests, the wettest are the tropical rainforests, where the temperature may exceed 18°C throughout the year, and rainfall may be as high as 200 cm, and in some cases 450 cm.

Rainforest vegetation layers usually include the emergents, canopy, understorey and forest floor layers, the first three being the most important for some forest eagles, which may hunt in these areas (Figure 7.7). Other eagles may kill prey on the ground. The emergent layer is composed of a few tall trees that grow above the canopy, sometimes over 50 meters. The canopy layer comprises most of the larger trees, around 30 to 45 meters, forming a continuous or variably broken carpet. The understorey layer is mostly smaller, shrub-sized plants between the canopy and the forest floor, where some prey species, such as monkeys may live. The forest floor is covered sparsely with shade-resistant plants and may have fauna such as rodents and other small mammals and reptiles, which may also be eagle prey species. Gaps may occur in the canopy (usually a result of human induced deforestation, forest fires or storm blowdowns, sunlight may reach the forest floor and new, shade intolerant shrubbery, grasses and saplings may grow, and new avian and mammal species may be present (Campbell, 1998; Cuervo, 2008).

The more open landcover clusters around grasses and shrubs, and in some scattered trees. For example, Biome 7, the tropical and subtropical grasslands, savannas, and shrublands, is an important habits type for eagles sub-Saharan Africa, North and South America and Asia. Grassland is land dominated by grasses, with savanna may include more shrubs and woodland patches. Scholes & Walker (1993) note that savannas cover about half the African continent and one fifth of the land surface of the world. Savanna landscapes may vary in the relative proportions of trees, shrubs and grasses, according to climate, soils,

A theoretical construct of forest layers.

A cross-section of the Amazon rainforest, showing emergents, a canopy layer and an understory and vegetated florest floor.

Reproduced with permission of Rômulo Ferreira.

A cross-section of the Congo forest in Central Africa, showing similar structure to the Amazon forest above.

Reproduced with permission of Andrey Gudkov.

Fig. 7.7. Forest layers

groundwater and human action (mostly deforestation and agricultural activities) (Campbell, 1998; Defries et al., 2010; Campbell, 2017, 2018). Savanna vegetation is thus very heterogeneous (Solbrig, 1993; Mistry, 2000; Campbell 2005; Parr & Andersen, 2006). For example, Harris (1980, p.3) argues that "the word 'savanna' has had an ambiguous career since it entered the English language in the 16th century." Scholes & Walker (1993, 3) nevertheless hit a key point in noting that "the central concept – a tropical mixed tree-grass community – is widely accepted, but the delimitation of the boundaries has always been a problem." Therefore, they consider savanna vegetation as including a "continuum of vegetation types which have trees and grasses as their main constituents", while "acknowledging that at the extremes, the distinction between savannas and woodlands is unavoidably arbitrary" (ibid. p. 3).

Despite the acknowledgment of the complexity of savanna vegetation and the difficulty of precise description, some definitions have been attempted. For example, Scholes & Walker (1993, 4) define savanna as a "tropical vegetation type in which ecological processes, such as primary production, hydrology and nutrient cycling, are strongly influenced by both woody plants and grasses, and only weakly influenced by plants of other growth forms", a definition based on a description from the International Responses of Savanna to Stresses and Disturbance (RSSD) Program (cited in Frost et al., 1986, in Scholes & Walker 1993, 4). Solbrig and Young (1993, 3) add that savanna "designates tropical grassland with scattered trees" and is the "most common tropical landscape unit", a point supported by Werner et al. (1990, 343) who argue that savannas are "delineated both structurally (specifying a wood/grass composition) and climatically (seasonality of water availability)", and also by Menaut et al.

Fig. 7.8. Deforestation in Central America [Author]

Fig. 7.9. Urban sprawl and deforestation in Central America [Author]

Fig. 7.10. Urban sprawl and deforestation by a Central American lake [Author]

(1990, 471) who referred to savanna vegetation as " generally expressed as the balance between trees and grasses, and more easily by tree density and distribution."

Shrubland refers to landcover where shrubs may dominate, at the expense of trees and grass, which may also be defined as tropical thicket (Scholes & Walker, 1993: 9). This tropical thicket is defined by small to medium height, multi-stemmed woody plants (either natural or stunted through regular trimming for firewood) with varying amounts of grass cover, which may be absent and hence the parent vegetation types may be "disqualified as savannas" (ibid). These may still be savannas, as the complexity of the compositional forms of savannas may involve transitions to thicket "in the absence of better understanding of the ecology of thickets, we treat them as a class of savannas rather than as miniature dry forests" (Scholes & Walker, 1993, 9). Cole (1986, 16) classifies "thicket and shrub" as savanna, and also by Menaut (1983, 128) who noted that thickets, which may be found in "a great diversity of environments, close to the rainforest as well as near the desert edge", are common aspects of coastal savannas.

The climate of savannas, intermediate between forests and deserts, may have average annual rainfall levels of 200 to 250 millimeters (Harris, 1980; Stott, 1994; Bucini & Lambin, 2002). In Africa, the driest savanna type is the Sahel or Sahelian savanna, which represents the zone of transition between the Sahara Desert to the north and the Sudan Savanna to the south. This type is mostly grassland, with a few shrubs and fewer trees. Grass species include *Cenchrus biflorus*, *Schoenefeldia gracilis*, and *Aristida stipoides*. Tree species include *Acacia* species *Commiphora africana*, *Balanites aegyptiaca*, *Faidherbia albida*, and *Boscia senegalensis*. Shrub species include *Panicum turgidum* and *Aristida sieberana*. Mammals include scimitar-horned oryx (*Oryx dammah*), dama gazelle (*Gazella dama*), Dorcas gazelle (*Gazella dorcas*), red-fronted gazelle (*Gazella rufifrons*), the giant prehistoric buffalo (*Pelorovis*) and Bubal Hartebeest (*Alcelaphus busephalus buselaphus*, lions, leopards, cheetahs and African wild dogs. These species may be rare due to competition with livestock, and due to human hunting.

The Sudanian Savanna, located in a belt south of the Sahel savanna, has more trees, and a longer rainy season. Common tree species are *Combretaceae* and *Caesalpinioideae*, with some *Acacia* species. Grasses include *Andropogoneae*, especially the genera *Andropogon* and *Hyparrhenia*. Vast areas of the Sudanian Savanna have been farmed, and culturally useful trees, such as shea, baobab, locust-bean tree and others are left standing, while farms of sorghum, maize, millet or other crops are cultivated. The Guinea Savanna or tree savanna is found south of the Sudan Savanna, and as the transition zone between the Sudan Savanna and the Dry Deciduous Forest, it has a longer rainy season than the Sudan Savanna and more trees. This region transitions to the Dry Deciduous Forest, which also transitions southwards into the tropical moist broadleaf forests of Equatorial Africa (Campbell, 2015).

The Temperate grasslands, savannas, and shrublands (Biome 8) include the grasslands and plains of southern Canada, the middle of the United States and Western Eurasia, the Prairie and Pacific Grasslands of North America, the Pampas of Argentina, Brazil and Uruguay, calcareous downland, and the steppes of Europe. It may interface with Biome 12. the Mediterranean forests, woodlands, and scrub or sclerophyll forests. The temperate grasslands are inhabited by large herbivores, such as American and European bison, antelopes and wild horses such as the Przewalski's horse. Carnivores include wolves, coyotes and leopards, while smaller herbivores include deer, prairie dogs, coyotes, snakes, foxes, owls and badgers, which may be the prey of eagles (Campbell, 2015, 2017, 2018).

The montane grasslands and shrublands (alpine or montane climate, Biome 10) are found mostly in the Asian Himalayan mountains and the South American Andes. Alpine tundra are those tundra vegetation types that are above the tree line, which is the maximum altitude at which trees can grow. Below the tree line, the tundra is referred to as subalpine and montane grasslands and shrublands. There are also stunted subalpine forests below the tree line, which are created due to the cold windy climate and poor soils. Examples are the Neotropic Páramo of the Andes Mountains and the montane steppes of the Tibetan Plateau, the latter being the largest and highest uplands in the world, comprising meadows, alpine shrub, and primitive forests (Lu et al., 2009). An example of the Neotropic Páramo of the Andes Mountains is the Páramo de Frontino, in the Western Andes of Colombia. Here Krabbe et al. (2006) recorded the Black-and-chestnut eagle and Solitary eagle in this area. The Asian or Himalayan Golden eagle (*Aquila chrysaetos daphanea* Severtzov, 1888) ranges through the Himalayan area, through central Kazakhstan, eastern Iran, and the easternmost Caucasus, and the Himalayas from northern Pakistan to Bhutan (Ferguson-Lees & Christie, 2001; Watson, 2010; Rasmussen & Anderton, 2012). Another example is the Australian Alps, located along the east coast of Australia from Tasmania to Cape York. The Wedge-tailed eagle ranges through this area (Mockrin, 2020).

Desert areas, classified as Biome13, are those areas drier than the savannas and grasslands. Between desert and grassland, there may be small clumps of grass, termed Xeric grassland. Deserts may be formed from long term climate change (such as the Sahara Desert), from the rain shadow effect from mountains (such as the American Southwest) or from human action that degrades grassland, a process termed desertification (Mazzucchelli, 1967; Mortimore, 1989; Mares, 1999; Marshak, 2009; Zeng & Yoon, 2009; McSweeney, 2019). Geist (2005) presents the widely accepted definition of desertification, from the Princeton University Dictionary which defines it as "the process of fertile land transforming into desert typically as a result of deforestation, drought or improper/inappropriate agriculture." According to Geist and Lambin (2004: 817) "The most authoritative definition of desertification remains that of the Convention to Combat Desertification" which is that it is "land degradation in arid, semi-arid and dry sub-humid areas resulting from various factors, including climatic variations and human activities" (see United Nations Environment Programme (UNEP), 1994). Geist & Lambin (2004) listed factors for desertification including: agricultural activities, comprising livestock production (nomadic/extensive grazing, intensive production), crop production (annuals, perennials); infrastructure extension including watering/irrigation (hydrotechnical installations, dams, canals, boreholes, etc.); transport (roads); human settlements; public/private companies (oil, gas, mining, quarrying); wood extraction and related activities such as harvesting of fuelwood or pole wood (from woodlands/forests), digging for medicinal herbs and other collection of plant or animal products; and demographic factors such as migration (in- and out-migration), natural increment (fertility, mortality), in population density and life-cycle features (Geist & Lambin, 2004, 819).

Eagles have been recorded in numerous deserts in the Americas, Africa and Eurasia. For example, Braham et al. (2015, 225) describe the foraging and presence parameters of Golden eagles in American Mojave Desert, especially near renewable energy projects. These authors found that Golden eagles ranged widely over large areas of desert to fulfil their foraging needs. Particularly, "eagles moved from desert heat to higher elevation cooler mountainous areas" and significantly "monthly home range patterns coincided with seasonal variation in breeding ecology, habitat associations, and temperature. The expanded home ranges in hot summer months included movements to cooler, prey-dense, mountainous areas characterized by forest, grasslands, and scrublands… Golden eagles in the Mojave Desert used more space and a wider range of habitat types than expected and renewable energy projects could affect a larger section of the regional population than was previously thought."

Kaczensky et al. (2004) examine the Great Gobi Strictly Protected Area (SPA) in Mongolia, which includes over 50,000 km^2 of "fragile desert steppe and desert habitat." Since 1991, the United Nations has established this as an international Biosphere Reserve, the largest in Asia. The Golden eagle and the Steppe eagle range over this semidesert and desert. The nearby Altai mountains are also part of the Golden eagle's range. In this area, Golden eagles may be used to hunt animals by falconers. Crowder (2015) describes this area as "one of the planet's most desolate and harsh terrains, the Altai Mountains which run from Siberia in Russia down to Mongolia's Gobi Desert" and further notes that "hunting with eagles is currently only practiced by a handful of Kyrgyz and Kazakhs… This form of falconry, the practice of hunting with the aid of birds of prey, can be traced back as far as 4,000 years in Central Asia."

7.5 Conclusions

Climate, relief and landcover variations are crucial for eagle distribution, foraging, ecology and migration. The evidence presented in this chapter indicates that eagles are very resourceful birds, that use landscape and land cover features to adapt to climate change. Some of the adaptation are evident at the intra- and interspecies levels. Common themes may be discerned between the eagles in the Americas, Eurasia and Africa. These include the ability of eagles to innovate when crossing water bodies during migration, and their flexible foraging habits in forest, open landcover and mixed vegetation. The next chapters will look at eagle relations with people, through historical cultural patterns across the world, and the impacts of human modifications of the eagle habitats and populations. People kill, venerate and conserve eagles, through conversion of forest, savanna and mixed landcover into urban areas, farmland and transportation zones, reforestation and reintroduction of eagles, worship of these species and legal protection and scientific study of eagles. Eagle adaptation and survival through these complex dynamics may be determined by their intra- and interspecies abilities and the cultural outlook of the peoples concerned. Part 3 and its composing chapters will examine these issues.

Part III

Eagles and Conservation Issues

This part looks at the relations between eagles and human actions and cultures, based on the interface of eagle ecology and human culture, politics, history, environmental policies and management. From this perspective, there are many relevant interdisciplinary, cross-disciplinary and multidisciplinary approaches. The introductory part of this Part 3 will look at the methodological and scientific background of these complex issues, and the succeeding chapters will examine the findings and conclusions of these studies relevant to the eagle-human relationship. Eagles are generally large, highly charismatic birds, and like lions, tigers, elephants and bears, they elicit strong human emotions and reactions, both positive and negative. There are five key issues to be covered: landcover change and impacts on eagles; geomatics as a tool for the documentation and analysis of the parameters of the human-eagle relationship; environmental chemistry and the impact of pollution on eagles; cultural perspectives of eagles, including falconry and representations of eagles; and eagle individualism and adaptation to human actions.

Eagles may range over very large areas, hence local, detailed ecological studies may not be sufficient to study the vagaries of their relations. Hence, research methods including geographical information systems (GIS), remote sensing and global positioning systems (GPS) and drones may be used. These may be used for habitat surveys and analyses, or as tracking technologies for tracing eagle movements. These can also study eagle movements within their foraging territories or track their migratory movements. Geomatics is mostly a North American term, the European equivalent being geoinformatics (especially in Europe), is defined variously, but there is consensus that it includes the fields of spatial data acquisition, assessment and analysis, and technologies used for these processes, and the application to geographical (natural and social) spatial and process issues. Geomatics usually includes the fields of remote sensing, geographical information science (GIS), geographical information systems (GIS), global positioning systems (GPS), geodesy, cartography and mapping, and applications to the natural and social sciences (Campbell, 2018).

Applied to the study of eagles, geomatics frequently intersects with the comparatively new field of conservation biology. Like geomatics, conservation biology is broad with various sometimes contested definitions. Conservation biology emerged from several disciplines including the biological sciences (especially ecology and biogeography), political paradigms associated with the conservation movement, environmental economics and sociology, political ecology, even history and anthropology. These disciplines share interpretive analysis in the application of their approaches to spatial socio-environmental issues, such as the factors for landcover change, habitat effects and the regionalization of ecological change.

At the root of the relationship between geomatics and conservation biology are four key concepts. The first concept is that of space (Campbell, 2017, 2018). This concept explains the areas where related entities are located. This may vary in extent, from a global scale, through continental, regional, national and local scales. For applications to issues relevant to eagle issues, these may be exemplified by migration patterns (global or continental), local juvenile dispersal (continental, regional and even national) and nesting, roosting and foraging selection (local or national). Of course, all these scales are linked to eagle ecology and also human actions that may affect habitats and consequently eagle ecology.

Another concept is pattern, defined as how entities, components and relational issues arrange in space (Jensen, 2007; Campbell, 2017, 2018). These could be physical entities such as habitat areas (forest trees, woodland or savanna) or foraging areas (open water or prey species breeding areas), relational components (for example, how water, vegetation, climate and air movements link together to create resources useful for eagle ecology) or wider relational issues such as how the areal differentiation of human actions impact on eagle ecosystem parameters. This may include human cultural variables, such as peoples and their cultures, and human-environment relations, with regional variations.

Space and pattern may combine into the third concept, that of place (Heywood et al., 2006; Campbell, 2017, 2020). This is defined as a localized point, where unique relationships or entities combine in particular (possibly unique) ways, from which patterned dynamics may emerge. Place may therefore refer to small local areas, such as the unique vegetation and prey patterns around eagle nests. These areas may be measured and analyzed using geomatics techniques such as drones, webcams, aerial photographs, satellite images, and/or point count surveys and statistical tables (Campbell, 2015, 2016, 2017, 2018). As documented in the earlier chapters of this book, there are several inter- and intra-species related variables that may vary according to place, such as frequencies of cliff or tree nesting, time and location of breeding season, and even time of migration.

The fourth concept, integrated analysis is based on the use of analytical tools from several disciplines or theoretical/methodological complexes to assess ecological entities or processes. This may combine image-based methods, GPS locations, point counts of animals and quantifiable parameters of vegetation, soils and climate, with the combine results put in into maps that may be integrated to classify the ecology of the studied areas. This, combined with the other concepts listed above, may create the fifth concept, analytical mapping. Analytical maps may be indispensable for conservation biology, as reference points, authoritative policy supports and derivable databases, and enable predictive modelling in conservation biology. This is especially relevant in cases where eagle prey species habitats are undergoing of multi-directional change (environmental changes that are unilinear but may include increases and reductions in species or habitat forms) (Campbell, 1998; Elliot & Campbell, 2002).

The relationship between people and eagles in changing landscapes is another important issue, and in many cases the most important form of landscape change is urbanization. Here, urbanization refers to both the increased number of people living in urban areas and the development of the physical infrastructure such as roads, parking lots, buildings, railways, parks, rubbish dumps and commercial centers. There is strong urbanization globally, especially in Asia and the Americas. Urbanization may remove the live food sources of eagles. In some cases, where eagles are scavengers, they may compete with urban avian scavengers such as vultures and crows and may even prey on urban mammals such as rats and urban waste may provide sources of urban foraging (Campbell, 2006; Gbogbo & Awotwe-Pratt, 2008). Foraging areas may comprise open sewers, rubbish dumps and open abattoirs, especially in tropical developing nations (Marzluff et al., 2001; Schochat et al., 2006). This habitat status may be based on the land cover configuration (parks, sidewalk vegetation, treed avenues, roofs, abandoned buildings and even gutters), human behavior (recreational food stands, garbage disposal sites, food markets and abattoirs), individual human reactions to birds in that particular area, and the levels of population density, food and cover availability, land cover heterogeneity and socio-cultural understanding (Campbell, 2006, 2015).

Avian species numbers and density (including those of eagles and other raptors) are much higher in the tropics than in temperate countries (Marzluff et al., 2001). Tropical cities in developing countries are often characterised by open food markets, large garbage disposal areas and in some cases poor sanitation that may facilitate avian colonisation (Mundy et al., 1992). Large, tropical avian scavengers are especially important, as they are relatively understudied (Borrow & Demey, 2001; Marzluff et al., 2001; Schochat et al., 2006). With the "expected increase in human populations and urbanization in the tropics and the rich biodiversity that characterizes this region", it may be surmised that "more studies are desperately needed to inform public policy so that the negative consequences of human development are mitigated" (Marzluff et al., 2001: 1; see also Schochat et al., 2006).

Although in most tropical urban areas, vultures and corvids are dominant foraging guilds (Borrow & Demey, 2001), eagles may exist, compete and even dominate the other scavengers (Lim & Sodhi, 2004; Simmons et al., 2007; Gbogbo & Awotwe-Pratt, 2008; Symes & Kruger, 2012; Campbell, 2015). In northern countries, Larus gulls and corvids are the dominant urban avian scavengers, with vultures mostly absent and eagles such as the Bald eagle foraging in peri-urban and coastal or riverine suburbs (Elliott et al., 2006; Sullivan, 2016; Campbell, 2015). Eagles, like other large avian scavengers, may affect human quality of life (Ma et al., 2010). The principal foraging areas for these birds are urban dumps, green spaces, car parks, roads, waterfronts, riverbanks, beaches, eating places, transport spaces, suburban farms). These areas are used by people for recreation, feeding, transport, and social interaction. Problems may concern threats to pet animals, nuisance scavenging, human life space intrusion, defecation, and obstruction of human communications and transport (Campbell, 2017; Camphuysen et al., 2011).

Urban avian scavengers and predators, including both eagles and vultures (which may be mistaken for each other), may be perceived positively or negatively, depending perceived actions and general knowledge (Adeola, 1992; Marzluff et al., 2001; Soewu, 2008). A common issue is the impact that large, charismatic animals may have on human quality of life (one basis for conservation policy), the latter issue defined variously and including social issues such as social structures, cultural changes and environmental dynamics (Campbell, 1998; Schalock et al., 2002). Definitions of human quality of life vary (Felce & Perry, 1995, 1996; Hughes & Hwang, 1996; Cummins, 1997a,b; Schalock, 1997, 2000). A consensus exists on the main components or entities for human quality of life: (a) emotional well-being, (b) interpersonal relationships, (c) material well-being, (d) personal development, (e) physical well-being, (f) self-determination, (g) social inclusion, and (h) rights (Schalock et al., 2002). These dimensions may vary with social variables such as age, culture and social perceptions (Elorriaga et al., 2000). Such large animals may both enhance and degrade human life, and this may emerge as the key issue in conservation policy (Campbell, 2017; 2018; Schaltegger & Beständig, 2012). The relevant environmental variables in this case are the physical spaces in urban areas (green spaces, car parks, river banks, beaches, airports, recreational facilities) and the biological actors (in this case raptors and scavengers).

A major issue for large raptors in urban areas concerns airplane collisions. Bird-Aircraft Strike Hazard (BASH) is an important part of avian and applied social studies (MacKinnon, 1999; Gard et al., 2007; Campbell, 2017, 2018; Camphuysen et al., 2011). For example, the U.S. Federal Aviation Administration recorded 57,702 bird-aircraft collision reports within the United States between 1990 and 2004, with over $600 million damage (Gard et al., 2007). "Large flocking birds and birds of large body size" are described as particularly susceptible to such collisions (United States Department of Agriculture, 2003). Factors include bird migration patterns and bird attraction to the prey sources in the grassed areas near the airports (Gard et al., 2007).

In rural areas, eagles may prey on livestock. For example, Avery and Cummings (2004, 58) note that "Golden eagle populations are increasing in western states with sheep production. It is unknown whether increased eagle numbers translate into increases in livestock depredations. It is important for livestock producers to understand that management techniques for Golden eagles are limited. The combination of human-like scarecrows, harassment and increased human activity is the most feasible means of protecting lambing bands from Golden eagles." They further point out that "Golden eagle predation on livestock has been documented in many areas of the western United States. Most depredations involve Golden eagles preying on young lambs and goats; depredations on domestic calves occur occasionally" (see also Phillips & Blom, 1988). Davies (1999, 67) referring to African eagles, also wrote that "Golden and Crowned eagles have killed prey (including domestic sheep) weighing 20-50 kg. Large eagles are, therefore, capable of killing domestic sheep up to about half adult size, so the issue appears to be not whether they kill domestic lambs, but to what extent does this occur and under what circumstances" (see also Lehti, 1947; Cooper, 1969; Bruns, 1970; Steyn, 1982; O'Gara et al., 1983, Bergo, 1987).

Another major interaction between eagles and people concerns the poisoning of birds with pesticides and other toxic chemicals (Campbell, 2015). For example, Elassar (2020) reports on the lead poisoning of Bald eagles in the United States, mostly from lead bullets in animal caracsses (see also Gibbens, 2017). The Canadian CTV News (2020) on Vancouver Island reported that "Eight bald eagles are fighting for their lives on Vancouver Island after feasting on an animal carcass that was loaded with sedative drugs." Other vultures may be poisoned by DDT (dichlorodiphenyl trichloroethane) and DDE (dichlorodiphenyl dichloroethylene) despite the ban on the agricultural use of DDT in the United States in 1972, and the global ban in the Stockholm Convention (1972). These issues will be explored at length and indicate the extent to which human action can interfere with ecology of eagles (Campbell, 2015).

The pollution chemistry of the environment, especially chemical element and compound accumulation at the upper levels of the food chain is a fundamental aspect of the study of the conservation biology of raptors (birds of prey) and waterbirds, with increasing relevance given the largely human-induced pollution of ecosystems and food chains (Campbell, 2015). Crucially, chemical element and compound intrusion is a factor behind local species extinction and the mitigation of species recovery efforts. Chemical pollutants are variously defined, but generally they are elements or compounds that may negatively affect the physiology and health of animals. The key points concern the contributory factors of industrial, agricultural and urban technology, and related human activities, with immediate or potential hazards to animals, plants and people, and animal body threshold levels for chemical pollutants. Such chemical elements and compounds include mercury, lead, copper, arsenic, chromium, rodenticides and insecticides, nonsteroidal anti-inflammatory drugs (NSAIDs) and polychlorinated biphenyls -PCBs. Raptors and waterbirds are particularly susceptible to chemical poisoning, as they eat animals that have ingested such poisons from infected materials, with accumulations at the upper levels of the food chain. This chapter takes a snapshot of the vast area of environmental chemical pollution and examines the factors for such events, the impacts and the ameliorative actions that have been attempted as part of conservation management. Raptors and waterbirds are taken as examples, due to their food chain status and the primary impact of pollutants on small mammals and fish, and carcasses of these and larger mammals. The methodologies employed in the studies are also examined. It is argued that the main culprits for such events, such as agricultural and industrial development, have been increasingly managed by technological advancements and increased ecological knowledge and more competent management. However, the increasing intensity of agricultural and industrial development, compromises the benefits derived from such efforts. Increasingly, fields of knowledge such as veterinary science and environmental management claim the interface between environmental chemistry and conservation biology.

Cultural perceptions of eagles are also important. Eagles have very powerful political and sociocultural significance and hence are very important for national symbols and heraldry. Several countries have eagle species as national birds: Albania, Germany, Indonesia, Mexico, Namibia, Panama, Philippines, Poland, Scotland, Serbia, the United States of America, Zambia and Zimbabwe. Historically, eagles have been symbols of heraldry, especially in royal Europe, including the two headed eagles of Ancient Greece, Rome, the Holy Roman Empire and the 19th to 20th century Austrian Empire. This portrayal of the eagle with a double head (probably a Golden eagle) is though by some to indicate power and dominion of a political entity (usually of an empire) over other entities. Eagles may also be used for hunting, as sport and for human food provision, This falconry was and still is an important interaction between people and eagles, especially in Asia and to a lesser extent Europe. This activity was a manifestation of the perception of eagles as the strong hunters that they are, and of the regal status that they confer on people who associated with them.

These complex relations between eagles and people can cover a vast amount of space. What is attempted in this section of the book is a summary of the various ways that people have perceived, interacted with, impacted on, and researched upon eagles through the ages. The evidence indicates that despite many negative impacts (e.g., persecution, indirect and direct killing, destruction of habitats, and trapping for feathers and zoos), there were also positive impacts, such as conservation policies and protections based on deity status. Although as noted in the above chapters the populations of several eagle species are declining, the charismatic status of the various species of eagles ensures that eagles may remain in the forefront of conservation policies in many countries.

8

Eagles and Cultural Landscapes

8.1 Introduction

This chapter examines the direct impact of human actions on eagles. This includes urbanization and rural environmental change, eagles and airplane collisions, collisions with wind turbines, eagles and electric power lines, hunting of eagle and the impacts of chemicals and poisons on eagles. These are crucial issues for eagle conservation: for example, Washburn et al. (2015, 196) referring to human/eagle conflicts in the United States argue that "most known eagle fatalities are anthropogenic; major contemporary challenges faced by eagle populations include wind energy development, electrocutions, collisions with vehicles (e.g., automobiles, aircraft), secondary effects of pesticides, and lead exposure and poisoning" (Stauber et al., 2010; McIntyre, 2012; Millsap et al., 2013; Pagel et al., 2013).

Urbanization is probably the most important issue, because it encompasses all the other issues in the project of supporting increased human populations in previously natural or pristine environments. Considering the extremely wide range of some eagle species, the global reach of urbanization may have serious impacts on eagle populations. In the process of urban development, most cities develop very large environmental 'footprints', sometimes with a near continent-wide impact. Here the footprint refers to the area of land that must be modified to create supporting resources for the human population of the urban area (e.g. water, crops, minerals, land space and raw materials) which may affect ecologies through environmental modification.

Three important concepts emerge: urbanization, urban ecology and environmental change. Urbanization may be defined as the increase in human population in urban areas, largely due to migration to rural areas, and the transitional changes that occur in a society as a result of this change. Through the process of urbanization, the most important changes concern greater modification of the landcover to support denser human populations, removal of natural ecological systems, replacement of natural vegetation with agricultural systems, possible elimination of dangerous animals, introduction of livestock and companion animals, and usually the fragmentation of animal habitats (Campbell, 2018). From the perspective of large raptors such as eagles, the main issues concern the increased proximity, in some cases the increased competition (at least for scavenging) from more human-tolerant species such as vultures, Larus gulls and feral dogs, loss of trees for nesting, disturbance of nesting sites (especially for the less human tolerant species), direct persecution (through beliefs that eagles kill companion animals and livestock), indirect threats such as poisoning of carcasses and pesticides in crops, which result in the chemicals moving up the food chain through rodents and other herbivores to eagles, and the killing or removal of prey species. Most urban areas also have airports, and as will be discussed later, large raptors may collide with airplanes (Campbell, 2015).

Urban ecology concerns the relationship between living creatures and their environment within the urban system. This includes their reactions and interactions with the green spaces in urban areas (parks, gardens, treed avenues, river banks, greenbelts, green hearts), the concrete structures (sidewalks, roof tops, balconies, car parks, gravel pits, rubbish dumps) and with people (scavenging in recreational parks, and crowded events) (Campbell, 2017, 2018). For eagles and large raptors, the issues would be either the possibilities for food sources (killing of companion animals or pests such as feral rodents and rival avian scavengers such as corvids or Larus gulls) or nesting (roof tops, abandoned buildings, urban trees and woodlots). Cities may vary in the area of green spaces in dense urban substrates (such as downtown areas) or dense areas such as the suburbs or peri urban areas (Clergeau et al., 1998; Marzluff et al., 2001).

Environmental change involves the difference between the pre-urban landscape and the current situation. A change from dense rainforest to a modern landscape may be extreme, to the point of the total disappearance of some bird species and forest eagle species. Rodrigues et al. (2018, 1) argue that "urbanization causes environment changes that directly affect biotic diversity and understanding the relationship between fauna and urban features is a key aspect of urban planning. Birds are particularly affected by urbanization. Noise levels, for instance, negatively affect birds' behavior and social communication, while the presence of green areas promotes bird diversity" (see also, Sick, 1997; Clergeau et al., 1998; Marzluff et al., 2001; Bino et al., 2008; Toledo et al., 2012; Chávez-Zichinelli et al., 2013; Taylor et al., 2013; Aronson et al., 2014; Puga-Caballero et al., 2014; Beninde et al.,2015). Aronson et al. (2014, 1) note that they "compiled the largest global dataset to date of two diverse taxa in cities: birds (54 cities) and plants (110 cities)" and that the "density of bird and plant species (the number of species per km^2) has declined substantially: only 8% of native bird and 25% of native plant species are currently present compared with estimates of non-urban density of species." Furthermore, "the current density of species in cities and the loss in density of species was best explained by anthropogenic features (landcover, city age) rather than by non-anthropogenic factors (geography, climate, topography)" (Aronson et al., 2014, 1).

The rest of this chapter will examine these issues in relation to eagle ecology. Case studies are cited from the five continents where eagles occur. It is evident that eagles, unlike smaller hawks, vultures, corvids and Larus gulls are less likely to forage in cities, including urban green spaces and denser urban substrates. Therefore, it may be surmised that urbanization will have a more serious impact on eagles than on other avian predator and scavengers. Solutions for eagle ecology may therefore include reforestation of areas known to be eagle habitats, and restriction of urban and agricultural expansion. Collisions between eagles and airplanes, wind turbines and electric power lines, and chemical poisoning are also serious global problems.

8.2 Eagles, Urbanization and Rural Environmental Change

Several eagle species are evidently not suited to urban environments. For example, Simmons et al. (2007) give the example of Verreaux's eagles nesting near the green urban areas of Johannesburg and Cape Town. They pose the question: "Can species, such as Verreaux's (Black) Eagles *Aquila verreauxii,* which live alongside man, continue to thrive in their mountain retreats?" They present evidence from both cities and conclude that urbanization has a serious impact on breeding and foraging eagles. Although tourists apparently have little impact on the eagles, the evidence indicates the main problem is the decline of prey species in the two study areas, and in two other cited areas (the Lesotho Highlands and the Matobo Hills in Zimbabwe, where the population of dassies, the local name for hyraxes, the main prey species declined). They conclude that "it seems probable that the lack of dassies in urbanised environments has caused the decline of Verreaux's eagles around South Africa's largest cities" and that there are "no other African cities where Verreaux's eagles are regular within the city limits" (Simmons et al., 2007).

Other eagles may nest in urban areas and may be considered more comfortable there, depending on the availability of prey. Figures 8.1, 8.2, 8.3 and 8.4 show urban and suburban areas in Victoria, Canada where there are frequent sightings of Bald eagles. Sullivan (2016) also gives the example of Bald eagles in Washington State in the United States, which have adapted to human presence. The authors point out that "while Bald eagles are still rare in large cities, researchers have documented eagle nests recently not just in the Washington area, which boasts the largest numbers on the East Coast, but also in New York — where in 2015 the first nesting pair was spotted in more than 100 years — Philadelphia, Pittsburgh and Miami, where a pair built a nest atop a cellphone tower." The increased urban presence may also be linked to the rise in population (from 487 in 1963 to 9,789 in 2006, and possibly 24,000 in 2016) the result of the ban in DDT, and various protective laws. Sullivan (2016) cites researchers who argue that "as eagle populations have expanded and the birds may no longer consider humans to be predators, suburbs and urban areas near waterways provide attractive nesting areas. In the Washington area, it's not just the Chesapeake Bay that provides a good home; in recent years nests have been seen near Ronald Reagan National Airport and the Beltway, and recently one couple has made a home at the National Arboretum in Northeast Washington."

In another example, McPherson et al (2015, 383) cite the presence of the Crowned eagle which is present within the urban greenspace mosaic of Durban and Pietermaritzburg, KwaZulu-Natal, South Africa. The rationale for this study was that the eagles may have been perceived as a threat to domestic stock and companion animals. The findings were that domestic stock comprised only 6% of the identifiable prey, and therefore these results were at variance with the common perception there were no domestic dogs (*Canis lupus familiaris*, Linnaeus, 1758) and few domestic cats (*Felis catus*, Linnaeus, 1758 or *Felis catus domesticus*, Erxleben, 1777) found in the nests of the breeding eagles in the study. Wild species dominated as prey. For example, Hadeda ibis (*Bostrychia hagedash*, Latham, 1790), which was common in the area was one of the commoner prey species. The Rock hyrax (*Procavia capensis*, Pallas, 1766) was the main prey and to a lesser extent vervet (*Chlorocebus pygerythrus*, F. Cuvier, 1821).

Some eagles may rely on urban landfills (Stalmaster, 1987; Gerrard & Bortolotti, 1988; Buehler, 2020), For example, Elliott et al. (2006) record Bald eagles foraging in landfills near Vancouver. The eagles primarily took red meat and bone refuse, and occasionally competed with and killed Glaucous gulls (*Larus hyperboreus*, Gunnerus, 1767). However, the landfills did not provide the main activity of the eagles. Elliott et al. (2006, 380) point out in their findings that "the Vancouver landfill was not a major energy source for eagles, in part because their foraging is inefficient due to the large number of potential pirates", "most eagles apparently used the landfill primarily as a site for resting during inclement weather (the landfill is protected from the wind, is slightly warmer than surrounding areas due to decomposing refuse and the surrounding conifer trees, and is relatively free of human activity)" and "a small population of largely subadult refuse specialists appeared to gain much or all of their energy from the landfill."

Turrin et al. (2015, 239) also describe Bald eagles foraging in the Chesapeake Bay area of the Mid-Atlantic region of the United States of America. They concluded that "landfill use varied significantly with age class, with hatch-year birds using landfills six times more often than adults and twice as often as third- and fourth-year birds" (Turrin et al., 2015, 239). This finding implied that the hatch-year birds frequented landfills more than expected, when considering the landfill area compared with the study area outside of landfills. The landfills most frequented by eagle where those within two kilometers of communal roosts, with less attendance at landfills more distant from communal roosts. The authors conclude that "if eagle presence at landfills is indicative of foraging at these sites, the results provide evidence that foraging strategies in Bald eagles change with age. Landfills may serve as important scavenging sites for hatch-year and second-year eagles, whereas older birds may be more successful obtaining higher quality prey elsewhere" (Turrin et al., 2015, 239).

Fig. 8.1. Esquimalt Victoria Bald eagle suburban habitat

8.3 Eagles and Aeroplane Collisions

Bird/airplane collisions are a serious issue globally. In the United States, such wildlife strikes cost civil aviation about $957 million annually (Dolbeer et al., 2013; Washburn et al., 2015). Several species may be involved, such as raptors, Larus gulls, large geese and other waterfowl, and smaller Passerines such as blackbirds and starlings (Dolbeer et al., 2000, Dolbeer & Wright, 2009; DeVault et al., 2011; DeVault et al., 2013). The incidents of bird/airplane collisions are not limited to eagles, but other species may be involved, such as vultures and Larus gulls that frequent airports near large cities (Lowney, 1999; Avery and Cummings, 2004; Blackwell and Wright, 2006; Campbell, 2015). Larus gulls and other seabirds (and in a few cases corvids) are more commonly killed in northern countries, as they are common social scavengers in these countries. For example, Satheesan and Satheesan (2000) report that from 1955 to 1999 more than 33 aircraft (27 military and other civil) were destroyed, and 21 lives were lost because of airplane/raptor collisions in several countries in Asia, Africa, Europe and North America, including United States, Spain,

Fig. 8.2. Sidney Victoria British Columbia Bald eagle suburban habitat with comorants and gulls

Fig. 8.3. Bald eagle suburban habitat, East Sooke, Victoria British Columbia

Fig. 8.4. Tofino area British Columbia – Bald eagle habitat

India, Pakistan, Kenya, Ethiopia and Tanzania. Collisions have occurred at high altitudes, such as one involving a Ruppell's Griffon and a commercial aircraft at 11,280 m over Abidjan, Cote d'Ivoire on 29 November 1973 (Laybourne, 1974).

Airplanes are increasingly colliding with eagles as well. Eagles, like vultures may be the largest bird may collide with in size and hence cause similar damage to aircraft (Buehler, 2000; Kochert et al., 2002; Dolbeer & Eschenfelder, 2003; Dolbeer & Wright, 2009; Campbell, 2015). The rate of collisions of raptors with airplanes represents a serious issue for eagle conservation especially in the so-called advanced nations (as there are more airports in these countries) (Campbell, 2015). For example, in Australia, a Boeing 737 airplane Virgin Australia Flight VA-319 departing Melbourne for Brisbane collided with an eagle carrying a rabbit and was forced to return to the departing airport (Morris, 2017). Washburn (2015, 193) cite the Bald and Golden eagles, which have sufficient weights (3.5 to 5.4 kilograms) that "far exceed the airworthiness standards for airframes, windshields, and engines set by the Federal Aviation Administration (FAA). Consequently, eagles pose a high risk of damage to aircraft and passenger or aircrew safety when eagle–aircraft collisions occur.

Washburn et al. (2015, 192) wrote that "Notably, the risk of collisions between eagles and aircraft is an increasing problem at civil airports and military airfields. Of the 234 eagle collisions with civil and military aircraft reported to the Federal Aviation Administration, the U.S. Air Force, and the U.S. Navy during 1990–2013, 52% caused damage to the aircraft. During this 24-yr time period, Bald Eagle–aircraft collisions increased by 2200% and Golden Eagle–aircraft collisions increased by 400%." Of the 234 eagle collisions with civil and military aircraft reported to the Federal Aviation Administration, 52% caused damage to the aircraft." The authors further acknowledge that "because of their large body size, eagles are capable of causing significant damage to aircraft" and that "to our knowledge, there has not been a comprehensive review of eagle–aircraft strikes for civil or military aircraft" (Washburn et al., 2015, 192). Over the 24-year period (1990–2013) studied, there were a total of 234 eagle–aircraft collisions, which involved 197 with civil aircraft, 26 with U.S.A.F. military aircraft, and 11 with Navy military aircraft (Washburn et al., 2015, 194). These collisions were mostly (88%) recorded during daytime, and mostly near the airfield (82.6% and 81.0% of the Bald eagle and Golden eagle collisions respectively occurred at or below 305 m altitude).

The airplane collisions are more important for Bald eagles than Golden eagles, and such collisions may increase as the Bald eagle populations increase (Washburn, 2015, 192). Bald eagles, although once endangered have increased in numbers over the last few decades (Buehler, 2000; Suckling & Hodges, 2007). Washburn (2015, 193) notes that "regional Bald eagle populations in the Pacific Northwest, Great Lakes, Chesapeake Bay, and Florida have increased five-fold during the past 20 years" (see also Watts et al., 2008; Elliott et al., 2011). The population of Bald eagles reached 11,000 pairs in the conterminous states in 2007, there were more than 11,000 nesting pairs of Bald eagles in the conterminous 48 states, with consequent increases in human/eagle conflicts possibly including airplane collisions (Suckling & Hodges, 2007; Stauber et al., 2010, Pagel et al., 2013; Washburn, 2015).

Fig. 8.5. Urbanization beginnings as farms and rural housing replace forest in Central America

Fig. 8.6. Birds attending the Vancouver Landfill: "City of Vancouver landfill Delta BC" dated January 23, 2009 and found on the internet at https://www.flickr.com/photos/stephen_rees/3220679823/

By contrast, the populations of Golden eagles have remained stable mostly in the western United States (Kochert et al., 2002, Katzner et al., 2012; Millsap et al., 2013). Therefore, there may be fewer Golden eagle deaths as a result of collisions, although both eagles are mostly killed through human actions and constructions (Kochert & Steenhof, 2002; Katzner et al., 2012; Pagel et al., 2013). Most Bald eagle incidents occur in areas of high species density (e.g. Alaska, Florida, and Chesapeake Bay) as do incidents involving Golden eagles (western North America, very few in the east, see Kochert et al., 2002; Katzner et al., 2012; Washburn et al., 2015). Washburn et al. (2015, 197) note that "given the relatively low incidence rate of Golden Eagle–aircraft collisions, we suggest Golden Eagles pose a low risk to civil and military aircraft relative to Bald Eagles and many other wildlife species" (see also Dolbeer & Wright, 2009, DeVault et al., 2011, Dolbeer et al., 2013).

The Associated Press (2016) also reported the collision of a small airplane (Cessna 172) and a Bald eagle near Anchorage Alaska, which was the "the first American civilian plane crash to result in deaths after an impact with a Bald eagle." It was further reported that the "remains of the eagle were found on the plane's tail structure." The Associated Press further noted that the "Alaska Department of Fish and Game's website says Alaska has the largest population of Bald eagles, which are found only in North America. It puts the Alaska Bald eagle population at about 30,000 birds." For all large raptors, the key factor for

Fig. 8.7. Airplane at 600 meters [Author]

Fig. 8.8. Airplane at 1500 meters [Author]

such collisions is individual or group thermal soaring near airports. Birds may be hit when ascending thermals or gliding between thermals. Larger raptors may be less maneuverable than smaller birds, and their heavy bodies may damage airplane engines. Faster airplanes, such as fighter jets may also cause more accidents than slower aircraft, especially at lower altitudes (e.g. below 200 meters). The legal, endangered status of eagles may prevent harsh ameliorative measures that have been used on other species. For example, shooting is commonly used to clear gulls from airports. From 1991 and 1997 shooters at JFK International Airport USA killed 52,235 gulls (47,601 Laughing Gulls, 4,634 others) (Dolbeer, 1998). Such shooting has also been practised in India, where vultures and some other species may also be shot (Satheesan & Satheesan, 2000). Sanitation management may also be practised at some airports in India, where the birds involved are frequently scavengers (Blokpoel, 1976; Ali & Grubh, 1984; Satheesan, 1992a, 1992b, 1994, 1996). Washburn et al. (2015, 198) concludes that for eagle such extreme measures may not be possible, admitting that "few mitigation tools and techniques are available to reduce eagle–aircraft collisions.

Nonlethal hazing (e.g., use of pyrotechnics), live-capture and translocation of problem individuals, nest removals, habitat modification, and other methods have the potential to reduce eagle use of airports and thus decrease the risk of eagle strikes, which must be "publicly acceptable" (Washburn et al., 2011; De Vault et al., 2013; Guerrant et al., 2013). The pressure for a solution is strong, as "eagles pose a higher risk of damage to aircraft and human safety relative to many other bird species due to their relatively large body mass. More than 40% of reported eagle strikes in this study caused physical damage to aircraft compared to 9% of all reported bird strike events (i.e., all species combined) with U.S. civil aircraft" Washburn et al., 2015, 196; see also Dolbeer et al., 2013). Costs go beyond aircraft damage, to legal penalties for airports where insufficient action has been taken to ameliorate the consequences of collisions (Dale, 2009; Washburn et al., 2015).

8.4 Eagles and Wind Turbines

Wind turbines and powerlines are another important source of conflict between eagles and human constructed structures, linked to urban development. In some cases, this conflict may be even more important than airplane collisions. The issue of bird fatalities on wind turbines is increasing (Hunt et al., 1998; Johnson et al., 2002; Langston & Pullan, 2003; Thelander et al., 2003; Barrios & Rodriguez, 2004; Smallwood & Thelander, 2005; Drewitt & Langston, 2006, 2008; Madders & Whitfield, 2006; DeLucas et al., 2008, Bevanger et al., 2010; Carrete et al., 2009; Campbell, 2015; Washburn et al., 2015). For example, Washburn et al. (2015, 197) compare the number of Golden eagles killed at the Altamont Pass Wind Resource Area in California (about 67 to 75, see Smallwood & Thelander, 2008) with the number of Golden eagles killed by aircraft collisions (usually one per). Wind energy developments or wind farms (mostly turbines) therefore rank among the main killers of eagles, most of these being anthropogenic (Moccia & Arapogianni, 2011; Martin et al., 2012; Washburn et al., 2015). Large raptors, especially eagles and vultures are very vulnerable to collisions with wind turbine blades and power lines, despite their sharp vision (Land & Nilsson, 2002). Collision incidents are important issues, discussed at international conferences, researched in science journals and acknowledged as factors for the decline of several raptor species (Drewitt & Langston, 2008; De Lucas et al., 2008; Carrete et al., 2009, Jenkins et al., 2010; Martin et al., 2012; Campbell, 2015).

The development of wind turbines is justified as an important aspect of the efforts towards the reduction of greenhouse gas emissions by using renewable sources. For example, Dahl et al. (2012) note that "climate change and global warming scenarios together with increased energy demands have made renewable energy production a key issue worldwide to reach the international targets on reducing CO_2 emissions. The potential for wind power as a source for electricity is substantial in most countries" (see also Hoogwijk et al., 2004; IEA, 2009; IPCC, 2007; Lu et al., 2009). In Europe, after the beginning of the 21st century, the target of the European Commission was that 20% of the energy generated by European Union (EU) should have renewable sources by 2020. Spain represents one of the nations strongly committed to the development of wind energy in the first decade of the 21st (AEE (AsociaciónEólica Española), 2009; Telleria, 2009; IDAE (Instituto para la Diversificación y el Ahorro de la Energía), 2010). In Spain, renewable sources soon generated 14% of energy production (IDAE, 2010), with about 16,842 turbines operating in 737 wind farms generating over 16 GW (gigawatts) of electricity in 2009 (AEE, 2009; Lucas et al., 2012).

Fig. 8.9. Windfarm USA [With permission from Nico – unsplash]

Fig. 8.10. Windfarm New Zealand [With permission from Wendy Schotsmans]

BirdLife South Africa (2017, 5) describes bird fatalities on wind turbines as a "well-documented risk associated with wind farms" (see also Drewitt & Langston, 2006; Hötker et al., 2006). Although there are variable reports of bird fatalities at wind farms (Erickson et al., 2001; Drewitt & Langston, 2006; Campbell, 2015), "most of the high-profile cases have involved large, soaring birds of prey (BirdLife South Africa, 2017, 5; see also Barrios & Rodríguez, 2004; Hötker et al., 2006; Smallwood & Thelander, 2008). May et al. (2010) point to fact that "large soaring birds of prey, such as the White-tailed eagle, are recognized to be perhaps the most vulnerable bird group regarding risk of collisions with turbines in wind-power plants." Examples of eagle mortalities include those of Golden eagles (e.g., Smallwood & Thelander, 2008; Smallwood, 2013), White-tailed eagles (Hötker et al., 2006; Dahl et al., 2012), Bald eagles (Pagel et al., 2013), White-bellied sea eagles (Smales & Muir, 2005) and Short-toed snake eagles (De Lucas et al., 2008).

Dahl et al. (2012, 79) also write that such wind energy developments have serious impacts on birds, especially large raptors (see also Hunt, 2000, 2002; Barrios & Rodriguez, 2004; Kingsley and Whittam, 2005; Drewitt and Langston, 2006, 2008; Smallwood & Thelander, 2008; Carrete et al., 2009; Pearce-Higgins et al., 2009). Wind turbines contribute significantly to raptor mortality (Campbell, 2015). Even low mortality rates for raptors may be important, as many, especially eagles have low reproductive rates (Orloff, 1992; Sæther & Bakke, 2000; Thelander & Rugge, 2000; Percival, 2003). Some studies have even indicated that some birds may leave areas (including nesting, foraging and fly-by areas) due to windfarms and relocate to suboptimal habitats (Frid and Dill, 2002; Langston & Pullan, 2003; Kingsley & Whittam, 2005; Drewitt & Langston, 2006, 2008; Pearce-Higgins et al., 2009). Dahl et al. (2012, 79) nevertheless note that "so far results are inconsistent, and the numbers of casualties recorded differ widely between wind-power plants and species" (see also Kuvlesky et al., 2007). One reason is that there are few BACI (before–after–control–impact assessments, and breeding raptors are generally sparsely distributed (Dahl et al., 2012, 79-80, see also Newton, 1979; Krebs,1999; Drewitt & Langston, 2006; Madders & Whitfield, 2006).

Barrios & Rodríguez (2004, 72) argue for the damage inflicted on raptor populations by windfarms, which is higher than that inflicted by electric power lines. Their study concluded that "bird vulnerability and mortality at wind power facilities reflect a combination of site-specific (wind–relief interaction), species-specific and seasonal factors" and "new wind installations must be preceded by detailed behavioural observation of soaring birds as well as careful mapping of migration routes" (Barrios & Rodríguez, 2004, 72). Other authors mention the flight patterns of large raptors as a factor for their possibly higher death rates. De Lucas et al. (2008, 1701) found that eagles (the Short-toed snake eagle) and vultures (the Griffon vulture) were more likely than smaller birds to be killed by turbines. They note that "it is inadequate to assume that collision mortality will increase with bird abundance. Rather, we propose that differences in mortality are equally or more likely to be related to species-specific flight behaviour and morphology, weather and topography around the wind farm." There is speculation that "high wing loading is associated with low manoeuvrability in flight and a low capability for powered flight is typical of some soaring birds like griffon vultures" and that "this increases their risk of collision with turbine blades compared with species that have a greater capability for powered flight."

Possibly due to their similarity in size and flight patterns, including high wing loading, reliance on thermals/winds and low flapping ability, raptors such as eagles and vultures are significantly affected by these wind turbines. For example, raptor mortality increased in several countries through the development of wind turbines. Examples for the general avian population, mostly raptors are in the Netherlands (Winkelman, 1990, 1992, 1995; Musters et al., 1996; Dirksen et al., 1998), USA (Orloff and Flannery, 1993; Howell, 1997; Morrison et al., 1998; Osborn et al., 2000; Erickson et al., 2002; Johnson et al., 2002; Thelander et al., 2003); Sweden (Peterson and Stalin, 2003) and Spain (Barrios & Rodríguez, 2004; Dorin et al., 2005; Telleria, 2009; De Lucas et al., 2012). Spanish Griffon vultures are known to be vulnerable to such wind power plants, especially when these plants are near their nesting, foraging and migratory presence (Telleria, 2009). Lekuona (2001) estimated a turbine generated death rate of eight griffons annually in near the Salajones wind plant, in Navarre, Southern Spain. Lekuona and Ursua (2007) also found that the Griffon vultures comprised 63.1 percent of bird fatalities near the wind plants of Navarre. DeLucas et al. (2012) also studied Griffon vulture mortality at 13 wind farms in Tarifa, Cadiz, Spain, before (2006–2007) and after (2008–2009) when selective turbine stopping programs were implemented for conservation factors and found 221 dead griffon vultures during the study. The implementation of the selective stopping technique reduced the vulture mortality rate by 50% with only a 0.07% reduction in total annual energy production of the wind farms.

For eagles, the situation is also bleak, as can be gleaned from international cases. Solutions can be found if critical assessments can be made of the behavioral habits of eagles in areas where the wind turbines are developed or where they may be developed in the future. For example, BirdLife South Africa (2017) assesses the impacts and possible impacts of wind farm developments on the Verreauxs' eagle in South Africa. Based on the findings of their research, they suggest that the ecology of eagles should be learned for planning less destructive locations of wind turbines. Basically, "the risk of collisions and displacement of eagles can be minimised if wind turbines are not placed within their core territory or areas heavily used by eagles (see also Fielding & Haworth, 2010; U.S. Fish and Wildlife Service, 2013). Additionally, monitoring of the impacts should go beyond the more usual assessment of the developable area, to include the whole territory of eagle pairs. It would be advantageous to extract data on the "relative importance of the wind farm site for the eagles, determine the potential significance of displacement, and point to areas that are more suitable for development" (see also Scottish Natural Heritage, 2010).

Taylor et al. (2015b) add that despite the potential threat from wind farms, "up until recently no wind farms had been constructed within the range of Verreauxs' eagle and there has been limited opportunity to document the impacts."One review of windfarms indicated Verreauxs' eagle mortalities at two farms, with four fatalities occurred at one wind farm (Smallie, 2015a). BirdLife South Africa (2017, 5) notes that "these preliminary data do suggest that this species is vulnerable to collisions, and points to the need for a precautionary approach for wind farms within the species' range... lessons may be learned from experiences with wind farms and other species of eagle, for example Golden eagle, Verreauxs' eagle's ecological counterpart in the northern hemisphere."

The U.S. Fish and Wildlife Service (2013) has evolved a constructive approach, defining "important eagle-use areas", which are – "eagle nest, foraging area, or communal roost site that eagles rely on for breeding, sheltering, or feeding, and the landscape features surrounding such nest, foraging area, or roost site that are essential for the continued viability of the site for breeding, feeding, or sheltering eagles." The hypothesis is that any wind energy sites that near these areas will form a risk to eagles (U.S. Fish and Wildlife Service, 2013). The U.S. Fish and Wildlife Service (2013) advocates the evaluation of the local ecological status of eagles and suggests the distance between eagle nests and the turbines should be half the mean inter-nest distance for eagles in the larger project area. BirdLife South Africa (2017) suggests that "a more precautionary approach would be to use the maximum inter-nest distance for an area, as this does not assume that the available foraging habitat is equally valuable and that there is an equitable allocation of resources amongst pairs."

Various studies have uncovered broadly similar effects of windfarms on eagles. For example, Dahl et al. (2012, 80) studied White tailed eagles in Norway's first large-scale windfarm on Smøla island, Central Norway, based on evidence of several birds found dead near the windfarm (39 dead eagles, 2005 to 2010, see also Bevanger et al., 2010; May et al. (2012). The Smøla Archipelago is an Important Bird Area, mostly because of the comparatively high breeding density of White-tailed Sea Eagles (BirdLife International, 2015). White-tailed eagles breed on Smøla; before the windfarm development, the current windfarm areas had the densest population of eagles, post development they moved (Follestad et al., 2007; Bevanger, et al., 2010). Dahl et al. (2012, 80) analysed a 10-year dataset of eagle territories using successful breeding as a response variable, and three predictors (distance to turbines, distance to roads and before/after turbine construction). The findings indicated that eagle territories within 500 m of the wind turbines had reduced breeding successes post windfarm construction, likely due to eagles vacating nests, or those killed by turbines not being replaced, or habitat loss due to windfarm development (the first factor was also found by Bevanger et al., 2010). Mortalities clustered in the spring, which was the early breeding season (Dahl et al., 2012) and affected mostly eagles breeding near the turbines (May et al., 2012). Such fatalities accounted for the majority of the deaths at Smøla. However, the population appeared to be stable, with some breeding success (May et al., 2012). Dahl et al. (2012, 82 - 84) concluded that the BACI (before–after–control–impact) approach is fundamental in assessing the impacts of windfarms on eagles, that windfarms must not be developed near central breeding areas for such species (at least one kilometer distant), windfarms may also recreational human activities, and "The cumulative impacts from several wind-power plants along the Norwegian coast and elsewhere could potentially have a substantial negative effect on the white-tailed eagle breeding population if they are located in important breeding areas, turning them from sources to sinks" (Dahl et al., 2012, 84).

Another study by Fielding and Haworth (2010) looked at the impact of windfarms on Golden eagles in Scotland. The windfarms impacted the Scottish Golden eagle population, through the displacement and disturbance of nearby breeding eagles, and eagle mortality. Displacement appeared to be the main factor for eagle absence from windfarm areas. The authors note that "to date, there has been no recorded golden eagle fatality at a Scottish wind farm", but "there is growing evidence of displacement of birds from wind farm sites." An important quantifier is however noted: despite the evidence for Golden eagle displacement, this assessment is "confounded by habitat management which may have, by itself, resulted in a shift in range use." (Fielding & Haworth, 2010). There were few flights near or over one of the windfarms, even on the windfarm's footprint. The authors conclude that the findings support the displacement hypothesis, and that "evidence from the impacts of conifer afforestation on Scottish Golden eagle productivity and range abandonment, suggests that the effects of displacement by windfarms are likely to be site-specific."

Golden eagles in the Altamont Pass in California have also been studied, where varying numbers of fatalities have been recorded (Smallwood & Thelander, 2008; Smallwood & Karas, 2009; Smallwood, 2013; BirdLife South Africa, 2017, 5). There is also inter-annual variation in fatality rates (Smallwood, 2013). Smallwood and Thelander (2008) tentatively estimated that 1127 raptors are killed in this area annually, this total including 67 Golden Eagles. Possibly, one factor for this large number is the number of turbines (5400 over 165 km^2) (Drewit & Langston, 2006; Smallwood & Thelander, 2008). Other authors blame the high fatalities on the high population of Golden eagles and other raptors, the large prey population, and the wind turbine design (e.g. lattice towers) and the comparatively closely spaced turbines (Hunt, 2002; Hunt & Hunt, 2006; Bright et al., 2006;

Smallwood & Thelander, 2008; Smallwood et al., 2009). BirdLife South Africa (2017, 5) notes that "Despite the high levels of mortality at Altamont Pass, the breeding population remains intact (all territories that were occupied in 2000 were still occupied in 2005). This suggests that even though the number of fatalities is high, there are enough 'floaters' to fill territories" (see also Hunt & Hunt, 2006; Smallwood & Thelander, 2008). The high rate of Golden eagle fatalities varies from other sites in the United States, but in general the fatalities in this country are higher than those of European sites, although the latter may lower populations of Golden eagles as a factor (Hötker et al., 2004; Hötker et al., 2006; Madders & Whitfield, 2006; Fielding & Haworth, 2010; Camiña, 2015).

Several models have been proposed for the study of eagle reactions to windfarms. For example, Whitfield (2009) suggests and employs Collision Risk Models (CRMs) specifically the Band' CRM (Band et al., 2007), which is cited as having "probably received the most utility in practice" and "potentially useful, albeit crude, tools in predicting the potential avian collision mortality rates which rotating turbine blades at operational windfarms may incur." The limitation of this model concerns assumptions that birds do not avoid collisions with turbines, a possibly incorrect position. Applied to a study of Golden eagles at four USA windfarms, the results indicated that the avoidance rate estimates for Golden eagles varied between 98.64% and 99.89%, varying by site and the calibration for biases and before and after mortality rates. (Band et al., 2007, 259–275) created "an overall 'worst case' estimate weighted by the scale of study, which was 99.33%" and a "mean unweighted 'worst case' (lowest) avoidance rate for the four wind farms was 99.19%." They recommended "a precautionary value of 99.0%" which would be recommended for use in predictive assessments of wind farm proposals" while suggested more research on the avoidance of biases which may be inherent in similar studies.

Another model is suggested by the United States Fish and Wildlife Service (USFWS) (Bay et al., 2016). This is a Bayesian modeling framework which could predict the annual Golden eagle fatality rate at a wind energy facility. This modeling approach defines the previous distributions for the collision rate and exposure rate from data derived from studies of existing wind projects. The collision rate refers to the number of collisions per exposure, where the exposure rate is a function of "minutes of eagle activity and survey effort" (Bay et al., 2016, 1000). Site-specific data was included update the prior distribution, which resulted in the posterior distribution. An expansion factor was used to adjust the fatality prediction, taking into account daylight hours and hazardous areas within a wind project footprint. The product of three constructs, the collision rate, posterior exposure rate and expansion factor created predicted annual fatality rate. An updated prior distribution for collision rate was created, by adding more contemporary information to the prior distribution for collision rate. The authors followed the USFWS directives, and "updated the prior distribution for collision rate from the USFWS baseline model with data from a site with modern specifications to obtain an updated prior distribution" and created other "prior distributions by estimating parameters for the distributions from data at 26 modern facilities only" (Bay et al., 2016, 1000). The predictions using the alternative prior distributions for collision rate were found to be about half the estimates using the original distribution, when using more recent data and a larger data set. They then collected data from studies conducted from 1994 to 2013 at 40 North American wind energy facilities. The authors conclude that this type of model would improve the uncertainty in eagle CRM results, provided the model used data collected with the same protocols, data collection methodologies, and the same species at the same type of wind energy facilities (Bay et al., 2016).

8.5 Eagles and Electric Power Lines

Eagles may be killed or seriously injured by electric power lines, either through electrocution or collisions (Alonso & Alonso, 1999; Manosa & Real, 2001; Mojica et al., 2009; Antal, 2010; Rollan et al., 2010; Campbell, 2015; Chevallier et al., 2015; Hernández-Matías et al., 2015; Watts et al., 2015; Allinson, 2017). Bird electrocutions have been a problem since the invention and mass production of aerial electric power lines in 1970s, when thousands, perhaps millions of birds were killed in North America (APLIC, 2006; Campbell, 2015). The problem was so serious that intensive research was conducted in the following years, to ameliorate the problem in North America, Western Europe and South Africa (Lehman et al., 2007; Haas et al., 2005; Manville, 2005). For large birds such as eagles, vultures or storks, such collisions are more common when power lines bisect foraging, nesting or migratory flight zones (Hunting, 2002). Such problems may increase in the absence of alternative perches such as tree branches, poles and cliffs, especially in deserts, open plains and intermontane basins (APLIC, 2006). The factors such accidents cover a broad range, on ecological, physical, and landscape issues, including the structure of the nearby vegetation, bird species and behavior, time, location, climate season and proximity of perching opportunities (Hunting, 2002; Lehman et al., 2007). The impacts may be severe. Rubolini et al. (2005, 132) note that "it has directly affected population structure and been among the causes of decline in White Stork *Ciconia ciconia*, Spanish Imperial eagle *Aquila adalberti* and Bonelli's eagle *Hieraaetus fasciatus* (see also Rieger & Winker, 1971; Fiedler & Wissner, 1980; Crivelli et al., 1988; Ferrer & Hiraldo, 1992; Mañosa & Real, 2001; Real et al., 2001).

Based on a copious review of the literature of raptor electrocution research, mitigation, and monitoring, Lehman et al. (2007) found that the majority of such research was focussed on North America, western Europe, and South Africa. The United States predominated in these studies, where after the 1970s there has been "extensive research, product testing, design standards development, and mitigation" (Lehman et al., 2007; see also Boeker & Nickerson, 1975; Nelson & Nelson, 1976; Olendorff et al., 1981). However, several studies have indicated that the bird deaths due to such collisions have not been substantially reduced (Franson et al., 1995; Melcher and Suazo, 1999; Harness and Wilson, 2001). Lehman et al. (2007) cites problems such as "negative media attention, increased scrutiny by regulatory agencies, and a landmark court conviction" (see also Melcher and Suazo, 1999; Williams, 2000; Suazo, 2000). Other problems include the impact of the electrocution problems, which may cost the energy suppliers billions of dollars annually, through power outages, equipment repairs, revenue losses and statutory compliance issues (Hunting, 2002).

Electrocution of birds requires contact on two energized components, such as two parallel wires, or an energized and an earthed (also called 'grounded') component of the pole structure. When a bird's body or wings bridge these two lines, a short circuit is created and an electric current, flows through the bird's body, resulting in a fatal electrocution. The conductors and ground/earth wires or earthed devices may be too far apart for smaller birds to touch simultaneously with their wings (Janss & Ferrer, 1999). Electrocution usually occurs in larger species, for example raptors and storks with wingspans over one meter, which may perch or nest on wires or poles (Olendorff et al., 1981; APLIC, 1996; Kruger, 1999). The closer the structures, such as the wires, the more dangerous for birds. Hence, low to medium voltage lines with closely spaced structures are the most dangerous. Most fatal incidents involve large raptors (especially eagles), vultures and storks during the breeding season (Boeker & Nickerson, 1975; Benson, 1981; Ledger & Annegarn, 1981; Ledger, 1984; Kruger, 1999; Harness & Wilson, 2001; Prinsen et al., 2011).

Birds may perch on power lines because there are no other perching structures, natural or artificial nearby (Olendorff et al., 1981; Janss & Ferrer, 1999a). Therefore, fewer forest birds are killed by electrocution (Switzer, 1977; O'Neil, 1988; Harness & Wilson, 2001). Ground-nesting

Fig. 8.11. Dangerous power lines [Author]

Fig. 8.12. Dangerous power lines [Author]

Fig. 8.13. Bald eagle and power lines [With permission from Sue Dietlein]

Fig. 8.14. Immature Bald eagles and power lines [With permission from Sue Dietlein

species are also only rarely killed (Pendleton, 1978; Benson, 1981). Vultures in Africa, mostly savanna or grassland species, such as the social Cape and African White-backed vultures may perch in large numbers on powerlines (Campbell, 2015). In the United States, more than 20 raptor species have been recorded as electrocution victims (Johnsgard, 1988, 1990; Campbell, 2015). Golden eagles have been cited as the commonest species killed through such electrocution, with 50 to 93% of deaths in some reports (Smith & Murphy, 1972; Boeker & Nickerson, 1975; Ansell & Smith, 1980; O'Neil, 1988; Harness & Wilson, 2001). Possibly, as most Golden eagles in the United States are commonest in the mostly treeless shrub/steppe regions in the Western inter-montane region, the eagles are more likely to perch on electric structures (Harlow & Bloom, 1989). Bald eagles mostly forage in forested areas with more natural perches (Stalmaster, 1987).

The topography, weather and climate of the relevant area is also important. For example, American studies have found that poles located on raised topography attract more raptors (and have more raptor electrocutions) than those on lower ground, possibly because raptors use poles for surveying food possibilities (Benton & Dickinson, 1966; Boeker & Nickerson, 1975; Nelson & Nelson, 1976; Benson, 1981; APLIC, 1996; Campbell, 2015). The daily weather and seasonal climate are also important factors for electrocutions. For Golden eagles in some western states of the United States, 80 percent of the deaths occurred during the winter (Benson, 1981). Some other species recorded losses during the nesting period (Harness & Wilson, 2001). Birds with wet feathers may be more vulnerable, with a ten-fold increase in the electrocution risk. Skin-to-skin contacts are also very conductive (Nelson, 1979, 1980). Wind direction relative to the crossarms, may also be important. Wind directions parallel or diagonal to the cross arms, increase the difficulty of landing relative to wind diections perpendicular to the crossarms. The greater difficulty in landing may result in more accidents, with the bird grasping two energized structures, or the wings touching two such structures. This is more likely to result in electrocutions (Nelson & Nelson, 1976; Benson, 1981).

Bird behavior is also important. For example, during fledging, the population increase may increase accidents, as juveniles and immature birds scatter into the surrounding habitats (Harness

& Wilson, 2001). Findings indicate that such young inexperienced birds are more vulnerable to electrocuted; in one case study at least 90% of Golden eagles killed in North America were immature or subadult birds (Boeker & Nickerson, 1975; Benson, 1981). In another case, young and female Spanish Imperial eagles (females due to their greater size) had more accidents than older birds and males (Ferrer & Hiraldo, 1992). Age is also important of smaller raptors (Fitzner, 1978; Dawson & Mannan, 1994). Bird physical contact during nest defence or mating may also link conductive components (Dickerman, 2003). Nests built on energized wires and components may create circuits capable of killing birds in the nest, if they connect two parallel lines (Hardy, 1970; Gillard, 1977; Switzer, 1977; Vanderburgh, 1993).

Fig. 8.15. Western Banded Snake eagle perching near power lines [With permission from Frank McClintock]

The design of the pole structures, power lines and the associated hardware is also important for electrocution risk. In the United States, most electrocutions have occurred on low-voltage distribution lines (<69 kV), which are mostly individual residential use. Before 1971, most of these lines had little or no insulation and narrow clearances between the energized components. These could cause electrocution when there is more possibility of links between energized components, e.g. with transformers or other auxiliary equipment (fused cutouts, capacitors, reclosers, jumper wires). Most poles are made of wood, which is nonconductive (Olendorff et al., 1981; APLIC, 1996; Harness and Wilson, 2001). The number of birds killed by electric power lines remained constant in the United States for the last decades of the 20th century (Lehman, 2001; see also Nelson and Nelson, 1976; Wildlife Management Institute, 1982; Phillips, 1986; Gauthereaux, 1993; Avian Power Line Interaction Committee [APLIC], 1996]). Lehman (2001) adds that there were more than 185 million wood distribution poles in North America in 2005 (see also American Iron and Steel Institute, 2005). The attempts to reduce bird deaths did result in some policy and design changes but the issues that led to the deaths were mostly ignored (Lehman, 2001).

In Europe, many poles are made of steel or steel-reinforced concrete rather than wood, which can kill birds of any size. Hence, bird size is less important (Bayle, 1999; Negro, 1999). These materials are conductive and grounded, so a bird on a cross can be electrocuted with one contact with a conductor (Bayle, 1999; Janss & Ferrer, 1999b; Janss, 2000). This results in much higher mortality levels than in North America, such as the 10,000 losses reported in the Slovak Republic (Adamec, 2004). The transfer of power lines in Europe underground, which end these events, and has been done in the Netherlands, Germany, the United Kingdom, Belgium, Denmark and Austria was undertaken partly for this reason (Bayle, 1999; Tucker et al., 2008). Collision with cable lines is also frequent, and the bird may be killed by the immediate impact or later through injuries. Collisions are particularly common with multiple vertical layered, high voltage lines. Up to 10,000 electrocutions and hundreds of thousands of collisions have been estimated to occur in the African-Eurasian region annually (Hoerschelman et al., 1988; Prinsen et al., 2011). Millions are estimated in Germany (Hoerschelman et al., 1988).

In European countries, where the lines are above ground, there are many accidents, affecting mostly large raptors such as Golden eagles, and storks. For example, 877 electrocuted birds of 46 species were found under one percent of the medium voltage electric poles in Hungary (6,500 poles) (Kovacs et al., 2008), and the annual dead bird count was estimated at over 30,000 (Demeter, 2004). One survey found the Golden eagles as the most affected species between 2003 and 2008. Other eagle species included the Imperial eagle and the White-tailed eagle (Horvath et al., 2008; Horvath et al., 2011). Other parts of Central Europe are also affected by such impacts (Lovaszi, 1998; Marti, 1998; Moritzi et al., 2001; Breuer, 2007; Schaub et al., 2010; Schürenberg et al., 2010). In Spain, some of the most electrocuted species are the Spanish Imperial eagle, Osprey, Bonelli's eagle and Golden eagle, which were among the most affected, probably due to "their morphologic traits, feeding habits and behavior" (Guil et al., 2015, 382; see also Janss, 2000; Frías et al., 2009). Another study found, using analysis of the 949 and 434 cases of mortality of Booted eagle and Short-toed Snake eagle respectively, recorded by wildlife rehabilitation centers in Spain over a 16-year period (1990–2006), that power lines (19.5 and 35.2%, respectively) and killing (32.5 and 22.9%, respectively) were the key causes of death (Martínez et al., 2015, 109).

In Eastern Europe, 45,000 kilometers of power lines were considered a risk to birds in Bulgaria (Stoychev & Karafeizov, 2003). Twenty-two species were recorded dead, mostly diurnal raptors, storks and crows (Demerdzhiev et al., 2009; Gerdzhikov and Demerdzhiev, 2009; Demerdzhiev, 2010). In Southern Europe, raptors were strongly affected by both electrocution and collisions. In France, one study showed that the deaths were mostly by electrocution (96.5%) and the rest due to collisions (Sériot & Rocamora, 1992 in Bayle, 1999; see also Schürenberg et al., 2010). The raptors affected included Bonelli's eagle and the Short-toed eagle, and smaller raptors (Cheylan et al., 1996; Bayle, 1999; Kabouche et al., 2006). BirdLife International (2017) notes that electrocutions are commoner for eagles and other raptors, while collisions are commoner for less maneuverable species such as cranes (Family *Gruidae*, Vigors, 1825), swans (Genus *Cygnus*, Garsault, 1764) and bustards (Family *Otididae,* Rafinesque, 1815).

In Asia, there are serious electrocution and collision problems. It is estimated that about 58,000 raptors are killed annually during spring migration along 9,478 kilometers of power lines (Karyakin, 2008). White-tailed eagles, Steppe eagles, Golden eagles, Greater Spotted eagles and Short-toed eagles were among the more severely affected in Kazakhstan (Karyakin et al., 2006; Karyakin & Novikova, 2006; Karyakin, 2008; Lasch et al., 2010). In Mongolia, more than 60% of the electrocuted birds were raptors, especially near poles with closely spaced electrical equipment. Species found dead near poles included Golden eagles and Steppe eagles (Harness & Gombobaatar, 2008; Harness et al., 2008). In Russia, aboveground power lines are estimated at 1,500,000 kilometers, with 0.5% equipped with isolated cables or modern facilities for bird protection, but it is estimated that 10 million birds of 100 species are killed annually (Matsyna & Matsyna, 2011).

In Africa, as in Europe, open landcover with few trees encouraged for bird use of power lines (Prinsen et al., 2011). In Ethiopia, the migratory species are at risk in the rift valley system and raptors were found dead near unsafe poles in open landcover (Haas, 2011). In Kenya, in the Magadi and Naivasha areas most pole designs posed an electrocution risk to many large birds, including the Martial eagle and several vulture, smaller raptor and large wading species (Virani, 2006; Smallie & Virani, 2010). In South Africa, electrical structures include 22-kV wooden T-structures, 88-kV steel kite transmission towers, terminal H-frame wood structures, and Delta suspension structures, and most of the electrocutions of different sized birds occur on the T-structures and terminal H-Frames (Kruger, 1999; Kruger et al., 2004). Birds killed include Martial and *Aquila* eagles, and smaller raptors. Larger species, including Martial eagles, Cape griffons and African White-backed vultures are often killed by the steel kite transmission towers and the Delta suspension structures (Lehman et al., 2007). The most commonly electrocuted birds, apart from the large eagles are vultures such as Cape griffons, Lappet-faced vultures and African White-backed vultures (Markus, 1972; Ledger & Annegarn, 1981; Ledger & Hobbs, 1999. Kruger et al., 2004; APLIC, 2006; Lehman et al., 2007; Smallie et al., 2009; Prinsen et al., 2011). In one study, between August 1996 and May 2011, 1,504 electrocuted eagles, Cape griffons, White-backed vultures, Lappet-faced vultures, corvids and other birds (Boshoff et al., 2011). Collisions may also kill some large raptors, and in some cases these birds survive for treatment (e.g., the De Wildt Cheetah and Wildlife Trust, the National Zoological Gardens of South Africa (NZG) and the Vulture Programme of the Rhino and Lion Wildlife Conservation NPO (R&L)) (Naidoo et al., 2011).

An important question concerns the inability of sharp-sighted species like eagles to see powerlines and wind turbines. For example, one study found the White-tailed eagle as "one of the most frequent casualties" of wind turbines on the island of Smøla, Norway (May et al., 2010). Halley et al. (2010) note "White-tailed Sea eagles show no tendency to avoid turbine blades, treating them as if they are not there." Yet eagles may have eyesight "perhaps 2.5 times higher than that of human eyes" (Martin, 2011, 243; see also Reymond, 1985). Martin (2011, 247) rates the visual fields of eagles as similar to those of bustards and cranes and notes that all these species are "highly vulnerable to collisions with artefacts."

There is evidence some birds, including eagles are restricted in their frontal vision. Martin (2011, 249) notes that for "most birds, slow flight, even for short periods, is not possible and this becomes more acute for birds with high wing loading and consequently higher average flight speeds", and "hence reducing speed to match the rate of gain of information to increasing perceptual challenges is unlikely to occur." This hints at problems when large birds are attempting to avoid a large structure. Also, when in many cases the eagle pitches its head forward and downward, possibly for foraging. Birds may pitch their heads forwards (about 60 degrees down) and may also sideways (yaw). Martin (2011, 250) concludes that bird collisions "may be the result of both visual and perceptual constraints", and "some birds may be blind to what is ahead of them; turning the head in both pitch and yaw to look downwards either with the binocular field or with the central part of an eye's visual field may not be unusual, and this may leave birds blind in the direction of travel", "frontal vision in birds is not high-resolution vision; high resolution occurs in the lateral fields of view", birds vision may be focussed on predators, prey and other environmental issues, rather than what lie ahead, they may not perceptually understand for human constructed structures such as wind turbines or power wires and cannot easily adjust flight speeds as related to information gain on the changing features ahead (see also Martin & Shaw, 2010; Martin & Portugal, 2011, Martin, 2011).

Although eagles and vultures may be slightly different in this regard, Martin et al. (2012, 626) remark that "the visual fields of vultures contain a small binocular region and large blind areas above, below and behind the head. Head positions typically adopted by foraging vultures suggest that these visual fields provide comprehensive visual coverage of the ground below, prohibit the eyes from imaging the sun and provide extensive visual coverage laterally. However, vultures will often be blind in the direction of travel." Therefore, "by erecting structures such as wind turbines, which extend into open airspace, humans have provided a perceptual challenge that the vision of foraging vultures cannot overcome" (Martin et al., 2012, 626).

8.6 Conclusions

The evidence presented in this chapter indicates the close relationship between eagles and large scale technological developments. This evidence further points to the eagles' need for uncontested space for development. Eagle ecologies are increasing inclusive of human constructed requirements. A major problem concerns the fact that some of these developments, such airplanes and wind turbines are irreplaceable, with major new inventions. Both the design of these machines and the behavior of eagles are contributory to the encountered problems. Wider ecological issues and greater attention to the requirements of foraging eagles are a starting point for the development of solutions.

9

Eagles and Environmental Chemistry

9.1 Introduction

Chemical poisoning is a major factor for and even determinant of avian and mammal health, distribution and numbers globally (Campbell, 2015). The ecology of wildlife is increasingly affected by the chemical changes resulting from industrial production and the consequences for environments and living organisms (Nighat et al., 2013). Eagles are particularly vulnerable to such chemical intrusions into the environment, because they sit at the top of the food chain and chemical pollutants tend to migrate up the food chain. As the chemical pollutants that infect eagles may be acquired from prey animals, any study of the accumulation of pollutants in eagles must primarily examine the acquisition of such animals in prey species. Other raptors need to be studied as well, as they may prey on some of the same species as eagles and may be more numerous and easier to study than the rarer, more distant eagles. All the species considered are either predators near the apex of their various food chains, as they kill other animals (including animals from live medium sized mammals, all sizes of carcasses, to insects, fish and worms) which may also feed on smaller species or on vegetable materials.

Raptors and waterbirds at the top and lower levels of the food chain are vulnerable to the chemicals such as heavy metals (a metal of relatively high density or atomic weight) ingested by prey species due to the processes of concentration and accumulation, both indirectly (e.g. lead from bullets fired at prey species, which are consumed by the predators, or the pollution of prey species foods from industrial pollutants) and deliberately (e.g. human poisoning of prey species carcasses to eliminate predators). Therefore, this chapter will examine the problem of chemical pollution in most birds of prey and waterbirds (Table 9.1) that feed on eagle prey species (small mammals, birds, fish and the carcasses of these and larger animals) (Cross, 2009; Campbell, 2015, 2018). Commoner families of these Orders are listed in Table 9.1. The list of species is not complete but selects eagles and other species that may share such processes, such as terrestrial, riverine and marine environments. Common chemical elements, such as lead, copper and zinc, and chemical compounds such as pesticides and veterinary chemicals are also examined and described for their status as pollutants. The research methods used for studying the impacts of these chemicals on ecosystems are also discussed. The research methods play the most important role in the assessment and prevention of the negative impacts on the selected birds and their related ecologies. These points are also related to the ecological relations for eagles described in the earlier chapters and the pointers on eagle conservation in the succeeding chapters.

Eagles, smaller raptors and waterbirds share similarly sized prey species and are usually the main obligate (full-time) and facultative (part-time) scavengers in terrestrial marine, riverine, vegetated and urban areas (Campbell, 2015). Fish-eating eagles such as the Bald eagle, White-tailed eagle and Steller's Sea eagle consume some of the same marine species as ospreys, gulls and waders. For example, Szumiło-Pilarska et al. (2017) mention the research on the White-tailed eagle, the great cormorant (*Phalacrocorax carbo*), herring gull (*Larus argentatus*), mallard (*Anas platyrhynchos*) and common merganser (*Mergus merganser*), on the Polish coast (see also Falandysz et al., 1988; Misztal-Szkudlińska et al., 2011; Szumiło et al., 2013; Szumiło-Pilarska et al., 2016; Kalisińska et al., 2013; 2014a, b). In some cases, eagles such as the Bald eagle may also prey on waterbirds, and hence suffer from any pollution effects that these species acquire (Middleton et al., 2018; Buehler, 2020). As noted in the earlier chapters, virtually all eagles prey on small mammals such as rodents, snakes and small birds. The concentration or mere existence of pollutants in the environments of raptors and waterbirds (i.e., plants, water, and prey species such as small fish, small mammals such as rodents, and small birds such as passerines and pigeons) begins direct and indirect paths to avian toxicity with dire consequences for raptor and waterbird pollutant levels (Szumiło-Pilarska et al., 2017; Campbell, 2018).

Table 9.1. Classification of common raptors and waterbirds

Orders	Families	English names
Birds of Prey or Raptors		
Accipitriformes	*Accipitridae*, Vieillot, 1816	Eagles, hawks, kites, vultures
	Pandionidae, Sclater & Salvin, 1873	Ospreys
Cathartiformes	*Cathartidae*, Lafresnaye, 1839	New World vultures
	Sagittariidae, Finsch & Hartlaub, 1870	Secretary birds
	Falconidae, Leach, 1820	Falcons, caracaras
Strigiformes	*Strigidae*, Leach, 1820	Typical owls
	Tytonidae, Ridgway, 1914	Barn owls
Waterbirds		
Charadriiformes	*Laridae*, Rafinesque, 1815	Gulls, terns
	Charadriidae, Leach, 1820	Plovers, lapwings
Anseriformes	*Anatidae*, Leach, 1820	Ducks, geese, swans
Ciconiiformes	*Ciconiidae*, J. E. Gray, 1840	Storks
Pelecaniformes	*Pelecanidae*, Rafinesque, 1815	Pelicans
	Ardeidae, Leach, 1820	Herons, egrets, bitterns
Suliformes	*Sulidae*, Reichenbach, 1849	Gannets, boobies
	Phalacrocoracidae, Reichenbach, 1850	Cormorants, shags
Gaviiformes	*Gaviidae*, Coues, 1903	Loons
Coraciiformes	*Alcedinidae*, Rafinesque, 1815	Kingfishers
Gruiformes	*Rallidae*, Rafinesque, 1815	Rails, crakes, coots, moorhens

Note that some classifications place the New World vultures in *Accipitriformes.* The waterbirds have nearly 20 families, so the commoner families are presented.

Research on the impacts of pollutants on birds is currently scattered varied over many species and ecosystems (Campbell, 2015). Kitowski et al. (2016) argue that "in spite of extensive surveys, many aspects of heavy metal accumulation and transmission in ecosystems remain unclear" and cite further studies by Helander et al. (2009), Stankovic et al. (2013) and Kitowski et al. (2014). Kitowski et al. (2016) also support the choice of raptors as good models for the study of chemical pollution of bird physiology, as they are apex predators, frequently sedentary, with wide distributions, and are consumers of other animals that in turn ingest contaminants from plants, water and other, yet smaller animals (also cited are Martin et al., 2008; Castro et al., 2011; Rajamani & Subramanian, 2015; Golden et al., 2016). These qualifications also apply to waterbirds, which may play the same role as raptors in riverine and coastal areas and have similar ecological interactions (Campbell, 2015). Basically, "metal concentrations in birds are good indicators of the level of heavy metal concentration in their diet, which is strongly associated with the degree of contamination in the exploited ecosystem" (Kitowski et al., 2016, 146, see also Pain et al., 1997; Castro et al., 2011).

Eagles and other raptors may ingest these chemicals from consumed prey, usually passerines, pigeons and small rodents, they may also acquire them by eating smaller species such as seabirds (Campbell, 2015). Waterbirds may acquire toxins, not only from the fish but also from the sea. Heavy metals such as mercury and cadmium are especially important because of their high toxicity. Kalisinska et al. (2014, 858) report that "research from the 1950s and later showed that among the various Hg species present in nature, MeHg is the main cause of neurological, neuromotorical, behavioral and reproductive disorders in warm-blooded vertebrates" (see also Clarkson & Magos, 2006; Rutkiewicz et al., 2011). They occur naturally in seawater, but levels can increase through natural causes and human-induced contamination. Fish-eating seabirds contain comparatively higher levels of mercury in their livers than other birds for this reason (Parslow & Jefferies, 1973). According to Kalisinska et al. (2014, 858) mercury (Hg) contamination in piscivorous birds, especially methylmercury (MeHg), has been drawing much attention worldwide in regard to its bioaccumulation and biomagnification in food chains.... As piscivorous species, the White-tailed eagle and osprey are exposed to chronic or acute poisoning with mercury (Hg), especially methylmercury (MeHg), in their food" (Holt et al., 1979).

Such chemicals may be ingested and accumulate in various organs, such as livers, kidneys, gastrointestinal (digestive) tract, nervous system, eggs, bones and even feathers of affected birds. The result for the bird may be increasingly severe illness, incapacity and even death, as the level of accumulation increases (Frantz et al., 2012, 23). The lethal toxicity is whichever leads to the death of the bird (Kramar et al., 2019). Heavy metal exposure may also negatively affect hormone, respiratory and reproduction systems and migration (Scheuhammer, 1987; Haschek et al., 2013; Williams et al., 2014; Kitowski et al., 2016). These threshold levels may vary according to the avian species, the body organ affected, the ecosystem of the bird and the toxicity of the chemical considered, but there may be generalizable levels (Campbell, 2015; Kramar et al., 2019). Heavy metals, such as lead, iron and zinc, and chemicals manufactured for agricultural and industrial use, such as anti-inflammatory drugs (NSAIDs), polychlorinated biphenyls (PCBs), rodenticides and insecticides are currently prevalent for birds (Campbell, 2018).

Other physical and behavioral impacts may occur with heavy metal ingestion at sublethal levels (Campbell, 2015). For example, for some species including the passerine prey of raptors, heavy metals ingestion may lead to reduced plumage intensity (especially copper, zinc, nickel and lead) (Eeva et al., 1998), genetic diversity (especially copper, zinc, cadmium and lead) (Eeva et al., 2006), fledging success (especially mercury, lead and cadmium) (Nam & Lee, 2006; Evers et al., 2008) and humoral immune responsiveness (especially lead and cadmium) (Snoeijs et al., 2004). Additional impacts may be "aberrant incubation behavior, lethargy and wing area asymmetry" (Frantz et al., 2012, 23) mostly as a result of mercury poison (Evers et al., 2008). Lead poisoning may also affect bone mineralization, which is "the process by which the organic bone matrix becomes filled with calcium phosphate nanocrystals" (Kuhn, 2001; Gangoso et al., 2009; Frantz et al., 2012). This is important for bone formation, which is "accomplished by coordinated multicellular activity", basically "a cell-mediated process involving the specific, highly ordered deposition of a unique form of calcium phosphate confined to precise locations within the organic matrix… essential for bone to meet its multifunctional property requirements" (Kuhn, 2001, 50).

9.2 Chemical Elements

The chemical elements to be considered in this chapter are listed in Table 9.2. These elements may combine with other elements to create compounds, which may have the same impacts as the elements. These are not exhaustive but represent some of the main chemical agents that compromise the health and existence of eagles, and along the food chain, smaller raptors and waterbirds, and the prey animals. Some studies have indicated that raptors may ingest and retain several elements in combination, so detecting the most dangerous element may be problematic. Intensive study may be required to determine which element or compound has the most impact on the bird's health and physical appearance (Campbell, 2015). Therefore, while some studies seek to discover the impacts of one element on the selected species, others seek to discover all the chemical elements in the studied birds.

As an example of the former (looking for one element), Kramar et al. (2019) looked at only mercury contamination in Bald eagles in both inland and coastal populations. The methodology used blood and feather samples from nestling Bald eagles in the Virginian coastal plain, piedmont, and western regions to find regional variations of Hg toxicity. The Hg levels were analyzed using a Milestone DMA-80. As an example of the latter (looking for several elements or compounds), Kitowski et al. (2017) looked at chemical impacts on White-tailed Sea eagles, using livers from 22 dead eagles in Eastern Poland. The birds were classified by sex, age (adults over 2 years old, immature less than 2 years old), based on iris color, plumage and gonad development (Helander et al., 1989, 2009; Forsman, 1999; Kitowski et al., 2017). After mineralization using laboratory techniques, and a solution of elements was derived, inductively coupled plasma optical emission spectrometry (ICP-OES) with a Thermo Scientific iCAP Series 6500 analyzer with a charge injection device (CID) was used for the identification of elements and toxic chemicals were identified.

Kitowski et al. (2017) judged the extracted chemicals through referral to published findings regarding the toxicity levels for birds. For example, lead is one of the more dangerous elements for birds. Lead hepatic (liver) concentrations above 6 mg/kg^{-1} dw (6 ppm) have resulted in subclinical toxicity (below the level for visible symptoms) (Pain et al., 1995; Helander et al., 2009). Lead levels equal to or above 10–15 mg/kg^{-1} dw (dry weight) (10–15 ppm) generally represent poisonous levels (Franson, 1996; Martin et al., 2008). These are based on the findings of research that links hepatic lead concentrations in raptors to the lead poisoning and subclinical effects (Pain et al., 1995; Franson, 1996; Martin et al., 2008; Kitowski et al., 2017). For cadmium, levels at $\geq$3 mg/kg^{-1} dw have been found to result from increased environmental exposure (Burgat, 1990; Battaglia et al., 2005). For mercury, Shore et al. (2011) argued that hepatic mercury concentrations greater than 6.7 mg/kg^{-1} dw (6.7 ppm) might affect avian reproduction, and mercury values above 67.0 mg/kg^{-1} dw might even kill non-marine birds. For selenium, Kitowski et al. (2017) argue that "selenium at the level of 4–10 mg/kg^{-1} dw is considered the background concentration in avian liver tissue" and cite a study by Ohlendorf (1989), where "a hepatic Se concentration above 10 mg/kg^{-1} dw was taken as exceeding the background level." For zinc, Kitowski et al. (2017) point to the study by Honda et al. (1990) in which 200 mg/kg^{-1} dw was defined as the background concentration in liver tissue (Kitowski et al., 2017).

Table 9.2. The main chemical elements and their impacts

Element	Symbol	Threshold	Impact on raptors, waterbirds and prey species (e.g. *Passerines*)
Lead	Pb	2-6 ppm dw[1, 5]	Gradual weight loss, polyuria/polydipsia, depression, weakness, anorexia, and/or rapid anemia[3]
Zinc	Zn	2 ppm[3]	Passive regurgitation, lethargy, weakness, weight loss, anemia, cyanosis, polyuria/polydipsia, diarrhea, regurgitation, neurologic signs, feather picking, anorexia, and/or hemoglobinuria[3]
Mercury	Hg	1+ ppm ww[2,5]	Reduced reproductive output, weak immune function, avoidance of high-energy behaviors, death with higher doses[4]
Cadmium	Cd	3 ppm dw[1]	Affects the endocrine system, kidneys, reproduction, moulting, haemoglobin formation and growth[5]
Arsenic	As	0.5 ppm ww[6]	Liver, kidney problems[6] Muscular incoordination, debility, slowness, jerkiness, falling hyperactivity, fluffed feathers, drooped eyelid, huddled position, unkempt appearance, loss of righting reflex, immobility, seizures[7]
Chromium	Cr	4 ppm dw[7]	Gastrointestinal and immune system, respiratory tract, male reproductive system damage, miscarriage, low birth weight, and development problems of the skeleton, reproductive system[8]
Copper	Cu	4 – 6 ppm[8]	Gastrointestinal tract damage, hyperplasia of the forestomach mucosa, liver and kidney damage, immune system damage, decreased embryo and fetal growth[8]
Selenium	Se	3 ppm ww[10]	Emaciation, breast muscle atrophy, absence of fat, lung fluids, excess body gelatinous or serous fluid, swollen livers, flaccid heart, congested and enlarged spleens, head feather loss[9,10]

The threshold represents the level at which the chemical begins to cause toxic reactions in the animals. Kitowski et al. (2016) suggest that the conversion from wet (ww) to dry weight (dw) 4.0.
Sources: 1. Pérez-López et al. (2008), Scheuhammer (1987), Nighat et al. (2013); 2. Weir et al. (2018); 3. Huang & Mayer (2020); 4. Whitney & Cristol (2017); 5. Gruz et al. (2019); 6. National Research Council, 1977; 7. Eisler (1988); 8. Agency for Toxic Substances and Disease Registry, 2019, 2020a; 9. Albers et al. (1996), O'Toole & Raisbeck (1997); 10. Spallholz & Hoffman (2002).
Note that, for example 10 mg/kg = 10 µg/g = 10 ppm.

In another study, Kitowski et al. (2016) examined concentrations of lead, cadmium, nickel, chromium and mercury in the livers of six species of raptors, represented by 58 specimens captured in Eastern Poland. The six species, in order of size were the White-tailed eagle *Haliaeetus albicilla*, Linnaeus, 1758 (2 specimens); Common buzzard *Buteo buteo*, Linnaeus, 1758 (31 specimens); Goshawk *Accipiter gentiles* Linnaeus, 1758 (9 specimens); Marsh harrier *Circus aeruginosus*, Linnaeus, 1758 (3 specimens), Common kestrel *Falco tinnunculus*, Linnaeus, 1758 (3 specimens) and Sparrowhawk *Accipiter nisus*, Linnaeus, 1758 (10 specimens). The findings indicated that higher mercury and chromium levels were correlated with passerine predation, while higher hepatic concentrations of lead and cadmium indicated links to small mammal predation. These results indicated that the passerine birds were more likely to have ingested mercury and chromium (possibly from seeds or invertebrates) and small mammals were more likely to have ingested lead and cadmium (possibly from seeds, leaves and small invertebrates). This illustrated the migration of elements through the food chain.

In this study, sparrowhawks livers differed from those of common buzzards in lead concentrations but were similar in the cadmium levels. Sparrowhawks had higher levels of mercury than goshawks (possibly due to the use of mercury in antifungal products for seed dressing, which eventually entered the bodies of small birds) despite both raptors being bird hunters (Kitowski et al., 2016). Water-sourced prey was not the main source of mercury source, as consumption of larger birds reduced mercury ingestion and accumulations (e.g. in goshawks), larger birds accumulated less mercury (e.g. the common buzzard and white-tailed eagle), cadmium and lead accumulations were linked to small and medium sized mammal consumption, and the study concluded that "prey preference has an effect on the heavy metal load in raptor species" (Kitowski et al., 2016, 5). There is evidence that white-tailed eagles respond to declining fish availability by switching to game carrion, increasing their ingestion of lead. This study showed that raptors may change their food preferences, and this may alter their vulnerability to metal poisoning and that human activities may alter these results (Helander et al., 2009; Nadjafzadeh et al., 2013).

Lead Toxicity and Research Methods

Lead toxicity is among the most dangerous forms of toxicity for birds (Cruz-Martinez, et al., 2012). Lead is a resource metal, but it is a neurotoxin for both animals and people (Campbell, 2015). Lead can accumulate in bones, flesh, and the nervous system, this impairing neurological functions and influencing brain damage and behavioral change. Lead in raptor digestive tracts may be absorbed into the blood, and may even play the role of calcium in chemical processes (Redig & Arent, 2008; Haig et al., 2014). This can disrupt the functions of calcium, affecting calcium homeostasis, signal transmissions, cholinergic nerve cells and the cerebellum, and creating anemia through blood cell impacts (Haig et al., 2014). Behavioral impacts may include impairment of flight, walking, foraging, digestion and strength (De Francisco et al., 2003). As pointed out by Wani et al. (2015, 55) "lead is highly persistent in the environment and because of its continuous use its levels rise in almost every country, posing serious threats" and "there is almost no function in the body which is not affected by lead. Lead toxicity disrupts the functions of the digestive system, nervous system, respiratory system, reproductive system, etc." (Wani et al., 2015, 62).

Lead toxicity has been identified as a neurotoxin for some North American raptors, such as Bald and Golden eagles, perhaps because of the high levels of pollution and use of lead in hunting (Jacobson et al., 1977; Kaiser et al., 1980; Pattee et al., 1981, 1983; Reichel et al., 1984; Frenzel & Anthony, 1989; Craig et al., 1990; Langelier et al., 1991; Elliott et al., 1992; Gill & Langelier, 1994; Wayland & Bollinger, 1999; Miller et al., 2000, 2001; Wayland et al., 2003; Campbell, 2015). For example, Cruz-Martinez et al. (2012) looked at Bald eagles admitted to a veterinary clinic over 40 years, noted "90% of the bald eagles received each year (120-130) for all types of problems have elevated lead residues in their blood…20–25% percent of these eagles have sufficiently high levels to cause clinical lead poisoning. Most of these birds die or are euthanized. In the last 24 years, over 500 eagles received or admitted to our clinic have either died or had to be euthanized due to lead poisoning." Spent lead ammunition from shotguns and rifles in waterfowl and other hunted birds was blamed, evidenced in the correlation between poisoning and the hunting seasons (Cruz-Martinez et al., 2012).

In 1986, a United States Federal Law banned lead ammunition for waterfowl hunting, due to the role of lead-poisoned waterfowl in the decline of eagle populations. This law was evaluated as ineffective (Kramer & Redig, 1997), manifested by the continued rate of poison fatalities, despite hunter compliance. This led to a search for another source of poison. This was found to be deer carcasses and deer gut piles left by hunters. Another study covering the period 1996-2009, evaluated the hypothesis that deer carcasses with lead bullets were the main contributory factor for lead poisoning of eagles. This study used four epidemiological parameters: (1) links between poisoned eagles and the deer hunting season in Minnesota, Wisconsin, and Iowa; (2) correlations between animal recovery locations within deer-hunting zones; (3) lead isotope ratio analysis of metal fragments in eagle gastrointestinal tracts and blood lead levels; and (4) comparison of eagle kidney copper concentrations in lead-exposed and non-exposed birds (Cruz-Martinez et al., 2012). Extracted livers were examined to explain the high lead levels in Bald eagles after the lead shot ban of white-tailed deer (*Odocoileus virginianus*, Zimmermann, 1780). Of the 1,277 Bald eagles in rehabilitation from January 1996 through December 2009, 334 had elevated lead levels. Copper concentrations in eagles with high lead were also assessed, as many high velocity rifle bullets used for deer hunting have copper jackets. Lead levels and poisoning were spatially and temporarily correlated with the hunting seasons, deer hunting rifle zones, and adult eagle age, with high levels of matching isotopic signatures in the lead samples (from ingestion of lead) and blood and fragment samples within the isotope ratio of ammunition (Cruz-Martinez et al., 2012; see also Church et al., 2006). Copper levels correlated with lead levels (Cruz-Martinez et al., 2012). Copper-jacketed bullets gave the high kidney copper concentrations in lead-exposed eagles (see the section on copper). Poisoning resulted from spent ammunition in deer and shotgun pellets in smaller pheasants, squirrels and rabbits, hence lead "is an important, if not the primary, source of lead exposure for Bald eagles" (Cruz-Martinez et al., 2012). These findings contributed to the larger public debate on lead poisoning and possible regulation of deer hunting activities (Campbell, 2015).

In another study, Manning et al. (2018) examined 93 severely lead poisoned Bald eagles and compared internal lesions with blood, liver, kidney, and/or bone lead concentrations. Findings indicated that gross lesions linked to lead toxicity were commonest within the heart (55%), detectable by "multifocal myocardial pallor and rounding of the apex" (Manning et al.,2018, 289). Gross lesions in the brain "included petechiae or hemorrhagic necrosis" (14%), while histologic lesions were in the heart (82%), brain (63%), and eyes (26%) (Manning et al., 2018, 289). The authors argued that "lead toxicity in bald eagles is characterized by fibrinoid necrosis of small- to medium-caliber arteries, most commonly affecting the heart, brain, and eyes"; additionally the gross and histologic lesions are "consistent with ischemia caused by a primary vascular injury." The findings also indicated that blood lead concentrations above 4 ppm, and higher liver lead concentrations were linked to more frequent heart lesions. They concluded that "the presence of fibrinoid arterial necrosis and parenchymal degeneration, necrosis, and/or hemorrhage within the heart, brain, and/or eyes is suggestive of lead toxicity in bald eagles and warrants evaluation of liver or bone lead concentrations" (Manning et al., 2018, 289).

In a Californian study, Kelly et al. (2014) looked at 48 carcases (21 Golden eagles, 23 Turkey vultures and 4 common ravens), which were either found alive (45) or dead (3, two Golden eagles and one Turkey vulture); of the 45 living birds 14 died at the centers and 31 were killed due to their condition. The methodology used blood samples and carcases, and the parameters for each bird were date of accession, extraction location, admission factor, age class, clinical presentation, treatments and outcome. Mortality factors were determined using the location and severity of lesions, radiography screening (for trauma and metallic opacities in the gastrointestinal tract, and heavy metal analysis), and "microscopic examination of tissues, including brain, spinal cord, peripheral nerves, bone, muscle, tongue, heart, trachea, lung, liver, kidney, gonads, spleen, adrenals, thyroid/parathyroid glands, oesophagus, ventriculus, proven triculus, pancreas, intestines and bursa of fabricius" (Kelly et al. (2014, 2).

Rodenticides, organophosphate pesticides, strychnine and zinc and other substances were also examined. Mortality was mostly due to lead poisoning (14% of the eagles and 17% of the turkey vultures) with some of the latter recoding pathological changes, elevated hepatic lead concentrations and anticoagulant rodenticide intoxication. Four turkey vultures (8% of the total sample) died of anticoagulant rodenticide intoxication (Kelly et al., 2014).

Canadian studies of eagles found similar results (Campbell, 2015). Owen (2019) cites the Golden eagles that are treated for lead poisoning at the Orphaned Wildlife Rehabilitation Society in Delta, B.C. Owen (2019) argues that "the Canadian government banned lead for hunting waterfowl two decades ago, and non-toxic shot is now required to hunt most migratory game birds across the country. Lead shot is also prohibited for all types of hunting in 55 national wildlife areas… but pellets and bullets containing lead are still widely used to hunt large game and upland game birds, such as grouse or pheasant." Furthermore, the "Canadian government estimates that 40 to 80 tonnes of lead are used every year for hunting activities in Canada" (Owen, 2019). A cited study in western Canada (1986 to 1998) found that 10 per cent of 546 eagles were lead-poisoned, and that Golden eagles were more affected than Bald eagles (Owen, 2019).

Vultures are also affected by lead poisoning. It may be important to compare them with eagles, as the smaller vultures (Turkey and Black vultures may be more common and hence may serve as indications of lead poisoning. Rarer species such as the California condor may also be more vulnerable than eagles, so a comparison may assist conservation assessments. For example, Bloom et al. (1989) wrote an article entitled "Lead contamination of golden eagles (*Aquila chrysaetos*) within the range of the Californian condor." Lead poisoning, from ammunition cartridges, is relevant to the survival of the California condor (Cade, 2007; Campbell, 2015). Cade (2007, 2125) argues that "the scientific evidence that California condors (*Gymnogyps californianus,* Shaw, 1797) are frequently sickened and killed by lead poisoning from spent ammunition supports the conclusion that current levels of lead exposure are too high to allow reintroduced condors to develop self-sustaining populations in the wild in Arizona and, by inference, in California." This lead-related mortality contributed to the reduction of the California condor population during the 1980s, and during the post-reintroduction period (Snyder & Snyder, 2000; Kelly & Johnson, 2011). Research findings have indicated lead accumulation in the blood of condors during and outside hunting seasons (Hall et al., 2007; Hunt et al., 2007; Sorensen et al., 2007). Such findings influenced the ban on lead ammunition in the condor's California range. Golden eagles in the condor's range were also found to have high lead levels (Pattee et al., 1990) as did ravens in the Greater Yellowstone Area (Craighead et al., 2009).

The New World Turkey vulture (*Cathartes aura*, Linnaeus, 1758), another common victim of the lead ammunition in animal carcasses, may suffer after ingestion from weakness and uncoordinated movements, including flight or death (Carpenter et al., 2003; Campbell, 2015). Turkey vultures in California have recently been recorded with higher lead poisoning levels, mostly correlated with higher blood lead concentrations during the deer hunting season and the intensity of wild pig hunting, this despite the possibly violated ban on lead ammunition in Californian hunting (Kelly & Johnson, 2011). Canada records similar issues (Martin et al., 2008). Clark and Scheuhammer (2003) reported that 184 carcasses of 16 species (including Golden eagles, Turkey vultures, Red-tailed hawks and Great horned owls) were found with evidence of lead ammunition poisoning from game bird and mammal consumption. Martin et al. (2008) add that the dissection of 225 carcasses from 19 species of raptors in southern Ontario, found that after bone, kidney and liver examination, Turkey vultures had the 'highest mean concentrations of lead in bone and kidney compared to other raptor species' (Martin et al., 2008: 96).

In Eurasia, Golden eagles (*Aquila chrysaetos*, Linnaeus, 1758) in the European Alps have also been studied (Madry et al., 2015). These authors note that the main issue concerns Golden eagles scavenging on either non-retrieved carcasses or offal, both of which are discarded by hunters. This study addressed an important question: did the three Golden eagles discovered with severe lead poisoning represent a one-off situation or a generalized trend (perhaps at sublethal levels) in the Alpine region? The methodology involved the measurement of body burdens (body burden is the term for the concentration (or amount) of chemical in the body at any given time, ScienceDirect, 2020) in Golden eagles (36 specimens) in relation to their prey and a comparison of the results with those of nearby similar eagle owls (19 specimens) in relation to their prey (marmots *Marmota marmota*, Linnaeus, 1758 and *Capra ibex*, Linnaeus, 1758) for eagles, and birds, rodents and snow hares (*Lepus timidus*, Linnaeus, 1758) for eagle owls. The study acknowledged that while eagles might feed on carcasses, eagle owls generally fed on live-caught prey (Madry et al., 2015).

Findings indicated that soft tissue lead levels were higher in Golden eagles than in eagle owls. Bone lead levels were higher in eagle bones than in owl bones. Prey species for both eagles and owls had low lead concentrations. The next question concerned the source of the sublethal lead in Golden eagles: from the general environment or from ammunition in the carcasses of prey species. Lead isotope ratios were examined, the findings indicating that the lead isotope signatures of the Golden eagle bones were like those of the ammunition but were different from the signatures of prey animals' bones, and those of eagle owls and soil. Additionally, the isotope signatures were unchanged with increasing bone lead concentration in any of the species examined. The conclusions were that eagles in the Alps generally pick up lead from spent ammunition in both carcasses and offal in sublethal and sometimes lethal quantities over several years (Madry et al., 2015).

White-tailed and Steller's Sea eagles have been seriously threatened by lead poisoning (Fisher et al., 2006; see for example, Falandysz et al., 1988; Kim et al., 1999; Kurosawa, 2000; Iwata et al., 2000; Ueta & Masterov, 2000; Kenntner et al., 2001; Krone et al., 2004; Kenntner et al., 2005). Fisher et al. (2006, 426) note that for the White-tailed and Steller's Sea eagles and the general topic of lead poisoning "much work on poisoning from bullets comes from Japan, where it is customary for hunters to leave whole and partial shot deer carcasses in the field. These then become a source of bullet fragments that are ingested by scavenging raptors during and after the hunting season, most notably Stellar's sea-eagles (*Haliaeetus pelagicus*) and White-tailed eagles (*Haliaeetus albicilla*) (see also Falandysz et al., 1988; Kim et al., 1999; Iwata et al., 2000). In one study, Kurosawa (2000) located 39 Stellar's and 15 White-tailed eagles in Hokkaido, which died from lead poisoning in winter from 1996 to 1998. Fisher et al. (2006, 426) further argue that "if this rate of mortality is an accurate measure of the proportion of deaths due to lead poisoning, then it represents a threefold increase in the known yearly eagle mortality." Other serious lead poisoning accidents have been reported through the range of the White-tailed eagle, for example in Germany and Austria (Kenntner et al., 2001, 2004). For example, Kenntner et al. (2005, 1831) found that 24% of 277 White-tailed eagles recovered "dead or moribund" in Germany from 1979 to 2005 had liver lead levels representative of acute poisoning.

In Africa there is serious problem concerning lead poisoning of eagles. However, it has been mooted that "vultures are more vulnerable to ingesting lead from ammunition than birds of prey such as falcons, hawks and eagles that often do their own hunting" (Verdoorn, G. in Torchia, 2018). But "Griffon vultures feed on carcasses often consumed by many other species, including threatened ones (e.g. Spanish Imperial eagle *Aquila adalberti*)" (Mateo-Tomás et al., 2016). Even though these species are mostly European, they may have migratory links with Africa (Campbell, 2015). The scavenging habits of eagles and the occasional link with vultures was illustrated when people may have poisoned (not lead) elephant carcasses in Botswana, and 537 vultures and two Tawny eagles died (BBC, 2019).

Lead poisoning on African vultures may therefore be a potential threat to eagles when they share carcasses with foraging vultures. Vultures are frequently affected by lead poisoning (García-Fernández et al., 2008; Cade, 2007; Dobrowolska & Melosik, 2008; Gangoso et al., 2009). Kruger & Amar (2018) discuss the lead poisoning of the Bearded vulture (*Gypaetus barbatus*, Linnaeus, 1758) in southern Africa, based on lead levels in bone and tissue samples, taken from carcasses with gun ammunition. The lead poisoning was examined in the context of the critically endangered status of this species. The methodology used blood lead levels in from six live birds (2017), five live birds (2009) and eight dead birds (with evidence that possibly lead exposure and accumulation was the factor for the deaths). The findings supported the conclusion that "the levels of lead may be detrimental to the survival and fecundity of this small and declining population" of Bearded vultures with recommendations being the banning of "hunting and culling with lead ammunition, which is the most likely source of this pollutant. Such actions may reduce the population's susceptibility to other threats, which may be compounded by high lead levels, and help ensure the success of planned reintroduction programs" (Kruger & Amar, 2018, 491).

In Europe, Spanish Greater Spotted eagles have been recorded with lead poisoning. Pérez-García et al. (2020, 224) found that this species fed mainly on birds, with 18 bird species accounting for about 73% of prey items, the rest being mainly mammals. The commonest prey species were the Common moorhen (*Gallinula chloropus* Linnaeus, 1758), rats (*Rattus* spp.) and Common Teal (*Anas crecca*, Linnaeus, 1758). A significant proportion of regurgitated pellets (42.3%, of which 64% contained lead shot) and prey remains (40.0%, of which 25% contain lead shot) contained ammunition. The authors concluded that "high lead shot presence in pellets and prey remains of wintering Greater Spotted eagles in southeast Spain warns of a high risk of lead poisoning. Factors such as feeding behavior, the large space–time overlap between the raptor presence and the waterbird hunting season and non-compliance with the ban on the use of lead ammunition are likely contributing to high lead exposure."

In Spain, as mentioned above Spanish Imperial eagles may feed on carcases also consumed by Griffon vultures (Mateo-Tomás et al., 2016). Hence, the lead poisoning of vultures for shared carcasses is relevant to eagle conservation. Rodriguez-Ramos et al. (2009: 235; see also Iñigo & Atienza, 2007) examined lead levels in Griffon and Cinereous vultures and argued that "this exposure may have increased after the ban on abandoning carcasses of domestic ruminants in the field due to the bovine spongiform encephalitis (BSE) crisis, both because the vultures consume hunting bag residues more frequently and because malnutrition may lead to mobilisation of lead stores." Lead toxicosis contributed to severe hypochromic anemia, disorientation, ataxia, impaired landing and posterior paresis.

Donázar et al. (2002: 89) also found high frequencies of lead poisoning in Egyptian vultures in the Canary Islands, which included sub-clinical and clinical intoxication levels, which likely resulted from the ingestion of lead shot. Around 16% of the surveyed birds had over 0.2 ppm of lead, which at higher levels could be sufficient for serious or fatal consequences. Sympathetic arguments have called for lead pellets to be replaced by steel or molybdenum/tungsten alloy bullets (Donázar et al., 2002). As Booted eagles (and more rarely Bonelli's and Short-toed eagles) also rarely occur on the Canary Islands (Lepage, 2020), this may be a serious threat, but is currently under research (Campbell, 2015).

Anglers using lead tackles may also affect eagles in North America (Bedrosian et al., 2012). Grade (2018, 155) notes that "ingested lead fishing tackle has been documented in 28 species of North American birds, as a risk factor for 75 species and as a leading cause of death for common loons" (*Gaviaimmer*, Brunnich, 1764) (see also Pokras & Chafel, 1992; Blus, 1994; United States Environmental Protection Agency, 1994; Scheuhammer & Norris, 1995; Anderson et al., 2000; Stone & Okoniewski, 2001; Scheuhammer et al., 2002; Franson et al., 2003; Sidor et al., 2003; Strom et al., 2009; Grade, 2011). Bald eagles may prey on young loons and hence are susceptible to any poisoning of loons (Evers et al., 2010). Grade (2018, 157) assessed the lead poisoning of loons using dead loon specimens and using the methodology of Sidor et al. (2003) and Franson et al. (2003). This study used measurements of lead tackle parts present in the gizzard and gastrointestinal tract, based on the parameters listed below. It also acknowledged a limitation of this type of methodology, as "ingested tackle erodes in the gizzard because of grinding action and stomach acids; and, in the case of jigs, hooks or other attachments break off and often dissolve or are passed prior to death" (Grade, 2018, 157; see also Pokras et al., 2009).

The cause of death was lead tackle toxicosis, based on five conditions: (1) "tissue, blood, or body fluids were tested for lead and exceeded thresholds at which clinical signs of lead poisoning have been observed" (Grade, 2018, 157; see also Sidor et al., 2003, Franson & Pain, 2011); (2) the necropsy form showed the death as due to lead toxicosis, with the discovery of lead tackle in the gastrointestinal tract, or evidence of clinical pathology identifiable as lead toxicosis; (3) a tackle item from the loon's gastrointestinal tract, revealed to be lead; (4) a radiograph revealing a tackle part, such as a lead sinker (a weight used with a fishing hook to increase its sinking rate, rate of sink, anchoring ability or casting distance) or jig (a hooked, lead sinker, covered to attract fish, jerked in vertical motion) in the loon's gastrointestinal tract; and (5) a report including evidence of lead toxicosis (Grade, 2018, 157). The methodology recorded the findings in the gizzard, using LeadCheckSwabsTM (3M, St. Paul, MN, USA), which can detect lead in metals at levels >0.2% (2,000 ppm; 3M 2011) (Grade, 2018).

Mercury Toxicity and Research Methods

Mercury is an extremely dangerous chemical element, largely due to its impact on people's health (Clarkson et al., 2003; Campbell, 2015). Henny and Elliott (2015, 337) note that mercury "toxicity depends on whether mercury is in the organic or inorganic form. Only a small percentage of inorganic mercury is absorbed, but almost all organic mercury is absorbed by the intestine... inorganic mercury produces methylmercury (MeHg), which fish accumulate from water and their diet; nearly all mercury in fish flesh is MeHg. MeHg can adversely affect developing neural tissue in birds, with fish-eating birds being especially vulnerable." In the United States human mercury exposure generally occurs through eating fish that contain methylmercury, a compound reacting from bacterial reactions with mercury in water, soil, or plants. Both elemental mercury and methylmercury affect the human nervous system, and even inhalation of mercury vapor results in severe symptoms and in some cases death. These results may occur in both people and animals (Brean, 2019).

Mercury has been considered a dangerous chemical for raptors for several decades and has increasingly been traced in raptors and waterbirds, these being tested through blood and feathers (Campbell, 2015; Henny & Elliott, 2018). An important problem in the 1960s and 1970s, was the use of alkyl mercury as a fungicide for seed dressing for agricultural use, which killed some raptors (Berg et al., 1966, Jenson et al., 1972). For example, Kitowski et al. (2016) argue that small birds, including passerines (about 20% of the prey of sparrowhawks) are a source of mercury for raptors, as they consume grains with high concentrations of mercury-based fungicides. Of course, many eagles may also kill small birds (Olendorff, 1976; Tjernberg, 1981; Houston, 1985a; Watson, 2010). Alkyl mercury was used during the 1940s and was later banned as a seed dressing in Sweden in 1966 (Johnels et al., 1979). Henny and Elliott (2015) described the main contemporary sources of mercury: these include the atmospheric deposition from coal-fired power plants, mostly from the Arctic, northeastern U.S. and southeastern Canada, which affect fish and fish-eaters; mining operations in the Amazon basin (see also Nriagu et al., 1992); and local mining around the world, either for mercury or gold and silver extraction.

The threat to eagles may also be assessed by the recorded poisoning of smaller raptors, if they eat similar prey (Ferguson-Lees & Christie, 2001). Golden eagles may also prey on peregrine falcons (Gainzaraian et al., 2000; Craig & Enderson, 2004; Sergio et al., 2004; Sergio & Hiraldo, 2008). Hays (1987) reported that a Peregrine falcon killed or critically injured a Golden eagle in the defence of its nest. Considering raptors, in a recent study, Barnes (2019, cited in Brean, 2019) tested 700 peregrines for mercury in Washington, Maryland, Southern Nevada, and the Gulf Coast of Texas, and found that all the birds tested positive for mercury. Peregrines are "one of the most impacted species when it comes to contamination," Barnes noted, because they feed on other birds that have been exposed to mercury. An identified contributory factor for the peregrines in the Lake Mead, Colorado River (Arizona and Nevada), was the eared grebe (*Podiceps nigricollis*, Brehm, 1831), a common prey species that might be exposed to mercury in Utah's Great Salt Lake. Barnes et al. (2019, 1) note that "peregrines have the potential to be a particularly useful biomonitor because they are globally distributed, nest in proximity to a diverse assortment of aquatic and terrestrial habitat types (e.g., freshwater rivers and lakes, marine systems, estuaries, wetlands, arctic tundra, temperate forests, and open desert), are long-lived with high breeding-site fidelity, and have an extremely diverse avian diet" (see also White et al., 2013).

White-tailed eagles have been recorded as having high and lower levels of mercury in their livers (Falandysz et al., 2001; Kalisinska et al., 2014). For raptors lower down on in the food chain, Kitowski et al. (2016) cite mercury poisoning for sparrowhawks in Poland, buzzards in Spain and the Czech Republic and Spanish and German goshawks (Houserova et al., 2005; Castro et al., 2011). In the study by Kitowski et al. (2016), a difference between goshawks and sparrowhawks was discerned: "the strong correlation between mercury and the consumption of small birds…implies that consumption of bigger avian prey by Goshawks allows them to avoid the accumulation of large amounts of mercury in the livers, in contrast to the Sparrowhawks specialized in catching small birds (Kitowski et al., 2016; see also Newton, 1986). Waterbodies and waterborne prey are also cited as sources of mercury poisoning in raptors (see also Kalisinska et al., 2014).

Weir et al. (2018) documented the impact of the rodenticide brodifacoum (formula $C_{31}H_{23}BrO_3$) and mercury and the effects of these on certain raptor species, including red-tailed hawks (*Buteo jamaicensis*, Gmelin, 1788), red-shouldered hawks (*Buteo lineatus*, Gmelin, 1788), barred owls (*Strix varia*, Barton, 1799), great horned owls (*Bubo virginianus*, Gmelin, 1788), and ospreys (*Pandion haliaetus*, Linnaeus, 1758). Liver samples from the raptor carcasses were analyzed for the total mercury residue (mg/kg, dry weight) and five anti-coagulant rodenticides (µg/kg wet weight). Mercury was analyzed using a direct mercury analyzer approach and rodenticides were quantified by liquid chromatography–mass spectrometry (LC-MS). The findings indicated that mercury residues were high in piscivorous (fish eating) birds (ospreys 15.09 mg/kg and great blue herons 6.93 mg/kg, dry weight) and "relatively high in red-shouldered hawks and one eastern screech owl tested" (Weir et al., 2018). Six samples (three ospreys, one each of great blue heron, red-shouldered hawk and eastern screech owl) were found to be above the health threshold of 1 mg/kg (wet weight). Brodifacoum, the only rodenticide consistently detected, exceeded 75% in barred owls, great horned owls, and red-shouldered hawks. The study concluded that brodifacoum was a more serious risk than mercury for most raptor species (possibly except for the osprey and to a lesser extent the red-shouldered hawk), as mercury ingestion was less.

Chromium Toxicity and Research Methods

Chromium toxicity occurs when an organism is exposed to certain types of chromium, for example hexavalent chromium (chromium (VI), usually toxic when inhaled or ingested, and is hemotoxic or haemotoxic (destructive of red blood cells, Merriam-Webster, 2020), genotoxic ("the process by which an agent produces a deleterious effect on DNA and other cellular targets that control the integrity of genetic material", Lopez et al., 2012), and carcinogenic (producing or tending to produce cancer, Merriam-Webster, 2020). Ingested hexavalent chromium affects blood cells by causing oxidation reactions, leading to hemolysis and possibly kidney and liver failure (Xie et al., 2017). Speer and Wise (2018) argue that "hexavalent chromium [Cr(VI)] is a valuable industrial resource but ubiquitous environmental contaminant. Humans can be exposed to Cr(VI) occupationally and through environmental sources. It is known Cr(VI) causes lung cancer and other health issues including reproductive and immune dysfunction."

In raptors, chromium has been detected as a factor for physiological problems. However, few studies record serious impacts or extremely high levels of chromium. For example, Falandysz et al. (2001) recorded mean Cr concentrations of 0.087 mg/kg dw in White-tailed eagles from the south Baltic, which was lower than the levels found in Kitowski et al. (2016). Manning et al. (2008) found a low chromium hepatic value for one Australian White-bellied sea-eagle (*Haliaeetus leucogaster*, Gmelin, 1788) of 0.04 mg/kg ww (about 0.16 mg/kg dw). In one non-eagle species, the white-backed vulture (*Gyps africanus*, Salvadori, 1865) from the Republic of South Africa, Van Wyk et al. (2001) estimated a higher average concentration of 19.57 mg/kg dw of hepatic Cr. Smaller raptors recorded low levels of chromium, especially when they prey on birds. Kitowski et al. (2016) note that in other studies, marsh harriers, kestrels, buzzards and sparrowhawks in southern Italy (from preying on passerines) had low hepatic concentrations of chromium, always less than 0.40 mg/kg dw (Zaccaroni et al., 2008). Tawny owls (*Strixaluco*, Linnaeus, 1758) had mean hepatic concentrations slightly over 0.30 mg/kg dw. In north Italy, the Little owls (*Athene noctua*, Scopoli, 1769) had a concentration of 0.29 mg/kg ww (about 1.16 mg/kg dw.) (Zaccaroni et al., 2003). Both owl species preyed in passerines, apparently an important source of chromium. This was similar to that of Sparrowhawks in their study (Kitowski et al., 2016).

Nickel Toxicity and Research Methods

Nickel is a heavy metal that occurs widely in the environment in small amounts (Das et al., 2018). Soils and volcanic emissions are common sources, generally bound with oxygen or sulfur, through oxides or sulfides. Industrial production and processing, recycling and disposal are also common sources of nickel, with emissions through mining, power plant operation and/or incinerators, and industrial production of rubber, plastic, nickel-cadmium battery and electroplating industries (Das et al., 2018). Mudd et al. (2010, 9) point out that nickel's uses, which include "stainless steel, alloys, electroplating and rechargeable batteries" are derived from either sulfide or to a lesser extent laterite-type ores, with recent extraction from laterite ores (especially in Sudbury in Canada and the Taimyr and Kola Peninsulas in Russia) (Mudd et al., 2010, 9). In terms of impacts on people and animals, Buxton et al. (2019) write that "notable human health toxicity effects identified from human and/or animal studies include respiratory cancer, non-cancer toxicity effects following inhalation, dermatitis, and reproductive effects. These effects have thresholds, with indirect genotoxic and epigenetic events underlying the threshold mode of action for nickel carcinogenicity."

DeForest et al. (2012, 107) examine the application of the European Union's Existing Substances regulation (EEC 793/93), which "was developed to assess the ecological risks posed by chemical substances such as Ni…" and also looked at "a generic framework for assessing secondary poisoning risks and prescribes the following terrestrial food chain: soil → earthworm → worm-eating bird or mammal." The Nickel Institute (2015) evaluates the impacts of nickel from marine, fresh water and soil. The methodology used a framework based on the ratio of dietary predicted exposure concentrations (PEC) and dietary predicted no effect concentrations (PNEC) (European Chemicals Bureau, 2003). The argument was that "the PEC in this assessment was the estimated Ni concentration in the diet of the representative predatory birds or mammals in each of the food chains evaluated, while the PNEC was the Ni concentration in bird and mammal diets at which no adverse effects are expected" (Nickel Institute, 2018). The ratio of the PEC to PNEC created Risk Characterization Ratio, or RCR, where an RCR <1 indicates low potential risk and an RCR >1 indicates a stronger risk. The formula is: a PEC-to-PNEC ratio, or RCR:

$$RCR = PEC \times RAF/PNEC, \text{ where RAF is relative bioavailability.}$$

Bioaccumulation of nickel from soil or water into organisms at the bottom of the food chain has been found to be low. The Nickel Institute (2015) argues that "there is no evidence that Ni biomagnifies in aquatic or terrestrial food chains (i.e., Ni concentrations do not increase with increasing trophic level). This contrasts with other chemicals, such as methyl mercury, which tends to biomagnify across multiple trophic levels, thereby resulting in higher trophic level organisms being potentially more susceptible to mercury poisoning." Nickel may even have a opposite reaction, termed "biodilution" when the nickel rather reduces as it moves up the food chain (references include, Campbell et al., 2005; Lapointe & Couture, 2006).

The study uses a tiered approach to estimate possible poisoning risk (Nickel Institute, 2015). In this the first tier represents more conservative assumptions, that may overestimate the potential risk of poisoning. The second tier represents greater refinement which may increase the possibility of poisoning, by assuming for example a wider diet for birds, rather than a diet of organisms that may have high nickel levels such as the common cockle (*Cerastoderma edule*, Linnaeus, 1758). For marine ecosystems, an example of oystercatchers (*Haematopus* spp, Linnaeus, 1758), in Tier 1, an assumption is that the diet is entirely on the *C. edule* diet, which gives a high risk, but in Tiers 2 and 3, the Risk Characterization Ratio is lower, when the ingestion rate of oystercatcher species and a mixed bivalve diet, are taken into account. For freshwater ecosystems, the diet of a representative mollusk-eating bird, subsequent Tiers failed to include other secondary poisoning risks, allowing for the food ingestion rate and body weight of mollusk-eating birds, which reduced the RCRs. For soil ecosystems, Tier 1, the analysis was based on a typical worm-eating bird, with subsequent Tiers adjusted for the food ingestion rate of the worm-eating bird and a wider diet including both worms and isopods (Nickel Institute, 2015).

For raptors, there is evidence of nickel ingestion. For example, white-tailed eagles in a study by Falandysz et al. (2001) in the Baltic sea area had liver-nickel concentrations 56 times higher (13.0 mg/kg dw) than those of the study by Kitowski et al. (2016) in east Poland, leading the latter researchers to speculate that "it seems that White-tailed eagles from east Poland have less opportunities to prey in highly mercury- and nickel-contaminated areas than raptors from the south Baltic Sea coast" (Kitowski et al. (2016, 11). For smaller raptors, Kitowski et al. (2016) looked at sparrowhawks and goshawks, which accumulated similar levels of nickel (Ni 0.3 mg/kg dw) through their diet of birds. Also cited are the low nickel levels in one peregrine falcon (0.107 mg/kg dw (Kalisinska et al., 2008) in northwest Poland.

Cadmium Toxicity and Research Methods

Kitowski et al. (2016) write that cadmium concentrations in birds are usually highest in kidneys, intermediate in muscle and lowest in the livers (see also García-Fernández et al., 1996; Battaglia et al., 2005; Binkowski et al., 2013). However, Kitowski et al. (2016) argue that it is still valid to use the liver to monitor Cd exposure, as this accumulates about 50% of the total cadmium and the body burden is stable in the liver, as this organ is resistant to cadmium toxic effects (Burgat, 1990; Battaglia et al., 2005). In their study, Kitowski et al. (2016) also note that the high concentration of cadmium in buzzard livers might be linked to their feeding on voles and other relevant animals, as the findings indicate a higher cadmium levels for small mammal consumption. This possibly indicates voles and other small mammals feeding on fields fertilized with chemicals including cadmium (Beyer, 2000; Louekari et al., 2000). Marine environments may also have high levels of cadmium. Coal combustion emissions are important cadmium sources especially in Eastern Europe (Seshadri et al., 2010; Pacyna et al., 2010).

For birds, cadmium can cause "sublethal behavioral effects at lower concentrations than lead and mercury" (Eisler, 1985). A study by Burger & Gochfeld (2009) found low, sublethal levels of cadmium in Bald eagles in the Aleutians. Falandysz et al. (2001) reported lower cadmium levels in the livers of Baltic White-tailed eagles than the goshawks in their study. For smaller raptors, there are also records. Cadmium levels in buzzards recorded in Northern Italy (Battaglia et al., 2005), Portugal (Carneiro et al., 2014) and Spain (Castro et al., 2011) were mostly lower than in East Poland (Kitowski et al., 2016). Other low levels of liver cadmium concentrations were recorded in goshawks in Germany (Kenntner et al., 2003) and buzzards in Spain (Castro et al.,

2011). Kitowski et al. (2016) note that the levels of cadmium for peregrines in Sweden (Ek et al., 2004) were similar to those of goshawks in their studies.

For possible Sea eagle prey species, Parslow and Jefferies (1973) point out that "gannets, in common with other seabirds which feed mainly in offshore waters, contain less cadmium in their livers (mean 4.7 ppm dry weight, seven adults) than do mainly pelagic feeders (e.g. Fulmar Fulmarus glacialis, mean 37 ppm, 11 individuals) or certain littoral and sublittoral mollusc feeders (e.g. Eurasian oystercatcher *Haematopus ostralegus*, mean 29 ppm… Eider *Somateria mollissima*, mean 10 ppm, eight adults). The argument is that the cadmium is derived from the invertebrates upon which the oceanic birds feed, which may be concentrators of cadmium. Rainbow (1997, 195) supports the position of marine invertebrates being possible concentrators of environmental metals: "Trace metals are accumulated by marine invertebrates to body concentrations higher, in many cases orders of magnitude higher, than the concentrations in an equivalent weight of the surrounding sea-water."

Copper Toxicity and Research Methods

Copper toxicity is generally less toxic for birds than lead toxicity (Campbell, 2015). Copper is "an essential cofactor for several oxidative stress-related enzymes including catalase, superoxide dismutase, peroxidase, cytochrome c oxidases, ferroxidases, monoamine oxidase, and dopamine β monooxygenase" (Tchounwou et al., 2012, 134, see also Harvey & McArdle, 2008; Stern, 2010; Agency for Toxic Substances and Disease Registry, 2020a). Manto (2014, 327) notes that "copper is a cofactor of proteins and enzymes (called cuproenzymes) involved in fundamental mechanisms, such as energy generation, oxygen transportation, hematopoiesis, cellular metabolism and signal transduction." Therefore, copper is not primarily a toxic element, being a nutrient within metalloenzymes in "hemoglobin formation, carbohydrate metabolism, catecholamine biosynthesis, and cross-linking of collagen, elastin, and hair keratin" (Tchounwou et al., 2012, 134). Toxicities may however be generated when copper transitions between the oxidized Cu(II) and a reduced states Cu(I), and possibly human cellular damage and Wilson disease (Chang et al., 1996; Harvey & McArdle, 2008; Tchounwou et al., 2008; Stern, 2010).

However, the use of copper jacketed bullets in hunting of potential prey animals and resultant raptor-accessed carcasses has fueled studies on possible negative copper impacts on raptors (Cruz-Martinez et al., 2012). Copper jacketed lead bullets are compared with lead bullets for impacts on bird health (Kallok, 2013). Kallok's (2013) study examined injured, sick and dead Bald eagles delivered to the University of Minnesota's Raptor Center, some of which die, and others recover and are released. The Minnesota State and Federal bans on lead shot for hunting in 1987 and 1991 resulted from assessments of possible impacts on eagles. In Canada, waterfowl hunters were required to use nontoxic shot on and after 1999. Lead shot in waterfowl was initially blamed, but later lead in deer gut piles was identified as a source of lead ingestion for birds. Copper jackets on soft lead bullets were hypothesized to be insufficient to prevent lead fragmentation into flesh after impact, poisoning eagles that ate the flesh. A Raptor Center study found measurable levels of copper in 63 out of 89 eagle carcasses. Suggested alternatives were based on copper bullets, which studies found did not fragment into flesh on impact (although lighter, copper bullets may penetrate the entire animal), unlike 12-gauge lead shotgun slugs (which left about 28 fragments per shot) and muzzleloader bullets (which left 82 to 141 lead fragments) all several centimeters from the channel of the wound (Kallok, 2013).

For smaller raptors, Franson et al. (2012) examined copper toxicity in American kestrels (*Falco sparverius*, Linnaeus, 1758). The study methodology used orally administered copper into birds, rather than use naturally abstracted copper sources. The 16 kestrels (there were also 10 m controls, which were sham gavaged – supplied with a feeding tube with no copper administered) were orally administered 5 mg Cu/g body mass copper pellets nine times during a 38-day period. Most birds retained the pellets for more than an hour, but 69% regurgitated the pellets in the 12-h monitoring period. The findings indicated that hepatic copper concentrations were higher in Cu-administered kestrels than in the controls, but no difference in treated and control bird's blood Cu concentrations.

Another important finding was that hepatic copper and the protein metallothionein concentrations were significantly correlated, but that metallothionein was higher in male birds that received copper than in controls and for female birds those that received copper were similar to controls. Hepatic copper and metallothionein concentrations in kestrels were significantly correlated. Crucially, there were no clinical signs: "no treatment effect on body mass; concentrations of Cu, hemoglobin, or methemoglobin in the blood; or Cu concentrations in kidney, plasma biochemistries, or hematocrit" (the percentage of the blood that consists of red blood cells) (Franson et al, 2012, 145). The conclusions were that ingested Cu pellets "pose little threat to American kestrels (and presumably phylogenetically related species), although the retention time of pellets in the stomach was of relatively short duration" (Franson et al., 2012, 145). Furthermore, "birds expected to regurgitate Cu fragments with a frequency similar to kestrels are not likely to be adversely affected by Cu ingestion, but the results of our study do not completely rule out the potential for toxicity in species that might retain Cu fragments for a longer time" (Franson et al., 2012, 145).

Selenium Toxicity and Research Methods

Selenium, a non-metal element, occurring frequently in sulfide ores (where it may replace sulfur, and be produced when these ores are processed) is generally evaluated as only moderately toxic in large amounts; in small amounts it is necessary for animal physiological health (Tchounwou et al., 2012). Selenium combines with other elements to form oxides (selenium dioxide, SeO_2; selenium trioxide, SeO_3, the former dissolving in water or nitric acid to form selenous acid, H_2SeO_3). Selenium is an essential trace element, needed for function of metabolism (Spallholz & Hoffman, 2002). Heinz and Hoffman (1998) also point out that interactions between selenium and mercury can reduce the toxicity of both metals in adult birds. However, large amounts are toxic to animals and humans. Selenium toxicity in humans induces gastrointestinal and behavorial disorders, and even neurological damage and even liver cirrhosis, pulmonary edema, or death (Agency for Toxic Substances and Disease Registry, 2020b).

Selenium studies focus on environmental, herbivore and carnivore impacts, in a food chain format. For example, Lemly (1996) gives such a procedure, applied to toxic levels of selenium as a threat to fish and wildlife. The justifying argument is that a "toxic threat is expressed as hazard and is based on the potential for food-chain bioaccumulation and reproductive impairment in fish and aquatic birds, which are the most sensitive biological responses for estimating ecosystem-level impacts of selenium contamination" (Lemly, 1996, 19). Five degrees of hazard are defended as possible, dependent on the variables of selenium concentrations in the target environments, possible bird and fish exposure and the possibilities for variable degrees of partial or total reproductive impairment. The five degrees are, on a 1 to 5 scale: no identifiable hazard, and then minimal, low, moderate and high hazard. The five ecosystem components; water, sediments, benthic macroinvertebrates, fish eggs, and aquatic bird eggs, each had a different separate hazard score. Environmental threats are assessed by totalling these scores. Therefore, the least hazardous environment would be calculated by giving each of the five ecosystem components a rating of one, and the highest hazard would be based on a rating of 5 for each of these components. Hence, based on the totals, a score of 5 would be no hazard, score of 6-8 would be minimal hazard, 9-11 would be low hazard, 12-15 would moderate hazard and 16 a high hazard (Lemly, 1996, 19). Supporting would be from ecosystem study, including selenium contamination and potential animal exposure. It is concluded that this method "is based on a large body of field and laboratory research data on selenium cycling and toxicity in aquatic ecosystems… a useful tool for assessing the toxic threat of selenium to fish and wildlife" (Lemly, 1996, 35).

Lemly (1993, 83) argues that "it is now possible to formulate diagnostic selenium concentrations in four distinct ecosystem-level components; water, food-chain, predatory fish (consuming fish or invertebrate prey), and aquatic birds." Furthermore, despite the low selenium toxicity, "waterborne selenium concentrations of 2 µg/l or greater (parts per billion; total recoverable basis in 0.45 µ filtered samples) should be considered hazardous to the health and long-term survival of fish and wildlife populations due to the high potential for food-chain bioaccumulation, dietary toxicity, and reproductive effects" (Lemly, 1993, 83). Bioaccumulation occurs, increasing the toxicity of waterborne concentrations below 1 µg/l. At the lower levels of the food chain, some organisms (e.g. zooplankton, benthic invertebrates, and some fishes) accumulate selenium with little effect up to 30 µg/g (even to 370 µg/g). Such levels are nevertheless dangerous at the higher levels of the food chain, where for many fish and other wildlife the tolerance level is only 3 µg/g. Lemly (1993, 83) further notes that the thresholds of selenium tolerance for freshwater and anadromous (fish that migrate up rivers from the sea to spawn) fish are: "whole body=4 µg/g; skeletal muscle=8 µg/g; liver=12 µg/g; ovaries and eggs=10 µg/g"… thresholds for aquatic birds are: liver=10 µg/g; eggs=3 µg/g." For impacts on the waterbirds that feed on these fish, measurements of selenium concentrations in gravid ovaries and eggs offer "the most precise way to evaluate potential reproductive impacts to adult fish and aquatic bird populations" (Lemly, 1993, 83). Therefore, "resource managers and aquatic biologists should obtain measurements of selenium concentrations present in water, food-chain organisms, and fish and wildlife tissues in order to formulate a comprehensive and conclusive assessment of the overall selenium status and health of aquatic ecosystems" (Lemly, 1993, 83).

Spallholz and Hoffman (2002) list details on methodologies for the measurement of selenium in animals, including the three forms of selenium toxicity. Firstly, there is the formation of CH_3Se- which may create free radicals that affect enzymes and proteins or create superoxide and oxidative stress. Secondly, where there is excess selenium in the form of selenocysteine, selenium methylation metabolism may be inhibited, which may result in the concentration of hydrogen selenide with possible toxic effects on animals. Thirdly, there is a possibility that excess selenium linked to sulfur-containing enzymes and structural proteins can relate to avian teratogenesis (possibly from L-selenomethionine or to a lesser extent L-selenocysteine, the most likely to be ingested by waterbirds). In this respect, a teratogen is a substance that can affect physiological development in young animals and people, especially affecting the fetus and embryo (teratology is the study of abnormalities of physiological development in the young) (Rogers & Kavlock, 1996; Heinz et al., 2011; Bastow & Holmes, 2016). Disputes exist as to the toxicity of L-selenomethionine to embryos, but Spallholz and Hoffman (2002, 27) note that "there is no suggestion from the literature that the levels of L-selenomethionine that would be expected to accumulate in eggs in the absence of environmental concentration of selenium pose harm to the developing embryo." This points to environmental concentration, rather than physiological accumulation as a possible factor for any toxic levels. Nevertheless, "for several species of aquatic birds, levels of Se as selenomethionine in the egg above 3 ppm on a wet weight basis result in reduced hatchability and deformed embryos" (Spallholz & Hoffman, 2002, 27).

Selenium toxicity in Bald eagles has been assessed in the San Francisco Estuary (Beckon & Maurer, 2008). Citing Lillebo et al. (1988), it is argued that an analysis of bioaccumulation dynamics and an estimated critical dietary threshold for toxicity of 3 µg/g, piscivorous birds such as the Bald eagle more at risk from selenium toxicity than waterbirds like mallards. Bald eagle could ingest selenium from ducks in the local ecosystem. Bowerman et al. (1994) studied selenium in association with mercury and detected selenium in all the feathers that were analyzed. They concluded that as "Se concentrations were not significantly different across regions or between adult and nestling feathers, and Se concentrations ranged from 0.8 to 3.2 mg/kg" and "no significant relationships between adult or nestling feather concentrations and either measure of reproduction, productivity or nesting success, was found for either Hg or Se", "neither Hg nor Se concentrations are currently affecting bald eagle reproduction in the Great Lakes region."

Selenium has also been studied in the context of water ecosystems and waterbird ecology. Selenium-polluted surface waters have an important role in waterbird toxicity (Hoffman, 2002). Physiological impacts included "mortality, impaired reproduction with teratogenesis, reduced growth, histopathological lesions and alterations in hepatic glutathione metabolism" (Hoffman, 2002, 11). This study looked at "adverse biological effects of Se in aquatic birds to altered glutathione metabolism and oxidative stress" and refers to "oxidative stress in different stages of the mallard (*Anas platyrhynchos*) life cycle"; consequences of increased Se include increases in plasma and hepatic GSH peroxidase" and "dose-dependent increases in the ratio of hepatic oxidized to reduced glutathione (GSSG:GSH) and ultimately hepatic lipid peroxidation measured as an increase in thiobarbituric acid reactive substances (TBARS)." Findings on mallard ducks (*Anas platyrhynchos*, Linnaeus, 1758) indicated links with teratogenesis (4.6 ppm wet weight Se in eggs), reduced duckling growth (15 ppm Se in liver), lowered immune function (5 ppm Se in liver) and histopathological lesions (29 ppm Se in liver) (Hoffman, 2002, 11).

Another case study examined the impact of selenium on mallard ducks (O'Toole & Raisbeck, 1997). These authors argue that "selenosis is thought to be a significant problem among waterfowl populations in selenium-contaminated wetlands in the western United States" (O'Toole & Raisbeck, 1997, 330). Two 150-day studies examined the possibility that morphological criteria for selenosis could supplement chemical analysis. The methodology involved the feeding of 48 young male mallard ducks, with proprietary waterfowl rations containing < 1, 10, 25, and 60 ppm selenium (seleno-L-methionine). Another 12 birds were fed twice daily with either a proprietary ration or a selenium-supplemented ration (120 µg/g) for one of the two daily feedings. Selenium in the blood increased from the baseline concentrations, which were < 0.4 µg/ml, to averages of 4.5 µg/ml (10 ppm group), 8.9 µg/ml (25 ppm group), and 16.0 µg/ml (60 ppm group). Ducks in the 60-ppm-group became emaciated and were either killed or died within 22 to 50 days of dietary exposure.

Impacts on this 60 ppm group, other than emaciation and general malaise, included "mild to moderate generalized hepatopathy with single-cell necrosis, karyomegaly of hepatocytes, hyperplastic bile duct epithelium, and/or iron accumulation in Kupffer cells" half of the birds. Those given lower dosages acquired lesions (integumentary structures of hard keratin), later developed into gross lesions of bilaterally symmetrical alopecia of the birds' scalp and the dorsal cervical midline, and also damaged digital nails, and "necrosis of the tip of the beak (maxillary nail)" (O'Toole & Raisbeck, 1997, 330). These increased with the increased dosages of selenium (for those with 10 ppm selenium, 0%; 25 ppm, 42%; 120 ppm 44%). Histologic lesions also occurred; in the digital and maxillary nails, single-cell to full-thickness necrosis of keratinocytes and multifocal parakeratosis in stratum corneum existed; in "alopecic skin (necrosis of the epidermal collar, inflammation of the feather pulp, and follicular keratosis) were mild" (O'Toole & Raisbeck, 1997, 330). As in some other studies of ingestion, the liver, kidneys and feathers had high selenium concentrations. The study records that "mean hepatic tissue concentrations were 14.5 µg/g (10 ppm group), 29.6 µg/g (25 ppm group), 60.6 µg/g (60 ppm group), 13.0 µg/g (120 ppm split-feed group), and 2.0 µg/g (controls)" (O'Toole & Raisbeck, 1997, 330). This study showed that selenium, perceived by some scholars requiring more studies (Lemly, 1993; Tchounwou et al., 2012) may present dangers at high concentrations and hence would pose risks due to accumulation effects, as recent studies appear to indicate (Ashbaugh et al., 2018).

Findings on other waterbird species indicate important Se impacts, especially concerning glutathione metabolism: growth reduction, possible immune function decline, increased hatchling liver to body weight and hepatic Glutathione disulfide (GSSG) to Glutathione (GSH) ratios in hatchings, when eggs contained 9 ppm Se (in American avocet, *Recurvirostra americana*, Gmelin, 1789); decrease in hepatic GSH (Black-necked stilts, *Himantopus mexicanus*, P.L.S. Müller, 1776); emaciation, increased feather losses, histopathological lesions, and increased hepatic GSH peroxidase, loss of hepatic protein-bound thiols and total thiols, and minor increases in GSH (American coot, *Fulica americana*, Gmelin, 1789). A thiol is defined as an organosulfur compound where sulfur replaces oxygen in the chemical structure. Other findings document a positive correlation between hepatic Se concentration and GSH peroxidase activity and a negative correlation between hepatic Se and GSH concentration in diving ducks (*Aythyini*spp), and high blood levels linked to increased plasma GSH peroxidase activity and decreased plasma GSSG reductase activity (Emperor geese, *Ansercanagicus*, Sevastianov 1802) (Hoffman, 2002, 11).

Selenium toxicity has been blamed for the decline of the snowy plover (*Charadrius nivosus*, Cassin, 1858) in the south western United States (Ashbaugh et al., 2018). In this study, selenium was the commonest metal, although vanadium, arsenic, cadmium and lead were also traced in the blood, feathers and potential prey of this species (e.g. of the tiger beetles, *Cicindela circumpicta*, Fitch, 1841, *Cicindela togata*, LaFerté-Sénectère, 1841). Ashbaugh et al. (2018, 703) note that in "samples greater than instrumentation detection limits, 98% of snowy plover blood and 22% of feather samples were greater than Se toxicity thresholds of 1 ppm ww for blood and 5 ppm dw for feathers (blood quantifiable range: 0.83-15.12 ppm; feathers quantifiable range: 1.90-27.47 ppm)." Variables considered included sex, individual body condition, local tiger beetle Se concentrations, state, and presence of body molt. Tiger beetle Se concentrations were related to individual study sites in Texas. The authors argue that "these results provide some of the first evidence of Se exposure risk for interior snowy plovers nesting in saline lake and alkali flat environments of the SGP" (Ashbaugh et al., 2018, 703).

Fish are also studied for their selenium storage. Selenium abstraction and poisoning in fish begins with the eggs, infected through the diet of the female fish. As the selenium (usually at concentrations above 10 µg/g) is metabolized by the growing fish, through the embryo stages, "biochemical functions may be disrupted, and teratogenic deformity and death may occur" (Lemly, 2002, 39). Lemly (2002) found that 20 fish species which were resident in Belews Lake, North Carolina were affected by selenium in wastewater from a coal-fired power plant during the mid-1970s, to the extent that 19 species were extirpated. Physical examination found "telangiectasia (swelling) of gill lamellae, elevated lymphocytes, reduced hematocrit and hemoglobin (anemia), corneal cataracts, exopthalmus (popeye), pathological alterations in liver, kidney, heart, and ovary (e.g. vacuolization of parenchymal hepatocytes, intracapillary proliferative glomerulonephritis, severe pericarditis and myocarditis, necrotic and ruptured mature egg follicles), reproductive failure (reduced production of viable eggs due to ovarian pathology, and post-hatch mortality due to bioaccumulation of selenium in eggs) and teratogenic deformities of the spine, head, mouth, and fins" (Lemly, 2002, 39). The conclusions highlighted the importance of these findings for the larger food chain: "Bioaccumulation in aquatic food chains causes otherwise harmless concentrations of selenium to reach toxic levels, and the selenium in contaminated sediments can be cycled into food chains for decades" (Lemly, 2002, 39).

Invertebrates, even lower on the food chain than fish are also important for selenium studies. Debruyn & Chapman (2007, 1766) argue that most studies of selenium pollution in freshwater ecosystems have been primarily on birds and fish, "with invertebrates most often considered only as dietary sources of Se to higher trophic levels." Their research targeted 29 macro-invertebrate species, reviewing links between toxic effects and 156 aqueous, dietary, or internal Se concentrations. The findings showed pronounced locational differences between the samples. Internal concentrations were the least variable, with about 30-fold variation. Concentrations in the diets varied about 100-fold and those among taxa varied over 1000-fold. There was also much less variation in sublethal effects (about 10-fold lower concentrations) than in lethal effects. Crucially, sublethal effects in the targeted invertebrates were found at 1-30 µg Se/g dry weight in invertebrate tissue. As this range covered some dietary toxicity thresholds for waterbirds and fish, Debruyn & Chapman (2007, 1766) argue that "Se may cause toxic effects in some invertebrate species at concentrations considered to be 'safe' for the organisms consuming them."

9.3 Chemical Compounds

Many chemical compounds also affect birds. In this sense, a chemical compound is defined as a substance which is composed of more than one element, with atoms from more than one element connected by chemical bonds. Common chemical compounds affecting birds are illustrated in Table 9.3. These compounds are usually artificial in that they are made for specific purposes: for example, to kill pests (animals defined as trouble to people, usually insects), termed pesticides; to kill rodents, these termed rodenticides; and veterinary compounds intended to benefit livestock. These compounds may contain chemicals that have different impacts on different species of raptors or waterbirds. These impacts are at least as serious as the impacts from the chemical elements described above. Some of the more dangerous chemical compounds have been banned, but in some cases the impacts prevail (Campbell, 2015).

DDT, DDE and Other Pesticides

Pesticides are generally composed of connected elements, an example being the banned pesticide dichlorodiphenyltrichloroethane (DDT) (chemical formula $C_{14}H_9Cl_5$). The inventor of this pesticide, Paul Hermann Müller, was awarded the Nobel Prize in Physiology or Medicine in 1948 "for his discovery of the high efficiency of DDT as a contact poison against several arthropods." DDT was used as a pesticide for insect control in the United States until 1972, when it was banned by the United States Environmental Protection Agency (EPA). The National Pesticide Information Center (NPIC) (2011) noted that "bald eagles, ospreys and peregrine falcons were brought to the brink of extinction because their food sources (fish and other birds) were contaminated with DDE, the breakdown product of the insecticide DDT." DDE (dichlorodiphenyldichloroethylene) is formed when hydrogen chloride (dehydrohalogenation) is lost from DDT (chemical formula $C_{14}H_8Cl_4$) (Walker et al., 1954). The change from DDT to DDE is shown in Figure 9.1.

Table 9.3. The main chemical compounds and their impacts

Compound	Formula	Impacts on Raptors and Waterbirds
DDT	$C_{14}H_9Cl_5$	Toxic to prey animals, reproductive degradation, shell thinning, endocrine problems.[1] Affects egg laying, pregnancy, lactation
DDE	$C_{14}H_8Cl_4$	Toxic to prey animals, shell thinning, embryonic and chick risk.[2] abnormal thyroid function, fertility decrease, hatching problems, immune function decline[2]
Dieldrin/ Aldrin	$C_{12}H_8Cl_6O$	Fatal to raptors, including juveniles and breeding adults, levels correlated to levels in prey animals[3]
PCB	$C_{12}H_{10-x}Cl_x$	Liver, immune, reproductive, system damage, anemia; skin damage conditions (chloracne); liver, stomach, and thyroid gland injuries (hepatocarcinoma thymocyte apoptosis[4]
Diclofenac	$C_{14}H_{11}C_{12}NO_2$	Renal and visceral gout (crystallization of uric acid in tissues), enteritis[5]
PAH e.g. di-methyl-naphthalene	$C_{12}H_{12}$	Toxic to embryos, young and adult birds, impacts including death, developmental abnormalities. Reduced egg production and hatching, increased clutch or brood abandonment, reduced growth. Trophic level accumulation unlikely[6]

Sources: 1. Colborn (1991); NPIC (1999); 2. Newton (1993); 3. Sibly, et al. (2000); 4. Fernie et al. (2001), Crinnion (2011); 5. Campbell (2015); 6. Fingas (2017), Albers (2006).

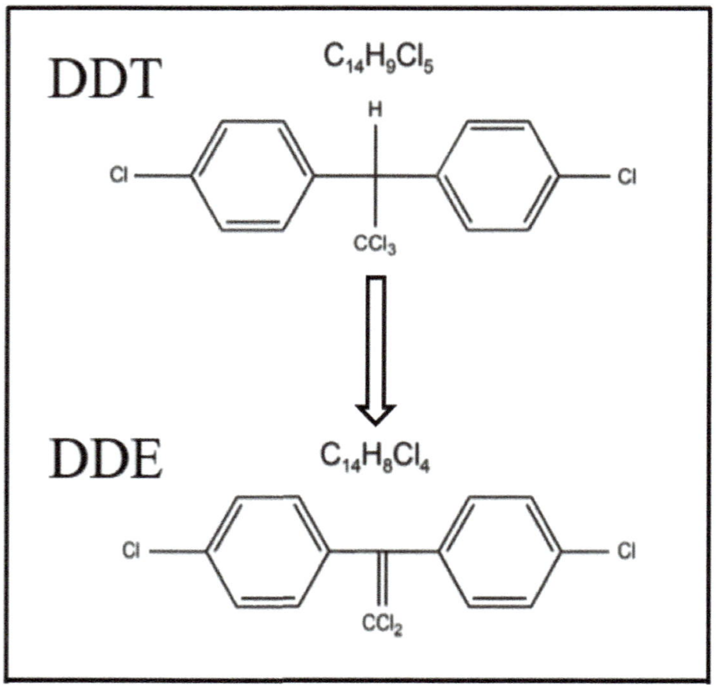

Fig. 9.1. Change from DDT to DDE

The National Pesticide Information Center (NPIC) (1999) describes DDT as "widely used… because it is effective, relatively inexpensive to manufacture, and lasts a long time in the environment." Furthermore, "DDT was canceled because it persists in the environment, accumulates in fatty tissues, and can cause adverse health effects on wildlife. In addition, resistance occurs in some insects (like the house fly) who develop the ability to quickly metabolize the DDT." Dieldrin was introduced as an alternative to DDT. This is an organochloride developed by 1948 by J. Hyman & Co, Shell Chemical, Denver, then AMVAC, Los Angles as an insecticide, related to aldrin (with a similar formula, $C_{12}H_8Cl_6$) which oxidizes after insect ingestion to form dieldrin (Jorgenson, 2001). It has strong environment persistence and biomagnifies as it ascends the food chain, a clear danger for avian predators and a factor for its abolition. PubChem (2020b) defines dieldrin as an organochlorine compound resulting from the epoxidation of the double bond of aldrin. It is the active metabolite of the proinsecticde aldrin. It has a role as a xenobiotic and a carcinogenic agent. It is an organochlorine insecticide, an organochlorine compound and an epoxide. It derives from an aldrin… aldrin and dieldrin were widely used pesticides for crops like corn and cotton. Because of concerns about damage to the environment and potentially to human health, EPA banned all uses of aldrin and dieldrin in 1974, except to control termites. In 1987, EPA banned all uses.

The impacts of DDT on animals included: negative effects on the nervous system, interfering with nerve impulses; hyperexcitability, tremors, incoordination, and convulsions in laboratory-tested animals; fatal doses resulting in liver lesions in the short term and liver changes over the longer term, increased risk of live tumors; long persistence in the environment, the soil half-life being 2 to 15 years and in an aquatic environment about 150 years; and moderately toxic to birds, but highly toxic to aquatic animals including fish and amphibians, affecting the heart, brain and other organs. The impacts on the animals at the bottom of the food chain are particularly important, especially for marine and aquatic animals preyed upon by waterbirds and relevant raptors such as ospreys and Bald eagles (Campbell, 2015).

DDT and DDE may be more environmentally persistent than some other pollutants, but in some cases such assessments cannot be clearly discerned (Campbell, 2015). For example, García-Fernández et al. (2008, 432) found that in Spain, despite many studies, few overall assessments have been published. García-Fernández et al. (2008, 432) conclude that "excepting DDT and DDE, most organochlorine-based pesticides in raptors from Spain have diminished over the last two decades. The concentrations of DDE found in the eggs of various species could in part explain problems in the reproductive success of raptors in Spain." Jürgens et al. (2016, 333) found high levels of DDT in fish in English ecosystems and noted that some fish had DDT levels that could be harmful to them or their predators. They compiled a list, based on relevant literature findings of the threshold levels of DDT for prey animals and associated raptors and waterbirds. These are: "based on egg shell thinning in ducks" 0.014 ppm (reference, Canadian Council of Ministers of the Environment, 1999); "dietary effects concentration" in ospreys, 0.09-0.19 ppm (reference, Hinck et al., 2009); "dietary effects concentration" in Bald eagles, 0.13 ppm (reference, Hinck et al., 2009); and "egg shell thinning in pelicans", 0.02 ppm (reference, Lazorchak et al., 2003).

British raptors and waterbirds were also affected during the peak years of DDT use in the late 19th century. For example, for raptors, Presst et al. (1970) found that ingestion levels for bird-feeding raptors the levels of PCBs in the liver were 70 ppm, which was similar to that of DDE. For fresh-water fish-feeding waterbirds in the same region, the figure was 900 ppm. The authors point out that in their study "most of the 559 specimens came from predatory birds obtained between April 1966 and August 1968. PCB was found in terrestrial species from most regions of Britain, in all the individual and bulked samples of seabird eggs examined from one west coast and two east coast colonies and in most of the freshwater species collected from the midlands, east and south of England." As DDE had similar levels, it appears both chemical compounds had similar dominance of ecosystems in the 1960s-70s period. Testing used gas-liquid chromatography on livers and kidneys (Presst et al., 1970). As such waterbirds are frequently taken by White-tailed eagles after its reintroduction to Britain, there is a real threat to eagles as well (Love, 1983).

In another example, Parslow et al. (1973) examined the impacts of DDT/DDE on gannets in British waters, and concluded that "measurable residues of polychlorinated biphenyls (PCB) and DDE (the chief breakdown product of DDT found in birds) are present in virtually all British seabirds; many contain residues of other organochlorine materials, such as dieldrin and hexachlorobenzene (HCB), usually in much smaller quantities; PCB and HCB derive chiefly from industrial use, the others mainly from agricultural insecticides." They also noted that as larger fish carry more pollutants, birds such as gannets that eat large fish are vulnerable (examples of such fish are pollack *Pollachius pollachius*, *P. virens* and haddock *Melanogrammus aeglefinus*). Toxicity and mortality are partly dependent on the level ingested and the area of deposition; i.e. the subcutaneous fat or the liver. In healthy birds there may be low liver deposits, with higher levels in the fat, but stress may mobilize pollutants when fat is utilized, and the result may affect the liver and endocrine system (Parslow et al., 1973). The study concluded PCBs occurred at higher levels in gannets and other British seabirds than DDE, gannets had higher mean liver levels (e.g. 200 ppm) than other seabirds (e.g. murres or guillemots *Uria aalge*, Pontoppidan, 1763, 116 ppm). In some dead gannets, PCB (also DDE) concentrations in the brain were a fraction of those in the liver. Residues of PCB and DDE also had a significant eggshell thickness (Parslow et al., 1973).

Raptors have also been directly targeted with pesticides and poisons. Generally, carcases are injected with poisons such as Carbofuran, and placed where it is likely that the predators would arrive for feeding. Pesticides may also be used. For example, Vandenbroucke et al. (2010, 266) write that in Belgium "in the last few years, corpses of birds of prey and other predators have frequently been suspected of malicious poisoning." The commonest poison detected was carbofuran, the others being aldicarb, carbaryl, strychnine and chlorophacinone (Vandenbroucke et al., 2010, 266). The chemical formulae of these compounds are displayed in Table 9.4. Another example of the poisoning of eagles was that of the Wedge-tailed eagle 'mass poisoning' in Australia (BBC, 2018). Ogada (2014, 10) describes the deliberate poison of several African eagles, the most targeted including the Martial, Crowned, Tawny and Verreaux's eagles (Brown, 1991a,b; Anderson, 2000; Odino & Ogada, 2008). For non-targeted eagles (e.g. the target of the poisoning may be mammal carnivores), common victims were Tawny and Bateleur eagles (Watson, 1986; Brown, 1991a,b; Mijele, 2009; Ogada & Kibuthu. 2012). Meyburg et al. (2020) also describe the poisoning of Greater-spotted eagles.

In Flanders (northern Belgium), 162 Common buzzards (*Buteo buteo*, Linnaeus, 1758) were examined (2003–2006) and 91 seemed to have been deliberately killed, by poisoning (n=79) or shooting (n=12). Other species were also killed by poison (Common kestrel, *Falco tinnunculus*, Linnaeus, 1758), Eurasian sparrowhawk (*Accipiter nisus,* Linnaeus, 1758), Red kite (*Milvus milvus*, Linnaeus, 1758), Eurasian magpie (*Pica pica*, Linnaeus, 1758), Carrion crow (*Corvus corone*, Linnaeus, 1758) and Common pheasant (*Phasianus colchicus*, Linnaeus, 1758) (Vandenbroucke et al., 2010, 266). Hernández and Margalida (2008) also describe the pesticide poisoning of the Cinereous vulture (*Aegypius monachus*, Linnaeus, 1766) in Spain from 1990 to 2006, involving 241 incidents and 464 vultures, using Carbofuran, aldicarb, and strychnine in 88% of the recorded incidents. They note about 98% were deliberate poisoning of the birds, and that 83% resulted in death (Hernández & Margalida, 2008).

Table 9.4. Pesticides toxic to birds

Compound	Formula	Research assessment for impacts on raptors/birds
Carbofuran	$C_{12}H_{15}NO_3$	Carbofuran is highly toxic to birds[1]
Aldicarb	$C_7H_{14}N_2O_2S$	Indeterminate impacts on birds[2]
Carbaryl	$C_{12}H_{11}NO_2$	Slight toxicity on birds[3]
Strychnine	$C_{21}H_{22}N_2O_2$	Highly toxic, but indeterminate results of poisoning[4]
Chlorophacinone	$C_{23}H_{15}ClO_3$	Serious possibilities for physiology[5]

Sources: 1. National Pesticide Information Center (1996); 2. Moore et al. (2010); 3. Walker et al. (1991); 4. Wobeser & Blakley (1987). Schmutz et al. (1989). 5. Rattner et al. (2015).

Poisoning, both deliberate and accidental, is an important contributory factor for the decline of hooded vultures across sub-Saharan Africa, especially in Cameroon, Uganda and Kenya and Southern Africa (Zimmerman et al., 1996; Mundy, 1997; Thiollay, 2001; Ogada et al., 2010; Ssemmanda & Pomeroy, 2010; Ogada & Buij, 2011; Virani et al., 2011). Vultures are sometimes poisoned in Botswana, as poachers believe that circling vultures will reveal illegal hunting kills (Hancock, 2009, 2010). Vultures may also be poisoned in Namibia, where the poisoning may target problem animals such as lions (*Panthera leo*, Linnaeus, 1758), hyenas (*Crocuta crocuta*, Erxleben, 1777) and jackals (*Canis mesomelas*, Schreber, 1775) (Bridgeford, 2004), or may target vultures (Pain et al., 2003) a situation existing elsewhere (Anderson, 1999; Verdoorn et al., 2004; Roche, 2006). Findings indicate that in places where vultures were not poisoned, their populations did not decline (Mundy et al., 1992). Multiple poisons have been recorded in such events. For example, Van Wyk et al. (2001a: 243) found 'the presence of quantifiable residues of 14 persistent chlorinated hydrocarbon pollutants', after collecting 'whole blood, clotted blood, heart, kidney, liver and muscle samples' from white-backed, Cape and lappet-faced vultures. Comparisons between wild adults and nestlings, and captive birds showed lower pesticide levels for captive birds (pesticides included gamma-BHC (lindane), alpha(cis)-chlordane and alphaendosulfan) (Van Wyk et al., 2001a, 243).

In the Americas, findings indicate that Turkey vultures are less affected by some chemical compounds, including potent rodenticides such as the compound 1080 (Snyder & Snyder, 1991). Serious threats to Turkey vultures include cyanide, strychnine (used to predators of livestock, such as coyotes) (Campbell, 2015) and pesticides such as dichlorodiphenyldichloroethylene (DDE) which induce eggshell thinning (Wilbur, 1978; Kiff et al., 1983). Burnett et al. (2013) also found eggshell thinning for California condors, when comparing a current cases study in central California with findings from more successful cases from southern California, where pesticide incidence was lower. Examining breeding, captive reared condors from 1997 through 2010, eggs or eggshell fragments were recovered from 12 nests. The mean thickness of shell fragments, without membrane, was found to be 0.46 mm, which was 34% lower than the average thickness of 0.70 mm of fragments recovered from nine successful nests in interior southern California, 2007–2009, and hatching success in the central California case was also lower (20–40%) than in southern California (70–80%). Furthermore, on the eggshells the outer crystalline layer was much thinner or absent similar to the southern Californian eggs during the 1960s (when DDE was more prevalent). Weight and water losses during incubation much higher than the normal rate representing successful hatching, and loss rates were correlated with decreasing shell thickness (Burnett et al., 2013).

Diclofenac

The Diclofenac Epidemic, highly published as an example of an avian and ecological catastrophe and human interference with the natural environment, was probably the most important poisoning episode of the twentieth century, an event termed the "global vulture crises" (Woodford et al., 2008; Olea & Mateo Thomas, 2009; Campbell, 2015). This disaster, initially defined as affecting Asian vultures, resulted in the near extinction of six of the 23 recognized vulture species (especially five Indian species), listed as such on the International Union for the Conservation of Nature (IUCN) Red List of Threatened Species. The affected species included the white-rumped vulture (*Gyps bengalensis*, Gmelin, 1788), slender-billed vulture (*Gyps tenuirostris*, Hodgson, 1844), Indian vulture (*Gyps indicus*, Scopoli, 1786) and red-headed vulture (*Sarcogyps calvus*, Scopoli, 1786)) (Houston, 1985b,c; Prakash 1999; Prakash & Rahmani, 1999; Virani et al., 2001; Prakash et al., 2003; Clements et al., 2012; Gilbert et al., 2002; Gilbert et al., 2004). Records indicate that 15 years (1992 to 2007) the white-rumped vulture population declined to one thousandth of the previous population, the red-headed vulture's population to below 10% and that of the Slender-billed vulture to about 1,000 birds (Cuthbert et al., 2006; Prakash et al., 2007). The Egyptian vulture (*Neophron percnopterus*) and red-headed vulture (*Sarcogyps calvus*) were also classified as Endangered and Critically Endangered respectively (Cuthbert et al., 2006; Campbell, 2015).

Symptoms of the Diclofenac poisoning were renal and visceral gout (crystallization of uric acid in the tissues) and enteritis (inflammation of the intestine, especially the small intestine) (Oaks et al., 2001; Gilbert et al., 2002; Pain et al., 2003). The factor for this was the tendency in the mid-1990s for Indian, Pakistani and Nepalese farmers to treat their cattle and water buffaloes with the non-steroidal anti-inflammatory drug (NSAID) diclofenac. This drug was extremely toxic to vultures that fed on the few carcasses at carcass dumps, which was the traditional method of livestock disposal in South Asia (Green et al., 2004; Oaks et al., 2004; Gross, 2006). Vulture deaths followed one to two days after ingestion, after kidney failure—severe renal dysfunction—led to a buildup of urates in the internal organs and increased blood uric acid levels, contributing to anorexia and emaciation and a chalky-white coating over the visceral organs, especially the liver and heart and in the body tissues.

Bans on diclofenac was passed during 2006 to 2010 in India, Nepal, Pakistan and Bangladesh, allowing a slow vulture population recovery in India and later in Pakistan and Nepal. For example, a breeding colony of long-billed vultures in Pakistan was regularly monitored before (2003–2006) and after (2007–2012), and the population was found to have declined (61%) pre-ban, and increased (55%) post-ban (Chaudhry et al., 2012). An alternative to diclofenac was meloxicam, which did not damage vulture kidneys (Swan et al., 2006). Studies concluded that "meloxicam is of low toxicity to Gyps vultures and that its use in place of diclofenac would reduce vulture mortality substantially in the Indian subcontinent. Meloxicam is already available for veterinary use in India" (Swan et al., 2006, 395). Asia was the main focus, rather than Europe and Africa, especially griffon vultures. The Americas were mostly ignored as *Cathartid* vultures were not affected by diclofenac ingestion at levels over 100 times over the dosage that killed Asian vultures, with no toxicity, visceral gout, renal necrosis, or elevate plasma uric acid being recorded (Rattner et al., 2008; Campbell, 2015).

More recent studies have indicated a risk to eagles as well. Sharma et al. (2014, 282) report that the Steppe eagle is susceptible. Although there is little evidence of "toxicity of diclofenac to the Accipitridae other than vultures… In this study, gross and microscopic lesions and diclofenac tissue levels in Steppe eagles (*Aquila nipalensis*) found at a cattle carcass dump in Rajasthan, India, showed evidence of the toxicity of diclofenac for this species. These findings suggest the possibility that diclofenac is toxic to other accipitrid raptors and is therefore a potential threat to much wider range of scavenging species in South Asia." Mackenzie (2014) also reports two dead Steppe eagles found near kidney failure. The author argues that if the Steppe eagle is vulnerable to diclofenac, then perhaps the other *Aquila* eagles may also be threatened.

Polychlorinated Biphenyls (PCBs) and Perfluoroalkyl Substances (PFASs)

Polychlorinated biphenyls (PCBs, molecular formula $C_{12}H_5Cl_5$) are another important environmental pollutant, which may seriously impact on eagle health. PubChem (2020a) notes:

> Polychlorinated biphenyls are mixtures of up to 209 individual chlorinated compounds (known as congeners). There are no known natural sources of PCBs. PCBs are either oily liquids or solids that are colorless to light yellow. Some PCBs can exist as a vapor in air. PCBs have no known smell or taste. Many commercial PCB mixtures are known in the U. S. by the trade name Aroclor. PCBs have been used as coolants and lubricants in transformers, capacitors, and other electrical equipment because they don't burn easily and are good insulators. The manufacture of PCBs was stopped in the U. S. in 1977 because of evidence they build up in the environment and can cause harmful health effects. Products made before 1977 that may contain PCBs include old fluorescent lighting fixtures and electrical devices containing PCB capacitors, and old microscope and hydraulic oils.

High PCB levels have been reported in eagles. Revkin (1997) reported that a young Bald eagle was found dead on the upper Hudson River with high concentrations of PCBs. Revkin states that this "finding, although limited to one eagle, is significant, the scientists said, because similar levels of PCBs in eagles or eagle eggs from polluted areas of the Great Lakes have been linked to reproductive problems and deformities in the birds." Revkin further notes that the recorded bird contained 71 parts per million of PCBs, which was higher than the average recorded in three deformed fledgling eagles from the Great Lakes area. This may also be significant because the Bald eagle appears to be returning to nest in the Hudson River area, after almost 100 years of rarity. PCB concentrations accumulate mostly in the brain and fat of the eagle. These concentrations start in fish and aquatic insects up the food chain and eventually reach the Bald eagles. As the recorded Bald eagle was young, the incident indicated a high accumulation in a relatively short period. PCBs have been banned for their effects on wildlife and people, but "the Hudson River is one of the worst PCB pollution sites in the world, with more than a million pounds of the chemicals released from two General Electric factories during a 30-year period ended in 1977" (Revkin, 1997).

Non-eagle avian species, which may be prey for eagles are also recorded with PCBs in their bodies. Zhang et al. (2011, 556) used a Chinese case study, on the impacts of polychlorinated biphenyls and organochlorinated pesticides on mostly waterbird species (*Rallidae*, *Ardeidae*) and doves which could be prey species (*Columbidae*) from a contaminated region in South China. This approach examined birds as indicators of the levels of organic pollutants, arguing that "there is exceedingly little data on organochlorinated pesticide (OCPs) and polychlorinated biphenyls (PCBs) residues in bird inhabiting in China." The findings described concentrations of PCBs, dichlorodiphenyltrichloroethane and its metabolites (DDTs) and hexachlorocyclohexanes (HCHs) in the muscle tissues, liver and kidney of birds recorded near an e-waste recycling site. The pectoral muscle, liver and kidney tissues were Soxhlet-extracted and dissolved in isooctane and the PCBs and OCPs were analyzed using gas chromatography-mass spectrometry. The subsamples for nitrogen stable isotope analysis were then lyophilized (i.e. preserved by removal of water, by freezing and then drying, in a vacuum, at low temperatures, allowing longer term storage) and then analyzed using a flash EA 112 series elemental analyzer (determines the weight percent composition of the studied elements) and an isotope ration mass spectrometer (used to determine the origins of substances, using the relative isotopic abundances of the elements which comprise the material (Zhang et al., 2011, 556). The findings indicated PCBs had higher concentrations than other compounds, with possible negative impacts on the target birds. Livers and kidneys had higher concentrations than muscle. Crucially, contaminant levels varied according to bird dietary composition and habitat, and PCB and DDT levels increased up the trophic levels. HCHs (hexachlorocyclohexane, formally benzene hexachloride, BHC) were level at the interspecies level (Zhang et al., 2011, 556).

In another study, Barghi et al. (2018, 220) contribute to the perceived gap in knowledge "on the accumulation and potential risks of persistent chlorinated and fluorinated contaminants in birds." Organohalogen contaminants (OHCs), including PCBs, organochlorine pesticides (OCPs), and perfluoroalkyl substances (PFASs), were measured, using the liver tissues of 69 Korean birds, members of 10 species. Christensen et al. (2019, 147) define perfluoroalkyl substances as "a class of contaminants used in many industrial applications and consumer products. Certain PFASs are regulated or voluntarily limited due to concern about environmental persistence and adverse health effects." As noted above, Jayaraj et al. (2016, 90) define Organochlorine pesticides (OCPs) as "synthetic pesticides" belonging "to the group of chlorinated hydrocarbon derivatives, which have vast application in the chemical industry and in agriculture…known for their high toxicity, slow degradation and bioaccumulation."

The findings were that PFASs had the highest concentrations, ranging from 0.0054 to 11.3 ppm (median: 0.294) wet weight. Median concentrations of OCPs and PCBs in the samples were 0.147 and 0.82 ppm lipid weight, respectively (Barghi et al., 2018). The concentrations of OHCs were higher in predatory than non-predatory birds, hence diet (on smaller, infected animals) and trophic position (the top of the food chain, allowing accumulation) were key issues. Further findings indicated concentrations of PCBs and DDTs were below toxic threshold levels, "PFOS concentrations in most of the predatory birds exceeded threshold values for adverse health effects…, implying potential health risks to Korean birds" (Barghi et al., 2018, 220). PFOs (perfluorooctane sulfonate) are a form of PFAS (National Institute of Environmental Health Sciences, 2020).

Another study by Luzardo et al. (2014) also looked at a larger sample with slightly different pollutants. These included 57 anthropogenic pollutants: 23 organochlorine pesticides (OCs), 18 polychlorinated biphenyls (PCBs) and 16 polycyclic aromatic hydrocarbons (PAHs), using liver samples from 102 specimens of six raptor species from Wildlife Recovery Centers of the Canary Islands from 2009 and 2012. The objectives were to measure contaminant levels and evaluate the six species as regional bioindicators. The findings indicated that the Eurasian sparrowhawk (*Accipiter nisus*, Linnaeus, 1758), Barbary falcon (*Falco peregrinus pelegrinoides*, Temminck, 1829) and kestrel (*Falco tinnunculus*, Linnaeus, 1758) were more contaminated than the other species, mostly by DDT and dieldrin and that these pollutants were comparatively high.

By contrast, the contamination by PCBs was comparatively low, possibly below threshold levels of toxicity for these contaminants. Polycyclic aromatic hydrocarbons (PAHs) were dependent on the feeding patterns but below threshold levels. The authors argue that their study "represents the first report of contamination by PAHs in all these species and is also the first report of PCB levels in barbary falcons" (Luzardo et al., 2014, 146). Polycyclic aromatic hydrocarbons (PAHs) are chemicals that occur naturally in coal, crude oil, and gasoline, also when these substances and others such as wood, tobacco and garbage are burned, forming small air particles (Environmental Protection Agency, 2013). Examples are petrogenic alkyl polycyclic aromatic hydrocarbons, which are those derived from petroleum and pyrogenic polycyclic aromatic hydrocarbons, which are derived from combustion and anthropogenic sources (Paruk et al., 2014).

9.4 Conclusions

This review of environmental toxicity in eagle and associated species ecology illustrates the impacts and dangers of environmental chemistry and toxicology for birds near the apex of the food web. The key issue emerging concerns the breadth of the sources of toxicity, and the requirement of the examination of many species, that may not readily be considered to be associated with eagles. Crucially, toxic elements and compounds may have wide source networks, some originating with or enabled by human action. Currently, there is insufficient knowledge of the status of toxicity from both elements and compounds. There are substantial variations at the inter- and intra-species levels in terms of the impact of such chemicals on birds, with pronounced differences in the chemical concentrations in prey animals and carcasses. Impacts of toxic chemicals vary and change through time and perhaps space, and this may vary depending on the sources and the species affected. Eagles may also vary their target prey, which may complicate the suitable ameliorative actions.

There may be inter- and intra-species variations in the reaction and resistance to toxic chemicals. There is a requirement for more research on these topics, from a spatial, case study approach. Some consequences were severe enough to require the attempted restriction or abolition of chemical compounds, such as lead, diclofenac, and DDT/DDE. Other elements, such as mercury may require similar restrictions. Other elements, such as arsenic or copper may require more research before continued use, restriction, or abolition. Some chemicals are useful at low concentrations, and only toxic at high concentrations. Determination of toxic levels requires more research at inter- and intraspecies levels, acknowledging that sublethal and lethal definitions may vary by the organs in which they accumulate, such as embryos, livers, kidney and the gastrointestinal tracts, and the toxic element or compound. The study of the impacts of toxic chemicals is an ongoing process, as understandings and biological/environmental parameters and assessments may change, as case studies and knowledge spread, chemicals may be outlawed and new chemical compounds enter the environment with industrial developments. Future scenarios may encourage the development of systems for detoxification of ingested chemicals, the continual and critical replacement of toxic chemicals with less dangerous substances, and more advanced detection methods. Industrial and agricultural developments may require more sensitive environmental impact assessments. This chapter used a relatively narrow case study base of eagle and non-eagle species within the complex and generally understudied reality of environmental change. Future research may focus on a wider range of species including prey and parallel bird species, mammals and research tools.

10

Eagle Ecology and Geomatics-Based Research

10.1 Introduction

Eagle ecology and movements, which may include nesting movements, breeding displays, foraging, dispersal and migration, may be recorded by basic techniques such as telescopes, binoculars, or even simple eye observation in point count surveys, which are techniques used by much older research. Recently, more sophisticated methods have been applied, starting with largely coarse resolution satellite images for the quantification of landcover character and change, and habitat distribution dispersal. Later techniques include the use of sharper resolution images and global positioning systems (GPS) technology for recording eagle behavior and movements and geographical information system (GIS) based tabulation and analyses of data. The umbrella terms geomatics (usually used in North America) and geoinformatics (used mostly in Europe) may be used to refer to these technologies, including most of the fields that deal with spatial data acquisition, assessment and analysis, the technologies used for these methods, the related natural and social science methodologies and theoretical frameworks (Decker et al., 2012; Goodfellow, 2014; Morzillo et al., 2014; Jackson & Fahrig, 2015). Geomatics is named from the French word Géomatique, coined by the French photogrammetrist Dubuisson) (Dubuisson, 1975; Gagnon & Coleman, 1990; Ehlers, 2008, 20). The field is very broad and is therefore well suited to the study of holistic concepts such as the ecosystem (Dempsey, 2011; Wedding et al., 2011; Goodfellow, 2014). This chapter looks at the development of remote sensing (RS), geographical information systems or science (GIS) and GPS, and how these have been applied to the study of eagles.

In the application of geomatics techniques to ecological studies, there are some basic structures or themes that underpin many studies of both animal behavior and movement tracking, and spatial analysis of habitats. Key concepts are space, pattern, place, integrated analysis and analytical mapping. Space may be defined as any area defined by the presence of the features or entities of interest are located (Campbell, 2017). This may exist at any scale, from continents to tree canopies, from mountain ranges to isolated lakes or even cliff ledges. In the context of eagles, with their extreme wide ranges, migratory patterns and foraging behavior, such spaces would be very large, but could also include spaces around nests, contested territories or carrion. Both eagle movements and habitats are included in such spaces, hence space is probably the most important theoretical and methodological entity for such studies. Pattern is the concept that describes how the entities of interest and related features are distributed in space, and the possibility of the distribution having a relationship among the components (Jensen, 2007). Geomatics techniques enable the detection, recording, analysis and prediction of patterns. For eagles, such patterns include nesting, mating, foraging, migratory and other movements, and how these link to patterns such as habitat characteristics, prey species distributions and more indirect ecological features such as climate and topography.

Place is a localized area, which describes certain unique characteristics distinguishing it from its surroundings (Heywood et al., 2006; Campbell, 2017). Places from the perspective of avian ecology could refer to areas where vegetation, wildlife and topography create an environment that may attract or repel some species, such as dense forest with large branches and arboreal mammal populations that may attract forest species such as the Harpy or Crowned eagles. Such places may be recorded and analyzed using images of varying resolution for landcover assessment, and GPS or tabulated statistical data to record species presence. Integrated analysis of such geographical and ecological entities allows assessments and predictions of their habitat suitability for wild animals. Such integrated analysis may utilize different geomatics-based data sources, including images with visible and non-visible (to the human eye) light bands, point counts from field surveys of animals, GPS-based measurements and measurable parameters of vegetation and climate. These may be analyzed in tabular or image form using GIS software, creating maps that document the full ecological relations of the animals under study. For eagles, such analysis is particularly useful for habitat suitability models, population trend analysis, predictions of eagle presence and conservation studies. This leads to the last concept, analytical mapping which maybe the quantitative basis of studies of conservation biology, lending authoritative support, scientific quantification, and predictive abilities to the models created by these methods. These may inform the conservation planning and policy and determine with more definitive analytical techniques the success or otherwise of the conservation strategies for eagle population sustainability. The rest of this chapter looks at the definitions and branches of geomatics, the possible applications to eagle ecology and conservation, and the current state of geomatics applications to research on eagles.

10.2 Definitions and Applications of Geomatics

Geotechnology, which includes the hardware and software fields related to geomatics, is described as one of the three "mega-technologies" of the 21st century (Gewin, 2004), "along with nanotechnology and biotechnology" (Boyd & Foody, 2011, 25). Boyd and Foody (2011, 25) further argue that "in the cusp of the second decade of the 21st century it is evident that recent developments in the geotechnologies of Geographic Information Science (GIS) and remote sensing have had a substantial impact on ecological research, providing spatial data and associated information to enable the further understanding of ecological systems" (see also Rundell et al., 2009).

The definitions of geomatics and those of the subfields such as remote sensing, GIS and GPS provide a starting point for the possible applications to the study of eagles and associated physiological attributes and ecological parameters (prey species, competitors, feeding behavior, foraging ranges, nesting sites, migration routes and mating behavior). Gomarasca (2010, 137) argues that geomatics as a term was initially defined at Laval University in Canada during the 1980s, based on the development of electronic computing and computer-based software applied to the geographical and related spatial sciences. From these beginnings, geomatics was "erected on the scientific framework of geodesy" ...using "terrestrial, marine, airborne, and satellite-based sensors to acquire spatial and other data" (Gomarasca, 2010, 137). As it advanced, geomatics incorporated several technological developments. These include the regulation of GeoSpatial Information or GeoInformation and Earth Observation data, international Spatial Data Infrastructures (SDIs), which may be the basis of a Global Spatial Data Infrastructure (GSDI), which may harmonize all the varying formats of SDIs at the regional, national, and continental levels. These harmonizing developments include the Global Earth Observation System of Systems (GEOSS) (February 16, 2005), and the Infrastructure for Spatial Information in the European Community (May 15, 2007) (Gomarasca, 2010).

Definitions of geomatics encompass its broad set of attributes and applications, and hence are generally long winded. Five such definitions are particularly instructive. One definition is that it is "a systemic, multidisciplinary, integrated approach to selecting the instruments and the appropriate techniques for collecting, storing, integrating, modeling, analyzing, retrieving at will, transforming, displaying, and distributing spatially georeferenced data from different sources with well-defined accuracy characteristics and continuity in a digital format" (Gomarasca, 2010, 137). The Canadian Council of Technicians and Technologist's (2008) definition is that geomatics is a "field of activities which, using a systematic approach, integrates all the means used to acquire and manage spatial data required as part of scientific, administrative, legal and technical operations involved in the process of the production and management of spatial information." This definition stresses the inclusion of all possible data acquisition methods; hence this means a total approach to gather all relevant data, the better to derive information for a conclusive result. Another definition by Natural Resources Canada (2018) stresses the role of data integration, another important characteristic if geomatics is applied to a broad study such as that of ecosystem appraisal and evaluation: geomatics is "the modern discipline which integrates the tasks of gathering, storing, processing, modeling, analyzing, and delivering spatially referenced or location information... surveying, hydrography, mapping, remote sensing (often called earth observation) and geographic information processing, often called geographic information systems (or GIS)."

The definition by the Journal of Remote Sensing and GIS (2018) points to the alternative names, applications and development of geomatics; geomatics may be termed as "geospatial technology or geomatics engineering, or geomatic engineering" and is "...the discipline of gathering, storing, processing, and delivering geographic information, or spatially referenced information. It consists of products, services and tools involved in the collection, integration and management of geographic data. The rapid progress and increased visibility of geomatics since the 1990s has been made possible by advances in computer hardware, computer science, and software engineering, as well as by airborne and space observation remote-sensing technologies" (Journal of Remote Sensing and GIS, 2018).

The British Columbia Institute of Technology (2018) also gives a definition that highlights the application and usefulness of geomatics: it is "the art and science of measuring, storing, managing and presenting of geospatial data. It has a rich history and an exciting future, where rapidly changing technology is applied directly to the sustainable management of the world's resources. Geomatics consists of a wide range of topics, including GPS technology, field surveying, cartography, geodesy, land use planning, photogrammetry, hydrography and cadastral studies." The last definition also points to the importance of both the technological components of geomatics and the perspective that dominate its application. It is both an art and science, based on human appraisal, judgement, presentation and understanding (hence an art), and scientific quantification, logic and rigor (hence a science). It also describes its numerous supporting subdisciplines: computer hardware, computer science, software engineering, space observation, GPS (global positioning systems) technology, field surveying, cartography, geodesy, land use planning, photogrammetry, hydrography and cadastral studies. Finally, it notes that geomatics is developing, and this advancement is relevant to changing world dynamics.

Geomatics is a term mostly used in North America. In Europe, especially Germany, the virtually synonymous term geoinformatics is frequently used. For example, Ehlers (2008, 17) defines geoinformatics "as a new scientific discipline designed for handling of geospatial information" .... not a part of Geography, Surveying, or Computer Science, but a new self-contained scientific discipline...and including "the acquisition and storing of geospatial data, the modelling and presentation of spatial information, geoscientific analyses and spatial planning, and the development of algorithms and geospatial database systems." This is like the definition of geomatics, as the difference is stylistic, although Ehlers (2008, 20) notes that "Geoinformatics... emphasizes a formal scientific approach to handle geoinformation which Geomatics does not imply."

Geomatics currently includes several disciplines and subdisciplines. In addition to remote sensing, global positioning systems (GPS) and geographic information systems or science (GIS), it also includes surveying, cartography, geodesy, hydrography and ocean mapping, photogrammetry, automated mapping/facilities management (AM/FM) and computer and communications science (Samuelsen, 2008). Gomarasca (2010) and some other authors (Heywood et al., 2006; Jensen, 2007), list up to 17 subjects within the ambit of geomatics. For this chapter, the most important disciplines are those that can record landcover, bird presence, behavior and movements, and analyze this data derived from images or field numerical records, using spatial and statistics software.

Developments in computer science are probably the most important issues for the advancement of geomatics. Current geomatics-based applications in the environmental sciences require massive databases and computing power, which would be impossible with older analytical methods. These developments are also linked to applications from the science of geodesy, which uses fundamental concepts related to geoid or the surface of reference of the Earth, the simplified ellipsoid, for spatial studies and reference points (Gomarasca, 2010). The study of topographical shape, point positions, point elevation above the geoid, and survey measurements are studied by the disciplines of topography, planimetry, altimetry, tachymetry and land surveying (Jensen, 2007). Cartography or map-making, and photogrammetry (the science of photograph analysis) are also important.

From the perspective of large animal biology and ecology, the most important disciplines within geomatics are RS, GIS and GPS. These three disciplines derive data on relevant objects (plants, animals, landscape, water, temperature and the associations and locations of these objects, subjects or components) which are relevant to the ecology of all species. These environmental features, subjects, objects and parameters may be recorded as points, lines, polygon, georeferenced features and images, tabular data or images of varying resolutions (Jensen, 2007; Campbell, 2017). For large, mobile species such as eagles the main measurable parameters are feeding, breeding and foraging behavior, migratory movements, similar parameters for prey and competing species, and habitat assessments linked to landcover change. From the perspective of geomatics, the most important issues concern resolution for image capture of landcover and bird presence (level of detail in images and data, through remote sensing), tracking detail (e.g. capability to measure individual bird movements, behavior and physiology, through GPS and remote sensing) and synthesis of ecological and physical data (e.g. statistical analysis of tabular data, representing ecological and other data, through GIS and geostatistics).

10.3 Remote Sensing Techniques

Remote sensing, termed as an earth-observing (EO) technique, refers to data acquisition representing a subject or object, without physical contact with that entity (Jensen, 2007). Jensen (2007, 4) promotes this simple definition, "Remote sensing is the noncontact recording of information from the ultraviolet, visible, infrared and microwave regions of the electromagnetic spectrum by means of instruments such as cameras, scanners, lasers, linear arrays, and/or area arrays located on platforms such as aircraft or spacecraft, and the analysis of acquired information by means of visual and digital image processing." This includes "virtually all remote sensing devices, including cameras, optical-mechanical scanners, linear and area arrays, lasers, radar systems, sonar, seismographs, gravimeters, magnetometers, and scintillation counters" (Jensen, 2007, 4). In the application of remote sensing tools to earth surface evaluation, the technology must work with human assessment, and in some cases subjectivity (Campbell, 2018; Bhatta, 2008).

A sensor may be either 'active' or 'passive.' An active sensor is "a radar instrument used for measuring signals transmitted by the sensor that were reflected, refracted or scattered by the Earth's surface or its atmosphere" (National Aeronautics and Space Administration, NASA, 2012). A passive sensor is "a microwave instrument designed to receive and to measure natural emissions produced by constituents of the Earth's surface and its atmosphere" (National Aeronautics and Space Administration, NASA, 2012). Therefore, the sensor may transmit its own emissions (active) or use those produced by the Earth (passive). As argued by McDermid et al. (2010) the active sensors have a wider range of operation than passive sensors as they operate without the requirement of solar illumination (the light and related energies from the sun) (Hamelin et al., 2014). They are more effective where there are obstructions such as cloud cover (Jensen, 2007). Synthetic aperture radar (SAR), which utilizes microwave wavelengths, is an example of an active sensor.

Sensors and scanned objects are the main components of remote sensing, the main connection between the two usually being the electromagnetic radiation (EMR) that is transmitted from the sensor to the target object (Campbell, 2018; Bhatta, 2008). Electromagnetic radiation is a stream of photons moving in a wave pattern at the speed of light. This is part of the electromagnetic spectrum, which includes visible light, infrared, ultra violet, radio, micro, gamma and X-rays (Livingston, 2015). The longest waves are radio waves (length <1 cm to 100 m +), then microwaves (1 mm to 30 cm), followed by infrared (from the visible region to about 1 mm), and the shorter visible bands, 400 and 700 billionths of a meter (400 to 700 nanometers, nm). The shortest are ultraviolet radiation (length 400 -10 nm), X-rays (10 nm – 10 picometers or 10 trillionths of a meter) and gamma rays (< 10 pm) (Jensen, 2007; National Aeronautical and Space Administration, NASA 2018a). The wavelengths are depicted below in tabular form (Table 10.1).

These remote sensing techniques are used by many disciplines, technologies and methodologies, such as field and aerial photography, photogrammetry, aspects of land surveying, laser-scanning systems (LS) including LIDAR (named from the combination of light and radar, and light detection and ranging, which measures the distance to a target by illuminating a target with rapid pulses of laser light and measuring the reflected pulses with a sensor) (Jensen, 2007; Gomarasca, 2010; Campbell, 2017a), and radio detection and ranging (RADAR) or interferometric synthetic aperture radar (IFSAR), which create advanced images and models such as digital elevation models (Jensen, 2007).

Table 10.1. Wavelengths of the electromagnetic spectrum

Name	Wavelength (m)
Radio	> 1 × 10-1
Microwave	1×10^{-3} - 1×10^{-1}
Infrared	7×10^{-7} - 1×10^{-3}
Optical	4×10^{-7} - 7×10^{-7}
Ultraviolet UV	1×10^{-8} - 4×10^{-7}
X-ray	1×10^{-11} - 1×10^{-8}
Gamma-ray	$< 1 \times 10^{-11}$

Source: Derived from Jensen (2007), NASA (2018) and Campbell (2018).

During the Modern Age (1946 +), satellite technology became the most important development in remote sensing (Falkner & Morgan, 2002; Bhatta, 2008; Campbell, 2017). Satellite images imparted the advantage of far greater aerial coverage, being taken from much higher altitudes, while being limited by coarser resolution; one pixel or picture element in the image could represent anything from 5 to 30 meters on the ground (Campbell, 2002). The Landsat program began on July 23, 1972, which included the launch of several satellites, starting with the Earth Resources Technology Satellite up to the Landsat 8, launched on February 11, 2013 (National Aeronautics and Space Administration NASA, 2018). After Landsat, there were more advanced satellites, such as the IKONOS series (1999 – 2015), Nimbus (seven satellites, launched August 28, 1964 to October 24, 1978, out of service 1994), RADARSAT (RADARSAT-1 1995–2013, RADARSAT-2 – 2007, by the Canadian Space Agency, CSA), Indian Space Research Organization IRS-1C (December 28, 1995 - 21 September, 2007), Upper Atmosphere Research Satellite (UARS) (15 September 1991 - 14 December 2005), EarlyBird-1 for Earth Watch Inc. (December 24, 1997), SPOT 1 – SPOT 7 and Earth Remote Observation System A (EROS-A) (Ramsay et al., 1998; Flett et al., 2001; Lindsey, 2005; Campbell, 2018; National Aeronautics and Space Administration NASA, 2005, 2018b). Tables 10.2 and 10.3 show the main satellites launched and the resolution of the images that these satellites produced.

There are also technological advancements in image enhancement techniques; examples were developed by research groups within and outside Silicon Valley, including NASA Ames Research Center and Electromagnetic Systems Laboratory Inc. (Ormsby, 1987; Showstack, 2004). IKONOS, which began as the Commercial Remote Sensing System (CRSS) satellite and operated by DigitalGlobe, Lockheed Corporation was the first commercial Earth observation satellite to produce high-resolution imagery (1 meter for panchromatic images, 4 meters for multispectral images) (Livingston, 2015). Panchromatic images present scenes like the visible appearance of the area captured. Some satellites, such as IKONOS, QuickBird and Cartosat, produce digital panchromatic imagery of a higher (spatial) resolution than multispectral imagery. For example, QuickBird (2001 – 2015) panchromatic image pixels are equivalent to an area 0.6 m × 0.6 m, but multispectral pixels are equivalent to 2.4 m × 2.4 m (Livingston, 2015). Multispectral images contain bands covering a wider range of the electromagnetic spectrum, i.e., the ultraviolet, visible and infrared bands. Such multispectral images may have up to ten bands. Landsat 5, for example, produces 7 band images, between wavelengths of 450 and 1250 nm (Ramsay et al., 1998; Flett et al., 2001; Livingston, 2015, 16). Examples of pixels and multispectral bands are shown in Figure 10.1.

Table 10.2. Recent satellite launched by the United States

Name	Nation	Active dates	Spatial resolution (m)
EarlyBird-1	USA	Dec. 24, 1997 - 28 Dec. 1997	3 m PAN, 15 m MS
Ikonos	USA	Sept. 24, 1999 - 31 Mar. 2015	82 cm PAN, 3.2 m MS
QuickBird	USA	Oct.18, 2001 - Dec. 17, 2014	60 cm PAN, 2.4 m MS
OrbView-3	USA	Jun. 2003 - March 13, 2011	1 m PAN, 4 m MS
WorldView-1	USA	Sept. 18, 2007 - 2018	0.5 m PAN
GeoEye-1	USA	Sep. 6, 2008 - 2018	0.41-46 m PAN, 1.65 m MS
WorldView-2	USA	Oct. 8, 2009 - 2018	46 cm PAN, 1.84 m MS
WorldView-3	USA	Aug. 13, 2014 - 2018	31 cm PAN, 1.24 m MS, 3.7 m SWIR, 30 m CAVIS

Sources: Livingston (2015, 15-16), Indian Space Research Organization (2018); NASA (2018d). NB: PAN refers to panchromatic, MS refers to multispectral and SAR to synthetic aperture radar. CAVIS refers to Clouds, Aerosols, Vapors, Ice and Snow. SW refers to short wave, IR to infrared, NIR to near infrared and SWIR to shortwave infrared imagery.

Table 10.3. Recent satellite launched by nations other than the United States

Name	Nation	Dates (Launch to retirement)	Spatial resolution (m)
RADARSAT-1	Canada	Nov. 4, 1995 - Mar. 28, 2013	10-100 SAR
IRS-1C	India	Dec. 28, 1995 - Sept. 21, 2007	6 m PAN
EROS-A	Israel	Dec. 5, 2000 - May 2016	1.8 meter
Spot 5	France	May 4, 2002 - Mar. 31, 2015	5 m PAN 10 m MS
Carosat-1	India	May 5, 2005 - 2018	2.5 m PAN
EROS-B	Israel	Apr. 25, 2006 - 2018	70 cm
Cartosat-2	India	Jan. 10, 2007 - 2018	<1 m
Cartosat-2	India	Jan. 10, 2007 - 2018	<1 m
Cosmo SkyMed-1	Italy	Jun. 8, 2007 - 2018	SAR
TerraSar-X	Germany	Jun. 15, 2007	Variable, to 1 m SAR
SAR-Lupe 1 -5	Germany	Dec.19, 2006 - Jul. 22, 2008 +	<1 m, SAR
Cosmo SkyMed-2	Italy	Dec. 9, 2007 - 2018	SAR
RADARSAT-2	Canada	Dec. 14, 2007 - 2018	3-100, <1 m SAR
Cartosat-2A	India	April 28, 2008	<1 m PAN
Cosmo SkyMed-3	Italy	Oct.25, 2008 - 2018	SAR
TanDEM-X	Germany	Jun. 21, 2010 - 2018	2 – 4 m
Cartosat-2B	India	Jul. 12, 2010 - 2018	<1 m PAN
Cosmo SkyMed-4	Italy	Nov. 6, 2010 - 2018	SAR
NigComSat-1R	Nigeria	Dec. 19, 2011 - 2018	2.5 m PAN, 5 m MS
Spot 6	France	Sept. 9, 2012 - 2018	PAN – 1.5 m, MS – 6 m, (BGR, NIR)
Spot 7	France	Jun. 30, 2014 - 2018	PAN – 1.5 m, MS – 6 m (BGR, NIR)
Pleiades-1A	France	Dec. 16, 2011 - 2018	PAN – 0.5 m, MS – 2 m
Pleiades-1B	France	Dec. 2, 2012 - 2018	PAN – 0.5 m, MS – 2 m

Sources: Livingston (2015, 15-16), Indian Space Research Organization (2018); NASA (2018d). NB: PAN refers to panchromatic, MS refers to multispectral and SAR to synthetic aperture radar. SW refers to short wave, IR to infrared and SWIR to shortwave infrared imagery. BGR refers to blue, green and red. Note for the Spot satellites, Spot 1 – 5 were launched 1986 – 2002, and deactivated 1990 – 2015. See also Campbell (2018).

LIDAR is a recent, high resolution technique for landcover assessment (National Ocean Service, 2018). It uses light pulses that generate three-dimensional data, recording Earth's surface features. LIDAR components usually include a laser, a scanner, and a global positioning system (GPS) receiver, based on an aircraft platform. The resolutions can be as high as 30-centimeters. LIDAR products are used for applications in many disciplines, including geography and forestry (Jensen, 2007). Tree heights, crown widths and diameter, and canopy volumes can be recorded (Wulder et al., 2008, 2012). LIDAR can record three-dimensional information for the multiple resource inventory, using measurements of the three-dimensional (3D) structure of, terrain, vegetation, and infrastructure. LIDAR and interferometric synthetic aperture radar (IFSAR or InSAR) can distinguish between vegetation and terrain at the 3D level (Næsett et al., 2004). LIDAR technology includes canopy height models (CHM) and canopy cover maps (Reutebuch et al., 2005, 291).

Figure 10.1 illustrates the pixel concept for a small area of landcover. The figure also shows small and large pixels representing a forested area, giving different resolutions and detail.

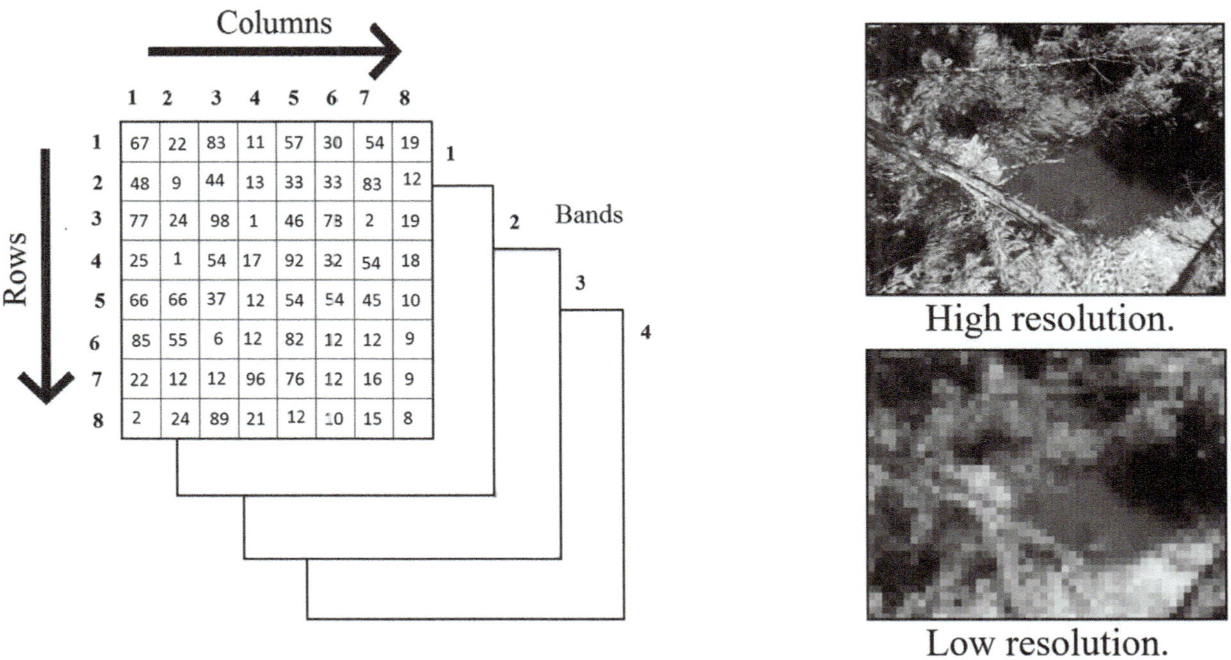

Fig. 10.1. Pixels in a multispectral image. The numbers represent digital numbers of the pixels, which illustrates the intensity of the feature the pixel represents. Higher numbers represent higher intensity for brightness. This example represents a five-band image. A multispectral IKONOS image has four bands: Blue, Green, Red and Near Infrared. A Landsat TM multispectral image has seven bands: blue, green, red, near-IR bands, two SWIR bands, and a thermal IR band.

10.4 Global Positioning Systems (GPS)

GPS is a form of remote sensing that employs technology to record locations, movements and behaviors of people or animals, which may complete other remotely sensed data such as images (Campbell, 2017). GPS systems may vary in accuracy and resolution, with the resolution usually varying between 100 meters (m) and 4 kilometers (km) (Campbell, 2018; Farve, 2018). Some sophisticated GPS technology systems may incorporate biotelemetry, which is defined as the "remote detection and measurement of a human or animal function, activity, or condition" (Merriam-Webster, 2018), "the instrumental technique for gaining and transmitting information from a living organism and its environment to a remote observer" (Slater, 1965) or the "remote measurement of the physiology, behavior and energetic status of free-living animals is made possible by a variety of techniques that we refer to collectively as 'biotelemetry'" (Cooke et al., 2004: 334). This includes the transmitters capable of sending signals over varying distances and to orbiting satellites and data devices. GPS may also include bio-logging, defined as the "on-animal deployment of miniaturised electronic data recorders' allowing "for the study of location, body position, and physiology of individuals throughout their ontogeny" (ibid.).

Technology includes "VHF (very high frequency) radio-tracking, GPS tracking, satellite tracking using the Argos system and a combination of the advantages of GPS and the Argos satellites into a GPS/satellite telemetry method" (Farve, 2018). Telemetry includes hand-held antenna to receive VHF transmissions from a VHF transmitter, which may be attached to the body of the observed animal. Triangulation of the transmissions from receivers and connections with GPS satellites orbiting the Earth determines the location of the subject under study. The satellites are the Argos Satellite system operated jointly by the United

States (National Oceanic and Atmospheric Administration, NOAA and National Aeronautics and Space Administration, NASA) and the France (French Space Agency) and data processing centers in Landover, Maryland and Toulouse, France (Tomkiewicz et al., 2010; Farve, 2018).

The GPS complements older research methods, bird ringing or banding, and field observations and even art (Campbell, 2017). Satellite images and aerial photographs may also require GPS derived information, as the derived images may be too coarse or infrequent to record the vagaries of avian foraging, feeding, migration, nesting and territorial conflicts. These applications are partly derived from the technological advancements during the First and Second World Wars, applied to the tracking and monitoring of military vehicles, aircraft and troop movements (Jensen, 2007). Early applications to bird studies include the combination of GPS technology with radar for the study of migrating birds during the 1940s (Russell & Gauthreaux, 1998). Acoustic telemetry using sonar technology (sonar is an acronym for sound navigation and ranging) also a product of the 1950s, for military submarine detection, and later for the monitoring of marine animals and avian predators (Cronkite et al., 2007; Hastie, 2012).

Tracking technology including GPS is currently a key method for the study of animal movements, including those related to feeding, energy expenditure, predatory behaviors, presence and absence and large-scale foraging and migration (Campbell, 2018). Neuman et al. (2015, 8) point out that "animal movement patterns in space and time are a central aspect of animal ecology... Remotely-sensed environmental indices can play a key role in understanding movement patterns by providing contiguous, relatively fine-scale data that link animal movements to their environment." These authors selected the use of remote sensing techniques for recording animal movements and reviewed the progress in this field over the period 2002 and 2013. They conclude that "greater documentation of the type and resolution of remotely-sensed products in ecological movement studies would enhance their usefulness" and "animal movement studies incorporate remotely-sensed products that provide time series of environmental response variables. This would facilitate wildlife management and conservation efforts, as well as the predictive ability of movement analyses."

Important progressions include movement path analyses (such as state-space models or step selection functions), which can track animal movements linked to resource selection over time (Patterson et al., 2010; Thurfjell et al., 2014; Neuman et al., 2015; Campbell, 2017). Such very detailed, fine scale data covering short term movements enable insights on some less understood and documented patterns, such as migratory corridors and stop-over sites (Takekawa et al., 2010; Morales et al., 2010; Buler & Moore, 2011; Börger & Fryxell, 2012; Cushman et al., 2011; Zeller et al., 2012). More advanced possibilities exist, as Neuman, et al. (2015, 8) note that recent sensor technology that can "link physiological data such as body temperature and heart rate to movement data, thereby directly linking animals' physiological conditions and movement behavior to landscape features" (Neuman et al., 2015, 8; see also Signer et al., 2010, 75).

Biotelemetry is one key, recent development in this vein. Biotelemetry is defined as the "remote detection and measurement of a human or animal function, activity, or condition" (Merriam-Webster, 2018). Another definition is that it is "the instrumental technique for gaining and transmitting information from a living organism and its environment to a remote observer" (Slater, 1965, 81). In justifying and promoting this field, Cooke et al. (2004: 334) note:

> Remote measurement of the physiology, behavior and energetic status of free-living animals is made possible by a variety of techniques that we refer to collectively as 'biotelemetry'. This set of tools ranges from transmitters that send their signals to receivers up to a few kilometers away to those that send data to orbiting satellites and, more frequently, to devices that log data.

The field is also on a incisive progression, as "the last major reviews on biotelemetry" including studies by Burggren (1987), and Butler (1989) have "concluded that, throughout the 1990s, we could anticipate major advances in our understanding of how animals interact with each other and their environment owing to these technologies" (Cooke et al., 2004: 334).

Currently, the main tracking methods are mostly "VHF (very high frequency) radio-tracking, GPS tracking, satellite tracking using the Argos system and a combination of the advantages of GPS and the Argos satellites into a GPS/satellite telemetry method" (Farve, 2018). VHF telemetry usually involves handling of a hand-held antenna to receive VHF transmissions from a VHF transmitter, which may be on a collar attached to the target animal. The locations of the transmitter and the animal are determined by triangulating the transmissions from receivers (Farve, 2018). Such GPS tracking telemetry uses GPS receivers connected through signals to GPS satellites in the Earth's orbit. Two dimensional locations (e.g. latitude and longitude) can be recorded using signals from three satellites, while signals from four satellites can locate a receiver in three dimensions (Campbell, 2018). Animal movement data in time and space are stored in the receiver and later retrieved by capturing the animal. Data may also be transmitted remotely to a receiver or to the Argos Satellite system (termed GPS/satellite telemetry, based on two Argos satellites, which are operated jointly by the United States (National Oceanic and Atmospheric Administration and National Aeronautics and Space Administration) and the French government (French Space Agency) and high technology centers in Landover, Maryland and Toulouse, France) (Tomkiewicz et al., 2010). Currently, GPS technology is quite accurate for monitoring animal presence, movements and behaviors. Presence may be determined within possible error, between 100 meters (m) and 4 kilometers (km). This enables the monitoring of animals of varying sizes and movement ranges (Campbell, 2017).

10.5 Drones and Remote Sensing Applications

Drones are defined as unmanned aerial vehicles or uncrewed aerial vehicles-UAV or remotely piloted aircraft-RPA, which are remotely, computer controlled, non-human occupied aircraft, usually incorporating GPS technology and attached (Campbell, 2018). Drones, like satellites and airplane aerial photography, started as military hardware, but later transposed into scientific and recreational technology, capable of accurate location recording, high resolution image capture and animal behavioral assessments. More sophisticated and advanced drones may incorporate an autopilot and computer system, a GPS with receivers and compasses, and even barometric altimeters, video cameras and photographic technology, and optical sensors (infrared and/or ultraviolet capable), physical sensors (measuring temperature, humidity etc.) and sound sensors (Ivošević et al., 2015; Ditmer et al., 2015a). Designs differ, from fixed wing machines to designs with one or more rotor blades similar to helicopters. These flexibilities have led to an increase in the number of drones used for variable tasks (Ditmer et al., 2015b; Franke, 2015; Kennedy & Rogers, 2015; Federal Aviation Administration FAA, 2018). The development of drone technology also parallels the applications for industry, including tracking, police work, fire assessment and search and rescue. These functions require increasingly "intelligent" designs that can react to the complex parameters that must be measured (DJI, 2019). Figures 10.2, 10.3, 10.4, 10.5 and 10.6 illustrate some of the recent developments of drone technology, quadcopters and hexacopters (drones with four or six rotors for lift and propulsion) (DJI, 2019).

Fig. 10.2. Matrice 200 Series V2. *Source*: Reproduced with permission of DJI (2019). This quadcopter drone, a cutting edge machine with a design that responds to the current needs for a versatile and maneuverable aircraft, is described by DJI (2019) as "the ultimate platform for aerial productivity" as it "combines a rugged design and simple configurability" with improved "intelligent control systems, flight performance, and add flight safety and data security features."

Fig. 10.3. Mavic 2 Enterprise. *Source*: Reproduced with permission of DJI (2019). This is also a foldable quadcopter drone, similarly hardy enough to be used for police work, fire response, infrastructure inspections and search and rescue. The Mavic 2 Enterprise Dual version adds thermal imagery. By extrapolation, this would also be good for ecological work.

Drones applied for ecological work may be termed "eco-drones" or "conservation drones" (Ivošević et al., 2015, 113). Ditmer et al. (2015a, 2278) argue that such "unmanned aerial vehicles (UAVs) have the potential to revolutionize the way research is conducted in many scientific fields" as drones "can access remote or difficult terrain, collect large amounts of data for lower cost than traditional aerial methods, and facilitate observations of species that are wary of human presence" (see also Anderson & Gaston, 2013; Marris, 2013; Vermeulen et al., 2013; Lucieer et al., 2014). GPS technology is merged with drone technology as the precise location of observed subjects usually mandatory for effective ecological studies (Campbell, 2018). Drones enable closer, more detailed image capture and video recordings than other technologies such as aerial photographs, satellite images and hand-held camera recordings. Drones also contribute frequency, such as patterns changing by the second, which may be required for ecological studies (Campbell, 2018).

Nowak et al. (2019: 56) examined the main weaknesses of older remote sensing tools, noting "remote sensing technology, which enables the observation of the Earth's surface and is currently very common in environmental research, has many limitations such as, insufficient spatial, spectral and temporal resolution, and a high cost of data acquisition." Anderson and Gaston (2013: 138) support these assessments: "to date, heavy reliance has been placed on obtaining such data from remote-sensing instruments mounted on spacecraft or manned aircraft, although the spatial and temporal resolutions of the data are often not suited to local-scale ecological investigations." Objectively, "flying low and slow, UAVs offer ecologists new opportunities for scale-appropriate measurements of ecological phenomena. Equipped with capable sensors, UAVs can deliver fine spatial resolution data at temporal resolutions defined by the end user" (Anderson & Gaston, 2013: 138). High resolution cameras are often added for critical photography and associated technological developments have enabled "improvements in navigation and

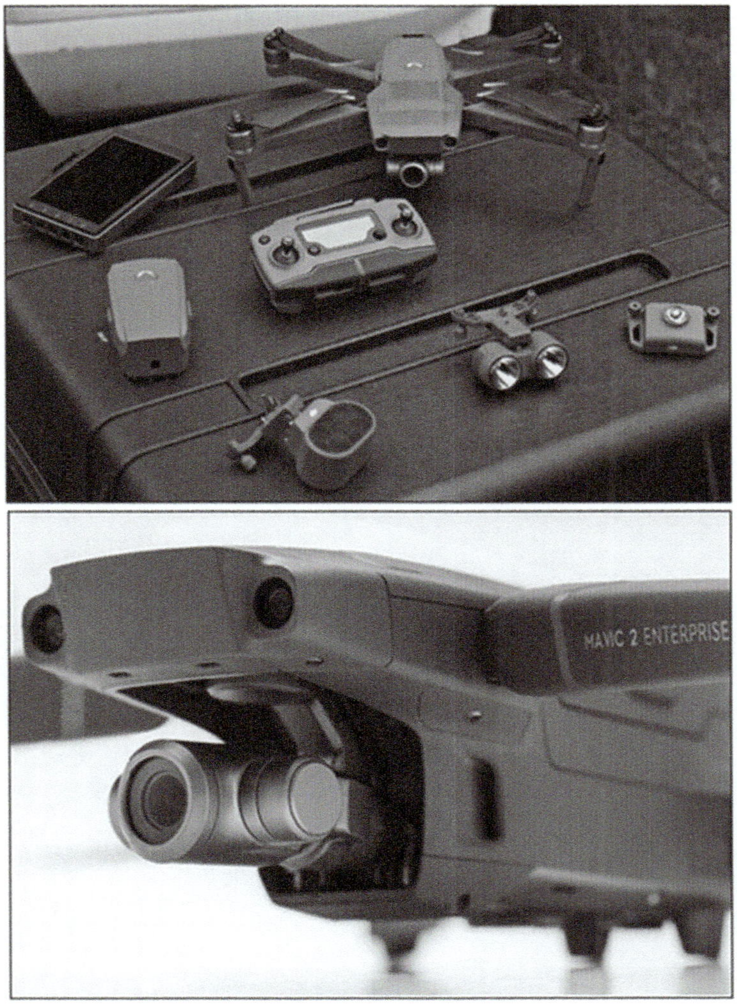

Fig. 10.4. Mavic 2 Enterprise. Accessories.
Source: Reproduced with permission of DJI (2019).

the miniaturization of measurement technologies, allowing the study of individual organisms and their spatiotemporal dynamics at close range" (Anderson & Gaston, 2013: 138).

Drones may also be used for the detailed georeferencing of required locations, when they are equipped with GPS tracking software, increasing the accuracy over other tracking methods for ecological assessments (Ditmer et al., 2015a; Ivošević et al., 2015; Hammill et al., 2017; Johnston et al., 2017; Hodgson et al., 2018; Lucas, 2018). Drones vary in size, from a few centimeters across to the size of airplanes, giving flexibility to operators in applications from photographing or filming insects or whole forest stands. Other applications include the observation and tracking of highly mobile wildlife such as large birds, their frequented foraging areas, prey species and nesting (Campbell et al., 2015). Drones may also participate in data acquisition for animal population estimation and poaching prevention (Koh & Wich, 2018; Barasona et al., 2014; Whitehead & Hugenholtz, 2014; Mulero-Pázmány et al., 2014; Ratcliffe et al., 2015). Drones may also be capable of short or long-range flight, from a few kilometers to over 500 km.

For these reasons, Hodgson et al. (2018, 1160) argue that "ecologists are increasingly using technologies, such as remotely piloted aircraft…unmanned aerial systems or unmanned aerial vehicles, for wildlife monitoring applications." Johnston et al. (2017, 178) support these arguments: "unoccupied aircraft systems (UAS) are now frequently used in wildlife research, including studies of marine species. Ivošević et al. (2015, 113) justify the increased interest in drones for environmental jobs: "a drone can provide a low-cost and low-impact solution to environmental managers working in a variety of ecosystems." Therefore, drones are "reliable tools for conducting population assessments… and establishing life history stages of animals. These new tools provide flexibility in operations and may reduce costs and human risk in some cases" (Johnston et al., 2017, 178). Thermal Dronesmount thermal infrared cameras that can capture images using the electromagnetic spectrum outside the visible light

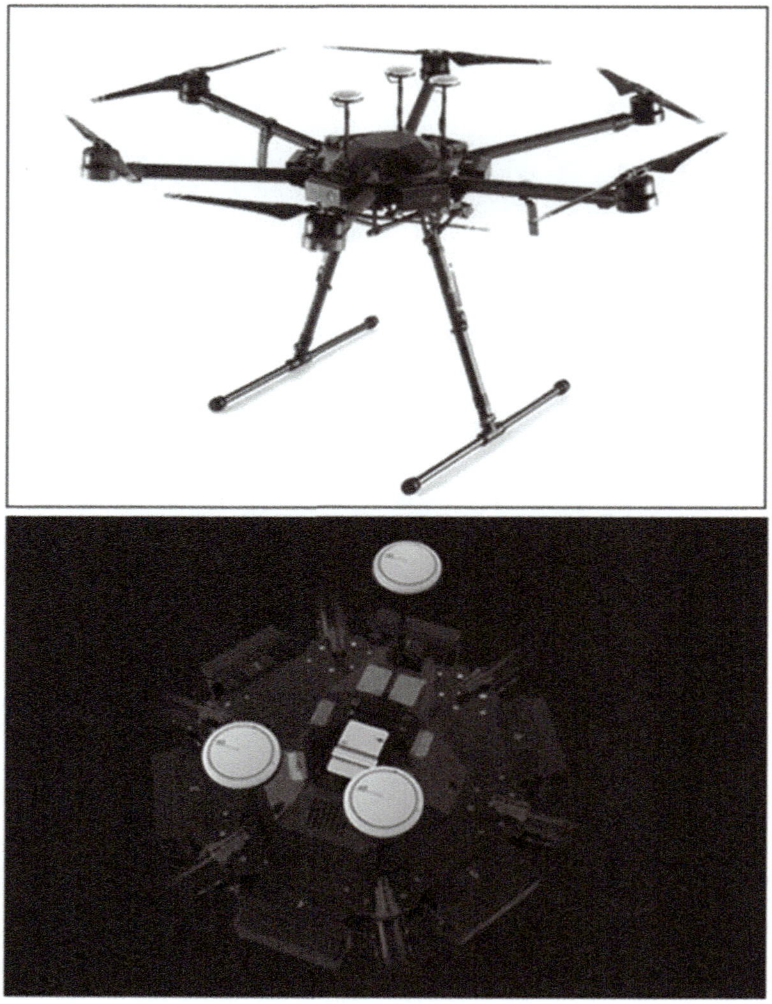

Fig. 10.5. Matrice 600 Pro. *Source*: Reproduced with permission of DJI (2019). This is a hexacopter or a drone with six rotor blades for lift and suspension. It can compare sensor data from three Global Navigation Satellite System (GNSS) units and has a dustfree dustproof propulsion system.

bands, based on the temperature of the subject animal in its cooler environment. This gives the capability to monitor animal movements in darkness, such as those by nesting and roosting raptors (Campbell, 2018; Witczuk, 2018; Spaan et al., 2019). Therefore, "the detection of heat radiation and the high contrast of thermal imaging cameras allow users to see the outlines of (warm-blooded) animals that camouflage well in their environments. It can also help distinguish animals (or people) in foggy or smoky conditions" (Fernando, 2017).

10.6 Remote Sensing Applications to Habitat Identification and Analysis

Remote sensing is increasingly relevant to ecological analysis, from primitive beginnings to sharper, more advanced space-based satellite technology, supported by advanced computer software and hardware. Currently, the development of remote sensing applications is concerned principally with developments that increase accuracy, resolution, speed of analysis, size of data capture, flexibility and breadth of syntheses. These enable better identification of animals and plants (smaller species, at greater distances, in more complex environments), recording of animal behavioral movements (e.g., movements, breeding, mating and feeding patterns), the finer aspects of landcover change (including temperature, moisture and species change, which may be relevant to the prey species of eagles), and the syntheses of such ecosystem components into the larger scenario of ecological change (Jensen, 2007; Campbell, 2018).

This technology has differing applications when applied to the ecology of large species such as eagles. As will be discussed later, the main points concern the identification, detection and mapping of landcover features relevant to eagle habitats and the

Fig. 10.6. Phantom 4 Pro. *Source*: Reproduced with permission of DJI (2019). This drone has a forward vision, backward vision, downward vision system, and an infrared sensing system.

recording of eagle presence and movements. As depicted in Tables 10.2 and 10.3, the resolutions of such images produced by the different sensors differ greatly. For the presence of eagles, a resolution of well below one meter may be necessary, as most eagles have wingspans under two meters and body lengths of under one meter. Even with a resolution of 30 cm, it would be difficult to detect an eagle in the pixel. Even if the eagle is detected in the image, differentiating it from other, similarly sized birds such as storks, herons, vultures or large hawks may be difficult. Therefore, the best function of aerial photographs and satellite images would be the recording of habitat features, such as forest canopy, mixed tree-grass savanna and grassland, and even large nests in trees or cliffs. The coarse resolution of most images would require intensive, supporting field surveys (Campbell, 2002). Maps created from such techniques may show landcover change, such as deforestation and desertification, which may be related to eagle habitat prediction (Campbell, 2017).

More advanced remote sensing techniques are extremely important for such wildlife studies, largely because it can identify both the target animal and its habitat, eradicating human error in identification and in some cases recording invisible bands (e.g. infrared) and moisture and temperature in landcover (Campbell, 2018). For example, McDermid et al. (2010, 193) wrote that "remote sensing has established a key role in modern wildlife ecology…For decades, remote sensing has been acknowledged as a critical data source of environmental information, theoretically capable of supporting a broad range of ecological applications." The development of applications to wildlife studies principally concerns the spectral, spatial, radiometric, and temporal resolutions of the remote sensing tools. McDermid et al. (2010, 193) further argue that remote sensing techniques enable "an impressive suite of wildlife ecology and habitat attributes to be modeled, predicted, and monitored through time, including land cover physiognomy, vegetation structure and condition, forage characteristics, specific nutrient concentrations, overall productivity, and biomass."

Habitat analysis is probably the most important application for remote sensing tools, considering that the resolution of many image products may be too coarse for species identification, and too infrequent (e.g. some time series images may have decade long, annual or monthly cycles) for regular, detailed assessment. For such habitat analysis, extensive, supporting 'ground truthing' or fieldwork may be required. As pixel sizes may vary from less than one meter to 30m x 30m in the older Landsat Thematic Mapper images, in some cases only large trees may be identified in the images, or particular textural colors linked to landcover combinations. A further complication is that the images with smaller pixel sizes (e.g.< 1 meter) may cover smaller areas, while those with larger pixels (e.g. = 30m x 30m) may cover larger parts of the Earth's surface (Jensen, 2007; Bhatta, 2008; McDermid et al., 2010).

For habitat analysis, especially where there are seasonal or rapid changes, the temporal resolution (the frequency at which images are recorded for an area) is vital. High resolutions refer to more frequent repetitions in time series images of an area. Lower resolutions may be suitable for records of long-term changes, including gradual deforestation, urbanization or spread of savanna landcover. Higher resolutions may be required to capture rapid landcover changes such as natural hazards (Campbell, 2017, 2018). The radiometric resolution (the precision of the radiance measurement in a data type, shown by the range of digital numbers in each pixel) is also important when detailed identification and analysis is required. Commenting of the detail of radiometric resolution, Natural Resources Canada (2016) notes "the radiometric resolution of an imaging system describes its ability to discriminate very slight differences in energy.... The finer the radiometric resolution of a sensor, the more sensitive it is to detecting small differences in reflected or emitted energy." This identifies small differences in the target object or subject, such as foliage type, for species level discrimination.

However, higher radiometric resolutions may encounter problems, as this may require lower spatial resolutions and increased energy detection. As noted by Natural Resources Canada (2018) "For high spatial resolution, the sensor has to have a small IFOV (Instantaneous Field of View).... this reduces the amount of energy that can be detected as the area of the ground resolution cell within the IFOV becomes smaller. This leads to reduced radiometric resolution – the ability to detect fine energy differences." Increasing the radiometric resolution through increased energy detected, may require either the reduction of the spatial resolution, or the broadening the wavelength range detected for the required channel or band. Therefore, higher radiometric resolutions may require coarse spatial resolutions (Natural Resources Canada, 2018). Increased temporal resolution may also require coarser spatial resolution (Campbell, 2018).

Several, mostly dated studies have used coarse resolution images, mostly for landcover and habitat assessment for non-eagle species and habitat studies such as deforestation trends. For example, Cannon et al. (1982) used Landsat MultiSpectral Scanner (MSS, 1978), with a spatial resolution of 80 m to assess Shinnery oak (*Quercus havardii*, Rydberg) dominated landcover in Western Oklahoma. Despite the coarse resolution, the findings were clear enough to indicate a positive correlation between percentage of grassland habitat and the presence of male lesser prairie chickens (*Tympanuchus pallidicinctus*, Ridgway, 1873). An example of a landcover study is that of Kamosuko and Amiya (2006) in Zimbabwe. This study used Landsat data for 1973, 1989 and 2000, comparing the trends of land use/cover changes over two periods (1973–1989 and 1989–2000). The study used a "hybrid supervised/unsupervised classification approach coupled with geographical information systems (GIS) analyses was employed to generate land use/cover maps with six classes; agriculture, woodland, mixed rangeland, settlement, bare land and water." The results of the study showed deforestation due largely to cultivation encroachment occurred in the land tenure systems, suggesting the needs for sustainable land use planning to reduce the loss of forested areas.

Another study is that of Peled and Gilichinsky (2010) which examined methods for classifying vegetation in remotely sensed satellite images using GIS methods. They argue that "valid classification of remotely sensed data is one of the most studied issues in the geoinformation science" which is an area of multivariate statistics, according to that deals with grouping a set of items by assigning them to several similar classes" (Peled & Gilichinsky, 2010, 218). This study integrated digital vector formats that correspond to the 1:40,000-scale air photographs from the Israeli National GIS database, and geo-referenced, radiometrically calibrated IKONOS images with four-meter spatial resolution dated 2005. The result was a landcover matrix classified into agricultural fields, natural forest, citrus orchards, fruit orchards and water reservoirs (Peled & Gilichinsky, 2010).

Another study is that of Rozenstein and Karnieli (2011) on the northern Negev, using a Landsat-5 TM image classified with ERDAS IMAGINE 2010, and a post-classification, a decision support system based on land-use databases using ArcGIS 9.3. The accuracies of these products were tested with STATISTICA 9.0. This study employed a Decision Support System (DSS), which used logical land-use trends, in which pixels in the images could be changed if there was evidence that the landcover it represents changed decisively. The result was a classification of landcover into urban or built-up land, agricultural fields, rangeland, forest, water bodies and barren land. The authors note that the accuracy of the classification was sufficient for planning and management, although not totally accurate (Rozenstein & Karnieli, 2011).

Rogan et al. (2003) used multitemporal Landsat TM data, aerial photographs and field studies to analyze land-cover change in San Diego County (1990–1996), USA. Palmeirim (1988) also used Landsat Thematic Mapper (TM) imagery, with 30 m spatial resolution to estimate the number of avian species in landcover classes in northeastern Kansas. Luoto et al. (2004) used similar data to predict the presence of bird species in five classes of landcover in the boreal forests of Finland. Miller & Conroy (1990) used SPOT High Resolution Visible (HRV) imagery (20 m spatial resolution) to develop a predictive model for the potential habitat for the Kirtland's warbler (*Setophaga kirtlandii*, Baird, 1852).

Older studies based on coarse resolution data include that of Wallin et al. (1992), who used 1-km NDVI data from the NOAA Advanced Very High Resolution Radiometer (AVHRR) sensor, and successfully related vegetation landcover variation to predict breeding habitat of the Red-billed quelea (*Quelea quelea*, Linnaeus, 1758) in East Africa. The authors concluded that coarse resolution satellite data can be used to monitor potential breeding habitat and provide an objective index of breeding site suitability (Wallin et al., 1992, 96). The authors acknowledge the coarse scale of the imagery, in that "data derived from the Advanced Very High Resolution Radiometer (AVHRR) on board the NOAA series of operational, polar orbiting, meteorological satellites have previously been shown to be quite useful for monitoring vegetation dynamics at scales ranging from regional (l04 km^2) to global" (Wallin et al., 1992, 87). Other studies based on 1-km NDVI imagery are Jørgensen and Nøhr (1996) on avian species presence and vegetative biomass in Senegal and Skidmore et al. (2003), Bonn et al. (2004), Amstrup et al. (2004), Stow et al. (2004), Nilsen et al. (2005), Rodriguez et al. (2005) and Van Bommel et al. (2006) on animal species presence and vegetation greenness.

Aspinall and Veitch (1993) describe a study using coarse Landsat TM Mapper imagery and digital elevation models (DEM) to document a non-eagle species (Eurasian curlews, *Numenius arquata,* Linnaeus, 1758) breeding and nesting patterns. In this case, the landcover classification was based on an analytical Bayesian probability method, namely the conditional probabilities for Curlew presence and absence using the spectral values in the image bands and DEM (representing landcover types). The final product was a map, with the same spatial resolution as the satellite image that predicts the presence and distribution of the specified species as occurrence probability. This study illustrates the possibilities for similar work on eagles, conditional on the greater complexity of the eagle's ecology and movements.

Resolution is therefore key for applications to landcover assessment. The sharper the resolution the better. Boyle et al. (2014) argue that the issue of sharper resolution may be resolved in the future. More recent technologies allow sharper resolutions, in some cases combining older image formats (e.g. Landsat TM or aerial photography) with newer image-based systems. For example, Imhoff et al. (1997) used aerial photography and field data with P-, L- and C-band SAR data from NASA's airborne AIRSAR system for the analysis of correlations between avian ecology and landscape variation at Kakadu National Park in Australia. The high-resolution SAR data recorded vegetation structures, which combined with the aerial photography created three-dimensional mapping of habitat dynamics (Imhoff et al., 1997).

Boyle et al. (2014) examined the hypothesis that high spatial resolution (≤5 m) imagery was different in applicability from lower-resolution imagery (≥30 m), taking a case study of the Interior Atlantic Forest in Paraguay and using 4 m multispectral IKONOS and 30 m multispectral Landsat imagery. The study found that higher-resolution imagery was more accurate for landcover identification than coarser images, especially for small patches of cover that varied from surrounding landcover. The cost and rarity of high-resolution imagery however precludes their universal use. For example, the authors examined articles in three conservation journals (Biological Conservation, Biotropica, Conservation Biology), and concluded that high-resolution imagery was used in only 26.8% of landcover studies that used satellite imagery, and only 10.4% of these used imagery (≤5 m resolution). The authors suggested the need for access to cheaper multiband, high-resolution satellite imagery.

Few studies have used remote sensing technology of this type to study eagle habitats and ecological relations. However, Nurfatimah et al. (2018) argue (see also Gairola et al., 2013) that finer-resolution and coarse-resolution methods are both useful for wildlife management, in this case for the study of the Javan Hawk eagle in the forests and modified landcover of Java, Indonesia. "Finer-resolution can provide higher accuracy in mapping the forest and predicting nesting site...But to provide the finer-resolution data will result in higher cost in human resources and technology... Therefore, the managers and conservationists can choose the most suitable image scale for managing, planning and monitoring based on the scale of application—site-scale or regional scale (Nurfatimah et al., 2018, 9).

Andrew and Shepard (2017) used very high-resolution (VHR) image data, including that from unmanned aerial vehicle (UAV) platforms, and developed semi-automated analyses to map White-bellied sea eagle nests in the Houtman Abrolhos Islands, Western Australia. They argue that "nest detection is complicated by high environmental heterogeneity at the scale of nests (~1–2 m), the presence of many features that resemble nests and the variability of nest size, shape and context" (Andrew & Shepard, 2017, 66). Very high-resolution (10 cm pixels) true color aerial photography was used, but even using such a sharp resolution there were still problems, as the landcover in which the eagle nests were situated was highly heterogenous, the materials from which the nests were constructed were similar to the surrounding landcover, the landcover was not wooded, hence the nests were constructed from the ground, low shrubs or the canopies of mangroves. The shape of the nests also varied, some modelled as towers, others as low matted, highly variable forms on sea cliffs. The nests were also not very common, and the habitats varied, creating a difficulty in obtaining suitable numbers of generalizable training class for the image classification. It was argued that these characteristics created greater difficulties than those in studies of nests in more uniform forest vegetation (Andrew & Shepard, 2017, 68).

10.7 GPS Applications to Animal Behavior, Presence and Movements

The high mobility and comparatively small size of birds (as opposed to mammals) test the applicability of many remote sensing and GPS tools. For this reason, many such techniques are more easily used for studies of mammals (Campbell, 2017). To adapt to the extreme mobility of animals, new developments of GPS adapt to the complexity of animal activities. Many such developments have occurred in private, competitive industry. Mostly based on the attachment of devices to the animal, these techniques are better suited to large mammals. Clark et al. (2006, 334) note that

> An improved global positioning system (GPS)–based animal tracking system is needed to meet quickly evolving demands of ecological research, range livestock production, and natural resource management. Commercially available tracking systems lack the data storage capacity needed to frequently collect animal location data (e.g., 15-minute intervals or less) over long-term deployment periods (e.g., 1 year or more). Some commercial systems have remote data–download capabilities, reducing the need to recapture tagged animals for data retrieval, but these systems download data via satellite (Argos), global system for mobile communications (GSM), cellular telephone, or telemetry radio frequencies.

One example of a recent, private developer is Du Toit (2018), sometimes associated with Animal Trackem which promotes the use of tracking and satellite collars, radio, GPS applied to endangered mammal species like cheetah, leopard or rhinoceros. The objective is the daily monitoring of the target animals, capturing records of possible inter- and intra-species interactions, movement patterns, population demographics, injuries and legal and illegal trapping (Wildlife Act, 2018). Collars are designed to emit signals whenever the target animal is stationary, hence possibly in a trap. There are also Rhino Ankle Collars, which replaced the less successful neck collars and horn implants and allowed for larger associated devices. Remote camera trapping is also used, in which movements trigger infrared or motion sensors to record animal presence, especially to detect rare or elusive animals (Wildlife Act, 2018).

Another recent development is the Clark GPS Animal Tracking System (Clark ATS), "designed to provide continual, real-time data access" (Clark et al., 2006, 334). The Clark ATS uses memory-card technology, with data storage of up to 8 gigabytes, which can be remotely downloaded by spectrum radio transceivers, without the need for narrow-band radio frequency allocations. This system also enables the monitoring of target animal behavior, including distribution patterns and responses to short-term disturbances (Clark et al., 2006, 334). Problems with these sophisticated systems include discomfort to the animals that must wear collars or tags, high cost of the technology and the lack of trained personnel to derive the full benefits (Tomkiewicz et al., 2010; Matthews et al., 2013).

The most important developments concern collection and analysis of ecological data, as well as "the physiological performance, energetics, foraging, migration, habitat selection, and sociality of wild animals" (Wilmers et al., 2015: 1741). Wilmers et al. (2015) note that:

> Great leaps forward in scientific understanding are often spurred by innovations in technology. The explosion of miniature sensors that are driving the boom in consumer electronics, such as smart phones, gaming platforms, and wearable fitness devices, are now becoming available to ecologists for remotely monitoring the activities of wild animals. While half a century ago researchers were attaching balloons to the backs of seals to measure their movement, today ecologists have access to an arsenal of sensors that can continuously measure most aspects of an animal's state (e.g., location, behavior, caloric expenditure, interactions with other animals) and external environment (e.g., temperature, salinity, depth).

These authors examine the recent technological advancements, which include smart phones, miniature sensors such as accelerometers, magnetometers, GPS and digital cameras. Wilmers et al. (2015) give examples, including tri-axial accelerometers, able to detect a fish moving its mouth (Viviant et al., 2014) and the number of calories burned by a stalking and attacking cougar (Williams et al., 2014). Some cameras can observe animals eating (Watanabe & Takahashi, 2013). Others can record the gaze direction of animals (Mitani et al., 2010). Sensors can measure the daily energy budgets, heart and stomach functions, metabolic rates, body temperatures, body and jaw movements, swimming speeds, use of oxygen when diving and oxygen stores, heat flux, stomach pH and temperatures, hunger and satiation indicators and reproductive states (Cooke et al., 2004, 2008, 2013; Costa & Sinervo, 2004; Ponganis, 2007; Rutz & Hays, 2009; Barron et al., 2010; White et al., 2013). Wilmers et al. (2015: 1741) argue that "this technology is revolutionizing our ability to study animal ecology by allowing researchers to (1) answer questions about the physiology, behavior, and ecology of wild animals that would have previously been limited to tests on model organisms in highly controlled settings; (2) study cryptic or wide-ranging animals that have previously evaded investigation; and (3) develop and test."

The "laboratory-to-field" method is an important methodology that enables the application of such technology (Wilmers et al., 2015, 1741). This involves the calibration of the sensors in laboratory conditions, to be later applied in field settings. So far, many or most of such programming efforts appear to be conducted on large mammals. For example, Williams et al. (2014) calibrated a GPS–accelerometer collar on caged cougars, which walked on level surfaces and treadmills, to calibrate the collar accelerometer signatures to the energetic costs of locomotion, applied to gaits, conditioned by the open-flow respirometry on the treadmill movements. The results were then used for the monitoring of wild ranging cougars, which revealed that the energy cost of kills by the cougars was positively correlated with the prey size. Low activity, including stalking, and ambush lowered the hunting energy cost. Such studies increased research interests in the energy expenditure of animal movement, the "energy landscape" (Shepard et al., 2013). Mosser et al. (2014) used a similar methodology to study the energy landscape of woodland caribou in northern Canada. Records included calorie expenditure in foraging in different ecological settings, such as variable snow depths, temperature levels and vegetation densities, and informed on the factors for the caribou population declines (Mosser et al., 2014; Wilmers et al., 2015).

Augliere (2015) presents a similar application to the ecology of wolves (*Canis lupus*, Linnaeus, 1758), based on the premise that "conventional collars show scientists where in the forest a wolf is, but new technology also tells you what the animal is doing hour by hour." The metabolic cost of wolves' activities is crucial, to assess the impact of stresses on the wolf ecological relations. Scientists at the University of California at Santa Cruz, used advanced, transmitting collars in a laboratory setting, like accelerometers used for smartphones and fitness trackers, recording calories burned during physical exercise. Correlations between calorie burnt and different exertion levels were recorded, creating an energetic "signature" for each activity. The findings were applied to wild ranging animals, including wolves, polar bears (*Ursus maritimus*, Phipps, 1774), narwhals (*Monodon Monoceros*, Linnaeus, 1758), and cougars (*Puma concolor*, Linnaeus, 1771).

Wilson et al. (2013) used combined GPS, accelerometer, and gyroscope collars on wild cheetahs (*Acinonyx jubatus*, Schreber 1775) in Botswana, the sampling data calibrated at up to 300 times a second, to replace older methods of assessment of cheetah foraging using speed cameras and videos (Weise et al., 2017). The advanced techniques measured the velocity, forward and lateral acceleration, and travel paths of 897 runs of five cheetahs over 17 months. The findings were that cheetah's maximum hunting speed was 25.9 m/s (58 mph), with an average top speed of 14.9 m/s for 1–2 seconds per chase. The rapid deceleration closing the chase was the key determinant for successful predation, over maximum speed, number of turns, or distance traveled, or vegetation cover. Studies of impacts of predator competition on cheetahs, by lions (*Panthera leo*, Linnaeus, 1758) and hyenas (*Crocuta crocuta*, Erxleben, 1777) included additional field data (feces collection and double-labeled water techniques for assessment of energy expenditure) (Scantlebury et al., 2014). The findings were that cheetahs hunted 1.1. more hours daily in response to theft levels of 25% by lions and hyenas, increasing the daily energy expenditure of the cheetahs by 12% (Wilmers et al., 2015).

Wilmers et al. (2015) also describe tilt sensors; accelerometers measuring the direction of the target animal's head (in terms of gravity) and devices using sound to measure activities such as chewing or silence to detect nearby predators. Lynch et al. (2014: 77) used acoustic data collection with preceding laboratory tests, where mule deer (*Odocoileus hemionus*, Rafinesque, 1817) were fitted with collar mounted acoustic recording devices (placed on the animal's throats) to record audible behaviors and related sounds. Sounds detected included those of vocalizations, foraging, footsteps, eating (including regurgitation, mastication, and swallowing), and respiration. Wild tests found that pauses between feeding and chewing had frequent links to awareness of predators and the surrounding landcover. These findings were similar to those of other studies, which used accelerometers on the jaws of marine predators, recording jaw movements in prey capture (Viviant et al., 2014) and accelerometers on bird wings recording wing beat frequencies related to body mass change and feeding. Lynch et al. (2014, 79) concluded:

> Our results demonstrated that pauses during mastication bouts (chewing) were structured by landscape properties that are expected to influence perceived risk and impede visual vigilance. Consistent with predictions, we found that deer paused more extensively where concealment cover abounded and thus where visual vigilance was likely to be less effective. In addition, deer allocated a larger proportion of time to pausing during mastication bouts at night, implying that auditory vigilance is an important defense mechanism when darkness reduces the effectiveness of visual scanning.

10.8 GPS and Drone Applications to Eagle Studies

Advanced GPS technology and drones are increasingly used for monitoring eagles, but the main applications appear to be simple tracking of eagles in foraging and migration. For example, Mojica (2012) documents the tracking of Bald eagles with 70g GSM/GPS transmitters, and argues that 'the Bald Eagle (*Haliaeetus leucocephalus,* Linnaeus, 1766) was one of the first wildlife species tracked with satellite telemetry" largely due to its large size which "can carry the 100+ g transmitter without impeding breeding or their 4,000 km distance migrations." This study acknowledged that Bald eagle were tracked using Platform Transmitter Terminals (PTT) over recent decades, especially during migration from eastern Canada to Florida, termed the Atlantic flyway. More recently, some transmitters are modified to use Global System for Mobile (GSM) communications technology. In the actual study, 70g GSM/GPS units were placed on two second-year Bald eagles in February 2011. These united gave an hourly GPS location, which at the time of writing recorded over 10,500 GPS locations over 473 days. The GSM/GPS units provided more data than the 70g solar Argos/GPS PTTs on 67 other Bald eagles.

The Raptor Resource Project (2016) documents similar technology, using PTT antennae placed on the backs of Bald eagles, which transmit signals to Argos satellites. The blog notes:

> The PTT and GPS units from Geo-Trak Inc. collect and encode data about location, heading, speed, time, activity, and transmitter and battery performance. The encoded data is supplied as energy to the antenna, which radiates it as UHF radio waves to the Argos satellite network orbiting 528 miles (850 kilometers) over our heads. From there, ground stations receive real time data from the satellites and retransmit it to regional processing centers where we can access it.

Such structures are argued to be harmless to the eagles, despite questions about radiation and size. Smaller transmitters such as some designed for implantation on insects have a much smaller range of transmission (100-500 m) and as such they are unsuitable for tracking eagles. The system parts are "light, safe, reliable, and trackable almost anywhere on earth" do not harm bald eagles or impact their social or reproductive interactions with eagles that aren't wearing transmitters" (Raptor Resource Project (2016).

Drones are useful, due the eagles' high mobility, frequently remote nesting sites, wide foraging and migratory ranges, and relevant, observable socially derived contexts (urban pollution, building damage, livestock attacks, collisions with aircraft). Drones have only recently been used for the study of raptors, so many of the possibilities are untested and applications to non-eagle species may be used to determine suitability for the study of eagles. Drone applications may be useful for the measurement of relevant parameters for eagle studies, including flight speeds altitudes, soaring duration, foraging flight patterns, courtship, nesting habits and locations, physical attacks on prey species, roosting and aggressive incidents, and human impact incidents. The usability of the drone technology will also depend on the accuracy, mobility and resolution of the mounted cameras, the precision of the GPS location measurements, the power of sound detection equipment and the speed, maneuverability, control and noise of the drone. The main contribution is increased accuracy in the documentation of avian behavior, within the limitations of animal avoidance, eagle attacks on drones, and human privacy issues (Anderson & Gaston, 2013; Fessenden, 2015; Rockwell, 2016).

Other issues that drones may be able to contribute to may include the quantification of inter- and intra-species differences (Rush et al., 2018), the automatic detection of species using spectral signatures which allows more accurate species detection (Fretwell et al., 2012, 2014; Grenzdörffer, 2013; Rush et al., 2018), more efficient surveys than aircraft, especially in mountain areas (Hausheer, 2016), the ability to maneuver to avoid detection, disturbance or subject reactions, enabling close observation and image capture which may be impossible with close human observation (Rush et al., 2018), and intra-species differences relevant to breeding patterns and adult/juvenile distinctions (Fretwell et al., 2012, 2014; Grenzdörffer, 2013; Rush et al., 2018). These are especially relevant when the drone's image capturing device has a much sharper resolution than the available satellite image (e.g., the Albris UAV with a resolution of 1 cm, compared with the Airbus Spot 6 satellite with a resolution of 1.5 meters (Martin, 2018). Junda et al. (2015, 234) argue that "investigating raptor nests with UAVs in urbanized areas may also raise some permit issues, arising from safety and privacy issues with the general public." These occurrences illustrate advantages and pitfalls of using drone-based technology for studying large, aggressive raptors. Issues to investigate include the detectability of the drone (based on its size, design, color, operating noise and speed) and interspecies and individual differences in drone awareness may determine the effectiveness of any drone-based study (see also Fretwell et al., 2012, 2014; Grenzdörffer, 2013; Campbell, 2015; Hausheer, 2016; Rush et al., 2018; Nowak et al., 2019).

Advantages of thermal imaging technology for ecological research include "the miniaturization of thermal imaging cameras" which "allowed them to be attached to an UAV and used for everything from monitoring wildfires, flying search and rescue missions, searching farm land for animals before mowing operations, and conducting wildlife surveys" (Bushaw et al., 2019; see also Haschberger et al., 1996; Ambrosia et al., 2003; Kastberger & Stachl, 2003; Collier et al., 2007; Rudol & Doherty, 2008; Israel, 2011). Relevant species for eagles are smaller mammals and birds, or the young of large mammals. Witczuk (2018, 5504) also used thermal infrared imaging tools on drones to study ungulates and concluded that such technology "may have great potential as a tool for ungulate surveys" enabling "safe operations at low flying altitudes, and at night – a time that often offers the optimal conditions for wildlife monitoring" and were able to demonstrate that "ungulate thermal signatures are visible both in leafless deciduous and in pine-dominated coniferous forests" especially at dawn, dusk and night. The resolution of approximately 10 cm enabled the identification of large mammals, but there were "difficulties in species identification due to relatively low resolution of TIR cameras, regulations limiting drone operations to visual line of sight, and high dependence on weather." Infrared capable drones are shown in Figure 10.7. Table 10.4 shows some of the applications of infrared imaging and filming. These are some of the mammals whose young may be taken by eagles. Hence their presence and foraging duing noctural movements may be helpful in predicting the movements of prey spcies for eagles.

Table 10.4.

No.	Animal	Drone	Camera
1.	Koala bear	S800 EVO Hexacopter	Forward-looking infrared (FLIR) camera, Tau 2-640 model
2.	Red deer, Roe deer	AVI-1 aircraft, Skywalker X8 Flying Wing	Thermal infrared (TIR) camera, IRMOD v640 model
3.	Monkeys, kinkajou	DJI Matrice 600Pro	DJI Zenmuse XT® V2.0 FLIR, TIR radiometric sensor, Canon EOS 5DS® RGB
4.	Bornean orangutan	Tarot X4	Dual thermal-visible 33 spectrum camera. Thermal Capture Fusion Zoom
5.	Meso-carnivore	DJI Inspire 1 quadcopter UAV	DJI Zenmuse XT2 R thermal imaging camera
6.	Spider monkey	Custom-built	TeAx Fusion Zoom dual vision camera, FLIR Tau2 640 core TIR, TAMRON RGB

Note: The numbered points in column 1 correspond to the following numbered references. 1. Gonzalez et al. (2016); 2. Witczuk et al. (2018); 3. Kays et al. (2019); 4. Burke et al. (2019); 5. Bushaw et al. (2019); 6. Spaan et al. (2019). The full scientific name for the koala is *Phascolarctos cinereus* (Goldfuss, 1817), red deer is *Cervus elaphus* L.1758, the roe deer is *Capreolus capreolus* (Linnaeus, 1758). The monkeys were Mantled howler monkey (*Alouatta palliata*, Gray, 1849), Geoffroy's spider monkey or Black-handed spider monkey (*Ateles geoffroyi*, Kuhl, 1820), and the kinkajous were *Potos flavus* (Schreber, 1774). Bornean orangutan is *Pongo pygmaeus* (Linnaeus, 1760). Spider monkey is *Ateles* spp. Meso-carnovore refers to medium sized carnivores such as the coyote (*Canis latrans*, Say, 1823), red fox (*Vulpes vulpes*, Linnaeus, 1758), striped skunk (*Mephitis mephitis*. Schreber, 1776) or raccoon (*Procyon lotor*, Linnaeus, 1758). Eagles that may take these species include the Wedge-tailed eagle for the Koala bear in Australia, Golden eagles for the red and Roe deer, and meso-carnivores in Eurasia and North America, Harpy eagles for the Spider monekys and Kinkajou and Black eagles for the orangutan.

Spaan et al. (2019) gives the example of advantages of the drone-based thermal infrared (TIR) camera compared with a ground survey, citing the former's lower cost and logistical difficulties. In a case study of spider monkeys (*Ateles geoffroyi*, Kuhl, 1820), it was possible to count larger numbers of monkeys in forested vegetation. Drone counts were approximately like ground counts, but larger numbers of monkeys were counted with the drone counts in 83% of the surveys, mostly due to the large ground coverage (see also Gonzalez et al., 2016; Witczuk et al., 2018; Burke et al., 2019; Kays et al., 2019). The authors concluded that "the use of a TIR camera is a feasible and valid method of obtaining data on the presence and number of spider monkeys at their sleeping sites in a closed-canopy forest" (Spaan et al., 2019).

There are several legal issues around the use of drone in areas near human habitation. This can be a serious limitation, as studies of eagles may be affected in such places, such as suburban, semi-rural and airport landcover (Lee, 2019). For example, in the United States, drone use may be restricted near airports. Shepardson (2019) notes "there has been rising concern about drones near U.S. airports." The Federal Aviation Administration (FAA) has informed U.S. airports that only authorized federal agencies could install systems for the detection and destruction of drones and airport authorities require federal consent to install such systems, especially as "they could pose an aviation safety risk by interfering with aircraft navigation and air navigation services" (Shepardson, 2019). In 2018, the US Congress went so far as to grant the Justice and Homeland Security departments authority to disable or destroy drones (Shepardson, 2019).

The National Park Service also banned drones from National Parks (Policy Memorandum 14–05 issued by the NPS on June 19, 2014), without a Special Use Permit, issued only for applications such as search and rescue, research and fire safety (Drone U, 2018). In the United States recreational and commercial uses are distinguished. Recreational rules include flight within sight, drones weighing less than 25 kg, stay 8 km from airports (without notification or permission) and avoid aircraft and emergency response efforts. Commercial rules include flight below 122 meters, drone weights less than 25 kg, flight during daylight to

Fig. 10.7. Phantom 4 Pro and Infrared. *Source*: Reproduced with permission of DJI (2019). This drone has a forward vision, backward vision, downward vision system, and an infrared sensing system.

twilight, less than about 162 km, avoid flight over people and operation of drones from a moving vehicle (unless in areas with low human presence) and yield to manned aircraft (UAV Coach, 2019).

In Canada, drones are banned within 5.6 kilometers from any airport listed as Certified ("Cert") in the Canada Flight Supplement and 1.9 kilometers heliports or aerodromes used by helicopters (Government of Canada, 2019a). Drones may not be used in Canadian National Parks except under Parks Canada supervision. This includes recreational flying of drones, while commercial drone use may be allowed with a permit. Such Restricted Activity Permits may be issued for the use of a drone for "natural or cultural resource management and protection, public safety, law enforcement, or park/site management purposes directly relating to park administrative purposes" (Government of Canada, 2019b).

New regulations were passed in Canada in 2019, applied to drones weighing "250 grams (g) up to and including 25 kilograms (kg) and are operated within the drone pilot's visual-line-of-sight" (Lee, 2019). The new legislation also gave equal treatment to recreational and commercial drone usage. Under this legislation, drone operators must register and acquire required certificates, and identify their usage as basic, which means the drone is flown in uncontrolled airspace, more than 30 meters horizontally from people and not over people. Drone operators must respect human privacy and be flown within sight, below 122 meters elevation, fly away from people, avoid forest fires and social events, stay 5.6 kilometers from airports and 1.9 kilometers from heliports and avoid other aircraft.

The GPS systems provide the three-dimensional position of fixed or moving objects, in space and time, over the entire surface of the Earth. When mounted on drones with aerial cameras, these enable location and images to be recorded, and provide vital information on bird ecology. (Campbell, 2018). López-López et al. (2016, 981–993) write that "fast-moving technological advances, such as satellite tracking technologies, are providing in-depth information of aspects of avian ecology hitherto unknown. In fact, detailed information about movement ecology and ranging behaviour of birds is important not only from the perspective of the basic ecology, but also from the conservation point of view."

Examples of these techniques have been recorded in the recent literature. For example, Krone et al. (2009b, 273–280) used GPS telemetry to measure the home range size and activity pattern of a White-tailed eagle. The eagle's positions were recorded three times daily, with the activity continuously recorded by two acceleration sensors, from July to January. The findings showed that the home range was 4.53 km^2 and minimum convex polygon of 8.22 km^2. The authors note that this was a "rather small area explained by an optimal habitat and by the much more precise location method used here compared to reports in the literature" (Krone et al., 2009b, 273). There was almost no night activity, with high day-activity which had no fixed daily pattern, the conclusions were that "the new technology of a combined GPS receiver and an acceleration sensor allows the automatic measuring of positions and activity of wild animals at a very precise level over prolonged periods which cannot be achieved by manpower" (Krone et al., 2009b, 273).

Another, similar study by Van Eeden et al. (2017) looked at the ranging behavior of the Martial eagle in South Africa. Van Eeden et al. (2017, 1) describe this study as "the first time that such movements have been quantified in adult Martial eagles." A problem is that Martial eagle is rare, widely ranging, and low-density forager and apex predator, that is both declining and little known in terms of these attributes. The technology was GPS tags fitted to Martial eagles, and the comparison of the results with known attributes of this species. The findings showed that the territorial eagles had home ranges averaging as large as 108 km^2. The authors note that these large home ranges were approximately what would be expected, using the large inter-nest distances, and the large home range sizes would make it difficult to include a large number of eagles in a conservation area. could constrain the carrying capacity of even the largest conservation areas (Van Eeden et al., 2017).

An important finding using the tracking technology was that a few eagles were non-territorial, and as such ranged over larger areas than the territorial individuals, around 44,000 km^2 and even beyond the Kruger National Park (KNP, which is South Africa's largest protected area covering 20,000 km^2) as floaters. A few territorial birds abandoned their territories and became floaters, ranging more widely than before. The authors speculate that "these unexpected movements after territory abandonment could indicate underlying environmental degradation" and found that these floaters had a much higher than average mortality rate due to anthropogenic causes (three of four), which is a serious conservation issue. This GPS technology was combined with digital elevation models, continuous fields of tree cover map, a HydroSHEDS river network layer, a roads layer, and a 72 class South African national land cover dataset that categorizes the national area into land classes at a 30 m resolution, with the main types including being vegetation type (e.g. grassland, open bush, dense bush) and artificial categories such as urban development. The constructed habitat preference models suggested Martial eagles frequented riverine areas and their nearby areas, with more tree cover, and were classed as dense bush rather than open bush or grassland. The authors conclude that this study "provides the first satellite-derived description of Martial eagle ranging behaviour and habitat preference" and hints at problems such large ranging behavior of territorial birds and floaters, low-density populations (such populations may be more vulnerable to stochastic events) and the need to conserve tall tree and dense shrubland losses (Van Eeden et al., 2017, 16).

Another similar study looked at the distribution and habitat selection of Golden eagles in the Mojave Desert of the United States, in relation to the development of energy projects in the Desert Renewable Energy Conservation Plan area (Braham et al., 2015, 225 - 232). This study used global positioning system–global system for mobile communications (GPS-GSM) telemetry to measure year-round movements of Golden eagles. It used two linear mixed-effects models to assess the relationship between size (the response variable) of monthly home ranges (the first model) and core ranges (the second model) related to month (the explanatory variable) (Braham et al., 2015, 227). This study also had a time series perspective and found that the average size of the Golden eagle home ranges was lowest and least variable from November through January and greatest in February to March and May to August. These patterns were related to the seasonal variations in breeding ecology, habitat associations, and temperature. The GPS measurements indicated that Golden eagles' greater ranging areas in the hotter summer months included ranging to the cooler, more biodiverse mountainous ecosystems, which comprised forest, grasslands, and scrublands. The breeding-season home ranges during October to May, were mostly lowland semi-desert and rock vegetation landcover. The overlap of these eagle home ranges and the focus areas for renewable energy development was most significant during the breeding season, when the eagle home ranges were smallest. The authors concluded on the major contribution of the GPS survey, and noted that the "Golden eagles in the Mojave Desert used more space and a wider range of habitat types than expected and renewable energy projects could affect a larger section of the regional population than was previously thought" (Braham et al., 2015, 225).

Another study examined the dispersal of Bonelli's eagles during their first winter (Cadahía et al., 2005, 415 - 419). This, termed the natal dispersal, includes three steps: the leaving of the natal area, the transitional dispersal stage and the selection of a new breeding location (Greenwood & Harvey, 1982; Clobert et al., 2001). The authors note that "only methods that provide information continuously on the location of an individual (mainly radiotracking) are adequate to separate different behaviours shown by individuals in this phase of dispersal" (Cadahía et al., 2005, 415). The most important development for this type of study is therefore use light-weight satellite telemetry, including platform transmitter terminals (PTTs), which allow the dispersal movements of wide-ranging birds (Fuller et al., 1995; Service Argos, 2000). The Bonelli's eagle has a complex range, mostly around the Mediterranean Sea, which underscores the usefulness of this type of research (Cramp & Simmons, 1979; Arroyo et al., 1992; Blanco & González, 1992; Rocamora, 1994; Real & Mañosa, 1997). This study tracked seven Spanish Bonelli's eagles by satellite from fledging to December 28, 2002 and included estimates of survival and including an estimate of survival and an analysis of their movement.

The authors found that the records using the satellite telemetry "concerning distances travelled and locations of dispersal areas agree with previous studies using visual sightings of wing-tagged birds and band recoveries, radiotracking and previous satellite tracking" (Cadahía et al., 2005, 417; see also Arroyo et al., 1992; Balbontín et al., 2000; Alcántara et al., 2001; Real & Mañosa, 2001). Another finding was that several of the birds died, one before leaving the natal area, and the distances travelled by the tracked birds were greater than those recorded using more conventional tracking methods (Cadahía et al., 2005). The distances from the nest ranged from 87 km to 663 km and the distances from dispersal areas to the nest ranged from 80 km to 536 km. Examples of the distances moved by the eagles in other studies include 32 km (Cheylan et al., 1996) and 100 km (Balbontín et al., 2000). Short distances have also been recorded (Real & Mañosa, 2001). Cadahía et al. (2005, 417) give three possible reasons why their findings may describe longer distances. Their method may be more advanced, as conventional radiotracking and sighting/recovery methods may fail to locate all the tracked birds, the shorter distances may reflect the close proximity of suitable habitat, and genetic factors may be relevant (see also Greenwood et al., 1979; Newton & Marquiss, 1983; Walls et al., 1999).

Another study on Bonelli's eagle by Pérez-Garcia et al. (2013) applied GPS satellite telemetry to the study of seven territorial Bonelli's eagles to ascertain their home range characteristics and annual and individual variations. The findings indicated that Bonelli's eagles had slight temporal changes in the areas of home ranges. There was a wide range in the eagles' use of space. Just over one quarter (27.3%) of the accumulated eagle home range was used during all the seasons. About 30% of the ranges was used only during a single season, illustrating considerable flexibility. The parts of the ranges near the nests and the common hunting and roosting areas were used regularly each year during breeding and non-breeding seasons. The evidence also revealed strong territorial instincts, during breeding and non-breeding seasons. Eagles also flew well outside these zones, up to trips 15 km. There was also a strong relationship between topographical landmarks (a line of mountainous areas and peaks and the Júcar river valley) that separated home ranges. The authors conclude that their contribution using GPS "reinforces the idea that combining information on season patterns, conspecific distribution and topography can help define both the shape and size of home ranges more realistically."

Another study looked at the movements of adult Verreaux's eagles from their nests using GPS tracking technology (Murgatroyd et al., 2016). The GPS technology is essential for the understanding of raptors spatial relations. Murgatroyd et al. (2016) acknowledge that "currently we have a limited understanding of the parameters of movement of territory-bound raptors, which are sensitive to environmental change." The GPS based study used a spatial resolution of 3 m and a time resolution of 3 mins to study the sample of Verreaux's eagle during the pre-laying period and the chick rearing period. This study looked at temporal (timing, duration, frequency and speed) and spatial (path length and maximum distance from nest) characteristics of the eagles' journeys from the nest. Habitat selection was also studied in the cultivated landcover of the Sandveld region. The results indicated that the home ranges were actually small, and mutually exclusive among the eagles, with most activities around the middle of the day, probably due to the need for thermals. An important finding was that the studied eagles preferred semi-modified environments (minimally or moderately modified landcover) to natural, unmodified or completely modified landcover. The authors therefore make the case that eagles may have benefitted from some of the agricultural developments, this having implications for the study of eagle ecology in the wider area (Murgatroyd et al., 2016).

A study by Väli & Bergmanis (2017, 104-107) compared the reliability of three tracking methods for the analysis of survival in adult Lesser Spotted eagles in the migratory and movement actions. This species is a Palearctic breeder that winters in southern Africa. Some of the Lesser Spotted eagles are philopatric (they return to the same breeding areas each year) allowing the recording of their annual survival rates (Väli & Bergmanis, 2017; see also Danko et al., 1996; Dravecký et al., 2013). Such tagging is facilitated by their foraging habits in open landcover. In this study, fifty-one adult eagles were marked at nest sites in Latvia and Estonia. Three types of tags were used; these were solar-powered satellite (Microwave Telemetry) or Global System for Mobile Communication-based (Ecotone) Global Positioning System (GPS)-transmitters fitted as a backpack on the eagle with teflon harnesses, plastic patagial wing-tags which were mounted by piercing and colored plastic leg rings (Varland et al., 2007; Dravecký et al., 2008). The results indicated that "colour rings and Global Positioning System transmitters indicated annual survival rates similar to other large raptors, but the rate suggested by wing-tags was significantly lower due to the higher rate of tag loss" (Väli & Bergmanis, 2017, 104). The survival rates were similar to those found for several other eagle species, and this showed the accuracy of the GPS methods.

Another study by Domenech et al. (2015, 429) looked at the possibility of using GPS technology to monitor the habitat requirements of the Golden eagles during the breeding and nonbreeding seasons. This study used PTT and GPS/PTT transmitter data from 14 adult, migratory Golden eagles, which were captured near the Rocky Mountain Front in Montana. The objective was to find the location and size of the Golden eagle's winter ranges and habitat use and selection within the chosen winter ranges. The findings indicated that the Golden eagles varied to a marked extent in the location and size of winter ranges in the western United States. Eagles showed high regularity to some wintering areas but varied more in choices of the annual range sizes. Features common in selected habitat types included perches, primary prey species, prey habitats, coniferous forests and riparian areas, usually in relatively low elevations, with possibilities for orographic uplift in mountainous areas. Avoided landcover types included urban substrates open grassland, agricultural developments and non-sagebrush-steppe habitat types (Domenech et al., 2015, 429). The authors suggest that these findings may support an effective conservation strategy for migrant wintering Golden eagles involving a "large geographic area with heterogeneous habitat allowing for adequate hunting perches and prey habitat, with little urban development or anthropogenic habitat conversion" (Domenech et al., 2015, 429).

López-López et al. (2016, 981–993) conducted a similar study, which was focussed on the foraging habits of Booted eagles. The Booted eagle is a forest raptor and may be used as a bioindicator of human-dominated agroforestry Mediterranean landscapes with traditional agricultural practices. The studied landscapes are changing, with increased landuse intensification, chemical and pesticide use, logging and forest fragmentation and biodiversity losses (Saunders et al., 1991; Stoate et al., 2001; Donald et al., 2002; García-Dios, 2009). The authors base their study on GPS telemetry and a large, spatially diverse population of marked birds over Spain, and promote this as the "first quantitative assessment of the home range size and space use of the Booted eagle by means of GPS satellite telemetry during the breeding season." The findings indicated that there were different levels of space use by eagles over the breeding season and these included long distance foraging movements from the nest, some up to 20 km throughout the breeding season. These findings implied that the eagles had larger home ranges than previously reported, these being scattered over large areas. The authors concluded that due to the wide ranges of eagle foraging, conservation management should avoid "setting restrictions around nesting sites using buffer areas of arbitrary radii" as these may result in "large areas of eagles' home ranges laying outside 'restriction' areas" and should therefore include all the agroforestry landcover (López-López et al., 2016, 981).

Another study of Booted eagles used GPS to look at migration patterns (Mellone et al., 2013, 409–413). In this case, the new information was gathered on the migration routes of the eagles. One important finding was that the fall routes were generally more eastern than spring routes, which revealed a loop migration. Average migration speeds were about 200 km/day, with generally no long stopovers. All the recorded eagles, except one individual used the same wintering area, which was located in Sub-Saharan Africa, about 2800–3500 km from their nests. The wide crossing at the Strait of Gibraltar forced long stopovers, up to 6 days, an average of 2.4 days for fall and 1.4 days for spring. In other cases, the eagles did stopovers in the Sahara Desert, and in Spain (Mellone et al., 2013, 409). In a similar study on migration, this using a Lesser Spotted eagle (ringed and fitted with two satellite transmitters) migrating from Slovakia as an example, Meyburg et al. (2004a, 1) found that the eagle migrated to South Africa, and in 2001, "it spent 43% of the year at its breeding site, 33% in its winter quarters, the remaining 24% being spent on migration." The fall and spring migrations took between 40 and 61 days, daily flight distance average of 178 km. The authors conclude that "the annual journeys flown, including movements around the wintering grounds, amounted in 2000-2001 to at least 20,396 km and in 2001-2002 to 19,041 km" (Meyburg et al., 2004a, 1).

Another study by Moss et al. (2014, 93-105) used GPS to record the home range sizes of Golden eagles in northern Sweden. The authors note that "this is the first detailed study on GPS-tracked adult Golden eagles in Fennoscandia. Our primary goal was to estimate home-range sizes for a Swedish Golden eagle population" (Moss et al., 2014, 93). The results indicated that the average home-range sizes for the eagles were among the largest reported by any study. An important finding was an "inverse relationship between home-range size and the percent of clear-cuts within the range" (ibid.). This hints that the Golden eagles may expand their range in compensation for their loss of habitat, which may be clear cut areas. The eagles also indulged in non-breeding movements during the breeding season, usually comparatively long distances. Some distances were very long, nearly 700 km north into northern Finland and Norway. There were also post-nesting movements, usually after nesting or breeding failure, which may be due to local food shortages (Moss et al., 2014, 93).

Another study by Weston et al. (2018), looked at the post-fledging period and the ranging of juvenile Golden eagles during this period, termed post-fledging dependence period (PFDP). The authors state that they used a "novel method" to determine this relationship, namely 35 juvenile Golden eagles were fitted with satellite transmitters, to determine their movements within the proximity or otherwise of the nesting area. The authors also note that "The period prior to an individual emigrating from its natal site and initiating dispersal is important for developing the skills that are ultimately required for surviving natal dispersal" (Weston et al., 2018). The results indicated that 29 of the 35 eagles showed increased mobility during the first phase, the ontogenic phase, which leveled off in a median number of 68 days. The next period, that of maintained mobility, ended at a median of 99 days, with a range of 24 to 176 days. The eagles that reached mobility faster had a shorter ontogenic period, but the variation in the length of the PFDP was mostly from the individual variations in the maintained phase. Other findings indicated that females developed at a slower pace and experienced a longer ontogenic phase. The males dispersed mostly in two modes, that of 89.5 days and 220 days, while females dispersed unimodally at 167.5 days. Most juveniles went through a maintained phase only after they were full mobile, even among the fast developers. Only six birds (mostly males) dispersed without a maintained period. According to the authors, this occurrence suggests that "the PFDP in Golden eagles is not simply a function of physical capacity for independence, but also a period when young eagles decide to remain, and are tolerated by parents, in their parental home range before dispersing.... We suggest that delaying dispersal may be a beneficial strategy for young raptors facing a competitive environment after PFDP (Weston et al., 2018).

This study is related to another similar research piece on dispersal by Golden eagles in Scotland (Weston et al., 2013). This study tracking data from 24 Golden eagles, which were fitted with durable GPS satellite transmitters as nestlings, and then followed through their first year. The findings of this study indicated Golden eagle dispersal occurred from 39 until 250 days after fledging. The methodology was based on two new methods, which to be contrasted with ten published methods. The first method was based on the "published ranging distances of parents as a measure of the natal home range, with the requirement that juveniles must exceed it for a minimum of 10 days (a literature-based measure of the maximum time that a juvenile can survive without food from its parents)" (Weston et al., 2013). The second method (Method 12) was based on "the biggest difference in the proportion of locations inside and outside of the natal home range smoothed over a 30-day period to assign the point of emigration:" this second method was then compared with the ten published methods (Weston et al., 2013). The method was used as the 'reference' or 'benchmark' against which to evaluate the other methods, on the justification that this was the most 'data intensive' method, due to the greater amount of local information on parental range use and the number of GPS fixes of the juvenile movements.

The other ten methods included seven that used a "derived distance-from-natal-nest threshold (methods 1-7) with or without an additional restriction of a minimum duration that this distance has to be exceeded before dispersal is deemed to have started" (Weston et al., 2013; see also Ferrer, 1993; Kenward et al., 1993a,b; Soutullo et al., 2006). The remaining methods (methods 8 – 11) utilised the "maximum rate of increase in distance from the nest as described by the coefficient of variation, the ratio of standard deviation to the mean over a set duration or number of locations" (Weston et al., 2013; see also Soutullo et al., 2006; Cadahía et al., 2008). The authors describe Method 12 as "novel to the present study and based on classifying an individual as inside or outside of its parents' home range in order to objectively identify when an individual becomes independent" (ibid.). The results indicated that Golden eagle dispersal in the study area occurred from 39 until 250 days after fledging (this was based on the use of method 12). This result was different from that found using the other published methods, as the latter gave estimates tended to assign the emigration point too soon. However, "by contrast the two methods we proposed provided very similar estimates for the point of emigration that under visual examination appeared to fit the definition of emigration much better" (Weston et al., 2013).

10.9 Definitions and Applications of Geographical Information Systems (GIS) to Eagle Studies

Geographical information systems (GIS) are defined as systems that record, store, transform and process spatial data, to produce information analyses of greater accuracy (Gomarasca, 2010). Like remote sensing, GIS is a broad field. One definition is that it is "an information system containing all spatial data of the atmosphere, the Earth's surface, and the lithosphere, allowing the systematic capture, update, manipulation, and analysis of such data, based on a standardized reference frame" (Göpfert, 1991). Another definition is "an information system with a database of observables of spatially distributed objects, activities, or events, which can be described by points, lines, or surfaces" (Dueker, 1979). It may also be "a system for decision support which integrates spatial data in a problem-solving environment" (Cowen, 1988); or "a comprehensive collection of tools for capturing, storing, retrieval, transformation, and visualization of spatial data of the real world for special applications (Burrough, 1979). It may also be a system or entity that undertakes the "capturing, storing, analysis, and visualization of data that describe a part of the Earth's surface, the technical and administrative entities, as well as findings of geoscience, economics, and ecological applications" (Bartelme, 2012, 145).

These definitions emphasize the work and capabilities of GIS. There are also definitions that look at the components of GIS. For example, Bartelme (2012, 148) argues that a GIS includes:

> A collection of spatial data plus the corresponding functions for storage and retrieval, a collection of algorithmic and functional tools, a set of hardware and software components needed for the handling of geospatial data, a special type of information technology, a gold mine for answers to geospatial questions, and a model of spatial relationships and spatial reconnaissance.

GIS, developed comparatively recently, when defined as the compilation and analysis of spatial data, with objective being the display of this data in spatial format. Mapping and the spatial display of field data have long existed within the fields of cartography and elementary surveying, but the added sophistications of GIS, which can incorporate remote sensing, GPS and computer technology is more recent. The GIS chronology generally starts with the federal Department of Forestry and Rural Development in Ottawa, Canada during 1960, which created the Canada Geographic Information System (CGIS), through the work of Roger Tomlinson (Tomlinson, 2003). This system was developed to collect, store and analyze data for the Canada Land Inventory, resulting in assessments of rural Canadian land capabilities linked to wildlife, waterfowl, forestry, agriculture and recreational landuse.

The CGIS was regarded as the originating GIS system largely because of its national coordinate system, integrated with a highly developed techniques for map and image overlays, measurements, digitizing and scanning, attribute and locational information. GIS structures with similar capabilities were developed later, such as Howard T. Fisher's Laboratory for Computer Graphics and Spatial Analysis at the Harvard Graduate School of Design (LCGSA 1965–1991). This complex conglomerate included SYMAP, GRID and ODYSSEY, which included sophisticated software codes and spatial data systems. This was one of the originators of the 1986 Mapping Display and Analysis System (MIDAS), later termed MapInfo for Windows, and the later web-based GIS programs on the World Wide Web (Fu & Sun, 2010; Goodchild, 2010; Maliene et al., 2011).

GIS has been strongly linked to various statistical technologies, both in terms of independent software such as SPSS, SAS and Microsoft Excel, and within GIS platforms such as IDRISI, ArcGIS, ERDAS Imaging and PCI Geomatica. These programs are at least partially linked to the science of geostatistics, defined as Olea (2012, 1) further defines geostatistics "as a collection of numerical techniques that deal with the characterization of spatial attributes, employing primarily random models in a manner similar to the way in which time series analysis characterizes temporal data" (Olea, 2012, 1). The task of geostatistics may generally in a larger sense be defined as the "study of phenomena that vary in space and/or time" (Deutsch, 2002).

Other definitions of geostatistics include those of Isaaks and Srivastava (1989): "Geostatistics offers a way of describing the spatial continuity of natural phenomena and provides adaptations of classical regression techniques to take advantage of this continuity"; and that of The Oxford American Dictionary (2018) which notes that geostatistics deals "with spatially autocorrelated data." Zhang (2009, 1) points out that geostatistics varies from general statistics because "classic statistics is generally devoted to the analysis and interpretation of uncertainties caused by limited sampling of a property under study… geostatistics however deviates from classic statistics in that geostatistics is not tied to a population distribution model that assumes, for example, all samples of a population are normally distributed and independent from one another." This focus on variation and uncertainty gives geostatistics an important link to the complexities of ecology (Campbell, 2017, 2018). The main function of geostatistics is the prediction of a subject's possible spatial distribution, through spatial expressions such as maps. There may be prediction based on simulation, where many possible maps are produced, which display the quantification of any uncertainties; or on estimation, where an estimate map is produced (Zhang, 2009, 1; Campbell, 2018).

Geology was the main birth science of geostatistics: "geostatistics evolved in the mining industry, it for a long time meant statistics applied to geology…geostatistical techniques have been successfully applied to applications in numerous fields, ranging from soil mapping, meteorology, ecology, oceanography, geochemistry, epidemiology, and human geography, to geomorphometry" (Hengl et al., 2009, 492). Currently, geostatistics has wide applicability, as "a branch of statistics that specializes in the analysis and interpretation of any spatially (and temporally) referenced data" (ibid.). For example, Bellier et al. (2010) published the results of geostatistics techniques applied to animal counts, using a hierarchical model of analysis for the spatial aspects of trends, patchiness and random fluctuations, recording the heterogeneity of wildlife populations and linked habitats. This was based on a hierarchical model with a Poisson distribution assumption (a Poisson distribution is discrete frequency distribution giving the probability of several independent events occurring in a fixed time). Their study used the example of fin whale (*Balaenoptera physalus*, Linnaeus, 1758) data, for mapping of animal densities with an assumption of non-stationary locations (Bellier et al., 2010).

GIS applications related to habitat studies use remote sensing images and other sources, such as field assessments and photographs, quadrats and samples, to derive supporting data, and computer-based software to classify the vegetation and landcover, and represent other features such as elevation, temperature and moisture. The ultimate result is maps and the spatial representation of statistical analysis. For elevation studies, a common tool is the Digital Elevation Model (DEM) (Campbell, 2018). This may be used for various landscape features, such as water, soil surfaces, and slopes. Others are the Digital Surface Models (DSM), Digital Terrain Models (DTM) and Triangular Irregular Networks (TIN) (GISGeography, 2018).

GISGeography (2018) defines the digital elevation model as "a bare-earth raster grid referenced to a vertical datum" which is created when the non-ground points, including cultural features (power lines, buildings, bridges, roads, etc.) and natural features (vegetation) are filtered out, resulting in a "smooth digital elevation model." This resultant "bare-earth elevation model is particularly useful in hydrology, soils and land use planning", as it may be used for hydrologic modelling (recording and measurement of water flow direction and accumulation, and watersheds); terrain stability studies (e.g., avalanche possibilities in high slope areas with little vegetation cover); and soil mapping (as soils are largely the result of elevation, geology, time and climate (GISGeography, 2018).

An example of a DEM-based hydrological study is that of Das et al. (2016, 1544) who define a DEM as "a regular gridded matrix representation of the continuous variation of relief over space." This study used digital elevation models (DEM) from Advanced Spaceborne Thermal Emission and Reflection Radiometer (ASTER V2), Shuttle Radar Topography Mission (SRTM V4, C-Band, 3 arc-second), Cartosat -1 (CartoDEM 1.0) and topographical maps (R.F. 1:250,000 and 1:50,000), to analyze topographical features (mostly relief, surface, size, shape and texture properties) of a drainage basin in the Indian Himalayas. The study found that DEMs derived from the 1:50,000 topographical map and ASTER GDEM V2 data were more accurate than the other DEM data products (Das et al., 2016, 1544).

An example of a Digital Surface Model (DSM) is the study by Laes et al. (2006). Here landcover data from aerial photographs is draped over a 3-D LIDAR digital surface model. This surface model allowed the mapping of the ground surface, and trees and other features to give a more accurate version of the landscape under investigation. In creating such a representation, two DEM, "a 'bare-earth' digital elevation model (DEM) and top-of-surface model depicting features above ground level such as vegetation." For the generation of the bare-earth DEM, the LIDAR pulses that represent vegetation or other nonground features are removed from the data. Laes et al. (2006) argue that "the ability to generate DEMs for both of the surfaces is the reason that lidar is a useful tool for forestry applications" and note that this, coupled with the use of terrain and surface models, detects and represents detailed landcover features.

An example of a Digital Terrain Model (DTM) is in the study of Durner et al. (2013). Here, a fine-grained digital elevation model was derived from Interferometric Synthetic Aperture Radar (IfSAR) to generate a predictive map of denning habitat in an area of polar bear (*Ursus maritimus*, Phipps, 1774) maternal denning, provided by the National Petroleum Reserve–Alaska (NPR-A) in northeastern Alaska. The objective was to locate potential denning habitat, to inform landuse, mostly petroleum activities. The authors tested the predictive accuracy of the model and found that the "final map correctly identified 82% of denning habitat estimated to be within the NPR-A" and concluded that although "mapping denning habitat with IfSAR data was as effective as mapping with the photogrammetric methods used for other regions of the Alaskan Arctic coastal plain, the use of GIS to analyze IfSAR data allowed greater objectivity and flexibility with less manual labor" and that "the use of IfSAR data to identify polar bear maternal denning habitat is a better management tool in the NPR-A (Durner et al., 2013, 197).

A triangulated irregular network (TIN) is a model of a continuous surface consisting of triangular faces. A study by Hazlett et al. (2008) gives the example of a TIN-based modelling exercise for forest-based research. This study used a TIN for slope assessment and a DEM for upslope areas and a topographic index, based in the Esker Lakes Research Area in northeastern Ontario. The focus of the study was on buffer zones as an aspect of forest-management strategies, due to the possibility that they reduce the effect of post-harvest nutrients on water quality, and the hypothesis that shallow slopes have longer residence times than steeper slopes. The implication of the last point is that the steep slopes are less able to moderate the quality of the water, and therefore such areas may need wider buffers. The findings were that the slopes were not correlated with nutrient (N and C) concentrations (except for dissolved organic carbon), while forest species and soil nitrogen were positively correlated with soil water N levels. The study concluded that slope is only a weak indicator for planning riparian buffers during harvesting (Hazlett et al., 2008). Although this study was not related to wildlife directly, the results may have implications for the management of habitats, as buffer zones may have different vegetation formations and densities than harvested areas (Campbell, 2017).

The recording of individual birds using GIS is largely done through point count records. Here a point count refers to a research method where a trained observer stands a specific, recorded locations and records all animal or single species sightings, or calls. The points at which such observations are made could be in a grid format, or at random locations. These sightings may then be recorded in tabular format in GIS or statistics software, as one column among others which may include dominant landcover features (e.g., coded numbers referring vegetation density or height, elevation, water proximity and coordinates such as latitudes or longitudes or UTM numbers). These columnar data can be used for multiple regression analyses in statistical packages or imported into GIS software and displayed as points on images or line and polygon-based vector maps.

Hostetler and Main (2001) list some of the main features of point surveys applied to birds. The point count in their study is defined as the observer standing in a location, counting all the individual birds (of each species) within a definite, consistent radius. The choice of radius is balanced between an extent limited enough to hear or see all the desired species, and large enough to maximize the required ecological information. The authors suggested a radius of 20 meters (65.5 ft.) "for most situations" (Hostetler & Main, 2001, 2). Also suggested were consistent count times (e.g. in this study, ten minutes), one to three counts per month, and the first three hours after sunrise, as this is the main activity period for birds: for nocturnal species, three hours after sunset.

During the count, the recorder counts all the birds seen or heard within the radius of the point count area. Those birds heard or sighted outside the radius of the study are also be counted, depending on the limitations of the study. Birds may be classified as FT (fly thrus), these including those that fly through the survey area below the tallest point in the area, but do not alight on any structure, and did not originate their flight from within the radius of the study area. They may also be classified FO (fly-overs), where they fly over the tallest structure in the census area. The count must also ensure that double counting of birds does not occur. Other features recorded include general climatic conditions, including wind, cloud cover and temperature.

An example of a point count survey is that of Volpato et al. (2009). In this study, undertaken in the southern Brazilian Atlantic forest (August 2003 to January 2006) there were 12 study sites, located in seasonal semideciduous forest (SSF), mixed ombrophilous forest (MOF), and dense ombrophilous forest (DOF). There were 12-point counts, with a radius of 100 m, a minimum of 200 m apart and 300 m from the forest edge. The counting lasted 15 minutes at six points per morning. The counts were surveyed once for each season; spring: September-November; summer: December-February; fall: March-May; winter: July-August), making a total of 576-point counts: 192 in SSF, 240 in MOF and 144 in DOF. The spring/summer (breeding season) and fall/winter (non-breeding season) were the main seasons. Statistical methods used included the Sorenson index, for the analysis of species numbers between breeding and non-breeding seasons; the Morisita-Horn Index to compare the number of contacts between the two seasons; and the Mann-Whitney test (U) for comparison of bird richness and abundance in breeding and non-breeding seasons, in the three forests classes and the three functional bird groups.

Another example of a point count survey is that of Applegate et al. (2011), located in Fort Campbell Military Reservation on the Tennessee–Kentucky state line, USA. This study used 110-point counts, at least 250 m apart, in hardwood forest, pine (*Pinus echinata*) plantations, grass and agricultural fields. The 10-minute counts were conducted from 0500 to 1000 hours during late May through June during 2003–2006 and 2008–2009, avoiding rain, heavy fog, or winds in excess of 6.5 km/h. Birds were located by both sounds and sightings. Data analyses used DISTANCE version 6.2 and SPSS version 17.0 (SPSS 2008), with point count data analyzed using cosine, simple polynomial, and hermite polynomial expansion (Thomas et al., 2010).

GIS applications to studies or eagles may emphasize the point counts flyovers, perching birds or may also use features such as nests. An example of the latter is the study by Letto et al. (2015), which used GIS analysis to evaluate the factors for Bald eagle population trends in Placentia Bay, Newfoundland during the 1990-2009 breeding seasons. This study tested the hypothesis that Bald eagle populations, in some cases recovering from the DDT bans, have experienced dense population surpluses of nonbreeding adults when the optimal habitat and food are restricted. The selected case study was of Placentia Bay, Newfoundland, which is one of the most densely populated Bald eagle populations in eastern North America. The methodology used field classifications of nest sites, GIS and statistical analyses. The nests were classified as active (including chicks, and incubating adults or vocalizing adults within 100 of the nest); inactive (no eagles nearby, or any observed flew away permanently); and occupied (at least one adult eagle nearby, with no vocalizations), which was neither active or inactive. Nest sites were classified as the area within which the nest was located, and nesting territories were within a 2.6 km^2 circular area, around the nest as the central point. This followed the methodology of Grier et al. (1983).

The GIS-based analysis rested on 163 pseudo-absence points (representing unused nest sites), which used Hawth's Tools in ArcGIS (version 9.3., ESRI, Redlands, California, USA), these points being to represent unused nest sites. The pseudoabsence points were all located at least 910 m apart and 5 m of the shoreline, the accuracy of these selections based on the well-surveyed location of the case study. Habitat parameters for the analysis nest site elevation, percentage of forest or water cover, and location on the mainland or islands, the last point relevant to previous research (Gerrard et al., 1975, Livingston et al., 1990), that found that Bald eagles preferred island nesting locations. Food resource parameters included the proximity of the nest/pseudo-absence site to the nearest relevant fish aggregations, including capelin (*Mallotusvillosus*, O.F. Müller, 1776) and Atlantic herring (*Clupea harengus*, Linnaeus, 1758), known to be an aspect of the eagles' diet (Gende et al., 1997). Using GIS, the circular 2.6 km^2 buffers encircling the observed and pseudo-absence nest sites denoted breeding territory, with calculations for elevation, landcover and proportion of water and forest, and the proximity between the nest sites and fish aggregations, and designations of 0 and 1 for mainland and island locations.

The supporting statistical tests used logistic regression, which included data partitioned into two equal sets (occupied nest sites and pseudo-absence data) comprising 20% of the data, termed the testing data and 80% as training data. The logistic regression analysis was used to determine the probability that a nest would be built within the habitat and food resource parameters. Linear models were used for occupied vs. pseudo-absence locations, and to determine the probability active or occupied nests (Letto et al., 2015). The findings indicated that there were no significant differences in habitat or feeding behaviors between the contrasted datasets (unoccupied versus occupied, high versus low activity). This indicated that there was no habitat saturation of Bald eagle populations in the study area and led the authors to suggest further studies to discover the possible factors for any high proportions of nonbreeding adults.

10.10 Conclusions

This chapter has examined the application of current geomatics techniques (remote sensing, drones, GPS, GIS) to the ecological relations surrounding eagles. This is obviously a burgeoning area of scientific inquiry, with new reams of research materials emerging annually. The key areas emerging concern the more accurate recording and documentation eagle reproductive, foraging and migration patterns. The study of eagles is highly complex and must include detailed assessments of the prey species and their favored habitats. Other parameters include those that measure the human modification of the environment (deforestation urbanization, agricultural expansion, pollution, numbers of prey species) which may be measured and analyzed with geomatics techniques. The results of geomatics-based studies indicate the impact on eagle numbers, distributions, migration and foraging patterns, and adaptive abilities. The adaptive abilities of eagles and their survival in the face of ecological change, are examined in the next chapter, along with the scientific paradigms evolved to examine and explain these relationships. These results may inform stakeholders on the actions required for eagle survival. The conservation actions and historic human assessments of eagles are covered in the last chapter.

11

Eagles as Individuals and Free Spatial Agents

11.1 Introduction

This chapter examines the theoretical concepts and physical research evidence for individualism in eagle behavior, including intra-specific behavioral differences and adaptation to the rapidly changing ecologies of eagle life spaces. In this it seeks to examine the intelligence of eagles in their survival strategies, and thus to hypothesize the possibilities for a future for eagles in an increasingly human dominated landscape. As has been noted through the first parts of this book, human action has greatly changed the environments that eagles occupy and there are inter-species differences in the ability of eagle species to adapt to the changes created directly and indirectly by people. Research findings also hint at intra-specific, individual adaptation to such ecological constraints. These must be matched with variable human behavior and explained as shared relations between people and animals in habitats and ecological contexts undergoing multidirectional change. As such, this is not an exercise in pure animal ecology, but must intertwine with urban and agricultural expansion, cultural perspectives on the relevant animals and the possible adaptive abilities of the relevant animals. Recently, some theoretical paradigms such as certain strands of animal geography have emerged that treat animals as at least partially active subjects, rather than only generalized objects. The contents of this chapter are summarized in Table 11.1.

Table 11.1.

Issue	Options	Examples	Refs.
Movements, reactions to technology	New routes	Bald eagles, four migration strategies	1
	Response to aircraft	Bald eagles' variable reactions to aircraft	2
	Collisions with aircraft	Bald eagle collisions with a aircraft	3
	New habitats	Bald eagle movements in urban areas	4
Foraging	New prey species	Martial, Crowned eagles' predation	5
	New foraging areas	Verreaux's eagle in suburbs	6
	Tracking prey species	Bald eagle tracking Mallard ducks	7
	Interspecific variation	Spanish Imperial eagle, kleptoparasitism	8
Breeding	New areas, e.g. urban	Bald Eagles in new urban areas	9
	Avoiding people	Varying nest height, distance from people	10
	Pairing	Indian Spotted eagle polygyny	11
	Dispersion	Martial eagles dispersing from territories	12
Inter-species relations	Conflict	Short-toed eagles vs Long-legged buzzards	13
	Kleptoparasitism	Bald eagle attacks on Peregrine falcons	14
	Dominance	Golden eagle dominating White-tailed eagle	15
	Avoidance	White-tailed eagle avoiding Golden eagle	16

Sources: 1. Wheat et al., 2017; 2. Grubb & Bowerman, 1997; 3. Lien, 2015; 4. Leahy, 2018; 5. Endangered Wildlife Trust, 2012; 6. Simmons et al., 2007; 7. Jorde & Lingle, 1988; 8. Margalida et al., 2017; 9. Goulet, 2009; 10. Watson, 2004; 11. Sant et al., 2017; 12. Van Eeden et al., 2017; 13. Friedemann et al., 2016; 14. Dekker et al., 2012, 2015; 15. Halley, 1998; 16. Halley & Gjershaug, 2008. These issues summarize some the issues of subjective eagle behavior discussed in detail below.

Therefore, this chapter will examine the development of such applied sciences and their usefulness in the study of animal and eagle behavior. This is rooted in the development of relevant strands of animal geography, as a complement and counterpoint for classical ecology and biogeography. The theory and development of zoogeography and animal ecology are examined, to explain and illustrate the differences between this discipline and that of animal ecology (Campbell, 2014a). Zoogeography, the study of the spatial ecology of animals is very important for the study of eagles, because much of eagle ecology is spatialized; migration and seasonal change, regional disparities in presence and extirpation, juvenile dispersal, local adaptation to deforestation and other environmental change, and spatial vagaries in foraging patterns all require spatial analyses, from which more critical understandings may be derived. Such spatial issues are strongly scaled, from global patterns to the micro-spaces of interaction between people, animals and ecologies. While large scale, generalizable studies may rely on data sources such as broad ranging observational samples, aerial photographs and satellite images covering large areas, smaller scale, more subjective studies may rely on critical, intensive long-term observation of individual animals, and data from drones. The latter approach offers a challenging approach, including the unpredictable, subjective, individualist behavior of animals to complement the more classical, large scale approach of zoogeography and animal ecology.

11.2 The Development of Zoogeography, 'Disequilibrium Ecology' and Animal Geography

Biogeography examines and analyzes the distribution of living species in geographic space and through geological time, and as such links ecology, evolutionary biology and palaeontology (McDonald, 2003). Branches of biogeography include ecological, historical, analytical and conservation biogeography, all linked to basic ecology and spatial concepts. Zoogeography, the biogeography of animals (as opposed to phytogeography, the geographical study of plants) studies the present and past geographic distribution of animal species, including the methodologies from both ecology, zoology and geography (McDonald, 2003; Campbell, 2017). Zoogeography evolved from a regional or global science; "the science of mapping out the surface of the earth with reference to its faunæ; the study of the distribution of animal life", this concerned with "local peculiarities in animals and plants" in the "various regions of the earth's surface" (Ingersoll, 1920). Various scholars have mooted that the lower, less studied status of zoogeography compared to phytogeography, as animals are a "paradox in biogeography"; although they are popular, but there is a "relative scarcity of texts in animal biogeography" (Ganderton & Coker, 2005: 123). Micro-scale interfaces between zoogeography and human geography may also be neglected, though these would be relevant to the survival of animals in human dominated landscapes (Campbell, 1998, 2017; McDonald, 2003). Two comparatively recent developments that encourage the integration of the global and regional analysis of traditional zoogeography and the more local micro-scale interactions between animals and human influenced landcover and ecological change are disequilibrium ecology and animal geography.

Disequilibrium ecology or the 'New' ecology contrasts with the classical ecology at the root of biogeography in that the latter was based on Newtonian physics, which has analyzed ecology as comprised of "deterministic dynamics", which may be determined by a "few simple laws" (Solbrig, 1994, 17), and applied to generalization, scientific universalism, vegetation climaxes, unilinear changes in animal populations, ecological succession and equilibrium, and human/nature separation (Pickett et al., 1992; Zimmerer, 1994; Begon et al., 2005). In disequilibrium ecology, the emphases shifted to continual, relational fluxes, unpredictable outcomes and trajectories, non-linear, non-equilibrium and multi-directional ecological dynamics, which may also be subject to "multiple interpretations" (Nicolis, 1994; Scoones, 1997, 162; Hobbs et al., 2009). Such trajectories are internally, rather than externally derived and are part of constantly changing "non-equilibrium or multi-state systems" (Young & Solbrig, 1993: 324).

This creates new scientific questions related to the ability and likelihood of animals adapting, through behavioral and coping strategies, with the continually changing environments that have become the modern reality (Campbell, 2017, 2018). This new, more flexible and possibly more relevant approach undergoes problems of construction, analysis and application. For example, Scoones (1999) noted the emergence of the new ecology encounters limitations in the study of socioenvironmental systems, which is possibly "a consequence of the lag times of cross-disciplinary communication: different languages, frames of reference and methodological approaches are evident across the disciplinary divides" (Scoones, 1999: 489). Scoones (1999) and other scholars point to three relevant engagements that the new ecology has facilitated for the study of nature-society relations, spatial ecological change and dynamic animal ecologies. One 'new' point was the increased focus on long-term and short-term environmental histories, including paleoecology and evolutionary ecology, and recent historical change within the engagements of nature-society relations (Williams et al., 1994; O'Connor, 1997; Campbell, 1998, 2017). Another recently developed point was the promotion of variability in both space and time, enabled the description of environmental dynamics to move "beyond the simple assumptions of equilibrial regulation to a wider appreciation of complex dynamics, uncertainty, and surprise" (Scoones,

1999: 483). Thirdly, more attention to scale, structure and agency, allowing the documentation of "nonlinear interactions", which concern unexpected results and adaptations (Scoones, 1999: 483) and the contribution of these to small and large-scale patterns, including human-animal relations (Allen & Starr, 1982; Turner, 2003; Campbell, 1998).

Concerning animal geography, during the 1960s, a dynamic period for paradigmatic changes in social and ecological studies, certain strands within the so-called 'animal geography' subdiscipline began to examine animals as subjects with active abilities (termed actancy, like human action) rather than as mechanistic objects (Wolch & Emel, 1998; Lorimer, 2007; Barua, 2013). Actancy in this sense is defined as encompassing behavioral reactions, which involve free, sometimes unpredictable choices are made in response to different events (Murdoch, 1998). Like the disequilibrium ecology paradigm, these approaches called for more flexible perspectives in the uncovering of subjectivities undetected by the more positivist science of zoogeography, possibly creating more effective links with the human geographies of wildlife (Anderson, 1997; Philo & Wilbert, 2000; Yarwood & Evans, 2000; Emel et al., 2002). There fundamental points were the more direct theoretical focus on animal behavioral subjectivity, and its engagement with 'cultural human geography.' Here there is a shared actancy between people and animals and the integration of these with quantitative zoogeographical field methods (Wolch & Emel, 1998; Johnston et al., 2000). The contribution of these to zoogeographical studies has been supported by recent studies (Campbell, 2017, 2018).

Therefore, for the study of birds like eagles with their complex environmental relations, animal geography and avian ecology/zoogeography would be two ways of looking at the relevant engagements; animal geography sees birds as actants, active agents creating relations with people as individuals and groups, while avian ecology/zoogeography focusses on statistical generalization, at inter- and intra-species levels (Campbell, 2017, 2018). The main question is the extent to which animal actancy may emerge and the methods available to determine and measure such behaviors. As pointed out by Johnston et al. (2000, 25) there are instances where "animals emerge as active agents fashioning the environment they inhabit or as experimental objects in the social relationships which bind them in different ways to people." In the case of eagles, such an issue would be particularly problematic, due to their generally huge ranges and capabilities for wide dispersal and in some cases aggressive intrusion into human life spaces.

11.3 Migration and Movements

Recent emphases on subjective behavior in eagles concern migration and movements. These may record unexpected behavior from eagles, in route, foraging or time selection, which deviate from 'normal' action. These may also describe individualized reactions to extremely dynamic circumstances, mostly due to human action. For example, Wheat et al. (2017) argue that "intraspecific variation in movement strategies may be particularly prevalent in environments in which resources are highly variable, as individuals that can learn when and where to access the best resources and alter their movement strategies accordingly will be those most likely to survive and reproduce" (see also Komers, 1997). The authors found evidence of what they term "four movement strategies" among some Bald eagles in southeastern Alaska and western Canada. These were sedentary breeding individuals; non-breeding, seasonal migratory individuals; non-breeding localized individuals; and non-breeding nomadic, irregularly moving individuals. There was some behavioral clustering in these activities, as males mostly traveled further than females and immatures more nomadic. The authors conclude that such intraspecific variation results in "different space use patterns among contingents, which has important implications for conservation and management" (Wheat et al., 2017). The authors note that there are many studies on interspecies differences, but only few on intraspecies variability of movement strategies (see also Komers, 1997). They also add that "intraspecific variation in movement strategies may be particularly prevalent in environments in which resources are highly variable, as individuals that can learn when and where to access the best resources and alter their movement strategies accordingly will be those most likely to survive and reproduce" (Wheat et al., 2017, 2).

Another issue concerns individual eagle choices for nesting areas, which appear to be based on their individual assessment of both tolerance of people and human presence. Leahy (2018) also reports on Bald eagle behavior and cite a respondent; "One of the interesting things about Bald eagles is that they're a very individualistic species and their behavior isn't the same across the board, so it's hard to generalize about them… For example, one eagle might learn to scavenge for food by the highway, while another bird may never even go near the roadside." Sullivan (2016) reports incidents of Bald eagles adapting to city living. This article quotes a respondent that argues about Bald eagles: "the main thing is they just don't really care as much about people anymore"" the result being that these Bald eagles have increasingly began to nest in urban areas. Based on protective laws, such as the Bald Eagle Protection Act (1940), the Bald and Golden Eagle Protection Act (1962), the ban on the pesticide DDT (1972), and other federal and state protection and recovery efforts, the Bald eagle population has increased, hence there are ecological problems. The population in the 48 contiguous states increased from 487 in 1963 to 9,789 in 2006, (U.S. Fish and Wildlife Service) and reached possibly 24,000 in 2016 (Sullivan, 2016). Nesting habits vary individually, within urban structures and substrates, some eagles avoid people, while others are unaffected by human presence.

Goulet (2009) also observed that “Bald eagles have begun invading suburban areas such as agricultural fields, landfill sites and city parks in large numbers” (see also Berry et al., 1988, Gerrard & Bortolotti, 1988, Buehler, 2000), while acknowledging that “overall, the ecology of Bald eagles in urban settings is poorly known… as well as the reasons leading them to establish in this new environment” and “how they differ from rural or wild counterparts.” Goulet (2009) notes that established knowledge hints that Bald eagle utilized mostly living Black cottonwood and Douglas fir, dominant strong trees and isolated trees for nest-building and dead trees are used for perching (see also for other areas, Gerrard et al., 1975; Stalmaster & Newman, 1979; Anthony et al., 1982; Swenson et al., 1986; Stalmaster, 1987; Anthony & Isaacs, 1989; Wood et al., 1989; Livingston et al., 1990; Curnutt & Robertson, 1994; Chandler et al., 1995; Watts & Byrd, 2007). However, Goulet (2009) found the eagles in their study neither selected the tallest or strongest trees for nesting or isolated trees (Stalmaster & Newman, 1979; Watts & Byrd, 2007). Goulet (2009) observed that “the fact that a third of the nests were built on residential grounds and that 75% of them were found within 230 m of buildings and roads or within 31 m from a potential source of disturbance illustrates the great adaptability of eagles to urban landscapes”: furthermore, “This contradicts earlier findings in some previous habitat selection studies which stated that eagles avoid manmade structures and that the distance of nests to human development in most populations averages >500 m” (Goulet, 2009; see also Andrew & Mosher, 1982; Fraser, 1985; Fraser et al., 1985; Anthony & Isaacs, 1989; Wood et al., 1989; Livingston et al., 1990; Bowerman et al., 1995).

11.4 Prey Foraging in New Environments

Prey foraging is another area where there is pronounced individualism among eagles. For example, Endangered Wildlife Trust (2012) report the three larger eagle species, Martial, Black and Crowned eagles vary in their attacks on small livestock, and in some cases “rogue” individual eagles choose predation of young lambs or goat kids. They emphasize that these rogue eagles are invariably inexperienced, young birds, which may be prevented from attacking young stock by the presence of adult eagles and less likely on an individual basis to attack stock if other prey species are present (Endangered Wildlife Trust, 2012). Margalida et al. (2017, 4192) also report in the juveniles of the Spanish Imperial eagle in scavenging, which has been presumed to be related the “influence of age, sex, breeding status, habitat quality, prey type, and landscape characteristics.” The findings indicated that, Spanish Imperial eagles accessed food through hunting in flight (42%), scavenging (30%), hunting from a perch (16%) and kleptoparasitism (12%). Scavenging and kleptoparasitism were mostly used by juveniles. There was no relationship between scavenging and the variables listed above, which suggested that “kleptoparasitism is an opportunistic behavior used sporadically… Scavenging is also independent of habitat quality and landscape characteristics.” Therefore, a lack of live prey was not found to be factor for scavenging habits in juveniles but might be individual judgement based on lack of hunting ability. The conservation implications were that the preadult individuals might risk mortality due the eating of poisoned carcasses. The authors note that “some individuals seem to use scavenging throughout their life regardless of habitat quality, suggesting that individual personality also could influence this behavior…. These individuals can learn to handle these less valuable prey as alternative prey balancing the biomass provided by other prey, as well the time and energy invested to obtain them with hunting techniques” (Margalida et al., 2017, 4197; see also Hughes, 1979),

In another prey foraging example, Friedemann et al. (2016) document the results of interactions between the long-legged buzzards and short-toed eagles (“two sympatric top-predators”), which had only recently (the previous two decades) become sympatric during their breeding season in the Judean Foothills, Israel. The study exemplified how “ecologically-similar species were found to develop specific strategies to partition their resources, leading to niche differentiation and divergence, in order to avoid interspecific competition” (Friedemann et al., 2016). The findings indicated adaptive behavior on the part of both species, the buzzards foraged closer to their nests than the eagles did to theirs, the buzzards foraged in natural vegetation, while the eagles frequented cultivated fields, the buzzards were active through the day, while the eagles were more active in the early afternoon and the buzzard took a wider range of prey and took mostly lizards, while the eagles favored snakes. Such adaptation ensured the spatial coexistence of these species. The buzzard was also found to be a more opportunistic predator in terms of prey selection (Friedemann et al., 2016, 9). The authors define this diet adaptability as “diet plasticity” which is found in some other species (Martin et al., 2014; Kajtoch et al., 2015). The Short-toed eagle widened its prey species from snake specialism to a broader diet in this new area, possibly as a reaction to the competition from the buzzard (Friedemann et al., 2016, 9).

In another example of improvised diets, Phillips & Blom (1988) examine Golden eagle predation on livestock. Opportunistic predation by Golden eagles on mammalian livestock and even larger than average wild mammals has been reported in other studies (Cooper, 1969; McEneaney & Jenkins, 1983; Love, 1989; Phillips et al., 1996; Nybakk et al., 1999), the animals killed by eagles including a Red deer calf, domestic sheep, semi-domestic reindeer and domestic calves. Phillips & Blom (1988) point to Golden eagles in this regard being more dangerous than Bald eagles in the predation on young sheep and goats. Specifically, the authors note that although “eagle numbers were reported to be increasing throughout the West, but livestock losses to eagles were staying at about the same level" (Phillips & Blom, 1988, 241). In this case, the issue was serious as the authors note that "the primary conclusion that can be drawn from past damage assessment studies and mail surveys is that eagle predation on lambs and kids can be locally severe and can have substantial economic impact on individual producers", this problem facing all western states except Washington and Oklahoma. Migrant eagles were blamed more than resident birds, and both Golden and Bald eagles were increasing in numbers at the time. Solutions tried included live trapping and relocation of eagles, which generally was ineffective (Phillips & Blom, 1988).

Bald eagles have also adapted to foraging in urban areas. Lien (2015) reports on accidents suffered by eagles in a case study of the Minneapolis and Saint Paul metropolitan area (frequently called the Twin Cities), arguing that "For an eagle, life in the suburbs can be dangerous. With pesticide use common, plenty of roadkill available and lots of power lines crisscrossing the Twin Cities, our national symbol must navigate a tricky set of obstacles here to stay healthy. Not all do." Individualistic scavenging is one factor for these occurrences, and the eagles have varying abilities to avoid traffic when eating road kills. Forty-two injured Bald eagles were captured and transported to the Raptor Center at the University of Minnesota in the period 2010 to 2015. Injuries indicate that these eagles were poisoned, or hit by vehicles, or injured by unknown factors. Injured eagles from the rest of the state brought to the center during that period numbered 279, and there were another 182 from other nearby states, especially Wisconsin. The author however cautions that such injured eagles are much more likely to be found in urban areas than in forested, rural areas. There were also likely to be poisonings from lead fragments from shot animals and pesticides, and collisions with power lines and airplane collisions (three over the past 20 years in the Twin Cities, out of the national total of 114 over the same period). Relocations of injured eagles may not be successful, as they are adaptable and may return, the evidence being the results of tracking with solar-powered GPS transmitters (Lien, 2015).

Other eagle species that are variably adapting to urban life include the Crowned eagle (Van der Meer et al., 2018). Van der Meer et al. (2018, 241) found that in the urban area of Durban, KwaZulu-Natal, South Africa, the main prey fed to nestlings were the Rock hyrax and the Hadeda Ibis. McPherson et al. (2016b) found similar findings, and that the expected high level of attacks on urban animals such as pet dogs and cats was lower than expected; such domestic stock was only 6% of the identifiable prey, with no dogs and few cats delivered to the nest. Additionally, however, "Juvenile and sub-adult eagles are most frequently identified at in situ attacks of pets, particularly toy dog breeds" (McPherson et al., 2016b, 383). Van der Meer et al. (2018, 241 stressed the importance of protecting the Durban Metropolitan Open Space System (D'MOSS) zones for the protection of the eagle and the prey species.

The Verreaux's eagle is another species adapting to urban expansion, exemplified by breeding pairs in the green urban areas of Johannesburg and Cape Town (Simmons et al., 2007, 42). Certain dynamics occurred for the two nesting pairs surveyed in the two cities. In one case, the population of the Rock hyrax declined under urban pressure, and the eagles resorted to preying on chickens. In this case, the study project organizers were obliged to feed the eagles with hyraxes, rabbits and birds to enable the eagles continual breeding and to alleviate the pressure from chicken owners. The authors pose the question, "Does human disturbance per se limit eagles breeding in urban areas?" (Simmons et al., 2007, 42). The conclusion based on the observation of a breeding pair under the constant scrutiny of tourists is that human presence may not necessarily be negative. The authors conclude that the actual factor is the decline of the Rock hyrax population under urban pressure, and the attempts of the eagles to improvise with bird prey, which may be insufficient to maintain a viable, breeding population of eagles (Simmons et al., 2007, 42). Despite the rareness of Crowned eagle attacks on companion animals, there is one record that details such an attack. Umraw (2015) documents is referred to as a "rare occurrence" when a "home yesterday morning, leaving a family in distress after losing their pooch to the large bird of prey… The incident caused hype on social media yesterday, but pet owners have been reassured that the incident was an isolated one and it is not a cause for panic."

Dispersal behavior is another individualized reaction of eagles to environmental change. For example, van Eeden et al. (2017, 1) examine the tendency of the Martial eagle to become a "floater" (a roaming, ranging bird with no territory). In this study, the home ranges averaged at 108 km^2 in study which acknowledges that "The habitat use and ranging behaviour of Martial Eagles is poorly understood." Machange et al. (2005) is cited as one of the few studies of the habitat preferences of the Martial eagle. Two of the tagged adult eagles in this study "apparently did not hold a territory, and accordingly ranged more widely (ca. 44,000 km^2), and beyond KNP boundaries as floaters. Another two territorial individuals abandoned their territories and joined the 'floater' population, and so ranged widely after leaving their territories." van Eeden et al. (2017, 1) hypothesize that such "unexpected" developments could be a reaction of some individuals to landcover and habitat degradation in their territories. They further argue that "these unexpected movements after territory abandonment could indicate underlying environmental degradation. Relatively high mortality of these wide-ranging 'floaters' due to anthropogenic causes (three of four) raises further concerns for the species' persistence." Even when floating, the habitat preferences Martial eagles appeared to remain landcover in near rivers, with more tree cover and denser bush rather than more open bush or grassland.

11.5 Species Reintroductions

Another area where individualism in eagles emerges is in species reintroductions, when such birds must fend for themselves for the first time. Morandini & Ferrer (2017) define "a successful reintroduction as one in which when the probability of extinction during two times the maximum live-span period for the species (20, 50, and 64 years respectively) was less than 0.001 (P<0.001)" and argue that "reintroductions have been increasingly used for species restoration and it seems that this conservation tool is going to be more used in the future." Muriel et al. (2011, 323-333) describe attempts to reintroduce the Spanish Imperial eagle into the Spanish province of Cadiz. Between 2002 and 2010, a total of 73 young eagles were released at four locations by means of hacking. Hacking is defined by the American Eagle Foundation as "the method used to simulate Eagle nesting and Eagle population recovery in a particular area by releasing fledgling eagles into the wild from an artificial nesting tower." Between 2002 and 2010, 58 eagles were released by means of hacking techniques in three neighboring locations in the La Janda area in Cadiz province and in the Doñana area. Most of the hacked birds were wild-hatched chicks and only a few came from wild-laid eggs hatched in captivity and most were rescued chicks highly vulnerable to disease, nest collapse, siblicide or parental desertion.

The authors argue that the procedures adopted, including supplementary feeding and consultation with nearby landowners have encouraged nestling survival and increased the productivity of low-quality territories in the past (González et al., 2006a; Consejería de Medio Ambiente, 2009; Madero & Pacheco, 2009). Nestlings were translocated when on average 47.8 ± 6.1 days old to the hacking facilities and spent 28.8 ± 6.2 days before fledging. Back-pack radio-transmitters (models TW-3, Biotrack Ltd., UK; and 5/XOB 17-04, Wagener Telemetrieanlagen, Germany) were attached to the nestlings using a harness before release for later flight and location monitoring. Eight more nestlings were tagged with satellite transmitters (Microwave Telemetry Inc., USA; and North Star Science and Technology, USA) (see Kenward, 2001, for the use of these methods).

The eagles were marked with a metal ring and a color ring for identification. The eagles adapted to their environment rapidly, with flight improvements, dispersal behavior and dispersed beyond the release areas, the result being 83% of hacked juveniles survived until six months old and 66.7% until their first year. The deaths were due to electrocution (79%) (females were 73.3% of electrocuted individuals, possibly due to their longer wings), poisoning (10.5%), fence collision (5.3%) and natural causes (5.3%). The survival rate was similar to the estimate for the global population during 1990-2001 (six months: 89.4%, one year: 66.9%; Ortego et al., 2009) and higher than the survival of the nearby Doñana subpopulation with similar dispersal areas, during 1986-1989 (six months: 39.2%; Ferrer, 2001). Muriel et al. (2011, 323- 333) point out that the lower death rate compared with the Doñana subpopulation could be due to "reduction of mortality factors over the last 20 years, especially the modification of dangerous powerlines in the breeding and dispersal areas" (see also López-López et al., 2011). Despite this development, electrocution was still the top killer of released juveniles (see also Ferrer & Hiraldo, 1992). This study showed that the released eagles were able to acquire skills and adapt to their environment as wild reared birds do and were an important contribution to the understanding of eagle ecology and adaptation.

In a similar study, Fernández et al. (2009, 17) followed the ranges and space use of eight radio-tracked Spanish Imperial eagles are described. The findings indicated that the annual mean breeding season range was 25,146 ha, while the non-breeding season range was 20,557 ha. The largest daily distance covered was 113.6 km. Males generally flew greater distances than females during the breeding season, breeding home ranges were negatively correlated with densities of Rabbits (*Oryctolagus cuniculus,* Linnaeus, 1758). Eagles adapted to rabbit populations by flying longer distances when their home range had fewer rabbits, sometimes flying as far as 16.2–28.9 km from their nests. The authors note that "our results suggest that home ranges varied with prey density and a bird's reproductive status. Larger breeding ranges are probably related to an increase in energy requirements, while habitat quality is probably a regulatory mechanism of space use."

Therefore, these eagles choose a wider foraging technique to achieve feeding requirements (Fernández et al., 2009, 17), unlike some other eagle species, which broadened their prey base to cope with the rarity of the main prey species (e.g. the Crowned eagle in the face of rarity of hyraxes, see Endangered Wildlife Trust, 2012; Lien, 2015; McPherson et al., 2016; Van der Meer et al., 2018). The distances for foraging Spanish Imperial eagles in such situations were longer than those of large eagles as the Golden eagle (Marzluff et al., 1997; McGrady et al., 2002). Fernández et al. (2009, 24) report in addition to male eagle flying further during the breeding as part of their role as food providers, that sone flights away from breeding zone were due to human disturbances, with some females virtually abandoning their nest to roost in the distance, due to human hunting near the nest (see also González et al., 2006c). The authors mention that the distant feeding grounds led to large overlaps feeding ranges and no overlaps in home ranges, which is similar to some results for other eagles (see for example, Marzluff et al., 1997). Furthermore, "Spanish Imperial eagles showed no territorial behaviour in distant feeding grounds, even though other eagles were regularly present. These results support the hypothesis that birds defend permanent territories where food is predictable and abundant whereas where food is very seasonal individuals do not defend any fixed areas" (Fernández et al., 2009, 24; see also Schoener, 1968). Also, these areas, unlike proximate feeding grounds, could serve as communal hunting grounds."

The study by Fernández et al. (2009) may be complemented by that of González et al. (1990, 311-320). In the latter study, the authors found that human presence is a factor for the absence of eagles from nesting and foraging areas. Eagles frequented forested areas, with lower levels of human land use, a typical Mediterranean climate with warm dry summers and temperate rainy winters, with more rabbits, but topographic differences were irrelevant. The best predictor of eagle presence was few or no irrigated farmlands. Therefore, unlike the Bald eagle and some of the eagles described above, human presence had serious impacts on eagle presence. González et al. (1990, 311) point out that "the rapid increase of people in areas previously depopulated, and human presence in fields where irrigated cultures have been developed, had a negative effect on the eagle's distribution." The fact that topography as a factor for distribution pointed to the ability of eagles to adapt to variable topographical situations, but their reaction to human presence would be to seek alternative breeding and foraging grounds.

11.6 Kleptoparasitism and Interspecific Competition

Another important activity where individualism in eagles emerges is in the practice of kleptoparasitism, where eagles may strategically and tactically use such methods to save energy, react to metabolic needs and enhance breeding areas. For example, an intricate study by Stalmaster & Gessaman (1984) found that Bald eagles adapt their energy requirements and energy relationships in a winter environment. The eagles' standard metabolic rate increased linearly with decreasing temperature. Metabolism also increased with higher rain levels. The energy and food needs of wild eagles were 10% more than for captive eagles. Many eagles saved energy by roosting in coniferous forest rather than in deciduous forest, despite the greater energy expenditure required to get to the coniferous forests. The justification was that when they roosted in coniferous trees they made "net daily energy savings of 61 kJ, or °5% of the daily energy budget after accounting for energy costs of flight to and from the roost" (Stalmaster & Gessaman, 1984, 407). The energy savings in the coniferous forest microclimate were created by the higher ambient temperature and long-wave radiation levels, and lower wind velocity. There was little impact from the reduced rainfall. Another strategy by the adult eagles was the kleptoparasitism, where adults took food from younger birds, which were less skilled at this method. This was the main means of food acquisition. Larger groups had more such interactions. The impact of such food deprivation on young birds was that they made less effective use of time and energy. Stalmaster & Gessaman (1984, 407) conclude that the findings "indicate that overwintering Bald Eagles are effective at exploiting and conserving energy. They maximize energy gain by foraging in groups, gorging, and assimilating more energy during cold stress. They minimize energy loss by becoming sedentary, seeking protective microclimates, reducing nocturnal body temperatures, and reducing foraging costs by living in groups."

Bald eagles also engaged in kleptoparasitism against other, smaller raptors, in a strategic response to dynamics in their ecological relations. For example, Jorde and Lingle (1988, 183) report Bald eagle kleptoparasitism by Bald eagles wintering in south-central Nebraska, along the North Platte and Platte rivers during the winters of 1978-1980. This was one of the most important ways in which Bald eagle obtained food. In this study, the Bald eagles took food mostly from the Ferruginous hawk (*Buteo regalis*, Gray, 1844), Red-tailed hawk (*Buteo jamaicensis*, Gmelin, 1788), Rough-legged hawk *(Buteo lagopus*, Pontoppidan, 1763), Golden eagle and other Bald eagles (see also Fischer, 1985). The findings indicated that Bald eagles stole prey from other birds to a greater extent during the severe winter of 1978-1979 (the stronger ice cover prevented eagles from feeding on fish) than during the milder winter of the following year (1979-1980). The recorded kleptoparasitism mostly occurred where Mallard ducks (*Anas platyrhynchos*, Linnaeus, 1758) foraging in agricultural habitats. During the winter when ice covered the rivers and prevented eagles from feeding on fish, Mallards appear to be the main prey item for Bald eagles (Lingle & Krapu 1986).

The authors noted that "subadults watched adults steal food and participated in food-stealing with adults, which indicated interspecific kleptoparasitism may be a learned behavior" and suggest factors that favor interspecific kleptoparasitism as a foraging strategy of Bald eagles in obtaining waterfowl prey during severe winters" (Jorde & Lingle, 1988, 184). During milder winters, mallards forage at the riverbank or cornfields near the rivers (Lingle & Krapu, 1986). Mallards changed their foraging strategies during the severe winters, as "cold weather and heavy snow accumulations on agricultural fields adjacent to the Platte River caused mallards to forage up to 19 km from river roosts, which resulted in kleptoparasitism by eagles at distant sites" (Jorde & Lingle, 1988, 184). Other factors that increased the likelihood of kleptoparasitism included the shortage of the main Bald eagle food, fish; the possibilities of secondary foods, waterbirds; predictable primary and secondary food sources, and the concentration of both the hosts and prey (see also Brockmann & Barnard, 1979, 492-493). The open habitats may also contribute to kleptoparasitism, possibly by increasing visibility (see also Paulson, 1985). Jorde & Lingle, (1988) point out that all of these factors were in the study area during 1978-1979. They also note that "Mallards were a large, visible alternate prey with predictable foraging behavior. They returned to the same fields for up to five consecutive days and usually foraged in groups of 2000-3000 birds. Hawks often perched on fence rows, power line poles, and trees along the border of cornfields. Hawk activity in the vicinity of cornfields used by mallards seemed to release interspecific kleptoparasitism by Bald eagles" (Jorde & Lingle, 1988, 186).

A particularly interesting aspect of the eagles kleptoparasitism was "beating" and "scavenging" (Jorde & Lingle, 1988, 187, see also Brockmann & Barnard, 1979). The first involved low flight to disturb the mallards into chaotic flight. In the study by Jorde & Lingle (1988), eagle behaviors included eagles flying to ground foraging or flying mallards, or crows; and flushing ground foraging mallards into flight, similar to beating "possibly to attract attention of other raptors… unless disturbed, foraging mallards remained on the ground and most likely were inconspicuous to hawks not in the immediate vicinity." The authors argue that such swooping over foraging mallards to flush them was not capture behavior as the eagles did not attempt to capture the mallards. Eagles also tried to take mallards just after a hawk had captured the duck, and where there were two or more eagles, they contested for the carcass, after the hawk had left the scene, a form of intraspecific kleptoparasitism. Jorde & Lingle (1988, 187) conclude that "although our small sample of data does not provide conclusive evidence, we hypothesize that the interspecific kleptoparasitism of bird and mammal prey by Bald eagles observed in south-central Nebraska could be an intentional foraging strategy rather than an opportunistic response and that such behavior occurs more frequently among Bald eagles than previously thought." This is partly because "Bald eagles seemed to learn which cornfields Mallards used" as "single eagles and groups of 2-4, often including at least one subadult, stood in cornfields waiting up to several hours before the Mallards and hawks arrived" (Jorde & Lingle, 1988, 187).

Another case of kleptoparasitism was recorded by Dekker et al. (2012, 290) who claim their study is the first to "examine the effect of kleptoparasitic Bald eagles on the kill rate of shorebird-hunting Peregrines and indirectly on a population of Dunlins (*Calidris alpine*, Linnaeus, 1758) wintering in coastal British Columbia." One factor was found to increase in the Bald eagle population increase over the period from October 2008 to January 2011. The Peregrines hunted ducks and Dunlins when eagles were scarce, but only Dunlins when eagles increased in numbers. Eagles and even Gyrfalcons (*Falcorusticolus*, Linnaeus, 1758) kleptoparasitized peregrine, this accounting 36% of the Peregrines' captures. The authors argue that "these results support the hypothesis that kleptoparasites had an indirect effect on a population of wintering Dunlins because Peregrines compensated for prey lost to kleptoparasites by increasing their kill rate."

Another study by Dekker et al. (2015, 159-164) which also looked at kleptoparasitism by Bald eagles, argued that these large raptors are a factor in reducing Peregrine falconpredation on Dunlins wintering in Boundary Bay in the Fraser River valley, British Columbia, Canada. The research involved observations between 1994 and 2014, in which there were 667 sightings of Peregrine Falcons, with 817 attacks on Dunlin and 120 captures. During this period, the number of wintering Bald eagles increased from approximately 200 in 1994 to 1800 in 2014, with a consequent increase in the number of kleptoparasitism incidents against peregrines and decline in January sightings of Peregrine falcons. This decrease in Peregrine falcon numbers may have resulted in a decline of predation risk for Dunlins. Eagles approached hunting peregrines 101 times, and succeeded in taking their prey 17 times, while the falcons evaded the eagles 48 times. In the other episodes the falcon may have been unsuccessful in catching the prey. The increase in the number of Bald eagles at the study site area was attributed to three main factors: the availability of food for eagles that kill waterfowl and kleptoparasitize gulls, a landfill nearby that attracts scavenging eagles and declining spawning runs of Chum salmon (*Oncorhynchus keta*, Walbaum, 1792) which have influenced eagles to move from inland rivers and coastal areas (Dekker, 1999; Elliott et al., 2006; Elliott et al., 2011).

Another interesting case study is the relationship between the White-tailed eagle and the Golden eagle in Norway and Scotland, where marked differences have emerged between the two locations. Halley (1998, 171-179) notes that during the eighteenth and nineteenth centuries, the two species of eagles occurred in Scotland, but the ranges there were segregated (allopatric), with the Golden eagle's distribution restricted by competition from the White-tailed eagle. The Golden eagles range expanded into the former Scottish range of White-tailed eagle after the latter's species extinction in the early twentieth century. This expansion suggested "that it had previously been excluded from these areas by competition from White-tailed eagles" (Halley, 1998, 171–179, see also Love, 1983; Watson, 1997). This possibly suggests that the earlier pattern of White-tailed eagle excluded the Golden eagle is re-emerging. More recent evidence is taken to suggest that Golden eagles have been displaced from their home ranges in western Scotland by these reintroduced White-tailed eagles (Watson et al., 1992a). This contrasts with the situation in Norway, where "the breeding ranges of the two species have apparently always overlapped in coastal areas."

The main difference between the two locations depends on the diet. In Scotland, the reintroduced White-tailed eagle and the Golden eagle eat mostly the same prey, namely sheep and deer carrion, and lagomorphs, mainly rabbits (Watson et al., 1992a, 1992b, 1993). The overlap in prey species is estimated at more than 90% (Watson et al., 1992a). In Norway, the diets were markedly different, as the White-tailed eagle relies more on the classic sea eagle diet, namely fish and waterbirds. The Golden eagle relies on hares (there are no wild rabbits in Norway), grouse (Subfamily *Tetraoninae*, Vigors, 1825) and other wild prey species. In summer there is little prey on carrion by either eagle, even when it is available (Halley & Gjershaug, 2008). The White-tailed eagle, being the larger heavier species, is presumed to be dominant when there is competition for food sources, possibly to the extent of the exclusion of the Golden eagle. Halley (1998) and Halley & Gjershaug (2008) found the reverse, as the Golden eagle is "strongly dominant" in food competition; in one case, a first-winter male displaced "seven White-tailed eagles in less than ten seconds, and on no occasion in two winters of observation was a White-tailed eagle able successfully to defend a carcase from an attacking Golden eagle; in those few cases where a White-tailed displaced a Golden eagle, the latter had invariably been feeding for some time beforehand" (Halley, 1998).

It must be noted that Golden eagles have physical advantages, including a wider talon span, longer legs and stronger flight that appear to compensate for their smaller size and weight (Willgohs, 1961; Cramp & Simmons, 1979). In Scotland, it appears that the factors for the White-tailed eagle excluding the Golden eagle is due to former bird's wider range of prey and the relative absence of the Golden eagle's prey set, such as grouse and hares. Hence Halley (1998, 171) concludes that "Although Golden eagles are wholly dominant in direct competition, White-tailed eagles are nevertheless able to outcompete them through having a wider spectrum of available prey. Increased competition and resulting competitive exclusion in western Scotland appear to be a consequence of long-term environmental degradation of the region through deforestation and overgrazing." For more reading on this see Fraser Darling, 1955; Willgohs, 1961; Love, 1983; Sydes & Miller, 1988; Watson et al., 1992a, b, 1993; Watson & Rae, 1993; Hunter, 1994; Gjershaug et al., 1994; Watson, 1997; Folkestad, 2003; Halley & Gjershaug, 2008.

The Indian Spotted eagle is also known for kleptoparasitism and even polygyny (Sant et al., 2017, 141-146). In this study two pairs of nesting eagles were assessed, two male eagles (named M1 and M2), and two females (named F1 and F2), which were nesting near each other. M2 and F2 lost their initial nest (N2) and moved their nesting location closer (N3) to that of M1 and F1 (N1). Both pairs successfully fledged chicks, but later M2 disappeared and F2 abandoned N3. The authors acknowledge identification issues for the four birds. M1 later mated with F2, with maintaining the paired chick rearing status with F1. Although F2 later abandoned N3, F1 eventually fledged a chick. F2 built a new nest (N4), and then again mated with M1. M1 engaged with both F1 and F2 in feeding and nest building (the nestling on N4 appeared to have died) and in some cases took some of F2's rodent kills to F1. In later cases, F1 attempted to stop M1 taking the food remains to the F1 in nest N1. Both nests eventually fledged successfully.

This type of polygyny and kleptoparasitism has yet to be explained in general terms. The authors (Sant et al., 2017) note that several attempts have been made to explain polygyny in raptors, such as a rich, local food supply and nomadic breeding dispersal (Korpimäki, 1988) (e.g. in the case study rodent prey was numerous, and the birds were nomadic, not residential birds with stronger pair bonds). Other authors have sought explanations (Orians, 1961; Verner & Willson, 1966; Weatherhead & Robertson, 1979; Alatalo et al., 1981). The authors contest other factors for such polygyny, such as the non-availability of male eagles in the area (the study area had many males Indian Spotted eagles), the non-availability of nesting sites (there were enough suitable trees) or attempts at deception (F1 often agitatedly loudly to M1 when he was with F2) or the immaturity of the second female (F2 was a full adult, having bred with M2 and raised two nestlings and had three seasons with M1). Sant et al. (2017, 146) note that "a possible reason could be that the territory of the male was particularly rich in food supply and/or that M1 was a healthier, and more successful individual than the other males in the area. But this is conjecture. However, it is interesting to note that F2 was unsuccessful in fledging any chick(s) for the first two years it mated with M1 and was successful in doing so only in the third year, making it seem strange for the female to continue with this arrangement, when there was limited success in raising chick(s)." The kleptoparasitism (taking food from one nest to another) was also highly unusual. They note that "as far as we are aware, this behaviour has not been reported or documented previously in a polygynous arrangement" (Sant et al., 2017, 146).

11.7 Eagles at Landfills

Landfills are another area where eagles may forage, at variance with their natural prey, and hence may need to adapt to new circumstances. Eagles may forage for meat-based discards in such landfills, and compete with other scavengers such as Larus gulls, corvids, vultures and feral dogs. In one case study, Elliott et al. (2006, 380–390) observed Bald eagles (foraging at the landfill in Vancouver, British Columbia, Canada, from 1994–1996 and 2001–2002. The eagles observed fed mostly on human refuse (95%), but the landfill food items comprised 10.3% of the eagles' daily energy needs. Other findings indicated that subadults foraged at the landfill more often than adults, and some of these subadults may be termed "refuse specialists" as the landfill may have been a larger portion of their food requirements. In terms of their time budgets they spent much of their time resting (91%), with smaller amounts of time spent drinking, scavenging and pirating (Elliott et al., 2006, 380). The authors note increases in wind speed resulted in increased resting times, and foraging efficiency declined with precipitation, these relationships agreeing with the hypothesis that the landfill is primarily a resting location during unsuitable weather. For two subadult eagles, their home ranges were partly comprised of the landfill. For the other eagles, the larger ranges were less likely to include the landfill area. The authors comment that "home-range size and percent of the home range that included the landfill were negatively correlated, suggesting that most eagles visited the landfill occasionally while a few spent most of their time there" (Elliott et al., 2006, 380).

The results were that the Vancouver landfill was a minor energy source, partly due to the numerous competitors, the landfill (mostly protected from the wind, warmer with low human activity) was primarily for resting during unsuitable weather and only a small population of largely subadult refuse specialists favored the area. There was a wide variation among the eagles in the use of the landfill. The younger eagles were more likely to become refuse specialists, as they foraged more while the older eagles rested more at the landfill, a possibility being the subadults are inferior hunters (Elliott et al., 2006, 380, see also Stalmaster & Gessaman, 1984; Brown, 1993; Bennetts & McClelland; 1997). The common pirating at the Vancouver landfill, was a possible reason for the low eagle presence, as foraging efficiency and the percent of birds foraging declined with an increase in bird numbers and pirating. Most pirating was between eagles, with few interactions with the competing gulls. Subadults eagle numbers increased over the winter, one possible reason being subadults acquired knowledge of food sources from adults and also as some breeding eagles returned to their territories in the fall (Knight & Knight, 1983; Bennetts & McClelland, 1997; Stalmaster & Kaiser, 1997; Restani et al., 2000).

11.8 Eagle Responses to Human Activities

Eagles differ in their tolerance of human activities and presence. In some cases, the variation may be due to the particular eagle species, while in other cases it may be due to the noise and threatening nature of the activity. Some eagle species may become more habituated to people than others (Goulet, 2009; Lien, 2015; Van der Meer et al., 2018). One way to test the adaptation of eagles to human activities is to examine the impacts of the noisier activities, and gauge eagle adaptive strategies. For example, Watson commented on research (1993, 171-178) that examined the impact of helicopter surveys on nesting Bald eagles in NW Washington. The findings were that the Bald eagles had higher response rates, defined by disturbance rates and flush distances, to survey helicopters than to fixed-wing aircraft at similar distances. Eagle responses to survey helicopters varied according to encounter distances and nest tenacity. Those nesting eagles with a brood and those closer to nests tolerated helicopters at closer range before flushing.

The findings also indicated that at encounter distances greater than 120 m eagles flushed at higher rates but then returned to their nests. There was no direct mortality of young or adults. As 53% of 270 eagles encountered were disturbed; 68% of disturbed eagles were flushed, and 32% were agitated, these findings indicated that 47% were not classified as disturbed, 32% were not flushed or were in another classification, and 68% were not flushed or belonged to another category. These results illustrate a high degree of individual reaction. The author acknowledges that interpretative analysis of eagle behavior is necessary. For example, concerning those eagles that appeared minimally or not disturbed by the helicopters approach, "what seems to be no disturbance and a lack of response to human activities closer to the nest might actually be a form of nest defense. Even though adults may not flush, regular disruption of nesting activities by aircraft or other human activities could result in reduced brooding and feeding of young, which might lead to reduced attentiveness and nest failure" (Watson, 1993, 176). The author concludes that to minimize disturbance of eagles and maximize cost efficiency, data reliability, and observer safety actions should include hovering for less than ten seconds at 60 m or greater from nests, and the surveying should be undertaken on calm, dry days, and as late in the season as possible to allow for minimal adult eagle presence (Watson, 1993).

Grubb & Bowerman (1997, 213-222) conducted a similar study concerning the variations of breeding Bald eagle responses to jets, light planes and helicopters in Arizona and Michigan, the reactions classified at three levels (none, alert, flight). Again, as in the study by Watson (1993) helicopters brought a greater frequency of response (47%), followed by jets (31%) and light planes (26%). Frequency of response (23-61%) and frequency of flight (2-13%) both increased through the nesting season from February to June. As in the earlier study there was high variability of response, revealed by these figures. The distances between the eagles and the aircraft, the duration of the overflight and the number of aircraft involved and the number of passes of the aircraft were the key factors for Bald eagle responses to the aircraft variables. Other findings supported the common result that "helicopters are the most disturbing type of aircraft" (see also Watson, 1993), and found that Bald eagles responses to aircraft varying towards the least response to aircraft near the start of the nesting season (usually remaining near the nests) probably due to the higher nest attendance requirements for eggs or more dependent nestlings, compared with later in the breeding season, when lower nest responsibilities might allow a flight response.

Grubb & Bowerman (1997, 221) noted that Watson (1993) found a similar result, the presence of young nestlings influenced a reduced adult response, such as flushing and flight, and eagles were more reactive when there were no nestlings In an earlier study by Grubb and King (1991), it was indicated that breeding Bald eagles in Arizona can be habituated to aircraft, and another study in Michigan found habituation at an eagle nest site close to a military air base (Grubb et al., 1992). Grubb & Bowerman (1997, 221) note that "if habituation occurs with repeated exposure, then our results may underestimate Bald eagle response at nest sites with limited air traffic and overestimate at sites with a high frequency of aircraft" and "whenever possible, surveys are better conducted with light planes, because they typically cause minimal disturbance to breeding Bald eagles" (see also Fraser et al., 1985). The authors concluded that considering the findings of variable reactions of Bald eagles' reactions to aircraft, "a categorical exclusion of aircraft within 600 m of nest sites would limit Bald Eagle response frequency to 19%" (Grubb & Bowerman, 1997, 213).

Watson (2004, 295-303) also looked at the impact of human pedestrian activities on nesting Bald eagles. The factors for eagle reactions included nest height, nest screening (coverage of the nest), pedestrian distance, and phenology (the study of phenomena or happenings). These responses were weaker where the nests were more than 40 m high and were well screened, when pedestrian distance from the nests increased from 60 to120 m, and during incubation rather than the brood period. Precipitation occurrence also reduced the time disturbed adults left eggs and young exposed. Watson (2004, 296) justify the basis of their study, because "human activities can change behavior of nesting eagles through auditory and visual disturbance... Such activities may cause reductions in incubation time, brooding time, feeding time, and other adult behaviors that affect attendance of eggs or young, and potentially survival of young. These changes in adult behavior may be possible indicators of impending nest failure of raptors" (see also Grier, 1969; Fraser et al., 1985; Anthony & Isaacs, 1989; Holthuijzen et al., 1989; Grubb & King, 1991; Steidl & Anthony, 1996).

Watson (2004, 300) concludes that "in non-pristine, forested environments, nest height, and vegetative screening are important factors for mitigating eagle responses to brief exposures to a single pedestrian." For nest height, those above 40 m elicited much less of an eagle response than nests closer within 72 m horizontal distance, showing that height was a stronger predictor than horizontal distance to pedestrians. For screening, eagle responses such as flushing were less when there was even partial screening (a reduction of response time by 32%) compared to open nests. The author concludes that "untrained observers may greatly underestimate their disturbance of nesting Bald eagles" (Watson, 2004, 302). The findings indicated that incubating and brooding eagles may react about 30% of the time when pedestrians are within 120 m of the nests, and the much of the eagle responsive behavior will involve passive perching, which dangerously expose the eggs and young to the threats. Crucially, the study by Watson (2004, 295) was located in "forested, non-pristine areas of residential development" where the eagles might be minimally habituated to human presence and activities. They acknowledge that "Bald eagles nesting in pristine environments may exhibit greater sensitivity to pedestrians than those in this study", being less acclimatized to people (Watson, 2004, 302).

11.9 Conclusions

The chapter has reviewed a few cases from the wide literature that documents the individual behavior of eagles, some of which may be the expected actions from generalized research findings on eagle behavior and some of which would be unexpected, mostly because the environments have changed under human influence, making the eagle behavioral changes examples of adaptation. In some case new, artificial environments were created, such as agricultural or urbanized landscapes. In other cases, primary forests might have been replaced with mixed natural and semi-artificial landcover. Some of the behavioral variations were at the intra-species level, while others were at the inter-species level. These studies can only grow in importance, as people continue to modify the landcover of eagle habitats and also change, remove or reintroduce both prey and competing species. These eagles, especially when monitored by geomatics and other more definitive research methods, may contribute to the growing body of scientific research on the dynamism and continually changing nature of ecology. Some of the human actions that contribute to eagle behavior, such as conservation policy and public attitudes to eagles, and human hunting foci are examined in the concluding chapter of this book.

12

Eagles in Social History and Conservation

12.1 Introduction

Eagles are the most charismatic birds in existence today, nearly matching vultures in size and outmatching them in killing and hunting abilities perceived, regal power and human respect (Campbell, 2015). As a consequence, eagles are much more likely to be associated with martial power (e.g., that of kings and generals) than even large vultures. The eagle is generally perceived as the 'King of Birds', a title that belongs not only to the Golden eagle in Europe, but also to the different species of eagle across the continents. Albornoz & Fernández (2014, 1) commenting on Eurasia, note that "the eagle is one of the most ancient symbols of humanity. The Greeks and the Persians consecrated it to the sun, as a symbol of elevation and of spirit associated with the supreme sky-god, representing the spiritual principle. For the Greeks, the Eagle was the emblem of Zeus. The Druids regarded it as the symbol of the Almighty who watches from higher spiritual states." Nebeský (2007) notes that "the lion is one of the two most popular symbols in European heraldry, the other being the eagle." Hull and Fergus (2009), commenting on the Americas, note that "among the apex predators of the air, eagles have always held special significance for both ancient and modern Mesoamerican societies. Eagles in general are common to oral lore, usually in a role exploiting their two chief characteristics – their strength and agility." In Africa, Ukpokolo (2011, 168) writes that "to the Igbo, the eagle bird symbolizes beauty, fortune and outstanding."

Therefore, a primary, either obvious or implied reason for the eagles' (whichever large raptor is deemed an eagle in that culture) status is its general morphology: extremely powerful beaks, vast wingspans, and powerful talons, and the hunting capabilities that emerge from these physical attributes. Another is their size, which gives them a conspicuous status above those of smaller birds of prey. Another issue concerns their large ranges and migratory habits, which ensure that every continent except Antarctica has at least one species of eagle. King (2017) notes that "the artistic record of man's interest in eagles goes back at least as far as ancient Mediterranean civilizations and cave dwellers in what is now Europe." This short chapter summarizes relations between people and eagles though the ages, examining eagles in falconry, eagles in religion, eagles in symbolism and heraldry, current public attitudes to eagles and current eagle-related conservation legislation policies. These topics are scattered on a regional basis, starting with Eurasia, followed by Africa, the Americas and Australia. In most cases the particular eagle concerned with the particular issue is identifiable, but in these cases the issues revolve around the general identification of any very large bird of prey.

The conservation of eagle species brings important disciplines such as political ecology, environmental sociology, conservation biology and applied ecology into the general sphere of ecology. These disciplines attempt to link the human mind and resultant action with the reality of the environment, to better understand nature-society relations and environmental change. In this case both the positive actions (conservation of eagles and their dependent environments) and negative actions and impacts (direct and indirect killing of eagles, habitat destruction and killing of prey species) are examined. The inclusion of social research methods means that some of the human-eagle interactions are subjective, based not on objective scientific research, but also on gossip, hearsay, crude observation, religious doctrines, cultural patterns and supports for political ambition. The common and prevailing mental association of eagles with powerful persons, political polities and ambitious projects has ensured that a copious literature exists covering these topics. There is a vast literature on these topics. Therefore, the chapter will select those literature pieces and discourses deemed to be central to the general and more specific human perceptions and actions towards eagles.

12.2 Eagles in Religion, Symbolism and Heraldry

Eurasia

The eagle is one of the most important religious and royal symbols in Eurasia. In this representation, the eagle may be either single or double headed. The figure of a double-headed bird is represented in many cultures around the globe, and in various time periods. "The double-headed eagle has been a royal symbol throughout its history until the present day" (Chariton, 2011; see also Deeds, 1935: 106). The history of the use of the double-headed eagle in various contexts since the Crusades has been well documented. It has been used as a heraldic emblem of many royal houses and is still a state symbol in some countries. However, ancient Near Eastern cultures had used the double-headed bird motif millennia before the Crusades. One striking example is that of the Hittites, for whom it was a "royal insignia" (Chariton, 2011; see also Collins, 2010: 60).

For the ancient Romans, the eagle was one of the most symbols or power and even military glory. The initial basis for the glorification of the eagle was its physical ability, which was acknowledged by the ancient Greeks of the pre-Roman period. These were keen eyesight, speed, strength, strong flight and strong talons (Greet, 2015). Other concepts identified with eagles include martial victory, freedom (exemplified with free, soaring flight), divine messaging (the eagle is linked to Zeus, the King of the Gods, the Roman Jupiter), a God or Goddess in bird form (for example, the Goddess Athena, the Roman Minerva changed to the form of an eagle in Homer's Odyssey) and communicator between Gods and people (Greet, 2015). For the Romans, possibly the most important role for the eagle was the development of the Roman military standard, the *Aquila*, which may have been linked to the eagle itself, at least in a theoretical sense. Eagles also appeared on coinage in Ancient Rome (linked to Mars, the God of war, and Minerva), and are debated to have appeared in the service of Jupiter (Greet, 2015).

In early Celtic Britain, the eagle was an important source of power perception. Puplett (2020) writes that "Few birds have come to symbolise majesty and splendour like the eagle… In Celtic mythology, the eagle is traditionally seen as one of the oldest of all creatures, surpassed only by the salmon in wisdom and age." Furthermore, in Wales "A lone eagle seen on a crag was seen as a sentry, warning of the arrival of an enemy, while a pair was a symbol of peace" (Puplett, 2020, see also Carr-Gomm, 1994; Matthews, 1995). Other nations across Europe idolized the eagle, not just as religious symbols, but as symbols of power, glory or nationhood. Examples are listed in Table 12.1. One of the main sources of religious symbolism for eagles is the Christian Bible. Table 12.2 lists some of the key states that exemplify representation of eagles as strong, and even representative of divine power, of the arrival of God or descriptive of human power, or the strength of an enemy of the recipients of the message refuse to accept God's edicts, and God as a protector (Wellman, 2015).

Table 12.1. Examples of eagles in symbolism in Europe

Eagle description	Users
Roman Eagle Standard	Roman Empire
Gold-and-red chequerboard eagle	German-speaking Moravians
Silver-and-red chequered eagle	Czech-speaking Moravians
Crowned black eagle, silver crescent on breast	Silesia
Reichsadler (Imperial Eagle), red beak, legs, gold field	Holy Roman Empire
Double-headed Eagle	Holy Roman Empire (1433)
Reichsadler (Imperial Eagle)	German politics (1871–1945)
Bundesadler (Federal Eagle)	Federal Republic of Germany
Double-headed Eagle	Palaiologi Emperors, Byzantine Empire
Double-headed Eagle	Tsars of Russian Empire
Double-headed Eagle	Teutonic State, Prussia (northern Germany)
Double-headed Eagle	Austrian Empire (1806)
Single-headed Eagle	Austria (1919)

Sources: 1. Nebeský (2007)

Table 12.2. Bible statements on eagles

Statement in the Bible	Reference in the Bible
"You yourselves have seen what I did to the Egyptians, and how I bore you on eagles' wings and brought you to myself"	Moses, Exodus 19:4
"Like an eagle that stirs up its nest, that flutters over its young, spreading out its wings, catching them, bearing them on its pinions"	Moses, Deuteronomy 32:11
God "who satisfies you with good so that your youth is renewed like the eagle's"	David, Psalm 103:5
"but they who wait for the Lord shall renew their strength; they shall mount up with wings like eagles; they shall run and not be weary; they shall walk and not faint"	Isaiah 40:31
"Behold, he comes up like clouds; his chariots like the whirlwind; his horses are swifter than eagles— woe to us, for we are ruined"	Jeremiah 4:13
"The Lord will bring a nation against you from far away, from the end of the earth, swooping down like the eagle, a nation whose language you do not understand"	Deuteronomy 28:49
These "pursuers were swifter than the eagles in the heavens; they chased us on the mountains; they lay in wait for us in the wilderness"	Lamentations 4:19
Their horses are swifter than leopards, more fierce than the evening wolves; their horsemen press proudly on. Their horsemen come from afar; they fly like an eagle swift to devour"	Habakkuk 1:8
"Behold, one shall mount up and fly swiftly like an eagle and spread his wings against Bozrah, and the heart of the warriors of Edom shall be in that day like the heart of a woman in her birth pains"	Jeremiah 49:22
"Though you soar aloft like the eagle, though your nest is set among the stars, from there I will bring you down, declares the Lord"	Obadiah 1:4
"For he will deliver you from the snare of the fowler and from the deadly pestilence. He will cover you with his pinions, and under his wings you will find refuge; his faithfulness is a shield and buckler."	Psalm 91:1-4

Sources: Bible; Wellman, 2015.

Africa

Africa also has a long tradition of deities linked to eagles, and this continent has no shortage of large, powerful eagles with majestic bearing. For example, among the Igbo people, an eagle's feather "signifies innocence and purity" (Udechukwu, 2019). The Mwari people of Great Zimbabwe venerated the magnificent Bateleur eagle as a spiritual messenger (Huffman, 1985). Huffman (1985, 68) writes about the "Soapstone Birds from Great Zimbabwe" which "Eight birds were found in the ruins of Great Zimbabwe... carved of soft green-gray soapstone." In the Shona culture of the area, birds are traditionally seen as messengers: "Eagles, being the largest and most powerful birds, are appropriate messengers for the most important people. Eagles also bring messages from ancestor spirits. The Shona believe that spirit people live at the bottom of certain deep pools" (Huffman, 1985, 68). These pools are seen as the entrance to the spirit world and the Bateleur eagle is an important messenger (see also Burbridge, 1925; Edwards, 1928; Franklin, 1933).

North America

This section mentions only a few of the numerous references to eagles among the Native American culture. Elder (2018) writes that "a great many of the Native American religious traditions regard the Bald eagle, along with the Golden eagle, as sacred, because they are the messengers of the Great Spirit, the Creator. Kovacs (2013, 73-74) points that among the Native American cultures that use eagle parts in their ceremonies "one unifying thread among those who require eagles is that the eagle is "considered a sacred messenger to the spirit world" (De Meo, supra note 23, at 774; see also ibid. at 775, "Eagles are, in essence, NativeAmerican prayer messengers"). For example, for the Hopi the Golden eagle "serves as the link between the spiritual world and the physical world . . . a connection that embodies the very essence of Hopi spirituality and belief" (Kovacs, 2013, 74, quoting Worthen, supra note 23, at 990, quoting Religious Ceremonial Collection of Golden Eaglets from Waputki National Monument, 66 Fed. Reg. 6516, 6517 (proposed Jan. 22, 2001)). The Hopi collect Golden eaglets soon after hatching, after which they send the eagles to what they believe is their spiritual home and use the feathers for other ceremonies. These ceremonies are believed to connect the physical and spiritual worlds for the next generation.

Some other tribes forbid the killing of eagles, allowing only the collection of molted feathers or plucking of feathers, provided this does not injure the eagle to the point of flightlessness. In these tribes, eagle feathers may be used in ceremonies celebrating naming, coming-of-age, marriage, burial, healing, and seasonal change (Kovacs, 2013). Kovacs (2013) cites the Northern Arapaho Tribe, where in the annual Sun Dance, the dance sponsor constructs an offering lodge and gets an eagle as part of the dance as a present to the god of creation. In this ceremony, the eagle "must be pure, meaning it has not died from poisoning, disease, electrocution, or automobile impact, and it has not been used in a prior SunDance (Kovacs, 2013, 74). Phillip (2015) argues that eagle feathers are key to the Native American cultures and cites the Lipan Apaches and many other Native American religions which "hold eagles to be sacred because of their physical and spiritual closeness to God. Eagle feathers and parts are considered sacred objects that are used in prayer, worship and for ceremonial cleansing."

According to Erdoes and Ortiz (1984, 140), a Zuni tradition details the association of an eagle and a coyote, who formed a hunting companionship, in which the eagle is the wiser: "Coyote is a bad hunter who never kills anything. Once he watched Eagle hunting rabbits, catching one after another – more rabbits than he could eat. Coyote thought, 'I'll team up with Eagle so I can have enough meat.' Coyote is always up to something." After some escapades, that came upon two boxes (a big box containing the Sun and a small box containing the moon). Taking the large box, the eagle carried the box while the coyote implored the eagle to let him carry it. After numerous rebuffs, the eagle gave the box to the coyote, which he did. The coyote being more deceitful, opened the box: "and Coyote opened the lid. Now, not only was the sun inside, but the moon also. Eagle had put them both together, thinking that it would be easier to carry one box than two... As soon as Coyote opened the lid, the moon escaped, flying high into the sky. At once all the plants shriveled up and turned brown. Just as quickly, all the leaves fell off the trees, and it was winter. Trying to catch the moon and put it back in the box, Coyote ran in pursuit as it skipped away from him. Meanwhile the sun flew out and rose into the sky. It drifted far away, and the peaches, squashes, and melons shriveled up with cold." This action brought the seasons, rather than the summer all year round (Erdoes & Ortiz, 1984, 142).

In another culture that of the Sioux, the eagle is seen as the originator of the people of this ethnic group. In this tradition, a water monster created a flood where all the people died: "Everyone was killed, and all the blood jelled, making one big pool. The blood turned to pipestone and created the pipestone quarry, the grave of those ancient ones. That is why the pipe, made of that red rock, is so sacred to us" as it represents the blood of the ancestors (Erdoes & Ortiz, 1984, 94). The water monster Unktehi turned to stone. One girl survived because a large spotted eagle, named Wanblee Galeshka, allowed her to take hold of his feet, and flew with her to his home, at top of a tall tree on the highest stone pinnacle in the Black Hills. That was the eagle's home, where he married the girl and she bore him a boy and a girl, whom the eagle placed on earth, telling them "Be a nation, become a great Nation – the Lakota Oyate." They had many children and "So we are descended from the eagle. We are an eagle nation. That is good, something to be proud of, because the eagle is the wisest of birds. He is the Great Spirit's messenger; he is a great warrior. That is why we always wore the eagle plume, and still wear it. We are a great nation. It is I, Lame Deer, who said this" (Told by Lame Deer in Winner, South Dakota, in 1969) (Erdoes & Ortiz, 1984, 95).

In the Sioux culture, there is another interesting narrative concerning eagles. After the world was created, a man and woman were created who bore children, "the red people, the Ikche Wichasha" the first people, and many other nations (Erdoes & Ortiz, 1984, 135). Their first children were two male twins, one of whom received a greater vision who had a son who was named "Ikche Wichasha" (the name of the Sioux people). When he climbed a hill, it turned into a nest, and when he took leave of his parents and grandparents, to fulfill his vision to take care of the winds of the universe, he turned into an eagle. He flew away with a gift from the Great Spirit, which were the four seasons. "They accompanied him in the shapes of the bald eagle, the spotted eagle, the golden eagle, and the northern eagle" (Erdoes & Ortiz, 1984, 135).

Australasia

The Australian aboriginal culture is based on many different cultures and includes eagles among some of the cultural symbols. The main eagle is the Wedge-tailed eagle. For example, the Kulin nation in central Victoria venerates Bunjil a creator deity, culture hero and ancestral being, who is often associated with a large eagle, possibly the wedge-tailed eagle (Mudrooroo, 1994; Everist, 2009; Fielding, 2019).

Illustrations of the symbols described above are shown in Figures 12.1 to 12.22.

12.3 Eagles in Falconry and Hunting

Falconry, defined as the training of raptors to hunt for game animals for sport or food, has long been an activity indulged with eagles, and in some cases smaller raptors such as falcons and hawks (Mitchell, 1971). Carroll (1996) defines falconry as "the capturing of quarry using trained birds of prey... hawking is used when a hawk or an eagle is used for the hunting, or even the broader term austringer is used for one who hunts with hawks, eagles, or even owls. The term falconry, some believe, should be used strictly for hunting with falcons" (see also Madden, 1924; Mitchell, 1962; Grossman & Hamlet, 1964; Bert, 1968; Gryndall, 1972; Glasier, 1986; Newton, 1990; Robinson, 1992; Hull, 1995). Schafer (1958, 293) referring to falconry as "hawking" writes that "the place of origin of the art of hawking is unknown. It is generally supposed to have been somewhere in 'Turkestan' or 'Central Asia'. From this conjectural center the craft is thought to have spread eastward into China, southward into India, and westward through Persia into the Mediterranean world and Europe. But Mongolia, for instance, might, with equal probability,

Fig. 12.1. Habsburger double-headed eagle, fascinating clock at the Schönbrunn Palace in Vienna, Austria [With permission from Arjan Richter]

Fig. 12.2. Two-headed eagle, Austria [With permission from Elrina753]

Fig. 12.3. Two-headed eagle [With permission from Werner Ustorf]

Fig. 12.4. Eagle cameo, onyx, Octavian's elevation to Augustos [With permission from Andrew McCabe]

Fig. 12.5. Philip I the Arab, 244-249 AD [With permission from Larry Frogel]

Fig. 12.6. Cassia Denarius. Romulus sceptre, Eagle thunderbolt jug lituus [With permission from Andrew McCabe]

Fig. 12.7. Caesar Denarius. Caesar, eagle, standard, plough, measuring rod, settlement type [With permission from Andrew McCabe]

Fig. 12.8. Galley Eagle Standards Praetorian Cohort type [With permission from Andrew McCabe]

Fig. 12.9. Jupiter Eagle thunderbolt Denarius [With permission from Andrew McCabe]

Fig. 12.10. Thrace Koson Brutus Accensus to gate figure lictor. Eagle with wreath [With permission from Andrew McCabe]

Fig. 12.11. Semilibral Sextans. Collateral series [With permission from Andrew McCabe]

Fig. 12.12. Petillia Denarius. Eagle on angled thunderbolt, temple [With permission from Andrew McCabe]

Fig. 12.13. Mars eagle [With permission from Andrew McCabe]

Fig. 12.14. Seleucid King Antiochus VII, 138-129 BC [With permission from Larry Frogel]

Fig. 12.15. Roman Republican Coinage [With permission from Larry Frogel]

Fig. 12.16. Roman Provincial Coinage Phillip 1 the Arab [With permission from Larry Frogel]

Fig. 12.17. Eagle Totem, Alaska [With permission from James Knitis]

Fig. 12.18. Eagle Totem, Alaska [With permission from James Knitis]

Fig. 12.19. Eagle Totem, Alaska [With permission from James Knitis

Fig. 12.20. Eagle Statue, British Columbia [Author]

Fig. 12.21. Eagle Totem, Alaska [With permission from James Knitis]

Fig. 12.22. Bald eagle feathers for Native American Warrior [With permission from James Knitis]

have been the region where the first hawk was tamed, especially if we rely on the intensity of the practice of falconry as a criterion. Or perhaps it was Manchuria."

Canby (2002, 167) adds that the "original home of falconry is not known. Perhaps it is not a coincidence that the earliest representation of what appears to be a falconer holding up his dead prey comes from the Syrian site of Tell Chudra in the third millennium BC." Canby (2002) argues that in one text the so-called "Protector God of the Countryside" is portrayed as standing on a hare with an eagle and a hare in the left hand and a bow and sword in the right hand. Canby (2002, 163) writes that "a bird on his fist who often holds dead prey looks and sounds like a falconer. If he is one, the date for the origin of the practice of falconry has to be pushed back at least to 2000 B.C." This argument is disputed and not all think that figure represents a falconer. Canby (2002, 163) further argues that, notwithstanding this possibility, "the earliest sure evidence for falconry is from the Han Dynasty in China 206 B.C.-A.D. 20" this assessment is based on the reference, Epstein (1943).

Mayor (2016, 1) notes that "the sight of a Eurasian nomad hunting with swift horse, loyal dog, and powerful golden eagle is a majestic scene with ancient roots. For thousands of years this was how the nomadic peoples of the northern steppes from the Caucasus to Manchuria hunted for survival." This is based on the Kazakh proverb "A fast horse and a soaring eagle are the wings of a nomad" (Mayor, 2016, 1). Areas touted for such origins include the Assyrians of the eighth century B.C., with later appearances in Han China around 2000 B.P. (or even earlier, around the 3rd century BC), and in Western Europe are no earlier about 400 to 500 years later (Schafer, 1958, 294-295).

The IAF (2018) gives evidence of the origins of falconry in many countries. That of the Asian nations are tabulated (Table 12.3). Concerning Europe, it appears falconry entered later. The IAF notes that "the earliest evidence of falconry in Europe is usually considered to be from the 5th century AD, written quotations from Paulinus of Pella and Sidonius Apollinaris in France and the famous mosaics in the Falconer's Villa in Argos (Greece)." Falconry remained popular for a thousand years, linked to high social status, and training in gentlemanly attributes (IAF, 2018).

Table 12.3. The origins of falconry in Asia

Region or country	Attributes
Mongolia	1000 B C. Evolved during the Mongol Empire. By the visit of Marco Polo (1270s to 1280s), over 10000 falconers and falconry workers
Korea	220 B.C. Associated with legal and military affairs, diplomacy and land colonisation, and high status
China	1000 to 700 B.C. Links to Imperial family, nobility and subjects. Images illustrate advanced techniques, like modern times.
Japan	355 A.D. Associated with the Imperial Household Ministry until the Second World War, when it declined with the decline of Imperial power.
Persia/Iran	8000 to 6000 B.C. disputed. Possible parallel evolution with Mongolian steppes, King Tahmooreth a falconer, possible 6000 BC.
India	600 B.C. Linked to the nobility and the Mughal dynasty. Endured socially for centuries. Popular, until the 1940s.
Arab	3500 B.C. For Kings, Sheikhs and cavalry. Spread eastwards to Central Asia and westwards to North Africa.
Russia	800 to 900 A.D. Possibly came to the Eastern Slavic tribes from the Turkic neighbours. Associated with nobility.
Turkey	1600 to 1200 B.C. Ottoman Empire flourished. Associated with royalty and nobility and declined with the decline of the Ottoman Empire.
Georgia	500 A D. Larger birds flown by higher ranking nobility, slow decline but legal now.
Kazakhstan and neighbors	5000 B.P. Oguz Khan tribes, forefathers of the Turkic people had falconry symbols. Subsistence necessity for food later.

Source: International Association for Falconry and Conservation of Birds of Prey (IAF) (2018).

It is generally accepted that the evolution of falconry appears to have begun in the most suitable habitats for this activity; grasslands and open savanna in Asia, commonly termed the steppe, including southwest Asia (commonly termed the Middle East) eastwards to Central Asia and Russia, as far east as Mongolia. Hence, many of the records of ancient falconry have been found in this vast area. Other, more recent records have been recorded from western Europe, including the Roman Empire and later, Britain and France (International Association for Falconry and Conservation of Birds of Prey IAF, 2018). Mayor (2016, 2) notes that the "earliest images of falconry appear in Assyrian and Hittite reliefs of the 9th and 8th centuries BC. Classical Greek and Roman authors Ctesias, Aristotle, Pliny, and Aelian described falconry"; later records come from Persia in the 10th century AD, Marco

Fig. 12.23. Falconry using Golden eagles in Central Asia [With permission from Julia Volk Prexels]

Fig. 12.24. Falconry using Golden eagles in Central Asia [With permission from Julia Volk Prexels]

Fig. 12.25. Crowned eagle in Modern Falconry [With permission from Picfair Licience Purchase

Fig. 12.26. African Fish eagle and Black-chested buzzard-eagle at the The Falconry Centre, Hagley UK [With permission from Steve Weaver

Fig. 12.27. Bald eagles at Hagley Bird of Prey Centre [With permission from Steve Weaver]

Fig. 12.28. Golden eagle at Hagley Bird of Prey Centre [With permission from Steve Weaver]

Fig. 12.29. Steller's Sea eagle at Royal Highland Show 2019 [With permission from Christine Sutcliffe]

Fig. 12.30. Steller's Sea eagle at Royal Highland Show 2019 [With permission from Christine Sutcliffe]

Polo on Central Asian nomads in AD 1270, Scythian (southern Siberian) sites, Kazakhstan, petroglyphs in the Altai region and Chinese stone reliefs (Mayor, 2016, 3). Canby (2002, 162) nevertheless disputes some of the records, especially the record "the often-quoted Assyrian relief of the eighth century BC from Khorsabad" which "does not, in my opinion, qualify as falconry (it appears to represent a man using an unwilling bird as decoy). Images depicting the falconry in different regions are shown in Figures 12.23 to 12.31.

Fig. 12.31. Verreaux's eagle in falconry [With permission from Jonathan Beilby]

12.4 Current Public Attitudes to Eagles and Eagle Conservation Legislation

Currently, there are many laws and policies that protect eagles. Some are nationally focussed, while others are based on international treaties, largely because of the acknowledgment of the vast ranges and migratory habits of eagles. For example, in the United States and Canada, laws may be Federal, State, Provincial or international. In Europe, they may be national, or at the level of the European Union (E.U.). Tables 12.4 and 12.5, list some of the more important laws that protect eagles. These give eagles various levels of protection, and also take account of some of the actions that may affect eagles, such as forestry, agriculture, urbanization and hunting. These laws evolved slowly, starting the late 19th century, with regular amendments as issues and vulnerable species changed. Some of the laws were successful, in that the target species (e.g., Bald and Golden eagles) increased in population. As noted above in the earlier chapters, such population increases create new problems for human-eagle conflicts, including tension with companion animal and livestock keepers, and conflicts between these groups and conservation groups.

Table 12.4. Legislation relevant to eagles in North America

Legislation	Date
Lacey Act (16 U.S.C. §§ 3371–3378	1900
Weeks-McLean Act	1913
Migratory Bird Treaty Act (replaced the Weeks-McLean Act).	1918
Convention on the International Trade in Endangered Species of Wild Fauna and Flora (CITES)*	1973
Convention on Wetlands (Ramsar)	1971
Convention on Biological Diversity, Convention on the Conservation of Migratory Species of Wild Animals	1979
Code of Federal Regulations Title 50 (50 C.F.R.)	1969+
Migratory Bird Treaty Act (Parts 10 and 21)	1918+
Bald and Golden Eagle Protection Act (Part 22)	1940+
Endangered Species Act (Parts 17 and 23).	1973
Endangered Species Preservation Act (ESPA; P.L. 89–669),	1966
Endangered Species Conservation Act (ESCA; P.L. 91–135),	1969
Endangered Species Act (ESA; 16 U.S.C. 1513–1543)	1973
Migratory Birds Convention Act (Canada) (from 1917)	1994
Criminal Code of Canada, Section 446, Cruelty to Animals (from 1892)	1892
Species at Risk Act (SARA) (Canada)	2003
Wild Animal and Plant Protection and Regulation of International and Inter-provincial Trade Act	1992
Canadian Council on Animal Care Committee	1968
Canada Wildlife Act (CWA)	1968

* Also termed the Washington Convention.
+ Indicate later additions or amendments.
Sources: Bond, 1974; Fisher & Cleva, 2000; Millsap et al., 2007. The Bald and Golden Eagle Protection Act was amended in 1962 and 1972.

Table 12.5. Legislation relevant to eagles in Eurasia and Africa

Legislation	Date
EEC Council Directive 79/409/EEC on the Conservation of Wild Birds ('Birds Directive')	1979
Convention on the Conservation of European Wildlife and Natural Habitats ('Berne Convention')	1979
EU on CITES, Council Regulation (EC) No. 338/97, Protection of Species of Wild Fauna and Flora	1984
Convention on the Conservation of Migratory Species ('Bonn Convention')	1979

Sources: Temple-Lang, 1982; Stroud, 2003

12.5 The Future of Eagles: Conclusions and Summary

The future of eagles clearly depends on the human actions on habitats, prey animals, environmental pollution, predatory competitors, direct killing of eagles, all ultimately dependent on the human will to protect the eagles and associated species and ecosystems. These issues are also linked to the biology and ecological relationships of eagles. Eagles are mostly large birds, all with predatory habits, strong enough to kill nearly all birds (with the possible exceptions of only the large flightless birds such as the ostrich and emu) and most smaller mammals (including even wolves, dogs, smaller herbivores and cats). Smaller prey species are mostly currently commoner than larger prey species, but eagles must compete for the former more than the latter, competitors for smaller prey including the commoner smaller raptors and cats. Additionally, the large size and obvious power of most eagles may encourage people to view them as threats to livestock, companion animals and even children. Eagles are also rather slow breeders; hence they recover slowly from population losses. Like the huge condors and vultures, eagles are sometimes argued to be relics from the past; huge birds that flourished at a time when there were vastly larger numbers of huge and large prey animals, these including the smaller herbivores they currently prefer for food and much larger species that provided the carcasses for which eagles successfully competed with vultures. Eagles are also relatively intolerant of human presence; unlike most medium-sized vultures and smaller raptors they are relatively absent from urbanized areas and some agricultural zones. Consequently, a pertinent question concerns the future sustainability of eagle populations. There is a balance between the adaptability of the eagle species, the tolerance and support from people, and the possibilities of negative environmental changes, from both natural factors and human action.

This chapter concludes by assessing the information presented in the previous chapters concerning the character, status and situation of eagles. The relevant issues are: (1) the adaptive capacity of eagles, in the face of the varied social and natural forces they encounter; (2) inter- and intra-species behavioral variations, especially the differences between the specialists and generalists in the adaptive abilities; (3) the needs of eagles in terms of habitats and prey species and whether these can be maintained in a changing world; (4) the public attitudes to eagles, which may result in either more conservation of eagles or their extermination; (5) the impact of technology, such as power lines, airplane collisions and wind turbines on eagle movements; and (6) the possible benefits of academic research on eagle ecology, if the findings of such research are implemented. These issues may ultimately depend more on the competitive and adaptive abilities of eagles than the environmental change itself, as some changes are inevitable due to human population increase, industrial development and urbanization. The evidence indicates that eagles can cope with all the issues mentioned, but considering the severe impacts of some changes, such as strong human actions with indirect effects, it is possible that eagles may survive with smaller populations than at present.

Eagle Adaptation

Eagles have been shown to be quite adaptive to environmental change, partially due to their size, which allows a generally wide prey range, their migratory habits over wide ranges and their superior abilities against competitors such as smaller raptors and the larger vultures. The evidence indicates that eagles are fairly generalized in their feeding and nesting habits, yet some specialisms exist. Eagles are either primarily bird eaters (such as the African hawk eagle), reptile killers (such as the Serpent and Snake eagles), fish eaters (such as the Sea eagles) or mammal killers (such as the *Aquila* and Harpy eagles). It remains to be seen if such specialisms can be changed, or if the loss of such prey may spell the doom of some species. All the prey mentioned have their own survival issues. For example, bird species preyed upon in the African savanna may be declining, due to the conversion of primal savanna to farmland, the declining stands of trees that some of these species rely on for nesting and human hunting of these species. Serious deforestation and habitat losses are also recorded in Central and South America, which has contributed to the decline and even local extirpation of many prey species. Large reptiles, such as snakes and large lizards are also vulnerable to human action and certain environmental changes such as deforestation and climate change. Strong fishing industries also reduce the food supplies of eagles such as the Bald eagle.

Eagles nest in large trees or in cliffs. The ability to shift between these two abodes is crucial in cases where human proximity may preclude cliff-nesting, or deforestation from agriculture, urbanization or forest fires prevents tree nesting. In the near-absence or rarity of natural nesting sites, eagles must find alternatives, a difficult proposition considering their large size and the huge dimensions of the nests that they build. Perhaps for this issue more than others will require human sympathy and help, either by conserving confirmed or potential nesting sites, or creating new structures that may be utilized for such purposes. Further studies would be required on possible intra- and inter-species differences on the likelihood of success for such experiments. This may require continual surveys using rudimentary observation and more advanced geomatics-based techniques to derive answers to such complex questions. Therefore, for these two points (foraging for prey and nesting) although eagle adaptation is key, human action may yet be the key issue.

Where eagles compete with vultures for carrion, an important issue concerns the increasing sanitation in urban areas of most countries. One of the important sources of food for vultures is urban waste, and in this activity, eagles have been recorded as dominant over most similarly sized vultures when in competition. However, the global trend has been the increased sanitation of cities, and this may affect the foraging abilities of urban scavengers. In most western countries, urban waste is either well controlled (for example, in covered bins, and collected regularly) or too piecemeal to attract birds as large as eagles (for example, urban gulls may be the main scavengers of pieces of pastry or fruit). Carrion is unacceptable in urban settings and would never be allowed to remain long enough for large avian scavengers to access it. As human welfare is increasingly defined as including a clean, disease-free environment, the future of urban waste as a food for avian scavengers may decline, even in the densely populated countries of the Global South, where large, open refuse dumps still exist. As described in this book, when human quality of life is defined as including physical, material, social, emotional well-being, eradication of the impacts of refuse such as disease and offensive smells become paramount.

One aspect of eagle adaptation, especially in North America and Europe, concerns their predation on livestock and companion animals. As documented in this book, this has become an important factor of conservation policy, as it influences public attitudes that may support conservation policy. This has influenced the killing of eagles in some western nations and Australia. In Africa, Asia and South America, the issue appears to be more on the eagles ingesting chemicals such as diclofenac although there is evidence that some people believe that eagles are killing smaller livestock and hence should be controlled. Future issues therefore concern the extent to which these predatory attacks may increase or decline.

Poisons and Impacts on Eagles

Current evidence indicates that poisons play an important role in the life span and habitat presence of eagles. With some research indications to the contrary, diclofenac poisoning appears to kill some eagles. This chemical compound has largely been replaced by less toxic substances, but eagles may be poisoned by other compounds that are yet to be eradicated. Lead from gunshots may have a near permanent presence, as guns are used for wars and hunting. In the United States and Africa, guns are unlikely to be banned, but hunting with guns may be restricted. This problem largely depends on the extent to eagles may access carrion, rather than live prey. In Europe, private hunting with guns is more likely to be banned entirely, but a global may be more difficult to achieve. Conservation efforts have gradually impressed on the public the negative aspects of hunting, so eventually a total ban may be possible.

Pesticides are still a serious problem for eagles, long after the 1972 American ban on DDT, which was strengthened by the global ban in the Stockholm Convention. DDT is still used in some places and has been replaced by more potent chemical compounds. The deadly compound carbofuran has killed many predators including eagles, and the bans of this substance have slowly made progress from Europe and North America, and have yet to be effectively banned in the Global South. Eagle deaths may move the public to influence policy on these matters, but such records require more detailed and sophisticated tracking methods. In Africa, desertification and reductions in landcover for livestock rearing encourages methods of carnivore removal with firepower and poisons. Poisons intended for jackals, hyenas, leopards and lions may also kill eagles, and guns purchased to kill such mammals may also target eagles when the birds attack small livestock like goats, sheep or calves. The rising human populations, coupled with more intensive landuse, may increasingly require the local eradication of large predators, and increase conflicts with conservation policies.

Although the shooting of eagles is currently generally in North America and Europe within the EU, in some other countries such as Australia culling is periodically allowed, and in the Global South there is little enforcement of protection laws. There are few detailed studies on the effectiveness of laws that protect eagles from illegal shootings, even in the more advanced countries. Problems encountered include the extremely wide range of most eagle species, their generally migratory or itinerant habits and the wide range of prey taken. Eagle movements may cross the borders of several countries, with different laws, resource exploitation patterns and interests in conservation policies. Comparative, cross border issues range across ecological, economic, socio-cultural and topographical factors, increasing the difficulties of monitoring, recording and conservation.

The amelioration of these problems clearly requires more detailed surveys of eagle foraging and movements, and accurate records of eagle deaths, population changes and actual prey species taken. This is currently very difficult, as the tracking methods lack sufficient detail and resolution. The new, more advanced geomatics methods such as GPS applications offer possibilities for meticulous examination of eagle movements and determination of their responsibility in the killing of particular species, changing foraging and nesting habits, and the impacts of landcover change on eagle populations. Such applications will nevertheless require heavy investments, considering the extensive foraging, wide ranges and migratory habits of most eagle species. Technological advances have lowered the costs of GPS technology, but the complexity of the subject is still daunting. Not only eagles, but their prey species and habitats must be monitored, as well as the factors (both natural and human induced) that affect these components of the eagles' existence.

Collisions and Electrocution

Modifications in the design, location and management of above ground structures such as power lines and wind turbines may affect the future of eagles. One important development is the trend of locating such power lines underground. It is however questionable if this is likely at a global scale. The location of underground lines for repairs, dangers for builders and other landusers who may accidently cut lines, and the equipment required to excavate trenches are some of the problems encountered in the laying of underground lines. Urban planning may require surface concrete structures, such as pavements or roads, that may cover the lines and render repairs difficult. For these reasons, in addition to the extra cost, most powerlines are still located above ground, supported by poles or taller pylon structures.

The key future problem concerns the growth of human populations, which results in strong urbanization and rural settlement and consequently urban and rural electrification. Electricity is touted as the energy of modernized regions, and most electric structures are similar: poles with electric wires closely spaced together, allowing the large birds such as eagles and vultures to touch two wires simultaneously, resulting in electrocution. The alternative designs suggested (including the wider spacing of wires) appear not to have taken hold, therefore any reductions in eagle deaths will depend largely on success or otherwise of this development. Underground wiring may be less likely, considering the contexts where rural electrification is developed, many of which are in areas of intensive cultivation and other landuse and underground cables may be considered dangerous for land users.

Wind turbines may also affect the future of eagles. This type of technology is largely limited to the northern nations and is unlikely to be developed in the Global South in the near future; however, some new developments are in the works in Asia (Japan, China, India, South Korea, Republic of the Philippines), in Chile and Brazil and some African countries (Kenya, South Africa and Morocco). If wind energy continues to develop in replacement of oil and gas sources, wind turbines may increasingly be built in open areas used by eagles for foraging and migration. The future impact of these structures on eagles may also depend on possible changes in the design (smaller turbines may be less lethal to high flying eagles), fence screens that would prevent birds from direct collisions, new methods to drive birds from the area, or relocation to area generally distant from eagle presence. Sea locations are useful, because most large heavy raptors avoid sea crossings during migration.

Airplane collisions are another serious future problem for eagles. The major problem of airport location, on flat land near urban areas, may not be changed in the near future. Unlike the other problems mentioned above, airports are vital and utterly non-negotiable; there is no possibility of relocating the airports. Future directions there must focus employing people to use technology to drive eagles and other species away from the airport areas, the redesign of airports with technology to do this job, removing any landcover or species that may attract eagles and/or attracting eagles away from the airports by any means possible. These are reliant on political will, and information exercises that would convince relevant stakeholders of the value of eagles. The objective would be to avoid the similar solution of shooting the eagles that may collide with airplanes, to save human lives. Problems apart from the political will include the increased number of airports and airplanes, which must keep up with the increased urban populations and the increased use of airports for passenger and cargo transit. Such problems may be even greater in the Global South, where there may be fewer airplanes but also less money, increased populations and transport needs, greater socio-economic and environmental disasters and less knowledge of the problems and solutions.

Academic Research

New research methods may be needed to study changes in eagle ecologies and populations in the future. Currently, common research methods are banding and capture, observation with telescopes or binoculars, or monitoring or nests and prey species. These studies document the general population and behavioral characteristics but may neglect some of the more individualist issues that may determine the fate of some species. Studies of eagle adaptability must necessarily include the micro-level behavioral variations of eagle species, that enable the birds to survive in extremely variable contexts. The evidence presented in this book indicates that eagles may be studied at more detailed behavioral levels. Certain strands of animal geography may provide the necessary theoretical background for such studies, using methodologies that examine individual eagle contrasts in reactions to variable human actions and ecological changes. Some studies have recorded variable individual eagle actions in terms of the killing of companion animals, livestock and scavenging in human life spaces. However, few studies have attempted to use such results to complement the species level studies documented in the first few chapters of this book.

Recent developments in geomatics techniques may allow the more detailed assessments of individual eagle feeding, nesting and mating behavior. This may be possible as smaller GPS devices are designed, that can be placed on the birds' bodies and can measure even physiological processes in reaction to life events. Such advancements would depend largely on technological developments and the use of these in research. Problems include the relative rarity of eagles, and their comparatively large size, aggressive behavior and high mobility, including cross-continental migration and remote, almost inaccessible habitats. These issues make eagles among the most difficult species to study.

Social science research will also be important, as eagle future are closely tied to human action. One of the most important research areas concerns impact of protective legislation on eagle populations. Currently, such research is largely limited to North America and Western Europe, with little knowledge on the impacts of the scanty legal protections in Asia, Africa and South America. Such studies in the non-Western continents will depend largely on the commitment of governments, private donors and the general publics of these countries. Increased environmental awareness, derived largely through the exposure of wider audiences to internet-based information. However, eagles are generally less studied than large mammals, this hampering knowledge on their status in protection efforts.

Studies of prey species and habitat variables from ecological and social perspectives may be important complements of eagle studies, especially where the eagles themselves are difficult to detect or to itinerant. Small mammal, medium-sized bird and coastal/riverine animal studies, linked to deforestation, forest-savanna boundary and coastal zone research are particularly relevant to eagle population assessments. Landcover studies are currently more easily done, as they can use the coarser resolutions of older, cheaper and readily available satellite images. Animal-based studies may require sharper resolutions, due to the small size and elusive behavior of most eagle prey species. The result of such complex and integrated studies should be habitat suitability maps that could predict or assess the suitability of landcover types and the locations for eagle presence and habitation.

Social science research would examine the human factors for landcover change and prey species change, and predict the future trajectories of such change. In addition to examining the impacts of legal developments, human population, agricultural, urban and industrial developments require examination. Currently, few such studies are examined in relation to eagles, which presents a unique opportunity to addition the existing bodies of knowledge. Well known eagle species, such the Bald, Golden, White-tailed and Steller's Sea eagles are well studied, and these could be examined for modelling the studies of rarer or less known species. Few comparative studies exist comparing the two groups of eagles and public awareness of the latter species is generally weak.

The development of interdisciplinary research, which began in the post-World War II era, may further develop to accommodate such complex research. Many of the issues concerning the welfare of eagles are outside the general ambit of ecological research. The broad fields of human (urban, agricultural, transportation, cultural and rural) and physical (geomorphology, climatology and biogeography) geography are all relevant to the study of most animals in modern times. Other multidisciplinary, transdisciplinary and interdisciplinary subjects, such as environmental law, environmental sociology and even cultural studies are also relevant. Many of these subjects developed after the significant paradigmatic shifts that occurred in the late twentieth century, when the themes of environmentalism and sustainable development, sometimes supported by Marxist positions on the methods of production, intruded into mainstream thought. Policy actions included the establishment of environmental departments, agencies, NGOs and/or committees, and legislative instruments in scores of countries, some including international treaties.

These issues feed into another issue, that of a coevolution of socio-economic developments and environmental (including animal and plant species, and landcover) change. As I have indicated elsewhere considering the conservation of vultures (Campbell, 2015), one possible framework is that of Norgaard (1994), which argues that nature and society are in permanent coevolution; society changes nature, and the changes done to nature affect society, which changes and further contributes to changes in nature and so on. This provides the foundation of a coevolutionary cosmology, with a framework comprising six basic principles. The application of this framework to the relationship between people and eagles is different from that of the human-vulture relationship, as the issue for eagles concerns perception of their predatory nature (killing livestock and companion animals, and believed by some to kill children) and their requirement of live prey (carrion may be an important side issue for some species), large foraging areas and shared preference for primal landcover (also required for urbanization, agriculture and recreation).

Considering the first principle of the coevolutionary cosmology, that people are part of the larger environment in which they live, serves as the foundation of most socio-environmental policies and philosophies, and justifies conservation policies that acknowledge the needs of animals, even those like eagles that are more difficult than most to conserve without some sacrifices on the part of society. The second principle, which states that "how we reason affects the social and environmental systems in which we evolve"; and "understandings must necessarily keep evolving, replacing themselves, to keep up with the cosmos that our understandings are changing" (Norgaard, 1994: 94-95) is key to the ecologies that must be investigated to conserve problematic species, supported by evolving technologies, the better to manage conservation for posterity. The third principle of the coevolutionary cosmology, that of conceptual pluralism, argues that there are different knowledge methods, conceptual frameworks and forms of application limitations, feeds directly into the multidisciplinary frameworks upon which all modern ecological studies and conservation applications must be based, especially for complex and extremely wide-ranging species such as eagles.

The fourth principle of the cosmology is that scientific knowledge is largely a socially derived process, the social issues being socially perceived wants and needs, revealing the "collective nature of understanding" (Norgaard, 1994: 97). In the context of the eagle-human relationship, this is particularly important, as many cultures and peoples have strong opinions and consequent policies on eagles. The size and power of eagles makes them possibly the most notice birds and enables them to intrude in a wide range of human experiences, from which even scientific knowledge may be derived and molded. The fifth principle of the coevolutionary cosmology (the relevance of history to changing conditions) is closely related to the sixth principle (the relevance of history to context) and both are relevant to the topic at hand, as the history of the relations between people and eagles (most the regal status of eagles, or their danger to human life spaces) still influences the policies towards eagles today.

Currently, academic research on eagles may soon feel the impacts of both technological and methodological advances. Possibly, the most important developments would be the use of new survey techniques, such as high-resolution imagery, precision GPS, and chemical analyses. These developments, which have been explored in this book require copious research but may not succeed alone. As one of the main issues is greater detail in the study of eagles, methodologies examining individualism in eagles, such as those included in the strands of animal geography are examined in this book. As eagle populations are reduced by various, differentiated and sometimes unrelated factors, and eagles must adapt to these issues, both inter- and intra-species issues must be examined. It is vital that these new research methods are equal to the task of evaluating the future of eagles at micro-levels.

12.6 Conclusions

The points raised in this chapter highlight the difficulty of assessing the global future of eagles. The issues in the history of eagles remain; they are important, relevant, symbolic and worth saving, and they cast light on human sensibilities. Legislation has been passed that may have some positive impacts; for example, it was noted that despite the fact that most eagle species have declining populations, Golden and Bald eagle populations in North America are increasing, and White-tailed eagle populations are expanding in Britain. The future of eagles is largely a human issue, as eagles without interference may easily survive in current contexts. This book has enumerated and analyzed the issues surrounding the unique but disputed group of birds called eagles. The most important points noted were the human attitudes and resultant actions towards eagles and their prey species and habitats, and the adaptive abilities of the eagles themselves. These issues contain within them the future of eagles, but the main job of the research that will give stakeholders knowledge of eagles' future, concern the development and application of technologies and methodologies that can examine in more detail the issues surrounding eagles, at intra- and inter-species, individual, group and species levels. With success in these developments, the future of eagles may be less bleak than is commonly assumed, subject of course to the human factors of continued human interest and support, habitat loss to human population growth and the ability of public policy to cope with these two last issues.

References

Aarts, G., MacKenzie, M., McConnell, B., Fedak, M. & Matthiopoulos, J. (2008). Estimating space-use and habitat preference from wildlife telemetry data. Ecography, 31(1), 140–160. https://doi.org/10.1111/j.2007.0906-7590.05236.x.

Abuladze, A., Kandaurov, A. & Eligulashvili, B. (2011). Seasonal migrations of birds of prey across Georgia: Results of long-term studies Materials of the International Conference, Birds of Prey and Owls in Caucasus. Tbilisi, Abastumani, Georgia, 26-29 October 2011, 3–4.

Abuladze, A. (2013). Birds of prey of Georgia. Institute of Zoology, Iliailia State University. [Tbilisi: Lasha Khvichia]

Abuladze, A. V. & Eligulashvili, V. E. (1996). White-tailed Sea eagle (Haliaeetus albicilla). Berlin, London & Paris: World Working Group on Birds of Prey (WWGBP). *In*: B.-U. Meyburg & R. D. Chancellor (Eds.), Eagle studies, 173–176.

Acharya, H. N. G. (1936). The White-bellied sea eagle (Haliaeetus leucogaster, Gmelin) in North Gujarat. Journal of the Bombay Natural History Society, 38(4), 828.

Acharya, R., Cuthbert, R., Baral, H. S. & Chaudhary, A. (2010). Rapid decline of the Bearded vulture Gypaetus barbatus in Upper Mustang, Nepal Forktail, 26, 117–120.

Acosta-Chaves, V., Granados, F. & Araya, D. (2012). Predation of long-tailed silky flycatcher (Ptilogonys caudatus) by Ornate hawk-eagle (Spizaetus ornatus) in a cloud forest of Costa Rica. Revista Brasileira de Ornitologia, 20(4), 451–452.

Adamec, M. (2004). Birds and power lines – Status in the Slovak Republic. In: Chancellor, R. D. & Meyburg, B.-U. (Eds.), Raptors, pp. 417–421 Worldwide. Berlin/Budapest: WWGBP/MME.

Adamian, M. S. & Klem, D. (1999). Handbook of the birds of Armenia. Oakland, CA: American University of Armenia.

Adamowicz, C. (1858). Énumération des travaux scientifiques littéraires publiés ou en manuscrits de C. Tyzenhauz. Revue et magasin de zoologie pure et appliquée, 2(9), 601–604.

Adeola, M. O. (1992). Importance of wild animals and their parts in the culture, religious festivals, and traditional medicine of Nigeria Environmental Conservation, 19(2), 125–134. https://doi.org/10.1017/S0376892900030605.

Adji, G. S. & Yuda, I. P. (2001). Population and distribution of raptors in central. *In*: Jawa, I. P. & Yuda, I. (Eds.), Report for Fauna and Flora International-Indonesia Programme and Kutilang Indonesia, 18. Yogyakarta, Indonesia: Yayasan Kutilang.

Adrianova, O. V. & Kashevarov, B. N. (2005). Some results of long-term raptor monitoring in the Kostomuksha nature reserve. Status of raptor populations in eastern Fennoscandia. Proceedings of the Workshop. Kostomuksha, Karelia, Russia.

AEE (Asociación Eólica Española). (2009). Observatorio Eólico 2009. Asociación Empresarial Eólica, Madrid, Spain.

Afianto, M. Y. (1999). Some ecological aspects of the Javan hawk-eagle Spizaetus bartelsi at G. Salak [Honors Thesis]. Bogor, Indonesia: Bogor Agriculture University.

Agency for Toxic Substances and Disease Registry (ATSDR). (2019). Toxicological profile for chromium. Retrieved from https://www.atsdr.cdc.gov/ToxProfiles/tp7.pdf

Agency for Toxic Substances and Disease Registry (ATSDR). (2020a). Toxicological profile for copper. Retrieved from https://www.atsdr.cdc.gov/toxprofiles/tp132.pdf

Agency for Toxic Substances and Disease Registry (ATSDR). (2020b). Selenium. Retrieved from https://www.atsdr.cdc.gov/toxprofiles/tp92-c3.pdf

Agostini, N. & Duchi, A. (1994). Water-Crossing behaviour of Black Kites (Milvus migrans) during migration. Bird Behavior, 10(1), 45–48. https://doi.org/10.3727/015613894791748935

Agostini, N., Malara, G., Neri, F., Mollicone, D. & Melotto, S. (1994a). Flight strategies of honey buzzards during spring migration across the central Mediterranean. Avocetta, 18, 73–76.

Agostini, N. & Malara, G. (1997). Entita delle popolazioni di alcune specie di rapaci Accipitriformi migranti, in primavera, sul Mediterraneo centrale. Rivista Italia di Ornitologia, 66, 174–176.

Agostini, N. & Logozzo, D. (1998). Primi dati sulla migrazione primaverile dei rapaci Accipitriformi sull'isola di Marettimo. Italian Ornithology, 68, 153–157.

Agostini, N. (2001). The island of Marettimo, a strategic point for surveying the migratory flow of Accipitriformes crossing the Channel of Sicily. Buteo, 12, 99–102.

Agostini, N. (2002). The migration of raptors in Italy. In: Brichetti, P. & Gariboldi, A. (Eds.), Manuale di ornitologia. Vol. III [Manual of Ornithology. Vol. III], 157–182. Bologna, Italy: Edagricole.

Agostini, N., Baghino, L., Coleiro, C., Corbi, F. & Premuda, G. (2002a). Circuitous autumn migration in the short-toed eagles (Circaetus gallicus). Journal of Raptor Research, 36, 111–114.

Agostini, N., Baghino, L., Panuccio, M. & Premuda, G. (2002b). A conservative strategy in migrating short-toed eagles (Circaetus gallicus). Ardeola, 49, 287–291.

Agostini, N., Baghino, L., Panuccio, M., Premuda, G. & Provenza, A. (2004). The autumn migration strategies of adult and juvenile short-toed eagles (Circaetus gallicus) in the central Mediterranean. Avocetta, 28, 37–40.

Agostini, N. (2005). Are earlier estimates of Accipitriformes crossing the Channel of Sicily (central Mediterranean) during spring migration accurate? Journal of Raptor Research, 39, 184–186.

Agostini, N., Panuccio, M., Mellone, U., Lucia, G., Wilson, S. & Ashton-Booth, J. (2007). Do migration counts reflect population trends? A case study of the honey buzzard Pernis apivorus. Ardeola, 54(2), 339–344.

Agostini, N. & Mellone, U. (2008). Does migration flyway of short toed snake-eagles breeding in Central Italy reflect the colonization history? Journal of Raptor Research, 42(2), 158–159. https://doi.org/10.3356/JRR-07-27.1

Agostini, N., Panuccio, M., Lucia, G., Liuzzi, C., Amato, P., Provenza, A., . . . Mellone, U. (2009). Evidence for age-dependent migration strategies in the short-toed eagle. British Birds, 102, 506–508.

Agostini, N., Panuccio, M. & Pasquaretta, C. (2015). Morphology, flight performance and water crossing tendencies of Afro-Palearctic raptors during migration. Current Zoology, 61(6), 951–958. https://doi.org/10.1093/czoolo/61.6.951

Aguiar-Silva, F. H. (2007). Dieta do gavião-real Harpia harpyja (Aves; Accipitridae). em florestas de terra firme de Parintins, Amazonas, Brasil [Thesis].

Aguiar-Silva, F. H., Sanaiotti, T. M., Jaudoin, O., Srbek-Araujo, A. C., Siqueira, G. & Banhos, A. (2012). Harpy eagle sightings, traces and nesting records at the "Reserva Natural Vale", Brazilian Atlantic Forest remnant in Espirito Santo, Brazil. Revista Brasileira de Ornitologia, 20, 148–155.

Aguiar-Silva, F. H., Junqueira, T. G., Sanaiotti, T. M., Guimarães, V. Y., Mathias, P. V. C. & Mendonça, C. V. (2015). Resource availability and diet in Harpy eagle breeding territories on the Xingu River, Brazilian Amazon. Brazilian Journal of Biology = Revista Brasleira de Biologia, 75(3)(Suppl. 1), 181–189. https://doi.org/10.1590/1519-6984.00914BM

Aguiar-Silva, F. H., Jaudoin, O., Sanaiotti, T. M., Seixas, G. H. F., Duleba, S. & Martins, F. D. (2017). Camera trapping at Harpy eagle nests: Interspecies interactions under predation risk. Journal of Raptor Research, 51(1), 72–78. https://doi.org/10.3356/JRR-15-58.1

Aguiar-Silva, F. H., Sanaiotti, T. M. & Luz, B. B. (2014). Food habits of the Harpy eagle, a top predator from the Amazonian rainforest canopy. Journal of Raptor Research, 48(1), 24–35. https://doi.org/10.3356/JRR-13-00017.1

Agustina, D. & Prawiradilaga, D. (2005). Committee for the 4th Symposium on Asian Raptors. Distribution and population status of Sulawesi hawk eagle Spizaetus lanceolatus (Temminck and Schlegel, 1844) at Rawa Aopa Watuymohai National Park, South-east Sulawesi, Indonesia. *In*: J. Abu et al. (Eds.). Proceedings of the 4th Symposium on Asian Raptors, 2005, pp. 217. Kuala Lumpur, Malaysia.

Ahrens, D. & Henson, R. (2019). Meteorology Today: An introduction to Weather, Climate, and the Environment (12th ed). Boston, MA: Cengage.

Ainslie, B., Alexander, N., Johnston, N., Bradley, J., Pomeroy, A. C., Jackson, P. L. & Otter, K. A. (2013). Predicting spatial patterns of eagle migration using a mesoscale atmospheric model: A case study associated with a mountain-ridge wind development. International Journal of Biometeorology, 58, 17–30. https://doi.org/10.1007/s00484-012-0620-0.

Airola, D. A. (2011). Dynamics of an urban turkey vulture roost in Sacramento, California. CVBC Bulletin. Retrieved from http://www.cvbirds.org/wp-content/themes/cvbirds/files/V.14no.1/ CVBC_ Vol14_No1_pp1-8.pdf, 14 (pp. 1–8).

Åkesson, S. & Hedenström, A. (2000). Wind selectivity of migratory flight departures in birds. Behavioral Ecology and Sociobiology, 47(3), 140–144. https://doi.org/10.1007/s002650050004

Åkesson, S., Karlsson, L., Walinder, G. & Alerstam, T. (1996). Bimodal orientation and the occurrence of temporary reverse bird migration during autumn in south Scandinavia. Behavioral Ecology and Sociobiology, 38(5), 293–302. https://doi.org/10.1007/s002650050245

Ákos, Z., Nagy, M. T. & Vicsek, T. S. (2008). Comparing bird and human soaring strategies. Proceedings of the National Academy of Sciences of the United States of America, 105(11), 4139–4143. https://doi.org/10.1073/pnas.0707711105

Ákos, Z., Nagy, M., Leven, S. & Vicsek, T. (2010). Thermal soaring flight of birds and unmanned aerial vehicles. Bioinspiration and Biomimetics, 5(4), 045003. https://doi.org/10.1088/1748-3182/5/4/045003, PubMed: 21098957

Alaska Department of Fish and Game. (2020). Bald eagle. Retrieved from https://www.adfg.alaska.gov/index.cfm?adfg=baldeagle.main

Alatalo, R. V., Carlson, A., Lundberg, A. & Ulfstrand, S. (1981). The conflict between male polygamy and female monogamy: The case of the pied flycatcher Ficedula hypoleuca. American Naturalist, 117(5), 738–753. https://doi.org/10.1086/283756

Alauda, J.A. (1960). The population of Imperial Eagles (Aquila heliaca adalberti) in the Guadalquivir salt marshes: Their evolution in the last century. Alauda, 28: 20–26. Sevilla, Valverde, Dominican Republic: CSIC.

Albano, C., Girão, W. & Pinto, T. (2007). First documented record for the Black hawk-eagle, Spizaetus tyrannus, for the state of Ceará, Brazil. Revista Brasileira de Ornitologia, 15, 123–124 (In Portuguese with English summary).

Albers, P. H. (2006). Birds and polycyclic aromatic hydrocarbons. Avian and Poultry Biology Reviews, 17(4), 125–140. https://doi.org/10.3184/147020606783438740

Albers, P. H., Green, D. E. & Sanderson, C. J. (1996). Diagnostic criteria for selenium toxicosis in aquatic birds: Dietary exposure, tissue concentrations, and macroscopic effects. Journal of Wildlife Diseases, 32(3), 468–485. https://doi.org/10.7589/0090-3558-32.3.468

Albornoz, M. A. & Fernández, M. A. (2014). The symbolism of the eagle. Retrieved from https://library.acropolis.org/the-symbolism-of-the-eagle/

Albuquerque, J. L. B. (1995). Observations of rare raptors in southern Atlantic rainforest of Brazil [Observaciónes sobre las Aves de Rapiña Raras en la Floresta Atlántica del Sureste de Brasil]. Journal of Field Ornithology, 363–369.

Albuquerque, J. L. B. (1986). Conservation and status of raptors in Southern Brazil. Birds of Prey Bulletin, 3, 88-94.

Albuquerque, J. L. B., Ghizoni, I. R. Jr., Silva, E. S., Trannini, G., Franz, I., Barcellos, A., Hassdenteufel, C. B., Arend, F. L. & Martins-Ferreira, C. (2006). Crowned Solitary-eagle (Harpyhaliaetus coronatus) and Crested eagle (Morphnus guianensis) in Santa Catarina and Rio Grande do Sul: Priorities and challenges for their conservation. Revista Brasileira de Ornitologia, 14, 411–415.

Albuquerque, J. L. B. (1979). On the presence of Harpyhaliaetus coronatus and Morphnus guianensis in southeast Brazil and recommendations for the conservation of the species by means of maintaining their natural habitat. Hornero Numero Especial, 70–73.

Alcántara, M., Ferreiro, E. & Gardiazábal, A. (2001). Dispersal of young Bonelli's eagle (Hieraaetus fasciatus), first results by satellite telemetry (LIFE project B4–3200/97/252 Conservation Plan of Bonelli's eagle in Sierra de Guara, Huesca, Aragón). *In*: Eurasian Congress on Raptors, IV. Seville, Spain: Estacion Biologica de Doñana.

Alderfer, J. (2006). Complete birds of North America. Washington, DC. National Geographic Society.

Alerstam, T. (1990). Bird Migration. Cambridge, UK: Cambridge University Press.

Alerstam, T. (2001). Detours in bird migration. Journal of Theoretical Biology, 209(3), 319–331. https://doi.org/10.1006/jtbi.2001.2266

Alerstam, T. (2011). Optimal bird migration revisited. Journal of Ornithology, 152(Suppl. 1), 5–23. https://doi.org/10.1007/s10336-011-0694-1

Alexander, R. M. (2003). Principles of Animal Location. Princeton: Princeton University Press.

Alexander, T. L. & Buxton, D. (Eds.). (1986). Management and diseases of deer: A handbook for the veterinary surgeon. Veterinary Deer Society.

Ali, R. (1986). Feeding ecology of the Bonnet Macaque at the Mundanthurai Sanctuary, Tamil Nadu. Journal of the Bombay Natural History Society, 83(1), 98–110.

Ali, S. (1993). The Book of Indian Birds. Bombay Natural History Society. Bombay, India.

Ali, S. & Ripley, S. D. (1978). Handbook of the Birds of India and Pakistan (2nd ed). Oxford University Press. New Delhi, India.

Ali, S. & Ripley, S. D. (1983). Handbook of the birds of India and Pakistan: Together with those of Bangladesh, Nepal, Bhutan and Sri Lanka. Oxford: Oxford University Press.

Aliskerov, S. V. (1996). Observations on Haliaeetus albicilla and Haliaeetus pelagicus on Lake Udil in the far east. *In*: B.-U. Meyburg & R. D. Chancellor (Eds.), Eagle studies, 177–179. Berlin: World Working Group on Birds of Prey and Owls.

Alivizatos, H., Papandropoulos, D. & Zogaris, S. (2020). Diet of the Lesser spotted eagle (Clanga pomarina) in Amvrakikos wetlands national park, Greece. Ecologica Montenegrina, 30, 68–76. https://doi.org/10.37828/em.2020.30.6

Allaby, M. (1992). The Concise Oxford Dictionary of Zoology. Oxford: Oxford University Press.

Allen, J. A. (1900). List of birds collected in the District of Santa Maria, Colombia, by Mr Herbert H. Smith. Bulletin of the American Museum of Natural History, 13, 117–183.

Allen, J. C. (1939). Ecology and management of Nelson's bighorn on the Nevada mountain ranges. Transactions of the Wildland Conference, 4, 253–256.

Allen, P. E., Goodrich, L. J. & Bildstein, K. L. (1996). Within—And among—Year effects of cold fronts on migrating raptors at hawk Mountain, Pennsylvania, 1934–1991. Auk, 113(2), 329–338. https://doi.org/10.2307/4088899

Allen, T. (2015). Ulgii Festival photographs, #No. taken Sept. (2014). Retrieved from http://www.allenfotowild.com/Mongolia/Eagle-Festival/i-xWVt8Fq40, 41 & 74.

Allen, T. F. H. & Starr, T. B. (1982). Hierarchy: Perspectives for Ecological Complexity. Chicago: University of Chicago Press.

Allendorf, F. W., Leary, R. F., Spruell, P. & Wenburg, J. K. (2001). The problems with hybrids: Setting conservation guidelines. Trends in Ecology and Evolution, 16(11), 613–622. https://doi.org/10.1016/S0169-5347(01)02290-X

Allinson, T. (2017). Introducing a new avian sensitivity mapping tool to support the siting of wind farms and power lines in the Middle East and Northeast Africa. *In*: Köppel, J. (Ed.), Wind Energy and Wildlife Interactions: Presentations from the CWW2015 Conference, pp. 207–218. https://doi.org/10.1007/978-3-319-51272-3_12. Cham, Germany: Springer International Publishing.

Allott, M., Allott, M. & Hatchett, N. (2006). The breeding cycle of a pair of Wedge-tailed eagles Aquila audax in southeast Queensland. Sunbird, 36, 37–41.

Alon, D., Granit, B., Shamoun-Baranes, J., Leshem, Y., Kirwan, G. & Shirihai, H. (2004). Soaring bird migration over Northern Israel in autumn. British Birds, 97, 160–182.

Alonso, J. A. & Alonso, J. C. (1999). Collision of birds with overhead transmission lines in Spain. *In*: M. Ferrer & G. F. E. Janss (Eds.), Birds and power lines: Collision, electrocution, and breeding, pp. 57–82. Madrid: Quercus.

Alonso, J. C., Gonzalez, L. M., Heredia, B. & Gonzalez, J. L. (1987). Parental care and the transition to independence of Spanish Imperial Eagles Aquila heliaca in Doñana National Park, southwest Spain. Ibis, 129, 212–224. https://doi.org/10.1111/j.1474-919X.1987.tb03202.x

Altalink. (2019). Avian electrocutions. Retrieved from http://www.altalink.ca/environment/avian-protection-plan/avian-fatalities.cfm

Altmann, J. (1974). Observational study of behavior sampling methods. Behaviour, 49(3), 227–267. https://doi.org/10.1163/156853974x00534

Alvarado, S. A. A., Figueroa, R. A., Faundez, P. V., Carrasco-Lagos, P. & Moreno, R. A. (2015). Aves rapaces de la Región Metropolitana de Santiago, Chile. Santiago: Universidad de Chile.

Alvarez-Cordero, E. (1996). Biology and conservation of the harpy eagle in Venezuela and Panama. PhD Thesis. Gainesville, FL: University of Florida.

Alvarez del Toro, M. (1980). The Birds of Chiapas (2nd ed). Universidad Autónoma de Chiapas, Tuxtla Gutierrez, Chiapas.

Alvarez, J. B., Jr. (1970). A report on the 1969 status of the Monkey-eating Eagle of the Philippines. Papers and Proceedings of the International Union of Conservation of National and Natural Resources, 11th Technical Meeting, New Delhi, India, November 25–28.

Alvarez, L. W., Alvarez, W., Asaro, F. & Michel, H. V. (1980). Extraterrestrial cause for the cretaceous-tertiary extinction. Science, 208(4448), 1095–108. https://doi.org/10.1126/science.208).4448).1095.

Amadon, D. (1949). Notes on Harpyhaliaetus. Auk, 66(1), 53–56. https://doi.org/10.2307/4080660.

Amadon, D. (1950). What is Spizaetus devillei Dubois? Auk, 67(2), 235–236. https://doi.org/10.2307/4081223
Amadon, D. (1953). Remarks on the Asiatic hawk-eagles of the genus Spizaetus. ibis, 95, 492–500.
Amadon, D. (1978). Remarks on the taxonomy of some Australasian raptors. Emu – Austral Ornithology, 78(3), 115–118. https://doi.org/10.1071/MU9780115
Amadon, D. (1982). A revision of the sub-buteonine hawks (Accipitridae, Aves). American Museum Novitates, 2741, 1-20.
Amadon, D. (1982). The genera of Booted Eagles: Aquila and relatives. Journal of the Yamashina Institute for Ornithology, 14(2–3), 108–121. https://doi.org/10.3312/jyio1952.14.108
Amadon, D. & Bull, J. (1988). Hawks and owls of the world: A distributional and taxonomic list. Proceedings of the Western Foundation of Vertebrate Zoology, 3, 295–357.
Amar, A., Buij, R., Suri, J., Sumasgutner, P. & Virani, M. Z. (2018). Conservation and ecology of African raptors. *In*: J. Sarasola, J. Grande & J. Negro (Eds.), Birds of prey. Cham, Germany: Springer. https://doi.org/10.1007/978-3-319-73745-4_18.
Amar, A. & Cloete, D. (2017). Quantifying the decline of the Martial eagle Polemaetus bellicosus in South Africa. Bird Conservation International, 1–12.
Amar, A., Cloete, D. & Whittington, M. (2016). Using independent nest survey data to validate changes in reporting rates of Martial Eagles between the Southern African Bird Atlas Project 1 and 2. Ostrich, 87(1), 1–5. https://doi.org/10.2989/00306525.2015.1089333
Amar, A., Krochuk, B. A., Weideman, E. A., Smyth, L. K. & Naude, V. V. N. (2019). Using web-sourced photography to explore the diet of a declining African raptor, the Martial eagle (Polemaetus bellicosus). Condor: Ornithological Applications, 121(1). https://doi.org/10.1093/condor/duy015
Amaral, F. S. R., Miller, M. J., Silveira, L. F., Bermingham, E. & Wajntal, A. (2006). Polyphyly of the hawk genera Leucopternis and Buteogallus (Aves, Accipitridae): Multiple habitat shifts during the Neotropical buteonine diversification. BMC Evolutionary Biology, 6, 10. https://doi.org/10.1186/1471-2148-6-10
Amaral, F. S. R. (2008). Molecular studies on the Accipitrids (Aves, Accipitridae): An evolutionary perspective. PhD Thesis. São Paulo, Brazil: Universidade de São Paulo.
Amaral, F. S. R. (2008). Phylogenetic inference in Buteonine hawks (Aves: Accipitridae) based on cranial Ostreological characters. Master's Thesis. São Paulo, Brazil: Universidade de São Paulo.
Amaral, F. S. R., Sheldon, F., Gamauf, A., Haring, E., Riesing, M., Silveira, L. & Wajntal, A. (2009). Patterns and processes of diversification in a widespread and ecologically diverse avian group, the buteonine hawks (Aves, Accipitridae). Molecular phylogenetics and evolution. 53, 703–715. https://doi.org/10.1016/j.ympev.2009.07.020.
Ambrosia, V. G., Wegener, S. S., Sullivan, D. V., Buechel, S. W., Dunagan, S. E., Brass, J. A., . . . Schoenung, S. M. (2003). Demonstrating UAV-acquired real-time thermal data over fires. Photogrammetric Engineering and Remote Sensing, 69(4), 391–402. https://doi.org/10.14358/PERS.69.4.391
American Bird Census. (2015). Golden Eagle. Retrieved from https://www.abcc-am.org/golden-eagle.html.
American Eagle Foundation. (2020). Eagle hacking. Retrieved from https://www.eagles.org/what-we-do/repopulate/eagle-hacking/#:~:text=Hacking%20is%20the%20method%%2020used, from%20an%20%20artificial%20nesting%20tower.
American Ornithologists' Union. (1998). Check-list of North American Birds (7th ed). Washington, DC: American Ornithologists' Union.
American Ornithologists' Union. (2008). The A.O.U. Check-list of North American Birds (7th ed.). Washington, D.C.: American Ornithologists' Union.
American Ornithologists' Union. (2009). Check-list of North American Birds, Tinamiformes to Falconiformes (7th ed). American Ornithologists' Union.
Amezian, M., Irizi, A., Errati, A., Loran, H., El Khamlichi, R., Morandini, V. & Garrido, J. R. (2015). Spanish Imperial eagles and other eagles found electrocuted in Morocco and proposition of correction measures. Figshare. https://doi.org/10.6084/m9.figshare.1613292
Ammann, K. & Ammann, K. (1989). The Hunters and the Hunted. Southern Book Publishers.
Amstrup, S. C., McDonald, T. L. & Durner, G. (2004). Using satellite radiotelemetry data to delineate and manage wildlife populations. Wildlife Society Bulletin, 32(3), 661–679.
Anderson, D. J. & Horwitz, R. J. (2008). Competitive interactions among vultures and their avian competitors. Jour name, 121(4), 505–509.
Anderson, D. L., Weidenfeld, D. A., Bechard, M. J. & Novak, S. J. (2004). Avian diversity in the Moskitia region of Honduras. Ornitologia Neotropical, 15, 447–482.
Anderson, K. (1997). A walk on the wild side: A critical geography of domestication. Progress in Human Geography, 21, 463–485.
Anderson, K. & Gaston, K. J. (2013). Lightweight unmanned aerial vehicles will revolutionize spatial ecology. Frontiers in Ecology and the Environment, 11(3), 138–146. https://doi.org/10.1890/120150
Anderson, M. D. (1999). Africa's hooded vulture: A dichotomy of lifestyle. Vulture News, 41, 3–5.
Anderson, M. D. (2000). Raptor conservation in the Northern Cape Province, South Africa. Ostrich, 71(1–2), 25–32. https://doi.org/10.1080/00306525.2000.9639859
Anderson, M. D., Maritz, A. W. A. & Oosthuysen, E. (1999). Raptors drowning in farm reservoirs in South Africa. Ostrich, 70(2), 139–144. https://doi.org/10.1080/00306525.1999.9634530
Anderson, M. D., Piper, S. E. & Boschoff, A. F. (Eds.). (2004) The vultures of Southern Africa—Quo vadis. Proceedings of the Workshop on Vulture Research and Conservation in Southern Africa. Kimberley. SA, Australia: McGregor Museum.
Anderson, M. & Hohne, P. (2007). An aerial survey of raptors along the Ghaap Plateau escarpment. Northern Cape Province, South Africa. Gabar, 18(1), 1–8.
Anderson, S. H. & Patterson, C. (1988). Characteristics of Bald eagle roost in Wyoming. Prairie Naturalist, 20, 147–152.

Anderson, W. L., Havera, S. P. & Zercher, B. W. (2000). Ingestion of lead and nontoxic shotgun pellets by ducks in the Mississippi Flyway. Journa of Wildlife Management, 64(3), 848–857. https://doi.org/10.2307/3802755
Andreenkov, O. V., Andreenkov, N. G., Zhimulev, E. I., Zhimulev, I. F., Sinkov, K. O., Chepurov, A. A. & Shnayder, E. P. (2009). Nesting of th White-tailed eagle in the Novosibirsk, Russia. Raptors Conservation, 17, 145–148.
Andrew, M. E. & Shephard, J. M. (2017). Semi-automated detection of eagle nests: An application of very high-resolution image data an advanced image analyses to wildlife surveys. Remote Sensing in Ecology and Conservation, 3(2), 66–80. https://doi.org/10.1002/rse2.38
Andrew, P. (1992). The Birds of Indonesia: A Checklist (Peter's sequence). Kukila Checklist no. 1. Jakarta: Indonesian Ornithological Society.
Andriamasimanana, R. H. & Cameron, A. (2013). Predicting the impacts of climate change on the distribution of threatened forest-restricted bird in Madagascar. Ecology and Evolution, 3(4), 763–769. https://doi.org/10.1002/ece3.497
Andrle, R. F. (1968). Raptors and other North American migrants in Mexico. Condor, 70(4), 393–395. https://doi.org/10.2307/1365945
Anfuso, J., Gil Suarez, M. V. & Chebez, J. C. (2008). New nesting record of the Harpy eagle (Harpia harpyja) in Misiones Province, Argentina an some considerations about its conservation. Notulas Faunistic, 21, 1–13.
Angehr, G. (2001). The Field Editor's report. Tucán, 27(5), 3–4.
Angehr, G. (2006). Annotated Checklist of the Birds of Panama. Panama City: Panama Audubon Society.
Angehr, G. R. (2003). Directory of important bird areas in Panama. [Sociedad Audubon de Panamá, Argentina]. *In*: Bildstein, K. L., Barber, D. R & Zimmerman, A. (Eds.), Neotropical Ornithology, p. 270. Orwigsburg: Hawk Mountain Sanctuary.
Angell, T. (1974). Owls. Seattle, WA: University of Washington Press.
Angerbjörn, A. & Flux, J. E. C. (1995). Lepus timidus. Mammalian Species, 495(495), 1–11. https://doi.org/10.2307/3504302
Ansell, A. R. & Smith, W. E. (1980). Raptor protection activities of the Idaho. Power Company. In: Howard, R. P. & Gore, J. F. (Eds.), Proceeding of a workshop on raptors and energy developments, pp. 56–70. Boise, Idaho: Idaho Chapter, The Wildlife Society.
Antal, M. (2010). Policy measures to address bird interactions with power lines: A comparative case study of four countries. Ostrich, 81(3) 217–223. https://doi.org/10.2989/00306525.2010.517921
Anthony, A. J. (1979). Patagial tagging of Wahlberg's eagles. Safring News, 8, 28–30. Retrieved from http://safring.birdmap.africa/papers safring_8_1_28.pdf.
Anthony, R. G., Knight, R. L., Allen, G. T., McClelland, B. R. & Hodges, J. I. (1982). Habitat use by nesting and roosting bald eagles in the Pacifi Northwest. Transactions of North American Wildlife and Natural Resource Conference, 47, 332–342.
Anthony, R. G. & Isaacs, F. B. (1989). Characteristics of Bald eagle nest sites in Oregon. Journal of Wildlife Management, 53(1), 148–159. https:/ doi.org/10.2307/3801322
Anthony, R. G., Estes, J. A., Ricca, M. A., Miles, A. K. & Forsman, E. D. (2008). Bald eagles and sea otters in the Aleutian archipelago: Indirec effects of trophic cascades. Ecology, 89(10), 2725–2735. https://doi.org/10.1890/07-1818.1
Anuari Ornitològic de les Balears. (2004). Retrieved from http://ibdigital.uib.es/greenstone/sites/ localsite/collect/anuariGob/index/assoc AOB_2004/v19pdf.dir/AOB_2004v19.pdf.
Aparecido Libório, R. & Moura Martins, M. (2013). Body size in predator–prey interactions: An investigation of Neotropical primates and thei predators. Studies on Neotropical Fauna and Environment, 48(1), 81–87. https://doi.org/10.1080/01650521.2013.789724
Aparicio, K. (2004). Using a Geographical Information System for the analysis of historical and current distribution of Harpy eagle (Harpi harpyja) in Panama. In: Rodriguez-Estrella, R. & Bojorquez-Tapia, L. A. (Eds.), Spatial analysis in raptor ecology and conservation, pp 27–37. La Paz, Baja California Sur, Mexico Centro de Investigaciones Biologicas del Noroeste, S.C.
APLIC (Avian Power Line Interaction Committee). (2006). Suggested practices for avian protection on power lines: The state of the art in 2006 Washington, D.C.: Edison Electric Institute.
Applegate, R. D., Kissell, R. E., Jr., Moss, E. D., Warr, E. L. & Kennedy, M. L. (2011). Problems with Avian point counts for estimating density of Northern bobwhite—A case study. Journal of Fish and Wildlife Management, 2(1), 117–121. https://doi.org/10.3996/092010-JFWM-033
Aranda, M., Burton, A., Iñigo-Elías, E. & Escalante, P. (2009). Registro del águila elegante (Spizaetus ornatus) en la Reserva de la Biosfera sierr de MAntlán, Jalisco–Colima, México. Revista Mexicana de Biodiversidad, 80(1), 265–268.
Aráoz, R. (2017). The first Black-and-Chestnut eagle (Spizaetus isidori) nest discovered in Argentina reveals potential human–predator conflicts Journal of Raptor Research, 51(1), 79–82. https://doi.org/10.3356/JRR-16-49.1
Ararat, K., Fadhil, O., Porter, R. F. & Salim, M. (2011). Breeding birds in Iraq: Important new discoveries. Sandgrouse, 33, 12–33.
Araújo, R. C., Silveira, L. F. & Da Luz, D. E. (2015). Rediscovery of the Crested eagle Morphnus guianensis (Daudin, 1800) in the fragmentec Atlantic Forest of Bahia, Brazil. Revista Brasileira de Ornitologia, 23(1), 25–28. https://doi.org/10.1007/BF03544286
Arent, L. A. (2007). Raptors in captivity. Blaine, WA: Hancock House.
Arkadevich, M. V., Yuryevna, P. Y., Vladimirovich, T. A. & Valerevna, M. J. (2014). Nesting of the White-tailed eagle (Haliaeetus albicilla) and th Long-legged buzzard (Buteo rufinus) in the Rostovsky Reserve. News of Higher Educational Institutions. North-Caucasian Region. Natura Sciences.
Armstrong, E. A. (1975). The Life and Lore of the Bird in Nature, Art, Myth, and Literature. New York: Crown Publishing Group.
Armstrong, R. H. (2008). The importance of fish to Bald eagles in Southeast Alaska: A review. *In*: Wright, B.A., & Schempf, P. (Eds.), Bald eagle in Alaska, pp. 54–67. Juneau, AK: Bald Eagle Research Institute, University of Alaska Southeast.
Arnold, K. E. & Owens, I. P. F. (1998). Cooperative breeding in birds: A comparative test of the life history hypothesis. Proceedings of the Roya Society of London. Series B, 265(1398), 739–745. https://doi.org/10.1098/rspb.1998.0355
Arnold, L. W. (1954). The Golden eagle and its economic status. U.S. Fish and Wildlife Service, 27, 35.
Arnold, M. L. (1992). Natural hybridisation as an evolutionary process. Annual Review of Ecology and Systematics, 23(1), 237–261. https://doi org/10.1146/annurev.es.23.110192.001321

Aronson, M. F. J., La Sorte, F. A., Nilon, C. H., Katti, M., Goddard, M. A., Lepczyk, C. A. & Winter, M. (2014). A global analysis of the impacts of urbanization on bird and plant diversity reveals key anthropogenic drivers. Proceedings of the Royal Society of London. Series B, 281(1780), 1–8. https://doi.org/10.1098/rspb.2013.3330

Arroyo, B., Bueno, J. M. & Pérez-Mellado, V. (1976). Biologia de reproducción de una pareja de Hieraaetus fasciatus en España Central. Doñana Acta Vertebrata, 3, 9–15.

Arroyo, B., Ferreiro, E. & Garza, V. (1992). Factores limitantes de la población de águila perdicera en España. Technical report ICONA. Madrid: Ministerio de Medio Ambiente.

Asai, S., Yamamoto, Y. & Yamagishi, S. (2006). Genetic diversity and extent of gene flow in the endangered Japanese population of Hodgson's hawk-eagle, Spizaetus nipalensis. Bird Conservation International, 16(2), 113–129. https://doi.org/10.1017/S0959270906000050

Ash, J. S. (1981). Ayres' eagle Hieraaetus dubius in Ethiopia and Sorealia. Scopus, 5, 54–56.

Ash, J. S. & Miskell, J. E. (1981). Birds of Somalia, their habitats and status: More than 50 additions of birds to the Somalia list, including two hybrids, together with notes from Ethiopia and Kenya. Scopus Special [Suppl. 1].

Ash, J. S. & Miskel, J. E. (1998). Birds of Somalia. Sussex, UK: Pica Press.

Ash, J. & Atkins, J. (2009). Birds of Ethiopia and Eritrea. London: Christopher Helm Publishers.

Ash, J. & Atkins, J. (2010). Birds of Ethiopia and Eritrea: An Atlas of Distribution. Bloomsbury Publishing.

Ash, J. & Atkins, J. (2000). Birds of Ethiopia and Eritrea. London: Christopher Helm Publishers, London.

Ashbaugh, H. M., Conway, W. C., Haukos, D. A., Collins, D. P., Comer, C. E. & French, A. D. (2018). Evidence for exposure to selenium by breeding interior snowy plovers (Charadrius nivosus) in saline systems of the Southern Great Plains. Ecotoxicology, 27(6), 703–718. https://doi.org/10.1007/s10646-018-1952-2

Asner, G. P. & Levick, S. R. (2012). Landscape-scale effects of herbivores on treefall in African savannas. Ecology Letters, 15(11), 1211–1217. https://doi.org/10.1111/j.1461-0248.2012.01842.x

Aspinall, R. & Veitch, N. (1993). Habitat mapping from satellite imagery and wildlife survey data using a Bayesian modeling procedure in a GIS. Photogrammetric Engineering and Remote Sensing, 59(4), 537–543.

Aspinall, S. & Hellyer, P. (2006). Important bird areas of the United Arab Emirates. British Birds, 99, 546–561.

Associated Press. (2016). Collision with Bald eagle caused plane to crash outside Alaska, killing 4 people on board. Retrieved from https://nationalpost.com/news/world/collision-with-bald-eagle-caused-plane-to-crash-outside-alaska-killing-4-people-on-board.

Associated Press. (2018). California mountain lion caught using thermal imaging drone. Retrieved from https://www.apnews.com/0254751f2e4c4027a1e7c94a24763b98.

Auburn, J. (1991). The timing of Wahlberg's eagles' migrations through Central Africa. Gabar, 6, 64–67.

Audet, A. M., Robbins, C. B. & Larivière, S. (2002). Alopex lagopus. Mammalian Species, 713, 1–10. https://doi.org/10.1644/1545-1410(2002)713<0001:AL>2.0.CO;2

Augliere, B. (2015). Beyond GPS: The Next High-Tech Frontier in Wild Animal Tracking. Retrieved https://blogs.scientificamerican.com/guest-blog/beyond-gps-the-next-high-tech-frontier-in-wild-animal-tracking/

Aumann, T. (1997). Sightings of diurnal raptors in relation to habitat and land use in the Matobo Hills, Zimbabwe. Journal of African Raptor Biology, 12, 2–8.

Avery, G., Robertson, A. S., Palmer, N. G. & Prins, A. J. (1985). Prey of giant eagle owls in the de Hoop Nature Reserve, Cape Province, Haiti and some observations on hunting strategy. Ostrich, 56(1–3), 117–122. https://doi.org/10.1080/00306525.1985.9639579

Avery, M. L. (2004a). Managing Depredation and Nuisance Problems Caused by Vultures. Washington: United States Department of Agriculture.

Avery, M. L. (2004b). Trends in North American vulture populations. *In*: R. M. Timm & W. P. Gorenzel (Eds.), Proceedings of the 21st Vertebrate Pest Conference, 1–4 March 2004, pp. 116–121. Visalia, CA: University of California, Davis, CA.

Avery, M. L. & Cummings, J. L. (2004). Livestock depredations by Black vultures and Golden eagles. Sheep and Goat Research Journal, 19, 58–63. Retrieved from https://www.aphis.usda.gov/wildlife_damage/nwrc/publications/04pubs/avery044.pdf

Avery, M. L., Humphrey, J. S., Daughtery, T. S., Fischer, J. W., Milleson, M. P., Tillman, E. A. & Walter, W. D. (2011). Vulture flight behavior and implications for aircraft safety. Journal of Wildlife Management, 75(7), 1581–1587. https://doi.org/10.1002/jwmg.205

Avery, M. L., Humphrey, J. S., Tillman, E. A., Phares, K. O. & Hatcher, J. E. (2002). Dispersing vulture roosts on communication towers. Journal of Raptor Research, 36, 45–50.

Avian Power Line Interaction Committee. (2006). Suggested practices for avian protection on power lines: The state of the art. *In*: Washington, DC: Edison Electric Institute.

Avise, J. C. (1994). Molecular Markers, Natural History and Evolution. New York and London: Chapman & Hall.

Avise, J. C., Nelson, W. S. & Sibley, C. G. (1994). DNA sequences support for a close phylogenetic relationship between some storks and New World vultures. Proceedings of the National Academy of Sciences of the United States of America, 91(11), 5173–5177. https://doi.org/10.1073/pnas.91.11.5173

Azpiroz, A. & Cortes Capano, G. D. (2014). On the status of Crowned eagle Buteogallus coronatus in Uruguay. Bulletin of the British Ornithologists' Club, 134(2), 110–115.

Azzopardi, J. (1978). The occurrence of the White-tailed eagle Haliaeetus albicilla in Malta. Il-Merill. 19, 12.

Babenko, V. G., Mazhulis, D. V., Ostapenko, V. A., Pererva, V. I. & Poyarkov, N. D. (1988). Distribution, number, and nesting ecology of Steller's Sea eagle (Haliaeetus pelagicus) in the area of Lower Amur River. Sbornik Trudov Zoologicheskogo Muzeya Moskovskogo Gosurdarstvennogo Universiteta, 26, 207–224.

Babenko, V. G., Mzhulis, D. V., Ostapenko, V. A., Pererva, V. I. & Poyarkov, N. D. (1988). Distribution, numbers and breeding ecology of the

Steller's Sea eagle in the Lower Amur. *In*: Litvinenko, N. M. (Ed.), Rare birds of the far east and their conservation, pp. 132–137. Vladivostok, Russian Federation: Nauka Publishers.
Baghino, L. (2003). L'importanza del ponente genovese per la migrazione del Biancone Circaetus gallicus nel Mediterraneo. Avocetta, 27, 67.
Baghino, L. (2005). Bollettino del Centro Ornitologico e di Educazione Ambientale di Case Vaccà. Parco Naturale regionale del Beigua, savona.
Baghino, L. & Premuda, G. (2005). Autunno 2004: Eccezionale invasio-ne di aquile minori. Quaderni di. Bird Watching, 13. Verona: Edizioni EBN Italia.
Baghino, L. & Premuda, G. (2005). Consistente migrazione preriproduttiva del biancone Circaetus gallicus lungo il versante tirrenico ligure-toscano. Avocetta, 29, 21.
Baghino, L. & Arenzano, G.E. (2006). La migrazione post-riproduttiva nel Parco del Beigua, Arenzano (GE). Centro Ornitologico e di Educazione Ambientale di Case Vaccà. Parco Naturale regionale del Beigua, savona. Baghino, L. In Info Migrans n.18, Parco Naturale Alpi Marittime, Valdieri.
Baghino, L. & Premuda, G. (2007). Nuovi dati sulla migrazione primaverile "a circuito" del biancone Circaetus gallicus in Italia. Avocetta, 31, 70–72.
Baghino, L., Bottero, M., Pedemonte, R. & Rapetti, C. (2006). La migrazione post-nuziale dell'Aquila minore ad Arenzano nel (2005). Biancone, 1, 17–23.
Baghino, L., Bottero, M., Pedemonte, R. & Rapetti, C. (2007). La migrazione post-nuziale dell'Aquila minore ad Arenzano nel (2006). Biancone, 2, 22–26.
Baghino, L., Premuda, G., Gustin, M., Corso, A., Mellone, U. & Cardelli, C. (2007). Exceptional wintering and spring migration of the Booted eagle Hieraaetus pennatus in Italy in 2004 and 2005. Avocetta, 31, 47–52.
Baghino, L., Campora, M. & Cattaneo G. (2009). Il Biancone – biologia e migrazione nell'Appennino ligure. Edizioni Il Piviere, Gavi.
Baghino, L., Premuda, G. & Giraudo, L. (2012). Nuove analisi sulla migrazione post-riproduttiva del biancone Circaetus gallicus nell'Italia nord-occidentale. Avocetta, 36, 107–111.
Baigorria, J., Barbar, F., Cavicchia, M., Gil Suárez, M. V. & Quaglia, A. (2008). New records for the Black hawk-eagle (Spizaetus tyrannus) for Misiones Province, Argentina. Revistã Nuestras Aves, 53, 36.
Bailey, K. M., Powers, D. M., Quattro, J. M., Villa, G., Nishimura, A., Traynor, J. J. & Walters, G. (1999). Population ecology and structural dynamics of Walleye pollock (Theragra chalcogramma). Dynamics of the Bering Sea, 581–614.
Bailey, S. F. (1992). Bird observations in lowland Madang Province. Muruk, 5, 111–135.
Bakaloudis, D. E., Iezekiel, S., Vlachos, C. G., Bontzorlos, V. A., Papakosta, M. & Birrer, S. (2012). Assessing bias in diet methods for the Long-legged buzzard Buteo rufinus. Journal of Arid Environments, 77, 59–65. https://doi.org/10.1016/j.jaridenv.2011.10.004
Baker, C. (2005). Field observations. Honeyguide, 51, 41–75.
Baker, E. C. S. (1914). Some notes on tame Serpent eagles. Avicultural Magazine, 5(5), 154–159.
Baker, E. C. S. (1918). Notes on the nidification of some Indian Falconidae. III. the genera Ictinaetus and Microhierax. Ibis, 60(1), 51–68. https://doi.org/10.1111/j.1474-919X.1918.tb00770.x.
Baker-Gabb, D. J. (1983). The breeding ecology of twelve species of diurnal raptor in north-western Victoria. Australian Wildlife Research, 10, 145–160.
Bakka, S. V. & Kiseleva, N. Y. (2009). Distribution and number of the Short-toed eagle in the N. Novgorod District, Russia. Raptors Conservation, 17, 121–124,
Bakka, S., Kiseleva, N. Y. & Karyakin, I. V. (2010). The Imperial eagle in the Nizhniy Novgorod District, Russia. Raptors Conservation, 20, 84–88.
Balbontín, J., Penteriani, V. & Ferrer, M. (2000). El águila perdicera en Andalucía: situaciónSituación actual y tendencias en las áreas de reproducción y de dispersión juvenil. Technical report. Consejería de Medio Ambiente, Junta de Andalucía.
Baldellou, M. & Henzi, S. P. (1992). Vigilance, predator detection and the presence of supernumerary males in vervet monkey troops. Animal Behaviour, 43(3), 451–461. https://doi.org/10.1016/S0003-3472(05)80104-6
Baldwin, W. P. (1940). Bald eagle robbing marsh hawk. Auk, 57(3), 413. https://doi.org/10.2307/4079007
Balen, S., Nijman, V. & Prins, H. (2000). The Javan hawk-eagle: Misconceptions about rareness and threat. Biological Conservation, 96, 297–304. https://doi.org/10.1016/S0006-3207(00)00092-6
Balme, G. A., Miller, J. R., Pitman, R. T. & Hunter, L. T. (2017). Catching reduces kleptoparasitism in a solitary, large field. Journal of Animal Ecology, 86(3), 634–644. https://doi.org/10.1111/1365-2656.12654
Balme, G. A., Slotow, R. & Hunter, L. T. B. (2010). Edge effects and the impact of nonprotected areas in carnivore conservation: Leopards in the Phinda–Mkhuze Complex, South Africa. Animal Conservation, 13(3), 315–323. https://doi.org/10.1111/j.1469-1795.2009.00342.x
Balmer, D. & Betton, K. (2007). Around the region. Sandgrouse, 29, 122–128.
Balmer, D. & Murdoch, D. (2010). Around the region. Sandgrouse, 32, 176–189.
Balotari-Chiebao, F., Brommer, J. E., Niinimäki, T. & Laaksonen, T. (2016). Proximity to wind-power plants reduces the breeding success of White-tailed eagle. Animal Conservation, 19(3), 265–272. https://doi.org/10.1111/acv.12238
Bamford, A. J., Razafindrajao, F., Young, R. P. & Hilton, G. M. (2017). Profound and pervasive degradation of Madagascar's freshwater wetlands and links with biodiversity. PloS One, 12(8), e0182673. https://doi.org/10.1371/journal.pone.0182673
Band, W., Madders, M. & Whitfield, D. P. (2007). Developing field and analytical methods to assess avian collision risk at wind farms. *In*: M. de Lucas, G. F. E. Janss & M. Ferrer (Eds.), Birds and wind farms: Risk assessment and mitigation, pp. 259–275. Madrid: Quercus.
Banfield, G. E. A. (1988). Long crested-eagle records around Bulawayo. Honeyguide, 34, 69–70.
Bank, L., Deme, T., Horvath, Z., Kalocsa, B. & Tamas, E. (2004). Population changes of the White-tailed eagle Haliaeetus albicilla in Hungary, with special attention to the lower Hungarian Danube-valley, 1987–2003. In: Chancellor, R. D. & Meyburg, B. (Eds.), Raptors worldwide, 529–536. Berlin and Budapest World Working Group on Birds of Prey/MME.

Banks, R. C., Cicero, C., Dunn, J. L., Kratter, A. W., Rasmussen, P. C., Remsen Jr., J. V. & Stotz, D. F. (2006). Forty-seventh supplement to the American Ornithologists' Union Check-List of North American Birds. Auk, 123(3), 926–936. https://doi.org/10.1093/auk/123.3.926

Banks, R. C., Chesser, R. T., Cicero, C., Dunn, J. L., Kratter, A. W., Lovette, I. J. & Stotz, D. F. (2007). Forty-eighth supplement to the American Ornithologists' Union Check-List of North American Birds. Auk., 2, 124(3), 1109–1115. https://doi.org/10.1642/0004-8038(2007)124[1109:FSTTAO]2.0.CO;2

Bannerman, D. A. (1951). The birds of tropical West Africa, with special reference to those of Gambia, Sierra Leone, the Gold Coast, and Nigeria. Vol. 8. London: Crown Agents for the Colonies.

Bannerman, D. A. (1959). The birds of the British isles, p. 8. Edinburgh, UK: Oliver & Boyd.

Barashkova, A. N., Smelansky, I. E., Tomilenko, A. A. & Akentiev, A. G. (2009). Some records of raptors in the East Kazakhstan. Raptors Conservation, 17, 131–144.

Barasona, J. A., Mulero-Pázmány, M., Acevedo, P., Negro, J. J., Torres, M. J., Gortázar, C. & Vicente, J. (2014). Unmanned aircraft systems for studying spatial abundance of ungulates: Relevance to spatial epidemiology. PloS One, 9(12), e115608. https://doi.org/10.1371/journal.pone.0115608

Barbar, F., Capdevielle, A. & Encabo, M. (2016). Direct persecution of Crowned eagles (Buteogallus coronatus) in Argentina: A new call for their conservation. Journal of Raptor Research, 50(1), 115–120. https://doi.org/10.3356/rapt-50-01-115-120.1

Barber, E. (1999). The mummies of Urumchi. WW Norton, London.

Barcellos, A. & Accordi, I. A. (2006). New records of the Crowned eagle, Harpyhaliaetus coronatus, in the state of Rio Grande do Sul, southern Brazil. Revista Brasileira de Ornitologia, 14, 345–349.

Barghi, M., Jin, X., Lee, S., Jeong, Y., Yu, J. P., Paek, W. K. & Moon, H. B. (2018). Accumulation and exposure assessment of persistent chlorinated and fluorinated contaminants in Korean birds. Science of the Total Environment, 645, 220–228. https://doi.org/10.1016/j.scitotenv.2018.07.040

Barkham, P. (2019). White-tailed eagles return to southern Britain after 240 years. The Guardian.

Barnes, J. G., Doney, G. E., Yates, M. A., Seegar, W. S. & Gerstenberger, S. L. (2019). A broadscale assessment of mercury contamination in peregrine falcons across the northern latitudes of North America. Journal of Raptor Research, 53(1), 1–13. https://doi.org/10.3356/JRR-18-0003

Barnes, K. N. (1998). The important bird areas of Southern Africa. Johannesburg, South Africa BirdLife South Africa.

Barnes, K. N. (2000a). Tawny eagle Aquila rapax. *In*: K. N. Barnes (Ed.), The Eskom Red Data Book of Birds of South Africa, Lesotho and Swaziland, 81–83. Randburg, Saudi Arabia: Birdlife South Africa.

Barnes, K. N. (2000b). Martial eagle Polemaetus bellicosus. *In*: K. N. Barnes (Ed.), The Eskom Red Data Book of Birds of South Africa, Lesotho and Swaziland, pp. 83–85. Johannesburg: Birdlife South Africa.

Barnes, K. N. (2000c). Bateleur Terathopius ecaudatus. *In*: Barnes, K. N. (Ed.), The Eskom Red Data Book of Birds of South Africa, Lesotho and Swaziland, pp. 86–88. Johannesburg, SA: BirdLife South Africa.

Barnes, R., Butchart, S., Clay, R., Davies, C. & Seddon, N. (1995). The conservation status of the Cordillera de Colán, northern Peru. Cotinga, 3, 6–7.

Barnett, A. A. (2015). Primate Predation by Black Hawk-Eagle (Spizaetus tyrannus) in Brazilian Amazonia. Journal of Raptor Research 49(1), 105-107. https://doi.org/10.3356/jrr-13-00090.1

Barnett, A., Andrade, E. S., Ferreira, M. C., Soares, J. B., da Silva, V. F. & de Oliveira, T. G. (2015). Primate predation by Black hawk-eagle (Spizaetus tyrannus) in Brazilian Amazonia. Journal of Raptor Research, 49(1), 105–107. https://doi.org/10.3356/jrr-13-00090.1.

Barnett, R., Barnes, I., Phillips, M., Martin, L. D., Harington, C. R., Leonard, J. & Cooper, A. (2005). Evolution of the extinct sabretooths (Smilodon and Homotherium) and American cheetahlike cat (Miracinonyx). Current Biology, 15, 589.

Barrett, G. W., Silcocks, A. F., Cunningham, R., Oliver, D. L., Weston, M. A. & Baker, J. (2007). Comparison of atlas data to determine the conservation status of bird species in New South Wales, with an emphasis on woodland-dependent species. Australian Zoologist, 34(1), 37–77. https://doi.org/10.7882/AZ 2007.003

Barrett, G., Silcocks, A., Barry, S., Cunningham, R. & Poulter, R. (2003). The New Atlas of Australian Birds. Melbourne: RAOU.

Barrett, S. W. (1980). Conservation in Amazonia. Biological Conservation, 18(3), 209–235. https://doi.org/10.1016/0006-3207(80)90034-8

Barrios, L. & Rodríguez, A. (2004). Behavioural and environmental correlates of soaring-bird mortality at on-shore wind turbines. Journal of Applied Ecology, 41(1), 72–81. https://doi.org/10.1111/j.1365-2664.2004.00876.x

Barrios, L. & Rodríguez, A. (2007). Spatiotemporal patterns of bird mortality at two wind farms of southern Spain. *In*: M. De Lucas, G. F. E. Janss & M. Ferrer (Eds.), Birds and wind farms: Risk assessment and mitigation, pp. 229–239. Madrid: Quercus.

Barron, D. G., Brawn, J. D. & Weatherhead, P. J. (2010). Meta-analysis of transmitter effects on avian behavior and ecology. Methods in Ecology and Evolution, 1, 180–187.

Barry, K. (2003). More on breeding in exotic gum plantations in Zimbabwe. Honeyguide, 49, 151–155.

Barry, R. E. & Shoshani, J. (2000). Heterohyrax brucei. Mammalian Species, 645, 1–7. https://doi.org/10.1644/1545-1410(2000)645<0001:HB>2.0.CO;2

Bartelme, N. (2012). Geographic Information Systems. Berlin, Heidelberg: Springer. https://doi.org/10.1007/978-3-540-72680-7_6

Barton, N. W. H. & Houston, D. C. (1993). A comparison of digestive efficiency in birds of prey. Ibis, 135(4), 363–371. https://doi.org/10.1111/j.1474-919X.1993.tb02107.x

Barua, M., Bhagwat, S. A. & Jadhav, S. (2013). The hidden dimensions of human-wildlife conflict: Health impacts, opportunity and transaction costs. Biological Conservation, 157, 309–316.

Bastille-Rousseau, G., Potts, J. R., Yackulic, C. B., Frair, J. L., Ellington, E. H. & Blake, S. (2016). Flexible characterization of animal movement pattern using net squared displacement and a latent state model. Movement Ecology, 4(1), 15. https://doi.org/10.1186/s40462-016-0080-y
Bastow, B. D. & Holmes, J. L. (2016). Teratology and drug use during pregnancy. Medscape. WebMD. Retrieved from https://emedicine.medscape.com/article/260725-overview
Bataamba, A. M. (1989). The ecology of raptors in and around the Impenetrable Forest, south western Uganda. Gabar, 4, 14–17.
Batchelor, D. M. (1958). A checklist of the birds seen at Asahan, Malacca. Part 1. Including Results for Malaysian Nature Journal, 12, 164–182.
Bates, J. M. (1993). Birds of Southwestern Brazil: Catalogue and guide to the birds of the Pantanal of Mato Grosso and its border areas balthasar Dubs. Auk, 110(3), 661. https://doi.org/10.2307/4088436
Bates, M. (1964). The Land and Wildlife of South America. Time: Life Books. New York.
Batschelet, E. (1981). Circular Statistics in Biology. London: Academic Press.
Battaglia, A., Ghidini, S., Campanini, G. & Spaggiari, R. (2005). Heavy metal contamination in little owl (Athene noctua) and common buzzard (Buteo buteo) from northern Italy. Ecotoxicology and Environmental Safety, 60(1), 61–66. https://doi.org/10.1016/j.ecoenv. (2003).
Bauer, S., Gienapp, P. & Madsen, J. (2008). The relevance of environmental conditions for departure decision changes enroute in migrating geese. Ecology, 89(7), 1953–1960. https://doi.org/10.1890/07-1101.1
Baum, A., De Pasquale, R. & Godek, M. L. (2014). Air mass frequency during precipitation events in the northern plains of the United States. Compass: Earth Science Journal of Sigma Gamma Epsilon, 86(1). Retrieved from https://digitalcommons.csbsju.edu/compass/vol86/iss1/1.
Baumgart, W. & Stephan, B. (1986). Ergebnisse ornithologischer Beobachtungen in der Syrischen Arabischen Republik. Mitteilungen aus dem Zoologischen Museum in Berlin 62. Suppl. Ann. Orn, 1: Non-Passeriformes, 10, 69–110.
Baumgart, W., Kasparek, M. & Stephan, B. (1995). Birds of Syria. Bedfordshire, UK: Max Kasparek Verlag and Ornithological Society of the Middle East.
Baumgart, W., Kasparek, M. & Stephan, B. (2003). Birds of Syria. Sandy, UK: OSME.
Baur, E. H. (2008). Structure of a Lowland Neotropical Galliform Bird Guild [PhD Thesis]. FL: University of Florida.
Bautista, L. M., García, J. T., Calmaestra, R. G., Palacín, C., Martín, C. A., Morales, M. B. & Viñuela, J. (2004). Effect of weekend road traffic on the use of space by raptors. Conservation Biology, 18(3), 726–732. https://doi.org/10.1111/j.1523-1739.2004.00499.x
Baxendale, D. (2014). Mongolia's lost secrets. Retrieved from https://www.lonelyplanet.com/mongolia/travel-tips-and-articles/mongolias-lost secrets-inpictures-the-golden-eagle-hunters. Golden Eagle, Hunters: Lonely Planet.
Bay, K., Nasman, K., Erickson, W., Taylor, K. & Kosciuch, K. (2016). Predicting eagle fatalities at wind facilities. Journal of Wildlife Management 80(6), 1000–1010. https://doi.org/10.1002/jwmg.21086
Bayle, P. (1999). Preventing birds of prey problems at transmission lines in Western Europe. Journal of Raptor Research, 33, 43–48. Retrieved from https://sora.unm.edu/sites/default/files/journals/jrr/v033n01/p00043-p00048.pdf
Baynard, O. E. (1909). Notes from Florida on Catharista urubu. Oologist, 26, 191–193. Retrieved from https://www.aphis.usda.gov/wildlife_damage/nwrc/publications/04pu bs/avery044.pdf
BBC. (2018). Wedge-tailed eagle "mass poisoning". prompts Australia arrest. Retrieved from https://www.bbc.com/news/world-australia-45504906
BBC. (2019). Critically endangered vultures poisoned en masse in Botswana. Retrieved from https://www.bbc.com/news/world-africa-48717018
BBC Earth. (2015). Coyote vs Bald Eagle. Retrieved from https://www.bbc.co.uk/ programmes/ p02y7n8x
Beaman, M. (1994). Palearctic Birds. A Checklist of the Birds of Europe, North Africa and Asia north of the foothills of the Himalayas. Stonyhurst Harrier.
Bearder, S. K. (2007). A comparison of calling patterns in two Nocturnal Primates, Otolemur crassicaudatus and Galago moholi as a guide to predation risk. *In*: Primate anti-predator strategies, pp. 206–221. NY: Springer US.
Beasley, J. C., Olson, Z. H. & DeVault, T. L. (2012). Carrion cycling in food webs: Comparisons among terrestrial and marine ecosystems. Oikos 121(7), 1021–1026. https://doi.org/10.1111/j.1600-0706.2012.20353.x
Beavers, R. A., Delaney, D. F., Leahy, C. W. & Oatman, G. F. (1991). New and noteworthy bird records from Petén, Guatemala, including Tika National Park. Bulletin of the British Ornithologists' Club, 111, 77–90.
Becker, C. D., Loughin, T. M. & Santander, T. T. (2008). Identifying forest-obligate birds in tropical moist cloud forest of Andean Ecuador. Journal of Field Ornithology, 79(3), 229–244. https://doi.org/10.1111/j.1557-9263.2008.00184.x
Beckford, R. (2017). Controlling predation on goats—Some ideas. Retrieved from http://lee.ifas.ufl.edu/AgNatRes/Pubs/Controlling_Predation_on_Goats.pdf. FL: Lee County Extension Service, University of Florida.
Bednarz, J. C., Klem, D. Jr., Goodrich, L. J. & Senner, S. E. (1990). Migration counts of raptors at Hawk Mountain, Pennsylvania, as indicators of population trends, 1934–1986. Auk, 107, 96–109.
Bedrosian, B., Craighead, D. & Crandall, R. (2012). Lead exposure in Bald eagles from Big Game Hunting, the continental implications and successful mitigation efforts. PLoS One, 7(12), e51978. https://doi.org/10.1371/journal.pone.0051978
Beecham, J. J. & Kochert, M. N. (1975). Breeding biology of the Golden eagle in southwestern Idaho. Wilson Bulletin, 506–513.
Beehler, B. M. (1978). Upland birds of northeastern New Guinea. Wau Ecology Institute Handbook 4.
Beehler, B. M., Pratt, T. K. & Zimmerman, D.A. (1986). Birds of New Guinea. Princeton University Press. Princeton, NJ.

Beehler, B. M., Crill, W., Jefferies, B. & Jefferies M. (1992). New Guinea Harpy eagle attempts to capture a monitor lizard. Emu, 92, 246–247.
Beehler, B. M., Angle, J. P., Gibbs, D., Hedemark, M. & Kuro, D. (2001). A field survey of the resident birds of southern New Ireland. In Beehler, B. M., Alonso, L. E. (Eds.), Southern New Ireland, Papua New Guinea: A biodiversity assessment. Washington, D.C. Conservation International Rapid Assessment Program.

Begon, M., Townsend, C. R. & Harper, J. L. (2005). Ecology: From Individuals to Ecosystems. Hoboken: Wiley-Blackwell.

Beier, P. (1991). Cougar attacks on humans in the United States and Canada. Wildlife Society Bulletin, 19, 403–412. Retrieved from https://www.jstor.org/stable/3782149.

Belding, L. & Ridgway, R. (1879). A partial list of the Birds of Central California. Proceedings of the United States National Museum, 1(56), 388–449. https://doi.org/10.5479/si.00963801.56.388

Bělka, T. & Horal, D. (2009). The White-tailed eagle (Haliaeetus albicilla) in the Czech Republic. Denisia, 27, 65–77.

Bellier, Edwige, Grégoire Certain, Benjamin Planque, Pascal Monestiez & Vincent Bretagnolle. (2010). Modelling habitat selection at multiple scales with multivariate geostatistics: An application to seabirds in open sea. Oikos, 119, 988–999. https://doi.org/10.1111/j.1600-0706.2009.17808).x

Bellocq, M. I., Bonaventura, S. M., Marcelino, F. N. & Sabatini, M. (1998). Habitat use of Crowned eagles (Harpyhaliaetus coronatus) in the southern limits of the species' range. Journal of Raptor Research, 32, 312–314.

Bellocq, M. I., Ramírez Llorens, P. & Filloy, J. (2002). Recent records of Crowned eagle (Harpyhaliaetus coronatus) from Argentina, 1981–2000. Journal of Raptor Research, 36, 206–212.

Belopolskiy, L. O. & Rogova, E. N. (1947). To the ornithofauna of the north-eastern part of the Kamchatka Peninsula. Bull. MOIP, Biol. Sec., 52(2), 39–50.

Belton, W. (1978). A list of birds of Rio Grande do Sul, Brazil. Iheringia Serie Zoologia, 52, 85–102.

Belton, W. (1984). Birds of Rio Grande do Sul, Brazil. Part 1. Rheidae through Furnariidae. Bulletin of the American Museum of Natural History, 178, 369–636.

Bencke, G. A., Fontana, C. S., Dias, R. A., Mauricio, G. N. & Mähler, J. F. K., Jr. (2003). Aves. *In*: C. S. Fontana Bencke, R. R. E. Ga (Eds.), Red Book of fauna threatened with extinction in Rio Grande do Sul, pp. 189–479. Porto Alegre, Brazil: EDIPURCRS (in Portuguese).

Beninde, J., Veith, M. & Hochkirch, A. (2015). Biodiversity in cities needs space: A meta-analysis of factors determining intra-urban biodiversity variation. Ecology Letters, 18(6), 581–592. https://doi.org/10.1111/ele.12427

Bennett Hennessey, A. B., Herzog, S. A., Kessler, M. & Robison, D. (2003). Avifauna of the Piló Lajas Biosphere Reserve and communal lands, Bolivia. Bird Conservation International, 13(4), 319–349. https://doi.org/10.1017/S0959270903003241

Bennett, A. F. (1982). The energetics of reptilian activity. Biology of the Reptilia, 13, 155–199.

Bennetts, R. E. & Mcclelland, B. R. (1997). Influence of age and prey availability on Bald eagle foraging behavior at Glacier National Park, Montana. Wilson Bulletin, 109(3), 393–409.

Bennike, O. & Feilberg, J. (1982). White-tailed eagle Haliaetus albicilla breeding at Disko Bugt, West Greenland. Dansk Ornitologisk Forenings Tidsskrift, 76, 73.

Bennitt, E., Bartlam-Brooks, H. L. A., Hubel, T. Y. & Wilson, A. M. (2019). Terrestrial mammalian wildlife responses to Unmanned Aerial Systems approaches. Scientific Reports, 9(1), 2142. https://doi.org/10.1038/s41598-019-38610-x

Benson, C. A. & Irwin, R. S. P. (1967). A contribution to the ornithology of Zambia. Zambia Museum Papers 1, pp. 1–139.

Benson, C. W. (1942). Additional notes on Nyasaland birds. Ibis, 299–337.

Benson, C. W. (1942). Additional notes on Nyasaland birds. Ibis, 6, 197–203.

Benson, C. W. & Benson, F. M. (1947). Some breeding records from southern Nyasaland 21, 1–3

Benson, C. W. & White, C. M. N. (1957). Checklist of the Birds of Northern Rhodesia. Rhodesian Department of Game & Tsetse Control. Lusaka: The Government Printer.

Benson, C. W. & Irwin, M. P. S. (1967). A contribution to the ornithology of Zambia. Zambia Museum Papers, 1, 1–139.

Benson, C. W., Brooke, R. K., Dowsett, R. J. & Irwin, M. P. S. (1970). Notes on the birds of Zambia: Part V. Arnoldia (Rhod.), 4(40), 1–59.

Benson, C. W., Brooke, R. K., Dowsett, R. J. & Irwin, M. P. S. (1971). The birds of Zambia. London: Collins.

Benson, C. W. & Benson, F. M. (1977). The Birds of Malawi. Limbe: Montfort Press.

Benson, P. C. (1981). Large raptor electrocution and power pole utilization: A study in six western states [PhD Thesis]. Provo, UT: Brigham Young University: Provo.

Bensusan, K. J., Garcia, E. F. J. & Cortes, J. E. (2007). Trends in abundance of migrating raptors at Gibraltar in spring. Ardea, 95(1), 83–90. https://doi.org/10.5253/078.095.0109

Bent, A. C. (1937). Life histories of North American birds of prey. Part 1. Smithsonian Institution National Museum. Bulletin of the United States National Museum, 167, i–409. https://doi.org/10.5479/si.03629236.167.i

Benton, A. H. & Dickinson, L. E. (1966). Wires, poles, and birds. In: Stefferud, A. (Ed.), Birds in our lives, pp. 390–395. Washington, D.C.: U.S. Bureau of Sport Fisheries and Wildlife.

Benton M. J. (1999). Early origins of modern birds and mammals: Molecules vs. morphology. BioEssays: News and Reviews in Molecular, Cellular and Developmental Biology, 21(12), 1043–1051. https://doi.org/10.1002/(SICI)1521-1878(199912)22:1<1043::AID-BIES8>3.0.CO;2-B

Berg, W., Johnels, A., Sjöstrand, B. & Westermark, T. (1966). Mercury content in feathers of Swedish birds from the past 100 years. Oikos, 17(1), 71–83. https://doi.org/10.2307/3564782

Bergallo, H. G., Rocha, C. F. D., Alves, M. A. S. & Van Sluys, M. (2003). Fauna ameacada de extincao no Rio de Jane. Rio de Janeiro, Brazil Ed. Universidade Estadual do Rio de Janeiro.

Berger, D. D. & Mueller, H. C. (1959). The Bal-Chatri: A trap for the birds of prey. Bird-Banding, 30(1), 18–26. https://doi.org/10.2307/4510726

Berger, K. M. & Gese, E. M. (2007). Does interference competition with wolves limit the distribution and abundance of coyotes? Journal of Animal Ecology, 76, 1075–1085. https://doi.org/10.1111/j.1365-2656.2007.01287.x

Berger, L. R. & Clarke, R. J. (1995). Eagle involvement in accumulation of the Taung child fauna. Journal of Human Evolution, 29(3), 275–299. https://doi.org/10.1006/jhev.1995.1060

Berger-Tal, O., Polak, T., Oron, A., Lubin, Y., Kotler, B. P. & Saltz, D. (2011). Integrating animal behavior and conservation biology: A conceptual framework. Behavioral Ecology, 22(2), 236–239. https://doi.org/10.1093/beheco/arq224

Bergmanis, U. (1989). How to identify the Spotted eagle Aquila clanga Pallas and the Lesser spotted eagle Aquila pomarina C. L. Brehm. Putni Daba, 2, 113–122.

Bergmanis, U. (1996). On the taxonomy of the Lesser spotted eagle Aquila pomarina and Greater spotted eagle A. Clanga. *In*: B.-U. Meyburg & R. D. Chancellor (Eds.), Eagle studies, pp. 199–207. World Working Group on Birds of Prey and Owls. Berlin.

Bergmanis, U. & Strazds, M. (2001). Another possible hybridization case of Lesser and Greater spotted eagle in Latvia. Putni Daba, 11(2), 6–7.

Bergmanis, U., Petrins, A., Strazds, A. M. & Krams, I. (2001). Probable case of hybridization of Greater Aquila clanga and Lesser Spotted Eagles A. pomarina in eastern Latvia. Acta Ornithoecologica, 4, 297–304.

Bergmanis, U., Petrins, A., Strazds, M. & Krams, I. (1997). Possible case of hybridization of the Lesser spotted eagle and the Greater spotted eagle in Eastern Latvia. Putni Daba, 6(3), 2–6.

Bergo, G. (1987). Eagles as predators on livestock and deer. Fauna Norvegica Series. C. 10: 95–102.

Berkelman, J., Fraser, J. D. & Watson, R. T. (1999a). Lake selection by Madagascar fish-eagles. The Auk, 116(4), 976–983. https://doi.org/10.2307/4089677

Berkelman, J., Fraser, J. D. & Watson, R. T. (1999b). Madagascar fish-eagle prey preference and foraging success. Wilson Bulletin, 111, 15–21.

Berkelman, J., Fraser, J. D. & Watson, R. T. (2002). Nesting and perching habitat use of the Madagascar fish eagle. Journal of Raptor Research, 36, 287–293.

Berkelman, J. (1997). Habitat requirements and foraging ecology of the Madagascar fish-eagle [PhD Thesis]. Virginia Polytechnic Institute & State University.

Berkunsky, I., Daniele, G., Kacoliris, F. P., Faegre, S. I. K., Gandoy, F. A., Gonzalez, L. & Luque, J. A. D. (2012). Records of the Crowned eagle in Moxos plains of Bolivia and observations about breeding behavior (Urubitinga Coronata). Revista Brasileira de Ornitologia, 20, 447–450.

Bernis, F. (1975). Migración de Falconiformes y Ciconia spp. por Gibraltar. Parte II, Análisis descriptivo del verano-otoño 1972. Ardeola, 21, 489–580.

Bernis, F. (1980). La migración de aves por el estrecho de Gibraltar, 1. Aves Planeadoras. Madrid: Universidad Complutense.

Berruti, A. (1997). Southern Banded Snake Eagle. In: Harrison, J. A., Allan, D., Underhill, L., Herremans, M., Tree, A. J. & Brown, C. J. (Eds.), The atlas of southern African birds, Vol. 1. Non-passerines, pp. 108–109. Johannesburg, SA: BirdLife South Africa.

Berry, M. E., Bock, C. E. & Haire, S. L. (1998). Abundance of diurnal raptors on open space grasslands in an urbanized landscape. Condor, 100, 601–608.

Bert, E. (1968). An Approved Treatise on Hawks and Hawking. Da Cap Press. London, New York.

Berthold, P. (2001). Bird migration. A general survey. Oxford University Press, Oxford.

Best, B. J. & Clarke, C. T. (1991). The threatened birds of the Sozoranga region, Southwest Ecuador. International Council for Bird Preservation. Cambridge, UK.

Best, B. J., Clarke, C. T., Checker, M., Broom, A. L., Thewlis, R. M., Duckworth, W. & McNab, A. (1993). Distributional records, natural history notes, and conservation of some poorly known birds from southwestern Ecuador and northwestern Peru. Bulletin of the British Ornithologists' Club, 113, 108–119.

Besten, J. W. D. (2004). Migration of Steppe eagles Aquila nipalensis and other raptors along the Himalayas past Dharamsala, India, in autumn 2001 and spring 2002. Forktail, 20, 9–13.

Bevanger, K., Berntsen, F., Clausen, S., Dahl, E. L., Flagstad, Ø. Follestad, A., Halley, D., Hanssen, F., Johnsen, L., Kvaløy, P., Lund-Hoel, P., May, R., Nygård, T., Pedersen, H. C., Reitan, O., Røskaft, E., Steinheim, Y., Stokke, B. & Vang, R. (2010). Pre- and post-construction studies of conflicts between birds and wind turbines in coastal Norway (BirdWind). Report on findings 2007–2010. NINA Report 620. Norwegian Institute for Nature Research.

Beyer, H. L., Haydon, D. T., Morales, J. M., Frair, J. L., Hebblewhite, M., Mitchell, M. et al. (2010). The interpretation of habitat preference metrics under use±availability designs. Philosophical Transactions of the Royal Society of London Series B, 365(1550), 2245–2254. https://doi.org/10.1098/rstb.2010.0083https://doi.org/10.1098/rstb.2010.0083.

Beyer, W. N. (2000). Hazards to wildlife for soil-borne cadmium reconsidered. Journal of Environmental Quality, 29(5), 1380–1384, https://doi.org/10.2134/jeq2000.00472425002 900050002x.

Bhatta, B. (2008). Remote Sensing and GIS. Oxford: Oxford University Press.

Bhola, N., Ogutu, J. O., Piepho, H-P., Said, M.Y., Reid, R. S., Hobbs, N. T. & Olff, H. 2012. Comparative changes in density and demography of large herbivores in the Maasai Maramara Reserve and its surrounding human-dominated pastoral ranches in Kenya. Biodiversity and Conservation, 21(6), 1509–1530.

Bidner, L. R. (2014). Primates on the menu: Direct and indirect effects of predation on primate communities. International Journal of Primatology, 35(6), 1164–1177. https://doi.org/10.1007/s10764-014-9797-x

Bierregaard, R. O. (1984). Observations of the nesting biology of the Guiana Crested Eagle (Morphnus guianensis). Wilson Bulletin, 96, 1–5.

Bierregaard, R. O. (1994). Crowned solitary eagle. *In*: Del Hoyo, J., Elliott, A. & Sargatal, J. (Eds.), Handbook of birds of the world. Vol. 2: New World vultures to guineafowl, pp. 175–176. Lynx Edicions. Barcelona, Spain.

Bierregaard, R. O. (1994). Neotropical Accipitridae (Hawks and Eagles). *In*: Del Hoyo, J., Elliott, A. & Sargatal, J. (Eds.), Handbook of the birds of the world, pp. 52–205. Lynx Edicions. Barcelona.

Bierregaard, R. O. (1994a). Black-chested buzzard-eagle. *In*: Del Hoyo, J., Elliott, A. & Sargatal, J. (Eds.), Handbook of birds of the world (Vol. 2: New World Vultures to Guineafowl). Lynx Edicions. Barcelona.

Bierregaard, R. O. (1994b). Black-and-white hawk-eagle. *In*: Del Hoyo, J., Elliott, A. & Sargatal, J. (Eds.), Handbook of birds of the world (Vol. 2: New World Vultures to Guineafowl), 205. Lynx Edicions. Barcelona.

Bierregaard, R. O., Jr. (1995). The biology and conservation status of Central and South American Falconiformes: A survey of current knowledge. Bird Conservation International, 5(2–3), 325–340. https://doi.org/10.1017/S0959270900001076

Bierregaard, R. O., Jr. (1995). The status of raptor conservation and our knowledge of the resident diurnal birds of prey of Mexico. Transactions of the North American Wildlife and Natural Resources Conference, 60, 203–213.

Bierregaard, R. O., Jr. (1998). Conservation status of birds of prey in the South American tropics. Journal of Raptor Research, 32, 19–27.

Bierregaard, R. O., Jr. & Kirwan, G. M. (2017). Crested eagle (Morphnus guianensis). In: del Hoyo, J., Elliott, A., Sargatal, J., Christie, D. A. & de Juana, E. (Eds.), Handbook of the Birds of the World. Barcelona, Spain: Lynx Edicions.

Bierregaard, R. O., Jr., Sharpe, C. J., Boesman, P. & Marks, J. S. (2018). Black-and-white hawk-eagle (Spizaetus melanoleucus). *In*: Del Hoyo, J., Elliott, A., Sargatal, J., Christie, D. A. & De Juana, E. (Eds.), Handbook of the birds of the world alive. Lynx Edicions. Barcelona.

Bierregaard, R. O., Kirwan, G. M., Boesman, P. F. D., Sharpe, C. J. & Marks, J. S. (2020). Black-and-Chestnut eagle (Spizaetus isidori), version 1.0. *In*: del Hoyo, J., Elliott, A., Sargatal, J., Christie, D. A. & de Juana, E. (Eds.), Birds of the world. Ithaca, NY: Cornell Lab of Ornithology. https://doi.org/10.2173/bow.baceag2.01

Bierregaard, R. O., Kirwan, G. M., Boesman, P. & Sharpe, C. J. (2014). Black-and-chestnut eagle (Oroaetus isidori). *In*: del Hoyo, J., Elliott, A., Sargatal, J., Christie, D. A. & De Juana, E. (Eds.), Handbook of the birds of the world. Lynx Edicions. Barcelona.

Biggs, H. C., Kemp, A. C., Mendelsohn, H. P. & Mendelsohn, J. M. (1979). Weights of South African raptors and owls. Durban Museum Novitates, 12, 73–81.

Bijleveld, M. F. (1974). Birds of Prey in Europe. Macmillan Press Ltd. London.

Bijleveld, M. (1986). Sur la migration de l'aigle botté, Hieraaetus pennatus, en Italie. Nos Oiseaux, 38, 377–378.

Bijlsma, R. G. (1987). Bottleneck Areas for Migratory Birds in the Mediterranean Region: An Assessment of the Problems and Recommendations for Action. International Council for Bird Preservation. Cambridge, UK.

Bijlsma, R. G. (2001). Trends and breeding success of raptors in The Netherlands in 2000. De Takkeling, 9, 12–52.

Bijlsma, R. G., Hustings, F. & Camphuysen, K. C. J. (2001). Avifauna of the Netherlands. 2. Common and scarce birds of the Netherlands. Haarlem/Utrecht, Netherlands GMB Uitgeverij/KNNV Uitgeverij.

Bijlsma, R. G. (2007). Behaviour of adult White-tailed eagle Haliaeetus albicilla towards mobbing raptors. Takkeling, 15, 206–209.

Bildstein, K. L. (1998). Long-term counts of migrating raptors: A role for volunteers in wildlife research. Journal of Wildlife Management, 62(2), 435–445. https://doi.org/10.2307/3802318

Bildstein, K. L. (1999). Racing with the sun: The forced migration of the Broad-winged hawk. *In*: K. P. Able (Ed.), Gatherings of angels, pp. 79–92. Comstock Press. Ithaca, NY.

Bildstein, K. L. & Zalles, J. I. (Eds.). (1995). Raptor Migration Watch-Site Manual. Hawk Mountain Sanctuary Association. Kempton, UK, PA.

Bildstein, K. L. & Zalles, J. I. (1998). Moving targets: The science and conservation of migrating raptors in the Western Hemisphere. Torgos, 28, 97–108.

Bildstein, K. L., Schelsky, W., Zalles, J. I. & Ellis, S. (1998). Conservation status of tropical raptors. Journal of Raptor Research, 32, 3–18.

Bildstein, K. L. & Saborio, M. (2000). Spring migration counts of raptors and New World vultures in Costa Rica. Ornitologia Neotropical, 11, 197–205.

Bildstein, K. L. (2006). Migrating Raptors of the World: Their Ecology and Conservation. Princeton University Press. Ithaca, NY.

Bildstein, K. L., Bechard, M. J., Farmer, C. & Newcomb, L. (2009). Narrow sea crossings present obstacles to migrating griffon vultures Gyps fulvus. Ibis, 151(2), 382–391. https://doi.org/10.1111/j.1474-919X.2009.00919.x

Bildstein, K. L., Farmer, C. J. & Yosef, R. (2009). Raptor population monitoring: Examples from migration watchsites in North America. Avocetta, 33, 43–51.

Bildstein, K. L., Smith, J. P., Ruelas, I. E. & Veit, R. R. (2008). State of North America's Birds of Prey. Nuttall Ornithological Club and American Ornithologists' Union. Cambridge, Washington, DC.

Bildstein, K., Schelsky, W., Zalles, J. & Ellis, S. (1998). Conservation status of tropical raptors. Journal of Raptor Research, 32, 3–18.

Binford, L. C. (1989). A distributional survey of the birds of the Mexican State of Oaxaca. Ornithological Monographs, 43(43), iii–418. https://doi.org/10.2307/40167673

Binkowski, Ł. J., Stawarz, R. M. & Zakrzewski, M. (2013). Concentrations of cadmium, copper and zinc in tissues of mallard and coot from southern Poland. Journal of Environmental Science and Health. Part. B, Pesticides, Food Contaminants, and Agricultural Wastes, 48(5), 410–415. https://doi.org/10.1080/03601234.2013.742725,

Bino, G., Levin, N., Darawshi, S., Van Der Hal, N., Reich-Solomon, A. & Kark, S. (2008). Accurate prediction of bird species richness patterns in an urban environment using Landsat-derived NDVI and spectral unmixing. International Journal of Remote Sensing, 29(13), 3675–3700. https://doi.org/10.1080/01431160701772534

Biology Dictionary. (2020). Paraphyletic. Retrieved from https://biologydictionary.net/paraphyletic/. 27, 289–301. https://doi.org/10.5751/ACE-00608-080212

Bird, D. M. (2004). The Bird Almanac: A Guide to Essential Facts and Figures of the World's Birds. ON, Firefly Books. Canada.

Bird, D. M. (Ed.). (1983). Biology and management of Bald Eagles and Ospreys [Ste-Anne-de-Bellevue, Quebec]. Proceedings of the 1st International Symposium on Bald Eagles and Ospreys, 1981. Harpell Press.

Bird, J. P. & Symes, A. J. (2009). Reviewing the conservation status of three Asian Aquila eagles. Birding ASIA, 12, 112–115.

Bird, J. P., Buchanan, J. M., Lees, A. C., Clay, R. P., Develey, P. F., Yépez, I. & Butchart, S. H. M. (2011). Integrating spatially explicit habitat

projections into extinction risk assessments: A reassessment of Amazonian avifauna incorporating projected deforestation. Diversity and Distributions, 18(3), 273–281. https://doi.org/10.1111/j.1472-4642.2011.00843.x
Bird, X. (2019). Got vulture problems? Retrieved from https://bird-x.com/interactive-problem-solver/vulture-control/
Birdlife International. (2000). Threatened Birds of the World. Lynx Edicions, Barcelona, and Birdlife International, Cambridge, UK.
Birdlife International. (2001). Threatened Birds of Asia: The Birdlife International Red Data Book. Birdlife International. Cambridge.
Birdlife International. (2002). Action plan for conservation of White-tailed sea eagle (Haliaeetus albicilla). Strasbourg, France.
Birdlife International. (2004). Birds in Europe. Populations estimates, trends and conservation status. BirdLife Conservations Series No. 12 Birdlife International. Cambridge, UK.
Birdlife International. (2004). Threatened birds of the world 2004 [CD-ROM]. Birdlife International. Cambridge, UK.
Birdlife International. (2009). Species factsheet: Aquila nipalensis. Retrieved from http://www.birdlife.org
Birdlife International. (2015). Aquila fasciata. IUCN Red List of Threatened Species, 2015: e.T22696076A60132541. International Union for Conservation of Nature.
Birdlife International. (2015). European Red List of Birds, Luxembourg. Office for Official Publications of the European Communities.
Birdlife International. (2015). Hieraaetus pennatus. IUCN Red List of Threatened Species, 2015: e.T22696092A60133199.
Birdlife International. (2015). Important bird Areas factsheet: Smøla archipelago. Retrieved from http://www.birdlife.org
Birdlife International. (2015). Species factsheet: Aquila verreauxii. Retrieved from http://www.birdlife.org
Birdlife International. (2016). Aquila gurneyi. IUCN Red List of Threatened Species, 2016: https://doi.org/10.2305/IUCN.UK.2016-3.RLTS. T22696056
Birdlife International. (2016). Circaetus fasciolatus. IUCN Red List of Threatened Species, 2016: https://doi.org/10.2305/IUCN.UK.2016-3. RLTS.T226 95277A93500599.en.
Birdlife International. (2016). Circaetus gallicus. IUCN Red List of Threatened Species, 2016. https://doi.org/10.2305/IUCN.UK.2016-3.RLTS. T22734216A95078150.en.
Birdlife International. (2016). Clanga hastata. IUCN Red List of Threatened Species, 2016: https://doi.org/10.2305/IUCN.UK.2016-3.RLTS. T22729779A95021573.en
Birdlife International. (2016). Clanga pomarina. IUCN Red List of Threatened Species, 2016: https://doi.org/10.2305/IUCN.UK.2016-3.RLTS. T22696022A, PubMed: 93539187.en.
Birdlife International. (2016). Dryotriorchis spectabilis. IUCN Red List of Threatened Species, 2016: https://doi.org/10.2305/IUCN.UK.2016-3. RLTS.T22695330A93502934.en.
Birdlife International. (2016). Eutriorchis astur. IUCN Red List of Threatened Species, 2016: https://doi.org/10.2305/IUCN.UK.2016-3.RLTS. T22695336A93503165.en.
Birdlife International. (2016). Haliaeetus albicilla. IUCN Red List of Threatened Species, 2016: https://doi.org/10.2305/IUCN.UK.2016-3.RLTS. T22695137A93491570.en.
Birdlife International. (2016). Haliaeetus leucogaster. IUCN Red List of Threatened Species, 2016: https://doi.org/10.2305/IUCN.UK.2016-3 RLTS.T22695097A93489471.en.
Birdlife International. (2016). Haliaeetus sanfordi. IUCN Red List of Threatened Species, 2016: https://doi.org/10.2305/IUCN.UK.2016-3.RLTS T22695105
Birdlife International. (2016). Haliaeetus vocifer. IUCN Red List of Threatened Species, 2016: https://doi.org/10.2305/IUCN.UK.2016-3.RLTS T22695115A93490143.en.
Birdlife International. (2016). Harpyopsis novaeguineae. IUCN Red List of Threatened Species, 2016: https://doi.org/10.2305/IUCN.UK.2016-3 RLTS.T22696007A93538251.en.
Birdlife International. (2016). Hieraaetus wahlbergi. IUCN Red List of Threatened Species, 2016: https://doi.org/10.2305/IUCN.UK.2016-3 RLTS.T22, PubMed: 696072
Birdlife International. (2016). Icthyophaga humilis. IUCN Red List of Threatened Species, 2016: https://doi.org/10.2305/IUCN.UK.2016-3.RLTS T2269 5156A93493324.en.
Birdlife International. (2016). Ictinaetus malaiensis. IUCN Red List of Threatened Species, 2016: https://doi.org/10.2305/IUCN.UK.2016-3 RLTS.T22, PubMed: 696019
Birdlife International. (2016). Lophaetus occipitalis. IUCN Red List of Threatened Species, 2016: https://doi.org/10.2305/IUCN.UK.2016-3 RLTS.T22696134A93546422.en.
Birdlife International. (2016). Nisaetus alboniger. IUCN Red List of Threatened Species, 2016: https://doi.org/10.2305/IUCN.UK.2016-3.RLTS T22696159
Birdlife International. (2016). Nisaetus cirrhatus. IUCN Red List of Threatened Species, 2016: https://doi.org/10.2305/IUCN.UK.2016-3.RLTS T22732090A95042127.en.
Birdlife International. (2016). Nisaetus lanceolatus. IUCN Red List of Threatened Species, 2016: https://doi.org/10.2305/IUCN.UK.2016-3.RLTS T22696 172A93547970.en.
Birdlife International. (2016). Nisaetus philippensis. IUCN Red List of Threatened Species, 2016: https://doi.org/10.2305/IUCN.UK.2016-3 RLTS.T4501 5567A95139313.en.
Birdlife International. (2016). Nisaetus pinskeri. IUCN Red List of Threatened Species, 2016: https://doi.org/10.2305/IUCN.UK.2016-3.RLTS T22734083A, PubMed: 950739
Birdlife International. (2016). Nisaetus pinskeri. IUCN Red List of Threatened Species, 2016: https://doi.org/10.2305/IUCN.UK.2016-3.RLTS T22734083 A950.

Birdlife International. (2016). Spilornis holospilus. IUCN Red List of Threatened Species, 2016https://doi.org/10.2305/IUCN.UK.2016-3.RLTS.T22695318A93502499.en.
Birdlife International. (2016). Spilornis klossi. IUCN Red List of Threatened Species, 2016: https://doi.org/10.2305/IUCN.UK.2016-3.RLTS.T2272946 5A9 5016577.en.
Birdlife International. (2016). Spizaetus isidori. IUCN Red List of Threatened Species, 2016: https://doi.org/10.2305/IUCN.UK.2016-3.RLTS.T22696207A9, PubMed: 3549661.en.
Birdlife International. (2016). Spizaetus isidori. IUCN Red List of Threatened Species, 2016: https://doi.org/10.2305/IUCN.UK.2016-3.RLTS.T22696207
Birdlife International. (2016). Spizaetus melanoleucus. IUCN Red List of Threatened Species, 2016: https://doi.org/10.2305/IUCN.UK.2016-3.RLTS.T2269612 0A93546145.en.
Birdlife International. (2016). Spizaetus ornatus. IUCN Red List of Threatened Species, 2016: https://doi.org/10.2305/IUCN.UK.2016-3.RLTS.T22696197A93548774.en.
Birdlife International. (2016). Spizaetus tyrannus. IUCN Red List of Threatened Species, 2016: https://doi.org/10.2305/IUCN.UK.2016-3.RLTS.T226961 93A9522 2363.en.
Birdlife International. (2016a). Buteogallus coronatus. IUCN Red List of Threatened Species, 2016 https://doi.org/10.2305/IUCN.UK.2016-3.RLTS.T22695855A93530845.en.
Birdlife International. (2016b). Buteogallus solitarius. IUCN Red List of Threatened Species, 2016 https://doi.org/10.2305/IUCN.UK.2016-3.RLTS.T22695849A93530532.en.
Birdlife International. (2017). Aquila africana. IUCN Red List of Threatened Species, 2017: https://doi.org/10.2305/IUCN.UK.2017-1.RLTS.T22696138 A111714549.en.
Birdlife International. (2017). Aquila rapax. IUCN Red List of Threatened Species, 2017: https://doi.org/10.2305/IUCN.UK.2017-3.RLTS.T22696033A119425859.en.
Birdlife International. (2017). Clanga clanga. IUCN Red List of Threatened Species, 2017: https://doi.org/10.2305/IUCN.UK.2017-1.RLTS.T22696027A110443604.en.
Birdlife International. (2017). Harpia harpyja. IUCN Red List of Threatened Species, 2017: https://doi.org/10.2305/IUCN.UK.2017-3.RLTS.T22695998A117357127.en.
Birdlife International. (2017). Icthyophaga Ichthyaetus (amended version of 2016 assessment). IUCN Red List of Threatened Species, 2017: https://doi.org/10.2305/IUCN.UK.2017-3.RLTS.T22695163A116996769.en.
Birdlife International. (2017). Nisaetus nanus. IUCN Red List of Threatened Species, 2017: https://doi.org/10.2305/IUCN.UK.2017-1.RLTS.T22696186A110049812.en.
Birdlife International. (2017). Polemaetus bellicosus. IUCN Red List of Threatened Species, 2017: https://dx.doi.org/10.2305/IUCN.UK.2020-3.RLTS.T22696116A172287822.en.
Birdlife International. (2017). Spilornis elgini. IUCN Red List of Threatened Species, 2017: https://doi.org/10.2305/IUCN.UK.2017-3.RLTS.T22695323A118207916.en.
Birdlife International. (2017). Spilornis kinabaluensis (amended version of 2016 assessment). IUCN Red List of Threatened Species, 2017: e.T22695306A110039784. https://doi.org/10.2305/IUCN.UK.2017-1.RLTS.T22695306A110039784.en.
Birdlife International. (2018). Aquila rapax. IUCN Red List of Threatened Species, 2018: https://doi.org/10.2305/IUCN.UK.2018-2.RLTS.T2269603 3A131671001.en.
Birdlife International. (2018). Haliaeetus leucoryphus. IUCN Red List of Threatened Species, 2018: https://doi.org/10.2305/IUCN.UK.2018-2.RLTS.T22695130A131934599.en.
Birdlife International. (2018). Haliaeetus vociferoides (amended version of 2016 assessment). IUCN Red List of Threatened Species, 2018: https://doi.org/10.2305/IUCN.UK.2016-3.RLTS.T22695121A125395004.en.
Birdlife International. (2018). Lophotriorchis kienerii. IUCN Red List of Threatened Species, 2018: https://doi.org/10.2305/IUCN.UK.2018-2.RLTS.T22696111A131375836.en.
Birdlife International. (2018). Nisaetus floris (amended version of 2017 assessment). IUCN Red List of Threatened Species, 2018: https://doi.org/10.2305/IUCN.UK.2017-3.RLTS.T22732096A125448523.en
Birdlife International. (2018). Pithecophaga jefferyi (amended version of 2017 assessment). IUCN Red List of Threatened Species, 2018: https://doi.org/10.2305/IUCN.UK.2017-3.RLTS.T22696012A129595746.en.
Birdlife International. (2018). Polemaetus bellicosus. IUCN Red List of Threatened Species, 2018: e.T22696116A129915349. https://doi.org/10.2305/IUCN.UK.2018-2.RLTS.T22696116A129915349.en.
Birdlife International. (2018). Species factsheet: Aquila chrysaetos. Retrieved from http://www.birdlife.org
Birdlife International. (2018). Species factsheet: Aquila nipalensis. Retrieved from http://www.birdlife.org
Birdlife International. (2018). Stephanoaetus coronatus. IUCN Red List of Threatened Species, 2018: https://doi.org/10.2305/IUCN.UK.2018-2.RLTS.T22696201A129914678.en.
Birdlife International. (2019). Aquila fasciata (amended version of 2016 assessment). IUCN Red List of Threatened Species, 2019: https://doi.org/10.2305/IUCN.UK.2019-3.RLTS.T22696076A155464015.en.
Birdlife International. (2019). Aquila heliaca (amended version of 2017 assessment). IUCN Red List of Threatened Species, 2019. https://doi.org/10.2305/IUCN.UK.2019-3.RLTS.T22696048A155464885.en
Birdlife International. (2019a). Turkey vulture Cathartes aura. Retrieved from http://datazone. Retrieved from http://birdlife.org/species/factsheet/22697627

Birdlife International. (2019b). American Black vulture Coragyps atratus. Retrieved from http://datazone.birdlife.org/species/factsheet/american-black-vulture-coragyps-atratus
Birdlife International. (2020) IUCN Red List for birds. Species factsheet: Haliaeetus vociferoides. Recommended citation for factsheets for more than one species. Retrieved from http://www.birdlife.org. Birdlife International.
Birdlife International. (2020) IUCN Red List for birds. Species factsheet: Icthyophaga Humilis. Recommended citation for factsheets for more than one species. Retrieved from http://www.birdlife.org. Birdlife International.
Birdlife International. (2020). Data Zone. Retrieved from http://datazone.birdlife.org/species/spcredcat.
Birdlife International. (2020). Species factsheet: Clanga hastata. IUCN Red List for birds. Retrieved from http://www.birdlife.org
Birdlife International. (2020). Species factsheet: Nisaetus floris. Retrieved from http://www.birdlife.org
Birdlife International. (2020). Species factsheet: Nisaetus kelaarti. Retrieved from http://www.birdlife.org
Birdlife International. (2020). Species factsheet: Nisaetus philippensis. Retrieved from http://www.birdlife.org
Birdlife International. (2020). Species factsheet: Nisaetus pinskeri. Retrieved from http://www.birdlife.org
Birdlife International. (2020). Species factsheet: Spizaetus isidori. Retrieved from http://www.birdlife.org
Birdlife International. (2020). Species factsheet: Terathopius ecaudatus. Retrieved from http://www.birdlife.org
Birdlife International/European Bird Census Council. (2000). European bird populations: Estimates and trends. *In*: M. Heath, C. Borggreve & N. Peet (Eds.), pp. 1–160. Birdlife International Royal Society for the Protection of Birds, UK. European Bird Census Council (EBCC). Birdlife International. Cambridge, UK.
BirdLife South Africa 2017. Verreaux's eagle and wind farms: Guidelines for impact assessment, monitoring, and mitigation. Retrieved from https://www.birdlife.org.za/wp-content/uploads/2020/03/BLSA-Guidelines-Verreauxs-Eagle-and-Wind.pdf
BirdwatchIreland. (2020). White-tailed eagle. Retrieved from https://birdwatchireland.ie/birds/white-tailed-eagle/
Bishop, K. D. (1999). Preliminary notes on some birds in Bhutan. Forktail, 15, 87–91.
Bishop, L. B. (1905). The Gray sea eagle (Haliæetus albicilla) in British Columbia. The Auk, 22(1), 79–80. https://doi.org/10.2307/4069884
Bisson, I. A., Ferrer, M. & Bird, D. M. (2002). Factors influencing nest-site selection by Spanish Imperial eagles. Journal of Field Ornithology, 73(3), 298–302. https://doi.org/10.1648/0273-8570-73.3.298
Bittner, L. (2010). Satellite telemetry takes bird banding to new heights. Wild News, 8, 1–5.
Blaber, S. J. M. (1990). A checklist and notes on the current status of birds of New Georgia, Western Province, Solomon Islands. Emu – Austral Ornithology, 90(4), 205–214. https://doi.org/10.1071/MU9900205
Blackwell, B. F. & Wright, S. E. (2006). Collisions of Red-tailed hawks (Buteo jamaicensis), Turkey vultures (Cathartes aura), and Black vultures (Coragyps atratus) with aircraft: Implications for bird strike reduction. Journal of Raptor Research, 40(1), 76–80. Retrieved from https://digitalcommons.unl.edu/cgi/viewcontent.cgi?article=1405&context=icwdm_usdanwrc. https://doi.org/10.3356/0892-1016(2006)40[76:CORHBJ]2.0.CO;2
Blake, E. R. (1958). Birds of volcán de Chiriquí, Panama. Fieldiana Zoology, 36, 499–577.
Blake, E. R. (1977). Manual of Neotropical birds, 1. Sphenisicdae (penguins) to Laridae (gulls and terns). Chicago: University of Chicago Press.
Blakers, M., Davies, S. J. J. F. & Reilly, P. N. (1984). The Atlas of Australian Birds. Royal Australasian Ornithologists' Union and Melbourne University Press. Melbourne.
Blanco Márquez, P. A. (2002). Rescue and rehabilitation of raptors in Venezuela [Rescate y rehabilitación de rapaces en Venezuela]. *In*: Fondo Peregrino-Panamá, Panamá City, Panamá Conference on Neotropical Raptorial Birds and Symposium on the Harpy Eagle, pp. 45–46.
Blanco, J. C. & González, J. L. (1992). Libro rojo de los vertebrados de España. [Technical report ICONA]. Ministerio de Agricultura, Pesca y Alimentación. Madrid.
Block, E. (2004). A species extinct in zoos – the Philippine eagle. International Zoo News, 51, 340–343.
Bloom, P. H. & Hawks, S. J. (1982). Food habits of nesting Golden eagles in northeast California and northwest Nevada. Raptor Research, 16, 110–115.
Bloom, P. H., Clark, W. S. & Kidd, J. W. (2007). Capture techniques. *In*: D. M. Bird, K. L. Bildstein, D. Barber, & R. A. Zimmerman (Eds.), Raptor research and management techniques. 193. Hancock House Publishers Ltd. Washington, DC.
Bloom, P. H., Scott, J. M., Pattee, O. H. & Smith, M. R. (1989). Lead contamination of Golden eagles (Aquila chrysaetos) within the range of the Californian condor (Gymnogyps californianus). *In*: Meyburg & R. D. Chancellor (Eds.), Raptors in the modern world. Proceedings of the 3rd World Conference on Birds of Prey and Owls, Eliat, Israel, 481–482.
Blume, I. E. (1947). One that got away. Yellowstone Nature Notes, 21, 70–71.
Blumstein, D. T., Rangchi, T. N., Briggs, T., De Andrade, F. S. & Natterson-Horowitz, B. (2017). A systematic review of carrion eaters' adaptations to avoid sickness. Journal of Wildlife Diseases, 53(3), 577–581. https://doi.org/10.7589/2016-07-162
Blus, L. J. (1994). A review of lead poisoning in swans. Comparative Biochemistry and Physiology Part C, 108(3), 259–267. https://doi.org/10.1016/0742-8413(94)00021-2
Blyth, E. (1845). Notes and descriptions of various new or little-known species of birds. Journal of the Asiatic Society of Bengal, 14, 173–212.
Boag, D. A. (1977). Summer food habits of Golden eagles in southwestern Alberta. Canadian Field-Naturalist, 91, 296–298.
Bobo, K. S., Williams, E., Anye, N. D., Njie, M. F., Fotso, R. C. & Languy, M. (2006). The birds of Mbam and Djerem National Park, Cameroon. Malimbus, 28, 90–106.
Bodio, S. J. (2012). Eternity of eagles: The human history of the most fascinating bird in the world. Lanham, MD: Rowman & Littlefield Publishing Group.
Boeker, E. L. & Ray, T. D. (1971). Golden eagle population studies in the Southwest. Condor, 73(4), 463–467. https://doi.org/10.2307/1366668
Boeker, E. L. & Nickerson, P. R. (1975). Raptor electrocutions. Wildlife Society Bulletin, 3, 79–81.

Boertmann, D. (1994). An annotated checklist to the birds of Greenland. Meddelelser.

Boesing, A. L., Menq, W. & Dos Anjos, L. (2012). First description of the reproductive biology of the grey-bellied hawk (Accipiter poliogaster). The Wilson Journal of Ornithology, 124(4), 767–774. https://doi.org/10.1676/1559-4491-124.4.767

Bogucki, Z. (1976). Status of the White-tailed eagle in Poland. *In*: Report of WWF Symposium on the White-tailed eagle, 31–32.

Bohrer, G., Brandes, D., Mandel, J. T., Bildstein, K. L., Miller, T. A., Lanzone, M., Katzner, T. E., Maisonneuve, C. & Tremblay, J. A. (2012). Estimating updraft velocity components over large spatial scales: Contrasting migration strategies of Golden eagles and Turkey vultures. Ecology Letters, 15(2), 96–103. http://www.https://doi.org/10.1111/j.1461-0248.2011.01713.x

Boinski, S., Kauffman, L., Westoll, A., Stickler, C. M., Cropp, S. & Ehmke, E. (2003). Are vigilance, risk from avian predators and group size consequences of habitat structure? A comparison of three species of squirrel monkey (Saimiri oerstedii, S. boliviensis, and S. sciureus). Behaviour, 140, 1421–1467.

Boitani, L. (1984). Simon and Schuster's guide to mammals. Simon & Schuster/Touchstone Books. New York.

Bolen, E. G. (1975). Eagles and sheep: A viewpoint. Journal of Range Management, 28(1), 11–17. https://doi.org/10.2307/3897571

Bollen, C. (1989). Notes on Little Eagles breeding on the southern Tablelands of New South Wales. Australian Bird Watcher, 13, 65–66.

Bonaparte, C. L. (1850). Revue générale de la classe des oiseaux. Première parte. Perroquets et oiseaux de proie. Revue et magasin de zoologie pure et appliquée, 2(2), 474–492.

Bonn, I., Yoshikawa, T. & Phan, P. (2004). Effects of board structure on firm performance: A comparison between Japan and Australia. Asian Business Management, 3, 105–125. https://doi.org/10.1057/palgrave.abm.9200068

Bonta, M. (2003). Seven names of the bellbird: Conservation geography in Honduras. Texas A and M University College Station, TX.

Bonta, M. & Anderson, D. L. (2002). Birding Honduras: A checklist and guide. Tegucigalpa, Honduras: EcoArte S. de R.L.

Borello, W. D. & Borello, R. M. (2004). Two incidents of talon-grappling and cartwheeling in the Tawny eagle Aquila rapax. Ostrich, 75(4), 320–321. https://doi.org/10.2989/00306520409485461

Borges, R. (1986). Predation attempt by Black eagle (Ictinaetus malayensis perniger) on Indian Giant Squirrel (Ratufa indica elphinstonii). Journal of the Bombay Natural History Society, 83, 203. [Suppl.] (Centenary issue).

Borges, S. E., Cohn-Haft, M., Carvalhaes, A. M. P., Henriques, L. M. P., Pacheco, J. F. & Whittaker, A. (2001). Birds of Jaú National Park, Brazilian Amazon: Species check-list, biogeography and conservation. Ornitologia Neotropical, 12, 109–140.

Borner, M. (1985). The rehabilitated chimpanzees of Rubondo Island. Oryx, 19(3), 151–154. https://doi.org/10.1017/S003060530002531X

Borrow, N. & Demey, R. (2001). Birds of Western Africa. Christopher Helm Publishers. London.

Borrow, N. & Demey, R. (2010). Birds of Ghana. Christopher Helm Publishers. London.

Borrow, N. & Demey, R. (2013). Field Guide to the Birds of Ghana. Bloomsbury Publishing.

Bortolotti, G. R. (1984). Age and sex size variation in Golden eagles. Journal of Field Ornithology, 55, 54–66.

Bortolotti, G. (1984a). Physical development of nestling Bald eagles with emphasis on the timing of growth events. Wilson Bulletin, 96(4), 524–542.

Bortolotti, G. R. (1984b). Sexual size dimorphism and age-related size variation in Bald eagles. The Journal of Wildlife Management, 48(1), 72–81. https://doi.org/10.2307/3808454

Bortolotti, G. R., González, L. M., Margalida, A., Sánchez, R. & Oria, J. (2008). Positive assortative pairing by plumage color in Spanish Imperial eagle. Behavioural Processes, 78(1), 100–107. https://doi.org/10.1016/j.beproc.2008.01.007

Bosch, J. & Meyburg, B.-U. (2012). The Lesser spotted eagle Aquila pomarina in Catalonia (Spain)—Breeding attempt and migration. die Vogelwelt, 133, 89–97.

Boshoff, A. (1993). Density, breeding performance and stability of Martial eagles Polemaetus bellicosus breeding on electricity pylons in the Nama-Karoo, South Africa. Cape Nature Conservation. Port Elizabeth, South Africa, 1(1), 7–19.

Boshoff, A. (1997). Martial eagle. *In*: J. A. Harrison et al. (Eds.), The atlas of southern African birds. 1. Birdlife South Africa. Johannesburg.

Boshoff, A. F. (1980). Some socio-economic aspects of a bird of prey questionnaire survey. South African Wildlife Research-24-month delayed open access, 10(2), 71–81.

Boshoff, A. F. & Palmer, N. G. (1980). Macro-analysis of prey remains from Martial eagle nests in the Cape Province. Ostrich, 51(1), 7–13. https://doi.org/10.1080/00306525.1980.9633535

Boshoff, A. F., Vernon, C. J. & Brooke, R. K. (1983). Historical atlas of the diurnal raptors of the Cape Province (Aves: Falconiformes). Annals of the Cape Provincial Museums (Natl Hist.), 14, 173–297.

Boshoff, A. F. (1986). Stability of a population of Martial eagles in a sheep farming area in the Great Karoo, South Africa: Project outline and early results. Gabar, 1, 9–13.

Boshoff, A. F. & Palmer, N. G. (1988). Black eagle nest spacing and estimated territory size in a Karoo habitat. South African Journal of Wildlife Research, 20, 57–68.

Boshoff, A. F., Palmer, N. G. & Avery, G. (1990). Regional variation in the diet of Martial eagles in the Cape Province, South Africa. South Africa Journal of Wildlife Research, 20, 57–68.

Boshoff, A. F., Palmer, N. G., Vernon, C. J. & Avery, G. (1994). Comparison of the diet of Crowned eagles in the savanna and forest biomes of south-eastern South Africa. South African Journal of Wildlife Research 24, 26–31.

Boshoff, A. F., Minnie, J. C., Tambling, C. J. & Michael, M. D. (2011). The impact of power line-related mortality on the Cape vulture Gyps coprotheres in a part of its range, with an emphasis on electrocution. Bird Conservation International, 21(3), 311–327. https://doi.org/10.1017/S095927091100013X

Boshoff, C. (1997). An experimental study of service recovery options. International Journal of Service Industry Management, 8(2), 110–130. https://doi.org/10.1108/09564239710166245

Bouchahdane, I., Boukhemza, M., Souttou, K. & Derridj, A. (2019). Breeding biology of Booted eagle Hieraaetus pennatus (Gmelin, J.F., 1788) The first study in North Africa. Ekológia (Bratislava), 38(4), 382–391. https://doi.org/10.2478/eko-2019-0028
Bouckaert, R. R. & Heled, J. (2014). DensiTree 2: Seeing trees through the forest. bioRxiv, 012401. Retrieved from https://doi.org/10.1101/01240
Bouglouan, N. (2020a). Steppe eagle Aquila nipalensis. Retrieved from http://www.oiseaux-birds.com/card-steppe-eagle.html
Bouglouan, N. (2020b). Eastern Imperial eagle Aquila heliaca. Accipitriforme order—Accipitridae Family. Retrieved from http://www.oiseaux birds.com/card-eastern-imperial-eagle.html
Bouglouan, N. (2020c). Wahlberg's eagle Hieraaetus wahlbergi. Accipitriformes Order—Accipitridae Family. Retrieved from http://www.oiseaux birds.com/card-wahlberg-s-eagle.html
Bowen, D. (2016). Halichoerus grypus. IUCN Red List of Threatened Species, 2016: e.T9660A45226042. https://doi.org/10.2305/IUCN UK.2016-1.RLTS.T9660A45226042.en
Bowen, W. D. (1980). Coyote-Golden eagle interactions at an ungulate carcass. Journal of Mammalogy, 61(2), 376–377. https://doi org/10.2307/1380075
Bowen, W. W. (1926). Catalogue of Sudan birds, based on the collection in the Sudan Government Museum (Natural History). Part I. Struthionida to Picidae. Sudan Government Museum (Natural History) Publication No. 1.
Bowerman, W. W., IV. (1991). Factors influencing breeding success of Bald eagles in upper Michigan [MAE Thesis]. Marquette, MI: Norther Michigan University.
Bowerman, W. W., Evans, E. D., Giesy, J. P. & Postupalsky, S. (1994). Using feathers to assess risk of mercury and selenium to Bald eagl reproduction in the Great Lakes region. Archives of Environmental Contamination and Toxicology, 27(3), 294–298. https://doi.org/10.1007 BF00213162
Bowerman, W. W., Giesy, J. P., Best, D. A. & Kramer, V. J. (1995). A review of factors affecting productivity of Bald eagles in the Great-Lake region – Implications for recovery. Environmental Health Perspectives, 103, 51–59.
Bowler, J. & Taylor, J. (1989). An annotated checklist of the birds of Manusela National Park, seram (birds recorded on the Operation Raleig Expedition). Kukila, 4, 3–29.
Bowman, J. L., Leopold, B. D., Vilella, F. J. & Gill, D. A. (2004). A spatially explicit model, derived from demographic variables, to predic attitudes toward black bear restoration. Journal of Wildlife Management. CO, 68(2), 223–232. https://doi.org/10.2193/0022-541X(2004)0 8[0223:ASEMDF]2.0.CO;2
Bowman, J. L., Leopold, B. D., Vilella, F. J., Gill, D. A. & Jacobson, H. A. (2001). Attitudes of landowners toward American black bear compare between areas of high and low bear populations. Ursus, 12, 153–160. Retrieved from https://www.jstor.org/stable/3873243
Bowns, J. E., Davenport, J. W., Workman, J. P., Nielson, D. B. & Dwyer, D. D. (1973). Determination of cause and magnitude of sheep losses i southwestern Utah. Utah Science, 34, 52.
Boycott, R. & Parker, V. (2003). Birds of the Malolotja Nature Reserve, Swaziland. Cape Town and Mbabane: Avian Demography Unit & th Conservation Trust of Swaziland.
Boyd, D. S. & Foody, G. M. (2011). An overview of recent remote sensing and GIS based research in ecological informatics. Ecological Informatics 6(1), 25–36. https://doi.org/10.1016/j.ecoinf.2010.07.007
Boyer, R. H. (1948). Fawns escape eagle's attack. Yosemite Nature Notes, 27, 129–130.
Boyle, S. A., Kennedy, C. M., Torres, J, Colman, K., Pérez-Estigarribia, P. E. & de la Sancha, N. U. (2014). High-resolution satellite imagery i an important yet underutilized resource in conservation biology. PLoS ONE, 9(1), e86908). https://doi.org/10.1371/journal.pone.0086908
Brabourne, L. & Chubb, C. (1912). The Birds of South America. R H Porter. London.
Bradley, D. W., Clark, R. G., Dunn, P. O., Laughlin, A. J., Taylor, C. M., Vleck, C..., Norris, D. R. (2014). Trans-Gulf of Mexico loop migration o tree swallows revealed by solar geolocation. Current Zoology, 60(5), 653–659. https://doi.org/10.1093/czoolo/60.5.653
Bragagnolo, L., Maceda, J., Reyes, M., Salvador, V., Santillan, M. & Sarasola, J. (2007). Characterization of the nesting habitat of the Crowne Solitary eagle (Harpyhaliaetus solitarius) in the province of la Pampa, Argentina. *In*: Bildstein, K. L., Barber, D. R. & Zimmerman, A. (Eds.) Neotropical raptors. Hawk Mountain Sanctuary. Kempton, UK.
Braham, M., Miller, T., Duerr, A. E., Lanzone, M., Fesnock, A., LaPre, L., Driscoll, D. & Katzner, T. (2015). Home in the heat: Dramatic seasona variation in home range of desert Golden eagles informs management for renewable energy development. Biological Conservation, 186 225–232. https://doi.org/10.1016/j.biocon.2015.03.020
Brandes, D. & Ombalski, D. (2004). Modeling raptor migration pathways using a fluid-flow analogy. Journal of Raptor Research, 38(3), 195–207
Brandl, R., Utschick, H. & Schmidtke, K. (1985). Raptors and land-use systems in southern Africa. African Journal of Ecology, 23(1), 11–20 https://doi.org/10.1111/j.1365-2028.1985.tb00707.x
Brathwaite, D. H. (1992). Notes on the weight, flying ability, habitat, and prey of Haast's eagle (Harpagornis moorei). Notornis. Ornithologica Society of New Zealand, 39(4), 239–247.
Braun, M. J., Finch, D. W., Robbins, M. B. & Schmidt, B. K. (2000). A field checklist to the birds of Guyana. Washington, DC, Brazil: Biologica Diversity Institute of the Guianas Program, National Museum of Natural History.
Brazil, M. (2009). Birds of East Asia. Princeton, China, Taiwan, Korea, Japan, and Princeton. London: Christopher Helm Publishers.
Brazil, M. (2010). The Whooper Swan. Bloomsbury Publishing.
Brazil, M. A. & Hanawa, S. (1991). The status and distribution of diurnal raptors in Japan. Birds of Prey Bulletin, 4, 175–238.
Brean, H. (2019). Peregrine falcons act as mercury bellwether, Nevada study shows. Las Vegas Review-Journal. Retrieved from https://www reviewjournal.com/local/local-nevada/peregrine-falcons-act-as-mercury-bellwether-nevada-study-shows-1640798/.
Brehm, A. E. (1856). Die Sippe der Seeadler. Naumannia, 6, 205–209.
Brehm, C. L. (1831). Handbuch der Naturgeschichte aller Vögel Deutschlands. Voigt, Ilmenau, 8 vols, 1085.

Brehm, C. L. (1855). Verzeichniss der europäischen Vögel nach den Species und Subspecies. Naumannia, 5, 266–300.

Brehm, C. L. (1861). Bericht 13 ueber die Versammlungen der allgemeinen Deutschen Ornithologen-Gesellschaft, 1860, 60–62.

Breuer, W. (2007). Stromopfer und Vogelschutz an Energiefreileitungen. Naturschutz und Landschaftsplanung, 39, 69–72.

Brichetti, P. & Fracasso, G. (2003). Ornitologia Italiana. Vol. 1. Gaviidae-Falconidae: Identificazione, Distribuzione, Consistenza e Movimenti degli Ucceli Italiani.

Bridge, E. S., Thorup, K., Bowlin, M. S., Chilson, P. B., Diehl, R. H., Fléron, R. W. et al. (2011). Technology on the move: Recent and forthcoming innovations for tracking migratory birds. BioScience. 61(9), 689–698. https://doi.org/10.1525/bio.2011.61.9.7

Bridgeford, P. (2004). Lappet-faced vulture Torgos tracheliotus. *In*: Monadjem, A., Anderson, M. D., Piper, S. E. & Boshoff, A. F. (Eds.), The vultures of Southern Africa—Quo vadis, pp. 28–33. Johannesburg: Birds of Prey Working Group.

Bridgeford, P. (2004). Status of vultures in Namibia. *In*: Monadjem, A., Anderson, M. D., Piper, S. E. & Boshoff, A. F. (Eds.), The vultures of Southern Africa—Quo vadis, pp. 144–147. Johannesburg: Birds of Prey Working Group.

Bright, J. A., Marugán-Lobón, J., Cobb, S. N. & Rayfield, E. J. (2016). The shapes of bird beaks are highly controlled by nondietary factors. Proceedings of the National Academy of Sciences of the United States of America, 113(19), 5352–5357. https://doi.org/10.1073/pnas.1602683113

Bright, J., Langston, R., Bullman, R., Evans, R. J., Gardner, S., Pearce-Higgins, J. & Wilson, E. (2006). Bird Sensitivity Map to provide locational guidance for onshore wind farms in Scotland. RSPB Research Report No. 20. Royal Society for the Protection of Birds.

Brinkley, E. S. & Byrne, A. M. (2013). A White-tailed eagle (Haliaeetus albicilla) at Derby Hill, New York. North American Birds, 66(4), 614–631.

Brisbin, Jr., I. L. & Mowbray, T. B. (2002). American coot (Fulica americana), version 1.0. *In*: A. F. Poole and Brisbin Jr., I. L. & Mowbray, T. B. (Ed.). American Coot (Fulica americana). Birds of North America Online. Ithaca, NY: Cornell Laboratory of Ornithology. https://doi.org/10.2173/bna.697a

British Birds. (2006). News and comment, 99(8), 441–445.

British Ornithological Union. (2020). Category D species. Retrieved from https://www.bou.org.uk/british-list/category-d-species/

Britton, P. L. (1980). Birds of East Africa: Their habitat, status and distribution. Nairobi, Kenya: East Africa Natural History Society.

Brockmann, H. J. & Barnard, C. J. (1979). Kleptoparasitism in birds. Animal Behaviour, 27, 487–514. https://doi.org/10.1016/0003-3472(79)90185-4

Brodeur, S., Bird, D. M. & Fuller, M. (1996). Complete migration cycle of Golden Eagles breeding in northern Quebec. Condor, 98(2), 293–299. https://doi.org/10.2307/1369147

Brokaw, N. V. L., Mallory, E. P. & Alcorn, P. W. (1990). Trees of Rio Bravo. A guide to the trees of the Rio Bravo Conservation and Management Area. Institution Report by the Plymouth, MA: Manomet Bird Observatory.

Broley, C. L. (1947). Migration and nesting of Florida Bald eagles. Wilson Bulletin, 59, 320.

Brooke, R. H., Grobler, J. H., Irwin, M. P. S. & Steyn, P. (1972). A study of the migratory eagles Aquila nipalensis and A. pomarina (Aves: Accipitridae) in southern Africa, with comparative notes on other large raptors. Occasional Papers of the National Museum of Rhodesia, 5, 61–114.

Brooke, R. K., Cannell, I. C. & Jeffery, R. D. (1971). New distributional records of raptors in western Angola. Bulletin of the British Ornithologists' Club, 91, 165–166.

Brooke, R. K., Grobler, J. H., Irwin, M. P. S. & Steyn, P. (1972). A study of the migratory eagles Aquila nipalensis and A. pomarina (Aves: Accipitridae) in southern Africa, with comparative notes on other large raptors. Occasional Papers of the Natural History Museum Series. *B.*, Natural Sciences, 5(2), 61–114.

Brooke, R. K. & Vernon, C. J. (1981). Early names and records of two small Hieraaetus eagles (Aves: Accipitridae) in the Cape Province of South Africa. Annals of the Cape Provincial Museums of Natural History, 13, 133–139.

Brooke, R. K. (1984). South African Red Data Book – Birds. South African National Scientific. Programmes Report No. 97.

Brooks, D. M., O'Neill, J. P., Foster, M. S., Mark, T., Dauphiné, N. & Franke, I. J. (2009). Avifauna of the pongos basin, Amazonas Department, Peru. Wilson Journal of Ornithology, 121(1), 54–74. https://doi.org/10.1676/07-160.1

Broomhall, L. S., Mills, M. G. L. & Du Toit, J. T. (2003). Home range and habitat use by cheetahs (Acinonyx jubatus) in the Kruger National Park. Journal of Zoology, 261(02), 119–128. https://doi.org/10.1017/S0952836903004059

Brown, A. & Grice, P. (2005). Birds in England. London: T. & A. D. Poyser.

Brown, B. T. (1993). Wintering foraging ecology of Bald eagles in Arizona condor, 95, 132–138.

Brown, C. J. (1988). Home ranges of Black eagles in the Natal Drakensberg. South African Journal of Wildlife Research, 18, 120–125.

Brown, C. J. (1991). Declining Martial Polemaetus bellicosus and Tawny Aquila rapax eagle populations and causes of mortality on farmlands in central Namibia. Biological Conservation, 56(1), 49–62. https://doi.org/10.1016/0006-3207(91)90088-Q

Brown, C. J. (1991). Declining martial and tawny eagle populations and causes of mortality on commercial farmlands in central Namibia. Biological Conservation, 56, 49–66

Brown, C. J. (1991a). An investigation into the decline of the bearded vulture in southern Africa. Biology Conservation, 57, 315–337.

Brown, C. J. & Hines, C. J. H. (1987). Western banded snake eagles in Namibia gabar. 2, 40–42.

Brown, D. L. & Couey, F. M. (1950). Eagle-game relationships in Montana. Third Annual Northwest Wildlife Conference.

Brown, J. E. (1976). Field criteria for predator damage assessment. Utah Science, 37, 26–30.

Brown, J. L. (1969). Territorial behavior and population regulation in birds: A review and re-evaluation. Wilson Bulletin, 293–329.

Brown, J. L. (1989). Birds and electricity transmission lines in South West Africa/Namibia. Madoqua, 16(1), 59–67.

Brown, J. W. & Mindell, D. P. (2009). Diurnal birds of prey (Falconiformes). *In*: S. B. Hedges & S. Kumar (Eds.), The timetree of life, 436–439. Oxford: Oxford University Press.

Brown, L. H. (1952). On the biology of the large birds of prey of the Embu District, Kenya colony. Ibis, 94, 577–620.

Brown, L. H. (1953). On the biology of the large birds of prey of the Embu District, Kenya Colony. Ibis, 95(1), 74–114.

Brown, L. H. (1955). Supplementary notes on the biology of large birds of prey of Embu District, Kenya Colony. Ibis, 97, 38–64.

Brown, L. H. (1963). Observations on East African birds of prey. African Journal of Ecology, 1(1), 5–17. https://doi.org/10.1111/j.1365-2028.1963.tb00174.x

Brown, L. H. (1966). Observations on some Kenya eagles. Ibis, 108(4), 531–572. https://doi.org/10.1111/j.1474-919X.1966.tb07208.x

Brown, L. H. (1970). African birds of prey. London: Collins.

Brown, L. H. (1970). Eagles. New York: Arco Publishing Company, Inc.

Brown, L. H. (1971). African Birds of Prey. Boston: Houghton Mifflin Company.

Brown, L. H. (1975). Brown on Golden eagles and sheep. *In*: J. R. Murphy et al. (Eds.), Population status of raptors. Research Report no. 3, pp. 129–133. Vermillion, SD: Proceedings of the Conference on Raptor Conservation Techniques, Raptor Research Foundation.

Brown, L. H. (1976). Eagles of the World. Exeter: David & Charles.

Brown, L. H. (1977). Eagles of the World. New York: Universe Books.

Brown, L. H. (1978). A note on the movements of Wahlberg's eagles Aquila wahlbergi in Kenya. Bokmakierie, 30, 107.

Brown, L. H. (1980). Scattered observations on the Long-crested eagle. Bokmakierie, 32, 117–120.

Brown, L. H. & Amadon, D. (1968). Hawks, Eagles, and Falcons of the World, 2. New York: McGraw-Hill Book Company.

Brown, L. H. & Britton, P. L. (1980). The Breeding Seasons of East African birds. Nairobi, Kenya: East African Natural History Society.

Brown, L. H., Urban, E. K. & Newman, K. (1982). The Birds of Africa, vol. 1. London: Academic Press.

Brown, L. H. & Davey, P. R. A. (1978). Natural longevity, as determined by plumage variation. *In*: Ayres' eagle Hieraaetus dubius. Bokmakierie, 30, 27–31.

Brown, L. H. & Pommery, D. E. (1984). The age structure of populations of wild birds in tropical Africa as demonstrated by plumage characters and marking techniques. *In*: Proceedings of the V Pan-African Ornithological Congress (pp. 97–119).

Brown, L. H., Urban, E. K. & Newman, K. (1982). The Birds of Africa. London: Academic Press.

Brown, L. & Amadon, D. (1986). Eagles, Hawks and Falcons of the World. New York: Wellfleet Press. ISBN 978-1555214722.

Bruce-Miller, I. (2008). Martial eagle Polemaetus bellicosus apparently killing White-headed vulture Trigonoceps occipitalis. Bulletin of the African Bird Club, 15, 102–103.

Bruderer, B., Underhill, L. & Liechti, F. (1995. Altitude choice of night migrants in a desert area predicted by meteorological factors. Ibis, 137, 44–45.

Bruns, E. H. (1970). Winter predation of Golden eagles and coyotes on pronghorn antelopes. Canadian Field-Naturalist, 84, 301–304.

Buchanan, K. (1899). Nesting of the Black eagle. Journal of the Bombay Natural History Society, 12(4), 776–777.

Bucher, E. & Huszar, P. (1999). Sustainable management of the Gran Chaco of South America: Ecological promise and economic constraints. Journal of Environmental Management, 57, 99–108. https://doi.org/10.1006/jema.1999.0290

Buchler, D. A. (2000). Bald eagle (Haliaeetus leucocephalus). *In*: A. Poole & F. Gill (Eds.), The birds of North America, 506. Philadelphia: Academy of Natural Sciences, and Washington, DC: American Ornithologists' Union.

Bucini, G. & Lambin, E. F. (2002). Fire impacts on vegetation in central Africa: A remote-sensing-based statistical analysis. Applied Geography, 22(1), 27–48). https://doi.org/10.1016/S0143-6228(01)00020-0

Bucini, G., Hanan, N. P., Boone, R. B., Smit, I. P. J., Saatchi Sassan, S., Lefsky, M. A. et al. (2010). Woody fractional cover in Kruger National Park, South Africa. Ecosystem Function in Savannas p. 219. Boca Raton: CRC Press.

Buckingham, D. L., Dutson, G. C. L. & Newman, J. L. (1995). Birds of Manus, Kolombangara and Makira (San Cristobal) with notes on mammals and records from other Solomon Islands.

Buckley, P. A., Foster, M. S., Morton, E. S., Ridgely, R. S. & Buckley, F. G. (1985). Neotropical ornithology. Ornithological Monographs. 36, vii–viii. https://doi.org/10.2307/40168271

Buckley, N. J. (1996). Food finding and the influence of information, local enhancement, and communal roosting on foraging success of North American vultures. Auk, 113(2), 473–488. https://doi.org/10.2307/4088913

Buckley, N. J. (1999). Black vulture (Coragyps atratus). *In*: A. Poole & F. Gil (Eds.), Birds of North America Online, 411. https://doi.org/10.2173/bna.411. Philadelphia: The Birds of North America, Inc.

Buehler, D. A. (2000). Bald eagle (Halieaeetus leucocephalus). Birds of North America, 564.

Buehler, D. A. (2020). Bald eagle (Haliaeetus leucocephalus), version 1.0. *In*: A. F. Poole and F. B. Gill (Eds), Birds of the World. Ithaca, NY: Cornell Laboratory of Ornithology. https://doi.org/10.2173/bow.baleag.01

Buehler, D. A., Fraser, J. D., Fuller, M. R., Mcallister, L. S. & Seegar, J. K. D. (1995). Captive and field-tested radio transmitter attachments for Bald eagles. Journal of Field Ornithology, 66, 173–180.

Buehler, D. A., Mersmann, T. J., Fraser, J. D. & Seegar, J. K. D. (1991). Differences in distribution of breeding, nonbreeding, and migrant bald eagles on the northern Chesapeake Bay. The Condor, 93(2), 399–408. https://doi.org/10.2307/1368956

Bueser, G. L. L., Bueser, K. G., Afan, D. S., Salvador, D. I., Grier, J. W., Kennedy, R. S. & Miranda, H. C., Jr. (2003). Distribution and nesting density of the Philippine eagle Pithecophaga jefferyi on Mindanao Island, Philippines: What do we know after 100 years? Ibis, 145(1), 130–135. https://doi.org/10.1046/j.1474-919X.2003.00131.x

Bueser, G. L., Afan, D. S., Balaquit, G. A., Gatil, K. M. & Ibañez, J. C. (1999). Distribution and conservation of the Philippine eagle (Pithecophaga jefferyi) in Davao Oriental and Davao Province. Sylvatop, 8, 114–119.

Buij, R. & Croes, B. M. (2014). Raptors in northern Cameroon, December 2005–December 2010. Bulletin of the African Bird Club, 21, 26–63.

Buij, R., Croes, B. M., Gort, G. & Komdeur, J. (2013). The role of breeding range, diet, mobility and body size in associations of raptor communities and land-use in a West African savanna. Biological Conservation, 166, 231–246. https://doi.org/10.1016/j.biocon.2013.06.028
Buij, R., Fox, E. A. & Wich, S. E. (2006). Birds of Gunung Leuser National Park, northern Sumatra. Kukila, 13, 23–47.
Bukreev, A., Boldbaatar, S. & Zvonov, B. M. (2010). The Imperial eagle in Mongolia. Raptors Conservation, 20, 186–194.
Bull, J. & Farrand, J., Jr. (1987). Audubon society field guide to North American birds: Eastern region (pp. 394–402). New York: Alfred A. Knopf. Bulletin 35.
Bulletin of the British Ornithologists' Club, 132: 70–72.
Bulut, Z., Bragin, E. A., DeWoody, J. A., Braham, M. A., Katzner, T. E. & Doyle, J. M. (2016). Use of noninvasive genetics to assess nest and space use by White-tailed eagles. Journal of Raptor Research, 50(4), 351–362. https://doi.org/10.3356/JRR-15-84.1
Bunce, M., Szulkin, M., Lerner, H. R. L., Barnes, I., Shapiro, B., Cooper, A. & Holdaway, R. N. (2005). Ancient DNA provides new insights into the evolutionary history of New Zealand's extinct giant eagle. PLOS Biology, 3(1), e9. https://doi.org/10.1371/journal.pbio.0030009
Bundy, G. (1976). The Birds of Libya. BOU Checklist 1. Pages 1–102.
Bundy, G. & Warr, E. (1980). A check-list of the birds of the Arabian Gulf states. Sandgrouse, 1, 4–49.
Bunnefeld, N., BoÈrger, L., van Moorter, B., Rolandsen, C. M., Dettki, H., Solberg, E. J., et al. (2011). A model-driven approach to quantify migration patterns: Individual, regional and yearly differences. Journal of Animal Ecology, 80(2), 466-476. https://doi.org/10.1111/j.1365-2656.2010.01776.x.
Burbridge, A. (1925). The witch doctor's power. NADA, 3, 22–31.
Bureau of Land Management Wyoming State Office, Collins, P. W., Guthrie, D. A., Rick, T. C. & Erlandson, J. M. (2005). Analysis of prey remains excavated from an historic Bald eagle nest site on San Miguel Island, California. *In*: Proceedings of the Sixth California Islands Symposium. Arcata, CA (pp. 103–120). Institute for Wildlife Studies.
Burgat, V. (1990). Un micropollutant: Le cadmium. Bulletin Mensuel de l'Office National de la Chasse, 146, 40–42.
Burger, A. E. & Shaffer, S. A. (2008). Application of tracking and data-logging technology in research and conservation of seabirds. The Auk, 125(2), 253–264. https://doi.org/10.1525/auk.2008.1408
Burger, J. & Gochfeld, M. (2009). Comparison of arsenic, cadmium, chromium, lead, manganese, mercury and selenium in feathers in Bald eagle (Haliaeetus leucocephalus), and comparison with common eider (Somateria mollissima), glaucous-winged gull (Larus glaucescens), pigeon guillemot (Cepphus columba), and tufted puffin (Fratercula cirrhata) from the Aleutian Chain of Alaska. Environmental Monitoring and Assessment, 152(1–4), 357–367. https://doi.org/10.1007/s10661-008-0321-7
Burgess, H. E. (1937). Eagles on the Nilgiris. Journal of the Bombay Natural History Society, 39(2), 399–403.
Burggren, W. W. (1987). Invasive and noninvasive methodologies in ecological physiology: A plea for integration. *In*: Feder, M. E. (Ed.), New directions in ecological physiology, 1987, pp. 251–274. Cambridge, UK: Cambridge University Press.
Burke, C., Rashman, M. F., Longmore, S. N., McAree, O., Glover-Kapfer, P., Ancrenaz, M. & Wich, S. A. (2019). Successful observation of orangutans in the wild with thermal-equipped drones. Journal of Unmanned Vehicle Systems, 7(3), 235–257. https://doi.org/10.1139/juvs-2018-0035
Burnett, L. J., Sorenson, K. J., Brandt, J., Sandhaus, E. A., Ciani, D., Clark, M. . . . Risebrough, R. W. (2013). Eggshell thinning and depressed hatching success of California Condors reintroduced to Central California. The Condor, 115(3), 477–491. https://doi.org/10.1525/cond.2013.110150
Burnett, S., Winter, J. & Russell, R. (1996). Successful foraging by the wedge-tailed eagle Aquila audax in tropical rainforest in North Queensland. Emu – Austral Ornithology, 96(4), 277–280. https://doi.org/10.1071/MU9960277
Burnham, K. P. & Anderson, D. R. (2002). Model selection and multimodel inference. A practical information – theoretic approach (2nd ed). New York: Springer.
Burnie, D. & Wilson, D. E. (Eds.). (2005). Animal: The definitive visual guide to the world's wildlife. DK adult.
Burrough, P. A. (1979). Principles of Geographical Information Systems for Land Resources Assessment, Oxford, UK: Clarendon Press.
Burt, W. H. (1943). Territoriality and home range concepts as applied to mammals. Journal of Mammalogy, 24(3), 346–352. https://doi.org/10.2307/1374834
Burton, A. M. (2006). Recent breeding records and sightings of Neotropical forest eagles in southeastern México, [S.C. and Comisión Nacional para el Conocimiento y Uso de la Biodiversidad, La Paz]. *In*: R. Rodríguez-Estrella (Ed.), Current raptor studies in Mexico, pp. 59–70. BCS, Mexico: Centro de Investigaciones Biológicas del Noreste.
Burton, P. (1989). Birds of Prey. New York: Gallery Press Books.
Bush, S. E., Clayton, D. G. & Fernandez, R. (2008). The avifauna of Mt. Kitanglad, Bukidnon Province, Mindanao, Philippines. Fieldiana Zoology, 110.
Bushaw, J. D., Ringelman, K. M. & Rohwer, F. C. (2019). Applications of unmanned aerial vehicles to survey Mesocarnivores. Drones, 3(1), 28. https://doi.org/10.3390/drones3010028
Butchart, S. H. M., Brooks, T. M., Davies, C. W. N., Dharmaputra, G., Dutson, G. C. L., Lowen, J. C. & Sahu, A. (1996). The conservation status of forest birds on Flores and Sumbawa, Indonesia. Bird Conservation International, 6(4), 335–370. https://doi.org/10.1017/S0959270900001817
Butler, P. J. (1989). Telemetric recording of physiological data from free-living animals. *In*: P. J. Grubb & J. B. Whittaker (Eds.), Towards a more exact ecology, pp. 63–84. Blackwell.
Butler, R. W. & Campbell, R. W. (1987). The Birds of the Fraser River Delta: Populations, Ecology, and International Significance. Canadian Wildlife Service Occasional Paper 65. Ottawa, Canada: Environment Canada.
Buxton, S., Garman, E., Heim, K. E., Lyons-Darden, T., Schlekat, C. E., Taylor, M. D. & Oller, A. R. (2019). Concise review of nickel human health toxicology and ecotoxicology. Inorganics, 7(7), 89. https://doi.org/10.3390/inorganics7070089

Byrd, G. V. & Day, R. H. (1986). The avifauna of Buldir Island, Aleutian Islands, Alaska. Arctic, 39, 109–118.
Byrne, M. St. Claire (Ed.). (1981). The Lisle Letters. Chicago: University of Chicago Press.
Cadahía, L., López-López, P., Urios, V. & Negro, J. J. (2008). Estimating the onset of dispersal in endangered Bonelli's eagles Hieraaetus fasciatu tracked by satellite telemetry: A comparison between methods. Ibis, 150(2), 416–420. https://doi.org/10.1111/j.1474-919X.2007.00781.x
Cadahía, L., Urios, V. & Negro, J. J. (2005). Survival and movements of satellite-tracked Bonelli"s eagles Hieraaetus fasciatus during their fir winter. Ibis, 147(2), 415–419. https://doi.org/10.1111/j.1474-919x.2005.00405.x
Cade, T. J. (1982). The Falcons of the World. Ithaca, NY: Cornell University Press.
Cade, T. J. (2007). Exposure of California Condors to lead from spent ammunition. Journal of Wildlife Management, 71(7), 2125–2133. https: doi.org/10.2193/2007-084
Cagnacci, F., Boitani, L., Powell, R. A. & Boyce, M. S. (2010). Animal ecology meets GPS-based radiotelemetry: A perfect storm of opportuniti and challenges. Philosophical Transactions of the Royal Society of London Series B, 365(1550), 2157–2162. https://doi.org/10.109 rstb.2010.0107
Cain, A. J. & Galbraith, I. C. J. (1956). Field notes on the birds of the eastern Solomon Islands. Ibis, 98, 100–134, 262–295.
Calder, W. A. & King, J. R. (1974). Thermal and caloric relations of birds. In: Farner, D. S. & King, J. R. (Eds.), Avian Biology, 4, 259–413. Ne York, NY: Academic Press.
Calderon, J., Castroviejo, J. & Garcia, L. (1987). El Aguilá Imperial (Aquila adalberti) en Donana: Algunos aspectos de su reproduccion. Ferre M. Alytes, 5, 47–72.
Calenge, C. (2006). The package 'adehabitat' for the R software: A tool for the analysis of space and habitat use by animals. Ecological Modellin 197(3), 516–519.
Calixto-Albarrán, L. I. & Osorno, J. L. (2000). The diet of the magnificent frigate bird during chick rearing. Condor, 102(3), 569–576. https://do org/10.1093/condor/102.3.569
Camargo, S./Conexão Planeta. (2020). Projeto Harpia: Saving the Amazon's largest raptor for more than 20 years. Mongabay Series. Amazo conservation. Retrieved from https://news.mongabay.com/2020/05/projeto-harpia-saving-the-amazons-largest-raptor-for-more-than-2C years/
Cambridge English Dictionary. (2020). Eagle. Retrieved from http://cambridge.org/dictionary/english/eagle.
Cameron, E. S. (1908). Observations on the Golden eagle in Montana. Auk, 25(3), 251–268. https://doi.org/10.2307/4070518
Camiña, A. (2015). Reduced effects of wind farms in the Golden eagle (Aquila).
Campbell, B. (1983). The dictionary of birds in color. New York, NY: Exeter Books.
Campbell, B. & Lack, E. (2013). A dictionary of birds. New York, NY: Bloomsburg Publishing.
Campbell, B. & Lack, E. (2013). Hawk. London: Bloomsbury Publishing.
Campbell, C. J., Fuentes, A., MacKinnon, K. C., Panger, M. & Bearder, S. K. (2011). Primates in perspective. NY: Springer USA.
Campbell, L. M., Norstrom, R. J., Hobson, K. A., Muir, D. C. G., Backus, S. & Fisk, A. T. (2005). Mercury and other trace elements in pelagic Arctic marine food web (Northwater Polynya, Baffin Bay). Science of the Total Environment, 247–263. https://doi.org/10.1016/ scitotenv.2005.02.043
Campbell, M. (1998). Interactions between biogeography and rural livelihoods in the coastal savanna of Ghana. Unpublished Doctoral Thesi University of London, Imperial College.
Campbell, M. (2006). Urban parks shared spaces? The utility of alert distances as indicators of avian tolerance of human in Stirling, Scotlan Area, 38(00695. x), 301–311. https://doi.org/10.1111/j.1475-4762.2006. 00695.x.
Campbell, M. (2014). The impact of urbanization and agricultural development on vultures in El Salvador. Vulture News, 66, 16–28.
Campbell, M. (2014). Urban planning for cougar presence in North America. *In*: M. P. Amado (Ed.), Urban planning: Practices, challenges an benefits, pp. 19–32. New York, NY: Nova Science Publishers.
Campbell, M. (2015). Vultures: Their evolution, ecology and conservation. Boca Raton, FL: CRC Press.
Campbell, M. (2016). Avian reactions towards human approaches in different urban greenery structures in Nanaimo. Urban Forestry and Urba Greening, 19, 47–55. https://doi.org/10.1016/j.ufug.2016.06.022
Campbell, M. (2017). Biological Conservation for the 21st Century: A Conservation Biology of Large Wildlife. New York, NY: Nova Scienc Publishers.
Campbell, M. (2018). Geomatics and conservation biology. New York, NY: Nova Science Publishers.
Campbell, M. O. (2012). The relevance of age and gender for public attitudes to brown bears (Ursus arctos), black bears (Ursus americanus), an cougars (Puma concolor) in Kamloops, British Columbia. Society and Animals, 21(4), 341–359. https://doi.org/10.1163/15685306-1234126
Campbell, M. O. & Torres Alvarado, M. E. (2011). Public perceptions of jaguars Panthera onca, pumas concolor and coyotes Canis latrans in E Salvador. Area, 43(3), 250–256. https://doi.org/10.1111/j.1475-4762.2011.00996.x
Campbell, M. & Lancaster, B. (2010). Public attitudes towards black bears Ursus americanus and cougars Puma concolor on Vancouver Island Society and Animals, 10(1), 47–51. https://doi.org/10.1163/106311110X12586086158448.
Camphuysen, K., Verheij, J. & Cremer, J. (2011). Risk assessment of bird strike hazards: Gulls Laridae. Oosterend, The Netherlands: Center fo the Study of Religion Consultancy.
Canadian Council of Ministers of the Environment. (1999). Canadian tissue Residue Guidelines for the Protection of Wildlife Consumers o Aquatic Biota! DDT. Excerpt from: Canadian environmental quality guidelines, Winnipeg. Retrieved from http://www.ccme.ca/publications ceqg_rcqe.html
Canadian Council of Technicians and Technologists (2008). Survey and Geomatics Technology. Retrieved from https://www.cctt.ca benchmarks/2014/technologist/survey-and-geomatics-technology

Canal, D., Roques, S., Negro, J. J. & Sarasola, J. H. (2017). Population genetics of the endangered Crowned Solitary eagle (Buteogallus coronatus) in South America. Conservation Genetics, 18(1), 235–240. https://doi.org/10.1007/s10592-016-0878-6

Canberra Ornithologists Group. (2008). Nomination of a Vulnerable Species to the ACT Flora and Fauna Committee—Little Eagle. Canberra Bird Notes, 33, 84–95.

Canby, J. V. (2002). Falconry (Hawking) in Hittite Lands. Journal of Near Eastern Studies, 61(3), 161–201. Retrieved from http://www.jstor.org/stable/3128920. https://doi.org/10.1086/469022

Cannon, R. W., Knopf, F. L. & Pettinger, L. R. (1982). Use of Landsat data to evaluate lesser prairie chicken habitat in western Oklahoma. Journal of Wildlife Management, 46, 915–922.

Cant, M. A., Nichols, H. J., Thompson, F. J. & Vitikainen, E. I. K. (2016). Banded mongooses: Demography, life history, and social behavior. Cooperative Breeding in Vertebrates, 318–337.

Canuto, M. (2008). Observations of two hawk-eagle species in a humid lowland tropical forest reserve in central Panama. Journal of Raptor Research, 42(4), 287–292. https://doi.org/10.3356/JRR-07-40.1

Canuto, M. (2008b). First description of the nest of the Black-and-white hawk-eagle (Spizaetus melanoleucus) in the Brazilian Atlantic Rainforest, southeast Brazil. Ornitología Neotropical, 19, 607–610.

Capdevielle, A., Barbar, F., Encabo, M. & Rodriguez, R. (2007a). Different hunting methods of Harpyhaliaetus coronatus during rehabilitation. *In*: K. L. Bildstein, D. R. Barber & A. Zimmerman (Eds.), Neotropical raptors, pp. 145–148. Orwigsburg: Hawk Mountain Sanctuary.

Capdevielle, A., Barbar, F., Encabo, M. & Rodriguez, R. (2007b). Recovery and release of a Crowned Solitary eagle (Harpyhaliaetus coronatus) in Sierra de las Quijadas National Park, San Luis, Argentina. *In*: K. L. Bildstein, D. R. Barber & A. Zimmerman (Eds.), Neotropical raptors, pp. 140–144. Orwigsburg: Hawk Mountain Sanctuary.

Cardoso da Silva, J. M. (1995). Birds of the Cerrado region, South America. Steensrupia, 21, 69–92.

Carey, G. J., Young, L., Chalmers, M. L., Diskin, D. A., Kennerley, P. R., Leader, P. J., Leven, M., Lewthwaite, R. W., Melville, D. S. & Turnbull, M. (2001). The avifauna of Hong Kong. Hong Kong Bird Watching Society.

Carlos, C. J. & Girao, W. (2006). The history of the Ornate Hawk-Eagle, Spizaetus ornatus, in the Atlantic forest of northeast Brazil. Revista Brasileira de Ornitologia, 14(4), 405–409.

Carneiro, M., Colaco, B., Brandão, R., Ferreira, C., Santos, N., Soeiro, V. . . . Lavín, S. (2014). Biomonitoring of heavy metals (Cd, Hg, and Pb) and metalloid (As) with the Portuguese common buzzard (Buteo). Environmental Monitoring and Assessment, 186(11), 7011–7021. https://doi.org/10.1007/s10661-014-3906-3.

Carnie, S. K. (1954). Food habits of Golden eagles in the coastal ranges of California. Condor, 56(1), 3–12. https://doi.org/10.2307/1364882

Caro, T. (2005). Antipredator defenses in birds and mammals. University of Chicago Press.

Caro, T. (2007). Behavior and conservation: A bridge too far? Trends in Ecology and Evolution, 22(8), 394–400. https://doi.org/10.1016/j.tree.2007.06.003

Caro, T. M. (1987). Cheetah mothers' vigilance: Looking out for prey or for predators? Behavioral Ecology and Sociobiology, 20(5), 351–361. https://doi.org/10.1007/BF00300681

Carpenter, J. W., Pattee, O. H., Fritts, S. H., Rattner, B. A., Wiemeyer, S. N., Royle, J. A. & Smith, M. N. M. R. (2003). Experimental lead poisoning in Turkey vultures (Cathartes aura). Journal of Wildlife Diseases, 39(1), 96–104. https://doi.org/10.7589/0090-3558-39.1.96.

Carpenter, P. A. & Bishop, P. C. (2009, December). The seventh mass extinction: Human-caused events contribute to a fatal consequence. Futures, 41(10), 715–722. https://doi.org/10.1016/j.futures.2009.07.008

Carr, A. (1964). The land and wildlife of Africa. Chicago, IL: Time, Incorporated.

Carrete, M., Lambertucci, S. A., Speziale, K., Ceballos, O., Travaini, A., Delibes, M. . . . Donázar, J. A. (2010). Winners and losers in human-made habitats: Interspecific competition outcomes in two Neotropical vultures. Animal Conservation, 13(4), 390–398. https://doi.org/10.1111/j.1469-1795.2010.00352.x

Carrete, M., Sánchez-Zapata, J. A., Benítez, J. R., Lobón, M. . . . Donázar, J. A. (2009). Large scale risk-assessment of windfarms on population viability of a globally endangered long-lived raptor. Biological Conservation, 142(12), 2954–2961. https://doi.org/10.1016/j.biocon.2009.07.027

Carrete, M., Tella, J. L., Blanco, G. & Bertellotti, M. (2009). Effects of habitat degradation on the abundance, richness and diversity of raptors across Neotropical biomes. Biological Conservation, 142(10), 2002–2011. https://doi.org/10.1016/j.biocon.2009.02.012

Carr-Gomm, P. S. (1994). The druid animal oracle. New York, NY: Fireside.

Carriker, M. A., Jr. 1910. An annotated list of the birds of Costa Rica including Cocos Island. Annals of the Carnegie Museum 6(7), 314–915.

Carroll, S. E. (1996). Ancient & Medieval Falconry: Origins & Functions in Medieval England. The Crown & Helm. Retrieved from http://www.r3.org/life/articles/falconry.html

Carswell, M., Pomeroy, D., Reynolds, J. & Tushabe, H. (2005). The bird atlas of Uganda. Oxford: British Ornithologists' Club and British Ornithologists' Union, Oxford, UK.

Cartron, J. L. E. (Ed.). (2010). Raptors of New Mexico. UNM Press.

Carvalho Filho, E. P. M., Zorzin, G., Canuto, M., Carvalho, C.E.A. & Carvalho, C. G. D. (2008). Diurnal raptors of the Rio Doce State Park, Minas Gerais, Brazil. MG Biota, 1, 4–43.

Carvalho, C. E. A., Carvalho Filho, E. P. M. & Chaves Filho, A. (2002). Observations of Crowned Solitary eagle nesting behavior (Harpyhaliaetus coronatus—Falconiformes: Accipitridae) in southeast Brazil. *In*: Program of Conference on Neotropical Raptors and Symposium on the Harpy Eagle. Fondo Peregrino—Panamá, 30, 89–90.

Carvalho, C. E. A. & Marini, M. (2007). Distribution patterns of diurnal raptors in open and forested habitats in south-eastern Brazil and the effects of urbanization. Bird Conservation International, 17(4), 367–380. https://doi.org/10.1017/S0959270907000822

Carvalho, J. C. M. (1949). Observações zoológicas no rio das Mortes e no alto Xingu. Publ. Avulsos Museu Nacional. Rio de Janeiro, 5, 7–19.
Casado, E., Suárez-seoane, S., Lamelin, J. & Ferrer, M. (2008). The regulation of brood reduction in Booted eagles Hieraaetus pennatus through habitat heterogeneity. Ibis, 150(4), 788–798. https://doi.org/10.1111/j.1474-919X.2008.00862.x
Casebeer, R. L., Rognrud, M. J. & Brandborg, S. (1950). The Rocky Mountain goat in Montana. Montana Fish and Game Commission, Wildlife Restoration Division Bulletin 5.
Cassin, J. (1865). Notes on some new and little-known rapacious birds. Proceedings of the Academy of Natural Sciences of Philadelphia, 17, 2–5.
Castelló, J. R. (2016). Bovids of the world: Antelopes, gazelles, cattle, goats, sheep, and relatives. Princeton, NJ: Princeton University Press.
Castelló, J. R. (2018). Canids of the world: Wolves, wild dogs, foxes, jackals, coyotes, and their relatives. Princeton, NJ: Princeton University Press.
Castro, I., Aboal, J. R., Fernandez, J. A. & Carballeira, A. (2011). Use of raptors for biomonitoring of heavy metals: Gender, age and tissue selection. Bulletin of Environmental Contamination and Toxicology, 86(3), 347–351. Retrieved from http://www.doi. https://doi.org/10.1007/s00128-011-0220-4
Catry, P., Lecoq, M. & Strange, I. J. (2008). Population growth and density, diet and breeding success of striated caracaras Phalcoboenus australis on New Island, Falkland Islands. Polar Biology, 31(10), 1167–1174. https://doi.org/10.1007/s00300-008-0454-8
Cavallini, P. (1992). Herpestes pulverulentas. Mammalian Species Archive, 409, 1–4.
Cawthorn Lang, K. (2001). Primate factsheets: Vervet (Chlorocebus) Taxonomy (On-line).
Ceballos, G. & Márquez-Valdelamar, L. (2000). Birds of México in danger of extinction. Mexico City, Mexico: CONABIO.
Ceballos, G., Ehrlich, P. R., Barnosky, A. D., García, A., Pringle, R. M. & Palmer, T. M. (2015). Accelerated modern human–induced species losses: Entering the sixth mass extinction. Science Advances, 1(5), e1400253. https://doi.org/10.1126/sciadv.1400253
Cederberg, H., Sandström, C., Rannug, U. & Helander, B. (2003). Genetic variation based on DNA fingerprinting in two White-tailed sea eagle populations in Sweden. *In*: B. Helander et al. (Eds.), Sea eagle 2000, 219–221. Swedish Society for Nature Conservation/SNF & Åtta. 45 Tryckeri AB, Stockholm.
Cerqueira, P. V., Gonçalves, G. S. R., de Araújo Sousa, S., Paz, R. L., Landim, A. S. & Santos, M. P. D. (2015). First record of the Ornate Hawk-eagle (Spizaetus ornatus) from the Brazilian Caatinga. The Wilson Journal of Ornithology, 127(1), 153–156. https://doi.org/10.1676/14-076.1
Chace, J. F., Walsh, J. J., Cruz, A., Prather, J. W. & Swanson, H. M. (2003). Spatial and temporal activity patterns of the brood parasitic brown-headed cowbird at an urban/wildland interface. Landscape and Urban Planning, 64(3), 179–190, https://doi.org/10.1016/S0169-2046(02)00220-7
Chakdar, B., Singha, H., Choudhury, M. R. & Deb, P. (2017). A nest of the Indian Spotted eagle Aquila hastata in Rajiv Gandhi Orang National Park, Assam. Indian Birds, 13(2), 42–43.
Chandler, S. K., Fraser, J. D., Buehler, D. A. & Janis, K. D. S. J. K. (1995). Perch trees and shoreline development as predictors of Bald eagle distribution on Chesapeake Bay. The Journal of Wildlife Management, 59(2), 325. https://doi.org/10.2307/3808946
Chang, L. W., Magos, L. & Suzuki, T. (Eds.). (1996). Toxicology of Metals. Boca Raton, FL: CRC Press.
Chapin, J. P. & Lang, H. (1953). The birds of the Belgian Congo. Part 3. by James P. Chapin. Bulletin of the AMNH, v. 75a.
Chapman, F. M. (1917). The distribution of bird-life in Colombia. Bulletin of the American Museum of Natural History, 36, 1–784.
Chapman, J. A. & Flux, J. E. (Eds.). (1990). Rabbits, Hares and Pikas: Status Survey and Conservation Action Plan. International Union for Conservation of Nature and Natural Resources.
Chariton, J. D. (2011). The mesopotamian origins of the Hittite double-headed eagle. UW-L Journal of Undergraduate Research, XIV, 1–13.
Chaudhary, A., Subedi, T. R., Giri, J. B., Baral, H. S., Bidari, B., Subedi, H. . . . Cuthbert, R. J. (2012). Population trends of critically endangered Gyps vultures in the lowlands of Nepal. Bird Conservation International, 22(3), 270–278. https://doi.org/10.1017/S0959270911000426
Chaudhry, M. J. I., Ogada, D. L., Malik, R. N., Virani, M. Z. & Giovanni, M. D. (2012). First evidence that populations of the critically endangered Long-billed vulture Gyps indicus in Pakistan have increased following the ban of the toxic veterinary drug diclofenac in South Asia. Bird Conservation International, 22(4), 389–397. https://doi.org/10.1017/S0959270912000445
Chávez-Zichinelli, C. A. MacGregor-Fors, I. M., Quesada, J., Rohana, P. T., Romano. M. C., Valdéz, R. & Schondube, J. E. (2013). How stressed are birds in an urbanizing landscape? Relationships between the physiology of birds and three levels of habitat alteration. Condor, 115(1), 84–92. https://doi.org/10.1525/cond.2013.110201.
Chavko, J. (2009). White-tailed eagle (Haliaeetus albicilla). Birds of Prey and Owls – Journal of Raptor Protection of Slovakia 5, 9.
Chebez, J. C. (1993). List of Argentine vertebrates in danger of extinction. Nuestras Aves, 10, 1.
Chebez, J. C., Seipke, S. & Rodríguez, L. (2008). Black-and-chestnut eagle. *In*: J. C. Chebez (Ed.). Los que se van: Fauna argentina amenazada, pp. 216–219. Albatros, Buenos Aires, Argentina.
Cheke, R. A. & Walsh, J. F. (1996). The birds of Togo: An annotated check-list 14. London: British Ornithologists' Union.
Cheney, D. L. & Seyfarth, R. M. (1981). Selective forces affecting the predator alarm calls of vervet monkeys. Behaviour, 76(1–2), 25–60. https://doi.org/10.1163/156853981X00022
Cherney, M. (2017). Bold eagles: Angry birds are ripping $80,000 drones out of the sky. Retrieved from https://www.wsj.com/articles/bold-eagles-angry-birds-are-ripping-80-000-drones-out-of-the-sky-1506701429
Chevallier, C., Hernández-Matías, A., Real, J., Vincent-Martin, N., Ravayrol, A. & Besnard, A. (2015). Retrofitting of power lines effectively reduces mortality by electrocution in large birds: An example with the endangered Bonelli's eagle. Journal of Applied Ecology, 52(6), 1465–1473. https://doi.org/10.1111/1365-2664.12476
Chevallier, D., Handrich, Y., Georges, J.-Y., Baillon, F., Brossault, P., Aurouet, A. . . . Massemin, S. (2010). Influence of weather conditions on the flight of migrating Black storks. Proceedings of the Royal Society B: Biological Sciences, 277(1695), 2755–2764. https://doi.org/10.1098/rspb.2010.0422.

Chevallier, D., Jiguet, F., Nore, T., Baillon, F. & Cavallin, P. (2010). Satellite tracking of a Booted eagle (Aquila pennata) during migration. Ring. Migr, 25, 62–64.

Cheylan, G., Ravayrol, A., Cugnasse, J.-M., Billet, J.-M. & Joulot, C. (1996). Dispersion des Aigles de Bonelli Hieraaetus fasciatus juvéniles bagués en France. Alauda, 64, 413–419.

Chia-hong Liu, C.-H. (2010). Diurnal activity pattern of Crested Serpent eagles Spilornis cheela hoya in Kenting, southern Taiwan (in Chinese) [Masters Thesis]. Archived from the original on 2012-04-02.

Chiaravalloti, R. M., Tomas, W. M., Tizianel, F. T. A. & Camilo, A. R. (2009). Aves, Accipitridae, Harpyhaliaetus coronatus: A documented record in the Pantanal wetland. Check List, 5(1), 89–91. https://doi.org/10.15560/5.1.89

Chiari, C. G. & Seeland, K. (2004). Are urban green spaces optimally distributed to act as places for social integration? Results of a geographical information system GIS approach for urban forestry research. Forest Policy and Economics, 6, 3–13. https://doi.org/10.1016/S1389-9341(02)00067-9

Chittenden, H. (2016). Roberts bird guide: Illustrating nearly 1,000 species in Southern Africa. Davies, Greg (Ornithologist), Weiersbye, Ingrid, John Voelcker Bird Book Fund. Percy FitzPatrick Institute of African Ornithology (Cape Town, South Africa) (Second ed.). Cape Town.

Chiu, S. C. & Lim, K. C. (2003). The first breeding record of the Black Eagle Ictinaetus malayensis in Malaysia. Suara Enggang, 11, 14–16.

Choi, C.-Y. & Park, J. G. (2012). Birds of prey in Korea. Korean Wild Birds Society & Shinan County, Seoul, South Korea.

Choijiljav, B. (2015). Golden eagle festival in Mongolia. Caters News. Retrieved from http://www.catersnews.com/stories/animals/golden-eagles-spectacularly-spread-their-wings-tocatch-their-prey/

Choudhury, A. (2012). Sudden decline of Grey-headed fish-eagles in Assam. Misnet, 13(4), 9–10.

Chow, J. (2017). Golden eagle Aquila chrysaetos. Retrieved from http://northwestwildlife.com/wp-content/uploads/2017/01/golden-eagle-new.pdf. Northwest Wildlife Preservation Society.

Christensen, J. (1979). Den grønlandske Havørns Haliaeetus albicilla groenlandicus Brehm: Ynglebiotop, redeplacering og rede the Greenland White-tailed eagle: Breeding habitat, nest placement, and nest. Dansk Ornitologisk Forenings Tidsskrift, 73, 131–156.

Christensen, K. Y., Raymond, M. & Meiman, J. (2019). Perfluoroalkyl substances and metabolic syndrome. International Journal of Hygiene and Environmental Health, 222(1), 147–153. https://doi.org/10.1016/j.ijheh.2018.08.014

Christensen, S. & Sorensen, U. G. (1989). A review of the migration and wintering of Aquila pomarina and Aquila nipalensis orientalis. *In*: Meyburg, B.-U. & R. D. Chancellor (Eds.), Raptors in the modern world, pp. 139–150. Berlin: World Working Group Birds of Prey.

Christiansen, P. & Harris, J. M. (2005). Body Size of Smilodon (Mammalia: Felidae). Journal of Morphology, 266(3), 369–384.

Christiansen, P., Steen, K. A., Jørgensen, R. N. & Karstoft, H. (2014). Automated detection and recognition of wildlife using thermal cameras. Sensors, 14(8), 13778–13793. https://doi.org/10.3390/s140813778

Christidis, L. & Boles, W. E. (2008). Systematics and Taxonomy of Australian Birds. Collingwood, Vic., Australia: CSIRO Press.

Church, M. E., Gwiazda, R., Risebrough, R. W., Sorenson, K., Chamberlain, C. P., Farry, S. . . . Smith, D. R. (2006). Ammunition is the principal source of lead accumulated by California condors re-introduced to the wild. Environmental Science and Technology, 40(19), 6143–6150. https://doi.org/10.1021/es060765s

Ciaccio, A. & Priolo, A. (1997). Avifauna della foce del Si-meto, del lago di Lentini e delle zone umide adiacenti (Sicilia, Italia). Il Naturalista Siciliano, 21, 309–413.

Cima, A., Farías, G., Zapotecos, U., de Juárez, C. & De Méndez, C. 2018. Enhancing our knowledge on the Ornate Hawk-Eagle (Spizaetus ornatus) through community-based monitoring records from Tropical Mexico. Revista Brasileira de Ornitologia, 26(3), 196-201.

Cincinnati Zoo & Botanical Garden. (2020). Steller's Sea Eagle (Haliaeetus pelagicus). Retrieved from http://cincinnatizoo.org/animals/stellers-Sea eagle/

Claessens, O. (2002). Diversity and guild structure of the Petit Saut bird community. Revue d'Ecologie (Terre et la Vie) Suppl., 8, 77–102.

Clancey, P. (1964). Birds of Natal and Zululand. London: Oliver & Boyd Ltd.

Clancey, P. A. (1985). The rare birds of southern Africa. Johannesburg, South Africa: Winchester Press.

Clancey, P. A. (1996). The birds of southern Mozambique. Westville, Kwazulu-Natal, South Africa African Bird Book Publishing.

Claramunt, S., Derryberry, E. P., Remsen, J. V. & Brumfield, R. T. (2012). High dispersal ability inhibits speciation in a continental radiation of passerine birds. Proceedings of the Royal Society of London B: Biological Sciences, 279(1733), 1567–1574. https://doi.org/10.1098/rspb.2011.1922

Clark, A. & Scheuhammer, A. (2003). Lead poisoning in upland-foraging birds of prey in Canada. Ecotoxicology, 12, 23–30. https://doi.org/10.1023/A:1022576510445

Clark, W. S. (1982). Turtles as a food source of nesting Bald eagles in the Chesapeake Bay region. Journal of Field Ornithology, 53. 49–51.

Clark, W. S. (1990). Spotted eagle with rufous nape patch. British Birds, 83, 397–399.

Clark, W. S. (1992). The taxonomy of Steppe and Tawny eagles, with criteria for separation of museum specimens and live eagles. Bulletin of the British Ornithoilogical Club, 12, 156–157.

Clark, W. S. (1994). Changeable hawk-eagle. *In*: Del Hoyo, J., et al. (Eds.), Handbook of birds of the world, 2. New world vultures to guineafowl. Barcelona: Lynx Edicions.

Clark, W. S. (1994). Crested serpent-eagle. *In*: del Hoyo, J., Elliott, A. & Sargatal, J. (Eds.), Handbook of birds of the world, 2. New World vultures to guineafowl, p. 133. Barcelona: Lynx Edicions.

Clark, W. S. (1994). Grey-headed fishing-eagle. *In*: Del Hoyo, J., et al. (Eds.), Handbook of birds of the world, 2. New World vultures to guineafowl, p. 123. Barcelona: Lynx Edicions.

Clark, W. S. (1994). Philippine eagle (Pithecophaga jefferyi). *In*: Del Hoyo, J., Elliott, A. & Sargatal, J. (Eds.), Handbook of the birds of the world, 2. New world vultures to guineafowl, p. 192. Barcelona: Lynx Edicions.

Clark, W. S. (1994). Philippine hawk-eagle. *In*: Del Hoyo, J., et al. (Eds.), Handbook of birds of the world, 2. New world vultures to guineafow p. 204. Barcelona: Lynx Edicions.

Clark, W. S. (1994). Species accounts for Asiatic raptors. *In*: Del Hoyo, J., Elliot, A. & Sargatal, J. (Eds.), Handbook of the birds of the world, 2 New World vultures to guineafowl, Lynx Ed. Barcelona.

Clark, W. S. (1994). Sulawesi hawk-eagle. *In*: Del Hoyo, J., et al. (Eds.), Handbook of birds of the world, 2. New World vultures to guineafow p. 204. Barcelona: Lynx Edicions.

Clark, W. S. (1994). Wallace's hawk-eagle. Handbook of birds of the world. *In*: del Hoyo, J., Elliott, A. & Sargatal, J. (Eds.), Handbook of bird of the world, 2. New World vultures to guineafowl, p. 204. Barcelona: Lynx Edicions.

Clark, W. S. (1999). Ayres' hawk-eagle Hieraaetus (dubius) ayresii sightings in Cameroon. African Bird Club Bulletin, 6, 115–116.

Clark, W. S. (1999). Plumage differences and taxonomic status of three similar Circaetus snake-eagles. Bulletin of the British Ornithologists' Club 119, 56–59.

Clark, W. S. (2007). Solitary eagle: Possible in Texas? Texas Birds Annual, 3, 16–18.

Clark, W. S. (2012). The eagle genus Hieraaetus is distinct from Aquila, with comments on the name Ayres' eagle. Bulletin of the Britis Ornithologists' Club, 132(4), 295–298. Retrieved from http://www.globalraptors.org/grin/researchers/uploads/155/hieraaetus_is_a_genus pdf

Clark, W. S., Jones, C. D. & Benesh, S. N. J. (2006). Field identification of the solitary eagle. Birding, 38, 66–74.

Clark, W. S. & Kirwan, G. M. (2020). Wallace's hawk-eagle (Nisaetus nanus), version 1.0. *In*: Del Hoyo, J., et al. (Eds.), Birds of the world. Ithaca NY: Cornell Laboratory of Ornithology. https://doi.org/10.2173/bow.walhae1.01.

Clark, W. S. & Schmitt, N. J. (1993). Field identification of the Rufous-bellied eagle Hieraaetus kienerii. Forktail, 7, 7–9.

Clark, W. S. & Schmitt, N. J. (1999). A Field Guide to the Raptors of Europe, the Middle East, and North Africa. Oxford: Oxford University Press

Clark, W. S., Fisher, D., Finch, B., Bruijn, B. D. & Shani, I. (2005). Status of Beaudouin's and Short-toed Eagle in Kenya. Bulletin African Bir Conservation, 12, 150–151.

Clark, W. S. & Schmitt, N. J. (2006). Melanistic specimen of Black-chested buzzard-eagle (Geranoaetus melanoleucus) with comments on th species' taxonomic position. Journal of Raptor Research, 40(1), 86–88. https://doi.org/10.3356/0892-1016(2006)40[86:MSOBBG]2.0.CO;

Clark, W. S., Jones, H. L., Benesh, C. D. & Schmitt, N. J. (2007). Field identification of the Montane Solitary eagle (Harpyhaliaetus solitarius). In Bildstein, K. L., Barber, D. R. & Zimmerman, A. (Eds.), Neotropical raptors, p. 281. Orwigsburg: Hawk Mountain Sanctuary.

Clark, W. S., Sharpe, C. J., Boesman, P. & Marks, J. S. (2015). Philippine eagle (Pithecophaga jefferyi). *In*: Del Hoyo, J., Elliott, A., Sargatal, J. Christie, D. A. & De Juana, E. (Eds.), Handbook of the birds of the world Alive. Barcelona: Lynx Edicions.

Clark, W. S. & Schmitt, N. J. (2017). Raptors of Mexico and Central America. Princeton, NJ: Princeton University Press.

Clark, W. S., Kirwan, G. M. & Christie, D. A. (2018). Lesser fish-eagle (Icthyophaga humilis). *In*: Del Hoyo, J., Elliott, A., Sargatal, J., Christie D. A. & De Juana, E. (Eds.), Handbook of the birds of the world alive. Retrieved from https://www.hbw.com/node/52989, Barcelona: Lyn Edicions.

Clark, W. S., Boesman, P., Kirwan, G. M. & Marks, J. S. (2019). Rufous-bellied eagle (Lophotriorchis kienerii). *In*: Del Hoyo, J., et al. (Eds.) Handbook of the birds of the world alive. Barcelona: Lynx Edicions, Barcelona.

Clark, W. S., Kirwan, G. M. & Christie, D. A. (2020). Nicobar serpent-eagle (Spilornis klossi), version 1.0. *In*: Del Hoyo, J., Elliott, A., Sargata J., Christie, D.A. & De Juana, E. (Eds.), Birds of the World. Ithaca, NY: Cornell Laboratory of Ornithology, USA. https://doi.org/10.2173 bow.nicsee1.01.

Clark, W. S. & Kirwan, G. M. (2020). Sulawesi Serpent-eagle (Spilornis rufipectus), version 1.0. *In*: Del Hoyo, J., et al. (Eds.), Handbook o birds of the world (Vol. 2: New World vultures to guineafowl). Ithaca, NY: Cornell Laboratory of Ornithology, Ithaca, NY, USA. https://do org/10.2173/bow.suseag1.01

Clark, W. S. & Marks, J. S. (2020). Philippine Serpent-eagle (Spilornis holospilus), version 1.0. *In*: Del Hoyo, J., et al. (Eds.), Birds of the World Ithaca, NY: Cornell Laboratory of Ornithology, Ithaca, NY, USA. https://doi.org/10.2173/bow.phseag1.01

Clark, W. S. and G. M. Kirwan (2020). Sulawesi Hawk-eagle (Nisaetus lanceolatus), version 1.0. In: del Hoyo, J., Elliott, A., Sargatal, J., Christie D. A. & de Juana, E. (Eds.), Birds of the world. Ithaca, NY: Cornell Lab of Ornithology. https://doi.org/10.2173/bow.sulhae1.01

Clark, W. S., Boesman, P. F. D., Kirwan, G. M. & Marks, J. S. (2020a). Rufous-bellied eagle (Lophotriorchis kienerii), version 1.0. *In*: del Hoyo J., Elliott, A., Sargatal, J., Christie, D. A. and de Juana, E. (Eds.), Birds of the world. Ithaca, NY: Cornell Lab of Ornithology. https://do org/10.2173/bow.rubeag2.01

Clark, W. S., Boesman, P. F. D., Kirwan, G. M. & Marks, J. S. (2020b). Legge's Hawk-eagle (Nisaetus kelaarti), version 1.0. In: Billerman, S M., Keeney, B. K., Rodewald, P. G. & Schulenberg, T. S. (Eds.), Birds of the world. Ithaca, NY: Cornell Lab of Ornithology. https://do org/10.2173/bow.mouhae2

Clark, W. S., Boesman, P. F. D., Marks, J. S. & Sharpe, C. J. (2020c). Javan Hawk-eagle (Nisaetus bartelsi), version 1.0. In: del Hoyo, J., Elliot A., Sargatal, J., Christie, D. A. and de Juana, E. (Eds.), Birds of the world. Ithaca, NY: Cornell Lab of Ornithology. https://doi.org/10.2173 bow.javhae1.01

Clark, W. S, Boesman, P. F. D. & Marks, J. S. (2020d). Blyth's Hawk-eagle (Nisaetus alboniger). In: del Hoyo, J., Elliott, A., Sargatal, J., Christie D. A. and de Juana, E. (Eds.), Birds of the world. Ithaca, NY: Cornell Lab of Ornithology. https://doi.org/10.2173/bow.blyhae1.01

Clark, W. S., Marks, J. S. & Kirwan, G. M. (2020e). Crested Serpent-eagle (Spilornis cheela), version 1.0. In: del Hoyo, J., Elliott, A., Sargatal, J Christie, D. A. and de Juana, E. (Eds.), Birds of the world. Ithaca, NY: Cornell Lab of Ornithology. https://doi.org/10.2173/bow.crseag1.01

Clark, W. S., Kirwan, G. M. & Christie, D. A. (2020f). Nicobar Serpent-eagle (Spilornis klossi), version 1.0. *In*: del Hoyo, J., Elliott, A., Sargata J., Christie, D. A. and de Juana, E. (Eds.), Birds of the World. Ithaca, NY: Cornell Lab of Ornithology. https://doi.org/10.2173/bow.nicsee1.0

Clark, W. S., Christie, D. A., Marks, J. S. & Sharpe, C. J. (2020g). Andaman Serpent-eagle (Spilornis elgini), version 1.0. *In*: del Hoyo, J., Elliott, A., Sargatal, J., Christie, D. A. and de Juana, E. (Eds.), Birds of the world. Cornell Lab of Ornithology, Ithaca, NY, USA. https://doi.org/10.2173/bow.anseag1.01

Clark, W. S., Kirwan, G. M. & Christie. D. A. (2020h). Lesser Fish-eagle (Haliaeetus humilis), version 1.0. *In*: del Hoyo, J., Elliott, A., Sargatal, J., Christie, D. A. and de Juana, E. (Eds.), Birds of the world. Cornell Lab of Ornithology, Ithaca, NY, USA. https://doi.org/10.2173/bow.lefeag1.01

Clarke, T. (2006). Field guide to the birds of the Atlantic Islands: Canary Islands, Madeira, Azores, Cape Verde. London: Christopher Helm Publishers.

Clarkson, T. W. & Magos, L. (2006). The toxicology of mercury and its chemical compounds. Critical Reviews in Toxicology, 36(8), 609–662. https://doi.org/10.1080/10408440600845619

Clarkson, T. W., Magos, L. & Myers, G. J. (2003). Human exposure to mercury: The three modern dilemmas. Journal of Trace Elements in Experimental Medicine, 16(4), 321–343. https://doi.org/10.1002/jtra.10050

Clawson, M. M. (1948). Golden eagle attacks young deer. Yosemite Nature Notes, 27, 107–109.

Clements Checklist Team. (2009). Updates and Corrections, September (2009). Cornell Laboratory of Ornithology, Little Eagle Hieraaetus morphnoides. *In*: Gjershaug, J. O., Lerner, H. R. L. & Diserud, O. H. (Eds.) (2009). Taxonomy and distribution of the Pygmy Eagle Aquila Hieraaetus weiskei (Accipitriformes: Accipitridae). Zootaxa number 2326: 24–38.

Clemens, J. (1907). Notes from the Philippines. Condor, 9(3), 92–93. https://doi.org/10.2307/1361140

Clements, J. F. (2007). The Clements checklist of birds of the world (6th ed.), pp. 47–48. Ithaca, NY. Cornell University Press.

Clements, J. F. & Shany, N. (2001). A field guide to the birds of Peru. Clements, Temecula, CA: Ibis Publishing Company. Retrieved from http://www.birds.cornell.edu/clementschecklist/

Clements, T., Gilbert, M, Rainey, H. J., Cuthbert, R., Eames, J. C., Bunnat, P., Teak, S., Chansocheat, S. & Setha, T. (2012). Vultures in Cambodia: Population, threats and conservation. Bird Conservation International, 1–18.

Clements, T., Gilbert, M., Rainey, H. J., Cuthbert, R., Eames, J. C., Bunnat, P., . . . Setha, T. 2013. Vultures in Cambodia: Population, threats and conservation. Bird Conservation International, 7–24. https://doi.org/10.1017/S0959270912000093.

Clergeau, P., Savard, J. L., Mennechez, G. & Falardeau, G. (1998). Bird abundance and diversity along an urban–rural gradient: A comparative study between two cities on different continents. Condor, 100(3), 413–425. https://doi.org/10.2307/1369707

Clinton-Eitniear, J. (1986). Status of the large forest eagles of Belize. Birds Prey Bull., 3, 107–110.

Clinton-Eitniear, J. (1991). The solitary eagle Harpyhaliaetus solitarius: A new threatened species. Birds of Prey Bulletin, 4, 81-85.

Clinton-Eitniear, J., Gartside, M. R. & Kainer, M. A. (1991). Ornate hawk-eagle feeding on green iguana. Journal of Raptor Research, 25(1), 19–20.

Clobert, J., Danchin, E., Dhont, A. A. & Nichols, J. (Eds.). (2001). Dispersal—Causes, consequences and mechanisms of dispersal at the individual, population and community level. Oxford: Oxford University Press.

Cloete, D. (2013). Investigating the decline of the Martial eagle (Polemaetus bellicosus) in South Africa [Thesis]. University of Cape Town.

Cloudsley-Thompson, J. L. (1990). Etosha and the Kaokoveld problems of conservation in Namibia. Environmental Conservation, 17(4), 351–354. https://doi.org/10.1017/S037689290003280X

Clouet, M., Barrau, C. & Goar, J. L. (1999). The Golden eagle (Aquila chrysaetos) in the Bale Mountains, Ethiopia. Journal of Raptor Research, 33, 102–109.

Clunie, P. (2003). Flora and fauna guarantee action statement: White-bellied sea eagle (Haliaeetus leucogaster). East Melbourne, Victoria. Retrieved from https://www.environment.vic.gov.au/__data/assets/pdf_file/0024/32892/White-bellied_Sea eagle_Haliaeetus_leucogaster.pdf. Department of Sustainability and Environment, State Government of Victoria.

Coates, B. J. (1977). Birds in Papua New Guinea. Port Moresby, Papua New Guinea Robert Brown and Associates.

Coates, B. J. (1985). The birds of Papua New Guinea, including the Bismarck Archipelago and Bougainville. Vol. I. Non-passerines. Alderley, Queensland, Australia: Dove Publications.

Coates, B. J. & Bishop, K. D. (1997). A guide to the birds of Wallacea, Sulawesi, the Moluccas and Lesser Sunda Islands, Indonesia. Alderley, UK: Dove Publications.

Coates, B. J. & Peckover, W. S. (2001). Birds of New Guinea and the Bismarck Archipelago: A photographic guide. Alderley, Queensland, Australia Dove Publications.

Coelho, C. M., de Melo, L. F. B., Sábato, M. A. L., Rizel, D. N. & Young, R. J. (2007). A note on the use of GPS collars to monitor wild maned wolves Chrysocyon brachyurus (Illiger 1815) (Mammalia, Canidae). Applied Animal Behaviour Science, 105(1), 259–264. https://doi.org/10.1016/j.applanim.2006.04.024

Colarusso, J. (2016). Nart Sagas: Ancient Myths and Legends of the Circassians and Abkhazians. Princeton, NJ: Princeton University Press.

Colborn, T. (1991). Epidemiology of great lakes Bald eagles. Journal of Toxicology and Environmental Health, 33(4), 395–453. https://doi.org/10.1080/15287399109531537

Colebrook-Robjent, J. F. R. & Aspinwall, D. R. (1986). The nest and eggs of Circaetus cinerascens and other snake-eagles in Zambia 106, 5–9.

Coleman, J. & Fraser, J. (1986). Predation on Black and Turkey vultures. Wilson Bulletin, 98(4), 600–601.

Collaerts, P., Collaerts, E., Verbelen, P. & Trainor, C. R. (2013). Discovery of the critically endangered flores hawk eagle Nisaetus flores on Alor Island, Indonesia. Birding ASIA, 19, 48–51.

Collar, N. J. (1986). Threatened raptors of the Americas: Work in progress from the ICBP/IUCN Red Data Book. Birds of Prey Bulletin, 3,13-20.

Collar, N. J. & Stuart, S. N. (1985). Threatened birds of Africa and related islands: The ICBP/IUCN Red Data Book. Part I (3rd edn.). Cambridge: ICBP.

Collar, N. J. & Andrew, P. (1988). Birds to watch: The ICBP world checklist of threatened birds. Cambridge, UK: International Council for Bird Preservation. Technical Publications, 8.

Collar, N. J., Gonzaga, L. P., Krabbe, N., Madronõ, A., Naraujo, L. G., Parker, T. A. & Wege, D. C. (1992). Threatened Birds of the Americas. Washington, DC: Smithsonian Institution Press.

Collar, N. J., Crosby, M. J. & Stattersfield, J. (1994). Birds to watch 2: The world list of threatened birds. BirdLife Conservation Series No 4. Cambridge: BirdLife International.

Collar, N. J., Dee, T. J. & Goriup, P. D. (1987). La conservation de la nature à Madagascar: Perspective du CIPO. Occasional Papers IUCN/. State (Corporation) Commission, 2 (pp. 97–108).

Collar, N. J., Gonzaga, L., Krabbe, N., Madroño Nieto, A., Naranjo, L., Parker, T. & Wege, D. (1992). Threatened birds of the America: The ICBP/IUCN Red Data Book. Cambridge: International Council for Bird Preservation.

Collar, N. J., Mallari, N. A. D. & Tabaranza, B. R. J. (1999). Threatened birds of the Philippines: The Haribon Foundation/Birdlife International Red Data Book. Makati City, Philippines: Bookmark.

Collette, B., Fernandes, P. & Heessen, H. (2015). Anarhichas lupus. IUCN Red List of Threatened Species, 2015: e.T18155993A44739312.

Collier, B. A., Ditchkoff, S. S., Raglin, J. B. & Smith, J. M. (2007). Detection probability and sources of variation in white-tailed deer spotlight surveys. Journal of Wildlife Management, 71(1), 277–281. Retrieved from http://www.jstor.org/stable/4495174http: //www. Retrieved from http://jstor.org/stable/4495174. https://doi.org/10.2193/2005-728

Collins, B. J. (2010). Animal mastery in Hittite art and texts. *In*: Counts, D. B. & Arnold, B. (Eds.), The master of animals in old world iconography, pp. 59–74. Main Series, no. 24. Budapest: Archaeolingua Foundation.

Collins, N. M., Sayer, J. A. & Whitmore, T. C. (1991). The Conservation Atlas of Tropical Forests – Asia and the pacific. London: Macmillan Press.

Collins, P. W., Guthrie, D. A., Rick, T. C. & Erlandson, J. (2005). Analysis of prey remains excavated from an historic Bald Eagle nest site on San Miguel Island, California. In: Garcelon, D. K. & Schwemm, C. A. (Eds.), Proceedings of the Sixth California Islands Symposium, pp. 103–120. Ventura, California, December 1–3, 2003, Arcata, CA: Institute for Wildlife Studies.

Collins, P. W. & Latta, B. C. (2009). Food habits of nesting Golden eagles (Aquila chrysaetos) on Santa Cruz and Santa Rosa islands, California. *In*: Damiani, C. C. & Garcelon, D. K. (Eds.), Proceedings of the 7th California Islands Symposium, pp. 255–268. Arcata, CA.

Collins, S. & Hines, M. (1999). Sea eagle predation of Wedge-tailed Shearwaters (NSW & QLD). Boobook, 19, 15.

Collinson, M. (2006). Splitting headaches? Recent taxonomic changes affecting the British and Western Palaearctic lists. British Birds, 99, 306–323.

Collopy, M. W. (1983). A comparison of direct observations and collections of prey remains in determining the diet of Golden eagles. Journal of Wildlife Management, 47(2), 360–368. https://doi.org/10.2307/3808508

Collopy, M. W. (1983). Foraging behavior and success of Golden eagles. Auk, 100(3), 747–749. https://doi.org/10.1093/auk/100.3.747

Collopy, M. W. (1984). Parental care and feeding ecology of Golden eagle nestling. Auk, 101(4), 753–760. https://doi.org/10.2307/4086902

Collopy, M. W. & Edwards, T. C. (1989). Territory size, activity budget, and role of undulating flight in nesting Golden eagles. Journal of Field Ornithology, 60, 43–51.

Colorado Springs, C. ETD collection for Texas State University. Retrieved from https://digital.library.txstate.edu/handle/10877/3004

Coltrain, J. B., Harris, J. M., Cerling, T. E., Ehleringer, J. R., Dearing, M.-D., Ward, J. & Allen, J. (2004). Rancho La Brea stable isotope biogeochemistry and its implications for the palaeoecology of late Pleistocene, coastal southern California. Palaeogeography, Palaeoclimatology, Palaeoecology, 205(3–4), 199–219.

Comte & Suzanne, S. (1978). Everyday life in the Middle Ages. *In*: David, D. & Macrae (Trans. by), Geneva editions. S.A.: Minerva.

Condon, H. T. & Amadon, D. (1954). Taxonomic notes on Australian hawks. Records of the South Australian Museum, 11, 189–246.

Cone, C. D. (1962). Thermal soaring of birds. American Science, 50, 180–209.

Contreras, J. R., Berry, L. M., Contreras, A. O., Bertonatti, C. C. & Utges, E. E. (1990). Cuadernos Tecnicos Felix de Azara. Ornithogeographical Atlas of Chaco Province, Republic of Argentina. Non-Passeriformes, 1, 1–165.

Cooke, S. J., Hinch, S. G., Wikelski, M., Andrews, R. D., Kuchel, L. J., Wolcott, T. G. et al. (2004). Biotelemetry: A mechanistic approach to ecology. Trends in Ecology and Evolution, 19(6), 334–343. https://doi.org/10.1016/j.tree.2004.04.003doi

Cooksey, B. F. 1962. A winter population of the Bald eagle (Haliaeetus leucocephalus) in northeastern Oklahoma [MSc Thesis]. KS State College, Pittsburg, KS.

Cooper, A. B. (1969). Golden eagle kills red deer calf. Journal of Zoology, 158(2), 215–216. https://doi.org/10.1111/j.1469-7998.1969.tb02140.x

Cooper, T. J. G. (2015). The effects of land use changes on the distribution of forest dependent bird species in South Africa. Stellenbosch University.

Corbett, L. & Hertog, T. (2011). Diet and breeding of White-bellied Sea eagles Haliaeetus leucogaster in subtropical river habitats in the Northern Territory, Australia. Corella, 35, 41–48.

Corbett, L. & Hertog, T. (2012). Impact of feral Water Buffalo Bubalus bubalis grazing on White-bellied Sea eagle Haliaeetus leucogaster breeding success in subtropical river habitat in the Northern Territory, Australia. Corella, 36, 24–26.

Cordeiro, N. & Githiru, M. (2000). Conservation evaluation for birds of Brachylaena woodland and mixed dry forest in north-east Tanzania. Bird Conservation International, 10, 47–65. https://doi.org/10.1017/S0959270900000058

Córdoba-Córdoba, S., Echeverry-Galvis, M. Á. & Estela, F. A. (2008). New distributional records for the Black-and-chestnut eagle (Spizaetus isidori) and the Black hawk-eagle (S. Tyrannus) for Colombia, with notes for field identification. Ornitología Colombiana, 7, 66–74.

Cornelis, J. & Hermy, M. (2004). Biodiversity relationships in urban and suburban parks in Flanders. Landscape and Urban Planning, 69(4), 385–401, https://doi.org/10.1016/j.landurbplan.2003.10.038

Corso, A. & Iapichino, C. (1998). I rapaci svernanti in Sicilia negli anni 1990–1997. Alula, 5, 125–130.

Corso, A. (2001). Raptor migration across the Strait of Messina, southern Italy. British Birds, 94, 196-202.
Cortes, J. E., Finlayson, J. C., Mosquera, M. A. & Garcia, E. F. J. (1980). The Birds of Gibraltar. Gibraltar, Spain Gibraltar Bookshop.
Cottam, G. & Curtis, J. T. (1956). The use of distance measures in phytosociological sampling. Ecology, 37(3), 451–460. https://doi.org/10.2307/1930167
Couey, F. M. (1944). Golden eagle vs. bighorn sheep. Montana Wildlife Bulletin, 1(1), 2–3.
Coulson, I. M. & Hailey, A. (2001). Low survival rate and high predation in the African hingeback tortoise Kinixys spekii. African Journal of Ecology, 39(4), 383–392. https://doi.org/10.1046/j.0141-6707.2001.00328.x
Counihan, J. (1975). Adiabatic atmospheric boundary layers: A review and analysis of data from the period 1880–1972. Atmospheric Environment, 9(10), 871–905, https://doi.org/10.1016/0004-6981(75)90088-8
Cowen, D. J. (1988). Cited in Rural Development (RD) RD-08. ICT and Geo-Informatics for Rural Development. Odisha: Odisha State Open University.
Crabtree, R. L. & Sheldon, J. W. (1999). The ecological role of coyotes on Yellowstone's northern range. Yellowstone Science, 7, 15–23.
Cracraft, J. & Rich, P. V. (1972). The systematics and evolution of the Cathartidae in the Old World Tertiary. Condor, 74(3), 272–283. https://doi.org/10.2307/1366589
Cracraft, J., Barker, F. K., Braun, M., Harshman, J., Dyke, G. J., Feinstein, J., Stanley, S., Cibois, A., Schikler, P., Beresford, P., García-Moreno, J., Sorenson, M.D., Yuri, T., . . . Mindell, D. P. (2004). Phylogenetic relationships among modern birds (Neornithes), Toward an avian tree of life. *In*: Cracraft, J. and & Donoghue, M. J. (Eds.), Assembling the tree of life, 468–489. New York: Oxford University Press.
Craig, R. C. & Enderson, J. H. (2004). Peregrine falcon biology and management in Colorado, 1973–2001. Technical Publication 43. Fort Collins, CO: Colorado Division of Wildlife.
Craig, T. H., Connelly, J. W., Craig, E. H. & Parker, T. L. (1990). Lead concentrations in Golden and Bald eagles. Wilson Bulletin, 102, 130–133.
Craighead, D., Bedrosian, B., Watson, R. T., Fuller, M., Pokras, M. & Hunt, W. (2009). A relationship between blood lead levels of common ravens and the hunting season in the southern Yellowstone ecosystem. Ingestion of lead from spent ammunition: Implications for wildlife and humans. Boise, ID: Peregrine Fund.
Cramp, S. & Simmons, K. E. L. (1979). Handbook of the birds of Europe, the Middle East, and North Africa: The birds of the western Palearctic, 2. Oxford: Oxford University Press.
Crawley, M. J. (2007). The R Book. London, UK: John Wiley and Sons.
Crease A. & Tepedino I. (2013). Observations at a nest of Crested eagle Morphnus guianensis in the southern Gran Sabana, Venezuela. Cotinga, 35, 90–93.
Crease, A. 2009. Avian range extensions from the southern headwaters of the Rio Caroni, Gran Sabana, Bolivar, Venezuela. Cotinga, 31, 5–19.
Cresswell, W. (1994). Flocking is an effective anti-predation strategy in redshanks, Tringa totanus. Animal Behaviour, 47(2), 433–442. https://doi.org/10.1006/anbe.1994.1057
Cresswell, W. (1996). Surprise as a winter hunting strategy in Sparrowhawks Accipiter nisus, Peregrines Falco peregrinus and Merlins F. columbarius. Ibis, 138(4), 684–692. https://doi.org/10.1111/j.1474-919X.1996.tb04770.x
Cresswell, W. (2004). Kleptoparasitism rates and aggressive interactions between raptors. *In*: Chancellor, R.D. & Meyburg, B. U. (Eds.), Proceedings of the 6th World Conference on Birds of Prey, 805–814. Berlin: World Working Group on Birds of Prey and Owls.
Cresswell, W. & Quinn, J. L. (2004). Faced with a choice, sparrowhawks more attack the more vulnerable prey group. Oikos, 104(1), 71–76. https://doi.org/10.1111/j.0030-1299.2004.12814.x
Cresswell, W. & Quinn, J. L. (2010). Attack frequency, attack success and choice of prey group size for two predators with contrasting hunting strategies. Animal Behaviour, 80(4), 643–648. https://doi.org/10.1016/j.anbehav.2010.06.024
Cresswell, W., Lind, J. & Quinn, J. L. (2010). Predator-hunting success and prey vulnerability: Quantifying the spatial scale over which lethal and non-lethal effects of predation occur. Journal of Animal Ecology, 79(3), 556–562. https://doi.org/10.1111/j.1365-2656.2010.01671.x
Crewe, T., Barry, K., Davidson, P. & Lepage, D. (2012). Coastal water-bird population trends in the Strait of Georgia 1999–2011: Results from the first 12 years of the British Columbia Coastal Waterbird Survey. British Columbia. Birds, 22, 8–35.
Crinnion, W. J. (2011). Polychlorinated Biphenyls: Persistent pollutants with immunological, neurological, and endocrinological consequences. Alternative Medicine Review, 16(1), 5–13.
Crivelli, A. J., Jerrentrup, H. & Mitchev, T. (1988). Electric power lines: A cause of mortality in Pelecanus crispus Bruch, a world endangered bird species, in Porto-Lago, Greece. Colonial Waterbirds, 11(2), 301–305. https://doi.org/10.2307/1521012
Cross, T. L. (2009). Waterbirds. New York, NY: W.W. Norton & Company.
Crowder, N. (2015). Hunting with eagles: This 4,000 year-old art is dying out in Mongolia. Retrieved from https://www.washingtonpost.com/news/in-sight/wp/2015/08/21/the-4000-year-old-art-of-falconry-is-dying-out-in-mongolia-see-some-of-the-last-eagle-hunters-in-the-world/
Cruysberghs, S. (2014). Ulgii Festival photographs, Sept. 21. Retrieved from http://www.scip.be/index.php?Page=PhotosMongolia&Lang=EN&ID=2941
Cruz-Martinez, L., Redig, P. T. & Deen, J. (2012). Lead from spent ammunition: A source of exposure and poisoning in Bald eagles. Human–Wildlife Interactions, 6(1), 94–104.
Csermely, D. & Rossi, O. (2006). Bird claws and bird of prey talons: Where is the difference? Italian Journal of Zoology, 73(1), 43–53. https://doi.org/10.1080/11250000500502368
CTV. (2020). Several Bald eagles poisoned by toxic carcass on Vancouver Island. Retrieved from https://vancouverisland.ctvnews.ca/several-bald-eagles-poisoned-by-toxic-carcass-on-vancouver-island-1.4815452
Cuervo, A. M., Pulgarín, P. C. & Calderón, D. 2008. New distributional bird data from the Cordillera Central of the Colombian Andes, with implications for the biogeography of northwestern South America. Condor, 110(3), 526–537. https://doi.org/10.1525/cond. 2008.8555

Cuervo, A. M., Pulgarín, P. C., Calderón, D., Ochoa-Quintero, J.M., Delgado-V., C.A., Palacio, A., Botero, J. M. & Múnera, W. A. (2008 Avifauna of the northern Cordillera Central of the Andes, Colombia. Ornitologia Neotropical, 19, 495–515.
Cullum, C. & Rogers, K. (2011) WRC Report no. TT 498/11. A framework for the classification of drainage networks in savanna landscape Centre for Water in the Environment. University of the Witwatersrand.
Cummins, R. A. (1997a). Assessing quality of life. *In*: Brown, R. J. (Ed.), Assessing quality of life for people with psychiatric disabilities, p 116–150. Cheltenham, UK: Stanley Thornes.
Cummins, R. A. (1997b). Comprehensive quality of life scale—Adult. Melbourne: School of Psychology. Melbourne: Deakin University.
Cupper, J. & Cupper, L. (1981). Hawks in focus. Mildura, Victoria: Jaclin.
Curry-Lindahl, K. (1971). Prey of the Martial eagle. Africana, 4(5), 8.
Cushing, R. & Washburn, B. E. (2014). Exploring the role of ospreys in education. Journal of Raptor Research, 48(4), 414–421. https://do org/10.3356/JRR-14-00011.1
Cuthbert, R., Green, D. R., Ranade, S., SaravAn, S., Pain, D. J., Prakash, V. & Cunningham, A. A. (2006). Rapid population declines of Egyptia vulture (Neophron percnopterus) and Red-headed vulture (Sarcogyps calvus) in India. Animal Conservation, 9(3), 349–354. https://do org/10.1111/j.1469-1795.2006.00041.x
Cyrus, D. & Robson, N. (1980). Bird atlas of Natal. Pietermaritzburg, South Africa: University of Natal Press.
Czechura, G. (1985. The raptors of the Blackall-Conondale Ranges and adjacent lowlands, south-eastern Queensland. Corella, 9, 49–54.
D.J.I. (2019). DJI. Retrieved from https://www.dji.com/ca.
D'Orbigny, A. D. (1835). Voyage of South America Executed between the years 1826 to 1833 by Alcide d'Orbigny. Birds. Paris: Pitois-Levraul
Dabbene, R. (1908). Ornitología Argentina: Systematic and descriptive catalogue of the birds of the Republic of Argentina. Vol. 1; 16. Anales d Museo Nacional de Buenos Aires, 1–513.
Dahl, E. L., Bevanger, K., Nygård, T., Roskaft, E. & Stokke, B. G. (2012). Reduced breeding success in White-tailed eagles at Smøla Windfarn western Norway, is caused by mortality and displacement. Biological Conservation, 145(1), 79–85. https://doi.org/10.1016/j. biocon.201 10.012.
Dahl, E. L., May, R., Hoel, P. L., Bevanger, K., Pedersen, H. C., Røskaft, E. & Stokke, B. G. (2013). White-tailed eagles (Haliaeetus albicilla) a the Smøla Wind-Power Plant, Central Norway: Lack Behavioral Flight Responses to Wind Turbines. Wildlife Society Bulletin, 37: 66–74.
Dal, S. K. (1954). Fauna of the Armenian SSR. Academy of Sciences Press, Armenian SSR, Yerevan, Republic of Armenia.
Dale, L. A. (2009). Personal and corporate liability in the aftermath of bird strikes: A costly consideration. Human-Wildlife Conflicts, 3, 216–22
Dall, S. R. X. (2002). Can information sharing explain recruitment to food from communal roosts? Behavioural Ecology, 13, 42–51.
Daly, W. M. (1899). Nesting of the Black eagle. Journal of the Bombay Natural History Society, 12 (3), 589.
Dänhardt, J. & Lindström, Å. (2001). Optimal departure decisions of songbirds from an experimental stopover site and the significance of weathe Animal Behaviour, 62, 235–243. https://doi.org/10.1006/anbe.2001.1749
Danko, Š., Meyburg, B.-U., Belka, T. & Karaska, D. (1996). Marking Lesser spotted Eagles Aquila pomarina: Methods, experiences and result *In*: Meyburg, B.-U. & Chancellor, R. D. (Eds.), Eagle studies, 209–243. Berlin: World Working Group on Birds of Prey.
Danko, S. (2009). Short-toed eagle (Circaetus gallicus). Birds of Prey and Owls – Journal of Raptor Protection of Slovakia, 5, 10.
Darawshi, S. (2009). The ecology of the Short-toed eagle (Circaetus gallicus) in the Judean Slopes Israel. Graduate Thesis. The Hebrew Universit of Jerusalem.
Das, D., Cuthbert, R. J., Jakati, R. D. & Prakash, V. (2011). Diclofenac is toxic to the Himalayan vulture Gyps himalayensis. Bird Conservatio International, 21(1), 72–75. https://doi.org/10.1017/S0959270910000171.
Das, K. K., Reddy, R. C., Bagoji, I. B., Das, S., Bagali, S., Mullur, L., Khodnapur, J.P. & Biradar, M. S. (2018). Primary concept of nickel toxicit – An overview. Journal of Basic and Clinical Physiology and Pharmacology, 30(2), 141–152. https://doi.org/10.1515/jbcpp-2017-0171.
Das, S., Patel, P. P. & Sengupta, S. (2016). Evaluation of different digital elevation models for analyzing drainage morphometric parameters i a mountainous terrain: A case study of the Supin–Upper Tons Basin, Indian Himalayas. Springer Plus, 5(1), 1544. https://doi.org/10.1186 s40064-016-3207-0
Daudin, F. M. (1800). Traitéélémentaire et completd'ornithologie, ou, Histoire naturelle des oiseaux., Tome 2. Paris: Bertrandet.
Davidar, P., Yoganand, K., Ganesh, T. & Geetha-Nayak, K. (2007). Avifauna of Andaman Islands: Preliminary inventory and distributiona patterns. Journal of the Bombay Natural History Society, 104(3), 35–47.
Davidar, P., Yoganand, K., Ganesh, T. & Joshi, N. (1996). An assessment of common and rare forest bird species of the Andaman Islands. Forktai 12, 99–105.
Davidson, H. E. (1993). The Lost Beliefs of Northern Europe. London: Routledge, London.
Davidson, I. (1987). Identification aids: Martial Eagle or Black-breasted Snake eagle? Witswatersrand Bird Club News, 137, 8–9.
Davidson, J. (1908). On the nesting habits of the Small Indian Spotted eagle (Aquila hastata). Journal of the Bombay Natural History Society 18(3), 682–683.
Davidson, M. E. M. (1934). Specimens of the Philippine monkey-eating eagle (Pithecophaga jefferyi). Auk, 51(3), 338–342. https://doi org/10.2307/4077661
Davies, C. W. N., Barnes, R., Butchart, S. H. M., Fernandez, M. & Seddon, N. (1997). The conservation status of birds on the Cordillera de Colán Peru. Bird Conservation International, 7(2), 181–195. https://doi.org/10.1017/S0959270900001490.
Davies, E. (2010). Dark Steller's sea eagle solves 100-year debate. BBC. Retrieved from http://news.bbc.co.uk/earth/hi/earth_news newsid_9036000/9036698.stm.
Davies, R. & Allan, D. G. (1997). Black eagle. *In*: Harrison, J. A., Allan, D. G., Underhill, L. G., Herremans, M., Tree, A. J., Parker, V. & Brown C.J. et al. (Eds.), The Atlas of Southern African Birds, 175–177. Johannesburg: BirdLife South Africa.

Davies, R. (1988). Problem eagles and conservation solutions. Gabar, 3, 60–66.
Davies, R. (1994). Black eagle Aquila verreauxii predation on Rock hyrax Procavia capensis and other prey in the Karoo. PhD. Thesis. Pretoria: University of Pretoria.
Davies, R. A. (1999). The extent, cost and control of livestock predation. Journal of Raptor Research, 33(1), 67–72.
Davis, R. M. (1973). The Ecology and Life History of the Vlei Rat, Otomysirroratus (Brants, 1827), on the Van Riebeck Nature Reserve, Pretoria. PhD. Thesis. Pretoria: University of Pretoria.
Dawe, N. K. (2016). Cooperative breeding by the Bald eagle (Haliaeetus leucocephalus) on Vancouver Island, British Columbia. British Columbia Birds, 26: 35–40.
Dawson, J. W. & Mannan, R. W. (1994). The ecology of Harris' hawks in urban environments. Unpublished Report, Arizona Game and Fish Department, Tucson, Arizona, USA. Available from the U.S. Geological Survey, Richard R. Olendorff Memorial Library, 970 Lusk St., Boise, ID 83706.
De Andrade, M. A. & de Andrade, M. V. G. (1998). Notes on the Crowned Solitary eagle (Harpyhaliaetus coronatus) and records of its occurrence in Minas Gerais. Atualidades Ornitológicas, 83, 11.
De Boer, L. E. M. & Sinoo, R. P. (1984). A karyological study of Accipitridae (Aves: Falconiformes), with karyotypic descriptions of 16 species new to cytology. Genetica, 65(1), 89–107. https://doi.org/10.1007/BF00056767.
De Boer, L. E. M. (1975). Karyological heterogeneity in the Falconiformes (Aves). Experientia, 31(10), 1138–1139. https://doi.org/10.1007/BF02326755.
De Buffrénil, V. & Hémery, G. (2002). Variation in longevity, growth, and morphology in exploited Nile monitors (Varanus niloticus) from Sahelian Africa. Journal of Herpetology, 36(3), 419–426. https://doi.org/10.1670/0022-1511(2002)0360419: VILGAM2.0.CO;2.
De Francisco, N., Troya, J. D. E. & Agüera, E. I. (2003). Lead and lead toxicity in domestic and free-living birds. Avian Pathology, 32(1), 3–13. https://doi.org/10.1080/ 0307945021000070 660.
De Vargas González, J. & Vargas, F. H. (2011) Nesting density of Harpy eagles in Darien with population size estimates for Panama. Journal of Raptor Research, 45(3), 199–210. https://doi.org/10.3356/JRR-10-57.1
De la Peña, M. R. (1978). Encyclopedia of Argentine birds. Fascicle II. Published by the author. Santa Fe, Argentina.
De la Peña, M. R. (1987). Nests and eggs of Argentina birds. Published by the author. Santa Fe, Argentina.
De la Peña, M. R. (1999). Argentine birds, list and distribution. Editorial L.O.L.A.
De la Peña, M. R. (2005). Reproducción de las aves argentinas (con descripción de pichones). Buenos Aires: Editorial L.O.L.A.
De la Peña, M. R. (2006). Photo guide to the nests, eggs, and young of Argentina birds. Editorial. Buenos Aires, Argentina: L.O.L.A.
De Labra, M. A., Escalante, P. & Monterrubio-Rico, T. (2013). Diurnal raptors in Los Tuxtlas Biosphere Reserve, Mexico: Current presence and relative abundance. Journal of Raptor Research, 47(4), 392–400. https://doi.org/10.3356/JRR-12-18.2.
De Lucas, M., Ferrer, M., Bechard, M. J. & Muñoz, A. R. (2012). Griffon vulture mortality at wind farms in southern Spain: Distribution of fatalities and active mitigation measures. Biological Conservation, 147(1), 184–189. https://doi.org/10.1016/j.biocon.2011.1 2.029.
De Lucas, M., Janss, G. F. E. & Ferrer, M. (2004). The effects of a wind farm on birds in a migration point: The Strait of Gibraltar. Biodiversity and Conservation, 13(2), 395–407. https://doi.org/10.1023/B: BIOC.0000006507.22024.93.
De Lucas, M., Janss, G. F. E. & Ferrer, M. (2007). Wind farm effects on birds in the Strait of Gibraltar. *In*: De Lucas, M., Janss, G. F. E. & Ferrer, M. (Eds.), Birds and wind farms: Risk assessment and mitigation, 219–228. Madrid: Quercus.
De Lucas, M., Janss, G. F. E., Whitfield. D. P. & Ferrer, M. (2008). Collision fatality of raptors in wind farms does not depend on raptor abundance. Journal of Applied Ecology, 45(6), 1695–1703. https://doi.org/10.1111/j.1365-2664.2008.01549.x.
De Lucas, M. & Ferrer, M., Bechard, M. & Muñoz, Antonio-Román. (2012). Griffon vulture mortality at wind farms in southern Spain: Distribution of fatalities and active mitigation measures. Biological Conservation, 147, 184–189. https://doi.org/10.1016/ j.biocon.2011.12.029.
De Lucca, E. R. (1991). Threatened raptors: The Black-and-chestnut eagle and the solitary eagle: State of current knowledge. Nuestras Aves, 25, 14–15.
De Lucca, E. R. (1992). The Crowned eagle, Harpyhaliaetus coronatus, in San Juan. Nuestras Aves, 26, 25.
De Lucca, E. R. (1993). Threatened raptors: The Crowned eagle. Nuestras Aves, 29, 14–17.
De Mendonça-Lima, A., Zilio, F., Joenck, C. M. & Barcellos, A. (2006). Novosregistros de Spizaetus ornatus (Accipitridae) no sul do Brasil. Revista Brasileira de Ornitologia, 14(3), 279–282.
De Roder, F. E. (1989). The migration of raptors south of Annapurna, Nepal, autumn 1985. Forktail, 4, 9–17.
De Schauensee, M. R. (1949). The birds of the Republic of Colombia. Part 2. Caldasia, 5, 381–644.
De Schauensee, M., R. & Phelps, W. H., Jr. (1978). A guide to the birds of Venezuela. Princeton, NJ: Princeton University Press.
De Schauensee, M., R. (1950). Colombian zoological survey. Part VII: A collection of birds from Bolivar, Colombia. Proceedings of the Academy of Natural Sciences of Philadelphia, 102, 111–139.
De Sparre, G. (1835). Description of Astur kienerii. Magasin de Zoologie, 5 (Classe, 2), pl. 35, unpaginated.
De, J., Vargas, G., Mosquera, J. & Watson, M. (2006). Crested eagle (Morphnus Guianensis) feeding a post-fledged young Harpy eagle (Harpia harpyja) in Panama. Ornitologia Neotropical, 17: 581–584. Retrieved from https://sora.unm.edu/sites/default/files/journals/ on/v017n04/ p0581-p0584.pdf.
Deacon, N. R. & Rockingham-Gill, D. V. (2009). Raptor survey in miombo woodland, Umfurudzi Safari area, Shamva District, Zimbabwe: Preliminary results. Gabar, 19, 2–9.
Dean, W. R. J. (1974). Breeding and distributional notes on some Angolan birds. Durban Museum Novitates, 10(8), 109–125.
Dean, W. R. J., Huntley, M. A., Huntley, B. J. & Vernon, C. J. (1988). Notes on some birds of Angola. Durban Museum Novitates, 14, 43–92.
Dean, W. R. J. (2000). The Birds of Angola. BOU Checklist no. 18. Tring, UK: British Ornithologists' Union.

Dean, W. R. J., Middlemiss, E., Cooper, J., Williams, A. J., Steyn, P., Kemp, A. C. & Tree, A. J. (1975). Martial eagles nesting on high tension pylons. Ostrich, 46(1), 116–117. https://doi.org/10.1080/00306525.1975.9639387.

Deblinger, R. D. & Alldredge, W. (1996). Golden eagle predation on pronghorns in Wyoming's Great Divide Basin. Journal of Raptor Research, 30, 157–159.

DeBruyn, A. M. H. & Chapman, P. M. (2007). Selenium toxicity to invertebrates: Will proposed thresholds for toxicity to fish and birds also protect their prey? Environmental Science & Technology, 41(5), 1766–1770. https://doi.org/10.1021/es062253j.

Debus, S. & Kirwan, G. M. (2020). White-bellied Sea eagle (Haliaeetus leucogaster), version 1.0. *In*: Del Hoyo, J., Elliott, A., Sargatal, J., Christie, D. A. & De Juana, E. (Eds.), Birds of the World. Ithaca, NY: Cornell Laboratory of Ornithology. https://doi.org/10.2173/bow.wbseag1.01.

Debus, S., Hatfield, T., Ley, A. & Rose, A. (2007). Breeding biology and diet of the Little eagle Hieraaetus morphnoides in the New England Region of New South Wales, 24.

Debus, S. (1998). The Birds of Prey of Australia: A Field Guide. Melbourne. UK: Oxford University Press.

Debus, S. (2006). Taxonomic status of Sanford's Sea eagle. Boobook, 24, 8.

Debus, S. (2008). Biology and Diet of the White-bellied Sea eagle Haliaeetus leucogaster Breeding in Northern Inland New South Wales. Australian Field Ornithology, 25(4), 165– 193.

Debus, S. (2008). Little Eagles in New England, NSW. Boobook, 26, 7.

Debus, S. (2017). Australasian Eagles and Eagle-like Birds. Clayton, Victoria, Australia: CSIRO Publishing.

Debus, S. J. S. & Ley, A. J. (2009). Aspects of the Breeding Cycle of the Little eagle Hieraaetus morphnoides. Australian Field Ornithology, 26, 76–99.

Debus, S. J. S. (1984). Biology of the Little eagle on the northern tablelands of New South Wales. Emu, 84(2), 87–92. https://doi.org/10.1071/MU9840087.

Debus, S. J. S. (1994). Gurney's eagle. *In*: Del Hoyo, J., Elliott, A. & Sargatal, J. (Eds.), Handbook of birds of the world, Vol. 2. New World vultures to guineafowl, p. 197. Barcelona: Lynx Edicions.

Debus, S. J. S. (1994). Sanford's sea eagle. *In*: Del Hoyo, J., Elliott, A. & Sargatal, J. (Eds.), Handbook of birds of the world, Vol. 2. New World vultures to guineafowl, p. 121. Barcelona: Lynx Edicions.

Debus, S. J. S. (1994). Wedge-tailed eagle. *In*: Del Hoyo, J., Elliott, A. &, Sargatal, J. (Eds.), Handbook of birds of the world, Vol. 2. New World vultures to guineafowl, p. 198. Barcelona: Lynx Edicions.

Debus, S. J. S. (1994). White-bellied sea eagle. *In*: Del Hoyo, J., Elliott, A. & Sargatal, J. (Eds.), Handbook of birds of the world, Vol. 2. New World vultures to guineafowl, p. 121. Barcelona: Lynx Edicions.

Debus, S. J. S. (2006). Sea eagle chasing duck. Boobook, 24, 19.

Debus, S. J. S. (2008). Biology and diet of the White-bellied Sea eagle Haliaeetus leucogaster breeding in northern inland of New South Wales. Australian Field Ornithology, 25, 165–193.

Debus, S. J. S. (2009). Aspects of the breeding cycle of the Little eagle. Australian Field Ornithology, 26, 76–99.

Debus, S. J. S. (2011). Parental time-budgets and breeding behaviour of the Little eagle Hieraaetus morphnoides in northern New South Wales. Corella, 35, 65–72.

Debus, S. J. S., Baker, G., Owner, D. & Nottidge, B. (2014). Response of White-bellied Sea eagles Haliaeetus leucogaster to encroaching human activities at nest sites. Corella, 38, 53–62.

Debus, S. J. S., Hatfield, T. S., Ley, A. J. & Rose, A. B. (2007b). Breeding biology and diet of the Little eagle Hieraaetus morphnoides in the New England region of New South Wales. Australian Field Ornithology, 24, 137–157.

Debus, S., Boesman, P. F. D. & Marks, J. S. (2020). Eagle: Little eagle (Hieraaetus morphnoides), version 1.0. *In*: Del Hoyo, J., Elliott, A., Sargatal, J., Christie, D. A. & De Juana, E. (Eds.), Birds of the World. Ithaca, NY: Cornell Laboratory of Ornithology. https://doi.org/10.2173/bow.liteag1.01.

Debus, S., Kirwan, G. M. & Christie, D.A. (2020). Sanford's Sea eagle (Haliaeetus sanfordi), version 1.0. *In*: Del Hoyo, J., Elliott, A., Sargatal, J., Christie, D. A. & De Juana, E. (Eds.), Birds of the World. Ithaca, NY: Cornell Laboratory of Ornithology. https://doi.org/10.2173/bow.solsee1.01.

Debus, S., Kirwan, G. M. & Marks, J. S. (2020). New Guinea eagle (Harpy opsis novaeguineae), version 1.0. *In*: Del Hoyo, J., Elliott, A., Sargatal, J., Christie, D. A. & De Juana, E. (Eds.) Birds of the World (Eds.). Ithaca, NY: Cornell Laboratory of Ornithology. https://doi.org/10.2173/bow.negeag1.01.

Debus, S. J. S. & Ley, A. J. (2009). Aspects of the breeding cycle of the Little eagle Hieraaetus morphnoides. Australian Field Ornithology, 26, 76–99.

Debus, S.J.S., Hatfield, T. S., Ley, A. J. & Rose, A. B. (2007a). Breeding biology and diet of the Wedge-tailed eagle Aquila audax in the New England region of New South Wales. Australian Field Ornithology, 24, 93–120.

DeCandido, R., Allen, D. & Bildstein, K. L. (2001). The migration of Steppe eagles (Aquila nipalensis) and other raptors in central Nepal, autumn 1999. Journal of Raptor Research, 35(1), 35–39.

DeCandido, R., Gurung, S., Subedi, T. & Allen, D. (2013). The east-west migration of Steppe Eagle Aquila nipalensis and other raptors in Nepal and India. BirdingASIA, 19, 18–25.

Dee, T. J. (1986). The Endemic Birds of Madagascar. Cambridge, UK: International Council for Bird Preservation.

Deedes, C. N. (1935). The Double-headed God. Folklore, 46(3), 194–243. https://doi.org/10.1080/0015587X.1935.9718603.

DeFries, R., Rudel, T., Uriarte, M. & Hansen, M. (2010). Deforestation driven by urban population growth and agricultural trade in the twenty-first century. Nature Geoscience, 3, 178–181. https://doi.org/10.1038/ngeo756

DeForest, D. K., Schlekat, C. E., Brix, K. V. & Fairbrother, A. (2012). Secondary poisoning risk assessment of terrestrial birds and mammals exposed to nickel. Integrated Environmental Assessment and Management, 8(1), 107–119. https://doi.org/10.1002/ieam.248.

Dekker, D. & Bogaert, L. (1997). Over-ocean hunting by Peregrine falcons in British Columbia. Journal of Raptor Research, 31, 381–383.

Dekker, D. & Court, G. (2003). Gyrfalcon predation on Mallards and the interaction of Bald eagles wintering in central Alberta. Journal of Raptor Research, 37, 161–163.

Dekker, D. (1980). Hunting success rates, foraging habits, and prey selection of Peregrine falcons migrating through central Alberta. Canadian Field-Naturalist, 94, 371–382.

Dekker, D. (1985). Hunting behavior of Golden eagles, Aquila chrysaetos, migrating in southwestern Alberta. Canadian Field-Naturalist, 99(3), 383–385.

Dekker, D. (1988). Peregrine falcon and Merlin predation on small shorebirds and passerines in Alberta. Canadian Journal of Zoology, 66(4), 925–928. https://doi.org/10.1139/z88-136.

Dekker, D. (1995). Prey capture by Peregrine falcons wintering on southern Vancouver Island, British Columbia. Journal of Raptor Research, 29, 26–29.

Dekker, D. (1999). Bolt from the Blue: Wild Peregrines on the Hunt. Hancock House Publishers, Surrey, British Columbia BC, Canada, and Blaine, Washington.

Dekker, D. (2003). Peregrine falcon predation on dunlins and ducks and kleptoparasitic interference from Bald eagles wintering in British Columbia. Journal of Raptor Research, 37, 91–97.

Dekker, D. (2009). Hunting tactics of Peregrines and other falcons. PhD Thesis. Wageningen, and Netherlands: Wageningen University, the Netherlands.

Dekker, D. (2013). High-tide flight by wintering dunlins (Calidris alpina): A weather dependent trade-off between energy loss and predation risk. Canadian Journal of Zoology, 91(1), 25–29. https://doi.org/10.1139/cjz-2012-0213.

Dekker, D., Out, M., Tabak, M. & Ydenberg, R. (2012). The effect of kleptoparasitic Bald eagles and Gyr falcons on the kill rate of Peregrine falcons hunting dunlins wintering in British Columbia. The Condor, 114(2), 290–294. https://doi.org/10.1525/ cond.2012.110110.

Dekker, D., Mark, C. & Drever, M. C. (2015). Kleptoparasitism by Bald eagles (Haliaeetus leucocephalus) as a factor in reducing Peregrine falcon (Falco peregrinus) predation on dunlin (Calidris alpina) wintering in British Columbia. Canadian Field-Naturalist, 129(2), 159–164.

Decker, D. J., Riley, S. J. & Siemer, W. F. (2012). Human Dimensions of Wildlife Management. Second edition. Baltimore: John Hopkins University Press.

del Castillo, H. & Clay, R. P. (2004). Annotated checklist of the birds of Paraguay. Associación Guyra Paraguay, Asunción, Paraguay.

del Castillo, H. & Clay, R. P. (2005). Atlas of the birds of Paraguay. Associación Guyra Paraguay. Asunción, Paraguay.

Del Hoyo, J., Elliott, A. & Sargatal, J. (1994). Handbook of birds of the world. Vol. 2. New World Vultures to Guineafowl. Barcelona, Spain: Lynx Edicions.

Del Hoyo, J., Collar, N. J., Christie, D. A., Elliott, A. & Fishpool, L. D. C. (2014). HBW and BirdLife International Illustrated Checklist of the Birds of the World. Volume 1: Non-Passerines. Barcelona: Lynx Edicions and Cambridge, UK: BirdLife International.

Del Hoyo, J., Collar, N., Marks, J. S. & Sharpe, C. J. (2018). Flores Hawk-eagle (Nisaetus floris). *In*: Del Hoyo, J., Elliott, A., Sargatal, J., Christie, D. A. & De Juana, E. (Eds.), Handbook of the Birds of the World Alive. Barcelona: Lynx Edicions.

Del Hoyo, J., Collar, N., Marks, J. S. & Sharpe, C. J. (2020). Flores Hawk eagle (Nisaetus floris), version 1.0. *In*: Del Hoyo, J., Elliott, A., Sargatal, J., Christie, D. A. & De Juana, E. (Eds.), Birds of the World. Ithaca, NY: Cornell Laboratory of Ornithology, Ithaca, NY. https://doi.org/10.2173/ bow.flohae1.01.

Del Hoyo, J., Collar, N. & Marks, J. S. (2020). Indian Spotted Eagle (Clangahastata), version 1.0. *In*: Del Hoyo, J., Elliott, A., Sargatal, J., Christie, D. A. & De Juana, E. (Eds.), Birds of the World. Ithaca, NY: Cornell Lab of Ornithology. https://doi.org/10.2173/bow.inseag1.01.

Del Hoyo, J., Elliott, A. & Sargatal, J. (1992). Handbook of the Birds of the World., Vol. 2. New World Vultures to Guinea Fowl. Barcelona: Lynx Edicions.

Del Hoyo, J., Elliot, A. & Sargatal, J. (1994). Handbook of the Birds of the World, 2. New World Vultures to Guinea Fowl. Barcelona: Lynx Edicions.

Del Hoyo, J., Elliot, A. & Sargatal, J. (2010). Handbook of the Birds of the World, Vol 2. Lynx Edicions.

Delacour, J. & Jabouille, P. (1931). Les oiseaux de I'Indochinefrangaise, 1-4. Paris: Exposition.

Delgado Maldonado, H. J. (2008). Influencia de los componentes y usosactuales del paisajesobre la comunidad de avesrapacesenvarios puntos del Valle de Magdalena. Bogotá, Colcmbia: Universidad Inca de Colombia, Bogotá, Colombia.

Delgado, O. (2007). The soya route in North West Argentina. The United Soy Republics: The truth about soy in South America, 132-159.

Delhey, R. (1992). Some new or poorly known birds from the region of Bahía Blanca. Nuestras Aves, 10, 31–32.

Delibes, M. (1978). Feeding ecology of the Spanish Imperial eagle (Aquila adalberti) during chick-rearing in the Coto of Doñana. Doñana Acta Vertebr., 5, 35–60.

Delong, D. C., Jr. (1990). Effects of food on Bald eagle distribution and abundance on the northern Chesapeake Bay: An experimental approach. Master's Thesis. The College of William and Mary in Virginia.

Dementavičius, D. (2007). White-tailed eagle (Haliaeetus albicilla) in Lithuania: Population numbers and trends 1900-2007. Acta Zoologica Lituanica, 17(4), 281–285. https://doi.org/10.1080/13921657.2007.10512845.

Dementiev, G. P. & Gladkov, N. A. (1951). Birds of the Soviet Union. Moscow: State Publishers, Soviet Science.

Dementiev, G. P., Gladkov, N. A., Ptushenko, E. S., Spangenberg, E. P. & Sudilovskaya, A. M. (1966). Birds of the Soviet Union. Vol. 1. Jerusalem: Israel Program for Scientific Translations, Jerusalem, Israel.

Demerdzhiev, D. A., Stoychev, S. A., Petrov, T. H., Angelov, I. D. & Nedyakov, N. P. (2009). Impact of power lines on bird mortality in Souther Bulgaria. Acta Zoologica Bulgarica, 61(2), 177–185.

Demey, R. (2006). Recent reports. Bulletin of the African Bird Club, 13, 219–230.

Demey, R. & Ossom, W. (2007). Rapid survey of the birds of the Atewa Range Forest Reserve, Ghana. *In*: McCullough, J., Alonso, L.E Naskrecki, P., Wright, H.E. & Osei-Owusu, Y. (Eds.), A Rapid Biological Assessment of the Atewa Range Forest Reserve, Eastern Ghana Rapid Assessment Program: Bulletin of Biological Assessment, 47, 84–89.

Den Besten, J. W. (2004). Migration of Steppe eagles Aquila nipalensis and other raptors along the Himalayas past Dharamsala, India, in autum 2001 and spring 2002. Forktail, 20, 9–13.

Dennis, T. E. & Lashmar, A. F. C. (1996). Distribution and abundance of White-bellied sea eagles in South Australia. Corella, 20, 93–102.

Dennis, T. E. (2004). Conservation status of the White-bellied Sea eagle, osprey and peregrine falcon on western Eyre Peninsula and adjacen offshore islands in South Australia. South Australian Ornithologist, 34, 222–228. Retrieved from https://birdssa.asn.au/images/saopdfs Volume34/2004V34P222.pdf.

Dennis, T. E. & Baxter, C. I. (2006). The status of the White-bellied Sea-eagle and Osprey on Kangaroo Island in 2005. South Australia Ornithologist, 35(1), 47–51.

Dennis, T. E. & Brittain, R. (2006). Attempted kleptoparasitism by White-bellied Sea eagles on fur-seal. South Australian Ornithologist, 35(1–2) 68.

Dennis, T. E., Detmar, S. A., Brooks, A. V. & Dennis, H. M. (2011a). Distribution and status of White-bellied Sea eagle, Haliaeetus leucogaste and Eastern Osprey, Pandion cristatus, populations in South Australia. South Australian Ornithologist, 37(1), 1–16.

Dennis, T. E., McIntosh, R. R. & Shaughnessy, P. D. (2011b). Effects of human disturbance on productivity of White-bellied Sea eagles (Haliaeetu leucogaster). Emu, 111(2), 179–185. https://doi.org/10.1071/MU10044.

Dennis, T., Fitzpatrick, G. & Brittain, R. (2012a). Phases and duration of the White-bellied Sea eagle Haliaeetus leucogaster breeding season i South Australia and the implications for habitat management. Corella, 36(3), 63–68.

Denyer, L. C. (1960). Aggressive martial eagle. Lammergeyer, 1(1), 43–44.

Derocher, A. E. & Wiig, Ø. (2002). Postnatal growth in body length and mass of polar bears (Ursus maritimus) at Svalbard. Journal of Zoology 256(3), 343–349. https://doi.org/10.1017/ S0952836902000377.

Des Murs, M. O. (1845). Description de quelquesespècesnouvelles d'Oiseaux. Revue Zoologique, 1845, 175–179.

Desholm, M. & Kahlert, J. (2005). Avian collision risk at an offshore wind farm. Biology Letters, 1(3), 296–298. https://doi.org/10.1098 rsbl.2005.0336.

Deutsch, C. V. (2002). Geostatistical Reservoir Modeling. Oxford: Oxford University Press.

DeVault, T. L. & Rhodes, O. E. (2002). Identification of vertebrate scavengers of small mammal carcasses in a forested landscape. Acta Theriologica 47(2), 185–192. https://doi.org/10.1007/BF03192458.

DeVault, T. L., Beasley, J. C., Olson, Z. H., Moleón, M., Carrete, M., Margalida, A. & Sánchez-Zapata, J. A. (2016). Ecosystem services provide by avian scavengers. *In*: Sekercioglu, C.H., Wenny, D. G. & Whelan, C. J. (Eds.), Ecosystem Services Provided by Birds, 235–270. Chicago IL: University of Chicago Press.

DeVault, T. L., Belant, J. L. & Blackwell, B. F. (Eds.). (2013). Wildlife in airport environments: Preventing animal-aircraft collisions throug science-based management. Bethesda, MD: Johns Hopkins University Press.

DeVault, T. L., Brisbin, Jr., I. L. & Rhodes, Jr., O. E. (2004). Factors influencing the acquisition of rodent carrion by vertebrate scavengers an decomposers. Canadian Journal of Zoology, 82(3), 502–509. https://doi.org/10.1139/z04-022.

DeVault, T. L., Belant, J. L., Blackwell, B. F. & Seamans, T. W. (2011). Interspecific variation in wildlife hazards to aircraft: Implications fo airport wildlife management. Wildlife Society.

DeVault, T. L., Reinhart, B. D., Brisbin, I. L. & Rhodes, O. E. (2010). Flight behavior of Black and Turkey vultures: Implications fo reducing bird-aircraft collisions. The Journal of Wildlife Management, 69(2), 601–608. https://doi.org/10.2193/0022-541X(2005 0690601:FBOBAT2.0.CO;2.

DeVault, T. L., Rhodes Jr., O. E. & Shivik, J. A. (2003). Scavenging by vertebrates: Behavioural, ecological and evolutionary perspectives on a important energy transfer pathway in terrestrial ecosystems. Oikos, 102(2), 225–234. https://doi.org/10.1034/j.1600-0706.2003.12378.x.

Dewhurst, C. F., van Someren, G. R. C., Allan, R. G. & Thomsett, S. (1988). Observation on the breeding ecology of Ayres' Hawk-eagle Hieraaetu ayresii at Karen, Nairobi, Kenya. Gabar, 3, 85–93.

Dharmakumarsinhji, K. S. (1985). The Black eagle Ictinaetus malayensis Temm. at Sawai Madhopur (Rajasthan). Journal of the Bombay Natura History Society, 82(3), 655.

Di Giacomo, A. G. (2005). Birds of the Bagual Reserve. *In*: Di Giacomo, A. G. & Krapovickas, S. F. (Eds.), Historia natural y paisaje de la Reserv El Bagual, Provincia de Formosa, Argentina: Inventario de la fauna de vertebrados y de la flora vascular de un area protegida del Chac Humedo. Temas de naturaleza y conservación, 4, 201–465. Aves Argentinas/Asociación del Plata. Buenos Aires, Argentina.

Diamond, J. M. (1972). Avifauna of the Eastern Highlands of New Guinea. Nuttall Ornithological Club Publications No. 12.

Diamond, J. M. (1975). Distributional ecology and habits of some Bougainville birds (Solomon Islands). Condor, 77, 14–23.

Dickerman, R. W. (2003). Talon-locking in the red-tailed hawk. Journal of Raptor Research, 37, 176. Retrieved from https://sora.unm.edu/sites default/files/journals/jrr/v037n02/ p00176-p00176.pdf

Dickey, D. R. & Van Rossem, A. J. (1938). The birds of El Salvador, Vol. 23. Chicago: Field Museum of Natural History.

Dickinson, E. C., Kennedy, R. S. & Parkes, K. C. (1991). The birds of the Philippines: An annotated Check-list, 12, 1–507.

Dickinson, E. C. (2005). The correct authorship of the name Astur kienerii (Rufous-bellied Hawk Eagle). Bulletin of the British Ornithologists Club, 125, 317–320.

Dickinson, E. C., Overstreet, L. K., Dowsett, R. J. & Bruce, M. D. (2011). Priority! The Dating of Scientific Names in Ornithology. A Directory to the Literature and its Reviewers. Northampton: Aves Press.

Dickinson, E. C. & Remsen, Jr., J.V. (Eds.) (2013). The Howard and Moore Complete Checklist of the Birds of the World. Fourth edition. Volume 1. Non-passerines. Eastbourne, UK: Aves Press.

Dickinson, E. D. (2003). The Howard and Moore Complete Checklist of the Birds of the World, 3rd edn. Princeton, NJ: Princeton University Press.

Dickinson, M. (1991). Field Guide to the Birds of North America. Washington, D.C.: The National Geographic Society.

Dickman, A. J., Macdonald, E. A. & Macdonald, D. W. (2011). A review of financial instruments to pay for predator conservation and encourage human–carnivore coexistence. Proceedings of the National Academy of Sciences of the United States of America, 108(34), 13937–13944. https://doi.org/10.1073/pnas.1012972108.

Dingle, H. (1996). Migration: The Biology of Life on the Move. New York: Oxford University Press.

Dirksen, S., Spaans, A. L., van der Winden, J. & van den Bergh, L. M. J. (1998). Nachtelijke vliegpatronen en vlieghoogtes van duikeenden in het IJsselmeergebied. Limosa, 71, 57–68.

Ditmer, M. A., Vincent, J. B., Werden, L. K., Tanner, J. C., Laske, T. G., Iaizzo, P. A., Garshelis, D. L. & Fieberg, J. R. (2015a). Bears show a physiological but limited behavioral response to unmanned aerial vehicles. Current Biology, 25(17), 2278–2283. https://doi.org/10. 1016/j.cub.2015.07.024.

Ditmer, M. A., Garshelis, D. L., Noyce, K. V., Laske, T. G., Iaizzo, P. A., Burk, T. E., Forester, J.D. & Fieberg, J. R. (2015b). Behavioral and physiological responses of American Black bears to landscape features within an agricultural region. Ecosphere, 6(3), 1–21. https://doi.org/10.1890/ES14-00199.1

DJI. (2019). DJI. Retrieved from https://www.dji.com/ca/

Do Amaral, F. R., Sheldon, F. H., Gamauf, A., Haring, E., Riesing, M., Silveira, L. F. & Wajntal, A. (2009). Patterns and processes of diversification in a widespread and ecologically diverse avian group, the Buteonine hawks (Aves, Accipitridae). Molecular Phylogenetics and Evolution, 53(3), 703–715. https://doi.org/10.1016/j.ympev.2009.07.020

Do Amaral, F.S.R., Miller, M.J., Silveira, L.F., Bermingham, E. & Wajntal, A. (2006). Polyphyly of the hawk genera Leucopternis and Buteogallus (Aves, Accipitridae), multiple habitat shifts during the Neotropical Buteonine diversification. BMC Evolutionary Biology, 6, 1–10. https://doi.org/10.1186/1471-2148-6-10.

Do, E. (2020). Chaco eagle (Buteogallus coronatus), version 1.0. *In*: Schulenberg, T.S. (Ed.), Birds of the World. Ithaca, NY: Cornell Lab of Ornithology. https://doi.org/10.2173/bow.croeag1.01

Dobrowolska, A. & Melosik, M. (2008). Bullet-derived lead in tissues of the wild boar (Sus scrofa) and red deer (Cervus elaphus). European Journal of Wildlife Research, 54(2), 231–235. https://doi.org/10.1007/s10344-007-0134-y.

Dolbeer, R. A. & Eschenfelder, P. E. (2003). Amplified bird-strike risks related to population increases of large birds in North America. Proceedings of the International Bird Strike Committee, 26, 49–67.

Dolbeer, R. A. & Wright, S. E. (2009). Safety management systems: How useful will the FAA National Wildlife Strike Database be? Human-Wildlife Conflicts, 3, 167–178.

Dolbeer, R. A., Wright, S. E. & Cleary, E. C. (2000). Ranking the hazard level of wildlife species to aviation. Wildlife Society Bulletin, 28, 372–378.

Dolbeer, R. A. (1998). Evaluation of shooting and falconry to reduce bird strikes with aircraft at John F. Kennedy International Airport. *In*: Lesna, S. (Ed.), Proceedings of the IBSC, 2, 145–153. Stara Lesna, Slovakia 14–18 September. WP13.

Dolbeer, R. A. (2009). Birds and aircraft – Fighting for airspace in ever more crowded skies. Human-Wildlife Conflicts, 3, 165–166.

Dolbeer, R. A., Wright, S. E., Weller, J. R. & Begier, M. J. (2013). Wildlife strikes to civil aircraft in the United States 1990–2012. Federal Aviation Administration, National Wildlife Strike Database. Serial Report Number 19: Washington, D.C.

Dolotovskaya, S., Flores Amasifuen, C., Haas, C. E., Nummert, F. & Heymann, E. W. (2019). Active anti-predator behaviour of red titi monkeys (Plecturocebus cupreus). Primate Biology, 6(1), 59–64. https://doi.org/10.5194/pb-6-59-2019.

Domashevsky, S. V. (2000). National Action Plan for the conservation of the Great Spotted eagle (Aquila clanga) and Lesser Spotted eagle (A. pomarina) in Ukraine. *In*: Mykytyuk, O. (Ed.), National action plans on conservation of globally threatened bird species, 122–132. Kyiv: SoftART.

Domashevsky, S. V. (2005). Towards ecology of Greater and Lesser Spotted eagles in the north of Ukraine. Berkut, 14(2), 180–188.

Dombrovski, V. C. (2002). Hybridization of Lesser and Greater Spotted eagles (Aquila pomarina et A. clanga). In Belarus: Rule or exception? Subbuteo, 5, 23–31.

Dombrovski, V. C., Tishechkin, A. K. & Ivanovski, V. V. (2002). Belarus' National Action Plan for Greater Spotted Eagle (Aquila clanga). Conservation Document. Minsk: A.P.B. – BirdLife-Belarus.

Domenech, R., Bedrosian, B. E., Crandall, R. H. & Slabe, V. A. (2015). Space use and habitat selection by adult migrant Golden eagles wintering in the western United States. Journal of Raptor Research, 49(4), 429–440. https://doi.org/10.3356/rapt-49-04-429-440.1.

Donald, C.H. (1923). Migration of eagles. Journal of the Bombay Natural History Society, 29, 1054–1055.

Donald, P. F., Pisano, G., Rayment, M. D. & Pain, D. J. (2002). The Common Agricultural Policy, EU enlargement and the conservation of Europe's farmland birds. Agriculture, Ecosystems & Environment, 89(3), 167–182. https://doi.org/10.1016/S0167-8809(01)00244-4.

Donázar, J. A., Palacios, C. J., Gangoso, L., Ceballos, O., González, M. J. & Hiraldo, F. (2002). Conservation status and limiting factors in the endangered population of Egyptian vulture (Neophron percnopterus) in the Canary Islands. Biological Conservation, 107(1), 89–97. https://doi.org/10.1016/S0006-3207(02)00049-6.

Donegan, T. M. & Dávalos, L. M. (1999). Ornithological observations from Reserva Natural Tambito, Cauca, south-west Colombia. Cotinga, 12, 48–55.

Donegan, T. M. (2012). Range extensions and other notes on the birds and conservation of the Serranía de San Lucas, an isolated mountain range in northern Colombia. Bulletin of the British Ornithological Club, 132, 140–161.

Dornas, T. & Pinheiro, R. T. (2007). Predação de Opisthocomushoazin por Spizaetus ornatus e de Bubulcus ibis por Bubo virginianus em Tocantins, Brasil. Revista Brasileira de Ornitologia, 15(4), 601–604.

Doughty, C., Day, N. & Plant, A. (1999). Birds of the Solomons, Vanuatu & New Caledonia. London: Christopher Helm Ltd.

Douglas, G. (2002). Appendix 6: Lakeshore bird survey. In: Cunliffe, R. (Ed.). Biodiversity and wilderness around Lake Cahora Bassa, Tete province, Mozambique. Consultancy report for DPAP Tete, pp. 157-180. Harare: Biodiversity Foundation for Africa & The Zambezi Society.

Douglas, J. T. D., Arne, F., Rowena, H. W. L., James, W. P.-H., Aleksi, L., Douglas, D. J. T., Follestad, A., Langston, R. H. W. & Pearce-Higgins, J. W. (2012). Modelled sensitivity of avian collision rate at wind turbines varies with number of hours of flight activity input data. Ibis, 154(4), 858–861. https://doi.org/10.1111/j.1474-919X.2012.01239.x.

Douthwaite, R. J. (1992). Effects of DDT on the fish eagle Haliaeetus vocifer population of Lake Kariba in Zimbabwe. Ibis, 134, 250–258.

Dowsett, R. J. (1982). On the supposed occurrence of the Eastern Banded Snake-eagle Circaetus fasciolatus in Malawi. Nyala, 8, 100–101.

Dowsett, R. J. & Dowsett-Lemaire, F. (1993). Comments on the taxonomy of some Afrotropical bird species. Tauraco Research Report, 45, 323–389.

Dowsett, R. J. & Forbes-Watson, A. D. (1993). Checklist of birds of the Afrotropical and Malagasy regions. Liège, Belgium: Tauraco Press.

Dowsett, R. J. & Leonard, P. M. (2001). Results from bird ringing in Zambia. Zambia Bird Report, 1999, 16–46.

Dowsett, R. J., Aspinwall, D. R. & Dowsett-Lemaire, F. (2008). The Birds of Zambia: An Atlas and Handbook. Liège, Belgium: Tauraco Press.

Dowsett-Lemaire, F. (2006). An annotated list and life history of the birds of Nyika National Park, Malawi-Zambia. Tauraco Research Report, 8, 1–64.

Dowsett-Lemaire, F. & Dowsett, R. J. (2006). The Birds of Malawi: An Atlas and Handbook. Liege, Belgium: Tauraco Press & Aves.

Dowsett-Lemaire, F. & Dowsett, R. J. (2008). Selected notes on birds of Gola Forest and surroundings, Sierra Leone, including three new species for the country. Bulletin of the African Bird Club, 15, 215–227.

Drake, A., Rock, C. A., Quinlan, S. P., Martin, M. & Green, D. J. (2014). Wind speed during migration influences the survival, timing of breeding, and productivity of a neotropical migrant Setophaga petechia. PLoS One, 9, e97112.

Dravecký, M. (2009). Lesser Spotted eagle (Aquila pomarina). The Annual Report (2008). Dravcea à Sovy, 5(1), 8.

Dravecký, M., Danko, Š., Obuch, J., Kicko, J., Maderič, B., Karaska, D., Vrána, J., Šreibr, O., Šotnár, K., Vrlík, P. & Bohačík, L. (2008). Diet of the Lesser Spotted eagle (Aquila pomarina) in Slovakia. Slovak Raptor Journal, 22(1), 1–18. https://doi.org/10.2478/v10262-012-0014-6.

Dravecký, M., Sellis, U., Bergmanis, U., Dombrovski, V., Lontkowski, J., Maciorowski, G., Maderič, B., Meyburg, B.-U., Mizera, T., Stój, M., Treinys, R. & Wójciak, J. (2008). Colour ringing of the Spotted eagles (Aquila pomarina, Aquila clanga and their hybrids) in Europe – A review. Slovak Raptor Journal, 2(1), 37–52. https://doi.org/10.2478/v10262-012-0017-3.

Dravecký, M., Danko, Š., Hrtan, E., Kicko, J., Maderič, B., Mihók, J., Balla, M., Belka, T. & Karaska, D. (2013). Colour ringing programme of the Lesser Spotted eagle (Aquila pomarina) population in Slovakia and its new results in the period 2009–2012. Slovak Raptor Journal, 7(1), 17–36. https://doi.org/10.2478/srj-2013-0008.

Dravecký, M., Maderič, B., Kicko, J., Danko, S.Š., Karaska, D., Mihók, J. & Guziová, Z. (2015). Reproductive success, selected nest characteristics and the effectiveness of establishing protection zones of the Lesser Spotted eagle (Aquila pomarina) population in Slovakia. Slovak Raptor Journal, 9(1), 127–145. https://doi.org/10.1515/srj-2015-0010.

Drever, M. C., Lemon, M. J. F., Butler, R. W. & Millikin, R. L. (2014). Monitoring populations of Western Sandpipers and Pacific Dunlins during northward migration on the Fraser River Delta, British Columbia, 1991–2013. Journal of Field Ornithology, 85, 10–22.

Drewitt, A. L. & Langston, R. H. W. (2006). Assessing the impacts of wind farms on birds. Ibis, 148(Suppl. 1), 29–42. https://doi.org/10.1111/j.1474-919X.2006.00516.x.

Drewitt, A. L. & Langston, R. H. W. (2008). Collision effects of wind-power generators and other obstacles on birds. Annals of the New York Academy of Sciences, 1134, 233–266. https://doi.org/10.1196/annals.1439.015.

Drone, U. (2018). What are the rules for flying your drone in national parks and national forests? Retrieved from https://medium.com/@askdroneu/what-are-the-rules-for-flying-your-drone-in-national-parks-and-national-forests-1d27554984ad.

Du Puy, D. J. & Moat, J. (1996). A refined classification of the primary vegetation of Madagascar based on the underlying geology: Using GIS to map its distribution and to assess its conservation status. *In*: Lourenço, W.R. (Ed.), Proceedings of the International Symposium on the Biogeography of Madagascar, 205–218. Paris: ORSTOM.

Dubois, A. (2008). Phylogenetic hypotheses, taxa and nomina in zoology. *In*: Zootaxa, A., Minelli, A., Bonato, L. & Fusco, G. (Eds.), Updating the Linnaean Heritage: Names as tools for thinking about animals and plants. Zootaxa, 1950(1), 51–86. https://doi.org/10.11646/zootaxa.1950.1.7.

Dubois, P. J., Le Maréchal, P., Olioso, G. & Yésou, P. (2001). Inventaire des Oiseaux de France. Avifaune de la France métropolitaine. Paris.

Dubs, B. (1992). Birds of Southwestern Brazil: Catalogue and guide to the birds of the Pantanal of Mato Grosso and its border areas. Küsnacht, Switzerland: Bertrona Verlag.

Dubuisson, B. (1975). Practique de la photogrammétrie et de moyens cartographiques derivés des ordinateurs. Paris: Editions Eyrolles. [Practical photogrammetry and cartographic means derived from computers]. Paris: Editiones Eyrolles.

Duchardt, K. (2012). Female Eagle Hunter Reaches New Heights. Reuters video, March 8. Retrieved from http://mobile.reuters.com/video/2012/03/08/female-eagle-hunter-reaches -newheights? videoVideo. IdID=231446109.

Duckett, J. E. (1985). Some general bird notes from the Fourth Division of Sarawak. Sarawak Museum Journal, 34, 145–159.

Duckworth, J. W., Inskipp, T. P., Pasquet, E., Rasmussen, P. C., Rice, N. H., Robson, C. R. & Russell, D. G. D. (2008). A re-evaluation of the status of Tawny eagle Aquila rapax in South-East Asia. Forktail, 24, 127–129.

Dueker, K. J. (1979). Cited in Rural Development (RD) RD-08). ICT and Geo-Informatics for Rural Development. Odisha: Odisha State Open University.
Duerr, A. E., Miller, T. A., Lanzone, M., Brandes, D., Cooper, J., O'Malley, K., Maisonneuve, C., Tremblay, J. & Katzner, T. (2012). Testing an emerging paradigm in migration ecology shows surprising differences in efficiency between flight modes. PLoS One, 7(4), e35548. https://doi.org/10. 1371/journal.pone.0035548.
Duff, D. (2006). Has the juvenile plumage of honey-buzzard evolved to mimic that of common buzzard? British Birds, 99, 118–128.
Duinker, P. N. & Greig, L. A. (2006). The impotence of cumulative effects assessment in Canada: Ailments and ideas for redeployment. Environmental Management, 37(2), 153–161. https://doi.org/10.1007/s00267-004-0240-5.
Dunne, P. (2017). Birds of Prey: Hawks, Eagles, Falcons, and Vultures of North America. Boston, MA: Houghton Mifflin Harcourt.
Dunning, J. S. (1982). South American land birds: A photographic aid to identification. Newtown Square, PA: Harrowood Books.
Dunning, J. B. (1992). CRC Handbook of Avian Body Masses. Boca Raton, FL: CRC Press.
Dunning, J. B. (1993). CRC Handbook of Avian Body Masses. Boca Raton, FL: CRC Press.
Dunning Jr., J. B. (Ed.). (2008). CRC Handbook of Avian Body Masses (2nd edn.). Boca Raton, FL: CRC Press.
Dupont, J. E. (1971). Philippine Birds. Wilmington, DE: Delaware Museum of Natural History.
Duquette, J. F., Ureña, L., Ortega, J., Cisneros, I., Moreno, R. & Flores, E. E. (2017). Coiban agouti (Dasyprocta coibae) density and temporal activity on Coiba Island, Veraguas, Panama. Mammal Study, 42(3), 153–161. https://doi.org/10.3106/041.042.0305.
Durner, G. M., Simac, K. & Amstrup, S. C. (2013). Mapping polar bear maternal denning habitat in the National Petroleum Reserve-Alaska with an IFSAR digital terrain model. Arctic, 66(2), 197–206.
Du Toit, M. (2018). Talking Endangered Species Conservation with Wildlife ACT's Jo Maree. Retrieved from https://blog.rhinoafrica.com/2018/04/17/endangered-species-conservation-wildlife-act/
Dutson, G. (2011). Birds of Melanesia: Bismarcks, Solomons, Vanuatu and New Caledonia. London: Christopher Helm Publishers, London.
Dwyer, J. F. & Cockwell, S. G. (2011). Social hierarchy of scavenging raptors on the Falkland Islands, Malvinas. Journal of Raptor Research, 45(3), 229–235. https://doi.org/10.3356/JRR-10-75.1.
Dyrcz, A., Okulewicz, J., Tomialoje, L. & Witkowski, J. (1972). Breeding avifauna of the Biebrza Marshes and adjacent territories. Acta Ornithologica, 13, 343–422.
Eagle Club. (2020). White-tailed Eagle. Retrieved from https://www.kotkas.ee/species/white-tailed-eagle.
Eagle Directory. (2012). Ayres's Hawk Eagle – Aquila ayresii. Retrieved from https://eagleencyclopedia.org/species/ayress_hawk_eagle.html.
Eagle Directory. (2012). Rufous-Bellied Eagle – Lophotriorchis kienerii. Retrieved from https://eagleencyclopedia.org/species/rufous_bellied_eagle.html.
Eagle Directory. (2017). Spanish Imperial Eagle – Aquila adalberti. Retrieved from https://www.eagledirectory.org/species/spanish_imperial_eagle.html.
Eagle Directory. (2017). Steppe Eagle – Aquila nipalensis https://www.eagledirectory.org/ species/steppe_eagle.html.
Eames, J. C. (2010). Little-known African bird: Madagascar Serpent eagle Eutriorchis astur. Bulletin of the African Bird Club, 17(1), 99–101.
Earth, B. C. (2015). Wily coyote tries to steal meal from three Bald eagles. Retrieved from https://www.bbc.com/earth/story/20150730-coyote-steals-meal-from-eagles.
Eason, P. (1989). Harpy Eagle attempts predation on adult Howler Monkey. Condor, 91, 469–470.
eBird. (2020). Black Hawk eagle. Retrieved from https://ebird.org/species/blheag1.
eBird. (2020). Blyth"s hawk-eagle. Retrieved from https://ebird.org/species/blyhae1.
eBird. (2020). Chaco Eagle. Retrieved from https://ebird.org/species/croeagl.
ebird. (2020). Indian Spotted Eagle. Retrieved from https://ebird.org/species/inseagl.
eBird. (2020). Wallace's hawk-eagle. Retrieved from https://ebird.org/species/walhae1.
eBird. (2021). Gray-headed Fish-Eagle. Retrieved from https://ebird.org/species/gyhfie1
Eckhardt, H., Van Wilgen, B. & Biggs, H. (2000). Trends in woody vegetation cover in the Kruger National Park, South Africa, between 1940 and 1998. African Journal of Ecology 38(2), 108–115.
Edgar, A. T. (1933). Notes on the nidification of some Perak birds. Bulletin of the Raffles Museum, 8, 121–162.
Eduardo, C., Carvalho, A. & Marini, M. Â. (2007). Distribution patterns of diurnal raptors in open and forested habitats in south-eastern Brazil and the effects of urbanization. Bird Conservation International, 17(4), 367–380. https://doi.org/10.1017/S0959270907000822
Edwards, C. C. (1969). Winter behavior and population dynamics of American eagles in Utah. Ph.D. Thesis. Provo, UT: Brigham Young University.
Edwards, S. V., Arctander, P. & Wilson, A. C. (1991). Mitochondrial resolution of a deep branch in the genealogical tree for perching birds. Proceedings of the Royal Society of London B. Biological Sciences, 243(1307), 99–107. https://doi.org/10.1098/rspb. 1991.0017.
Edwards, W. (1928). Sacred Places. NADA, 6, 23–27.
Eeva, T., Belskii, E. & Kuranov, B. (2006). Environmental pollution affects genetic diversity in wild bird populations. Mutation Research/Genetic Toxicology and Environmental Mutagenesis, 608(1), 8–15. https://doi.org/10.1016/j.mrgentox.2006.04.021.
Eeva, T., Lehikoinen, E. & Ronkarönkä, M. (1998). Air pollution fades the plumage of the Great Tit. Functional Ecology 12(4), 607–612. https://doi.org/10.1046/j.1365-2435.1998.00221.x.
Egan. (2018). Evaluating the potential utility of drones to deter birds from areas of human-wildlife conflict. Master of Science Thesis. North Dakota State University of Agriculture and Applied Science.
Ehlers, M. (2008). Geoinformatics and digital earth initiatives: A German perspective. International Journal of Digital Earth, 1(1), 17–30. https://doi.org/10.1080/17538940701781975

Ehmsen, E., Pedersen, L., Meltofte, H., Clausen, T. & Nyegaard, T. (2011). The occurrence and reestablishment of White-tailed eagle and Golde eagle as breeding birds in Denmark. Dansk Ornitologisk Forenings Tidsskrift, 105(2), 13.
Einoder, L. L. D. & Richardson, A. M. (2007). Aspects of the hind limb morphology of some Australian birds of prey: A comparative an quantitative study. The Auk, 124(3), 773–788. https://doi.org/10.1093/auk/124.3.773.
Eisenmann, E. (1952). Annotated list of birds of Barro Colorado Island, Panama Canal Zone. Smithsonian Miscellaneous Collections, 117, 1–62
Eisermann, K. & Avendaño, C. (2006). Diversidad de aves en Guatemala, con una lista bibliográfica. In: Cano, E. (Ed.), Biodiversidad d Guatemala, Vol. 1, 525–623. Guatemala: Universidad del Valle de Guatemala.
Eisermann, K. & Avendaño, C. (2007). Annotated checklist of the birds of Guatemala. Barcelona: Lynx Edicions.
Eisermann, K. (2005). Noteworthy bird observations in Alta Verapaz, Guatemala. Bulletin of the British Ornithologists' Club, 125, 3–11.
Eisler, R. (1985). Cadmium hazards to fish, wildlife and invertebrates: A synoptic review. Biological Report, 1.2, Vol. 85, p. 2. Washington, DC United States Fish and Wildlife Service.
Eisler, R. (1988). Arsenic hazards to fish, wildlife, and invertebrates: A synoptic review. Biological Report 85 (1.12). Contaminant Hazard Review January 1988 Report No. 12. Laurel, MD: United States Fish and Wildlife Service, Patuxent Wildlife Research Center.
Eitniear, C. J. (1986). Status of the large forest eagles of Belize. Birds of Prey Bulletin, 3, 107– 110.
Eitniear, C. J., Gartside, M. R. & Kainer, M. A. (1991). Ornate Hawk eagle feeding on Green Iguana. Journal of Raptor Research, 25, 19–20.
Ek, K. H., Morrison, G. M., Lindberg, P. & Rauch, S. (2004). Comparative tissue distribution of metals in birds in Sweden using ICP-MS and lase ablation ICP-MS. Bulletin of Environmental Contamination and Toxicology, 47(2), 259–269. https://doi.org/10.1007/ s00244-004-3138-6.
Ekblad, C. (2004). The sea eagle's breeding season nesting choices in different archipelago zones on Åland. Doktorsavhandling. Institutionen fö Ekologioch Systematik. Avdelningen för populations biologi. Februari.
Ekstrom, J. M. & Isherwood, I. (2000). New and distribution bird records from the Island of Seram, Maluku. Kukila, 11, 51–57.
Elassar, A. (2020). Bald eagles across the United States are dying from lead poisoning. Retrieved from https://www.cnn.com/2020/02/16/us/bald eagles-dying-lead-poisoning-trnd/index. html.
Elassar, A. (2020). A bald eagle has been released back into the wild after recovering from lead poisoning. https://www.azfamily.com/news/us_ world_news/a-bald-eagle-has-been-released-back-into-the-wild-after-recovering-from-lead-poisoning/article_80cc3aed-fa8b-51a9-8c05 103c536f52ec.amp.html
Elder, G. (2018). Professing faith: The eagle is a religious symbol as well as a national one. Redlands Daily Facts.
Elderkin, M. (2014). Eagles: Too much of a good thing? Nova Scotian Herald.
Elgood, J. H., Heigham, J. B., Moore, A. M., Nason, A. M., Sharland, R. E. & Skinner, N. J. (1994). The Birds of Nigeria: An Annotated Check-list BOU Check-list No. 4 (2nd ed). Tring, UK: Hertsfordshire: British Ornithologists' Union.
Elliott, C. C. H. (1984). Martial eagle Polemaetus bellicosus attacking Hartebeest calf Alcephalus buselaphus. Scopus, 8, 45.
Elliott, J. E., Langelier, K. M., Scheuhammer, A. M., Sinclair, P. H. & Whitehead, P. E. (1992). Incidence of lead poisoning in Bald eagles an lead shot in waterfowl Gizzards from British Columbia, 1988–91. CWS Prog. Note 200. QC, Canada: Canadian Wildlife Service, Quebec.
Elliott, K. H., Duffe, J., Lee, S. L., Mineau, P. & Elliott, J. E. (2006). Foraging ecology of Bald eagles at an urban landfill. Wilson Journal o Ornithology, 118(3), 380–390. https://doi.org/10.1676/04-126.1.
Elliott, K. H., Elliott, J. E., Wilson, L. K., Jones, I. & Stenerson, K. (2011). Density-dependence in the survival and reproduction of bald eagles Linkages to chum salmon. Journal of Wildlife Management, 75(8), 1688–1699. https://doi.org/10.1002/jwmg.233.
Elliott, K.H., Duffe, J., Lee, S.L., Mineau, P. & Elliott, J.E. (2006). Foraging ecology of Bald eagles at an urban landfill. The Wilson Journal o Ornithology, 118(3), 380–390.
Ellis, D. H., Bednarz, J. C., Smith, D. G. & Flemming, S. P. (1993). Social foraging classes in raptorial birds: Highly developed cooperativ hunting may be important for many raptors. Bioscience, 43(1), 14–20. https://doi.org/10.2307/1312102.
Ellis, D. H., Moon, S. L. & Robinson, J. W. (2001). Annual movements of a Steppe eagle (Aquila nipalensis) summering in Mongolia an wintering in Tibet. Journal of the Bombay Natural History Society, 98, 335–340.
Ellis, D. H., Tsengeg, P., Whitlock, P. & Ellis, M. H. (2000). Predators as prey at a Golden Eagle Aquila chrysaetos eyrie in Mongolia. Ibis, 142(1) 139–141. https://doi.org/10.1111/j.1474-919X.2000.tb07694.x.
Eloff, F. C. (1980). Cub mortality in the Kalahari lion Panthera leovernayi (Roberts, 1948). Koedoe, 23(1), 163–170. https://doi.org/10.4102 koedoe.v23i1.643.
Elorriaga, J., Garcia, L., Martinez, J. & Unamunzaga, E. (2000). Quality of life of persons with mental retardation in Spain. *In*: Keith, K.D. & Schlock, R.L. (Eds.), Cross-cultural perspectives on quality of life. Washington, DC: American Association on Mental Retardation.
Eltringham, S. K. (1975). Territory size and distribution in the African Fish Eagle. Journal of Zoology, 175(1), 1–13. https://doi org/10.1111/j.1469-7998).1975.tb01386.x
Emlen, S. T. 1982. The evolution of helping. I: An ecological constraints model. The American Naturalist, 119, 29–39.
Encyclopaedia Britannica. (2020). Sea eagle. Retrieved from https://www.britannica.com/animal/Sea eagle.
Encyclopaedia Britannica. (2020a). Gray wolf. Retrieved from https://www.britannica.com/ animal/gray-wolf.
Encyclopaedia Britannica. (2020b). Coyote. Retrieved from https://www.britannica.com/ animal/coyote-mammal
Endangered Wildlife Trust. (2012). Eagles and Farmers. Retrieved from https://www.ewt.org.za/eBooks/booklets/Eagles%20and%20Farmers%2(booklet.pdf
Engelbach, P. (1932). Les oiseaux du Laos meridiona!. L'Oiseau et R.f. O. 2, 439–498.
Environmental Protection Agency. (2013). Polycyclic Aromatic Hydrocarbons (PAHs). Retrieved from https://www.epa.gov/sites/production files/2014-03/documents/pahs_f actsheet_cdc _2013.pdf.

Epstein, H. J. (1943). The origin and earliest history of falconry. International Review of the History of Science. Ibis, 34(6), 497–509. https://doi.org/10.1086/347874.
Erdoes, R. & Ortiz, A. (1984). American Indian Myth and Legends. New York, NY: Pantheon Books. Retrieved from https://www.gusd.net/cms/lib/ CA01000648/ Centricity/ Domain/2027/ AmericanIndianMythsAndLegends.pdf. New York: Pantheon Books.
Erickson, W. P., Johnson, G. D., Strickland, M. D., Young, D. P. Jr., Sernka, K. J. & Good, R. E. (2001). Avian collisions with wind turbines: A summary of existing studies and comparisons to other sources of avian collision mortality in the United States. National Wind Coordinating Committee, c/o RESOLVE, Inc., Washington, D.C
Erickson, W. P., Johnson, G. D., Young, D. P., Strickland, M. D., Good, R. E., Bourassa, M. et al. (2002). Synthesis and comparison of baseline avian and bat use, raptor nesting and mortality information from proposed and existing wind developments. Final. Edited by Western EcoSystems Technology Inc. Bonneville Power Administration. Cheyenne, WY.
Ericson, P. G. P., Anderson, C. L., Britton, T., Elżanowski, A., Johansson, U. S., Kallersjö, M., Ohlson, J. I., Parsons, T. J., Zuccon, D. & Mayr, G. (2006). Diversification of Neoaves: Integration of molecular sequence data and fossils. Biology Letters, 2(4), 543–547. https://doi.org/10.1098/rsbl.2006.0523.
Eriksen, J., Sargaent, D.E. & Victor, R. (2003). Oman bird list: The official list of the birds of the Sultanate of Oman 6. Oman Centre for Environmental Studies and Research, Sultan Qaboos University.
Erni, B., Liechti, F. & Bruderer, B. (2005). The role of wind in passerine autumn migration between Europe and Africa. Behavioral Ecology, 16, 732–740.
Erskine, A. J. (1968). Encounters between Bald eagles and other birds in winter. Auk, 85(4), 681–683. https://doi.org/10.2307/4083378.
Estes, R. (1992). The behavior guide to African mammals: Including hoofed mammals, carnivores, primates, 398–404. Berkeley, CA: University of California Press.
Estes, R. D. (2004). Cheetah Acinonyx jubatus. *In*: The behavior guide to African mammals: Including hoofed mammals, carnivores, primates (4th ed.), 377–383. Berkeley, CA: University of California Press.
Etheridge, B., Holling, M., Riley, H. T., Wernham, C. V. & Thompson, D. B. A. (2008). Scottish Raptor Monitoring Scheme Report (2006). Aberlady, UK: Scottish Ornithologists' Club.
European Chemicals Bureau. (2003). Technical guidance document in support of Commission. Directive 93/67/EEC on risk assessment for new notified substances, Commission Regulation (EC) 1488/94 on risk assessment for existing substances and Directive 98/8/EC of the European Parliament and of the Council concerning the placing of biocidal products on the market. Ispra, Italy: European Commission, Joint Research Centre, European Chemicals Bureau.
Evans, K. (2018). Return of the lost birds. New Zealand Geographic, 154, 30. Retrieved from https://www.nzgeo.com/stories/return-of-the-lost-birds/.
Evans, R. J., Pearce-Higgins, J., Whitfield, D. P., Grant, J. R., MacLennan, A. & Reid, R. (2010). Comparative nest habitat characteristics of sympatric White-tailed Haliaeetus albicilla and Golden eagles Aquila chrysaetos in western Scotland. Bird Study, 57(4), 473–482. https://doi.org/10.1080/00063657.2010.489317.
Evans, R. J., Wilson, J. D., Amar, A., Douse, A., MacLennan, A., Ratcliffe, N. & Whitfield, D. P. (2009). Growth and demography of a re-introduced population of White-tailed eagles Haliaeetus albicilla. Ibis, 151(2), 244–254. https://doi.org/10.1111/j.1474-919X.2009.00 908.x.
Evans, R. M. & Knopf, F. L. (1993). American White pelican (Pelecanus erythrorhynchos). *In*: Poole, A. & Gill, F. (Eds.), The Birds of North America, No. 57. Philadelphia: The Academy of Natural Sciences; Washington, D.C.: American Ornithologists' Union.
Evans, R.J., Pearce-Higgins, J., Whitfield, D.P., Grant, J.R., MacLennan, A. & Reid, R. (2010). Comparative nest habitat characteristics of sympatric White-tailed Haliaeetus albicilla and Golden eagles Aquila chrysaetos in western Scotland. Bird Study, 57(4), 473–482.
Everist, R. (2009). Traveller's Guide to the Goldfields: History and natural heritage trails through Central and Western Victoria. Australia: Best Shot.
Evers, D. C., Paruk, J. D., McIntyre, J. W. & Barr, J. F. (2010). Common Loon (Gaviaimmer). *In*: Poole, A.F. (Eds.), The Birds of North America. Ithaca: Cornell Laboratory of Ornithology.
Evers, D. C., Savoy, L. J., DeSorbo, C. R., Yates, D. E., Hanson, W., Taylor, K. M., Siegel, L.S., Cooley, J.H. Jr., Bank, M.S., Major, A., Munney, K., Mower, B.F., Vogel, H.S., Schoch, N., Pokras, M., Goodale, M.W. & Fair, J. (2008). Adverse effects from environmental mercury loads on breeding common loons. Ecotoxicology, 17(2), 69–81. https://doi.org/10.1007/s10646-007-0168-7.
Fagan, J. & Komar, O. (2016). Peterson field guide to birds of northern Central America. Houghton Mifflin Harcourt.
Fain, M. G. & Houde, P. (2004). Parallel radiations in the primary clades of birds. International Journal of Organic Evolution, 58(11), 2558–2573. https://doi.org/10.1111/j.0014-3820.2004.tb00884.x.
Fairbairn, D., Blanckenhorn, W. & Szekely, T. (2007). Sex, size and gender roles: Evolutionary studies of sexual size dimorphism. Oxford: Oxford University Press.
Falandysz, J., Ichihashi, H., Szymczak, K., Yamasaki, S. & Mizera, T. (2001). Metallic elements and metal poisoning among White-tailed sea eagles from the Baltic Sea. Marine Pollution Bulletin, 42, 1190–1193. https://doi.org/10.1016/s0025-326x(01)00217-x.
Falandysz, J., Jakuczun, B. & Mizera, T. (1988). Metals and organochlorines in four female White-tailed eagles. Marine Pollution Bulletin, 19(10), 521–526. https://doi.org/10.1016/0025-326X(88)90542-5 0542-5.
Fam, S. D. & Nijman, V. (2011). Spizaetus hawk-eagles as predators of arboreal colobines. Primates; Journal of Primatology, 52(2), 105–110. https://doi.org/10.1007/s10329-011-0240-z.
Fantin, G. (1974). L'Aquila minore (Hieraaetus pennatus). Rivista Italiana di Ornitologia, 44: 241–249.
Farmer, C. J. & Smith, J. P. (2010). Seasonal differences in migration counts of raptors: Utility of spring counts for population monitoring. Journal of Raptor Research, 44(2), 101–112. doi: 10.3356/JRR-09-31.1, PubMed: 101112.

Farmer, C. J., Goodrich, L. J., Ruelas, I. E. & Smith, J. P. (2008). Conservation status of North America's Birds of Prey. *In*: Bildstein, K. L., Smith, J. P., Ruelas, I. E. & Veit, R. R. (Eds.), State of North America's Birds of Prey. Cambridge, Washington, D.C.: Nuttall Ornithological Club and American Ornithologists' Union.

Farve, R. (2018). Demonstration of Satellite/GPS Telemetry for Monitoring Fine-Scale Movements of Lesser Prairie-Chickens. Retrieved from https://www.fs.fed.us/t-/programs/im/satellite _gps_telemetry/ wildlifetrackingtelementry.htm.

Faulkner, S. (2011). Eagle Hunters. Retrieved from http://samfaulkner.co.uk/work/eagle-hunters/4/caption.

Favaloro, N. (1944). The White-breasted Sea eagle along the Murray Valley. Emu-Austral Ornithology, 43(4), 233–242. https://doi.org/10.1071/MU943233.

Fawkner, J. (1991). More on Wedge-tailed eagles hunting in forests. Australasian Raptor Association News, 12, 25–26.

Federal Aviation Administration. (2018). FAA Aerospace Forecast. Retrieved from https://www.faa.gov/data_research/aviation/aerospace_forecasts/media/FY2019-9_FAA_Aerospace_Forecast.pdf

Federal Aviation Administration (FAA). (2004). Airplane flying handbook. Washington, DC: United States Department of Transportation. Federal Aviation Administration, Flight Standards Service.

Fedjuschin, A. W. & Dolbik, M. S. (1967). The Birds of White Russia. Minsk: Raptors in the New Millennium.

Feduccia, A. (1999). The Origin and Evolution of Birds (2nd ed.). New Haven: Yale University Press.

Felce, D. & Perry, J. (1995). Quality of life: Its definition and measurement. Research in Developmental Disabilities, 16(1), 51–74. https://doi.org/10.1016/0891-4222(94)00028-8.

Felce, D. & Perry, J. (1996). Assessment of quality of life. *In*: Schalock, R. L. & Siperstein, G. N. (Eds.), Quality of life, volume I: Conceptualization and measurement, pp. 63–70. Washington, DC: American Association on Mental Retardation.

Feldhamer, G. A., Thompson, B. C. & Chapman, J. A. (2003). Wild mammals of North America: Biology, management, and conservation. Baltimore, MD: Johns Hopkins University Press.

Fennec, R. S. (2005). Growth rate and duration of growth in the adult canine of Smilodon gracilis and inferences on diet through stable isotope analysis. Feranec Bull FLMNH, 45(4), 369–377.

Ferguson-Lees, J. & Christie, D. A. (2001). Raptors of the world. London: Christopher Helm Publishers.

Fernández, M., Oria, J., Sánchez, R., Garcia, L. & Margalida, A. (2009). Space use of adult Spanish Imperial eagles Aquila adalberti. Acta Ornithologica, 44, 17–26. https://doi.org/10.3161/00016 4509X464849.

Fernandez-de-Simon, J., Díaz-Ruiz, F., Cirilli, F., Sánchez, F., Villafuerte, R., Delibes-Mateos, M. & Ferreras, P. (2011). Towards a standardized index of European rabbit abundance in Iberian Mediterranean habitats. European Journal of Wildlife Research, 57(5), 1091–1100. https://doi.org/10.1007/s10344-011-0524-z.

Fernando, R. H. S. S., Palihawadana, A. & Gunasekara, S. (2001). First site record of a Black eagle Ictinaetus malayensis nest in Sri Lanka. Sri Lanka Naturalist, 4(03), 49.

Fernando, S. (2017). Seeing in the dark and more: Facts and FAQs about thermal imaging. Retrieved from https://news.mongabay.com/2017/03/facts-and-faqs-about-thermal-imaging/

Fernie, K. J., Smits, J. E., Bortolotti, G. R. & Bird, D. M. (2001). Reproduction success of American Kestrels exposed to dietary polychlorinated biphenyls. Environmental Toxicology and Chemistry, 20(4), 776–781. https://doi.org/10.1002/etc.5620200412doi

Ferrari, S. F. & Ferrari, M. A. L. (1990). Predator avoidance behaviour in the buffy-headed marmoset, Callithrix flaviceps. Primates, 31(3), 323–338. https://doi.org/10.1007/BF02381104

Ferrer, M. (1993). The Spanish Imperial eagle. (El Águila imperial ibérica). Madrid (Ed.): Quercus.

Ferrer, M. (1993). Juvenile dispersal behaviour and natal philopatry of a long-lived raptor, the Spanish Imperial eagle Aquila adalberti. Ibis, 135(2), 132–138. https://doi.org/10.1111/j.1474-919X.1993.tb02824.x

Ferrer, M. (2001). The Spanish Imperial eagle. Barcelona: Lynx Edicions.

Ferrer, M. & Hiraldo, F. (1992). Man-induced sex-biased mortality in the Spanish Imperial eagle. Biological Conservation, 60(1), 57–60. https://doi.org/10.1016/0006-3207(92)90799-S

Ferrer, M. & Janss, G. F. (Eds.). (1999). Birds and power lines: Collision, electrocution, and breeding. Madrid: Quercus.

Ferrer, M. & Negro, J. J. (2004). The near extinction of two large European predators: Super specialists pay a price. Conservation Biology, 18(2), 344–349. https://doi.org/10.1111/j.1523-1739.2004.00096.x

Ferrer, M., de Lucas, M., Janss, G. F. E., Casado, E., Muñoz, A. R., Bechard, M. J. & Calabuig, C. P. (2012). Weak relationship between risk assessment studies and recorded mortality in wind farms. Journal of Applied Ecology, 49(1), 38–46. https://doi.org/10.1111/j.1365-2664.2011.02054.x

Ferrer, M., Morandini, V. & Newton, I. (2015). Floater interference reflects territory quality in the Spanish Imperial eagle Aquila adalberti: A test of a density-dependent mechanism. Ibis, 157(4), 849–859. https://doi.org/10.1111/ibi.12289

Fessenden, M. (2015). A drone encounters two eagles, and the birds win. Retrieved from https://www.smithsonianmag.com/smart-news/drone-encounter-two-eagles-didnt-go-planned-180957336/

Ffrench, R., O'Neill, J. P. & Eckelberry, D. R. (1991). Guide to the birds of Trinidad and Tobago. Illustrated by John P. O'Neill. Ithaca, NY: Cornell University Press.

Fiedler, G. & Wissner, A. (1980). Freileitungen als tödliche Gefahr für Störche Ciconia ciconia. Okol. Vögel, 2 (Sonderheft), 59–109.

Fielding, A. H., Whitfield, D. P. & McLeod, D. R. A. (2006). Spatial association as an indicator of the potential for future interactions between wind energy developments and Golden eagles Aquila chrysaetos in Scotland. Biological Conservation, 131(3), 359–369. https://doi.org/10.1016/j.biocon.2006.02.011

Fielding, A. & Haworth, P. (2010). Golden eagles and wind farms. Report to Scottish Natural Heritage.

Fielding, M. (2019). Australia's birds are embedded with Aboriginal Australian Culture. Retrieved from https://www.deep-group.com/post/australia-s-birds-are-embedded-within-aboriginal-australian-culture.

Figueiredo, A. M., Valente, A. M., Barros, T., Carvalho, J., Silva, D. A. M., Fonseca, C. & Torres, R. T. (2020). What does the wolf eat? Assessing the diet of the endangered Iberian wolf (Canis lupus signatus) in northeast Portugal, 15(3), e0230433. https://doi.org/10.1371/journal.pone.0230433

Filho, R. C. B., Roos, A. L., Lacerda, R. C. A., Costa, J. A., Freitas, G. L. & Arantes, M. S. (2007). Report of the Brazilian national banding system (1966–1995). Ornithologica, 2, 119–169.

Finch-Davies, C. G. (1919). Some notes on Hieraaetus ayresi Gurney Sen. [Lophotriorchis lucani Sharpe et auctorum]. Ibis, 11(1), 167–179.

Fingas, M. (2017). The fate of PAHs resulting from in-situ oil burns. International Oil Spill Conference Proceedings, 2017(1), 1041–1056. https://doi.org/10.7901/2169-3358-2017.1.1041

Finlayson, C. (1992). Birds of the Strait of Gibraltar. London: T. & A. D. Poyser.

Firouz, E. (2005). The complete fauna of Iran. New York. NY: I. B. Tauris.

Fischer, D. L. (1985). Piracy behavior of wintering Bald eagles. Condor, 87(2), 246–251. https://doi.org/10.2307/1366889

Fischer, W. (1970). Die Seeadler. A. Ziemsen Verlag.

Fischer, W. (1982). Die Seeadler. Die Neueneue Brehm-Bücherei, A. Wittenberg, Lutherstadt: Ziemsen Verlag.

Fisher, I. J., Pain, D. J. & Thomas, V. G. (2006). A review of lead poisoning from ammunition sources in terrestrial birds. Biological Conservation, 131(3), 421–432. https://doi.org/10.1016/j.biocon.2006.02.018

Fitzherbert, K. (1992). Wedge-tailed eagle (Tasmanian subspecies). *In*: S. T. Garnett (Ed.). Threatened and extinct birds of Australia, 36–37. RAOU Report 82.

Fitzner, R. E. (1978). Behavioral ecology of the Swainson's hawk Buteo swainsoni in southeastern Washington. Ph.D. Thesis. Washington State University, Pullman.

Fjeldså, J. & Krabbe, N. (1990). Birds of the high Andes. Svendborg, Denmark: Zoological Museum: University of Copenhagen and Apollo Books.

Fjeldså, J. (2020). Pinsker's hawk-eagle (Nisaetus pinskeri), version 1.0. *In*: Del Hoyo, J. Elliott et al. (Eds.), Birds of the World. Ithaca, NY: Cornell Laboratory of Ornithology. https://doi.org/10.2173/bow.pinhae1.01doi: 10.2173/bow.pinhae1.01.

Flatten, C. J., Madrid, J. A., Hernandez, A. & Gerhardt, R. (1989). Observations at the nest of an Ornate hawk-eagle. The Peregrine Fund Maya Project Progress Report II: 69–79. Boise, ID: Peregrine Fund, Inc.

Flatten, C. J., Madrid, J. A., Madrid, H. D., Funes, S. H., Hernandez, A. E. & Botzoc, R. (1990). Biology of the ornate hawk-eagle (Spizaetus ornatus). Maya Project Progress Report III: 129–144. Boise, ID: Peregrine Fund, Inc.

Fleay, D. (1948). Notes on the White-breasted sea eagle. Emu–Austral Ornithology, 48(1), 20–31. https://doi.org/10.1071/MU948020

Fleischer, R. C., Olson, S. L., James, H. F. & Cooper, A. C. (2000). Identification of the extinct Hawaiian eagle (Haliaeetus) by mtDNA sequence analysis. Auk, 117(4), 1051–1056. https://doi.org/10.1093/auk/117.4.1051

Fleming, R. L., Jr. (1983). An east–west Aquila migration in the Himalayas. Journal of the Bombay Natural History Society, 80, 58–62.

Flerov, A. I. (1970). The ecology of the White-tailed eagle of the Kandalaksha Bay. Proceeding of the Kandalaksha State Nature Reserve, VII (pp. 215–232).

Flett, D. G., Manore, M. J., Ramsay, B. R. & Falkingham, J. C. (2001). Preparing for operational use of RADARSAT-2 data at the Canadian Ice Service. IGARSS 2001. Scanning the Present and Resolving the Future. Proceedings. IEEE 2001 International Geoscience and Remote Sensing Symposium (Cat. No.01CH37217), 2001, vol. 1, 493–495. https://doi.org/10.1109/IGARSS.2001.976200

Flint, V. E., Boehme, R. L., Kostin, Y. V. & Kuznetsov, A. A. (1984). A field guide to birds of the USSR. Princeton, NJ: Princeton University Press.

Flores-Puebla, L., Montejo, D. J. & Burton, A. (2002). Analysis of the status of five species of Neotropical eagles in the Montes Azules and El Ocote Biosphere Reserves, Chiapas, México. Eagle: Program of the Conference on Neotropical Raptors and Symposium on the Harpy. Fondo Peregrino – Panamá, Panamá City, Brazil: Fondo Peregrino.

Folkestad, A. O. (1998). Prosjekt Havorn: Organisering, bestandforhold, populasjonsdynamikk, forvaltningsproblematikk. [Report of Direktoratet for Naturforvaltning, Trondheim, Norway. Fraser Darlingdarling, F], 1955. Oxford: West Highland Survey.

Folkestad, A. O. (2003). Status of the White-tailed sea eagle in Norway. *In*: Helander, B., Marquiss, M. & Bowerman, W. (Eds.), Sea Eagle. Proceedings of the International Sea Conference in Björkö, Sweden, 13–17 September 2000, 51–55. Stockholm: SNF & Åtta. 45 Tryckeri AB.

Follestad, A. R., Flagstad, Ø., Nygård, T., Reitan, O. & Schulze, J. (2007). Vindkraft og fugl på Smøla 2003–2006 p. 78s. In Nina Rapport. Trondheim: NINA.

Follestad, A., Reitan, O., Pedersen, H. C., Brøseth, H. & Bevanger, K. (1999). Vindkraftverk på Smøla: Mulige konsekvenser for "rødlistede" fuglearter. *In*: NINAN. Oppdragsmelding. Trondheim, p. 48. Trondheim: Norsk Instituut for Naturforskning.

Fombonnat, J. (2004). lL'Aigleaigle botté Hieraaetus pennatus. *In*: Thiollay, I. & Bretagnolle, J. (Eds.), Delachaux et Niestlé, V. apaces nicheurs de France, distribution, effectifs et conservation, pp. 100-103.

Fonturbel, F. E. & Simonetti, J. A. (2011). Translocations and human-carnivore conflicts: Problem solving or problem creating? Wildlife Biology, 17(2), 217–224. https://doi.org/10.2981/10-091

Ford, H. S. & Alcorn, J. R. (1964). Observations of Golden eagle attacks on coyotes. Condor, 66(1), 76–77. https://doi.org/10.1093/condor/66.1.76adoi: 10.1093/condor/66.1.76a.

Forero, M. G., Donázar, J. A., Blas, J. & Hiraldo, F. (1999). Causes and consequences of territory change and breeding dispersal distance in the Black kite. Ecology, 80(4), 1298–1310. https://doi.org/10.1890/0012-9658(1999)0801298: CACOTC2.0.

Forney, J. M. & Aragón, R. (2009). Notes on Oaxacan birds. Huitzil, 10, 38–47.
Forrester, R. & Andrews, I. J. (2007). The birds of Scotland. Aberlady. UK: Scottish Ornithologists Club.
Forsman, D. (1991). Die Bestimmung von schell – Aquila clanga, Schrei – A. pomarina und Steppenadler A. nipalensis. Limicola, 5, 145–185.
Forsman, D. (1999). The raptors of Europe and the Middle East. A handbook of field identification. London: Poyser, London.
Fowler, D. W., Freedman, E. A. & Scannella, J. B. (2009). Predatory functional morphology in raptors: Interdigital variation in talon size is relate to prey restraint and immobilisation technique. PLoS One, 4(11), e7999. https://doi.org/10.1371/journal.pone.0007999
Fowler, J. M. & Cope, J. B. (1964). Notes on the Harpy eagle in British Guiana. The Auk, 81(3), 257–273. https://doi.org/10.2307/4082683.
Fox, R., Lehmkuhle, S. W. & Westendorf, D. H. (1976). Falcon visual acuity. Science, 192(4236), 263–265. https://doi.org/10.1126/science.125776
Fra, E. A., Salinas, R. S., Rodríguez, P. R. & Barrioneuvo, C. (2008). Presence and habitat of the Black-and-chestnut eagle (Oroaetus isidori De Murs, 1845) in Catamarca Province, Argentine Republic. Nuestras Aves, 24, 6–9.
Fraga, R. M. & Clark, R. (1999). Notes on the avifauna of the upper Bermejo River (Argentina and Bolivia) with a new species for Argentina Cotinga, 12, 77–78.
François, J. (1975). 'Contribution à la connaissance de l'avifauned'Afrique du Nord.' [Contribution to the knowledge of the North Africar Avifauna]. Alauda, 43, 279–293.
Frank, R. A. & Lutz, R. S. (1999). Productivity and survival of Great horned owls exposed to dieldrin. The Condor, 101(2), 331–339. https://doi org/10.2307/1369996
Franke, U. E. (2015). Civilian drones: Fixing an image problem? International relations and Security Network. Retrieved from https://isnblog.ethz ch/security/civilian-drones-fixing-an-image-problem
Franklin, H. (1933). Vakaranga superstitions. NADA, 11, 122–124.
Franklin, J. (1831). Exhibition of a collection of birds formed by Major James Franklin. Proceedings of the Committee of Science anc Correspondence of the Zoological Society of London, Part 1 (pp. 114–125).
Franson, J. C. (1996). Interpretation of tissue lead residues in birds other than waterfowl. *In*: W. N. Beyer et al. (Eds.), Environmental contaminant in wildlife: Interpreting tissue concentrations, pp. 265–279. Boca Raton, FL: CRC Press, Lewis Publishers.
Franson, J. C. & Pain, D. J. (2011). Lead in birds. *In*: W. N. Beyer & J. P. Meador (Eds.), Environmental contaminants in biota: Interpreting tissu concentrations (2nd ed.), pp. 563–593). Boca Raton, FL: CRC Press.
Franson, J. C., Hansen, S. P., Creekmore, T. E., Brand, C. J., Evers, D. C., Duerr, A. E. & DeStefano, S. (2003). Lead fishing weights and othe fishing tackle in selected waterbirds. Waterbirds, 26(3), 345–352. https://doi.org/10.1675/1524-4695(2003)026[0345:LFWAOF]2.0.CO;2.
Franson, J. C., Lahner, L. L., Meteyer, C. U. & Rattner, B. A. (2012). Copper pellets simulating oral exposure to copper ammunition: Absence o toxicity in American kestrels (Falco sparverius). Archives of Environmental Contamination and Toxicology, 62(1), 145–153. Retrieved fror http://www.doi.org/10.1007/s00244-011-9671-1
Franson, J. C., Siloe, L. & Thomas, J. J. (1995). Causes of eagle deaths. *In*: E. T. LaRoe, G. S. Farris, C. E. Puckett & P. D. Doran (Eds.), Ou living resources: A report to the nation on the distribution, abundance, and health of U. S. Plants, animals, and ecosystems, 68. Washington DC: National Biological Service.
Frantz, A., Pottier, M. A., Karimi, B., Corbel, H., Aubry, E., Haussy, C., . . . Castrec-Rouelle, M. (2012). Contrasting levels of heavy metals in th feathers of urban pigeons from close habitats suggest limited movements at a restricted scale. Environmental Pollution, 168, 23–28. https:/ doi.org/10.1016/j.envpol.2012.04.003
Fraser Darling, F. (1955). Highland Survey. Oxford: West.
Fraser, J. D. (1985). The impact of human activities on Bald eagle populations – A review. *In*: Gerrard, J. M. & Ingram, T. M. (Eds.), The Bal Eagle in Canada, pp. 68–84. Headingley, Manitoba: White Horse Plains Publications.
Fraser, J. D., Frenzel, L. D. & Mathisen, J. E. (1985). The impact of human activities on breeding Bald eagles in north-central Minnesota. Journa of Wildlife Management, 49(3), 585–592. https://doi.org/10.2307/3801676
Fraser, J. D., Frenzel, L. D., Mathisen, J. E. & Shough, M. E. (1983). Three adult Bald eagles at an active nest. Journal of Raptor Research, 1(7) 29–30.
Freile, J. F. & Chaves, J. A. (2004). Interesting distributional records and notes on the biology of bird species from a cloud forest reserve in North West Ecuador. Bulletin of the British Ornithologists' Club, 124, 6–16.
Freitas, M. A., Lima, D. M. & Gomes, F. B. R. (2014). Slaughter record of Harpy harpyja (Accipitridae) for human consumption in Maranhão Brazil. Ornithological News, 178, 12–15.
French, W. (1957). Birds of the Solomon Islands. Ibis, 99, 126–130.
Frenzel, R. W. & Anthony, R. G. (1989). Relationship of diets and environmental contaminants in wintering Bald eagles. Journal of Wildlif Management, 53(3), 792–802. https://doi.org/10.2307/3809214
Fretwell, P. T., LaRue, M. A., Morin, P., Kooyman, G. L., Wienecke, B., Ratcliffe, N., . . . Trathan, P. N. (2012). An Emperor penguin populatio estimate: The first global, synoptic survey of a species from space. PLoS ONE, 7(4), e33751. https://doi.org/10.1371/journal.pone.0033751
Fretwell, P. T., Trathan, P. N., Wienecke, B. & Kooyman, G. L. (2014). Emperor penguins breeding on ice shelves. PLoS One, 9(1), e85285. https:/ doi.org/10.1371/journal.pone.0085285
Fretwell, P. T., Staniland, I. J. & Forcada, J. (2014). Whales from space: Counting southern Right Whales by satellite. PLoS ONE, 9(2), e88655 https://doi.org/10.1371/journal.pone.0088655
Frías, O., Serradilla, A. I. & Escudero, E. (2009). Informe de actividades de la Centralde Anillamientode Aves "ICONA". año2007. Ecología, 22 137–229. Retrieved from http://gob.es/ministerio/pags/Biblioteca/Revistas/pdf_REPN%2FECO_2008_22_137_229.pdf
Frid, A. & Dill, L. M. (2002). Human-caused disturbance stimuli as a form of predation risk. Conservation Ecology, 6(1), 11–27. https://doi org/10.5751/ES-00404-060111

Friedemann, G., Leshem, Y., Kerem, L., Shacham, B., Bar-Massada, A., McClain, K. M., . . . Izhaki, I. (2016). Multidimensional differentiation in foraging resource use during breeding of two sympatric top predators. Scientific Reports, 6, 35031. https://doi.org/10.1038/srep35031

Friedmann, H. (1935). A new race of the Crested hawk-eagle Spizaetus ornatus. Journal of the Washington Academy of Sciences, 25(10), 450–451.

Friedmann, H. (1950). Birds of North and Middle America, Pt. 2. Bulletin of the United States National Museum, 50.

Friedmann, H. & Foster, D. S. (1950). A contribution to the ornithology of northeastern Venezuela. Proceedings of the United States National Museum, 100(3268), 411–538, plates 10. https://doi.org/10.5479/si.00963801.100-3268.411

Friedmann, H., Griscom, L. & Moore, R. T. (1950). Distributional check-list of the birds of Mexico. Part I. Pacific Coast Avifauna, 29, 1–202.

Friedmann, H. & Smith, F. D., Jr. (1955). A further contribution to the ornithology of northeastern Venezuela. Proceedings of the United States National Museum, 104(3345), 463–524. https://doi.org/10.5479/si.00963801.104-3345.463

Friend, M. (1981). Waterfowl diseases—Changing perspectives for the future. Int. Waterfowl Symp., 4, 189–196.

Frost, P. G. H., Menaut, J. C., Walker, B. H., Medina, E., Solbrig, O. & Swift, M. (1986). Responses of savannas to stress and disturbance. Biology International, Special Issue, 10, 1–78.

Fu, P. & Sun, J. (2010). WebGIS: Principles and Applications, ESRI Press.

Fuchs, J., Cibois, A., Duckworth, J. W., Eve, R., Robichaud, W. G., Tizard, T. & Van Gansberghe, D. (2007). Birds of Phongsaly Province and the Nam Ou River, Laos. Forktail, 23, 22–86.

Fuller, M. R., Seegar, W. S. & Howey, P. W. (1995). The use of satellite systems for the study of bird migration. Israel Journal of Zoology, 41, 243–252.

Fuller, M. R., Seegar, W. S. & Schueck, L. S. (1998). Routes and travel rates of migrating Peregrine falcons (Falco peregrinus) and Swainson's hawks (Buteo swainsoni) in the Western Hemisphere. J. Journal of Avian Biology, 29(4), 433–440. https://doi.org/10.2307/3677162

Fülöp, A., Kovács, I., Baltag, E., Daróczi, S. J., Dehelean, A. S., Dehelean, L. A. (2014). Autumn migration of soaring birds at Bosporus: Validating a new survey station design. Bird Study, 61(2), 264–269. https://doi.org/10.1080/00063657.2014.907236

Funes, S. H., Lopez, J. & Lopez, G. (1992). Reproductive biology, food habits, and behavior of the Black Hawk eagle in Tikal National Park. *In*: D. F. Whitacre & R. K. Thorstrom (Eds.), Maya Project progress report V, pp. 173–178. Boise, ID: Peregrine Fund, Inc.

Furness, R. W. (1987). Kleptoparasitism in seabirds. *In*: J. P. Croxall (Ed.), Seabirds: Feeding ecology and role in marine ecosystems, pp. 77–100. Cambridge University.

Gábor, M. (1998). Annotated list of the birds of Hungary. KTM Termeszetvedelmi.

Gadgil, M. (2004). Karnataka State of Environment Report and Action Plan: Biodiversity Sectors. ENVIS technical report no. 16. ENVIS, CES, IISc, Bangalore. Retrieved from http://iisc.ernet.in/hpg/envis/ETR/CES_ETR/etr16.pdf.

Gaffney, R. F. & Mooney, N. J. (1997). The Wedge-tailed eagle recovery plan: Management phase. Hobart, Tasmania, Australia: Department of Parks, Wildlife and Heritage.

Gagnon, P. & Coleman, D. J. (1990). Geomatics: An integrated, systemic approach to meet the needs for spatial information. Canadian Institute of Surveying and Mapping Journal, 44(4), 377–382.

Gairola, S., Procheş, Ş. & Rocchini, D. (2013). High-resolution satellite remote sensing: A new frontier for biodiversity exploration in Indian Himalayan forests. International Journal of Remote Sensing, 34(6), 2006–2022. https://doi.org/10.1080/01431161.2012.730161

Galetti, M., Martuscelli, P., Pizo, M. A. & Simao, I. (1997). Records of Harpy and Crested Eagles in the Brazilian Atlantic Forest, 117, 27–31.

Galetti, M. & de Carvalho, O. (2000). Sloths in the diet of a Harpy eagle nestling in eastern Amazon. Wilson Bulletin, 112(4), 535–536, SITDOA2.0. https://doi.org/10.1676/0043-5643(2000)1120535

Gallardo del Ángel, J. C. & Montejo, J. E. (2002). Notes about the tolerance of some Neotropical Forest raptors to altered environments in Veracruz, Mexico. In: Program of Neotropical Raptor Conference and Harpy Eagle. Symposium, 39, 99–100. Fondo Peregrino – Panamá, Panamá City, Panamá.

Galmes, M. (2009). Ecology and conservation of Crowned eagle Harpyhaliaetus coronatus in central Argentina. The Peregrine Fund Rep., 34, 1–33.

Galushin, V. M. (1962). The Greater Spotted eagle of the Oka Valley and its influence on the numbers of some birds. Moscow. Pedag. Inst. im. Lenina, Uch. Zapiski, 186, 115–151 (in Russian).

Galushin, V. M. (1980). Lesnaya Promyshlennost, Moscow, USSR (in Russian). Birds of prey of the Wood.

Galushin, V. M. (1982). Role of the birds of prey in the ecosystems. Results of the Sciences and Technique. Ser. Zoology of Vertebrates, 11: 158–239.

Gamauf, A., Winkler, H. & Preleuthner, M. (1998). Philippine birds of prey: Interrelations among habitat, morphology, and behavior. The Auk, 115(3): 713–726. https://doi.org/10.2307/4089419._

Gamauf, A., Preleuthner, M. & Pinsker, W. (1998). Distribution and field identification of Philippine birds of prey: 1. Philippine hawk eagle (Spizaetus philippensis) and Changeable hawk eagle (Spizaetus cirrhatus). Forktail, 14, 1–11.

Gamauf, A., Nittinger, F., Haring, E. & Pinsker, W. (2005). Committee for the 4th Symposium on Asian Raptors. Phylogenetic re-examination of Asian raptor species based on molecular studies and its consequences for conservation. *In*: J. Abu et al. (Eds.), Proceedings of the 4th Symposium on Asian Raptors, pp. 209–210. Kuala Lumpur, Malaysia.

Gamauf, A., Gjershaug, J. O., Kvaløy, K., Røv, N. & Haring, E. (2005a). Molecular phylogeny of the hawk-eagles (genus Spizaetus). Zoologische Mededelingen (Leiden), 79(3), 179–180.

Gamauf, A., Gjershaug, J. O., Røv, N., Kvaløy, K. & Haring, E. (2005b). Species or subspecies? The dilemma of taxonomic ranking of some South-East Asian hawk-eagles (genus Spizaetus). Bird Conservation International, 15(1), 99–117. https://doi.org/10.1017/S0959270905000080

Ganderton, P. & Coker, P. (2005). Environmental Biogeography. Harlow: Pearson/Prentice Hall.

Gangoso, L., Alvarez-Lloret, P., Rodríguez-Navarro, A. A., Mateo, R., Hiraldo, F. & Donázar, J. A. (2009). Long-term effects of lead poisoning

on bone mineralization in vultures exposed to ammunition sources. Environmental Pollution, 157(2), 569–574. https://doi.org/10.1016/j.envpol.2008.09.015
Ganusevich, S. (2003). Status, monitoring and colour ringing of the White-tailed Sea eagle population of the Kola Peninsula, Russia. In: Helander, B. (Ed.), Sea eagle 2000: Proceedings from the International Sea Eagle Conference in Bjorko, Sweden, 13–17 September, 20, 113–119. Stockholm: Sweden Proceedings of the Swedish Society for Nature Conservation, SNF.
Garber, D. P. (1972). Breeding ecology of Ospreys in Lassen and Las Plumas counties, California. Unpublished M.S. Thesis, California State University.
Garcelon, D. K. & Slater, G. L. (1995). Cooperative nesting by a trio of Bald eagles. Journal of Raptor Research, 29, 210–213.
Garcia, E. (2006). Birds in Gibraltar 2005. Gibraltar Bird Report 5, 9–45.
Garcia, E. (2007). Birds of Gibraltar 2006. Gibraltar Bird Report 6, 8–44.
Garcia, E. (2008). Birds in Gibraltar 2007. Gibraltar Bird Report 7, 8–46
Garcia, J. T. & Arroyo, B. E. (2005). Food-niche differentiation in sympatric Hen Circus cyaneus and Montagu's Harriers Circus pygargus. Ibis, 147(1), 144–154. https://doi.org/10.1111/j.1474-919x.2004.00377.x
García, V., Moreno-Opo, R. & Tintó, A. (2013). Sex differentiation of Bonelli's eagle Aquila fasciata in Western Europe using morphometrics and Plumage Colour Patterns. Ardeola, 60(2), 261–277. https://doi.org/10.13157/arla.60.2.2013.261
Garcıá-Dios, I. S. (2009). Aguililla calzada—Hieraaetus pennatus. *In*: A. Salvador & L. M. Bautista (Eds.). Retrieved from http://www.vertebradosibericos.org/, Enciclopedia Virtual de los Vertebrados Españoles. Madrid: Museo Nacional de Ciencias Naturales.
García-Fernández, A. J., Calvo, J. F., Martínez-López, E., María-Mojica, P. & Martínez, J. E. (2008). Raptor ecotoxicology in Spain: A review on persistent environmental contaminants. AMBIO: A Journal of the Human Environment. 37(6), 432–439, REISAR2.0.CO;2 https://doi.org/10.1579/0044-7447(2008)37432
García-Fernández, A. J., Sanchez-Garcia, J. A., Gomez-Zapata, M. & Luna, A. (1996). Distribution of cadmium in blood and tissues of wild birds. Archives of Environmental Contamination and Toxicology, 30(2), 252–258. https://doi.org/10.1007/BF00215805
Garcia-González, L. M. (2008). Origin and formation of the Spanish Imperial eagle (Aquila adalberti). Journal of Ornithology, 149(2), 151–159. https://doi.org/10.1007/s10336-007-0252-z
García-Ripollés, C., López-López, P. & Urios, V. (2010). First description of migration and wintering of adult Egyptian vultures Neophron percnopterus tracked by GPS satellite telemetry. Bird Study, 57(2), 261–265. https://doi.org/10.1080/00063650903505762
Gard, K., Groszos, M. S., Brevik, E. C. & Lee, G. W. (2007). Spatial analysis of bird-aircraft strike hazard for Moody Air Force Base aircraft in the State of Georgia. Georgia Journal of Science, 65(4), 160–169.
Gardner, C. J. (2016). Birds of Madagascar and the Indian Ocean Islands, Seychelles, Comoros, Mauritius. Reunion and Rodrigues. The Condor, 118(4), 864–865. https://doi.org/10.1650/CONDOR-16-139.1
Gargett, V. (1990). A study. Eagle: Black. Acorn Books.
Gargett, V., Gargett, E. & Damania, D. (1995). The influence of rainfall on Black eagle breeding over 31 years in the Matobo Hills, Zimbabwe. Ostrich, 66(4), 114–121. https://doi.org/10.1080/00306525.1995.9632715
Garnett, S. (1987). An Australian record of Gurney's eagle Aquila gurneyi. Australian Bird Watcher, 12, 134–135.
Garnett, S. (1992). Threatened and extinct birds of Australia. RAOU Report no. 82. Melbourne: Royal Australasian Ornithologists' Union.
Garnett, S. T. & Crowley, G. M. (2000). The action plan for Australian birds (2000). Environment Australia, Canberra, Australia.
Garrett, K. & Dunn, J. (1981). Birds of Southern California: Status and distribution. Los Angeles Audubon Society.
Garrod, A. H. (1973). On certain muscles in the thigh of birds and their value for classification. Proceedings of the Zoological Society of London, 1873, 626–644.
Garzon, J. (1977). Birds of prey in Spain, the present situation. *In*: R. D. Chancellor (Ed.), Proceedings of the World Conference on Birds of Prey, pp. 159–170. Cambridge: International Council for Bird Preservation.
Gaston, K. J. & Spicer, J. I. (2001). The relationship between range size and niche breadth: A test using five species of Gammarus (Amphipoda). Global Ecology and Biogeography, 10(2), 179–188. https://doi.org/10.1046/j.1466-822x.2001.00225.x
Gates, A. (2002). A gathering of eagle Huntershunters. Austringer. Journal of the Welsh Hawking Club. Eagle Falconry. Retrieved from http://www.eaglefalconer.com/about.html.
Gatter, W. (1988). The birds of Liberia. Africa West: A preliminary list with status and open questions. Verhandlungen der Ornithologischen Gesellschaft in Bayern, 689–723.
Gatter, W. (1997). Birds of Liberia. Robertsbridge, UK: Pica Press.
Gatter, W. & Klump, G. (1979). Ausgeprägte fälle von Zugumkehr bei Eichelhäher (Garrulus glandarius) und Tannenhäher (Nucifraga caryocatactes). Schütt, R. Vogelwarte, 30, 101–107.
Gauthereaux, S. A. (1993). Avian interactions with utility structures: Background and milestones. In: Huckabee, J. W. (Ed.), Proceedings of the Avian Interactions with Utility Structures International Workshop, 13–16 September 1992, 1–6. Miami, Florida: Electric Power Research Institute, Palo Alto, CA.
Gavrilov, E. I. & Gavrilov, A. E. (2005). The birds of Kazakhstan. Tethys Ornithological Research, 2, 1(1/2), 1–16. Retrieved from http://www.wesca.net/Podoces/Podoces1/PODOCES_1
Gavrilyuk, M. & Grishchenko, V. (2003). Current status of the White-tailed Eagle in Ukraine. In: Helander, B. (Ed.), Sea eagle 2000: Proceedings from the International Sea Eagle Conference in Bjorko, Sweden, 13–17 September 20, Pages 129–132. Stockholm, Sweden Proceedings of the Swedish Society for Nature Conservation, SNF.
Gbogbo, F. & Awotwe-Pratt, V. (2008). Waste management and Hooded Vultures on the Legon Campus of the University of Ghana in Accra, Ghana, West Africa. Vulture News, 58, 16–22.

Gedeon, K. & Stubbe, M. (1991). Tagesrhythmik, Raumnutzung und Jagdverhalten des Schreiadlers Aquila pomarina Brehm. Populationsökologie von Greifvogel- und Eulenarten 2. Martin: Luther University Halle-Wittenberg & Wiss. Beitr (pp. 107–129).
Geist, H. J. & Lambin, E. F. (2004). Dynamic causal patterns of desertification. BioScience, 54.9(9), 817, DCPOD2.0. https://doi.org/10.1641/0006-3568(2004)0540817
Geist, H. (2005). The Causes and Progression of Desertification. Abingdon-on-Thames, UK: Routledge.
Geller, S. (1983). Changes in the raptors' population in the north of the Upper Galilee since the establishment of the state of Israel. Torgos, 3(6), 17–27, 123–124.
Gende, S. M. (2008). Perspectives on the breeding biology of Bald eagles in Southeast Alaska. *In*: B. A. Wright & P. Schempf (Eds.), Southeast, Vietnam: University of Alaska Southeast.
Gende, S. M., Wilson, M. F., & Jacobsen, M. (1997). Reproductive success of Bald eagles (Haliaeetus leucocephalus) and its association with habitat or landscape features and weather in southeast Alaska. Canadian Journal of Zoology, 75(10), 1595–1604. https://doi.org/10.1139/z97-786
Genetics Home Reference. (2020). What is a Gemome? Retrieved from https://ghr.nlm.nih.gov/primer/hgp/genome.
Gensbol, B. (1992). Guida ai rapaci diurni d'Europa, Nord Africa e Medio Oriente. Bologna, Italy: Zanichelli, Bologna.
Génsbøl, B. (2003). The re-colonization of the Sea Eagle in Denmark. In: Helander, B. (Ed.), Sea Eagle (2000). Proceedings from the International Sea Eagle Conference in Björkö, Sweden, 13–17 September 2000, 67–69. Stockholm: Swedish Society for Nature Conservation.
GeoNet. (2019). Drones for snow leopard conservation: Detecting and counting prey for an endangered species in Mongolia. Retrieved from https://community.esri.com/docs/DOC-13613-drones-for-snow-leopard-conservation
GeoTerra Image. (2013). G. T. (GTI) Pty Ltd SA. GTI SA National Land-Cover.
Gerdzhikov, G. P. & Demerdzhiev, D. A. (2009). Data on bird mortality in 'Sakar' IBA (BG021), caused by hazardous power lines. Ecologia Balkanica, 1, 67–77.
Gerrard, J. M., Gerrard, P., Maher, W. J. & Whitfield, D. W. A. (1975). Factors influencing nest site selection of Bald eagles in northern Saskatchewan and Manitoba. Blue Jay, 33, 169–176.
Gerrard, J. M., Wilitfield, D. W. A. & Mailer, W. J. (1976). Osprey-Bald eagle relationships in Saskatchewan. Blue Jay, 34, 240–246.
Gerrard, J. M. & Ingram, T. M. (Eds.). (1983). White Horse Plains Publications. The Bald eagle in Canada. Proceedings of the Bald Eagle Days. Headingley, Manitoba.
Gerrard, J. M. & Ingram, T. N. (1985). The Bald Eagle in Canada: Proceedings of Bald Eagle Days, Winnipeg, August 18–20, 1982. Headingley, Manitoba, Canada and Apple River, IL: White Horse Plains and The Eagle Foundation.
Gerrard, J. M. & Bortolotti, G. R. (1988). The Bald eagle: haunts and habits of a wilderness monarch. Washington, DC: Smithsonian Institution Press.
Gerrard, J. M. & Bortolotti, G. R. (1988). The Bald Eagle. Saskatoon, Canada: Western Producer Prairie Books.
Gertenbach, W. P. D. (1983). Landscapes of the Kruger National Park. Koedoe, 26(1), 121. https://doi.org/10.4102/koedoe.v26i1.591
Gese, E. M., Ruff R. L. & Crabtree, R. L. (1996). Foraging ecology of coyotes (Canis latrans): The influence of extrinsic factors and a dominance hierarchy. Canadian Journal of Zoology, 74, 769–783.
Getz, W. M. (2011). Biomass transformation webs provide a unified approach to consumer–resource modelling. Ecology Letters, 14(2), 113–124. https://doi.org/10.1111/j.1461-0248.2010.01566.x
Gewin, V. (2004). Mapping opportunities. Nature, 427(6972), 376–377. https://doi.org/10.1038/nj6972-376a
Gibb, G. C., Kardailsky, O., Kimball, R. T., Braun, E. L. & Penny, D. (2007). Mitochondrial genomes and avian phylogeny: Complex characters and resolvability without explosive radiations. Molecular Biology Evolution, 24(1), 269–280. https://doi.org/10.1093/molbev/msl158
Gibbens, S. (2017). Poisoned Bald eagle highlights lead bullet controversy. Retrieved from https://www.nationalgeographic.com/news/2017/12/eagle-death-lead-poisoning-ammunition-bullets-spd/
Gibson, D. D. & Byrd, G. V. (2007). Birds of the Aleutian Islands, Alaska. Cambridge, MA, & Washington, DC: Nuttall Ornithological Club & The American Ornithologists' Union.
Gibson-Hill, C. A. (1950). A checklist of the birds of Singapore Island. Bulletin of the Raffles Museum, 21, 132–183.
Gil, G., Haene, E. & Chebez, J. C. (1995). Notes on the avifauna of the Sierra de las Quijadas. Nuestras Aves, 13, 26–28.
Gilbert, C. C., McGraw, W. S. & Delson, E. (2009). Brief communication: Plio-Pleistocene eagle predation on fossil cercopithecids from the Humpata Plateau, southern Angola. American Journal of Physical Anthropology, 139(3), 421–429. https://doi.org/10.1002/ajpa.21004
Gilbert, K. A. (2000). Attempted predation on a White-faced Saki in the central Amazon. Neotropical Primates, 8, 103–104.
Gilbert, M. & Gombobataar, S. (2009). The status and distribution of Pallas's Fish eagle in Mongolia: A report on field surveys June–August (2009).
Gilbert, M., Oaks, J. L., Virani, M. Z., Watson, R. T., Ahmed, S., Chaudhry, J., . . . Khan, A. A. (2004). The status and decline of vultures in the provinces of Punjab and Sind, Pakistan: A 2003 update. *In*: R. C. Chancellor & Meyburg (Eds.), Raptors worldwide. Proceedings of the 6th World Conference on Birds of Prey and Owls, pp. 221–234. Berlin and Budapest: WWGBP and MME/Birdlife Hungary.
Gilbert, M., Tingay, R., Losolmaa, J., Sureda, N., Gilbert, C., Batmunkh, D. & Gombobaatar, S. (2014). Distribution and status of the Pallas's Fish eagle Haliaeetus leucoryphus in Mongolia: A cause for conservation concern. Bird Conservation International, 24(3), 379–388. https://doi.org/10.1017/S0959270913000543
Gilbert, M., Virani, M. Z., Watson, R. T., Oaks, J. L., Benson, P. C., Khan, A. A., . . . Shah Ali Shah, Q. A. (2002). Breeding and mortality of Oriental White-backed vulture Gyps bengalensis in Punjab Province, Pakistan. Bird Conservation International, 12(4), 311–326. https://doi.org/10.1017/S0959270902002198.

Gilbert, M., Watson, R. T., Virani, M. Z., Oaks, J. L., Ahmed, S., Chaudhry, M. J. I., . . . Khan, A. A. (2006). Rapid population declines an mortality clusters in three Oriental White-backed vulture Gyps bengalensis colonies in Pakistan due to diclofenac poisoning. Oryx, 40(4 388–399. https://doi.org/10.1017/S0030605306001347

Gilbert, S. L. (2016). Bald eagle predation on Sitka Black-tailed deer fawns. Northwestern Naturalist, 97(1), 66–69. https://doi.org/10.1898/1051 1733-97.1.66

Gill, C. E. & Langelier, K. M. (1994). British Columbia. Acute lead poisoning in a Bald eagle secondary to bullet ingestion. Canadian Veterinar [Journal] = la Revue Veterinaire Canadienne, 35(5), 303–304.

Gill, F. B. L. L. & Wolf, L. L. (1975). Economics of feeding territoriality in the golden-winged sunbird. Ecology, 56(2), 333–345. https://do org/10.2307/1934964

Gill, F. & Donsker, D. (2018). IOC World Bird List (v 8.1), https://www.worldbirdnames.org/ioc-lists/crossref/

Gill, F. & Brown, L. H. (2020). Falconiform. Retrieved from https://www.britannica.com/animal/falconiform

Gill, F., Donsker, D. & Rasmussen, P. (Eds.). (2020). IOC world bird list. Retrieved from http://www.worldbirdnames.org/, https://doi.org/10.14344 IOC.ML.10.2.

Gill, F. & Donsker, D. (Eds.). (2019). New World vultures, secretary bird, kites, hawks and eagles. World bird list version 9.2. Baton Rouge, LA International Ornithologists' Union.

Gill, F. & Donsker, D. (Eds.). (2016). IOC world bird names (version 6.1). Retrieved from http://www.world.birdnames.org

Gill, F. & Donsker, D. (Eds.). (2014). New World vultures, secretary bird, kites, hawks and eagles. IOC world bird list, version 4.2. Internationa Ornithologists' Union. https://doi.org/10.14344/IOC.ML.4.2.

Gill, F. & Wright, M. (2006). Birds of the world: Recommended English names. Princeton, NJ: Princeton University Press.

Gill, J. A., Norris, K. & Sutherland, W. J. (2001). Why behavioural responses may not reflect the population consequences of human disturbanc Biological Conservation, 97(2), 265–268. https://doi.org/10.1016/S0006-3207(00)00002-1

Gill, Jr., R. E. & Kincheloe, K. L. (1993). Are Bald eagles important predators of Emperor geese? Journal of Raptor Research, 27(1), 34–36.

Gillard, R. (1977). Unnecessary electrocution of owls. Blue Jay, 35, 259.

Gilliard, E. T. (1958). Living birds of the world. New York: Doubleday Publishing.

Gilliard, E. T. & LeCroy, M. (1966). Birds of the Middle Sepik Region, New Guinea: Results of the American Museum of Natural Histor Expedition to New Guinea 1953–1954. Bulletin of the American Museum of Natural History, 132, 245–276.

Ginn, P. J., Mcilleron, W. G. & Milstein, P. Le S. (1989). The complete book of South African birds. Cape Town, SA: Struik Winchester Publishers

Giordano, A. (1991). The migration of birds of prey and storks in the Straits of Messina. Birds of Prey Bulletin, 4, 239–249.

GIS Geography. (2018). DEM, DSM and DTM differences—A look at elevation models in GIS. Retrieved from https://gisgeography.com/dem dsm-dtm-differences/

Giudice, R. (2007). First nesting report of the Ornate hawk-eagle (Spizaetus ornatus) in Peru. *In*: K. L. Bildstein, D. R. Barber & A. Zimmerma (Eds.), Neotropical raptors, pp. 9–13. Orwigsburg, PA: Hawk Mountain Sanctuary.

Giudice, R., Piana, R. & Williams, M. (2007). Tree architecture as a determinant factor for nest-tree selection by Harpy Eagles. *In*: Bildstein, K L., Bechard, M. J., Porras-Peñaranda, P., Campo, E. & Farmer, C. J. (Eds.), Neotropical raptors, pp. 14–22. Orwigsburg, PA: Hawk Mountai Sanctuary.

Gjershaug, J. O. (1981). Hekkeokologi hos kongeorn Aquila chrysaetos i Atae og Romsdal [MSc. Thesis]. University of Trondheim.

Gjershaug, J. O. (2006). Taxonomy and conservation status of hawk-eagles (Genus Nisaetus) in South-East Asia. PhD Thesis. Trondheim Norwegian University of Science and Technology.

Gjershaug, J. O., Diserud, O. H., Rasmussen, P. C. & Warakagoda, D. (2008). An overlooked threatened species of eagle: Legge's hawk eagl Nisaetus kelaarti (Aves: Accipitriformes) (PDF). Zootaxa, 1792(1), 54–66. https://doi.org/10.11646/zootaxa.1792.1.4

Gjershaug, J. O., Kvaløy, K., Røv, N., Prawiradilaga, D.M., Suparman, U. & Rahman. Z. 2004. Taxonomic Status of Flores Hawk Eagle spizaetu Floris. Forktail, 20, 55–62.

Gjershaug, J. O., Lerner, H. R. L. & Diserud, O. H. (2009). Taxonomic status and distribution of the Pygmy eagle Aquila weiskei. Zootaxa 2326(1), 24–38. https://doi.org/10.11646/zootaxa.2326.1.2

Gjershaug, J. O., Rov, N., Nygard, T., Prawiradilaga, D. M., Afianto, M. Y., Hapsoro & Supriatna, A. (2004). Home-range size of the Javan Hawk eagle (Spizaetus bartelsi) estimated from direct observations and radiotelemetry. Journal of Raptor Research, 38(4), 343–349.

Gjershaug, J. O., Thingstad, P.-G., Eldoy, S. & Byrkjeland, S. (Eds.). (1994). Norsk Fugleatlas. Norsk Ornitologisk Forening, Klsebu.

Gjershaug, J.-O., Lerner, H. R. L. & Diserud, O. A. (2009). Taxonomy and distribution of the Pygmy eagle Aquila (Hieraaetus) weiske (Acciptriformes: Accipitridae). Zootaxa, 2316, 24–28.

Glasier, P. (1979). Charles T. Branford Co. MA: Newton Center. Falconry and Hawking. MA: Newton Center.

Glasier, P. (1986). Guide to Falconry and Hawking, London: Batsford.

Glazener, W. C. (1964). Note on the feeding habits of the caracara in south Texas. Condor, 66, 162.

Gleason, T. M. & Norconk, M. A. (2002). Predation risk and antipredator strategies of White-faced Saki, Pithecia. *In*: Eat or be eaten: Predato sensitive foraging among primates. Cambridge: Cambridge University Press.

Global Health Group and National Vector Borne Disease Control Program. (2013). Eliminating malaria in the Solomon Islands. Retrieved fron https://s116768.gridserver.com/ sites/default/files/content/ghg/country-briefings/asia-pacific/Solomon-Islands_2013.pdf

Global Raptor Information Network. (2021a). Search Results for: "Accipitridae." Retrieved from http://www.globalraptors.org/grin/SpeciesLookup asp

Global Raptor Information Network. (2021b). Species account: Golden Eagle Aquila chrysaetos. Retrieved from http://www.globalraptors.org

Global Raptor Information Network. (2021c). Species account: Tawny Eagle Aquila rapax. Retrieved from http://www.globalraptors.org

Global Raptor Information Network. (2021d). Species account: Steppe Eagle Aquila nipalensis. Retrieved from http://www.globalraptors.org

Global Raptor Information Network. (2021aa). Species account: Blyth's Hawk-eagle Nisaetus alboniger. Retrieved from http://www.globalraptors.org

Global Raptor Information Network. (2021ab). Species account: Wallace's Hawk-eagle Nisaetus nanus. Retrieved from http://www.globalraptors.org

Global Raptor Information Network. (2021ac). Species account: Black Hawk-eagle Spizaetus tyrannus. Retrieved from http://www.globalraptors.org

Global Raptor Information Network. (2021ad). Species account: Black-and-white Hawk-eagle Spizaetus melanoleucus. Retrieved from http://www.globalraptors.org

Global Raptor Information Network. (2021ae). Species account: Ornate Hawk-eagle Spizaetus ornatus. Retrieved from http://www.globalraptors.org

Global Raptor Information Network. (2021af). Species account: Black-and-chestnut Eagle Spizaetus isidori. Retrieved from http://www.globalraptors.org

Global Raptor Information Network. (2021ag). Species account: Greater Spotted Eagle Clanga clanga. Retrieved from http://www.globalraptors.org

Global Raptor Information Network. (2021ah). Species account: Lesser Spotted Eagle Lophaetus pomarinus. Retrieved from http://www.globalraptors.org

Global Raptor Information Network. (2021ai). Species account: Indian Spotted Eagle Clanga hastata. Retrieved from http://www.globalraptors.org

Global Raptor Information Network. (2021aj). Species account: Martial Eagle Polemaetus bellicosus. Retrieved from http://www.globalraptors.org

Global Raptor Information Network. (2021ak). Species account: Crowned Hawk-eagle Stephanoaetus coronatus. Retrieved from http://www.globalraptors.org

Global Raptor Information Network. (2021al). Species account: Long-crested Eagle Lophaetus occipitalis. Retrieved from http://www.globalraptors.org

Global Raptor Information Network. (2021am). Species account: Indian Black Eagle Ictinaetus malayensis. Retrieved from http://www.globalraptors.org

Global Raptor Information Network. (2021an). Species account: Short-toed Snake Eagle Circaetus gallicus. Retrieved from http://www.globalraptors.org

Global Raptor Information Network. (2021ao). Species account: Black-breasted Snake Eagle Circaetus pectoralis. Retrieved from http://www.globalraptors.org

Global Raptor Information Network. (2021ap). Species account: Beaudouin's Snake Eagle Circaetus beaudouini. Retrieved from http://www.globalraptors.org

Global Raptor Information Network. (2021aq). Species account: Brown Snake Eagle Circaetus cinereus. Retrieved from http://www.globalraptors.org

Global Raptor Information Network. (2021ar). Species account: Southern Banded Snake Eagle Circaetus fasciolatus. Retrieved from http://www.globalraptors.org

Global Raptor Information Network. (2021as). Species account: Western Banded Snake Eagle Circaetus cinerascens. Retrieved from http://www.globalraptors.org

Global Raptor Information Network. (2021at). Species account: Bateleur Terathopius ecaudatus. Retrieved from http://www.globalraptors.org

Global Raptor Information Network. (2021au). Species account: Philippine Serpent Eagle Spilornis holospilus. Retrieved from http://www.globalraptors.org

Global Raptor Information Network. (2021av). Species account: Sulawesi Serpent Eagle Spilornis rufipectus. Retrieved from http://www.globalraptors.org.

Global Raptor Information Network. (2021aw). Species account: Crested Serpent Eagle Spilornis cheela. Retrieved from http://www.globalraptors.org

Global Raptor Information Network. (2021ax). Species account: Great Nicobar Serpent Eagle Spilornis klossi. Retrieved from http://www.globalraptors.org

Global Raptor Information Network. (2021ay). Species account: Andaman Serpent Eagle Spilornis elgini. Retrieved from http://www.globalraptors.org

Global Raptor Information Network. (2021az). Species account: Congo Serpent Eagle Dryotriorchis spectabilis. Retrieved from http://www.globalraptors.org

Global Raptor Information Network. (2021ba). Species account: Madagascar Serpent Eagle Eutriorchis astur. Retrieved from http://www.globalraptors.org

Global Raptor Information Network. (2021bb). Species account: Philippine Eagle Pithecophaga jefferyi. Retrieved from http://www.globalraptors.org

Global Raptor Information Network. (2021bc). Species account: Steller's Sea Eagle Haliaeetus pelagicus. Retrieved from http://www.globalraptors.org

Global Raptor Information Network. (2021bd). Species account: White-tailed Eagle Haliaeetus albicilla. Retrieved from http://www.globalraptors.org
Global Raptor Information Network. (2021be). Species account: Bald Eagle Haliaeetus leucocephalus. Retrieved from http://www.globalraptors.org
Global Raptor Information Network. (2021bf). Species account: Pallas''s Fish Eagle Haliaeetus leucoryphus. Retrieved from http://www.globalraptors.org
Global Raptor Information Network. (2021bg). Species account: Madagascar Fish Eagle Haliaeetus vociferoides. Retrieved from http://www.globalraptors.org
Global Raptor Information Network. (2021bh). Species account: African Fish Eagle Haliaeetus vocifer. Retrieved from http://www.globalraptors.org
Global Raptor Information Network. (2021bi). Species account: White-bellied Sea Eagle Haliaeetus leucogaster. Retrieved from http://www.globalraptors.org
Global Raptor Information Network. (2021bj). Species account: Sanford's Sea Eagle Haliaeetus sanfordi. Retrieved from http://www.globalraptors.org
Global Raptor Information Network. (2021bk). Species account: Lesser Fish Eagle Ichthyophaga humilis. Retrieved from http://www.globalraptors.org
Global Raptor Information Network. (2021bl). Species account: Grey-headed Fish Eagle Ichthyophaga ichthyaetus. Retrieved from http://www.globalraptors.org
Global Raptor Information Network. (2021bm). Species account: Black Solitary Eagle Buteogallus solitarius. Retrieved from http://www.globalraptors.org
Global Raptor Information Network. (2021bn). Species account: Crested Eagle Morphnus guianensis. Retrieved from http://www.globalraptors.org
Global Raptor Information Network. (2021bo). Species account: Harpy Eagle Harpia harpyja. Retrieved from http://www.globalraptors.org
Global Raptor Information Network. (2021bp). Species account: New Guinea Harpy Eagle Harpyopsis novaeguineae. Retrieved from http://www.globalraptors.org
Global Raptor Information Network. (2021bq). Species account: Crowned Solitary Eagle Buteogallus coronatus. Retrieved from http://www.globalraptors.org
Global Raptor Information Network. (2021e). Species account: Spanish Imperial Eagle Aquila adalberti. Retrieved from http://www.globalraptors.org
Global Raptor Information Network. (2021f). Species account: Eastern Imperial Eagle Aquila heliaca. Retreived from http://www.globalraptors.org
Global Raptor Information Network. (2021g). Species account: Wahlberg's Eagle Hieraaetus wahlbergi. Retrieved from http://www.globalraptors.org
Global Raptor Information Network. (2021h). Species account: Gurney's Eagle Aquila gurneyi. Retrieved from http://www.globalraptors.org
Global Raptor Information Network. (2021i). Species account: Wedge-tailed Eagle Aquila audax. Retrieved from http://www.globalraptors.org
Global Raptor Information Network. (2021j). Species account: Verreaux's Eagle Aquila verreauxii. Retrieved from http://www.globalraptors.org/
Global Raptor Information Network. (2021k). Species account: Bonelli's Eagle Aquila fasciata. Retrieved from http://www.globalraptors.org
Global Raptor Information Network. (2021l). Species account: African Hawk-eagle Aquila spilogaster. Retrieved from http://www.globalraptors.org
Global Raptor Information Network. (2021m). Species account: Cassin's Hawk-eagle Aquila africana. Retrieved from http://www.globalraptors.org
Global Raptor Information Network. (2021n). Species account: Booted Eagle Hieraaetus pennatus. Retrieved from http://www.globalraptors.org
Global Raptor Information Network. (2021o). Species account: Little Eagle Aquila morphnoides. Retrieved from http://www.globalraptors.org
Global Raptor Information Network. (2021p). Species account: Pygmy Eagle Aquila weiskei. Retrieved from http://www.globalraptors.org
Global Raptor Information Network. (2021q). Species account: Ayres's Hawk-eagle Aquila ayresii. Retrieved from http://www.globalraptors.org
Global Raptor Information Network. (2021r). Species account: Rufous-bellied Eagle Lophotriorchis kienerii. Retrieved from http://www.globalraptors.org
Global Raptor Information Network. (2021s). Species account: Crested Hawk-eagle Nisaetus cirrhatus. Retrieved from http://www.globalraptors.org
Global Raptor Information Network. (2021t). Species account: Flores Hawk-eagle Nisaetus floris. Retrieved from http://www.globalraptors.org
Global Raptor Information Network. (2021u). Species account: Mountain Hawk-eagle Nisaetus nipalensis. Retrieved from http://www.globalraptors.org
Global Raptor Information Network. (2021v). Species account: Legge's Hawk-eagle Nisaetus kelaarti. Retrieved from http://www.globalraptors.org
Global Raptor Information Network. (2021w). Species account: Javan Hawk-eagle Nisaetus bartelsi. Retrieved from http://www.globalraptors.org
Global Raptor Information Network. (2021x). Species account: Sulawesi Hawk-eagle Nisaetus lanceolatus. Retrieved from http://www.globalraptors.org
Global Raptor Information Network. (2021y). Species account: Northern Philippine Hawk-eagle Nisaetus philippensis. Retrieved from http://www.globalraptors.org

Global Raptor Information Network. (2021z). Species account: Southern Philippine (Pinsker's) Hawk-eagle Nisaetus pinskeri. Retrieved from http://www.globalraptors.org

Glotov, I. N. (1959). Material on the biology of the Greater Spotted eagle, Aquila clanga Pall. Trudy Biol. Inst. Sib. Otd. Akad. Nauk Novosibirsk, 5, 167–170 (in Russian).

Glover, F. A. & Heugly, L.G. (1970). Golden eagle ecology in West Texas. [final report] National Audubon Society, Colorado Coop.: Wildlife Research Unit. CO.

Glutz von Blotzheim, U. N., Bauer, K. M. & Bezzel, E. (1971). Handbuch der Vogel Mitteleuropas. Falconiformes. Verlagsgesellschaft, Frankfurt am Main, 4, pp. 620–637. Akadem.

Glutz von Blotzheim, U. N., Bauer, K. M. & Bezzel, E. (1989). Falconiformes. Handbuch der VoVol.¨gel Mitteleuropas. Aula. Aula, Wiesbaden (22nd Aufl.). Wiesbaden, Germany: Band 4 Falconiformes.

Gmelin, J. F. (1788). Systema naturae per regna tria naturae, secundum Classes, Ordines, Genera, Species, cum Characteribus, Differentiis, Synonymis, Locis. Tomus I. [Editio Decima Tertia, Aucta, Reformata]. Beer: Lipsiae.

Gober, D. R. & Lockhart, J. M. (1979). Factors affecting Golden eagle predation on livestock in Trans Pecos, Texas. Paper presented at the 1979 annual meeting of the Texas chapter of The Wildlife Soc., April 6, Alpine, TexTX.

Godfrey, W. E. (1986). The birds of Canada. (Revised edition). Natural Science, Ottawa. Ottawa: National Museum of Natural Science.

Godoi, M. N., Morante-Filho, J. C., Faxina, C., Modena, E. S., Pivatto, M. A. C., Manço, D. D. G., . . . Stavis, V. K. (2012). Aves de rapina raras no estado de Mato Grosso do Sul, Brasil. Atual Ornitol, 170, 41–47.

Gokula, V. (2012). Breeding ecology of the crested serpent eagle Spilornis cheela (Latham, 1790) (Aves: Accipitri formes: Accipitridae) in Kolli Hills, Tamil Nadu, India. TAPROBANICA, 4(2), 77–82. https://doi.org/10.4038/tapro.v4i2.5059

Golden, N. H., Warner, S. E. & Coffey, M. J. (2016). A review and assessment of spent lead ammunition and its exposure and effects to scavenging birds in the United States. Reviews of Environmental Contamination and Toxicology, 237, 123–191. https://doi.org/10.1007/978-3-319-23573-8_6

Goloduschko, B. Z. (1958). Material on Lesser Spotted eagle ecology in Bialowiecza Forest Reserve. The First Zoology Conference on Belarus SSR [Thesis] of reports, pp. 34–35 (in Russian), Minsk.

Goloduschko, B. Z. (1961). On feeding relations of the birds of prey of the Bialowiecza Forest Reserve. Fauna y ekologiya nazemnykh pozvono chnykh Belorussii, 143–149, Minsk (in Russian).

Golubova, E. Yu. & Pleshenko, S. V. (1997). Colonies of sea birds in the northern part of the Sea of Okhotsk. *In*: Kondratiev, A. Ya. (Ed.), Species diversity and state of the populations of Waterfowl Birds in the North-East of Asia. A Collection of Study Reports, 141–162. Magadan: Institute of Biological Problems of the North. In Russian.

Gomarasca, M. (2010). Basics of Geomatics. Applied Geomatics, 2, 137–146. https://doi.org/10.1007/s12518-010-0029-6

Gomes, F. B. R. (2014). Distribuição e ecologia do Uiraçu-falso (Morphnus guianensis, Daudin 1800). Ph.D. Thesis. Manaus: INPA.

Gomes, F. B. R. & Sanaiotti, T. M. (2015). A review of the distribution of the Crested eagle, Morphnus guianensis (Daudin, 1800) (Accipitridae: Harpiinae), including range extensions. Revista Brasileira de Ornitologia, 23(1), 36–63. https://doi.org/10.1007/BF03544289

Gómez de Silva, H. (2005). Mexico. North American Birds, 59, 500–505.

Gómez de Silva, H. (2006). Mexico. North American Birds, 60, 290–295.

Gómez de Silva, H. (2007). Mexico. North American Birds, 61, 150–154.

Gómez de Silva, H. (2008). Mexico. North American Birds, 62, 159–163.

Gómez de Silva Garza, H. (2008). Mexico. North American Birds, 62, 310–313.

Gómez de Silva, H., González Garcia, F. & Casillas-Trejo, M. P. (1999). Birds of the upper cloud forest of Elel Triunfo, Chiapas, Mexico. Ornitologia Neotropical, 10, 1–26.

Gomez, L. G., Houston, D. C., Cotton, P., Tye, A., Luis, G., Houston, D. C., . . . Tye, A. (1994). The role of greater yellow-headed vultures Cathartes melambrotus as scavengers in neotropical forest. Ibis, 136(2), 193–196. https://doi.org/10.1111/j.1474-919X.1994.tb01084.x

Gone, B. B. (2019). Vulture buzzard control. Retrieved from https://www.birdbgone.com/bird/vulture-buzzard-control/

Gonnett, J. M. & Blendinger, P. G. (1998). New distributional records for the Crowned eagle (Harpyhaliaetus coronatus) in western Argentina. Hornero, 15, 39–42.

Gonzales, R. B. (1968). A study of the breeding biology and ecology of the monkey-eating eagle. Silliman Journal, 15, 461–491.

Gonzales, R. B. (1969). Monkey-eating eagle survey, Mindanao Island, Philippines. *In*: F. Vollmer (Ed.), World Wildlife Fund Yearbook, pp. 116–119. Morges, Switzerland: World Wildlife Fund.

Gonzales, R. B. (1971). Report on the 1969 status of the monkey-eating eagle on Mindanao Island, Philippines. Bulletin of the International Council for Bird Preservation, 11, 154–168.

Gonzalez, L. M., Oria, J., Sánchez, R., Margalida, A., Aranda, A., Prada, L., Caldera, J., Mariano González, L. & Ignacio Molina, J. (2008). Status and habitat changes in the endangered Spanish Imperial eagle Aquila adalberti population during 1974–2004: Implications for its recovery. Bird Conservation International, 18(3), 242–259. https://doi.org/10.1017/S0959270908000245.

González, L. M., Margalida, A., Oria, J., Sánchez, R. & González, L. M. (2006). Supplementary feeding as an effective tool to improving breeding success in the Spanish Imperial eagle (Aquila adalberti). Biological Conservation, 129(4), 477–486. https://doi.org/10.1016/j.biocon.2005.11.014.

Gonzalez, C. A., Guevara, E. P., Serrano, K. P., Soria, H. L. & Corona, R. S. (2004). A record of the Ornate Hawk-eagle (Spizaetus ornatus) in Nayarit, Mexico. Journal of Raptor Research, 38(2), 193–194.

Gonzalez, L. F., Montes, G. A., Puig, E., Johnson, S., Mengersen, K. & Gaston, K. J. (2016). Unmanned Aerial Vehicles (UAVs) and artificial intelligence revolutionizing wildlife monitoring and conservation. Sensors, 16(1), 97. https://doi.org/10.3390/s16010097

Gonzalez, L. M. (1991). Historia natural del Aguila Imperial Iberica (Aquila adalberti Brehm, 1861). Madrid: Coleccio´n Tecnica, ICONA Ministerio de Agricultura.
González, L. M. (1991). Historia natural del Águila Imperial Ibérica (Aquila adalberti Brehm, 1861). Taxonomía, población, análisis de l distribución geográfica, alimentación, reproducción y conservación. Madrid: Servicio de Vida Silvestre, ICONA.
González, L. M. (1996). Action plan for the Spanish Imperial eagle (Aquila adalberti). *In*: B. Heredia, L. Rose & M. Painter (Eds.), Globall threatened birds in Europe action plans, pp. 175–189. Strasbourg, France: Council of Europe and Birdlife International.
González, L. M. & Margalida, A. (2008). Conservation biology of the Spanish Imperial eagle (Aquila adalberti). [Organismo autónomo parque nacionales]. Madrid: Ministerio de medio ambiente y medio marino y rural.
Gonzalez, L. M., Margalida, A., Sánchez, R. & Oria, J. (2006a). Cooperative breeding in the Spanish Imperial eagle Aquila adalberti: A case o polyandry with reversed sexual behaviour? Ibis, 148, 159–163.
Gonzalez, L. M., Arroyo, B. E., Margalida, A., Oria, J., . . . (2006c). Effect of human activities on behaviour and success of breeding Spanis Imperial eagles Aquila adalberti. Animal Conservation, 9, 85–93.
González, L. M. & Oria, J. (2003). Águila Imperial Ibérica, Aquila adalberti. *In*: R. Martí & J. C. del Moral (Eds.), Atlas de las aves reproductora de España, pp. 186–187. Madrid: Dirección General de Conservación de la Naturaleza-Sociedad Española de Ornitología, Madrid, Spain.
Gonzalez, L. M., Heredia, B., González, J. L. & Alonso, J. C. (1989). Juvenile dispersal of Spanish imperial eagles. Journal of Field. Ornithology 60, 369–379.
Gonzalez, L. M., Hiraldo, F., Delibes, M. & Calderon, J. (1989). Zoogeographic support to consider the Spanish Imperial eagle as a distinc species. Bulletin of the British Ornithological Club, 109, 86–93.
González, L. M., Margalida, A., Mañosa, S., Sánchez, R., Oria, J., Molina, J. I. & Prada, L. (2007). Causes and spatio-temporal variations o non-natural mortality in the globally endangered Spanish Imperial eagle (Aquila adalberti) during a recovery period. Oryx, 41(4), 495–502 https://doi.org/10.1017/S0030605307414119
González, L. M., Oria, J., Margalida, A., Sánchez, R., Prada, L., Caldera, J., . . . Molina, J. I. (2006). Effective natal dispersal and age o maturity in the threatened Spanish Imperial eagle Aquila adalberti: Conservation implications. Bird Study, 53(3), 285–293. https://doi org/10.1080/00063650609461444
Gonzalez, L. M., Oria, J., Sánchez, R. & Margalida, A. (2008). Status and habitat changes in the endangered Spanish Imperial eagle Aquil adalberti population during 1974–2004: Implications for its recovery. BirdLife Conservation International, 18(3), 242–259. https://doi org/10.1017/S0959270908, PubMed: 000245
González, L. M., Bustamante, J. & Hiraldo, F. (1990). Factors influencing the present distribution of the Spanish Imperial eagle (Aquila adalberti) Biological Conservation, 51(4), 311–319. https://doi.org/10.1016/0006-3207(90)90116-7
Good, T. P. (2020). Great Black-backed Gull (Larus marinus), version 1.0. *In*: S. M. Billerman MCh (Ed.), Birds of the World. Ithaca, NY: Cornel Laboratory of Ornithology, Ithaca, USA. https://doi.org/10.2173/bow.gbbgul.01
Goodchild, M. (2010). Twenty years of progress: GIScience in 2010. Journal of Spatial Information Science, 1. https://doi.org/10.5311 JOSIS.2010.1.2
Goodfellow, D. (2014). How GIS is Being Used in Conservation Biology. GIS Lounge. Retrieved from https://www.gislounge.com/gis-used conservation-biology/
Goodman, S. M. & Meininger, P. L. (1989). The birds of Egypt. Oxford: Oxford University Press.
Goodman, S. M. (1994). The enigma of antipredator behavior in lemurs: evidence of a large extinct eagle on Madagascar. International Journal o Primatology, 15(1), 129–134, Berlin: Springer. https://doi.org/10.1007/BF02735238
Goodman, S. S. M. (1994). Description of a new species of subfossil eagle from Madagascar: Stephanoaetus (Aves; Falconiformes) from th deposits of Ampasambazimba. Proceedings of the Biological Society of Washington, 107, 421–428.
Goodrich, L. J. & Smith, J. P. (2008). (2020). Eagle. Raptor migration in North America. *In*: K. L. Bildstein, Smith, J. P., Ruelas, I. E., Veit, R. R (Eds.). Retrieved from https://www.lexico.com/en/definition/eagle
Göpfert, W. (1991). In: Kresse, W. & Danko, D.M. (Eds.), Springer handbook of geographic information, 147. Berlin, Germany: Springer Scienc & Business Media.
Gopi, G. V. & Pandav, B. (2006). White-bellied Sea eagle Haliaeetus leucogaster preying on salt-water crocodile Crocodylus porosus. India Birds, 2(6), 171.
Gorbet, G. B. & Harris, S. (1991). The Handbook of British Mammals (3rd Edition). Oxford: Blackwell Publishing Science Publications.
Gore, M. E. J. & Won, P. (1971). Birds of Korea. Seoul, Korea: Royal Asiatic Society, Korean Branch.
Gore, M. E. J. (1990). Birds of the Gambia: An annotated check-list. 2nd (revised) ed. Vol. 3. Series: BOU Checklists. London: Britis Ornithologists' Union.
Gore, M. L., Knuth, B. A., Curtis, P. D. & Shanahan, J. E. (2006). Education programs for reducing black bear-human conflicts: Indicators o success? Ursus., 17(1), 75–80, EPFRAB2.0. https://doi.org/10.2192/1537-6176(2006)1775
Goriup, P., Baboianu, G. & Chernichko, J. (2007). The Danube Delta: Europe's remarkable wetland. British Birds, 100(4), 194–213.
Gould, J. (1841). Descriptions of new birds from Australia. Aquila morphnoides. Proceedings of the Zoological Society of London,1840, 161–162
Gould, J. (1841). Exhibition of new Australian birds. Proceedings of the Zoological Society of London, 1840, 159–165.
Gould, J. (1863). The birds of Asia, 1. Published by the author. London, unpaginated.
Gould, J. & Rutgers, A. (1966). Birds of Europe, 1. London: Methuen.
Goulet, R. (2009). Aspects of the ecology of urban-nesting Bald eagles (Haliaeetus leucocephalus) in south-coastal British Columbia [MS Thesis]. Quebec, Canada. Montréal: McGill University.

Gove, B., Langston, R., McCluskie, A., Pullan, J. & Scrase, I. (2013). Wind farms and birds: An updated analysis of the effects of wind farms on birds, and best practice guidance on integrated planning and impact assessment. Highway Mail. Zimbabwe.
Govender, B. (2015). Waterfall pets fall prey to eagles. Highway Mail. Zimbabwe.
Government of Canada. (2019). Weather and meteorology glossary. Retrieved from https://www.canada.ca/en/environment-climate-change/services/weather-general-tools-resources/glossary.html.
Government of Canada. (2019a). Where to fly your drone. Retrieved from https://www.tc.gc.ca/en/services/aviation/drone-safety/where-fly-drone.html.
Government of Canada. (2019b). Use of Unmanned Aerial Vehicles (UAV) or drones in Canada's national parks. Retrieved from https://www.pc.gc.ca/en/pn-np/ab/banff/info/permis-permit/drone.
Grade, T. J. (2011). Effects of lead fishing tackle on common loons (Gavia immer) in New Hampshire, pp. 1989–2010 [Thesis]. Madison: University of Wisconsin.
Grade, T. J., Pokras, M. A., Laflamme, E. M. & Vogel, H. S. (2018). Population-level effects of lead fishing tackle on common loons. Journal of Wildlife Management, 82(1), 155–164. https://doi.org/10.1002/jwmg.21348, PubMed: 20172018
Graham, K., Beckerman, A. P. & Thirgood, S. (2005). Human–predator–prey conflicts: Ecological correlates, prey losses, and patterns of management. Biological Conservation, 122(2), 159–171. https://doi.org/10.1016/j.biocon.2004.06.006
Grainger Hunt, W. (2015). C. B. Moffat's anticipation of twenty-first century bird population dynamics theory. 157(4), 888–891. https://doi.org/10.1111/ibi.12283
Grambo, R. L. (2003). Eagles. Minneapolis: Voyageur Press.
Grant, P. R. & Grant, B. R. (1992). Hybridization of bird species. Science, 256(5054), 193–197. https://doi.org/10.1126/science.256.5054.193
Granzinolli, M. A. M. & Motta-Junior, J. C. (2007). Raptors in the heterogeneous landscape of central São Paulo State, Brazil. *In*: K. L. Bildstein, D. R. Barber & A. Zimmerman (Eds.), Neotropical raptors, pp. 302–303. Orwigsburg: Hawk Mountain Sanctuary.
Granzinolli, M. A. M., Pereira, R. J. G. & Motta-Junior, J. C. (2006). The Crowned Solitary-eagle Harpyhaliaetus coronatus (Accipitridae) in the Cerrado of Estação Ecologica de Itirapina, southeast Brazil. Revista Brasileira de Ornitologia, 14, 429–432.
Graszynski, K., Komischke, B. & Meyburg, B.-U. (2002). On the biology of the Greater Spotted eagle (Aquila clanga, Pallas 1811). *In*: R. Yosef, M. L. Miller & D. Pepler (Eds.), Raptors in the new Millennium, pp. 62–75. Eilat, Israel: International Birding & Research Center.
Graves, G. R. (1992). Greater Yellow-headed vulture (Cathartes melambrotus) locates food by olfaction. Journal of Raptor Research, 26, 38–39.
Gray, G. R. (1840). A list of the genera of birds, with an indication of the typical species of each genus. London: Taylor.
Gray, G. R. (1860). List of birds collected by Mr. Wallace at the Molucca Islands, with descriptions of new species, etc. Proceedings of the Zoological Society of London, 28(3), 341–366.
Greco, M. V., Andrade, M. A., Carvalho, G. D. M., Carvalho-Filho, E. P. M. & Carvalho, C. E. (2004). Callithrix penicillata na dieta de Spizaetus ornatus (Aves; Accipitridae) em área de cerrado no estado de Minas Gerais. A Primatologia no Brasil, 8, 155–160.
Green, C. R. & Sanford, W. R. (1986). The peregrine falcon. Mankato, MN: Crestwood House.
Green, M., Alerstam, T., Gudmundsson, G. A., Hedenströöm, A. & Piersma, T. (2004). Do Arctic waders use adaptive wind drift? Journal of Avian Biology, 35(4), 305–315. https://doi.org/10.1111/j.0908-8857.2004.03181.x
Green, R. E., Donázar, J. A., Sánchez-Zapata, J. A. & Margalida, A. (2016). The threat to Eurasian Griffon vultures Gyps fulvus in Spain from veterinary use of the drug diclofenac. Journal of Applied Ecology, 53(4), 993–1003. https://doi.org/10.1111/1365-2664.12663
Green, R. E., Newton, I., Shultz, S., Cunningham, A. A., Gilbert, M., Pain, D. J. & Prakash, V. (2004). Diclofenac poisoning as a cause of vulture population declines across the Indian subcontinent. Journal of Applied Ecology, 41(5), 793–800. https://doi.org/10.1111/j.0021-8901.2004.00954.x
Green, R. E., Pienkowski, M. W. & Love, J. A. (1996). Long-term viability of the re-introduced population of the White-tailed eagle Haliaeetus albicilla in Scotland. Journal of Applied Ecology, 33(2), 357–368. https://doi.org/10.2307/2404757
Greeney, H. F., Gelis, R. A. & White, R. (2004). Notes on breeding birds from an Ecuadorian lowland forest. Bulletin of the British Ornithologists' Club, 124(1), 28–37.
Greenwood, P. J., Harvey, P. H. & Perrins, C. M. (1979). The role of dispersal in the great tit (Parus major), The causes, consequences and heritability of dispersal. Journal of Animal Ecology, 48, 123–142.
Greet, B. J. R. (2015). The Roman eagle: A symbol and its evolution. PhD. Thesis. Leeds: University of Leeds.
Gregory, G. (2005). The birds of the State of Kuwait. Stegness, UK: Published by the author.
Gregory, M., Gordon, A. & Moss, R. (2003). Impact of nest-trapping and radiotagging on breeding Golden eagles Aquila chrysaetos in Argyll, Scotland. Ibis, 145, 113–119.
Gregory, P. (2008). Birds of New Guinea and associated islands. A checklist. Kuranda. Australia: Sicklebill Publications.
Gregory, S. M. S. & Dickinson, E. C. (2012). Clanga has priority over Aquiloides (or how to drop a clanger). Bulletin of British Ornithology, 132, 135–136.
Gregory, V. (2018). Infrared imaging helps save Yellowstone's wolves. Retrieved from https://www.wired.com/2012/05/st-photo-wolves/
Greig-Smith, P. W. (1976). The composition and habitat preferences of the avifauna of Mole National Park, Ghana. Bulletin Nigerian Ornithological Society, 12, 49–66.
Grenzdörffer, G. J. (2013). UAS-based automatic bird count of a common gull colony. International Archives of the Photogrammetry, Remote Sensing and Spatial Information Sciences, XL-1/W2. 169–174. https://doi.org/10.5194/isprsarchives-XL-1-W2-169-2013
Grewal, B., Harvey, B. & Pfister, O. (2002). A photographic guide to the birds of India and the Indian subcontinent, including Pakistan, Nepal, Bhutan, Bangladesh, Sri Lanka and the Maldives. Princeton, NJ: Princeton University Press.

Grier, J. W. (1969). Bald eagle behavior and productivity responses to climbing nests. Journal of Wildlife Management, 41, 438–443.

Grier, J. W. (1982). Ban of DDT and subsequent recovery of reproduction in Bald eagles. Science, 218(4578), 1232–1235. https://doi.org/10.1126/science.7146905

Grier, J. W., Elder, J. B., Gramlich, F. J., Green, N. F. . . . (1983). Northern states Bald eagle recovery plan. Denver, CO: United States Fish and Wildlife Service, Denver, Colorado, USA.

Griffin, C. R. (1981). Interactive behavior among Bald eagles wintering in north-central Missouri. Wilson Bulletin, 93, 259–264.

Griffin, C. R., Baskett, T. S. & Sparrowe, R. D. (1982). Ecology of Bald eagles wintering near a waterfowl concentration Special Scientific Report—Wildlife 247. U.S. Fish and Wildlife Service.

Griffiths, C. S. (1994). Monophyly of the Falconi formes based on syringeal morphology. Auk, 111(4), 787–805. https://doi.org/10.2307/4088811

Griffiths, C. S., Barrowclough, G. F., Groth, J. G. & Mertz, L. A. (2007). Phylogeny, diversity, and classification of the Accipitridae based on DNA sequences of the RAG-1 exon. Journal of Avian Biology, 38(5), 587–602. https://doi.org/10.1111/j.0908-8857.2007.03971.x

Griffiths, R., Double, M. C., Orr, K. & Dawson, R. J. G. (1998). A DNA test to sex most birds. Molecular Ecology, 7(8), 1071–1075. https://doi.org/10.1046/j.1365-294x.1998.00389.x

Grimes, L. G. (1987). The birds of Ghana: An annotated checklist. B.O.U. Checklist. London: British Ornithologists' Union, London.

Grimmett, R., Inskipp, C. & Inskipp, T. (1998). Birds of the Indian subcontinent. London: Christopher Helm Publishers.

Grimmett, R., Inskipp, C. & Inskjpp, T. (2011). Birds of the Indian Subcontinent. (2nd edn.). 528 pp. London: Oxford University Press.

Grimmett, R., Inskipp, C., Inskipp, T. & Baral, H. S. (2016). Birds of Nepal: (Revised Edition). Bloomsbury Publishing.

Grimmett, R., Inskipp, C., Inskipp, T. & Byers, C. (1999). Birds of India, Pakistan, Nepal, Bangladesh, Bhutan, Sri Lanka, and the Maldives. Princeton, NJ: Princeton University Press.

Griscom, L. (1932). The distribution of bird-life in Guatemala. Bulletin of the American Museum of Natural History, 64, 1–439.

Griscom, L. (1935). The ornithology of the Republic of Panama. Bulletin of the Museum of Comparative Zoology, 78, 261–382.

Gron, K. (2007). Primate Factsheets: Dusky titi (Callicebus moloch) Taxonomy, Morphology & Ecology. Grønland Biosci. 38: 1–63.

Gross, L. (2006). Switching drugs for livestock may help save critically endangered Asian vultures. PLoS Biology, 4(3), e61. https://doi.org/10.1371/journal.pbio.0040061

Grosselet, M. & Gutierres-Carbonell, D. (2007). Primera observación confirmada del Águila crestada Morphnus guianensis para México. Cotinga, 28, 74–75.

Grossman, M. L. & Hamlet, J. (1964). Birds of prey of the world. New York: Bonanza Books.

Grossman, M. L. & Hamlet, J. (1964). Birds of Prey of the World. New York: Clarkson N. Group on Birds of Prey and Owls, Berlin, pp. 291–295.

Grubac, B. R. (1988). The Golden eagle in Southeastern Yugoslavia. Lotus, 38/39, 95–135.

Grubb, T. G. (1995). Food habits of Bald eagles breeding in the Arizona desert. Wilson Bulletin, 107(2), 258–274.

Grubb, T. G. & Bowerman, W. W. (1997). Variations in breeding Bald eagle responses to jets, light planes and helicopters. Journal of Raptor Research, 31(3), 213–222.

Grubb, T. G. & Coffey, M. A. (1982). Evidence of Bald eagles feeding on freshwater mussels. Wilson Bulletin, 94(1), 84–85.

Grubb, T. G. & King, R. M. (1991). Assessing human disturbance of breeding Bald eagles with classification tree models. Journal of Wildlife Management, 55(3), 500–511. https://doi.org/10.2307/3808982

Grubb, T. G., Bowerman, W. W. & Howey, P. H. (1994). Tracking local and seasonal movements of wintering Bald eagles Haliaeetus leucocephalus from Arizona and Michigan with satellite telemetry. *In*: Meyburg, B.-U. & Chancellor, R. D. (Eds.), Raptor conservation today, pp. 347–358. Moose Jaw, Saskatchewan: Pica Press.

Grubb, T. G., Bowerman, W. W., Giesy, J. P. & Dawson, G. A. (1992). Responses of breeding Bald eagles to human activities in northcentral Michigan. Canadian Field-Naturalist, 106, 443–453.

Grumbine, R. E. (1990). Viable populations, reserve size, and federal lands management: A critique. Conservation Biology, 4(2), 127–134. https://doi.org/10.1111/j.1523-1739.1990.tb00101.x

Grussu, M. & Corso, A. (1997). Sardegna e Sicilia: Aree privilegiate in Italia per lo svernamento dell'avifauna. Atti IX CIO. Avocetta, 21, 34.

Grúz, A., Mackle, O., Bartha, A., Szabó, R., Déri, J., Budai, P. & Lehel, J. (2019). Biomonitoring of toxic metals in feathers of predatory birds from eastern regions of Hungary. Environmental Science and Pollution Research International, 26(25), 26324–26331. https://doi.org/10.1007/s11356-019-05723-9

Gryndall, W. (1972). Hawking, Hunting, Fouling and Fishing p. 1596. New York: Da Cap Press. London.

Guedim, Z. (2017). How drones are helping with waste management. Retrieved from https://edgy.app/drones-waste-management-china-dubai

Guerrant, T. L., Pullins, C. K., Beckerman, S. F. & Washburn, B. E. (2013). Managing raptors to reduce wildlife strikes at Chicago's O'Hare International Airport. Proceedings of the Wildlife Damage Management Conference, 15, 63–68.

Guerrero Ayuso, J. & Arambiza, A. Segundo (2001). Preliminary list of the birds of kaa-iya del Gran Chaco E Izozog National Park, Santa Cruz, Bolivia. Santa Cruz, Bolivia: Wildlife Conservation Society.

Guggisberg, C. A. W. (1988). Birds of East Africa, vol, 1. Nairobi, Kenya: Mount Kenya Sundries Ltd.

Guil, F., Colomer, M. A., Moreno-Opo, R. & Margalida, A. (2015). Space-time trends in Spanish bird electrocution rates from alternative information sources. Global Ecology and Conservation, 3, 379–388. https://doi.org/10.1016/j.gecco.2015.01.005.

Guillemard, F. H. H. (1885). Report on the collection of birds obtained during the Voyage of the Yacht "Marchesa". Celebes Part IV. Proceedings of the Zoological Society of London, 3, 542–561. https://doi.org/10.1111/j.1469-7998.1885.tb07862.x, PubMed: 188553

Guillosson, T. & Garcia, F. (2006). Rétromigration d'Aigleaigle bottés (Hieraaetus pennatus) dans le midi de la France à l'automne (2004). Jardin, M. Ornithos, 13, 2–11.

Gunawardena, K. (2006). A nest of the Black eagle Ictinaetus malayensis in Sri Lanka. Birding ASIA, 6, 23–27.

Gunawardene, K. (2002). The first record of the nesting of the Black eagle. Loris, 23(1&2), 14–19.

Gurney, J. H. (1879). Notes on a "Catalogue of the Accipitres in the British Museum". Ibis, 21(4), 464–470.

Gurney, J. H. (1862). A fourth additional list of birds received from Natal. Ibis, 4(2), 149–158. https://doi.org/10.1111/j.1474-919X.1862.tb07483.x.

Gustafsson, R., Hjort, C., Ottosson, U. & Hall, P. (2003). Birds at Lake Chad and in the Sahel of NE Nigeria 1997–2000. Degerhamn, Sweden: Ottenby Bird Observatory.

Gutiéérrez, R. & Villa, S. (2002). A possible hybrid Spotted eagle × Lesser spotted eagle in Spain. Birding World, 15, 104–105.

Guyra Paraguay. (2004). Guyra Paraguay, Asunción, Paraguay.

Haagner, A. (2009). Observations of immature Southern Banded Snake-eagle Circaetus fasciolatus at nest in Zululand. Gabar, 19, 15–17.

Haas, D. (2011). Electrocution of birds. Some further aspects of international high significance. Presentation at International Conference on Power Lines and Bird Mortality in Europe, Budapest.

Haas, M. (2012). Extremely rare birds in the Western Palearctic. Barcelona, Spain: Lynx Edicions.

Haast, J. (1872). Notes on Harpagornis moorei, an extinct gigantic bird of prey, containing description of femur, ungual phalanges, and rib. Transactions and Proceedings of the New Zealand Institute, 4, 192–196.

Hackett, S. J., Kimball, R. T., Reddy, S., Bowie, R. C. K., Braun, E. L., Braun, M. J., . . . Yuri, T. (2008). A phylogenomic study of birds reveals their evolutionary history. Science, 320(5884), 1763–1768. https://doi.org/10.1126/science.1157704

Hadarics, T. & Zalai, T. (2008). An Annotated List of the Birds of Hungary. [Magyar Madártani és Természetvédelmi Egyesület, Budapest, Hungary].

Hadden, D. (2004). Birds and Bird Lore of Bougainville and the North Solomons. Alderley, Queensland, UK: Dove Publications.

Hadi, F. (2001). Food composition, hunting technique and breeding behaviour of the Javan Hawk-eagle (Spizaetus bartelsi, Stresemann, 1924) [Honors Thesis]. West Java, Indonesia: University of Padjadjaran, Jatinangor.

Hadi, F. (2003). Observations on the breeding behaviour of Javan Hawk-eagle (Spizaetus bartelsi). *In*: D. M. Prawiradilaga (Ed.), Proceedings of the Second Symposium on Raptor of Asia, pp. 145–154. The Indonesian Committee for the Second Symposium of Asian Raptor Research & Conservation Network. Bogor, Indonesia.

Hagar, J. A. (1988). Swainson's hawk (migration). *In*: R. S. Palmer (Ed.), Handbook of North American Birds. 5: Diurnal raptors (pt. 2),pp. 56–64. New Haven, CT: Yale University Press.

Hagen, Y. (1976). Havørn og kongeørn i Norge. Viltrapport, 1: 93 pg.

Haig, S. M., D'Elia, J., Eagles-Smith, C., Fair, J. M., Gervais, J., Herring, G., . . . Schulz, J. H. (2014). The persistent problem of lead poisoning in birds from ammunition and fishing tackle. The Condor, 116(3), 408–428. https://doi.org/10.1650/condor-14-36.1

Hailer, F., Helander, B., Folkestad, A. O., Ganusevich, S. A., Garstad, S., Hauff, P., . . . Ellegren, H. (2006). Bottlenecked but long-lived: High genetic diversity retained in White-tailed eagles upon recovery from population decline. Biology Letters, 2(2), 316–319. https://doi.org/10.1098/rsbl.2006.0453

Hailer, F., Helander, B., Folkestad, A. O., Ganusevich, S. A., Garstad, S., Hauff, P., Koren, C., Masterov, V. B., Nygård, B. T., Rudnick, J. A., Shiraki , S., Skarphedinsson, K., Volke, V., Wille, F. & Vilà, C. 2007. Phylogeography of the White-tailed eagle, a generalist with large dispersal capacity. Journal of Biogeography, 34(7), 1193–1206. https://doi.org/10.1111/j.1365-2699.2007.01697.x.

Hailer, F., James, H. F., Olson, S. L. & Fleischer, R. C. (2015). Distinct and extinct: Genetic differentiation of the Hawaiian eagle. Molecular Phylogenetics and Evolution, 83, 40–43. Online publication date: 1-Feb-2015. https://doi.org/10.1016/j.ympev.2014.11.005

Haldane, J. B. S. (1922). Sex-ratio and unisexual sterility in hybrid animals. Journal of Genetics, 12(2), 101–109. https://doi.org/10.1007/BF02983075

Hall, B. P. (Ed.) (1974). Birds of the Harold Hall Australian Expeditions, 1962–70. London: British Museum of Natural History.

Hall, D. (1979). Food of the Long-crested eagle. Ostrich, 48, 256–257.

Hall, D. (1979). Records of Long-crested eagles rearing two young. Ostrich, 50, 187.

Hall, D. (1984). Long-crested eaglet hooked on twigs on its first flight. Bokmakierie, 36, 72.

Hall, D. G. (1979). Observations at three Long-crested eagle nests in the Nelspruit District. Bokmakierie, 31, 65–72.

Hall, D. G. (1982). High reproductive output by a pair of Long-crested Eagles. Ostrich, 53(4), 227. https://doi.org/10.1080/00306525.1982.9634578

Hall, D. G. (1991). Long-crested eagles' legs, an aid to identifying the sexes. Gabar, 6, 29.

Hall, D. G. (1992). A 16-year study of Long-crested eagles in the Nelspruit District, South Africa. Gabar, 7, 15–20.

Hall, K. R. L. (1966). Behaviour and ecology of the wild patas monkey, Erythrocebus patas, in Uganda. Journal of Zoology, 148(1), 15–87. https://doi.org/10.1111/j.1469-7998.1966.tb02942.x

Hall, M., Grantham, J., Posey, R. & Mee, A. (2007). Lead exposure among reintroduced California condors in southern California. California condors in the 21st century. Series in Ornithology, 2, 163–184. Cambridge, MA: Nuttall Ornithological Club and American Ornithologists' Union.

Halley, D. J. (1998). Golden and White-tailed eagle in Scotland and Norway: Co-existence, competition and environmental degradation. Breast British Birds, 91, 171–179.

Halley, D. J. & Gjershaug, J. O. (1996). Inter- and intra-specific dominance relationships and feeding behaviour of Golden eagles (Aquila chrysaetos) and Sea eagles (Haliaeetus albicilla) at carcasses. Ibis, 140, 295–301.

Halley, D. J., Bevanger, K., Berntsen, F. E. H., Clausen, S. M., Dahl, E. L., Flagstad, Ø., . . . Vang, R. (2010). The effect of wind turbines on birds at the Smøla Wind Power Plant, Norway: Collisions, displacement, and avoidance. Poster presented at the Conference Renewable Energy, 2010, Yokohama, Japan. Trondheim: Norsk Institutt for Naturforskning (NINA), 27 June–2 July.

Hallmann, B. (1985). Status and conservation problems of birds of prey in Greece. *In*: I. Newton & R. D. Chancellor (Eds.), Proceedings of th ICBP World Conference on Birds of Prey, pp. 55–59. ICBP Technical Publication No 5. Thessaloniki, Greece.
Hallmann, B. C. G. (1989). Status and distribution of the genus Aquila in Greece. Biologia Gallo-hellenica, 15, 171–176.
Halse, S. A. & Skead, D. M. (1983). Wing moult, body measurements and condition indices of spur-winged geese. Wildfowl, 34(34), 108–114.
Hamer, T. E., Forsman, E. D., Fuchs, A. D. & Walters, M. L. (1994). Hybridization between barred and spotted owls. Auk, 111(2), 487–492. https: doi.org/10.2307/4088616
Hamilton, S. G. (2004). Pelage colouration of the endemic Solomons Flying-fox Pteropus rayneri (Pteropodidae; Chiroptera), responses t predation by the endemic Sanford's eagle Haliaeetus sanfordi in the Melanesian Archipelago: Solomon Islands and Bougainville. Australasia Bat Society Newsletter, 23, 14–16.
Hamilton, W. J. & Heppner, F. (1967). Radiant solar energy and the function of black homeotherm pigmentation: An hypothesis. Science, 15 196–197.
Hammerschlag, N., Bell, I., Fitzpatrick, R., Gallagher, A. J., Hawkes, L. A., . . . Barnett, A. (2016). Behavioral evidence suggests facultativ scavenging by a marine apex predator during a food pulse. Behavioral Ecology and Sociobiology, 70(10), 1777–1788. https://doi.org/10.1007 s00265-016-2183-2
Hammill, M. O., Dale, J., Stenson, G. B., den Heyer, C., Gosselin, J.-F., Leblanc, P. & Johnston, D. W. (2017). Comparison of methods to estima grey seal pup production at different colonies. DFO Canadian Science Advisory Secretariat (CSAS) Research Document/041.
Hanafusa, T., Lee, C. B. & Lo, A. K. (1986). Dependence of the exponent in power law wind profiles on stability and height interval. Atmospheri Environment, 20(10), 2059–2066, https://doi.org/10.1016/0004-6981(86)90348-3
Hancock, D. (1964). Bald eagles wintering in the Southern Gulf Islands, British Columbia. Wilson Bulletin, 76, 111–120.
Hancock, D. (2003). The Bald eagle of Alaska, BC and Washington. Surrey, British Columbia. Canada: Hancock House Publishers.
Hancock, P. (2009). Poisons devastate vultures throughout Africa. African Raptors Newsletter, 2 (November).
Hancock, P. (2009a). Botswana—Major poisoning incidents. African Raptors, 10–11.
Hancock, P., Muller, M. & Tyler, S. J. (2007). Inventory of birds of the Okavango Delta Ramsar site. Babbler, 49, 3–38.
Hancock, P. (2010). Vulture poisoning at Khutse. Birds and People, 27, 7. Retrieved from http://hdl.handle.net/1854/LU-1039067.
Hancock, P. & Weiersbye, I. (2015). Birds of Botswana. Princeton, NJ: Princeton University Press.
Handbook of the birds of the World. (2017). Golden eagle (Aquila chrysaetos). Retrieved from https://www.hbw.com/species/golden-eagle-aquila chrysaetos
Handbook of the birds of the World. (2018). Bald eagle (Haliaeetus leucocephalus). Retrieved from https://www.hbw.com/species/bald-eagle haliaeetus-leucocephalus
Handrinos, G. (1987). The Golden eagle in Greece. Actes. 1er Coll. [Intern]. Aigle royal en Europe.
Handrinos, G. & Akriotis, T. (1997). The birds of Greece. London: A & C Black.
Handschuh, M., van Zalinge, R. N., Olsson, U., Samphos, P., Chamnan, H. & Evans, T. D. (2011). First confirmed record and first breeding recor of Indian Spotted Eagle Aquila hastata in Indochina. Bulletin of the British Ornithologists' Club, 131, 118–122.
Hanmer, D. B. (1977). Man-induced changes in the Lower Shire valley with special reference to the avifauna. Nyala, 3(1), 33–37.
Hansen, K. (1979). Population status for the Greenland white-tailed eagle Haliaeetus albicilla groenlandicus Brehm covering the years 1972–74 Dansk orn. Foren. Tidsskr., 73, 107–130.
Hantge, E. (1980). Untersuchungen über den Jagderfolg mehrerer Europäischer Greifvögel. Journal of Ornithology, 121(2), 200–207. https://do org/10.1007/BF01642933
Hapsoro, Lesmana, D., Kartiwa, H., Mulyadi, Valentinus, A. & Purba, C. (1998). Future from the high frontier: Final report of Javan Hawk-eagl surveys in West Java. Bogor, Indonesia: Telapak Indonesia Foundation, Bogor.
Harder, M. (2000). Diet and breeding biology of the Wedge-tailed Eagle Aquila audax at three nests in northeastern New South Wales. Corella 24, 1–5.
Hardey, J., Crick, H., Wernham, C., Riley, H., Etheridge, B. & Thompson, D. (2009). Raptors: A field guide for surveys and monitoring [Mitteleuropäische Gesellschaft zur Erhaltung der Greifvögel (MEGEG)]. Central European Society for Raptor Protection.
Harford, H. (2012). Sea eagles catching flying-foxes in NSW. Boobook, 30, 49.
Haring, E., Kvaløy, K., Gjershaug, J.-O., Røv, N. & Gamauf, A. (2007). Convergent evolution and paraphyly of the hawk-eagles of the genu Spizaetus (Aves, Accipitridae), Phylogenetic analyses based on mitochondrial markers. Journal of Zoological Systematics and Evolutionar Research, 45(4), 353–365. https://doi.org/10.1111/j.1439-0469.2007.00410.x
Harlow, D. L. & Bloom, P. H. (1989). Buteos and the golden eagle. In: Pendleton, B. G. (Ed.), Proceedings of the Western Raptor Managemen Symposium and Workshop, 102–110. Washington, D.C.: National Wildlife Federation Scientific and Technical Series 12.
Harmata, A. R., Toepfer, J. E. & Gerrard, J. M. (1985). Fall migration of Bald Eagles produced in northern Saskatchewan. Blue Jay, 43, 232–237
Harmata, A. R. (2002). Vernal migration of Bald Eagles from a southern Colorado wintering area. Journal of Raptor Research, 36, 256–264.
Harness, R. & Gombobaatar, D. R. S. (2008). Raptor electrocutions in the Mongolia Steppe. Winging It, 20, 1–6.
Harness, R., Gombobaatar, D. R. S. & Yosef, R. (2008). Mongolian Distribution of Power Lines and Raptor Electrocutions. Rural Electric Powe Conference, Charleston, South Caroline.
Harness, R. E. & Wilson, K. R. (2001). Utility structures associated with raptor electrocutions in rural areas. Wildlife Society Bulletin, 29(2) 612–623. Retrieved from https://www.jstor.org/stable/3784188.
Harris, D. R. (1980). Human Ecology in Savanna Environments. London: Academic.
Harris, L. N., Tallman, R. F. & Ottawa, O. N. (2010). Information to support the assessment of Arctic Char, Salvelinus alpinus, from the Isuitu River system (Canada), [Nunavut]. Department of Fisheries and Oceans. Ottawa, Canada: Canadian Science Advisory Secretariat.

Harrison, E. N. & Kiff, L. F. (1977). The nest and egg of the Black Solitary eagle. Condor, 79(1), 132–133. https://doi.org/10.2307/1367547
Harrison, J. (1999). A Field Guide to the Birds of Sri Lanka. Oxford: Oxford University Press.
Harrison, J. (2011). A Field Guide to the Birds of Sri Lanka. Oxford: Oxford University Press.
Harrison, J. A., Allan, D. G., Underhill, L. G., Herremans, M., Tree, A. J., Parker, V. & Brown, C. J. (Eds.) (1997). The atlas of southern African birds, 1. Non-passerines. Johannesburg: Birdlife South Africa.
Hartert, E. (1898). On the birds collected by Mr. Everett in South Flores. Novitates Zoologicae, 5, 42–50.
Hartley, R. (1998). Raptor migration and conservation in Zimbabwe. Torgos, 28, 135–150.
Hartley, R. R. & Verdoorn, G. H. (2000). Lophaetus occipitalis Long-crested Eagle. *In*: Verdoorn, G. H., Bildstein, K. L. & Ellis, S. (Eds.), Selected African Falconiformes conservation assessment and management plan, pp. 54–56. Apple Valley, MN IUCN/SSC Conservation Breeding Specialist Group.
Hartley, R. R. & Verdoorn, G. H. (compilers) (2000). Lophaetus occipitalis Long-crested eagle. *In*: G. H. Verdoorn, K. L. Bildstein & S. Ellis (Eds.), Selected African Falconiformes conservation assessment and management plan, pp. 54–56. Apple Valley, MN: International Union for Conservation of Nature and Natural Resources. SSC Conservation Breeding Specialist Group.
Hartley, R. R., Bramford, D. & Fenn, T. (2005). Diversity and ecology of raptors in the Save Conservancy. Honeyguide, 48, 153–166.
Hartley, R. (1998). Raptor migration and conservation in Zimbabwe. Torgos, 28, 135–150.
Hartley, R. & Mundy, P. (2003). Morphometrics and status of Ayres's Hawk-eagle in Zimbabwe. Journal of Raptor Research, 37, 44–54.
Hartshorn, G. S., Nicolait, L., Hartshorn, L., Bevier, G., Brightman, R., Cal, J., . . . Wright, C. (1984). Belize Country Environmental Profile: A Field Study. No. 505-0000-C-00-3001-00. 151 pp. USAID Contract.
Harvey, L. J. & McArdle, H. J. (2008). Biomarkers of copper status: A brief update. British Journal of Nutrition, 99(Suppl. 3), S10-S13. https://doi.org/10.1017/S0007114508006806
Harvey, W. G. & Harrison. I. D. (1970). The birds of the Mole game reserve. Bulletin of the Nigerian Ornithologists Society, 7, 43–52, 63–75.
Haschberger, P., Bundschuh, M. & Tank, V. (1996). Infrared sensor for the detection and protection of wildlife. Optical Engineering, 35(3), 882–889. https://doi.org/10.1117/1.601005
Haschek, W. M., Rousseaux, C. G. & Wallig, M. A. (2013). Haschek and Rousseaux's Handbook of Toxicologic Pathology (3rd ed.). San Diego: Academic Press [Imprint]. Elsevier Science & Technology Books.
Hastie, G. D. (2012). Tracking marine mammals around marine renewable energy devices using active sonar. SMRU Ltd report number SMRUL-DEC-2012-002 to the Department of Energy and Climate Change, 93. St Andrews: SMRU Ltd.
Hatch, D. R. M. (1968). Golden eagle hunting tactics. Blue Jay, 26(2), 78–80. https://doi.org/10.29173/bluejay3140
Hatchwell, B. J. & Komdeur, J. (2000). Ecological constraints, life history traits and the evolution of cooperative breeding. Animal Behaviour, 59(6), 1079–1086. https://doi.org/10.1006/anbe.2000.1394
Hatfield, R. (2018). Diet and space use of the Martial eagle (Polemaetus bellicosus) in the Maasai Mara region of Kenya. Theses and Dissertations—Forestry and Natural Resources.
Hauff, P. (2003). Sea-eagles in Germany and their population growth in the 20th century. In: Helander B. (Ed.), Sea eagle 2000: Proceedings from the International Sea Eagle Conference in Bjorko, Sweden, 13–17 September, 20. 71–72. Stockholm, Sweden Proceedings of the Swedish Society for Nature Conservation, SNF.
Hauff, P. & Mizera, T. (2006). Distribution and density of White-tailed Sea Eagles Haliaeetus albicilla in Germany and Poland: A current atlas-map. Vogelwarte, 44, 134–136.
Hauff, P. (2009). Nesting sites of White-tailed eagles Haliaeetus albicilla. die Vogelwelt, 130, 67–76.
Hauser, M. D. (1996). The Evolution of Communication. MIT Press. Cambridge, MA: MIT Press.
Hausheer, J. E. (2016). Can drones help monitor vultures on Mongolia's eastern Steppe? Retrieved from https://blog.nature.org/science/2016/03/09/can-drones-help-monitor-vultures-on-mongolias-eastern-steppe/
Haverschmidt, F. (1968). The birds of Surinam Suriname. Edinburgh, UK: Oliver & Boyd.
Haverschmidt, F. & Mees, G. F. (1994). Birds of Suriname. Paramaribo, Surinam: VACO.
Haworth, P. F., McGrady, M. J., Whitfield, D. P., Fielding, A. H. & McLeod, D. R. A. (2006). Ranging distance of resident Golden eagles Aquila chrysaetos in western Scotland according to season and breeding status. Bird Study, 53(3), 265–273. https://doi.org/10.1080/00063650609461442
Hayes, F. E. (1995). Status, distribution and biogeography of the birds of Paraguay. Monograph of Field Ornithology, 1.
Hays, L. L. (1987). Peregrine falcon nest defense against a Golden eagle. Journal of Raptor Research, 21(2), 67.
Hayward, J. J. L., Galusha, J. J. G. & Henson, S. S. M. (2010). Foraging-related activity of Bald eagles at a Washington Seabird colony and Seal rookery. Journal of Raptor Research, 44(1), 19–29. https://doi.org/10.3356/JRR-08-107.1.
Hazlett, P. W., Broad, K., Gordon, A. M., Sibley, P. K., Buttle, J. M. & Larmer, D. (2008). The importance of catchment slope to soil water N and C concentrations in riparian zones: Implications for riparian buffer width. Canadian Journal of Forest Research, 38(1), 16–30. https://doi.org/10.1139/X07-146
Healey, C. (1990). Maring hunters and traders: Production and exchange in the Papua New Guinea Highlands. University of California Press, Berkeley, CA.
Heath, M. F. & Evans, M. I. (2000). Important bird areas in Europe: Priority sites for conservation, 2 p. 8. Cambridge: Birdlife International, BirdLife Conservation Series.
Hedenström, A. (1993). Migration by soaring or flapping flight in birds: The relative importance of energy cost and speed. Philosophical Transactions of the Royal Society of London. Series B, 342(1302), 353–361. https://doi.org/10.1098/rstb.1993.0164
Hedges, S. B. & Sibley, C. G. (1994). Molecules vs. morphology in avian evolution. The case of the pelecaniform birds. Proceedings of the National Academy of Sciences of the United States of America, 91(21), 9861–9865. https://doi.org/10.1073/pnas.91.21.9861

Heglund, P. J. & Reiswig, B. (1980). Raptor survey, the breeding Bald eagle population, Amchitka Island, Alaska, U.S.D.I. AK: Fish and Wildlife Service, Aleutian Islands National Wildlife Refuge, Alaska.

Heidrich, P., Koenig, C. & Wink, M. (1995). Molecular phylogeny of the South American Screech Owls of the Otus atricapillus complex (Aves Strigidae) inferred from nucleotide sequences of the mitochondrial cytochrome b gene. Naturforsch, Z. Society Société, 294–302.

Heim de Balsac, H. & Mayaud, N. (1962). The birds of North-east Africa: Distribution, ecology, migration, reproduction. Encyclopedie Ornithologiqu, 10, 1–486.

Heinsohn, T. (2000). Predation by the White-breasted sea eagle Haliaeetus leucogaster on phalangerid possums in New Ireland, Papua New Guinea. Emu – Austral Ornithology, 100(3), 245–246. https://doi.org/10.1071/MU00913.

Heinrich, B. (1988). Winter foraging at carcasses by 3 sympatric corvids, with emphasis on recruitment by the raven, Corvus corax. Behavioral Ecology and Sociobiology, 23, 141–156.

Heinrich, G. (1958). Zur Verbreitung und Lebensweise der Vogel von Angola. Ornithol., 99, 399–421.

Heinz, G. H. & Hoffman, D. J. (1998). Methylmercury chloride and selenomethionine interactions on health and reproduction in mallards Environmental Toxicology and Chemistry, 17(2), 139–145. https://doi.org/10.1002/etc.5620170202

Heinz, G. H., Hoffman, D. J., Klimstra, J. D., Stebbins, K. R., Kondrad, S. L. & Erwin, C. A. (2011). Teratogenic effects of injected methylmercury on avian embryos. Environmental Toxicology and Chemistry, 30(7), 1593–1598. https://doi.org/10.1002/etc.530

Heintzelman, D. S. (1975). Autumn hawk flights. New Brunswick, NJ: Rutgers University Press.

Helander, B. (1985). Colour-ringing of White-tailed sea eagles in northern Europe. Conservation studies on raptors. ICBP Technical Publication, (5), 401–407.

Helander, B. (1988). Conservation and management of the White-tailed Sea-eagle Haliaeetus albicilla in Sweden. Torgos, 7, 86–90.

Helander, B., Ekman, B., Hagerroth, J.-E., Hagerroth, P.-A. & Israelsson, J. (1989). Age-specific field characteristics of the white-tailed sea eagle, Haliaeetus albicilla. Var Vagelvarld, 48, 319–334.

Helander, B. (2003). The international colour-ringing programme – Adult survival, homing, and the expansion of the White-tailed Sea Eagle in Sweden. *In*: Helander, B., Marquiss, M. & Bowermann, W. (Eds.), Sea Eagle 2000. Proceedings from an International Conference at Björko, Sweden, 13-17 September 2000. Swedish Society for Nature Conservation/SNF & Åtta.45. Tryckeri AB. Stockholm, 145–167.

Helander, B. & Stjernberg, T. (2002). Action plan for the conservation of White-tailed sea eagle (Haliaeetus albicilla). *In*: BirdLife International and Conseil de l'Europe, Strasbourg.

Helander, B., Axelsson, J., Borg, H., Holm, K. & Bignert, A. (2009). Ingestion of lead from ammunition and lead concentrations in White-tailed sea eagles (Haliaeetus albicilla) in Sweden. Science of the Total Environment, 407(21), 5555–5563. https://doi.org/10.1016/j.scitotenv.2009.07.027

Helander, B., Bignert, A. & Asplund, L. (2008). Using raptors as environmental sentinels: Monitoring the White-tailed sea eagle Haliaeetus albicilla in Sweden. Ambio – A Journal of the Human Environment, 37(6), 425–431. https://doi.org/10.1579/0044-7447(2008)37425.

Helbig, A. J. (2000). Was ist eine Vogel-"Art"? Ein Beitrag zur aktuellen Diskussion um Artkonzepte in der Ornithologie - Teil I. Limicola, 14, 57–79, 172–184.

Helbig, A. J., Kocum, A., Seibold, I. & Braun, M. J. (2004). A multi-gene phylogeny of aquiline eagles (Aves: Accipitriformes) reveals extensive paraphyly at the genus level. Molecular Phylogenetics and Evolution, 35(1), 147–164. https://doi.org/10.1016/j.ympev.2004.10. 003

Helbig, A. J., Salomon, M., Bensch, S. & Seibold, I. (2001). Male-biased gene flow across an avian hybrid zone: Evidence from mitochondrial and microsatellite DNA. Journal of Evolutionary Biology, 14(2), 277–287. https://doi.org/10.1046/j.1420-9101.2001.00273.x

Helbig, A. J., Seibold, I., Kocum, A., Liebers, D., Irwin, J. & Bensch, S. (2005). Genetic differentiation and hybridization between Greater and Lesser spotted eagles (Accipitriformes: Aquila clanga, A. pomarina). Journal of Ornithology, 146(3), 226–234. https://doi.org/10.1007/s10336-005-0083-8

Hellebrekers, W. P.h.J. & Hoogerwerf, A. (1967). A further contribution to our knowledge of the island of Java (Indonesia). Zoölogische Verhandelingen (Leiden), 88, 1–164.

Hellmayr, C. E. & Conover, B. (1949). Catalog of the birds of the Americas and the adjacent islands. Field Museum of Natural History, Part I, Zoological Series, 13(4), 1–358.

Helm-Bychowski, K. & Cracraft, J. (1993). Recovering a phylogenetic signal from DNA sequences. Relationships within the Corvine assemblage (Class Aves) as inferred from complete sequence of the mitochondrial DNA cytochrome-b-gene. Molecular Biology and Evolution, 10(6), 1196–1214. https://doi.org/10.1093/oxfordjournals.molbev.a040072

Helms, O. (1926). The birds of Angmagssalik. Meddelelser om Grønland, 58(4), 205–274.

Hemson, G., Johnson, P., South, A., Kenward, R., Ripley, R. & Macdonald, D. (2005). Are kernels the mustard? Data from global positioning system (GPS) collars suggests problems for kernel home-range analyses with least-squares cross-validation. Journal of Animal Ecology, 74(3), 455–463. https://doi.org/10.1111/j.1365-2656.2005.00944.x

Hendarsah, G. (2003). Flight skill development, home range and movements of Javan Hawk-eagle (Spizaetus bartelsi) during post-fledging period (at 56–243 days) in Ciasem, Gunung Tangkuban Perahu, Gunung T. Nature Reserve. *In*: Prawiradilaga, D. M. (Ed.), Proceedings of the Second Symposium on Raptor of Asia, pp. 131–138. Indonesian Committee for the Second Symposium.

Hengl, T., Minasny, B. & Gould, M. (2009). A geostatistical analysis of geostatistics. Scientometrics, 80, 491–514. https://doi.org/10.1007/s11192-009-0073-3

Hennessey, A. B. & Gomez, M. I. (2003). Four bird species new to Bolivia: An ornithological survey of the Yungas site Tokoaque, Madidi National Park. Cotinga, 19, 25–33.

Hennessey, B., Herzog, S. K. & Sagot, F. (2003). Annotated list of the birds of Bolivia. Asociación Armonia, Santa Cruz de la Sierra, Bolivia, Asociación Armonia.

Henny, C. J. & Elliott, J. E. (2015). Toxicology. *In*: Bird, D.M. & Bildstein, K. L. (Eds.), Raptor Research and Management Techniques Manual (2nd edn.), pp. 329–350. Surrey: Hancock House Publishing, Ltd.
Henry, G. M. (1933). Flight of the Black eagle Ictinaetus malayensis perniger, Hodg. Nature, 132(3335), 516–517. https://doi.org/10.1038/132516c0
Heptner, V. G. & Naumov, N. P. (1998). Mammals of the Soviet Union, II. Part 1a, Sirenia and Carnivora (Sea cows; Wolves and Bears). Boca Raton, FL: Science Publishers, Inc.
Herholdt, J. J. & De Villiers, D. J. (1991). Breeding success and population density of the Bateleur Terathopius ecaudatus in the Kalahari Gemsbok National Park. Gabar, 6, 3–6.
Herholdt, J. J., Kemp, A. C. & du-Plessis, D. (1996). Aspects of the breeding status and ecology of the Bateleur and Tawny Eagle in the Kalahari Gemsbok National Park, South Africa. Ostrich, 67, 126–137.
Herholdt, J. J. & Kemp, A. C. (1997). Breeding status and ecology of the Martial eagle in the Kalahari Gemsbok National Park, South Africa. Ostrich, 68(2–4), 80–85. https://doi.org/10.1080/00306525.1997.9639717
Herholdt, J. J. (1998). Survival threats and conservation management of raptors in the Kgalagadi Transfrontier Park. Transactions of the Royal Society of South Africa, 53(2), 201–218. https://doi.org/10.1080/00359199809520387
Herklots, G. A. C. (1953). Hong Kong birds. South China Morning Post, Hong Kong.
Herklots, G. A. C. (1961). The birds of Trinidad and Tobago. London: Collins, London.
Hernández, D. A. & Zook, J. R. (1993). Northward migration of Peregrine Falcons along the Caribbean coast of Costa Rica. Journal of Raptor Research, 27, 123–125.
Hernández, M. & Margalida, A. (2008). Pesticide abuse in Europe: Effects on the Cinereous vulture (Aegypius monachus) population in Spain. Ecotoxicology, 17(4), 264–272. https://doi.org/10.1007/s10646-008-0193-1
Hernández, M. & Margalida, A. (2009). Poison-related mortality effects in the endangered Egyptian vulture (Neophron percnopterus) population in Spain: Conservation measures. European Journal of Wildlife Research, 55(4), 415–423. https://doi.org/10.1007/s10344-009-0255-6
Hernandez-Matias, A., Real, J., Pradel, R., Ravayrol, A., Vincent-Martin, N., Bosca, F. & Cheylan, G. (2010). Determinants of territorial recruitment in Bonelli's eagle (Aquila fasciata) populations. Auk, 127(1), 173–184. https://doi.org/10.1525/auk.2009.09143
Hernández-Matías, A., Real, J. & Pradel, R. (2011). Quick methods for evaluating survival of age-characterizable long-lived territorial birds. Journal of Wildlife Management, 75(4), 856–866. https://doi.org/10.1002/jwmg.102
Hernández-Matías, A., Real, J., Moleón, M., Palma, L., Sánchez-Zapata, J., Pradel, R., . . . García, J. (2013). From local monitoring to a broad-scale viability assessment: A case study for the Bonelli's eagle in western Europe. Ecological Monographs, 83(2), 239-261
Hernández-Matías, A., Real, J., Parés, F. & Pradel, R. (2015). Electrocution threatens the viability of populations of the endangered Bonelli's eagle (Aquila fasciata) in Southern Europe. Biological Conservation, 191, 110–116. https://doi.org/10.1016/j.biocon.2015.06.028
Hernandez-Pliego, J., Rodríguez, C. & Bustamante, J. (2014). Gone with wind: Seasonal trends in foraging movement directions for a central-place forager. Current Zoology, 60(5), 604–615. https://doi.org/10.1093/czoolo/60.5.604
Herremans, M. (1997). Western Banded Snake Eagle. Harrison, J. A. et al. (Eds.), The atlas of South African birds. Volume 1, Non-passerines, 201. Johannesburg, SA: BirdLife South Africa and Avian Demography Unit.
Herremans, M. & Herremans-Tonnoeyr, D. (2000). Land use and the conservation status of raptors in Botswana. Biological Conservation, 94(1), 31–41. https://doi.org/10.1016/S0006-3207(99)00166-4
Herrero, S. & Higgins, A. (1999). Human injuries inflicted by bears in British Columbia: 1960–1997. Ursus, 11, 209–218. Retrieved from https://www.jstor.org/stable/3873003
Herrmann, C., Krone, O. & Helander, B. (2020). Population development of Baltic bird species: White-tailed Sea eagle (Haliaeetus albicilla). HELCOM Baltic Sea Environment Fact Sheet. Retrieved from https://helcom.fi/wp-content/uploads/2020/06/BSEFS-White_tailed-Eagle.pdf.
Hertel, F. (1995). Ecomorphological indicators of feeding behavior in recent and fossil raptors. The Auk, 112(4), 890–903. https://doi.org/10.2307/4089021
Herzog, S. K. & Kessler, M. (2002). Biogeography and composition of dry forest bird communities in Bolivia. Journal für Ornithologie, 143(2), 171–204. https://doi.org/10.1046/j.1439-0361.2002.02004.x
Herzog, S. K., Kessler, M., Maijer, S. & Hohnwald, S. (1997). Distributional notes on birds of Andean dry forests in Bolivia. Bulletin of the British Ornithologists' Club, 117, 223–235.
Heuck, C., Herrmann, C., Schabo, D. G., Brandl, R. & Albrecht, J. (2017). Density-dependent effects on reproductive performance in a recovering population of White-tailed eagles Haliaeetus albicilla. Ibis, 159(2), 297–310. https://doi.org/10.1111/ibi.12444
Heuglin, M. T. (1856). Haliaëtos cinereus. *In*: Systematische Übersicht der Vögel Nordost-Afrika's mit Einschluss der arabischen Küste, des rothen Meeres und der Nil-Quellen-Länder südwärts bis zum 4. Grade nördl. Breite. Sitzungsberichte der kaiserlichen Akademie der Wissenschaften (math.-nat. Classe) Wissenschaften (math.-nat. Classe). Wien: 225–324.
Heymann, E. W. (1990). Reactions of wild tamarins, Saguinus mystax and Saguinus fuscicollis to avian predators. International Journal of Primatology, 11(4), 327–337. https://doi.org/10.1007/BF02193004
Heywood, I., Cornelius, S. & Carver, S. (2006). An Introduction to Geographical Information Systems. Third Edition. Harlow: Pearson Education.
Hidalgo, C., Sánchez, J., Sánchez, C. & Saborío, M. T. (1995). Migración de Falconiformes en Costa Rica. [Hawk Migration Association of North America, hawk]. Migration Studies, 21(1), 10–13.
Hilgerloh, G., Weinbecker, J. & Weiss, I. (2009). The timing of spring passage of soaring birds at Zait Bay, Egypt. Sandgrouse, 31, 26–35.
Hill, J. R. & Neto, P. S. (1991). Black vultures nesting on skyscrapers in southern Brazil. Journal of Field Ornithology, 62(2), 173–176. Retrieved from https://sora.unm.edu/sites/default/files/ journals/jfo/v062n02/p0173-p0176.pdf
Hillis, D. M. & Moritz, C. (1990). Molecular Systematics. Sinauer Press, Sundaland.

Hilty, S. L. (1977). A guide to the birds of Colombia. Princeton, NJ: Princeton University Press.

Hilty, S. L. (1985). Distributional changes in the Colombia avifauna: A preliminary Blue List. *In*: Buckley, P. A., Foster, M. S., Morton, E. S Ridgely, R. S. & Buckley, F. G. (Eds.), Neotropical Ornithology, pp. 1000–1012. Washington, DC: American Ornithologists' Union.

Hilty, S. L. (2002). Birds of Venezuela (2nd ed). Princeton, NJ: Princeton University Press.

Hilty, S. L. (2003). Birds of Venezuela. London: A&C Black, London.

Hilty, S. L. & Brown, W. L. (1986). A guide to the birds of Colombia. Princeton, NJ: Princeton University Press.

Hilty, S. S. L. (2003). Birds of Venezuela. London: Christopher Helm Publishers.

Hinck, J. E., Norstrom, R. J., Orazio, C. E., Schmitt, C. J. & Tillitt, D. E. (2009). Persistence of organochlorine chemical residues in fish from th Tombigbee River (Alabama, USA), Continuing risk to wildlife from a former DDT manufacturing facility. Environmental Pollution, 157(2 582–591. https://doi.org/10.1016/j.envpol.2008.08.021

Hipfner, J. M., Blight, L. K., Lowe, R. W., Wilhelm, S. I., Robertson, G. J., Barrett, R. T., Anker-Nilssen, T. & Good, T. P. 2012. Unintende consequences: How the recovery of Sea eagle Haliaeetus spp. populations in the northern hemisphere is affecting seabirds. Marine Ornitholog 40, 39–52.

Hiraldo, F., Delibes, M. & Calderon, J. (1976). The Spanish Imperial eagle. Donana Acta Vertebrata, 3/2, 171–182.

Hiraldo, F. & Donázar, J. (2009). Foraging time in the Cinereous Vulture Aegypius monachus: Seasonal and local variations and influence c weather. Bird Study, 37, 128–132. https://doi.org/10.1080/00063659009477048

Hobbs, R. J., Higgs, E. & Harris, J. A. (2009). Novel ecosystems: Implications for conservation and restoration. Trends in Ecology and Evolution 24, 599–605.

Hockey, P. A. R., Dean, W. R. J. & Ryan, P. G. (Eds.) (2005). Roberts' birds of Southern Africa. Cape Town, SA: Trustees of the John Voelke Bird Book Fund.

Hockey, P. A. R., Dean, W. R. J. & Ryan, P. G. (2016). Roberts VII Birds of Southern Africa. John Voelcker Book Fund.

Hodgson, B. H. (1833). On a species of Aquila, Circætus, and Dicrurus. Asiatic Researches, 18(2), 13–26.

Hodgson, B. H. (1836). Summary description of some new species of Falconidae. Journal of the Asiatic Society of Bengal, 5(1), 227–231.

Hodgson, J. C., Mott, R., Baylis, S. M., Pham, T. T., Wotherspoon, S., Kilpatrick, A. D., Segaran, R. R., Reid, I., Terauds, A. & Koh, L. P. (2018) Drones count wildlife more accurately and precisely than humans. Methods in Ecology and Evolution, 9(5), 1160–1167. Retrieved fror https://doi.org/10.1111/2041-210X.12974

Hoeck, H. (2001). Hyraxes. *In*: Macdonald, D. & Norris S. (Eds.), The New Encyclopedia of Mammals, pp. 448–451. Oxford: Oxford Universit Press.

Hoerschelmann, H. von, Haack, A. & Wohlgemuth, F. (1988). Bird casualties and bird behaviour at a 380-kV-power line (in German with Englis summary). İkologie der Vogel, 10, 85–103.

Hoffman, D. J. (2002). Role of selenium toxicity and oxidative stress in aquatic birds. Aquatic Toxicology, 57(1–2), 11–26. https://doi.org/10.1016 S0166-445X(01)00263-6

Hoffmann, M., Mech, D. & Sillero-Zubiri, C. (Eds.) (2004). Canids: Foxes, wolves, jackals and dogs: Status survey and conservation action plar World Conservation Union.

Höfle, U., Blanco, J. M., Crespo, E., Naranjo, V., Jiménez-Clavero, M. A., Sanchez, A., . . . Gortazar, C. (2008). West Nile virus in the endangere Spanish Imperial eagle. Veterinary Microbiology, 129(1–2), 171–178. https://doi.org/10.1016/j.vetmic.2007.11.006

Holdaway, R. N. (1994). An exploratory phylogenetic analysis of the genera of the Accipitridae, with notes on the biogeography of the family *In*: Meyburg, B.-U. & Chancellor, R. D. (Eds.), Raptor Conservation Today, 601–649. London: World Working Group on Birds of Prey an Owls.

Holland, A. E., Byrne, M. E., Bryan, A. L., DeVault, T. L., Rhodes, O. E. & Beasley, J. C. (2017). Fine-scale assessment of home ranges an activity patterns for resident black vultures (Coragyps atratus) and turkey vultures (Cathartes aura). PloS One, 12(7), e0179819. https://doi org/10.1371/journal.pone.0179819

Hollands, D. (1984). Eagles, hawks and falcons of Australia. Nelson, Melbourne, Australia: Nelson Publishing.

Hollands, D. (2003). Eagles, hawks and falcons of Australia. Melbourne, Victoria: Bloomings Books.

Holmes, D. A. (1996). Sumatra Bird Report. Kukila, 8, 9–56.

Holt, G., Frøslie, A. & Norheim, G. (1979). Acta Veterinaria Scandinavica. Mercury. France: DDE, and PCB in the avian fauna in Norway, 70 1–28, 1965–1976.

Holthuijzen, A. M. A., Eastland, W. G., Ansell, A. R., Kochert, M. N., Williams, R. D. & Young, L. S. (1989). Effects of blasting on behavior an productivity of nesting Prairie falcons. Wildlife Society Bulletin, 18, 270–281.

Honda, K., Marcovecchio, J. E., Kan, S., Tatsukawa, R. & Ogi, H. (1990). Metal concentrations in pelagic seabirds from the North Pacific Ocean Archives of Environmental Contamination and Toxicology, 19(5), 704–711. https://doi.org/10.1007/BF01183988

Honnen, A.-C., Hailer, F., Kenntner, N., Literák, I., Dubská, L. & Zachos, F. E. (2010). Mitochondrial DNA and nuclear microsatellites revea high diversity and genetic structure in an avian top predator, the White-tailed sea eagle, in central Europe. Biological Journal of the Linnea Society, 99(4), 727–737. https://doi.org/10.1111/j.1095-8312.2010.01373.x

Hooge, P. N. B. & Wichenlaub. (1997). Animal movement extension to Arc view. Ver 1.1. Anchorage, AK: Alaska Biological Science Cente United States Geological Survey.

Hoogerwerf, A. (1949). Een Bijdrage tot de Oölogie van het Eiland Java. [Koninklijke Plantetuin van Indonesië, Buitenzorg].

Hoogwijk, M., de Vries, B. & Turkenburg, W. (2004). Assessment of the global and regional geographical, technical and economic potential o onshore wind energy. Energy Economics, 26(5), 889–919. https://doi.org/10.1016/j.eneco.2004.04.016

Hoover, S. L. & Morrison, M. L. (2005). Behavior of red-tailed hawks in a wind turbine development. Journal of Wildlife Management, 69(1), 150–159. https://doi.org/10.2193/0022-541X(2005)069<0150:BORHIA>2.0.CO;2

Hopkins, D. A., Mersereau, G. S. & Nordell, A. C. (1999). A third adult again assists at a Connecticut Bald eagle nest. Connecticut Warbler, 19, 26–27.

Hopkins, D. A., Mersereau, G. S. & O'Leary, M. J. (1993). A third adult Bald eagle takes an active part in raising young eagles in Connecticut. Connecticut Warbler, 13, 114–116.

Hopkins, D. A., Mersereau, G. S. & O'Leary, M. J. (1995). A third adult Bald eagle replaces a nesting eagle. Connecticut Warbler, 15, 5–11.

Hornocker, M. G. & Hash, H. S. (1981). Ecology of the wolverine in northwestern Montana. Canadian Journal of Zoology, 59(7), 1286–1301. https://doi.org/10.1139/z81-181

Horsfield, T. (1821). Systematic arrangement and description of birds from the island of Java. Transactions of the Linnean Society of London, 13(1), 133–200. https://doi.org/10.1111/j.1095-8339.1821.tb00061.x.

Horvath, M., Nagy, K., Papp, F., Kovacs, A., Demeter, I., Szugyi, K. & Halmos, G. (2008). The evaluation of the Hungarian medium-voltage electricity network from a bird conservation perspective (in Hungarian). Budapest: Magyar Madártani és Természetvédelmi Egyesület.

Horvath, M., Nagy, K., Demeter, I., Kovacs, A., Bagyura, J., Toth, P., Solt, S. & Halmos, G. (2011). Birds and power lines in Hungary: Mitigation planning, monitoring and research. Presentation at International Conference on Power Lines and Bird Mortality in Europe, Budapest.

Horvath, M., Karyakin, I., Perlman, Y. & Vyas, V. (2017). Aquila nipalensis. The IUCN Red List of threatened species. Retrieved from http://www.iucnredlist.org/details/ 22696038/0

Horvath, Z. & Pinter, T. (2005). Nest site selection of Hungarian White-tailed eagles (Haliaeetus albicilla) according to a survey made in 2000. Aquila, 112, 23–32.

Horváth, Z. (2009). White-tailed eagle (Haliaeetus albicilla) populations in Hungary between 1987–2007. Denisia, 27, 85–95. Retrieved from https://www.zobodat.at/pdf/DENISIA_0027_0085-0095.pdf

Horváth, Z. (2009). White-tailed eagle Haliaeetus albicilla (Linnaeus, 1758). *In*: T. Csörgõ et al. (Eds.), Hungarian bird migration atlas, pp. 204–205. Budapest, Hungary: Kossuth Kiadó.

Hostetler, M. E. & Main, M. B. (2001). Florida monitoring program: Point count method to survey birds. Retrieved from http://edis.ifas.ufl.edu/uw140

Hostetler, M. & Knowles-Yanez, K. (2003). Land use, scale, and bird distributions in the Phoenix metropolitan area. Landscape and Urban Planning, 62(2), 55–68, https://doi.org/10.1016/S0169-2046(02)00096-8

Hötker, H. (2008). Birds of prey and wind farms: Analysis of problems and possible solutions October (2008). *In*: Documentation of an international workshop in Berlin, October 21 and 22, pp. 1–77.

Hötker, H., Thomsen, K.-M. & Köster, H. (2004). Auswirkungen regenerativer Energiegewinnung auf die biologische Vielfalt am Beispiel der Vögel und der Fliedermäuse—Fakten, Wissenslücken, Anforderungen an die Forschung, ornithologische Kriterien zum Ausbau von regenerativen Energiegewinnungsformen. Nabu / Bundesamt für Naturschutz, Bonn, Germany. Förd. Nr Z1.3-684 11-5/03. http://bergenhusen.nabu.de/bericht/VoegelRegEnergien.pdf.

Hötker, H., Thomsen, K.-M. & Jeromin, H. (2006). Impacts on biodiversity of exploitation of renewable energy sources: The example of birds and bats – Facts, gaps in knowledge, demands for further research, and ornithological guidelines for the development of renewable energy exploitation. Bergenhusen, Germany.

Houde, P. (1988). Paleognathous birds from the early tertiary of the Northern Hemisphere. Publications of the Nuttall Ornithological Club, 22, 34–35.

Houserova, P., Hedbavny, J., Matejicek, D., Kracmar, S., Sitko, J. & Kuban, V. (2005). Determination of total mercury in muscle, intestines, liver and kidney tissues of cormorant (Phalacrocorax carbo), great crested grebe (Podiceps cristatus) and Eurasian buzzard (Buteo buteo). Veterinarni Medicina, 50, 61–68. https://doi.org/10.17221/5597-VETMED

Houston, C. S. (1985a). Golden eagle banding in the South Saskatchewan River Valley. North American Bird Bander, 10, 110–114.

Houston, C. S., Terry, B., Blom, M. & Stoffel, M. J. (2007). Turkey vulture nest success in abandoned houses in Saskatchewan. Wilson Journal of Ornithology, 119(4), 742–747. https://doi.org/10.1676/06-055.1

Houston, D. C. (1975). Ecological isolation of African scavenging birds. Ardea, 63, 55–64.

Houston, D. C. (1979). The adaptations of scavengers. *In*: Sinclair, A.R.E. & Griffiths, M.N. (Eds.), Serengeti: Dynamics of an ecosystem, 263–286. Chicago, IL: University of Chicago.

Houston, D. C. (1983). The adaptive radiation of the griffon vultures. *In*: Wilbur, S. R. & Jackson, J. A. (Eds.), Vulture biology and management, pp. 135–152. Berkeley, CA: University of California Press.

Houston, D. C. (1985b). Evolutionary ecology of Afrotropical and Neotropical vultures in forests. Ornithological Monographs, 36(36), 856–864. https://doi.org/10.2307/40168321

Houston, D. C. (1985c). Indian White-backed vulture (G. bengalensis). *In*: Newton, I. & Chancellor, R. D. (Eds.), Conservation Studies on Raptors Technical Publication, pp. 465–466. Cambridge: International Council for Bird Preservation.

Houston, D. C. (1986). Scavenging efficiency of Turkey vultures in tropical forest. The Condor, 88(3), 318–323. https://doi.org/10.2307/1368878

Houston, D. C. (1988). Competition for food between Neotropical vultures in forest. Ibis, 130, 402–417.

Houston, D. C. (1994). Cathartidae (New World Vultures). *In*: Del Hoyo, J., Elliott, A. & Sargatal, J. (Eds.), Handbook of the birds of the world, pp. 24–41. Barcelona: Lynx Edicions.

Houston, D. C. & Maddox, J. G. (1974). Causes of mortality among young Scottish Blackface lambs. Veterinary Record, 95(25–26), 575. https://doi.org/10.1136/vr.95.25-26.575

Howard, H. (1935). A new species of eagle from a Quaternary cave deposit in eastern Nevada. Condor, 37(4), 206–209. https://doi.org/10.2307/1363473
Howard, H. (1947). A preliminary survey of trends in avian evolution from Pleistocene to recent time. The Condor, 49(1), 10–13. https://doi.org/10.2307/1364422
Howard, H. (1962). Bird remains from a prehistoric cave deposit in Grant County, New Mexico. Condor, 64(3), 241–242.
Howard, R. & Moore, A. (1991). A Complete Checklist of the Birds of the World. Second edn. London, UK: Academic Press.
Howell, A. H. (1932). Florida Bird Life. Tallahassee: Florida Dept. Game and Freshwater Fish.
Howell, S. N. G. & Webb, S. (1995). A guide to the birds of Mexico and northern Central America. New York: Oxford University Press.
Howell, S. N., Lewington, I. & Russell, W. (2014). Rare birds of North America. Princeton, NJ: Princeton University Press.
Howells, W. W. (1982). Martial eagle preys on Klipspringer. Honeyguide, 111/112, 61–62.
Höy, G. (1969). Additions to the Salteña avifauna. Hornero, 11, 53–56. https://www.lightrocket.com/wolfgangkaehler/search/results?skeywords=Ulgii
Huang, J. & Mayer, J. (2020). Lead and zinc toxicity in birds. Retrieved from https://todaysveterinarypractice.com/lead-and-zinc-toxicity-in-birds/
Hudec, K. (1977). Research on the occurrence of birds in the Czechoslovak Socialist Republic and the problems of faunal work. Cs. Společ. Zool., 10–12, 30–31.
Hudson, W. H. (1872). On the birds of the Río Negro of Patagonia, with notes by P. L. Sclater. Proceedings of the Zoological Society of London, 1872, 534–550.
Huelsenbeck, J. P. & Ronquist, F. (2001). MRBAYES: Bayesian inference of phylogeny. Bioinformatics, 17(8), 754–755. https://doi.org/10.1093/bioinformatics/17.8.754.
Huffman, T. N. (1985). The soapstone birds from Great Zimbabwe. African Arts, 18(3), 68–100. https://doi.org/10.2307/3336358
Hughes, C. & Hwang, B. (1996). Attempts to conceptualize and measure quality of life. *In*: Schalock, R. L. (Ed.), Quality of Life, Vol. I: Conceptualization and Measurement, pp. 51–61. Washington, DC: American Association on Mental Retardation.
Hughes, R. N. (1979). Optimal diets under the energy maximization premise: The effects of recognition time and learning. American Naturalist, 113(2), 209–221. https://doi.org/10.1086/283380
Hull, J. (1995). The raptors of the world. The Atlantic Monthly, 124–128.
Hull, K. & Fergus, R. (2009). Eagles in Mesoamerican thought and mythology. Reitaku Review, 15, 83–134.
Hume, A. O. (1876). The Laccadives and the West Coast. Stray Feathers, 4, 413–483.
Hume, J. J. P., Walters & Michael, M. (2012). Extinct birds, 217 (pp. 81–82). London. A & C Black (Poyser Imprint).
Hunt, G. (1995). A pilot Golden eagle population study in the Altamont Pass Wind Resource Area: Population trend analysis (pp. 1994–1997). https://doi.org/10.2172/86813. Golden, CO: Prepared by Predatory Bird Research Group, University of California, Santa Cruz, National Renewable Energy Laboratory, Golden, Colorado, USA.
Hunt, G. (2002). Golden eagles in a perilous landscape: Predicting the effects of mitigation for wind turbine blade–strike mortality. Sacramento: Report to the California Energy Commission.
Hunt, G. & Hunt, T. (2006). The trend of Golden eagle territory occupancy in the vicinity of the Altamont Pass Wind Resource Area, 2005 Survey. Report of the California Energy Commission.
Hunt, W. G. (1998). Raptor floaters at Moffat's equilibrium. Oikos, 82(1), 191–197. https://doi.org/10.2307/3546929
Hunt, W. G., Jackman, R. E., Hunt, T. L., Driscoll, D. E. & Culp, L. (1998). A population study of golden eagles in the Altamont Pass Wind Resource Area: A population trend analysis (1994–1997). National Resource Energy Laboratory SR-500-26092, Golden, Colorado.
Hunt, W. G. (2000). A population study of Golden eagles in the Altamont pass wind resource area: Population trend analysis (1994–1997): Executive summary. *In*: Richardson, W. J. & Harris, R. E. (Eds.), Proceedings of the National Avian – Wind Power Planning Meeting III, pp. 15–17. San Diego, CA: Avian Subcommittee of the National Wind Coordinating Committee.
Hunt, W. G., Parish, C. N., Farry, S. C., Lord, T. G. & Sieg, R. (2007). Movements of introduced California Condors in Arizona in relation to lead exposure. *In*: Mee, A., Hall, L.S. & Grantham, J. (Eds.), California Condors in the 21st Century, pp. 79–96. Cambridge, MA: Nuttall Ornithological Club, American Ornithologists' Union, McLean.
Hunt, W. G., Wiens, J. D., Law, P. R., Fuller, M. R., Hunt, T.L., Driscoll, D.E. & Jackman, R. E. (2017). Quantifying the demographic cost of human related mortality to a raptor population. PLoS One, 12, 1–23.
Hunter, N. D. (1977). 'Nyala' records. Birds: Eastern Banded Snake-eagle Circaetus fasciolatus. Nyala, 3, 108–109.
Hunter, J. (1994). From West Highland Survey to SNH: The beginnings of a sustainable development strategy for the highlands, the islands, and their surrounding seas. *In*: Usher, M. B. & Baxter, J. M. (Eds.), The Islands of Scotland: A living marine heritage. Edinburgh.
Hunter, J. S. (2002). Adaptations to intraguild competition in Mesocarnivores [PhD. Thesis]. Seattle, WA: Seattle University of Washington.
Hunter, L. (2011). Carnivores of the world. Princeton, NJ: Princeton University Press.
Hunter, L. (2015). Wild cats of the world. Bloomsbury Publishing.
Hunting, K. (2002). A road map for PIER research on avian power line electrocution in California. Sacramento: California Energy Commission.
Hüppop, O. & Hüppop, K. (2003). North Atlantic Oscillation and timing of spring migration in birds. Proceedings of the Royal Society of London B, 270, 233–240.
Hurlbert, S. H. (1978). The measurements of niche overlap and some relatives. Ecology, 59(1), 67–77. https://doi.org/10.2307/1936632
Hurrell, S. (2014). The king of the birds. World Birdwatch, 36(3), 16–19.
Hustler, K. (1988). Martial eagle lays two-egg clutch. Honeyguide, 34, 70.
Hustler, K. & Howells, W. W. (1987). Breeding periodicity, productivity and conservation of the Martial eagle. Ostrich, 58(3), 135–138. https://doi.org/10.1080/00306525.1987.9633687

Hustler, K. & Howells, W. W. (1990). The influence of primary production on a raptor community in Hwange National Park, Zimbabwe. Journal of Tropical Ecology, 6(3), 343–354. https://doi.org/10.1017/S0266467400004624
Iankov, P. (2008). Atlas of breeding birds in Bulgaria. Sofia, Bulgaria: Bulgarian Society for the Protection of Birds.
Iapichino, C. (1993). Rapporto ornitologico Sicilia 1987–89. Naturalista Siciliano, 17, 149–168.
Iapichino, C. & Massa, B. (1989). The birds of Sicily. Dorchester, UK: Dorset Press.
Ibañez, J. C. (2007). Philippine eagle Pithecophaga jefferyi breeding biology, diet, behavior, nest characteristics, and longevity estimate in Mindanao Island [Master of Science Thesis] Ateneo de Davao University.
Ibañez, J. C., Miranda, H. C., Jr., Balaquit-Ibañez, G., Afan, D. S. & Kennedy, R. S. (2003). Observations on the breeding behavior of a Philippine eagle pair at Mount Sinaka, Central Mindanao. Wilson Bulletin, 115(3), 333–336. https://doi.org/10.1676/01-054
ICZN. (1999). International code of zoological nomenclature (4th ed). International Trust for Zoological Nomenclature, London.
IDEA (Instituto para la Diversificación y el Ahorro de la Energía). (2010). Plan de Acción Nacional de Energías Renovables 2010–2020. Madrid, Spain.*
Igl, L. D. & Peterson, S. L. (2010). Repeated use of an abandoned vehicle by nesting Turkey vultures (Cathartes aura). Journal of Raptor Research, 44(1), 73–75. https://doi.org/10.3356/JRR-09-02.1
Iliff, M. (2020). Ornate hawk-eagle (Spizaetus ornatus). *In*: Schulenberg, T. S. (Ed.), Neotropical birds online. Ithaca: Cornell Laboratory of Ornithology. https://doi.org/10.2173/nb.orheag1.01
Imhoff, M. L., Lawrence, W. T., Stutzer, D. C. & Elvidge, C. D. (1997). A technique for using composite DMSO/OLS 'city lights' satellite data to map urban area. Remote Sensing of Environment, 61, 361–370.
Imler, R. H. & Kalmbach, E. R. (1955). The Bald eagle and its economic status. U.S. Fish and Wildlife Service circula.
Incropera, F. P. & DeWitt, D. P. (1990). Fundamentals of Heat and Mass Transfer (3rd ed.). Chichester, UK: John Wiley & Sons.
India Biodiversity Portal. (2020). Hieraaetus pennatus (J. F. Gmelin, 1788). Retrieved from https://indiabiodiversity.org/biodiv/species/show/239188.
Indrawan, M., Somadikarta, S., Supriatna, J., Bruce, M. D., Djanubudiman, S. & Djanubudiman, G. (2006). The birds of the Togian Islands, central Sulawesi, Indonesia. Forktail, 22, 7–22.
Ingels, J., Cleere, N. & Pelletier, V. (2003). Noteworthy observations on some French Guianan birds. Alauda, 71, 59–67.
Ingersoll, E. (1920). Zoogeography. In: Encyclopedia Americana. New York: Grolier.
Ingólfsson, A. (1961). The distribution and breeding ecology of the White-tailed eagle Haliaeetus albicilla (L.) in Iceland [BSc (Hons) Thesis]. Aberdeen: University of Aberdeen.
Inigo Elias, E. E. (1987). Feeding habits and ingestion of synthetic products in a Black vulture population from Chiapas, Mexico. Acta Zoologica Mexicana, 22, 1–15. https://doi.org/10.21829/azm.1987.19222040
Iñigo, A. & Atienza, J. C. (2007). Efectos del Reglamento 1774/2002 y las decisions adoptadas por la Comisión Europea en, 2003 y 2005 sobre las aves necrófagas en la peninsula Ibérica y sus posibles soluciones. Official report to the European Commission. SEO/Birdlife 15-06-2007.
Iñigo, E. E. I., Ramos, M. & González, F. (1987). Two recent records of Neotropical eagles in Southern Veracruz, Mexico. Condor, 89(3), 671–672. https://doi.org/10.2307/1368659_
Iñigo-Elias, E. (1991). Ecological correlates of forest fragmentation and habitat alteration on a tropical raptor community in the Selva Lacandona region of Chiapas, Mexico. PhD Thesis. Gainesville, FL: University of Florida.
Iñigo-Elias, E. (2000). Solitary eagle. *In*: Ceballos, G. & Márquez-Valdelamar, L. (Eds.), Birds of México in danger of extinction, pp. 110–111. Mexico City, Mexico: CONABIO.
Iñigo-Elias, E. (2000). Black Hawk-eagle. *In*: Ceballos, G. & Márquez-Valdelamar, L. (Eds.), Birds of México in danger of extinction, pp. 125-126. Mexico City: CONABIO.
Iñigo-Elias, E., Ramos, E. & Gonzáles, F. (1989). Some ecological aspects of two primary evergreen forest raptor communities compared with cultivated tropical areas in southern Mexico. *In*: Meyburg, B.-U. & Chancellor, R. D. (Eds.), Raptors in the modern world, pp. 529–543. Berlin, London and Paris: WWGBP.
Inskip, C. & Zimmermann, A. (2009). Human–felid conflict: A review of patterns and priorities worldwide. Oryx, 43(1), 18–34. https://doi.org/10.1017/S003060530899030X
Inskipp, C. & Inskipp, T. (1991). A guide to the birds of Nepal (2nd edn). Washington: DC. Smithsonian Institution Press.
Inskipp, T., Lindsey, N. & Duckworth, W. (1996). An annotated checklist of birds of the Oriental region. Sandy, UK: Oriental Bird Club.
Integrated Taxonomic Information System. (2020). Haliaeetus leucocephalus (Linnaeus, 1766), Retrieved from ITIS Report. (2018). Ichthyophaga ichthyaetus (Horsfield, 1821) https://www.itis.gov/servlet/SingleRpt/SingleRpt?search_topic=TSN&search_value=175420#null.
Intel. (2017). From polar bears to whales, Intel pushes the boundaries of wildlife research with drone and artificial intelligence. Retrieved from https://newsroom.intel.com/news-releases/from-polar-bears-to-whales-intel-pushes-the-boundaries-of-wildlife-research-with-drone-and-artificial-intelligence/#gs.x9oo3y.
International Association for Falconry and Conservation of Birds of Prey (IAF). (2018). A History of Falconry. Retrieved from https://iaf.org/a-history-of-falconry/.
International Union for Conservation of Nature and Natural Resources. (2001). Red List categories and criteria version 3.1. Gland, Switzerland. IUCN. International Union for Conservation of Nature and Natural Resources.
International Union for Conservation of Nature and Natural Resources. (2016). The IUCN Red List of threatened species. Retrieved from http://www.iucnredlist.org.
International Union for Conservation of Nature and Natural Resources. (2017). The IUCN Red List of threatened species. Retrieved from http://www.iucnredlist.org.

International Union for Conservation of Nature and Natural Resources. (2018). The IUCN Red List of threatened species. Retrieved from http: www.iucnredlist.org.

Iqbal, M., Mulyawati, D., Fujita, M. S., Hua, F. & Zetra, B. (2011). A breeding record of the Rufous-bellied eagle Lophotriorchis (Hieraaetus kienerii in Sumatra. Kukila, 15, 75–79.

Irham, M. (2012). Avifauna diversity at central Halmahera North Maluku, Indonesia. Zoo Indonesia, 21(1), 17–31. Retrieved from https://(journal.biologi.lipi.go.id/ index.php/zoo_indonesia/article/view/2355/2021

Irwin, M. P. S. (1981). The birds of Zimbabwe. Salisbury, Zimbabwe: Quest Publishing.

Isaacs, F. B. & Silvosky, G. (1981). Bald eagle management on the Fremont National Forest, Lakeview, Oregon. Lakeview, OR: USDA Fores Service, Fremont National Forest, Supervisor's Office.

Isbell, L. A. & Enstam, K. L. (2002). Predator (in) sensitive foraging in sympatric female vervets (Cercopithecus aeihiops) and patas monkey (Erythrocebus patas), a test of ecological models of group dispersion. Eat or be eaten: Predator sensitive foraging among primates, 154.

Isenmann, P. (1993). Oiseaux de Camargue. Société Française d'Ornithologieornithologie, Paris.

Isenmann, P. & Moali, A. (2000). Birds of Algeria. Paris.

Isenmann, P., Gaultier, T., Hili, A. E., Azafzaf, H., Dlensi, H. & Smart, M. (2005). Birds of Tunisia. Paris: Société d'Etudes Ornithologiques d France.

Isenmann, P., Benmergui, M., Browne, P., Ba, A. D., Diagana, C. H., Diawara, Y. & Eine El Abidine ould Sidaty, Z. E. A. (2010). Birds c Mauritania. Paris: SEOF.

Israel, M. A. (2011). A UAV-based roe deer fawn detection system. ISPRS – International Archives of the Photogrammetry, Remote Sensing an Spatial Information Sciences, XXXVIII–1/C22, 51–55. https://doi.org/10.5194/isprsarchives-XXXVIII-1-C22-51-2011

Issam, B., Boukhemza, M., Souttou, K., Arezki, D., Bouchahdane, I., Boukhemza, M., Souttou, K. & Derridj (2019). Breeding biology of Boote eagle Hieraaetus pennatus (Gmelin, 1788), the first study in North Africa. Ekológia (Bratislava), 38(4), 382–391. https://doi.org/10.247 eko-2019-0028._

ITIS Report. (2018). Ichthyophaga ichthyaetus (Horsfield, 1821). https://www.itis.gov/servlet/SingleRpt/SingleRpt?search_topic=TSN&search value=175475#null.

Ivanovsky, V. V. (2010). White-tailed eagle Haliaeetus albicilla in the Byelorussian Poozerie: Materials on the biology of the species within th range. Russian Ornithological Journal, 19, 1876–1887.

Ivanovsky, V. (1996). Current status of the White-tailed Sea Eagle Haliaeetus albicilla in Byelorussia. In: Meyburg, B. & Chancellor, R. D. (Eds. Eagle studies, pp. 137–139. Berlin: World Working Group on Birds of Prey.

Ivanowsky, V. (1996a). Notes on the breeding biology of Spotted Eagles Aquila clanga and Aquila pomarina in Byelorussia. *In*: Meyburg, B.-U & Chancellor, R. D. (Eds.), Eagle studies, pp. 297–299. Berlin: World Working Group Birds of Prey.

Ivanowsky, V. (1996b). Vergleichende brutökologische Angaben von Schelladler und chreiadler, Aquila clanga, Aquila pomarina in Weißrusslanc Ornithologische Mitteilungen, 48, 72–75.

Ivanowsky, V., Bashkirov, I. V. & Shamovich, D. J. (1999). Der Schreiadler in Weißrussland. Ornithologische Mitteilungen, 51, 260–264.

Ivanovsky, V. (2003). The White-tailed Sea eagle in northern Belarus. In: Helander, B. (Ed.), Sea eagle 2000: Proceedings from the Internationa Sea Eagle Conference in Bjorko, Sweden, 13–17 September 20, 127–132. Stockholm: Sweden Proceedings of the Swedish Society for Natur Conservation, SNF.

Ivošević, B., Han, Y., Cho, Y. & Kwon, O. (2015). The use of conservation drones in ecology and wildlife research. Journal of Ecology an Environment, 38(1), 113–118. https://doi.org/10.5141/Ecoenv.2015.012

Iwata, H., Watanabe, M., Kim, E., Gotoh, R., Yasunaga, G., Tanabe, S., Masuda, Y. & Fujita, S. (2000). Contamination by chlorinated hydrocarbon and lead in Steller's Sea eagle and White-tailed Sea eagle from Hokkaido, Japan. *In*: Ueta, M. & Mcgrady, M. J. (Eds.), First Symposium o Steller's and White-tailed Sea eagles in East Asia, 91–106. Tokyo Japan: Wild Bird Society of Japan.

Izor, R. J. (1985). Sloths and other mammalian prey of the Harpy eagle. *In*: G. G. Montgomery (Ed.), The evolution and ecology of armadillos sloths, and vermilinguas, 343–346. Washington: DC. Smithsonian Institution.

Jackman, B. & Scott, J. (2012). The marsh lions: The story of an African pride. Bradt Travel Guides.

Jackman, R. E., Hunt, W. G., Jenkins, J. M. & Detrich, P. J. (1999). Prey of nesting Bald eagles in northern California. Journal of Raptor Research 33, 87–96.

Jackson, F. J. (1938). The birds of Kenya Colony and the Uganda Protectorate, Vols. 1–3. London: Gurney & Jackson.

Jackson, H. B. & Fahrig, L. (2015). Are ecologists conducting research at the optimal scale? Global Ecology and Biogeography, 24, 52–63.

Jackson, J. A. (1983). Nesting phenology, nest site selection, and reproductive success of Black and Turkey vultures. *In*: Wilbur, A. & Jackson, J A. (Eds.), Vulture Biology and Management, pp. 245–270. Berkeley, CA: University of California Press.

Jackson, J. A. (1988a). American Black vulture. *In*: Palmer, R. S. (Ed.). Handbook of North American birds, Research Series, 11–24. New Haven Yale University Press.

Jackson, J. A. (1988b). Turkey vulture. *In*: Palmer, R. S. (Ed.). Handbook of North American birds, Research Series, 25–42. New Haven: Yal University Press.

Jackson, P. M. M. (1985). Martial eagle breeding in Gutu district. Honeyguide, 31, 213.

Jackson, P. M. M. (1987). Further on Martial eagle breeding in Gutu district. Honeyguide, 33, 59.

Jacobson, E., Carpenter, J. W. & Novilla, M. (1977). Suspected lead toxicosis in a Bald eagle. Journal of the American Veterinary Medica Association, 171(9), 952–954.

Jais, M. (2017). Spanish Imperial eagle, Aquila adalberti. Retrieved from http://www.europeanraptors.org/raptors/spanish_imperial_eagle.html.

Jakubiec, Z. & Peterson, U. (2005). Relationship between White-tailed eagle Haliaeetus albicilla and White stork Ciconia ciconia. Buteo, 14, 51–52.
James, H. F. (1995). Prehistoric extinctions and ecological changes on oceanic islands. *In*: Vitousek, P. et al. (Eds.), Ecological studies, 115, pp. 88–102. Islands, Éire: Biological Diversity and Ecosystem Function.
James, H. F. & Olson, S. L. (1991). Descriptions of thirty-two new species of birds from the Hawaiian Islands. Part II. Passeriformes. Ornithological Monographs, 46(46), 1–88. https://doi.org/10.2307/40166713
James, H. W. (1925). Birds observed in the Somerset East district, Cape Province, Union of South Africa. Ibis, 12(1), 626.
Janson, C., Baldovino, M. C. & Di Bitetti, M. (2012). The group life cycle and demography of brown capuchin monkeys (Cebus apella nigritus) in Iguazú National Park, Argentina. *In*: Long-term field studies of primates, pp. 185–212. Berlin, Heidelberg: Springer.
Janss, G. F. E. & Ferrer, M. (1999). Mitigation of raptor electrocution on steel power poles. Wildlife Society Bulletin, 27, 263–273.
Janss, G. F. E. (2000). Avian mortality from power lines: A morphologic approach of a species-specific mortality. Biological Conservation, 95(3), 353–359. Retrieved from http://www.atclave.es/english/publicaciones/descargas/pub_avifauna/47_avian_mortality.pdf. https://doi.org/10.1016/S0006-3207(00)00021-5
Jarvis, A., Robertson, A. J., Brown, C. J. & Simmons, R. E. (2001). Namibian avifaunal database. Windhoek, Namibia: Biodiversity Programme.
Jarvis, M. J. F. (1978). Notes on Long-crested eagles in Rhodesia. *In*: Proceedings of the Symposium on African Predatory Birds, pp. 17–24. Pretoria, South Africa: Northern Transvalle Ornithological Society, Pretoria, South Africa.
Jayaneththi, H. B. (2010). Some observations of nesting behavior of endangered Rufous-Bellied Hawk eagle (Hieraaetus Kienerii) in Udagama-Kegalle Mountains of Central Sri Lanka. Tiger Paper, 38(2), 1–4.
Jayaneththi, H. B. (2010). Some observations of nesting behavior of endangered Rufous-bellied Hawk eagle (Hieraaetus kienerii) in Udagama-Kegalle mountains of Central Sri Lanka. Tiger Paper, 33(2), 1–3.
Jayaraj, R., Megha, P. & Sreedev, P. (2016). Organochlorine pesticides, their toxic effects on living organisms and their fate in the environment. Interdisciplinary Toxicology, 9(3–4), 90–100. https://doi.org/10.1515/intox-2016-0012
Jenkins, A. (2008). WEIG visits the roof of Africa – Surveying power lines for vulture casualties in the Lesotho Highlands. Newsletter of the Wildlife and Energy Interaction Group (WEIG), 3–5.
Jenkins, A., Smallie, J. & Diamond, M. (2010). Avian collisions with power lines: A global review of causes and mitigation with a South African perspective. Bird Conservation International, 20(3), 263–278. https://doi.org/10.1017/S0959270910000122
Jenkins, A. R. (1997). Long-crested eagle. *In*: J. A. Harrison et al. (Eds.), The atlas of South African birds. 1: Non-passerines, pp. 190–191. Johannesburg: BirdLife South Africa and Avian Demography Unit.
Jenkins, A. R. (2000). Hunting mode and success of African Peregrines Falco peregrinus minor: Does nesting habitat quality affect foraging efficiency? Ibis, 142(2), 235–246. https://doi.org/10.1111/j.1474-919X.2000.tb04863.x
Jenkins, A. R., van Rooyen, C. S., Smallie, J. J., Harrison, J. A., Diamond, M., Smit-Robinson, H. A. & Ralston-Paton, S. (2015). Birds and Wind-Energy Best Practice Guidelines: Best practice guidelines for avian monitoring and impact mitigation at proposed wind energy development sites in southern Africa (3rd edn). Johannesburg: Endangered Wildlife Trust/BirdLife South Africa, Johannesburg.
Jenkins, J. M. & Jackman, R. E. (1994). Field experiments in prey selection by resident Bald eagles in the breeding and non-breeding season. Journal of Field Ornithology, 65(4), 441–446.
Jennings, M. C. (1981). Birds of the Arabian Gulf. London: George Allen and Unwin Ltd.
Jennings, M. C. (1981). The birds of Saudi Arabia: A check-list. Cambridge, UK: M.C. Jennings.
Jennings, M. C. (1996). Summary report of ABBA Survey 19 to south central Saudi Arabia; March and April 1996. Phoenix, 13, 21–23.
Jensen, F. P., Christensen, K. D. & Petersen, B. S. (2008). The avifauna of southeast Niger. Malimbus, 30, 30–54.
Jensen, J. R. (2007). Remote Sensing of the Environment: An Earth Resource Perspective, 2nd Edition. Upper Saddle River: Pearson Education Inc.
Jensen, R. A. C. (1972). The Steppe eagle Aquila nipalensis and other termite-eating raptors in South West Africa. Madoqua, 1, 73–76.
Jenson, S., Johnels, A. G., Olsson, M. & Westermark, T. (1972). The avifauna of Sweden as indicators of environmental contamination with mercury and chlorinated hydrocarbons. Proceedings of the International Ornithological Congress, 15, 455–465.
Jeyarajasingam, A. & Pearson, A. (1999). A Field Guide to the Birds of West Malaysia and Singapore. [Oxford: Oxford University Press.
Jiang, Z. G. & Luo, Z. H. (2012). Assessing species endangerment status: Progress in research and an example from China. Biodiversity Science, 20, 612–622.
Jiang, Z. G., Luo, Z. H., Tang, S. H., Fang, H. X., Li, C. W. & Meng, Z. B. (2013). Assessment report on threatened status of terrestrial vertebrates in China. Beijing, China: Environmental Science.
Jiménez-Franco, M. V., Martõanez, J. E. & Calvo, J. F. (2011). Territorial occupancy dynamics in a forest raptor community. Oecologia, 166(2), 507–516. https://doi.org/10.1007/s00442-010-1857-0doi: 10.1007/s00442-010-1857-0.
Jobling, J. A. (2010). The Helm dictionary of scientific bird names. London: Christopher Helm Publishers.
Joenck, C. M. (2006). Records of Spizaetus tyrannus (Accipitridae) in the Centro de Pesquisa e Conservação da Natureza Pró-Meta (CPCN Pró-Mata) in northeastern Rio Grande do Sul, Brazil. Revista Brasileira de Ornitologia, 14, 427–428.
Joenck, C. M., Zilio, F. & Ad, M.-L. (2007). Raptor species richness of the Barra Grande Hydroelectric Power Plant. *In*: Bildstein, K. L., Barber, D. R. & Zimmerman, A. (Eds.), Neotropical raptors Pelota River Basin at the boundary between Rio Grande do Sul and Santa Catarina States, Brazil p. 305. Orwigsburg: Hawk Mountain Sanctuary.
Joenck, C. M., Zilio, F. & de Mendonça-Lima, A. (2011). First record of breeding of the Ornate Hawk-eagle (Spizaetus ornatus) in Southern Brazil. El Hornero, 26(2), 163–166.

Joenck, C. M., Zilio, F. & de Mendonca-Lima, A. (2013). Successful translocation of a nestling Ornate Hawk-eagle (Spizaetus ornatus) in Souther Brazil. Revista Brasileira de Ornitologia, 21(2), 136–140.
Johnels, A., Tyler, G. & Westermark, T. (1979). A history of mercury levels in Swedish fauna. Ambio, 8(4), 160–168.
Johnsen, P. (1953). Birds and mammals of Peary Land in North Greenland. Meddelelser om Grønland, 128(6), 6–135.
Johnsgard, P. (1990). Hawks, eagles & falcons of North America. Washington, DC: Smithsonian Institution Press, Washington, D.C.
Johnsgard, P. A. (1988). North American owls: Biology and natural history. Washington, DC: Smithsonian Institute Press.
Johnsgard, P. A. (1990). Hawks, eagles, and falcons of North America. Washington, DC: Smithsonian Institute Press.
Johnsingh, A. J. T. & Manjrekar, N. (Eds.) (2013). Mammals of South Asia. Himayatnagar, India: Universities Press.
Johnson, J. A., Lerner, H. R. L., Rasmussen, P. C. & Mindell, D. P. (2006). Systematics within Gyps vultures: A clade at risk. BMC Evolutionary Biology, 6(1), 65. https://doi.org/10.1186/1471-2148-6-65
Johnson, J. A., Tingay, R. E., Culver, M., Hailer, F., Clarke, M. L. & Mindell, D. P. (2009). Long-term survival despite low genetic diversity in the critically endangered Madagascar Fish-eagle. Molecular Ecology, 18(1), 54–63. https://doi.org/10.1111/j.1365-294X.2008.04012.x
Johnson, J. B. & Omland, K. S. (2004). Model selection in ecology and evolution. Trends in Ecology and Evolution, 19(2), 101–108. https://doi org/10.1016/j.tree.2003.10.013
Johnston, D. W., Dale, J., Murray, K. T., Josephson, E., Newton, E. & Wood, S. (2017). Comparing occupied and unoccupied aircraft surveys o wildlife populations: Assessing the gray seal (Halichoerus grypus) breeding colony on Muskeget Island, USA. Journal of Unmanned Vehicle Systems, 5(4), 178–191. https://doi.org/10.1139/juvs-2017-0012.
Johnston, N. N., Bradley, J. E., Pomeroy, A. C. & Otter, K. A. (2013). Flight paths of migrating Golden eagles and the risk associated with wind energy development in the Rocky Mountains. Avian Conservation and Ecology, 8(2), 12. https://doi.org/10.5751/ACE-00608-080212
Johnston, R. J., Gregory, D., Pratt, G. & Watts, M. (2000). Human Geography (4th ed.). Malden: Blackwell Publishing.
Johnstone, R. E. & Storr, G. M. (1998). Handbook of Western Australian Birds, vol. 1. Perth: WA Museum.
Jollie, M. (1957). Notes on the bird genus Aquila and its occurrence in New Guinea and Australia. Nova Guinea, 8, 179–181.
Jollie, M. (1976). A contribution to morphology and phylogeny of the Falconiformes. Evolutionary Theory, 1, 285–298.
Jollie, M. (1977a). A contribution to morphology and phylogeny of the Falconiformes. Evolutionary Theory, 2, 115–300.
Jollie, M. (1977b). A contribution to the morphology and phylogeny of the Falconiformes, Part 2. Evolutionary Theory, 2, 285–298.
Jollie, M. (1977c). A contribution to the morphology and phylogeny of the Falconiformes, Part 3. Evolutionary Theory, 3, 1–141.
Jollie, M. (1977d). A contribution to the morphology and phylogeny of the Falconiformes, Part 3. Evolutionary Theory, 2, 209–300.
Jollie, M. T. (1977). A contribution to the morphology and phylogeny of the Falconiformes, Parts 2–3. Evolutionary Theory, 2, 115–130.
Jones, H. L. (2002). Central America. North American Birds, 56, 232–234.
Jones, H. L. (2003). Birds of Belize. Austin, TX: University of Texas Press.
Jones, H. L. (2004). Central America. North American Birds, 58, 155–157.
Jones, H. L. (2005). Central America. North American Birds, 59, 337–340.
Jones, H. L. (2005). Central America. North American Birds, 59, 506–510.
Jones, H. L. & Komar, O. (2006a). Central America. North American Birds, 60, 152–156.
Jones, H. L. & Komar, O. (2006b). Central America. North American Birds, 63, 507–510.
Jones, H. L. & Komar, O. (2007). Central America. North American Birds, 61, 155–159.
Jones, H. L. & Komar, O. (2008). Central America. North American Birds, 62, 163–170.
Jones, H. L. & Komar, O. (2009). Central America. North American Birds, 63, 167–172.
Jones, H. L. & Komar, O. (2011). Central America. North American Birds, 65, 174–180.
Jones, H. L. & Vallely, A. C. (2001). Annotated checklist of the birds of Belize. Barcelona, Spain: Lynx Edicions.
Jones, H. L., McRae, E., Meadows, M. & Howell, S. N. G. (2000). Status updates for selected bird species in Belize, including several species previously undocumented from the country. Cotinga, 13, 17–31.
Jones, T., Laurent, S., Mselewa, F. & Mtui, A. (2006). Sanje mangabey Cercocebus sanjei kills an African crowned eagle Stephanoaetus coronatus. Folia primatologica; International Journal of Primatology, 77(5), 359–363. https://doi.org/10.1159/000093701
Jonesand, S. E. I. & Dorward, L. J. (2014). Possible scavenging behavior in Ornate Hawk-eagle (Spizaetus ornatus) in Amazonas, Brazil. Revista Brasileira de Ornitologia, 22(1), 27–31. https://doi.org/10.1007/BF03544228
Jonzén, N., Lindén, A., Ergon, T., Knudsen, E., Vik, J. O., Rubolini, D., Piacentini, D., Brinch, C., Spina, F., Karlsson, L., Stervander, M., Andersson, A., Waldenström, J., Lehikoinen, A., Edvardsen, E., Solvang, R. & Stenseth, N. C. (2006). Rapid advance of spring arrival dates in long-distance migratory birds. Science, 312(5782), 1959–1961. https://doi.org/10.1126/science.1126119
Jorde, D. G. & Lingle, G. R. (1988). Kleptoparasitism by Bald eagles wintering in south-central Nebraska. Journal of Field Ornithology, 59, 183–188.
Jorde, D. G., Krapu, G. L. & Crawford, R. D. (1983). Feeding ecology of Mallards wintering in Nebraska. Journal of Wildlife Management, 47(4), 1044–1053. https://doi.org/10.2307/3808163
Jordheim, S. (1980). Eagle-coyote incident. Blue Jay, 38, 47–48.
Jørgensen, A. F. & Nøhr, H. (1996). The use of satellite images for mapping of landscape and biological diversity in Sahel. International Journal of Remote Sensing, 17, 91–109.
Jorgenson, J. L. (2001). Aldrin and dieldrin: A review of research on their production, environmental deposition and fate, bioaccumulation, toxicology, and epidemiology in the United States. Environmental Health Perspectives, 109 (Suppl. 1), 113–139. https://doi.org/10.1289/ ehp.01109s1113

Joubert, S. C. J. (1972). Territorial behaviour of the tsessebe (Damaliscus lunatus lunatus Burchell) in the Kruger National Park. African Zoology Zoologica Africana, 7(1), 141–156. https://doi.org/10.1080/00445096.1972.11447435

Juhant, M. (2007). Habitat use of diurnal raptors in Calilegua National Park, Argentina. *In*: K. L. Bildstein, D. R. Barber & A. Zimmerman (Eds.), Neotropical raptors, pp. 306–307. Orwigsburg: Hawk Mountain Sanctuary.

Julio, A., Madrid, M., Héctor, D., Sixto, H., Funes, A., Lopez, J., . . . Ramos, A. (1991). Reproductive biology and behavior of the Ornate Hawk-eagle (Spizaetus ornatus) in Tikal National Park. Maya Project: Use of raptors and other fauna as environmental indices for design and management of protected areas for building local capacity for conservation in Latin America, progress report IV (pp. 93–13). Boise, ID: Peregrine Fund, Inc.

Julliot, C. (1994). Predation of a young Spider Monkey (Ateles paniscus) by a Crested Eagle (Morphnus guianensis). Folia Primatologica, 63, 75–77.

Junda, J., Greene, E. & Bird, D. M. (2015). Proper flight technique for using a small rotary-winged drone aircraft to safely, quickly, and accurately survey raptor nests. Journal of Unmanned Vehicle Systems, 3(4), 222–236. https://doi.org/10.1139/juvs-2015-0003

Jung, T. S., Everatt, K. T. & Andresen-Everatt, L. M. (2009). Kleptoparasitism of a coyote (Canis latrans) by a Golden eagle (Aquila chrysaetos) in Northwestern Canada. Northwestern Naturalist, 90(1), 53–55. https://doi.org/10.1898/1051-1733-90.1.53

Junge, G. C. A. (1936). Fauna simalurensis—Aves. Temminckia, 1, 1–76.

Jürgens, M. D., Crosse, J., Hamilton, P. B., Johnson, A. C. & Jones, K. C. (2016). The long shadow of our chemical past—High DDT concentrations in fish near a former agrochemicals factory in England. Chemosphere, 162, 333–344. https://doi.org/10.1016/j.chemosphere.2016.07.078

Jyoti, A. (2019). Anti-poaching drones face shooting down by DGCA. Pioneer. Retrieved from https://www.dailypioneer.com/2019/pioneer-exclusive/anti-poaching-drones-face-shooting-down-by-dgca.html

Kabouche, B., Bayeul, J., Zimmermann, L. & Bayle, P. (2006). Bird mortality in aerial power lines: Challenges and prospects in Provence—Alpes-Cote d'Azur (in French). Report DIREN PACA–LPO PACA, Hyères.

Kaczensky, P., Walzer, C. & Steinhauer-Burkart, B. (2004). Great Gobi B Strictly Protected Area – A Wild Horse Refuge. Oberaula, Germany: ECO Nature Edition Steinhauer-Burkart OHG.

Kaehler, W. (2015). Amanbol and Aisholpan at Ulgii festival, October (2015). Retrieved from https://websites.lightrocket.com/wolfgangkaehler/

Kaisanlahti-Jokimäki, M. & Jokimäki, J. (2008). Territory occupancy and breeding success of the Golden eagle (Aquila chrysaetos) around tourist destinations in northern Finland. Ornis Fennica, 85, 2–12.

Kaiser, M. (2010). A living specimen of the dark form of Steller's Sea eagle, Haliaeetus pelagicus in captivity. Journal of Ornithology, 152(1), 207–208. https://doi.org/10.1007/s10336-010-0580-2.

Kaiser, T. E., Reichel, W. L., Locke, L. N., Cromartie, E., Krynitsky, A. J., Lamont, T. G., . . . Swineford, D. M. (1980b). Organochlorine pesticide, PCB, and PBB residues and necropsy data for Bald eagles from 29 states—1975–77. Pesticides Monitoring Journal, 4, 145–149.

Kajtoch, Ł., Żmihorski, M. & Wieczorek, P. (2015). Habitat displacement effect between two competing owl species in fragmented forests. Population Ecology, 57(3), 517–527. https://doi.org/10.1007/s10144-015-0497-y

Kalina, J. & Butynski, T. M. (1994). Natural deaths of two Crowned Eagles in Uganda. Gabar, 9, 28–31.

Kalisinska, E., Lisowski, P., Czernomysy-Furowicz, D. & Kavetska, K. (2008). Serratospiculiasis, mycosis, and haemosiderosis in wild peregrine falcon from Poland. A case report. Bulletin of the Veterinary Institute in Pulawy, 52, 75–79.

Kalisińska, E., Kosik-Bogacka, D. D. I., Lisowski, P., Lanocha, N. & Jackowski, A. (2013). Mercury in the body of the most commonly occurring European game duck, the mallard (Anas platyrhynchos, L. 1758) from Northwestern Poland. Archives of Environmental Contamination and Toxicology, 64(4), 538–593. https://doi.org/10.1007/s00244-012-9860-6doi: 10.1007/s00244-012-9860-6.

Kalisinska, E., Gorecki, J., Lanocha, N., Okonska, A., Melgarejo, J. B., Budis, H., Rzad, I. & Golas, J. (2014a). Total and methylmercury in soft tissues of White-tailed eagle (Haliaeetus albicilla) and osprey (Pandion haliaetus) collected in Poland. Ambio – A Journal of the Human Environment, 43(7), 858–870. https://doi.org/10.1007/s13280-014-0533-8

Kalisinska, E., Gorecki, J., Okonska, A., Pilarczyk, B., Tomza-Marciniak, A., Budis, H., Lanocha, N., Kosik-Bogacka, D. I., Kavetska, K. M., Macherzynski, M. & Golas, J. M. (2014b). Mercury and selenium in the muscle of piscivorous common mergansers (Mergus merganser) from a selenium-deficient European country. Ecotoxicology and Environmental Safety, 101, 107–115. https://doi.org/10.1016/j.ecoenv.2013.12.019

Kallok, M. A. (2013). The case for copper: Research uncovers problems with lead bullets for deer hunting. Minnesota Conservation Volunteer. Retrieved from https://www.dnr.state.mn.us/mcvmagazine/issues/2013/sep-oct/copper.html.

Kalra, M., Rahmani, A. R., Choudhury, B. C. & Qureshi, Q. (2006). Development of desert national park as biosphere reserve. Final Report, p. 73. Mumbai, India: Bombay Natural History Society.

Kalyakin, M. & Voltzit, O. V. (2006). Birds of Moscow city and the Moscow region. Sofia, Bulgaria: Pensoft Publishers.

Kamarauskaitė, A., Dementavičius, D., Skuja, S., Dagys, M. & Treinys, R. (2020). Interaction between the White-tailed eagle and Common buzzard estimated by diet analysis and brood defence behaviour/Mita ravintoanalyysi ja poikasten puolustaminen paljastavat merikotkan ja hiirihaukan vuorovaikutuksesta? Ornis Fennica, 97(1), 26–37.

Kampp, K. & Wille, F. (1979). Fodevaner hos den Gronlandske Havorn Haliaeetus albicilla groenlandicus Brehm. (Food habits of the Greenland White-tailed eagle). Dansk Ornithologisk Forenings Tidsskrift, 73, 157–164.

Kampp, K. & Wille, F. (1990). Bestanden af Havørn i Grønland, 1985–1989 (Status of White-tailed eagle in Greenland, 1985–1989). Dansk Ornitologisk Forenings Tidsskrift, 84, 37–44.

Kamstra, J. (1987). An ecological study of the Cockscomb Basin, Belize. MSc Thesis. York, ON, Canada: York University.

Kamusoko, C. & Aniya, M. (2009). Hybrid classification of Landsat data and GIS for land use/cover change analysis of the Bindura district, Zimbabwe. International Journal of Remote Sensing, 30(1), 97–115, https://doi.org/10.1080/01431160802244268

Kane, A., Jackson, A. L., Ogada, D. L., Monadjem, A. & McNally, L. (2014). Vultures acquire information on carcass location from scavengin eagles. Proceedings of the Royal Society of London Series B, 281, https://doi.org/10.1098/rspb.2014.1072doi: 10.1098/rspb.2014.1072.

Kaneda, H., Prawiradilage, D. M. & Yamagishi, S. (2009). Home range and habitat use of an individual of the Javan Hawk-eagle in Telaga Warr Forest Reserve, West Java, based on radio tracking and direct observation. Journal of Raptor Research, 41(1), 68–71, doi: 10.3356/089. 1016(2007)4168.

Kansky, R., Kidd, M. & Knight, A. T. (2014). Meta-analysis of attitudes toward damage causing mammalian wildlife. Conservation Biolog 28(4), 924–938. Retrieved from https://doi.org/10.1111/cobi.12275.

Karanth, K. U. & Stith, B. M. (1999). Prey depletion as a critical determinant of tiger population viability. *In*: J. Seidensticker et al. (Eds.), Ridin the tiger: Tiger conservation in human-dominated landscapes, pp. 100–113. Cambridge: Cambridge University Press.

Karanth, K. U. & Chellam, R. (2009). Carnivore conservation at the crossroads. Oryx, 43(1), 1–2. https://doi.org/10.1017/S003060530843106X

Karyakin, I. V. & Novikova, L. M. (2006). The Steppe Eagle and power lines in Western Kazakhstan. Does coexistence have any chance? Rapto Conservation, 6, 48–57.

Karyakin, I. V., Kovalenko, A. V. & Novikova, L. M. (2006). The Imperial eagle in the Volga-Ural Sands: Results of researches in 2006. Rapto Conservation, 6, 39–47

Karyakin, I. V. (2008). Lines-killers continue to harvest the mortal crop in Kazakhstan. Raptors Conservation, 11, 14–21.

Karyakin, I. V., Nikolenko, E. N., Levin, A. S. & Kovalenko, A. V. (2011). Eastern Imperial eagle in Russia and Kazakhstan: Population status an trends. Acta Zoologica Bulgarica Suppl., 3, 95–104.

Karyakin, I. V. (2013). Review of the modern population status of the Steppe eagle in the world and in Russia. Raptors Conservation, 26, 22–43

Karyakin, I. V., Zinevich, L. S., Schepetov, D. M. & Sorokina, S. Y. (2016a). Population structure within range of Steppe eagle and preliminar data of the genetic diversity of the population and the status of the subspecies. Raptors Conservation, 32, 67–88.

Karyakin, I. V., Kovalenko, A. V., Barashkova, A. N., Smelansky, I. E. & Nikolenko, E. G. (2016b). Strategy of the Steppe eagle conservation i the Russian Federation. Moscow.

Kas'yanova, L. (2014). Kyrgystan: Hunting with birds of prey. Retrieved from http://www.uzbekjourneys.com/2014/01/kyrgyzstan-hunting-with eagles.html

Kastberger, G. & Stachl, R. (2003). Infrared imaging technology and biological applications. Behavior Research Methods, Instruments, an Computers, 35(3), 429–439. Retrieved from https://doi.org/10.3758/BF03195520

Katdare, V. & Mone, R. (2003). Status of the White-bellied Sea eagle in Ratnagiri District, Maharashtra. Journal of the Bombay Natural Histor Society, 100(1), 113–116.

Katdare, V., Mone, R. & Joshi, P. (2004). Status of White-bellied Sea eagle Haliaeetus leucogaster in Sindhudurg District, Maharashtra. Journal c the Bombay Natural History Society, 101(2), 314–316.

Katovai, E., Edwards, W. & Laurance, W. F. (2015). Dynamics of logging in Solomon Islands: The need for restoration and conservatio alternatives. Tropical Conservation Science, 8(3), 718–731. https://doi.org/10.1177/194008291500800309

Katzenberger, J., Tate, G., Koeslag, A. & Amar, A. (2015). Black Sparrowhawk brooding behaviour in relation to chick age and weather variatio in the recently colonised Cape Peninsula, South Africa. Journal of Ornithology, 156(4), 903–913. https://doi.org/10.1007/s10336-015-1199-

Katzner, T. E., Brandes, D., Miller, T., Lanzone, M., Maisonneuve, C., Tremblay, J. A., . . . Merovich, G. T. (2012). Topography drives migrator flight altitude of Golden eagles: Implications for on-shore wind energy development. Journal of Applied Ecology, 49(5), 1178–1186. https:. doi.org/10.1111/j.1365-2664.2012.02185.x

Katzner, T., Brandes, D., Lanzone, M., Miller, T. & Ombalski, D. (2008). Raptors and wind energy development in the Central Appalachian Where we stand on the issue. Pittsburgh, PA: National Aviary. Retrieved from http://docs.wind-watch.org/windenergyraptorswhitepaper.pd

Katzner, T., Smith, B. W., Miller, T. A., Brandes, D., . . . (2012). Status, biology, and conservation priorities for North America's eastern Golde eagle (Aquila chrysaetos) population. Auk, 129(1), 168–176. https://doi.org/10.1525/auk.2011.11078

Kaufman, K. (1996). Lives of North American Birds. Boston, MA: Houghton Mifflin Field Guides.

Kaup, J. J. (1844). Classification der Säugetiere und Vögel. Leske, Darmstadt, Germany.

Kaup, J. J. (1847). Monographien der Genera der Falconidae. Isis von Oken, Leipzigleipzig.

Kays, R., Sheppard, J., Mclean, K., Welch, C., Paunescu, C., Wang, V., . . . Crofoot, M. (2019). Hot monkey, cold reality: Surveying rainfore canopy mammals using drone-mounted thermal infrared sensors. International Journal of Remote Sensing, 40(2), 407–419. https://doi.org/ 0.1080/01431161.2018.1523580

Kayser, Y, Girard, C., Massez, G., Ch6rain, Y., Cohez, D., Hafner, H., Johnson, A., Sadoul, N., Tamisier, A. & Isenmann, P. (2003). Compte-rend ornithologique camarguais pour les annes (1995–2000). Revue d'Ecologie (Terre Vie), 58, 20.

Kellert, S. R., Black, M., Rush, C. R. & Bath, A. J. (1996). Human culture and large carnivore conservation in North America. Conservatio Biology, 10(4), 977–990. https://doi.org/10.1046/j.1523-1739.1996.10040977.x

Kelly, T. R. & Johnson, C. K. (2011). Lead exposure in free-flying Turkey vultures is associated with big game hunting in California. PLoS On 6(4), e15350. https://doi.org/10.1371/journal.pone.0015350._

Kelly, T. R., Poppenga, R. H., Woods, L. A., Hernandez, Y. Z., Boyce, W. M., Samaniego, F. J., . . . Johnson, C. K. (2014). Causes of mortalit and unintentional poisoning in predatory and scavenging birds in California. Veterinary Record Open, 1(1), e000028. https://doi.org/10.113 vropen-2014-000028.

Kemp, A. C. (1994). Cassin's hawk-eagle. *In*: del Hoyo, J., Elliott, A. & Sargatal, J. (Eds.), Handbook of birds of the world, 2. New World vulture to guineafowl. Barcelona: Lynx Edicions.

Kemp, A. C. (1994). Short-toed snake-eagle. *In*: del Hoyo, J., Elliott, A. & Sargatal, J. (Eds.), Handbook of birds of the world, 2. New Worl vultures to guineafowl, p. 131. Barcelona: Lynx Edicions.

Kemp, A. C. (1994a). Wahlberg's eagle. *In*: del Hoyo, J., Elliott, A. & Sargatal, J. (Eds). Handbook of birds of the world. Vol. 2. New World vultures to guineafowl, p. 197. Barcelona, Spain: Lynx Edicions.
Kemp, A. C. (1994c). African Hawk-eagle. In: del Hoyo, J., Elliott, A. & Sargatal, J. (Eds.), Handbook of birds of the world. Vol. 2. New world vultures to guineafowl, 199. Barcelona, Spain: Lynx Edicions.
Kemp, A. C. (1994d). Crowned Hawk-eagle (Stephanoaetus coronatus). In: del Hoyo, J., Elliott, A. & Sargatal, J. (Eds.), Handbook of birds of the world. Vol. 2. New World vultures to guineafowl, 205. Barcelona, Spain: Lynx Edicions.
Kemp, A. C. (1994f). Southern Banded Snake-eagle. Handbook of birds of the world. Vol. 2. New world vultures to guineafowl, 132. Barcelona, Spain: Lynx Edicions.
Kemp, A. C. (1994g). Bateleur. Handbook of birds of the world. Vol. 2. New world vultures to guineafowl, 132. Barcelona, Spain: Lynx Edicions.
Kemp, A. C. & Kemp, M. I. (1980). The biology of the Southern Ground Hornbill Bucorvus leadbeateri (Vigors) (Aves; Bucerotidae). Annals of the Transvaal Museum, 32, 65–100. https://doi.org/10.1017/S0959270905000237_
Kemp, A. C., Herholdt, J. J., Whyte, I. & Harrison, J. (2001). Birds of the two largest national parks in South Africa: A method to generate estimates of population size for all species and assess their conservation ecology. South African Journal of Science, 97, 393–403.
Kemp, A. C. & Kemp, M. I. (2005). Birds of prey of Africa and its islands. Cape Town, S.A.: Struik Publishers.
Kemp, A. C. & Boesman, P. (2013). Ayres's Hawk-eagle (Hieraaetus ayresii). In: del Hoyo, J., Elliott, A., Sargatal, J., Christie, D. A. and de Juana, E. (Eds.), Handbook of the birds of the world alive. Barcelona: Lynx Edicions.
Kemp, A. C. & Kirwan, G. M. (2014). Madagascar Fish-eagle (Haliaeetus vociferoides). *In*: Del Hoyo, J., Elliott, A., Sargatal, J., Christie, D.A. & De Juana, E. (Eds.), Handbook of the birds of the world alive. Barcelona: Lynx Edicions.
Kemp, A. C. & Christie, D. A. (2020). Madagascar Serpent-eagle (Eutriorchis astur), version 1.0. In: del Hoyo, J., Elliott, A., Sargatal, J., Christie, D. A. & de Juana, E. (Eds.), Handbook of the birds of the world. Cornell Lab of Ornithology, Ithaca, NY, USA: Cornell Lab of Ornithology. https://doi.org/10.2173/bow.maseag1.01
Kemp, A. C. & Kirwan, G. M. (2020a). Cassin's Hawk-eagle (Aquila africana), version 1.0. *In*: Del Hoyo, J., Elliott, A., Sargatal, J., Christie, D.A. & De Juana, E. (Eds.), Birds of the World. Ithaca, NY: Cornell Laboratory of Ornithology. https://doi.org/10.2173/bow.cashae1.01._
Kemp, A. C., Boesman, P. F. D. & Marks, J. S. (2020a). Ayres's Hawk-eagle (Hieraaetus ayresii), version 1.0. *In*: Del Hoyo, J., Elliott, A., Sargatal, J., Christie, D.A. & De Juana, E. (Eds.), Birds of the World. Ithaca, NY: Cornell Laboratory of Ornithology. https://doi.org/10.2173/bow.ayheag1.01._
Kemp, A. C. & Kirwan, G. M. (2020b). Banded Snake-eagle (Circaetus cinerascens), version 1.0. *In*: del Hoyo, J., Elliott, A., Sargatal, J., Christie, D. A. & De Juana, E. (Eds.), Birds of the World. Ithaca, NY: Cornell Laboratory of Ornithology. https://doi.org/10.2173/bow.baseag1.01._
Kemp, A. C., Boesman, P. F. D. & Marks, J. S. (2020b). Martial eagle (Polemaetus bellicosus), version 1.0. *In*: Del Hoyo, J., Elliott, A., Sargatal, J., Christie, D.A. & De Juana, E. (Eds.), Birds of the World. Ithaca, NY: Cornell Laboratory of Ornithology. https://doi.org/10.2173/bow.mareag1.01.
Kemp, A. C. & Kirwan, G. M. (2020c). Congo Serpent-eagle (Dryotriorchis spectabilis), version 1.0. *In*: del Hoyo, J., Elliott, A., Sargatal, J., Christie, D.A. & De Juana, E. (Eds.), Birds of the World. Ithaca, NY: Cornell Laboratory of Ornithology. https://doi.org/10.2173/bow.coseag1.01
Kemp, A. C., Kirwan, G. M. and Marks, J. S. (2020c). Crowned Eagle (Stephanoaetus coronatus), version 1.0. In: del Hoyo, J., Elliott, A., Sargatal, J., Christie, D. A. & de Juana, E. (Eds.), Birds of the world. Ithaca, NY, USA: Cornell Lab of Ornithology. https://doi.org/10.2173/bow.crheag1.0
Kemp, A. C., Boesman, P. F. D. & Marks, J. S. (2020d). Long-crested eagle (Lophaetus occipitalis), version 1.0. *In*: Del Hoyo, J., Elliott, A., Sargatal, J., Christie, D.A. & De Juana, E. (Eds.), Birds of the World. Ithaca, NY: Cornell Laboratory of Ornithology. https://doi.org/10.2173/bow.loceag1.01.
Kemp, A. C., Kirwan, G. M., Christie, D. A. & Garcia, E. F. J. (2020e). Beaudouin's Snake-Eagle (Circaetus beaudouini), version 1.0. *In*: Del Hoyo, J., Elliott, A., Sargatal, J., Christie, D.A. & De Juana, E. (Eds.), Birds of the World. Cornell Lab of Ornithology, Ithaca, NY, USA. https://doi.org/10.2173/bow.beasne1.01
Kemp, A. C., Christie, D. A. & Marks, J. S. (2020f). Fasciated snake-eagle (Circaetus fasciolatus), version 1.0. *In*: Del Hoyo, J., Elliott, A., Sargatal, J., Christie, D.A. & De Juana, E. (Eds.), Birds of the World. Ithaca, NY: Cornell Laboratory of Ornithology. https://doi.org/10.2173/bow.faseag1.01
Kemp, A. C. & Marks, J. S. (2020). African fish-eagle (Haliaeetus vocifer), version 1.0. *In*: Del Hoyo, J. et al. (Eds.), Birds of the World. Ithaca, NY: Cornell Laboratory of Ornithology. USA. https://doi.org/10.2173/bow.affeag1.01.
Kemp, A. C., Kirwan, G. M. & Christie, D. A. (2020g). Madagascar Fish-eagle (Haliaeetus vociferoides), version 1.0. *In*: Del Hoyo, J., Elliott, A., Sargatal, J., Christie, D.A. & De Juana, E. (Eds.) Birds of the World. Ithaca, NY: Cornell Laboratory of Ornithology. https://doi.org/10.2173/bow.mafeag1.01.
Kemp, A. & Kemp, M. (1998). Sasol birds of prey: Of Africa and its islands. New Holland, UK.
Kemp, A. & Kemp, M. (2005). Birds of prey of Africa and its Islands. Johannesburg: Struik Publishers.
Kemp, A. & Newton, I. (2003). New World Vultures. *In*: C. Perrins (Ed.), The Firefly Encyclopedia of birds. Firefly Books.
Kemp, M. I. (1980). The biology of the Southern ground hornbill Bucorvus leadbeateri (Vigors) (Aves; Bucerotidae). Annals of the Transvaal Museum, 32(4), 35–247.
Kemp, M. U., Shamoun-Baranes, J., Van Gasteren, H., Bouten, W. & Van Loon, E. E. (2010). Can wind help explain seasonal differences in avian migration speed? Journal of Avian Biology, 41(6), 672–677. https://doi.org/10.1111/j.1600-048X.2010.05053.x
Kemp, M. U., Van Loon, E. E., Shamoun-Baranes, J. & Bouten, W. (2012). RNCEP: Global weather and climate data at your fingertips. Methods in Ecology and Evolution, 3(1), 65–70. https://doi.org/10.1111/j.2041-210X.2011.00138.x

Kendall, C. J. (2013). Alternative strategies in avian scavengers: How subordinate species foil the despotic distribution. Behavioral Ecology and Sociobiology, 67(3), 383–393. https://doi.org/10.1007/s00265-012-1458-5.

Kendall, C. J. (2014). The early bird gets the carcass: Temporal segregation and its effects on foraging success in avian scavengers. The Auk 131(1), 12–19.

Kenefick, M. (2005). Third report of the Trinidad and Tobago Rare Birds Committee: Rare birds in Trinidad and Tobago in 2001–2003. Living World, Journal of the Trinidad and Tobago Field Naturalists' Club, 2005, 56–61.

Kenefick, M. (2007). Fourth report of the Trinidad and Tobago Rare Bird Committee: Rare birds in Trinidad and Tobago in 2004–2005. Living World, Journal of the Trinidad and Tobago Field Naturalists' Club, 2007, 79–81.

Kennedy, C. A. (1948). Golden eagle kills bighorn lamb. Journal of Mammalogy, 29(1), 68–69. https://doi.org/10.1093/jmammal/29.1.68-a

Kennedy, C. & Rogers, J. I. (2015). Virtuous drones? The International Journal of Human Rights, 19(2), 211–227. https://doi.org/10.1080/13642987.2014.991217

Kennedy, R. S. (1977). Notes on the biology and population status of the monkey-eating eagle of the Philippines. Wilson Bulletin, 89(1), 1–20.

Kennedy, R. S. (1985). Conservation research of the Philippine Eagle. National Geographic Society Research Report, 18, 401–414.

Kennedy, R. S., Gonzales, P. C., Dickinson, E. C., Miranda, H. C., Jr. & Fisher, T. H. (2000). A guide to the birds of the Philippines. New York Oxford University Press.

Kenntner, N., Tataruch, F. & Krone, O. (2001). Heavy metals in soft tissue of White-tailed eagles found dead or moribund in Germany and Austria from 1993 to 2000. Environmental Toxicology and Chemistry, 20(8), 1831–1837. https://doi.org/10.1002/etc.5620200829

Kenntner, N., Krone, O., Altenkamp, R. & Tataruch, F. (2003). Environmental contaminants in liver and kidney of free-ranging northern Goshawks (Accipiter gentilis) from three regions of Germany. Archives of environmental contamination and toxicology, 45, 128–135. https://doi.org/10.1007/s00244-002-2100-8

Kenntner, N., Crettenand, Y., Fünfstück, H., Janovsky, M. & Tataruch, F. (2004). Lead poisoning and heavy metal exposure of Golden eagles (Aquila chrysaetos) from the European Alps. Journal of Ornithology, 148(2), 173–177: https://doi.org/10.1007/s10336-006-0115-zdoi

Kenntner, N., Tataruch, F. & Krone, O. (2005). Risk assessment of environmental contaminants in White-tailed Sea eagles (Haliaeetus albicilla) from Germany. *In*: K. Pohlmeyer (Ed.), Extended Abstracts of the XXVIIth Congress of the International Union of Game Biologists, Hanover 125–127. Hamburg: DSV Verlag.

Kenward, R. E. (1992). Quantity versus quality: Programmed collection and analysis of radio-tracking data. *In*: Pried, I. G. & Swift, S. M. (Eds.) Wildlife telemetry: Remote and tracking of animals, pp. 231–246. New York: Ellis Horwood.

Kenward, R. E. (2001). A manual for wildlife radio tagging. London: Academic Press.

Kenward, R. E. (2006). The goshawk. London: A. D. Poyser.

Kenward, R. E. & Hodder, K. H. (1996). Ranges V: An analysis system for biological location data. Wareham, UK: Institute of Terrestial Ecology Wareham.

Kenward, R. E., Clarke, R. T., Hodder, K. H. & Walls, S. S. (2001). Density and linkage estimators of home range nearest-neighbor clustering defines multinuclear cores. Ecology, 82(7), 1905–1920. https://doi.org/10.1890/0012-9658(2001)082[1905:DALEOH]2.0.CO;2

Kenward, R. E., Marcström, V. & Karlbom, M. (1993). Post-nestling behaviour in goshawks, Accipiter gentilis: I. The causes of dispersal. Animal Behaviour, 46(2), 365–370. https://doi.org/10.1006/anbe.1993.1198.

Kenward, R. E., Marcström, V. & Karlbom, M. (1993). Post-nestling behaviour in goshawks, Accipiter gentilis: II. Sex differences in sociality and nest-switching. Animal Behaviour, 46(2), 371–378. https://doi.org/10.1006/anbe.1993.1199.

Kenyon, K. W. (1961). Birds of Amchitka Island, Alaska. Auk, 73, 304–326.

Kerley, L. L. & Slaght, J. C. (2013). First documented predation of Sika deer (Cervus nippon) by Golden eagle (Aquila chrysaetos) in Russian far east. Journal of Raptor Research, 47(3), 328–330. https://doi.org/10.3356/JRR-12-00008.1

Kerlinger, P. & Moore, F. R. (1989). Atmospheric structure and avian migration. *In*: Power, D. M. (Ed.), Current Ornithology, 6(109–142). New York: Plenum Press. https://doi.org/10.1007/978-1-4757-9918-7_3.

Kerlinger, P. (1985). Water-crossing behaviour of raptors during migration. Wilson Bulletin, 97, 109–113.

Kerlinger, P. (1989). Flight Strategies of Migrating Hawks. Chicago: University of Chicago Press.

Kerr, J. G. (1892). On the avifauna of the lower Pilcomayo. Ibis, 34(1), 120–152. https://doi.org/10.1111/j.1474-919X.1892.tb01191.x

Kertson, B. N., Spencer, R. D. & Grue, C. E. 2013. Demographic influences on cougar residential use and interactions with people in western Washington. Journal of Mammalogy, 94(2), 269–281. https://doi.org/10.1644/12-MAMM-A-051.1

Kertson, B. N., Spencer, R. D., Marzluff, J. M., Hepinstall-Cymerman, J. & Grue, C. E. (2011). Cougar space use and movements in the wildland–urban landscape of western Washington. Ecological Applications, 21(8), 2866–2881. https://doi.org/10.1890/11-0947.1

Kesteloo, H. (2018). A thermal drone shows the Nkuhuma Pride of lions hunt at night in Kruger National Park. Retrieved from https://dronedj.com/2018/07/23/a-thermal-drone-shows-the-nkuhuma-pride-of-lions-hunt-at-night-in-kruger-national-park/

Khacher, L. (2003). The birds of Gujarat – A Salim Ali centenary year overview. In: Daniel, J. C. & Ugra, G. W. (Eds.), Petronia: Fifty years of post-independence ornithology in India, a centenary dedication to Dr. Salim Ali, 1896–1996, pp. 104–154. Mumbai, India: Bombay Natural History Society and Oxford University Press.

Khan, A. N. (2015). Sighting of a juvenile White-tailed eagle at Jorjor Beed carcass dump, Bikaner, Rajasthan, India. Journal of the Bombay Natural History Society, 112(2), 92–93. https://doi.org/10.17087/jbnhs/2015/v112i2/104934

Kidd, E. & Beales, R. (1977). A study of lowland rainforest birds in Brunei. Brunei Museum Journal, 4(1), 197–225.

Kiff, L. F. (2000). The current status of North American vultures. *In*: Chancellor, R. D. & Meyburg, B.-U. (Eds.), Raptors at Risk, 175–189. Berlin and Surrey: WWGBP/Hancock House.

Kiff, L. F. & Cunningham, M. (1980). The egg of the Ornate Hawk eagle (Spizaetus ornatus). Journal of Raptor Research, 14, 51.

Kiff, L. F., Peakall, D. B., Morrison, M. L. & Wilbur, S. R. (1983). Eggshell thickness and DDE residue levels in vulture eggs. *In*: Wilbur, A. & Jackson, J. A. (Eds.), Vulture biology and management, pp. 440–458. Berkeley, CA: University of California Press.

Kiff, L. F., Wallace, M. P. & Gale, N. B. (1989). Eggs of captive Crested eagles (Morphnus guianensis). Journal of Raptor Research, 23, 107–108.

Kilham, L. (1978). Alarm call of Crested guan when attacked by Ornate hawk-eagle. The Condor, 80(3), 347–348. https://doi.org/10.2307/1368050.

Kim, E.-Y., Goto, R., Iwata, H., Masuda, Y., Tanabe, H. & Fujita, S. (1999). Preliminary survey of lead poisoning of Stellar's sea eagle (Haliaeetus pelagicus) and White-tailed eagle (Haliaeetus albicilla) in Hokkaido, Japan. Environmental Toxicology and Chemistry, 18(3), 448–451. https://doi.org/10.1002/etc.5620180312.

Kimball, R. T., Parker, P. G. & Bednarz, J. C. (2003). Occurrence and evolution of cooperative breeding among the diurnal raptors (Accipitridae and Falconidae). The Auk, 120(3), 717–729. https://doi.org/10.2307/4090102.

King, B. F. (1990). Little eagle Hieraaetus morphnoides on Halmahera. Kukila, 5, 77.

King, H. (2006). Bahrain Systematic List 2006. In: Babbington, J. & Roberts, P. (Eds.), Greater spotted eagles Aquila clanga summering in Saudi Arabia. Retrieved from https://osme.org/wp-content/uploads/2019/10/F_Babbington-Roberts_Sandgrouse34-1_compr.pdf

King, J. (2020). The Bald Eagle in American Culture. https://www.researchgate.net/publication/268339930_The_Bald_Eagle_in_American_Culture

King, J. G. (2017). The Bald Eagle in American Culture. Juneau, AK: Fish and Wildlife Service.

Kingdon, J. (1988). East African Mammals: An Atlas of Evolution in Africa, 3. Part A: Carnivores (Vol. 4). University of Chicago Press.

Kingdon, J. & Hoffman, M. (2013). Mammals of Africa, V (pp. 39–45). London: Bloomsbury Publishing.

Kingdon, J., Happold, D., Hoffman, M., Butynski, T., Happold, M. & Kalina, J. (2013). Mammals of Africa, 6 vols. London: A & C Black.

Kingsley, A. & Whittam, B. (2005). Wind Turbines and Birds: A Background Review for Environmental Assessment. Gatineau, QC: Canadian Wildlife Service.

Kinloch, A. M. (1907). The nesting of the Rufous-bellied Hawk-eagle (Lophotriorchis kieneri). Journal of the Bombay Natural History Society, 17(4), 1027–1028.

Kipp, F. A. (1959). Der Handflügel-Index als flugbiologisches Maβ. Vogelwarte, 20, 77–86.

Kirk, D. A. & Mossman, M. J. (2020). Turkey Vulture (Cathartes aura), version 1.0. *In*: Poole, A. F. & Gill, F. B. (Eds.), Birds of the World. Ithaca, NY: Cornell Lab of Ornithology. https://doi.org/10.2173/bow.turvul.01

Kirmse, W. (1998). Morphometric features characterizing flight properties of Palearctic eagles. *In*: Chancellor, R. D., Meyburg, B.U. & Ferrero, J. J. (Eds.), Holarctic birds of prey, pp. 339–348. Merida and Berlin: ADENEX-WWGBP.

Kirwan, G. M. & Marlow, T. (1996). A review of avifaunal records from Mindo, Pichincha Province, north-western Ecuador. Cotinga, 6, 47–57.

Kirwan, G. M., Boyla, K., Castell, P., Demirci, B., Özen, M., Welch, H. & Marlow, T. (2008). The Birds of Turkey. London: Christopher Helm Publishers.

Kislev, L. (2020). Tazpit: Birding Tours and Bird Watching in Israel. Retrieved from http://www.tatzpit.com/Site/en/pages/inPage.asp?catID=9&subID=12

Kitowski, I. (2003). Age-related differences in foraging behaviour of Montagu's harrier Circus pygargus males in south-east Poland. Acta Ethologica, 6(1), 35–38. https://doi.org/10.1007/s10211-003-0078-5

Kitowski, I., Sujak, A., Wiacek, D., Strobel, W. & Rymarz, M. (2014). Trace element residues in eggshells of Grey heron (Ardea cinerea) from colonies of East Poland. North-West Journal of Zoology, 10, 346–354. https://doi.org/10.1016/j.foodchem.2005.06.036

Kitowski, I., Sujak, A., Wiącek, D., Strobel, W., Komosa, A. & Stobiński, M. (2016). Heavy metals in livers of raptors from Eastern Poland – The importance of diet composition. Belgian Journal of Zoology, 146(1), 3–13. https://doi.org/10.26496/bjz.2016.34

Kitowski, I., Jakubas, D., Wiącek, D. & Sujak, A. (2017). Concentrations of lead and other elements in the liver of the White-tailed eagle (Haliaeetus albicilla), a European flagship species, wintering in Eastern Poland. Ambio – A Journal of the Human Environment, 46(8), 825–841. https://doi.org/10.1007/s13280-017-0929-3

Kjeldsen, J. P. (2005). Aves del municipio Rio Prinzapolka, un inventarío de base. Wani Revista Caribe Nicaragüense, 41, 31–64.

Kjellén, N. (1992). Differential timing of autumn migration between sex and age groups in raptors at Falsterbo, Sweden. Ornis Scandinavica, 23(4), 420–434. https://doi.org/10.2307/3676673

Kjellén, N. & Roos, G. (2000). Population trends in Swedish raptors demonstrated by migration counts at Falsterbo, Sweden 1942–97. Bird Study, 47(2), 195–211. https://doi.org/10.1080/00063650009461175

Klaassen, R. H. G., Hake, M., Strandberg, R. & Alerstam, T. (2011). Geographical and temporal flexibility in the response to crosswinds by migrating raptors. Proceedings. Biological Sciences, 278(1710), 1339–1346. https://doi.org/10.1098/rspb.2010.2106.

Klaassen, R. H. G., Strandberg, R., Hake, M., Olofsson, P., Tøttrup, A. P. & Alerstam, T. (2010). Loop migration in adult marsh harriers Circus aeruginosus as revealed by satellite telemetry. Journal of Avian Biology, 41(2), 200–207. https://doi.org/10.1111/j.1600-048X.2010.05058.x

Klebenow, D. A. & McAdoo, K. (1976) Predation on Domestic Sheep in Northeastern Nevada. Journal of Range Management, 29(2), 96–100. https://doi.org/10.2307/3897400.

Klein, B. C., Harper, L. H., Bierregaard, R. O. & Powell, G. V. (1988). The nesting and feeding behavior of the Ornate hawk-eagle near Manaus, Brazil. The Condor, 90(1), 239–241. https://doi.org/10.2307/1368455

Kleiven, J., Bjerke, T. & Kaltenborn, B. P. (2004). Factors influencing the social acceptability of large carnivore behaviours. Biodiversity and Conservation, 13(9), 1647–1658. https://doi.org/10.1023/B:BIOC.0000029328.81255.38

Knight, S. K. & Knight, R. L. (1983). Aspects of food finding by wintering Bald eagles. Auk, 100(2), 477–484. https://doi.org/10.1093/auk/100.2.477

Knudsen, K. (1935). Noter om Lindenowfjords fuglefauna. [Dansk Ornithologisk Forenings]. Tidsskrift, 29, 37–41.

Knystautas, A. (1993). Birds of Russia. London: Harper Collins Publishers.

Kocher, T. D., Thomas, W. K., Meyer, A., Edwards, S. V., Paabo, S., Villablanqt, F. X. & Wilson, A. C. (1989). Dynamics of mitochondrial DN evolution in animals, amplification and sequencing with conserved primers. Proceedings of the National Academy of Sciences, 86(16 6196–6200. https://doi.org/10.1073/pnas.86.16.6196

Kochert, M. N., Mcintyre, C. L. & Craig, E. H. (2002). Golden eagle (Aquila chrysaetos). *In*: A. Poole & F. Gill (Eds.), The Birds of Nort America, 684. Philadelphia: Academy of Natural Sciences, U.S.A. and the American Ornithologists' Union, Washington, D.C.

Kochert, M. N. & Steenhof, K. (2002). Golden eagles in the U.S. and Canada: Status, trends, and conservation challenges. Journal of Rapto Research, 36(1), 32–40.

Kocum, A. (2006). Phylogeny of the Accipitriformes (birds of prey) based on different nuclear and mitochondrial DNA sequences. PhD Thesi Ernst-, Moritz-Arndt-Universität Greifswald. Greifswald, Germany.

Kocum, A., Huddleston, C. J., Helbig, A. J., Holznagel, W. E., Seibold, I., Krosby, M. & Braun, M. J. (2009). Evolutionary history of diurnal bird of prey (Aves; Accipitriformes). Genbank Direct Submission.

Koford, C. B. (1953). The California Condor. Toronto: General Publishing Company.

Koh, L. P. & Wich, S. A. (2018). Conservation drones: Mapping and monitoring biodiversity. Oxford: Oxford University Press.

Kohler, G. & Rezinni, J. A. (2013). Occasional observations on the courtship behavior of Black-and-white Hawk-eagle, Spizaetus melanoleucu at Santa Catarina state, Southern Brazil. Atualidades Ornitológicas, 171, 26.

Komar, O. (1998). Avian diversity in El Salvador. Wilson Bulletin, 110, 511–533.

Komar, O. (2002). Birds of Montecristo National Park, El Salvador. Ornitologia Neotropical, 13, 167–193.

Komar, O. (2002). Priority conservation areas for birds in El Salvador. Animal Conservation, 5(3), 173–183. https://doi.org/10.1017 S1367943002002238

Komar, O. & Domínguez, J. P. (2001). Lista de aves de El Salvador. San Salvador: Fundación Ecológica de El Salvador—SalvaNATURA.

Komers, P. E. (1997). Behavioral plasticity in variable environments. Canadian Journal of Zoology, 75(2), 161–169. https://doi.org/10.1139/z97 023

Kondratiev, A. Ya., Zubakin, V. A., Golubova, E. Yu., Kondratieva, L. F., Kharitonov, S. P. & Kitayskiy, A. S. (1992). Fauna of terrestrial vertebrate of Talan Island. *In*: A. Ya. Kondratiev (Ed.), Coastal Ecosystems of the Northern Okhotsk Sea – Talan Island, pp. 72–108. Magadan, Russia Russian Academy of Sciences.

König, C. (1973). Zum Sozialverhalten von Geiern am Futterplatz in Spanien. Ornithologische Mitteilungen, 25, 1–3.

König, C. (1974). On the behavior of Spanish vultures on carcasses. Zum Verhalten spanischer Geier an Kadavern. Journal of Ornithology, 115(3 289–320. https://doi.org/10.1007/BF01644326.

König, C. (1982). On the systematic position of the New World vultures (Cathartidae). Journal für Ornithologie, 123, 259–267.

König, C. (1983). Interspecific and intraspecific competition for food among Old World vultures. *In*: Jackson, J. A. & Wilbur, S. R. (Ed.), Vultur Biology and Management, pp. 153–171. Berkeley, CA: University of California Press.

Konijnendijk, C. C. (2017). Cities and nature: Urban forestry for greater biocultural diversity. *In*: M. Campbell (Ed.), Biological Conservation i the 21st Century: A Conservation Biology of Large Wildlife, pp. 15–29. New York: Nova Science Publishers.

Koparde, P. & Manchi, S. S. (2013). Avifaunal records from Chalis Ek, North Andaman Island: Insights into distribution of some Andaman Islan birds. Check List, 9(1), 34–41. https://doi.org/10.15560/9.1.34

Kornegay, J. R., Kocher, T. D., Williams, L. A. & Wilson, A. C. (1993). Pathways of lysozyme evolution inferred from the sequences of cytochrom b in birds. Journal of Molecular Evolution, 37(4), 367–379. https://doi.org/10.1007/BF00178867

Korpimäki, E. (1988). Factors promoting polygyny in European birds of prey – a hypothesis. Oecologia, 77(2), 278–285. https://doi.org/10.1007 BF00379199

Korpimäki, E., Norrdahl, K. & Korpimaki, E. (1989). Avian predation on mustelids in Europe 1: Occurrence and effects on body size variation an life traits. Oikos, 55(2), 205–215. https://doi.org/10.2307/3565424

Korsman, J. C., Schipper, A. M., Lenders, H. J. R., Foppen, R. P. B. & Hendriks, A. J. (2012). Modelling the impact of toxic and disturbance stres on White-tailed eagle (Haliaeetus albicilla) populations. Ecotoxicology, 21(1), 27–36. https://doi.org/10.1007/s10646-011-0760-8

Kortland, K., Evans, R., Douse, A. & Patterson, G. (2011). Managing forests for White-tailed eagles. Mitteleuropäische Gesellschaft zur Erhaltun der Greifvögel (MEGEG). Central European Society for Raptor Protection.

Kovacs, K. E. (2013). Eagles, Indian Tribes, and the Free Exercise of Religion. 47 Loy. L.A. L. Rev. 53, 73–74. Available at: https://digitalcommons lmu.edu/llr/vol47/iss1/.

Kovalenko, A. V. (2009). Modern records of the Pallas' Fish eagle in Kazakhstan. Raptors Conservation, 15, 134–139.

Kozie, K. D. & Anderson, R. K. (1991). Productivity, diet, and environmental contaminants in Bald eagles nesting near the Wisconsin shoreline o Lake Superior. Archives of Environmental Contamination and Toxicology, 20(1), 41–48. https://doi.org/10.1007/BF01065326

Krabbe, N. (2007). Birds collected by P. W. Lund and J. T. Reinhardt in south-eastern Brazil between 1825 and 1855, with notes of P. W. Lund' travels in Rio de Janeiro. Revista Brasileira de Ornitologia, 15, 331–357.

Krabbe, N. & Fjeldså, J. (1990). Birds of the High Andes. Newport News, VA: Apollo Press.

Krabbe, N., Flórez, P., Suárez, G., Castaño, J., Arango, J. D. & Duque, A. (2006). The birds of Páramo de Frontino, Western Andes of Colombia Ornitología Colombiana, 4, 39–50. Retrieved from https://asociacioncolombianadeornitologia.org/wp-content/uploads/revista/oc4/Frontinc pdf

Kramar, D. E., Carstensen, B., Prisley, S. & Campbell, J. (2019). Mercury concentrations in blood and feathers of nestling Bald eagles in coasta and inland Virginia. Avian Research, 10(1), 3. https://doi.org/10.1186/s40657-019-0142-0

Kramer, J. & Redig, P. T. (1997). Sixteen years of lead poisoning in eagles, 1980–1995: An epizootiologic view. Journal of Raptor Research, 31(4 327–332.

Kratter, A. W., Sillett, T. S., Chesser, T. R., O'Neill, J. P., Parker, T. A., III & Castillo, A. (1993). Avifauna of a chaco locality in Bolivia. Wilson Bulletin, 105, 114–141.
Krebs, C. J. (1999). Ecological Methodology (2nd ed). Menlo Park, CA. Benjamin-Cummings.
Krebs, J. R. (1971). Territory and breeding density in the Great Tit, Parus major L. Ecology, 52(1), 2–22. https://doi.org/10.2307/1934734
Kremer, W. (2014). A 13-year-old eagle Huntress in Mongolia. BBC News Magazine, April 14. Retrieved from http://www.bbc.com/news/magazine-26969150
Kren, J. (2000). Birds of the Czech Republic. London: Christopher Helm Publishers Ltd.
Kreuger, R. (1963). Details of three previously undescribed South American raptor eggs. Oologists. Record, 43, 5–6.
Krone, O. & Scharnweber, C. (2003). Two White-tailed Sea eagles (Haliaeetus albicilla) collide with wind generators in Northern Germany. Journal of Raptor Research, 37, 174–176.
Krone, O., Kenntner, N., Trinogga, A., Nadjafzadeh, M., Scholz, F., Sulawa, J., Totschek, K., Schuck-Wersig, P. & Ziehschank, R. (2009a). Lead poisoning in White-tailed Sea eagles: Causes and approaches to solutions in Germany. Retreived from https://science.peregrinefund.org/legacy-sites/conference-lead/PDF/0207%20Krone.pdf
Krone, O., Berger, A. & Schulte, R. (2009b). Recording movement and activity pattern of a White-tailed Sea eagle (Haliaeetus albicilla) by a GPS datalogger. Journal of Ornithology, 150(1), 273–280. https://doi.org/10.1007/s10336-008-0347-1
Krone, O., Nadjafzadeh, M. & Berger, A. (2013). White-tailed Sea eagles (Haliaeetus albicilla) defend small home ranges in north-east Germany throughout the year. Journal of Ornithology, 154(3), 827–835. https://doi.org/10.1007/s10336-013-0951-6
Krone, O., Stjernberg, T., Kenntner, N., Tataruch, F., Koivusaari, J. & Nuuja, I. (2006). Mortality Factors, Helminth Burden, and Contaminant Residues in White-tailed Sea eagles (Haliaeetus albicilla) from Finland. Ambio – A Journal of the Human Environment, 35(3), 98–104. https://doi.org/10.1579/0044-7447(2006)3598
Krone, O., Wille, F., Kenntner, N., Boertmann, D. & Tataruch, F. (2004). Mortality factors, environmental contaminants, and parasites of White-tailed Sea eagles from Greenland. Avian Diseases, 48(2), 417–424. https://doi.org/10.1637/7095
Krueger, O. (1997). Population density and intra- and interspecific competition of the African Fish Eagle Haliaeetus vocifer in Kyambura Game Reserve, southwest Uganda. Ibis, 139, 19–24.
Krüger, O., Grünkorn, T. & Struwe-Juhl, B. (2010). The return of the White-tailed eagle (Haliaeetus albicilla) to northern Germany: Modelling the past to predict the future. Biological Conservation, 143(3), 710–721. https://doi.org/10.1016/j.biocon.2009.12.010
Kruger, R. (1999). Towards solving raptor electrocutions on Eskom distribution structures in South Africa. Mini-Thesis. University of the Orange Free State, Bloemfontein.
Kruger, R., Maritz, A. & van Rooyen, C. (2004). Vulture electrocutions on vertically configured medium voltage structures in the Northern Cape Province South Africa. pp. 437–441. In: Chancellor, R. D. & Meyburg, B.-U. (Eds.), Raptors Worldwide. Berlin: WWGBP/MME.
Krüger, S. C. & Amar, A. (2018). Lead exposure in the critically endangered Bearded vulture (Gypaetus barbatus) population in Southern Africa. Journal of Raptor Research, 52(4), 491–499. https://doi.org/10.3356/JRR-17-86.1_
Krüger, S., Reid, T. & Amar, A. (2014). Differential Range Use between Age Classes of Southern African Bearded Vultures Gypaetus barbatus. PLoS One 9(12), e114920. https://doi.org/10.1371/journal.pone.0114920
Kruger, T. L. (2010). Long term prospects for the persistence of breeding Verreaux's eagles (Aquila verreauxii) at the Walter Sisulu National Botanical Garden, Johannesburg. PhD Thesis.
Kruuk, H. (1967). Competition for food between vultures in East Africa. Ardea, 5, 171–192.
Kruuk, H. (1972). The Spotted Hyena: A study of predation and social behaviour. Berkeley, CA: University of California Press.
Kuhn, L. T. (2001). Bone mineralization. *In*: Martin, J. W. (Ed.), Encyclopedia of materials: Science and technology (2nd ed.), pp. 47–54. Retrieved from https://epdf.pub/queue/ concise-encyclopedia-of-the-structure-of-materials.html.
Kühn, M. (2003). Greenbelt and greenheart: Separating and integrating landscapes in European city regions. Landscape and Urban Planning, 64(1–2), 19–27. https://doi.org/10.1016/S0169-2046(02)00198-6
Kuijper, D. P. J., Sahlén, E., Elmhagen, B., Chamaillé-Jammes, S., Sand, H., Lone, K. & Cromsigt, J. P. G. M. (2016). Paws without claws? Ecological effects of large carnivores in anthropogenic landscapes. Biological Sciences. Proceedings of the Royal Society of Biology, 283 (1841). https://doi.org/10.1098/rspb.2016.1625
Kumar, S., Tamura, K. & Nei, M. (2004). MEGA3: Integrated software for molecular evolutionary genetics analysis and sequence alignment. Briefings in Bioinformatics, 5(2), 150–163. https://doi.org/10.1093/bib/5.2.150
Kuniy, A. A., Viana, M. A., Zahn, T. M. G., Lazarinos, D. E. A. & Portes, C. E. B. (2015). Translocation of Harpy eagle (Harpia harpyja) (Linnaeus, 1758) nest. Brazilian Journal of Biology = Revista Brasleira de Biologia, 75(4), 1035–1036. https://doi.org/10.1590/1519-6984.24714
Kurosawa, N. (1998). Lead poisoning of Steller's Sea eagles and White-tailed eagles in Hokkaido. Journal of the Hokkaido Veterinary Medicine Association, 336–338.
Kurosawa, N. (2000). Lead poisoning in Steller's Sea eagles and White-tailed sea eagles. *In*: Ueta, M. & McGrady, M. J. (Eds.), First Symposium on Stellar's and White-tailed Sea eagles in East Asia, 107–109. Tokyo, Japan: Wild Bird Society of Japan.
Kuswandono, D., Ekawati, S., Mulyati, M. T., Inoue, T. & Sakaguchi, N. (2003). Javan Hawk-eagle Spizaetus bartelsi conservation in Gunung Halimun National Park. Research on endangered species in GHNP. Research and Conservation of Biodiversity in Indonesia, 9, 62–70.
Kuvlesky, W. P., Brennan, L. A., Morrison, M. L., Boydston, K. K., Ballard, B. M. & Bryant, F. C. (2007). Wind energy development and wildlife conservation: Challenges and opportunities. Journal of Wildlife Management, 71(8), 2487–2498. https://doi.org/10.2193/2007-248
Kysar, K. (2015). Visit to Mongolia's Altai Mountains Takes Flight with Eagle Hunting. Report on 2104 Ulgii festival. Minneapolis. Star Tribune. Retrieved from http://www.startribune.com/mongolian-visit-takes-flight-with-eagle-hunting/2880728 81/#1.
Lack, D. (1971). Ecological Isolation in Birds. Oxford: Blackwell Publishing.

Ladigin, A. V., Lobkov, E. G. & Ladigna, O. N., (1991). The largest wintering of the White-shouldered eagles at Kuril Lake (South Kamchatka) Byull. Mosk. Ova Ispyt. Prix. Otd. Biol., 96, 48–57.

Ladygin, A. (1992). Kleptoparasitism among Steller's Sea eagle on Kamtchatka Peninnsula. Proceedings of the Raptor Research Foundation Annual Meeting. November 11–15, 1992, Bellevue p. 117.

Ladygin, A. (1992b). Morpho-ecological adaptations of the Steller's Sea eagle and White-tailed Sea eagle. *In*: Kurochkin, E. (Ed.), Modern Ornithology, pp. 56–59. Moscow: Nauka Publishers.

Ladygin, A. (1994). Relationships of raptors wintering on the places of salmon spawning on the Kurilsky Lake (South Kamtchatka). Current Ornithology, 1, 96–106.

Ladyguin, A. (1996). Foraging strategy of three raptors in salmon spawning ground in Kamtchatka Peninsula, Russian Far East. Proceedings of the Second International Conference on Raptors, Urbino, Italy.

Ladyguin, A. (1997). Group behaviour of Steller's Sea eagle on Kamtchatka. Zoological Journal, 76(1), 83–93.

Ladyguin, A.A. (2000). The morphology of the bill apparatus in the Steller's Sea eagle. *In*: Ueta, M. & McGrady, M. J. (Eds.), First Symposium on Steller's and White-tailed Sea eagles in East Asia, pp. 1–10. Wild Bird Society of Japan.

Laes, D., Warnick, R. & Goetz, W. (2006). Lidar applications for forestry and geosciences. Retrieved from https://static1.squarespace.com/static/59c944de59cc68469d159b28/t/5a21e5b453450aa90cada3bd/1512170934265/lidar-overview.pdf

Lambertucci, S. & Ruggiero, A. (2013). Cliffs used as communal roosts by Andean Condors protect the birds from weather and predators. PLOS ONE, 8(6), e67304. https://doi.org/10.1371/ journal.pone.0067304

Lambertucci, S. A., Alarcón, P. A. E., Hiraldo, F., Sanchez-Zapata, J. A., Blanco, G. & Donázar, J. A. (2014). Apex scavenger movements call for transboundary conservation policies. Biological Conservation, 170, 145–150. https://doi.org/10.1016/j.biocon.2013.12.041

Lamprey, R. H. & Reid, R. S. (2004). Special Paper: Expansion of human settlement in Kenya's Maasai Mara: What future for pastoralism and wildlife? Journal of Biogeography, 31(6), 997–1032. https://doi.org/10.1111/j.1365-2699.2004.01062.x

Land, H. C. (1970). Birds of Guatemala. Wynnewood: Livingstone.

Land, M. F. & Nilsson, D.-E. (2002). Animal Eyes. Oxford: Oxford University Press.

Lande, R. (1998). Anthropogenic, ecological and genetic factors in extinction and conservation. Researches on Population Ecology, 40(3), 259–269. https://doi.org/10.1007/BF02763457

Lanfear, R., Calcott, B., Ho, S. Y. W. & Guindon, S. (2012). Partition Finder: Combined selection of partitioning schemes and substitution models for phylogenetic analyses. Molecular Biology Evolution, 29(6), 1695–1701. https://doi.org/10.1093/molbev/mss020

Lang, A., Bishop, M. A. & Le Sueur, A. (2007). An annotated list of birds wintering in the Lhasa river watershed and Yamzho Yumco, Tibet Autonomous Region, China. Forktail, 23, 1–11.

Langelier, K. M., Andress, C. E., Grey, T. K., Wooldridge, C., Lewis, R. J. & Marchetti, R. (1991). Lead poisoning in Bald eagles in British Columbia. Canadian Veterinary Journal = la Revue Veterinaire Canadienne, 32(2), 108–109.

Langgemach, T., Blohm, T. & Frey, T. (2000). Zur Habitatstruktur Des Schreiadlers (Aquila pomarina) an seinem westlichen Arealrand. Untersuchungen aus dem Land Brandenburg. Acta Ornithoecologica, 4, 237–267.

Langguth, T., Honnen, A.-C., Hailer, F., Mizera, T., Skoric, S., Väli, Ü. & Zachos, F. E. (2013). Genetic structure and phylogeography of a European flagship species, the White-tailed Sea eagle Haliaeetus albicilla. Journal of Avian Biology, 44(3), 263–271. https://doi.org/10.1111/j.1600-048X.2012.00075.x

Langrand, O. (1987). Distribution, status and conservation of the Madagascar Fish eagle Haliaeetus vociferoides (Desmurs 1845). Biological Conservation, 42(1), 73–77. https://doi.org/10.1016/0006-3207(87)90053-X

Langrand, O. (1990). Guide to the birds of Madagascar. New Haven, CT: Yale University Press.

Langrand, O. & Meyburg, B.-U. (1984). Birds of prey and owls in Madagascar: Their distribution, status and conservation. *In*: Mendelsohn, J. M. & Sapsford, C. W. (Eds.), Proceedings of the III World Conference on Birds of Prey and Owls, 22–27 March 1987, pp. 3–13. London, UK: World Working Group on Birds of Prey and Owls.

Langrand, O. & Meyburg, B.-U. (1989). Range, status and biology of the Madagascar Sea eagle Haliaeetus vociferoides. *In*: Meyburg, B.-U. & Chancellor, R. D. (Eds.), Raptors in the Modern World, pp. 269–277. Berlin, London & Paris: WWGBP.

Langston, R. H. W. & Pullan, J. D. (2003). Wind farms and birds: An analysis of the effects of windfarms on birds, and guidance on environmental assessment criteria and site selection issues. *In*: Secretariat Memorandum of the Standing Committee, Convention on the conservation of European wildlife and natural habitats, pp. 1–58. Strasbourg, France: Birdlife International, Royal Society for the Protection of Birds.

Lanszki, J., Heltai, M., Kövér, G. & Zalewski, A. (2019). Non-linear relationship between body size of terrestrial carnivores and their trophic niche breadth and overlap. Basic and Applied Ecology, 38, 36–46. https://doi.org/10.1016/j.baae.2019.06.004

Lanzone, M.J., Miller, T.A., Turk, P., Brandes, D., Halverson, C., Maisonneuve, C., Tremblay, J., Cooper, J., O'Malley, K., Brooks, R. P. & Katzner, T. (Eds.) (2012). Flight responses by a migratory soaring raptor to changing meteorological conditions. Biology Letters, 8(5), 710-713. https://doi.org/10.1098/rsbl.2012.0359.

Lapointe, D. & Couture, P. (2006). Importance of the route of exposure in accumulation and subcellular partitioning of nickel in fathead minnows (Pimephales promelas). Presentation P484 at the 27th Annual Meeting of the Society for Environmental Toxicology and Chemistry, Montreal, QC, November.

Larivière, S. & Pasitschniak-Arts, M. (1996). Vulpes. Mammalian Species, 537(537), 1–11. https://doi.org/10.2307/3504236

Larkin, C., Jenkins, R., McDonald, P. G. & Debus, S. J. S. (2020). Breeding habitat, nest-site characteristics and productivity of the Little eagle (Hieraaetus morphnoides) near Armidale, New South Wales. Pacific Conservation Biology, 26(3). https://doi.org/10.1071/PC19033

Larson, J. (2010). Wind energy sector in British Columbia. Victoria, BC, Canada: Ministry of Energy, Mines and Petroleum Resources, Victoria, British Columbia. Retrieved from http://www.empr.gov.bc.ca/EAED/InvestmentInfo/Documents/Wind27May2010.pdf.

Lasch, U., Zerbe S. & Lenk, M. (2010). Electrocution of raptors at power lines in Central Kazakhstan. Waldökologie, Landschaftsforschung und Naturschutz, 9, 95–100.
Latham, J. (1801). Supplementum indicis ornithologici sive systematis ornithologiae. London: Leigh & Sotheby.
Laver, P. N. & Kelly, M. J. (2008). A critical review of home range studies. Journal of Wildlife Management, 72(1), 290–298. https://doi.org/10.2193/2005-589
Laverde-R., O., Stiles, F. G. & Múnera-R., C. (2005). New records and updated inventory of the avifauna of the Serranía de las Quinchas, an Important Bird Area (IBA) in Colombia. Caldasia, 27, 247–265.
Layard, E. L. (1873). Notes on birds observed at Para, with descriptions of two new species. Ibis, 15, 374–396.
Laybourne, R. C. (1974). Collision between a vulture and an aircraft at an altitude of 37,000 feet. Wilson Bulletin, 86, 461–462.
Lazorchak, J. M., McCormick, F. H., Henry, T. R. & Herlihy, A. T. (2003). Contamination of fish in streams of the mid-Atlantic region: An approach to regional indicator selection and wildlife assessment. Environmental Toxicology and Chemistry, 22, 545–553. https://doi.org/10.1002/etc.5620220312
Leahy, C. (2018). Thriving Wisconsin Bald eagle population good news for bird watchers. Retrieved from https://www.wpr.org/thriving-wisconsin-bald-eagle-population-good-news-birdwatchers.
Leal, I. R., Da Silva, J. M. C. D., Tabarelli, M. & Lacher, T. E. (2005). Changing the course of biodiversity conservation in the caatinga of Northeastern Brazil. Conservation Biology, 19(3), 701–706. https://doi.org/10.1111/j.1523-1739.2005.00703.x
Lebossé, A. & Bour, R. (2011). The first twenty livraisons of Lesles Planches Coloriées d'Oiseaux of Temminck & Laugier (1820–1839), The ten wrappers now known. Zoological Bibliography, 1(4), 141–150.
Ledger, J. A. & Annegarn, H. J. (1981). Electrocution Hazards to the Cape Vulture (Gyps coprotheres) in South Africa. Biol. Conservation, 20, 15–24.
Ledger, J. A. & Hobbs, J. C. A. (1999). Raptor use and abuse of powerlines in southern Africa. Journal of Raptor Research, 33(1), 49–52.
LeDuc, P. V. (1970). Red-tailed Hawk attacks Bald Eagle. Auk, 87, 58.
Lee, I. (2019). Canada announces new drone laws that allow recreational and commercial pilots to operate under same rules. Plans to implement June 2019. Retrieved from https://uavcoach.com/canada-new-drone-laws/
Lee, W. S., Koo, T. H. & Park, J. Y. (2005). A field guide to the birds of Korea. Seoul, South Korea: LG Evergreen Foundation.
Lees, A. C. & Peres, C. A. (2006). Rapid avifaunal collapse along the Amazonian deforestation frontier. Biological Conservation, 133(2), 198–211. https://doi.org/10.1016/j.biocon.2006.06.005
LeFranc, M. N., Jr. & Clark, W. S. (1983). Working bibliography of the golden eagle and the genus AQUILA. National Wildlife Federation Scientific & Technical Series No. 7, Washington, D.C.
Legg, C. & Jewell, N. (1995). A 1:50,000 scale forest map of Sri Lanka: The basis for a National Forest Geographic Information System. The Sri Lanka Forester. Remote Sensing, Special Issue, 3–24.
Legge, W. V. (1878). Remarks on birds of Ceylon. The Ibis, 4(2), 201–204.
Legra, L. A. T. (2005). Nest-site Selection and Behavioural Biology of the New Guinea Harpy Eagle, Harpyopsis novaeguineae. Retrieved from http://assets.peregrinefund.org/docs/pdf/research-library/2005/2005-Legra-New-Guinea-Harpy-Eagle.pdf
Lehman, R. N. (1979). A survey of selected habitat features of 95 Bald eagle nest sites in California. California Department of Fish and Game Wildlife Management Branch Administrative Report, 1–23.
Lehman, R. N. (2001). Raptor electrocution on power lines: Current issues and outlook. Wildlife Society Bulletin, 29, 804–813.
Lehman, R. N., Kennedy, P. L. & Savidge, J. A. (2007). The state of the art in raptor electrocution research: A global review. Biological Conservation, 136(2), 159–174. https://doi.org/10.1016/j.biocon.2006.09.015
Lehmann, F. C. (1959). Contributions to the study of the Colombian fauna XIV. New observations on Oroaetus isidori (desd. Murs). Novedades Colombianas, 1, 168–195.
Lehmann, V. F. C. (1943). El genero Morphnus the genus Morphnus. Caldasia, 2, 165–179.
Lehmann, V. F. C. (1944). Notes on Colombian birds of prey. Caldasia, 2, 411–415.
Lehner, B., Verdin, K. & Jarvis, A. (2006). Hydro SHEDS technical documentation, version 1.0. Washington, DC: World Wildlife Fund United States.
Lehti, R. W. (1947). The Golden eagle attacking antelope. Journal of Wildlife Management, 11(4), 348–349. https://doi.org/10.2307/3796216
Lei, Z., Xiao-Nong, Y., Guang, H., Qin, H., Tian-Tian, L., Zi-Yue, D., Qian, W., Lewthwaite, R. W. & Yue-Hua, S. (2014). A review of the distribution of Black eagle Ictinaetus malaiensis in mainland China. Forktail, 30, 45–49.
Lei, Z., Xiao-Nong, Y., Guang, H., Tian-Tian, L., Zi-Yue, D. & Yue-Hua, S. (2015). Notes on the diet of the Black eagle Ictinaetus malaiensis. BirdingASIA, 23, 100–101.
Lekuona, J. M. (2001). Uso del espacio por la avifauna y control de la mortalidad de aves en los parques eólicos de Navarra. Gobierno de Navarra, Pamplona.
Lekuona, J. M. & Ursua, C. (2007). Avian mortality in wind power plants of Navarra (Northern Spain). *In*: de Lucas, M., Janss, G. F. E. & Ferrer, M. (Eds.), Birds and wind farms: Risk assessment and mitigation, pp. 177–192. Madrid: Quercus.
Lemke, T., Mack, J. A. & Houston, D. B. (1998). Winter range expansion by the Northern Yellowstone elk herd. Intermountain Journal of Sciences, 4, 1–9.
Lemly, A. D. (1993). Guidelines for evaluating selenium data from aquatic monitoring and assessment studies. Environmental Monitoring and Assessment, 28(1), 83–100. https://doi.org/10.1007/BF00547213
Lemly, A. D. (1996). Assessing the toxic threat of selenium to fish and aquatic birds. Environmental Monitoring and Assessment, 43(1), 19–35. Retrieved from http://www.doi. https://doi.org/10.1007/BF00399568

Lemly, A. D. (2002). Symptoms and implications of selenium toxicity in fish: The Belews Lake case example. Aquatic Toxicology, 57(1–2) 39–49. https://doi.org/10.1016/S0166-445X(01)doi

Lemon, W. C. (1991). Foraging behavior of a guild of Neotropical vultures. Wilson Bulletin, 103(4), 698–702.

Lendrum, A. L. (1982). Ayres' Hawk-eagle in Bulawayo, Zimbabwe. Honeyguide, 110, 15–22.

Leonard, J. (1995). White-bellied sea eagle takes shearwater from sea. Canberra Bird Notes, 20, 10.

Lepage, D. (2020). Avibase – Bird Checklists of the World: Canary Islands. Retrieved from https://avibase.bsc-eoc.org/checklis jsp?region=esic&list=howard moore

Lepage, D. (2020). Booted eagle (Hieraaetus pennatus) (Gmelin, 1788). Avibase. Retrieved from https://avibase.bsc-eoc.org/species.jsp?avibas id=4B8CC2853542FFED.

Lerner, H. R. L. (2007). Molecular Phylogenetics of Diurnal Birds of Prey in the Avian Accipitridae Family. PhD. Thesis. Ann Arbor, M University of Michigan.

Lerner, H. R. L. & Mindell, D. P. (2005). Phylogeny of eagles, Old World vultures, and other Accipitridae based on nuclear and mitochondri DNA. Molecular Phylogenetics and Evolution, 37(2), 327–346. https://doi.org/10.1016/j.ympev.2005.04.010

Lerner, H. R. L., Klaver, M. C. & Mindell, D. P. (2008). Molecular phylogenetics of the buteonine birds of prey (Accipitridae). Auk, 125(2 304–315. https://doi.org/10.1525/auk.2008.06161

Lerner, H., Christidis, L. E. S., Gamauf, A., Griffiths, C., Haring, E., Huddleston, C. J., . . . Jan, O. (2017). Phylogeny and new taxonomy of th Booted eagles (Accipitriformes: Aquilinae). Zootaxa, 4216(4), 301. https://doi.org/10.11646/zootaxa.4216.4.1

Les Christidis, L. & Boles, W. E. (2008). Systematics and Taxonomy of Australian birds. CSIRO Publishing.

Leshem, Y. (1977). Bonelli's Eagle. Israel Land and Nature, 3, 9–15.

Leshem, Y. (1985). Israel: An international axis of raptor migration, ICBP. ICBP Technical Publication No. 5, 243–251. http://www.raptors international.org/book/conservation_studies_on_raptors_1985/Leshem_1985_243-250.pdf_

Leshem, Y. & Yom-Tov, Y. (1998). Routes of migrating soaring birds. Ibis, 140, 41–52. https://doi.org/10.1111/j.1474-919X.1998).tb04539.x

Lesson, R. P. (1831a). Oiseaux. *In*: C. F. Bélanger "1834". Voyage aux Indes-Orientales par le Nord de l'Europe, les provinces du Caucase, l Géorgie, l'Arménie et la Perse, suivi de détails topographiques, statistiques et autres sur le Pégou, les iles de Java, de Maurice et de Bourbo sur le Cap-de-Bonne-Espérance et Sainte-Hélène, pendant les années 1825, 1826, 1827, Zoologie. Livraisons, 3 & 4, 161–288. Paris: Arthu Bertrand.

Lesson, R. P. (1831b). Century zoologique, ou choix d'animaux rares, nouveaux ou imparfaitement connus. Livraisons 4–12, 49–192. Strasbour France: Levrault.

Lestiyanto, T. (2005). Population and habitat of Javan Hawk-eagle Spizaetus bartelsi (Stresemann, 1924) at G. Sanggabuana, Karawang district west Java. Honors Thesis, University As-Syafi'iyah-Jakarta. Jakarta.

Lethaby, N. (2005). The occurrence of Lesser Fish eagle Ichthyophaga humilis on the Cauvery River, Karnataka, India and some notes on th identification of this species. BirdingASIA, 4, 33–38.

Letto, K. R., Wiersma, Y. F., Brazil, J. & Rodrigues, B. (2015). Bald eagle (Haliaeetus leucocephalus) population increases in Placentia Ba Newfoundland: Evidence for habitat saturation? Avian Conservation and Ecology, 10(1), 4. https://doi.org/10.5751/ACE-00729-100104

Levenson, H. (1976). Behavior and energetics of Ospreys nesting in northern California. Unpublished M.S. Thesis, California State University.

Lévêque, R. (1964). Notes on Ecuadorian birds. Ibis, 106(1), 52–62. https://doi.org/10.1111/j.1474-919X.1964.tb03680.x

Levin, A. S. & Kurkin, G. A. (2013). The scope of death of eagles on power lines in Western Kazakhstan. Raptors Conservation, 27, 240–244.

Levy, G. (2011). White-tailed Eagles in Ireland. Scottish Birds, 31, 362.

Lewington, I., Alström, P. & Colston, P. (1991). A Field Guide to the birds of Britain and Europe. London: Harper Collins.

Lewis, A. & Pomeroy, D. (1989). A bird atlas of Kenya. Boca Raton, FL: CRC Press.

Lewis, S. S. B., Fuller, M. M. R. & Titus, K. (2004). A comparison of 3 methods for assessing raptor diet during the breeding season. Wildlif Society Bulletin, 32(2), 373–385. https://doi.org/10.2193/0091-7648(2004)32[373:ACOMFA]2.0.CO;2.

Lewis, S. & Gould, W. R. (2000). Survey effort effects on power to detect trends in raptor migration counts. Wildlife Soc. Società B, 28(2 317–329.

Li, W., Buitenwerf, R., Munk, M., Amoke, I., Bøcher, P. K. & Svenning, J.-C. (2020). Accelerating savanna degradation threatens the Maasai Mar socio-ecological system. Global Environmental Change. PubMed, 60. https://doi.org/10.1016/j.gloenvcha.2019.102030

Lian, X., Zhang, T., Cao, Y., Su, J. & Thirgood, S. (2011). Road proximity and traffic flow perceived as potential predation risks: Evidence fro the Tibetan antelope in the Kekexili National Nature Reserve, China. Wildlife Research, 38(2), 141–146. https://doi.org/10.1071/WR1015

Liechti, F. (2006). Birds: Blowin' by the wind? Journal of Ornithology, 147, 202–211. https://doi.org/10.1007/s10336-006-0061-9

Lien, D. (2015). As eagle population soars, so do conflicts with humans. Retrieved from https://www.twincities.com/2009/06/22/as-eagl population-soars-so-do-conflicts-with-humans/

Ligon, J. D. (1967). Relationships of cathartid vultures. Occasional Papers of the Museum of Zoology University of Michigan, 651, 1–26.

Liguori, J. (2004). How to age Golden eagles: Techniques for birds observed in flight. Birding, 36, 278–283. online. URL. Retrieved from http: www.aba.org/birding/v36n3p278.pdf.

Lillebo, H. P., Shaner, S., Carlson, D., Richard, N. & DuBowy, P. (1988). Water quality criteria for selenium and other trace elements for protecti of aquatic life and its uses in the San Joaquin Valley. SWRCB Order(W).Q., 85–1 . Technical Committee Report, Appendix D. Sacrament CA: California State Water Resources Control Board.

Lim, H. C. & Sodhi, N. S. (2004) Responses of avian guilds to urbanisation in a tropical city. Landscape and Urban Planning, 66(4), 199–21 https://doi.org/10.1016/S0169-2046(03)00111-7

Lim, W. S. & Hut, T. S. S. (2005). Observation of Black eagle Ictinaetus malayensis in Plawangan-Turgo Nature Reserve, Gunung Merapi mountain, Indonesia. In: Abu, J., Chong, M. H. N., Sebastian, A. C. & Aik, Y. C. (Eds.), Proceedings of the 4th Symposium on Asian Raptors, 48–56. Kuala Lumpur: Committee for the 4th Symposium on Asian Raptors.

Limiñana, R., Romero, M., Mellone, U. & Urios, V. (2012a). Mapping the migratory routes and wintering areas of Lesser kestrels Falco naumanni: New insights from satellite telemetry. Ibis, 154(2), 389–399. https://doi.org/10.1111/j.1474-919X.2011.01210.x

Limiñana, R., Soutullo, A., Urios, V. & Reig-Ferrer, A. (2012b). Migration and wintering areas of adult Montagu's harriers Circus pygargus breeding in Spain. Journal of Ornithology, 153(1), 85–93. https://doi.org/10.1007/s10336-011-0698-x

Limiñana, R., Romero, M., Mellone, U. & Urios, V. (2013). Is there a different response to winds during migration between soaring and flapping raptors? An example with the Montagu's harrier and the Lesser kestrel. Behavioral Ecology and Sociobiology, 67(5), 823–835. https://doi.org/10.1007/s00265-013-1506-9

Lin, W. (2005). Habitat use and breeding ecology of Black eagle Ictinaetus malayensis in northern Taiwan. In: Abu, J., Chong, M. H. N., Sebastian, A. C. & Aik, Y. C. (Eds.), Proceedings of the 4th Symposium on Asian Raptors, 35–41. Kuala Lumpur: Committee for the 4th Symposium on Asian Raptors.

Lin, W. H. & Lin W. C. (2010). Diet survey of the Indian Black Eagle during the breeding season: A case study in northeast Taiwan. In: Severinghaus, L. L. & Lin, S. M. (Eds.), The 4th Symposium on Ecology of Raptors in Taiwan: Program and Abstracts, 13. Taipei: Raptor Research Group of Taiwan.

Lindgren, E. (1972). Courtship display of the White-breasted Sea Eagle Haliaeetus leucogaster. Australian Bird Watcher, 4, 132.

Lindsey, R. (2005). Nimbus: 40th Anniversary. Retrieved from https://earthobservatory.nasa.gov/ Features/Nimbus/

Lingle, G. & Krapu, G. 1988. Ingestion of Lead Shot and Aluminum Bands by Bald Eagles During Winter in Nebraska. USGS Northern Prairie Wildlife Research Center, 51. https://digitalcommons.unl.edu/usgsnpwrc/51

Lingle, G. R. & Krapu, G. L. (1986). Winter ecology of Bald eagles in south central Nebraska. Prairie Naturalist, 18, 65–78.

Linnaeus, C. (1758). Systema naturæ per regna tria naturæ, secundum classes, ordines, genera, species, cum Characteribus, Differentiis, Synonymis, Locis. Tomus I. Editio decima, reformata, Salviae, Holmiae.

Linnell, J. D. C., Swenson, J. E. & Anderson, R. (2001). Predators and people: Conservation of large carnivores is possible at high human densities if management policy is favorable. Animal Conservation, 4(4), 345–349. https://doi.org/10.1017/S1367943001001408

Linthicum, J., Jackman, R. E., Latta, B. C., Koshear, J. & Smith, M. (2007). Annual migrations of Bald eagles to and from California. Journal of Raptor Research, 41, 106–112.

Lipsbergs, J. & Bergmanis, U. (2003). Recent population status and conservation measures for the White-tailed eagle in Latvia. In: Helander, B. (Ed.), Sea eagle 2000: Proceedings from the International Sea Eagle Conference in Bjorko, Sweden, 13–17 September 20. 91–95. Stockholm: Sweden Proceedings of the Swedish Society for Nature Conservation, SNF.

Lish, J. W. (1973). Status and ecology of Bald eagles and nesting Golden eagles in Oklahoma. MSc Thesis. Stillwater, OK: Oklahoma State University.

Lish, J. W., Domenech, R., Bedrosian, B. E., Ellis, D. H. & Payton, M. (2016). Wing loading in North American Golden eagles (Aquila chrysaetos). Journal of Raptor Research, 50(1), 70–75. https://doi.org/10.3356/rapt-50-01-70-75.1

Littauer, G. A. & White, R. J. (1961). Sheep and lamb losses in New Mexico. Las Cruces, New Mexico: New Mexico Department of Agriculture Unpublished Report.

Liu, C.-H. (2010). Diurnal activity pattern of Crested Serpent eagles Spilornis cheela hoya in Kenting, southern Taiwan (in Chinese) [Masters Thesis]. Archived from the original on 2012-04-02.

Livezey, B. C. & Zusi, R. L. (2007) Higher-order phylogeny of modern birds (Theropoda, Aves: Neornithes) based on comparative anatomy. II. Analysis and discussion. Zoological Journal of the Linnean Society, 149(1), 1–95. https://doi.org/10.1111/j.1096-3642.2006.00293.x

Livingston, S. (2015). Commercial remote sensing satellites and the regulation of violence in areas of limited statehood. CGCS Occasional Paper Series on ICTs, State building, and Peacebuilding in Africa No. 5. Philadelphia, PA: University of Pennsylvania.

Livingston, S. A., Todd, C. S., Krohn, W. B. & Owen, R. B. Jr. (1990). Habitat models for nesting Bald eagles in Maine. Journal of Wildlife Management, 54(4), 644–653. https://doi.org/10.2307/3809363

Lloyd, D. A. (1998). Predation on the Tent tortoise Psammobates tentorius: A whodunit with the honey badger Mellivora capensis as prime suspect. African Zoology, 33(4), 200–202.

Lo Valvo, M., Massa, B. & Sarà, M. (1993). Uccelli e paesaggio in Sicilia alle soglie del terzo millennio. Il Naturalista siciliano, 17 (suppl.), 1–373.

Lobkov, E. (1991). The population of Steller's Sea eagles Haliaeetus pelagicus (Pallas) on Kamtschatka and problems with their protection. Martin Luther Universität Halle Wittenberg Wissenschaftliche Beiträge, 1991, 141–152.

Lobkov, E. G. (1978). Steller's Sea eagle (Haliaeetus pelagicus) in Kamchatka. Zoologicheskii Zhurnal, 57, 1048–1053.

Lobkov, E. G. & Neufeldt, I. A. (1986). Distribution and biology of the Steller's Sea eagle (Haliaeetus pelagicus, Pall.). Proceedings of the Zoological Institute, 150, 107–146). Academy of Science.

Lobley, G. (2007). Wintering of Greater Spotted eagle Aquila clanga and Eastern Imperial eagle A. heliaca in the Arabian Peninsula. Sandgrouse, 29, 177–182.

Lobos, R. P. & Alvarado, O. S. (2006). Mutualism between the Crowned Solitary eagle (Harpyhaliaetus coronatus) and the Monk parakeet (Myiopsitta monacha) during nest building in the Telteca Natural Forest Reserve, Department of Lavalle, Mendoza Province, Argentina. VIII Jornadas de Etologia. Pages 1-1.

Lockhart, J. (2020). Eagle goes head to head with jackal for food. Retrieved from https://roaring.earth/eagle-with-jackal/

Lockie, J. D. & Stephen, D. (1959). Eagles, lambs and land management on Lewis. Journal of Animal Ecology, 28(1), 43–50. https://doi.org/10.2307/2012

Lockwood, M. W. & Freeman, B. (2004). The Texas Ornithological Society Handbook of Texas birds. College Station, TX Texas A&M University Press.

Löe, J. & Röskaft, E. (2004). Large carnivores and human safety: A review. Ambio – A Journal of the Human Environment, 33(6), 283–288. https://doi.org/10.1579/0044-7447-33.6.283

Loehr, V. J. (2012). Activity of the greater paedoloper, Homopus femoralis (Testudinidae), in relation to rainfall. African Zoology, 47(2), 294–300. https://doi.org/10.3377/004.047.0212.

Løfaldli, L. (1994). Småspove Numenius phaeopus. *In*: J. O. Gjershaug et al. (Eds.), Norsk fugleatlas, 204. Klæbu: Norsk Ornitologisk Forening.

Lohmus, A. (1998). Numbers of the Greater Spotted eagle and the Lesser Spotted eagle in Estonia. Hirundo, 11, 24–34.

Lõhmus, A. & Väli, Ü. (2001). Interbreeding of the Greater Aquila clanga and Lesser spotted eagle A. pomarina. Acta Ornithoecologica, 4, 377–384.

Long, S. (1998). Tree-nesting crowned crane and response to potential predators. Honeyguide, 44, 143.

Longman's Dictionary of Contemporary. (2014). Eagle (English 6th ed). Retrieved from https://www.ldoceonline.com/dictionary/eagle

Lopes, L. E., Batista de Pinho, J., Bernardon, B., Ficagna de Oliveira, F., . . . Albonette de Nóbrega, P. F. (2009). Aves da Chapada dos Guimarães, Mato Grosso, Brasil: Uma síntese histórica do conhecimento. Papéis Avulsos de Zoologia. MT, Brazil: Birds of Chapada dos Guimaráes: A synthesis of historical knowledge, 49(2), 9–47. https://doi.org/10.1590/S0031-10492009000200001.

Lopez, R. R., Silvy, N. J., Peterson, M. J., Brooks, D. M., Baur, E. H. & González-García, F. (2019). Curassows, Guans, and Chachalacas in Mexico. Wildlife Ecology and Management in Mexico, 77.

López, S. L., Aiassa, D., Benítez-Leite, S., Lajmanovich, R., Mañas, F., Poletta, G., . . . Carrasco, A. E. (2012). Chapter Two. Pesticides used in South American GMO-based agriculture: A review of their effects on humans and animal models. Advances in Molecular Toxicology, 6, 41–75. https://doi.org/10.1016/B978-0-444-59389-4.00002-1

López-Jurado, C. (2004). Homologació Dede Rareses Ornitològiques a Balears. Informe Dede (2004). Anuari Ornitolňgic de les Balears, 19, 109–125.

López-Lanús, B., Salaman, P. G. W., Cowley, T. P., Arango, S., . . . (2000). The threatened birds of the Río Toche, Cordillera Central, Colombia. Cotinga, 14, 17–23.

López-López, P., Ferrer, M., Madero, A., Casado, E. & Mcgrady, M. (2011). Solving man-induced large-scale conservation problems: The Spanish imperial eagle and power lines. PLoS ONE, 6, e17196. https://doi.org/10.1371/journal.pone.0017196

López-López, P., Benavent-Corai, J., Garcíía-Ripolléés, C. & Urios, V. (2013). Scavengers on the move: Behavioural changes in foraging search patterns during the annual cycle. PLoS One, 8(1), e54352. https://doi.org/10.1371/journal.pone.0054352

López-López, P., De la Puente, J., Mellone, U., Bermejo, A. & Urios, V. (2016). Spatial ecology and habitat use of adult Booted eagles (Aquila pennata) during the breeding season: Implications for conservation. Journal of Ornithology, 157(4), 981–993. https://doi.org/10.1007/s10336-016-1357-z

López-López, P., García-Ripollés, C. & Urios, V. (2014). Individual repeatability in timing and spatial flexibility of migration routes of trans-Saharan migratory raptors. Current Zoology, 60(5), 642–652. https://doi.org/10.1093/czoolo/60.5.642

Lorimer, J. (2007). Nonhuman charisma. Environment and Planning D: Society and Space, 25, 911–932.

Louekari, K., Mäkelä-Kurtto, R., Pasanen, J., Virtanen, V., Sippola, J. & Malm, J. (2000). Cadmium in Fertilizers: Risks to human health and the environment. Ministry of Agriculture and Forestry in Finland.

Lovaszi, P. (1998). Status of the White Stork (Ciconia ciconia) in Hungary: Results of national censuses between 1941 and 1994. Ornis Hungarica, 8, 1–8.

Love, J. A. (1983). The return of the Sea eagle. Cambridge: Cambridge University Press.

Love, J. A. (1988). The reintroduction of the White-tailed Sea eagle to Scotland, 1975–1987. Nature Publishing Conservancy Council, Royal Society for the Protection of Birds, Institute of Terrestrial Ecology.

Love, J. A. (1989). Eagles. London: Whittet Books.

Lovell, C. D. & Dolbeer, R. A. (1999). Validation of the U.S. Air Force bird avoidance model. Wildlife and Society Bulletin, 27. https://www.jstor.org/stable/3783954.

Loveridge, A. (1922). Notes on East African birds (chiefly nesting habits and stomach contents) collected 1915–1919. Proceedings of the Zoological Society of London, 92(4), 837–862. https://doi.org/10.1111/j.1469-7998.1922.tb07084.x

Loveridge, A. (1923). Notes on East African birds (chiefly nesting habits and endo-parasites) collected 1920–1923. Proceedings of the Zoological Society of London, 93(4), 899–921. https://doi.org/10.1111/j.1096-3642.1923.tb02213.x

Loveridge, A. J. & Macdonald, D. W. 2002. Habitat Ecology of two Sympatric Species of Jackals in Zimbabwe. Journal of Mammalogy, 83(2), 599–607. https://doi.org/10.1644/1545-1542(2002)083<0599:HEOTSS>2.0.CO;2

Lovette, I. J. (2004). Mitochondrial dating and mixed support for the 2% rule in birds. Auk, 121, 1–6.

Løvschal, M., Bøcher, P. K., Pilgaard, J., Amoke, I., . . . (2017). Fencing bodes a rapid collapse of the unique Greater Mara ecosystem. Scientific Reports, 7, 1–841450. https://doi.org/10.1038/srep41450

Lowen, J. C., Bartrina, L., Brooks, T. M., Clay, R. P., Tobias, J. & Project YACUTINGA. (1995). Bird surveys and conservation priorities in eastern Paraguay. Cotinga. 1996, 5, 14–19.

Lowery, G. H. & Dalquest, W. W. (1951). Birds from the State of Veracruz, Mexico. University of Kansas Publications, 3 (pp. 531–649). Museum of Natural History. Retrieved from https://www.biodiversitylibrary.org/item/21915#page/563/mode/1up.

Lowry, M. (2006). Courage and love. Gabar, 17, 29–30.

Lowney, M. S. (1999). Damage by Black and Turkey vultures in Virginia, 1990–1996. Wildllife Society Bulletin, 27(3), 715–719. Retrieved from https://www.jstor.org/stable/3784093.

Lu, X., McElroy, M. B. & Kiviluoma, J. (2009). Global potential for wind-generated electricity. Proceedings of the National Academy of Sciences of the United States of America, 106(27), 10933–10938. https://doi.org/10.1073/pnas.0904101106

Lubkov, E. G. & Neufeldt, I. A. (1986). Distribution and biology of the Steller's sea eagle Haliaeetus pelagicus pelagicus (Pallas). Proceedings of the Zoological Institute of the Academy of Science, 150, 107–145.

Lucas, M. D., Ferrer, M., Bechard, M. J. & Muñoz, A. R. (2012). Griffon vulture mortality at wind farms in southern Spain: Distribution of fatalities and active mitigation measures. Biological Conservation, 147, 184–189. https://doi.org/10.1016/j.biocon.2011.12.029

Lucas, T. (2017). New Project Uses Phones and Drones to Monitor Endangered Species. Retrieved from https://nicholas.duke.edu/about/news/new-project-uses-phones-and-drones-monitor-endangered-species.

Lucas, T. (2018). Drones prove reliable, cost-efficient tool for wildlife surveys. Retrieved from https://nicholas.duke.edu/about/news/drones-prove-reliable-cost-efficient-tool-wildlife-surveys

Lucieer, A., Turner, D., King, D. H. & Robinson, S. A. (2014). Using an unmanned aerial vehicle (UAV) to capture micro-topography of Antarctic moss beds. International Journal of Applied Earth Observation and Geoinformation, 27(A), 53–62. https://doi.org/10.1016/j.jag.2013.05.011

Luna, L. G. (1972). Manual of histologic staining methods of Armed Forces Institute of Pathology (3rd edition). Philadelphia, PA: W. B. Saunders.

Luoto, M., Virkkala, R., Heikkinen, R. K. & Rainio, K. (2004). Predicting bird species richness using remote sensing in boreal agricultural-forest mosaics. Ecological Applications, 14(6), 1946–1962. https://doi.org/10.1890/02-5176

Luz, B. B. (2005). Características de árvores emergentes utilizadas por Gavião-real (Harpia harpyja) para nidificação no centro e leste da Amazônia Brasileira. [MSc Dissertation]. Manaus, Brazil: INPA/UFAM.

Luzardo, O. P., Ruiz-Suárez, N., Henríquez-Hernández, L. A., Valerón, P. F., Camacho, M., Zumbado, M. & Boada, L. D. (2014). Assessment of the exposure to organochlorine pesticides, PCBs and PAHs in six species of predatory birds of the Canary Islands, Spain. Science of the Total Environment, 15(472), 146–153. HTTPS://DOI.ORG/10.1016/j.scitotenv.2013.11.021.

Lyly, M. S., Villers, A., Koivisto, E., Helle, P., Ollila, T. & Korpimäki, E. (2015). Avian top predator and the landscape of fear: Responses of mammalian mesopredators to risk imposed by the Golden eagle. Ecology and Evolution, 5(2), 503–514. https://doi.org/10.1002/ece3.1370

Lynch, E., Northrup, J. M., McKenna, M. F., Anderson, C. R. Jr., Angeloni, L. & Wittemyer, G. (2014). Landscape and anthropogenic features influence the use of auditory vigilance by mule deer. Behavioral Ecology, 26, 75–82. https://doi.org/10.1093/beheco/aru158

Lyon, B. & Kuhnigk, A. (1985). Observations on nesting Ornate hawk-eagles in Guatemala. The Wilson Bulletin, 97(2), 141–147.

Ma, Z. J., Cheng, Y. X., Wang, J. Y. & Fu, X. H. (2012). The rapid development of birdwatching in mainland China: A new force for bird study and conservation. Bird Conservation International, 13, 259–269.

Maala, C. P. (2001). Endangered Philippine wildlife species with special reference to the Philippine eagle (Pithecophaga jefferyi) and Tamaraw (Bubalus mindorensis) [Review]. https://doi.org/10.15027/14360

Mabie, D. W., Merendino, M. T. & Reid, D. H. (1995). Prey of nesting Bald eagles in Texas. Journal of Raptor Research 29(1), 10–14. Retrieved from https://sora.unm.edu/sites/ default/files/journals/jrr/v029n01/p00010-p00014.pdf

MacArthur, R. H. & Pianka, E. R. (1966). On optimal use of a patchy environment. The American Naturalist, 100(916), 603–609. https://doi.org/10.1086/282454

Macdonald, D. W. & Norris, S. (2001). Encyclopedia of mammals. Oxford: Oxford University Press.

MacDonald, G. (2003). Biogeography: Introduction to Space, Time, and Life. 55(2), 283–285. https://doi.org/10.1111/0033-0124.5502018

MacDonald, J. (1994). Bald eagle attacks adult Osprey. Journal of Raptor Research, 28(2), 122.

Maceda, J. J. (2007a). Biology and conservation of the Crowned Solitary eagle (Harpyhaliaetus coronatus) in Argentina. Hornero, 22, 159–171.

Maceda, J. J., Sarasola, J. H. & Pessino, M. E. M. (2003). Prey consumed by the Crowned eagle (Harpyhaliaetus coronatus) in the southern limits of its range in Argentina. Ornitologia Neotropical, 14, 419–422.

Maceda, J., Salvador, V., Santillán, M., Reyes, M., Bragagnolo, L., Lanusse, A., . . . Bechard, M. J. (2007). Behavior and monitoring of nests of the Crowned Solitary eagle (Harpyhaliaetus coronatus) in central Argentina using a video camera. *In*: Bildstein, K. L., Barber, D. R. & Zimmerman, A. (Eds.), Neotropical ornithology, 311. Orwigsburg, PA: Hawk Mountain Sanctuary.

Machado, A. B. M., Fonseca, G. A. B., Machado, R. B., Aguiar, L. M. S. & Lins, L. V. (1998). Red Book of the threatened and extinct species of the Minas Gerais fauna. Belo Horizonte, Brazil: Fundação Biodiversitas. [In Portuguese]

Machange, R. W., Jenkins, A. R. & Navarro, R. A. (2005). Eagles as indicators of ecosystem health: Is the distribution of Martial eagle nests in the Karoo, South Africa, influenced by variations in land-use and rangeland quality? Journal of Arid Environments, 63(1), 223–243. https://doi.org/10.1016/j.jaridenv.2005.03.031

Maciorowski, G. & Mizera, T. (2010). Conservations and studies on Greater Spotted eagle in Poland—LIFE Project [Studia imateriały CEPL w Rogowie], 25, 181–190.

Mack, A. L. & Wright, D. W. (1996). Birds of the Crater Mountain Biological Research Station, Papua New Guinea. Emu, 96, 89–101.

Mackal, R. P. (1987). A Living Dinosaur? In Search of Mokele-Mbembe. Brill Archive.

MacKay, R. (1980). A list of birds of the Baiyer River Sanctuary and adjacent areas. Papua New Guinea Bird Society Newsletter, 167–168, 24–38.

Mackay, R. D. (1988). Gurney's eagle Aquila gurneyi in the highlands. Muruk, 3, 56.

Mackay, R. D. (1993). New distribution records—Lihir Island. Muruk, 6, 10.

Mackenzie, D. (2014). Europe's eagles under threat from vulture-killing drug. Retrieved from https://www.newscientist.com/article/dn25650-europes-eagles-under-threat-from-vulture-killing-drug/

MacKinnon, J. (1990). Field guide to the birds of Java and Bali. Yogyakarta, Indonesia: Gadjah Mada University Press.

MacKinnon, J. & Phillips, K. (1993). A Field Guide to the Birds of Borneo, Sumatra, Java, and Bali, the Greater Sunda Islands. Oxford: Oxford University Press.

MacKinnon, J. R., MacKinnon, J., Phillips, K. & He, F. Q. (2000). A Field Guide to the Birds of China. Oxford: Oxford University Press.
Mackinnon, J. & Hicks, N. (2017). Birds of China. Bloomsbury Publishing.
Maclean, G. (1988). Robert's Birds of Southern Africa. London: New Holland Publishers Ltd.
Maclean, G. L. (1969). The breeding seasons of birds in the south-western Kalahari. Ostrich, 40(Suppl. 1), 179–192. https://doi.org/10.1080/00 06525.1969.9639120
Maclean, G. L. (1973). The sociable weaver, Part 4: Predators, parasites and symbionts. Ostrich, 44(3–4), 241–253. https://doi.org/10.1080/003 6525.1973.9639161
Maclean, G. L. (1993). Roberts' Birds of Southern Africa (6th Ed.). Cape Town, S.A.: Trustees John Voelcker Bird Book Fund.
MacLeod, J. G. R., Murray, E. M. & Murray, C. D. (1952). The birds of the Hottentots Holland (Part II). Ostrich 23(1), 16–25. https://doi.or 10.1080/00306525.1952.9632598.
Macleod, R., Ewing, S. K., Herzog, S. K., Bryce, R., Evans, K. L. & Maccormick, A. (2005). First ornithological inventory and conservatic assessment for the Yungas forests of the Cordilleras Cocapata and Mosetenes, Cochabamba, Bolivia. Bird Conservation International, 15(4 361–382. https://doi.org/10.1017/S095927090500064X
MacWilliam, I. (2004). Kazakhstan's Eagle Hunters. BBC News. December 16. Retrieved from http://news.bbc.co.uk/2/hi/asia-pacific/410188 stm
Madden, D. H. (1924). A chapter of mediæval history. Port Washington, WI, NY: Kennikat Press.
Madders, M. & Whitfield, D. P. (2006). Upland raptors and the assessment of wind farm impacts. Ibis 148: 43–56. https://doi.org/10.1111/j.147 919X.2006.00506.x
Madero, A. & Pacheco, S. (2009). Criterios para la obtención de pollos silvestres para proyectos de reintroducción y reforzamiento de águil imperial ibérica (Aquila adalberti) en Andalucía. I Congreso ibérico de reintroducción de especies silvestres. 5–8 Noviembre, Jerez d la Frontera, Cádiz. www.zoobotanicojerez.com/fileadmin/documentos/2009/Documento_entre gado_a_los_participantes_Congreso Reintroducciones__DEFINITIVO_01.pdf
Madhyastha, N. A. (2004). A paradise for White-bellied Sea eagle: Netrani Island. Newsletter for Birdwatchers, 44(1), 14.
Madoc, G. C. (1956). An Introduction to Malayan Birds 3rd. Kuala Lumpur, Malaysia: Malayan Nature Society.
Madrid, M. H. D., Madrid, M. J. A., Cruz, E. J. R., Cordova, A. J., Castillo, R. M., Martinez, A. W. & Ramos, C. A. (1992). Behavior and breedin biology of the Ornate Hawk-eagle. *In*: Whitacre, D.F. & Thorstrom, R.K. (Eds.), Progress Report V, 1992. Maya Project: Use of rapto and other fauna as environmental indicators for design, management, and monitoring of protected areas and for building local capacity fc conservation in Latin America, pp. 179–182. Boise, ID: The Peregrine Fund.
Madroño-Nieto, A., Clay, R. P., Robbins, M. B., Rice, N. H., Faucett, R. C. & Lowen, J. C. (1997). An avifaunal survey of the vanishing interic Atlantic Forest of San Rafael National Park. Departments Itapúa/Caazapá, Paraguay. Cotinga, 7, 45–53.
Madry, M. M., Kraemer, T., Kupper, J., Naegeli, H., Jenny, H., Jenni, L. & Jenny, D. (2015). Excessive lead burden among Golden eagles in th Swiss Alps. Environmental Research Letters. PubMed, 10(3). https://doi.org/10.1088/1748-9326/10/3/034003
Maehr, D. S. & Kale, H. W. (2005). Florida's birds: A Field Guide and Reference. Pineapple Press, Inc.
Mahoney, S. A. (1983). Heat production and heat loss in vultures: Rest and exercise. In: Wilbur, S. R. & Jackson, J. A. (Eds.), Vulture Biology an Management, 330–342. University of California Press, Berkeley.
Maindonald, J. & Braun, J. W. (2010). Data analysis and graphics using R: An example-based approach. Retrieved from http://www.stats.uwo.c DAAG. Cambridge, UK: Cambridge University Press Online. (3rd Edition).
Makatsch, W. (1950). Die Vogelwelt Macedoniens. Leipzig, Germany.
Malan, A. R. (1996). Territory and nest defence in polyandrous pale chanting goshawks: Do co-breeders help? African Zoology, 31(4), 170–176
Malan, G. & Crowe, T. M. (1996). The diet and conservation of monogamous and polyandrous pale chanting goshawks in the Little Karoo, Sout Africa. South African Wildlife Research-24-month delayed open access, 26(1), 1–10.
Malan, G. (2009). Raptor Survey and Monitoring: A Field Guide for African Birds of Prey. Pretoria, South Africa: Briza Publications.
Maliene, V., Grigonis, V., Palevičius, V. & Griffiths, S. (2011). Geographic information system: Old principles with new capabilities. Urban Desig International, 16. https://doi.org/10.1057/udi.2010.25
Mallalieu, M. (2007). Greater Spotted eagles Aquila clanga in central Thailand. Forktail, 23, 167–170.
Mallalieu, M. (2008). Kathmandu Bird Report 2004–2006. Kathmandu: Bird Conservation Nepal.
Mallet-Rodrigues, F. & Noronha, M. L. M. (2009). Birds in the Parque Estadual dos TrêsPicos, Rio de Janeiro state, south-east Brazil. Coting 31, 61–71.
Malmiga, G., Nilsson, C., Bääckman, J. & Alerstam, T. (2014). Interspecific comparison of the flight perfomance between sparrowhawk and common buzzards migrating at the Falsterbo peninsula: A radar study. Current Zoology, 60(5), 670–679. https://doi.org/10.109 czoolo/60.5.670
Mandel, J. T. & Bildstein, K. L. (2007). Turkey vultures use anthropogenic thermals to extend their daily activity period. Wilson Journal c Ornithology 119(1), 102–105. https://doi.org/10.1676/05-154.1
Mandel, J. T., Bildstein, K. L., Bohrer, G. & Winkler, D. W. (2008). Movement ecology of migration in Turkey vultures. Proceedings of th National Academy of Sciences of the United States of America, 105(49), 19102–19107. https://doi.org/10.1073/pnas.0801789105
Manning, L. K., Wünschmann, A., Armién, A. G., Willette, M., MacAulay, K., Bender, J. B., . . . Redig, P. (2018). First lead intoxication in free ranging Bald eagles (Haliaeetus leucocephalus). Veterinary Pathology, 56(2), 289–299. https://doi.org/10.1177/0300985818813099.
Manning, T., Ross, G. A. & Symons, R. (2008). Environmental contaminants in White-bellied sea eagles (Haliaeetus leucogaster) found in Sydne Australia. Australian Journal Ecotoxicology, 14: 21–30.

Mañosa, S. & Real, J. (2001). Potential negative effects of collisions with transmission lines on a Bonelli's eagle population. Journal of Raptor Research, 35, 247–252.

Mañosa, S., Mateos, E., Pedrocchi, V. & de Campos Martins, F. (2002). Birds of prey survey (Aves: Cathartiformes and Accipitriformes) in the Paranapiacaba Forest fragment. *In*: Mateos, E. J., Guix, C., Serra, A. & Pisciotta, K. (Eds.), Censuses of vertebrates in a Brazilian Atlantic rainforest area: The Paranapiacaba fragment, pp. 165–179. Barcelona: Centre de Recursos de Biodiversitat Animal, Universitat de Barcelona. Barcelona: Lynx Edicions.

Mañosa, S., Real, J. & Codina, J. (1995). Age estimation and growth patterns in nestling Bonelli's eagles. Journal of Raptor Research, 29, 273–275.

Manson, A. J. (1982). Predation of wood owl by Long Crested eagle. Honeyguide, 110, 46.

Manto, M. (2014). Abnormal copper homeostasis: Mechanisms and roles in neurodegeneration. Toxics, 2(2), 327–345. https://doi.org/10.3390/toxics2020327

Manuel, H. (2020). Nesting Martial eagle (Polemaetus bellicocus) diet: Influence of prey delivery and diversity in two Kenyan ecosystems. FULLTEXT04. Retrieved from https://www.diva-portal.org/smash/get/diva2. PubMed: 1442659

Maphalala, M. I., Monadjem, A., Bildstein, K. L., McPherson, S., Hoffman, B. & Downs, C. T. (2020). Ranging behaviour of Long-crested eagles Lophaetus occipitalis in human-modified landscapes of KwaZulu-Natal, South Africa. Ostrich, 91(3), 221–227. https://doi.org/10.2989/00306525.2020.1770888

MAPS Media Institute. (2019). Golden eagles versus coyotes: Episode 1, Bitterroot Valley Winter Eagle Project Series. Retrieved from https://www.youtube.com/ watch?v=xcgWSOocNSE.

Mara Raptor Project. (2018). Annual Report. Conserving Raptors in the Maasai Mara Ecosystem.

Maransky, B. P., Goodrich, L. J. & Bildstein, K. L. (1997). Seasonal shifts in the effects of weather on the visible migration of Red-Tailed Hawks at Hawk Mountain, Pennsylvania, 1992–1994. The Wilson Journal of Ornithology, 109, 246–252.

Marchant, S. & Higgins, P. J. (Eds.). (1993). Handbook of Australian, New Zealand and Antarctic birds, 2: Raptors to Lapwings. Melbourne, Victoria: Oxford University Press.

Marchant, S. & Higgins, P. J. (Eds.). (1993). Handbook of Australian, New Zealand and Antarctic birds, 2: Raptors to Pratincoles. Melbourne, UK: Oxford University Press.

Mares, M. (Ed.). (1999). Encyclopedia of Deserts. University of Oklahoma Press.

Margalida, A. (2012). Baits, budget cuts: A deadly mix. Science, 338(6104), 192. https://doi.org/10.1126/science.338.6104.192-a

Margalida, A., Bertran, J. & Heredia, R. (2009). Diet and food preferences of the endangered Bearded vulture Gypaetus barbatus: A basis for their conservation. Ibis, 151(2), 235–243. https://doi.org/10.1111/j.1474-919X.2009.00904.x

Margalida, A., Bogliani, G., Bowden, C. G., Donázar, J. A., Genero, F., Gilbert, M., . . . Green. (2014). One health approach to the use of veterinary pharmaceuticals. Science, 346(6215), 1296–1298. https://doi.org/10.1126/science.1260260

Margalida, A., Colomer, M., Sánchez, R., Sánchez, F. J., Oria, J. & González, L. M. (2017). Behavioral evidence of hunting and foraging techniques by a top predator suggests the importance of scavenging for preadults. Ecology and Evolution, 7(12), 4192–4199. https://doi.org/10.1002/ece3.2944

Margalida, A., González, L. M., Sánchez, R., Oria, J., Prada, L., Caldera, J., . . . Molina, J. I. (2007). A long-term large-scale study of the breeding biology of the Spanish Imperial eagle (Aquila adalberti). Journal of Ornithology, 148(3), 309–322. https://doi.org/10.1007/s10336-007-0133-5

Margalida, A., González, L. M., Sánchez, R., Oria, J. & Prada, L. (2007a). Parental behaviour of Spanish Imperial eagles (Aquila adalberti), Sexual differences in a moderately dimorphic raptor. Bird Study, 54(1), 112–119. https://doi.org/10.1080/00063650709461462

Margalida, A., Mañosa, S., Mariano González, L. M., Ortega, E., Sánchez, R. & Oria, J. (2008). Breeding of non-adults and effects of age on productivity in the Spanish Imperial eagle Aquila adalberti. Ardea, 96(2), 173–180. https://doi.org/10.5253/078.096.0203

Mariano González, L., Oria, J., Sánchez, R., Margalida, A., Aranda, A., Prada, L., Caldera, J., Molina, J. I. (2008). Status and habitat changes in the endangered Spanish Imperial eagle Aquila adalberti population during 1974–2004: Implications for its recovery. Bird Conservation International, 18(3), 242–259. https://doi.org/10.1017/S0959270908000245

Mark, T., Augustine, L., Barrio, J., Flanagan, J., . . . (2008). New records of birds from the northern Cordillera Central of Peru in a historical perspective. Cotinga, 29, 108–125.

Marks, J. & Doremus, J. (1988). Breeding-season diet of Northern saw-whet owls in Southwestern Idaho. Wilson Bulletin, 100(4), 690–694.

Markus, M.B. (1972). Mortality of vultures caused by electrocution. Nature, 238, 228.

Marques, A. T., Batalha, H., Rodrigues, S., Costa, H., Pereira, M. J. R., Fonseca, C., . . . Bernardino, J. (2014). Understanding bird collisions at wind farms: An updated review on the causes and possible mitigation strategies. Biological Conservation, 179, 40–52. https://doi.org/10.1016/j.biocon.2014.08.017

Márques Reyes, C., Zalles, J., Stubbs, L., Bildstein, K. L. & Ellis, S. (2000). Latin American Falconiformes Conservation Assessment and Management Plan. Conservation Breeding Specialist Group, Species Survival Commission, IUCN, Apple Valley.

Marquez, C. & Delgado, H. (2010). Feeding ecology and conservation of Isidor's eagle (Spizaetus isidori) in Colombia. Boise, ID: The Peregrine Fund Reports. The Peregrine Fund.

Márquez, C., Gast, F., Vanegas, V. H. & Bechard, M. (2005). Diurnal raptors of Colombia (1st ed.). Instituto Investigación de Recursos Biológicos Alexander von Humboldt. Bogotá, DC, Colombia.

Marquiss, M., Madders, M. & Carss, D. N. (2003). White-tailed eagles (Haliaeetus albicilla) and Lambs (Ovisaries). *In:* Thompson, D.B.A., Redpath, S.M., Fielding, A.H., Marquiss, M. and Galbraith, C.A. (Eds.), Birds of prey in a changing environment, pp. 471–479. Edinburgh: The Stationery Office.

Marquiss, M., Madders, M., Irvine, J. & Carss, D. N. (2004). The impact of White-tailed eagles on sheep farming on mull. Edinburgh: Scottish Executive.

Marquiss, M. & Newton, I. A radio-tracking study of the ranging behaviour and dispersion of European sparrow hawks Accipiter nisus. Journal of Animal Ecology, 51(1), 111–133. https://doi.org/10.2307/4314.

Marris, E. (2013). Drones in science: Fly and bring me data. Nature, 498(7453), 156–158. Retrieved from https://doi.org/10.1038/498156a

Marshak, S. (2009). Essentials of Geology (3rd ed). New York: W.W. Norton & Company.

Martell, M. S., Henny, C. J., Nye, P. E. & Solensky, M. J. (2001). Fall migration routes, timing, and wintering sites of North American Ospreys as determined by satellite telemetry. Condor, 103(4), 715–724. https://doi.org/10.1093/condor/103.4.715

Marti, C. (1998). Effects of power lines on birds: Documentation (in German with English summary). Schriftenreihe Umwelt Nr. 292. Bundesamt für Umwelt, Wald und Landschaft (BUWAL), Bern.

Marti, C. D. (1987). Raptor food habits studies. *In*: Giron Pendleton, B. A., Millsap, B. A., Cline, K. W. & Bird, D. M. (Eds.), Raptor management techniques manual, pp. 67-80. Washington, D.C.: National Wildlife Federation.

Marti, C. D., Korpimäki, E. & Jaksić, F. M. (1993). Trophic structure of raptor communities: A three-continent comparison and synthesis. *In* Current ornithology, pp. 47–137. Boston, MA: Springer.

Martí, R. & Del Moral, J. C. (2003). Atlas de las avesreproductoras de España. Madrid: Dirección General de Conservación de la Naturaleza SEO/BirdLife.

Martín, B., Onrubia, A., De la Cruz, A. & Ferrer, M. (2016). Trends of autumn counts at Iberian migration bottlenecks as a tool for monitoring continental populations of soaring birds in Europe. Biodiversity and Conservation, 25(2), 295–309. https://doi.org/10.1007/s10531-016-1047-4

Martin, G. (2018). UAVs a useful tool in wildlife conservation. Retrieved from https://www.defenceweb.co.za/aerospace/unmanned-aerial-vehicles/uavs-a-useful-tool-in-wildlife-conservation/.

Martin, G. R. & Shaw, J. M. (2010). Bird collisions with power lines: Failing to see the way ahead? Biological Conservation, 143(11), 2695–2702 https://doi.org/10.1016/j.biocon.2010.07.014

Martin, G. R. (2011). Understanding bird collisions with man-made objects: A sensory ecology approach. Ibis, 153(2), 239–254. https://doi.org/10.1111/j.1474-919X.2011.01117.x

Martin, G. R. & Portugal, S. J. (2011). Differences in foraging ecology determine variation in visual fields in ibises and spoonbills (Threskiornithidae) Ibis, 153(4), 662–671. https://doi.org/10.1111/j.1474-919X.2011.01151.x

Martin, G. R., Portugal, S. J. & Murn, C. P. (2012). Visual fields, foraging and collision vulnerability in Gyps vultures. Ibis, 154(3), 626–631 https://doi.org/10.1111/j.1474-919X.2012.

Martin, G. R. (2012). Through birds' eyes: Insights into avian sensory ecology. Journal of Ornithology, 153, 23–48. https://doi.org/10.1007/s10336-011-0771-5.

Martin, L. D. (1971). An early Pleistocene eagle from Nebraska. The Condor, 73(2), 248–250. https://doi.org/10.2307/1365848

Martin, L. D. (1975). A new species of Spizaetus from the Pliocene of Nebraska. Wilson Bulletin 87(3), 413–416.

Martin, L. D. (1989). Fossil history of the terrestrial carnivora. *In*: Gittleman, J. (Ed.), Carnivore behavior, ecology and evolution, pp. 535–568 London: Chapman and Hall.

Martin, P. A., Campbell, D., Hughes, K. & McDaniel, T. (2008). Lead in the tissues of terrestrial raptors in southern Ontario, Canada, 1995–2001 Science of the Total Environment, 391(1), 96–103. https://doi.org/10.1016/j.scitotenv.2007.11.012

Martin, R., Pepler, D. & Hamerse, P. (1987). Martial eagles nesting on a boulder. Bokmakierie, 39, 91.

Martin, R. E., Wehmiller, J. F., Harris, M. S. & Liddell, W. D. (1996). Comparative taphonomy of foraminifera and bivalves in Holocene shallow-water carbonate and siliciclastic regimes: Taphonomic grades and temporal resolution. Paleobiology, 22(1), 80–90.

Martin, R. O., Sebele, L., Koeslag, A., Curtis, O., Abadi, F. & Amar, A. (2014). Phenological shifts assist colonisation of a novel environment in a range-expanding raptor. Oikos, 123(12), 1457–1468. https://doi.org/10.1111/oik.01058

Martinez, J. E. & Sanchez-Zapata, J. A. (1999). Invernada de aguilillacalzada (Hieraaetus pennatus) y culebreraeuropea (Circaetus gallicus) en España. Ardeola, 46, 93–96.

Martínez, J. E., Zuberogoitia, I., Gómez, G., Escarabajal, J. M., Cerezo, E., Jiménez-Franco, J. V. & Calvo, J. F. (2014). Attack success in Bonelli's eagle Aquila fasciata. Ornis Fennica, 91, 67–78.

Martínez, J. E., Zuberogoitia, I., Jiménez-Franco, M. V., Mañosa, S. & Calvo, J. F. (2016). Spatio-temporal variations in mortality causes of two migratory forest raptors in Spain. European Journal of Wildlife Research, 62(1), 109–118. https://doi.org/10.1007/s10344-015-0981-x.

Martínez-Abraín, A., Oro, D., Jiménez, J., Stewart, G. & Pullin, A. (2010). A systematic review of the effects of recreational activities on nesting birds of prey. Basic and Applied Ecology, 11(4), 312–319. https://doi.org/10.1016/j.baae.2009.12.011.

Martinez-Cruz, B. & Godoy, J. A. (2007). Genetic evidence for a recent divergence and subsequent gene flow between Spanish and Eastern Imperial eagles. BMC Evolutionary Biology, 7, 170. https://doi.org/10.1186/1471-2148-7-170.

Martinez-Gomez, J. E. (1992). Raptor conservation in Veracruz, Mexico. Journal of Raptor Research, 26, 184–188.

Martinez-Sanchez, J. C. (1986). Causes affecting the survival of birds of prey in Nicaragua. Birds of Prey Bulletin, 3: 43–47.

Martínez-Sánchez, J. C. (2000). Check-list of the birds of Nicaragua. Fundación Cocibolca.

Martínez-Sánchez, J. C. & Will, T. (Eds.). (2010). Thomas R. Howell's Check-list of the birds of Nicaragua as of 1993. Ornithological Monographs 68(1), x–xvi. https://doi.org/10.1525/om.2010.68.1.x

Martín, B., Onrubia, A. & Ferrer, M. (2014). Effects of climate change on the migration behavior of the Common buzzard (Buteo buteo). Climate Research, 60(3), 187–197. https://doi.org/10.3354/cr01233

Martõ Ânez, J. A., Serrano, D. & Zuberogoitia, I. (2003). Predictive models of habitat preferences for the Eurasian eagle owl Bubo: A multiscale approach. Ecography, 26(1), 21–28. https://doi.org/10.1034/j.1600-0587.2003.03368.x

Marx, J. T. & van Staden, M. (1989). A survey of the Black and Martial eagles in the Orange Free State. Gabar, 4(2), 26.

Marzluff, J. M., Heinrich, B. & Marzluff, C. S. (1996). Raven roosts are mobile information centres. Animal Behaviour, 51, 89–103.

Marzluff, J. M., Bowman, R. & Donnelly, R. (2001). Avian Ecology and Conservation: An Urbanizing World. Boston: Kluwer Academic Publishers.

Marzluff, J. M., Knick, S. T., Vekasy, M. S., Schueck, L. S. & Thomas, J. Z. (1997). Spatial use and habitat selection of Golden eagles in southwestern Idaho. The Auk, 114(4), 673–687. https://doi.org/10.2307/4089287

Marzluff, J., Vekasy, M., Kochert, M. & Steenhof, K. (1997). Productivity of Golden eagles wearing backpack radio transmitters. Journal of Raptor Research, 31(3), 224–227.

Mason, J. R. (2000). Golden eagle attacks and kills adult male coyote. Journal of Raptor Research, 34(3), 244–245. Retrieved from https://sora.unm.edu/sites/default/files/ journals/jrr/v034n03/p002 44-p00245.pdf

Mason, N. W. H., de Bello, F., Doležal, J. & Lepš, J. (2011). Niche overlap reveals the effects of competition, disturbance and contrasting assembly processes in experimental grassland communities. Journal of Ecology, 99(3), 788–796. https://doi.org/10.1111/j.1365-2745.2011.01801.x

Masterov, V. (1992). Ecological energetics and interspecies relationships of Sea eagles H. albicilla and H. pelagicus at Lower Amur and Sakhalin. PhD. Thesis. Moscow State University.

Masterov, V. B. (2000). Postnatal development of Steller's Sea eagles in East Asia. *In*: First symposium on Steller's and White-tailed Sea eagles in East Asia, pp. 17–28. Tokyo, Japan: Wild Bird Society of Japan, Tokyo.

Masterov, V. B. & Zykov, V. B. (1992). The distribution and abundance of Steller's Sea eagles (Haliaeetus pelagicus, Pall.) at the northeast coast of Sakhalin Island. Biologicheske Nauki, 1992, 71–76.

Matchett, M. R. & O'Gara, B. W. (1987). Methods of controlling Golden eagle depredation on domestic sheep in southwestern Montana. Journal of Raptor Research, 21(3), 85–94.

Mateos-Rodríguez, M. & Liechti, F. (2012). How do diurnal long-distance migrants select flight altitude in relation to wind? Behavioral Ecology, 23(2), 403–409. https://doi.org/10.1093/beheco/arr204

Mateo-Tomás, P., Olea, P. P., Sánchez-Barbudo, I. S. & Mateo, R. (2012). Alleviating human-wildlife conflicts: Identifying the causes and mapping the risk of illegal poisoning of wild fauna. Journal of Applied Ecology, 49(2), 376–385. https://doi.org/10.1111/j.1365-2664.2012.02119.x

Mateo-Tomás, P., Olea, P., Jiménez-Moreno, M., Camarero, P. R., Sánchez-Barbudo, I. S., Martín-Doimeadios, R. C. R. & Mateo, R. (2016). Mapping the spatio-temporal risk of lead exposure in apex species for more effective mitigation. Proceedings of the Royal Society of London Series B. https://doi.org/10.1098/rspb.2016.0662.

Matola, S. (2006). The Harpy Eagle (Harpia harpyja) restored to former Central American range. Oryx, 40, 12–17.

Matsyna, A. I. & Matsyna, E. L. (2011). Protection of birds on the power lines in Russia. Poster presented on International Conference on Power Lines and Bird Mortality in Europe, Budapest.

Matthews, A., Ruykys, L., Ellis, W., Fitzgibbon, S., Lunney, D., Crowther, M., Glen, A., Purcell, B., Moseby, K., Stott, J., Fletcher, D., Wimpenny, C., Allen, B., van Bommel, L., Roberts, M., Davies, N., Green, K., Newsome, T., Ballard, G. & Wiggins, N. (2013). The success of GPS collar deployments on mammals in Australia. Australian Mammalogy, 35, 65–83. 10.1071/AM12021. https://doi.org/10.1071/AM12021

Matthews, J. C. (1995). British and Irish Mythology. London: Diamond Books.

MauMara (2018). Serengeti (MaMaSe) Sustainable Water Initiative. Map of Maasai Mara Ecosystem.

Maurer, G., Russell, D. G. D., Woog, F. & Cassey, P. (2010). The eggs of the extinct Egyptian population of White-tailed eagle Haliaeetus albicilla. Bulletin of the British Ornithologists' Club, 130(3), 208–214.

May, R. F., Hoel, P. L., Langston, R., Dahl, E. L., Bevanger, K., Reitan, O. & Stokke, B. G. (2010). Collision risk in White-tailed eagles. Modelling collision risk using vantage point observations in Smøla wind-power plant. Report 639. Trondheim: Norsk Institutt for Naturforskning (NINA).

May, R. F., Nygard, T., Dahl, E. L., Reitan, O. & Bevanger, K. M. (2011). Collision risk in White-tailed eagles. Modelling kernel-based collision risk using satellite telemtry data in Smøla wind power plant. [Mitteleuropäische Gesellschaft zurErhaltung der Greifvögel (MEGEG)]. Central European Society for Raptor Protection.

May, R., Bevanger, K., van Dijk, J., Petrin, Z. &Brende, H. (2012). Renewable energy respecting nature. A synthesis of knowledge on environmental impacts of renewable energy financed by the Research Council of Norway [Report].

May, R., Nygård, T., Dahl, E. L. & Bevanger, K. (2013). Habitat utilization in White-tailed eagles (Haliaeetus albicilla) and the displacement impact of the Smøla wind-power plant. Wildlife Society Bulletin, 37(1), 75–83. https://doi.org/10.1002/wsb.264

Maynard, M. C. J. (1881). The Birds of Eastern North America. Newtonville: C. J. Maynard and Company.

Mayor, A. (2014). The Amazons: Lives and Legends of Warrior Women across the Ancient World. Princeton, NJ: Princeton University Press.

Mayor, A. (2016). The eagle huntress: Ancient traditions and new generations. Retrieved from https://web.stanford.edu/dept/HPS/EagleHuntress 2016long.pdf

Mayr, E. (1936). New subspecies of birds from the New Guinea region. American Museum Novitates, 869, 1–7.

Mayr, E. & Gilliard, E. T. (1954). Birds of central New Guinea: Results of the American Museum of Natural History Expedition to New Guinea in 1950 and 1952. Bulletin of the American Museum of Natural History, 103, 317–374.

Mayr, E. & Cottrell, G. W. (Eds.) (1979). Check-list of Birds of the World (2nd ed.), pp. 384–385. Cambridge, MA: Museum of Comparative Zoology.

Mazák, V. (1981). Panthera tigris. Mammalian Species, 152, 1–8. https://doi.org/10.2307/3504004

Mazar Barnett, J. & Kirwan, G. M. (1999). Neotropical notebook. Cotinga, 11, 96–105.

Mazar Barnett, J. & Pearman, M. (2001). Annotated Checklist of the Birds of Argentina. Barcelona: Lynx Edicions.

Mazzucchelli, V. G. (1967). The southern limits of the Mohave Desert, California. California Geographer, 8: 127–133.

McADoo, J. K. & Klebenow, D. A. (1978). Predation on range sheep with no predator control. Journal of Range Management, 31(2), 111–114. https://doi.org/10.2307/3897655

McClelland, B. R., Young, L. S., McClelland, P. T., Crenshaw, J. G., Allen, H. L. & Shea, D. S. (1994). Migration ecology of Bald eagles fro autumn concentrations in Glacier National Park, Montana. Wildlife Monographs 125.
McCollough, M. A., Todd, C. S. & Owen, R. B., Jr. (1994). Supplemental feeding program for wintering Bald eagles in Maine. Wildlife Societ Bulletin, 22(2), 147–154.
McCrary, J. K., Arendt, W. J., Chavarría, L. et al. (2009). A contribution to Nicaraguan ornithology, with a focus on the pine-oak region. Coting 31, 72–78.
McDermid, G. J., Coops, N. C., Wulder, M. A., Franklin, S. E. & Seitz, N. E. (2010). Critical remote sensing contributions to spatial wildli ecological knowledge and management. In: Cushman, S. A., Huettmann, F. (Eds.), Spatial complexity, informatics, and wildlife conservatio Springer, Tokyo. https://doi.org/10.1007/978-4-431-87771-4_11
McEneaney, T. P. & Jenkins, M. A. (1983). Bald eagle predation on domestic sheep. Wilson Bulletin, 95, 694–695.
McEwan, L. C. & Hirth, D. H. (1980). Food habits of the Bald eagle in north-central Florida. The Condor, 82(2), 229–231. Retrieved from http: www.jstor.org/stable/1367485. https://doi.org/10.2307/1367485
McFarland, D. (1987). The Oxford Companion of Animal Behaviour. Oxford: Oxford University Press.
McGahan, J. (1967). Quantified estimates of predation by a Golden eagle population. Journal of Wildlife Management, 31(3), 496–501. https: doi.org/10.2307/3798130
McGahan, J. (1973). Gliding flight of the Andean condor in nature. Journal of Experimental Biology, 58, 225–237.
Mcgarigal, K., Anthony, R. G. & Isaacs, F. B. (1991). Interactions of humans and Bald eagles on the Columbia River Estuary. Wildlife Monograph 115, 3–47.
McGrady, M. J., Grant, J. R., Bainbridge, I. P. & McLeod, D. R. A. (2002). A model of Golden eagle (Aquila chrysaetos) ranging behaviou Journal of Raptor Research, 36, 62–69.
McGrady, M. J., Maechtle, T. L., Vargas, J. J., Seegar, W. S. & Porras Peña, M. C. (2002). Migration and ranging of Peregrine Falcons winterin on the Gulf of Mexico coast, Tamaulipas, Mexico. Condor, 104(1), 39–48. https://doi.org/10.1093/condor/104.1.39
McGrady, M. J., Ueta, M., Potapov, E. R., Utekhina, I., Masterov, V., Ladyguine, A., Zykov, V., Cibor, J., Fuller, M. & Seegar, W. S. (2003 Movements by juvenile and immature Steller's Sea eagles Haliaeetus pelagicus tracked by satellite. Ibis, 145(2), 318–328. https://do org/10.1046/j.1474-919X.2003.00153.x
McGraw, W. S. & Berger, L. R. (2013). Raptors and primate evolution. Evolutionary Anthropology, 22(6), 280–293. https://doi.org/10.100 evan.21378
McGraw, W. S., Cooke, C. & Shultz, S. (2006). Primate remains from African Crowned eagle (Stephanoaetus coronatus) nests in Ivory Coast Tai Forest: Implications for primate predation and early hominid taphonomy in South Africa. American Journal of Physical Anthropolog 131(2), 151–165. https://doi.org/10.1002/ajpa.20420
McGregor, R. C. (1907). Notes on specimens of the Monkey-eating eagle (Pithecophaga jefferyi Grant) from Mindanao and Luzon. Philippin Journal of Science, 2, 297.
McGregor, R. C. (1918). New or noteworthy Philippine birds, II. Philippine Journal of Science, 13, 1–20.
McGregor, R. C. (1921). New or noteworthy Philippine birds, IV. Philippine Journal of Science, 19, 691–703.
McGregor, R. C. (1927). New or noteworthy Philippine birds, V. Philippine Journal of Science, 32, 513–528.
McIntyre, C. L. (2012). Quantifying sources of mortality and wintering ranges of Golden eagles from interior Alaska using banding and satelli tracking. Journal of Raptor Research, 46(1), 129–134. https://doi.org/10.3356/JRR-10-96.1
McIntyre, C. L., Douglas, D. C. & Collopy, M. W. (2008). Movements of Golden eagles (Aquila chrysaetos) from interior Alaska during their fir year of independence. Auk, 125(1), 214–224. https://doi.org/10.1525/auk.2008.125.1.214
McLachlan, G. R. & Liversidge, R. (1978). Roberts birds of South Africa. Cape Town, S.A.: The Trustees of the John Voelcker Bird Book Fund
Mclean, D. D. (1925). The Golden eagle as a deer enemy. Yosemite Nature Notes, 4, 27.
McLeod, D. R. A., Whitfield, D. P. & McGrady, M. J. (2002). Improving prediction of Golden eagle (Aquila chrysaetos) ranging in wester Scotland using GIS and terrain modeling. Journal of Raptor Research, 36, 70–77.
McMaster, M. K. & Downs, C. T. (2009). Home range and daily movement of Leopard tortoises (Stigmochelys pardalis) in the Nama-Karo South Africa. Journal of Herpetology, 43(4), 561–569. https://doi.org/10.1670/07-078.1
McNary, D. (2016). Fox Animation Buys Remake Rights to Daisy Ridley's Documentary "Eagle Huntress" Variety. Retrieved from http://variet com/2016/film/news/daisy-ridley-eaglehuntress-remake-fox-animation-1201750222/
McPherson, S. (2015). Urban ecology of the Crowned eagle Stephanoaetus coronatus in KwaZulu-Natal, South Africa [PhD Thesis]. South Afric University of KwaZulu-Natal.
McPherson, S. C., Brown, M. & Downs, C. T. (2016a). Diet of the crowned eagle (Stephanoaetus coronatus) in an urban landscape: Potential fc human-wildlife conflict? Urban Ecosystems, 19(1), 383–396. https://doi.org/10.1007/s11252-015-0500-6
McPherson, S. C., Brown, M. & Downs, C. T. (2016b). Crowned eagle nest sites in an urban landscape: Requirements of a large eagle in th Durban metropolitan Open Space System. Landscape and Urban Planning, 146, 43–50. https://doi.org/10.1016/j.landurbplan.2015.10.004
McSweeney, R. (2019). Explainer: 'Desertification' and the role of climate change. Retrieved from https://www.carbonbrief.org/explaine desertification-and-the-role-of-climate-change.
Mearns, E. A. (1905). Note on a specimen of Pithecophaga jefferyi Ogilvie-Grant. Proceedings of the Biological Society of Washington, 18, 76–7
Mebs, T. & Schmidt, D. (2006). Die Greifvögel Europas, [Nordafrikas und Vorderasiens]. Kosmos Verlag.
Mech, L. D., Smith, D. W., Murphy, K. M. & MacNulty, D. R. (2001). Winter severity and wolf predation on a formerly wolf-free elk herd. Journ of Wildlife Management, 65, 998–1003.

Mecionis, R. & Jusys, V. (1994). The White-tailed eagle Haliaeetus albicilla at Kursiu Lagoon. Acta Ornithologica Lituanica, 9–10.
Mecionis, R. & Dementavicius, D. (2003). Monitoring of White-tailed eagle in Lithuania (1994–1999). In: Helander, B. (Eds.), Sea eagle 2000: Proceedings from the International Sea Eagle Conference in Bjorko, Sweden, 13–17 September 20. 85–89. Stockholm, Sweden Proceedings of the Swedish Society for Nature Conservation, SNF.
Mee, A., Ohlson, J., Stewart, I., Wilson, M., Örn, P. & Diaz Ferreyra, J. (2002). The Cerros del Sira revisited: Birds of submontane and montane forest. Cotinga, 18, 46–57.
Meensk, C. (1974b). Comparative ecological studies of some East African birds of prey. Bibliotheck 4p. Ardea, Italy: Nederl., da. 1974, 62(1-2), 1–97.
Mees, G. F. (1965). The avifauna of Misool. Nova, Guinea. Zoology, 31, 139–203.
Meinertzhagen, R. (1930). Nicoll's birds of Egypt. London: Hugh Rees Ltd.
Meinertzhagen, R. (1940). How do larger raptorial birds hunt their prey? Ibis, 4, 530–535.
Meinertzhagen, R. (1959). Pirates and Predators: The Piratical and Predatory Habits of Birds. Edinburgh, UK: Oliver and Boyd.
Meir, B. (1988). Nesting survey and breeding success of Short-toed eagles (Circaetus gallicus) in the Judean Plain during 1983–1985. Torgos, 7, 37–50.
Meiri, S. (2010). Length-weight allometries in lizards. Journal of Zoology. https://doi.org/10.1111/j.1469-7998.2010.00696.x
Meiri, S. & Dayan, T. (2003). On the validity of Bergmann's rule. Journal of Biogeography, 30(3), 331–351. https://doi.org/10.1046/j.1365-2699.2003.00837.x
Melcher, C. & Suazo, L. (1999). Raptor electrocutions: The un-necessary losses continue. Journal of the Colorado Field Ornithologists, 33, 221–224. Retrieved from https://birdconservancy.org/wp-content/uploads/2018/10/CO_birds1999v33_news_from_ field.pdf.
Mellone, U., Limiñana, R., Mallia, E. & Urios, V. (2011a). Extremely detoured migration in an inexperienced bird: Interplay of transport costs and social interactions. Journal of Avian Biology, 42(5), 468–472. https://doi.org/10.1111/j.1600-048X.2011.05454.x
Mellone, U., López-López, P., Limiñana, R. & Urios, V. (2011b) Weather conditions promote route flexibility during open ocean crossing in a long-distance migratory raptor. International Journal of Biometeorology, 55(4), 463–468. https://doi.org/10.1007/s00484-010-0368-3
Mellone, U., Yáñez, B., Limiñana, R., Muñoz, A. -R., Pavón, D., González, J.-M. & Ferrer, M. (2011). Summer staging areas of non-breeding Short-toed Snake eagles Circaetus gallicus. Bird Study, 58(4), 516–521. https://doi.org/10.1080/00063657.2011.598914
Mellone, U., Klaassen, R. H. G., Garcia-Ripolléés, C., Limiñana, R., López-López, P., Pavón, D. & Alerstam, T. (2012). Interspecific comparison of the performance of soaring migrants in relation to morphology, meteorological conditions and migration strategies. PLoS One, 7(7), e39833. https://doi.org/10.1371/journal.pone.0039833
Mellone, U., De La Puente, J., López-López, P., Limiñana, R., Bermejo, A. & Urios, V. (2013). Migration routes and wintering areas of Booted eagles Aquila pennata breeding in Spain. Bird Study, 60(3), 409–413. https://doi.org/10.1080/00063657.2013.781113
Mellone, U., De la Puente, J., López-López, P., Limiñana, R., Bermejo, A. & Urios, V. (2015). Seasonal differences in migration patterns of a soaring bird in relation to environmental conditions: A multiscale approach. Behavioral Ecology and Sociobiology, 69(1), 75–82. https://doi.org/10.1007/s00265-014-1818-4.
Mellone, U., Limiñana, R., López-López, P. & Urios, V. (2015). Regional and age-dependent differences in the effect of wind on the migratory routes of Eleonora's falcon. Current Zoology, 61(3), 428–434. https://doi.org/10.1093/czoolo/61.3.428
Mellone, U. (2016). Resolving the puzzle of Short-toed eagle migration. Retrieved from https://www.bou.org.uk/mellone-short-toed-eagle-migration/
Melnikov, V. N., Solovkov, D. A., Kostin, A. B., Egorova, N. A. & Bogomolov, D. V. (2001). Nesting of Lesser Spotted eagle (Aquila pomarina) in Ivanovo Region (Russia). Acta Ornithoecologica 4(2–4), 287–289.
Meloro, C. & Slater, G. (2012). Covariation in the skull modules of cats: The challenge of growing saber-like canines. Journal of Vertebrate Paleontology, 32, 677–685. https://doi.org/10.1080/02724634.2012.649328
Melville, D. (1980). Some observations on birds in Irian Jaya, New Guinea. Emu – Austral Ornithology, 80(2), 89–91. https://doi.org/10.1071/MU9800089a
Menaut, J. C., Gignoux, J., Prado, C. & Clobert, J. (1990). Tree Community Dynamics in a Humid Savanna of the Cote de Ivoire: Modelling the effects of fire and competition with grass and neighbours. Journal of Biogeography, 17, 471–481.
Mendelsohn, H. P. (1975). A ring recovery of a Lesser Spotted eagle Aquila pomarina (Brehm.), Koedoe, 18(1), 193. https://doi.org/10.4102/koedoe.v18i1.922
Mendelsohn, J. (1990). Recovery of colour-marked Martial eagle. Safring News, 19, 16.
Mendelsohn, J. M., Kemp, A. C., Biggs, H. C., Biggs, R. & Brown, C. J. (1989). Wing areas, wing loadings and wing spans of 66 species of African raptors ostrich, 60(1), 35–60. https://doi.org/10.1080/00306525.1989.9634503
Menq, W. (2013). Mimicry in Birds of Prey. Aves de Rapina Brasil.
Menq, W. (2015). Observaçõescomportamentais do gavião-pato Spizaetus melanoleucus Accipitriformes, Accipitridae) no Estado do Paraná, Brasil. Arquivos de Ciências Veterinárias e zoologia da UNIPAR, 18, 175–178.
Meretsky, V. J., Snyder, N. F. R., Beissinger, S. R., Clendenen, D. A. & Wiley, J. W. (2000). Demography of the California Condor: Implications for Reestablishment. Conservation Biology, 14(4), 957–967. https://doi.org/10.1046/j.1523-1739.2000.99113.x
Merom, H. (1960). Birds of Israel. Tel Aviv: Hakibbutz Hameuchad.
Merriam-Webster. (2017). Eagle. Retrieved from https://www.merriam-webster.com/ dictionary/eagle.
Merriam-Webster. (2018). Biotelemetry. Retrieved from https://www.merriam-webster.com/ dictionary/biotelemetry.
Merriam Webster. (2020). Medical Dictionary. Carcinogenic. Retrieved from https://www.merriam-webster.com/dictionary/carcinogenic

Merriam Webster. (2020). Medical Dictionary. Hemotoxic. Retrieved from https://www.merriam-webster.com/medical/hemotoxic.

Mersmann, T. J. (1989). Foraging ecology of Bald eagles on the northern Chesapeake Bay with an examination of techniques used in the study o: Bald eagle food habits. Master's Thesis. Blacksburg, VA: Virginia Polytechnic Institute and State University.

Meyburg, B. & Meyburg, C. (2005). Tracking the endangered Greater Spotted Eagle. Tracker News.

Meyburg, B.-U. (1970). On the biology of the Lesser Spotted eagle (Aquila pomarina). Jahrbuch Deutscher Falkenorden, 1969, 32–66.

Meyburg, B.-U. (1973). Studies of less familiar birds: Lesser Spotted eagle. British Birds, 66, 439–447.

Meyburg, B.-U. (1974). On the breeding biology and taxonomic position of the spotted eagles. Falke, 21, 126–134.

Meyburg, B.-U. (1974). Sibling aggression and mortality among nestling eagles. Ibis, 116(2), 224–228. https://doi.org/10.1111/j.1474-919X.1974 tb00243.x

Meyburg, B. U. (1975). On the biology of the Spanish Imperial eagle (Aquila heliacaadalberti). Ardeola, 21, 245–283.

Meyburg, B. U. & Langrand, O. (1985). Conservation problems regarding birds of prey in Madagascar. Berlin: World Working Group on Birds of Prey.

Meyburg, B.-U. (1986). Threatened and near-threatened diurnal birds of prey of the world. Birds of Prey Bulletin, 3, 1–12.

Meyburg, B.-U. & Meyburg, C. (1988). The Lesser Spotted eagle (Aquila pomarina) as an alpine hunter in the Caucasus. Journal für Ornithologie 129(1), 102–103. https://doi.org/10.1007/BF01641537

Meyburg, B.-U., van Balen, S., Thiollay, J. M. & Meybrug, C. (1989). Observations on the endangered Java hawk eagle Spizaetus bartelsi. *In* Meyburg, B.-U. & Chancellor, R. D. (Eds.), Raptors in the modern world. Proceedings of the 3rd World Conference on Birds of Prey anc Owls, pp. 279–300. Sussex: Pica Press.

Meyburg, B.-U. (1991). Der Schreiadler (Aquila pomarina), Bisherige und zukünftige Bemühungen um seine Erforschung und seinem Schutz Populationsökologie Greifvogel- und Eulenarten, 2, 89–105.

Meyburg, B.-U. & Pielowski, Z. (1991). Cainism in the Greater Spotted eagle Aquila clanga. Birds of Prey Bulletin 4, 143–148.

Meyburg, B.-U., Scheller, W. & Meyburg, C. (1993). Satelliten Telemetriebeieinemjuvenilen Schreiadler (Aquila pomarina) auf dem Herbstzug Journal of Ornithology, 134(2), 173–179. https://doi.org/10.1007/BF01640086

Meyburg, B.-U., Scheller, W. & Meyburg, C. (1993). Zug und Überwinterung des Schreiadlers Aquila pomarina: Satellitentelemetrische Untersuchungen. Journal of Ornithology, 136, 401–422.

Meyburg, B.-U. (1994). Aquila clanga. *In*: Del Hoyo, J., Elliot, A. & Sargatal, J. (Eds.), Handbook of the birds of the world, 2. New World Vultures to Guinea Fowl. Barcelona: Lynx Edicions.

Meyburg, B.-U. (1994). Steppe eagle Aquila nipalensis. *In*: Del Hoyo, J., Elliott, A. & Sargatal, J. (Eds.), Handbook of the birds of the world, 2 p. 194. Barcelona: Lynx Edicions.

Meyburg, B. U. (1994). Imperial eagles. *In*: Del Hoyo, J., Elliott, A. & Sargatal, J. (Eds.), Handbook of birds of the world, 2: New World Vultures to Guineafowl: 194–195. Barcelona: Lynx Edicions.

Meyburg, B. U. (1994). Lesser Spotted eagle. *In*: Del Hoyo, J., Elliott, A. & Sargatal, J. (Eds.), Handbook of birds of the world, 2. New Worlc Vultures to Guineafowl, 192–193. Barcelona: Lynx Edicions.

Meyburg, B.-U., Eichaker, X., Meyburg, C. & Paillat, P. (1995). Migrations of an adult Spotted eagle tracked by satellite. British Birds, 88 357–361.

Meyburg, B.-U., Mendelsohn, J. M., Ellis, D. H., Smith, D. G., Meyburg, C. & Kemp. A. C. (1995). Year-round movements of a Wahlberg's eagle Aquila wahlbergi tracked by satellite. Ostrich, 66(4), 135–140. https://doi.org/10.1080/00306525.1995.9632720

Meyburg, B.-U. (1996). The Lesser Spotted eagle Aquila pomarina: Population status and present state of research. *In*: Chancellor, R. D. & Meyburg, B.-U (Eds.), Eagle studies, 377–387. Berlin. World Working Group on Birds of Prey.

Meyburg, B.-U., Meyburg, C. & Pacteau, C. (1996). Migration automnale d'un Circaète Jean-Le-Blanc Circaetus gallicus suivi par satellite Alauda, 64, 339–344.

Meyburg, B.-U., Haraszthy, L., Strazds, M. & Schäffer, N. (1997). European Union species action plan for Greater Spotted eagle (Aquila clanga) *In*: Species Action Plans for 8 European Threatened Bird species. Sandy, Bedfordshire: RSPB.

Meyburg, B.-U., Meyburg, C. & Barbraud, J. C. (1998). Migration strategies of an adult Short-toed eagle Circaetus gallicus tracked by satellite Alauda, 66(1), 39–48.

Meyburg, B.-U., Meyburg, C., Mizera, T., Maciorowski, G. & Kowalski, J. (1998). Greater Spotted eagle wintering in Zambia. Africa-Birds anc Birding, 3, 62–68.

Meyburg, B.-U., Haraszthy, L., Strazds, M. & Schäffer, N. (1999). European species action plan for Greater Spotted eagle (Aquila clanga). *In* Schäffer, N. & Gallo-Orsi, U. (Eds.), European Union action plans for eight priority bird species, 1–24. Luxembourg: Office for Officia Publications of the European Communities.

Meyburg, B.-U. & Meyburg, C. (2000). Greifvogel-Monitoring mittels Satelliten-Telemetrie. Populationsökologie Greifvogel- und Eulenarten, 4 33–49. Martin: Luther University Halle-Wittenberg &Wiss. Beitr, 4/2000, 33–49.

Meyburg, B.-U., Scheller, W. & Meyburg, C. (2000). Migration and wintering of the Lesser Spotted eagle Aquila pomarina: A study by means o: satellite telemetry. Global Environmental Research, 4, 183–193.

Meyburg, B.-U., Haraszthy, L., Strazds, M. & Schäffer, N. (2001). European species action plan for Lesser Spotted eagle (Aquila pomarina). *In* Schäffer, N. & Gallo-Orsi, U. (Eds.), European Union action plans for eight priority bird species, 1–24. Luxembourg: Office for Officia Publications of the European Communities.

Meyburg, B.-U., Haraszthy, L. & Strazds, M. (2001). European species action plan for Greater Spotted eagle. *In*: Schääffer, N. & Gallo-Orsi, U (Eds.), European Union action plans for eight priority bird species, European Commission, Luxembourg.

Meyburg, B., Ellis, D. H., Meyburg, C., Mendelsohn, J. M. & Scheller, W. (2001). Satellite tracking of two Lesser Spotted Eagles, Aquila pomarina, migrating from Namibia. Ostrich, 72(1–2), 35–40. https://doi.org/10.2989/00306520109485281

Meyburg, B.-U., Matthes, J. & Meyburg, C. (2002). Satellite-tracked Lesser Spotted eagle avoids crossing water at the Gulf of Suez. British Birds, 95, 372–376.

Meyburg, B.-U., Paillat, P. & Meyburg, C. (2003). Migration routes of Steppe Eagles between Asia and Africa: A study by means of satellite telemetry. Condor, 105(2), 219–227. https://doi.org/10.1093/condor/105.2.219

Meyburg, B.-U., Scheller, W. & Bergmanis, U. (2004). Home range size, habitat utilisation, hunting and time budgets of Lesser Spotted eagles Aquila pomarina with regard to disturbance and landscape fragmentation. *In*: Chancellor, R. D. & Meyburg, B.-U. (Eds.), Raptors Worldwide, 615–635. Berlin and Budapest: World Working Group on Birds of Prey/MME-BirdLife Hungary.

Meyburg, B.-U., Meyburg, C., Bělka, T., Šreibr, O. & Vrana, J. (2004a). Migration, wintering and breeding of a Lesser Spotted eagle (Aquila pomarina) from Slovakia tracked by satellite. Journal of Ornithology, 145(1), 1–7. https://doi.org/10.1007/s10336-003-0014-5

Meyburg, B.-U., Mizera, T., Matthes, J., Graszynski, K., Scwanbeck, J. P., Maciorowski, G. & Scheller, W. (2005). Hybridisation of Greater (Aquila clanga) and Lesser Spotted eagle (A. pomarina) in Poland and Germany. *In*: Mizera, T. & Meyburg, B.-U. (Eds.), Proceedings of an International Symposium, International Meeting on Spotted Eagles. Osowiec: Biebrza National Park, 115-118.

Meyburg, B.-U., Meyburg, C., Matthes, J. & Matthes, H. (2006). GPS-Satelliten-Telemetriebem Schreiadler Aquila pomarina: Aktionsraum und Territorialverhaltenim Brutgebiet. die Vogelwelt, 127, 127–144.

Meyburg, B.-U., Meyburg, C., Matthes, J. & Matthes, H. (2007). Spring migration, late arrival, temporary change of partner and breeding success in the Lesser Spotted eagle Aquila pomarina. die Vogelwelt, 128, 21–31.

Meyburg, B.-U. & Meyburg, C. (2007). Post-fledging behavior and outward migration of a hybrid Greater × Lesser Spotted eagle (Aquila clanga × A. pomarina). Journal of Raptor Research 41(2), 165–170. https://doi.org/10.3356/0892-1016(2007)41165

Meyburg, B.-U., Meyburg, C. & Franck-Neumann, F. (2007). Why do female Lesser Spotted Eagles (Aquila pomarina) visit strange nests remote from their own? Journal of Ornithology, 148(2), 157–166. https://doi.org/10.1007/s10336-006-0113-1

Meyburg, B.-U., Graszynski, K., Langgemach, T., Sömmer, P. & Bergmanis, U. (2008). Cainism, nestling management in Germany in 2004–2007 and satellite tracking of juveniles in the Lesser Spotted eagle Aquila pomarina. Slovak Raptors Journals, 2, 53–72.

Meyburg, B.-U., Bergmanis, U., Graszynski, K., Langgemach, T., . . . P. (2011). Experience with the hacking method for support of the Lesser Spotted eagle (Aquila pomarina) population in Brandenburg (Germany). Poster on 8th Conference of the European Ornithologists' Union, Riga, Latvia.

Meyburg, B. U. & Boesman, P. (2013). Steppe eagle (Aquila nipalensis). *In*: Del Hoyo, J., Elliott, A., Sargatal, J. Christie, D. A. & De Juana, E. (Eds.), Handbook of the birds of the world alive. Barcelona: Lynx Edicions.

Meyburg, B. U., Boesman, P. & Marks, J. S. (2014). Lesser Spotted eagle (Clanga pomarina). *In*: Del Hoyo, J., Elliott, A., Sargatal, J., Christie, D. A. & De Juana, E. (Eds.), Handbook of the birds of the world alive. Barcelona: Lynx Edicions.

Meyburg, B. U., Boesman, P., Marks, J. S. & Kirwan, G. M. (2016). Lesser Spotted eagle (Clanga pomarina). *In*: Del Hoyo, J., Elliott, A., Sargatal, J., Christie, D. A. & De Juana, E. (Eds.), Handbook of the birds of the world alive. Barcelona: Lynx Edicions.

Meyburg, B.-U., Boesman, P. F. D., Marks, J. S. & Kirwan, G. M. (2020). Lesser Spotted eagle (Clanga pomarina), version 1.0. *In*: Del Hoyo, J., Elliott, A., Sargatal, J., Christie, D. A. and De Juana, E. (Eds.), Birds of the World. Ithaca, NY: Cornell Laboratory of Ornithology. https://doi.org/10.2173/bow.leseag1.01.

Meyburg, B.-U., Kirwan, G. M. & Garcia, E. F. J. (2020). Greater Spotted eagle (Clanga clanga), version 1.0. *In*: Del Hoyo et al. (Eds.), Birds of the World. Ithaca, NY: Cornell Laboratory of Ornithology. https://doi.org/10.2173/bow.grseag1.01.

Meyer de Schauensee, R. (1966). The species of birds of South America and their distribution. Narberth, Edinburgh, PA: Livingstone Publishing Company.

Meyer de Schauensee, R. & Phelps, W. H., Jr. (1978). A guide to the birds of Venezuela. Princeton, NJ: Princeton University Press, Princeton, NJ.

Meyer, A. (1994). Shortcomings of the cytochrome b gene as a molecular marker. Trends in Ecology and Evolution, 9(8), 278–280. https://doi.org/10.1016/0169-5347(94)90028-0

Meyer, A. B. & Wigglesworth, L. W. (1898). The birds of Celebes and neighbouring islands. Two volumes. Berlin: Friedländer.

Meyer de Schauensee, R. (1949). The birds of the republic of Colombia: Segunda entrega (accipitridae-picidae).

Meyer, S. K., Spaar, R. & Bruderer, B. (2000). To cross the sea or to follow the coast? Flight directions and behaviour of migrating raptors approaching the Mediterranean Sea in autumn. Behaviour, 137(3), 379–399. https://doi.org/10.1163/156853900502132.

Michev, T. & Petrov. (1979). On the distribution of the Imperial eagle (Aquila heliaca Savigny) in Bulgaria. Journal of the Museums of South Bulgaria, 5, 65–77.

Middleton, H. A., Butler, R. W. & Davidson, P. (2018). Waterbirds alter their distribution and behavior in the presence of Bald eagles (Haliaeetus leucocephalus). Northwestern Naturalist, 99(1), 21–30. https://doi.org/10.1898/NWN16-21.1.

Mijele, D. (2009). Incidences of poisoning of vultures and lions in the Masai Maramara National Reserve. Kenya Wildlife Service, Masai Mara Veterinary Report. Nairobi.

Mikami, O. K. & Kawata, M. (2004). Does interspecific territoriality reflect the intensity of ecological interactions? A theoretical model for interspecific territoriality. Evolutionary Ecological Research, 6, 765–775.

Mikhelson, K. A. (1982). Steppe eagle, Aquila nipalensis. *In*: Mikhelson, K. A., Viksne, Y. A. (Eds.), Migrations of birds of eastern Europe and northern Asia, 92–95. Falconiformes-Gruiformes. Moscow: Nauka Press.

Mikoyan, L. (2004). Vegetation study in the home range of Javan Hawk-eagle at TelagaWarna Nature Reserve, Tugu Village, Bogor, West Java Honors Thesis. Jakarta: Faculty of Biology. Jakarta, Indonesia: National University.

Mikula, P., Morelli, F., Lučan, R. K., Jones, D. N. & Tryjanowski, P. (2016). Bats as prey of diurnal birds: A global perspective. Mammal Reviev 46(3), 160–174. https://doi.org/10.1111/mam.12060
Miller, B. W. & Miller, C. M. (1998). Ornithology in Belize since 1960. Wilson Bulletin, 110, 544–558.
Miller, K. V. & Conroy, M. J. (1990). SPOT satellite imagery for mapping Kirtland's Warbler wintering habitat in the Bahamas. Wildlife Societ Bulletin, 18, 252–257.
Miller, M. J. R., Wayland, M. E., Dzus, E. H. & Bortolotti, G. R. (2000). Availability and ingestion of lead shotshell pellets by migrant Bald eagle in Saskatchewan. Journal of Raptor Research, 34, 167–174.
Miller, M. J. R., Wayland, M. E. & Bortolotti, G. R. (2001) Exposure of migrant Bald eagles to lead in prairie Canada. Environmental Pollutio 112(2), 153–162. https://doi.org/10.1016/S0269-7491(00)00122-6
Miller, M., Pfeiffer, W. & Schwartz, T. (2010). Creating the CIPRES Science Gateway for inference of large phylogenetic trees. Gatewa Computing Environments Workshop (GCE), 14 November 2010, 1–7. https://doi.org/10.1109/GCE.2010.5676129.
Mills, G. & Hofer, H. (1998). Hyaenas. Gland: IUCN/SSC Hyaena Specialist Group.
Mills, M. G. L. & Biggs, H. C. (1993). Prey apportionment and related ecological relationships between large carnivores in Kruger National Par *In*: Symposia of the Zoological Society of London, 65.
Millsap, B. A., Zimmerman, G. S., Sauer, J. R., Nielson, R. M. Y. A. . . . (2013). Golden eagle population trends in the western United State 1968–2010. Journal of Wildlife Management, 77(7), 1436–1448. https://doi.org/10.1002/jwmg.588
Millsap, B., Breen, T., Mcconnell, E., Steffer, T., Phillips, L., Douglass, N. & Taylor, S. (2004). Comparative fecundity and survival of Bald eagle fledged from suburban and rural natal areas in Florida. Journal of Wildlife Management, 68(4), 1018–1031. https://doi.org/10.2193/002 541X(2004)0681018
Milon, P., Petter, P. & Randrianasolo, J.-J. (1973). Faune de madagascar, 35. Oiseaux. [TAnarive and Paris: ORSTOM and CRNS].
Ministerio de Medio Ambiente MMA. (2004). National Forestry Inventory. Dirección General para la Biodiversidad. Madrid: Ministerio de Medi Ambiente, Madrid.
Ministerio de Medio Ambiente MMA. (2001). National strategy for the conservation of the Spanish Imperial eagle (Aquila adalberti). Direcció General para la Biodiversidad. Madrid: Ministerio de Medio Ambiente, Madrid.
Miranda, E. B. P. (2018). Prey composition of Harpy eagles (Harpia harpyja) in Raleighvallen, Suriname. Tropical Conservation Science, 11, 1– https://doi.org/10.1177/1940082918800789.
Miranda, E. B. P., Campbell-Thompson, E., Muela, A. & Vargas, F. H. (2018). Sex and breeding status affect prey composition of Harpy eagle Harpia harpyja. Journal of Ornithology, 159(1), 141–150. https://doi.org/10.1007/s10336-017-1482-3
Miranda, E. P. B. (2015). Conservation implications of Harpy eagle Harpia harpyja predation patterns. Endangered Species Research, 29(1 69–79. https://doi.org/10.3354/esr00700
Miranda, H. C., Jr., Salvador, D. I., Ibanez, J. C. & Ibanez, G. B. (2000). Summary of Philippine eagle Pithecophaga jefferyi breeding succes 1978–1998. Journal of Raptor Research, 34, 37–41.
Miranda, H. C., Jr., Salvador, D. J. I. & Bueser, G. L. L. (2008). Updates on the Nesting Biology of Population Status of the Philippine Eag Pithecophaga jefferyi. Davao City.
Miranda, J. M. D., Bernardi, I. P., Moro-Rios, R. F. & Passos, F. C. (2006). Antipredator behavior of Brown Howlers attacked by Black Hawk eagl in southern Brazil. International Journal of Primatology, 27(4), 1097–1101. https://doi.org/10.1007/s10764-006-9062-z
Mirski, P. (2012). European Greater Spotted eagle Species Action Plan and the need for an update. Conservation of the Greater Spotted Eagl Goniądz, Poland, 23.
Misztal-Szkudlińska, M., Szefer, P., Konieczka, P. & Namieśnik, J. (2011). Biomagnification of mercury in trophic relation of Great Cormorar (Phalacrocorax carbo) and fish in the Vistula Lagoon, Poland. Environmental Monitoring and Assessment, 176(1–4), 439–449. https://do org/10.1007/s10661-010-1595-0
Mistry, J. (2000). World Savannas: Ecology and Human Use. London: Prentice Hall.
Mitani, J. C., Sanders, W. J., Lwanga, J. S. & Windfelder, T. L. (2001). Predatory behavior of Crowned hawk-eagles (Stephanoaetus coronatus) i Kibale National Park, Uganda. Behavioral Ecology and Sociobiology, 49(2–3), 187–195. https://doi.org/10.1007/s002650000283
Mitchell, E. B. (1962). The Art and Practice of Hawking. Boston, MA: Charles E. Branford.
Mitchell, J. C. M. (1965). Predation on large mammals in the Kafue National Park, Zambia. African Zoology, 1(2), 297–318.
Mittermeier, J. C., Cottee-Jones, H. E. W., Purba, E. C., Ashuri, N. M., Hesdianti, E. & Supriatna, J. (2013). A survey of the avifauna of Obi islan north Moluccas, Indonesia. Forktail, 29, 128–137.
Mizera, T., Kasperczyk, B. & Bobrowicz, D. (1990). Clutch of three in a White-tailed eagle (Haliaeetus albicilla). Ornithologische Mitteilunger 42, 12–15.
Mizera, T. (2003). White-tailed eagle in Poland. In: Helander, B. (Ed.), Sea eagle 2000: Proceedings from the International Sea Eagle Conferenc in Bjorko, Sweden, 13–17 September 20, 79–83. Stockholm, Sweden Proceedings of the Swedish Society for Nature Conservation, SN.
Mizera, T. & Meyburg, B.-U. (2005). Biebrza National Park, Osowiec, Poland. Proceedings of the International Symposium. International Meetin on Spotted eagles.
Mizera, T. & Sielicki, J. (2009). Breeding status of the Peregrine falcon in Poland during the pre- and post-DDT era (pp. 153–168). Poznań: Turu Poznań University of Life Sciences Press, Warsaw.
Mlíkovský, J. (2009). The food of the White-tailed Sea eagle (Haliaeetus albicilla) at Lake Baikal, East Siberia. Slovak Raptor Journal, 3(1 35–39. https://doi.org/10.2478/v10262-012-0031-5
Mo, M., Wrightson, C., Waterhouse, D. R., Hayler, P. & Hayler, A. (2017). Predation by White-bellied sea eagles Haliaeetus leucogaster on bird Victorian Naturalist, 134(2), 43–47.

Moccia, J. & Arapogianni, A. (2011). Pure Power: Wind Energy Targets for 2020 and 2030. Brussels: European Wind Energy Association.

Mock, D. W. & Parker, G. A. (1997). The evolution of sibling rivalry. Oxford: Oxford University Press.

Mockrin, M. (2020). Southeastern Australia. Retrieved from https://www.worldwildlife.org/ecoregions/aa1001

Modha, K. L. & Eltringham, S. K. (1976) Population ecology of the Uganda kob (Adenotakob thomasi, Neumann) in relation to the territorial system in the Rwenzori National Park, Uganda. Journal of Applied Ecology, 13(2), 453–473. https://doi.org/10.2307/2401795

Moe, S. R., Rutina, L. P. & Du Toit, J. T. (2007). Trade-off between resource seasonality and predation risk explains reproductive chronology in impala. Journal of Zoology, 273(3), 237–243. https://doi.org/10.1111/j.1469-7998.2007.00319.x

Moehrenschlager, A., List, R. & Macdonald, D. W. (2007). Escaping Intraguild predation: Mexican Kit foxes survive while coyotes and Golden eagles kill Canadian swift foxes. Journal of Mammalogy, 88(4), 1029–1039. https://doi.org/10.1644/06-MAMM-A-159R.1.

Mojica, E. K., Meyers, J. M., Millsap, B. A. & Haley, K. L. (2008). Migration of Florida sub-adult Bald eagles. The Wilson Journal of Ornithology, 120(2), 304–310. https://doi.org/10.1676/07-079.1

Mojica, E. K., Watts, B. D., Paul, J. T., Voss, S. T. & Pottie, J. (2009). Factors contributing to Bald eagle electrocutions and line collisions on Aberdeen Proving Ground, Maryland MD. Journal of Raptor Research 43(1), 57–61. https://doi.org/10.3356/JRR-07-60.1

Mojica, E. K., Watts, B. D. & Turrin, C. L. (2016). Utilization probability map for migrating Bald eagles in Northeastern North America: A tool for siting wind energy facilities and other flight hazards. PLoS One, 11(6), e0157807. https://doi.org/10.1371/journal.pone.0157807

Mojica, L. (2012). Tracking the Bald eagle with 70 g GSM/GPS. Transmitters. Retrieved from http://www.microwavetelemetry.com/Spring2012_newsletter_website_Page_4.pdf

Moleón, M., Sánchez-Zapata, J. A., Selva, N., Donázar, J. A. & Owen-Smith, N. (2014). Inter-specific interactions linking predation and scavenging in terrestrial vertebrate assemblages. Biological Reviews of the Cambridge Philosophical Society, 89(4), 1042–1054. https://doi.org/10.1111/brv.12097

Mollhagen, T. R., Wiley, R. W. & Packard, R. L. (1972). Prey remains in Golden eagle nests: Texas and New Mexico. Journal of Wildlife Management, 36(3), 784–792. https://doi.org/10.2307/3799431

Monadjem, A., Boycott, R., Parker, V. & Culverwell, J. (2003). Threatened Vertebrates of Swaziland. Swaziland Red Data Book: Fishes, Amphibians, Reptiles, Birds and Mammals. Environmental Communication. Mbabane, Swaziland: Ministry of Tourism.

Monadjem, A., Anderson, M. D., Piper, S. E. & Boshoff, A. F. (Eds.). (2004). The vultures of Southern Africa—Ouovadis? 28–33. Johannesburg, SA: Birds of Prey Working Group.

Monadjem, A. & Rasmussen, M. W. (2008). Nest distribution and conservation status of eagles, selected hawks and owls in Swaziland. Gabar, 19, 1–22.

Monirul, M. & Khan H. (2009). Sundarbans: Bangladesh's unique birding hotspot. BirdingASIA, 12, 46–52.

Monroe, B. L., Jr. (1968). A distributional survey of the birds of Honduras. Ornithological Monographs, 7(7), 1–458. https://doi.org/10.2307/40168043

Monroy-Ojeda, A., Wolf, K., Martínez-Leyva, E., Goodrich, L., Mesa, R. R., Domínguez, I., Lara, G. R., Hoffman, S. W. & Rojas, E. P. (2016). Neotropical records of Nearctic Raptors: Observational data from Veracruz, Mexico. The Wilson Journal of Ornithology, 128, 631–637. https://doi.org/10.1676/1559-4491-128).3.631

Monsalvo, J. A. B., Heming, N. M. & Marini, M. Â. (2018). Breeding biology of Neotropical Accipitriformes: Current knowledge and research priorities. Revista Brasileira de Ornitologia, 26(2), 151–186. https://doi.org/10.1007/BF03544426

Montalvo, C. I., Fernández, F. J., Galmes, M. A., Santillán, M. A. & Cereghetti, J. (2015). Crowned Solitary eagle (Buteogallus coronatus) as accumulator of armadillo osteoderms in the archaeological record? An actualistic taphonomic study for central Argentina. Quaternary International, 391, 90–99. https://doi.org/10.1016/j.quaint.2015.08.017

Montejo Díaz, J. & Ruelas Inzunza, E. (1997). Notes on spring migration in the Pacific and Atlantic slopes of Guatemala, with observations on Wood Storks and Laughing Gulls. Hawk Migration Association of North America, Hawk. Migration Studies, 22(2), 6–8.

Mooney, N. (1988). Tasmania area co-ordinators report for 1987. Australasian Raptor Association News, 9, 24.

Mooney, N. J. (1997). The Wedge-tailed eagle recovery plan [Tasmania, Australia]. Hobart, Tasmania, Australia: Tasmanian Parks and Wildlife Service, 1998–2003.

Moore, D. R., Teed, R. S., Rodney, S. I., Thompson, R. P. & Fischer, D. L. (2010). Refined avian risk assessment for aldicarb in the United States. Integrated Environmental Assessment and Management, 6(1), 83–101. https://doi.org/10.1897/IEAM_2009-022.1

Moores, N., Ha, J.-M. & Seo, H.-M. (2018). The Birds, Korea Checklist. Yongdangdong: Birds, Korea.

Mooring, M. S. (1993). Predation on a newborn impala by a Martial eagle. Ostrich, 64(4), 185–189. https://doi.org/10.1080/00306525.1993.9632664

Morales, J. M., Moorcroft, P. R., Matthiopoulos, J., Frair, J. L., Kie, J. G., Powell, R. A. et al. (2010). Building the bridge between animal movement and population dynamics. Philosophical Transactions of the Royal Society of London Series B, 365(1550), 2289–2301. https://doi.org/10.1098/rstb.2010.0082

Morandini, V. & Ferrer, M. (2017). How to plan reintroductions of long-lived birds. PloS One, 12(4), e0174186. https://doi.org/10.1371/journal.pone.0174186

Morandini, V. & Ferrer, M. (2015). Sibling aggression and brood reduction: A review. Ethology, Ecology and Evolution, 27(1), 2–16. https://doi.org/10.1080/03949370.2014.880161

Moreno, F. P. (1997). Viaje a la Patagonia Austral. Editorial Elefante Blanco, Buenos Aires.

Moreno-Opo, R., Trujillano, A. & Margalida, A. (2016). Behavioral coexistence and feeding efficiency drive niche partitioning in European avian scavengers. Behavioral Ecology, 27(4), 1041–1052. https://doi.org/10.1093/beheco/arw010

Moreno-Opo, R., Trujillano, A., Arredondo, Á, González, L. M. & Margalida, A. (2015). Manipulating size, amount and appearance of food inputs to optimize supplementary feeding programs for European vultures. Biological Conservation, 181, 27–35. https://doi.org/10.1016/j.biocon.2014.10.022

Mori, S. (1980). Breeding biology of the White-tailed eagle Haliaeetus albicilla in Hokkaido, Japan. Japanese Journal of Ornithology, 29(2–3), 47–68. https://doi.org/10.3838/jjo1915.29.47

Morin, P. J. (1999). Community ecology. Malden, MA: Blackwell Publishing Science, Inc.

Morris, H. (2017). Plane makes emergency landing after freak collision with both an eagle and a rabbit. Retrieved from https://www.telegraph.co.uk/travel/news/plane-hits-eagle-carrying-rabbit-strike/

Morris, P. & Hawkins, F. (1998). Birds of Madagascar: A photographic guide. Robertsbridge, UK: Pica Press.

Mortimore, M. (1989). Adapting to drought: Farmers, famines, and desertification in West Africa. Cambridge, UK: Cambridge University Press.

Morzillo, A. T., De Beurs, K. M. & Martin-Mikle, C. J. (2014). A conceptual framework to evaluate human-wildlife interactions within coupled human and natural systems. Ecology and Society, 19(3), 44. Retrieved from http://dx.doi.org/10.5751/ES-06883-190344

Mosbergen, D. (2014). This 13-year-old Girl Hunts with Golden Eagles. April 17–21. Huffington Post. Retrieved from http://www.huffingtonpost.com/2014/04/17/13-year-old-eagle-huntressmongolia_n_5162623.html.

Moss, E. H. R., Hipkiss, T., Ecke, F., Dettki, H., Sandström, P. & Bloom, P. H. (2014). Home-range size and examples of post-nesting movements for adult Golden eagles (Aquila chrysaetos) in Boreal Sweden. Journal of Raptor Research, 48(2), 93–105. https://doi.org/10.3356/JRR-13-00044.1

Mountfort, G. (1958). Portrait of a Wilderness. London: Hutchinson and Co.

Moura, N. G., Lees, A. C., Aleixo, A., Barlow, J., Dantas, S. M., Ferreira, J., . . . Gardner, T. A. (2014). Two hundred years of local avian extinctions in Eastern Amazonia. Conservation Biology, 28(5), 1271–1281. https://doi.org/10.1111/cobi.12300

Mourer-Chauvire, C. (1982). Les oiseauxfossiles des phosphorites du Quercy. [Implications paleobiogeographiques]. Geobios Memoires Speciale, 6, 413–426.

Mu, M. (2015). Multi-model inference. R package version 1.15. 1. United States USGS Geological Survey. Global Land Survey Digital Elevation Model (GLSDEM). College Park, Maryland 2008.

Muchtar, M. & Setiawan, I. (2005). Raptor research and conservation in Indonesia: Recent initiatives. *In*: J. Abu et al. (Eds.), Proceedings of the 4th Symposium on Asian Raptors, pp. 121–134. Committee for the 4th Symposium on Asian Raptors. Kuala Lumpur, Malaysia.

Mucina, L. & Rutherford, M. C. (2006). The Vegetation of South Africa, Lesotho and Swaziland. Pretoria: South African National Biodiversity Institute.

Mudd, G. M. (2010). Global trends and environmental issues in nickel mining: Sulfides versus laterites. Ore Geology Reviews, 38(1–2), 9–26. https://doi.org/10.1016/j.oregeorev.2010.05.003

Mudrooroo. (1994). Aboriginal mythology: An A-Z spanning the history of the Australian Aboriginal people from the earliest legends to the present day. London: Thorsons.

Muela, A., Curti, M. & Campbell, E. (2007). Time to independence, diet, and dispersion patterns of captive-bred and released Harpy eagles (Harpia harpyja). In: Bildstein, K. L., Barber, D. R. & Zimmerman, A. (Eds.), Neotropical raptors, pp. 316–317. Orwigsburg: PA Hawk Mountain Sanctuary.

Mueller, H. C. & Berger, D. D. (1967). Fall migration of Sharp-shinned Hawks. Wilson Bulletin, 79, 397–415.

Mulero-Pázmány, M., Stolper, R., van Essen, L. D., Negro, J. J. & Sassen, T. (2014). Remotely piloted aircraft systems as a rhinoceros anti-poaching tool in Africa. PloS One, 9(1), e83873. https://doi.org/10.1371/journal.pone.0083873

Mullarney, K., Svensson, L., Zetterstrom, D. & Grant, P. J. (1999). Birds of Europe. Princeton, NJ: Princeton University Press.

Müller, K., Altenkamp, R. & Brunnberg, L. (2007). Morbidity of free-ranging White-tailed Sea Eagles (Haliaeetus albicilla) in Germany. Journal of Avian Medicine and Surgery, 21(4), 265–274. https://doi.org/10.1647/2007-001R.1

Muller-Schwarze, D. (2006). Chemical ecology of vertebrates. Cambridge, UK: Cambridge University Press.

Múnera-Roldán, C., Cody, M., Schiele-Zavala, R., Sigel, B., Woltmann, S. & Kjeldsen, J. (2007). New and noteworthy records of birds from southeastern Nicaragua. Bulletin of the British Ornithologists' Club, 127, 152–161.

Mundy, P. J. (1992). The vultures of Africa. Acorn Books.

Mundy, P. J. (1997). Hooded vulture Necrosyrtes monachus. *In*: Harrison, J. A. et al. (Eds.), The atlas of southern African birds, 1: Non-passerines, pp. 156–157. Johannesburg: Birdlife South Africa.

Mundy, P. J., Robertson, A. S., Komen, J. & O'Connor, T. G. (1986). Attacks by Black eagles on vultures. Journal of Raptor Research, 20(2), 61–64.

Mundy, P., Butchart, D., Ledger, J. & Piper, S. (1992). The vultures of Africa. London: Academic Press.

Muniz-Lopez, R. (2005). Discovery of the first Harpy eagle (Harpia harpyja) nest on the western side of the Andean Cordillera. Newsl. Neotrop Raptor Network, 2, 1–2.

Muñiz-López, R. (2007). Ecology, biology and habitat of the Harpy eagle (Harpia harpyja). *In*: P. Tufiño (Ed.), C. Pindo, conservation of the Harpy eagle in Ecuador, pp. 191–244. Quito, Ecuador: SIMBIOE (in Spanish).

Muñiz-López, R. (2008). Revisión de la situación del águilaharpía Harpia harpyjaen Ecuador. Cotinga, 29, 42–47.

Muñiz-López, R. (2016). Biología y conservación del águilaharpía (Harpia harpyja) en Ecuador [Tesis Doctoral]. Universidad de Alicante.

Muñiz-López, R., Criollo, O. & Mendúa, A. (2007). Results of five years of the Harpy eagle (Harpia harpyja) Research Program in the Ecuadorian tropical forest. *In*: K. L. Bildstein et al. (Eds.), Neotropical raptors, pp. 23–32. Orwigsburg, PA: Hawk Mountain Sanctuary.

Muñiz-López, R., Limiñana, R., Cortés, G. D. & Urios, V. (2012). Movements of Harpy eagles Harpia harpyja during their first two years after hatching. Bird Study, 59(4), 509–514. https://doi.org/10.1080/00063657.2012.722190

Muñoz, A. R., Toxopeus, B., Elorriaga, J., Gonzalez, J. M. & Yáñez, B. (2010). First record of a communal roost of Short-toed eagles Circaetus gallicus. Ibis, 152(1), 173–175. https://doi.org/10.1111/j.1474-919X.2009.00974.x

Murdoch, J. (1998). The spaces of actor network theory. Geoforum, 29, 357–374.

Murgatroyd, M., Underhill, L. G., Bouten, W. & Amar, A. (2016). Ranging behaviour of Verreaux's eagles during the pre-breeding period determined through the use of high temporal resolution tracking. PLoS One, 11(10), e0163378. https://doi.org/10.1371/journal.pone.0163378

Murgatroyd, M., Underhill, L. G., Rodrigues, L. & Amar, A. (2016a). The influence of agricultural transformation on the breeding performance of a top predator: Verreauxs' eagles in contrasting land use areas. Condor, 118(2), 238–252. https://doi.org/10.1650/CONDOR-15-142.1

Murgatroyd, M., Avery, G., Underhill, L. & Amar, A. (2016b). Adaptability of a specialist predator: The effects of land use on diet diversification and breeding performance of Verreaux's eagles. Journal of Avian Biology, 47, 001–012.

Murgatroyd, T. M. M., Photopoulou, T., Underhill, L. G., Bouten, W. & Amar, A. (2018). Where eagles soar: Fine-resolution tracking reveals the spatiotemporal use of differential soaring modes in a large raptor. Ecology and Evolution, 8(13), 6788–6799. https://doi.org/10.1002/ece3.4189

Murie, A. (1940). Ecology of the Coyote in the Yellowstone. US National Park Service Fauna Series No. 4. Washington, D.C.

Murie, O. J. (1959). Fauna of the Aleutian Islands and Alaska peninsula. North American Fauna, 61, 1–364. https://doi.org/10.3996/nafa.61.0001

Muriel, R., Ferrer, M., Casado, E., Madero, A. & Calabuig, C. P. (2011). Settlement and successful breeding of reintroduced Spanish Imperial eagles Aquila adalberti in the province of Cadiz (Spain). Ardeola, 58(2), 323–333. https://doi.org/10.13157/arla.58.2.2011.323

Murn, C., Betchley, P. & Robert, C. (2009). Talon-locking and cartwheeling as a prelude to copulation in Tawny eagles Aquila rapa. Gabar, 20(2), 12–14.

Murphy, T. & Hope, C. (2006). Bald eagles in South Carolina. Department of Natural Resources of SC. Retrieved from https://www.dnr.sc.gov/cwcs/pdf/Baldeagle.pdf

Murphy, W. L. (2004). A birdwatchers' guide to Trinidad and Tobago. Norfolk: Prion Ltd.

Musindo, P. T. (2012). Morphological variation in bills and claws in relation to prey type in Southern African Birds of Prey (Orders Falconiformes and Strigiformes) [Thesis]. University of Zimbabwe.

Musters, C. J. M., Noordervliet, M. A. W. & ter Keurs, W. J. (1996). Bird casualties caused by a wind energy project in an estuary. Bird Study, 43, 124–126.

Myers, N. (1975). The cheetah Acinonyx jubatus in Africa: Report of a survey in Africa from the Sahara southwards. International Union for Conservation of Nature and Natural Resources.

Nadjafzadeh, M., Hofer, H. & Krone, O. (2013). The link between feeding ecology and lead poisoning in White-tailed eagles. Journal of Wildlife Management, 77(1), 48–57. https://doi.org/10.1002/jwmg.440

Nadjafzadeh, M., Hofer, H. & Krone, O. (2015). Lead exposure and food processing in white-tailed eagles and other scavengers: An experimental approach to simulate lead uptake at shot mammalian carcasses. European Journal of Wildlife Research, 61, 763–774. https://doi.org/10.1007/s10344-015-0953-1

Nadjafzadeh, M., Hofer, H. & Krone, O. (2015a). Sit-and-wait for large prey: Foraging strategy and prey choice of White-tailed eagles. Journal of Ornithology, 157(1), 165–178. https://doi.org/10.1007/s10336-015-1264-8

Nadjafzadeh, M., Voigt, C. C. & Krone, O. (2015b). Spatial, seasonal and individual variation in the diet of White-tailed eagles Haliaeetus albicilla assessed using stable isotope ratios. Ibis, 158(1), 1–15. https://doi.org/10.1111/ibi.12311

Næsset, E., Gobakken, T., Holmgren, J., Hyyppa, H., Hyyppa, J., Maltamo, M., Nilsson, M., Olsson, H., Persson, A. & Soderman, U. (2004). Laser scanning of forest resources: The Nordic experience. Scandinavian Journal of Forest Research, 19(6), 482–499.

Nagy, K. A., Girard, I. A. & Brown, T. K. (1999). Energetics of free-ranging mammals, reptiles and birds. Annual Review of Nutrition, 19, 247–277. https://doi.org/10.1146/annurev.nutr.19.1.247

Naidoo, V., Duncan, N., Bekker, L. & Swan, G. (2007). Validating the domestic fowl as a model to investigate the pathophysiology of diclofenac in Gyps vultures. Environmental Toxicology and Pharmacology, 24(3), 260–266. https://doi.org/10.1016/j.etap.2007.06.003

Naidoo, V., Wolter, K., Cuthbert, R. & Duncan, N. (2009). Veterinary diclofenac threatens Africa's endangered vulture species. Regulatory Toxicology and Pharmacology, 53(3), 205–208. https://doi.org/10.1016/j.yrtph.2009.01.010

Naidoo, V., Wolter, K., Cromarty, D., Diekmann, M., Duncan, N., Meharg, A. A. & Cuthbert, R. (2010). Toxicity of NSAIDs to Gyps vultures: A new threat from ketoprofen. Biology Letters, 6(3), 339–341. https://doi.org/10.1098/rsbl.2009.0818

Naidoo, V., Mompati, K. F., Duncan, N. & Taggart, M. A. (2011). The pied crow (Corvus albus) is insensitive to diclofenac at concentrations present in carrion. Journal of Wildlife Diseases, 47(4), 936–944. https://doi.org/10.7589/0090-3558-47.4.936

Nakagawa, H., Lobkov, E. G. & Fujimaki, Y. (1987). Winter censuses on Haliaeetus pelagicus in the Kamchatka and northern Japan in 1985. Strix, 6, 14–19.

Nam, D. H. & Lee, D. P. (2006b). Reproductive effects of heavy metal accumulation on breeding feral pigeons (Columba livia). Science of the Total Environment, 366(2–3), 682–687. https://doi.org/10.1016/j.scitotenv.2006.02.004

Nankinov, D. (1992). Check list of bird species and subspecies in Bulgaria. Avocetta, 16, 1–17.

Nankinov, D., Simeonov, S., Michev, T. & Ivanov, B. (1997). Fauna of Bulgaria, 26, Aves, Part II [Sofia, BAS, 426 p].

Nankinov, D. (1998). First addition to list of birds in Bulgaria. Rivista Italiana di Ornitologia, 68(2), 183–187.

Nankinov, D. (2003). Second addition to the list of birds in Bulgaria. Rivista Italiana di Ornitologia, 73(2), 157–163.

Nankinov, D., Kirov, D., Tilova, E., Nikolov, I., Ivanov, I., Dilchev, K., Popov, K., Karaivanov, N., Todoov, N., Shurulinkov, P., Stanchev, R., Dutsov, A., Aleksov, R., Tsonev, R., Dalakchieva, S., Ivanov, S., Marin, S., Stajkov, S., Nikolov, B., Borisov, B., Stoyanov, G., Gradev, G., Georgiev, D., Popov, D. & Domuschiev, D. (2004). Breeding Totals of the Ornithofauna Bulgaria. Green Balkans, Plovdiv.

Naoroji, R. (1997). Contamination in eggshells of Himalayan Grey headed Fishing-eagle Ichthyophaga nana plumbea in Corbett National Park India. Journal of the Bombay Natural History Society, 94, 398–400.
Naoroji, R. (2006). Birds of Prey of the Indian Subcontinent. London: Christopher Helm Publishers, A&C Black Publishers Ltd.
Naoroji, R. (2007). Birds of Prey of the Indian Subcontinent. New Delhi: Om Books International.
Naoroji, R. K. & Monga, S. G. (1983). Observations on the Crested serpent eagle (Spilornischeela) in Rajpipla forests—South Gujarat. Journal c the Bombay Natural History Society, 80(2), 273–285.
Naoroji, R. & Schmitt, N. J. (2007). Birds of prey of the Indian subcontinent. Uttar Pradesh: Om Books International.
National Aeronautics and Space Administration, NASA (2012). What are passive and active sensors? Retrieved from https://www.nasa.gov directorates/heo/scan/communications/ outreach/funfacts/txt_passive_active.html
National Aeronautics and Space Administration National Astronautics and Space Administration. (2018). Regions of the Electromagneti Spectrum. Retrieved from https://imagine.gsfc.nasa.gov/science/toolbox/spectrum_ chart.html.
National Eagle Center. (2020). Eagle Nesting and Young. Retrieved from https://www.nationaleaglecenter.org/eagle-nesting-young/
National Geographic. (2002). Field guide to the birds of North America (4th ed.). Washington, DC: National Geographic.
National Geographic. (2015). Steller's Sea eagle. Retrieved from https://www.nationalgeographic.com/animals/birds/s/stellers-eagle/
National Institute of Biological Resources. (2011). Red Data Book of endangered birds in Korea. Seoul, South Korea: National Institute c Biological Resources.
National Institute of Environmental Health Sciences. (2020). Perfluoroalkyl and Polyfluoroalkyl Substances (PFAS). Retrieved from https://www niehs.nih.gov/health/topics/agents/pfc/index.cfm
National Research Council. (1977). (US) Committee on Medical and Biological Effects of Environmental Pollutants. Retrieved from https://www ncbi.nlm.nih.gov/books/NBK231025/
Naude, V. N., Smyth, L. K., Weideman, E. A., Krochuk, B. A. & Amar, A. (2019). Using web-sourced photography to explore the diet of declining African raptor, the Martial eagle (Polemaetus bellicosus). Condor, 121(1), 1–9. https://doi.org/10.1093/condor/duy015
Naumburg, E. M. B. (1930). The birds of Matto Grosso, Brazil: A report on the birds secured by the Roosevelt-Rondon Expedition. Bulletin of th American Museum of Natural History, 40, 1–432.
Naveda-Rodríguez, A. (2002). Description of ecological aspects of large birds of prey in Henri Pittier. National Parks, Venezuela, 28, 8 Conferencia sobre Aves Rapaces Neotropicales y Simposio del Águila Arpía. Fondo Peregrino – Panamá, Panamá City.
Naveda-Rodríguez, A. (2004). Contribution to the natural history of Ornate hawk-eagle Spizaetus ornatus (Daudin, 1801) and Great black-haw Buteogallus urubitinga (Gmelin, 1788). Revistã de Ecologia Latino-America, 11, 23–26.
Naveda-Rodríguez, A. (2007). Distribution and habitat of hawk-eagles in Venezuela. *In*: A. Zimmerman (Ed.), Neotropical raptors, pp. 99–10 Orwigsburg, PA: Hawk Mountain Sanctuary.
Neale, J. C. C., Sacks, B. N., Jaeger, M. M. & McCullough, D. R. (1998). A comparison of bobcat and coyote predation on lambs in north-coast California. The Journal of Wildlife Management, 62(2), 700–706. https://doi.org/10.2307/3802346
Nebeský, J. J. K. (2007). Czech Heraldry through the centuries. Retrieved from http: //historie. hranet.cz/heraldika/pdf/heraldika_AN.pdf
Negro, J. J. (1999). Past and future research on wildlife interaction with power lines. *In*: Ferrer, M. & Janss, G. F. (Eds.), Birds and power line Collision, electrocution, and breeding, 21–28. Madrid, Spain: Quercus.
Negro, J. J. (2008). Two aberrant serpent-eagles may be visual mimics of bird-eating raptors. Ibis, 150(2), 307–314. https://doi.org/10.1111/j.147 919X.2007.00782.x
Nelson, M. W. & Nelson, P. (1976). Power lines and birds of prey. Idaho Wildlife Review, 28, 3–7.
Nelson, M. W. (1979). Power lines progress report on eagle protection research. Boise, ID: Idaho Power Company.
Nelson, M. W. (1980). Update on eagle protection practices. Boise, ID: Idaho Power Company.
Nemesházi, E., Kövér, S., Zachos, F. E., Horváth, Z., Tihanyi, G., Mórocz, A., . . . Szabó, K. (2016). Natural and anthropogenic influences on th population structure of White-tailed eagles in the Carpathian Basin and central Europe. Journal of Avian Biology, 47(6), 795–805. https: doi.org/10.1111/jav.00938
Nesse, G. E., Longhurst, W. M. & Howard, W. E. (1976). Predation and the sheep industry in California, 1972–1974. Division of Agricultur Science Bulletin, 1878. CA: University of California, Davis.
Neto, P. S. [in Portuguese]. Birds of Paraná. Published by the author. Nilópolis, Brazil: Rio de Janerio.
Nette, T., Burles, D. & Hoefs, M. (1984). Observation of Golden eagle Aquila chrysaetos predation Dall sheep Ovis dalli dalli lambs. Canadia Field-Naturalist, 98, 252–254.
Neumann, W., Martinuzzi, S., Estes, A. B., Pidgeon, A. M., Dettki, H., Ericsson, G. & Radeloff, V. C. (2015). Opportunities for the applicatio of advanced remotely-sensed data in ecological studies of terrestrial animal movement. Movement Ecology, 3(1). https://doi.org/10.118 s40462-015-0036-7
Newby, J., Grettenberger, J. & Watkins, J. (1987). The birds of Northern Aïr, Niger. Malimbus, 9, 4–16.
Newman, K. (1998). Newman's birds of Southern Africa. Halfway House, UK: Southern Book Publishers.
Newsome, S. D., Collins, P. W. & Sharpe, P. (2015). Foraging ecology of a reintroduced population of breeding Bald eagles on the Chann Islands, California, USA, inferred from prey remains and stable isotope analysis. The Condor, 117(3), 396–413. https://doi.org/10.165 CONDOR-14-213.1
Newton, I. (1979). Population ecology of Raptors. Hertfordshire, UK: T & A D Poyser Ltd., Hertfordshire, United.
Newton, I. (1986). The Sparrow hawk. Calton, UK: T and A D Poyser Ltd.
Newton, I. (Ed.) (1990). Birds of Prey. New York: Facts on File.

Newton, I. (1991). L'Àguilaperdiguera Hieraaetus fasciatus a Catalunya: Estatus, ecologiatròfica, biologiareproductora I demografia [PhD Dissertation]. Barcelona: Universitat de Barcelona. Lynx Edicions.

Newton, I. (1993). Causes of breeding failure in wild raptors: A review. *In*: P. Y. Redig (Ed.), Raptor, biomedicine, pp. 62–70. Minneapolis, MN: University of Minnesota Press.

Newton, I. (1998). Population Limitation in Birds. London: Academic Press.

Newton, I. (2004). Aguila azor-perdicera, Hieraaetus fasciatus. *In*: J. C. Atienza (Ed.), Libro Rojo de las aves de España, pp. 154–157. Madrid: Dirección General para la Biodiversidad-SEO/BirdLife, Madrid, Spain.

Newton, I. & Marquiss, M. (1983). Dispersal of Sparrow hawks between birthplace and breeding place. Journal of Animal Ecology, 52(2), 463–477. https://doi.org/10.2307/4566

Nickel Institute (2015). Secondary Poisoning Risk Assessment of Birds and Mammals Exposed to Nickel in their Diets. No. 6. Retrieved from https://nickelinstitute.org/media/3719/eu-ni-ra-fact-sheet-6-2015-july.pdf

Nickens, T. E. (2008). Vultures take over suburbia. Retrieved from https://www.audubon.org/magazine/november-december-2008/vultures-take-over-suburbia

Nicolis, G. (1994). Dynamical systems, biological complexity and global change. In: Solbrig, O. T., Van Emden, H. M. & Oordt, P. G. W. J. (Eds.), Biodiversity and global change, 21–31, Wallingford: CAB International.

Nighat, S., Iqbal, S., Nadeem, M. S., Mahood, T. & Shah, S. I. (2013). Estimation of heavy metal residues from the feathers of Falconidae, Accipitridae, and Strigidae in Punjab, Pakistan. Turkish Journal of Zoology, 37, 488–500. Retrieved from http://www.doi. https://doi.org/10.3906/zoo-1112-1

Nijman, V. & Sözer, R. (1996). Sexual dimorphism in the Javan Hawk-eagle Spizaetus bartelsi. Bulletin of the British Ornithological Club, 116, 251–253.

Nijman, V. & Sözer, R. (1998). Field identification of the Javan hawk eagle Spizaetus bartelsi. Forktail, 14, 13–16. Retrieved from http://vincentnijman.org/files/a9_nijmansozer_ sbartelsi_forktail_1997.pdf

Nijman, V. & Prawiradilaga, D. M. (2003). Tropical forest raptors as indicators in biodiversity monitoring programmes and conservation area selection: Minimal area requirements for the raptor community of the Sundaic region. *In*: Nijman, V. & Prawiradilaga, D. M. (Eds.), Quarterly report prepared for the Indonesian Institute of Sciences (LIPI), 32. Amsterdam: University of Amsterdam & PuslitbangBiologi-LIPI, Bogor.

Nijman, V. & Van Balen, S. B. S. (2003). Wandering stars: Age-related habitat use and dispersal of Javan Hawk-eagles (Spizaetus bartelsi). Journal für Ornithologie, 144(4), 451–458. https://doi.org/10.1111/j.1439-0361.2003.03044.x

Nijman, V. & van Balen, S. S. (1999). Breeding biology of Javan Hawk-eagle Spizaetus bartelsi in West Java, Indonesia. Emu–Austral Ornithology, 100(2), 125–132. https://doi.org/10.1071/MU9826

Nijman, V., Shepherd, C. R. & van Balen, S. (2009). Declaration of the Javan Hawk eagle Spizaetus bartelsi as Indonesia's National Rare Animal impedes conservation of the species. Oryx, 43(1), 122–128. https://doi.org/10.1017/S0030605307001081

Nikolaus, G. (1987). Distribution atlas of Sudan's birds with notes on habitat and status. Zoologisches Forschungsinstitut und Museum Alexander Koenig (Bonn).

Nilsson, M. (2014). Movement ecology of the Golden eagle Aquila chrysaetos and the semidomesticated reindeer Rangifer tarandus – Synchronous movements in a boreal ecosystem. Umeå, Sweden: Swedish University of Agricultural Sciences.

Nobel Prize in Physiology or Medicine (1948). Retrieved from https://www.nobelprize.org/prizes/medicine/1948/summary/

Nolet, B. A., Fuld, V. N. & Van Rijswijk, M. E. C. (2006). Foraging costs and accessibility as determinants of giving-up densities in a swan-pondweed system. Oikos, 112(2), 353–362. https://doi.org/10.1111/j.0030-1299.2006.13463.x

Nores, M., Barker, K. & Remsen, V. (2011). Proposal, 491 to South American Classification Committee: Change linear sequence of orders for Falconiformes, Psittaciformes, and Cariamiformes. Retrieved from http://www.museum.lsu.edu/~Remsen/SACCprop491. html, 491.

Nores, M., Yzurieta D. & Miatello, R. (1983). List and distribution of the birds of Cordoba, Argentina. Boletin de la Academia Nacional de Ciencias (Cordoba), Argentina, 56, 1–114.

North, M. E. W. (1939). Field notes on certain raptorials and water-birds in Kenya Colony—Part II. Ibis, 3(4), 617–643.

Norton, P. (1980). The habitat and feeding ecology of the klipspringer Oreotragus (Zimmerman, 1783) in two areas of the Cape Province. Zoological Society of Southern Africa Occasional Bulletin [MSc Thesis]. Pretoria: University of Pretoria (pp. 1–209).

Novaes, W. G. & Cintra, R. (2013). Factors influencing the selection of communal roost sites by the Black vulture Coragypsatratus (Aves: Cathartidae) in an urban area in Central Amazon. Zoologia, 30(6), 607–614. https://doi.org/10.1590/S1984-46702013005000014

Novy, S. A. & Van Putte, R. D. (2016). Behavioral notes and nesting of the Black Solitary eagle (Buteogallus solitarius) in Belize. Transactions of the Illinois State Academy of Science, 109(1), 29–33.

Nowak, M., Bogawski, P. & Dzióbk, K. (2019). Unmanned aerial vehicles (UAVs) in environmental biology: A review. European Journal of Ecology, 4(2), 56–74. https://doi.org/10.2478/eje-2018-0012

Nowak, R. M. (2005). Walker's Carnivores of the World. Baltimore, MD: Johns Hopkins University Press.

Nowell, K. & Jackson, P. (1996). Wild Cats. Gland, Switzerland: International Union for Conservation of Nature and Natural Resources.

NPIC. (1996). Carbofuran. Retrieved from http:extoxnet. orst.edu/pips/carbofur.htm.

NPIC. (1999). DDT Drug Discovery. Retrieved from http://npic.orst.edu/ factsheets/ddtgen.pdf

NPIC. (2011). Ecotoxicology. Retrieved from http://npic.orst.edu/factsheets/ecotox.html

Nriagu, J. O., Pfeiffer, W. C., Malm, O., Magalhaes De Souza, C. M. & Mierle, G. (1992). Mercury pollution in Brazil. Nature, 356(6368), 389. https://doi.org/10.1038/356389a0

Nurfatimah, C., Syartinilia & Mulyani, Y. A. (2018). GIS-based approach for quantifying landscape connectivity of Javan Hawk-Eagle habitat. IOP Conference Series: Earth and Environmental Science, 149, 012017. https://doi.org/10.1088/1755-1315/149/1/012017

Nurwantha, P. F. & Rachman, Z. (2000). Status and distribution of Sulawesi Hawk-eagle Spizaetus lanceolatus in South. *In*: Proceedings of the Asian Raptor Research Conservation: Therapeutic Second Symposium on Raptors of Asia, Grand Aquila, Bandung, Indonesia, 25-27 July 2000, 203. S. Sulawesi. D. I. Prawiradilaga (Ed.). Indonesian Committee for the Second Symposium of Asian Raptor Research and Conservation, Cibinong, Indonesia.

Nurwantha, P. F. & Rachman, Z. (2000). Raptor notes from South. Proceedings of the Asian Raptor Research Conservation: Therapeutic Second Symposium on Raptors of Asia, Grand Aquila, Bandung, Indonesia, 25-27 July 2000, 204. S. Sulawesi. D. I. Prawiradilaga (Ed.). Indonesian Committee for the Second Symposium of Asian Raptor Research and Conservation, Cibinong, Indonesia.

Nurwatha, P. F., Rahman, Z. & Raharjaningtrah, W. (2000). Distribution and population of Sulawesi hawk-eagle Spizaetus lanceolatus in south and Central Sulawesi. Y.P.A.L. Jakarta, Indonesia: Pro Natura Fund.

Nurwatha, P. F., Rahman, Z. & Faridah, I. (2004). Project garuda: The conservation effort of the Javan hawk-eagle and its habitat at Mt GnBurangrang—Tangkubanperahu nature Reserve, West Java, Indonesia [Final report]. Bandung, Indonesia: YPAL.

Nybakk, K., Kjelvik, O. & Kvam, T. (1999). Golden eagle predation on semi-domestic reindeer. Wildlife Society Bulletin, 27, 1038–1042.

Nye, P. E. (1994). Polygamy in Bald eagles. *In*: Abstracts of the presentations made at the annual Journal of Raptor Research Meeting of the Raptor Research Foundation, Inc., held at Charlotte, North Carolina, on 3–7 November 1993, 28: 45–71.

Nye, P. G. (1998). Bald eagle (Haliaeetus leucocephalus). *In*: E. Levine (Ed.), Bull's Birds of New York State. Ithaca: Cornell University Press.

Nygård, T., Bevanger, K., Dahl, E. L., Flagstad, Ø., Follestad, A., Lund Hoel, P. & Reitan, O. (2010). A study of White-tailed eagle Haliaeetus albicilla movements and mortality at a wind farm in Norway. *In*: Climate Change and Birds, BOU Conference, pp. 1–4. British Ornithologists Union.

Oaks, L., Rideout, B. A., Gilbert, M., Watson, R., Virani, M. & Ahmed Khan, A. (2001). Summary of diagnostic investigation into vulture mortality: Punjab Province, Pakistan, 2000–2001. Reports from the workshop on Indian Gyps vultures. In: Katzner, T. & Parry- Jones, J (Eds.), Proceedings of the 4th Eurasian Congress on Raptors, 12–13. Seville: Estación Biológica Donaña, Raptor Research Foundation.

Oatley T. B., Oschadleus, H. D., Navarro, R. A. & Underhill, L. G. (1998). Review of ring recoveries of birds of prey in southern Africa, 1948–1998. Johannesburg, South Africa Endangered Wildlife Trust.

Obrien, M. & Lacey, G. (2016). White-bellied Sea-eagle breeding in the Western Port area, Victoria. Australian Field Ornithology, 33, 46–56.

O'Brien, T. G. & Kinnaird, M. F. (2008). A picture is worth a thousand words: The application of camera trapping to the study of birds. Bird Conservation International, 18(Suppl. 1), S144–S162. https://doi.org/10.1017/S0959270908000348

O'Connor & Rebecca, R. (2015). Lauren McGough Apprentices wtih eagle Faconers in Mongolia. Retrieved from http://www.primaoutdoor.com/lauren-mcgough-apprentices-with-eagle-falconers-in-mongolia/.

O'Connor, R. J. (1984). The Growth and Development of Birds. Hoboken, NJ: John Wiley & Sons.

O'Connor, J. (1997). What is environmental history? Why environmental history? Capitalism, Nature, Socialism, 8, 3–29.

O'Daniel, D. (1995). Raising Cain . . . and Abel. Living Bird, 14: 30–35.

O'Donnell, W. B. & Debus, S. J. S. (2012). Nest-sites and foraging of the White-bellied sea eagle Haliaeetus leucogaster on the subtropical eastern coast of Australia. Australian Field Ornithology, 29: 149–159.

O'Donoghue, F. (2002). Long-crested eagle breeds at Karoi. Honeyguide, 48: 250.

O'Donoghue, F. (2002). Yellow-billed kite and Long-crested eagle nesting close together. Honeyguide, 48: 250.

O'French, R., Neill, J. P. & Eckelberry, D. R. (1991). Guide to the birds of Trinidad and Tobago (2nd ed.). Ithaca, NY: Cornell University Press.

O'Gara, B. (1994). Bald and Golden eagles damage management and predation control. Cooperative Extension Division, Institute of Agriculture and Natural Resources, University of Nebraska.

O'Gara, B. (1994). Eagles. The Handbook: Prevention and Control of Wildlife Damage. 61. Retrieved from https://digitalcommons.unl.edu/icwdmhandbook/61/.

O'Gara, B. W. (1978). Sheep predation by Golden eagles in Montana. Proceedings of the Veterinary. Pest Control Conference, 8: 206–213.

O'Gara, B. W. (1978). Differential characteristics of predator kills. In: Proceedings 8th Biennial Pronghorn Antelope Workshop, Jasper, Alberta 8, 380–393.

O'Gara, B. W. (1981). Predation by Golden Eagles on domestic lambs in Montana. *In*: Peek, J. M. & Dalke, P. D. (Eds.), 1982 Wildlife-Livestock Relationships Symposium: Proc. 10. April 20–22, 1981, 345–358. University of Idaho, Forestry, Wildlife and. Range Experiment Station Moscow, Idaho.

O'Gara, B. W. (1983). Predation on domestic sheep on a western Montana ranch. *In*: Brawley, K. C., Munoz, J. R. & Henne, D. R. (Eds.), Wildlife Society Bulletin, 11: 253–264.

O'Gara, B. W. (2005). Eagles. Retrieved from http://icwdm.org/handbook/birds/Eagles.asp.

O'Gara, B. W. & Getz, D. C. (1986). Capturing Golden eagles using a helicopter and net gun. Wildlife Society Bulletin, 14: 400–402.

O'Shea, T. (1991). Xerus rutilus. The American Society of Mammalogists, Mammalian Species, 370: 1–5.

O'Toole, D. & Raisbeck, M. F. (1997). Experimentally induced selenosis of adult mallard ducks: Clinical signs, lesions, and toxicology. Veterinary Pathology, 34(4), 330–340. https://doi.org/10.1177/030098589703400409

Oaks, J. L., Gilbert, M., Virani, M. Z., Watson, R. T., Meteyer, C. U., Rideout, B. A., Shivaprasad, H. L., Ahmed, S., Jamshed, M., Chaudhry, I. Arshad, M., Mahmood, S., Ali, A. & Khan, A. A. (2004). Diclofenac residues as the cause of vulture population decline in Pakistan. Nature 427(6975), 630–633. https://doi.org/10.1038/nature02317

Oatley, T. B., Oschadleus, H. D., Navarro, R. A. & Underhill, L. G. (1998). Review of ring recoveries of birds of prey in southern Africa, 1948–1998. Johannesburg: Endangered Wildlife Trust.

Oberholser, H. C. in Bent, A. C. (1937). Life histories of North American birds of prey. Part 1. Smithsonian Institution National Museum Bulletin 167.

Odino, M. & Ogada, D. L. (2008). Furadan use in Kenya and its impacts on birds and other wildlife: A survey of the regulatory agency, distributors, and end-users of this highly toxic pesticide. Report to the Bird Committee of Nature, Kenya.

Odum, E. P. & Kuenzler, E. J. (1955). Measurement of territory and home range size in birds. Auk, 72(2), 128–137. https://doi.org/10.2307/4081419

Oehme, G. (1961). Die Bestandsentwicklung Des Seeadlers, Haliaeetus albicilla (L.). Deutschland mit Untersuchungenzur Wahl der Brutbiotope. *In*: H. Schildmacher, Beiträgezur Kenntnis Deutscher Vögel, Jena, 1–61.

Oehme, G. (2003). On the methods, terminology and criteria in population studies of the White-tailed eagle. *In*: B. Helander et al. (Eds.), Sea eagle (2000). Proceedings of the Therapeutic International Sea Eagle Conference in Björkö, Sweden, September 13–17 (2000), 25–27. Stockholm: SNF & Åtta, 45 Tryckeri, A. B. & Stockholm.

Ofelt, C. H. (1975). Food habits of nesting Bald eagles in Southeast Alaska. Condor, 77(3), 337–338. https://doi.org/10.2307/1366232

Ogada, D. L., Gadd, M. E., Ostfeld, R. S., Young, T. P. & Keesing, F. (2008). Impacts of large herbivorous mammals on bird diversity and abundance in an African savanna. Oecologia, 156(2), 387–397. https://doi.org/10.1007/s00442-008-0994-1

Ogada, D. L., Thomsett, S., Virani, M. Z., Kendall, C. & Odino, M. (2010). Raptor road counts in Kenya: With emphasis on vultures. Report to the Royal Society for the Protection of Birds.

Ogada, D. L. & Buij, R. (2011). Large declines of the Hooded vulture Necrosyrtes monachus across its African range. Ostrich, 82(2), 101–113. https://doi.org/10.2989/00306525.2011.603464

Ogada, D. L., Torchin, M. E., Kinnaird, M. F. & Ezenwa, V. O. (2012). Effects of vulture declines on facultative scavengers and potential implications for mammalian disease transmission. Conservation Biology, 26(3), 453–460. https://doi.org/10.1111/j.1523-1739.2012.01827.x

Ogada, D. L. & Kibuthu, P. M. (2012). Breeding ecology of Mackinder's eagle owls Bubo capensis mackinderi in farmlands of central Kenya. Journal of Raptor Research, 46(4), 327–335. https://doi.org/10.3356/JRR-11-90.1

Ogada, D. L. (2014). The power of poison: Pesticide poisoning of Africas wildlife. Annals of the New York Academy of Sciences, 1322, 1–20. https://doi.org/10.1111/nyas.12405

Ogden, J. C. (1975). Effects of Bald eagle territoriality on nesting Ospreys. Wilson. Bulletin, 87, 496–505.

Ohishi, I., Sakaguchi, G., Riemann, H., Behymer, D. & Hurvell, B. (1979). Antibodies to Clostridium botulinum toxins in free-living birds and mammals. Journal of Wildlife Diseases, 15(1), 3–9. https://doi.org/10.7589/0090-3558-15.1.3

Ohlendorf, H. M. (1989). Bioaccumulation and effects of selenium in wildlife. *In*: Jacobs, L. W. (Ed.), Selenium in agriculture and the environment, 133–177. Madison: Soil Science Society of America.

Oindo, B. O., Skidmore, A. K. & de Salvo, P. (2003). Mapping habitat and biological diversity in the Maasai Mara ecosystem. International Journal of Remote Sensing, 24(5), 1053–1069. https://doi.org/10.1080/01431160210144552

Olea, P. P. & Mateo-Tomás, P. (2009). The role of traditional farming practices in ecosystem conservation: The case of transhumance and vultures. Biological Conservation, 142(8), 1844–1853. https://doi.org/10.1016/j.biocon.2009.03.024

Olendorff, R. R. (1972). Eagles, sheep, and power lines. Colorado Outdoors, 21, 3–11.

Olendorff, R. R. (1976). The food habits of North American Golden eagles. American Midland Naturalist, 95(1), 231–236. https://doi.org/10.2307/2424254

Olendorff, R. R., Miller, A. D. & Lehman, R. N. (1981). Suggested practices for raptor protection on power lines—The state-of-the-art in 1981. Raptor Research Report No. 4. St. Paul, MN. Raptor Research Foundation, Inc.

Olioso, G. (1991). L'aiglebotté. *In*: Yeatman-Berthelot, D. (Ed.), Atlas des oiseaux de France en Hiver, SOF (pp. 170–171).

Oliveira, P. S. & Marquis, R. J. (2002). The cerrados of Brazil: Ecology and natural history of a Neotropical savanna. New York, NY: Columbia University Press.

Ollington, R. F. & Loh, E. (1992). 1991 Singapore Bird Report. Unpublished Report. Nature Society (Singapore), Singapore.

Olmos, F., Pacheco, J. F. & Silveira, L.F. (2006). Notes on Brazilian birds of prey (Cathartidae, Accipitridae, and Falconidae). Revista Brasileira de Ornitologia, 14, 401–404 (In Portuguese with English summary).

Olmos, F. & Brito, G. R. R. (2007). Birds from the Boa Esperança Dam region, middle Parnaíba River, Brazil. Revista Brasileira de Ornitologia, 15, 37–52.

Olmos, F. & Turshak, L. G. (2009). A survey of birds in Omo Forest Reserve, south-western Nigeria. Bulletin of the African Bird Club, 16, 184–196.

Olrog, C. C. (1985). Status of wet forest raptors in northern Argentina. *In*: Newton, I. & Chancellor, R. D. (Eds.), Conservation studies on raptors. 5, 191–197. International Council for Bird Preservation Technical Publication, Cambridge, UK: ICBP.

Olsen, J. (1994). Some Time with Eagles and Falcons. Canberra: University of Canberra.

Olsen, J. (1997). Notes on Sanford's sea eagle Haliaeetus sanfordi and other raptors in the Solomon Islands. Australian Bird Watcher, 17(2), 81–86.

Olsen, J. & Fuentes, E. (2005). Collapse in numbers of breeding Little Eagles in the Australian Capital Territory. Canberra Bird Notes, 30, 141–145.

Olsen, J., Fuentes, E. & Rose, A. B. (2006). Trophic relationships between neighbouring White-bellied sea eagles (Haliaeetus leucogaster) and Wedge-tailed eagles (Aquila audax) breeding on rivers and dams near Canberra. Emu – Austral Ornithology, 106(3), 193–201. https://doi.org/10.1071/MU05046

Olsen, J., Fuentes, E., Rose, A. B. & Trost, S. (2006a). Food and hunting of eight breeding raptors near Canberra, 1990–1994. Australian Field Ornithology, 23, 77–95.

Olsen, J. & Trost, S. (2007). Diurnal raptors on the Island of Sumba, Indonesia, in June/July and December/January 2001–2002. Australian Field Ornithology, 24, 158–166.

Olsen, J., Osgood, M., Maconachie, M. & Dabb, G. (2008). Numbers of breeding Little eagles Hieraaetus morphnoides in the Australian Capital Territory in 2007. Canberra Bird Notes, 33, 77–80.

Olsen, J., Osgood, M., Maconachie, M., Dabb, G. & Butterfield, M. (2009). Numbers of breeding Little eagles Hieraaetus morphnoides in th Australian Capital Territory in 2008. Canberra Bird Notes, 34: 81–85.
Olsen, J., Judge, D., Fuentes, E., Rose, A. B. & Debus, S. J. S. (2010). Diets of Wedge-tailed eagles (Aquila audax) and Little eagles (Hieraaetu morphnoides) breeding near Canberra, Australia. Journal of Raptor Research, 44(1), 50–61. https://doi.org/10.3356/JRR-09-28.1
Olsen, J., Debus, S. J. S., Rose, A. B. & Judge, D. (2013). Diets of White-bellied sea eagles Haliaeetus leucogaster and Whistling kites Haliastu sphenurus breeding near Canberra, 2003–2008. Corella, 37, 13–18.
Olsen, P., Crome, F. & Olsen, J. (1983). Birds of Prey & Ground Birds of Australia. Sydney, Australia: Angus & Robertson.
Olsen, P., Fuller, P. & Marples, T. T. G. (1993). Pesticide-related eggshell thinning in Australian raptors. Emu – Austral Ornithology, 93(1), 1–1 https://doi.org/10.1071/MU9930001
Olsen, P. (1995). Australian Birds of Prey. Baltimore, MD: Johns Hopkins University Press.
Olsen, P. (2005). Wedge-tailed Eagle. Collingwood, Victoria: CSIRO Publishing Press.
Olsen, P. D., Cunningham, R. B. & Donnelly, C. F. (1994). Avian egg morphometrics -Allometric models of egg volume, clutch volume and shap Australian Journal of Zoology, 42(3), 307–321. https://doi.org/10.1071/ZO9940307
Olson, D. M., Dinerstein, E., Wikramanayake, E. D., Burgess, N. D., Powell, G. V. N., Underwood, E. C., D'Amico, J. A., Itoua, I., Strand, H. E Morrison, J. C., Loucks, C. J., Allnutt, T. F., Ricketts, T. H., Kura, Y., Lamoreux, J. F., Wettengel, W. W., Hedao, P. & Kassem, K. R. (2001 Terrestrial ecoregions of the world: A new map of life on earth. BioScience, 51(11), 933–938.
Olson, S. L. & James, H. F. (1982). Prodromus of the fossil avifauna of the Hawaiian Islands. Smithsonian Contributions to Zoology, 365, 1–5 https://doi.org/10.5479/si.00810282.365
Olson, S. L. (1982). The distribution of fused phalanges of the inner toe in the Accipitridae. Bulletin of the British Ornithological Club, 102, 8–1
Olson, S. L. (1994). Cranial osteology of Tawny and Steppe eagles Aquila rapax and A. nipalensis. Bulletin of the British Ornithologists' Clu 114, 264–267.
Olson, S. L. & James, H. F. (1991). Descriptions of thirty-two new species of birds from the Hawaiian Islands. Part I. Non-Passeriforme Ornithological Monographs, 45, 1–88.
Omland, K. S. & Hoffman, S. W. (1996). Seasonal, diel, and spatial dispersion patterns of Golden eagle autumn migration in southwester Montana. Condor, 98(3), 633–636. https://doi.org/10.2307/1369577
Ong, C. D. (2000). The home range of the Martial eagle Polemaetus bellicosus determined with the use of radio telemetry. Leicester: Universit of Leicester.
Onrubia, A., Muñoz, A.-R., Arroyo, G. M., Ramírez, J., de la Cruz, A., Barrios, L., Meyburg, B.-U., Meyburg, C. & Langgemach, T. (2011 Autumn migration of Lesser Spotted eagle Aquila pomarina in the Strait of Gibraltar: Accidental or regular? Ardea, 99(1), 113–116. https: doi.org/10.5253/078.099.0113.
Onrubia, A., Muñoz, A. R. & de la Cruz, A. (2011b). The magnitude and timing of migration by soaring storks and raptors in the Straits of Gibralta An updated study. Proceedings of the British Ornithologists Union Conference on Migratory Birds: Their Ecology and Conservation.
Ontiveros, D., Pleguezuelos, J. M. & Caro, J. (2005). Prey density, prey detectability and food habits: The case of Bonelli's eagle and th conservation measures. Biological Conservation, 123(1), 19–25. https://doi.org/10.1016/j.biocon.2004.10.004
Ordano, M. A. & Bosisio, A. C. (1994). Catalogue of Birds of the Museo Provincial de Ciencias Naturales "Florentino Ameghino". Santa F Argentina.
Ordano, M. & Bosisio, A. (1997). Historical records of threatened and near threatened Argentinian birds from Museo Provincial de Cienci Naturales "Florentino Ameghino" of Santa Fe, Argentina. Bulletin of the British Ornithologists' Club, 117, 77–79.
Orejuela Gartner, J. E. (1987). The "La Planada" Natural Reserve and Andean Biogeography. Humboldtia, 1, 117–148.
Orians, G. H. (1961). The ecology of blackbird (Agelaius) social systems. Ecological Monographs, 31(3), 285–312. https://doi.org/10.2307/194855
Oring, L.W. (1986). Avian polyandry. *In*: Johnston, R. J. (Ed.), Current Ornithology, 3, 309–351. New York, NY: Plenum Press.
Orloff, S. (1992). Tehachapi wind resource area avian collision baseline study. BioSystems Analysis, Inc. Prepared for California Energ Commission, 40. Sacramento.
Orloff, S. & Flannery, A. (1992). Wind turbine effects on avian activity, habitat use, and mortality in Altamont Pass and Solano County Win Resource Areas, 1989–1991. Final report by BioSystems Analysis to the California Energy Commission, 700-92-001. Sacramento, CA California Energy Commission.
Orloff, S. G. and Flannery, A. W. (1993). Wind turbine effects on avian activity, habitat use, and mortality in the Altamont Pass and Solan County Wind Resource Areas. *In*: Huckabee J. W. (Ed.), Avian interactions with utility structures, pp. 1–14. Avian Power Line Interactior Committee (APLIC). Electric Power Research Institute, Palo Alto, California.
Ormsby, J. P. (1987). Whitetail deer food availability maps from thematic mapper data. Photogrammetric Engineering and Remote Sensing, 53(8 1081–1085.
Orta, J. (1994). Golden eagle. *In*: Del Hoyo, J., Elliott, A. & Sargatal, J. (Eds.), Handbook of birds of the World 2. New World Vultures t Guineafowl, 197–198. Barcelona: Lynx Edicions.
Orta, J., Kirwan, G. M., Christie, D. A. and Boesman, P. (2013). White-tailed Sea-eagle (Haliaeetus albicilla). *In*: del Hoyo, J., Elliott, A., Sargata J., Christie, D. A. & de Juana, E. (Eds.), Handbook of the birds of the world. Barcelona: Lynx Edicions.
Orta, J., Kirwan, G. M., Christie, D. A. & Boesman, P. (2016). Bonelli's Eagle. *In*: Del Hoyo, J., Elliott, A., Sargatal, J., Christie, D.A. and d Juana, E. (Eds.), Handbook of the Birds of the World Alive. Barcelona: Lynx Edicions.
Orta, J., Christie, D. A., Kirwan, G. M. & Sharpe, C. J. (2020). Pallas's fish-eagle (Haliaeetus leucoryphus), version 1.0. *In*: Del Hoyo, J., Ellio A. & Sargatal J. (Eds.). Birds of the World. Ithaca, NY: Cornell Laboratory of Ornithology. https://doi.org/10.2173/bow.pafeag1.01

Orta, J., Kirwan, G. M. & Garcia, E. F. J. (2020). Short-toed snake-eagle (Circaetus gallicus), version 1.0. *In*: Del Hoyo, J., Elliott, A., Sargatal, J., Christie, D.A. & de Juana, E. (Eds.), Birds of the World. Ithaca, NY: Cornell Laboratory of Ornithology. https://doi.org/10.2173/bow.shteag1.01

Orta, J., Kirwan, G. M., Christie, D. A., Boesman, P. F. D & Marks, J. S. (2020). White-tailed eagle (Haliaeetus albicilla), version 1.0. *In*: Del Hoyo, J., Elliott, A., Sargatal, J., Christie, D.A. and de Juana, E. (Eds.), Birds of the World. Ithaca, NY: Cornell Laboratory of Ornithology. https://doi.org/10.2173/bow.whteag.01doi

Ortego, B., Gregory, C., Mabi, D., Mitchell, M. & Schmidt, D. (2009). Texas Bald Eagles. Bulletin of the Texas Ornithological Society, 42, 1–17.

Ortiz, F., Greenfield, P. & Matheus, J. C. (1990). Birds of continental Ecuador and Galapagos archipelago. Quito, Ecuador: Fundacion Ecuatoriana de Promocion Turistica/CECIA.

Ortiz-Crespo, F. I. (1986). Notes on the status of diurnal raptor populations in Ecuador. Birds of Prey Bulletin, 3, 71–78.

Osborn, R. G., Higgins, K. F., Usgaard, R. E., Dieter, C. D. & Neiger, R. G. (2000). Bird mortality associated with wind turbines at the Buffalo Ridge Wind Resource Area, Minnesota. American Midland Naturalist, 143, 41–52.

Osborne, T. O. & Colebrook-Robjent, J. F. R. (1975). Wahlberg's eagle breeding success in the Southern Province of Zambia in 1974. Bulletin of the Zambian Ornithological Society, 7, 30.

Osborne, T. O. & Osborne, L. (1991). A short week in Papua New Guinea. Muruk, 5, 21–23.

O'Shea, B. (2005). Notes on Birds of the Sipaliwini Savanna and Other Localities in Southern Suriname, with Six New Species for the Country. Ornitologia Neotropical, 16(3), 361–370.

Osman, S. M. (1973). The Changeable Hawk eagle Spizaetus cirrhatus (Gmelin). Journal of the Bombay Natural History Society, 70, 417–425.

Oustalet, E. (1899–1903). Les oiseaux du Cambodge, du Laos, de l'Annamannam et du Tonkin. Nouv. Arch. Mus. Hist. Nat. National Paris, 4(1), 221–296.

Oversluijs Vasquez, M. R. & Heymann, E. W. (2001). Crested eagle (Morphnus guianensis) predation on Infant Tamarins (Saguinusmystax and Saguinusfuscicollis, Callitrichinae). Folia Primatologica International Journal of Primatology, 72(5), 301–303. https://doi.org/10.1159/000049952

Owen, B. (2019). Switch ammo to stop lead poisoning in birds, B.C. wildlife experts tell hunters. Canadian Press. Retrieved from https://www.cbc.ca/news/canada/british-columbia/switch-ammo-lead-poisoning-1.5361718

Owen, W. F. W. (1833). Narrative of voyages to explore the shores of Africa, Arabia and Madagascar, II. London: Richard Bentley.

Owen-Smith, N. & Mills, M. G. L. (2008). Predator–prey size relationships in an African large-mammal food web. Journal of Animal Ecology, 77(1), 173–183. https://doi.org/10.1111/j.1365-2656.2007.01314.x

Oxford dictionaries online 2020. Eagle. Retrieved from https://www.lexico.com/definition/eagle

Paakinen, M., Rantanen, J. & Rantanen, P. (1981). Linturetkiaulkomaille I. Staijaamassa Gruusiassa. Ornis Karelica, 7, 68–72.

Pacheco, J. F., Kirwan, G. M., Aleixo, A., Whitney, B. M., Whittaker, A., Minns, J. & Oren, D. C. (2007). An avifaunal inventory of the CVRD Serra dos Carajás project, Pará, Brazil. Cotinga, 27, 15–30.

Pacheco, J. F., Parrini, R., Lopes, L. E. & de Vasconcelos, M. F. (2008). On the avifauna of Parque Estadual do Ibitipoca Park and adjacent areas, Minas Gerais, Brasil, with a critical revision of previous records and comments on biogeography and conservation. Cotinga, 30, 16–32.

Pacyna, E. G., Pacyna, J. M., Sundseth, K., Munthe, J., Kindbom, K., Wilson, S. & Maxson, P. (2010). Global emission of mercury to the atmosphere from anthropogenic sources in 2005 and projections to 2020. Atmospheric Environment, 44(20), 2487–2499. https://doi.org/10.1016/j.atmosenv.2009.06.009

Padilla, J. A., Martinez-Trancón, M., Rabasco, A. & Fernández-García, J. L. (1999). The karyotype of the Iberian Imperial eagle (Aquila adalberti) analyzed by classical and DNA replication banding. Cytogenetics and Cell Genetics, 84(1–2), 61–66. https://doi.org/10.1159/000015216

Pagel, J. E., Kritz, K. J., Millsap, B. A., Murphy, R. K., Kershner, E. L. & Covington, S. (2013). Bald eagle and Golden eagle mortalities at wind energy facilities in the contiguous United States. Journal of Raptor Research, 47(3), 311–315. https://doi.org/10.3356/JRR-12-00019.1

Paijmans, D. (2016). New longevity record of the Martial eagle. Retreived from http://safring.adu.org.za/long_news.php 2016.

Pain, D. J., Bavoux, C. & Burneleau, G. (1997). Seasonal blood lead concentration in marsh harrier Circus aeruginosus from Charente-Maritime, France: Relationship with the hunting season. Biological Conservation, 81(1–2), 1–7. https://doi.org/10.1016/S0006-3207(96)00132-2

Pain, D. J., Sears, J. & Newton, I. (1995). Lead concentrations in birds of prey in Britain. Environmental Pollution, 87(2), 173–180. https://doi.org/10.1016/0269-7491(94)P2604-8

Palasthy, J. & Meyburg, B.-U. (1973). On the food of the Lesser Spotted eagle in Eastern Slovakia under atypic climatic conditions. Ornithologische Mitteilgungen, 25, 61–72.

Palásthy, J. & Meyburg, B.-U. (1973). Zur Ernährungdes Des Schreiadlers (Aquila pomarina) in der Ostslowakeiunteratypischenklimatischen Bedingungen. Ornithologische Mitteilungen, 25, 63–72. Raptors in the new millennium 75.

Pallas, P. S. (1811). Zoographia Rosso-Asiatica, sistens omnium animalium in extenso Imperio Rossico et adjacentibus maribus observatorum recensionem, domicilia, mores et descriptiones, anatomen atque icones plurimorum. Petropoli. 3 vols. Caes. Academiae Scientiarum, St. Petersburg, 1, 1.

Palleser, T. (2007). Rare birds in 2006. Wingspan, 17, 43.

Palmer, R. S. (1962). Handbook of North American birds. Vol. 1. Loons through flamingos, New Haven, CT Yale University Press.

Palmeirim, J. M. (1988). Automatic mapping of avian species habitat using satellite imagery. Oikos, 52(1), 59–68. https://doi.org/10.2307/3565982

Palmer, R. S. (1988). Golden eagle: Aquila chrysaetos. *In*: Palmer, R. S. (Ed.), Handbook of North American birds, 5, 180–231. New Haven, CT: Yale University Press.

Palomares, F. (2001). Comparison of three methods to estimate rabbit abundance in a Mediterranean environment. Wildlife Society Bulletin, 29, 578–585.

Palomino, D. & Carrascal, L. M. (2007). Habitat associations of a raptor community in a mosaic landscape of central Spain under urban development. Landscape and Urban Planning, 83, 268–274.

Pande, S., Sant, N. & Pednekar, S. (2009). Avifaunal survey of South- and Middle-Andamans, November, 2009. Indian Birds, 7(4), 104–106.

Pande, S., Tambe, S., Francis, C. & Sant, N. (2003). Birds of Western Ghats, Kokan & Malabar. Mumbai: Bombay Natural History Society/Oxford University Press.

Panov, E. N. (1989). Natural hybridisation and ethological isolation in birds. Moscow: Nauka.

Panuccio, M., Agostini, N. & Massa, B. (2004). Spring raptor migration over Ustica, Southern Italy.

Panuccio, M., Agostini, N., Lucia, G., Mellone, U., Ashton-Boot, J. et al. (2010). Local weather conditions affect migration strategies of adult Western honey buzzards Pernis apivorus through an isthmus area. Zoological Studies, 49, 651–656.

Panuccio, M., Agostini, N. & Premuda, G. (2012). Ecological barriers promote risk minimization and social learning in migrating short-toed snake eagles. Ethology, Ecology and Evolution, 24(1), 74–80. https://doi.org/10.1080/03949370.2011.583692

Panuccio, M., Agostini, N., Baghino, L. & Bogliani, G. (2013). Visible migration of Short-toed snake-eagles: Interplay of weather and topographical features. Journal of Raptor Research, 47(1), 60–68. https://doi.org/10.3356/JRR-12-23.1

Panuccio, M., Stanzione, V., Catoni, C., Santini, M. & Dell'Omo, G. (2016). Radar tracking reveals influence of crosswinds and topography on migratory behavior of European honey buzzards. Journal Ethology, 34, 73–77.

Panuccio, M., Martín, B., Morganti, M., Onrubia, A. & Ferrer, M. (2017). Long-term changes in autumn migration dates at the Strait of Gibraltar reflect population trends of soaring birds. Ibis, 159(1), 55–65. https://doi.org/10.1111/ibi.12420, PubMed: 20162017

Paraguay, G. (2005). Atlas de las aves de Paraguay. Guyra Paraguay, Asunción.

Parasharya, B. M. (2010). Black eagle Ictinaetus malayensis at Narmada Dam, Gujarat. Indian Birds, 6, 95–96.

Parco Del Beigua, P. (2005). Bollettino del Centro Ornitologico e di Educazione Ambientale di Case Vaccà. Savona. Parco Naturale Regionale del Beigua. Retrieved from http://www.parcobeigua.it.

Park, C. R. & Lee, W. S. (2000). Relationship between species composition and area in breeding birds of urban woods in Seoul, Korea. Landscape and Urban Planning, 51(1), 29–36, https://doi.org/10.1016/S0169-2046(00)00094-3

Parker, T. A., III. (1997). Bird species recorded at three sites on the northern and western slopes of the Cordillera del Cóndor. In T. S. Schulenberg & K. Awbrey (Eds.), The Cordillera del Cóndor region of Ecuador and Peru, 168–179. Washington, DC: Conservation International.

Parker, T. A., III, Parker, S. A. & Plenge, M. A. (1982). An annotated checklist of Peruvian birds. Vermillion, SD: Buteo Books.

Parker, T. A., III, Castillo, A., Gell-Mann, M. & Rocha, O. (1991). Records of new and unusual birds from northern Bolivia. Bulletin of the British Ornithologists' Club, 111, 120–138.

Parker, T. A., Stotz, D. F. & Fitzpatrick, J. W. (1996). Ecological and distributional databases. *In*: Stotz, D. F. et al. (Eds.), Neotropical bird ecology and conservation, 113–436. Chicago: University of Chicago Press.

Parker, T. A., III. (1997). Bird species recorded at three sites on the northern and western slopes of the Cordillera del Cóndor. *In*: Schulenberg, T. S. & Awbrey, K. (Eds.), The Cordillera del Cóndor region of Ecuador and Peru, pp. 168–179. Washington, DC: Conservation International.

Parker, T. A., III, Schulenberg, T. S., Graves, G. R. & Braun, M. J. (1985). The avifauna of the Huancabamba region, northern Peru. *In*: Buckley P. A., Foster, M. S., Morton, E. S., Ridgely, R. S. & Buckley, F. G. (Eds.), Ornithological Monographs, 36(36), 169–197. https://doi.org/10.2307/40168282

Parker, V. (1994). Swaziland bird atlas 1985–1991. Mbabane, Swaziland: Websters.

Parker, V. (1995). The current status of some birds of prey in southern Mozambique. Gabar, 10, 2–3.

Parker, V. (1999). The atlas of the birds of Sul do Save, Southern Mozambique. Avian Demography Unit and endangered Wildlife Trust, Cape Town and Johannesburg. Retrieved from http://www.adu.org.za/docs/Southern_Mozabique_Atlas.pdf

Parker, V. (2005). The Atlas of the birds of Central Mozambique. Johannesburg, SA: Endangered Wildlife Trust.

Parker, V. (2005a). Birds of the Niassa Reserve, Mozambique. Bright Continent Guide 4. Cape Town, South Africa: Avian Demography Unit.

Parker, V. (2005b). The atlas of the birds of central Mozambique. Johannesburg, SA. South Africa Endangered Wildlife Trust.

Parkes, D. A. (1998). Western Banded Snake-eagle nesting on the Middleveld. Honeyguide, 44, 25–27.

Parr, C. L. & Andersen, A. N. (2006). Patch mosaic burning for biodiversity conservation: A critique of the Pyrodiversity Paradigm. Conservation Biology: The Journal of the Society for Conservation Biology, 20(6), 1610–1619. https://doi.org/10.1111/j.1523-1739.2006.00492.x

Parr, J. W. K., Eames, J. C., Sun Hean, S., Chamnan Hong, C., Som Han, S., Vi La Pich, V. & Kim Hout Seng, K. H. (1996). Biological and socio-economic aspects of water-bird exploitation and natural resource utilization at Prek Toal, Tonle Sap Lake, Cambodia. Bangkok: IUCN International Union for Conservation of Nature and Natural Resources Species Survival Commission.

Parrini, R., Raposo, M. A., Pacheco, J. F., Carvalhaes, A.M.P., Melo Junior, T.A., Fonseca, P.S.M. & Minnset , J. C. (1999). Birds of the Chapada Diamantina, Bahia, Brazil. Cotinga, 11, 86–95.

Parry, S. J. (2001). The Booted eagles (Aves; Accipitridae), perspectives in evolutionary biology. PhD Thesis. University College, London.

Parry, S. J., Clark, W. S. & Prakash, V. (2002). On the taxonomic status of the Indian Spotted eagle Aquila hastata. Ibis, 144(4), 665–675. https://doi.org/10.1046/j.1474-919X.2002.00109.x

Parry-Jones, J. (1997). Eagle. Dorling Kindersley Ltd.

Parslow, J. L. F., Jefferies, D. J. & Hanson, H. M. (1973). Gannet mortality incidents in 1972. Marine Pollution Bulletin, 4(3), 41–43, https://doi.org/10.1016/0025-326X(73)90303-2

Partridge, W. H. (1956). Notes on the Brazilian Merganser in Argentina. Auk, 73, 473–488.

Paruk, J. D., Long, D., Perkins, C., East, A., Sigel, B. J. & Evers, D. C. (2014). Polycyclic aromatic hydrocarbons detected in common loons (Gaviaimmer) wintering off coastal Louisiana. Waterbirds, 37: 85–93. https://doi.org/10.1675/063.037.sp111

Patrikeev, M. (2004). The Birds of Azerbaijan. Pensoft Series Faunistica. Sofia, Bulgaria: Pensoft Publishers, 38.

Pattee, O. H. & Hennes, S. K. (1983). Bald eagles and waterfowl: The lead shot connection. Transactions of the North American Wildlife and Natural Resources Conference, 48, 230–237.

Pattee, O. H., Bloom, P. H., Scott, J. M. & Smith, M. R. (1990). Lead hazards within the range of the California Condor. Condor, 92(4), 931–937. https://doi.org/10.2307/1368729https://doi.org/10.2307/1368729.

Pattee, O. H., Wiemeyer, S. N., Mulhern, B. M., Sileo, L. & Carpenter, J. W. (1981). Experimental lead-shot poisoning in Bald eagles. Journal of Wildlife Management, 45(3), 806–810. https://doi.org/10.2307/3808728

Patterson, D. A., McClelland, B. R., Shea, D. S. & McClelland, P. T. (1998). Size variation of migrant Bald eagles at Glacier National Park, Montana. Journal of Raptor Research, 32(2), 120–125.

Patterson, I. J. (1980). Territorial behaviour and the limitation of population density. Ardea, 68, 53–62.

Patterson, T. A., McConnell, B. J., Fedak, M. A., Bravington, M. V. & Hindell, M. A. (2010). Using GPS data to evaluate the accuracy of state-space methods for correction of Argos satellite telemetry error. Ecology, 91(1), 273–285. https://doi.org/10.1890/08-1480.1

Paulson, D. R. (1985). The importance of open habitat to the occurrence of kleptoparasitism. Auk, 102(3), 637–639. https://doi.org/10.1093/auk/102.3.637

Pautasso, A. A. (2003). Aprovechamiento de la fauna silvestre por pobladores rurales en la fracción norte de los bajos submeridionales de la Provincia de Santa Fe, Argentina. Comunicaciones del Museo Provincial de Ciencias Naturales "Florentino Ameghino", 8(2), 1–66.

Pautasso, A. A., de la Peña, M. R. & Mastropaolo, J. M. (2003). New records of the Crowned Solitary eagle (Harypyhaliaetus coronatus) for Santa Fe Province, Argentina. Nuestras Aves, 19, 29–32.

Pavés, H. J., Schlatter, R. P. & Espinoza, C.I. 2008. Scavenging and Predation by Black vultures Coragyps atratus at a South American sea lion breeding colony. Vulture News, 58, 4–15.

Pavón, D., Limiñana, R., Urios, V., Izquierdo, A., Yáñez, B., Ferrer, M. & Vega, Adl. (2010). Autumn migration of juvenile Short-toed eagles Circaetus gallicus from Southeastern Spain. Ardea, 98(1), 113–117. https://doi.org/10.5253/078.098.0114

Paynter, R. A. (1955). The ornithogeography of the Yucatan Peninsula. Bulletin of the Peabody Museum of Natural History, 9, 1–347.

Pearce-Higgins, J. W., Stephen, L., Langston, R. H. W., Bainbridge, I. P. & Bullman, R. (2009). The distribution of breeding birds around upland wind farms. Journal of Applied Ecology, 46, 1323–1331. https://doi.org/10.1111/j.1365-2664.2009.01715.x

Peârez-Garcõâa, J. M., Margalida, A., Afonso, I., Ferreiro, E., Gardiazaâbal, A., Botella, F. & Sánchez-Zapata, J. A. (2013). Interannual home range variation, territoriality and overlap in breeding Bonelli's eagles (Aquila fasciata) tracked by GPS satellite telemetry. Journal of Ornithology, 154(1), 63–71. https://doi.org/10.1007/s10336-012-0871-x.

Pearson, D. L. (1972). A study of the birds of Limoncocha, Napo Province, Ecuador. Boletin Inform. Cient. Nac., Quito, 13, 1–14.

Peckover, W. S. & Filewood, L. W. C. (1976). Birds of New Guinea and Tropical Australia. Terrey Hills, Australia A.H. & A.W. Reed.

Pedrocchi, V., da Silva, C. R. & da Silva, A. (2002). Check list of birds and mammals in the Paranapiacaba Forest fragment. *In*: Mateos, E., Guix, J. C., Serra, A. & Pisciotta, K. (Eds.), Censuses of vertebrates in a Brazilian Atlantic rainforest area: The Paranapiacaba fragment, 183–204. Centre de Recursos de Biodiversitat Animal, Universitat de Barcelona, Barcelona.

Peled, A. & Gilichinsky, M. (2010). Knowledge-based classification of land cover for the quality assessment of GIS databases. International Archives of the Photogrammetry, Remote Sensing and Spatial Information Sciences – ISPRS Archives, 38.

Pendlebury, H. M., Chasen, F. F. & Nutter, N. (1932). A zoological expedition to Mount Kinabalu, British North Borneo (1929), Singapore.

Pendleton, E. (1978). To save raptors from electrocution. Defenders, 53, 18–21.

Pennycuick, C. J. (1972). Soaring behaviour and performance of some East African birds, observed from a motor-glider. Ibis, 114(2), 178–218. https://doi.org/10.1111/j.1474-919X.1972.tb02603.x

Pennycuick, C. J. (1975). Mechanics of flight. *In*: Farner, D. S. & King, J. R. (Eds.), Avian biology, 5, 1–75. London: Academic Press.

Pennycuick, C. J., Alerstam, T. & Larsson, B. (1979). Soaring migration of the Common Crane Grus grus observed by radar and from an aircraft. Ornis Scand., 10, 241–251. https://doi.org/10.2307/3676068

Pennycuick, C. J. (1983). Thermal soaring compared in three dissimilar tropical bird species, Fregata magnificens, Pelecanus occidentals and Coragyps atratus. Journal of Experimental Biology, 102, 307–325.

Pennycuick, C. J. & Scholey, K. D. (1984). Flight behavior of Andean condors Vultur gyphys and Turkey vultures Cathartes aura around the Pacas Peninsula, Peru. Ibis, 126, 253–256. https://doi.org/10.1111/j.1474-919X.1984.tb08005.x

Pennycuick, C. J. (2008). Modelling the flying bird. Burlington: Elsevier.

Penry, H. (1994). Bird atlas of Botswana. Pietermaritzburg, NHBS, South Africa: University of Natal Press.

Penteriani, V., Otalora, F. & Ferrer, M. (2005). Floater survival affects population persistence. The role of prey availability and environmental stochasticity. Oikos, 108(3), 523–534. https://doi.org/10.1111/j.0030-1299.2005.13514.x

Penteriani, V., Ferrer, M. & Delgado, M. M. (2011). Floater strategies and dynamics in birds, and their importance in conservation biology: Towards an understanding of nonbreeders in avian populations. Animal Conservation, 14(3), 233–241. https://doi.org/10.1111/j.1469-1795.2010.00433.x

Penteriani, V., Rutz, C. & Kenward, R. (2013). Hunting behaviour and breeding performance of northern Goshawks Accipiter gentilis, in relation to resource availability, sex, age and morphology. Naturwissenschaften, 100(10), 935–942. https://doi.org/10.1007/s00114-013-1093-7

Pequeño, T., Salazar, E. & Aucca, C. (2001). Birds of the southern Vilcabamba region, Peru. *In*: Alonso, L. E., Alonso, A., Schulenberg, T. S. & Dallmeier, F. (Eds.), Biological and social assessments of the Cordillera de Vilcabamba, pp. 98–104. Washington, DC: Conservation International.

Percival, S. (2003). Birds and windfarms in Ireland: A review of potential issues and impact assessment (pp. 1–25). Durham: Ecology Consulting.

Percival, S. (2013). Witberg Wind Farm: Ornithological Collision Risk Modelling (2013). Report prepared for Witberg Wind Power (Pty) Ltd.

Peregrine Fund. (2009). Plan d'aménagement et de gestion de la nouvelle aire protégée de Mandrozo. Boise, ID: The Peregrine Fund.

Pereira, A. D. M. M. & Salzo, I. (2006). First nesting record for Harpia harpyja in the State of Mato Grosso do Sul, Brazil. Revista Brasileira d Ornitologia, 14.

Pereira, H. (2002). Forecasting bird extinctions in Costa Rica, 14(1), pp. 8–9. Center for Conservation Biology Update.

Pereira, L. M., Owen-Smith, N. & Moleón, M. (2014). Facultative predation and scavenging by mammalian carnivores: Seasonal, regional, an intraguild comparisons. Mammal Review, 44(1), 44–55. https://doi.org/10.1111/mam.12005

Pereira, M. E. (1988). Effects of age and sex on intra-group spacing behaviour in juvenile savannah baboons, Papio cynocephalus. Anima Behaviour, 36(1), 184–204. https://doi.org/10.1016/S0003-3472(88)80262-8

Peres, C. A. (1993). Anti-predation benefits in a mixed-species group of Amazonian tamarins. Folia Primatologica; International Journal c Primatology, 61(2), 61–76. https://doi.org/10.1159/000156732

Pereyra Lobos, R. (2004). New records of Crowned Solitary eagle (Harpyhaliaetus coronatus) for Mendoza Province, Argentina. Nuestras Ave 20, 25–26.

Pereyra Lobos, R. P., Muñoz, L. & Rodriguez, R. (2007). First record of natural mortality in a nestling Crowned Solitary eagle (Harpyhaliaetu coronatus) in the desert of Lavalle, Mendoza Province, Argentina. *In*: Bildstein, K. L., Barber, D. R. & Zimmerman, A. (Eds.), Neotropica ornithology, p. 327. Orwigsburg: Hawk Mountain Sanctuary.

Pereyra Lobos, R. P., Muñoz, L., Mayorga, S., Martinez, F. & Yacante, R. (2007). Use of local observers as a strategy for studying the Crowne Solitary eagle (Harpyhaliaetus coronatus) at the Nature Reserve Telteca Forests, Department of Lavalle, Mendoze Province, Argentina. *I* Bildstein, K. L., Barber D. R. & Zimmerman, A. (Eds.), Neotropical ornithology, pp. 325–326. Orwigsburg: Hawk Mountain Sanctuary.

Pereyra Lobos, R., López, H. & Natale, E. (2007). Sites of interest for the conservation of the Crowned Solitary eagle (Harpyhaliaetus coronatu in Parque Nacional and the Reserva Provincial Sierra de las Quijadas, Provincia de San Luis, Argentina. *In*: Bildstein, K. L., Barber, D. R. & Zimmerman, A. (Eds.), Neotropical raptors, p. 234. Orwigsburg, PA: Hawk Mountain Sanctuary.

Pereyra Lobos, R., Muñoz, L., Rodríguez, R. & Mayorga, S. (2008). First record of anthropogenic mortality of a Crowned Solitary eag (Harpyhaliaetus coronatus) chick in the desert of Lavalle, Mendoze Provinca, Argentina. Nuestras Aves, 24, 27–28.

Pereyra Lobos, R. P., Santander, F. J., Orellana, S. A., Ramírez, P. A., Muñoz, L. & Bellón, D. F. (2011). Diet of the Crowned eagle (Harpyhaliaetu coronatus) during the breeding season in the Monte Desert, Mendoza, Argentina. Journal of Raptor Research, 45(2), 180–183. https://do org/10.3356/JRR-10-36.1

Pérez-García, J. M., Marco-Tresserras, J. & Orihuela-Torres, A. (2020). Winter diet and lead poisoning risk of Greater Spotted eagles Clang clanga in southeast Spain. Bird Study, 67(2), 224–231. https://doi.org/10.1080/00063657.2020.1810206

Pérez-Garcia, J. M., Margalida, A., Afonso, I., Ferreiro, E., Gardiazábal, A., Botella, F. & Sánchez-Zapata, J. A. (2013). Interannual home rang variation, territoriality and overlap in breeding Bonelli's eagles (Aquila fasciata) tracked by GPS satellite telemetry. Journal of Ornitholog 154(1), 63–71. https://doi.org/10.1007/s10336-012-0871-x

Pérez-León, R. A. (2002). Local distribution of diurnal raptorial birds en el Impossible National Park, El Salvador. Licenciado en Biologia. E Salvador, Philippines: Universidad de El Salvador, Ciudad Universitaria.

Pérez-López, M., De Mendoza, M. H., López-Beceiro, A. & Rodriguez, F. S. (2008). Heavy metal (Cd, Pb, Zn) and metalloid (As) content in rapto species from Galicia (NW Spain). Ecotoxicology and Environmental Safety, 70(1), 154–162. https://doi.org/10.1016/j.ecoenv. 2007.04.01

Pergantis, F. & Handrinos, G. (1992). A preliminary check-list of the birds of the Messolonghi wetlands (1860–1986) with seasonal occurrenc and population data. Biologia Gallo-Hellenica, 19, 89–112.

Perlman, Y. & Granit, B. (2012). Wintering Raptors in Israel 2010/11. Tel Aviv: Society for the Protection of Nature in Israel.

Péron, G., Fleming, C. H., Duriez, O., Fluhr, J., Itty, C., Lambertucci, S. & Calabrese, J. M. (2017). The energy landscape predicts flight heigh and wind turbine collision hazard in three species of large soaring raptor. Journal of Applied Ecology, 54(6), 1895–1906. https://do org/10.1111/1365-2664.12909

Perry, A., Kessler, M. & Helme, N. (1997). Birds of the Central Río Tuichi Valley, with emphasis on dry forest. *In*: Remsen, J. V., Jr. (Ed.), Studie in neotropical ornithology honoring Ted Parker, Ornithological monographs, 48. Parque N Madidi, pp. 557–576. Washington, DC: America Ornithologists' Union.

Perry, P. (2007). Lappetfaced vulture attacking Golden jackal at Gnu carcass. Locality Serengeti, Tanzania. Retrieved from http://ibc.lynxeds.com photo/lappet-faced-vulture-torgos-tracheliotus/lappetfaced-vulture-attacking-golden-jackal-gnu-carca.

Peters, J. L. (1931). Check-list of birds of the world (1st ed.); Vol I. Cambridge, MA: Harvard University Press.

Petersen, J. & Stalin, T. (2003). Influence of offshore windmills on migration birds – In southeast coast of Sweden. https://www.osti.gov/etdewe servlets/purl/20501066

Peterson, A. (1986). Habitat suitability index models: Bald eagle (breeding season). Biology of Reproduction. Washington, DC: United State Department of the Interior, Fish and Wildlife Service, 82, 10(126).

Peterson, A. T., Brooks, T., Gamauf, A., Gonzalez, J. C. T., Mallari, N. A. D., Dutson, G. & Fernandez, R. (2008). The avifauna of Mt. Kitangla Bukidnon Province, Mindanao, Philippines. Fieldiana Zoology, CO;2 No. 110, 114(1), TAOMKB2.0. https://doi.org/10.3158/001 0754(2008)1141

Peterson, A. T., Navarro-Sigüenza, A. G., Hernández-Baños, B. E., Escalona-Segura, G., Rebón-Gallardo, F., Rodríguez-Ayala, E., Figuero Esquivel, E. M. & Cabrera-García, L. (2003). The Chimalapas region, Oaxaca, Mexico: A high-priority region for bird conservation i Mesoamerica. Bird Conservation International, 13(3), 227–253. https://doi.org/10.1017/S0959270903003186

Peterson, C. (1987). Wrong-Way Eagle Jets Back Home from Ireland. Retrieved from https://www.washingtonpost.com/archive/politics/1987/12/2 wrong-way-eagle-jets-back-home-from-ireland/d98dd9a2-ef40-4684-9356-0865c549f921/

Peterson, R. T. (1936). The hawk migrations at Cape May Point, New Jersey. Auk, 53(4), 393–404. https://doi.org/10.2307/4078257

Peterson, R. T. (1969). The Osprey: Endangered world citizen. Natl. Geographic (July), 52–66.

Petersson, L., Brooks, T., Gamauf, A., Gonzalez, J. C. T., Mallari, N. A. D. & Dutson, G. (2016). First breeding record of Sanford's sea eagle Haliaeetus sanfordi, a Solomon Islands endemic. Australian Field Ornithology, 33, 222–223. https://doi.org/10.20938/afo33222223

Petrides, G. A. (1959). Competition for food between five species of East African vultures. Auk, 76(1), 104–106. https://doi.org/10.2307/4081860

Pettersson, J. (1977). White-tailed eagle, Haliaeetus albicilla, robbing Golden eagle, Aquila chrysaetos, of its prey. Vår Fågelvärld, 36, 54.

Petty, S. J. (1998). Ecology and Conservation of Raptors in Forests. London: HMSO Publications Centre.

Pfeifer, A. M., Verhelst, J. C. & Botero, J. E. (2001). State of conservation of birds of the Natural National Park Los Nevados and its buffer zone. Boletín Sao, 12(22), 21–41.

Philippine Eagle Foundation. (2008). Annual report. Davao City, Philippines: PEF.

Phillip, A. (2015). Why eagle feathers could land this Native American pastor in prison? Retrieved from https://www.washingtonpost.com/news/acts-of-faith/wp/2015/05/ 26/why-eagle-feathers-could-land-this-native-American-pastor-in-prison/

Phillips, R. & Martinez, R. (2013). Solitary Eagle Project Report. Campbell, CA: Belize Raptor Research Institute.

Phillips, R., Martinez, R., Sanchez M., Cordova, M., Bradshaw, A. & Britt, C. (2014). Nest and nest-site characteristics and prey of the Solitary eagle Buteogallus solitarius. Corpus Christi: Raptor Research Foundation Annual Conference.

Phillips, R. A. & Hatten, C. J. (2012). Nest observations on the Ornate Hawk-eagle (Spizaetus ornatus) in Belize, Central America. Boletin SAO, 21(1), 1–9.

Phillips, J. A. (1995). Movement patterns and density of Varanus albigularis. Journal of Herpetology, 29(3), 407–416. https://doi.org/10.2307/1564991

Phillips, Q. & Phillips, K. (2011). Phillips' Field guide to the birds of Borneo. Oxford, UK: John Beaufoy Publishing.

Phillips, R. (2009). Studying hawk-eagles in Belize. Neotropical Raptor Network Newsletter, 7(1), 10–11. Retrieved from http://www.globalraptors.org/grin/researchers/ uploads/245/nrn_ boletin_ english_june_

Phillips, R. (2012). An active nest of the rare Solitary eagle Harpyhaliaetus solitarius discovered in Belize. Spizaetus, 13, 2–8.

Phillips, R. A. & Hatten, C. J. (2012). Nest observations on the Ornate Hawk-eagle (Spizaetus ornatus) in Belize, Central America. Boletín Sao, 22(1 & 2), 1–9.

Phillips, R. A., Xavier, J. C. & Croxall, J. P. (2003). Effects of satellite transmitters on albatrosses and petrels. Auk, CO, 120(4), 1082–1090. https://doi.org/10.1642/0004-8038(2003)120 [1082:EOSTOA]2.0.CO;2

Phillips, R. L., McEneaney, T. P. & Beske, A. E. (1984). Population densities of breeding Golden eagles in Wyoming. Wildlife Society Bulletin, 12. 269–273.

Phillips, R. L. (1986). Current issues concerning the management of Golden eagles in Western U.S.A. *In*: Chancellor, R. D. & Meyburg, B.-U. (Eds.), Birds of Prey Bulletin No. 3, 149–156. Berlin, Germany: Proceedings of the Western Hemisphere Meeting of the World Working Group on Birds of Prey.

Phillips, R. L. & Blom, F. S. (1988). Distribution and magnitude of eagle/Livestock conflicts in the Western United States. *In*: Crabb, A. A. & Marsh, R. E. (Eds.), Proceedings of the Thirteenth Vertebrate Pest Conference, 13, 241–244. Davis, CA: University of California.

Phillps, R. L., Cummings, J. L. & Berry, J. D. (1991). Responses of breeding Golden eagles to relocation. Wildlife Society Bulletin, 19, 430-434.

Phillips, R. L., Cummings, J. L., Notah, G. & Mullis, C. (1996). Golden eagle predation on domestic calves. Wildlife Society Bulletin, 24, 468–470.

Philo, C. & Wilbert, C. (Eds.) (2000). Animal Spaces Beastly Spaces: New Geographies of Human–Animal Relations. London: Routledge.

Piana, R. (2007). Nesting and diet of Harpia harpyja (Linnaeus) in the native community of Infierno, Madre de Dios, Peru. Revista Peruana de Biología, 14, 135–138.

Piana, R. P. & Arpia, E. A. (2002). Harpia harpyja en la comunidad nativa de Infierno. Madre de Dios, Peru, Panamá, Brazil: Conferencia sobre Aves Rapaces Neotropicales y Simposio del Águila Arpía.

Piana, R. P., Crespo, S., Angulo, F., Ormaeche, E. & Alzamora, M. (2010). Grey hawk Buteo nitidus and Ornate hawk-eagle Spizaetus ornatus in north-west Peru. Cotinga, 32, 106–108.

Pienaar, U. D. V. (1969). Predator-prey relationships amongst the larger mammals of the Kruger National Park. Koedoe, 12(1). https://doi.org/10.4102/koedoe.v12i1.753

Pikacha, P., Sirikolo, M., Boseto, D. & Filardi, C. (2012). Ecological observations on Sanford's sea eagle Haliaeetus (leucogaster) sanfordi. Australian Field Ornithology, 29, 143–148.

Pimentel, L. & Olmos, F. (2011). The birds of Reserva Ecológica Guapiau (REGUA), Rio de Janeiro, Brazil. Cotinga, 33, 6–22.

Pineau, O. (1986). Observations d'Aiglesaiglesbottésen migration dans l'Aude. Guêpied, 4, 69.

Pinto, A. A. (1983). Ornithologia de angola. Lisbon: Instituto de Investigacao Cientifica Tropical.

Pinto, O. (1947). Contribution to the ornithology of the Lower Amazon: Critical study of a bird collection from Pará State. Arquivos de Zoologia do Estado de São Paulo, 5(6), 311–482.

Pinto, O. M. O. (1938). Catalogue of the birds of Brazil and list of the examples represented in the Museu Paulista. Revista do Museu Paulista, 22, 1–566.

Pinto, O. M. O. (1978). New catalogue of the birds of Brazil. First part. Non-Passeriformes and Oscine Passeriformes, excluding the family Tyrannidae. Published by the author. São Paulo.

Piper, R. (2007). Extraordinary Animals: An Encyclopedia of Curious and Unusual Animals. Greenwood Press.

Pitman, R. S. (1960). An unusual case of predation by Aquila verreauxi. Bulletin of the British Ornitological Club, 80, 67.

Planet of Birds. (2013). Lesser fish eagle [Icthyophaga humilis]. Retrieved from http://www.planetofbirds.com/accipitriformes-accipitridae-lesser-fish-eagle-icthyophaga-humilis

Platt, J. B. (1976). Bald eagles wintering in a Utah desert. American Birds, 30, 783–788.

Poirazidis, K., Bontzorlos, V., Schindler, S. & Vasilakis, D. (2019). Lesser Spotted eagle Clanga pomarina (Brehm, 1831) (Aves; Accipitridae) i Dadia-Lefkimi-Soufli national park, Greece: Population trends and spatial use in respect to continuous landscape changes during the last 3 years. Acta Zoologica Bulgarica Supplement, 14, 7–14.

Pokras, M. A. & Chafel, R. (1992). Lead toxicosis from ingested fishing sinkers in adult Common loons (Gaviaimmer) in New England. Journa of Zoo and Wildlife Medicine, 23, 92–97. Retrieved from http://www.jstor.org/stable/20460274.

Pokras, M., Kneeland, M., Ludi, A., Golden, E., Major, A., Miconi, R. & Poppenga, R. H. (2009). Lead objects ingested by Common loons in Nev England. Northeastern Naturalist, 16(2), 177–182. https://doi.org/10.1656/045.016.0202

Polite, C. & Pratt, J. (1990). Life History Account for Golden Eagle. California Department of Fish and Game, California Interagency Wildlif Task Group. Retrieved from http://nrm.dfg.ca.gov/FileHandler.ashx?DocumentVersionID=17730.

Pollock, C. & Chow, C. (2009). Heavy metal poisoning in birds. Retrieved from https://lafeber.com/vet/heavy-metal-poisoning-in-birds/

Ponder, J. B. & Willette, M. M. (2015). Chapter 23. Strigiformes. *In*: Fowler's zoo and wild animal medicine, pp. 189–198. https://doi.org/10.1016 B978-1-4557-7397-8.00023-2.

Ponganis, P. J. (2007). Bio-logging of physiological parameters in higher marine vertebrates. Deep Sea Research Part II: Topical Studies i Oceanography, 54, 18.

Ponnikas, S., Kvist, L., Ollila, T., Stjernberg, T. & Orell, M. (2013). Genetic structure of an endangered raptor at individual and population levels Conservation Genetics, 14(6), 1135–1147. https://doi.org/10.1007/s10592-013-0501-z

Poole, C. (2005). Tonle Sap: The Heart of Cambodia's Natural Heritage. Bangkok, Thailand: River Books Ltd.

Pooley, A. C. (1970). Powerful eagle – Martial eagle kills red duiker. Lammegeyer, 11, 82.

Porter, R. F. (1981). Flight Identification of European Raptors. London: A&C Black.

Porter, R. F., Christensen, S. & Schiermacker-Hansen, P. (1996). A Field Guide to the Birds of the Middle East. London: T. & A.D. Poyser.

Porter, R. & Aspinall, S. (2010). Birds of the Middle East. London: Christopher Helm Publishers.

Potapov, E., Utekhina, I., McGrady, M. & Rimlinger, D. (2010). Low breeding success of Steller's Sea eagles in Magadan District (Russia) i 2009: Start of a decline? Raptors Conservation, 18, 163–165.

Potapov, E., Utekhina, I. & McGrady, M. (2012). Source-sink populations of the Steller's Sea eagles (Haliaetus pelagicus) in the northern part o the sea of Okhotsk: Ecological traps and their conservation implications. Ornis Mongolica, 1, 20–25.

Potapov, E., Utekhina, I. & McGrady, M. (2015). Steller's sea eagle Haliaeetus pelagicus: Display and copulation in early spring. Birding ASIA 23, 77–83.

Pounds, J. A., Fogden, M. P. L. & Campbell, J. H. (1999). Biological response to climate change on a tropical mountain. Nature, 398(6728) 611–615. https://doi.org/10.1038/19297

Powell, G. V. N., Barborak, J. & Rodriguez, M. S. (2000). Assessing representativeness of protected areas in Costa Rica for conserving biodiversity A preliminary gap analysis. Biological Conservation, 93, 35–41.

Prakash, V. (1988). Lesser Spotted eagle (Aquila pomarinahastata) nesting in Keoladeo National Park, Bharatpur. Journal of the Bombay Natura History Society, 85(3), 614.

Prakash, V. (1996). Status, distribution and breeding biology of Lesser Spotted eagle Aquila pomarinahastata in Keoladeo National Park. *In* Meyburg, B. U. & Chancellor, R. D. (Eds.), Eagle studies, pp. 357–375. Berlin, London and Paris: World Working Group on Birds of Prey.

Prakash, V. (1999). Status of vultures in Keoladeo National Park, Bharatpur, Rajasthan with special reference to population crash in Gyps species Journal of the Bombay Natural History Society, 96, 365–378.

Prakash, V. & Rahmani, A. R. (1999). Notes about the decline of Indian vultures, with particular reference to Keoladeo National Park. Vulture' News, 41, 6–13.

Prakash, V., Green, R., Pain, D., Ranade, S., Saravanan, S., Prakash, N., Venkitachalam, R., Cuthbert, R., Rahmani, A. & Cunningham, A. (2007) Recent changes in populations of resident Gyps vultures in India. Journal of the Bombay Natural History Society, 104, 129-135.

Prakasha, V., Pain, D. J., Cunningham, A. A., Donald, P. F., Prakash, N., Verma, A., Gargia, R., Sivakumar, S. & Rahmania, A. R. (2003) Catastrophic collapse of Indian White-backed Gyps bengalensis and Long-billed Gyps indicus vulture populations. Biological Conservation 109(3), 381–390. https://doi.org/10.1016/S0006-3207(02)00164-7

Prater, S. H. (1940). The Indian Black eagle (Ictinaetus malayanus perniger Hodgs.) in Salsette. Journal of the Bombay Natural History Society 41(4), 899.

Pratt, T. K. & Beehler, B. M. (2015). Birds of New Guinea. Princeton, NJ: Princeton University Press.

Prawiradilaga, D. (2004). Javan hawk-eagle (Spizaetus bartelsi). Fauna and Flora, 6, 29.

Prawiradilaga, D. M. (1999). Javan Hawk-eagle: An endangered species. *In*: Prawiradilaga, D. M. (Ed.), Educational series on biodiversit conservation, 11. Bogor, Indonesia: Biodiversity Conservation Project LIPI-JICA-PKA.

Prawiradilaga, D. M. (2000). Conservation studies of Javan Hawk-eagle Spizaetus bartelsi: An overview. *In*: Ichinose, H. et al. (Eds.), Asia raptor research and conservation: Proceedings of the First Symposium on Raptors of Asia, pp. 118–127. Shiga, Japan: The Committee for th Symposium on Raptors of South East Asia.

Prawiradilaga, D. M. (2001). Diversity and distribution of raptors at Gunung Halimun National Park with particular reference to the Javan Hawk eagle. Berita Biologi, 5, 649–657.

Prawiradilaga, D. M. (2006). Ecology and conservation of endangered Javan hawk-eagle Spizaetus bartelsi. Ornithological Science, 5(2), 177–186, EACOEJ2.0.CO;2, https://doi.org/10.2326/1347-0558(2006)5177

Prawiradilaga, D. M., Røv, N., Gjershaug, J. O., Hapsoro & Supriatna, A. (2000). Feeding ecology of Javan hawk eagle (Spizaetus bartelsi) durin nestling period. *In*: Chancellor, R. D. & Meyburg, B.-U. (Eds.), Raptor at risk, pp. 559–563. Washington: World Working Group on Birds o Prey and Owls.

Prawiradilaga, D. M., Widyaningrum, I. K., Suparman, U., Noni, F. T., Ridwan, I., Kosar & Sola, E. (2005). Raptor survey at G. Endut and adjacent areas: The corridor of Halimun-Salak. *In*: Prawiradilaga, D. M. (Ed.), Report for JICA G. Halimun-Salak Management Project, 8. Bogor, Indonesia: Team Endangered Species JICA-GHSNPMP.

Preleuthner, M. & Gamauf, A. (1998). A possible new subspecies of the Philippine hawk-eagle (Spizaetus philippensis) and its future prospects. Journal of Raptor Research, 32(2), 126–135.

Premuda, G. (2002). Primidatisullamigrazione post-riproduttiva del Biancone, Circaetus gallicus, nelleAlpiApuane. Riv.rivistaital.Italia di ornitol. ornitologia, 71, 181–186.

Premuda, G. (2003). La migrazionedell'Aquilaminore Hieraaetus pennatus in Italia. Quaderni di birdwatching Birdwatching, 9.

Premuda, G. (2004a). La migrazioneautunnaledell'Aquilaminore, Hieraaetus pennatus, attraverso la PenisolaItaliana. And &Baghino, L. Rivista Italiana di Ornitologia, 74, 103–115.

Premuda, G. (2004b). First observations on the circuitous spring migration of Short-toed eagle, Cicaetus gallicus, in the Apuan Alps (Tuscany, Italy). Rivista Italiana di Ornitologia, 74, 119–124.

Premuda, G. (2004c). La migrazione del Biancone Circaetus gallicus in Italia: Statodelleconoscenzeattuali. *In*: Atti del Convegno, Rapaci in volo verso l'Appennino. CorpoForestaleDelloStato, C. F. Gestione ex ASFD di Lucca. RiservaNaturaleStataledell'Orecchiella, 21–24. La graficapisana, Bientina, 21–24. Cambridge, UK.

Premuda, G. (2006). La migrazionedeirapacisulleAlpiApuane (LU), campiCampiautunno (2006). *In*: Info Migrans n.18. Parco Naturale N. Valdieri, Italy: AlpiMarittime.

Premuda, G. (2007). L'importanzadelle Alpi Apuane per lo studio dellamigrazionedeirapaci. Acta Apuana, VI, 47–52.

Premuda, G., Baghino, L., Guillosson, T., Jardin, M., Tirado Bernat, M. & Esteller, V. (2007). A remarkable case of circuitous autumn migration of the booted eagle Hieraaetus pennatus through the western and central Mediterranean. Ardeola. 54, 349–357.

Premuda, G., Baghino, L., Gustin, M. & Borioni, M. (2010a). Data on spring migration of immature Short-toed eagles Circaetus gallicus through the Central Mediterranean route (Italy, Tunisia). Avocetta, 34, 65–68.

Premuda, G., Belosi, A., Viviani, F. & Franchini, M. (2015). Short-toed eagle Circaetus gallicus population monitoring at the Apuane Alps migration watch-site (Tuscany). Avocetta, 39, 5–12.

Premuda, G., Gustin, M., Pandolfi, M., Sonet, L. & Cento, M. (2008). Spring raptor migration along the Adriatic coast (Italy). A comparative study over three sites. Avocetta, 32, 13–20.

Premuda, G., Ricci, U. & Viviani, F. (2010b). RapacidelleAlpiApuane. Parco DelleAlpiApuane: PaciniEditore, Pisa.

Prestt, I., Jefferies, D. J. & Moore, N. W. (1970). Polychlorinated biphenyls in wild birds in Britain and their avian toxicity. Environmental Pollution, 1(1), 3–26, https://doi.org/10.1016/0013-9327(70)90003-0

Prieta, J. (2012). First White-tailed eagle for Extremadura. Retrieved from http://birds-extremadura.blogspot.com/2012/12/first-white-tailed-eagle-for-extremadura.html.

Prigogine, A. (1971). Les Oiseaux de l'ltombwe et de son Hinterland, 1. Musée Royal de l'Afrique centrale. Annales Série 8, Sciences Zoologiques, 185.

Prinsen, H. A. M., Boere, G. C., Píres, N. & Smallie, J. J. (2011). Review of the conflict between migratory birds and electricity power grids in the African-Eurasian region. CMS Technical Series No. XX, AEWA Technical Series No. XX Bonn.

Prinsen, H. A. M., Boere, G. C., Smallie, J. & Píres, N. (2011). Review of the conflict between migratory birds and the electricity power grids in the African-Eurasian Region. AEWA/CMS Technical Series No. 42, Bonn, Germany.

Prior, K. A. & Weatherhead, P. J. (1991). Competition at the carcass: Opportunities for social foraging by Turkey vultures in southern Ontario. Canadian Journal of Zoology, 69(6), 1550–1556. https://doi.org/10.1139/z91-218

Probst, R. (2009). Der Seeadler (Haliaeetus albicilla) in Österreich: Das WWF Österreich Seeadlerprojekt. Denisia, 27, 29–50.

Probst, R. & Gaborik, A. (2012). Action Plan for the conservation of the White-tailed Sea eagle (Haliaeetus albicilla) along the Danube. [Mitteleuropäische Gesellschaft zurErhaltung der Greifvögel (MEGEG)]. Central European Society for Raptor Protection.

Probst, R. & Peter, H. (2009). Der Seeadler (Haliaeetus albicilla) in Österreich: Eine Revision historischer Daten. Denisia, 27, 19–28.

Probst, R., Bogdea, L., Bandacu, D. S., Bohuš, M., Cheshmedzhiev, S., Gáborik, Á., Geissler, S., Hodor, C. V., Ionescu, D. T., Koev, V., Mikuska, T., Nagy, Z., Parrag, T., Rožac, V., Schmidt, M., Schneider, T., Šćiban, M., Tatai, S., Todorov, E., Tomik, A., Tucakov, M., Váczi, M. & Frank, G. (2014). The first comprehensive estimate of the winter population of the White-tailed eagle Haliaeetus albicilla along the Danube/Prva celostnaocenazimskepopulacijebelorepca Haliaeetus albicilla vzdolž Donave. Acrocephalus, 35(162–163), 115–123. https://doi.org/10.1515/acro-2014-0006

Probst, R., Kohler, B., Krone, O., Ranner, A. & Rössler, M. (2009). Schutzanforderungenfür den Seeadlerim Herzen Europas - Ergebnisse des Workshops der WWF-ÖsterreichTagung in Illmitz, 18, November (2007). Denisia 27: 147–157.

ProgramaMigres. (2009). Seguimento de la migracion de las-avesen el Estrecho de Gibraltar: Resultados del ProgramaMigres (2008). Migres, 1, 83–101.

Prout-Jones, D. V. & Kemp, A. G. (1997). Moult, plumage sequence and maintenance behaviour of a captive male and female Crowned eagle, Stephanoaetus coronatus (Aves; Accipitridae). Annals of the Transvaal Museum, 36(19).

Ptschelinzev, V. (2003). The White-tailed eagle in the northwest of Russia. In: Helander, B. (Ed.), Sea eagle 2000: Proceedings from the International Sea Eagle Conference in Bjorko, Sweden, 13–17 September 20, 121–126. Stockholm, Sweden Proceedings of the Swedish Society for Nature Conservation, SNF.

PubChem. (2020a). Aroclor, 1254. Retrieved from https://pubchem.ncbi.nlm.nih.gov/ compound/40470.

PubChem. (2020b). Dieldrin. Retrieved from https://pubchem.ncbi.nlm.nih.gov/compound/ Dieldrin

Pueblo-Olivares, F., Rodriguez-Ayala, E., Hernández-Baños, B. E. & Navarro S. A. G. (2002). Status and conservation of the avifauna of th Yaxchilán Natural Monument, Chiapas, Mexico. Ornitologia Neotropical, 13, 381–396.

Puga-Caballero, A., MacGregor-Fors, I. & Ortega-Álvarez, R. (2014). Birds at the urban fringe: Avian community shifts in different peri-urba ecotones of a megacity. Ecological Research, 29(4), 619–628. https://doi.org/10.1007/s11284-014-1145-2

Pulido Jiménez, J. F. & Scheidegger Soboll, D. (2002). Display enhancement impact on raptors in captivity at the Brasilia Zoo, Brazil. p. S44. 10 *In*: Program of Neotropical raptor conference and Harpy eagle symposium. Panamá, Panamá: Fondo Peregrino.

Puplett, D. (2020). Eagle mythology and folklore. Trees for Life. Retrieved from https://treesforlife.org.uk/into-the-forest/trees-plants-animal birds/eagle/eagle-mythology-and-folklore/

Purandare, K. (2002). Attempt by the Crested serpent eagle Spilornischeela to seize the Indian cobra Najanaja. Journal of the Bombay Natur History Society, 99(2), 299.

Purroy, F. J. (1997). Atlas de las aves de Espanã (1975–1995). Barcelona: Lynx Editions.

Puscariu, V. & Filipascu, A. (1975). The situation of birds of prey in Romania in 1970–1974. *In*: World Conference Birds of Prey. Vienna, 148–15

Pyke, G. H., Pulliam, H. R. & Charnov, E. L. (1977). Optimal foraging: A selective review of theory and tests. The Quarterly Review of Biolog 52(2), 137–154. https://doi.org/10.1086/409852

Pyle, R. L. & Pyle, P. (2009). The birds of the Hawaiian Islands: Occurrence, History, Distribution, and Status. Honolulu, HI: BP Global Bishc Museum.

QGIS Development Team. (2015). QGIS Geographic Information System. Open Source Geospatial Foundation.

Quick, D. (2018). Astronomy and ecology come together in drone designed to tackle poaching problem. Retrieved from https://newatlas.con animal-poaching-astro-ecology-drone/54066/.

Quinn, J. L. & Cresswell, W. (2004). Predator hunting behaviour and prey vulnerability. Journal of Animal Ecology, 73(1), 143–154. https://do org/10.1046/j.0021-8790.2004.00787.x

Quintero, I. & Jácome, A. (2020). Black hawk eagle (Spizaetus tyrannus), version 1.0. *In*: Schulenberg, T. S. (Ed.), Birds of the World. Ithaca, N Cornell Laboratory of Ornithology. https://doi.org/10.2173/bow.blheag1.01

Quiroga Durán, S. P. (2008). Aves rapacesdiurnascomoherramienta de indicación de calidadambientalentrestiposgenerales de ecosistema (bosque basal, bosque subandino y bosque andino) bajo la jurisdicción de la CorporaciónAutónoma de Coporpochivor, Boyacá. Bogotá, DC Colombia: Universidad Distrital Francisco José Morales Sánchez, Bogotá D.C., Colombia.

Rabarisoa, R. (1999). Nouveaux éléments sur la distribution du Pygargue de Madagascar Haliaeetus vociferoides dans la région du sud-ouest d Madagascar. *In*: Andrianarimisa, A. (Ed.), Projet de conservation des zones humides de Madagascar. 110. Conservation de la biodiversit versune gestion communautaire du complexelacustre de la Forêt de Tsimembo. Rapport d'avancement, V, 1997 and 1998, 110–11 AntAnarivo, Madagascar: The Peregrine Fund's Madagascar Project.

Rabarisoa, R., Watson, R. T., Thorstrom, R. & Berkelman, J. (1997). Status of the Madagascar Fish eagle Haliaeetus vociferoides in 1995. Ostric 68, 8–12.

Rabarisoa, R., Rafanomezantsoa, S. A. & Watson, R.T. (2003). Falconiformes; Haliaeetus vociferoides, Madagascar Fish eagle, Ankoay. *I* Goodman, S. M. & Benstead, J. P. (Eds.), The Natural History of Madagascar. Chicago, United States and London, UK: Chicago Universi Press.

Rabearivony, J., Fanameha, E., Mampiandra, J. & Thorstrom, R. (2008). Taboos and social contracts: Tools for ecosystem management - Lessor from the Manambolomaty Lakes RAMSAR site, western Madagascar. Madagascar Conservation & Development, 3, 7–16.

Rabor, D. S. (1965). The present status of the Monkey-eating eagle, Pithecophaga jefferyi Ogilvie-Grant, of the Philippines. *In*: Talbot, L. M. Talbot, M. H. (Eds.), Conservation in tropical Southeast Asia. pp. 312–314. Morges, Switzerland International Union for Conservation Nature and Natural Resources.

Rabor, D. S. (1971). The present status of conservation of the Monkey-eating eagle of the Philippines. Philippine Geographical Journal, 1 90–103.

Radovic, A. & Mikuska, T. (2009). Population size, distribution and habitat selection of the White-tailed eagle Haliaeetus albicilla in the alluvi wetlands of Croatia. Biologia, 64, 1–9. https://bib.irb.hr/datoteka/390711.haliaeetus_croatia.pdf

Radović, A. & Mikuska, T. (2009a). Population size, distribution and habitat selection of the White-tailed eagle Haliaeetus albicilla in the alluvi wetlands of Croatia. Biologia, 64(1), 156–164. https://doi.org/10.2478/s11756-009-0011-0

Radović, A. & Mikuska, T. (2009b). Testing the effect of persecution and permanent dispersion of sub-adult birds in long-term sustainability White Tailed Eagles (Haliaeetus albicilla L.) population at different management options in Croatia. Acta Zoologica Academiae Scientiaru Hungaricae, 55, 395–407.

Radović, D., Kralj, J., Tutiš, V. & Ciković, D. (2003). Red Book of Endangered Birds in Croatia. Ministry of Environmental Protection an Physical Planning.

Rafanomezantsoa, S. (1997). Behavior and natal dispersal of fledgling Madagascar Fish eagles. *In*: Watson, R. T. (Ed.), Madagascar wetlan conservation project: Developing community-based conservation in a proposed wetland Biosphere Reserve in Madagascar. Progress Rep III, 1995–1996, 403–412. Boise, ID: The Peregrine Fund.

Rafanomezantsoa, S. A. (1998). Contribution to the study of the behavior and the dispersal of young Madagascar Fish-eagles in the three lak complex of Antsalova. Madagascar: Antananarivo.

Raharjaningtrah, W. & Rahman, Z. (2004). Study on the distribution, habitat and ecology of Flores Hawk-eagle Spizaetus cirrhatus floris Lombok, Sumbawa, Flores, Komodo, and Rinca Islands, Nusa Tenggara, Indonesia. Annual Report of Pro Natura Fund, 13, 177–191.

Rainbow, P. S. (1997). Trace metal accumulation in marine invertebrates: Marine biology or marine chemistry? Journal of the Marine Biologic Association of the United Kingdom, 77(1), 195–210. https://doi.org/10.1017/S0025315400033877

Raine, A. F. (2007). Breeding bird records from the Tambopata-Candamo Reserve Zone, Madre de Dios, South-east Peru. Cotinga, 28, 53–58.
Rainey, H. J., Ikonga, J. M., Vernon, R. & Kingm, T. (2009). Additions to the avifauna of Congo-Brazzaville. Bulletin of the African Bird Club, 16, 53–600.
Rainier of Monaco & Bourne, GH. (Eds.). (1977). Primate Conservation. New York: Academic Press.
Rajamani, J. & Subramanian, M. (2015). Toxicity assessment on the levels of select metals in the critically endangered Indian White-backed Vulture, Gyps bengalensis, in India. Bulletin of Environmental Contamination and Toxicology, 94(6), 722–726. https://doi.org/10.1007/s00128-015-1548-y
Rajchard, J. & Prochaka. (2009). Restoration of sea eagle population: A review. Current Zoology, 55, s315–318.
Rajeshkumar, S., Raghunathan, C. & Venkataraman, K. (2015). Avifaunal community in Neil Island: A conservation approach in the evergreen forest ecosystem. Proceedings of the National Conference on Biodiversity for Sustainable Development, Chapter 20, 138–147. Lucknow, India: Uttar Pradesh State Biodiversity Board.
Ralls, K. & White, P. J. (1995). Predation on endangered San Joaquin kit foxes by larger canids. Journal of Mammalogy, 76(3), 723–729. https://doi.org/10.2307/1382743
Ramadan-Jaradi, G. & Ramadan-Jaradi, M. (1999). An updated checklist of the birds of Lebanon. Sandgrouse, 21, 132–170.
Ramadan-Jaradi, G., Bara, T. & Ramadan-Jaradi, M. (2008). Revised checklist of the birds of Lebanon (1999–2007).
Ramarao, D., Karuthedathu, D., Mohanram, K., Prakash, H. L., Raju, A. K., Sreekumar, H. & Das, V. (2011). On the breeding of Lesser Fish-eagle Ichthyophaga humilis in Cauvery Wildlife Sanctuary, Karnataka. Indian Birds, 7(1), 9–13.
Rambaut, A. (2012). FigTree v1.4.0. Oxford, UK: University of Oxford.
Rambaut, A. & Drummond, A. J. (2012). Tracer, MC Trace Analysis Tool, v.1.5.0. Oxford, UK: University of Oxford.
Ramírez-Bastida, P. & Navarro-Sigüenza, A. G. (2006). Current curatorial and bibliographic knowledge of Mexican Falconiformes. [Centro de Investigaciones Biológicas del Noreste, S.C. and Comisión Nacional para el Conocimiento y Uso de la Biodiversidad]. *In*: Rodríguez-Estrella, R. (Ed.), Current raptor studies in México, 231–253. La Paz: Baja Book Company.
Ramos, M. A. (1986). Birds in peril in Mexico: The diurnal raptors. Birds of Prey Bulletin, 3, 26-42.
Ramsay, B., Manore, M., Weir, L., Wilson, K. & Bradley, D. (1998). Use of RADARSAT data in the Canadian Ice Service. Canadian Journal of Remote Sensing, 24(1), 36–42.
Ramsay, S. L. (2002). Decapitation of the tortoise Chersinaangulata: Is the large grey mongoose a headhunter?: Short communication. African Zoology, 37(1), 114–116. https://doi.org/10.1080/15627020.2002.11657164
Ranada, P. (2015). First active Philippine eagle nest in Luzon found in Apayao. Rappler. Retrieved from http://www.rappler.com/science-nature/environment/99306-first-active-philippine-eagle-nest-luzon-apayao.
Rand, A. L. (1936). The distribution and habits of Madagascar birds. Bulletin of the American Museum of Natural History, 72, 143–500.
Rand, A. L. & Gilliard, E. T. (1967). Handbook of New Guinea birds. New York, NY: Natural History Press.
Rand, A. L. & Gilliard, E. T. (1967). Handbook of New Guinea birds. London: Weidenfeld & Nicolson.
Rangel-Salazar, J. L. & Enriquez-Rocha, P. L. (1993). Nest record and dietary items for the Black hawk-eagle (Spizaetus tyrannus) from the Yucatan Peninsula. Journal of Raptor Research, 27(2), 121–122.
Rao, D. V., Chandra, K. & Devi, K. (2013). Endemic animals of Andaman and Nicobar Islands. Zoological Survey of India, Kolkata, India.
Rao, V. U. S. (1968). The Black eagle Ictinaetus malayensis perniger within Bombay limits. Newsletter for Birdwatchers, 8(12), 6–7.
Raptor Resource Project. (2016). Eagle tracking: Can you do something about the tracking antenna? Retrieved from https://raptorresource.blogspot.com/2016/08/eagle-tracking-can-you-do-something.html
Raptor View Research Institute. (2020). Golden eagle research projects. Retrieved from http://raptorview.org/golden-eagle-research-projects.html
Rasamoelina, A. D. (2001). Impact of fishing on the piscivorous resources of the Madagascar Fish-eagle population of the lake complex of Manambolomaty, Antsalova, Madagascar. Ostrich Supplement, 15, 210–214.
Rasmussen, J. F., Rahbek, C., Poulsen, B. O., Poulsen, M. K. & Loch, H. (1996). Distributional records and natural history notes on threatened and little-known birds of southern Ecuador. Bulletin of the British Ornithologists' Club, 116, 26–46.
Rasmussen, P. C. & Anderton, J. C. (2012). Birds of South Asia: The Ripley Guide, 2. Attributes and status. Smithsonian Institution. Washington, D.C. and Barcelona: Lynx Edicions.
Ratcliffe, N., Guihen, D., Robst, J., Crofts, S., Stanworth, A. & Enderlein, P. (2015). A protocol for the aerial survey of penguin colonies using UAVs. Journal of Unmanned Vehicle Systems, 3(3), 95–101. https://doi.org/10.1139/juvs-2015-0006
Rattner, B. A., Horak, K. E., Lazarus, R. S., Schultz, S. L., Knowles, S., Abbo, B. G. & Volker, S. F. (2015). Toxicity reference values for chlorophacinone and their application for assessing anticoagulant rodenticide risk to raptors. Ecotoxicology, 24(4), 720–734. https://doi.org/10.1007/s10646-015-1418-8
Rattner, B. A., Whitehead, M. A., Gasper, G., Meteyer, C. U., Link, W. A., Taggart, M. A., Meharg, A. A., Pattee, O. H. & Pain, D. J. (2008). Apparent tolerance of Turkey vultures (Cathartes aura) to the non-steroidal anti-inflammatory drug diclofenac. Environmental Toxicology and Chemistry, 27(11), 2341–2345. https://doi.org/10.1897/08-123.1
Razafimanjato, G., de Roland, L.-A. R., Rakotondratsima, M., Razakaratrimo, S., Pruvot, Y. Z. M., RazafimAntsoa, T. M. &. Thorstrom, R. (2017). Madagascar fish eagle productivity in the Tsimembo-Manambolomaty Protected Area and surrounding habitat of western Madagascar. https://doi.org/10.2989/00306525.2017.1350887
Razafimanjato, G., Sam, T. S. & Thorstrom, R. (2007). Waterbird monitoring in the Antsalova region, Western Madagascar. Waterbirds, 30(3), 441–447, WMITAR2.0. CO;2, https://doi.org/10.1675/1524-4695(2007)0300441
Razafimanjato, G., Sam, T. S., Rakotondratsima, M., Rene De Roland, L. & Thorstrom, R. (2014). Population status of the Madagascar fish eagle Haliaeetus vociferoides in 2005–2006. Bird Conservation International, 24(1), 88–99. https://doi.org/10.1017/S0959270913000038

Razafindramana, S. (1995). Contribution à l'étude de la biologie de Haliaeetus vociferoides, Desmurs 1845 (Pygargue de Madagascar). Reproduction et domaine vital. Mémoire de Diplômed'EtudesApprofondies de Sciences Biologiques Appliqués. Antanarivo. Madagascar: Universitéd'AntAnarivo.
Razandrizanakanirina, D. & Kalavaha, L. (1997). Community-based wetland conservation project. *In*: Watson, R. T. (Ed.), Madagascar wetlands conservation project: Developing community-based conservation in a proposed wetland Biosphere Reserve in Madagascar. Progress Report III, 1995–1996, Mission notes, 1996, 105–120. Boise, ID: The Peregrine Fund.
Razgour, O., Hanmer, J. & Jones, G. (2011). Using multi-scale modelling to predict habitat suitability for species of conservation concern: The grey long-eared bat as a case study. Biological Conservation, 144(12), 2922–2930. https://doi.org/10.1016/j.biocon.2011.08.010
Rea, A. M. (1983). Cathartid affinities: A brief overview. *In*: Wilbur, S. R. & Jackson, J. A. (Eds.), Vulture Biology and Management, 26–54. CA: University of California Berkeley.
Read, J. L. (2013). The birds of Tetepare Island, Solomon Islands. Australian Field Ornithology, 30, 67–78.
Real, J. (1987). Evolucióncronológica del régimenalimenticio de una población de Hieraaetus fasciatus en Catalunya: Factorescausantes adaptación y efectos. Rapaci Mediterranei, 3, 185–205.
Real, J. (1996). Biases in diet study methods in the Bonelli"s eagle. Journal of Wildlife Management, 60(3), 632–638. https://doi.org/10.2307/3802082
Real, J. & Mañosa, S. (1997). Demography and conservation of Western European Bonelli's eagle Hieraaetus fasciatus populations. Biological Conservation, 79(1), 59–66. https://doi.org/10.1016/S0006-3207(96)00100-0
Real, J. & Mañosa, S. (2001). Dispersal of juvenile and immature Bonelli's eagles in northeastern Spain. Journal of Raptor Research, 35, 9–14.
Real, J., Grande, J. M., Mañosa, S. & Sánchez-Zapata, J. A. (2001). Causes of death in different areas for Bonelli's eagle Hieraaetus fasciatus in Spain. Bird Study, 48(2), 221–228. https://doi.org/10.1080/00063650109461221
Redig, P. T. & Arent, L. R. (2008). Raptor toxicology. Veterinary Clinics of North America. Exotic Animal Practice, 11(2), 261–282. https://doi.org/10.1016/j.cvex.2007.12.004
Redman, N., Stevenson, T. & Fanshawe, J. (2010). Birds of the Horn of Africa, Ethiopia. Djibouti, Eritrea, Somalia, and Socotra: Bloomsbury Publishing.
Redpath, S. M., Clarke, R., Madders, M. & Thirgood, S. J. (2001). Assessing raptor diet: Comparing pellets, prey remains, and observational data at Hen Harrier nests. Condor, 103(1), 184–188. https://doi.org/10.1093/condor/103.1.184
Redpath, S., Amar, A., Madders, M., Leckie, F. & Thirgood, S. (2002). Hen harrier foraging success in relation to land use in Scotland. Animal Conservation, 5(2), 113–118. https://doi.org/10.1017/S1367943002002159
Reeve, A. H., Haryoko, T., Poulsen, M. K., Fabre, P.-H. & Jonsson, K. A. (2014). New ornithological records from Buru and Seram, south Maluku Indonesia, 1995–2012. Forktail, 30, 10–22.
Reichel, W. L., Schmeling, S. K., Cromartie, E., Kaiser, T. E., Krynitsky, A. J., Lamont, T. G. & Swineford, D. M. (1984). Pesticide, PCB and lead residues and necropsy data for Bald eagles from 32 states—1978–1981. Environmental Monitoring and Assessment, 4(4), 395–403. https://doi.org/10.1007/BF00394177
Reichenow, A. 1900. Neue Artenvomsüdöstlichen Neuguinea und Nordqueensland. Ornithologische Monatsberichte, 8, 185–188.
Reid, D. & Reid, S. (1999). Sea eagle takes a magpie, terrorizes an island. *In*: NSW. Boobook, 19: 17.
Reid, T., Krüeger, S., Whitfield, D. P. & Amar, A. (2015). Using spatial analyses of Bearded vulture movements in southern Africa to inform wind turbine placement. Journal of Applied Ecology, 52(4), 881–892. https://doi.org/10.1111/1365-2664.12468
Reilly, J. R. & Reilly, R. J. (2009). Bet-hedging and the orientation of juvenile passerines in fall migration. Journal of Animal Ecology, 78, 990–1001. https://doi.org/10.1111/j.1365-2656.2009.01576.x
Reiser, O. (1905). MaterialenzueinerOmisBalcanica; III Griechenland und die GriechischenInseln. Wien.
Remsen, J. V. & Ridgely, R. S. (1980). Additions to the avifauna of Bolivia. Condor, 82(1), 69–75. https://doi.org/10.2307/1366786
Remsen, J. V., Jr. & Traylor, M. A., Jr. (1989). An annotated list of the birds of Bolivia. Vermillion, SD: Buteo Books.
Remsen, J. V., Jr. & Parker, T. A., III. (1995). Bolivia has the opportunity to create the planet's richest park for terrestrial biota. Bird Conservation International, 5(2–3), 181–199. https://doi.org/10.1017/S0959270900001003
Remsen, J. V., Jr., Cadena, C. D., Jaramillo, A., Nores, M., Pacheco, J. F., Robbins, M. B., Schulenberg, T. S., Stiles, F. G., Stotz, D. F. & Zimmer, K. J. (2008). A classification of the bird species of South America (section Accipitridae (Hawks) 3 note 1). Version 11 December (2008). American Ornithologists' Union.
Remsen, J. V., Jr., Cadena, C. D., Jaramillo, A., Nores, M., Pacheco, J. F., Robbins, M. B., Schulenberg, T. S., Stiles, F. G., Stotz, D. F. & Zimmer, K. J. (2009). A classification of the bird species of South America. Version 1 September (2009). American Ornithologists' Union. http://www.museum.lsu.edu/-Remsen/SACC Base line.html
Remsen, J. V., Jr., Areta, J. I., Cadena, C. D., Jaramillo, A., Nores, M., Pacheco, J. F., Perez-Emen, J., Robbins, M. B., Stiles, F. G., Stotz, D. F. & Zimmer, K. J. (2015). A Classification of the Bird Species of South America. American Ornithologists' Union. Version 13 May (2015). Retrieved from: http: //www.museum. lsu.edu/~Remsen/SACCBaseline05.htm
Remsen, V. (2008). Proposal (383) to South American Classification Committee: Separate Accipitriformes from Falconiformes. Retrieved from https://web.archive.org/web/2010 0628113348/ http://www.museum.lsu.edu/~Remsen/SACCprop383.html
Renjifo, L. M., Franco, A. M., Amaya-Espinel, J. D., Kattán, G. H. & López-Lanús, B. (Eds.) (2002). Libro rojo de aves de Colombia. Serie Libros Rojos de Especies Amenazadas de Colombia. Bogotá, Colombia: Instituto de Investigación de Recursos Biológicos Alexander von Humboldt, Ministerio del Medio Ambiente.
Renjifo, L. M., Servat, G. P., Goerck, J. M., Loiselle, B. A. & Blake, J. G. (1997). Patterns of species composition and endemism in the northern Neotropics: A case for conservation of montane avifaunas. *In*: Remsen, Jr., J. V. (Ed.), Studies in neotropical ornithology honoring Ted Parker, 577–594. Washington, DC: American Ornithologists Union.

Restall, R., Rodner, C. and Lentino, M. (2006). Birds of northern South America: An identification guide. Volume 1: Species accounts. London: Christopher Helm.

Restani, M., Harmata, A. R. & Madden, E. M. (2000). Numerical and functional responses of migrant Bald Eagles exploiting a seasonally concentrated food source. Condor, 102(3), 561–568. https://doi.org/10.1093/condor/102.3.561

Restrepo-Cardona, J. S., Márquez, C., Echeverry-Galvis, M. Á., Vargas, F., Sánchez-Bellaizá, D. M. & Renjifo, L. M. (2019). Deforestation may trigger Black-and-Chestnut eagle (Spizaetus isidori) predation on domestic fowl. Tropical Conservation Science, 12. https://doi.org/10.1177/1940082919831838.

Restrepo-Cardona, J. S., Echeverry-Galvis, M. Á., Maya, D. L., Vargas, F. H., Tapasco, O. & Renjifo, L. M. (2020). Human-raptor conflict in rural settlements of Colombia. PLoS One, 15(1), e0227704. https://doi.org/10.1371/journal.pone.0227704

Retana, S. A. V., Araya, H. D. & García, R. (2016). Record of depredation by Spizaetus ornatus on Nyctibius grandis in the Selva biological Station, Sarapiquí, Heredia, Costa Rica. Spizaetus, 22, 14–19.

Retief, E., Diamond, M., Anderson, M. D., Smit, D. H. A., Jenkins, D. A. & Brooks, M. (2012). Avian wind farm sensitivity map for South Africa. Johannesburg. Retrieved from http://hdl.handle.net/1854/LU-1039067.

Rettig, N. & Hayes, K. 1995. Remote world of the Harpy eagle. National Geographic, 187(2), 40–49.

Rettig, N. (1978). Breeding behavior of the Harpy eagle (Harpia harpyja). Auk, 95(4), 629–643. Retrieved from https://sora.unm.edu/sites/default/files/journals/auk/v095n04/p0.pdf.

Rettig, N. (2002). Informacion adicional sobre La conducta de anidacion del Aguila Arpia. Panamá, Brazil: Conferencia sobre Aves Rapaces Neotropicales y Simposio del Águila Arpía.

Reutebuch, S. E., Andersen, H. E. & McGaughey, R. J. (2005). Light detection and ranging (LIDAR): An emerging tool for multiple resource inventory. Journal of Forestry, 103(6), 286–292.

Revkin, A. C. (1997). High PCB Level is Found in a Hudson Bald Eagle. Retrieved from https://www.nytimes.com/1997/09/17/nyregion/high-pcb-level-is-found-in-a-hudson-bald-eagle.html

Reyes-Velarde, A. (2018). Young Mountain Lion Found Lounging in Pleasanton is Released to the Wild. Retrieved from https://www.latimes.com/local/lanow/la-me-ln-mountain-lion-pleasanton-20181030-story.html

Reymond, L. (1985). Spatial visual acuity of the eagle Aquila audax: A behavioural, optical and anatomical investigation. Vision Research, 25, 1477–1491. https://doi.org/10.1016/0042-6989(85) 90226-3

Reynolds, R. T. & Meslow, E. C. (1984). Partitioning of food and niche characteristics of coexisting Accipiter during breeding. The Auk, 101(4), 761–779. https://doi.org/10.2307/4086903

Rheindt, F. E. & Hutchinson, R. O. (2007). New island records and new elevational records of birds from South Maluku, Indonesia. Forktail, 23, 158–161.

Rheindt, F. E., Prawiradilaga, D. M., Suparno, Ashari, H. & Wilton, P. R. (2014). New and significant island records, range extensions and elevational extensions of birds in Eastern Sulawesi, its nearby satellites, and ternate. Treubia, 41, 67–97.

Rhymer, J. M. & Simberloff, D. (1996). Extinction by hybridization and introgression. Annual Review of Ecology and Systematics, 27(1), 83–109. https://doi.org/10.1146/nnurev.ecolsys 27.1.83

Ribon, R., Simon, J. E. & Theodoro De Mattos, G. (2003). Bird extinctions in Atlantic forest fragments of the Viçosa region, southeastern Brazil. Conservation Biology, 17(6), 1827–1839. https://doi.org/10.1111/j.1523-1739.2003.00377.x

Ricard, F., Berhongaray, G. & Viglizzo, E. (2019). The Argentine Pampas: A novel ecosystem at the crossroad. *In*: Elias, S. (Ed.), Module in earth systems and environmental sciences. https://doi.org/10.1016/B978-0-12-409548-9.12060-3

Richardson, C. (1990). The birds of the United Arab Emirates. Freehold, NJ: Hobby Publications.

Richardson, W. J. (1978). Timing and amount of bird migration in relation to weather: A review. Oikos, 30(2), 224–272. https://doi.org/10.2307/3543482

Richman, A. D. & Price, T. (1992). Evolution of ecological differences in the Old World leaf warblers. Nature, 355(6363), 817–821. https://doi.org/10.1038/355817a0

Rideout, B. A., Stalis, I., Papendick, R., Pessier, A., Puschner, B., Finkelstein, M. E. & Grantham, J. (2012). Patterns of mortality in free-ranging California condors (Gymnogyps californianus). Journal of Wildlife Diseases, 48(1), 95–112. https://doi.org/10.7589/0090-3558-48.1.95

Ridgely, R. S. (1980). Notes on some rare or previously unrecorded birds in Ecuador with additional information on identification and habitat. American Birds, 34, 242–248.

Ridgely, R. S. & Gwynne, J. A., Jr. (1989). A guide to the birds of Panama with Costa Rica, Nicaragua, and Honduras (2nd ed). Princeton, NJ: Princeton University Press.

Ridgely, R. S. & Greenfield, P. J. (2001). The birds of Ecuador, Volume I: Status, distribution, and taxonomy. Ithaca, NY: Comstock Publishing Associates.

Ridgely, R. S. & Greenfield, P. J. (2001). The birds of Ecuador, Volume II: Field Guide. Ithaca, NY: Cornell University Press.

Rieger, M. & Winkel, W. (1971). Über Todesursachenbeim Weisstorch (Ciconia ciconia) an Hand von Ringfundangaben. Vogelwarte, 26, 128–135.

Riesing, M. J., Kruckenhauser, L., Gamauf, A. & Haring, E. (2003). Molecular phylogeny of the genus Buteo (Aves: Accipitridae) based on mitochondrial marker sequences. Molecular Phylogenetics and Evolution, 27(2), 328–342. https://doi.org/10.1016/s1055-7903(02)00450-5

Rijn, S. V., Zijlstra, M. & Bijlsma, R. G. (2010). Wintering White-tailed eagles Haliaeetus albicilla in The Netherlands: Aspects of habitat scale and quality. Ardea, 98(3), 373–382. https://doi.org/10.5253/078.098.0311

Riley, J., Hunowu, Y., Mole, J. & Wangko, M. F. (2003). Noteworthy records of birds from the Panua Nature Reserve, North Sulawesi. Kukila, 12, 17–26.

Riney, T. (1951). Relationships between birds and deer. Condor, 53(4), 178–185. https://doi.org/10.2307/1364874
Ripley, S. D. (1982). A synopsis of the birds of India and Pakistan: Together with those of Nepal, Bhutan, Bangladesh, and Sri Lanka. Bomba India: Bombay Natural History Society.
Ripple, W. J., Estes, J. A., Beschta, R. L., Wilmers, C. C., Ritchie, E. G., Hebblewhite, M., Berger, J., Elmhagen, B., Letnic, M., Nelson, I Schmitz, O. J., Smith, D. W., Wallach, A. D. & Wirsing, A. J. (2014). Status and ecological effects of the world's largest carnivores. Scienc 343(6167), 1241484. https://doi.org/10.1126/science.1241484
Ristow, D., Wink, C. & Wink, M. (1982). Biology of Eleonora's falcon (Falco eleonorae), Individual and social defense behavior. Journal c Raptor Research, 16, 65–70.
Ritter, L. V. (1983). Growth, development, and behavior of nestling Turkey vultures in Central California. *In*: Wilbur, S. R. & Jackson, J. A. (Eds. Vulture biology and management, 287–302. Berkeley, CA: University of California Press.
Rivers, J. W., Johnson, J. M., Haig, S. M., Schwarz, C. J., Glendening, J. W., Burnett, L. J., George, D. & Grantham, J. (2014). Resource selectior by the California Condor (Gymnogyps californianus) relative to terrestrial-based habitats and meteorological conditions. PLoS One, 9(2 e88430. https://doi.org/10.1371/journal.pone.0088430
Riviera, G. & Yiu, K. (2013). Kyrgyzstan's Golden Eagle Hunters: In Search of a Legend. https://abcnews.go.com/International/kyrgyzstan golden-eagle-hunters-search-legend/story?id=18782383.
Robards, F. C. & Hodges, J. I. (1976). Observations from 2,760 Bald eagle nests in Southeast Alaska, progress report (1969–1976). Juneau, AK Fish and Wildlife Service.
Robbins, M. B. & Ridgely, R. S. (1990). The avifauna of an upper tropical cloud forest in southwestern Ecuador. Proceedings of the Academy c Natural Sciences of Philadelphia, 142, 59–71.
Robbins, M. B., Braun, M. J., Milensky, C. M., Schmidt, B. K., Prince, W., Rice, N. H., Finch, D. W. & O'Shea, B. J. (2007). Avifauna of the uppe Essequibo River and Acary Mountains, southern Guyana. Ornitologia Neotropical, 18, 339–368.
Robbins, M. B., Faucett, R. C. & Rice, N. H. (1999). Avifauna of a Paraguayan cerrado Locality: Parque Nacional Serranía San Luis, Dept Concepción. Wilson Bulletin, 111, 216–212.
Roberts, A. (1940). The Birds of South Africa. London: H. F. & G. Witherby.
Roberts, J. (2000). Romania: A bird watching and wildlife guide. Published by the author. Somerset, UK: Burton Expeditions.
Roberts, S. C. & Dunbar, R. I. M. (1991). Climatic influences on the behavioural ecology of Chanter's mountain reedbuck in Kenya. Africa Journal of Ecology, 29(4), 316–329. https://doi.org/10.1111/j.1365-2028.1991.tb00469.x
Robertson, A., Jarvis, A. M., Brown, C. J. & Simmons, R. E. (1998). Avian diversity and endemism in Namibia: Patterns from the Southern Africa Bird Atlas Project. Biodiversity & Conservation, 7(4), 495–511. https://doi.org/10.1023/A: 1008827628666.
Robinson, D. & Seywald, G. (1987). The birth of Blyth's hawk eagle. Malayan Naturalist, 412, 16–18.
Robinson, H. C. (1927). The Birds of the Malay Peninsular. Vol. I: The Commoner Birds. London: Houston Foundation & G. Whiterby.
Robinson, J. B. (1992). Hunting with Hawks. Field and Stream, 45, 86–88.
Robinson G. P. (1973). African Hawk Eagle (Hieraaetus spilogaster) eating a squirrel. Bulletin of the Zambian Ornithological Society, 5, 77.
Robinson, J. G. (1988). Group size in wedge-capped capuchin monkeys Cebus olivaceus and the reproductive success of males and female Behavioral Ecology and Sociobiology, 23(3), 187-197. https://doi.org/10.1007/BF00300353
Robinson, J. G. & Bennett, E. L. (2000). Hunting for Sustainability in Tropical Forests. Irvinton, NY: Columbia University Press.
Robinson, S. K. (1994). Habitat selection and foraging ecology of raptors in Amazonian Peru. Biotropica, 26(4), 443–458. https://do org/10.2307/2389239.
Robinson, W. D. (1999). Long-term changes in the avifauna of Barro Colorado Island, Panama, a tropical forest isolate. Conservation Biolog 13(1), 85–97. https://doi.org/10.1046/j.1523-1739.1999.97492.x
Robson, C. & Allen, R. (2005). Birds of Southeast Asia: Thailand, Peninsular Malaysia, Singapore, Vietnam, Cambodia, Laos, Myanma Princeton, NJ: Princeton University Press.
Robson, C. (2008). A Field Guide to the Birds of South-East Asia. London: New Holland.
Rocamora, G. (1994). Bonelli's eagle Hieraaetus fasciatus. *In*: Tucker, G. M. & Heath, M. F. (Eds.), Birds in Europe: Their conservation statu pp. 184–185. Cambridge: Birdlife International.
Roche, C. (2006). Breeding records and nest site preference of Hooded vultures in the Greater Kruger National Park. Ostrich, 77(1–2), 99–10 https://doi.org/10.2989/00306520609485515
Rockingharm-Gill, D. V. (2001). Field observations. Honeyguide, 47, 101–105.
Rockwell, A. (2016). Flying the Phantom 3 Pro with Turkey Vultures – Nearly Attacked. Retrieved from https://www.youtube.con watch?v=nb3Gfo_0iys
Rodwell, S. P., Sauvage, A., Rumsey, S. J. R. & Bräunlich, A. (1996). An annotated check-list of birds occurring at the Parc National des Oiseau du Djoudj in Senegal, 1984–1994. Malimbus, 18, 74–111.
Roda, S. A. & Carlos, C. J. (2003). New records for some poorly known birds of Atlantic forest of northeastern Brazil. Cotinga, 20, 17–20.
Roda, S. A. & Pereira, G. A. (2006). Recent distribution and conservation of the forest raptor species of central Pernambuco. Revista Brasilei de Ornitologia, 14, 331–344.
Roderick, M. (2014). Observation of a White-bellied Sea eagle taking submerged seabird prey. The Whistler, 8, 56–57.
Rodrigues, A. G., Borges-Martins, M. & Zilio, F. (2018). Bird diversity in an urban ecosystem: The role of local habitats in understanding th effects of urbanization. Iheringia, Série Zoologia, Vol. 108. https://doi.org/10.1590/1678-4766e2018017
Rodriguez-Ayala, E., Hernández-Baños, B. E., Navarro, A. G. & Pueblo-Olivares, F. (2002). Status and conservation of the avifauna of th Yaxchilán Natural Monument, Chiapas, Mexico. Ornitologia Neotropical, 13, 381–396.

Rodriguez-Estella, R. & Rivera-Rodriguez, L. (1992). Kleptoparasitism and other interactions of Crested Caracara in the Cape Region, Baja California, Mexico. Journal of Field Ornithology, 63(2), 177–180.

Rodríguez-Estrella, R. & Rivera-Rodríguez, L. B. (2006). Raptor studies in México: An overview. *In*: Rodríguez-Estrella, R. (Ed.), Current raptor studies in México. Centro de Investigaciones Biológicas del Noreste, S.C. and Comisión Nacional para el Conocimiento y Uso de la Biodiversidad, La Paz, Baja California Sur, Mexico (1–32).

Rodziewicz, M. (1996). The status, range and breeding success of the Lesser Spotted eagle Aquila pomarina in Poland. *In*: Meyburg, B.-U. & Chancellor, R. D. (Eds.), Eagle studies. World Working Group on Birds of Prey and Owls.

Roemer, G. W., Donlan, C. J. & Courchamp, F. (2002). Golden eagles, feral pigs, and insular carnivores: How exotic species turn native predators into prey. Proceedings of the National Academy of Sciences of the United States of America, 99(2), 791–796. https://doi.org/10.1073/pnas.012422499

Roesch, L. F. W., Vieira, F. C. B., Pereira, V. A., Schünemann, A. L., Teixeira, I. F., Senna, A. J. T. & Stefenon, V. M. (2009). The Brazilian Pampa: A Fragile Biome. Diversity, 1, 182–198. doi.: 10.3390/d1020182

Roesler, I. & Juhant, M. (2004). Raptors of Calilegua National Park, Jujuy, Argentina. *In*: Proceedings of the First Argentine Symposium on Investigation and Conservation of Raptors - SAICR I (pp. 3–4). Museo de la Plata, Buenos Aires, Argentina: Museo de La Plata.

Rogers, J. M. & Kavlock, R. J. 1996. Developmental toxicology. *In*: Klaassen, C. D. (Ed.), Casarett & Doull's Toxicology (5th ed.), 301–331. New York: McGraw-Hill.

Rogers, M. J. & Rarities Committee (1980). Report on rare birds in Great Britain in 1980. British Birds, 73(11), 491–534.

Rohner, C. & Krebs, C. J. (1998). Response of Great horned owls to experimental "Hot Spots" of Snowshoe hare density. The Auk, 115(3), 694–705. https://doi.org/10.2307/4089417

Rollan, À., Real, J., Bosch, R., Tintó, A. & Hernández-Matías, A. (2010). Modelling the risk of collision with power lines in Bonelli's eagle Hieraaetus fasciatus and its conservation implications. Bird Conservation International, 20(3), 279–294. https://doi.org/10.1017/ S0959270910000250

Rondeau, G. & Thiollay, J.-M. (2004). West African vulture decline. Vulture News, 51, 13–33.

Rood, J. P. (1983). Banded mongoose rescues pack member from eagle. Animal Behaviour, 31(4), 1261–1262. https://doi.org/10.1016/S0003-3472(83)80036-0

Rood, J. P. (1986). Ecology and social evolution in the mongooses. Ecological Aspects of Social Evolution, 131, 152.

Rose, A. B. (1973). Food of some Australian birds. Emu – Austral Ornithology, 73(4), 177–183. https://doi.org/10.1071/MU973177

Rosenfield, R. N., Sonsthagen, S. A., Stout, W. E. & Talbot, S. L. (2015). High frequency of extra-pair paternity in an urban population of Cooper's Hawks. Journal of Field Ornithology, 86(2), 144–152. https://doi.org/10.1111/jofo.12097

Rosenvald, R. & Lõhmus, A. (2003). Nesting of the Black stork (Ciconia nigra) and White-tailed eagle (Haliaeetus albicilla) in relation to forest management. Forest Ecology and Management, 185(3), 217–223. https://doi.org/10.1016/S0378-1127(03)00216-0

Røskaft, H. C. & Stokke, B. G. (2010). Collision risk in White-tailed eagles. Modelling collision risk using vantage point observations in Smøla wind-power plant. NINA Report 639.

Rosko, L. (1948). Losses of sheep from predatory animals on summer range in Iron County, Utah. Utah Coop Wildlife Research Unit Report.

Rotenberg, A. J., Marlin, J., Meacham, S. & Tolfree, S. (2000). An integrated community-based Harpy eagle and avian conservation program for the Maya Mountains Massif, Belize. Proceedings of the Fourth International Partners in Flight Conference: Tundra to Tropics, pp. 493–507.

Rotenberg, J. A. & Marlin, J. A., Pop. L. & Garcia, W. (2012). First record of a Harpy eagle (Harpia harpyja) nest in Belize. Wilson Journal of Ornithology, 124(2), 292–297. https://doi.org/10.1676/11-156.1

Roulin, A. & Wink, M. (2004). Predator-prey relationships and the evolution of colour polymorphism: A comparative analysis in diurnal raptors. Biological Journal of the Linnean Society, 81(4), 565–578. https://doi.org/10.1111/j.1095-8312.2004.00308.x

Round, P. D. (1988). Resident forest birds in Thailand: Their status and conservation. International Council for Bird Preservation Monographs 2, 1–211.

Round, P. D. & Jukmongkol, R. (2003). Recent Reports, late October 2002–February 2003. BCST Bulletin, 20(4), 10–20.

Røv, N. & Gjershaug, J. O. (2000). Population density, territory size and habitat use of Gurney's eagle Aquila gurneyi in the North Moluccas, Indonesia. *In*: Chancellor, R. D. & Meyburg, B.-U. (Eds.), Raptors at Risk, 677–683. Berlin: World Working Group on Birds of Prey and Owls.

Roxburgh, L., Bruce-Miller, E. & Wood, C. (2009). Lesser Spotted eagle recovery in Zambia. Gabar, 19: 31–33.

Rozenstein, O. & Karnieli, A. (2011). Comparison of methods for land-use classification incorporating remote sensing and GIS inputs. Applied Geography, 31(2), 533–544. https://doi.org/10.1016/j.apgeog.2010.11.006

RSPB. (2006). Cited in BBC, Wind farm 'hits eagle numbers'. Retrieved from http://news.bbc.co.uk/2/hi/europe/5108666.stm

Ruddock, M. & Whitfield, D. (2007). A review of disturbance distances in selected bird species. Report to Scottish Natural Heritage.

Rudebeck, G. (1950). The choice of prey and modes of hunting of predatory birds with special reference to their selective effect. Oikos, 2(1), 65–88. https://doi.org/10.2307/3564663

Rudol, P. & Doherty, P. (2008). Human body detection and geolocalization for UAV search and rescue missions using color and thermal imagery. *In*: Proceedings of the 2008 IEEE Aerospace Conference, 1–8. Big Sky, MT.

Ruggieri, L. (Ed.). (2005). Annuario (2004). Verona: Edizioni EBN Italiaa.

Ruggieri, L., Premuda, G., Baghino, L. & Giraudo, L. (2006). A large-scale survey of the autumn migration of Short-toed eagles Circaetus gallicus across Italy and the central Mediterranean. Avocetta, 30, 76–80.

Rundell, P. W., Graham, E. A., Allen, M. F., Fisher, J. C. & Harmon, T. C. (2009). Environmental sensor networks in ecological research. New Phytologist, 182(3), 589–607. https://doi.org/10.1111/j.1469-8137.2009.02811.x

Runge, C. A., Martin, T. G., Possingham, H. P., Willis, S. G. & Fuller, R. A. (2014). Conserving mobile species. Frontiers in Ecology and th Environment, 12(7), 395–402. https://doi.org/10.1890/130237

Rush, G. P., Clarke, L. E., Stone, M. & Wood, M. J. (2018). Can drones count gulls? Minimal disturbance and semiautomated image processing with an unmanned aerial vehicle for colony-nesting seabirds. Ecology and Evolution, 8(24), 12322–12334. https://doi.org/10.1002/ece3.449

Rushworth, I. & Krüger, S. (2014). Wind farms threaten southern Africa's cliffnesting vultures. Ostrich, 85(1), 1–11.

Russell, K. R. & Gauthreaux, Jr., S. (1998). A. Use of weather radar to characterize movements of roosting purple martins. Wildlife Society Bulletin, 26, 5–16.

Russell, S. M. (1964). A distributional study of the birds of British Honduras. Ornithological Monographs, 1, 1–195. https://doi.org/10.2307/4016666

Russell, S. M. & Monson, G. (1998). The birds of Sonora. Tucson, AZ: University of Arizona Press.

Russian Raptor Research and Conservation Network. (2018). White-tailed eagle was chosen the bird of 2013 in Russia. Russian Raptor Research and Conservation Network 2018.

Rutkiewicz, J., Nam, D. H., Cooley, T., Neumann, K., Padilla, I. B., Route, W., Strom, S. & Basu, N. (2011). Mercury exposure and neurochemica impacts in Bald eagles across several Great Lakes states. Ecotoxicology, 20(7), 1669–1676. https://doi.org/10.1007/s10646-011-0730-1

Rutledge, L. Y., White, B. N., Row, J. R. & Patterson, B. R. (2012). Intense harvesting of eastern wolves facilitated hybridization with coyotes Ecology and Evolution, 2(1), 19–33. https://doi.org/10.1002/ece3.61

Rutz, C. & Bijlsma, R. G. (2006). Food-limitation in a generalist predator. Proceedings of the Royal Society of London. Series B: Biologica Sciences, 273(1597), 2069–2076. https://doi.org/10.1098/rspb.2006.3507

Rutz, C. & Hays, G. C. (2009). New frontiers in biologging science. Biology Letters, 5(3), 289–292. https://doi.org/10.1098/rsbl.2009.0089

Ruxton, G. D. & Houston, D. C. (2004). Obligate vertebrate scavengers must be large soaring fliers. Journal of Theoretical Biology, 228, 431–436 https://doi.org/10.1016/j.jtbi.2004.02.005

Ruzicka, V. A. (2006). The Influence of Predation on the Nesting Ecology of Diamondback Terrapins (Malaclemys terrapin) in the Lowe Chesapeake Bay. Dissertations, Theses, and Master's Projects. Journal of Field Ornithology, 53, 49–51.

Rydell, J., Engström, H., Hedenström, A., Larsen, J. K., Pettersson, J. & Green, M. (2012). The effect of wind power on birds and bats – A synthesis. Stockholm, Sweden: Swedish Environmental Protection Agency.

Rylands, A. B. (Ed.) (1993). Marmosets and tamarins: Systematics, behaviour, and ecology. Oxford: Oxford University Press.

Ryser, F. A. & Ryser, F. A., Jr. (1985). Birds of the Great Basin: A Natural History. University of Nevada Press.

Ryves, V. W. (1938). Notes. Cited in Global Raptor Information Network. 2020. Species account: Crested hawk-eagle Nisaetus cirrhatus. Retrieve from http: //www.globalraptors.org on 5 Nov. 2020.

Sabine, N. & Gardner, K. (1987). Agonistic encounters between Bald eagles and other raptors wintering in west central Utah. Journal of Rapto Research, 21, 118–120.

Sadvi, K. M. (2009). Occurrence of the White-bellied Sea eagle Haliaeetus leucogaster in inland waters. Indian Birds, 5(2), 50. Retrieved from http: //indianbirds.in/pdfs/IB.5.2_50.pdf.

Sæther, B. E. & Bakke, Ø. (2000). Avian life history variation and contribution of demographic traits to the population growth rate. Ecology, 81(3) 642–653. https://doi.org/10.1890/0012-9658(2000)0810642: ALHVAC2.0.CO;2.

Safford, R. J. & Duckworth, J. W. (Eds.) (1990). A wildlife survey of Marojejy nature reserve, Madagascar. International Council for Bir Preservation Study Rep., 40, 1–184.

Safford, R. J. & Hawkins, A. F. A. (2013). The Birds of Africa. Volume VIII: The Malagasy Region. London: Christopher Helm Publishers.

Safi, K., Kranstauber, B., Weinzierl, R., Griffin, L., Rees, E. C., Cabot, D. & Bohrer, G. (2013). Flying with the wind: Scale dependency of spee and direction measurements in modelling wind support in avian flight. Movement Ecology, 1(1), 4. https://doi.org/10.1186/2051-3933-1-4.

Saggese, M. D., Quaglia, A., Lambertucci, S. A. et al. (2009). Survey of lead toxicosis in free-ranging raptors from central Argentina. *In*: Watson R. T., Fuller, M., Pokras, M. & Hunt, G. (Eds.), Ingestion of lead from spent ammunition: Implications for wildlife and humans. The peregrin Fund, pp. 223–231. Boise, ID.

Saini, M., Taggart, M. A., Knopp, D., Upreti, S., Swarup, D., Das, A., Gupta, P. K., Niessner, R., Prakash, V., Mateo, R. & Cuthbert, R. J. (2012) Detecting diclofenac in livestock carcasses in India with an ELISA: A tool to prevent widespread vulture poisoning. Environmental Policy and Law, 60, 11–16.

Saito, K. (2009). Lead poisoning of Steller's Sea eagle (Haliaeetus pelagicus) and White-tailed eagle (Haliaeetus albicilla) caused by the ingestio of lead bullets and slugs in Hokkaido, Japan. *In*: Watson, R. T., Fuller, M., Pokras, M. & Hunt, W. G. (Eds.), Ingestion of lead from spen ammunition: Implications for wildlife and humans. Boise, ID: The Peregrine Fund.

Saito, K. (2017). Changing Environment of the Steller's Sea eagles. Retrieved from http://www.foejapan.org/ en/aid/jbic02/sakhalin ChangingEnvironment.pdf

Salaman, P. G. W., Donegan, T. M. & Cuervo, A. M. (1999). Ornithological surveys in Serranía de los Churumbelos, southern Colombia. Cotinga 12, 29–39.

Salaman, P., Cuadros, T., Jaramillo, J. G. & Weber, W. H. (2001). Checklist of the birds of Colombia. Sociedad Antioqueña de Ornitologia Medellin, Colombia.

Salaman, P., Donegan, T. M., Cuervo, A. M., Davison, D. & Ochoa, J. M. (2007). Birds of Serranía de los Churumbelos, their conservation an elevational distribution. Conservación Columbiana, 3, 29–58.

Salomonsen, F. (1967). The Seabirds of Greenland. Copenhagen: Rhodos.

Salomonsen, F. (1979). Ornithological and ecological studies in SW Greenland. Meddelelser om Groenland, 204(6). Nyt Nordisk Forlag Arnol Busck, Copenhagen, 66–78.

Salvador, D. & Quiblatt, T. (2009). Philippine Eagle Foundation annual report (2008). Davao City, Philippines: Philippine Eagle Foundation.

Salvador, D. J. I. & Ibanez, J. C. (2006). Ecology and conservation of Philippine eagles. Ornithological Science, 5(2), 171–176. https://doi.org/10.2326/1347-0558(2006)5171: EAC OPE2.0.CO;2.

Salvador, L. (2008). Raptor survey in the upper Rio Doce Valley, southeast Brazil. Neotropical Raptor Network Newsletter, 6, 4–5.

Salvador, L. F., Jr. & Silva, F. A. (2009). Diurnal raptors in a fragmented Atlantic Forest landscape in the upper Doce River, Minas Gerais, Brazil. Boletim do Museu de Biologia Mello Leitão (Nuevo Serie), 25, 53–65.

Salvador, S. A. & Eroles, P. G. (1994). Notes on birds of Santiago del Estero. Nuestras Aves, 12, 24–25.

Salvador, V., Santillán, M., Pereyra Lobos, R., Maceda, J. J., Bregagnolo, L., Galmes, M., Peretti, V., Reyes, M., Sarasola, J. H. & Negro, J.J. (2007). Mortality of large eagles caused by powerlines in central Argentina. *In*: Bildstein, K. L., Barber, D. R. & Zimmerman, A. (Eds.), Neotropical ornithology, 338. Orwigsburg: Hawk Mountain Sanctuary.

Salvador-Jr., L. F., Canuto, M., Carvalho, C. E. A. & Zorzin, G. (2011). Aves, Accipitridae, Spizaetus tyrannus (Wied, 1820), New records in the Quadrilátero Ferrífero region, Minas Gerais, Brazil. Check List, 7(1), 032–036. https://doi.org/10.15560/7.1.32

Samago, L. (1995). White-tailed eagle – Haliaeetus albicilla L. Propagation of the Urals and adjacent territories to poly-terrestrial sources, 107–122.

Samant, J. S., Prakash, V. & Naoroji, R. (1995). Ecology and behaviour of resident raptors with special reference to endangered species: Final Report 1990-1993. Bombay, India: Bombay Natural History Society.

Samarawickrama, P., Ariyasiri, K., Menike, U., Samarasingha, N. & Meegaskumbura, M. (2011). Observations of Black Eagle (Ictinaetus malayensis) Nesting in Sri Lanka, 38(2), 1–4.

Samuelsen, R. (2008). The Field of Geomatics: Scientific, Technological, Business and Infrastructural Aspects. NF-working Paper 1027/2008.

Sanaiotti, T. M. (2002). Ubicacion geografica y monitoreo de los nidos del Aguila Arpia en la Amazonia brasilena. Panamá, Brazil: Conferencia sobre Aves Rapaces Neotropicales y Simposio del Águila Arpía.

Sanaiotti, T. M., Junqueira, T. G., Palhares, V., Aguiar-Silva, F. H., Henriques, L. M., Oliveira, G., Guimaraes, V. Y., Castro, V., Mota, D., Trombin, D. F., Villar, D. N. A., Lara, K. M., Fernandes, D., Castilho, L., Yosheno, E., Alencar, R. M., Cesca, L., Dantas, S. M., Laranjeiras, T. O., Mathias, P. C. & Mendonça, C. V. (2015). Abundance of Harpy and Crested eagles from a reservoir-impact area in the low- and mid-Xingu River. Brazilian Journal of Biology = Revista Brasleira de Biologia, 75(3)(Suppl. 1), 190–204. https://doi.org/10.1590/1519-6984.00614BM

Sanaiotti, T. M., Rittl Filho, C., Luz, B. B. & Soler, I. G. P. (2001). Dieta de filhotes de Harpia harpyja na região de. *In*: Straube, F.C. (Ed.), Ornitologia sem fronteiras, 349–350. Curitiba, Brazil: Pontifícia Universidade Católica do Paraná.

Sánchez, R., Margalida, A., González, L. M. & Oria, J. (2008). Biases in diet sampling methods in the Spanish Imperial eagle Aquila adalberti. Ornis Fennica, 85(3), 82-89.

Sánchez, R., Margalida, A., González, L. M. & Oria, J. (2009). Temporal and spatial differences in the feeding ecology of the Spanish Imperial eagle Aquila adalberti during the non-breeding season: Effects of the rabbit population's crash. Acta Ornitologica, 44(1), 53–58. https://doi.org/10.3161/000164509X464885

Sanchez-Lalinde, C., Velez-Garcia, F., Cornelio, A. C., Silveira, L. F. & Alvarez, M. R. (2011). Records of the Harpy eagle (Harpia harpyja) in the Serra Bonita reserves complex, Camacan, Bahia, with evidence of breeding. Revista Brasileira de Ornitologia, 19, 436–438.

Sánchez-Zapata, J. A., Eguía, S., Blázquez, M., Moleón, M. & Botella, F. (2010). Unexpected role of ungulate carcasses in the diet of Golden eagles Aquila chrysaetos in Mediterranean mountains. Bird Study, 57(3), 352–360. https://doi.org/10.1080/00063651003674946

Sanderfoot, O. V. & Holloway, T. (2017). Air pollution impacts on avian species via inhalation exposure and associated outcomes. Environmental Research Letters, 12(8), 1–16. https://doi.org/10.1088/1748-9326/aa8051

Sandor, A. D., Alexe, V., Marinov, M., Dorosencu, A., Domsa, C. & Kiss, B. J. (2015). Nest-site selection, breeding success, and diet of White-tailed eagles (Haliaeetus albicilla) in the Danube Delta, Romania. Turkish Journal of Zoology, 39(2), 300–307. https://doi.org/10.3906/zoo-1401-64

Sangster, G., Knox, A. G., Helbig, A. J. & Parkin, D. T. (2002). Taxonomic recommendations for European birds. Ibis, 144(1), 153–159. https://doi.org/10.1046/j.0019-1019.2001.00026.x

Sangster, G., Collinson, J. M., Helbig, A. J., Knox, A. G. & Parkin, D. T. (2005). Taxonomic recommendations for British birds: Third report. Ibis, 147(4), 821–826. https://doi.org/10.1111/j.1474-919X.2005.00483.x

Sant, N. R., Shelke, V. D. & Shelke, S. D. (2017). Polygyny and kleptoparasitism in the Indian Spotted eagle Clanga hastata in Belgaum, Karnataka, India. Indian Birds, 13(6), 141–146.

Sant, N., Prabhukhanolkar, R., Shelke, V. & Shelke, S. (2020). Breeding ecology of the Indian Spotted Eagle Clanga hastata around Belgaum, India. Indian Birds, 16(6), 176–184.

Santillán, M. A., Carpintero, D. L., Galmes, M. A. & Sarasola, J. H. (2009). Presence of cimicid bugs (Hemiptera: Cimicidae) on a Crowned eagle (Harpyhaliaetus coronatus) nestling. Journal of Raptor Research, 43(3), 255–256. https://doi.org/10.3356/JRR-08-57.1

Santos, J. (2015). Eagle Hunters and Summer Migration. Retrieved from http://joelsantos.net/mongoliaeagle-hunters-and-summer-migration/

Santos, J. C., Leal, I. R., Almeida-Cortez, J. S., Fernandes, G. W. & Tabarelli, M. (2011). Caatinga: The scientific negligence experienced by a dry tropical forest. Tropical Conservation Science, 4(3), 276–286.

Santos, M. P. D. & Cardosa de Silva, J. M. (2007). On the birds of the Roraima savannas. Revista Brasileira de Ornitologia, 15, 189.

Sarà, M. (2003). The colonization of Sicily by the Black kite (Milvus migrans). Journal of Raptor Research, 37, 167–172.

Sarasola, J. H. & Maceda, J. J. (2006). Past and current evidence of persecution of the endangered Crowned eagle Harpyhaliaetus coronatus in Argentina. Oryx, 40, 347–350. https://doi.org/10.1017/S0030605306001013

Sarasola, J. H., Santillán, M. Á. & Galmes, M. A. (2010). Crowned eagles rarely prey on livestock in central Argentina: Persecution is not justified. Endangered Species Research, 11, 207–213. https://doi.org/10.3354/esr00280

Satheesan. (1992a). An updated list of birds and bat species involved in collision with aircraft in India. Journal of the Bombay Natural Histor Society, 89(1), 129–132.
Satheesan. (1992b). Bird and bat collisions with aircraft. Journal of the Bombay Natural History Society, 89(3), 379–380.
Satheesan. (1994). The more serious vulture hits to Military Aircraft in India between 1980 and 1994. *In*: Proc. 22 BSCE Meeting Vienna, Austr 29 August–2 September, WP 23, 163–168.
Satheesan. (1996). Raptors associated with airports and aircraft in raptors in human landscapes. *In*: Bird, D., Varland, D. & Negro, J. (Eds. Raptors in human landscapes: Adaptations to built and cultivated environments, Chapter 30, pp. 315–323. Academic Press, U.K. and Rapt Research Foundation.
Satheesan, S. M. & Satheesan, M. (2000). Serious Vulture Hits to Aircraft Over the World. International Bird Strike Committee IBSC25/WP-SA Amsterdam: IBSC.
Saunders, D. A. & Ingram, J. A. (1995). Birds of Southwestern Australia. An Atlas of Changes in Distribution and Abundance of the Wheatbe Avifauna, 1–296. Chipping Norton: Surrey Beatty & Sons.
Saunders, D. A., Hobbs, R. J. & Margules, C. R. (1991). Biological consequences of ecosystem fragmentation: A review. Conservation Biolog 5(1), 18–32. https://doi.org/10.1111/j.1523-1739.1991.tb00384.x
Savage, C. (1989). Eagles of North America. Saskatoon: Western Producer Prairie Books.
Savage, C. (1992). Peregrine Falcons. San Francisco: Sierra Club Books.
Savard, J. P. L., Clergeau, P. & Mennechez, G. (2000). Biodiversity concepts and urban ecosystems. Landscape and Urban Planning, 48(3–4 131–142. https://doi.org/10.1016/S0169-2046(00)00037-2
Savigny, J.-C. (1809). Système des oiseaux de l'Egyptel'Égypte et de la Syrie. Description de l'Egypte, Histoire naturelle, Tome Remier, 62–11
Savigny, M. J. C. (1809). Description de l'Égyptel'Égypte: Histoire naturelle, Volume 1 (in French). Paris: Imprimerie impériale, 68, 85.
Sayad, A. (2007). Causes of extinction and regression in Morocco's birds. Ostrich, 78(2), 170–170.
Scantlebury, D. M., Mills, M. E. J., Wilson, R., Wilson, J., Mills, M. G. L., Durant, S., Bennett, N. C., Bradford, P., Marks, N. J. & Speakman, (2014). Mammalian energetics. Flexible energetics of cheetah hunting strategies provide resistance against kleptoparasitism. Science (Ne York), 346, 79–81. https://doi.org/10.1126/science.1256424
Schafer, E. E. H. (1958). Falconry in T'ang Times. T'oung Pao, 46(3/5), Second Series, 46(1), 293–338. https://doi.org/10.1163/156853258X001C
Schäffer, N. (1996). Narew und Biebrza, Natur – Reiseführer, Naturerbe-Verlag Jürgen Resch, Überlingen.
Schaldach, W. J., Jr. & Escalante-Pliego, B. P. (1997). Lista de aves. *In*: González-Soriano, E., Dirzo, R. & Vogt, R. C. (Eds.), Natural history Los Tuxtlas, pp. 571–588. Mexico City, Mexico: Universidad Nacional Autónoma de México.
Schalock, R. L. (1997). The concept of quality of life in 21st century disability programs. *In*: Brown, R. I. (Eds.), Quality of life for people wi disabilities: Models, research and practice, pp. 327–340. Cheltenham, UK: Stanley Thornes.
Schalock, R. L. (2000). Three Decades of Quality of Life. Focus on Autism and Other Developmental Disabilities, 15, 116–127. https://do org/10.1177/108835760001500207
Schalock, R. L., Alonso, M. A. V. & Braddock, D. L. (2002). Handbook on Quality of Life for Human Service Practitioners. Washington, DC American Association on Mental Retardation.
Schaltegger, S. & Beständig, U. (2012). Corporate Biodiversity Management Handbook: A Guide for Practical Implementation. Berlin: Feder Ministry of the Environment.
Schaub, M., Liechti, F. & Jenni, L. (2004). Departure of migrating European robins, Erithacus rubecula, from a stopover site in relation to wir and rain. Animal Behaviour, 67, 229–237. https://doi.org/10.1016/j.anbehav.2003.03.011
Schaub, M., Aebischer, A., Gimenez, O., Berger, S. & Arlettaz, R. (2010). Massive immigration balances high anthropogenic mortality in a stab eagle owl population: Lessons for conservation. Biological Conservation, 143, 1911–1918.
Scheller, W. & Meyburg, B.-U. (1996). A study of the breeding and feeding biology of the Lesser Spotted eagle Aquila pomarina by means of remote control video camera. *In*: Meyburg, B.-U. & Chancellor, R. D. (Eds.), Eagle studies, pp. 245–256. Berlin: World Working Group c Bird of Prey.
Scheller, W. & Meyburg, B.-U. (1996). Observations on cainism in the Lesser Spotted eagle Aquila pomarina by means of a remote-controlle video camera. Populationsökologie Greifvogel- und Eulenarten, 3, 177–184.
Scheller, W., Bergmanis, U., Meyburg, B.-U., Furkert, B., Knack, A. & Röper, S. (2000). Untersuchungen zum Raum-Zeit-Verhalten vc Schreiadlern Aquila pomarina. Acta Ornithoecologica, 4, 75–236.
Scheller, W., Bergmanis, U., Meyburg, B.-U., Furkert, B., Knack, A. & Röper, S. (2001). Raum-Zeit-Verhalten des Schreiadlers (Aquila pomarin (Home range size, habitat utilization and time budgets of Lesser Steppe eagles Aquila pomarina). Acta Ornithoecologica, 4, 75–236.
Scherer Neto, P., Kajiwara, D., Carrano, E. & Abe, L. M. (1997). New records of the Black Hawk-eagle Spizaetus tyrannus (Wied, 1820) for t States of Parana and Santa Catarina. Atualidades Ornitológicas, 79, 11.
Scherer-Neto, P. & Straube, F. C. (1995). Birds of Paraná: History, annotated list and bibliography. Curitiba, Brazil: Logos Press.
Scheuhammer, A. M. (1987). The chronic toxicity of aluminium, cadmium, mercury, and lead in birds: A review. Environmental Pollution, 46(4 263–295. Retrieved from https://doi.org/10.1016/0269-7491(87)90173-4.
Scheuhammer, A. M. & Norris, S. L. (1995). A review of the environmental impacts of lead shotshell ammunition and lead fishing weights Canada. Ottawa: Canadian Wildlife Service Occasional Paper 88.
Scheuhammer, A. M., Money, S. L. & Kirk, D. (2002). Lead fishing sinkers and jigs in Canada: Review of their use patterns and toxic impacts c wildlife. Ottawa: Canadian Wildlife Service Occasional Paper 108.
Schlee, M. (2005). King vultures (Sarcoramphus papa) forage in Moriche and Cucurit palm stands. Journal of Raptor Research, 39(4), 458–46

Schlegel, H. & Pollen, F. P. L. (1868). Recherches sur la faune de Madagascar et des ses dépendances, d'après les découvertes de François P. L. Pollen et D. C. van Dam 2ème partie. Mammifères et oiseaux. Leyden, The Netherlands.

Schmaljohann, H., Liechti, F. & Bruderer, B. (2009). Trans-Sahara migrants select flight altitudes to minimize energy costs rather than water loss. Behavioral Ecology and Sociobiology, 63, 1609–1619. https://doi.org/10.1007/s00265-009-0758-x

Schmaljohann, H., Fox, J. W. & Bairlein, F. (2012). Phenotypic response to environmental cues, orientation and migration costs in songbirds flying halfway around the world. Animal Behaviour, 84(3), 623–640. https://doi.org/10.1016/j.anbehav.2012.06.018

Schmidt, G. D. & Kuntz, R. R. E. (1971). Nematode Parasites of Oceanica. XI. Madelinema angelae gen. et sp. n., and Inglisonema mawsonae sp. n. (Heterakoidea: Inglisonematidae). Journal of Parasitology, 57(3), 479–484. https://doi.org/10.2307/3277897

Schmutz, J. K., Rose, K. A. & Johnson, R. G. (1989). Hazards to raptors from strychnine poisoned ground squirrels. Journal of Raptor Research, 23(4), 147–151. Retrieved from https://sora.unm.edu/sites/default/files/journals/jrr/v023n04/p00147-p00151.pdf

Schneider-Jacoby, M., Mohl, A. & Schwarz, U. (2003). The White-tailed eagle in the Danube River Basin. In: Helander, B. (Ed.), Sea eagle 2000: Proceedings from the International Sea Eagle Conference in Bjorko, Sweden, 13–17 September 20, 133–140. Stockholm: Sweden Proceedings of the Swedish Society for Nature Conservation, SNF.

Schoener, T. W. (1968). Sizes of feeding territories among birds. Ecology, 49(1), 123–141. https://doi.org/10.2307/1933567

Schoener, T. W. (1971). Theory of feeding strategies. Annual Review of Ecology and Systematics, 2(1), 369–404. https://doi.org/10.1146/annurev.es.02.110171.002101

Schofield, G., Bishop, C. M., MacLean, G., Brown, P., Baker, M., Katselidis, K. A., et al. (2007). Novel GPS tracking of sea turtles as a tool for conservation management. Journal of Experimental Marine Biology and Ecology, 347(1), 58–68. https://doi.org/10.1016/j.jembe.2007.03.009

Scholes, R. J. & Walker, B. H. (1993). An African Savanna: Synthesis of the Nylsvley Study. Cambridge: Cambridge University Press.

Schoultz, E. (2011). Eagle Man's Vision. Falconers and Raptor Conservation Magazine.

Schreiber, A. & Weitzel, T. (1995). Biochemical systematics of sea eagles (genus Haliaeetus, Savigny 1809), with a note on allozyme differentiation between black and red kites (Genus Milvus L. 1758). Biochemical Systematics and Ecology 23(3), 235–244. https://doi.org/10.1016/0305-1978(95)00001-B

Schulenberg, T. (2015). Proposal (681) to South American Classification Committee: Change the English name of Buteogallus coronatus. Retrieved from https://www.museum.lsu.edu/~Remsen/SACC prop681.htm

Schulenberg, T. S. & Parker, T. A., III. (1981). Status and distribution of some northwest Peruvian birds. Condor, 83(3), 209–216. https://doi.org/10.2307/1367309

Schulenberg, T. S. & Wust, W. H. (1997). Birds of the Upper Río Comainas, Cordillera del Cóndor. *In*: Schulenberg, T. S. & Awbrey, K. (Eds.), The Cordillera del Cóndor region of Ecuador and Peru, 180–187. Washington, DC: Conservation International.

Schulenberg, T. S. (Ed.). (2010). Greater Yellow-headed vulture (Cathartes melambrotus), Neotropical Birds Online. Ithaca: Cornell Laboratory of Ornithology. Retrieved from http://neotropical.birds.cornell.edu/portal/species/overview?p_p_spp=118556

Schulenberg, T. S., Stotz, D. F., Lane, D. F., O'Neill, J. P. & Parker, III, T. A. (2010). Birds of Peru: Revised and updated edition, 63. Princeton, NJ: Princeton University Press.

Schulenberg, T. S. (2020). Harpy Eagle (Harpia harpyja), version 1.0. *In*: Schulenberg, T. S. (Ed.), Birds of the World. Ithaca, NY: Cornell Laboratory of Ornithology. https://doi.org/10.2173/bow.hareag1.01

Schultz, R. (2005). The Eagle Hunters. Saudi Aramco World, 56, 1 (January- February). Kazakh eagle hunters in Xinjiang, China. Retrieved from http://archive.aramcoworld.com/issue/200501/the. eagle.hunters.htm1

Schulze, A. & Dingler, K. H. (2003). Die Vogelstimmen Europas, Nordafrikas und Vorderasiens.

ScienceDirect. (2020). Body Burden. Retrieved from https://www.sciencedirect.com/topics/ medicine-and-dentistry/body-burden

Sclater, P. L. (1873). On the birds of eastern Peru. Proceedings of the Zoological Society of London, 1873, 252–311.

Sclater, P. L. & Salvin, O. (1959). On the ornithology of Central America. Part III. Ibis, 1(3), 213–234. https://doi.org/10.1111/j.1474-919X.1859.tb06204.x

Sclater, W. L. (1922). Notes on the nomenclature and taxonomy of African birds. Bulletin of the British Ornithologists' Club, 42, 72–77.

Sclater, W. L. & Moreau, R. E. (1932). Taxonomic and field notes on some birds of north-eastern Tanganyika Territory. Ibis, 13(2), 500–501.

Scoones, I. (1997). The dynamics of soil fertility change: Historical perspectives on environmental transformation from Zimbabwe. The Geographical Journal, 63(2), 161–169.

Scoones, I. (1999). New ecology and social sciences: What prospects for a fruitful engagement? Annual Review of Anthropology, 28, 479–507.

Scott, D. A. & Adhami, A. (2006). An updated checklist of the birds of Iran. Podoces, 1(1/2), 1–16.

Scott, J. (1986). Predation of Black Sparrowhawks by Long-crested eagles. Honeyguide, 32, 151–152.

Scott, J. & Scott, A. (2013). The Leopard's Tale: Featuring Half-Tail and Zawadi, Stars of Big Cat Diary. Bradt Travel Narratives.

Scott, J. A. (1983). Further notes on the breeding success of Long crested eagles. Ostrich, 54, 183–184.

Scott, J. A. & Scot, J. P. (1995). Notes on the breeding biology of the Southern Banded Snake-eagle. Honeyguide, 41, 156–160.

Scottish Natural Heritage. (2010). Survey methods for use in assessing the impacts of onshore windfarms on bird communities. Inverness, Scotland: Scottish Natural Heritage.

Seamans, T. W. (2004). Response of roosting Turkey vultures to a vulture effigy. Ohio Journal of Sciences, 104, 136–138. Retrieved from https://digitalcommons.unl.edu/icwdm_ usdanwrc/385/

Secretaría de Medio Ambiente y Recursos Naturales. (2002). NOM-059-ECOL-2001: Official Mexican List. Environmental protection, Native species of flora and wildlife of Mexico, Categories of risk and specifications for inclusion, exclusion or change, List of species at risk. Diario Oficial de la Federación, Mexico City, Distrito Federal, Mexico.

Segal, M. & Pielke, R. A. (1988). The extrapolation of vertical profiles of wind speed within the marine atmospheric surface layer using the p formula. Journal of Applied Meteorology, 27(2), 174–181. https://doi.org/10.1175/15120-0450(1988)027<0174: TEOVPO>2.0.CO;2.

Seguin, J. F., Thibault, J. C., Torre, J., Bayle, P. & Vigne, J. D. (2001). The diet of young Golden eagles Aquila chrysaetos in Corsica: Foraging in a man-made mammal fauna. Ardea, 89, 527–535.

Seibold, I., Helbig, A. J. & Wink, M. (1993). Molecular systematics of falcons (family Falconidae). Naturwis Senschaften, 80, 87–90.

Seibold, I. (1994). Untersuchungen zur molekularen Phy-logenie der Greifvogel anhand von DNA-Sequenzen des mitochondriellen Cytochron b-Gens. PhD Dissertation, University of Heidelberg, Heidelberg.

Seibold, I. & Helbig, A. J. (1995). Evolutionary history of new and old world vultures inferred from nucleotide sequences of the mitochondria cytochrome b gene. Philosophical Transactions of the Royal Society B: Biological Sciences, 350(1332), 163–178.

Seibold, I. & Helbig, A. J. (1996). Phylogenetic relationships of the Sea eagles (genus Haliaeetus), reconstructions based on morphology allozymes and mitochondrial DNA sequences. Journal of Zoological Systematics and Evolutionary Research, 34(2), 103–112. https://doi org/10.1111/j.1439-0469.1996.tb00815.x

Seibold, I., Helbig, A. J., Meyburg, B.-U., Negro, J. J. & Wink, M. (1996). Genetic differentiation and molecular phylogeny of European Aquila eagles according to cytochrome b nucleotide sequences. *In*: Meyburg, B.-U. & Chancellor, R. D. (Eds.), Eagle studies, 1–15. Berlin: World Working Group on Birds of Prey.

Seipke, S. H. (2004). Field identification of raptors threatened in Argentina. *In*: Proceedings of the First Argentine Symposium on Investigation and Conservation of Raptors—SAICR I. Buenos Aires, Argentina (pp. 5–6). Museo de La Plata (in Spanish).

Sekhar, P. S., Kumar, P. K., Anil, K. & Babu, A. S. (2004). Sighting of the White-bellied Sea eagle (Haliaeetus leucogaster) in Andhra University Campus, Visakhapatnam, Andhra Pradesh. Zoos' Print Journal, 19(5), 20. Retrieved from https://www.zoosprint. zooreach.org/index.php/zp article/ view/3085

Selås, V. & Rafoss, T. (1999). Ranging behaviour and foraging habitats of breeding Sparrowhawks Accipiter nisus in a continuous forested area in Norway. Ibis, 141(2), 269–276. https://doi.org/10.1111/j.1474-919X.1999.tb07549.x

Selva, N. & Fortuna, M. A. (2007). The nested structure of a scavenger community. Proceedings of the Royal Society B Biological Sciences 274(1613), 1101–1108. https://doi.org/10.1098/rspb.2006.0232

Selva, N., Jedrzejewska, B., Jedrzejewski, W. & Wajrak, A. (2003). Scavenging on European bison carcasses in Bialowieza primeval fores (eastern Poland). Ecoscience, 10(3), 303–311. https://doi.org/10.1080/11956860.2003.11682778

Seminario, Y., Phillips, R. & Curti, M. (2011). Observations of the post-fledging behavior and prey of the Solitary eagle (Harpyhaliaetus solitarius). Journal of Raptor Research, 45, 261–264. https://doi.org/10.3356/JRR-11-08.1

Sergio, F., Marchesi, L. & Pedrini, P. (2004). Integrating individual habitat choices and regional distribution of a biodiversity indicator and top predator. Journal of Biogeography, 31(4), 619–628). https://doi.org/10.1046/j.1365-2699.2003.01002.x

Sergio, F., Marchesi, L., Pedrini, P. & Penteriani, V. (2007). Coexistence of Tawny owls with their intraguild predator: Distance-sensitive o habitat-mediated avoidance? Animal Behavior, 74(6), 1607–1616. https://doi.org/10.1016/j.anbehav.2006.10.022

Sergio, F. & Hiraldo, F. (2008). Intraguild predation in raptor assemblages: A review. Ibis, 150, 132–145. https://doi.org/10.1111/j.1474 919X.2008.00786.x

Sergio, F., Newton, I. & Marchesi, L. (2008). Top predators and biodiversity: Much debate, few data. Journal of Applied Ecology, 45, 992–999.

Serhal, A. A. & Khatib, B. C. (2014). State of Lebanon's Birds and IBAs. Beirut, Lebanon: Dar Bilal for Printing and Publishing. Retrieved from http://datazone.birdlife.org/userfiles/ file/sowb/countries/LebanonReport.pdf.

Serhal, A.A., Khatib, B.C. & Khaddaj, V. (2014). Field Guide to the Soaring Birds in Lebanon. Beirut: Dar Bilal for Printing and Publishing.

Serna, H. M. A. (1992). Notes on some species of Colombian birds, some of which are potentially threatened. Boletín Sao, 3, 9–23.

Servheen, C. W. (1975). Ecology of the wintering Bald eagles on the Skagit River, Washington. M.S.Thesis. Seattle, WA: University of Washington

Seshadri, B., Bolan, N. S., Naidu, R. & Brodie, K. J. (2010). The role of coal combustion products in managing the bioavailability of nutrients and heavy metals in soils. Journal of Soil Science and Plant Nutrition, 10(3), 378–398. https://doi.org/10.4067/S0718

Seth-Smith, D. (1910). On the Monkey-eating Eagle of the Philippines (Pithecophaga jefferyi). Ibis, 52(2), 285–290. https://doi.org/10.1111 j.1474-919X.1910.tb07905.x

Setiadi, A. P. Z., Rahman, P. F., Nurwatha & Muchtar, M. (1999). Status distribution ecology and conservation of Javan Hawk-eagle (Spizaetu bartelsi, Stresemann,1924) in southern west Java. Bandung: YPAL & HIMBIO UNPAD.

Seton, O. P. (1937). The King of Birds. Natural History, 39, 353–356.

Sexton, J. O., Song, X.- P., Feng, M., Noojipady, P., Huang, C., et al. (2013). Global, 30-m resolution continuous fields of tree cover: Landsat-base rescaling of MODIS vegetation continuous fields with lidar based estimates of error. International Journal of Digital Earth, 6(5), 427–448 https://doi.org/10.1080/17538947.2013.786146

Seyfarth, R. M., Cheney, D. L. & Marler, P. (1980). Monkey responses to three different alarm calls: Evidence of predator classification and semantic communication. Science, 210(4471), 801–803. https://doi.org/10.1126/science.7433999

Seymour, A. S., Hatherley, G., Contreras, F. J., Aldred, J. & Beeley, F. (2010). Hatching synchrony, green branch collecting, and prey use by nesting Harpy eagles (Harpia harpyja). Wilson Journal of Ornithology, 122(4), 792–795. https://doi.org/10.1676/10-060.1

Sferco, G. D. & Nores, M. (2003). An annotated check-list of the birds of Chancaní Natural Reserve, Córdoba, Argentina. Hornero, 18, 21–29.

Shäfer, E. & Phelps, W. H. (1954). The birds of "Henri Pittier" (Rancho Grande) National Park and their ecological functions. Boletín de l Sociedad Venezolana de Ciencias Naturales, 83, 1–171.

Shamoun-Baranes, J., Baharad, A., Alpert, P., Berthold, P., Yom-Tov, Y., Dvir, Y. & Leshem, Y. (2003). The effect of wind, season and latitude o the migration speed of White storks Ciconia ciconia along the eastern migration route. Journal of Avian Biology, 34(1), 97–104. https://doi org/10.1034/j.1600-048X.2003.03079.x hamoun-Baranes, J., Liechti, O., Yom-Tov, Y. & Leshem, Y. (2003). Using a convection model to predict altitudes of White stork migration over central Israel. Boundary-Layer Meteorology, 107(3), 673–681. https://doi.org/10.1023/A: 1022824008388

hamoun-Baranes, J., van Loon, E., van Gasterenvan Gastere, H., van Belle, J., Bouten, W. & Buurma, L. (2006). A comparative analysis of the influence of weather on the flight altitudes of birds. Bulletin of the American Meteorological Society, 87(1), 47–62. https://doi.org/10.1175/BAMS-87-1-47

hamoun-Baranes, J., Bouten, W. & van Loon, E. E. (2010). Integrating meteorology into research on migration. Integrative and Comparative Biology, 50(3), 280–292. https://doi.org/10.1093/icb/ icq011

hannon, H. D., Young, G. S., Yates, M. A., Fuller, M. R. & Seegar, W. S. (2002). American White pelican soaring flight times and altitudes relative to changes in thermal depth and intensity. The Condor, 104(3), 679–683. https://doi.org/10.1650/0010-5422(2002)1040679: AWPSFT2.0.CO;2.

harma, A. K. & Singh, K. R. (1989). Tal Chhapar Blackbuck Sanctuary. Sanctuary Asia, 9(2), 36–45.

harma, A. K., Saini, M., Singh, S. D., Prakash, V., Das, A., Dasan, R. B., Pandey, S., Bohara, D., Galligan, T. H., Green, R. E., Knopp, D. & Cuthbert, R. J. (2014). Diclofenac is toxic to the Steppe eagle Aquila nipalensis: Widening the diversity of raptors threatened by NSAID misuse in South Asia. Bird Conservation International, 24, 282–286. https://doi.org/10.1017/S095927 0913000609

harma, P. & Gopi Sunder, K. S. (2009). Counts of Steppe eagles (Aquila nipalensis) at a carcass dump in Jorbeer, Rajasthan, India. Forktail, 25, 160–163.

harma, S. C. & Chanda, J. (2010). Indian Spotted eagle Aquila hastata nesting in Sonepat, Haryana, India. Indian Birds, 6(1), 18.

harpe, C. F., Ascanio-Echeverria, D. & Rodriguez, G. A. (2001). Further range extensions and noteworthy records for Venezuelan birds. Bulletin of the British Ornithologists' Club, 121, 50–62.

harpe, R. B. (1874). Catalogue of the Accipitres, or diurnal birds of prey, in the collection of the British Museum. Catalogue of the Birds of the British Museum, Vol. 1. London: Trustees of the British Museum.

harpe, R. B. (1879). On Contributions to the Ornithology of Borneo. Part IV. Ibis, 233–272.

harpe, R. B. (1888). List of a collection of birds made by Mr. L. Wray in the main range of mountains of the Malay Peninsula, Perak. Proceedings of the Zoological Society of London, 268–281.

harpe, R. B. (1889). On the Ornithology of Northern Borneo. With Notes by John Whitehead. Idem. Part I, 03-85.

healer, D. A., Spendelow, J. A., Hatfield, J. S. & Nisbet, I. C. T. (2005). The adaptive significance of stealing in a marine bird and its relationship to parental quality. Behavioral Ecology, 16(2), 371–376. https://doi.org/10.1093/beheco/ari008

heard, C., Neate-Clegg, M. H. C., Alioravainen, N., Jones, S. E. I., Vincent, C., MacGregor, H. E. A., Bregman, T. P., Claramunt, S. & Tobias, J. A. (2020). Ecological drivers of global gradients in avian dispersal inferred from wing morphology. Nature Communications, 11(1), 2463. https://doi.org/10.1038/s41467-020-16313-6

heldon, B. C. & Duckworth, J. W. (1990). Rediscovery of the Madagascar Serpent-eagle Eutriorchis astur. Bulletin of the British Ornithologists' Club, 110, 126–130.

heldon, F. H., Moyle, R. G. & Kennard, J. (2001). Ornithology of Sabah: History, Gazetteer, Annotated Checklist, and Bibliography. Washington, DC: American Ornithologists' Union: Ornithological Monographs, 52, 1–285. https://doi.org/10.2307/40166890

helley, G. E. (1874). Note on Dryotriorchis, a new genus of Harrier eagles from West Africa. The Ibis, IV(3), 90–91.

hepard, E. L. C., Lambertucci, S. A., Vallmitjana, D. & Wilson, R. P. (2013). Energy beyond food: Foraging theory informs time spent in thermals by a large soaring bird. PLoS ONE, 6(11), e27375. https://doi.org/10.1371/journal.pone.0027375

hepard, E. L. C., Wilson, R. P., Gareth Rees, W., Grundy, E., Lambertucci, S. A. & Vosper, S. B. (2013). Energy landscapes shape animal movement ecology. The American Naturalist, 182(3), 298–312. https://doi.org/10.1086/671257

hepardson, D. (2019). FAA to U.S. airports: Do Not Install Drone Countermeasures On Your Own. Retrieved from https://www.reuters.com/article/us-usa-drones-airports/faa-to-u-s-airports-do-not-install-drone-countermeasures-on-your-own-idUSKCN1SE2QZ

hephard, J. M., Catterall, C. P. & Hughes, J. M. (2004). Discrimination of sex in the White-bellied Sea eagle, Haliaeetus leucogaster, using genetic and morphometric techniques. Emu – Austral Ornithology, 104(1), 83–87. https://doi.org/10.1071/MU03043

heridan, S. (2002). The redevelopment of a weather-type classification scheme for North America. International Journal of Climatology, 22, 51–68. https://doi.org/10.1002/joc.709

heridan, S. C. (2013). Spatial Synoptic Classification. Retrieved from http://sheridan.geog.kent. edu/ssc.html

herman, P. T. (1991). Harpy Eagle predation on a Red Howler Monkey. Folia Primatologica, 56, 53–56.

herrington, P. (2003). Trends in a migratory population of Golden eagle (Aquila chrysaetos) in the Canadian Rocky Mountains. Bird Trends Canada, 9, 34–39.

herrington, P. (2008). The road to the place of the eagles. Birdwatch Canada, 43, 4–7.

herrod, S. K. (1978). Diets of North American Falconiformes. Journal of Raptor Research, 12(3/4), 49–121.

herrod, S. K., White, C. M. & Williamson, F. S. L. (1976). Biology of the Bald eagle on Amchitka Island, Alaska. Living Bird, 15, 143–182.

hibnev, Yu. B. (1981). Wintering of large birds of prey in Prymorie. *In*: Litvinenko, N. M. (Ed.), Rare Birds of the Far East, pp. 100–107. Vladivostok.

himelis, A., Sande, E., Evans, S. & Mundy, P. (Eds.) (2005). International Species Action Plan for the Lappet-faced vulture, Torgos tracheliotus. Nairobi: BirdLife International, Sandy, UK: Royal Society for the Protection of Birds.

hiraki, S. (1994). Characteristics of White-tailed sea eagle nest sites in Hokkaido, Japan. Condor, 96(4), 1003–1008. https://doi.org/10.2307/1369109

hiraki, S. (1996). Distributions of Steller's sea and White-tailed eagles in the rivers of Hokkaido. *In*: Survey of the Status and Habitat Conditions of Threatened Species, 15–27. Tokyo: Environment Agency.

hiraki, S. (2001). Foraging habitats of Steller's Sea eagles during the wintering season in Hokkaido, Japan. Journal of Raptor Research, 35, 91–97.

Shirihai, H. (1994). Separation of Tawny eagle from Steppe eagle in Israel. British Birds, 87, 396.

Shirihai, H. (1996). The Birds of Israel. San Diego, CA: Academic Press.

Shirihai, H., Yosef, R., Alon, D., Kirwan, G. M. & Spaar, R. (2000). Raptor migration in Israel and the Middle East. A summary of 30 years of fie research. Eilat, Israel: International Birding and Research Center.

Shivakoti, S., Ito, H., Otsuki, K. & Ito, T. (2010). Characterization of H5N1 highly pathogenic avian influenza virus isolated from a Mounta hawk eagle in Japan. Journal of Veterinary Medical Science, 72(4), 459–463. https://doi.org/10.1292/jvms.09-0478

Shivprakash, A., Das, K. R. K., Shiv, T., Girija, T. & Sharath, A. (2006). Notes on the breeding of the Indian Spotted eagle Aquila hastata. Indi Birds, 2(1), 2–4.

Shlaer, R. (1972). An eagle's eye: Quality of the retinal image. Science, 176(4037), 920–922. https://doi.org/10.1126/science.176.4037.920

Shobrak, M. (1996). Ecology of lappet-faced vulture Torgos tracheliotus in Saudi Arabia. PhD Thesis, Glasgow: University of Glasgow.

Shochat, E., Warren, P., Faeth, S., McIntyre, N. & Hope, D. (2006). From patterns to emerging processes in mechanistic urban ecology. Trends Ecology & Evolution, 21. https://doi.org/10.1016/j.tree.2005.11.019

Shore, R. F., Pereira, M. G., Walker, L. A. & Thompson, D. R. (2011). Mercury in nonmarine birds and mammals. *In*: Beyer, W. N. & Meador, P. (Eds.), Environmental contaminants in biota, pp. 609–642. Boca Raton, FL: CRC Press.

Showstack, R. (2004). New exploration focus will not diminish Earth science agenda, NASA says. Eos. American Geophysical Union, 85(5), 4 https://doi.org/10.1029/2004EO050003

Shufeldt, R. W. (1919). Osteological and other notes on the monkey-eating eagle of the Philippines, Pithecophaga jefferyi Grant. Philippi Journal of Science, 15, 31–58.

Shultz, S. (2002). The flying leopard. Wildlife Conservation, 105, 42–43.

Shultz, S. (2019). African Crowned eagle. *In*: Tingay, R. & Katzner, T. (Eds.), The Eagle Watchers, pp. 119–125. Ithaca: Cornell University Pres https://doi.org/10.7591/9780801459382-015

Shultz, S., Baral, H. S., Charman, S., Cunningham, A. A., Das, D., Ghalsasi, G. R., Goudar, M.S., Green, R. E., Jones, A., Nighot, P., Pain, D. J. Prakash, V. (2004). Diclofenac poisoning is widespread in declining vulture populations across the Indian subcontinent. Proceedings of t Royal Society of London B Biological Sciences, 271(Suppl. 6), S458–S460. https://doi.org/10.1098/rsbl.2004.0223

Shultz, S., Noë, R., McGraw, W. S. & Dunbar, R. I. M. (2004). A community-level evaluation of the impact of prey behavioural and ecologic characteristics on predator diet composition. Proceedings of the Royal Society of London B Biological Sciences, 271 (1540), 725–73 https://doi.org/10.1098/rspb.2003.2626

Shurpali, S. (2008). White-bellied Sea-eagle in Bhadra Reservoir. *In*: Sadvi, K. M. (2009). Occurrence of the White-bellied Sea-eagle Haliaeet leucogaster in inland waters. Indian Birds, 5(2), 50.

Si, X. F., Ding, P., Xingfeng, S. & Ping, D. (2011). History, status of monitoring land birds in Europe and America and countermeasures of Chin Biodiversity Science, 19(3), 303–310. https://doi.org/10.3724/SP.J.1003.2011.08314

Siberian Times. (2014). On Eagle Hunters in Siberia, Oct. 3. Retrieved from http://siberiantimes.com /ecology/casestudy/features/a-dying-forn of-hunting-with-eagles-couldbe-resurrected-in-siberia-despite-opposition-from-conservationists/.

Sibley, C. G. (1951). Notes on the birds of New Georgia, Central Solomon Islands. Condor, 53, 81–92.

Sibley, C. G. & Monroe, B. L. (1990). Distribution and taxonomy of birds of the world. New Haven, USA: Yale University Press.

Sibley, C. G. & Monroe, B. L. (1993). A supplement to "Distribution and Taxonomy of Birds of the World.' New Haven: Yale University Press.

Sibley, C. G. (1994). On the phylogeny and classification of living birds. Journal of Avian Biology, 25(2), 87–92. https://doi.org/10.2307/367702

Sibley, D. (2000). The Sibley Guide to Birds. New York, NY: National Audubon Society.

Sibley, D. (2003). The Sibley Field Guide to Birds of Eastern North America. New York: Alfred A. Knopf.

Sibley, S. G. & Ahlquist, J. E. (1990). Phylogeny and Classification of Birds. A Study in Molecular Evolution. New Haven: Yale University Pres

Sibley-Ahlquist Taxonomy, S.-A. (2013). Sibley-Ahlquist Taxonomy. Retrieved from http: //www.birding.in/sibley-ahlquist_taxonomy.htm

Sibly, R. M., Newton, I. & Walker, C. H. (2000). Effects of dieldrin on population growth rates of sparrowhawks 1963–1986. Journal of Applie Ecology, 37(3), 540–546. https://doi.org/10.1046/j.1365-2664.2000.00516.x

Sick, H. & Teixeira, D. M. (1977). The egg of the Crowned Solitary eagle, Harpyhaliaetus coronatus. The Condor, 79, 133.

Sick, H. & Teixeira, D. M. (1979). Notes on Brazilian birds rare or threatened with extinction. Publicações Avulsas do Museu Nacional, 62, 1–3

Sick, H. (1993). Birds in Brazil: A Natural History. Princeton, NJ: Princeton University Press.

Sick, H. (1997). Brazilian Ornithology. Rio de Janeiro: Nova Press Fronteira.

Sick, H., Rosário, L. A. & Azevedo, T. R. (1981). Birds of the State of Santa Catarina: Systematic list based on a bibliography, museum materia and field observations. Sellówia Série Zoologia, 1, 1–51.

Sidle, W. B., Suring, L. H. & Hodges, Jr., J. I. (1986). The Bald Eagle in Southeast Alaska. Wildlife and Fisheries Habitat Management Note Juneau, AK: United States Forest Service.

Sidor, I. F., Pokras, M. A., Major, A. R., Poppenga, R. H., Taylor, K. M. & Miconi, R. M. (2003). Mortality of common loons in New Englan 1987 to 2000. Journal of Wildlife Diseases, 39(2), 306–315. https://doi.org/10.7589/0090-3558-39.2.306

Sidorovich, V. E., Rotenko, I. I. & Krasko, D. A. (2011). Badger Meles meles spatial structure and diet in an area of low earthworm biomass and hig predation risk. *In*: Annales Zoologici Fennici, 48(1), 1–16. Finnish Zoological and Botanical Publishing. https://doi.org/10.5735/086.048.01C

Siebold, I. & Helbig, A. J. (1995). Evolutionary history of New World and Old World vultures inferred from nucleotide sequences of t mitochondrial cytochrome b gene. Philosophical Transactions of the Royal Society of London, 350B, 163–178.

Siebold, I., Helbig, A. J. & Wink, M. (1993). Molecular systematics of falcons (family Falconidae). Naturwissenschaften, 80(2), 87–90. https: doi.org/10.1007/BF01140425

Siegfried, W. R. (1963). A preliminary report on Black and Martial eagles in the Laingsburg and Philipstown divisions. Cape Province Department of Conservation Investigational Report No, 5, 1–15.

Sievi, J. R. (1998). Double brooding in the Long-crested eagle. Honeyguide, 44, 217.

Sigfusson, L. (2017). Drones Are Keeping Watch on the Arctic's Polar Bear Population. Retrieved from http://blogs.discovermagazine.com/drone360/2017/10/04/polar-bear-drone/#.XVo4s OhKi70

Sigrist, T. (2006). Birds of Brazil: An artistic view (2nd ed). Avis Brasilis. São Paulo, Brazil.

Sigrist, T. (2013). Ornitologia Brasileira. Vinhedo, Brazil: Avis Brasilis.

Sillero-Zubiri, C., Hoffmann, M. & Macdonald, D. W. (Eds.). (2004). Canids: Foxes, wolves, jackals and dogs: Status survey and conservation action plan (pp. x–430). IUCN.

Silveira, L. F., Mendonca, F. & d'Horta. (2002). On the avifauna of the Vila Bela da Santíssima Trinidade region, Mato Grosso. Papeis Avulsos de Zoologia (São Paulo), 42, 265–286.

Silverstein, A., Silverstein, V. & Silverstein, R. (1995). The Peregrine Falcon. Brookfield, CT: The Millbrook Press.

Simmons, R. (1988). Offspring quality and the evolution of cainism. Ibis, 130, 339–357.

Simmons, R. (1990). Africa's best kept secret – where Wahlberg's eagles winter. Gabar, 5, 31–32.

Simmons, R. (1991). The Wahlberg's eagle migration mystery. Gabar, 6, 13–15.

Simmons, R. (2005). Verreauxs' Eagle. *In*: Hockey, P. A. R., Dean, W. R. J. & Ryan, P. G. (Eds.), Roberts' Birds of Southern Africa (7th Ed.). Cape Town, SA.

Simmons, R. & Brown, C. (2006). Birds to Watch in Nambia: Red, Rare, and Endemic Species. Windhoek, Namibia: National Biodiversity Programme.

Simmons, R. & Barnard, P. (2007). Breeding Booted Eagles at Brukkaros, Namibia. Gabar, 18, 9–15.

Simmons, R. & Brown, C. (2004). RED Data Status of the Martial eagle in Namibia. Gabar, 15(1), 5–8.

Simmons, R. E. & Brown., C. J. (1997). Bateleur Terathopius ecaudatus. *In*: Harrison, J. A. et al. (Eds.), The atlas of South African birds. Volume 1: Non-passerines. Johannesburg, SA: BirdLife South Africa and Avian Demography Unit.

Simmons, R. E. & Brown, C. J. (2004). The conservation status of the Martial eagle Polemaetus bellicosus in Namibia. Gabar, 15, 12–15.

Simmons. R. E. (2005). African Crowned Eagle Stephanoaetus coronatus. *In*: Hockey, P. A. R., Dean, W. R. J. & Ryan, P. G. (Eds.), Roberts Birds of Southern Africa. VIIth edition, 541–542. Cape Town, SA: Trustees of the John Voelcker Bird Book Fund.

Simmons, R. E. & Brown, C. J. (2008). Birds to watch in Namibia: Red, rare and endemic species. Windhoek: National Biodiversity Programme and Namibia Nature Foundation.

Simmons, R. E. & Mendelsohn, J. M. (1993). A critical review of cartwheeling flights of raptors. Ostrich, 64(1), 13–24. https://doi.org/10.1080/00306525.1993.9634190

Simmons, R. E. (1997). Why don't all siblicidal eagles lay insurance eggs? The egg quality hypothesis. Behavioral Ecology, 8(5), 544–550. https://doi.org/10.1093/beheco/8.5.544

Simmons, R. E. (1997a). Tawny eagle. *In*: Harrison, J. A., Allan, D. G., Underhill, L. G., Herremans, M., Tree, A. J., Parker, V. & Brown, C. J., et al. (Eds.), The Atlas of South African birds. Volume 1: Non-passerines, pp. 178–179. Johannesburg, SA: BirdLife South Africa and Avian Demography Unit.

Simmons, R. E. (1997b). Steppe eagle. *In*: Harrison, J. A., Allan, D. G., Underhill, L. G., Herremans, M., Tree, A. J., Parker, V. & Brown, C. J., et al. (Eds.), The Atlas of South African Birds. Vol. 1: Non-passerines, p. 180. Johannesburg, BirdLife South Africa.

Simmons, R. E. (1997c). Wahlberg's eagle. *In*: Harrison, J.A., et al. (Eds.), The Atlas of South African Birds. Volume 1: Non-passerines. Johannesburg, SA: BirdLife South Africa and Avian Demography Unit.

Simmons, R. E. (1997d). African Hawk-eagle. *In*: Harrison, J. A. et al. (Eds.), The Atlas of South African Birds. Volume 1: Non-passerines. pp. 188–189. Johannesburg, SA: BirdLife South Africa and Avian Demography Unit.

Simmons, R. E. (1997e). Lesser Spotted eagle. *In*: Harrison, J. A. et al. (Eds.), The Atlas of South African Birds. Volume 1: Non-passerines, p. 181. Johannesburg, BirdLife South Africa and Avian Demography Unit.

Simmons, R. E. (1997f). Bateleur eagle. *In*: Harrison, J. A. et al. (Eds.), The Atlas of South African Birds. Volume 1: Non-passerines, pp. 188–189. Johannesburg, SA: BirdLife South Africa and Avian Demography Unit.

Simmons, R. E. (2005). Martial Eagle Polemaetus bellicosus. *In*: Hockey, P. A. R., Dean, W. R. J. & Ryan, P. G. (Eds.), Roberts' Birds of southern Africa (7th ed.), 538–539. Cape Town, SA: Trustees of the John Voelcker Bird Book Fund.

Simmons, R. E. (2005). African Hawk-Eagle Aquila spilogaster. *In*: Hockey, P. A. R., Dean, W. R. J. & Ryan, P. G. (Eds.), Roberts' Birds of southern Africa (7th ed.), 533–534. Cape Town, South Africa: Trustees of the John Voelcker Bird Book Fund.

Simmons, R. E. (2005). Long-crested Eagle Lophaetus occipitalis. *In*: Hockey, P. A. R., Dean, W. R. J. & Ryan, P. G. (Eds.), Roberts' Birds of southern Africa (7th ed.), 539-540. Cape Town, SA: Trustees of the John Voelcker Bird Book Fund.

Simmons, R. E. & Brown, C. J. (2009). Birds to watch in Namibia: Red, rare and endemic species. Windhoek: National Biodiversity Programme.

Simmons, R. E. (2015). Tawny Eagle Aquila rapax. *In*: Simmons, R. E., Brown, C. J. & Kemper, J. (Eds.), Birds to watch in Namibia: Red, rare and endemic species, 127–129. Windhoek: Ministry of Environment and Tourism and Namibia Nature Foundation.

Simmons, R. E. (2015). Martial eagle Polemaetus bellicosus. In: Simmons, R. E., Brown, C. J. & Kemper, J. (Eds.), Birds to watch in Namibia: Red, rare and endemic species, pp. 135–137. Ministry of Environment and Tourism, Namibia Nature Foundation.

Simmons, R., Rodrigues, L., Woodcock, L., Steyn, P. & Jenkins, A. (2007). City Limits: Can the Verreaux's eagle survive urbanisation? Birds and Birding, 12(3), 42–46.

Simon, J. E. & S. Pacheco, S. (2005). On the standardization of nest descriptions of Neotropical birds. Revista Brasileira de Ornitologia, 13, 143–151.

Simon, N. & Geroudet, P. (1970). Last Survivors: Natural History of 48 Animals Danger of Extinction. Sierra Nevada. Reno, NV: Thrift Books.
Simpson, M. R., Gauthier, J., Benoît, H. P., MacDonald, D., Hedges, K., Collins, R., Mello, L. & Miri, C. (2016). A pre-COSEWIC assessment c the Common Lumpfish (Cyclopterus lumpus, Linnaeus 1758) in Canadian – Atlantic and Arctic waters. Canadian Science Adversary.
Sinclair, A. R. E. & Norton-Griffiths, M. (Eds.) (1979). Serengeti: Dynamics of an Ecosystem. Chicago: University of Chicago Press.
Sinclair, I., Hockey, P. & Tarboton, W. (2002). Birds of Southern Africa. Struik Publishers, Cape Town.
Sinclair, I. & Ryan, P. (2011). Birds of Africa Souths of the Sahara. Princeton, NJ, Oxford, UK: Princeton University Press and Oxford Universit Press.
Sinclair, K. & De George, E. (2016). Wind energy industry eagle detection and deterrents: Research gaps and solutions. Workshop summar report. National Renewable Energy Laboratory.
Sing, T. S. O. & Beerenwinkel, N. (2005). Lengauer T. ROCR: Visualizing classifier performance in R. Bioinformatics, 21(20), 7881.
Sixto, H. F. (1992). Reproductive biology, food habits, and behavior of the Black Hawk eagle in Tikal National Park. Maya Project: Use of raptor and other fauna as environmental indices for design and management of protected areas for building local capacity for conservation in Lati America, Progress Report V, 173–178. Boise, ID: The Peregrine Fund, Inc.
Skorupa, J. P. (1989). Crowned eagles Strephanoaetus coronatus in rainforest: Observations on breeding chronology and diet at a nest in Uganda Ibis, 131(2), 294–298.
Skutch, A. (1945). The migration of Swainson's and Broad-winged hawks through Costa Rica. Northwest Science, 19, 80–89.
Skutch, A. F. (1960). Life histories of Central American birds II. Pacific Coast Avifauna, 34, 1–593.
Skutch, A. F. (1981). New studies of tropical American birds. Publications of the Nuttall Ornithological Club, 19, 1–281.
Slack, K. E., Delsuc, F., McClenachan, P. A., Arnason, U. & Penny, D. (2007). Resolving the root of the avian mitogenic tree by breaking up lon branches. Molecular Phylogenetics and Evolution, 42(1), 1–13. https://doi.org/10.1016/j.ympev.2006.06.002
Sládek, J. (1959). About the nutrition of the Lesser Spotted eagle in Slowakia. Zoologicke Listy, 8, 105–113.
Slagsvold, T. & Sonerud, G. A. (2007). Prey size and ingestion rate in raptors: Importance for sex roles and reversed sexual size dimorphism Journal of Avian Biology, 38(6), 650–661. https://doi.org/10.1111/j.2007.0908-8857.04022.x
Slater, G. J. & Van Valkenburgh, B. (2009). Allometry and performance: The evolution of skull form and function in felid. Jour 22(11), 2278–2287 https://doi.org/10.1111/j.1420-9101.2009.01845.x
Slater, L. E. (1965). Introduction (special issue on telemetry). BioScience, 15(2), 81–82.
Slobodchikoff, C. N., Paseka, A. & Verdolin, J. L. (2009). Prairie dog alarm calls encode labels about predator colors. Animal Cognition, 12(3) 435–439. https://doi.org/10.1007/s10071-008-0203-y
Slud, P. (1960). The birds of Finca "La Selva," Costa Rica: A tropical wet forest locality. Bulletin of the American Museum of Natural History 121, 49–148.
Slud, P. (1964). The Birds of Costa Rica: Distribution and ecology. Las aves de Costa Rica: Distribución y ecología. Bulletin of the America Museum of Natural History, 128, 1–430.
Smales, I. & Muir, S. (2005). Modelled cumulative impacts on the Tasmanian Wedge-tailed eagle of wind farms across the species' range. Biosi Research Pty. Ltd. for Australian Department of the Environment and Heritage.
Smallie, J., Diamond, M. & Jenkins, A. (2009). Lighting up the African continent – What does it mean for our birds? In: Harebottle, D. M., Craig A. J. F. K., Anderson, M. D., Rakotomanana, H. & Muchai (Eds.), Proceedings of the 12th Pan-African Ornithological Congress, 2008. Cap Town, Animal Demography Unit.
Smallie, J. (2015a). Operational phase bird monitoring at wind farms in SA. Birds and Renewable Energy Forum, Cape Town (Presentation).
Smallie, J. (2015b). Verreauxs' eagle Aquila verreauxii wind turbine collision fatalities. Short note.
Smallwood, K. S. & Thelander, C. (2004). Developing methods to reduce bird mortality in the Altamont Pass Wind Resource Area. Final repor by BioResource Consultants to the California Energy Commission, Public Interest Energy Research-Environmental Area, Contract No. 500 01-019. Sacramento, CA: California Energy Commission.
Smallwood, K. S. & Thelander, C. G. (2005). Bird mortality at the Altamont Pass Wind Resource Area: March 1998–September 2001. Nationa Renewable Energy Laboratory NREL/SR-500-36973, Golden, Colorado, USA.
Smallwood, K. S. (2007). Estimating wind turbine-caused bird mortality. Journal of Wildlife Management, 71(8), 2781–2791. https://doi. org/10.2193/2007-006
Smallwood, K. S. & Thelander, C. (2008). Bird mortality in the Altamont Pass Wind Resource Area, California. Journal of Wildlife Management 72(1), 215–223. https://doi.org/10.2193/2007-032
Smallwood, K. S. & Neher, L. (2009). Map-based repowering of the Altamont Pass Wind Resource Area based on Burrowing Owl burrows, rapto flights, and collisions with wind turbines. Final Report, CEC-500-2009-065. Sacramento, CA: California Energy Commission, PIER Energy Related Environmental Research Program.
Smallwood, K. S., Rugge, L. & Morrison, M. L. (2009). Influence of behavior on bird mortality in wind energy developments. Journal of Wildlif Management, 73(7), 1082–1098. https://doi.org/10.2193/2008-555
Smallwood, K. S. & Karas, B. (2009). Avian and bat fatality rates at old-generation and repowered wind turbines in California. The Journal o Wildlife Management, 73(7), 1062–1071. https://doi.org/10.2193/2008-464
Smallwood, K. S. (2013). Comparing bird and bat fatality-rate estimates among North American wind-energy projects. Wildlife Society Bulletin 37(1), 19–33. https://doi.org/10.1002/wsb.260
Smárason, H. (2011). World's largest White-tailed eagle found at the bottom of Sermitsiaq fjord in Greenland. Raving Ravens Blog, March, 201
Smeenk, C. (1974). Comparative-ecological studies of some East African birds of prey. Ardea, 62(1–2), 1–97.
Smiddy, P. & O'Sullivan, O. (1996). Forty-third Irish Bird Report, (1995). Irish Birds, 5, 445–474.

›mit, II. P. J. & Asner, G. P. (2012). Roads increase woody cover under varying geological, rainfall and fire regimes in African savanna. Journal of Arid Environments, 80, 74. https://doi.org/10.1016/j.jaridenv.2011.11.026

›mith, C. (2007). White-tailed Eagles return to eastern Scotland. Scottish Bird News, 86, 12–13.

›mith, D. G. & Murphy, J. R. (1972). Unusual causes of raptor mortality. Raptor Research, 6, 4–5.

›mith, D. G. & Murphy, J. R. (1973). Breeding ecology of raptors in the eastern Great Basin of Utah. Brigham Young University Science Bulletin, Biological Series, 18, 1–76.

›mith, D. G., Ellis, D. H. & Johnson, T. H. (1988). Raptors and aircraft. *In*: Glinski, R. L. Giron-Pendleton, B., Moss, M. B., LeFranc, Jr., M. N., Millsap, B. A., Hoffman, S. W. et al. (Eds.), Proceedings of the Southwest Raptor Management Symposium and Workshop, 360–367. Washington, DC: National Wildlife Federation.

›mith, G. C. (1985). An analysis of prey remnants from Osprey Pandion haliaetus and White-bellied Sea-eagle Haliaeetus leucogaster feeding roosts. Emu, 85, 198–200.

›mith, G. (1992). Gurney's eagle Aquila gurneyi at Lae's city centre. Muruk, 5, 92.

›mith, J. W. (2012). Crested Eagle (Morphnus guianensis), version 1.0. *In*: Schulenberg, T. S. (Ed.), Neotropical Birds Online. Ithaca, NY: Cornell Laboratory of Ornithology.

›mith, K. D. (1947). Notes on some birds of the Eritrean coastline. Red Sea. Ibis, 89(2), 355–356.

›mith, K. D. (1955). The winter breeding season of land birds in eastern Eritrea. Ibis, 97(3), 450–507.

›mith, K. D. (1957). An annotated check list of the birds of Eritrea. Ibis, 99(2), 307–337.

›mith, K. D. (1960). The passage of Palearctic migrants through Eritrea. Ibis, 102, 536–544.

›mith, N. G. (1970). Nesting of King vulture and Black Hawk-eagle in Panama. Condor, 72(2), 247–248. https://doi.org/10.2307/1366655

›mith, N. G. (1980). Hawk and vulture migrations in the neotropics. *In*: Keast, A. & Morton, E. S. (Eds.), Migrant birds in the Neotropics, pp. 51–65. Washington, DC: Smithsonian Institution Press.

›mith, N. G. (1985a). Dynamics of the trans-isthmian migration of raptors between Central and South America. *In*: Newton, I. & Chancellor (Eds.), Conservation studies on raptors, pp. 271–290. ICBP Tech, No. 5. Cambridge, UK: International Council for Bird Preservation.

›mith, N. G. (1985b). Thermals, cloud streets, trade winds, and tropical storms: How migrating raptors make the most of atmospheric energy in Central America. *In*: Harwood, M. (Ed.), Proceedings of the Hawk Migration Conference IV, pp. 51–65. Lynchburg, VA: Hawk Migration Association of North America.

›mith, N. R., Hess, T. J. & Afton, A. D. (2016). History and nesting population of Bald eagles in Louisiana. Southeastern Naturalist, 15(1), 12–25. https://doi.org/10.1656/058.015.0102

›mith, T. B. (1982). Nests and young of two rare raptors from Mexico. Biotropica, 14(1), 79–80. https://doi.org/10.2307/2387765

›mythies, B.E. (1968). The Birds of Borneo (2nd ed.). Edinburgh, UK: Oliver and Boyd.

›mythies, B. E. (1981). The Birds of Borneo (3rd ed.). The Sabah Society with the Malayan Nature Society.

›mythies, B. E. (1986). The Birds of Burma (3rd). Ontario, Canada: Nimrod Press/Silvio Mattacchione & Co.

›mythies, B. E. & Davison, G. W. H. (1999). The Birds of Borneo. Kota Kinabalu, Sabah, Malaysia: Natural History Publications.

›noeijs, T., Dauwe, T., Pinxten, R., Vandesande, F. & Eens, M. (2004). Heavy metal exposure affects the humoral immune response in a free-living small songbird, the Great Tittit (Parus major). Archives of Environmental Contamination and Toxicology, 46(3), 399–404. https://doi.org/10.1007/s00244-003-2195-6

›now, C. (1973). Habitat management series for endangered species, Report No. 7: Golden eagle, Aquila chrysaetos. Grand Junction, CO: Bureau of Land Management, U.S. Dept. of the Interior.

›now, D. W. (Ed.). (1978). Atlas of Speciation in African Non-Passerine Birds. British Museum (Natural History).

›now, D. W. & Perrins, C. M. (1998). The Birds of the Western Palearctic, Volume 1: Non-Passerines. Oxford: Oxford University Press.

›nyder, N. & Snyder, H. (2000). The California Condor: A Saga of Natural History and Conservation. San Diego, CA: Academic Press.

›nyder, N. & Schmitt, J. (2002). California Condor (Gymnogyps californianus). *In*: Poole, A. & Gill, F. (Eds.), The Birds of North America, No. 610. Philadelphia, PA: The Birds of North American, Inc.

›nyder, N. F. R. & Snyder, H. A. (1991). Birds of Prey: Natural History of North American Raptors. Stillwater, MN: Voyageur Press.

›nyder, N. F. R. & Snyder, H. (2006). Raptors of North America: Natural History and Conservation. Beverly, MA: Voyageur Press.

›oares-Filho, B. S., Nepstad, D. C., Curran, L. M., Cerqueira, G. C., Garcia, R. A., Ramos, C. A., Voll, E., McDonald, A., Lefebvre, P. & Schlesinger, P. (2006). Modelling conservation in the Amazon Basin. Nature, 440(7083), 520–523. https://doi.org/10.1038/nature04389

›ojamo, E. (1982). Lintutretkia ulkomaille IV. Gruusiassa kevaalla 1982. Observing raptor migration in Batumi in the spring of 1982. Ornis Karelica, 8, 58–62.

›ojamo, E. (1983). Gruusiassa kevaalla 1983. Observations of raptor migration in Batumi, Georgian SSR, in the spring 1983. Ornis Karelica, 9, 69–73.

›olbrig, O. T. (1993). Ecological constraints to savanna land use. *In*: Young, M. D. & Solbrig, O. T. (Eds.), The world's savannas: Economic driving forces, ecological constraints and policy options for sustainable land use, pp. 21–47. Paris: The Pantheon Publishing Group.

›olbrig, O. T. (1994) Biodiversity: An introduction. *In*: Solbrig, O. T., van Emden, H. M. & van Oordt, P. G. W. (Eds.), Biodiversity and global change, pp. 13–17. Wallingford: CAB International.

›oleha, D. (2000). Reproductive behavior of Black eagle (Ictinaetus malayensis) in the southern slope of Mount Merapi, Yogyakarta. In: Prawiradilaga, D. (Ed.), Proceedings Asian Raptor Research Conservation: The Second Symposium on Raptors of Asia, Grand Aquila, 208. Bandung – Indonesia, 25–27 July 2000.

›oma, T. (2013). Ethnoarchaeology of Ancient Falconry in East Asia. Asian Conference on Asian Studies, International Academic Forum.

Somers, M. J. & Penzhorn, B. L. (1992). Reproduction in a reintroduced warthog population in the eastern Cape Province. South African Journ of Wildlife Research, 22, 57–60.

Sonerud, G. A., Steen, R., Løw, L. M., Røed, L. T., Skar, K., Selås, V. & Slagsvold, T. (2014). Evolution of parental roles in raptors: Prey ty determines role asymmetry in the Eurasian kestrel. Animal Behaviour, 96, 31–38. https://doi.org/10.1016/j.anbehav.2014.07.011

Soon, C. K. (2008). Recent sightings. Suara Enggang, 16, 26–33.

Sorenson, K., Burnett, J., Mee, A. & Hall, L. (2007). Lead concentrations in the blood of Big Sur California condors. California condors in t 21st century. Series in Ornithology, No. 2, pp. 185–195. Cambridge, MA: Nuttall Ornithological Club and American Ornithologists' Unio

Sorley, C. S. & Andersen, D. E. (1994). Raptor abundance in south-central Kenya in relation to land-use patterns. African Journal of Ecolog 32(1), 30–38. https://doi.org/10.1111/j.1365-2028. 1994.tb00552.x

Soroczynski, M. J. (1992). An account of a Martial eagle pair breeding in the Jwaneng area. Babbler, 23, 28–32.

Soroczynski, M. (1994). A breeding account of a Martial eagle Polemaetus bellicosus in the Jwaneng Game Park in September 1993. Babbl 26/27: 27–29.

Sourav, M. S. H., Ahmed, B. & Thompson, P. (2011). Pallas's fish eagle Haliaeetus leucoryphus in Bangladesh. BirdingASIA, 16, 101–105.

Sousa, G. L., Oliveira, C., Procopio, J. et al. (2015). Harpy eagle (Harpia harpyja) monitoring project in São Geraldo do Araguaia, south-easte Brazilian Amazon. Spizaetus, 20, 2–10.

Southern, W. E. (1963). Winter populations, behavior, and seasonal dispersal of Bald eagles in northwestern Illinois. Wilson Bulletin, 75, 186–18

Soutullo, A., Urios, V., Ferrer, M. & Peñarrubia, S. G. (2006). Post-fledging behaviour in Golden eagles Aquila chrysaetos: Onset of juveni dispersal and progressive distancing from the nest. Ibis, 148(2), 307–312. https://doi.org/10.1111/j.1474-919X.2006.00530.x

Sovada, M. A., Roy, C. C., Bright, J. B. & Gillis, J. R. (1998). Causes and rates of mortality of swift foxes in western Kansas. Journal of Wildli Management, 62(4), 1300–1306. https://doi.org/10.2307/3801994

Sözer, R. & Nijman, V. (1995a). Behavioural ecology, distribution and conservation of the Javan Hawk-eagle Spizaetus bartelsi (Streseman 1924). Verslagen en Technische Gegevens, 62, 1–122.

Sözer, R. & Nijman, V. (1995b). The Javan Hawk-eagle: New information on its distribution in central Java and notes on its threats. Tropic Biodiversity, 3, 49–55.

Spaan, D., Burke, C., McAree, O., Aureli, F., Rangel-Rivera, C. E., Hutschenreiter, A., Longmore, S. N., McWhirter, P. R. & Wich, S. (2019). Thermal infrared imaging from drones offers a major advance for spider monkey surveys. Drones, 32, 34. https://doi.org/10.339 drones3020034

Spaar, R. & Bruderer, B. (1996). Soaring migration of Steppe eagles (Aquila nipalensis) in southern Israel: Flight behaviour under various wi and thermal conditions. Journal of Avian Biology, 27(4), 289–301. https://doi.org/10.2307/3677260

Spaar, R. & Bruderer, B. (1997). Migration by flapping or soaring: Flight strategies of Marsh, Montagu's and Pallid harriers in southern Isra Condor, 99(2), 458–469. https://doi.org/10.2307/1369952

Spaar, R. (1997). Flight strategies of migrating raptors: A comparative study of interspecific variation in flight characteristics. Ibis, 139(3), 52 535. https://doi.org/10.1111/j.1474-919X.1997. tb04669.x

Spallholz, J. E. & Hoffman, D. J. (2002). Selenium toxicity: Cause and effects in aquatic birds. Aquatic Toxicology, 57(1–2), 27–37. https://d org/10.1016/s0166-445x(01)00268-5

Spasov, S., Arkumarev, V., Dobrev, D. & Dobrev, V. (2012). An overview of monitoring for raptors in Bulgaria. Acrocephalus, 33(154–15 181–189. https://doi.org/10.2478/v10100-012-0005-9

Speer, R. & Wise, J. P. (2018). Current status on chromium research and its implications for health and risk assessment. *In*: Reference Module Chemistry, Molecular Sciences and Chemical Engineering. https://doi.org/10.1016/B978-0-12-409547-2.14283-0

Spencer, J. A. & Lynch, T. P. (2005). Patterns in the abundance of White-bellied Sea-eagles (Haliaeetus leucogaster) in Jervis Bay, South-centr Australia. Emu, 105, 211–216.

Spencer, R. D., Beausoleil, R. A. & Martorello, D. A. (2007). How agencies respond to human black bear conflicts: A survey of wildlife agenci in North America. Ursus, 18(2), 217–229. https://doi.org/10.2192/1537-6176(2007)18217:HARTHB2.0.CO;2

SPI Infrared. (2014). Thermal Cameras on Drones to Spot Hibernating Bears in Study. Retrieved from https://www.x20.org/thermal-cameras-o drones-bears/

Spiegel, O., Getz, W. M. & Nathan, R. (2013). Factors influencing foraging search efficiency: Why do scarce Lappet-Faced vultures outperfor ubiquitous White-backed vultures? American Naturalist, 181(5), e102–e115. https://doi.org/10.1086/670009

Spierenburg, P. (2005). Birds in Bhutan: Status and Distribution. Bedford, U.K.: Oriental Bird Club.

Spittle, R. J. (1952). Feeding habits of some Singapore birds. Malayan Nature Journal, 7, 23–32.

Springer, M. T., Nielsen, C. K., Carver, A. D. & Correa, N. J. (2011). Harpy eagle (Harpia harpyja) feeding behavior on a Brown-throated thre toed Sloth (Bradypus variegatus). Journal of Raptor Research, 45, 101–103.

Sriram, M., Sriram, T. & Anand, M. O. (2008). Sighting of the White-bellied Sea eagle Haliaeetus leucogaster at the Lakkavalli Dam. Indian Bird 4(2), 71.

Ssemmanda, R. & Pomeroy, D. (2010). Scavenging birds of Kampala: 1973–2009. Scopus, 30, 26–31.

Stace, T. (2015). Sea eagle catches Wood duck. Boobook, 33, 49.

Stager, K. E. (1954). Birds of the Barranca de Cobre region of southwestern Chihuahua, Mexico. Condor, 56(1), 21–32. https://d org/10.2307/1364884

Stager, K. E. (1964). The role of olfaction in food location by the Turkey vulture and other cathartids. Los Angeles County Museum Contributio in Science, 81, 1–63.

Stagg, A. J. (1991). Birds of the Riyadh region: An annotated checklist (2nd Ed.). Riyadh: National Commission for Wildlife Conservation.
Stalmaster, M. V. & Newman, J. R. (1979). Perch site preferences of wintering Bald eagles in northwest Washington. Journal of Wildlife Management, 43(1), 221–224. https://doi.org/10.2307/3800658
Stalmaster, M. V. & Gessaman, J. A. (1982). Food consumption and energy requirements of captive Bald eagles. Journal of Wildlife Management, 46(3), 646–654. https://doi.org/10.2307/3808555
Stalmaster, M. V. (1983). An energetics simulation model for managing wintering Bald eagles. Journal of Wildlife Management, 47(2), 349–359. https://doi.org/10.2307/3808507
Stalmaster, M. V. & Gessaman, J. A. (1984). Ecological energetics and foraging behavior of overwintering Bald eagles. Ecological Monographs, 54(4), 407–428. https://doi.org/10.2307/1942594
Stalmaster, M. V. (1987). The Bald Eagle. New York: Universe Books.
Stalmaster, M. V. & Plettner, R. G. (1992). Diets and foraging effectiveness of Bald eagles during extreme winter weather in Nebraska. Journal of Wildlife Management, 56, 355–367.
Stalmaster, M. V. & Kaiser, J. L. (1997). Winter ecology of Bald eagles in the Nisqually River drainage, Washington. Northwest Science, 71, 214–223.
Stanford, C. B., Wallis, J., Matama, H. & Goodall, J. (1994). Patterns of predation by chimpanzees on red colobus monkeys in Gombe National Park, 1982–1991. American Journal of Physical Anthropology, 94(2), 213–228.
Stankovic, D. & Rakovic, M. (2008). Ninth report of the Center for Animal Marking. Ciconia, 17, 119–136.
Stankovic, S., Kalaba, P. & Stankovic, A. R. (2014). Biota as toxic metal indicators. Environmental Chemistry Letters, 12(1), 63–84. https://doi.org/10.1007/s10311-013-0430-6.
State of Victoria Department of Environment, Land, Water and Planning. (2018). Australia's largest raptor is the Wedge-tailed eagle, which is truly a magnificent bird. Our Wildlife Fact Sheet: Wedge-tailed eagle. Retrieved from https://www.wildlife.vic.gov.au/__data/ assets/pdf_file/0017/115343/Wedge-tailed-Eagle.pdf.
Stattersfield, A. J., Crosby, M. J., Long, A. J. & Wege, D. C. (1998). Endemic Bird Areas of the World: Priorities for biodiversity conservation. Cambridge, UK: BirdLife International.
Stauber, E., Finch, N., Talcott, P. A. & Gay, J. M. (2010). Lead poisoning of Bald (Haliaeetus leucocephalus) and Golden (Aquila chrysaetos) eagles in the US Island Pacific Northwest Region—An 18-year retrospective study: 1991-2008. Journal of Avian Medicine and Surgery, 24(4), 279–287. https://doi.org/10.1647/2009-006.1
Steele, M. L. (2017). Where in the World are Pallas's Fish Eagles? Migration and Ecology of Haliaeetus leucoryphus in Asia. Theses and Dissertations (2015). Retrieved from http://scholarworks.uark.edu/etd/2015http: //scholarworks.uark.edu/etd/2015
Steele, W. (1998). Birds of prey and the rabbit calicivirus. Wingspan, 8(3), 23.
Steenhof, K. & Kochert, M. N. (1988). Dietary responses of three raptor species to changing prey densities in a natural environment. Journal of Animal Ecology, 57(1), 37–48. https://doi.org/10.2307/4761
Steenhof, K. (1984). Use of an interspecific communal roost by wintering Ferruginous Hawks. Wilson Bulletin, 95, 137–138.
Steenhof, K., Kochert, M. N. & Roppe, J. A. (1993). Nesting by raptors and common ravens on electrical transmission line towers. The Journal of Wildlife Management, 57(2), 271–281. https://doi.org/10.2307/3809424
Steenhof, K. & Newton, I. (2007). Assessing nesting success and productivity. *In*: Bird, D. M., Steidl, R. J. & Anthony, R. G. (Eds.), Responses of Bald eagles to human activity during the summer in interior Alaska. Ecological Applications, 6, 482–491.
Steenhof, K. & Newton, I. (2007). Assessing raptor nest success and productivity. *In*: Bird, D. M. & Bildstein, K. L. (Eds.), Raptor Research and Management Techniques, 181-192. Surrey, Canada: Hancock House.
Steidl, R. J. & Anthony, R. G. (1996). Responses of Bald eagles to human activity during the summer in interior Alaska. Ecological Applications, 6(2), 482–491. https://doi.org/10.2307/2269385
Steinitz, C., Faris, R., Flaxman, M., Karish, K., Mellinger, A. D., Canfield, T. & Sucre, L. (2005). A delicate balance: Conservation and development scenarios for Panama's Coiba National Park. Environment: Science and Policy for Sustainable Development, 47(5), 24–39. https://doi.org/10.3200/ENVT.47.5.24-43
Stenchenko, A. M. (1974). Steller's Sea eagle in Kamchatka. Materials of VI-All-Union Ornithological Conference, Moscow, Part 2, 139–140.
Stepanyan, L. S. (1990). Conspectus of the Ornithological Fauna of the USSR. Moscow: Nauka Press.
Stephens, W. B. & Blackwood, V. (1993). Wahlberg's eagle catching bats in Malawi. Ostrich, 54, 25. https://doi.org/10.1080/00306525.1983.9634439
Stern, B. R. (2010). Essentiality and toxicity in copper health risk assessment: Overview, update and regulatory considerations. Journal of Toxicology and Environmental Health. Part A, 73(2), 114–127. https://doi.org/10.1080/15287390903337100
Stevenson, A. & Evans, R. (2015). White-tailed eagle. Mitteleuropäische Gesellschaft zur Erhaltung der Greifvögel (MEGEG). Hamilton, Bermuda: Central European Society for Raptor Protection (CESRP).
Stevenson-Hamilton, J. (1954). Wild life in South Africa. London: Cassell and Co.
Stevenson, T. & Fanshawe, J. (2002). Field guide to the birds of East Africa. London: T & A D Poyser.
Stewart, P. A. (1978). Behavioral interactions and niche separation in Black and Turkey vultures. Living Bird, 17, 79–84.
Stewart, R. (2002). Kyrgyzstan. Hong Kong: Odyssey Books & Guides.
Steyn, P. (1962). Observations on Wahlberg's Eagle. Bokmakierie, 14, 7–14.
Steyn, P. (1970). Prey of the Martiall Eagle. African Wild Life, 24, 172.
Steyn, P. (1980). Notes on the prey and breeding success of the Martial eagle. Ostrich, 51, 115–116.
Steyn, P. (1982). Birds of Prey in Southern Africa. Cape Town, SA: David Philip Publishers.

Steyn, P. (1983). Birds of Prey of Southern Africa: Their identification and life histories. Beckenham (UK), Croom Helm.
Steyn, P. (1987). Martial eagle kills Woolly-necked stork. African Wildlife, 41, 147.
Stiles, F. G. (1985). Conservation of forest birds in Costa Rica: Problems and perspectives. *In*: Diamond, A. W. & Lovejoy, T. E. (Eds.), Conservation of tropical forest birds. International Council for Bird Preservation Technical Publication, 4 (pp. 141–168). Cambridge, UK: ICBP.
Stiles, F. G. & Skutch, A. (1989). A Guide to the Birds of Costa Rica. Ithaca, NY: Cornell University Press.
Still, C. J., Foster, P. N. & Schneider, S. H. (1999). Simulating the effects of climate change on tropical montane cloud forests. Nature, 398(6728), 608–610. https://doi.org/10.1038/19293
Stillman, J. D. B. (1967). The Gold Rush Letters with an Introduction by Kenneth Johnson. Lewis: Osborne: Palo Alto, California.
Stjernberg, T. (1981). Projekt havsörn i Finland – I. Stjernberg, T. (Ed.), Projekt havsörn i Finland och Sverige på Tvärminne Zoologiska station, Finland. Förhandlingar från ett havsörnssymposium, 8–9.1(1), 1979 på Tvärminne Zoologiska station, Finland.
Stjernberg, T., Koivusaari, J. & Hogmander, J. (2003). Population trends and breeding success of the White-tailed Sea eagle in Finland (1970–2000). *In*: Helander, B. (Ed.), Sea eagle 2000: Proceedings from the International Sea Eagle Conference in Bjorko, Sweden, 13–17 September 20, 103–112. Stockholm: Sweden Proceedings of the Swedish Society for Nature Conservation, SNF.
Stjernstedt, B. (1975). Eagle attack. Black Lechwe, 12, 18–22.
Stoate, C., Boatman, N. D., Borralho, R. J., Carvalho, C. R., De Snoo, G. R. & Eden, P. (2001). Ecological impacts of arable intensification in Europe. Journal of Environmental Management, 63(4), 337–365. https://doi.org/10.1006/jema.2001.0473
Stolen, E. D. (1996). Black and Turkey vulture interactions with Bald eagles in Florida. Florida Field Naturalist, 24(2), 43–45.
Stone, W. B. & Okoniewski, J. C. (2001). Necropsy findings and environmental contaminants in common loons from New York. Journal of Wildlife Diseases, 37, 178–184. https://doi.org/10.7589/0090-3558-37.1.178
Stone, W. & Roberts, H. R. (1934). Zoological results of the Matto Grosso Expedition to Brazil in 1931 – III. Birds. Proceedings of the Academy of Natural Sciences of Philadelphia, 86, 363–397.
Storr, G. M. (1991). Birds of the south-west division of Western Australia. Records of the Western Australian Museum Supplement, No. 35, 1–150.
Stoychev, S. & Karafeisov, T. (2003). Power Line Design and Raptor Protection in Bulgaria. Budapest: Sixth World Conference on Birds of Prey and Owls.
Strandberg, R., Klaassen, R. H. G., Hake, M. & Alerstam, T. (2010). How hazardous is the Sahara Desert crossing for migratory birds? Indications from satellite tracking of raptors. Biology Letters, 6(3), 297–300. https://doi.org/10.1098/rsbl.2009.0785
Strange, I. J. (1996). The Striated Caracara Phalcoboenus australis in the Falkland Islands. Author's Edition, Falkland Islands.
Strange, M. (2000). Photographic Guide to the Birds of Southeast Asia. Singapore: Periplus.
Straube, F. C. (1993). Revision of the itinerary of the Natterer Expedition to the State of Paraná (Brazil). Acta Biol Leopoldensia, 15, 5–20.
Straube, F. C., Urben-Filho, A. & Kajiwara, D. (2004). Aves. *In*: Mikich, S. B. & Bérnils, R. S. (Eds.), Red Book of Threatened Fauna in the State of Paraná, pp. 143–146. Curitiba, Brazil: Instituto Ambiental do Paraná.
Strauch, J. G., Jr. (1975). Observations at a nest of the Black-and-white Hawk-eagle. Condor, 77(4), 512. https://doi.org/10.2307/1366111
Stresemann, E. (1924). Raubvogelstudien II. Die Formenkreise Spizaëtus nipalensis (Hodgson) und Spizaëtus cirrhatus (Gmelin). Journal für Ornithologie, 72, 430–432.
Stresemann, E. (1932). Ein neuer Habichtsadler aus Sumbawa: Hieraaëtus fasciatus renschi subsp. nova. Ornithologische Monatsberichte, 40, 78–80.
Stresemann, E. & Amadon, D. (1979). Order Falconiformes. *In*: Mayr, E. & Cottrell, G. W. (Eds.), Check-list of birds of the world. Vol. I. (2nd ed.), pp. 271–425. Cambridge, MA: Museum of Comparative Zoology.
Strewe, R. (1999). Notes on the distribution and nesting of the Black-and-chestnut eagle, Oroaetus isidori, in Nariño. Boletín SAO, 10, 45–52.
Strewe, R. & Navarro, C. (2003). New distributional records and conservation importance of the San Salvador Valley, Sierra Nevada de Santa Maria, northern Colombia. Ornitología Colombiana, 1, 29–41.
Strewe, R. & Navarro, C. (2004). The threatened birds of the Rio Frio Valley, Sierra Nevada de Santa Marta, Colombia. Cotinga, 22, 47–55.
Strickland, M. D., Arnett, E. B., Erickson, W. P., Johnson, D. H., Johnson, G. D., Morrison, M. L., Shaffer, J. A. & Warren-Hicks, W. (2011). Comprehensive Guide to Studying Wind Energy/Wildlife Interactions. Prepared for the National Wind Coordinating Collaborative, Washington, D.C.
Strix. (2012). Developing and testing the methodology for assessing and mapping the sensitivity of migratory birds to wind energy development. Cambridge: BirdLife International.
Strom, S. M., Langenberg, J. A., Businga, N. K. & Batten, J. K. (2009). Lead exposure in Wisconsin birds. *In*: Watson, R. T., Fuller, M., Pokras, M. & Hunt, W. G. (Eds.), Ingestion of lead from spent ammunition: Implications for wildlife and humans, pp. 1–8. Boise, Idaho ID: The Peregrine Fund.
Stuart, C. T. (1982). Aspects of the biology of the caracal (Felis caracal, Schreber, 1776) in the Cape Province, South Africa. Pietermaritzburg: University of KwaZulu Natal.
Sturgis, B. B. (1928). Field book of birds of the Panama Canal Zone: A description on the habits, call notes and songs of the birds of the Panama Canal Zone for the purpose of identifying them. Many of these birds are also common in Central and South America. New York: G. P. Putnam's Sons.
Suana, I. W., Ahyadi, H., Amin, S., Kalih, L. A. T. T. W. S. & Hadiprayitno, G. (2016). Birdwatching in Kerandangan Natural Park, Lombok, Nusa Tenggara Baratbarat, Indonesia. BirdingASIA, 26, 8–16.
Suárez, S., Balbontín, J. & Ferrer, M. (2000). Nesting habitat selection by Booted eagles Hieraaetus pennatus and implications for management. Journal of Applied Ecology, 37(2), 215–223. https://doi.org/10.1046/j.1365-2664.2000.00503.x

ubedi, T. R. (2014). East to west migration of Steppe eagle (Aquila nipalensis) and other raptors in Nepal; abundance, timing and age class determination. https://doi.org/10.13140/RG.2.2.26813.46565.

uckling, K. &. Hodges, W. (2007). Status of the Bald eagle in the lower 48 states and the District of Columbia: 1963–2007. Retrieved from http://www.biologicaldiversity.org/ species/ birds/bald_eagle/ report/

ulawa, J., Robert, A., Köppen, U., Hauff, P. & Krone, O. (2010). Recovery dynamics and viability of the White-tailed eagle (Haliaeetus albicilla) in Germany. Biodiversity and Conservation, 19(1), 97–112. https://doi.org/10.1007/s10531-009-9705-4

ulkava, S., Tornberg, R. & Koivusaari, J. (1997). Diet of the White-tailed eagle Haliaeetus albicilla in Finland. Ornis Fennica, 74(2), 65–67.

ullivan, G. L. (2016). Bald eagles are adapting to city living. Retrieved from https://www.washingtonpost.com/national/health-science/bald-eagles-are-adapting-to-city-living/2016/02/29/97df78be-d5ad-11e5-be55-2cc3c1e4b76b_story.html.

umasgutner, P., Zuna-Kratky, T. & Krenn, H. W. (2010). Influence of the vegetation structure on the nest-site selection of birds of prey in the March floodplain forests/Lower Austria. Vogelwarte, 48, 81–95.

umasgutner, P., Tate, G. J., Koeslag, A. & Amar, A. (2016). Seasonal patterns in space use of Black sparrowhawks Accipiter melanoleucus in an urban environment. Bird Study, 63(3), 430–435. https://doi.org/10.1080/00063657.2016.1214814

undev, G., Yosef, R., Birazana, O. & Damdin, S. (2012). Breeding ecology of the Steppe eagle (Aquila nipalensis) in Mongolia. Ornis Mongolica, 1, 13.e19.

undevall, C. J. (1850). Foglar från Södra Afrika. Öfversigt af Konglig Vetenskaps-Akademiens Forhandlingar, 7, 96–110.

unquist, M. & Sunquist, F. (2002a). Snow leopard Uncia uncia (Schreber, 1775). Wild Cats of the World, 377–394. Chicago: University of Chicago Press.

unquist, M. & Sunquist, F. (2002b). Clouded leopard Neofelis nebulosa (Griffiths, 1821). Wild Cats of the World, pp. 278–284. Chicago: University of Chicago Press.

unyer, J. R. & Viñuela, J. (1996). Invernada de rapaces (O. Falconiformes) en España Peninsular e Islas Baleares (O. Falconiformes) en España Peninsular e Islas Baleares. *In*: Muntaner, J. & Mayol, J. (Eds.), Biología y Conservación de las Rapaces Mediterráneas, 1994.

uparman, U. (2002). Surveys on protected birds: Javan Hawk-eagle and other raptors in southern Cianjur District. Final report for FFI-Indonesia. KPB. CIBA, Cianjur, Indonesia: Ciba.

uring, L. H. (2008). Habitat relationships of Bald eagles in Alaska. *In*: Wright, B. A. & Schempf, P. (Eds.), Bald Eagles in Alaska, pp. 106–116. Juneau, AK: Bald Eagle Research Institute, University of Alaska Southeast.

urve, S. S., Patil, P., Khan, N. & Narwade, S. (2014). Sighting of Indian Spotted eagle Clanga hastata from Taltal Chhapar Sanctuary, Rajasthan, India. Journal of the Bombay Natural History Society, 111(3), 220–221.

uryawanshi, K. R., Bhatia, S., Bhatnagar, Y. V., Redpath, S. & Mishra, C. (2014). Multiscale factors affecting human attitudes toward snow leopards and wolves. Conservation Biclogy, 28(6), 1657–1666. doi.org/10.1111/cobi.12320

ustaita, D. & Hertel, F. (2010). In vivo bite and grip forces, morphology and prey-killing behavior of North American accipiters (Accipitridae) and falcons (Falconidae). The Journal of Experimental Biology, 213(15), 2617–2628. https://doi.org/10.1242/jeb.041731

utton, A. E. (2015). Leadership and management influence the outcome of wildlife reintroduction programs: Findings from the Sea Eagle Recovery Project. Peer J, 3, e1012. https://doi.org/10.7717/peerj.1012

utton, C. & Sutton, P. (1999). River of raptors: Exploring and enjoying Pronatura Veracruz's raptor conservation project. Birding, 31, 229–236.

utton, C., Dunne, P. & Sibley, D. (1989). Hawks in Flight: The Flight Identification of North American Migrant Raptors. Boston: Houghton Mifflin Harcourt.

utton, G. M. (1951). Mexican birds: First impressions, based upon an ornithological expeditions to Tamaulipas, Nuevo León, and Coahuila, with an appendix briefly describing all Mexican birds. Norman, OK: University of Oklahoma Press.

vidensky, A. (2013). The Eagle Huntress Story. Retrieved from http://www.svidensky.com/ projects/the-eagle-huntress

vidensky, A. (2013). Interview on Mongolian TV, Part 1. Retrieved from https://www.youtube.com/watch?v=nEPMUb-aeCQ&feature=youtu.be

vidensky, A. (2014). Heroes as Stories. TEDX Talk at Ulaanbaatar, Mongolia, Nov. 21. Retrieved from https://www.youtube.com/watch?v=DhhXh9YMQ0M

wan, G., Naidoo, V., Cuthbert, R., Green, R. E., Pain, D. J., Swarup, D., Prakash, V., Taggart, M., Bekker, L., Das, D., Diekmann, J., Diekmann, M., Killian, E., Meharg, A., Chandra Patra, R., Saini, M. & Wolter, K. (2006). Removing the threat of diclofenac to critically endangered Asian vultures. PLoS Biology, 4(3), 395–402. https://doi.org/10.1371/journal.pbio.0040066

wan, G., Cuthbert, R., Quevedo, M., Green, R., Pain, D., Bartels, P., Cunningham, A., Duncan, N., Meharg, A., Oaks, J., Parry-Jones, J., Shultz, S., Taggart, M., Verdoorn, G. & Wolter, K. (2006). Toxicity of diclofenac to Gyps vultures. Biology Letters, 2, 279–282. http://dx.doi.org/10.1098/rsbl.2005.0425.

wann, H. K. (1921). Notes on a collection of Accipitres from the Mérida District, W. Venezuela. Auk, 38(3), 357–364. https://doi.org/10.2307/4073763

wart, J. M., Richardson, P. R. K. & Ferguson, J. W. H. (1999). Ecological factors affecting the feeding behaviour of pangolins (Manis temminckii). Journal of Zoology, 247(3), 281–292. https://doi.org/10.1111/j.1469-7998.1999.tb00992.x

watridge, C. J., Monadjem, A., Steyn, D. J., Batchelor, G. R. & Hardy, I. C. W. (2014). Factors affecting diet, habitat selection and breeding success of the African Crowned eagle Stephanoaetus coronatus in a fragmented landscape. Ostrich, 85(1), 47–55. https://doi.org/10.2989/00306525.2014.896832

wenson, J. E., Adamič, M., Huber, D. & Stokke, S. (2007). Brown bear body mass and growth in northern and southern Europe. Oecologia, 153(1), 37–47. https://doi.org/10.1007/s00442-007-0715-1

wenson, J. E., Alt, K. L. & Eng, R. L. (1985). Ecology of Bald eagles in the Greater Yellowstone ecosystem. Wildlife Monogram, 95(95), 3–46.

wets, J. A. (1988). Measuring the accuracy of diagnostic systems. Science, 240(4857), 1285–1293. https://doi.org/10.1126/science.3287615

Swihart, R. K. & Slade, N. A. (1985). Testing for independence of observations in animal movements. Ecology, 66(4), 1176–1184. https://dc org/10.2307/1939170

Switzer, F. (1977). Saskatchewan Power's experience. Blue Jay, 35, 259–260.

Syartinilia & Tsuyuki, S. (2008). GIS-based modelling of Javan Hawk-eagle distribution using logistic and autologistic regression model Biological Conservation, 141(3), 756–769. https://doi.org/10.1016/j.biocon.2007.12.030

Sydes, C. & Miller, G. R. (1988). Range management and nature conservation in the British uplands. *In*: Usher, M. B. & Thompson, D. B. (Eds.), Ecological Change in the Uplands, pp. 323–337. British Ecological Society Special Publications Series No. 7. Oxford: Blackwell.

Symes, C. T. & Antonites, A. R. (2014). Notes on African Crowned eagle Stephanoaetus coronatus diet in savanna and forest in KwaZulu-Nat South Africa. Ostrich, 85(1), 85–88. https://doi.org/10.2989/00306525.2014.892036

Symes, C. C. T. & Kruger, T. T. L. (2012). The persistence of an apex avian predator, Verreaux's eagle, in a rapidly urbanizing environment. Sou African Journal of Wildlife Research, 42(1), 45–53. https://doi.org/10.3957/056.042.0109

Szumiło, E., Szubska, M., Meissner, W., Bełdowska, M. & Falkowska, L. (2013). Mercury in immature and adults Herring gulls (Larus argentatu wintering on the Gulf of Gdańsk area. Oceanological and Hydrobiological Studies, 42(3), 260–267. https://doi.org/10.2478/s13545-01 0082-y

Szumiło-Pilarska, E., Grajewska, A., Falkowska, L., Hajdrych, J., Meissner, W., Frączek, T., Bełdowska, M. & Bzoma, S. (2016). Speci differences in total mercury concentration in gulls from the Gulf of Gdansk (Southern Baltic). Journal of Trace Elements in Medicine a Biology: Organ of the Society for Minerals and Trace Elements, 33, 100–109. https://doi.org/10.1016/j.jtemb.2015.09.005

Szumiło-Pilarska, E., Falkowska, L., Grajewska, A. & Meissner, W. (2017). Mercury in feathers and blood of gulls from the Southern Baltic Coa Poland. Water, Air and Soil Pollution, 228(4), 138. https://doi.org/10.1007/s11270-017-3308-6

Tabaranza, Blas B. R., Jr. (2005). The Largest Eagle in the World. Haribon Foundation.

Taberlet, P., Meyer, A. & Bouvet, J. (1992). Unusual mitochondrial DNA polymorphism in two local populations of blue tit Parus caeruleu Molecular Ecology, 1(1), 27–36. https://doi.org/10.1111/j.1365-294X.1992.tb00152.x

Tago, K., Suzuki, Y., Shirai, A. & Yamagishi, S. (2015). The breeding success of Mountain Hawk-eagles in relation to environmental facto Japanese Journal of Ornithology, 64(2), 195–206. https://doi.org/10.3838/jjo.64.195

Takekawa, J. Y., Newman, S. H., Xiao, X., Prosser, D. J., Spragens, K. A., Palm, E. C., Yan, B., Li, T., Lei, F., Zhao, D., Douglas, D. C., Muzaff S. B. & Ji, W. (2010). Migration of waterfowl in the East Asian flyway and spatial relationship to HPAI H5N1 outbreaks. Avian Disease 54(1 Suppl), 466–476. https://doi.org/10.1637/8914-043009-Reg.1

Tamisier, A. & Isenmann, P. (2003). Compte rendu ornithologique camarguais pour les années 1995–2000. Revue d'Ecologie (Terre et Vie), 5 5–76.

Tamura, K., Peterson, D., Peterson, N., Stecher, G., Nei, M. & Kumar, S. (2011). MEGA5: Molecular evolutionary genetics analysis usi maximum likelihood, evolutionary distance, and maximum parsimony methods. Molecular Biology and Evolution, 28(10), 2731–273 https://doi.org/10.1093/molbev/msr121

Tannerfeldt, M. & Angerbjörn, A. (1996). Life history strategies in a fluctuating environment: Establishment and reproductive success in the arc fox. Ecography, 19(3), 209–220. http://www.jstor.org/stable/3682906

Tappen, M. & Wrangham, R. (2000). Recognizing hominoid-modified bones: The taphonomy of colobus bones partially digested by free-rangi chimpanzees in the Kibale Forest, Uganda. American Journal of Physical Anthropology, 113(2), 217–234. https://doi.org/10.1002/109 8644(200010)113: 2<217:: AID-AJPA6>3.0.CO;2-G

Tarboton, W. (1976). Martial eagles: An unusual breeding episode. Bokmakierie, 28, 29–32.

Tarboton, W. R. & Allan, D. G. (1984). The Status and Conservation of Birds of Prey in the Transvaal. Transvaal Museum Monograph No. Pretoria, 3.

Tarboton, W. R. (1990). African Birds of Prey. New York: Ithaca Cornell University Press.

Tate, A. R. (2020). Black-and-white Hawk-Eagle (Spizaetus melanoleucus), version 1.0. *In*: Schulenberg, T.S. (Ed.), Birds of the World. Ithac NY: Cornell Laboratory of Ornithology. https://doi.org/10.2173/bow.bawhae1.01

Taylor, L., Taylor, C. & Davis, A. (2013). The impact of urbanization on avian species: The inextricable link between people and birds. Urb Ecosystems, 16(3), 481–498. https://doi.org/10.1007/s11252-012-0283-y

Taylor, M. (1933). Further notes on the White-bellied Sea eagle Haliaeetus leucogaster. Hong Kong Naturalist, 4, 11–12.

Taylor, M. R. (2015a). Martial Eagle Polemaetus bellicosus. *In*: Taylor, M. R., Peacock, F. & Wanless, R. M. (Eds.), The Eskom Red Data Bo of Birds of South Africa, Lesotho and Swaziland, pp. 113–115. Johannesburg, SA: Birdlife South Africa.

Taylor, M. R. (2015b). Verreauxs' Eagle. The Eskom Red Data Book of Birds of South Africa, Lesotho and Swaziland. Johannesburg: BirdLi South Africa.

Tchounwou, P., Newsome, C., Williams, J. & Glass, K. (2008). Copper-induced cytotoxicity and transcriptional activation of stress genes in hum liver carcinoma cells. Metal Ions in Biology and Medicine, 10, 285–290.

Tchounwou, P. B., Yedjou, C. G., Patlolla, A. K. & Sutton, D. J. (2012). Heavy metals toxicity and the environment. *In*: Luch, A. (Ed.), Molecul Clinical and Environmental Toxicology. Experientia Supplementum, 101, 133–164. Basel: Springer. https://doi.org/10.1007/978-3-764 8340-4_6

Tellería, J. L. (2009). Wind power plants and the conservation of birds and bats in Spain: A geographical assessment. Biodiversity and Conservatio 18(7), 1781–1791. https://doi.org/10.1007/s10531-008-9558-2

Temeles, E. J. (1985). Sexual size dimorphism of bird-eating hawks: The effect of prey vulnerability. The American Naturalist, 125(4), 485–49 https://doi.org/10.1086/284357

Temminck, C. J. & Schlegel, H. (1844). Aves. *In*: Von Siebold, P. F. (Ed.), Fauna Japonica, sive, Descriptio animalium, quae in itinere per Japoniam, jussu et auspiciis, superiorum, qui summum in India Batava imperium tenant, suscepto, annis 1823–1830. Leiden, The Netherlands: Arnz & Company.

Temminck, C. J. (1822). Nouveau Recueil de Planches coloriées d'Oiseaux, pour server de suite et de complément aux Planches Enluminées de Buffon. Published by the author, Paris.

Tennesen, M. (1992). Flight of the Falcon. Toronto, ON: Key Porter Books.

Terborgh, J. & Weske, J. S. (1969). Colonization of secondary habitats by Peruvian birds. Ecology, 50(5), 765–782. https://doi.org/10.2307/1933691

Terborgh, J. W., Fitzpatrick, J. W. & Emmons, L. (1984). Annotated checklist of bird and mammal species of Cocha Cashu Biological Station, Manu National Park, Peru. Fieldiana Zoology, New series, 21, 1–29.

Terres, J. K. & National Audubon Society (1991). The Audubon Society Encyclopedia of North American Birds. New York, NY: National Audubon Society.

Thaler, E., Maschler, S. & Steinkellner, V. (1986). Vergleichende Studien zur Postembryonalentwicklung Dreier Altweltgeier. Annalen Naturhistorisches Museum Wien 881898, 361–376.

Thelander, C. G. & Rugge, L. (2000). Bird risk behaviours and fatalities at the Altamont wind resource area. In: Proceedings of National Avian – Wind Power Planning Meeting III. San Diego, California, May (1998). Prepared for the Avian Subcommittee of the National Wind Coordinating Committee by LGL Ltd., King City, Ontario.

Thelander, C. & Rugge, L. (2001). Examining relationships between bird risk behaviors and fatalities at the Altamont Pass Wind Resource Area: A second year's progress report. *In*: Proceedings of the National Avian-Wind Planning Meeting IV. The Avian Subcommittee of the National Wind Coordinating Committee, 5–14. Carmel, CA.

Thelander, C. G., Smallwood, K. S. & Rugge, L. (2003). Bird risk behaviors and fatalities at the Altamont Pass Wind Resource Area. Period of performance: March 1998–December 2000. Golden, CO: National Renewable Energy Laboratory.

Thévenot, M., Vernon, R. & Bergier, P. (2003). The Birds of Morocco. An Annotated Checklist. Tring, UK: BOU Checklist (pp. 141–143).

Thewlis, R. M., Duckworth, J. W., Evans, T. D. & Timmins, R. J. (1998). The conservation status of birds in Laos: A review of key species. Bird Conservation International, 8 (Suppl. 1), 1–159. https://doi.org/10.1017/S09 59270900002197

Thiollay, J. (2000). Dryotriorhis spectabilis Congo Serpent-eagle. *In*: Verdoorn, G. H., Bildstein, K. L. & Ellis, S. (Eds.), Selected African Falconiformes conservation assessment and management plan, pp. 46–47. Apple Valley, MN IUCN/SSC Conservation Breeding Specialist Group.

Thiollay, J. (2006). The decline of raptors in West Africa: Long-term assessment and the role of protected areas. Ibis, 148(2), 240–254. https://doi.org/10.1111/j.1474-919X.2006.00531.x

Thiollay, J. (2007). Raptor declines in West Africa: Comparisons between protected, buffer and cultivated areas. Oryx, 41(3), 322–329. https://doi.org/10.1017/S0030605307000809

Thiollay, J.-M. (1975). Migrations de rapaces africains en Ouganda et au Rwanda. Oiseau Rev. Fr. Orn.Ornithol., 45, 192–194.

Thiollay, J.-M. (1976). Les rapaces diurnes dans l'Ouestouest africaine: Analyse d'un peuplement de savane préforestière et recherches sur les migrations saisonnières. Thèse de Doctorat. Université de Paris.

Thiollay, J.-M. (1977). Distribution saisonnière des rapaces diurnes en Afrique occidentale. Oiseau Rev. Fr. Ornithol., 47, 253–285.

Thiollay, J.-M. (1980). Spring hawk migration in eastern Mexico. Journal of Raptor Research, 14, 13–19.

Thiollay, J.-M. (1985). Birds of Prey in French Guiana—A preliminary review. Birds of Prey Bulletin, 2, 11–15.

Thiollay, J.-M. (1985). The West African forest avifauna: A review. *In*: Diamond, A. W. & Lovejoy, T. E. (Eds.), pp. 171–186. Conservation of Tropical Forest Birds, Cambridge, UK: International Council for Bird Preservation.

Thiollay, J.-M. (1985a). Falconiformes of tropical rainforests: A review. *In*: Newton, I. & Chancellor, R. D. (Eds.), Conservation Studies on Raptors, pp. 155–165. Cambridge, UK: International Council for Bird Preservation Technical Publication No. 5. ICBP.

Thiollay, J.-M. (1985b). Species diversity and comparative ecology of rainforest falconiforms on three continents. *In*: Newton, I. & Chancellor, R. D. (Eds.), Conservation Studies on Raptors, pp. 167–174. Cambridge, UK: International Council for Bird Preservation.

Thiollay, J.-M. (1991). Altitudinal distribution and conservation of raptors in southwestern Colombia. Journal of Raptor Research, 25, 1–8.

Thiollay, J-M. (1993). Response of a raptor community to shrinking area and degradation of tropical rain forest in the South Western Ghats (India). Ecography 16(2), 97–110. https://doi.org/10.1111/j.1600-0587.1993.tb00062.x

Thiollay, J.-M. (1998). Current status and conservation of Falconiformes in Tropical Asia. Journal of Raptor Research, 32, 40–55.

Thiollay, J.-M. (2006a). The decline of raptors in West Africa: Long-term assessment and the role of protected areas. Ibis, 148, 240–254.

Thiollay, J.-M. (2006b). Severe decline of large birds in the northern Sahel of West Africa: A long-term assessment. Bird Conservation International, 16(4), 353–365. https://doi.org/10.1017/S0959270 906000487

Thiollay, J.-M. (2007). Raptor communities in French Guiana: Distribution, habitat selection, and conservation. Journal of Raptor Research, 41(2), 90–105. https://doi.org/10.3356/08921016(2007)4190: RCIFGD2.0.CO;2

Thiollay, J.-M. (2007a). Raptor population decline in West Africa. Ostrich, 78(2), 405–413. https://doi.org/10.2989/OSTRICH.2007.78.2.46.126

Thiollay, J.-M. (2007b). Raptor declines in West Africa: Comparisons between protected, buffer and cultivated areas. Oryx, 41(3), 322–329.

Thiollay, J. M. & Meyburg, B.-U. (1981). Remarques sur l'organisation d'un peuplement insulaire de rapaces: Madagascar. Alauda, 49, 216–226.

Thiollay, J. M. (1994). Family Accipitridae (Hawks and Eagles), *In*: Del Hoyo, J., Elliott, A. & Sargatal, J. (Eds.), Handbook of the Birds of the World, Vol. 2. New World Vultures to Guineafowl, pp. 52–205. Barcelona: Lynx Edicions.

Thiollay, J. M. (1994). Structure, density and rarity in an Amazonian rainforest bird community. Journal of Tropical Ecology, 10(4), 449–481. https://doi.org/10.1017/S0266467400008154

Thiollay, J. M. (2001). Long-term changes of raptor populations in northern Cameroon. Journal of Raptor Research, 35, 173–186.

Thiollay, J. M. (2006). Large bird declines with increasing human pressure in savanna woodlands (Burkina Faso). Biodiversity and Conservatior 15(7), 2085–2108. https://doi.org/10.1007/s10531-004-6684-3

Thiollay, J.-M. & Meyburg, B.-U. (1988). Forest fragmentation and the conservation of raptors: A survey on the island of Java. Biologica Conservation, 44(4), 229–250. https://doi.org/10.1016/0006-3207(88)90018-3

Thiollay, J. M., Del Hoyo, J., Elliott, A. & Sargatal, J. (1994). Handbook of the Birds of the World, vol. 2, New World Vultures to Guineafow Barcelona: Lynx Edicions.

Thiollay, J. M. & Rahman, Z. (2002). The raptor community of Central Sulawesi: Habitat selection and conservation status. Biological Conservatior 107(1), 111–122. https://doi.org/10.1016/ S0006-3207(02)00051-4

Thirgood, S., Woodroffe, R. & Rabinowitz, A. (2005). The impact of human-wildlife conflict on human lives and livelihoods. Conservatio Biology Series (Cambridge), 9, 13.

Thomas, L., Buckland, S. T., Rexstad, E. A., Laake, J. L., Strindberg, S., Hedley, S. L., Bishop, J. R. B., Marques, T. A. & Burnham, K. P. (2010) Distance software: Design and analysis of distance sampling surveys for estimating population size. Journal of Applied Ecology, 47, 5–14.

Thompson, J. D., Higgins, D. G. & Gibson, T. J. (1994). Improving the sensitivity of progressive multiple sequence alignment through sequenc weighting, position-specific gap penalties and weight matrix choice. Nucleic Acids Research, 22(22), 4673–4680. https://doi.org/10.1093 nar/22.22.4673

Thompson, N. (1999). Coot scoopin' Sea eagle. Boobook, 19, 16.

Thompson, P. M. & Johnson, D. L. (2003). Further notable bird records from Bangladesh. Forktail, 19, 85–102.

Thomsett, S. (2011). Simon Thomsett on the African Crowned Eagle. Retrieved from http://www.africanraptors.org/simon-thomsett-on-the african-crowned-eagle-part-1/

Thomson, A. (2015). A Year After Her Photo Went Viral, How Has Teenage Eagle Hunter's Life Changed? New York Times, Women in the Worlc Oct. 16. Retrieved from http: //nytlive.nytimes.com/womenintheworld/2015/10/16/ashol-pan-eagle-hunter/

Thomson, S. (2014). Two young eagles preying on cats in Campbell River. Retrieved from http://www.timescolonist.com/news/local/two-young eagles-preying-on-cats-in-campbell-river-1.981479

Thorn, D. & Peacock, J. (1962). Martial Eagle. Honeyguide, 39, 27.

Thorstrom, R., Watson, R. T., Damary, B., Toto, E., Baba, M. & Baba, V. (1995). Repeated sightings and first capture of a live Madagascar Serpen eagle Eutriorchis astur. Bulletin of the British Ornithologists' Club, 115, 40–45.

Thorstrom, R. & Watson, R. T. (1997). Avian inventory and key species of the Masoala Peninsula, Madagascar. Bird Conservation Internationa 7(2), 99–115. https://doi.org/10.1017/S0959270900 001453

Thorstrom, R. & René de Roland, L.-A. (2000). First nest description, breeding behaviour and distribution of the Madagascar Serpent-eagl Eutriorchis astur. Ibis, 142(2), 217–224. https://doi.org/10.1111/j.1474-919X.2000.tb04861.x

Thorstrom, R., Rene de Roland, L. & Watson, R. T. (2003). Falconiformes and strigiformes: Ecology and status of raptors. In: Goodman, S. M & Benstead, J. P. (Eds.), The natural history of Madagascar, Falconiformes and Strigiformes: Ecology and status of raptors, pp. 1080–108 Chicago: Chicago University Press.

Thorup, K., Alerstam, T., Hake, M. & Kjellén, N. (2003). Bird orientation: Compensation for wind drift in migrating raptors is age dependen Proceedings of the Royal Society of London, Series B, 270(Suppl. 1), 8–11.

Thorup, K., Alerstam, T., Hake, M. & Kjellén, N. (2006). Traveling or stopping of migrating birds in relation to wind: An illustration for the osprey Behavioral Ecology, 17(3), 497–502. https://doi.org/10.1093/beheco/arj054

Threatened Species Section, DPIW. (2006). Threatened Tasmanian Eagles Recovery Plan 2006–2010. Hobart, Tasmania, Australia: Departme of Primary Industries and Water, Tasmanian Government. Retrieved from http://www.environment.gov.au/resource/threatened-tasmanian eagles-recovery-plan-2006-2010.

Thurber, W. A., Serrano, J. F., Sermeño, A. & Benítez, M. (1987). Status of uncommon and previously unreported birds of El Salvador. Proceeding of the Western Foundation of Vertebrate Zoology, 3, 109–293.

Tigner, J. R. & Larson, G. E. (1977). Sheep losses on selected ranches in southern Wyoming. Journal of Range Management, 30(4), 244–25 https://doi.org/10.2307/3897297

Tigner, J. R. & Larson, G. E. (1981). Golden eagles – Scavengers and predators on domestic lambs. *In*: Peck, J. M. & Dalke, P. D. (Eds.), Wildlife livestock relationships symposium, pp. 359–361. University of Idaho, Forestry, Wildlife and Range Experiment Station, Moscow, Idaho.

Tikhomirov, E. A. (1966). Some data on distribution and biology of Largha Seal in the Sea of Okhotsk in summer-autumn and organisation of i commercial use. Proceedings of the Icthiological Commission (12).

Tilly, F. C., Hoffman, S. W. & Tilly, R. C. (1990). Spring hawk migration in southern Mexico, 1989. Hawk Migration Association of Nort America, Hawk Migration Studies, 15(2), 21–29.

Tingay, R. E. (2000). Sex, lies and dominance: Paternity and behaviour of extra-pair Madagascar Fish eagle Haliaeetus vociferoides. MSc Thesi Nottingham, UK: University of Nottingham.

Tingay, R. E. (2005). Historical distribution, contemporary status and cooperative breeding in the Madagascar Fish eagle: Implications fc conservation. PhD Thesis. Nottingham, UK: University of Nottingham.

Tingay, R. E., Nicoll, M. A. C. & Visal, S. (2006). Status and distribution of the Grey-headed Fish-eagle (Ichthyophaga ichthyaetus) i the Prek Toal core area of Tonle Sap Lake, Cambodia. Journal of Raptor Research, 40(4), 277–283. https://doi.org/10.3356/089 1016(2006)40277:SADOTG2.0.CO;2

Tingay, R. E., Nicoll, M. A. C., Whitfield, D. P., Visal, S. & McLeod, D. R. A. (2010). Nesting ecology of the Grey-headed Fish-eagle at Prek Toa Tonle Sap lake, Cambodia. Journal of Raptor Research, 44(3), 165–174. https://doi.org/10.3356/JRR-09-04.1

'ingay, R. E. & Katzner, T. E. (Eds.) (2010). The Eagle Watchers: Observing and Conserving Raptors Around the World. Cornell University Press.
'ingay, R., Visal, S. & Nicoll, M. (2012). Ecology and conservation of the Grey-headed Fish-eagle (Ichthyophaga ichthyaetus) at Prek Toal, Tonle Sap Lake, Cambodia. Ornis Mongolica, 1(1), 26–29.
'jernberg, M. (1981). Diet of the Golden eagle Aquila chrysaetos during the breeding season in Sweden. Holarctic Ecology. Nordic Society Oikos, 4(1), 12–19. https://doi.org/10.1111/j.1600-0587.1981.tb00975.x
·obish, Jr., T. G. & Balch, L. G. (1987). First North American nesting and occurrence of Haliaeetus albicilla on Attu Island, Alaska. The Condor, 89(2), 433–434. https://doi.org/10.2307/1368501
'odorov, E., Daskalova, G. & Shurulinkov, P. (2015). Current breeding distribution and conservation of White-tailed eagle, Haliaeetus albicilla (L.) in Bulgaria. Acta Zoologica Bulgarica, 67, 3–10.
'oland, B. (1986). Hunting success of some Missouri raptors. Wilson Bulletin, 98, 116–125.
'oledo, M. C. B., Donatelli, R. J. & Batista, G. T. (2012). Relation between green spaces and bird community structure in an urban area in southeast Brazil. Urban Ecosystems, 15(1), 111–131. https://doi.org/10.1007/s11252-011-0195-2
'omkiewicz, S. M., Fuller, M. R., Kie, J. G. & Bates, K. K. (2010). Global positioning system and associated technologies in animal behaviour and ecological research. Philosophical Transactions of the Royal Society of London Series B: Biological Sciences, 365(1550), 2163–2176. https://doi.org/10.1098/rstb.2010.0090
'omlinson, R. F. (2003). Thinking about GIS: Geographic Information System Planning for Managers. Redlands, CA: ESRI Press.
'orchia, C. (2018). A new concern for African vultures: Bullet lead in carrion. Retrieved from https://apnews.com/article/e3d078ac03b3491c98012ec2b0743b46
·ordoff, A. W. (2002). Raptor migration at Hoang Lien Nature Reserve, northern Vietnam. Forktail, 18, 45–48.
·ordoff, A. W., Timmins, R. J., Maxwell, A., Huy, K., Lic, V. & Khou Eang, H. (2005). Biological assessment of the Lower Mekong dry forests ecoregion. Phnom Penh: World Wildlife Fund Cambodia.
'orrelio, S. (2014). Kissaki Films' Otto Bell to Direct Ashol-Pan Documentary 'Eagle Huntress'. Retrieved from http: //variety.com/2014/film/news/kissaki-films-otto-bell-to-direct-ashol-pandocumentary-.eagle-huntress-1201263741/
·ostain, O., Dujardin, J.-L., Érard, C. & Thiollay, J.-M. (1992). Birds of French Guiana. Sociéte de'Etudes Ornithologiques. Brunoy, France.
'ouchton, J. M., Yu-Cheng, H. & Palleroni, A. (2002). Foraging ecology of reintroduced captive-bred subadult harpy eagles (Harpia harpiya) on Barro Colorado Island, Panama. Ornitologia Neotropical (The Neotropical Ornithological Society) 13.
'ownsend, C. W. (1937). Life histories of North American birds of prey. Part 1. Smithsonian Institution National Museum Bulletin 167. Toxicologic Pathology. San Diego, CA: Academic Press.
'racewski, L. Ł, Butchart, S. H. M., Di Marco, M., Ficetola, G. F., Rondinini, C., Symes, A., Wheatley, H., Beresford, A. E. & Buchan, G. M. (2016). Toward quantification of the impact of 21st-century deforestation on the extinction risk of terrestrial vertebrates. Conservation Biology, 30(5), 1070–1079. https://doi.org/10.1111/cobi.12715
'rail, P. W. (1987). Predation and antipredator behavior at Guianan cock-of-the-rock leks. The Auk, 104(3), 496–507. https://doi.org/10.2307/4087549
'rail, P. W. (2017). Identifying Bald Versus Golden Eagle Bones: A Primer for Wildlife Biologists and Law Enforcement Officers. Journal of Fish and Wildlife Management, 8(2), 596–610. https://doi.org/10.3996/042017-JFWM-035
'rainor, C. & Lesmana, D. (2000). Exploding volcanoes, unique birds, gigantic rats and elegant ikat: Identifying sites of international biodiversity significance on Flores, Nusa Tenggara. Bogor: PKA/BirdLife International/World Wildlife Fund.
'rainor, C. R. (2002). Birds of Gunung Tambora, Sumbawa, Indonesia: Effects of altitude, the 1815 cataclysmic volcanic eruption and trade. Forktail, 18, 49–61.
'rainor, C. R., Debus, S. J. S., Olsen, J., Norman, J. A. & Christidis, L. (2013). Bonelli's eagle Aquila fasciata renschi in the Lesser Sundas, Wallacea: Distribution taxonomic status, likely origins and conservation status. Forktail, 29, 100–106.
'ransehe, N. (1942). Das Vogelleben des Angerschen Sees. Korrespondenzblatt Naturforscher-Vereins Riga, 64, 101–124.
'ravaini, A., Rodriguez, A., Ceballos, O., Donázar, J. A. & Hiraldo, F. (1995). Roadside raptor surveys in central Argentina. Hornero, 14, 64–66.
'ravsky, A. & Beauvais, G. P. (2004). Species Assessment for Bald Eagle (Haliaeetus leucocephalus) in Wyoming. Cheyenne, Wyoming: United States Department of the Interior.
'raylor, M. A. (1963). Check-list of Angolan Birds. Lisbon, Portugal: Diamang.
'reinys, R. (2005). The Greater Spotted eagle (Aquila clanga), previous and current status and hybridisation in Lithuania. Acta. Zoologica Lituanica, 15(1), 31–38. https://doi.org/10.1080/13921657.2005.10512606
'reinys, R., Dementavičius, D., Rumbutis, S., Švažas, S., Butkauskas, D., Sruoga, A. & Dagys, M. (2016). Settlement, habitat preference, reproduction, and genetic diversity in recovering the White-tailed eagle Haliaeetus albicilla population. Journal of Ornithology, 157(1), 311–323. https://doi.org/10.1007/s10336-015-1280-8.
'reves, A. & Karanth, K. U. (2003). Human-carnivore conflict and perspectives on carnivore management worldwide. Conservation Biology, 17(6), 1491–1499. https://doi.org/10.1111/ j.1523-1739.2003.00059.x
'reves, A., Drescher, A. & Ingrisano, N. (2001). Vigilance and aggregation in Black howler monkeys (Alouatta pigra). Behavioral Ecology and Sociobiology, 50(1), 90–95. https://doi.org/10.1007/s002650100328
'rinca, C. T., Ferrari, S. F. & Lees, A. C. (2008). Curiosity killed the bird: Arbitrary hunting of Harpy eagles Harpia harpyja on an agricultural frontier in southern Brazilian Amazonia. Cotinga, 30, 12–15.
'ucker, G. M. & Heath, M. F. (1994) . Birds in Europe: Their conservation status. Cambridge: BirdLife International.
'ucker, G., Bassi, S., Anderson, J., Chiavari, J., Casper, K. & Fergusson, M. (2008). Provision of evidence of the conservation impacts of energy production. London: Institute for European Environmental Policy (IEEP).

Turelli, M. (1998). The causes of Haldane's rule. Science, 282(5390), 889–891. https://doi.org/10.1126/ science.282.5390.889
Turrin, C., Watts, B. D. & Mojica, E. K. (2015). Landfill use by Bald eagles in the Chesapeake Bay region. The Journal of Raptor Research, 49(3 239–249. https://doi.org/10.3356/JRR-14-50.1.
Tuvia, J. & Välia, Ü. (2007). The impact of the White-tailed eagle Haliaeetus albicilla and the Osprey Pandion haliaetus on Estonian Common ca Cyprinus carpio production: How large is the economic loss? Estonian Journal of Ecology, 56(3).
Tytler, R. C. (1865). Description of a new species of Spizaetus. Proceedings of the Asiatic Society of Bengal, 112.
U.S. Fish and Wildlife Service (1996). California Condor Recovery Plan, Third revision. Portland, OR: United States Fish and Wildlife Service
UAV Coach, U. (2019). Drone Laws in the United States of America. Retrieved from https://uavcoach.com/drone-laws-in-united-states-of-americ
Ubaid, F. K., Ferreira, L. P., Oliveira, S. B. & Antas, P. T. Z. (2011). Primeiro registro de Harpia harpyja para o bioma Pantanal, com dados sob atividade reprodutiva. Revista Brasileira de Ornitologia, 19, 88–92.
Udechukwu, G. I. (2019). The significance and use of cultural symbols in the contemporary African society: Igbo symbols as a paradig Mgbakoigba, Journal of African Studies, 8(1), 110–116.
Ueta, M. & Minton, J. S. (1996). Habitat preference of Crested Serpent eagles in southern Japan. Journal of Raptor Research, 30(2), 99–100.
Ueta, M., Sato, F., Lobkov, E. G. & Mitza, N. (1998). Migration route of White-tailed Sea eagles Haliaeetus albicilla in northeastern Asia. Ib 140: 684–686.
Ueta, M., Kotta, M. & Fukui, K. (1999). The relationship between the autumn distributions of salmon and of Steller's and White-tailed eagles Hokkaido, Japan. Strix, 17, 25–29.
Ueta, M. & Masterov, V. B. (2000). Estimation by a computer simulation of population trend of Steller's Sea Eagles. *In*: Ueta, M. & McGrady, M J. (Eds.), First Symposium on Steller's and White-tailed Sea Eagles in East Asia, pp. 111–116. Tokyo: Wild Bird Society of Japan.
Ueta, M., Sato, F., Nakagawa, H. & Mita, N. (2000). Migration routes and differences of migration schedule between adult and young Steller's S eagles Haliaeetus pelagicus. Ibis, 142(1), 35–39. https://doi.org/10.1111/j.1474-919X.2000.tb07681.x
Ueta, M., McGrady, M. J., Nakagawa, H., Sato, F. & Masterov, V. B. (2003). Seasonal change in habitat use in Steller's sea eagles. Oryx, 37(110–124. https://doi.org/10.1017/S003060 530300019X
Uhlig, R. (1996) . Zur Verbreitung des Schreiadlers Aquila pomarina in Rumänien. *In:* Meyburg, B.-U. & Chancellor, R. D. (Ed.), Eagle studie pp. 313–318. Berlin, London & Paris: World Working Group on Birds of Prey.
Uitamo, E. (1999). Modelling deforestation caused by the expansion of subsistence farming in the Philippines. Journal of Forest Economics, 5(99–122.
Ukpokolo, C. (2011). Gender, symbols and traditional peacemaking among the Nanka-Igbo of south-eastern Nigeria. Human Affairs, 21(163–183. https://doi.org/10.2478/s13374-011-0018-4.
Ulfstrand, S. (1960). Migration of African kites, Milvus migrants and Wahlberg's eagle Aquila wahlbergi at Lake Tanganyika. Förh. K. fysio Sällsk. Lund, 30, 31–39.
Umraw, A. (2015). Eagle attack 'unusual'. Retrieved from https://www.news24.com/SouthAfrica/News/Eagle-attack-unusual-20150729
Unck, C. E., Waterman, J. M., Verburgt, L. & Bateman, P. W. (2009). Quantity versus quality. How does level of predation threat affect Ca ground squirrel vigilance? Animal Behaviour, 78(3), 625–632. https://doi.org/10.1016/j.anbehav.2009.05.028
Underhill, L. G. (2012). Martial eagles spiral downwards. Retrieved from http://www.adu.org.za/news.php? id=2384.
United Nations Environment Programme (UNEP). (1994). United Nations Convention to Combat Desertification. Nairobi (Kenya).
United States Department of Agriculture (U.S.D.A) (2003). Fact Sheet: Managing Vulture Damage. Pittstown: United States Department Agriculture, Animal and Plant Health Inspection Service, Wildlife Services.
United States Department of Agriculture (U.S.D.A.) (2005). Managing Wildlife Hazards at Airports. U.S.D.A. Washington, DC: Animal and Pla Health Inspection Service, Wildlife Services.
United States Department of Agriculture (U.S.D.A.). (2016). Managing Vulture Damage. Retrieved from https://www.aphis.usda.gov/publicatio wildlife_damage/content/ printable_version/fs_vulture_damage_man.pdf
United States Environmental Protection Agency (U.S.E.P.A.). (1994). Lead fishing sinkers: Response to citizens' petition and proposed ban. CFR Part 745. Federal Register, 59, 11121–11143.
United States Fish and Wildlife Service. (1996). California Condor Recovery Plan. USFWS, Washington DC.
United States Fish & Wildlife Service (U.S.F.W.S.). (2009). Final environmental assessment: Proposal to permit take under the Bald and Gold Eagle Protection Act. Washington, DC: United States Fish and Wildlife Service.
United States Fish and Wildlife Service. (2011). Golden Eagles Status Fact Sheet. Retrieved from https://www.fws.gov/migratorybirds/pc education/educational-activities/golden-eagle-fact-sheet.pdf
United States Fish and Wildlife Service. (2012). U.S. Fish and Wildlife Service land-based wind energy guidelines. Arlington, United States.
United States Fish and Wildlife Service. (2013). Eagle conservation plan guidance. Module (2nd version), 1 – Land-based wind energy (versi 2). Retrieved from https://www.ecfr.gov/current/title-50/chapter-I/subchapter-B/part-22
Unitt, P. (1984). The birds of San Diego County. San Diego Society of Natural History Memoir 13.
Urbani, B. (2005). The targeted monkey: A re-evaluation of predation on New World primates. Journal of Anthropological Sciences, 83, 89–10
Urbina Torres, F. (1996). Birds of prey of Mexico. Centro de Investigaciones Biologicas UAEM, Cuernavaca, Morelos, Mexico.
Urios, V., Donat-Torres, M. P., Bechard, M. & Ferrer, M. (2014). Movements of a juvenile Crowned eagle (Harpyhaliaetus coronatus) tracked satellite telemetry in central Argentina. Journal of Biological Research, 21(1), 12. https://doi.org/10.1186/2241-5793-21-12
Usher, M. B. & Thompson, D. B. A. (Eds.), Ecological Change in the Uplands. https://doi.org/10.2172/86813
Usher, M. B. & Baxter, J. M. (Eds.) (1994). The Islands of Scotland: A living marine heritage: 201–212. Edinburgh: Her Majesty's Statione Office.

Ussher, R. J. & Warren, R. (1900). Birds of Ireland. London: Gurney.

Utami, B. D. (2001). Study on potential prey of Javan Hawk-eagle Spizaetus bartelsi (Stresemann, 1924) at G. Salak. Honors Thesis. Bogor: Bogor Agriculture University.

Utekhina, I., Potapov, I. E. & McGrady, M. J. (2000). Diet of the Steller's Sea eagles in the northern Sea of Okhotsk. *In:* Ueta, M. & McGrady, M. J. (Eds.), First Symposium on Steller's and White-tailed Sea eagles in East Asia, pp. 71–92. Tokyo, Japan. Wild Bird Society of Japan.

Uttendorfer, O. (1939). Die Ernahrungder deutschen Raubvogel und Eulen. Stuttgart.

Uttendörfer, O. (1952). Neue Ergebnisse überdie Ernährung der Greifvögel und Eulen. Stuttgart & Ludwigsburg: Eugen Ulmer.

Vaasen, A. & Aykurt, M. A. (2003). Status and conservation of vultures in Anatolia. Vulture News, 48, 60.

Vaassen, E. W. A. M. & Aykurt, M. A. (2000). An alternative spring migration route of Lesser Spotted eagles (Aquila pomarina) in southern Turkey. International Hawkwatcher, 1, 4–7.

Valdez, U. (2001). Isidor's Eagles: Owners of the cloud forest. The Peregrine Fund. Newsletter, 32, 15–17.

Valdez, U. & Osborn, S. (2004). Observations on the ecology of the Black-and-chestnut eagle (Oroaetus isidori) in a montane forest of southeastern Peru. Ornitologia Neotropical, 15, 31–40.

Valencia-Herverth, J. & Valencia-Herverth, R. (2009). New records of Elanus leucurus and Spizaetus tyrannus in Hidalgo, México, and interesting avifaunal records. Cotinga, 31, 124–125.

Valero, S. O. P. (2008). White-tailed eagle hunting behaviour and food choice in Kvismaren. Sveriges Ornitologiska Förening.

Väli, U. & Lõhmus, A. (2000). The Greater Spotted eagle and its conservation in Estonia. Hirundo Supplment, 3, 1–50.

Väli, Ü., Treinys, R. & Poirazidis, K. (2001). Genetic structure of Greater Spotted eagle Aquila clanga and Lesser Spotted eagle A. pomarina populations: Implications for phylogeography and conservation. *In:* Chancellor, R. D. & Meyburg, B.-U. (Eds.), Raptors worldwide, pp. 473–482. Berlin, Budapest: World Working Group on Birds of Prey, MME/BirdLife Hungary.

Väli, Ü. (2002). Mitochondrial pseudo-control region in Old World eagles' genus Aquila. Molecular Ecology, 11(10), 2189–2194. https://doi.org/10.1046/j.1365-294x.2002.01578.x

Väli, Ü. (2003). The Lesser Spotted eagle and its conservation in Estonia. Hirundo Suppl., 6, 1–64.

Väli, Ü. & Lõhmus, A. (2004). Nestling characteristics and identification in the Greater Spotted eagle Aquila clanga, Lesser Spotted Eagle A. pomarina and their hybrids. Journal of Ornithology, 145, 256–263. https://doi.org/10.1007/s10336-004-0028-7

Väli, Ü. (2004). Sex ratio of Lesser Spotted eagle Aquila pomarine nestlings in good and poor breeding years. Bird Study, 51(2), 189–191. https://doi.org/10.1080/00063650409461352

Väli, U. (2006). Mitochondrial DNA sequences support species status for the Indian Spotted Eagle Aquila hastata. Bulletin of the British Ornithologists' Club, 126(3), 238–242.

Vâli, Ü., Belik, V. P. & Babkin, I. G. (2009). The Lesser Spotted eagle Aquila pomarina in the North Caucasus, Russian Federation: Taxonomic status, genetic diversity, breeding density and nest site characteristics. Sandgrouse, 31, 122–127.

Väli, Ü., Dombrovski, V., Treinys, R., Bergmanis, U., Daróczi, S. S. J., Dravecky, M., Ivanovsky, V., Lontkowski, J., Maciorowski, G., Meyburg, B.U., Mizera, T., Zeitz, R. & Ellegren, H. (2010). Wide-spread hybridization between the Greater Spotted eagle Aquila clanga and the Lesser Spotted eagle Aquila pomarina (Aves: Accipitriformes) in Europe. Biological Journal of the Linnean Society, 100(3), 725–736. https://doi.org/10.1111/j.1095-8312.2010.01455.x.

Väli, Ü. (2011). Numbers and hybridization of spotted eagles in Estonia as revealed by country-wide field observations and genetic analysis. Estonian Journal of Ecology, 60(2), 143–154. https://doi.org/10.3176/eco.2011.2.05

Väli, Ü & Bergmanis, U. (2017). Apparent survival rates of adult Lesser Spotted eagle Clanga pomarina estimated by GPS-tracking, colour rings and wing-tags. Bird Study, 64(1), 104–107. https://doi.org/10.1080/00063657.2016.1271395.

Vallely, A. C. & Dyer, D. (2018). Birds of Central America: Belize, Guatemala, Honduras, El Salvador, Nicaragua, Costa Rica Confederación Unitaria de Trabajadores, and Panama (Vol. 136). Princeton, NJ: Princeton University Press.

Valverde, J. A. (1960). Vertebrates of the Guadalquivir Marshes. Archivos del Instituto de Aclimatación de Almería., Vol. IX.

Valverde, J. A. (1967). Structure of a land vertebrate community. Monografia N° 1 Estación Biológica de Doñanadoñana. CSIC, Sevilla. Valverde.

Van Balen, S. (1991). Ecology and conservation of endangered Javan Hawk-eagle. *In:* Chancellor, R. & Meyburg, B.-U. (Eds.), Birds of prey, Bulletin 4, 33–39. Berlin, London & Paris: World Working Group on Birds of Prey and Owls.

Van Balen, S. (1993). Special review: A Field Guide to the Birds of Borneo, Sumatra, Java and Bali. Bulletin of the Oriental Bird Club, 18: 48–51.

Van Balen, S., Dewi, D. & Jepson, P. R. (1994). Observations at a Java hawk-eagle nest. Tropical Biodiversity, 2, 329–331.

Van Balen, S. & Meyburg, B.-U. (1994). The Java Hawk Eagle Spizaetus bartelsi: Results of recent research on distribution, status and ecology. *In:* Meyburg, B.-U. & Chancellor, R. D. (Eds.), Raptor conservation today, pp. 245-254. Proceedings of the 4th World Conference on Birds of Prey and Owls. Sussex: Pica Press.

Van Balen, S. (1998). Tropical forest raptors in Indonesia: Recent information on distribution status, and conservation. Journal of Raptor Research, 32(1), 56–63. Retrieved from https://sora.unm.edu/sites/default/files/journals/jrr/v032n01/p00056-p00063.pdf.

Van Balen, S., Sozer, R. & Nijman, V. (1999). Distribution and conservation of the endemic Javan hawk-eagle Spizaetus bartelsi. Bird Conservation International, 9, 333–349.

Van Balen, S., Nijman, V. & Prins, H. T. (2000). The Javan hawk-eagle: Misconceptions about rareness and threat. Biological Conservation, 96(3), 297–304. https://doi.org/10.1016/S0006-3207(00)00092-6

Van Belle, J., Shamoun-Baranes, J., van Loon, E. E. & Bouten, W. (2007). An operational model predicting autumn bird migration intensities for flight safety. Journal of Applied Ecology, 44, 864–874. https://doi.org/10.1111/j.1365-2664.2007.01322.x

Van Bommel, F., Heitkonig, I., Epema, G., Ringrose, S., Bonyongo, M. & Veenendaal, E. (2006). Remotely sensed habitat indicators for predicting distribution of impala (Aepyceros melampus) in the Okavango Delta, Botswana. Journal of Tropical Ecology, 22, 101–110.

Van De Weghe, J.-P. (1979). L'aigle de Wahlberg, Aquila wahlbergi, au Rwanda. Gerfaut, 69, 461–474.
Van den Berg, A. B. & Haas, M. (2007). WP reports late May-early July (2007). Dutch Birding, 29(4), 239–252.
Van den Hoek Ostende, L. W., Morlo, M. & Nagel, D. (2006). Fossils explained 52 Majestic killers: The sabre-toothed cats. Geology Today, 22(4) 150.
Van der Meer, T., McPherson, S. & Downs, C. (2018). Temporal changes in prey composition and biomass delivery to African Crowned eagl nestlings in urban areas of KwaZulu-Natal, South Africa. Ostrich, 89: 3(3), 241–250. https://doi.org/10.2989/00306525.2018.1449027
Van Eeden, R. (2016). Understanding the decline of Martial eagles Polemaetus bellicosus in the Kruger National Park, South Africa. University o Cape Town. Retrieved from http://ac.za/ pdf/Van_Eeden_Rowen_2016_PhD_thesis.pdf. University of Cape Town.
Van Eeden, R., Whitfield, D. P., Botha, A. & Amar, A. (2017). Ranging behaviour and habitat preferences of the Martial eagle: Implications for th conservation of a declining apex predator. PLoS One, 12(3), e0173956. https://doi.org/10.1371/journal.pone.0173956
Van Jaarsveld, J. (1985). The ecology of the Martial eagle in the Kruger National Park. Quagga, 9, 10–11.
Van Kleef, H. & Bustamante, J. (1999). First recorded polygynous mating in the Red kite (Milvus milvus). Journal of Raptor Research, 33 254–257.
Van Maanen, E., Goradze, I., Gavashelishvili, A. & Rezo, G. (2001). Trapping and hunting of migratory raptors in western Georgia. Bir Conservation International, 11(2), 77–92. https://doi.org/10.1017/S095927090100017X
Van Niekerk, J. & Van Ginkel, C. (2003). Notes on the behavioural ecology of Coqui francolin in the Rustenburg district, South Africa. Sout African Journal of Wildlife Research, 33(1), 59–62.
Van Niekerk, J. H. (2015). Vocal behavior of Crested guineafowl (Guttera edouardi) based on visual and sound playback surveys in the Umhlang Lagoon Nature Reserve, KwaZulu-Natal Province, South Africa. Avian Research, 6(1), 13. https://doi.org/10.1186/s40657-015-0022-1
Van Niekerk, J. J. H. (2011). Habitat-use and range contraction of Swainson's spurfowl at the Krugersdorp Game Reserve, Gauteng Province South Africa. Ostrich. 82(1), 43–47. https://doi.org/10.2989/00306525.2010.515398
Van Oosten, J. R. & Wyant, D. (1999). Interim report of the bird life in the Komarindi catchment conservation area. Solomon Islands avifaun survey and conservation project. Avian Ark Foundation Report.
Van Perlo, B. (2009). A Field Guide to the Birds of Brazil. Oxford: Oxford University Press.
Van Rijn, S.Sv, Zijlstra, M. & Bijlsma, R. G. (2010). Wintering White-tailed eagles Haliaeetus albicilla in the Netherlands: Aspects of habitat scal and quality. Ardea, 98(3), 373–382. https://doi.org/10.5253/078.098.0311
Van Rooyen, C. & Ledger, J. (1999). Birds and utility structures: Developments in southern Africa. *In:* Ferrer, M. & Janns, G. F. M. (Eds.), Bird and power lines. Madrid: Quercus.
Van Rossem, A. J. (1948). A race of Urubitornis solitaria from northwestern Mexico. Proceedings. Biological Society of Washington, 61, 67–68
van Someren, V. D. (1947). Field notes on some Madagascar birds. Ibis, 89, 235–267.
Van Someren, V. G. L. (1932). Birds of Kenya and Uganda, being addenda and corrigenda to my previous paper in Novit. Zool. 37, 29, 252–38
Van Someren, V. G. L. & Van Someren, G. R. C. (1949). The birds of Bwamba. Uganda J., 13(Special Suppl.), 1–111.
Van Valkenburgh, B. & Hertel, F. (1993). Tough times at La Brea: Tooth breakage in large carnivores of the late Pleistocene. Science, 261(5120) 456–459.
Van Valkenburgh, B. (2007). Déjà vu: The evolution of feeding morphologies in the Carnivora, Integrative and Comparative Biology, 47(1) 147–163. https://doi.org/10.1093/icb/icm016
Van Wyk, E., Bouwman, H., van der Bank, H., Verdoorn, G. H. & Hofmann, D. (2001). Persistent organochlorine pesticides detected in bloo and tissue samples of vultures from different localities in South Africa. Comparative Biochemistry and Physiology. Part C: Toxicology & Pharmacology 129(3), 243–264. https://doi.org/10.1016/S1532-0456(01)90201-7
Van Wyk, E., Van der Bank, F. H. &, Verdoorn, G. H. & Hofmann, D. (2001). Selected mineral and heavy metal concentrations in blood an tissues of vultures in different regions of South Africa. South African Journal of Animal Science, 31(2), 57–64. https://doi.org/10.4314/saja v31i2.3831
Vanak, A. T., Shannon, G., Thaker, M., Page, B., Grant, R. & Slotow, R. (2012). Biocomplexity in large tree mortality: Interactions betwee elephant, fire and landscape in an African savanna. Ecography, 35(4), 315–321. https://doi.org/10.1111/j.1600-0587.2011.07213.x
Vandenbroucke, V., Van Pelt, H., De Backer, P. & Croubels, S. (2010). Animal poisonings in Belgium: A review of the past decade. Vlaam Diergeneeskundig Tijdschrift, 79, 259–268.
Vanderburgh, D. C. (1993). Manitoba Hydro accommodates osprey activity. Blue Jay, 51(3), 173–177. https://doi.org/10.29173/bluejay5430
Vannini, J. P. (1989). Neotropical raptors and deforestation: Notes on diurnal raptors at Finca El Faro, Quetzaltenango, Guatemala. Journal c Raptor Research, 23, 27–38.
Vansteelant, W. M. G., Bouten, W., Klaassen, R. H. G., Koks, B. J., Schlaich, A. E., van Diermen, J. & Shamoun-Baranes, J. (2015). Region and seasonal flight speeds of soaring migrants and the role of weather conditions at hourly and daily scales. Journal of Avian Biology, 46(1 25–39. https://doi.org/10.1111/jav.00457
Vardanis, Y., Klaassen, R. H., Strandberg, R. & Alerstam, T. (2011). Individuality in bird migration: Routes and timing. Biology Letters, 7(4 502–505. https://doi.org/10.1098/rsbl.2010.1180
Vargas-Gonzalez, J. J., Mosquera, R. & Watson, M. (2006a). Crested eagle feeding a post-fledged young Harpy eagle (Harpia harpyja) in Panam Ornitologia Neotropical, 17, 581–584.
Vargas-Gonzalez, J. J., Whitacre, D., Mosquera, R. et al. (2006b). Estado y distribucion actual del Aguila Arpia (Harpia harpyja) en Centro y Su America. Ornitologia Neotropical, 17, 39–55.
Vargas-Gonzalez, J. D. & Vargas, F. H. (2011). Nesting density of Harpy Eagles in Darien with population size estimates for Panama. Journal c Raptor Research, 45, 199–210.

'argas-Gonzalez, J. J., Vargas, F. H., Carpio, D. & McClure, C. J. W. (2014). Caracteristicas de la vegetacion en sitios de anidacion del Aguila Arpia (Harpia harpyja) en Darien, Panama. Ornitologia Neotropical, 25, 207–218.
'arland, D. E., Smallwood, J. A., Young, L. S. & Kocert, M. N. (2007). Marking techniques. *In:* Bird, D. M. & Bildstein, K. L. (Eds.), Raptor research and management techniques, pp. 221–236. Surrey: Hancock House.
'as, E., Lescroël, A., Duriez, O., Boguszewski, G. & Grémillet, D. (2015). Approaching birds with drones: First experiments and ethical guidelines. Biology Letters, 11(2), 20140754. https://doi.org/10.1098/rsbl.2014.0754
'asconcelos, M. F., Pacheco, J. F. & Parrini, R. (2007). Establishment and conservation of the avifauna of the urban zone of Marabá, Pará, Brazil. Cotinga, 28, 45–52.
'asconcelos, M. Fd. & D'Angelo Neto, S. (2009). First assessment of the avifauna of Araucaria forests and other habitats from extreme southern Minas Gerais, Serra da Mantiqueira, Brazil, with notes on biogeography and conservation. Papéis Avulsos de Zoologia, 49(3), 49–71. https://doi.org/10.1590/S0031-10492009000300001
'aughan, C. (2011). Change in dense forest habitat for endangered wildlife species in Costa Rica from 1940 to 1977. UNED Research Journal. Cuadernos de Investigación UNED, 3(1), 99–161.
'aurie, C. (1965). The birds of the Palearctic fauna, a systematic reference. Non-Passeriformes. London: H. F. and G. Witherby.
'azhov, S. V. (2009). Notes on the White-tailed eagle breeding in the Ob' River floodlands, Altai Kray, Russia. Raptors Conservation, 17, 151–152.
'een, T., Borge, T., Griffith, S. C., Saetre, G. P., Bures, S., Gustafsson, L. & Sheldon, B. C. (2001). Hybridization and adaptive mate choice in flycatchers. Nature, 411(6833), 45–50. https://doi.org/10.1038/35075000
'enables, W. N. & Ripley, B. D. (2002). Modern Applied Statistics (4th ed.). New York, NY: Springer. https://doi.org/10.1007/978-0-387-21706-2
'erbelen, P. (2014). Persecution and killing of critically endangered Flores hawk eagle Nisaetus floris on Alor, Lesser Sundas, Indonesia. BirdingASIA, 21, 18.
'erdoorn, G. H. & Roth, L. A. (2000). Polemaetus bellicosus Martial Eagle. *In:* Verdoorn, G. H., Bildstein, K. L. & Ellis, S. (Eds.), Selected African Falconiformes conservation assessment and management plan, pp. 68–69. Apple Valley, MN: International Union for Conservation of Nature and Natural Resources/SSC Conservation Breeding Specialist Group.
'erdoorn, G. H., van Zijl, N., Snow, T. V., Komen, L. & Marais, E. W. (2004). Vulture poisoning in southern Africa. *In:* Monadjem, A., Anderson, M. D., Piper, S. E. & Boshoff, A. F. (Eds.). The Vultures of Southern Africa—Quo Vadis? 195–201. Johannesburg: Birds of Prey Working Group.
'erheijen, J. A. (1961). Some notes on the birds of the Island of Paloé, Flores, observed from 13 April to 5 May 1960. Ardea, 49, 183–187.
'erheijen, J. A. (1964). Breeding season on the Island of Flores. Ardea, 52, 194–201.
'erhelst, B., Jansen, J. & Vansteelant, W. (2011). South West Georgia: An important bottleneck for raptor migration during autumn. Ardea, 99(2), 137–146. https://doi.org/10.5253/078.099.0203
'erhoeye, J. & Holmes, D. A. (1998). The birds of the islands of Flores – A review. Kukila, 10, 3–59.
'ermeulen, C., Lejeune, P., Lisein, J., Sawadogo, P. & Bouché, P. (2013). Unmanned aerial survey of elephants. PLoS One, 8(2), e54700. https://doi.org/10.1371/journal.pone.0054700
'erner, J. & Willson, M. F. (1966). The influence of habitats on mating systems of North American passerine birds. Ecology, 47(1), 143–147. https://doi.org/10.2307/1935753
'idal-Mateo, J., Mellone, U., López-López, P., De La Puente, J., García-Ripollés, C., Bermejo, A. & Urios, V. (2015). Wind effects on the migration routes of trans-Saharan soaring raptors: Geographical, seasonal, and interspecific variation. Current Zoology, 62(2), 89–97, https://doi.org/10.1093/cz/zow008
'idoz, J. Q., Jahn, A. E. & Mamani, A. M. (2010). The avifauna of Estación Biológica Caparú, Bolivia. Cotinga, 32, 5–22.
'ieillot, L. P. (1816a). Analyse d'une Nouvelle Ornithologie Elementaire. A. Belin, Deterville, Paris.
'ieillot, L. P. (1816b). Nouveau Dictionnaire d'Histoirehistoire naturelle. Tome 4. Deterville, Paris, 4.
'ieillot, L. P. (1822). Sur une nouvelle espèce d'aigle découverte en France, et décrite par M. Vieillot (I), un des members honoraires de la Société Linnéenne de Paris. Mémoires de la Société Linnéenne de Paris, 2 (Part 2), 152–154.
'ijayan, L. (2009). Conservation of birds of the Andaman & Nicobar Islands. Indian Birds, 5(4), 103–107.
'ilà, C., Walker, C., Sundqvist, A. K., Flagstad, Ø., Andersone, Z., Casulli, A., Kojola, I., Valdmann, H., Halverson, J. & Ellegren, H. (2003). Combined use of maternal, parental and bi-parental genetic markers for the identification of wolf-dog hybrids. Heredity, 90(1), 17–24. https://doi.org/10.1038/sj.hdy.6800175
'illa-Bonilla, B., Rojas-Soto, O. R., Colodner-Chamudis, A. & Tejeda-Cruz, C. (2008). Municipal-level inventories of the avifauna and its application to conservation: The case of Zacapoaxtla, Puebla, Mexico. Ornitologia Neotropical, 19, 531–551.
'irani, M. & Watson, R. T. (1998). Raptors in the east African tropics and west Indian Ocean islands: State of ecological knowledge and conservation status. Journal of Raptor Research, 32, 28–39.
'irani, M., Gilbert, M., Watson, R., Oaks, L., Benson, P., Kham, A. A. & Baral, H.-S. (2001). Asian vulture crisis project: Field results from Pakistan and Nepal for the 2000–2001 field season. Reports from the workshop on Indian Gyps vultures. *In:* Katzner, T. & Parry-Jones, J. (Eds.). Proceedings of the 4th Eurasian Congress on Raptors, pp. 461–469. Seville: Estación Biológica Donaña, Raptor Research Foundation.
'irani, M. M. Z., Kendall, C., Njoroge, P. & Thomsett, S. (2011). Major declines in the abundance of vultures and other scavenging raptors in and around the Masai Mara ecosystem, Kenya. Biological Conservation, 144(2), 746–752. https://doi.org/10.1016/j.biocon.2010.10.024
'isal, S. & Mahood, S. (2015). Wildlife monitoring at Prek Toal Ramsar site, Tonle Sap Great Lake 2013 and 2014. Wildlife Conservation Society, Cambodia Program, Phnom Penh.
'iviant, M., Monestiez, P. & Guinet, C. (2014). Can we predict foraging success in a marine predator from dive patterns only? Validation with prey capture attempt data. PLoS ONE, 9(3), e88503. https://doi.org/10.1371/journal.pone.0088503

Vlachos, C. G. (1989). The ecology of the Lesser Spotted eagle (Aquila pomarina) in the forest of Dadia, Evros Prefecture. PhD Thesis. Thessalonik Greece: Department of Forestry and Natural Environment, School of Geotechnical Sciences, Aristotle University of Thessaloniki, Thessalonik

Vlachos, C. G. & Papageorgiou, N. K. (1994). Diet, breeding success, and nest-site selection of the Short-toed eagle (Circaetus gallicus) northeastern Greece. Journal of Raptor Research, 28(1), 39–42.

Vlachos, C. G. & Papageorgiou, N. K. (1996). Breeding biology and feeding of the Lesser Spotted eagle Aquila pomarina in Dadia Forest, Nort Eastern Greece. *In:* Meyburg, B.-U. & Chancellor, R. D. (Eds.), Eagle studies, pp. 337–347. Berlin, London, Paris: World Working Gro on Birds of Prey and Owls.

Vlietstra, L. S. & Paruk, J. A. (1997). Predation attempts on incubating Common loons, Gavia immer, and the significance of shoreline nestin Canadian Field-Naturalist, 111, 654–655.

Volke, V. & Randla, T. (2003). Population trends of the White-tailed eagle in Estonia in the 1990s. *In*: Helander, B. (Eds.), Sea eagle 200 Proceedings from the International Sea Eagle Conference in Bjorko, Sweden, 13–17 September 20, 97–101. Stockholm: Sweden Proceedin of the Swedish Society for Nature Conservation, SNF.

Volpato, G. H., Lopes, E. V., Mendonça, L. B., Boçon, R., Bisheimer, M. V., Serafini, P. P. & Anjos, L. D. (2009). The use of the point count meth for bird survey in the Atlantic Forest. Zoologia, 26(1), 74–78. https://doi.org/10.1590/S1984-46702009000100012

Von Blotzheim, U. N. G., Bauer, K. & Bezzel, E. (1993). Handbuch der vögel mitteleuropas. Akademische Verlagsgesellschaft.

Voous, K. H. & Wijsman, H. J. W. (1964). Notes on African hawk-eagles. Biologisch Jaarboek, 32, 151–163.

Voous, K. H. (1964). Wood owls of the genera Strix and Ciccaba. Zoologische Mededelingen (Leiden), 39, 471–478.

Voous, K. H. (1969). Predation potential in birds of prey from Suriname. Ardea, 57, 117–148.

Voous, K. H. (1973). List of recent Holarctic bird species Non-Passerines. Ibis, 115(4), 612–638. https://doi.org/10.1111/j.1474-919X.197 tb02004.x

Voous, K. H. (1977). List of recent Holarctic bird species passerines. Ibis, 119(3), 376–406. https://doi.org/10.1111/j.1474-919X.1977.tb08262.

Voous, K. H. (1982). Taxonomy and conservation. Birds of Prey Bulletin, 1, 50-57.

Vrezec, A. (2016). Prehranska niša orla belorepca (Haliaeetus albicilla) z idejnim predlogom za izvedbo dodatnega krmljenja vrste v širši okoli Reškega jezera pri Kočevski reki. Poročilo v okviru projekta LIFE Kočevsko (LIFE13 NAT/SI/000314). Ljubljana.

Vrijdagh, J. M. (1949). Observations ornithologiques enrégion occidentale du Lac Albert et principalementde la plaine d'Ishwa. Gerfaut, 3 1–115.

Vuilleumier, F. & Ewert, D. N. (1978). The distribution of birds in Venezuelan páramos. Bulletin of the American Museum of Natural Histor 162, 47–90.

Vuong, Q. H. (1989). Likelihood ratio tests for model selection of non-nested hypotheses. Econometrica, 57(2), 307–333. https://d org/10.2307/1912557

Wade, M. (2017). When Birds Attack! – Experienced Drone Operators Discuss This (Surprisingly Common) Mission Threat. Retrieved fro https://waypoint.sensefly.com/ bird-drone-attacks-avoid-threat/

Waithaka, J. (2004). Maasai Mara – An ecosystem under siege: An African case study on the societal dimension of rangeland conservation. Afric Journal of Range & Forage Science, 21(2), 79–88. https://doi.org/10.2989/10220110409485838

Walker, B. (2002). A field guide to the birds of the Macchu Picchu Historical Sanctuary, Peru. Lima, Peru: PFORONANPE y Programa Picchu

Walker, C. H., Brealey, C. J., Mackness, M. M. I. & Johnston, G. (1991). Toxicity of pesticides to birds: The enzymic factor. Biochemical Socie Transactions, 19(3), 741–745. https://doi.org/10.1042/bst0190741

Walker, D., McGrady, M., McCluskie, A., Madders, M. & McLeod, D. R. A. (2005). Resident Golden eagle ranging behaviour before and aft construction of a windfarm in Argyll. Scottish Birds, 25, 24–40.

Walker, K. C., Goette, M. B. & Batchelor, G. S. (1954). Pesticide residues in foods, Dichlorodiphenyltrichloroethane a Dichlorodiphenyldichloroethylene content in prepared foods. Journal of Agricultural and Food Chemistry, 2(20), 1034–1037. https://d org/10.1021/jf600 40a006

Walker, L. & Walker, M. (1939). From egg to eaglehood. Natural History, 43, 284–289, 302.

Wallace, A. R. (1868). On the raptorial birds of the Malay Archipelago. The Ibis, New Series, 13, 1–27.

Wallace, D. (1987). Rains: Life in the Balance. New York: Harcourt.

Wallace, M. P. & Temple, S. A. (1987). Competitive interactions within and between species in a guild of avian scavengers. Auk, 104(2), 290–29 https://doi.org/10.1093/auk/104.2.290

Wallach, A. D., Izhaki, I., Toms, J. D., Ripple, W. J. & Shanas, U. (2015). What is an apex predator? Oikos, 124(11), 1453–1461. https://d org/10.1111/oik.01977

Wallin, D., Elliott, C., Shugart, H., Tucker, C. & Wilhelmi, F. (1992). Satellite remote sensing of breeding habitat for an African weaver-bi Landscape Ecology, 7, 87–99. https://doi.org/10.1007/BF02418940

Walls, S. S., Mañosa, S., Fuller, R. M., Hodder, K. H. & Kenward, R. E. (1999). Is early dispersal enterprise or exile? Evidence from radio-tagg buzzards. Journal of Avian Biology, 30(4), 407–415. https://doi.org/10.2307/3677013

Walpole, P. (2010). Figuring the Forest Figures: Understanding Forest Cover Data in the Philippines and where we might be proceedi Environmental Science for Social Change, Quezon City, Philippines. Available at: essc.org.ph.

Wang, S. & Xie, Y. (2009). China Species Red List, 2. Beijing: Higher Education Press.

Wani, A. L., Ara, A. & Usmani, J. A. (2015). Lead toxicity: A review. Interdisciplinary toxicology, 8(2), 55–64. https://doi.org/10.151 intox-2015-0009

Washburn, B. E., Begier, M. J. & Wright, S. E. (2015). Collisions between eagles and aircraft: An increasing problem in the airport environme Journal of Raptor Research, 49(2), 192–200. https://doi.org/10.3356/rapt-49-02-192-200.1

Vashburn, B. E., Bernhardt, G. E. & Kutschbachbrohl, L. A. (2011). Using dietary analysis to reduce the risk of wildlife-aircraft collisions. Human-Wildlife Interactions, 5, 204–209.
Wassink, A. & Oreel, G. J. (2007). The Birds of Kazakhstan. Texel, Netherlands: De Cocksdorp.
Wassink, A. (2009). Birds of Kazakhstan: New and interesting data, Part 2. Dutch Birding, 31, 101–110.
Watling, D. (1983). Ornithological notes from Sulawesi. Emu – Austral Ornithology, 83(4), 247–261. https://doi.org/10.1071/ MU9830247
Watson, J., Garrett, M. & Anthony, R. (1991). Foraging Ecology of Bald Eagles in the Columbia River Estuary. Oregon State University.
Watson, J., Leitch, A. F. & Broad, R. A. (1992). The diet of the Sea eagle Haliaeetus albicilla and Golden eagle Aquila chrysaetos in western Scotland. Ibis, 134(1), 27–31. https://doi.org/10.1111/j.14 74-919X.1992.tb07225.x
Watson, J., Rae, S. & Stillman, R. (1992b). Nesting density and breeding success of Golden eagles in relation to food supply in scotland. Journal of Animal Ecology, 61(3), 543–550. https://doi.org/10.2307/5609
Watson, J. (1992). Golden Eagle Aquila chrysaetos breeding success and afforestation in Argyll. Bird Study, 39, 203–206.
Watson, J. (1993). Responses of nesting Bald eagles to helicopter surveys. Wildlife Society Bulletin, 21(2), 171–178.
Watson, J., Davison, M. & Leschner, L. (1993). Bald eagles rear Red-tailed hawks. Journal of Raptor Research, 27(2), 126–127.
Watson, J. & Rae, S. R. (1993). The diet of Golden eagles Aquila chrysaetos in Scotland. Ibis, 135(4), 387–393. https://doi.org/10.1111/j.1474-919X.1993.tb02110.x
Watson, J. (1997). The Golden Eagle. London, U.K: T & A D Poyser.
Watson, J. (2010). The Golden Eagle. London: A&C Black.
Watson, J. W., Garrett, M. G. & Anthony, R. G. (1991). Foraging Ecology of Bald Eagles in the Columbia River Estuary. The Journal of Wildlife Management, 55(3), 492–499. https://doi.org/10.2307/3808981
Watson, J. W. (1993). Responses of nesting Bald eagles to helicopter surveys. Wildlife Society Bulletin, 21(2): 171–178.
Watson, J. W. (2004). Responses of nesting Bald eagles to experimental pedestrian activity. The Journal of Raptor Research, 38(4), 295–303.
Watson, J. W., Duff, A. A. & Davies, R. W. (2014). Home range and resource selection by GPS-monitored adult Golden eagles in the Columbia Plateau Ecoregion: Implications for wind power development. The Journal of Wildlife Management, 78(6), 1012–1021. https://doi.org/10.1002/jwmg.745
Watson, M. & Asoyama, S. (2001). Dispersion, habitat use, hunting behavior, vocalizations, and conservation status of the New Guinea Harpy eagle (Harpyopsis novaeguineae). Journal of Raptor Research, 355(3), 235–239.
Watson, R. (1998). The plight of the fish eagle: People, eagles and wetlands' conservation in Madagascar. Africa – Birds & Birding 3(4), 34–41.
Watson, R. A. & Maritz, A. W. A. (2000). Terathopius ecaudatus Bateleur. *In*: Verdoorn, G. H., Bildstein, K. L. & Ellis, S. (Eds.), Selected African Falconiformes conservation assessment and management plan, pp. 74–76. Apple Valley, MN IUCN/SSC Conservation Breeding Specialist Group.
Watson, R. T. (1986). The ecology, biology and population dynamics of the Bateleur eagle Terathopius ecaudatus. PhD Thesis. University of Witwatersrand, Johannesburg.
Watson, R. T. & Watson, C. (1987). Interspecific piracy between Tawny eagles and Bateleurs: How common is it? Gabar, 2, 9–11.
Watson, R. T. (1988). The influence of nestling predation on nest site selection and behaviour of the Bateleur. South African Journal of Zoology, 23, 143–149.
Watson, R. T. (1990). Breeding biology of the Bateleur. Ostrich, 61(1–2), 13–23. https://doi.org/10.1080/00306525.1990.9633933
Watson, R. T., Thomsett, S., O'Daniel, D. Lewis L. & R. (1996). Breeding, growth, development, and management of the Madagascar Fish eagle (Haliaeetus vociferoides). Journal of Raptor Research, 30: 21–27.
Watson, R. T., Razafindramanana, S., Thorstrom, R. and Rafanomezantsoa, S. (1997). Breeding biology, trios, productivity, siblicide and conservation of the Madagascar Fish-eagle. *In:* Watson, R. T. (Eds.), Madagascar wetlands conservation project. Progress report III, 1995–1996, pp. 289–306. Boise: The Peregrine Fund.
Watson, R. T. (1997a). Madagascar Fish eagle conservation. *In:* Watson, R. T. (Ed.), Madagascar wetlands conservation project: Developing community-based conservation in a proposed wetland Biosphere Reserve in Madagascar. Progress Report III, 1995–1996, pp. 45–50. Boise, USA, ID: The Peregrine Fund.
Watson, R. T. (1997b). Madagascar Fish eagle sibling rescue in 1995 and 1996. *In:* Watson, R. T. (Ed.), Madagascar wetlands conservation project: Developing community-based conservation in a proposed wetland Biosphere Reserve in Madagascar. Progress Report III, 1995–1996, pp. 283–288. Boise, USA, ID: The Peregrine Fund.
Watson, R. T. (1997c). Captive breeding and display aviary for Madagascar Fish eagles in Parc Tsimbazaza: A multiple conservation opportunity. *In:* Watson, R. T. (Ed.), Developing community-based conservation in a proposed wetland Biosphere Reserve in Madagascar. Progress report, III, 1995–1996, pp. 413–416. Boise, USA, ID: The Peregrine Fund.
Watson, R. T., Razafindramanana, S., Thorstrom, R. & Rafanomezantsoa, S. (1999). Breeding biology, extra-pair birds, productivity, siblicide and conservation of the Madagascar Fish-eagle. Ostrich, 70, 105–111.
Watson, R. T., Andrianarimisa, A., Razandrizanakanirina, D. & Kalavaha, L. (2000). Madagascar Fish-eagle Haliaeetus vociferoides and community based wetland conservation. *In*: Chancellor, R. D. & Meyburg, B. (Eds.), Raptors at risk, pp. 333–342. Berlin: WWGBP/Hancock House.
Watson, R. T., Rene R., de Roland, L.-A., Rabearivony, J. & Thorstrom, R. (2007). Community-based wetland conservation protects endangered species in Madagascar: Lessons from science and conservation. Banwa, 4, 81–95.
Watson, R. T., McClure, C. J. W., Vargas, F. H. & Jenny, J. P. (2016). Trial restoration of the Harpy eagle, a large, long-lived, tropical forest raptor, in Panama and Belize. Journal of Raptor Research, 50(1), 3–22. https://doi.org/10.3356/rapt-50-01-3-22.1
Watson, R.T. & Rabarisoa, R. (2000). Sakalava fishermen and Madagascar Fish eagles: Enhancing traditional conservation rules to control resource abuse that threatens a key breeding area for an endangered eagle. Ostrich, 71(1-2), 2–10. https://doi.org/10.1080/00306525.2000.9639854

Wattel, J. (1973). Geographical differentiation in the genus Accipiter. Nuttall Ornithologists Club Publication, No. 13. Cambridge, MA.
Watts, B. D. & Byrd, M. A. (2007). Virginia Bald eagle nest and productivity survey: Year 2007 Report. CCB Technical Report, 368. Retrieve from https://scholarworks.wm.edu/ ccb_reports/368
Watts, B. D., Therres, G. D. & Byrd, M. A. (2007). Status, distribution, and the future of Bald eagles in the Chesapeake Bay area. Waterbird 30(sp1), 25–38. https://doi.org/10.1675/1524-4695(2007)0300025: SDATFO2.0.CO;2
Watts, B. D., Therres, G. D. & Byrd, M. A. (2008). Recovery of the Chesapeake Bay Bald eagle nesting population. Journal of Wildlife Managemen 72(1), 152–158. https://doi.org/10.2193/2005-616
Watts, B. D., Mojica, E. K. & Paxton, B. J. (2015). Using Brownian bridges to assess potential interactions between Bald eagles and electric hazards within the upper Chesapeake Bay. Journal of Wildlife Management, 79(3), 435–445. https://doi.org/10.1002/jwmg.853https://do org/10.1002/ jwmg.853
Wayland, M. & Bollinger, T. (1999). Lead exposure and poisoning in Bald eagles and Golden eagles in the Canadian prairie province Environmental Pollution, 104(3), 341–350. https://doi.org/10.1016/S0269-7491(98)00201-2
Wayland, M., Wilson, L. K., Elliott, J. R., Miller, J. E., Bollinger, T., McAdie, M., Langelier, K., Keating, J. & Froese, J. M. W. (2003). Mortalit morbidity, and lead poisoning of eagles in western Canada 1986–1998. Journal of Raptor Research, 37, 8–18.
Weatherhead, P. J. & Robertson, R. J. (1979). Offspring quality and the polygyny threshold: The Sexy Son Hypothesis. The American Naturalis 113(2), 201–208. https://doi.org/10.1086/283379
Webb, H. P. (1992). Field observations of the birds of Santa Isabel, Solomon Islands. Emu – Austral Ornithology, 92(1), 52–57. https://do org/10.1071/MU9920052
Webb, S. & Howell, S. (1993). New records of Hawk-eagles from Guerrero, México. Euphonia, 2, 19–21.
Wedding, L. M., Lepczyk, C. A., Pittman, S. J., Friedlander, A. M. & Jorgensen, S. (2011). Quantifying seascape structure: Extending terrestri spatial pattern metrics to the marine realm. Marine Ecology Progress Series, 427, 219–232.
Weenink, R., van Duivendijk, N. & Ebels, E. B. (2011). Spanish Imperial Eagle at Loozerheide in May 2007. Dutch Birding, 33, 94–102.
Wege, D. C. & Long, A. J. (1995). Key areas for threatened birds in the neotropics. Birdlife Conservation, 311 (series 5).
Weick, F. & Brown, L. H. (1980). Birds of Prey of the World. London: Collins.
Weick, F. (1980). Birds of Prey of the World. Hamburg und Berlin: Paul Parey.
Weidensaul, S. (1996). Raptors. New York: Lyons and Burford.
Weidensaul, S. (2000). The Raptor Almanac: A Comprehensive Guide to Eagles, Hawks, Falcons, and Vultures. Guilford, CT: Lyons Press.
Weidensaul, S. (2004). The Raptor Almanac: A Comprehensive Guide to Eagles, Hawks, Falcons, and Vultures. New York, NY: Lyons Press.
Weimerskirch, H., Guionnet, T., Martin, J., Shaffer, S. A. & Costa, D. P. 2000. Fast and fuel efficient? Optimal use of wind by flying albatrosse Proceedings of the Royal Society B: Biological Sciences, 267, 1869–1874. https://doi.org/10.1098/rspb.2000.1223
Weir, S.M., Thomas, J. F. & Blauch, D. N. 2018. Investigating spatial patterns of mercury and rodenticide residues in raptors collected near th Charlotte, NC, USA, metropolitan area. Environmental Science and Pollution Research, (33), 33153–33161. https://doi.org/10.1007/s1135 018-3229-y
Weizhi Ji, W. (2006). Birds in Yunnan. China: China Forestry Publishing House.
Welch, G. & Welch, H. (1991). The autumn migration of the Steppe eagle Aquila nipalensis. Sandgrouse, 13, 24–33.
Welch, G. & Welch, H. (1998). Raptor migration Bab al Mandab, Yemen – Spring 1998. Phoenix, 15, 11–12.
Welch, G. R., Welch, H. J., Coghlan, S. M. & Denton, M. L. (1986). Djibouti II, Autumn '85.
Welch. G. R. & Welch, H.J. (1992). Djibouti III, migrant raptor count. Minsmere Reserve, Suffolk, UK: Privately published.
Wellman, J. (2015). What Does the Eagle Represent in the Bible? A Christian Study. Retrieved from https://www.patheos.com/blog christiancrier/2015/07/17/what-does-the-eagle-represent-in-the-bible-a-christian-study/
Wells, D. R. & Inskipp, T. P. (2012). A proposed new genus of Booted eagles (tribe Aquilini). Bulletin of the British Ornithologists' Club, 13 70–72.
Wells, D. R. (1999). The Birds of the Thai-Malay Peninsula, covering Burma and Thailand south of the eleventh parallel, Peninsular Malaysia an Singapore. Vol. 1. Non-passerines. San Diego, CA: Academic Press.
Wells, M. C. & Bekoff, M. (1978). Coyote-Bald eagle interactions at Carrion. Journal of Mammalogy, 59(4), 886–887, https://do org/10.2307/1380170
Wendland, V. (1971). Aquila clanga (Pallas 1811), Schelladler. Handbuch der Vögel Mitteleuropas. *In:* von Blotzheim, U. N. G., Bauer, K. M. Bezzel, E. (Eds.), Handbuch der Vögel Mitteleuropas Bd., 4, Akad. Verl.-Ges., Frankfurt/Main, Germany.
Werner, P. A. (1990). Ecological determinants of savannas: Abiotic and biotic. Journal of Biogeography, 17, 401–402.
West, J. N. (1988). Raptors of El Imposible Forest, El Salvador, C.A. M.Sc. Thesis, Central Washington University, Ellensburg, WA.
West, P. M. & Packer, C. (2013). Panthera P. leol. Lion. *In:* Kingdon, J., Happold, D., Butynski, T., Hoffmann, M., Happold, M. & Kalina, J. et a (Eds.). Mammals of Africa, 150–159. London, UK: Bloomsbury Publishing.
Weston, E. D., Whitfield, D. P., Travis, J. M. J. & Lambin, X. (2018). The contribution of flight capability to the post-fledging dependence perio of Golden eagles. Journal of Avian Biology, 49(1), e01265. https://doi.org/10.1111/jav.01265
Weston, Ewan, Whitfield, D., Travis, J. & Lambin, X. (2013). When do young birds disperse? Tests from studies of Golden eagles in Scotlan BMC Ecology, X, 13, 42. https://doi.org/10.1186/1472-6785-13-42
Wetmore, A. (1965). The Birds of the Republic of Panama. Part I: Tinamidae (Tinamous) to Rynchopidae (Skimmers). Washington: Smithsonia Miscellaneous Collections, 150, 1–483. Retrieved from https://archive.org/details/smithsonianmisce15011965smit/page/n5/mode/2up
Weyer, D. (1985). Diurnal birds of prey of Belize. Hawk Trust Annual Report, 14, 22–39.
What does that mean Dictionary (2020). Definition of eagle. Retrieved from http: //www.whatdoesthatmean.com/ dictionary/E/eagle.html

/heat, R. E., Lewis, S. B., Wang, Y., Levi, T. & Wilmers, C. C. (2017). To migrate, stay put, or wander? Varied movement strategies in Bald eagles (Haliaeetus leucocephalus). Movement Ecology, 5, 9. https://doi.org/10.1186/s40462-017-0102-4
/heeler, B. C. (2008). Selfish or altruistic? An analysis of alarm call function in wild capuchin monkeys, Cebus apella nigritus. Animal Behaviour, 76(5), 1465–1475. https://doi.org/10.1016/ j.anbehav.2008.06.023
'heeler, B. C. (2010). Production and perception of situationally variable alarm calls in wild tufted capuchin monkeys (Cebus apella nigritus). Behavioral Ecology and Sociobiology, 54(6), 989–1000. https://doi.org/10.1007/s00265-010-0914-3
'heeler, B. K. (2003). Raptors of Eastern North America. Princeton, NJ: Princeton University Press.
'hitacre, D., Bjork, R. D., Aguirre, B. O., Cordova, A. & Lopez, A. (2002). Current status and distribution of the Harpy Eagle in Guatemala (Estado actual y distribucion del Aguila Harpia en Guatema). Program of Neotropical Raptor Conference and Harpy Eagle Symposium, pp. 17–18.
/hitacre, D. F., Lopes-Avila, J. & Lopez-Avila, G. (2002). Behavioral and physical development of a nestling Crested eagle (Morphnus guianensis). Journal of Raptor Research, 36, 77–81.
/hitacre, D. F., Lopez, J. & Lopez, G. (2012). Crested Eagle. *In*: Whitacre, D. F. (Ed.), Neotropical birds of prey: Biology and ecology of a forest raptor community, pp. 164–184. Ithaca, NY: Comstock Publishing Associates.
/hitacre, D. F., López, J., López, G., Funes, S. H., Flatten, C. J. & Madrid, J. A. (2012a). Black hawk-eagle. *In:* Whitacre, D. F. (Ed.), Neotropical birds of prey: Biology and ecology of a forest raptor community, pp. 185–202. Ithaca, NY: Cornell University Press. https://doi.org/10.1525/bio.2013.63.8.11
/hitacre, D. F., Madrid, J. A., Madrid, H. D., Cruz, R., Flatten, C. J. & Funes, S. H. (2012b). Ornate hawk-eagle. *In:* Whitacre, D. F. (Ed.), Neotropical birds of prey: Biology and ecology of a forest raptor community, pp. 203–233. Ithaca, NY: Cornell University Press.
/hitacre, D. F. & Jenny, J. P. (2013). Neotropical birds of prey: Biology and ecology of a forest raptor community. Ithaca: Cornell University Press.
/hitacre, D. F., López, J. & López, G. (2017). Crested Eagle: Biology and Ecology of a Forest Raptor Community. https://doi.org/10.7591/9780801464287-016
/hite, C. M., Emison, W. B. & Williamson, F. S. L. (1971). Dynamics of raptor populations on Amchitka Island, Alaska. BioScience, 21, 623–627.
/hite, C. M. N. & Bruce, M. D. (1986). The Birds of Wallacea (Suluwesi, the Moluccas and Lesser Sunda Islands, Indonesia), An annotated check-list. British Ornithologists' Union Check-list no. 7.
'hite, C. M., Clum, N. J., Cade, T. J. & Hunt, W. G. (2002). Peregrine Falcon (Falco peregrinus). *In:* Poole, A. & Gill, F. (Eds.), The Birds of North America, 660. Philadelphia, PA, Washington, DC: Academy of Natural Sciences, and the American Ornithologists' Union.
/hite, C. M., Cade, T. J. & Enderson, J. H. (2013). Peregrine Falcons of the World. Barcelona: Lynx Edicions.
/hite, G. & Hayes, F. E. (2002). Second report of Trinidad and Tobago Rare Bird Committee. Living World. Journal of the Trinidad and Tobago Field Naturalists' Club, 51–56.
/hite, G., Kenefick, M. & Murphy, W. L. (2007). The status and abundance of birds in Trinidad and Tobago. Living World. Journal of the Trinidad and Tobago Field Naturalists' Club, 57–68.
/hite, J. G., Antos, M. J., Fitzsimons, J. A. & Palmer, G. C. (2005). Non-uniform bird assemblages in urban environments: The influence of street vegetation. Landscape and Urban Planning, 71(2–4), 123–135. https://doi.org/10.1016/j.landurbplan.2004.02.006
/hitehead, K. & Hugenholtz, C. H. (2014). Remote sensing of the environment with small, unmanned aircraft systems (UASs), Part 1: A review of progress and challenges. Journal of Unmanned Vehicle Systems, 202(3), 69–85. https://doi.org/10.1139/juvs-2014-0007
/hitfield, D. P. (2009). Collision avoidance of Golden eagles at wind farms under the 'Band' collision risk model. Natural Research, Banchury, UK. Retrieved from http://www.snh.org.uk/pdfs/strategy/renewables/B362718.pdf
'hitfield, D. P., Douse, A., Evans, R. J., Grant, J., Love, J., McLeod, D. R. A., Reid, R. & Wilson, J. D. (2009a). Natal and breeding dispersal in a reintroduced population of White-tailed eagles Haliaeetus albicilla. Bird Study, 56(2), 177–186. https://doi.org/10.1080/00063650902792023
'hitfield, D. P., Duffy, K., McLeod, D. R. A., Evans, R. J., MacLennan, A. M., Reid, R., Sexton, D., Wilson, J. D. & Douse, A. (2009b). Juvenile dispersal of White-tailed eagles in Western Scotland. Journal of Raptor Research, 43(2), 110–120. https://doi.org/10.3356/JRR-08-54.1
/hitfield, D. P., Marquiss, M., Reid, R., Grant, J., Tingay, R. & Evans, R. J. (2013). Breeding season diets of sympatric White-tailed eagles and Golden eagles in Scotland: No evidence for competitive effects. Bird Study, 60: 1(1), 67–76. https://doi.org/10.1080/00063657.2012.742997
'hitney, M. C. & Cristol, D. A. (2017). Impacts of sublethal mercury exposure on birds: A detailed review. Reviews of Environmental Contamination and Toxicology, 244, 113–163. https://doi.org/10.1007/398_2017_4
'hittaker, A. (2009). Pousada Rio Roosevelt: A provisional avifaunal inventory in south-western Amazonian Brazil, with information on life history, new distributional data and comments on taxonomy. Cotinga, 31, 23–46.
/hittaker, D. G. & Burns, A. G. (2001). Black bear status in western North America: Summary of western state and province bear status report surveys. Western, Fiji: Black Bear Workshop, 7, 32–55.
/hittington-Jones, C., West, S., Neser, W., Davidson, I., Saunders, P. & Saunders, N. (2013). The status and breeding success of Verreaux's eagles Aquila verreauxii in the Magaliesberg, South Africa. Gabar, 24(1), 7–15.
/hymper, S. L. (1907). Curious behaviour of Aquila hastata, the Lesser Spotted eagle. Journal of the Bombay Natural History Society, 18(1), 187–188.
/ichmann, G. & Kohler, B. (2009). Prenasledovanie dravocov v dolnom rakusku. Birds of Prey and Owls – Journal of Raptor Protection of Slovakia, 5, 4–5.
/ichmann, M. C., Dean, W. R. J. & Jeltsch, F. (2004). Global change challenges the Tawny eagle (Aquila rapax), modelling extinction risk with respect to predicted climate and land use changes. Ostrich, 75(4), 204–210. https://doi.org/10.2989/00306520409485446

Wichmann, M. C., Dean, W. R. J. & Jeltsch, F. (2006). Predicting the breeding success of large raptors in arid southern Africa: A first assessme Ostrich-Journal of African Ornithology, 77(1–2), 22–27. https://doi.org/10.2989/00306520609485503
Widodo, T. (2004). Population and home range of Javan Hawk-eagle (Spizaetus bartelsi, Stresemann, 1924) at G. Kendeng, Cikaniki-G. Halim National Park. Honors Thesis. Bogor, Indonesia: Bogor Agriculture University.
Wiedenfeld, D. A., Schulenberg, T. S. & Robbins, M. B. (1985). Birds of a tropical deciduous forest in extreme northwestern Peru. Ornithologic Monographs, 36, 305–315. https://doi.org/10.2307/40168288
Wikipedia (2020). Eagle. Retrieved from https://en.wikipedia.org/wiki/Eagle
Wilbur, S. R. (1978). The California Condor, 1966–1976: A Look at its Past and Future. Washington, DC: Fish and Wildlife Service.
Wilcove, D. S. (1999). The Condor's Shadow: The Loss and Recovery of Wildlife in America. New York: W.H. Freeman.
Wild Andalucia (2020). The Bird Migration in Numbers. Retrieved from http://www.wildandalucia. com/bird-migration-strait-of-gibraltar/
Wilde, H. (1980). Ornithologische Beobachtungen auf Kreta. Ornithologische Mitteilungen, 32, 13–16.
Wiley, R. W. & Bolen, E. G. (1971). Eagle-livestock relationships: Livestock carcass census and wound characteristics. Southwestern Naturali 16(2), 151–169. https://doi.org/10.2307/3670496
Wilkinson, A. (2016). The Eagle Huntress (review). Eye for Film, March 25. Retrieved from http://www.eyeforfilm.co.uk/review/the-eag huntress-2016-film-review-by-amber-wilkinson
Willard, D. E., Foster, M. S., Barrowclough, G. F., Dickerman, R.W., Cannell, P.F., Coats, S.L., Cracraft, J. L. & O'Neill, J. P. (1991). The bir of Cerro de la Neblina, Territorio Federal Amazonas, Venezuela. Fieldiana Zoology, New series, 65, 1–80.
Wille, F. & Kampp, K. (1983). Food of the White-tailed eagle Haliaeetus albicilla in Greenland. Holarctic Ecology, 6(1), 81–88. http://www.jst org/stable/3682720
Wille, F. (2003). Status of the White-tailed Sea eagles in Greenland. *In:* Helander, B., Marquiss, M. & Bowermann, W. (Eds.), Sea eagle, p 27–29. Proceedings of the International Conference at Björko, Sweden, 13-17 September 2000. Swedish Society for Nature Conservatic SNF & Åtta. 45 Tryckeri AB. Stockholm.
Willgohs, J. F. (1961). The White-tailed Eagle Haliaëtus albicilla albicilla (Linné) in Norway. Norwegian Universities Press.
Williams, P. L., James, R. C. & Roberts, S. M. (2014). Principles of toxicology: Environmental and industrial applications. New York: John Wil & Sons.
Williams, T. (2000). Zapped! Audubon, 102, 32–44.
Williamson, D. F. (2002). Status, Management, and Trade of the American Black Bear (Ursus americanus) in North America. Washington, D.C World Wildlife Fund.
Willis, E. O. & Eisenmann, E. (1979). A revised list of birds of Barro Colorado Island, Panama. Smithsonian Contributions to Zoology, 291(29 1–30. https://doi.org/10.5479/si.00810282.291
Willis, E. O. (1988). A hunting technique of the Black-and-white Hawk-eagle (Spizastur melanoleucus). Wilson Bulletin, 100, 672–675.
Willis, E. O. & Oniki, Y. (2002). Birds of Santa Teresa, Espírito Santo, Brazil: Do humans add or substract species? Papeis Avulsos de Zoolog (São Paulo), 42, 193–264.
Willow, G. (2020). Strange Visitors from the North: Irruptive Migrants. NY: New York City Audubon.
Wilmers, C. C., Stahler, D. R., Crabtree, R. L., Smith, D. W. & Getz, W. M. (2003). Resource dispersion and consumer dominance: Scavengi at wolf- and hunter-killed carcasses in Greater Yellowstone, USA. Ecology Letters, 6(11), 996–1003. https://doi.org/10.1046/ j.146 0248.2003.00522.x
Wilmers, C. C., Nickel, B., Bryce, C. M., Smith, J. A., Wheat, R. E. & Yovovich, V. (2015). The golden age of bio-logging: How animal-bor sensors are advancing the frontiers of ecology. Ecology, 96, 1741–1753.
Wilson, E. E. & Wolkovich, E. M. (2011). Scavenging: How carnivores and carrion structure communities. Trends in Ecology and Evolutic 26(3), 129–135. https://doi.org/10.1016/j.tree. 2010.12.011
Wilson, E. O. (1975). Sociobiology. Cambridge, MA: Harvard University Press.
Wilson, R. P., Shepard, E. & Liebsch, N. (2008). Prying into the intimate details of animal lives: Use of a daily diary on animals. Endanger Species Research, 4(1–2), 123–137. https://doi.org/10.3354/esr00064
Wilson, V. J. (1993). A zoological survey of Mole National Park north-western Ghana. Part II. Birds (dry season list). Unpubl. Report to For Resource Management Programme, GWD/IUCN Project 9786, Accra.
Wimarna, T. (2002). The presence of Javan Hawk-eagle Spizaetus bartelsi at G. Simpang and G. Tilu Nature Reserves, southern West Java. Hon Thesis. Bandung, Indonesia: University of Pajaran.
Wink, M., Heidrich, P., Kahl, U., Witt, H. & Ristow, D. (1993a). Inter- and intraspecific variation of the nucleotide sequence for cytochrome in Cory's shearwater (Calonectris diomedea), Manx Shearwater (Puffinus puffinus) and Fulmar (Fulmarus glacialis). Z. Naturforsch., 48 504–508.
Wink, M., Heidrich, P. & Ristow, D. (1993b). Genetic evidence for speciation of the Manx shearwater (Puffinus puffinus) and the Mediterrane shearwater (P. yelkouan). die Vogelwelt, 114, 226–232.
Wink, M. (1994). PCR in der Evolutionsforschung. *In:* Wink, M. & Wehrle, H. (Eds.), PCR im medizinischen und biologischen labor, pp. 16 184, GIT-Verlag, Darmstadt, Germany.
Wink, M. (1995). Phylogeny of Old and New World vultures (Aves: Accipitridae and Cathartidae) inferred from nucleotide sequences of t mitochondrial cytochrome b gene. Verlag der Zeitschrift für Naturforschung, 50, 868–882.
Wink, M., Heidrich, P. & Fentzloff, C. (1996). A mtDNA phylogeny of sea eagles (genus Haliaeetus) based on nucleotide sequences of t cytochrome b-gene. Biochemical Systematics and Ecology, 24(7–8), 783–791.

Wink, M. (2000). Advances in DNA studies of diurnal and nocturnal raptors. *In*: Chancellor, R. D. & Meyburg, B.-U. (Eds.), Raptors at risk, 831–844. Towcester, UK/Surrey, BC:WWGBP/Hancock House.

Wink, M., Heidrich, P. & Fentzloff, C. (1996). A mtDNA phylogeny of Sea eagles (genus Haliaeetus) based on nucleotide sequences of the cytochrome b-gene. Biochemical Systematics and Ecology, 24(7–8), 783–791. https://doi.org/10.1016/S0305-1978(96)00049-X

Wink, M. & Seibold, I. (1996). Molecular phylogeny of Mediterranean raptors (families Accipitridae and Falconidae). *In:* Muntaner, J. & Mayol, J. (Eds.), Biology and conservation of Mediterranean raptors, pp. 335–344. Madrid, Spain. SEO BirdLife.

Wink, M. 2000. Advances in DNA studies of diurnal and nocturnal raptors. *In*: Chancellor, R. D. & Meyburg, B.-U. (Eds.), Raptors at risk, pp. 831–844. Towcester, UK/Surrey, BC:WWGBP/Hancock House.

Wink, M. & Sauer-Gürth, H. (2000). Advances in the molecular systematics of African raptors. *In:* Chancellor, R. D. & Meyburg, B.-U. (Eds.), Raptors at risk, pp. 135–147. Vancouver: Hancock House.

Wink, M. & Sauer-Gürth, H. (2004). Phylogenetic relationships in diurnal raptors based on nucleotide sequences of mitochondrial and nuclear marker genes. *In:* Chancellor, R. D. & Meyburg, B.-U. (Eds.), Raptors at risk, pp. 483–498. Berlin: World Working Group on Birds of Prey and Owls.

Winkelman, J. E. (1990). Verstoring van vogels door des Sep-proefwindcentrale te Oosterbierum (Fr.) tijdens boufwase in half-operationale situaties (1984–1989). Rijksinstitut voor Natuurbeheer, RIN-rapport 9/157, Arnhem.

Winkelman, J. E. (1992). The impact of the Sep wind park near Oosterbierum (Fr.), The Netherlands, on birds, 4: disturbance. DLO-Instituut voor Bos-en Natuuronderzoek.

Winker, K., Oehlenschlager, R. J., Ramos, M. A., Zink, R. M., Rappole, J. H. & Warner, D. W. (1992). Avian distribution and abundance records for the Sierra de los Tuxtlas, Veracruz, Mexico. Wilson Bulletin, 104, 699–718.

Winker, K. (1997). Introduction to the birds of Los Tuxtlas. *In:* González-Soriano, E., Dirzo, R. & Vogt, R. C. (Eds.), Natural history of Los Tuxtlas, pp. 535–543. Mexico: Universidad Nacional Autónoma de México.

Winkler, D. W. (1987). A general model for parental care. American Naturalist, 130, 526–543.

Winkler, D. W., Billerman, S. M. & Lovette, I. J. (2020). Hawks, eagles, and kites (Accipitridae), version 1.0. *In*: Billerman, S. M., Keeney, B. K., Rodewald, P. G. & Schulenberg, T. S. (Eds.), Birds of the world. Ithaca, NY: Cornell Lab of Ornithology. https://doi.org/10.2173/bow.accipi1.01

Witczuk, J., Pagacz, S., Zmarz, A. & Cypel, M. (2018). Exploring the feasibility of unmanned aerial vehicles and thermal imaging for ungulate surveys in forests – Preliminary results. International Journal of Remote Sensing, 39, 15-16, 5504–5521. https://doi.org/10.1080/01431161.2017.1390621

Witherby, H. F., Jourdain, F. C. R., Ticehurst, N. F. & Tucker, B. W. (1943). Handbook of British Birds. London: H.F. & G. Witherby.

Wobeser, G. & Blakley, B. R. (1987). Strychnine poisoning of aquatic birds. Journal of Wildlife Diseases, 23(2), 341–343. https://doi.org/10.7589/0090-3558-23.2.341

Woelfl, M. & Woelfl, S. (1994). Evidence of Bobcats, Lynx rufus, in southeastern Alberta. Canadian Field-Naturalist, 108(4), 495–496.

Wolch, J. & Emel, J. (1998). Animal Geographies: Place Politics and Identity in the Nature-Culture Borderlands. London: Verso.

Wolley, J. (1907). Ootheca Wolleyana: An Illustrated Catalogue of the Collection of Birds' Eggs. Repressed Publishing LLC.

Wolters, H. E. (1975–1982). Die Vogelarten der Erde. Eine systematische Liste mit Verbreitungsangaben sowie deutschen und englischen Namen. Hamburg: Paul, Cape Verde: Parey.

Wood, D. T. (1946). Eye-witness account of Golden eagle killing calf. Condor, 48, 143.

Wood, G. (1983). The Guinness Book of Animal Facts and Feats (3rd Revised edition). Sterling Publishers Co., Inc.

Wood, P. B., Edwards, T. C. & Collopy, M. W. (1989). Characteristics of Bald eagle nesting habitat in Florida. Journal of Wildlife Management, 53, 441–449.

Wood, P. B., Nesbitt, S. A. & Steffer A. (1993). Bald Eagles prey on Sandhill Cranes in Florida. Journal of Raptor Research, 27, 164–165.

Wood, P. B. & Collopy, M. W. (1994). Population ecology of sub-adult southern Bald eagles in Florida: Post-fledging ecology, migration patterns, habitat use, and survival. Final Report. Nongame Wildlife Program Project NG87-026. Florida Game and Fresh Water Fish Commission, Tallahassee, USA.

Wood, P. B., Collopy, M. W. & Sekerak, C. M. (1998). Postfledging nest dependence period for Bald eagles in Florida. Journal of Wildlife Management, 62, 333–339.

Woodford, M. H., Bowden, C. G. R. & Shah, N. (2008). Diclofenac in Asia and Africa –Repeating the Same Mistake? Harmonisation and Improvement of Registration and Quality Control of Veterinary Medicinal Products in Africa. Paris: OIE World Organisation for Animal Health.

Woodroffe, R. & Ginsberg, J. R. (1998). Edge effects and the extinction of populations inside protected areas. Science, 280(5372), 2126–2128. https://doi.org/10.1126/science.280.5372.2126

Woodroffe, R., Thirgood, S. & Rabinowitz, A. (2005). The impact of human-wildlife conflict on natural systems. *In:* Woodroffe, R., Thirgood, S. & Rabinowitz, A. (Eds.), People and wildlife conflict or coexistence, 1–12. Cambridge, U.K.: Cambridge University Press.

Working Group for White-tailed Eagles and Steller's Sea Eagles. (1996). Wintering status of Steller's Sea eagles and White-tailed eagles in northern Japan. *In:* Survey of the status and habitat conditions of threatened species, pp. 1–9. Tokyo, Japan: Environment Agency.

World Meteorological Organization. (2019). What is Climate? Retrieved from https://www.wmo.int/pages/prog/wcp/ccl/faq/faq_doc_en.htmlHTML; and https://www.wmo#: ~:text=The%20classical%20period%20is%2030,description%2C%20of%20the%20climate%20system

World Wildlife Fund. (2016). New Anti-Poaching Technology Leads to Dozens of Arrests of Wildlife Criminals in Africa. Retrieved from https://www.worldwildlife.org/stories/new-anti-poaching-technology-leads-to-dozens-of-arrests-of-wildlife-criminals-in-africa.

World Wildlife Fund. (2018). Terrestrial Ecoregions. Retrieved from https://www.worldwildlife. org/biome-categories/terrestrial-ecoregions.
Worthy, T. H. & Holdaway, R. N. (2002). The Lost World of the Moamoa: Prehistoric Life of New Zealand. Indiana University Press.
Worton, B. J. (1989). Kernel methods for estimating the utilization distribution in home-range studies. Ecology, 70(1), 164–168: https://do org/10.2307/1938423
Worton, B. J. (1995). Using Monte Carlo simulation to evaluate kernel-based home range estimators. Journal of Wildlife Management, 59(4 794–800. https://doi.org/10.2307/3801959
Wright, A. (2008). Aquila the Eagle. *In:* http: //www.constellationsofwords.com/Constellations/ Aquila.htm
Wright, B. S. (1953). The relation of Bald eagles to breeding ducks in New Brunswick. The Journal of Wildlife Management, 17(1), 55–62. https: doi.org/10.2307/3796805
Wright, B.A. & Schempf, P. (Eds.). 2005. Bald Eagles in Alaska. Bald Eagle Research Institute. http://www.hancockhouse.com/products/akba htm
Wright, K. R. (2016). Count trends for migratory Bald eagles reveal differences between two populations at a spring site along the Lake Ontari shoreline. Peer J, 4, e1986. https://doi.org/10.7717/peerj.1986
Wright, P. C. (1998). Impact of predation risk on the behaviour of Propithecus diadema edwardsi in the rainforest of Madagascar. Behaviou 135(4), 483–512. Brill Publishers. https://doi.org/10.1163/156853998793066186
Wu, W. W., Xu, H. G., Wu, J., Cao, M. C., Weiwei, W., Haigen, X., Jun, W. & Mingchang, C. (2012). The impact of climate change on birds: review. Biodiversity Science, 20(1), 108–115. https://doi.org/10.3724/SP.J.1003.2012.08152
Wulder, M. A., Bater, C. W., Coops, N. C., Hilker, T. & White, J. C. (2008). The role of LiDAR in sustainable forest management. The Forestr Chronicle, 84(6), 807–826.
Wulder, M. A., White, J. C., Nelson, R. F., Næsset, E., Ørka, H. O., Coops, N. C., et al. (2012). Lidar sampling for large-area forest characterizatio A review. Remote Sensing of Environment, 121, 196–209.
Wunderlich, K. (1980). Zum Nahrungsspektrum freilebender Riesenseeadler, Haliaeetus pelagicus (Pallas), im Fernen Osten//Mitu., 5(1–2).
Wuxgohs, J. F. (1961). The White-tailed eagle Haliaeetus albicilla (L.) in Norway. Mat. Naturv. Serie, No. 12. Bergen and Oslo: Norwegia University Press.
Xie, Y., Holmgren, S., Andrews, D. M. K. & Wolfe, M. S. (2017). Evaluating the impact of the U.S. National Toxicology Program: A case stud on Hexavalent Chromium. Environmental Health Perspectives, 125(2), 181–188. https://doi.org/10.1289/ehp21
Yahoo! (2020). Eagle. Definition. Retrieved from https://ca.search.yahoo.com/search?p=eagle+ definition&fr=yfp-t&fp=1&toggle=1&cop mss&ei=UTF-8
Yalden, D. W. (2007). The older history of the White-tailed eagle in Britain. British Birds, 100, 471–480.
Yarwood, R. & Evans, N. (2000). Taking stock of farm animals and rurality. *In*: Philo, C. & Wilbert, C. (Eds.), Animal spaces, beastly spaces: Ne geographies of human-animal relations, pp. 98–114. London: Routledge.
Yates, R. E., McClelland, B. R., McClelland, P. T., Key, C. H. & Bennetts, R. E. (2001). The influence of weather on golden Eagle migration i northwestern Montana. Journal of Raptor Research, 35, 81–90.
Ydenberg, R. C., Dekker, D., Kaiser, G., Shepherd, P. C. F., Ogden, L. E., Richards, K. & Lank, D. B. (2010). Winter body mass and over-ocea flocking as components of danger management by Pacific Dunlins. BMC Ecology, 10, 1. https://doi.org/10.1186/1472-6785-10-1.
Yom-tov, Y. (1984). On the difference between the spring and autumn migrations in Eilat, southern Israel. Ringing and Migration, 5(3), 141–14 https://doi.org/10.1080/03078698.1984. 9673845
Yom-tov, Y. (1988). Bird Migration in Israel. *In:* Yom-Tov, Y. & Tchernov, E. (Eds.), The Zoogeography of Israel, pp. 497–514. Dordrech Netherlands: Dr W. Junk Publishers.
Yosef, R., Verdoorn, G., Helbig, A. & Seibold, I. (2000). A new subspecies of the Booted eagle from Southern Africa inferred from biometrics an mitochondrial DNA. *In*: Chancellor, R. D. & Meyburg, B.-U. (Eds.), Raptors at risk, pp. 43–49. Publisher: WWGBP and Hancock House.
Yosef, R. (2003). Raptor visible-migration monitoring, banding and conservation at Eilat on the westernmost Asiatic flyway. Asian Rapt Research and Conservation Network. 3rd Symposium on Asian Raptors, Kenting.
Yosef, R. & Fornasari, L. (2004). Simultaneous decline in Steppe eagle (Aquila nipalensis) populations and Levant Sparrow hawk (Accipit brevipes) reproductive success: Coincidence of a Chernobyl legacy? Ostrich, 75(1–2), 20–24. https://doi.org/10.2989/0030652040 948540
Yoshino, T., Shingake, T., Onuma, M., Kinjo, T., Yanai, T., Fukushi, H., Kuwana, T. & Asakawa, M. (2009). Parasitic helminths and arthropods the Crested Serpent Eagle Spilornis cheela perplexus Swann, 1922 from the Yaeyama. Journal of the Yamashina Institute for Ornithology, 4 55–61. https://doi.org/10.3312/jyio.41.55
Young, M. D. & Solbrig, O. T. (1993). Providing an environmentally sustainable, economically profitable and socially equitable future for th world's savannas. *In*: Young, M. D. & Solbrig, O. T. (Eds.), The world's savannas: Economic driving forces, ecological constraints and polic options for sustainable land use, pp. 321–344. Paris: Pantheon Publishing Group.
Yuda, I. P., Nurcahyo, A. & Atmojo, H. (2003). Javan Hawk eagle at Mount Merapi, Yogyakarta, Indonesia. *In:* Prawiradilaga, D. M. (Ed. Proceedings of the Second Symposium on Raptor of Asia, 205. Bogor: The Indonesian Committee for the Secound Symposium of Asi Raptor Research & Conservation Network.
Yurko, V. V. (2016). Diet of the White-tailed eagle during the breeding season in the Polesie State Radiation-Ecological Reserve, Belarus. Rapt Conservation, 32, 21–31.
Yves Prévost, Y. (1979). Osprey-Bald eagle interactions at a common foraging site. Auk, 96(2), 413–414. https://doi.org/10.1093/auk/96.2.414
Zaccaroni, A., Amorena, M., Naso, B., Castellani, G., Lucisano, A. & Stracciari, G. L. (2003). Cadmium, chromium and lead contaminati of Athene noctua, the Little owl, of Bologna and Parma, Italy. Chemosphere, 52(7), 1251–1258. https://doi.org/10.1016/s0045-6535(0 00363-1

Zaccaroni, A., Andreani, G., Ferrante, M. C., Carpene, E., Isani, G., Lusiano, A., … Gloria, I. (2008). Metal concentrations in the liver and kidney of raptor species from the Calabria region, Italy. Acta Veterinaria (Beograd), 58(4), 315–324. https://doi.org/10.2298/AVB0804315Z

Zachos, F. E. (2004). Critical remarks on taxonomy and species delineation, the Aquila (Accipitridae) taxon as an example – Or: How many species of eagles are there? Ornithologische Mitteilungen, 56, 166–172.

Zahavi, A. (1957). The breeding birds of Huleh Swamp and lake. Ibis, 99, 600–607.

Zahm, S. (2015). Haliaeetus leucogaster (Online). Animal Diversity Web. Retrieved from https://animaldiversity.org/accounts/Haliaeetus_leucogaster/

Zakrajsek, E. J. & Bissonette, J. A. (2005). Ranking risk of wildlife species hazardous to military aircraft. Wildlife Society Bulletin, 33(1), 258–264. https://doi.org/10.2193/0091-7648(2005)33258: RTROWS2.0.CO;2

Zalles, J. I. & Bildstein, K. L. (Eds.) (2000). Raptor Watch: A Global Directory of Raptor Migration Sites. Cambridge, UK: Bird-Life International; and Kempton, PA, USA: Hawk Mountain Sanctuary (BirdLife Conservation Series No. 9).

Zaun, B. J. (2009). First modern record of the White-tailed eagle in Hawaii. Western Birds, 40(1), 35–38.

Zawadzka, D. (1999). Feeding habits of the Black kite Milvus migrans, Red kite Milvus milvus, White-tailed eagle Haliaeetus albicilla and Lesser Spotted eagle Aquila pomarina in Wigry National Park NE Poland. Acta Ornithologica, 34(1), 65–75.

Zeng, N. & Yoon, J. (2009). Expansion of the world's deserts due to vegetation-albedo feedback under global warming. Geophysical Research Letters, 36(17), L17401. https://doi.org/10.1029/2009 GL039699

Zhang, J., Zhang, J. & Yao, N. (2009). Geostatistics for spatial uncertainty characterization, Geo-spatial Information Science, 12(1), 7–12. https://doi.org/10.1007/s11806-009-0191-3

Zhang, X. L., Luo, X. J., Liu, J., Luo, Y., Chen, S. J. & Mai, B. X. (2011). Polychlorinated biphenyls and organochlorinated pesticides in birds from a contaminated region in South China: Association with trophic level, tissue distribution and risk assessment. Environmental Science and Pollution Research International, 18(4), 556–565. https://doi.org/10.1007/s11356-010-0397-9

Zheng, G. M. & Wang, Q. S. (1998). China Red Data Book of endangered animals: Aves. Beijing, People's Republic of China: Science Press.

Zhengjie, Z. (2001). Birds of China, Vol. 1 (Non-Sparrow-Shaped). Jilin Science and Technology Press, 275–276.

Zhezherin, V. P. (1969). On taxonomic interrelation of Aquila clanga Pall. and Aquila pomarina Brehm. Zbirn Prats Zool Muz, 33, 91–97.

Zhu Lei, Z., Yang Xiao-Nong, Y., Hao Guang, H., Huang Qin, H., Liu Tian-Tian, L., Zizi-Yue, D., Qian, W., Lewthwaite, R. L. & Yue-Hua, S. (2014). A review of the distribution of Black eagle Ictinaetus malaiensis in mainland China. Forktail, 30, 45–49.

Zilio, F. (2017). Breeding biology and conservation of Hawk-eagles (Spizaetus spp.) (Aves, Accipitridae) in southern Atlantic Forest, Brazil. Iheringia Série Zoologia, 107. https://doi.org/10.1590/1678-4766e2017037

Zimbelmann, F. (1992). Ein vergleichend morphologischer Beitrag zur Phylogenetik der Seeadler (Haiaeetus Savigny 1809). Mitt. Zool. Mus. Berlin. Ann. Orn., 68(16), 61–114.

Zimmerer, K. S. (1994). Human geography and the 'New Ecology': The prospect of integration. Annals of the Association of American Geographers, 84(1), 108–125.

Zimmerman, D. A., Turner, D. A. & Pearson, D. J. (1996). Birds of Kenya and Northern Tanzania. London: Christopher Helm Publishers.

Zorzin, G., Carvalho, E. A., Carvalho Filho, E. P. M. & Canuto, M. (2006). New records of rare and threatened Falconiformes for the State of Minas Gerais. Revista Brasileira de Ornitologia, 14(4), 416–421.

Zoumakis, N. M. (1993). The dependence of the power-law exponent on surface roughness and stability in a neutrally and stably stratified surface boundary layer. Atmósfera, 6, 79–83.

Zu Wied-Neuwied, M. A. P. (1820). Reise nach Brasilien in den Jahren 1815 bis 1817. Erster Band. Frankfurt: Brönner.

Zu-Aretz, S. & Leshem, Y. (1983). The sea – A trap for gliding birds. Torgos, 5, 16–17.

Zuluaga, S. & Echeverry-Galvis, M. Á. (2016). Domestic fowl in the diet of the Black- and Chestnut eagle (Spizaetus isidori) in the eastern Andes of Colombia: A potential conflict with humans? Ornitología Neotropical, 27, 113–120.

Zuna-Kratky, T. & Kürthy, A. (1999). Long-term study of birds of prey in the lower March-Thaya-valley along the Austrian-Slovakian border. Egretta, 42, 17–29.

Zuskin, J., Zuskinova, M. & Kornan, I. J. (2003). Štátna ochrana prírody Slovenskej republiky Banská Bystrica Správa Tatranského národného parku: Program Záchrany; orla skalného (Aquila chrysaetos Linnaeus, 1758). Liptovský Mikuláš, Slovakia.

Zuur, A. F., Ieno, E. N., Walker, N. J., Saveliev, A. A. & Smith, G. M. (2009). Zero-truncated and zero-inflated models for count data. *In:* Mixed effects models and extensions in ecology with R. Statistics for Biology and Health. Springer, New York, 261–293. https://doi.org/10.1007/978-0-387-87458-6_11

Zylo, M. T. (2012). Bald eagles (Haliaeetus leucocephalus) wintering in northern Arizona select perches based on food availability, visibility and cover. PhD Thesis, Northern Arizona University.

Index

D

E

F

Bald Eagle Perching on Totem Pole. [With permission from Ignacio Beya Marshal].

For Product Safety Concerns and Information please contact our EU representative GPSR@taylorandfrancis.com Taylor & Francis Verlag GmbH, Kaufingerstraße 24, 80331 München, Germany

Batch number: 10723926

Printed by Printforce, the Netherlands